Classical and Contemporary Sociological Theory

Text and Readings

Edition

Scott Appelrouth / Laura Desfor Edles

California State University, Northridge

SAGE | PINE FORGE

Los Angeles | London | New Delhi
Singapore | Washington DC

Los Angeles | London | New Delhi
Singapore | Washington DC

FOR INFORMATION:

Pine Forge Press
An Imprint of SAGE Publications, Inc.
2455 Teller Road
Thousand Oaks, California 91320
E-mail: order@sagepub.com

SAGE Publications Ltd.
1 Oliver's Yard
55 City Road
London EC1Y 1SP
United Kingdom

SAGE Publications India Pvt. Ltd.
B 1/I 1 Mohan Cooperative Industrial Area
Mathura Road, New Delhi 110 044
India

SAGE Publications Asia-Pacific Pte. Ltd.
33 Pekin Street #02-01
Far East Square
Singapore 048763

Acquisitions Editor: Jerry Westby
Editorial Assistant: Erim Sarbuland
Production Editor: Brittany Bauhaus
Typesetter: C&M Digitals (P) Ltd.
Proofreader: Wendy Jo Dymond
Indexer: Sylvia Coates
Cover Designer: Candice Harman
Marketing Manager: Erica DeLuca
Permissions Editor: Karen Ehrmann

Printed in the United States of America

*A catalog record of this book is available from the Library of
Congress.*

9781412992336

This book is printed on acid-free paper.

11 12 13 14 15 10 9 8 7 6 5 4 3 2

BRIEF CONTENTS

DETAILED CONTENTS

LIST OF ILLUSTRATIONS AND PHOTOS

LIST OF FIGURES AND TABLES

Figures

Tables

ABOUT THE AUTHORS

Scott Appelrouth (PhD, New York University, 2000) is Professor of Sociology at California State University, Northridge. His interests include social theory, cultural sociology, and social movements. He teaches classical and contemporary theory at both the graduate and undergraduate levels and has published articles in research and teaching journals.

Laura Desfor Edles (PhD, UCLA, 1990) is Professor of Sociology at California State University, Northridge. She is the author of *Symbol and Ritual in the New Spain* (Cambridge University Press, 1998), *Cultural Sociology in Practice* (Blackwell Publishers, 2002), and various articles on social theory and culture.

PREFACE

Every semester we begin our sociological theory courses by telling students that we love sociological theory and that one of our goals is to get each and every one of them to love theory, too. This challenge we set for ourselves makes teaching sociological theory exciting. If you teach "sexy" topics like the sociology of drugs, crime, or sex, students come into class expecting the course to be titillating. By contrast, when you teach sociological theory, students tend to come into class expecting the course to be abstract, dry, and irrelevant to their lives. The fun in teaching sociological theory is in proving students wrong. The thrill in teaching sociological theory is in getting students to see that sociological theory is absolutely central to their everyday lives—and fascinating as well. What a reward it is to have students who adamantly insisted that they "hated" theory at the beginning of the semester "converted" into theorists by the end!

In teaching sociological theory, we use original texts. We rely on original texts in part because every time we read these works we derive new meaning. Core sociological works tend to become "core" precisely for this reason. However, using original readings requires that the professor spend lots of time and energy explaining issues and material that is unexplained or taken for granted by the theorist. This book was born of this process—teaching from original works and explaining them to our students. Hence, this book includes the original readings we use in our courses, as well as our interpretation and explanation of them.

Thus, this book is distinct in that it is both a reader *and* a text. It is unlike existing readers in several ways, however. First and foremost, this book is not just a collection of seemingly disconnected readings. Rather, in this book we provide an overarching theoretical framework with which to understand, compare, and contrast these selections. In our experience, this overarching theoretical framework is essential in explaining the relevance and excitement of sociological theory. In addition, we discuss the social and intellectual milieus in which the selections were written, as well as their contemporary relevance. Thus, we connect these seemingly disparate works not only via theory but also via concrete applications to today's world.

Finally, this book is unique in that we provide a variety of visuals and pedagogical devices—historical and contemporary photographs, diagrams, and charts illuminating core theoretical concepts and comparing specific ideas—to enhance student understanding. Our thinking is, Why should only introductory-level textbooks have visual images and pedagogical aids? Almost everyone, and not just the youngest audiences, enjoys—and learns from—visuals.

As is often the case in book projects, this turned out to be a much bigger and thornier project than either of us first imagined. And, in the process of writing this book, we have accrued many intellectual and social debts. First, we especially thank Jerry Westby of SAGE/Pine Forge Press for helping us get this project started. Jerry literally walked into our offices at California State University, Northridge, and turned what had been a nebulous, long-standing idea into a concrete plan. In the several years since then, Jerry has been a continuously enthusiastic and conscientious ally, and we are grateful for his hard work and unflagging support. On the production of this second edition, we especially thank Nichole O'Grady, who organized and collected the many copyright permissions essential to the production of this book, and we thank the following reviewers for their comments:

For the First Edition

John Bartkowski
Mississippi State University

Robert Bausch
Cameron University

Matthew Bond
Oxford University

Leslie Cintron
Washington and Lee University

Harry Dahms
University of Tennessee, Knoxville

Mathieu Deflem
University of Southern California

Stephen B. Groce
Western Kentucky University

Neil Gross
Harvard University

Melanie Hildebrandt
Indiana University

Laurel Holland
University of West Georgia

Joy Crissey Honea
Montana State University–Billings

Chris Hunter
Grinnell College

Amanda Kennedy
Ohio State University

Paul-Brian McInerney
Indiana University–South Bend

Elizabeth Mitchell
Rutgers, the State University of New Jersey

John Murray
Manhattanville College

Darek Niklas
Rhode Island College

William Nye
Hollins University

William Outhwaite
University of Sussex

Brian Rich
Transylvania University

Scott Schaffer
Millersville University of Pennsylvania

David Schweingruber
Iowa State University

Anne Szopa
Indiana University East

For the Second Edition

Jordan Brown
Florida State University

Daniel Egan
University of Massachusetts Medical School

Stephen Groce
Western Kentucky University

Angela Henderson
University of Northern Colorado

Gabriel Ignatow
University of North Texas

Stan Knapp
Brigham Young University

Peter Venturelli
Valparaiso University

Ingrid Whitaker
Old Dominion University

At California State University, Northridge, we are especially indebted to our friend and colleague David Boyns, who helped us in writing the first edition of this book when things got tight. This book is much better because of David's pivotal contributions to Chapter 8 and Chapter 9.

1 INTRODUCTION

Key Concepts

- Theory
- Order
  - Collective/Individual
- Action
  - Rational/Nonrational
- Enlightenment
- Counter-Enlightenment

"But I'm not a serpent, I tell you!" said Alice. "I'm a—I'm a—"

"Well! What are you?" said the Pigeon. "I can see you're trying to invent something!"

"I—I'm a little girl," said Alice, rather doubtfully, as she remembered the number of changes she had gone through that day.

"A likely story indeed!" said the Pigeon, in a tone of the deepest contempt. "I've seen a good many little girls in my time, but never one with such a neck as that! No, no! You're a serpent; and there's no use denying it. I suppose you'll be telling me next that you never tasted an egg!"

"I have tasted eggs, certainly," said Alice, who was a very truthful child; "but little girls eat eggs quite as much as serpents do, you know."

"I don't believe it," said the Pigeon; "but if they do, why, then they're a kind of serpent: that's all I can say."

—Lewis Carroll, Alice's Adventures in Wonderland

In the previous passage, the Pigeon had a theory—Alice is a serpent because she has a long neck and eats eggs. Alice, however, had a different theory—that she was a little girl. Yet, it was not the "facts" that were disputed in the passage. Alice freely admitted that she had a long neck and ate eggs. So why did Alice and the Pigeon come to such different conclusions? Why didn't the facts "speak for themselves"?

Alice and the Pigeon both *interpreted* the question (what *is* Alice?) using the categories, concepts, and assumptions with which each was familiar. It was these unarticulated concepts, assumptions, and categories that led the Pigeon and Alice to have such different conclusions.

Likewise, social life can be perplexing and complex. It is hard enough to know "the facts"—let alone to know *why* things are as they seem. In this regard, theory is vital to making sense of social life because it holds assorted observations and facts together (as it did for Alice and the Pigeon). Facts make sense only because we interpret them using preexisting categories and assumptions, that is, "theories." The point is that even so-called facts are based on implicit assumptions and unacknowledged presuppositions. Whether we are consciously aware of them or not, our everyday life is filled with theories as we seek to understand the world around us. The importance of formal sociological theorizing is that it makes assumptions and categories explicit, hence open to examination, scrutiny, and reformulation.

To be sure, some students find sociological theory as befuddling as the conversation between Alice and the Pigeon in *Alice's Adventures in Wonderland.* Some students find it difficult to understand and interpret what sociological theorists are saying. Moreover, some students wonder why they have to read works written over a century ago or why they have to study classical sociological theory at all. After all, classical sociological theory is abstract and dry and has "nothing to do with my life." So why not just study contemporary stuff and leave the old, classical theories behind?

In this book, we seek to demonstrate the continuing relevance of classical as well as contemporary sociological theory. By "classical" sociological theory, we mean the era during which sociology first emerged as a discipline and was then institutionalized in universities—the mid-19th to early 20th centuries. We argue that the classical theorists whose work you will read in this book are vital, first, because they helped chart the course of the discipline of sociology from its inception until the present time and, second, because their concepts and theories still resonate with contemporary concerns. These theoretical concerns include the nature of capitalism, the basis of social solidarity or cohesion, the role of authority in social life, the benefits and dangers posed by modern bureaucracies, the dynamics of gender and racial oppression, and the nature of the "self," to name but a few.

"Contemporary" sociological theory can be periodized roughly from 1935 to the present. However, the dividing line between "classical" and "contemporary" theory is not set in stone, and a few classical thinkers, such as W. E. B. Du Bois, wrote from the late 1800s right up until the 1960s! In identifying core contemporary theorists, we consider the extent to which a writer extends and expands on the theoretical issues at the heart of sociology. To a person, these thinkers all talk back to, revise, and reformulate the ideas of the "founding" theorists of sociology while taking up important issues raised by the social context in which they were/are writing and by the human condition itself.

Yet, the purpose of this book is to provide students not only with both core classical and contemporary sociological readings but also a framework for comprehending them. In this introductory chapter we discuss (1) *what* sociological theory is, (2) *who* the "core" theorists in sociological theory are, and (3) *how* students can develop a more critical and gratifying understanding of some of the most important ideas advanced by these theorists.

▪▪ WHAT IS SOCIOLOGICAL THEORY?

Theory is a system of generalized statements or propositions about phenomena. However, there are two additional features that, together, distinguish scientific theories from other idea systems, such as those found in religion or philosophy. "Scientific" theories

1. explain and predict the phenomena in question, and

2. produce testable and thus falsifiable hypotheses.

Universal laws are intended to explain and predict events occurring in the natural or physical world. For instance, Isaac Newton established three laws of motion. The first law, the law of inertia, states that objects in motion will remain in motion, while objects at rest will remain at rest unless are acted on by another force. In its explanation and predictions regarding the movement of objects, this law extends beyond the boundaries of time and space. For their part, sociologists seek to develop or refine general statements about some aspect of *social* life. For example, a long-standing (though not uncontested) sociological theory predicts that as a society becomes more modern, the salience of religion will decline. Similar to Newton's law of inertia, the secularization theory, as it is called, is not restricted in its scope to any one period or population. Instead, it is an abstract proposition that can be tested in any society once the key concepts making up the theory, "modern" and "religion," are defined and observable measures are specified.

Thus, sociological theories share certain characteristics with theories developed in other branches of science. However, there are significant differences between social and other scientific theories (i.e., theories in the social sciences as opposed to the natural sciences) as well. First, sociological theories tend to be more evaluative and critical than theories in the natural sciences. Sociological theories are often rooted in implicit moral assumptions, which contrasts with traditional notions of scientific objectivity. In other words, it is often supposed that the pursuit of scientific knowledge should be free from value judgments or moral assessments; that is, the first and foremost concern of science is to uncover what *is,* not what *ought* to be. Indeed, such objectivity is often cast as a defining feature of science, one that separates it from other forms of knowledge based on tradition, religion, or philosophy. While some sociologists adopt this model of scientific inquiry, others tend to be interested not only in an objective understanding of the workings of society but also in realizing a more just or equitable social order. As you will see, the work of many theorists is shaped in important respects by their own moral sensibilities regarding the condition of modern societies and what the future may bring. Thus, sociological theorizing at times falls short of the "ideal" science practiced more closely (though still imperfectly) by "hard" sciences like physics, biology, or chemistry. The failure to consistently conform to the ideals of either science or philosophy is, for some observers, a primary reason for the discipline's troublesome identity crisis and "ugly duckling" status within the world of academics. For others, it represents the opportunity to develop a unique understanding of social life.

A second difference between sociological theories and those found in other scientific disciplines stems from the nature of their respective subjects. Societies are always in the process of change, while the changes themselves can be spurred by any number of causes, including internal conflicts, wars with other countries (whether ideological or through direct invasion), scientific or technological advances, or through the expansion of economic markets that in turn spreads foreign cultures and goods. As a result, it is more difficult to fashion universal laws to explain societal dynamics. Moreover, we must also bear in mind that humans, unlike most other animals or naturally occurring elements in the physical world, are motivated to act by a complex array of social and psychological forces. Our behaviors are not the product of any one principle; instead, they can be driven by self-interest, altruism, loyalty, passion, tradition, or habit, to name but a few factors. From these remarks you can see the difficulties inherent in developing universal laws of societal development and individual behavior, this despite our earlier example of the secularization theory and other efforts to forge such laws.

These two aspects of sociological theory (the significance of moral assumptions and the nature of the subject matter) are responsible, in part, for the form in which much sociological theory is written. While some theorists construct formal propositions or laws to explain and predict social events and individual actions, more often theories are developed through storylike narratives. Thus, few of the original readings included in this volume contain explicitly stated propositions. One of the intellectual challenges you will face in studying the selections is to uncover the general propositions that are embedded in the texts. Regardless of the style in which they are presented, however, the theories (or narratives, if you prefer) you will explore in this text answer the most central social questions while uncovering taken-for-granted truths and encouraging you to examine who you are and where we as a society are headed.

The Enlightenment

Many of the seeds of the debate as to the nature of sociology were first planted in the **Enlightenment**, a period of remarkable intellectual development that occurred in Europe during the late seventeenth and early eighteenth centuries. During the Enlightenment, a number of long-standing ideas and beliefs on social life were turned upside down. The development of **civil society** (open spaces of debate relatively free from government control) and the rapid pace of the modern world enabled a critical mass of literate citizens to think about the economic, political, and cultural conditions that shaped society. Before that, explanations of the conditions of existence were so taken for granted that there was no institutionalized discipline examining them (Lemert 1993; Seidman 1994). Enlightenment intellectuals advocated rule by rational, impersonal laws and the end to arbitrary, despotic governments. They sought to define the rights and responsibilities of free citizens. In so doing, Enlighteners called into question the authority of kings whose rule was justified by divine right.

However, the Enlightenment was not so much a fixed set of ideas, but a new attitude, a new method of thought. One of the most important aspects of this new attitude was an emphasis on *reason*. Central to this new attitude was questioning and reexamining received ideas and values.

The Enlightenment emphasis on reason was part and parcel of the rise of science. Scientific thought had begun to emerge in the fifteenth century through the efforts of astronomers and physicists such as Copernicus, Galileo, and Newton. Enlightenment intellectuals developed an approach to the world based on methodical observations. Rather than seeing the universe as divinely created and hierarchically ordered, Enlighteners insisted that the universe was a mechanical system composed of matter in motion that obeyed natural laws. Moreover, they argued that these laws could be uncovered by means of science and empirical research. In advocating the triumph of reasoned investigation and systematic observation of phenomena over religious faith and common-sense ways of understanding, Enlightenment intellectuals rebuked existing knowledge as fraught with prejudice and mindless tradition (Seidman 1994:20– 21). Not surprisingly, such views were dangerous, for they challenged the authority of religious beliefs and those charged with advancing them. Indeed, some Enlighteners were tortured and imprisoned, or their work was burned for being heretical.

The rise of science and empiricism would give birth to sociology in the mid-nineteenth century. The central idea behind the emerging discipline was that society could be the subject of scientific examination in the same manner as biological organisms or the physical properties of material objects. Indeed, the French intellectual **Auguste Comte** (1798–1857), who coined the term *sociology* in 1839, also used the term *social physics* to refer to this new discipline and his organic conceptualization of society. The term *social physics* reflects the Enlightenment view that the discipline of sociology parallels other natural sciences. Comte argued that like natural scientists, sociologists should rationally and scientifically uncover the laws of the social world.[1] For Enlighteners, the main difference between scientific knowledge and either theological explanation or mere conjecture is that scientific knowledge can be tested. Thus, for Comte, the new science of society—sociology—involved (1) the analysis of the central elements and functions of social systems using (2) concrete historical and comparative methods in order to (3) establish testable generalizations about them (Fletcher 1966:14).[2]

---

[1]Physics is often considered the most scientific and rational of all the natural sciences because it focuses on the basic elements of matter and energy and their interactions.

[2]Of course, the scientists of the Enlightenment were not uninfluenced by subjectivity or morality. Rather, as Seidman (1994:30–31) points out, paradoxically the Enlighteners sacralized science, progress, and reason; they deified the creators of science, such as Galileo and Newton, and fervently believed that "science" could resolve all social problems and restore social order, which is itself a type of "faith."

However, it was the French theorist **Émile Durkheim** (1858–1917) who arguably was most instrumental in laying the groundwork for the emerging discipline of sociology. Durkheim emphasized that while the primary domain of psychology is to understand processes internal to the individual (for example, personality or instincts), the primary domain of sociology is "social facts," that is, conditions and circumstances external to the individual that, nevertheless, determine one's course of action. As a scientist, Durkheim advocated a systematic and methodical examination of social facts and their impact on individuals.

Yet interestingly, sociology reflects a complex mix of Enlightenment and **counter-Enlightenment** ideas (Seidman 1994). In the late eighteenth century, a conservative reaction to the Enlightenment took place. Under the influence of Jean-Jacques Rousseau (1712–1778), the unabashed embrace of rationality, technology, and progress was challenged. Against the emphasis on reason, counter-Enlighteners highlighted the significance of nonrational factors, such as tradition, emotions, ritual, and ceremony. Most importantly, counter-Enlighteners were concerned that the accelerating pace of industrialization and urbanization and growing pervasiveness of bureaucratization were producing profoundly disorganizing effects. In one of his most important works, *The Social Contract* (1762), Rousseau argued that in order to have a free and equal society, there must be a genuine social contract in which everyone participates in creating laws for the good of society. Thus, rather than being oppressed by impersonal bureaucracy and laws imposed from above, people would willingly obey the laws because they helped make them. Rousseau's challenge of the age of reason echoed Pascal's view that the heart has reasons that reason does not know. When left to themselves, our rational faculties leave us lifeless and cold, uncertain and unsure (see McMahon 2001:35).

In a parallel way, Durkheim was interested in objective or external social facts and the more subjective elements of society, such as feelings of solidarity or commitment to a moral code. Akin to Rousseau, Durkheim felt that it was these subjective elements that ultimately held societies together. Similarly, **Karl Marx** (1818–1883), who is another of sociology's core classical figures (though he saw himself as an economist and social critic), fashioned an economic philosophy that was at once rooted in science and humanist prophecy. Marx analyzed not only the economic dynamics of capitalism but also the social and moral problems inherent to the capitalist system. So, too, did the third of sociology's core classical theorists, **Max Weber** (1864–1920), combine a methodical, scientific approach with a concern about both the material conditions and idea systems of modern societies.

Economic and Political Revolutions

Thus far we have discussed how the discipline of sociology emerged within a specific intellectual environment. But, of course, the Enlightenment and counter-Enlightenment were both the cause and the effect of a whole host of political and social developments, which also affected the newly emerging discipline of sociology. Tremendous economic, political, and religious transformations had been taking place in western Europe since the sixteenth century. The new discipline of sociology sought to scientifically explain both the causes and the effects of such extraordinary social change.

One of the most important of these changes was the **Industrial Revolution**, which began in England in the eighteenth century. The Industrial Revolution refers to the application of power-driven machinery to manufacturing. Though industrialization began in remote times and continues today, this process completely transformed Europe in the eighteenth century. It turned Europe from a predominantly agricultural to a predominantly industrial society. It not only radically altered how goods were produced and distributed, it galvanized the system of capitalism as well.

Specifically, with the Industrial Revolution, large numbers of people began to leave farms and agricultural work to become wage earners in factories located in the rapidly growing cities. Indeed, though most of the world's population was rural before the Industrial Revolution, by the mid-nineteenth century, half of the population of England lived in the cities, and by the end of the nineteenth century so did half of the population of Europe. Moreover, while there were scarcely 25 cities in Europe with a population of 100,000 in 1800, there were more than 150 cities this size a century later. At the same time, factories were transformed by a long series of technological changes. Ever more efficient machines were adopted, and tasks were routinized. Thus, for instance, with the introduction of the power loom in the textile industry, an unskilled worker could produce three and a half times as much as the best handloom weaver.

However, this rise in efficiency came at a tremendous human cost. Mechanized production reduced both the number of jobs available and the technical skills needed for work in the factory. A few profited enormously, but most worked long hours for low wages. Accidents were frequent and often quite serious. Workers were harshly punished and/or their wages were docked for even the slightest mistakes. Women and children worked alongside men in noisy, unsafe conditions. Most factories were dirty, poorly venti-lated and lit, and dangerous.

As you will read in Chapter 2, Karl Marx was particularly concerned about the economic changes and disorganizing social effects that followed in the wake of the Industrial Revolution. Marx not only wrote articles and books on the harsh conditions faced by workers under capitalism; he also was a political activist who helped organize revolutionary labor movements to provoke broad social change.

As you will read in Chapter 4, Max Weber also explored the profound social transformations taking place in European society in the eighteenth and nineteenth centuries. Akin to Marx, Weber was concerned about the social consequences wrought by such profound structural change. However, in contrast to Marx, Weber argued that it was not only modern economic structures (e.g., capitalism) but also organizational structures—mostly importantly bureaucracies—that profoundly affected social relations. Indeed, in one of the most famous metaphors in all of sociology, Weber compares modern society to an "iron cage." Even more importantly, in contrast to Marx, Weber also examined the particular systems of meaning, or ideas, that both induced and resulted from such profound structural change.

The eighteenth century was a time of not only tremendous economic but also political transformation. One of the most significant political events of that time was the French Revolution, which shook France between 1787 and 1799 and toppled the ancien régime. Inspired in large part by Rousseau's *Social Contract,* the basic principle of the French Revolution as contained in its primary manifesto, *The Declaration of Rights of Man and of the Citizen,* was that "all men are born free and equal in rights." The French revolutionaries called for "liberty, fraternity, and equality." They sought to substitute reason for tradition and equal rights for privilege. Because the revolutionaries sought to rebuild government from the bottom up, the French Revolution stimulated profound political rethinking about the nature of government from its inception and set the stage for democratic uprisings throughout Europe.

However, the French Revolution sparked a bloody aftermath, making it clear that even democratic revolutions involve tremendous social disruption and that heinous deeds can be done in the name of free-dom. During the Reign of Terror led by Maximilien Robespierre, radical democrats rounded up and executed anyone—whether on the left or the right—suspected of opposing the revolution. In the months between September 1793 (when Robespierre took power) and July 1794 (when Robespierre was over-thrown), revolutionary zealots arrested about 300,000 people, executed some 17,000, and imprisoned thousands more. It was during this radical period of the Republic that the guillotine, adopted as an effi-cient and merciful method of execution, became the symbol of the Terror.[3]

▪▪ WHO ARE SOCIOLOGY'S CORE THEORISTS?

Thus far we have argued that the central figures at the heart of classical sociological theory all sought to explain the extraordinary economic, political, and social transformations taking place in Europe in the late nineteenth century. Yet, concerns about the nature of social bonds and how these bonds can be maintained

[3]R. W. Connell (1997) notes that sociology was born during a decisive period of European colonial expansion. In turn, much of the discipline was devoted to collecting information about the colonizers' encounters with "primitive Others." Early sociologists' views on progress, human evolution, and racial hierarchies, however, were largely mar-ginalized as the process of canon formation began during the 1930s. This had the effect of purging the discourse of imperialism from the history of the discipline. See "Why Is Classical Theory Classical?" *American Journal of Sociology* 102(6):1511–57.

in the face of extant social change existed long before the eighteenth century and in many places, not only Western Europe. Indeed, in the late fourteenth century, Abdel Rahman Ibn-Khaldun (1332–1406), born in Tunis, North Africa, wrote extensively on subjects that have much in common with contemporary sociology (Martindale 1981:134–36; Ritzer 2000a:10). And long before the fourteenth century, Plato (circa 428–circa 347 B.C.), Aristotle (384–322 B.C.), and Thucydides (circa 460–circa 400 B.C.) wrote about the nature of war, the origins of the family and the state, and the relationship between religion and the government—topics that have since become central to sociology (Seidman 1994:19). Aristotle, for example, emphasized that human beings were naturally political animals (*zoon politikon;* Martin 1999:157), and he sought to identify the "essence" that made a stone a stone or a society a society (Ashe 1999:89). For that matter, well before Aristotle's time, Confucius (551–479 B.C.) developed a theory for understanding Chinese society. Akin to Aristotle, Confucius maintained that government is the center of people's lives and that all other considerations derive from it. According to Confucius, a good government must be concerned with three things: sufficient food, a sufficient army, and the confidence of the people (Jaspers 1957:47).

Yet, these premodern thinkers are better understood as philosophers, not sociologists. Both Aristotle and Confucius were less concerned with explaining social dynamics than with prescribing a perfected, moral social world. As a result, their ideas are guided less by a scientific pursuit of knowledge than by an ideological commitment to a specific set of values. Moreover, in contrast to modern sociologists, premodern thinkers tended to see the universe as a static, hierarchical order in which all beings, human and otherwise, have a more or less fixed and proper place and purpose, and they sought to identify the "natural" moral structure of the universe (Seidman 1994:19).

Our key point here is that while the ideas of Marx, Durkheim, and Weber are today at the heart of the classical sociological theoretical canon—and deemed "founding figures" in Part I—this does not mean that they are inherently "better" or more original than those of other intellectuals who wrote before or after them. Rather, it is to say that, for specific historical, social, and cultural as well as intellectual reasons, their works have helped define the discipline of sociology and that sociologists refine, rework, and challenge their ideas in different ways to this day.

For that matter, Marx, Weber, and Durkheim have not always been considered the core theorists in sociology. On the contrary, until 1940, Marx, Weber, and Durkheim were not especially adulated by American sociologists (Bierstedt 1981); up to this time, discussions of their work are largely absent from texts. For that matter, Marx was not included in the canon until the 1960s. Meanwhile, even a cursory look at mid-century sociological theory textbooks reveals an array of important "core figures," including Sumner, Sorokin, Sorel, Pareto, Le Play, Ammon, Veblen, De Tocqueville, Cooley, Spencer, Tönnies, and Martineau. Though an extended discussion of all these theorists is outside the scope of this volume, we provide a brief look at some of these scholars in the "Significant Others" section of the chapters that follow.

In Part II of this book, we focus on several classical writers who for social and/or cultural reasons were underappreciated as sociologists in their day. **Charlotte Perkins Gilman** (1860–1935), for example, was well known as a writer and radical feminist in her time but not as a sociologist (Degler 1966:vii). It was not until the 1960s that there was a formalized sociological area called "feminist theory." Gilman sought to explain the basis of gender inequality in modern industrial society. She explored the fundamental questions that would become the heart of feminist social theory some 50 years later, when writers such as Simone de Beauvoir and Betty Friedan popularized these same concerns.

Georg Simmel (1868–1963), a German sociologist, wrote works that would later become pivotal in sociology, though his career was consistently stymied both because of the unusual breadth and content of his work and because of his Jewish background.[4] Simmel sought to uncover the basic *forms* of social interaction, such

---

[4]Durkheim was also Jewish (indeed, he was the son of a rabbi). But anti-Semitism did not significantly impede Durkheim's career. In fact, it was Durkheim's eloquent article "Individualism and Intellectuals" (1898) on the Dreyfus affair (a political scandal that emerged after a Jewish staff officer named Captain Alfred Dreyfus was erroneously court-martialed for selling secrets to the German embassy in Paris) that shot him to prominence and eventually brought Durkheim his first academic appointment in Paris. In sum, German anti-Semitism was much more harmful to Simmel than French anti-Semitism was to Durkheim.

as "exchange," "conflict," and "domination," that take place between individuals. Above all, Simmel underscored the contradictions of modern life; for instance, he emphasized how individuals strive to conform to social groups and, at the same time, to distinguish themselves from others. Simmel's provocative work is gaining more and more relevance in today's world where contradictions and ironies abound.

While anti-Semitism prevented Simmel from receiving his full due, and sexism impeded Gilman (as well as other women scholars) from achieving hers, the forces of racism in the United States forestalled the sociological career of the African American intellectual **W. E. B. Du Bois** (1868–1963). Not surprisingly, it was this very racism that would become Du Bois's most pressing scholarly concern. Du Bois sought to develop a sociological theory about the interpenetration of race and class in America at a time when most sociologists ignored or glossed over the issue of racism. Though underappreciated in his day, Du Bois's insights are at the heart of contemporary sociological theories of race relations.

We conclude our discussion of classical sociology with the work of social philosopher **George Herbert Mead** (1863–1931). Mead laid the foundation of symbolic interactionism, which, as you will see in Part III, has been one of the major perspectives in sociological theory since the middle of the twentieth century. Mead challenged prevailing psychological theories about the mind by highlighting the social basis of thinking and communication. Mead's provocative work on the emergent, symbolic dimensions of human interaction continue to shape virtually all social psychological and symbolic interactionist work today.

Contemporary Sociological Theory

This brings us to contemporary sociological theory, which, as indicated previously, can be periodized roughly from 1935 to the present. If ascertaining who sociology's core classical theorists are was difficult, determining who sociology's core contemporary theorists are is even thornier. There are myriad possibilities, and contemporary sociologists disagree not only as to who is a core theorist and who is not but even as to the major genres or categories of contemporary theory. For that matter, even defining what theory "is" or should be is a far-from-settled issue. Tied to this state of affairs is the increasing fragmentation of sociological theory over the past 25 years. During this period, sociology has become both increasingly specialized (breaking into such subspecialties as sociology of emotions and world-systems theory) and increasingly broad as sociologists have built new bridges between sociology and other academic fields (e.g., anthropology, psychology, biology, political science, and literary studies), further contributing to the diversity of the discipline.

That said, in this book we take a broad, historical perspective, prioritizing individuals who have significantly influenced others—and the discipline—from the mid-twentieth century until today. In the end, however, determining the "ins and outs" of contemporary theory is a contentious matter, and as such, the writers whose work we feature in this volume are by no means unanimously "core." As per the classical theorists we discussed earlier, we address this issue within the space constraints of this book by providing a briefer look at a number of important theorists in the "Significant Others" section of the chapters that follow.

In Part III, we focus on several major perspectives that have emerged in contemporary sociological theory. We begin with the tradition of **structural functionalism** and the work of **Talcott Parsons** and one of his most prolific students, **Robert Merton**. From the 1930s through the 1970s, functionalism was the dominant theoretical approach in American sociology. A major emphasis of this approach lies in analyzing the societal forces that sustain or disrupt the stability of existing conditions. Functionalists introduced central concepts, such as "role," "norm," and "social system," into the discipline of sociology. They also coined several concepts (such as "role model" and "self-fulfilling prophecy") that are in widespread colloquial and academic use today.

Chapter 10 examines the **Frankfurt School of critical theory**, particularly the work of **Max Horkheimer, Theodor Adorno**, and **Herbert Marcuse**. Due in large measure to the dominance of the functionalist paradigm, the ideas expressed within this perspective would not find wide dissemination in the United States until the 1960s when the sweeping social and cultural changes occurring in the broader society demanded a radically different theoretical approach to their explanation. Rather than emphasizing societal cohesion or consensus (as functionalist typically did), critical theorists underscore the divisive aspects of the social order. Drawing particularly from the works of Karl Marx and Max Weber, the theorists presented in this section seek to expose the oppressive and alienating conditions that are said to characterize modern societies.

As you will see, one of the most important characteristics of both functionalism and critical theory is their collectivist or "macro" approach to social order. (This point will be further explained.) However, a variety of more individualistic perspectives focusing more on the "micro" dimension of the social order were developing alongside these two theoretical camps. In Chapter 11 we examine two of the most important of these perspectives: **exchange theory** and **rational choice theory**. Instead of looking to social systems or institutions for explanations of social life, exchange theorists emphasize individual behavior. Moreover, they consider individuals to be strategic actors whose behavior is guided by exchanges of benefits and costs. Based on rational calculations, individuals use the resources they have at their disposal in an effort to optimize their rewards. We focus especially on the work of two renowned exchange theorists: **George Homans**, who draws principally from behavioral psychology and neoclassical economics, and **Peter Blau**, who, while sympathetic to economics, evinces a greater indebtedness to the German sociologist Georg Simmel (see Chapter 6). In addition, in Chapter 4, we examine the work of **James S. Coleman**, who is one of the central figures within rational choice theory. While both exchange and rational choice theorists view the actors as a purposive agent motivated by maximizing rewards, exchange theorists focus on the strategic decision making of individuals and how such decisions affect social relationships. For their part, rational choice theorists emphasize how group dynamics themselves shape individuals' decisions.

In Chapter 12, **dramaturgical theory** and **symbolic interactionism,** we continue our discussion of analyses of everyday life by examining the work of **Erving Goffman** and **Arlie Russell Hochschild**. As the leading proponent of dramaturgy, Erving Goffman occupies a unique place in the pantheon of contemporary theorists. While rooted in part in a symbolic interactionist approach, Goffman also drew from the work of Émile Durkheim and Georg Simmel. In doing so, he developed a fascinating account of the commonplace rituals that pervade daily interaction and their significance in constructing and presenting an individual's self. Arlie Hochschild's work bears the imprint of Goffman but incorporates a focus on a crucial, though often neglected, aspect of social life: emotions. Additionally, she brings within her purview an examination of gender and family dynamics in contemporary capitalist society.

In Chapter 13, we discuss **phenomenology**: a perspective that, akin to exchange theory and symbolic interactionism, focuses not on political, economic, and social institutions at the collectivist level but on the everyday world of the individual. However, in contrast to exchange theorists (who emphasize the *strategic* calculation of rewards and costs in everyday life), phenomenologists such as **Alfred Schutz** and **Peter Berger** and **Thomas Luckmann** emphasize the subjective categories behind and within which everyday life revolves. They are interested in how people actively produce and sustain *meaning*.

Hochschild's integration of questions of gender into symbolic interactionism and dramaturgical theory brings us to an obvious but all-too-often overlooked point: Sociological theory has traditionally been written by men from the perspective of men. In Chapter 14, we focus on a tradition that takes seriously both the dearth of female voices in sociological theorizing and the distinct social situation of men and women in society: **feminist and gender theories**. As you will see, feminist theory is very diverse. Feminist theorists all address a specific topic—gender equality (or the lack thereof)—but they examine this issue from a number of theoretical perspectives. Indeed, in this chapter you will read selections from the works of the institutional ethnographer **Dorothy Smith**, who extends and integrates the seemingly disparate traditions of phenomenology and Marxism, and neo-Marxist feminist **Patricia Hill-Collins**, whose Black feminist thought speaks to the particular situation of African American women. The final two theorists whose works are featured in this chapter—Australian sociologist **Raewyn Connell** and American postmodernist philosopher **Judith Butler**; both challenge the prevailing nation that "sex" is "the *biological* fact, the difference between the male and the female human animal" and "gender" is the "*social* fact," the difference between masculine and feminine roles, or men's and women's personalities" (Connell 2002:33; emphasis added). Indeed, Butler (1990/1999:145) rejects the very idea that "women" can be understood as a concrete category at all and, instead, construes gender as unstable "fictions" (1990, p. 145).

Judith Butler's postmodern approach to gender brings us to the topic of Chapter 8: **poststructuralism** and **postmodernism.** One of the greatest challenges to theory in the twentieth century has come from poststructuralism/postmodernism. The theorists whose work you will read in Chapter 15, **Michel Foucault** and **Jean Baudrillard**, both critically engage the meaning of modernity by emphasizing how all knowledge, including science, is a representation of reality—not "reality" itself. Baudrillard goes the furthest here, contending that in contemporary society "reality" has completely given way to a simulation of reality, or *hyperreality,* as simulated experience has replaced the "real."[5]

In Chapter 16, we present the work of three leading contemporary theorists: **Pierre Bourdieu, Jürgen Habermas,** and **Anthony Giddens**. Each of these theorists has been involved in a similar project, namely, developing a multidimensional approach to social life that integrates elements from distinct theoretical orientations. In articulating their perspectives, each has emphasized a different theme. Bourdieu develops his project through an emphasis on the reproduction of class relations; Habermas's approach addresses the prospects for democracy in the modern world. For his part, Giddens explores the effects of modernity on trust, risk, and the self.

We conclude this book with an examination of various theories pertaining to contemporary global society. As you will see, although the works of **Immanuel Wallerstein, George Ritzer,** and **Edward Said** are quite distinct, these three theorists focus not on the dynamics of interpersonal interaction (à la symbolic interactionism; see Chapter 12) or the forces that give form to a single society per se (à la functionalism; see Chapter 9) but, rather, on how such aspects of social life are themselves embedded in a global context and how what happens in any given country (or geographical zone) is a function of its interconnections with other geographical regions. Indeed, these theorists both underscore that, given the increasingly unrestricted flow of economic capital and cultural images across countries, the nation-state—a self-governing territory demarcated by recognized spatial boundaries—can no longer serve as the dominant unit of analysis today.

▪▪ HOW Can We Navigate Sociological Theory?

This brief overview of the topics and perspectives covered in this book clearly reveals that contemporary sociological theory is not an easy subject to master. This task is made even more difficult by the fact that sociological theorists oftentimes develop their own terminologies and implicitly "talk back" to a wide variety of thinkers whose ideas may or may not be explained to the reader. As a result, some professors (and students) contend that original theoretical works are just too hard to decipher. These professors use secondary textbooks that interpret and simplify the ideas of core contemporary sociological theorists. Their argument is that students' attention simply can't be captured using original works, and, because students must be engaged in order to understand, secondary texts ultimately lead to a better grasp of the covered theories.

Primary Versus Secondary Sources

There is an important problem with reading only secondary interpretations of original works, however: The secondary and the original text are not the same thing. Secondary texts do not merely translate what the theorist wrote into simpler terms; rather, in order to simplify, the authors of secondary texts must revise what the original writer said.

The problems that can arise from even the most faithfully produced interpretations can be illustrated by the "telephone game." Recall that childhood game in which you sit in a circle and one person thinks of a message. He or she then whispers the message to the next person, and then that person passes the message

---

[5]Thus, for instance, by the time they are school age, many American children will have watched more hours of television than the total number of hours they will spend in classroom instruction (Lemert 1997, p. 27).

on to the next person, until the last person in the circle announces the message aloud. Usually, everyone roars with laughter because the message at the end typically is not the same as the one circulated at the beginning. This is because the message inadvertently gets misinterpreted and changed as it goes around.

In the telephone game the goal is to repeat exactly what has been said to you. Nevertheless, misinterpretations and modifications are commonplace. Consider now a secondary text in which the goal is not to restate exactly what was originally written but to take the original source—that is by nature open to multiple interpretations—and make it "easier" to understand. While this process of simplification perhaps allows students to understand the secondary text, they are at least one step removed from what the original author actually wrote.[6] At the same time, you have no way of knowing what was written in the original works. Moreover, when you start thinking and writing about the material presented in the secondary reading, you are not one—but *two*—steps removed from the original text. If the objective of a course in sociological theory is to grapple with the ideas that preoccupied the core figures of the field—the ideas and analyses that currently shape the direction of sociology—then studying original works must be a cornerstone.

To this end, we provide lengthy excerpts from the original writings of those we consider to be sociology's core classical and contemporary theorists. We believe that if students are to learn Marx, they must read Marx and not a simplified interpretation of his ideas. They must learn to study for themselves what the leading theorists have said about some of the most fundamental social issues, the relevance of which are timeless.

Nevertheless, in this book we also provide a secondary interpretation of the theorists' overall frameworks and the selected readings. Our intent is to provide a guide (albeit simplified) for understanding the original works. The secondary interpretation will help you navigate the different writing styles often resulting from the particular historical, contextual, and geographical locations in which the theorists were and are rooted. Perhaps even more important than the secondary explanations that this book provides, however, is the analytical frame or "map" that we use to explore, compare, and contrast the work of each theorist. It is to this vital tool for comprehension and analysis that we now turn.

The Questions of "Order" and "Action"

Our analytical frame or map revolves around two central questions that social theorists and philosophers have grappled with since well before the establishment of sociology as an institutionalized discipline: the questions of *order* and *action* (Alexander 1987). Indeed, these two questions have been a cornerstone in social thought at least since the time of the ancient Greek philosophers. The first question, that of order, asks what accounts for the patterns and/or predictability of behavior that lead us to experience social life as routine. Or, expressed somewhat differently, how do we explain the fact that social life is not random, chaotic, or disconnected but instead demonstrates the existence of an ordered social universe? The second question, that of action, considers the factors that motivate individuals or groups to act. The question of action, then, turns our attention to the forces that are held responsible for steering individual or group behavior in a particular direction.

Similar to how the north-south, east-west coordinates allow you to orient yourself to the details on a street map, our analytical map is anchored by four "coordinates" that assist in navigating the details of the theories presented in this volume. In this case, the coordinates situate the answers to the two questions. Thus, to the question of order, one answer is that the patterns of social life are the product of structural arrangements or historical conditions that confront individuals or groups. As such, preexisting social arrangements produce the apparent orderliness of social life as individuals and groups are pursuing trajectories that, in a sense, are not of their own making. Society is thus pictured as an overarching system that works *down* on individuals and groups to determine the shape of the social order. Society is understood as a reality "sui generis" that operates according to its own logic distinct from the will of individuals. This orientation has assumed many different names—macro, holistic, objectivist, structuralist, and the label we use here, **collectivist** (see Figure 1.1).

---

[6]Further complicating the matter is that many of the original works that make up the core of sociological theory were written in a language other than English. Language translation is itself an imperfect exercise.

Figure 1.1   Basic Theoretical Continuum as to the Nature of Social Order

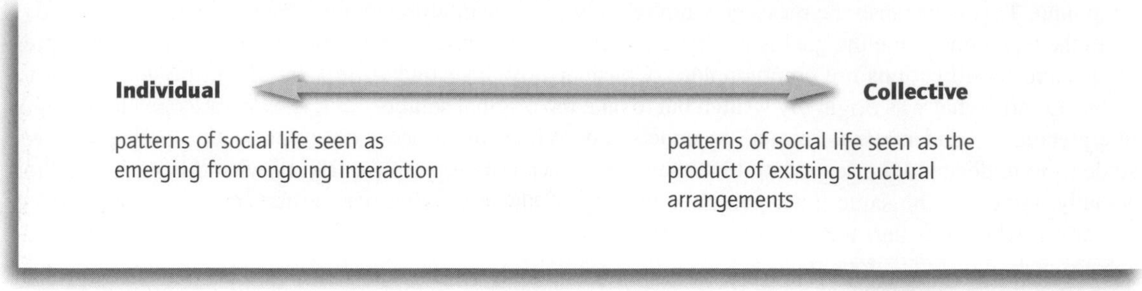

Individual ⟵⟶ **Collective**

patterns of social life seen as
emerging from ongoing interaction

patterns of social life seen as the
product of existing structural
arrangements

By contrast, the other answer to the question of order is that social order is a product of ongoing inter-actions between individuals and groups. Here, it is individuals and groups creating, re-creating, or altering the social order that works *up* to produce society. This position grants more autonomy to actors, as they are seen as relatively free to reproduce the patterns and routines of social life (i.e., the social order) or transform them. Over time, this orientation has earned several names as well—micro, elementarism, subjectivist, and the term we adopt, **individualist** (see Figure 1.1).

Turning to the question of action, we again find two answers labeled here as **nonrational** and **rational**.[7] Specifically, action is primarily nonrational when it is guided by values, morals, norms, tradi-tions, the quest for meaning, unconscious desires, and/or emotional states. While the nonrationalist ori-entation is relatively broad in capturing a number of motivating forces, the rationalist orientation is far less encompassing. It contends that individual and group actions are motivated primarily by the attempt to maximize rewards while minimizing costs. Here, individuals and groups are viewed as essentially calculating and strategic as they seek to achieve the "selfish" goal of improving their position. In short, interests—not values—motivate action (see Figure 1.2).

Intersecting the two questions and their answers, we can create a four-celled map on which we are able to plot the basic theoretical orientation of some of the core classical theorists (see Figure 1.3) and the major contemporary perspectives (see Figure 1.4) discussed in this book. The four cells are identified as individual-nonrational, individual-rational, collective-nonrational, and collective-rational. Yet, we cannot overemphasize that these four coordinates are "ideal types"; theorists and theories are never "pure." Implicitly and/or explicitly, theorists inevitably incorporate more than one orientation in their work. This is even truer today than in the past, as today's theorists explicitly attempt to bridge the theo-retical gaps and dilemmas left by earlier thinkers. Thus, these coordinates (or cells in the table) are best understood as endpoints on a continuum on which theories typically occupy a position somewhere between the extremes. This multidimensionality and ambiguity is reflected in our maps by the lack of fixed points.

In addition, it is important to note that this map is something *you* apply to the theories under consideration. Though all of theorists address the questions of order and action, they generally do not use these terms in their writing. For that matter, their approaches to order and action tend to be implicit, rather than explicit, in their work. Thus, at times, you will have to read between the lines to determine a theorist's position on these fun-damental questions. While this may pose some challenges, it also expands the opportunities for learning.

---

[7]The terms *rational* and *nonrational* are problematic in that they have a commonsensical usage that is at odds with how theorists use these terms. By "rational" we do not mean "good and smart," and by "nonrational" we do not mean irrational, nonsensical, or stupid (Alexander 1987:11). Despite these problems, however, we continue to use the terms rational and nonrational.

Figure 1.2　Basic Theoretical Continuum as to the Nature of Social Action

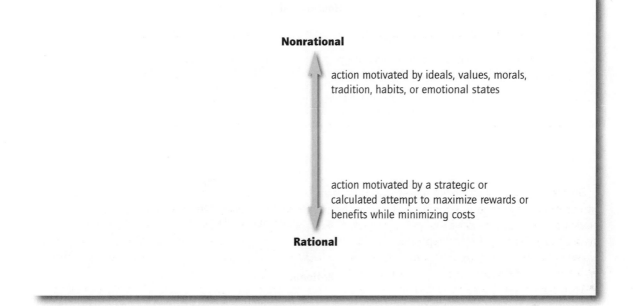

Figure 1.3　Core Classical Theorists' Basic Orientation

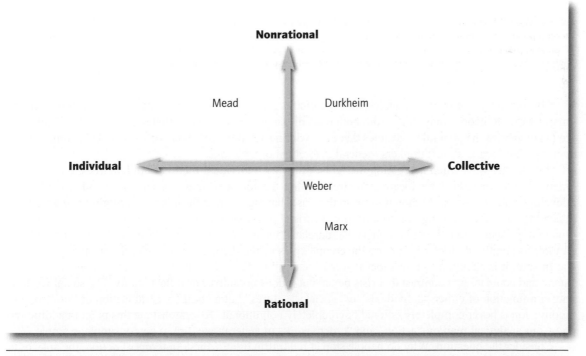

NOTE: This diagram reflects the basic theoretical orientation of a few core classical sociological theorists: George Herbert Mead, Émile Durkheim, Max Weber, and Karl Marx. However, each of these theorists—as well as every theorist in this volume—is far more nuanced and multidimensional than this simple figure lets on. The point is not to "fix" each theorist in a particular box, but rather to provide a means for illuminating and discussing each theorist's orientation relative to other theorists, and within their own works.

Figure 1.4   Basic Orientation of Core Perspectives in Contemporary Sociological Theory

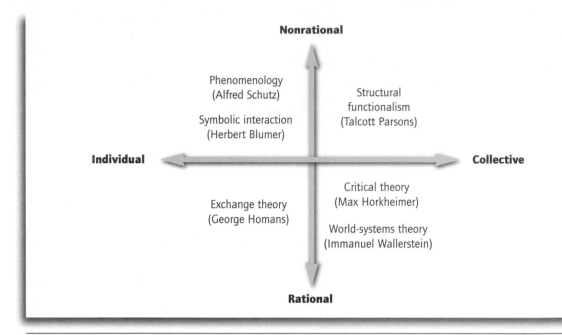

NOTE: This simplified diagram is intended to serve as a guide to comparing and contrasting the theoretical orientations underlying several contemporary theoretical perspectives and the work of authors who are aligned with them. For the sake of visual clarity, we include only the names of those theorists who arguably are most commonly associated with a given perspective, although, as the contents alone of this book suggest, any number of theorists are aligned with a given approach. The point is not to "fix" each theorist in a particular box; nor do all of the works associated with a given perspective fit neatly in a given quadrant. Indeed, each of the theorists and perspectives in this diagram is far more nuanced and multidimensional than this simple figure suggests. Moreover, several perspectives discussed in this book—for instance, postmodernism and feminist theory—are not included in this diagram. Postmodern theory deliberately challenges the very idea of fixed categories such as those that form the basis of this figure, while feminist theories draw from, extend, and fuse a wide variety of perspectives and traditions.

Consequently, not everyone views each theorist in exactly the same light. Moreover, even within one major work, a theorist may draw from both ends of the continuum. Nevertheless, these maps enable you to (1) recognize the general tendencies that exist within each theorist's body of work and (2) compare and contrast (and argue about) thinkers' general theoretical orientations.

Put another way, when navigating the forest of theory, individual theorists are like trees. Our analytic map is a tool or device for locating the trees within the forest so that you can enter and leave having developed a better sense of direction or, in this case, having learned far more than might have otherwise been the case. By enabling you to compare theorists' positions on two crucial issues, their work is less likely to be seen as a collection of separate, unrelated ideas. Bear in mind, however, that the map is only a tool. Its simplicity does not capture the complexities of the theories or of social life itself.

In sum, it is essential to remember that this four-cell table is an analytical device that helps us understand and compare and contrast theorists better, but it does not mirror or reflect reality. The social world is never a function of either "individuals" or "social structures" but a complex combination of both; so, too, motivation is never completely rational or completely nonrational. To demonstrate this point in addition to how our analytical map on "action" and "order" works in general, we turn to a very simple example.

Consider the question, Why do people stop at red traffic lights? First, in terms of *action,* the answer to this question resides on a continuum with rational and nonrational orientations serving as the endpoints. On one hand, you might say that people stop at red traffic lights because it's in their best interest to avoid getting a ticket or into an accident. This answer reflects a rationalist response; the action (stopping at a red light) is rooted in minimizing costs (see Table 1.1).

A nonrationalist answer to this question is that people stop at red traffic lights because they believe that it is good and right to follow the law. Here the individual takes his or her bearings from internalized morals or values. Interestingly, if this moral or normative imperative is the only motivation for action, the individual will stop at the traffic light even if there is no police car or oncoming cars in sight. External circumstances, such as whether or not the individual will get hit or caught if he or she go through the red light, are irrelevant. By contrast, if one's only motivation for action is rationalist, and there are absolutely no visible dangers (i.e., no other cars in sight and hence no possibility of getting a ticket or getting into an accident), the driver will *not* stop at the red light. Rather, on the basis of a calculated appraisal of the relevant conditions, she will go.

Another nonrationalist answer to the question "Why do people stop at red traffic lights?" involves "habits" (see Table 1.1). By definition, habits are relatively unconscious; that is, we don't think about them. They come "automatically," not from strategic calculations of interests or a concern for consequences; that is why they are typically considered nonrationalist. Interestingly, habits may or may not have their roots in morality. Some habits are "folkways," or routinized ways people do things in a particular society (paying your bills by mail rather than in person; driving on the right side of the road), while other habits are attached to sacred values (putting your hand over your heart when you salute the flag). Getting back to our example, let's say you are driving your car on a deserted road at 2:00 in the morning, and you automatically stop at a red traffic light out of habit. Your friend riding with you might say, "Why are you stopping? There's not a car in sight." If your action were motivated simply from habit and not a moral imperative to follow the law, you might say, "Hey, you're right!" and run through the red light.

Of course, actions often have—indeed, they usually have—both rational and nonrational dimensions. For instance, in this last example, you may have interpreted your friend's question, "Why are you stopping? There's not a car in sight" to mean "Don't be a goody-goody—let's go!" In other words, you may have succumbed to peer pressure even though you knew it was wrong. If such was the case, you may have wittingly or unwittingly felt that your ego, or "sense of self," was on the line. Thus, it was not so much that rational trumped nonrational motivation; rather, you acted out of a complex combination of your assessment of the traffic conditions, pressure from your friend to do the "cool" thing, and your desire to be the particular type of person you want to be.

Table 1.1   Why Do People Stop at Red Traffic Lights? Basic Approaches to Order and Action

		ORDER	
		Individual	**Collective**
ACTION	**Nonrational**	Value fidelity: Individual believes it is good and right to follow the law. Habit: Individual stops without thinking.	Hegemonic moral order: Society teaches it is wrong to disobey the law. "Red" means "stop" and "green" means "go" in hegemonic symbolic system.
	Rational	Instrumentality: Individual does not want to get a traffic ticket. Individual does not want to get into an accident.	Hegemonic legal structure: Society punishes those who break the law.

Indeed, a basic premise of this book is that because social life is extremely complex, a complete social theory must account for multiple sources of action and levels of social order. Theorists must be able to account for the wide variety of components (individual predispositions, personality and emotions, social and symbolic structures) constitutive of this world. Thus, for instance, our rationalist response to the question of why people stop at red traffic lights—that people stop simply because they don't want to get a ticket or get into an accident—is, in fact, incomplete. It is undercut by a series of unacknowledged nonrational motivations. There is a whole host of information that undergirds the very ability of an individual to make this choice. For example, before one can even begin to make the decision as to whether to stop for the red light or not, one must know that normally (and legally) "red" *means* "stop" and "green" *means "go."* That we know and take for granted that "red" means "stop" and "green" means "go" and then consciously think about and decide to override that cultural knowledge (and norm) indicates that even at our most rationalist moments we are still using the tools of a largely taken-for-granted, symbolic or nonrational realm (see Table 1.1).

Now let's turn to the issue of *order.*

If we say that people stop at red lights because they don't want to get a ticket, this can reflect a collectivist approach to order if we are emphasizing that there is a coercive state apparatus (e.g., the law, police) that hems in behavior. If such is the case, we are emphasizing that external social structures precede and shape individual choice. This collectivist approach to order (and rationalist approach to action) is illustrated in Table 1.1.

If we say that people stop because they believe it is good and right to follow the law, we would be taking a collectivist approach to order as well. Here we assume that individuals are socialized to obey the law. We emphasize that socially imposed collective morals and norms are internalized by individuals and reproduced in their everyday behavior. Similarly, if we emphasize that it is only because of the preexisting symbolic code in which red means stop and green means go that individuals can then decide what to do, we would be taking a collectivist approach. These versions of order and action are illustrated in Table 1.1.

On the other hand, that people stop at red traffic lights because they don't want to get into an accident or get a ticket also might reflect an individualist approach to order, if the assumption is that the individual determines his action using his own free will, and from this the traffic system is born. At the same time, another important individualist, albeit nonrationalist, answer to this question emphasizes the role of emotions. For instance, one might fear getting a ticket or into an accident, and to the extent that the fear comes from within the individual, rather than from a set of laws or socialization into a preexisting symbolic code, we can say that this represents an individualist explanation for the patterning of social life.

Sociological theorists hold a variety of views on the action/order continua even within their own work. Overall, however, each theorist can be said to have a basic or general theoretical orientation. For instance, in terms of the classical theorists discussed earlier, Marx was most interested in the collectivist and rationalist conditions behind and within order and action, while Durkheim, especially in his later work, was most interested in the collectivist and nonrationalist realms. Thus, juxtaposing Figure 1.3 and Table 1.1, you can see that if we were to resurrect Marx and Durkheim from their graves and ask them the hypothetical question, "Why do people stop at red traffic lights?" it would be more likely that Marx would emphasize the rationalist motivation behind this act (they seek to avoid getting a ticket), while Durkheim would emphasize the nonrational motivation (they consider it the "right" thing to do)—though both would emphasize that these seemingly individualist acts are actually rooted in collectivist social and cultural structures (that it is the law with its coercive and moral force that undergirds individual behavior). Meanwhile, at the more individualist end of the continuum, Mead (see Chapter 8) would probably emphasize the immediate ideational process in which individuals interpret the meanings for and consequences of each possible action. (Naturally, each of these theorists' work is far more complex and multidimensional than this simple example lets on.)

Of course, the purpose of this book is not to examine the work of sociological theorists in order to figure out how they might answer a hypothetical question about traffic lights. Rather, the purpose of this book is to examine the central issues core classical and contemporary theorists themselves raise and analyze the particular theoretical stance they take as they explore these concerns. These tasks are particularly challenging because the contemporary theorists and perspectives you will encounter in this book tend to be even more theoretically complex than sociology's classical founding figures. This is because contemporary theorists are not only drawing from and extending the classical theorists' ideas; they are also seeking to better them. For instance, Jürgen Habermas, Pierre Bourdieu, and Anthony Giddens, whose works are discussed in Chapter 16, have each set out to develop a theoretical model that explicitly synthesizes and bridges nonrationalist and rationalist and individualist and collectivist concerns and ideas. However, all of the contemporary theorists whose works you will read in this book are well aware of, and seek to correct in some way, the theoretical dilemmas posed by sociology's founding figures. Some thinkers, for instance those aligned with exchange theory, symbolic interactionism, and phenomenology (see Chapters 11, 12 and 13), look to address more fully the individualist realm that Marx, Weber, and Durkheim underemphasized (see Figure 1.4). Other theorists, such as the structural functionalist Talcott Parsons (see Chapter 9) and critical theorist Herbert Marcuse (see Chapter 10), meld the collectivist focus of Marx, Durkheim, and Weber with the ideas of the psychoanalyst Sigmund Freud and thus incorporate an individualist (and nonrational) component to their respective arguments.

Notwithstanding their attempts to construct multidimensional theories, contemporary theorists and the perspectives with which they are aligned generally evince a "basic" theoretical orientation, many of which are illustrated in Figure 1.4. For instance, as you will see in Chapter 13, phenomenology focuses, above all, on how individuals apprehend social life on the basis of taken-for-granted assumptions. Instead of positing an overarching, objective social order that establishes behavioral codes according to which individuals are more or less compelled to act (reflecting a collectivist approach to order), they see behavior as patterned or predictable only to the extent that individuals rely on commonplace schemes of understanding to navigate their everyday life. This reflects an emphasis on the individualist/nonrationalist realm. Here, social life is pictured as an intricate panoply of interaction as individuals go about the process of making sense of the situations they face. As we discuss in Chapter 11, exchange theory and rational choice theory, on the other hand, posit that individual conduct is not motivated by attempts to construct meaning or by an intersubjectivity that enables actors to coordinate their behavior. Instead, this perspective argues that individuals are motivated by conscious attempts to satisfy their interests, which reflects an individualist/rationalist theoretical orientation. Moreover, society itself is seen as an accumulation of individual efforts to maximize rewards that have the effect of producing and sustaining institutional structures. Thus, while an exchange theorist would recognize that individuals act within existing institutions, his focus would be how individuals maneuver within a given institution in order to maximize their self-interests.

Unlike the individualist perspectives just outlined, collectivist approaches argue that individual and group conduct is largely shaped by external forces. For instance, as you will see in Chapter 9, structural functionalism posits that societies are self-contained systems that possess their own needs necessary to their survival. It is the existence of such societal needs that in large measure accounts for patterns of individual and group of behavior. For example, because all societies must ensure some measure of peaceful coexistence between its members, a system of shared values and morals must be developed in order to establish the basis for consensual relations. As depicted in Figure 1.4, these assumptions reflect an emphasis on the collective/nonrational realm. For its part, world-systems theory, discussed in Chapter 17, explores the historical dynamics that have created the modern capitalist economy, an economy whose reach spans the globe. Far from studying the routines of everyday interaction, or the consciousness of individuals, world-systems theory explores how distinct regions of the world are tied to one another by relations of domination and subordination that in turn affect economic and social dynamics within a given country. These regions have developed according to a strategic, profit-driven logic that has produced the world's winners and losers, its colonizers and colonized. This argument reflects a collectivist/rationalist orientation.

Yet, it cannot be overemphasized that the point is not to "fix" each theorist or tradition in a particular box. All of the theorists and traditions presented in this book are far more complex than these simple figures let on. As indicated previously, many theorists featured in this book explicitly seek to develop a multidimensional framework, incorporating distinct traditions into mulifaced theoretical paradigms. In addition to the explicitly synthetical theorists discussed previously (Pierre Bourdieu, Jürgen Habermas, and Anthony Giddens, discussed in Chapter 16), dramaturgical theorists (Erving Goffman and Arlie Hochschild, featured in Chapter 12) and gender theorists (Dorothy Smith, Patricia Hill Collins, and Raewyn Connell, featured in Chapter 14) clearly fall into this camp. Moreover, as you will see postmodern thinkers, such as Judith Butler (Chapter 14) and Jean Baudrillard (Chapter 15), generally speaking, dismiss—rather than—advance overarching theoretical frameworks as "essentializing" and misguided. These theorists are probably best viewed not as exemplifying a specific "quadrant" of our model or even bridging quadrants but as rejecting the model altogether. Postmodern theorists are an important exception to our assertion that one of the main goals of contemporary theory is to achieve theoretical synthesis and/or multidimensionality. Throughout Part IV, but especially in Chapter 15, we explore the ideas of these provocative thinkers who challenge some of sociology's central tenets and concerns.

Discussion Questions

1. Explain the difference between "primary" and "secondary" theoretical sources. What are the advantages and disadvantages of reading each type of work?

2. Using Table 1.1 as a reference, devise your own question, and then give hypothetical answers that reflect the four basic theoretical orientations: individual/rational, individual/nonrational, collective/rational, and collective/nonrational. For instance, why do 16-year-olds stay in (or drop out of) school? Why might a man or woman stay in a situation of domestic violence? What are possible explanations for gender inequality? Why are you reading this book?

3. Numerous works of fiction speak to the social conditions that early sociologists were examining. For instance, Charles Dickens's *Hard Times* portrays the hardships of the Industrial Revolution, while Victor Hugo's *Les Miserables* addresses the political and social dynamics of the French Revolution. Read (or watch the play) either of these works, and discuss the tremendous social changes they highlight.

4. One's answers to the questions of order and action have methodological as well as theoretical implications. Theories, after all, should be testable through the use of empirical data. Particularly with regard to the question of order, the perspective one adopts will have important bearing on what counts as evidence and how to collect it. Consider both an individualist and collectivist perspective: How might you design a research project studying the causes and effects of job outsourcing or studying the causes and effects of affirmative action? How about a study of the causes and effects of the rising costs of college tuition or the causes and effects of drug and alcohol abuse? What types of questions or data would be most relevant for each approach? How would you collect the answers to these questions? What are some of the strengths and weaknesses associated with each approach?

Part I

Foundations of Classical Sociological Theory

2 KARL MARX (1818–1883)

Karl Marx

Key Concepts

- Class
- Bourgeoisie
- Proletariat
- Forces and relations of production
- Capital
- Surplus value
- Alienation
- Labor theory of value
- Exploitation
- Class consciousness

The history of all hitherto existing society is the history of class struggles.

—Marx and Engels ([1848] 1978:473)

Have you ever worked at a job that left you feeling empty inside? Perhaps you have worked as a telemarketer, reading a script and selling a product that, in all likelihood, you had never seen or used. Or perhaps you have worked in a fast-food restaurant, or in a large factory or corporation. Sometimes we have jobs that make us feel like we are "just a number," that even though we do our job, we might be easily replaced. This is precisely the type of situation that greatly concerned Karl Marx. Marx sought to explain the nature of the capitalist economies that came to the fore in western Europe in the eighteenth and nineteenth centuries. He maintained that the economic deficiencies and social injustices inherent to capitalism would ultimately lead to the breakdown of capitalist societies. Yet Marx was not an academic writing in an "ivory tower": he was an activist, a revolutionary committed to the overthrow of capitalism. And as you will see shortly, Marx paid a personal price for his revolutionary activities.

Though Marx's prediction that capitalism would be replaced by communism has not come true (some would say, "not yet"), his critique of capitalism continues to resonate with contemporary society.

His discussions regarding the concentration of wealth, the growth of monopoly capitalism, business's unscrupulous pursuit of profit (demonstrated, for instance, by the recent scandals surrounding WorldCom, Enron, Countrywide, Bear Stearns, Merrill Lynch, and Bernard Madoff, to name but a few), the relationship between government economic policy and the interests of the capitalist class, and the alienation experienced in the workplace all speak to concerns that affect almost everyone, even today. Indeed, who has not felt at one time or another that his job was solely a means to an end—a paycheck, money—instead of a forum for fulfilling his aspirations or cultivating his talents? Who has not felt as though she were an expendable "commodity," a means or tool in the production of a good or the provision of a service where even her emotions must be manufactured for the sake of the job? Clearly, Marx's ideas are as relevant today as they were more than a century ago.

A Biographical Sketch ▪▪

Karl Marx was born on May 5, 1818, in Trier, a commercial city in southwestern Germany's Rhineland.[1] Descended from a line of rabbis on both sides of his family, Marx's father, Heinrich, was a secularly educated lawyer. Though Heinrich did not actively practice Judaism, he was subject to anti-Semitism. With France's ceding of the Rhineland to Prussia after the defeat of Napoleon, Jews living in the region were faced with a repeal of the civil rights granted under French rule. In order to keep his legal practice, Heinrich converted to Lutheranism in 1817. As a result, Karl was afforded the comforts of a middle-class home.

Following in his father's footsteps, Marx pursued a secular education. He enrolled as a law student at the University of Bonn in 1835, then transferred the following year to the University of Berlin. In addition to studying law, Marx devoted himself to the study of history and philosophy. While in Berlin, Marx also joined the Young Hegelians, a group of radical thinkers who developed a powerful critique of the philosophy of **Georg W. F. Hegel** (1770–1831), the dominant German intellectual figure of the day and one of the most influential thinkers of the nineteenth century. Marx constructed the basis of his theoretical system, historical materialism, by inverting Hegel's philosophy of social change. (See pp. 32–33 for a brief sketch of Hegel's philosophy and its relation to Marx's theory.)

In 1841, Marx earned a doctorate in philosophy from the University of Jena. However, his ambitions for an academic career ended when the Berlin ministry of education blacklisted him for his radical views.[2] Having established little in the way of career prospects during his student years, Marx accepted an offer to write for the *Rheinische Zeitung,* a liberal newspaper published in Cologne.

Marx soon worked his way up to become editor of the newspaper. Writing on the social conditions in Prussia, Marx criticized the government's treatment of the poor and exposed the harsh conditions of peasants working in the Moselle wine-producing region. However, Marx's condemnation of the authorities brought on the censors, and he was forced to resign his post.

Soon after, Marx married his childhood love, Jenny Von Westphalen, the daughter of a Prussian baron. The two moved to Paris in the fall of 1843. At the time, Paris was the center of European intellectual and political movements. While there, Marx became acquainted with a number of leading socialist writers and revolutionaries. Of particular importance to his intellectual development were the works of the French philosopher **Henri de Saint-Simon** (1760–1825) and his followers. Saint-Simon's ideas led to the creation of Christian Socialism, a movement that sought to organize modern industrial society according to

---

[1]Prussia was a former kingdom in eastern Europe established in 1701 that included present-day Germany and Poland. It was dissolved following World War II.

[2]Marx's mentor and colleague, Bruno Bauer, had promised him a faculty position at the University of Bonn. But when Bauer was dismissed from the university for advocating leftist, antireligious views, Marx was effectively shut off from pursuing an academic career.

the social principles espoused by Christianity. In their efforts to counter the exploitation and egoistic competition that accompany industrial capitalism, Saint-Simonians advocated that industry and commerce be guided according to an ethic of brotherhood and cooperation. By instituting common ownership of society's productive forces and an end to rights of inheritance, they believed that the powers of science and industry could be marshaled to create a more just society free from poverty.

Marx also studied the work of the seminal political economists **Adam Smith** (1723–1790) and **David Ricardo** (1772–1823). Smith's book *An Inquiry into the Nature and Causes of the Wealth of Nations* (1776) represents the first systematic examination of the relationship between government policy and a nation's economic growth. As such, it played a central role in defining the field of political economy. (See p. 24 for summary remarks on Smith's views.) For his part, Ricardo, building on Smith's earlier works, would further refine the study of economics. He wrote on a number of subjects, including the condition of wages, the source of value, taxation, and the production and distribution of goods. A leading economist in his day, Ricardo's writings were influential in shaping England's economic policies. It was from his critique of these writers that Marx would develop his humanist philosophy and economic theories.

During his time in Paris, Marx also began what would become a lifelong collaboration and friendship with **Friedrich Engels**, whom he met while serving as editor of the *Zeitung.* Marx's stay in France was short-lived, however, and again it was his journalism that sparked the ire of government authorities. In January 1845, he was expelled from the country at the request of the Prussian government for his antiroyalist articles. Unable to return to his home country (Prussia), Marx renounced his Prussian citizenship and settled in Brussels, where he lived with his family until 1848. In Brussels, Marx extended his ties to revolutionary working-class movements through associations with members of the League of the Just and the Communist League. Moreover, it was while living in Brussels that Marx and Engels produced two of their most important early works, *The German Ideology* and *The Communist Manifesto* (see selections at the end of this chapter). In 1848, workers and peasants began staging revolts throughout much of Europe. As the revolution spread, Marx and Engels left Brussels and headed for Cologne to serve as coeditors of the radical *Neue Rheinische Zeitung,* a paper devoted to furthering the revolutionary cause. For his part in the protests, Marx was charged with inciting rebellion and defaming the Prussian royal family. Though acquitted, Marx was forced to leave the country. He returned to Paris, but soon was pressured by the French government to leave the country as well, so Marx and his family moved to London in 1849.

In London, Marx turned his attention more fully to the study of economics. Spending some 60 hours per week in the British Museum, Marx produced a number of important works, including *Capital* (see below), considered a masterpiece critique of capitalist economic principles and their human costs. Marx also continued his political activism.

From 1851–1862, he was a regular contributor to the *New York Daily Tribune,* writing on such issues as political upheavals in France, the civil war in the United States, Britain's colonization of India, and the hidden causes of war.[3] In 1864, Marx helped found and direct the *International Working Men's Association,* a socialist movement committed to ending the inequities and alienation or "loss of self" experienced under capitalism. The *International* had branches across the European continent and the United States, and Marx's popular writing and activism gave him an international audience for his ideas.

Yet, the revolutionary workers' movements were floundering. In 1876, the *International* disintegrated, and Marx was barely able to support himself and his wife as they struggled against failing health. Jenny died on December 2, 1881, and Marx himself died on March 14, 1883.

---

[3]A number of articles attributed to Marx were actually written by Engels, whose assistance allowed Marx to continue to collect a wage from the newspaper. Engels, whose father owned textile mills in Germany and England (that he would later inherit), also provided Marx with financial support throughout his years in London. The depth of Engels's devotion even led him to support an out-of-wedlock child fathered by Marx.

The revolutionary spirit that inflamed Marx's work cannot be understood outside the backdrop of the sweeping economic and social changes occurring during this period. By the middle of the nineteenth century, the Industrial Revolution that began in Britain 100 years earlier was spreading throughout western Europe. Technological advances in transportation, communication, and manufacturing spurred an explosion in commercial markets for goods. The result was the birth of modern capitalism and the rise of middle-class owners of capital, or the **bourgeoisie**, to economic and political power. In the wake of these changes came a radical reorganization of both work and domestic life. With the rapid expansion of industry, agricultural work declined, forcing families to move from rural areas to the growing urban centers. It would not take long for the size of the manufacturing labor force to rival and then surpass the numbers working in agriculture.

Nowhere were the disorganizing effects of the Industrial Revolution and the growth of capitalism more readily apparent than in Manchester, England. In the first half of the nineteenth century Manchester's population exploded by 1,000 percent as it rapidly became a major industrial city.[4] The excessive rate of population growth meant that families had to live in makeshift housing without heat or light and in dismal sanitary conditions that fueled the spread of disease. The conditions in the mechanized factories were no better. The factories were poorly ventilated and lit and often dangerous, and factory owners disciplined workers to the monotonous rhythms of mass production. A 70-hour workweek was not uncommon for men and women, and children as young as six often worked as much as 50 hours a week. Yet, the wages earned by laborers left families on the brink of beggary. The appalling living and work standards led Engels to describe Manchester as "Hell upon Earth."

It was in reaction to such dire economic and social conditions that Marx sought to forge a theoretical model intended not only to *interpret* the world, but also to *change* it. In doing so, he centered his analysis on economic **classes**. For Marx, classes are groups of individuals who share a common position in relation to the means or **forces of production**. These refer to the raw materials, technology, machines, factories, and land that are necessary in the production of goods. Each class is distinguished by what it owns with regard to the means of production. Marx argued, "wage labourers,

Photo 2.1 Sordid Factory Conditions: A Young Girl Working as a Spinner in a U.S. Textile Mill, Circa 1910.

Photo 2.2 Sadly, for some factory workers, little has changed over the past century. Here, 16-year-old girls are assembling Keds sneakers at the Kunshan Sun Hwa Footwear Company, in China. The girls apply the toxic glue with their bare hands. At the end of the day, they must line up and leave single file. The factory is surrounded by a 15-foot wall topped with barbed wire.

---

[4]Manchester was also the site of Engels's urban ethnography, *The Condition of the Working Class in England* (1845), and the location of one of his family's textile mills. It was Engels's work that early on helped to crystallize Marx's conception of the proletariat as the revolutionary force in modern industrial society.

capitalists and landowners constitute [the] three big classes of modern society based upon the capitalist mode of production." Thus, under capitalism, there are "the owners merely of labour-power, owners of capital, and landowners, whose respective sources of income are wages, profit, and ground-rent" ([1867] 1978:441).[5]

Private ownership of the means of production leads to class relations based on domination and subordination. While wage earners are free to quit or refuse a particular job, they nevertheless must sell their labor power to someone in the capitalist class in order to live. This is because laborers have only their ability to work to exchange for money that can then be used to purchase the goods necessary for their survival. However, the amount of wages paid is far exceeded by the profits reaped by those who control the productive forces. As a result, classes are pitted against each other in a struggle to control the means of production, the distribution of resources, and the profits.

For Marx, this class struggle is the catalyst for social change and the prime mover of history. This is because any mode of production based on private property (e.g., slavery, feudalism, capitalism) bears the seeds of its own destruction by igniting ongoing economic conflicts that inevitably will sweep away existing social arrangements and give birth to new classes of oppressors and the oppressed. Indeed, as Marx states in one of the most famous passages in *The Communist Manifesto*, "The history of all hitherto existing society is the history of class struggles" (Marx and Engels [1848] 1978:473; see below).

Marx developed his theory in reaction to laissez-faire capitalism, an economic system based on individual competition for markets. It emerged out of the destruction of feudalism, in which peasant agricultural production was based on subsistence standards in the service to lords, and the collapse of merchant and craft guilds, where all aspects of commerce and industry were tightly controlled by monopolistic professional organizations. The basic premise behind this form of capitalism, as outlined by Adam Smith, is that any and all should be free to enter and compete in the marketplace of goods and services. Under the guiding force of the "invisible hand," the best products at the lowest prices will prevail, and a "universal opulence [will] extend itself to the lowest ranks of the people" (A. Smith [1776] 1990:6). Without the interference of regulations that artificially distort supply and demand and disturb the "natural" adjusting of prices, the economy will be controlled by those in the best position to dictate its course of development: consumers and producers. Exchanges between buyers and sellers are rooted not in appeals to the others' "humanity but to their self-love . . . [by showing] them that it is for their own advantage to do for him what he requires of them" (ibid.:8). The potentially destructive drive for selfishly bettering one's lot is checked, however, by a rationally controlled competition for markets that discourages deceptive business practices, because whatever gains a seller can win through illicit means will be nullified as soon as the "market" uncovers them. According to Smith, a "system of perfect liberty" is thus created that both generates greater wealth for all and promotes the general well-being of society.

Marx shared much of Smith's analysis of economics. For instance, both viewed history as unfolding through evolutionary stages in economic organization and understood the central role of governments to be protecting the privilege of the wealthy through upholding the right to private property. Nevertheless, important differences separate the two theories. Most notable is Marx's insistence that, far from establishing a system of perfect liberty, private ownership of the means of production necessarily leads to the alienation of workers. They sell not only their labor power but also their souls. They have no control over the product they are producing, while their work is devoid of any redeeming human qualities. Although capitalism produces self-betterment for owners of capital, it necessarily prevents workers from realizing their essential human capacity to engage in creative labor.

Indeed, in highly mechanized factories, a worker's task might be so mundane and repetitive (e.g., "insert bolt A into widget B") that she seems to become part of the machine itself. For example, a student once said she worked in a job in which she had a scanner attached to her arm. Her job was simply to stand by a conveyer belt in which boxes of various sizes came by. She stuck her arm out and "read" the boxes with her scanner arm. Her individual human potential was completely irrelevant to her job. She was just

---

[5]Marx was not entirely consistent when discussing the number and types of classes that compose capitalist societies. Most often, however, he described such societies as consisting of two antagonistic classes: the bourgeoisie and the proletariat.

a "cog in a wheel" of mechanization. Marx maintained that when human actions are no different from those of a machine, the individual is dehumanized.

Moreover, according to Marx, capitalism is inherently exploitative. It is the labor power of workers that produces the products to be sold by the owners of businesses. Workers mine the raw materials, tend to the machines, and assemble the products. Yet, it is the owner who takes for himself the profits generated by the sale of goods. Meanwhile, workers' wages hover around subsistence levels, allowing them to purchase only the necessities—sold at a profit by capitalists to ensure their return to work the next day. One of Marx's near contemporaries, Thorstein Veblen (1857–1929), an American sociologist and economist, held a similar view on the nature of the relationship between owners and workers. (See the Significant Others box that follows.)

From the point of view of the business owner, capitalism is a "dog-eat-dog" system in which business owners must always watch the "bottom line" in

Photo 2.3   Many of Charlie Chaplin's silent films during the 1920s and 1930s offered a comedic—and quite critical—look at the industrial order. Here, in a scene from *Modern Times* (1936), Chaplin is literally a "cog in a machine."

order to compete for market dominance. Business owners can never rest on their laurels—because someone can always come along and create either a better or newer product, or the same product at a lower price. Thus, a business owner must constantly think strategically and work to improve her product or reduce her costs, or both. Cutting costs can increase a business owner's profit either directly (as she keeps more money for herself) or indirectly (by enabling the business owner to lower the price and sell more of her products).

While competition between capitalists may lead to greater levels of productivity, it also results in a concentration of wealth into fewer and fewer hands. One of the basic truths of capitalism is that it takes money to make money, and the more money a business owner has at his or her disposal, the more ability the business owner has to generate profit-making schemes. For instance, a wealthy capitalist might temporarily underprice a product (i.e., sell it below the cost of its production) in order to force his or her competitors out of business. Once the competition is eliminated and a monopoly is established, the product can be priced as high as the market will bear.

Significant Others

Thorstein Veblen (1857–1929): The Leisure Class and Conspicuous Consumption

While Karl Marx's ideas would remain largely on the periphery of sociology until the 1960s, his ideas, nevertheless, inspired a legion of scholars even before his death. One early student of Marx's theories was Thorstein Veblen. Veblen was born in Wisconsin, the son of Norwegian immigrants. His parents, like so many others of that time and place, were poor tenant farmers who came to America seeking to better their lives. Fortunately, after a number of years of hardship and thrift, the Veblens were able to attain a modest lifestyle working as family farmers. Thorstein's humble upbringing, however, contrasted sharply with the vast fortunes being reaped by America's robber barons, who ruthlessly dominated the nation's budding industrial economy.

(Continued)

(Continued)

Veblen's cognizance of the nation's gross inequities of wealth found expression in his writings, most notably *The Theory of the Leisure Class* (1899) and *The Theory of Business Enterprise* (1904). As a sociologist and economist, Veblen, in his scholarly analyses, did not pretend to value the neutrality often associated with scientific endeavors. Instead, his work presents a highly critical picture of modern capitalism and the well-to-do, the "leisure class," who benefit most from the economic system built on "waste." Though the efficiency of mechanized production is capable of creating a surplus of goods that could in turn provide a decent standard of living for all, Veblen argued that "parasitic" business leaders "sabotaged" the industrial system in their quest for personal profit.

Though Veblen by no means embraced Marxist models of society and social change in their entirety, his work nevertheless contains important parallels with some of Marx's key ideas. For instance, his assertion that the state of a society's technological development forms the foundation for its "schemes of thought" bears a pronounced resemblance to Marx's distinction between the economic base and superstructure. Additionally, Veblen's analysis of the modern-day conflict between "business" (those who make money) and "industry" (those who make "things") recalls Marx's own two-class model of capitalist society and its attendant moral critique of the exploitation of workers and the clash between the forces and relations of production. However, it was his twin notions of "conspicuous consumption" and "conspicuous leisure" that would come to have the greatest impact on sociology.

Veblen here calls our attention to the "waste" of both money and time that individuals of all social classes engage in as a means for improving their self-esteem and elevating their status in the community. Whether it's purchasing expensive cars or clothes when inexpensive brands will suffice, or dedicating oneself to learning the finer points of golf or dining etiquette, such practices signal an underlying competitive attempt to best others and secure one's position in the status order.

The business owners who are unable to compete successfully for a share of the market find themselves joining the swelling ranks of propertyless wage earners: the **proletariat**. This adds to the revolutionary potential of the working-class movement in two ways. First, the proletariat is transformed into an overwhelming majority of the population, making its class interests an irresistible force for change. Second, as Marx points out, the former capitalists bring with them a level of education not possessed by the typical wage laborer. This breeds political consequences as the former members of the bourgeoisie translate their economic resentment into a radicalization of the proletariat by educating the workers with regard to both the nature of capitalist accumulation and the workers' essential role in overthrowing the system of their oppression.

This was precisely the purpose of Marx's political activities: He sought to generate **class consciousness**: an awareness on the part of the working class of its common relationship to the means of production. Marx believed that this awareness was a vital key for sparking a revolution that would create a "dictatorship of the proletariat," transforming it from a wage-earning, propertyless mass into the ruling class. Unlike all previous class-based revolutions, however, this one would be fought in the interests of a vast majority of the population and not for the benefit of a few as the particular class interests of the proletariat had come to represent the universal interests of humanity. The epoch of capitalism was a necessary stage in this evolution—and the last historical period rooted in class conflict (see Figure 2.1). Capitalism, with its unleashing of immense economic productivity, had created the capital and technology needed to sustain a communist society, the final stage of history.

Using the power of the state to abolish private ownership of the means of production, the proletariat would wrest control of society's productive forces from the hands of the bourgeoisie and create

Figure 2.1 Marx's Model of Social Change: The Communist Revolution

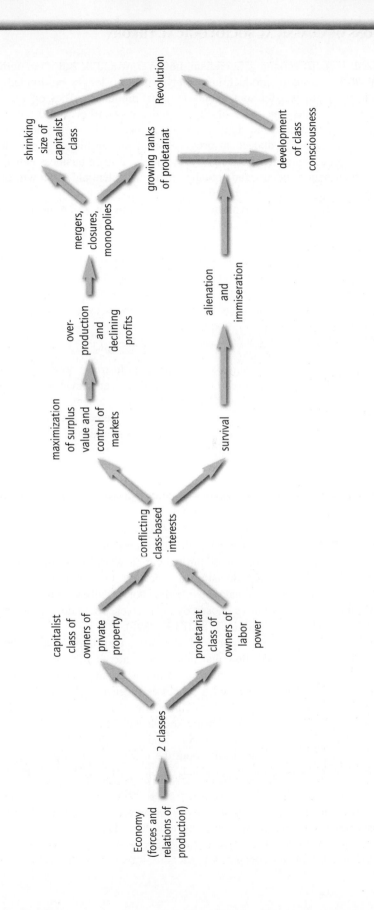

a centralized, socialist economy. Socialism, however, would be but a temporary phase. Without private ownership of the means of production, society would no longer be divided along class lines; without antagonistic class interests, the social conditions that produce conflict, exploitation, and alienation would no longer exist. The disappearance of classes and class conflict would render obsolete the state whose primary charge is to secure the right to private property. Finally, without class conflict—the fuel that ignites social change—the dialectical progression of human history comes to a utopian end. With the production of goods controlled collectively and not by private business elites, individuals would be free to cultivate their natural talents and actualize their full potential.[6] (You will read more about this, in the excerpt from *The Communist Manifesto*.)

As indicated previously, this evolutionary type of thinking was typical of Enlightenment intellectuals. Today, however, many consider Marx's "end of prehistory" vision of communism as the least viable part of his theory. While the internal contradictions of capitalism are real, they have been checked by a number of practices, including ongoing government intervention in the economy, the continued expansion of markets (i.e., Western-dominated globalization), and cost-saving advances in production and organizational technologies.

■■ MARX'S THEORETICAL ORIENTATION

In terms of our metatheoretical framework, Figure 2.2 illustrates how Marx's work is predominantly collectivist and rationalist in orientation. Of course, as discussed previously, the action/order dimensions are intended to serve as heuristic devices. Certainly, there are elements of Marx's theory that do not fit neatly into this particular "box." Nevertheless, Marx pursued themes that, taken as a whole, underscored his vision of a social order shaped by broad historical transitions and classes of actors (collectivist) pitted against one another in a struggle to realize their economic interests (rationalist).

Regarding the question of order, Marx saw human societies as evolving toward an ultimate, utopian end—a process spurred by *class* conflict. It is the struggle to control the forces of production and the distribution of resources and profits they create that leads classes—not individuals—to become the prime movers of history.

Of course, one might counter that it is *individuals* who "make history." Is it not individuals who make up classes, join labor unions, manage factories, merge corporations, and devise industry strategies? Though this is perhaps true on one level, throughout his work Marx emphasized the structural parameters that inhibit and shape individual decisions and actions. On this point Marx stated in one of his most famous passages, although "men make their own history . . . they do not make it just as they please; they do not make it under circumstances chosen by themselves, but under circumstances directly found, given and transmitted from the past" (Marx [1852] 1978:595).

The circumstance of greatest import in this regard is that individuals are born into societies where the **forces** and **relations of production** that make up "material life"—classes and property relations—are already established independent of their will. From this existing economic base is born a "superstructure" or "the social, political, and intellectual life processes in general" (Marx [1859] 1978:4). The superstructure, in short, consists of everything noneconomic in nature such as a society's legal, political, and educational systems, as well as its stock of commonsense knowledge. As a result, an individual's very consciousness—how she views the world, develops aspirations, and defines her interests—is not determined by the individual's own subjectivity. Instead, ideas about the world and one's place in it are structured by, or built into, the *objective* class position an individual occupies. And while there are capitalists and laborers who

---

[6]By no means have modern communist societies—for instance, the former Soviet Union, China, and North Vietnam—resembled the type of free and creative society envisioned by Marx.

Figure 2.2 Marx's Basic Theoretical Orientation

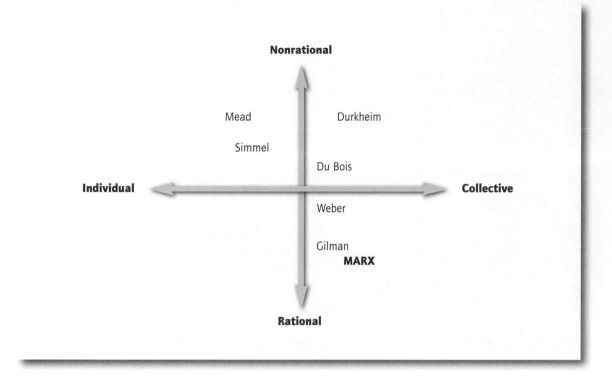

seemingly do not pursue their antagonistic class interests, such exceptions to the rule do not disprove it: "It is not the consciousness of men that determines their being, but, on the contrary, their social being that determines their consciousness" (Marx [1859] 1978:4).

In terms of the motivation for action, Marx's work is primarily rationalist. This tendency is most clearly reflected in his emphasis on class-based *interests.* According to Marx, humans are separated from other animals due to our innate need to realize our full potential through engaging in creative labor. It is through freely developed "conscious life-activity" (Marx [1844] 1978:76) that women and men are able to develop their "true" selves and forge meaningful relationships with others. It is in the process of production and in the objects that result that women and men realize themselves and their significance in a world that they create. (The corruption of the link between labor and self-realization by capitalism is addressed most fully in the selection "Alienated Labour.")

Because self-fulfillment is derived through labor, it is in the individual's interest to control the production process that is so vital to meeting this most basic of human needs. The crucial arena of this struggle is the network of economic relationships: the manufacturing, distribution, and sale of goods, and the selling of one's labor. Even if individuals are unaware of their true class interests, they will still be moved by them. Recall that interests are a reflection of one's objective position in relation to the process of production; they are not spawned by one's subjective disposition. The essential point here is that Marx's model *presupposes* that our actions are driven by our attempts to maximize our interests. (See Figure 2.3.) Of course, whether we are truly as rationalistic as Marx maintains is a point of great theoretical debate.

Figure 2.3 Marx's Core Concepts

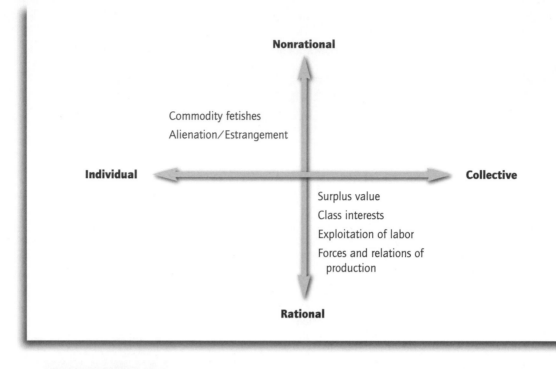

Antonio Gramsci (1891–1937): Hegemony and the Ruling Ideas

Antonio Gramsci was an Italian philosopher, journalist, and political activist who spent much of his adult life ardently supporting the revolutionary cause of the working class. His foray into politics began in earnest in 1915 when he became a member of the Italian Socialist Party (Partito Socialista Italiano; PSI) and published critical essays in the Party's official paper, *L'Avanti*. In 1919, he cofounded the periodical *The New Order: A Weekly Review of Socialist Culture*. Covering political events across Europe, the United States, and the Soviet Union, the paper was widely influential among Italy's radical Left. After a split within the PSI in 1921, Gramsci became a prominent member of the Italian Communist Party (Partito Comunista Italiano; PCI), serving first in the Party's central committee and then as a delegate to the Communist International in Moscow. He would go on to be elected to the Party's Chamber of Deputies and later rise to the position of general secretary.

Gramsci would pay a heavy cost for his political activism. His sympathies with the Bolshevik revolution and its leaders, and his alliance with his country's workers' movements, made him an enemy of Italy's newly formed fascist government. In 1926, Gramsci was arrested for his political activities and was sentenced to 20 years in prison. He would serve only 11 years in prison, however, before dying of a brain hemorrhage in April 1937.

Despite the harsh conditions of his imprisonment and his fragile health, Gramsci produced 29 notebooks—some 3,000 pages—of political and philosophical analysis while serving his sentence. The notebooks were smuggled out, but none was published until several years after the end of World War II. It would be another 20 years before the notebooks were compiled and published in

English, under the title *Prison Notebooks*. The notebooks reveal one of Gramsci's central concerns: to explain why Europe's working-class failed to spearhead a socialist revolution, and how, in Italy and elsewhere, it could act against its own class interests by supporting a fascist regime. In addressing these issues, Gramsci confronted an oft-noted weakness in Marx's historical materialism: the role of ideas in preventing or advancing revolutionary change. Asserting, "the ruling ideas are the ideas of the ruling class," Marx nevertheless portended that the proletariat, with its numbers increasing, would come to recognize its class interests and unite to overthrow the bourgeoisie and the conditions of alienation and exploitation that serve its narrow ambitions for profit. Yet, despite the fact that the material or economic conditions were ripe for a revolutionary movement across much of Europe, no successful challenge to the ruling powers was mounted.

To account for the lack of revolutionary foment on the part of the working class, Gramsci emphasized the role of ideas in establishing "hegemony," or domination, over subaltern (oppressed) classes. For Gramsci, the bourgeoisie maintained its dominance not primarily through force or coercion, but through the willing, "spontaneous" consent of the ruled. This consent was the outgrowth of the proletariat adopting as its own the values, beliefs, and attitudes that serve the interests of the ruling class. In other words, the working class is socialized (particularly through the educational system) into accepting a bourgeois ideology as an unquestioned or commonsense view of the world and their place in it. As a result, the working class aligns itself with the status quo, thus granting legitimacy to social and economic arrangements that perpetuate their own exploitation.

Recognizing that economic crises alone could not spark a socialist revolution, Gramsci was convinced that in order for the proletariat to unmask the real sources of its oppression and generate a unified, popular revolt, it must first develop its own "organic" consciousness, or counter-hegemony. This counter hegemony would articulate the real interests and needs of the masses. Moreover, he insisted that this counter ideology must originate from within the masses; to be effective in provoking revolutionary change, it cannot be imposed on them by bourgeois "traditional" intellectuals who remain detached from the everyday realities of working-class life. Declaring, "all men are intellectuals," Gramsci sought to encourage the development of "organic" intellectuals from within the ranks of the working class through his political journalism and active participation in the workers' movement. Such individuals are intellectuals not in the sense of their profession or social function, but in terms of their "directing the ideas and aspirations of the class to which they organically belong" (Gramsci 1971:3). In this way, the factory worker and truck driver, the financial accountant and government bureaucrat, are all potential intellectuals. Indeed, the intellectuals most capable of contributing to progressive social change were not those of the "traditional" or professional type—writers, artists, scientists, philosophers—but rather those who engage in "praxis," connecting theoretical insights to an active attempt to fashion a more just society. For Gramsci, this was the "new intellectual" drawn from the working class:

> In the modern world, technical education, closely bound to industrial labour even at the most primitive and unqualified level, must form the basis of the new type of intellectual. The mode of being of the new intellectual can no longer consist in eloquence, which is an exterior and momentary mover of feelings and passions, but in active participation in practical life, as constructor, organiser, "permanent persuader" and not just a simple orator . . . One of the most important characteristics of any group that is developing towards dominance is its struggle to assimilate and to conquer "ideologically" the traditional intellectuals, but this assimilation and conquest is made quicker and more efficacious the more the group in question succeeds in simultaneously elaborating its own organic intellectuals. (Gramsci 1971:10)

Readings

Marx's writings included here are divided into four sections. The first section centers on his "materialist conception of history," developed in reaction to the works of the German idealist philosopher Georg W. F. Hegel (1770–1831). The second section offers his critique of the human costs of capitalism. The third section contains Marx's call for the inevitable communist revolution that will usher in the "end of prehistory" and, with it, the end of alienation, private property, and oppressive government. In the fourth set of readings, we move from Marx's prophecy of emancipation to his theory of economics. Here you will read his analyses of the sources of value and the nature of commodities.

Introduction to *The German Ideology*

Written in 1845–1846, *The German Ideology* presents the most detailed account of Marx's theory of history. In it, Marx set out to reformulate the work of the eminent German philosopher Georg W. F. Hegel. In contrast to previous philosophers who focused on explaining the roots of stability in the physical and social worlds (i.e., why things seemingly stayed the same), Hegel saw change as the motor of history. For Hegel, change was driven by a dialectical process in which a given state of being or idea contains within it the seeds of an opposing state of being or opposing idea. The resolution of the conflict produces yet a new state of being or idea. This synthesis, in turn, forms the basis of a new contradiction, thus continuing the process of change.

As an example, consider the division of gender roles. Traditionally, women and the roles they perform have been devalued relative to the positions occupied by men. Thus, the notion that a "woman's place is in the home" serves as a justification for male dominance in economic and political affairs. Out of the ideas that sustained the oppression of women was born the opposing view that women are in fact superior to men. From this vantage, it is argued, for instance, that women's "innate" compassion or empathy better qualifies them for positions of leadership compared to the "innate" aggression said to characterize men. From the clash of these opposing ideas, a synthesis or state of being has evolved in which neither women nor men are considered superior to the other, but instead are viewed as equals. Thus, women have entered into roles formerly reserved for men, while men have begun to perform more traditionally "feminine" tasks.

Is society thus faced with a never-ending challenging of ideas as one "truth" replaces another in the evolution of history? Hegel's answer is a definitive "no." He expressed, instead, a belief in the ultimate perfectibility of the consciousness of humankind. Such perfection occurs through the progressive realization of "Absolute Idea" as revealed by God. In other words, every idea (thesis) is a distorted expression of an all-embracing "Spirit" or "Mind" (God) that produces an opposite idea (antithesis). The two contradictory ideas are unified to form a synthesis that in turn becomes the basis for a new idea (thesis). Progress and history itself come to an end as the contradictions between our ideas about reality and the "Truth" of reality as designed by God are finally resolved. In arguing that the evolution of human history proceeds purposively according to an immanent or predestined design, Hegel offers a teleological vision shared with both Christian theology and Enlightenment philosophy. (As you will read below, Marx, too, fashioned a teleological theory, but one that casts communism as the end toward which history progresses.)

If this seems abstract, it is because it is! Perhaps we can clarify Hegel's dialectic idealism a bit further. The essence of reality lies in thought or ideas because it is only in and through the concepts that order our experiences that experiences, as such, are known. Reality is a product of our conceptual categories or consciousness and thus has no existence independent of our own construction of it. As our ideas or knowledge changes, so does reality. The stages of history or reality are then defined by progressive stages in the negation of the prevailing conceptual ordering of experience. The utopian aspect of this development is found in the assertion that humankind's knowledge will reach the perfected state of "Pure Reason" or "Absolute Idea" in which freedom takes the form of self-knowledge.

In contending that history is marked by a distortion of "Truth" or "Pure Reason," it follows that our consciousness is alienated from Spirit (God). The condition of alienation thus stems from a religiously grounded misunderstanding of reality. At its core, this misunderstanding comes from the failure to recognize that man and Spirit are one. Instead, man exists as an "unhappy soul," placing in God all that is good and righteous, while seeing in himself only that which is base and sinful. God becomes an alien, all-knowing, powerful force separated from ignorant, powerless man. Yet, as consciousness evolves through the historical dialectic, it advances closer to utopia in the form of an absolute self-knowledge that recognizes that reality is a product of the human spirit and not an alien force. No longer plagued by the irrationality that comes from a distorted view of the essence of mankind, man, in unity with Mind, can order the world in a rational way.[1]

The German Ideology reflects both Marx's indebtedness to and break from Hegel's philosophy. On one hand, akin to Hegel, Marx depicts the unfolding of history as a progressive, dialectical process that culminates in a utopia of freedom and self-realization. In other words, like Hegel, Marx argues that each successive period in societal evolution is a necessary consequence of the preceding stage; and Marx projects a millennial significance onto the process itself, claiming that social development ends in a "necessary" utopia free of conflict and exploitation.

However, Marx breaks decisively from Hegel by insisting that it is *material* existence—not consciousness—that fuels historical change. Thus, Marx sought to take Hegel's idealism, which had the evolution of history "standing on its head," and "turn it right side up" in order to discover the real basis of the progression of human societies. Theoretically, this inversion is of utmost significance because it reflects a shift from a nonrationalist to a rationalist theoretical orientation.

The German Ideology is a pivotal writing because it offers the fullest treatment of Marx's materialist conception of history. It is in Marx's theory of historical materialism that we find one of his most important philosophical contributions, namely his conviction that ideas or interests have no existence independent of physical reality. In numerous passages, you will see Marx's rejection of Hegel's notion that ideas determine experience in favor of the materialist view that experience determines ideas. For instance, Marx asserts, "Consciousness can never be anything else than conscious existence, and the existence of men is their actual life-process" (Marx and Engels [1846] 1978:154). And again, "Life is not determined by consciousness, but consciousness by life" (ibid.:155). In short, Marx argues that the essence of individuals, what they truly are and how they see the world, is determined by their material, economic conditions—"both with *what* they produce and with *how* they produce"—in which they live out their very existence (ibid.:150; emphasis in original).

Moreover, to argue that experience determines consciousness yields a radical conclusion: "The ideas of the ruling class are in every epoch the ruling ideas" (Marx and Engels [1846] 1978:172). In other words, Marx maintains that the dominant economic class controls not only a society's means of material production, but the production of ideas as well. To illustrate this point, consider, for instance, the *idea* of individual equality. From where did it spring? The notion of equality is by no means universal. Not only do some contemporary societies reject the concept of equality, but even those societies that do guarantee such rights (the United States, France, England, to name but a few) have not always done so. How are we then to account for the development of this principle? The answer, in short, lies in the development of capitalism. As an economic system, capitalism is based on the notion of "freedom"—workers are "free" to find work or to quit their job. Entrepreneurs are "free" to open or close their businesses. In order for

---

[1]Hegel's notion of alienation would play a central role in Marx's work. Marx, however, argued that alienation was not a consequence of distorted consciousness but, rather, that it resulted from the material conditions of production. Marx takes up this issue in his essay "Alienated Labour," the next excerpt.

competitive capitalism to develop to its fullest productive capacities, individuals must be able to move, work, and invest their capital freely. This ability is expressed through the idea of individual equality. Thus, the concept of equality is born out of the capitalist mode of production and the nature of the social relationships it demands. It is an idea advanced by the bourgeoisie to sanction individualism that, in turn, justifies and sustains the economic conditions in which they themselves are the dominant force. In short, it serves the economic and political interests of the ruling class.

The German Ideology (1845–1846)

Karl Marx and Friedrich Engels

The premises from which we begin are not arbitrary ones, not dogmas, but real premises from which abstraction can only be made in the imagination. They are the real individuals, their activity and the material conditions under which they live, both those which they find already existing and those produced by their activity. The premises can thus be verified in a purely empirical way.

The first premise of all human history is, of course, the existence of living human individuals. Thus the first fact to be established is the physical organisation of these individuals and their consequent relation to the rest of nature. Of course, we cannot here go either into the actual physical nature of man, or into the natural conditions in which man finds himself—geological, orohydrographical, climatic and so on. The writing of history must always set out from these natural bases and their modification in the course of history through the action of men.

Men can be distinguished from animals by consciousness, by religion or anything else you like. They themselves begin to distinguish themselves from animals as soon as they begin to produce their means of subsistence, a step which is conditioned by their physical organisation. By producing their means of subsistence men are indirectly producing their actual material life.

The way in which men produce their means of subsistence depends first of all on the nature of the actual means of subsistence they find in existence and have to reproduce. This mode of production must not be considered simply as being the reproduction of the physical existence of the individuals. Rather it is a definite form of activity of these individuals, a definite form of expressing their life, a definite *mode of life* on their part. As individuals express their life, so they are. What they

are, therefore, coincides with their production, both with *what* they produce and with *how* they produce. The nature of individuals thus depends on the material conditions determining their production. . . .

The relations of different nations among themselves depend upon the extent to which each has developed its productive forces, the division of labour and internal intercourse. This statement is generally recognised. But not only the relation of one nation to others, but also the whole internal structure of the nation itself depends on the stage of development reached by its production and its internal and external intercourse. How far the productive forces of a nation are developed is shown most manifestly by the degree to which the division of labour has been carried. Each new productive force, insofar as it is not merely a quantitative of productive forces already known (for instance the bringing into cultivation of fresh land), causes a further development of the division of labour.

The division of labour inside a nation leads at first to the separation of industrial and commercial from agricultural labour, and hence to the separation of *town* and *country* and to the conflict of their interests. Its further development leads to the separation of commercial from industrial labour. At the same time through the division of labour inside these various branches there develop various divisions among the individuals co-operating in definite kinds of labour. The relative position of these individual groups is determined by the methods employed in agriculture, industry and commerce (patriarchalism, slavery, estates, classes). These same conditions are to be seen (given a more developed intercourse) in the relations of different nations to one another.

The various stages of development in the division of labour are just so many different forms of ownership, i.e., the existing stage in the division of labour determines also the relations of individuals to one another with reference to the material, instrument, and product of labour.

The first form of ownership is tribal [*Stammeigentum*] ownership. It corresponds to the undeveloped stage of production, at which a people lives by hunting and fishing, by the rearing of beasts or, in the highest stage, agriculture. In the latter case it pre-supposes a great mass of uncultivated stretches of land. The division of labour is at this stage still very elementary and is confined to a further extension of the natural division of labour existing in the family. The social structure is, therefore, limited to an extension of the family; patriarchal family chieftains, below them the members of the tribe, finally slaves. The slavery latent in the family only develops gradually with the increase of population, the growth of wants, and with the extension of external relations, both of war and of barter.

The second form is the ancient communal and State ownership which proceeds especially from the union of several tribes into a city by agreement or by conquest, and which is still accompanied by slavery. Beside communal ownership we already find movable, and later also immovable, private property developing, but as an abnormal form subordinate to communal ownership. The citizens hold power over their labouring slaves only in their community, and on this account alone, therefore, they are bound to the form of communal ownership. It is the communal private property which compels the active citizens to remain in this spontaneously derived form of association over against their slaves. For this reason the whole structure of society based on this communal ownership, and with it the power of the people, decays in the same measure as, in particular, immovable private property evolves. The division of labour is already more developed. We already find the antagonism of town and country; later the antagonism between those states which represent town interests and those which represent country interests, and inside the towns themselves the antagonism between industry and maritime commerce. The class relation between citizens and slaves is now completely developed. . . .

The third form of ownership is feudal or estate property. If antiquity started out from the town and its little territory, the Middle Ages started out from the *country.* This different starting-point was determined by the sparseness of the population at that time, which was scattered over a large area and which received no large increase from the conquerors. In contrast to Greece and Rome, feudal development at the outset, therefore, extends over a much wider territory, prepared by the Roman conquests and the spread of agriculture at first associated with them. The last centuries of the declining Roman Empire and its conquest by the barbarians destroyed a number of productive forces; agriculture had declined, industry had decayed for want of a market, trade had died out or been violently suspended, the rural and urban population had decreased. From these conditions and the mode of organisation of the conquest determined by them, feudal property developed under the influence of the Germanic military constitution. Like tribal and communal ownership, it is based again on a community; but the directly producing class standing over against it is not, as in the case of the ancient community, the slaves, but the enserfed small peasantry. As soon as feudalism is fully developed, there also arises antagonism to the towns. The hierarchical structure of landownership, and the armed bodies of retainers associated with it, gave the nobility power over the serfs. This feudal organisation was, just as much as the ancient communal ownership, an association against a subjected producing class; but the form of association and the relation to the direct producers were different because of the different conditions of production.

This feudal system of landownership had its counterpart in the *towns* in the shape of corporative property, the feudal organisation of trades. Here property consisted chiefly in the labour of each individual person. The necessity for association against the organised robber nobility, the need for communal covered markets in an age when the industrialist was at the same time a merchant, the growing competition of the escaped serfs swarming into the rising towns, the feudal structure of the whole country: these combined to bring about the *guilds.* The gradually accumulated small capital of individual craftsmen and their stable numbers, as against the growing population, evolved the relation of journeyman and apprentice, which brought into being in the towns a hierarchy similar to that in the country.

Thus the chief form of property during the feudal epoch consisted on the one hand of landed property with serf labour chained to it, and on the other of the labour of the individual with small capital commanding the labour of journeymen. The organisation of both was determined by the restricted conditions of production— the small-scale and primitive cultivation of the land, and the craft type of industry. There was little division of labour in the heyday of feudalism. Each country bore in itself the antithesis of town and country; the division into estates was certainly strongly marked; but apart from the differentiation of princes, nobility, clergy and peasants in the country, and masters, journeymen, apprentices and soon also the rabble of casual labourers in the towns, no division of importance took place. In agriculture it was rendered difficult by the strip-system, beside which the cottage industry of the peasants themselves emerged. In industry there was no division of labour at all in the individual trades themselves, and very little between them. The separation of industry and commerce was found already in existence in older towns; in the newer it only developed later, when the towns entered into mutual relations. . . .

The fact is, therefore, that definite individuals who are productively active in a definite way enter into these definite social and political relations. Empirical observation must in each separate instance bring out empirically, and without any mystification and speculation, the connection of the social and political structure with production. The social structure and the State are continually evolving out of the life process of definite individuals, but of individuals, not as they may appear in their own or other people's imagination, but as they *really* are; i.e., as they operate, produce materially, and hence as they work under definite material limits, presuppositions and conditions independent of their will.

The production of ideas, of conceptions, of consciousness, is at first directly interwoven with the material activity and the material intercourse of men, the language of real life. Conceiving, thinking, the mental intercourse of men, appear at this stage as the direct efflux of their material behaviour. The same applies to mental production as expressed in the language of politics, laws, morality, religion, metaphysics, etc., of a people. Men are the producers of their conceptions, ideas, etc.—real, active men, as they are conditioned by a definite development of their productive forces and of the intercourse corresponding to these, up to its furthest forms. Consciousness can never be anything else than conscious existence, and the existence of men is their actual life-process. If in all ideology men and their circumstances appear upside-down as in a *camera obscura,* this phenomenon arises just as much from their historical life-process as the inversion of objects on the retina does from their physical life-process.

In direct contrast to German philosophy which descends from heaven to earth, here we ascend from earth to heaven. That is to say, we do not set out from what men say, imagine, conceive, nor from men as narrated, thought of, imagined, conceived, in order to arrive at men in the flesh. We set out from real, active men, and on the basis of their real life-process we demonstrate the development of the ideological reflexes and echoes of this life-process. The phantoms formed in the human brain are also, necessarily, sublimates of their material life-process, which is empirically verifiable and bound to material premises. Morality, religion, metaphysics, all the rest of ideology and their corresponding forms of consciousness, thus no longer retain the semblance of independence. They have no history, no development; but men, developing their material production and their material intercourse, alter, along with this their real existence, their thinking and the products of their thinking. Life is not determined by consciousness, but consciousness by life. In the first method of approach the starting-point is consciousness taken as the living individual; in the second method, which conforms to real life, it is the real living individuals themselves, and consciousness is considered solely as *their* consciousness. . . .

The production of life, both of one's own in labour and of fresh life in procreation, now appears as a double relationship: on the one hand as a natural, on the other as a social relationship. By social we understand the co-operation of several individuals, no matter under what conditions, in what manner and to what end. It follows from this that a certain mode of production, or industrial stage, is always combined with a certain mode of co-operation, or social stage, and this mode of co-operation is itself a "productive force." Further, that the multitude of productive forces accessible to men determines the nature of society, hence, that the "history of humanity" must always be studied and treated in relation to the history of industry and exchange. . . . Thus it is quite obvious from the start that there exists a materialistic connection of men with one another, which is determined by their needs and their mode of production, and which is as old as men themselves. This connection is ever taking on new forms, and thus presents a "history" independently of the existence of any political or religious nonsense which would especially hold men together.

Only now, after having considered four moments, four aspects of the primary historical relationships, do we find that man also possesses "consciousness";[i] but,

---

[i]Marginal note by Marx: "Men have history because they must produce their life, and because they must produce it moreover in a certain way: this is determined by their physical organisation: their consciousness is determined in just the same way."

even so, not inherent, not "pure" consciousness. From the start the "spirit" is afflicted with the curse of being "burdened" with matter, which here makes its appearance in the form of agitated layers of air, sounds, in short, of language. Language is as old as consciousness, language is practical consciousness that exists also for other men, and for that reason alone it really exists for me personally as well; language, like consciousness, only arises from the need, the necessity, of intercourse with other men. Where there exists a relationship, it exists for me: the animal does not enter into *"relations"* with anything, it does not enter into any relation at all. For the animal, its relation to others does not exist as a relation. Consciousness is, therefore, from the very beginning a social product, and remains so as long as men exist at all. Consciousness is at first, of course, merely consciousness concerning the *immediate* sensuous environment and consciousness of the limited connection with other persons and things outside the individual who is growing self-conscious. At the same time it is consciousness of nature, which first appears to men as a completely alien, all-powerful and unassailable force, with which men's relations are purely animal and by which they are overawed like beasts; it is thus a purely animal consciousness of nature (natural religion).

We see here immediately: this natural religion or this particular relation of men to nature is determined by the form of society and vice versa. Here, as everywhere, the identity of nature and man appears in such a way that the restricted relation of men to nature determines their restricted relation to one another, and their restricted relation to one another determines men's restricted relation to nature, just because nature is as yet hardly modified historically; and, on the other hand, man's consciousness of the necessity of associating with the individuals around him is the beginning of the consciousness that he is living in society at all. This beginning is as animal as social life itself at this stage. It is mere herd-consciousness, and at this point man is only distinguished from sheep by the fact that with him consciousness takes the place of instinct or that his instinct is a conscious one. This sheep-like or tribal consciousness receives its further development and extension through increased productivity, the increase of needs, and, what is fundamental to both of these, the increase of population. With these there develops the division of labour, which was originally nothing but the division of labour in the sexual act, then that division of labour which develops

spontaneously or "naturally" by virtue of natural predisposition (e.g., physical strength), needs, accidents, etc., etc. Division of labour only becomes truly such from the moment when a division of material and mental labour appears.[ii] From this moment onwards consciousness *can* really flatter itself that it is something other than consciousness of existing practice, that it *really* represents something without representing something real; from now on consciousness is in a position to emancipate itself from the world and to proceed to the formation of "pure" theory, theology, philosophy, ethics, etc. But even if this theory, theology, philosophy, ethics, etc., comes into contradiction with the existing relations, this can only occur because existing social relations have come into contradiction with existing forces of production. . . .

With the division of labour, in which all these contradictions are implicit, and which in its turn is based on the natural division of labour in the family and the separation of society into individual families opposed to one another, is given simultaneously the *distribution,* and indeed the *unequal* distribution, both quantitative and qualitative, of labour and its products, hence property: the nucleus, the first form, of which lies in the family, where wife and children are the slaves of the husband. This latent slavery in the family, though still very crude, is the first property, but even at this early stage it corresponds perfectly to the definition of modern economists who call it the power of disposing of the labour-power of others. Division of labour and private property are, moreover, identical expressions: in the one the same thing is affirmed with reference to activity as is affirmed in the other with reference to the product of the activity.

Further, the division of labour implies the contradiction between the interest of the separate individual or the individual family and the communal interest of all individuals who have intercourse with one another. And indeed, this communal interest does not exist merely in the imagination, as the "general interest," but first of all in reality, as the mutual interdependence of the individuals among whom the labour is divided. And finally, the division of labour offers us the first example of how, as long as man remains in natural society, that is, as long as a cleavage exists between the particular and the common interest, as long, therefore, as activity is not voluntarily, but naturally, divided, man's own deed becomes an alien power opposed to him, which enslaves him instead of being controlled by him. For as soon as the distribution of labour comes into being, each man

---

[ii]Marginal note by Marx: "The first form of ideologists, priests, is concurrent."

has a particular, exclusive sphere of activity, which is forced upon him and from which he cannot escape. He is a hunter, a fisherman, a shepherd, or a critical critic, and must remain so if he does not want to lose his means of livelihood: while in communist society, where nobody has one exclusive sphere of activity but each can become accomplished in any branch he wishes, society regulates the general production and thus makes it possible for me to do one thing today and another tomorrow, to hunt in the morning, fish in the afternoon, rear cattle in the evening, criticise after dinner, just as I have a mind, without ever becoming hunter, fisherman, shepherd or critic. This fixation of social activity, this consolidation of what we ourselves produce into an objective power above us, growing out of our control, thwarting our expectations, bringing to naught our calculations, is one of the chief factors in historical development up till now.

And out of this very contradiction between the interest of the individual and that of the community the latter takes an independent form as the *State,* divorced from the real interests of individual and community, and at the same time as an illusory communal life, always based, however, on the real ties existing in every family and tribal conglomeration—such as flesh and blood, language, division of labour on a larger scale, and other interests—and especially, as we shall enlarge upon later, on the classes, already determined by the division of labour, which in every such mass of men separate out, and of which one dominates all the others. It follows from this that all struggles within the State, the struggle between democracy, aristocracy, and monarchy, the struggle for the franchise, etc., etc., are merely the illusory forms in which the real struggles of the different classes are fought out among one another. . . . Further, it follows that every class which is struggling for mastery, even when its domination, as is the case with the proletariat, postulates the abolition of the old form of society in its entirety and of domination itself, must first conquer for itself political power in order to represent its interest in turn as the general interest, which in the first moment it is forced to do. Just because individuals seek only their particular interest, which for them does not coincide with their communal interest (in fact the general is the illusory form of communal life), the latter will be imposed on them as an interest "alien" to them, and "independent" of them, as in its turn a particular, peculiar "general" interest; or they themselves must remain within this discord, as in democracy. On the other hand, too, the *practical* struggle of these particular interests, which constantly *really* run counter to the communal and illusory communal interests, makes *practical* intervention and control necessary through the illusory "general" interest in the form of the State. The social power, i.e., the multiplied productive force, which arises through the co-operation of different individuals as it is determined by the division of labour, appears to these individuals, since their co-operation is not voluntary but has come about naturally, not as their own united power, but as an alien force existing outside them, of the origin and goal of which they are ignorant, which they thus cannot control, which on the contrary passes through a peculiar series of phases and stages independent of the will and the action of man, nay even being the prime governor of these.

This *"estrangement"* (to use a term which will be comprehensible to the philosophers) can, of course, only be abolished given two *practical* premises. For it to become an "intolerable" power, i.e., a power against which men make a revolution, it must necessarily have rendered the great mass of humanity "propertyless," and produced, at the same time, the contradiction of an existing world of wealth and culture, both of which conditions presuppose a great increase in productive power, a high degree of its development. And, on the other hand, this development of productive forces (which itself implies the actual empirical existence of men in their *world-historical,* instead of local, being) is an absolutely necessary practical premise because without it *want* is merely made general, and with *destitution* the struggle for necessities and all the old filthy business would necessarily be reproduced; and furthermore, because only with this universal development of productive forces is a universal intercourse between men established, which produces in all nations simultaneously the phenomenon of the "propertyless" mass (universal competition), makes each nation dependent on the revolutions of the others, and finally has put *world-historical,* empirically universal individuals in place of local ones. Without this, (1) communism could only exist as a local event; (2) the *forces* of intercourse themselves could not have developed as *universal,* hence intolerable powers: they would have remained home-bred conditions surrounded by superstition; and (3) each extension of intercourse would abolish local communism. Empirically, communism is only possible as the act of the dominant peoples "all at once" and simultaneously, which presupposes the universal development of productive forces and the world intercourse bound up with communism. How otherwise could for instance property have had a history at all, have taken on different forms, and landed property, for example, according to the different premises given, have proceeded in France from parcellation to centralisation in the hands of a few, in England from centralisation in the hands of a few to parcellation, as is actually the case today? Or how does it happen that trade, which after all is nothing more than the exchange of products of various individuals and countries, rules the whole world through the relation of supply and demand—a relation which, as an English economist says, hovers over the earth like the fate of the ancients, and with invisible hand allots fortune and misfortune to men, sets up empires and overthrows empires, causes nations to rise and to disappear—while with the abolition of the basis of private property, with the communistic regulation of production (and, implicit in this, the destruction of the alien relation between men and what they themselves produce), the power of the relation of supply and demand is dissolved into nothing, and men get exchange, production, the mode of their mutual relation, under their own control again?

Communism is for us not a *state of affairs* which is to be established, an *ideal* to which reality [will] have to adjust itself. We call communism the *real* movement which abolishes the present state of things. The conditions of this movement result from the premises now in existence. Moreover, the mass of *propertyless* workers— the utterly precarious position of labour-power on a mass scale cut off from capital or from even a limited satisfaction and, therefore, no longer merely temporarily deprived of work itself as a secure source of life— presupposes the *world market* through competition. The proletariat can thus only exist *world-historically,* just as communism, its activity, can only have a "world-historical" existence. World-historical existence of individuals, i.e., existence of individuals which is directly linked up with world history.

The form of intercourse determined by the existing productive forces at all previous historical stages, and in its turn determining these, is *civil society.* The latter, as is clear from what we have said above, has as its premises and basis the simple family and the multiple, the so-called tribe, and the more precise determinants of this society are enumerated in our remarks above. Already here we see how this civil society is the true source and theatre of all history, and how absurd is the conception of history held hitherto, which neglects the real relationships and confines itself to high-sounding dramas of princes and states.

Civil society embraces the whole material intercourse of individuals within a definite stage of the development of productive forces. It embraces the whole commercial and industrial life of a given stage and, insofar, transcends the State and the nation, though, on the other hand again, it must assert itself in its foreign relations as nationality, and inwardly must organise itself as State. The term "civil society" [*bürgerliche Gesellschaft*][iii] emerged in the eighteenth century, when property relationships had already extricated themselves from the ancient and medieval communal society. Civil society as such only develops with the bourgeoisie; the social organisation evolving directly out of production and commerce, which in all ages forms the basis of the State and of the rest of the idealistic superstructure, has, however, always been designated by the same name. . . .

This conception of history depends on our ability to expound the real process of production, starting out from the material production of itself, and to comprehend the form of intercourse connected with this and created by this mode of production (i.e., civil society in its various stages), as the basis of all history; and to show it in its action as State, to explain all the different theoretical products and forms of consciousness, religion, philosophy, ethics, etc., etc., and trace their origins and growth from that basis; by which means, of course, the whole thing can be depicted in its totality (and therefore, too, the reciprocal action of these various sides on one another). It has not, like the idealistic view of history, in every period to look for a category, but remains constantly on the real *ground* of history; it does not explain practice from the idea but explains the formation of ideas from material practice; and accordingly it comes to the conclusion that all forms and products of consciousness cannot be dissolved by mental criticism, by resolution into "self-consciousness" or transformation into "apparitions," "spectres," "fancies," etc., but only by the practical overthrow of the actual social relations which gave rise to this idealistic humbug; that not criticism but revolution is the driving force of history, also of religion, of philosophy and all other types of theory. It shows that history does not end by being resolved into "self-consciousness" as "spirit of the spirit," but that in it at each stage there is found a

---

iiiBurgerliche Gesellschaft can mean either "bourgeois society" or "civil society."

material result: a sum of productive forces, a historically created relation of individuals to nature and to one another, which is handed down to each generation from its predecessor; a mass of productive forces, capital funds and conditions, which on the one hand, is indeed modified by the new generation, but also on the other prescribes for it its conditions of life and gives it a definite development, a special character. It shows that circumstances make men just as much as men make circumstances. This sum of productive forces, capital funds and social forms of intercourse, which every individual and generation finds in existence as something given, is the real basis of what the philosophers have conceived as "substance" and "essence of man," and what they have deified and attacked: a real basis which is not in the least disturbed, in its effect and influence on the development of men, by the fact that these philosophers revolt against it as "self-consciousness" and the "Unique." These conditions of life, which different generations find in existence, decide also whether or not the periodically recurring revolutionary convulsion will be strong enough to overthrow the basis of the entire existing system. And if these material elements of a complete revolution are not present (namely, on the one hand the existing productive forces, on the other the formation of a revolutionary mass, which revolts not only against separate conditions of society up till then, but against the very "production of life" till then, the "total activity" on which it was based), then, as far as practical development is concerned, it is absolutely immaterial whether the idea of this revolution has been expressed a hundred times already, as the history of communism proves. . . .

The ideas of the ruling class are in every epoch the ruling ideas: i.e., the class which is the ruling *material* force of society, is at the same time its ruling *intellectual* force. The class which has the means of material production at its disposal, has control at the same time over the means of mental production, so that thereby, generally speaking, the ideas of those who lack the means of mental production are subject to it. The ruling ideas are nothing more than the ideal expression of the dominant material relationships, the dominant material relationships grasped as ideas; hence of the relationships which make the one class the ruling one, therefore, the ideas of its dominance. The individuals composing the ruling class possess among other things consciousness, and therefore think. Insofar, therefore, as they rule as a class and determine the extent and compass of an epoch, it is self-evident that they do this

in its whole range, hence among other things rule also as thinkers, as producers of ideas, and regulate the production and distribution of the ideas of their age: thus their ideas are the ruling ideas of the epoch. For instance, in an age and in a country where royal power, aristocracy and bourgeoisie are contending for mastery and where, therefore, mastery is shared, the doctrine of the separation of powers proves to be the dominant idea and is expressed as an "eternal law."

The division of labour, which we have already seen above as one of the chief forces of history up till now, manifests itself also in the ruling class as the division of mental and material labour, so that inside this class one part appears as the thinkers of the class (its active, conceptive ideologists, who make the perfecting of the illusion of the class about itself their chief source of livelihood), while the others' attitude to these ideas and illusions is more passive and receptive because they are in reality the active members of this class and have less time to make up illusions and ideas about themselves. Within this class this cleavage can even develop into a certain opposition and hostility between the two parts, which, however, in the case of a practical collision, in which the class itself is endangered, automatically comes to nothing, in which case there also vanishes the semblance that the ruling ideas were not the ideas of the ruling class and had a power distinct from the power of this class. The existence of revolutionary ideas in a particular period presupposes the existence of a revolutionary class; about the premises for the latter sufficient has already been said above.

If now in considering the course of history we detach the ideas of the ruling class from the ruling class itself and attribute to them an independent existence, if we confine ourselves to saying that these or those ideas were dominant at a given time, without bothering ourselves about the conditions of production and the producers of these ideas, if we thus ignore the individuals and world conditions which are the source of the ideas, we can say, for instance, that during the time that the aristocracy was dominant, the concepts honour, loyalty, etc., were dominant, during the dominance of the bourgeoisie the concepts freedom, equality, etc. The ruling class itself on the whole imagines this to be so. This conception of history, which is common to all historians, particularly since the eighteenth century, will necessarily come up against the phenomenon that increasingly abstract ideas hold sway, i.e., ideas which increasingly take on the form of universality. For each new class which puts itself in the place of one ruling before it, is compelled, merely in order to carry through its aim, to represent its interest as the common interest of all the members of society, that is, expressed in ideal form: it has to give its ideas the form of universality, and represent them as the only rational, universally valid ones. The class making a revolution appears from the very start, if only because it is opposed to a *class,* not as a class but as the representative of the whole of society; it appears as the whole mass of society confronting the one ruling class.[iv] It can do this because, to start with, its interest really is more connected with the common interest of all other non-ruling classes, because under the pressure of hitherto existing conditions its interest has not yet been able to develop as the particular interest of a particular class. Its victory, therefore, benefits also many individuals of the other classes which are not winning a dominant position, but only insofar as it now puts these individuals in a position to raise themselves into the ruling class. When the French bourgeoisie overthrew the power of the aristocracy, it thereby made it possible for many proletarians to raise themselves above the proletariat, but only insofar as they became bourgeois. Every new class, therefore, achieves its hegemony only on a broader basis than that of the class ruling previously, whereas the opposition of the non-ruling class against the new ruling class later develops all the more sharply and profoundly. Both these things determine the fact that the struggle to be waged against this new ruling class, in its turn, aims at a more decided and radical negation of the previous conditions of society than could all previous classes which sought to rule.

This whole semblance, that the rule of a certain class is only the rule of certain ideas, comes to a natural end, of course, as soon as class rule in general ceases to be the form in which society is organised, that is to say, as soon as it is no longer necessary to represent a particular interest as general or the "general interest" as ruling.

▦

Introduction to *Economic and Philosophic Manuscripts of 1844*

In the essay "Alienated Labour" (taken from the *Economic and Philosophic Manuscripts of 1844*), Marx examines the condition of **alienation** or estrangement. For Marx, alienation is inherent in capitalism, because the process of production and the results of our labor confront us as a dominating power. It stems not from religiously rooted errors of consciousness, as Hegel argued, but from the material conditions in which we apply our essential productive capacities. For, contrary to Hegel's assertion, God does not create man and his ideas. Instead, it is man who creates the idea of God.

How is it that alienation is a necessary feature of capitalism? For the wage earner, work is alienating because it serves solely to provide the means (i.e., money) for maintaining her physical existence. Instead of labor representing an end in itself—an activity that expresses our capacity to shape our lives and our relationships with others—private ownership of the means of production reduces the role of the worker to that of a cog in a machine. The worker is an expendable object that performs routinized tasks. Put in another way, for Marx, working just for money—and not for the creative potential of labor itself—is akin to selling your soul.

The wage earner has little, if any, control over the production *process*. The types of materials or machines to be used, how to divide the necessary tasks, and the rate at which goods are to be manufactured are all determined by the owner of the factory or business. The worker is thus subject to the demands of the production process; it confronts her as an alienating power that controls her labor. Because the worker is alienated in her role as producer, she can only be but alienated from that which the process of her labor produces. In turn, the *product* opposes the worker as an object over which she has no control.

[iv]Marginal note by Marx: "Universality corresponds to (1) the class versus the estate, (2) the competition, world-wide intercourse, etc., (3) the great numerical strength of the ruling class, (4) the illusion of the common interests (in the beginning this illusion is true), (5) the delusion of the ideologists and the division of labour."

The questions of where and how it is sold and how much to charge are determined by the capitalist. More profoundly, the worker is dependent on the object for her very existence. It is only for her labor expended in producing the object that she earns a wage and is thus able to survive. If the object disappears—when the factory closes or technology renders the worker's labor obsolete—through no fault of her own she is left clinging to survival.

Because the worker is alienated from the process of production as well as the product of his labor, he becomes inescapably alienated from *himself.* The wage earner spends two-thirds of his waking hours engaged in a meaningless activity, save its providing him with the means of subsistence. Torn away from the object of his labor, he is unable to realize the essence of his creative nature or "species being" through his work. Finally, the worker is alienated from the rest of humanity, and becomes just another commodity to be bought and sold. To himself and others he is more like an animal or a machine than a human. Tragically, Marx asserts that the worker is free only in the performance of his "animal functions—eating, drinking, procreating . . . and in his human functions [labor] he no longer feels himself to be anything but an animal" (Marx [1844] 1978:74).

In "The Power of Money in Bourgeois Society" (also taken from the *Economic and Philosophic Manuscripts*), Marx extends his critique of capitalist production to money itself. Here he describes how the possessor of money can be transformed into anything money can buy; how one's individuality is determined not by his own characteristics or capacities, but by the power of money to transform what he wants to be into what he *is*. Money is a medium capable of being exchanged not only for a specific good or service, but also for traits such as beauty, talent, or honesty. It is not simply something that we earn, spend, or save—rather, it *does* things, it makes us who and what we are. Money is "the alienating *ability of mankind*" (Marx [1844] 1978:104, emphasis in the original) that bonds us to life itself and to our relationships with others, not through our innate qualities, but through what we have the power to buy.

Significantly, this concern with the subjective consequences of the capitalist system reflects a nonrationalist dimension to Marx's argument that contrasts with his overall rationalist theoretical orientation. In "Alienated Labour," Marx does not focus on the nature of class interests and the struggle to realize them (though it certainly would be in our interest to reform, if not abolish, the productive arrangements he describes). Rather, he describes a "way of being," a sensibility imposed on workers and capitalists alike by the properties inherent to capitalism. Indeed, the nonrationalist logic of this essay is highlighted further by the fact that Marx is constructing a moral critique as much as a scientific argument concerning the degradation wreaked by capitalism.

─────── *Economic and Philosophic Manuscripts of 1844* ───────

Karl Marx

ALIENATED LABOUR

We have proceeded from the premises of political economy. We have accepted its language and its laws. We presupposed private property, the separation of labour, capital and land, and of wages, profit of capital and rent of land—likewise division of labour, competition, the concept of exchange-value, etc. On the basis of political economy itself, in its own words, we have shown that the worker sinks to the level of a commodity and becomes indeed the most wretched of commodities; that the wretchedness of the worker is in inverse proportion to the power and magnitude of his production; that the necessary result of competition is the accumulation of capital in a few hands, and thus the restoration of monopoly in a more terrible form; that finally the distinction between capitalist and land-rentier, like that between the tiller of the soil and the factory-worker,

SOURCE: Excerpts from *Economic and Philosophic Manuscripts of 1844,* first published in German, 1932, translated by Robert Tucker from *The Marx-Engels Reader,* Second Edition, by Karl Marx and Friedrich Engels, edited by Robert C. Tucker. Copyright © 1978, 1972 by W. W. Norton & Company, Inc. Used by permission of W. W. Norton & Company, Inc.

disappears and that the whole of society must fall apart into the two classes—the property-*owners* and the propertyless *workers*. . . .

Now, therefore, we have to grasp the essential connection between private property, avarice, and the separation of labour, capital and landed property; between exchange and competition, value and the devaluation of men, monopoly and competition, etc.; the connection between this whole estrangement and the *money*-system.

Do not let us go back to a fictitious primordial condition as the political economist does, when he tries to explain. Such a primordial condition explains nothing. He merely pushes the question away into a grey nebulous distance. He assumes in the form of fact, of an event, what he is supposed to deduce—namely, the necessary relationship between two things—between, for example, division of labour and exchange. Theology in the same way explains the origin of evil by the fall of man: that is, it assumes as a fact, in historical form, what has to be explained.

We proceed from an *actual* economic fact.

The worker becomes all the poorer the more wealth he produces, the more his production increases in power and range. The worker becomes an ever cheaper commodity the more commodities he creates. With the *increasing value* of the world of things proceeds in direct proportion the *devaluation* of the world of men. Labour produces not only commodities; it produces itself and the worker as a *commodity*—and does so in the proportion in which it produces commodities generally.

This fact expresses merely that the object which labour produces—labour's product—confronts it as *something alien,* as a *power independent* of the producer. The product of labour is labour which has been congealed in an object, which has become material: it is the *objectification* of labour. Labour's realization is its objectification. In the conditions dealt with by political economy this realization of labour appears as *loss of reality* for the workers; objectification as *loss of the object* and *object-bondage;* appropriation as *estrangement,* as *alienation.*

So much does labour's realization appear as loss of reality that the worker loses reality to the point of starving to death. So much does objectification appear as loss of the object that the worker is robbed of the objects most necessary not only for his life but for his work. Indeed, labour itself becomes an object which he can get hold of only with the greatest effort and with the most irregular interruptions. So much does the appropriation of the object appear as estrangement that the more objects the worker produces the fewer can he possess and the more he falls under the dominion of his product, capital.

All these consequences are contained in the definition that the worker is related to the *product of his labour* as to an *alien* object. For on this premise it is clear that the more the worker spends himself, the more powerful the alien objective world becomes which he creates over-against himself, the poorer he himself—his inner world—becomes, the less belongs to him as his own. It is the same in religion. The more man puts into God, the less he retains in himself. The worker puts his life into the object; but now his life no longer belongs to him but to the object. Hence, the greater this activity, the greater is the worker's lack of objects. Whatever the product of his labour is, he is not. Therefore the greater this product, the less is he himself. The *alienation* of the worker in his product means not only that his labour becomes an object, an *external* existence, but that it exists *outside him,* independently, as something alien to him, and that it becomes a power of its own confronting him; it means that the life which he has conferred on the object confronts him as something hostile and alien.

Let us now look more closely at the *objectification,* at the production of the worker; and therein at the *estrangement,* the *loss* of the object, his product.

The worker can create nothing without *nature,* without the *sensuous external world.* It is the material on which his labor is manifested, in which it is active, from which and by means of which it produces.

But just as nature provides labor with the *means of life* in the sense that labour cannot *live* without objects on which to operate, on the other hand, it also provides the *means of life* in the more restricted sense—i.e., the means for the physical subsistence of the *worker* himself.

Thus the more the worker by his labour *appropriates* the external world, sensuous nature, the more he deprives himself of *means of life* in the double respect: first, that the sensuous external world more and more ceases to be an object belonging to his labour—to be his labour's *means of life;* and secondly, that it more and more ceases to be *means of life* in the immediate sense, means for the physical subsistence of the worker.

Thus in this double respect the worker becomes a slave of his object, first, in that he receives an *object of labour,* i.e., in that he receives *work;* and secondly, in that he receives *means of subsistence.* Therefore, it enables him to exist, first, as a *worker;* and, second, as a *physical subject.* The extremity of this bondage is that it is only as a *worker* that he continues to maintain himself as a *physical subject,* and that it is only as a *physical subject* that he is a worker.

(The laws of political economy express the estrangement of the worker in his object thus: the more the worker produces, the less he has to consume; the more values he creates, the more valueless, the more unworthy he becomes; the better formed his product, the more deformed becomes the worker; the more civilized his object, the more barbarous becomes the worker; the mightier labour becomes, the more powerless becomes the worker; the more ingenious labour becomes, the duller becomes the worker and the more he becomes nature's bondsman.)

Political economy conceals the estrangement inherent in the nature of labour by not considering the direct relationship between the worker (labour) *and production.* It is true that labour produces for the rich wonderful things—but for the worker it produces privation. It produces palaces—but for the worker, hovels. It produces beauty—but for the worker, deformity. It replaces labour by machines—but some of the workers it throws back to a barbarous type of labour, and the other workers it turns into machines. It produces intelligence—but for the worker idiocy, cretinism.

The direct relationship of labour to its produce is the relationship of the worker to the objects of his production. The relationship of the man of means to the objects of production and to production itself is only a *consequence* of this first relationship—and confirms it. We shall consider this other aspect later.

When we ask, then, what is the essential relationship of labour we are asking about the relationship of the *worker* to production.

Till now we have been considering the estrangement, the alienation of the worker only in one of its aspects, i.e., the worker's *relationship to the products of his labour.* But the estrangement is manifested not only in the result but in the *act of production*—within the *producing activity* itself. How would the worker come to face the product of his activity as a stranger, were it not that in the very act of production he was estranging himself from himself? The product is after all but the summary of the activity of production. If then the product of labour is alienation, production itself must be active alienation, the alienation of activity, the activity of alienation. In the estrangement of the object of labour is merely summarized the estrangement, the alienation, in the activity of labour itself.

What, then, constitutes the alienation of labour?

First, the fact that labour is *external* to the worker, i.e., it does not belong to his essential being; that in his work, therefore, he does not affirm himself but denies himself, does not feel content but unhappy, does not develop freely his physical and mental energy but mortifies his body and ruins his mind. The worker therefore only feels himself outside his work, and in his work feels outside himself. He is at home when he is not working, and when he is working he is not at home. His labour is therefore not voluntary, but coerced; it is *forced labour.* It is therefore not the satisfaction of a need; it is merely a *means* to satisfy needs external to it. Its alien character emerges clearly in the fact that as soon as no physical or other compulsion exists, labour is shunned like the plague. External labour, labour in which man alienates himself, is a labour of self-sacrifice, of mortification. Lastly, the external character of labour for the worker appears in the fact that it is not his own, but someone else's, that it does not belong to him, that in it he belongs, not to himself, but to another. Just as in religion the spontaneous activity of the human imagination, of the human brain and the human heart, operates independently of the individual—that is, operates on him as an alien, divine or diabolical activity—in the same way the worker's activity is not his spontaneous activity. It belongs to another; it is the loss of his self.

As a result, therefore, man (the worker) no longer feels himself to be freely active in any but his animal functions—eating, drinking, procreating, or at most in his dwelling and in dressing-up, etc.; and in his human functions he no longer feels himself to be anything but an animal. What is animal becomes human and what is human becomes animal.

Certainly eating, drinking, procreating, etc., are also genuinely human functions. But in the abstraction which separates them from the sphere of all other human activity and turns them into sole and ultimate ends, they are animal.

We have considered the act of estranging practical human activity, labour, in two of its aspects.

1. The relation of the worker to the *product of labour* as an alien object exercising power over him. This relation is at the same time the relation to the sensuous external world, to the objects of nature as an alien world antagonistically opposed to him.

2. The relation of labour to the *act of production* within the *labour* process. This relation is the relation of the worker to his own activity as an alien activity not belonging to him; it is activity

as suffering, strength as weakness, begetting as emasculating, the worker's *own* physical and mental energy, his personal life or what is life other than activity—as an activity which is turned against him, neither depends on nor belongs to him. Here we have *self-estrangement,* as we had previously the estrangement of the *thing.*

We have yet a third aspect of *estranged labour* to deduce from the two already considered.

Man is a species being, not only because in practice and in theory he adopts the species as his object (his own as well as those of other things), but—and this is only another way of expressing it—but also because he treats himself as the actual, living species; because he treats himself as a *universal* and therefore a free being.

The life of the species, both in man and in animals, consists physically in the fact that man (like the animal) lives on inorganic nature; and the more universal man is compared with an animal, the more universal is the sphere of inorganic nature on which he lives. Just as plants, animals, stones, the air, light, etc., constitute a part of human consciousness in the realm of theory, partly as objects of natural science, partly as objects of art—his spiritual inorganic nature, spiritual nourishment which he must first prepare to make it palatable and digestible—so too in the realm of practice they constitute a part of human life and human activity. Physically man lives only on these products of nature, whether they appear in the form of food, heating, clothes, a dwelling, or whatever it may be. The universality of man is in practice manifested precisely in the universality which makes all nature his *inorganic body*—both inasmuch as nature is (1) his direct means of life, and (2) the material, the object, and the instrument of his life-activity. Nature is man's *inorganic body*—nature, that is, in so far as it is not itself the human body. Man *lives* on nature—means that nature is his *body,* with which he must remain in continuous intercourse if he is not to die. That man's physical and spiritual life is linked to nature means simply that nature is linked to itself, for man is a part of nature.

In estranging from man (1) nature, and (2) himself, his own active functions, his life-activity, estranged labour estranges the *species* from man. It turns for him the *life of the species* into a means of individual life. First it estranges the life of the species and individual life, and secondly it makes individual life in its abstract form the purpose of the life of the species, likewise in its abstract and estranged form.

For in the first place labour, *life-activity, productive life* itself, appears to man merely as a *means* of satisfying a need—the need to maintain the physical existence. Yet the productive life is the life of the species. It is life-engendering life. The whole character of a species—its species character—is contained in the character of its life-activity; and free, conscious activity is man's species character. Life itself appears only as *a means to life.*

The animal is immediately identical with its life-activity. It does not distinguish itself from it. It is its *life-activity.* Man makes his life-activity itself the object of his will and of his consciousness. He has conscious life-activity. It is not a determination with which he directly merges. Conscious life-activity directly distinguishes man from animal life-activity. It is just because of this that he is a species being. Or it is only because he is a species being that he is a Conscious Being, i.e., that his own life is an object for him. Only because of that is his activity free activity. Estranged labour reverses this relationship, so that it is just because man is a conscious being that he makes his life-activity, his *essential* being, a mere means to his *existence.*

In creating an *objective world* by his practical activity, in *working-up* inorganic nature, man proves himself a conscious species being, i.e., as a being that treats the species as its own essential being, or that treats itself as a species being. Admittedly animals also produce. They build themselves nests, dwellings, like the bees, beavers, ants, etc. But an animal only produces what it immediately needs for itself or its young. It produces one-sidedly, whilst man produces universally. It produces only under the dominion of immediate physical need, whilst man produces even when he is free from physical need and only truly produces in freedom therefrom. An animal produces only itself, whilst man reproduces the whole of nature. An animal's product belongs immediately to its physical body, whilst man freely confronts his product. An animal forms things in accordance with the standard and the need of the species to which it belongs, whilst man knows how to produce in accordance with the standard of every species, and knows how to apply everywhere the inherent standard to the object. Man therefore also forms things in accordance with the laws of beauty.

It is just in the working-up of the objective world, therefore, that man first really proves himself to be a *species being.* This production is his active species life. Through and because of this production, nature appears as *his* work and his reality. The object of labour is, therefore, the *objectification of man's species life:* for he duplicates himself not only, as in consciousness, intellectually, but also actively, in reality, and therefore he contemplates himself in a world that he has created. In tearing away from man the object of his production, therefore, estranged labour tears from him his *species life,* his real species objectivity, and transforms his advantage over animals into the disadvantage that his inorganic body, nature, is taken from him.

Similarly, in degrading spontaneous activity, free activity, to a means, estranged labour makes man's species life a means to his physical existence.

The consciousness which man has of his species is thus transformed by estrangement in such a way that the species life becomes for him a means.

Estranged labour turns thus:

3. *Man's species being,* both nature and his spiritual species property, into a being *alien* to him, into a *means* to his *individual existence.* It estranges man's own body from him, as it does external nature and his spiritual essence, his *human* being.

4. An immediate consequence of the fact that man is estranged from the product of his labour, from his life-activity, from his species being is the *estrangement of man* from *man.* If a man is confronted by himself, he is confronted by the *other* man. What applies to a man's relation to his work, to the product of his labour and to himself, also holds of a man's relation to the other man, and to the other man's labour and object of labour.

In fact, the proposition that man's species nature is estranged from him means that one man is estranged from the other, as each of them is from man's essential nature.

The estrangement of man, and in fact every relationship in which man stands to himself, is first realized and expressed in the relationship in which a man stands to other men.

Hence within the relationship of estranged labour each man views the other in accordance with the standard and the position in which he finds himself as a worker.

We took our departure from a fact of political economy—the estrangement of the worker and his production. We have formulated the concept of this fact—*estranged, alienated* labour. We have analysed this concept—hence analysing merely a fact of political economy.

Let us now see, further, how in real life the concept of estranged, alienated labour must express and present itself.

If the product of labour is alien to me, if it confronts me as an alien power, to whom, then, does it belong?

If my own activity does not belong to me, if it is an alien, a coerced activity, to whom, then, does it belong?

To a being *other* than me.

Who is this being?

The *gods?* To be sure, in the earliest times the principal production (for example, the building of temples, etc., in Egypt, India and Mexico) appears to be in the service of the gods, and the product belongs to the gods. However, the gods on their own were never the lords of labour. No more was *nature.* And what a contradiction it would be if, the more man subjugated nature by his labour and the more the miracles of the gods were rendered superfluous by the miracles of industry, the more man were to renounce the joy of production and the enjoyment of the produce in favour of these powers.

The *alien* being, to whom labour and the produce of labour belongs, in whose service labour is done and for whose benefit the produce of labour is provided, can only be *man* himself.

If the product of labour does not belong to the worker, if it confronts him as an alien power, this can only be because it belongs to some *other man than the worker.* If the worker's activity is a torment to him, to another it must be *delight* and his life's joy. Not the gods, not nature, but only man himself can be this alien power over man.

We must bear in mind the above-stated proposition that man's relation to himself only becomes *objective* and *real* for him through his relation to the other man. Thus, if the product of his labour, his labour *objectified,* is for him an *alien,* hostile, powerful object independent of him, then his position towards it is such that someone else is master of this object, someone who is alien, hostile, powerful, and independent of him. If his own activity is to him an unfree activity, then he is treating it as activity performed in the service, under the dominion, the coercion and the yoke of another man.

Every self-estrangement of man from himself and from nature appears in the relation in which he places himself and nature to men other than and differentiated from himself. For this reason religious self-estrangement necessarily appears in the relationship of the layman to the priest, or again to a mediator, etc., since we are here dealing with the intellectual world. In the real practical world self-estrangement can only become manifest through the real practical relationship to other men. The medium through which estrangement takes place is itself *practical.* Thus through estranged labour man not only engenders his relationship to the object and to the act of production as to powers that are alien and hostile to him; he also engenders the relationship in which other men stand to his production and to his product, and the relationship in which he stands to these other men. Just as he begets his own production as the loss of his reality, as his punishment; just as he begets his own product as a loss, as a product not belonging to him; so he begets the dominion of the one who does not produce over production and over the product. Just as he estranges from himself his own activity, so he confers to the stranger activity which is not his own.

Till now we have only considered this relationship from the standpoint of the worker and later we shall be considering it also from the standpoint of the non-worker.

Through *estranged, alienated labour,* then, the worker produces the relationship to this labour of a man alien to labour and standing outside it. The relationship of the worker to labour engenders the relation to it of the capitalist, or whatever one chooses to call the master of labour. *Private property* is thus the product, the result, the necessary consequence, of *alienated labour,* of the external relation of the worker to nature and to himself.

Private property thus results by analysis from the concept of *alienated labour*—i.e., of *alienated man,* of estranged labour, of estranged life, of *estranged* man.

True, it is a result of the *movement of private property* that we have obtained the concept of *alienated labour (of alienated life)* from political economy. But on analysis of this concept it becomes clear that though private property appears to be the source, the cause of alienated labour, it is really its consequence, just as the gods *in the beginning* are not the cause but the effect of man's intellectual confusion. Later this relationship becomes reciprocal.

Only at the very culmination of the development of private property does this, its secret, re-emerge, namely, that on the one hand it is the *product* of alienated labour, and that secondly it is the *means* by which labour alienates itself, the *realization of this alienation.*

This exposition immediately sheds light on various hitherto unsolved conflicts.

1. Political economy starts from labour as the real soul of production; yet to labour it gives nothing, and to

private property everything. From this contradiction Proudhon has concluded in favour of labour and against private property. We understand, however, that this apparent contradiction is the contradiction of *estranged labour* with itself, and that political economy has merely formulated the laws of estranged labour.

We also understand, therefore, that *wages* and *private property* are identical: where the product, the object of labour pays for labour itself, the wage is but a necessary consequence of labour's estrangement, for after all in the wage of labour, labour does not appear as an end in itself but as the servant of the wage. We shall develop this point later, and meanwhile will only deduce some conclusions.

A *forcing-up of wages* (disregarding all other difficulties, including the fact that it would only be by force, too, that the higher wages, being an anomaly, could be maintained) would therefore be nothing but *better payment for the slave,* and would not conquer either for the worker or for labour their human status and dignity.

Indeed, even the *equality of wages* demanded by Proudhon only transforms the relationship of the present-day worker to his labour into the relationship of all men to labour. Society is then conceived as an abstract capitalist.

Wages are a direct consequence of estranged labour, and estranged labour is the direct cause of private property. The downfall of the one aspect must therefore mean the downfall of the other.

2. From the relationship of estranged labour to private property it further follows that the emancipation of society from private property, etc., from servitude, is expressed in the *political* form of the *emancipation of the workers;* not that *their* emancipation alone was at stake but because the emancipation of the workers contains universal human emancipation—and it contains this, because the whole of human servitude is involved in the relation of the worker to production, and every relation of servitude is but a modification and consequence of this relation. . . .

THE POWER OF MONEY IN BOURGEOIS SOCIETY

If man's *feelings,* passions, etc., are not merely anthropological phenomena in the [narrower] sense, but truly *ontological* affirmations of essential being (of nature), and if they are only really affirmed because their *object* exists for them as an object of *sense,* then it is clear:

1. That they have by no means merely one mode of affirmation, but rather that the distinctive character of their existence, of their life, is constituted by the distinctive mode of their affirmation. In what manner the object exists for them, is the characteristic mode of their *gratification.*

2. Whenever the sensuous affirmation is the direct annulment of the object in its independent form (as in eating, drinking, working up of the object, etc.), this is the affirmation of the object.

3. In so far as man, and hence also his feeling, etc., are *human,* the affirmation of the object by another is likewise his own enjoyment.

4. Only through developed industry—i.e., through the medium of private property—does the ontological essence of human passion come to be both in its totality and in its humanity; the science of man is therefore itself a product of man's establishment of himself by practical activity.

5. The meaning of private property—liberated from its estrangement—is the *existence of essential objects* for man, both as objects of enjoyment and as objects of activity.

By possessing the *property* of buying everything, by possessing the property of appropriating all objects, *money* is thus the *object* of eminent possession. The universality of its *property* is the omnipotence of its being. It therefore functions as the almighty being. Money is the *pimp* between man's need and the object, between his life and his means of life. But that which mediates *my* life for me, also *mediates* the existence of other people *for me.* For me it is the *other* person.

"What, man! confound it, hands and feet
And head and backside, all are yours!
And what we take while life is sweet,
Is that to be declared not ours?
Six stallions, say, I can afford.
Is not their strength my property?

I tear along, a sporting lord,
As if their legs belonged to me."
                    (Mephistopheles, in *Faust*)[i]

Shakespeare in *Timon of Athens:*

"Gold? Yellow, glittering, precious gold?
No, Gods, I am no idle votarist! . . .
Thus much of this will make black white, foul fair,
Wrong right, base noble, old young, coward valiant.
. . . Why, this
Will lug your priests and servants from your sides,
Pluck stout men's pillows from below their heads:
This yellow *slave*
Will knit and break religions, bless the accursed;
Make the hoar leprosy adored, place thieves
And give them title, knee and approbation
With senators on the bench: This is it
That makes the wappen'd widow wed again;
She, whom the spital-house and ulcerous sores
Would cast the gorge at, this embalms and spices
To the April day again. . . . Damned earth,
Thou common whore of mankind, that putt'st odds
Among the rout of nations."[ii]

And also later:

"O thou sweet king-killer, and dear divorce
Twixt natural son and sire! thou bright defiler
Of Hymen's purest bed! thou valiant Mars!
Thou ever young, fresh, loved and delicate wooer,
Whose blush doth thaw the consecrated snow
That lies on Dian's lap! Thou *visible God!*
That solder'st *close impossibilities,*
And mak'st them kiss! That speak'st with every tongue,
To every purpose! O thou touch of hearts!
Think thy slave man rebels, and by thy virtue

---

[i]Goethe, Faust (Part I–Faust's Study, III), translated by Philip Wayne (Penguin, 1949), p. 91.

[ii]Shakespeare, *Timon of Athens,* Act 4, Scene 3. Marx quotes the Schlegel-Tieck German translation. (Marx's emphasis.)

Set them into confounding odds, that beasts
May have the world in empire!"[iii]

Shakespeare excellently depicts the real nature of *money*. To understand him, let us begin, first of all, by expounding the passage from Goethe.

That which is for me through the medium of *money*—that for which I can pay (i.e., which money can buy)—that am I, the possessor of the money. The extent of the power of money is the extent of my power. Money's properties are my properties and essential powers—the properties and powers of its possessor. Thus, what I *am* and *am capable* of is by no means determined by my individuality. I am ugly, but I can buy for myself the most *beautiful* of women. Therefore I am not *ugly,* for the effect of *ugliness*—its deterrent power—is nullified by money. I, in my character as an individual, am *lame,* but money furnishes me with twenty-four feet. Therefore I am not lame. I am bad, dishonest, unscrupulous, stupid; but money is honoured, and therefore so is its possessor. Money is the supreme good, therefore its possessor is good. Money, besides, saves me the trouble of being dishonest: I am therefore presumed honest. I am *stupid,* but money is the *real mind* of all things and how then should its possessor be stupid? Besides, he can buy talented people for himself, and is he who has power over the talented not more talented than the talented? Do not I, who thanks to money am capable of *all* that the human heart longs for, possess all human capacities? Does not my money therefore transform all my incapacities into their contrary?

If *money* is the bond binding me to *human* life, binding society to me, binding me and nature and man, is not money the bond of all *bonds?* Can it not dissolve and bind all ties? Is it not, therefore, the universal *agent of divorce?* It is the true *agent of divorce* as well as the true *binding agent*—the [universal][iv] *galvano-chemical* power of Society.

Shakespeare stresses especially two properties of money:

1. It is the visible divinity—the transformation of all human and natural properties into their contraries, the universal confounding and overturning of things: it makes brothers of impossibilities.

2. It is the common whore, the common pimp of people and nations.

The overturning and confounding of all human and natural qualities, the fraternization of impossibilities—the *divine* power of money—lies in its *character* as men's estranged, alienating and self-disposing *species-nature.* Money is the alienated *ability of mankind.*

That which I am unable to do as a *man,* and of which therefore all my individual essential powers are incapable, I am able to do by means of *money.* Money thus turns each of these powers into something which in itself it is not—turns it, that is, into its *contrary.*

If I long for a particular dish or want to take the mail-coach because I am not strong enough to go by foot, money fetches me the dish and the mail-coach: that is, it converts my wishes from something in the realm of imagination, translates them from their meditated, imagined or willed existence into their *sensuous, actual* existence—from imagination to life, from imagined being into real being. In effecting this mediation, money is the *truly creative* power.

No doubt *demand* also exists for him who has no money, but his demand is a mere thing of the imagination without effect or existence for me, for a third party, for the others, and which therefore remains for me *unreal* and *objectless.* The difference between effective demand based on money and ineffective demand based on my need, my passion, my wish, etc., is the difference between being and *thinking,* between the imagined which *exists* merely within me and the imagined as it is for me outside me as a *real object.*

If I have no money for travel, I have no *need*—that is, no real and self-realizing need—to travel. If I have the *vocation* for study but *no* money for it, I have no vocation for study—that is, no *effective,* no *true* vocation. On the other hand, if I have really *no* vocation for study but have the will *and* the money for it, I have an *effective* vocation for it. Being the external, common

---

[iii]Ibid.

[iv]An end of the page is torn out of the manuscript [Trans.].

medium and *faculty* for turning an *image* into *reality* and *reality* into a mere *image* (a faculty not springing from man as man or from human society as society), *money* transforms the *real essential powers of man and nature* into what are merely abstract conceits and therefore *imperfections*—into tormenting chimeras—just as it transforms *real imperfections and chimeras*—essential powers which are really impotent, which exist only in the imagination of the individual—into *real powers* and *faculties.*

In the light of this characteristic alone, money is thus the general overturning of *individualities* which turns them into their contrary and adds contradictory attributes to their attributes.

Money, then, appears as this *overturning* power both against the individual and against the bonds of society, etc., which claim to be *essences* in themselves. It transforms fidelity into infidelity, love into hate, hate into love, virtue into vice, vice into virtue, servant into master, master into servant, idiocy into intelligence and intelligence into idiocy.

Since money, as the existing and active concept of value, confounds and exchanges all things, it is the general *confounding* and *compounding* of all things—the world upside-down—the confounding and compounding of all natural and human qualities.

He who can buy bravery is brave, though a coward. As money is not exchanged for any one specific quality, for any one specific thing, or for any particular human essential power, but for the entire objective world of man and nature, from the standpoint of its possessor it therefore serves to exchange every property for every other, even contradictory, property and object: it is the fraternization of impossibilities. It makes contradictions embrace.

Assume *man* to be *man* and his relationship to the world to be a human one: then you can exchange love only for love, trust for trust, etc. If you want to enjoy art, you must be an artistically cultivated person; if you want to exercise influence over other people, you must be a person with a stimulating and encouraging effect on other people. Every one of your relations to man and to nature must be a *specific expression,* corresponding to the object of your will, of your *real individual* life. If you love without evoking love in return—that is, if your loving as loving does not produce reciprocal love; if through a *living expression* of yourself as a loving person you do not make yourself a *loved person,* then your love is impotent—a misfortune.

───────────────── ■■ ─────────────────

Introduction to *The Communist Manifesto*

In 1847, the Communist League, an association formed by radical workers in 1836, commissioned Marx and Engels to write a political tract outlining the organization's program. The result was the now-famous *Communist Manifesto* (also called *The Manifesto of the Communist Party*). In contrast to other readings in this volume, the *Manifesto* is a deliberately adversarial work intended to inspire allegiance to the movement's cause. Though it had only modest impact at the time of its publication in 1848, shortly afterward workers and peasants staged revolts throughout much of Europe including France, Germany, and Italy.

Notwithstanding its origins as a political tract, *The Communist Manifesto* is of great theoretical significance. In it, you will again encounter Marx's theory of historical materialism and his inversion of Hegel's idealism. You will also see Marx's commitment to the Enlightenment belief in the perfectibility of humanity, which in his view will be realized through an inevitable communist revolution. The *Manifesto* also describes the economic processes that led to the ascendancy of the capitalist class and that eventually will produce to its own "grave-diggers"—a class-conscious proletariat.

Indeed, much of the *Manifesto* is a "scientific prophecy" detailing the downfall of the capitalist class and the rise of the proletariat. As such, it represents a penetrating theory of social change. The eventual collapse of capitalism will occur much in the way as previous economic systems: the social **relations of production** (how productive activity is organized and the laws governing property ownership) will become a "fetter" or obstacle to the continued development of the means of production (i.e., machinery, technology). The result is an "epidemic of overproduction" (Marx and Engels [1848] 1978:478) in which the bourgeoisie "chokes" on the overabundance of goods produced by ever-increasing industrial efficiency.

The final crisis of capitalism is thus a necessary consequence of the technological progress that was itself spurred by the capitalist class's private ownership of the means of production and the goods produced.

As an example of this process, consider the debates on music file sharing over the Internet. Though by no means spelling the doom of capitalism, the controversy nevertheless highlights the contradictions that arise between the forces and relations of production. Technology—the forces of production—advances more quickly than changes in the laws governing the relations of production—that is, ownership of property. Computer and communication technologies have been developed that enable a virtually infinite number of users to simultaneously share data stored on their hard drives. To avail themselves of this capability, users must first connect to a central terminal that serves as a temporary holding station. Napster was such a central terminal for sharing music files. The company itself did not own or control the files; it simply provided a conduit for the individuals who did. However, because the company provided a singular, "tangible" site for the free exchange of music, lawyers for the record companies were able to successfully argue that *Napster*—and not the individual users—circumvented copyright regulations despite the fact that it did not "steal" the files for its own use.

This advancement in computer technology undermines current laws governing copyright ownership and the rights that accompany proprietorship. Internet users can download and thus possess music with unparalleled ease and speed without having to pay for it, making it impossible for the owners of the copyrights to fully control the distribution of their property. (While cassette tapes and albums have been copied for decades, the convenience with which such "pirated" versions are made and the scope of their distribution pales in comparison to that of file sharing on the Internet.) This, of course, runs completely counter to a legal cornerstone of capitalism, namely, that owners must be compensated for the use of their "private" property. However, the social relations of production—the laws of ownership—do not prevent *individuals* from sharing the products that they have purchased and thus rightfully own, and someone at some point purchased the music that now is stored on their hard drive. However, to the extent that current laws are enforced or rewritten in an effort to combat the infringement of property rights, the social relations of production become a "fetter" to the full development of advances in technology.

Returning to our theoretical discussion of the dynamics of capitalism, capitalists must forever seek to eliminate their competitors, create new markets, destroy some of their products, or cut back their productive capacity in order to minimize the oversupply of goods that results from increasingly sophisticated means of production. If production is reduced, however, capitalists, in turn, will be forced to reduce their work force and, with it, their source of profit as well as the size of the market able to purchase their goods. Yet, the bourgeoisie is confronted not only with these economic realities of capitalism, but also with political consequences, as competition creates an obstacle to class unity and to the ability to implement coherent economic policies that will ensure its dominance. And so the cycle continues.

Meanwhile, factory conditions themselves facilitate the development of a revolutionary class consciousness through which workers come to realize the true source of their alienation and the possibility of breaking free from the chains of their enslavement. Placed side by side in their performance of tedious, monotonous tasks, the physical settings of factories increase the contact between the workers, making it easier to communicate and spread allegiance to the proletariat's cause. Urging "WORKING MEN OF ALL COUNTRIES, UNITE!" Marx warns that the Communists

> openly declare that their ends can be attained only by the forcible overthrow of all existing social conditions. Let the ruling class tremble at a Communistic revolution. The proletarians have nothing to lose but their chains. They have a world to win. (Marx and Engels [1848] 1978:500)

Yet, the question remains: Why would the establishment of a communist economy create a more humane society? At the risk of oversimplifying the matter, the communist utopia hinges on the abolishment of private property. Marx maintains that once the means of production becomes collectively owned, exploitation of the worker is no longer possible. This is because the surplus value (i.e., profit) produced by the worker is not appropriated or siphoned off by an individual owner. Instead, it is distributed among the workers themselves.

Alienation is also ended because the worker, now a part owner of the enterprise, is able to direct the production process and maintain control over the products she creates. In turn, the worker is no longer estranged from herself and the species being. Finally, the competition for profit that characterizes bourgeois capitalism is brought to a close and, with it, recurring economic crises. Periods of "boom or bust" and their accompanying disruptions to employment are replaced by a more stable form of economic planning that produces according to the needs of the population and not the whims of an unpredictable market. "In place of the old bourgeois society, with its classes and class antagonisms, we shall have an association, in which the free development of each is the condition for the free development of all" (Marx and Engels [1848] 1978:491).

The economic crisis currently unfolding is a textbook example of the continuing relevance and prescience of Marx's ideas. Today's crisis is America's worst since the Great Depression. A record high of nearly 14 million workers are unemployed; those still employed are left with no choice but to accept steep cuts in pay and benefits. The United States, however, is by no means alone in experiencing the dramatic downturn. Just as Marx predicted, the spread of capitalism has ensured that the ever-worsening economic conditions cannot be confined to any one country's borders, but necessarily must reach across the entire globe. With sales of commodities plummeting worldwide, capitalists are beginning to "choke" on their supplies as warehouses are filling to capacity with unshipped goods. To compensate, stores are slashing prices—and losing profits—in order to sell their products. But what of the workers, the proletariat? Millions of people are looking for jobs, struggling to meet their basic needs (as are many millions of the employed). Nevertheless, production across all sectors of the economy is slowing to a virtual halt, but not because the machines are broken or somehow malfunctioning, or because there are not enough skilled laborers available to carry out the required tasks. Production has been stopped artificially by capitalists, and they must do so in order to prevent glutting the market with their goods while preserving whatever profits they are still able to earn. The relations of production—private ownership and its accompanying drive for private profit—have become a fetter to the forces of production, despite the fact that millions are living in increasingly desperate conditions.

While the causes of the current crisis are complex, many analysts have pointed to the dominant role played by the bundling of individual home loans into mortgage-backed securities that were then sold to investors. When the housing bubble that made investment in these financial instruments profitable burst, banks and investment companies around the world were left holding assets with rapidly declining values. However, the very corporations who invented and sold this new form of security are unable to root out the problems caused by these "troubled assets" because the originally bundled securities have been rebundled and traded so frequently that it has become impossible to determine the value of the securities as well as who actually owns a specific asset. Capitalists, like a "sorcerer who is no longer able to control the powers of the nether world whom he has called upon by his spells" (Marx and Engels [1848] 1978:478), created a financial instrument that they are incapable of controlling and that has metastasized to the point where it threatens the stability of the global capitalist economy.

To stem the tide of the fallout, governments are intensifying their intervention in their respective economies. In the United States, intervention to this point has taken the form of giving billions of taxpayer dollars to the very financial institutions that are largely responsible for creating the crisis with little oversight or accountability for how the funds will be used. And should the government decide to use public funds to purchase the troubled assets from the banks and investment companies, it will be impossible for taxpayers to know whether or not they are paying a fair price for them, because the value of the assets cannot be determined. At the same time, the government has provided comparatively little funds for the increasingly distressed auto industry—one of the few remaining manufacturing industries in the country—prompting some observers to claim that the government is concerned only with the well-being of Wall Street and not Main Street. In rescuing the "moneyed interests" while letting drown those blue-collar workers who *make* things, a ring of truth is sounded in Marx's assertion, "The executive of the modern State is but a committee for managing the common affairs of the whole bourgeoisie" (Marx and Engels [1848] 1978:475). Yet, to avoid a complete economic collapse, the capitalists and the state have no choice but to appeal to the public—the proletariat—"to ask for its help, and thus drag it into the political arena" (ibid.:481), in turn supplying it with a political and intellectual education that will later be used as a weapon against them.

The Communist Manifesto (1848)

Karl Marx and Friedrich Engels

A spectre is haunting Europe—the spectre of communism. All the powers of old Europe have entered into a holy alliance to exorcise this spectre: Pope and Tsar, Metternich and Guizot, French Radicals and German police-spies.

Where is the party in opposition that has not been decried as communistic by its opponents in power? Where the opposition that has not hurled back the branding reproach of Communism, against the more advanced opposition parties, as well as against its reactionary adversaries?

Two things result from this fact.

I. Communism is already acknowledged by all European powers to be itself a power.

II. It is high time that Communists should openly, in the face of the whole world, publish their views, their aims, their tendencies, and meet this nursery tale of the spectre of communism with a manifesto of the party itself.

To this end, Communists of various nationalities have assembled in London, and sketched the following manifesto, to be published in the English, French, German, Italian, Flemish and Danish languages.

Bourgeois and Proletarians[i]

The history of all hitherto existing society[ii] is the history of class struggles.

Freeman and slave, patrician and plebeian, lord and serf, guild-master[iii] and journeyman, in a word, oppressor and oppressed, stood in constant opposition to one another, carried on an uninterrupted, now hidden, now open fight, a fight that each time ended, either in a revolutionary re-constitution of society at large, or in the common ruin of the contending classes.

In the earlier epochs of history, we find almost everywhere a complicated arrangement of society into various orders, a manifold gradation of social rank. In ancient Rome we have patricians, knights, plebeians, slaves; in the Middle Ages, feudal lords, vassals, guild-masters, journeymen, apprentices, serfs; in almost all of these classes, again, subordinate gradations.

The modern bourgeois society that has sprouted from the ruins of feudal society has not done away with class antagonisms. It has but established new classes, new conditions of oppression, new forms of struggle in place of the old ones.

Our epoch, the epoch of the bourgeoisie, possesses, however, this distinct feature: it has simplified class antagonisms: Society as a whole is more and more splitting up into two great hostile camps, into two great classes directly facing each other: bourgeoisie and proletariat.

From the serfs of the Middle Ages sprang the chartered burghers of the earliest towns. From these burgesses the first elements of the bourgeoisie were developed.

The discovery of America, the rounding of the Cape, opened up fresh ground for the rising bourgeoisie.

SOURCE: Marx/Engels Internet Archive.

[i]By bourgeoisie is meant the class of modern Capitalists, owners of the means of social production and employers of wage-labour. By proletariat, the class of modern wage-labourers who, having no means of production of their own, are reduced to selling their labour-power in order to live. [Engels, English edition of 1888]

[ii]That is, all written history. In 1847, the pre-history of society, the social organisation existing previous to recorded history, was all but unknown. Since then, Haxthausen discovered common ownership of land in Russia, Maurer proved it to be the social foundation from which all Teutonic races started in history, and by and by village communities were found to be, or to have been the primitive form of society everywhere from India to Ireland. The inner organisation of this primitive Communistic society was laid bare, in its typical form, by Morgan's crowning discovery of the true nature of the gens and its relation to the tribe. With the dissolution of these primaeval communities society begins to be differentiated into separate and finally antagonistic classes. I have attempted to retrace this process of dissolution in: "Der Ursprung der Familie, des Privateigenthums und des Staats" [The Origin of the Family, Private Property and the State], 2nd edition, Stuttgart 1886. [Engels, English edition of 1888]

[iii]Guild-master, that is, a full member of a guild, a master within, not a head of a guild. [Engels, English edition of 1888]

The East-Indian and Chinese markets, the colonisation of America, trade with the colonies, the increase in the means of exchange and in commodities generally, gave to commerce, to navigation, to industry, an impulse never before known, and thereby, to the revolutionary element in the tottering feudal society, a rapid development.

The feudal system of industry, in which industrial production was monopolized by closed guilds, now no longer suffices for the growing wants of the new markets. The manufacturing system took its place. The guild-masters were pushed aside by the manufacturing middle class; division of labor between the different corporate guilds vanished in the face of division of labor in each single workshop.

Meantime the markets kept ever growing, the demand ever rising. Even manufacturers no longer sufficed. Thereupon, steam and machinery revolutionized industrial production. The place of manufacture was taken by the giant, Modern Industry, the place of the industrial middle class, by industrial millionaires, the leaders of the whole industrial armies, the modern bourgeois.

Modern industry has established the world-market, for which the discovery of America paved the way. This market has given an immense development to commere, to navigation, to communication by land. This development has, in turn, reacted on the extension of industry; and in proportion as industry, commerce, navigation, railways extended, in the same proportion the bourgeoisie developed, increased its capital, and pushed into the background every class handed down from the Middle Ages.

We see, therefore, how the modern bourgeoisie is itself the product of a long course of development, of a series of revolutions in the modes of production and of exchange.

Each step in the development of the bourgeoisie was accompanied by a corresponding political advance in that class. An oppressed class under the sway of the feudal nobility, an armed and self-governing association in the medieval commune;[iv] here independent urban republic (as in Italy and Germany), there taxable "third estate" of the monarchy (as in France), afterward, in the period of manufacturing proper, serving either the semi-feudal or the absolute monarchy as a counterpoise against the nobility, and, in fact, corner-stone of the great monarchies in general, the bourgeoisie has at last, since the establishment of Modern Industry and of the world-market, conquered for itself, in the modern representative state, exclusive political sway. The executive of the modern state is but a committee for managing the common affairs of the whole bourgeoisie.

The bourgeoisie, historically, has played a most revolutionary part.

The bourgeoisie, wherever it has got the upper hand, has put an end to all feudal, patriarchal, idyllic relations. It has pitilessly torn asunder the motley feudal ties that bound man to his "natural superiors," and has left no other nexus between man and man than naked self-interest, than callous "cash payment." It has drowned out the most heavenly ecstasies of religious fervour, of chivalrous enthusiasm, of philistine sentimentalism, in the icy water of egotistical calculation. It has resolved personal worth into exchange value, and in place of the numberless indefeasible chartered freedoms, has set up that single, unconscionable freedom—Free Trade. In one word, for exploitation, veiled by religious and political illusions, it has substituted naked, shameless, direct, brutal exploitation.

The bourgeoisie has stripped of its halo every occupation hitherto honored and looked up to with reverent awe. It has converted the physician, the lawyer, the priest, the poet, the man of science, into its paid wage-laborers.

The bourgeoisie has torn away from the family its sentimental veil, and has reduced the family relation into a mere money relation.

The bourgeoisie has disclosed how it came to pass that the brutal display of vigour in the Middle Ages, which reactionaries so much admire, found its fitting complement in the most slothful indolence. It has been the first to show what man's activity can bring about. It has accomplished wonders far surpassing Egyptian pyramids, Roman aqueducts, and Gothic cathedrals; it has conducted expeditions that put in the shade all former exoduses of nations and crusades.

The bourgeoisie cannot exist without constantly revolutionizing the instruments of production, and thereby

---

[iv]"Commune" was the name taken, in France, by the nascent towns even before they had conquered from their feudal lords and masters local self-government and political rights as the "Third Estate." Generally speaking, for the economical development of the bourgeoisie, England is here taken as the typical country; for its political development, France. [Engels, English edition of 1888] This was the name given their urban communities by the townsmen of Italy and France, after they had purchased or wrested their initial rights of self-government from their feudal lords. [Engels, German edition of 1890]

the relations of production, and with them the whole relations of society. Conservation of the old modes of production in unaltered form, was, on the contrary, the first condition of existence for all earlier industrial classes. Constant revolutionizing of production, uninterrupted disturbance of all social conditions, everlasting uncertainty and agitation distinguish the bourgeois epoch from all earlier ones. All fixed, fast-frozen relations, with their train of ancient and venerable prejudices and opinions, are swept away, all new-formed ones become antiquated before they can ossify. All that is solid melts into air, all that is holy is profaned, and man is at last compelled to face with sober senses, his real condition of life, and his relations with his kind.

The need of a constantly expanding market for its products chases the bourgeoisie over the entire surface of the globe. It must nestle everywhere, settle everywhere, establish connections everywhere.

The bourgeoisie has through its exploitation of the world-market given a cosmopolitan character to production and consumption in every country. To the great chagrin of reactionaries, it has drawn from under the feet of industry the national ground on which it stood. All old-established national industries have been destroyed or are daily being destroyed. They are dislodged by new industries, whose introduction becomes a life and death question for all civilized nations, by industries that no longer work up indigenous raw material, but raw material drawn from the remotest zones; industries whose products are consumed, not only at home, but in every quarter of the globe. In place of the old wants, satisfied by the production of the country, we find new wants, requiring for their satisfaction the products of distant lands and climes. In place of the old local and national seclusion and self-sufficiency, we have intercourse in every direction, universal inter-dependence of nations. And as in material, so also in intellectual production. The intellectual creations of individual nations become common property. National one-sidedness and narrow-mindedness become more and more impossible, and from the numerous national and local literatures, there arises a world literature.

The bourgeoisie, by the rapid improvement of all instruments of production, by the immensely facilitated means of communication, draws all, even the most barbarian, nations into civilization. The cheap prices of commodities are the heavy artillery with which it batters down all Chinese walls, with which it forces the barbarians' intensely obstinate hatred of foreigners to capitulate. It compels all nations, on pain of extinction,

to adopt the bourgeois mode of production; it compels them to introduce what it calls civilization into their midst, *i.e.,* to become bourgeois themselves. In one word, it creates a world after its own image.

The bourgeoisie has subjected the country to the rule of the towns. It has created enormous cities, has greatly increased the urban population as compared with the rural, and has thus rescued a considerable part of the population from the idiocy of rural life. Just as it has made the country dependent on the towns, so it has made barbarian and semi-barbarian countries dependent on the civilized ones, nations of peasants on nations of bourgeois, the East on the West.

The bourgeoisie keeps more and more doing away with the scattered state of the population, of the means of production, and of property. It has agglomerated population, centralized means of production, and has concentrated property in a few hands. The necessary consequence of this was political centralization. Independent, or but loosely connected provinces, with separate interests, laws, governments and systems of taxation, became lumped together into one nation, with one government, one code of laws, one national class-interest, one frontier and one customs-tariff.

The bourgeoisie, during its rule of scarce one hundred years, has created more massive and more colossal productive forces than have all preceding generations together. Subjection of nature's forces to man, machinery, application of chemistry to industry and agriculture, steam-navigation, railways, electric telegraphs, clearing of whole continents for cultivation, canalization of rivers, whole populations conjured out of the ground—what earlier century had even a presentiment that such productive forces slumbered in the lap of social labor?

We see then: the means of production and of exchange, on whose foundation the bourgeoisie built itself up, were generated in feudal society. At a certain stage in the development of these means of production and of exchange, the conditions under which feudal society produced and exchanged, the feudal organization of agriculture and manufacturing industry, in one word, the feudal relations of property became no longer compatible with the already developed productive forces; they became so many fetters. They had to be burst asunder; they were burst asunder.

Into their place stepped free competition, accompanied by a social and political constitution adapted in it, and the economic and political sway of the bourgeois class.

A similar movement is going on before our own eyes. Modern bourgeois society with its relations of production, of exchange and of property, a society that has conjured up such gigantic means of production and of exchange, is like the sorcerer, who is no longer able to control the powers of the nether world whom he has called up by his spells. For many a decade past the history of industry and commerce is but the history of the revolt of modern productive forces against modern conditions of production, against the property relations that are the conditions for the existence of the bourgeois and of its rule. It is enough to mention the commercial crises that, by their periodical return, put the existence of the entire bourgeois society on its trial, each time more threateningly. In these crises a great part not only of the existing products, but also of the previously created productive forces, are periodically destroyed. In these crises there breaks out an epidemic that, in all earlier epochs, would have seemed an absurdity—the epidemic of over-production. Society suddenly finds itself put back into a state of momentary barbarism; it appears as if a famine, a universal war of devastation had cut off the supply of every means of subsistence; industry and commerce seem to be destroyed. And why? Because there is too much civilization, too much means of subsistence, too much industry, too much commerce. The productive forces at the disposal of society no longer tend to further the development of the conditions of bourgeois property; on the contrary, they have become too powerful for these conditions, by which they are fettered, and so soon as they overcome these fetters, they bring disorder into the whole of bourgeois society, endanger the existence of bourgeois property. The conditions of bourgeois society are too narrow to comprise the wealth created by them. And how does the bourgeoisie get over these crises? On the one hand by enforced destruction of a mass of productive forces; on the other, by the conquest of new markets, and by the more thorough exploitation of the old ones. That is to say, by paving the way for more extensive and more destructive crises, and by diminishing the means whereby crises are prevented.

The weapons with which the bourgeoisie felled feudalism to the ground are now turned against the bourgeoisie itself.

But not only has the bourgeoisie forged the weapons that bring death to itself; it has also called into existence the men who are to wield those weapons—the modern working class—the proletarians.

In proportion as the bourgeoisie, *i.e.,* capital, is developed, in the same proportion is the proletariat, the modern working class, developed—a class of laborers, who live only so long as they find work, and who find work only so long as their labor increases capital. These laborers, who must sell themselves piece-meal, are a commodity, like every other article of commerce, and are consequently exposed to all the vicissitudes of competition, to all the fluctuations of the market.

Owing to the extensive use of machinery and to the division of labor, the work of the proletarians has lost all individual character, and consequently, all charm for the workman. He becomes an appendage of the machine, and it is only the most simple, most monotonous, and most easily acquired knack, that is required of him. Hence, the cost of production of a workman is restricted, almost entirely, to the means of subsistence that he requires for maintenance, and for the propagation of his race. But the price of a commodity, and therefore also of labor,[v] is equal to its cost of production. In proportion, therefore, as the repulsiveness of the work increases, the wage decreases. What is more, in proportion as the use of machinery and division of labor increases, in the same proportion the burden of toil also increases, whether by prolongation of the working hours, by the increase of the work exacted in a given time or by increased speed of machinery, etc.

Modern Industry has converted the little workshop of the patriarchal master into the great factory of the industrial capitalist. Masses of laborers, crowded into the factory, are organized like soldiers. As privates of the industrial army they are placed under the command of a perfect hierarchy of officers and sergeants. Not only are they slaves of the bourgeois class, and of the bourgeois state; they are daily and hourly enslaved by the machine, by the over-looker, and, above all, in the individual bourgeois manufacturer himself. The more openly this despotism proclaims gain to be its end and aim, the more petty, the more hateful and the more embittering it is.

The less the skill and exertion of strength implied in manual labor, in other words, the more modern industry becomes developed, the more is the labor of men superseded by that of women. Differences of age and sex have no longer any distinctive social validity for the

---

[v]Subsequently Marx pointed out that the worker sells not his labor but his labor power.

working class. All are instruments of labor, more or less expensive to use, according to their age and sex.

No sooner is the exploitation of the laborer by the manufacturer, so far, at an end, that he receives his wages in cash, than he is set upon by the other portion of the bourgeoisie, the landlord, the shopkeeper, the pawnbroker, etc.

The lower strata of the middle class—the small tradespeople, shopkeepers, and retired tradesmen generally, the handicraftsmen and peasants—all these sink gradually into the proletariat, partly because their diminutive capital does not suffice for the scale on which Modern Industry is carried on, and is swamped in the competition with the large capitalists, partly because their specialized skill is rendered worthless by new methods of production. Thus the proletariat is recruited from all classes of the population.

The proletariat goes through various stages of development. With its birth begins its struggle with the bourgeoisie. At first the contest is carried on by individual laborers, then by the work of people of a factory, then by the operative of one trade, in one locality, against the individual bourgeois who directly exploits them. They direct their attacks not against the bourgeois condition of production, but against the instruments of production themselves; they destroy imported wares that compete with their labor, they smash to pieces machinery, they set factories ablaze, they seek to restore by force the vanished status of the workman of the Middle Ages.

At this stage the laborers still form an incoherent mass scattered over the whole country, and broken up by their mutual competition. If anywhere they unite to form more compact bodies, this is not yet the consequence of their own active union, but of the union of the bourgeoisie, which class, in order to attain its own political ends, is compelled to set the whole proletariat in motion, and is moreover yet, for a time, able to do so. At this stage, therefore, the proletarians do not fight their enemies, but the enemies of their enemies, the remnants of absolute monarchy, the landowners, the non-industrial bourgeois, the petty bourgeois. Thus the whole historical movement is concentrated in the hands of the bourgeoisie; every victory so obtained is a victory for the bourgeoisie.

But with the development of industry the proletariat not only increases in number; it becomes concentrated in greater masses, its strength grows, and it feels that strength more. The various interests and conditions of life within the ranks of the proletariat are more and more equalized, in proportion as machinery obliterates all distinctions of labor, and nearly everywhere reduces wages to the same low level. The growing competition among the bourgeois, and the resulting commercial crises, make the wages of the workers ever more fluctuating. The increasing improvement of machinery, ever more rapidly developing, makes their livelihood more and more precarious; the collisions between individual workmen and individual bourgeois take more and more the character of collisions between two classes. Thereupon the workers begin to form combinations (trade unions) against the bourgeois; they club together in order to keep up the rate of wages; they found permanent associations in order to make provision beforehand for these occasional revolts. Here and there the contest breaks out into riots.

Now and then the workers are victorious, but only for a time. The real fruit of their battles lie, not in the immediate result, but in the ever-expanding union of the workers. This union is helped on by the improved means of communication that are created by Modern Industry and that place the workers of different localities in contact with one another. It was just this contact that was needed to centralize the numerous local struggles, all of the same character, into one national struggle between classes. But every class struggle is a political struggle. And that union, to attain which the burghers of the Middle Ages, with their miserable highways, required centuries, the modern proletarians, thanks to railways, achieve in a few years.

This organization of the proletarians into a class, and consequently into a political party, is continually being upset again by the competition between the workers themselves. But it ever rises up again, stronger, firmer, mightier. It compels legislative recognition of particular interests of the workers, by taking advantage of the divisions among the bourgeoisie itself. Thus the Ten-Hours Bill in England was carried.

Altogether collisions between the classes of the old society further, in many ways, the course of development of the proletariat. The bourgeoisie finds itself involved in a constant battle. At first with the aristocracy; later on, with those portions of the bourgeoisie itself, whose interests have become antagonistic to the progress of industry; at all time, with the bourgeoisie of foreign countries. In all these battles it sees itself compelled to appeal to the proletariat, to ask for its help, and thus, to drag it into the political arena. The bourgeoisie itself, therefore, supplies the proletariat with its own elements of political and general education, in other words, it furnishes the proletariat with weapons for fighting the bourgeoisie.

Further, as we have already seen, entire sections of the ruling class are, by the advance of industry, precipitated into the proletariat, or are at least threatened in their conditions of existence. These also supply the proletariat with fresh elements of enlightenment and progress.

Finally, in times when the class struggle nears the decisive hour, the progress of dissolution going on within the ruling class, in fact within the whole range of old society, assumes such a violent, glaring character, that a small section of the ruling class cuts itself adrift, and joins the revolutionary class, the class that holds the future in its hands. Just as, therefore, at an earlier period, a section of the nobility went over to the bourgeoisie, so now a portion of the bourgeoisie goes over to the proletariat, and in particular, a portion of the bourgeois ideologists, who have raised themselves to the level of comprehending theoretically the historical movement as a whole.

Of all the classes that stand face to face with the bourgeoisie today, the proletariat alone is a genuinely revolutionary class. The other classes decay and finally disappear in the face of Modern Industry; the proletariat is its special and essential product.

The lower middle class, the small manufacturer, the shopkeeper, the artisan, the peasant, all these fight against the bourgeoisie, to save from extinction their existence as fractions of the middle class. They are therefore not revolutionary, but conservative. Nay more, they are reactionary, for they try to roll back the wheel of history. If by chance they are revolutionary, they are so only in view of their impending transfer into the proletariat, they thus defend not their present, but their future interests, they desert their own standpoint to place themselves at that of the proletariat.

The "dangerous class," the social scum, that passively rotting mass thrown off by the lowest layers of the old society, may, here and there, be swept into the movement by a proletarian revolution; its conditions of life, however, prepare it far more for the part of a bribed tool of reactionary intrigue.

In the condition of the proletariat, those of old society at large are already virtually swamped. The proletarian is without property; his relation to his wife and children has no longer anything in common with the bourgeois family-relations; modern industry labor, modern subjection to capital, the same in England as in France, in America as in Germany, has stripped him of every trace of national character. Law, morality, religion, are to him so many bourgeois prejudices, behind which lurk in ambush just as many bourgeois interests.

All the preceding classes that got the upper hand, sought to fortify their already acquired status by subjecting society at large to their conditions of appropriation. The proletarians cannot become masters of the productive forces of society, except by abolishing their own previous mode of appropriation, and thereby also every other previous mode of appropriation. They have nothing of their own to secure and to fortify; their mission is to destroy all previous securities for, and insurances of, individual property.

All previous historical movements were movements of minorities, or in the interest of minorities. The proletarian movement is the self-conscious, independent movement of the immense majority, in the interest of the immense majority. The proletariat, the lowest stratum of our present society, cannot stir, cannot raise itself up, without the whole superincumbent strata of official society being sprung into the air.

Though not in substance, yet in form, the struggle of the proletariat with the bourgeoisie is at first a national struggle. The proletariat of each country must, of course, first of all settle matters with its own bourgeoisie.

In depicting the most general phases of the development of the proletariat, we traced the more or less veiled civil war, raging within existing society, up to the point where that war breaks out into open revolution, and where the violent overthrow of the bourgeoisie lays the foundation for the sway of the proletariat.

Hitherto, every form of society has been based, as we have already seen, on the antagonism of oppressing and oppressed classes. But in order to oppress a class, certain conditions must be assured to it under which it can, at least, continue its slavish existence. The serf, in the period of serfdom, raised himself to membership in the commune, just as the petty bourgeois, under the yoke of the feudal absolutism, managed to develop into a bourgeois. The modern laborer, on the contrary, instead of rising with the process of industry, sinks deeper and deeper below the conditions of existence of his own class. He becomes a pauper, and pauperism develops more rapidly than population and wealth. And here it becomes evident, that the bourgeoisie is unfit any longer to be the ruling class in society, and to impose its conditions of existence upon society as an over-riding law. It is unfit to rule because it is incompetent to assure an existence to its slave within his slavery, because it cannot help letting him sink into such a state, that it has to feed him, instead of being fed by him. Society can no longer live under this bourgeoisie, in other words, its existence is no longer compatible with society.

The essential conditions for the existence, and for the sway of the bourgeois class, is the formation and augmentation of capital; the condition for capital is wage-labor. Wage-labor rests exclusively on competition between the labourers. The advance of industry, whose involuntary promoter is the bourgeoisie, replaces the isolation of the laborers, due to competition, by the revolutionary combination, due to association. The development of Modern Industry, therefore, cuts from under its feet the very foundation on which the bourgeoisie produces and appropriates products. What the bourgeoisie, therefore, produces, above all, is its own grave-diggers. Its fall and the victory of the proletariat are equally inevitable.

PROLETARIANS AND COMMUNISTS

In what relation do the Communists stand to the proletarians as a whole?

The Communists do not form a separate party opposed to other working-class parties.

They have no interests separate and apart from those of the proletariat as a whole.

They do not set up any sectarian principles of their own, by which to shape and mold the proletarian movement.

The Communists are distinguished from the other working-class parties by this only: (1) In the national struggles of the proletarians of the different countries, they point out and bring to the front the common interests of the entire proletariat, independently of all nationality. (2) In the various stages of development which the struggle of the working class against the bourgeoisie has to pass through, they always and everywhere represent the interests of the movement as a whole.

The Communists, therefore, are on the one hand, practically, the most advanced and resolute section of the working-class parties of every country, that section which pushes forward all others; on the other hand, theoretically, they have over the great mass of the proletariat the advantage of clearly understanding the lines of march, the conditions, and the ultimate general results of the proletarian movement.

The immediate aim of the Communists is the same as that of all other proletarian parties: Formation of the proletariat into a class, overthrow of the bourgeois supremacy, conquest of political power by the proletariat.

The theoretical conclusions of the Communists are in no way based on ideas or principles that have been invented, or discovered, by this or that would-be universal reformer.

They merely express, in general terms, actual relations springing from an existing class struggle, from a historical movement going on under our very eyes. The abolition of existing property relations is not at all a distinctive feature of communism.

All property relations in the past have continually been subject to historical change consequent upon the change in historical conditions.

The French Revolution, for example, abolished feudal property in favor of bourgeois property.

The distinguishing feature of communism is not the abolition of property, but the abolition of bourgeois property generally, but modern bourgeois private property is the final and most complete expression of the system of producing and appropriating products, that is based on class antagonisms, on the exploitation of the many by the few.

In this sense, the theory of the Communists may be summed up in the single sentence: Abolition of private property.

We Communists have been reproached with the desire of abolishing the right of personally acquiring property as the fruit of a man's own labor, which property is alleged to be the groundwork of all personal freedom, activity and independence.

Hard-won, self-acquired, self-earned property! Do you mean the property of petty artisan and of the small peasant, a form of property that preceded the bourgeois form? There is no need to abolish that; the development of industry has to a great extent already destroyed it, and is still destroying it daily.

Or do you mean the modern bourgeois private property?

But does wage-labour create any property for the laborer? Not a bit. It creates capital, i.e., that kind of property which exploits wage-labor, and which cannot increase except upon conditions of begetting a new supply of wage-labor for fresh exploitation. Property, in its present form, is based on the antagonism of capital and wage-labor. Let us examine both sides of this antagonism.

To be a capitalist, is to have not only a purely personal, but a social *status* in production. Capital is a collective product, and only by the united action of many members, nay, in the last resort, only by the united action of all members of society, can it be set in motion.

Capital is, therefore, not only personal; it is a social power.

When, therefore, capital is converted into common property, into the property of all members of society, personal property is not thereby transformed into social property. It is only the social character of the property that is changed. It loses its class-character.

Let us now take wage-labor.

The average price of wage-labor is the minimum wage, i.e., that quantum of the means of subsistence, which is absolutely requisite to keep the laborer in bare existence as a laborer. What, therefore, the wage-laborer appropriates by means of his labour, merely suffices to prolong and reproduce a bare existence. We by no means intend to abolish this personal appropriation of the products of labor, an appropriation that is made for the maintenance and reproduction of human life, and that leaves no surplus wherewith to command the labour of others. All that we want to do away with, is the miserable character of this appropriation, under which the laborer lives merely to increase capital, and is allowed to live only in so far as the interest of the ruling class requires it.

In bourgeois society, living labor is but a means to increase accumulated labor. In communist society, accumulated labor is but a means to widen, to enrich, to promote the existence of the laborer.

In bourgeois society, therefore, the past dominates the present; in communist society, the present dominates the past. In bourgeois society capital is independent and has individuality, while the living person is dependent and has no individuality.

And the abolition of this state of things is called by the bourgeois, abolition of individuality and freedom! And rightly so. The abolition of bourgeois individuality, bourgeois independence, and bourgeois freedom is undoubtedly aimed at.

By freedom is meant, under the present bourgeois conditions of production, free trade, free selling and buying.

But if selling and buying disappears, free selling and buying disappears also. This talk about free selling and buying, and all the other "brave words" of our bourgeois about freedom in general, have a meaning, if any, only in contrast with restricted selling and buying, with the fettered traders of the Middle Ages, but have no meaning when opposed to the communistic abolition of buying and selling, or the bourgeois conditions of production, and of the bourgeoisie itself.

You are horrified at our intending to do away with private property. But in your existing society, private property is already done away with for nine-tenths of the population; its existence for the few is solely due to its non-existence in the hands of those nine-tenths. You reproach us, therefore, with intending to do away with a form of property, the necessary condition for whose existence is the non-existence of any property for the immense majority of society.

In one word, you reproach us with intending to do away with your property. Precisely so; that is just what we intend.

From the moment when labor can no longer be converted into capital, money, or rent, into a social power capable of being monopolized, i.e., from the moment when individual property can no longer be transformed into bourgeois property, into capital, from that moment, you say, individuality vanishes.

You must, therefore, confess that by "individual" you mean no other person than the bourgeois, than the middle-class owner of property. This person must, indeed, be swept out of the way, and made impossible.

Communism deprives no man of the power to appropriate the products of society; all that it does is to deprive him of the power to subjugate the labor of others by means of such appropriation.

It has been objected that upon the abolition of private property all work will cease, and universal laziness will overtake us.

According to this, bourgeois society ought long ago to have gone to the dogs through sheer idleness; for those who acquire anything, do not work. The whole of this objection is but another expression of the tautology: There can no longer be any wage-labor when there is no longer any capital.

All objections urged against the communistic mode of producing and appropriating material products have, in the same way, been urged against the communistic modes of producing and appropriating intellectual products. Just as, to the bourgeois, the disappearance of class property is the disappearance of production itself, so the disappearance of class culture is to him identical with the disappearance of all culture.

That culture, the loss of which he laments, is, for the enormous majority, a mere training to act as a machine.

But don't wrangle with us so long as you apply, to our intended abolition of bourgeois property, the standard of your bourgeois notions of freedom, culture, law, etc. Your very ideas are but the outgrowth of the conditions of your bourgeois production and bourgeois property, just as your jurisprudence is but the will of your class made into a law for all, a will, whose essential character and direction are determined by the economical conditions of existence of your class.

The selfish misconception that induces you to transform into eternal laws of nature and of reason, the social forms springing from your present mode of production and form of property—historical relations that rise and disappear in the progress of production—this misconception you share with every ruling class that has preceded you. What you see clearly in the case of ancient property, what you admit in the case of feudal property, you are of course forbidden to admit in the case of your own bourgeois form of property.

Abolition of the family! Even the most radical flare up at this infamous proposal of the Communists.

On what foundation is the present family, the bourgeois family, based? On capital, on private gain. In its completely developed form this family exists only among the bourgeoisie. But this state of things finds its complement in the practical absence of the family among proletarians, and in public prostitution.

The bourgeois family will vanish as a matter of course when its complement vanishes, and both will vanish with the vanishing of capital.

Do you charge us with wanting to stop the exploitation of children by their parents? To this crime we plead guilty.

But, you say, we destroy the most hallowed of relations, when we replace home education by social.

And your education! Is not that also social, and determined by the social conditions under which you educate, by the intervention, direct or indirect, of society, by means of schools, etc.? The Communists have not intended the intervention of society in education; they do but seek to alter the character of that intervention, and to rescue education from the influence of the ruling class.

The bourgeois clap-trap about the family and education, about the hallowed co-relation of parents and child, becomes all the more disgusting, the more, by the action of Modern Industry, all the family ties among the proletarians are torn asunder, and their children transformed into simple articles of commerce and instruments of labor.

But you Communists would introduce community of women, screams the bourgeoisie in chorus.

The bourgeois sees his wife a mere instrument of production. He hears that the instruments of production are to be exploited in common, and, naturally, can come to no other conclusion [than] that the lot of being common to all will likewise fall to the women.

He has not even a suspicion that the real point aimed at is to do away with the status of women as mere instruments of production.

For the rest, nothing is more ridiculous than the virtuous indignation of our bourgeois at the community of women which, they pretend, is to be openly and officially established by the Communists. The Communists have no need to introduce free love; it has existed almost from time immemorial.

Our bourgeois, not content with having wives and daughters of their proletarians at their disposal, not to speak of common prostitutes, take the greatest pleasure in seducing each other's wives.

Bourgeois marriage is in reality a system of wives in common and thus, at the most, what the Communists might possibly be reproached with, is that they desire to introduce, in substitution for a hypocritically concealed, an openly legalized system of free love. For the rest, it is self-evident that the abolition of the present system of production must bring with it the abolition of free love springing from that system, i.e., of prostitution both public and private.

The Communists are further reproached with desiring to abolish countries and nationality.

The working men have no country. We cannot take from them what they have not got. Since the proletariat must first of all acquire political supremacy, must rise to be the leading class of *the* nation, must constitute itself the nation, it is, so far, itself national, though not in the bourgeois sense of the word.

National differences and antagonism between peoples are daily more and more vanishing, owing to the development of the bourgeoisie, to freedom of commerce, to the world-market, to uniformity in the mode of production and in the conditions of life corresponding thereto.

The supremacy of the proletariat will cause them to vanish still faster. United action, of the leading civilized countries at least, is one of the first conditions for the emancipation of the proletariat.

In proportion as the exploitation of one individual by another will also be put an end to, the exploitation of one nation by another will also be put an end to. In proportion as the antagonism between classes within the nation vanishes, the hostility of one nation to another will come to an end.

The charges against communism made from a religious, a philosophical, and, generally, from an ideological standpoint, are not deserving of serious examination.

Does it require deep intuition to comprehend that man's ideas, views and conceptions, in one word, man's consciousness, changes with every change in the conditions of his material existence, in his social relations and in his social life?

What else does the history of ideas prove, than that intellectual production changes its character in proportion as material production is changed? The ruling ideas of each age have ever been the ideas of its ruling class.

When people speak of the ideas that revolutionize society, they do but express that fact, that within the old society, the elements of a new one have been created, and that the dissolution of the old ideas keeps even pace with the dissolution of the old conditions of existence.

When the ancient world was in its last throes, the ancient religions were overcome by Christianity. When Christian ideas succumbed in the eighteenth century to rationalist ideas, feudal society fought its death battle with the then revolutionary bourgeoisie. The ideas of religious liberty and freedom of conscience merely gave expression to the sway of free competition within the domain of knowledge.

"Undoubtedly," it will be said, "religious, moral, philosophical and juridicial ideas have been modified in the course of historical development. But religion, morality, philosophy, political science, and law, constantly survived this change."

"There are, besides, eternal truths, such as Freedom, Justice, etc., that are common to all states of society. But communism abolishes eternal truths, it abolishes all religion, and all morality, instead of constituting them on a new basis; it therefore acts in contradiction to all past historical experience."

What does this accusation reduce itself to? The history of all past society has consisted in the development of class antagonisms, antagonisms that assumed different forms at different epochs.

But whatever form they may have taken, one fact is common to all past ages, *viz.,* the exploitation of one part of society by the other. No wonder, then, that the social consciousness of past ages, despite all the multiplicity and variety it displays, moves within certain common forms, or general ideas, which cannot completely vanish except with the total disappearance of class antagonisms.

The communist revolution is the most radical rupture with traditional property relations; no wonder that its development involved the most radical rupture with traditional ideas.

But let us have done with the bourgeois objections to communism.

We have seen above, that the first step in the revolution by the working class, is to raise the proletariat to the position of ruling class, to win the battle of democracy.

The proletariat will use its political supremacy to wrest, by degrees, all capital from the bourgeoisie, to centralize all instruments of production in the hands of the state, *i.e.,* of the proletariat organized as the ruling class; and to increase the total productive forces as rapidly as possible.

Of course, in the beginning, this cannot be effected except by means of despotic inroads on the rights of property, and on the conditions of bourgeois production; by means of measures, therefore, which appear economically insufficient and untenable, but which, in the course of the movement, outstrip themselves, necessitate further inroads upon the old social order, and are unavoidable as a means of entirely revolutionizing the mode of production.

These measures will of course be different in different countries.

Nevertheless in most advanced countries, the following will be pretty generally applicable.

1. Abolition of property in land and application of all rents of land to public purposes.

2. A heavy progressive or graduated income tax.

3. Abolition of all right of inheritance.

4. Confiscation of the property of all emigrants and rebels.

5. Centralization of credit in the hands of the state, by means of a national bank with state capital and an exclusive monopoly.

6. Centralization of the means of communication and transport in the banks of the state.

7. Extension of factories and instruments of production owned by the state; the bringing into cultivation of waste-lands, and the improvement of the soil generally in accordance with a common plan.

8. Equal obligation of all to work. Establishment of industrial armies, especially for agriculture.

9. Combination of agriculture with manufacturing industries; gradual abolition of all the distinction between town and country, by a more equable distribution of the populace over the country.

10. Free education for all children in public schools. Abolition of children's factory labor in its present form. Combination of education with industrial production, etc.

When, in the course of development, class distinctions have disappeared, and all production has been concentrated in the hands of a vast association of the whole nation, the public power will lose its political character. Political power, properly so called, is merely the organised power of one class for oppressing another. If the proletariat during its contest with the bourgeoisie is compelled, by the force of circumstances, to organise itself as a class, if, by means of a revolution, it makes itself the ruling class, and, as such, sweeps away by force the old conditions of production, then it will, along with these conditions, have swept away the conditions for the existence of class antagonisms and of classes generally, and will thereby have abolished its own supremacy as a class.

In place of the old bourgeois society, with its classes and class antagonisms, we shall have an association, in which the free development of each is the condition for the free development of all.

▦

Introduction to *Capital*

In this section, we turn to what many consider Marx's masterpiece of economic analysis: *Capital.* Here, we provide excerpts from two chapters: "Commodities" and "The General Formula for Capital."

In "Commodities," Marx explores the sources of "value" by asking what determines the worth or price of goods bought and sold on the market. In answering this question, Marx again borrowed from the work of Adam Smith to draw a distinction between "use-value" and "exchange-value." Use-value refers to the utility of a commodity or its ability to satisfy wants.[1] A commodity has use-value only if it is consumed or otherwise put to use. For instance, a one-legged stool cannot readily satisfy a person's desire to sit; therefore, it has no use-value for most individuals. The use-value of a commodity, however, does not determine its actual price; although the usefulness of a commodity may differ between individuals (maybe you really do prefer sitting on a one-legged stool), the cost of the good does not likewise change (we'll all pay the same price for it). Moreover, because use-value refers to the *qualities* of commodities—what they do—it cannot establish a *quantifiable* standard for measuring the price of goods. After all, how can one quantify and compare the usefulness of a lightbulb with that of a fork?

Exchange-value, on the other hand, does express equivalencies—how much of a given commodity (e.g., corn) it takes to equal the value of another commodity (e.g., iron). Because exchange-value is derived from trade, it cannot be a property inherent in the commodity itself. Instead, it is dependent on what goods are being exchanged. For instance, one DVD player might be exchanged fairly for one guitar, two jackets, or three CD burners. Thus, a DVD player has not one, but many exchange-values. But if different quantities of different commodities can nevertheless be equal in exchange-value, then the value of the commodities must be determined by something else separate from yet common to the commodities themselves.

---

[1]Marx explicitly excluded questions concerning the origins of "wants" as well as how commodities actually satisfied them. Some Marxist-inspired theorists, most notably those associated with the Frankfurt School, would later turn their attention to precisely such questions—that is, how the continued expansion of capitalism requires the production of "false" needs.

For Marx, this common "something else" is labor. In Marx's **labor theory of value** (which he appropriated from Adam Smith and David Ricardo), the value of an object is determined ultimately by the amount of labor time (hours, weeks, months, etc.) that it took to produce it. "Commodities, therefore, in which equal quantities of labour are embodied, or which can be produced in the same time, have the same value. . . . As values, all commodities are only definite masses of congealed labour-time" (Marx [1867] 1978:306). By equating the value of goods with labor time, Marx not only outlined the economic principles that purportedly guide exchange; he also unmasked the root source of exploitation inherent in capitalist production.

In a capitalist economy, those who do not own the means of production have no choice but to sell their labor power in order to survive. The worker's labor power is thus treated as a commodity exchanged, in this case, for a wage. But at what rate is the worker paid? What determines the exchange-value of labor? Like all other commodities, the value of labor power is a function of the amount of labor time necessary to produce itself. In other words, the value of labor power is equivalent to the costs incurred by the worker for food, clothing, shelter, training, and other goods necessary to ensure both the survival of his family and his return to work the next day.

However, the length of the working day exceeds the time needed on the job in order for the worker to reproduce his labor power. Say, for instance, that in six hours of work a laborer is able to produce for the capitalist the equivalent value of what he needs in order to support his family and return to work. Because the worker's wage is equal to the value of the goods necessary for his family's survival, he is paid, in this case, for six hours worth of labor. Yet, the capitalist employs the worker for a longer duration, say 12 hours a day. During these additional six hours, the worker produces surplus value for the capitalist. **Surplus value** is the difference between what workers earn for their labor and the price or value of the goods that they produce. Surplus value is thus the source of the capitalist's profit: the capitalist pays the worker less than the value of what she actually produces. Human labor is thus the one commodity that is exchanged for its value while being capable of producing more than its value.

To illustrate this concept more clearly, consider a simplified example of a furniture manufacturing plant employing 100 workers. A worker paid $10.00 an hour to assemble tables would earn $400 for a 40-hour workweek. Annually, the worker would earn $20,800. This annual wage would barely keep a family of four out of poverty, to say nothing of attaining the "American Dream." On the other hand, let's assume the worker assembles 100 tables over the course of a year, each sold on the market for $300. The worker thus generates $30,000 for the owner of the plant. The nearly $10,000 difference between wages earned and money generated is appropriated by the capitalist both to reinvest in her business and to support her own family. While this may not seem like a significant difference, recall that the plant employs 100 workers, each of whose labor produces roughly $30,000 in sales. Now the owner is appropriating nearly $1 million in surplus value over the course of only one year, while the workers, whose labor produced the goods sold on the market for a profit, cling with their families to a near-poverty existence.

Additionally, private ownership of the means of the production allows the owner to control the production process and appropriate the products, thus enabling him to take this profit solely for himself. In turn, surplus value is also the source of the capitalists' **exploitation** of the worker because the worker gives more than is given in return without having any voice in this relationship of exchange.

In his effort to increase his profit and market share, the capitalist has two principal means at his disposal: increasing "absolute" or increasing "relative" surplus value. He can increase his *absolute surplus value* by extending the working day. The increase in hours on the job, in turn, increases the productivity of his workforce. With wages remaining constant, greater productivity yields higher profits for the capitalist. During Marx's time, 12- and 14-hour working days were not uncommon, and capitalists routinely opposed legislation aimed at reducing laborers' hours.

Capitalists can also increase their *relative surplus value*. This stems from increasing the productivity of labor by instituting timesaving procedures. With a decrease in the time and thus the cost of production, a capitalist is able to undersell competitors and capture a larger share of the market. For instance, production efficiency can be improved as capitalists specialize their labor force by reorganizing workers and the allocation of tasks. Specialization simplifies a worker's role in the production process so that, rather than performing a variety of tasks, his contribution is reduced to one or two operations. Often this entails adopting an assembly line system of manufacturing such as Henry Ford did when he revolutionized the automobile industry in the early twentieth century. However, although specialization increases efficiency by enabling more products to be produced in less time, it also leads to the routinization of labor and the workers' loss of self-fulfillment.

Similarly, in their competition for markets, capitalists can turn to more-sophisticated machines and technology to enable laborers to produce more goods in less time. To the extent that mechanized production decreases the necessary labor time, surplus value is increased, along with the level of worker alienation and exploitation.

Although a machine may be able to run 24 hours a day (and does not need insurance or bathroom breaks), mechanized production has its costs. In the short run, it can lead to a reduction in profits, despite the higher volume of productivity, as machines take the place of workers who are the capitalist's source of surplus value. Increasing productivity as a means for selling commodities more cheaply than one's competitors sell also *compels* a capitalist to sell more products and dominate a larger share of the market. Without selling more commodities, the capitalist cannot offset the lower selling price and the expense of adopting more costly machines, to say nothing of turning a profit. Moreover, as the capitalist's competitors begin to make use of the new technology, she is forced to seek—and pay for—ever-newer and more-efficient machines, lest she suffer the very fate she intends to inflict on others.

The competition for markets and the need to increase productivity bear long-run costs, as well. Specialization and mechanization force more workers into unstable employment and a marginal existence. Needed to perform only the most monotonous of unskilled tasks, workers become easily replaceable and expendable. Indeed, "it is the absolute interest of every capitalist to press a given quantity of labour out of a smaller, rather than a greater number of labourers," because doing so increases their relative surplus value and accumulation of capital (Marx [1867] 1978:425). As a result, an "industrial reserve army" of unemployed and underemployed laborers is created, the ranks of which swell as the employed segments of the proletariat are overworked. Thus, despite the increasing levels of productivity and growth in the amount of wealth controlled by the capitalists, the market for their products begins to shrink as a growing "relative surplus-population" of laborers is left unable to afford little more than the necessities for survival. At the same time, the increasing competition for jobs due to the expanding industrial reserve army combines with the marginalization of skills to decrease the wages of those fortunate enough to be employed. Meanwhile, competition between capitalists forever breeds greater specialization and mechanization, and all that follows in their wake. Recurring crises of overproduction and "boom or bust" are thus endemic to the capitalist system, while economic recessions and depressed wages become more severe.[2]

In this chapter, Marx also reworks his earlier analysis of alienation in the form of the "fetishism of commodities." Recall that alienation, according to Marx, is a dehumanizing consequence of the worker's estrangement or separation from the means of production and the goods produced (see our

---

[2]Though Marx contended that the continuing expansion of the industrial reserve army operates as "a law of population peculiar to the capitalist mode of production" (Marx [1867] 1978:423), it is clear that rising rates of unemployment are not inevitable, nor are fluctuations in rates of unemployment due entirely to changing levels of production. Instead, unemployment rates are as much a product of government policy as they are of general economic conditions. Nevertheless, a recent (2006) report issued by the International Labour Organization revealed that the number of people unemployed worldwide reached an all-time high of 191.8 million in 2005, an increase of 34.4 million (21 percent) since 1995. Additionally, of the more than 2.8 billion workers in the world, 1.4 billion earned less than $2 dollars per day.

discussion of "Alienated Labour" earlier). Similarly, commodity fetishism refers to the distorted relationship existing between individuals and the production and consumption of goods. However, in fetishizing commodities, Marx argues that we treat the goods we buy as if they have "magical" powers. We lose sight of the fact that *we* create commodities and, in doing so, grant them a power over us that in reality they do not hold.

Perhaps you can think of how products directed at our personal appearance are marketed. Advertisements for shampoos, lotions, deodorants, toothpastes, and the like routinely convey the message that interpersonal "success" is dependent on our using these products. Boy gets girl because he buys a specific brand of mouthwash. Girl gets boy because she uses a toothpaste that "whitens" her teeth. Likewise, driving a particular type of car or drinking a particular brand of soft drink or beer magically transforms us into the "type" of person who uses the products. In each instance, our accomplishments and failures are derived not from who we are as individuals, but magically from what we buy as consumers. As a result, our social interactions as well as our sense of self are mediated through or steered by products, not by our individual qualities. When we fetishize commodities, we relate to things, not people. (Compare Marx's argument here with the one made earlier in the excerpt from "The Power of Money in Bourgeois Society.")

Not only are commodities fetishized, but so too is the process of commodity production. When we blame machines for our dissatisfaction, we endow them with human qualities of conscious intent or will. In turn, we fail to recognize that it is the owner of the means of production who is responsible for transforming the production process, not the machines. Thus, if the introduction of new technology increases the speed of the labor process or alters how that process is organized among workers, fetishizing commodity production prevents laborers from holding capitalists accountable for their growing dissatisfaction. Instead, workers will assign the source of their increasing exploitation not to the capitalists who benefit from it, but to the new technology. This carries with it important political consequences, because the intrinsically social nature of the production process is veiled, making workers less able to effectively press their class-based interests for change. The Luddites were one such group of handicraft workers who in early-nineteenth-century England destroyed the textile machines that rendered their skilled labor obsolete, displacing them with cheap, unskilled laborers. Their protests were met with repressive government actions that included hangings and imprisonment in exile.

Finally, in "The General Formula for Capital," Marx describes the cycle or circulation of commodities peculiar to capitalism. Unlike other economic arrangements, production under capitalism is driven by the quest for increasing profits and capital for reinvestment, not toward simply fulfilling needs or wants established through tradition. Guiding the profit motive is a cycle of exchange Marx labeled "M-C-M." By definition, the capitalist enters into economic exchange already possessing **capital** (raw materials, machinery for production) or, more generally, money (M). Seeking to expand her business and profits, the capitalist converts her money into a commodity (C) by purchasing additional machinery, raw materials, or labor. The capitalist then uses these commodities to produce other commodities that are then sold for money (M). Hence, the meaning of the slogan "It takes money to make money."

For the proletariat, the cycle of exchange takes an inverse path. Take a typical wage earner, for example. The worker enters into the labor market possessing only his labor power, which he sells as a commodity (C). His commodity, labor, is then exchanged for money (M) or a wage. The worker then takes the money and spends it on the commodities (C) necessary to his survival. The circulation of commodities here follows the pattern C-M-C. The worker sells his one commodity in order to purchase goods he does not otherwise possess. Such a pattern of exchange cannot generate a profit. Instead, it is a cycle of economic activity that provides solely for the satisfaction of basic needs and a subsistence level of existence. Moreover, this cycle must be repeated daily as the commodities bought by the worker—food, fuel, clothing, shelter—tied as they are to survival, are more or less immediately consumed or in need of continual replacement. Rent is paid not once, but monthly. Clothes are bought not once, but regularly, when worn out or outgrown.

Capital (1867)

Karl Marx

COMMODITIES

The Two Factors of a Commodity:
Use-Value and Value (The Substance
of Value and the Magnitude of Value)

The wealth of those societies in which the capitalist mode of production prevails, presents itself as "an immense accumulation of commodities," its unit being a single commodity. Our investigation must therefore begin with the analysis of a commodity.

A commodity is, in the first place, an object outside us, a thing that by its properties satisfies human wants of some sort or another. The nature of such wants, whether, for instance, they spring from the stomach or from fancy, makes no difference. Neither are we here concerned to know how the object satisfies these wants, whether directly as means of subsistence, or indirectly as means of production.

Every useful thing, as iron, paper, etc., may be looked at from the two points of view of quality and quantity. It is an assemblage of many properties, and may therefore be of use in various ways. To discover the various uses of things is the work of history. So also is the establishment of socially recognized standards of measure for the quantities of these useful objects. The diversity of these measures has its origin partly in the diverse nature of the objects to be measured, partly in convention.

The utility of a thing makes it a use-value. But this utility is not a thing of air. Being limited by the physical properties of the commodity, it has no existence apart from that commodity. A commodity, such as iron, corn, or a diamond, is therefore, so far as it is a material thing, a use-value, something useful. This property of a commodity is independent of the amount of labour required to appropriate its useful qualities. When treating of use-value, we always assume to be dealing with definite quantities, such as dozens of watches, yards of linen, or tons of iron. The use-values of commodities furnish the material for a special study, that of the commercial knowledge of commodities.[i] Use-values become a reality only by use or consumption: they also constitute the substance of all wealth, whatever may be the social form of that wealth. In the form of society we are about to consider, they are, in addition, the material depositories of exchange-value.

Exchange-value, at first sight, presents itself as a quantitative relation, as the proportion in which values in use of one sort are exchanged for those of another sort, a relation constantly changing with time and place. Hence exchange-value appears to be something accidental and purely relative, and consequently an intrinsic value, *i.e.,* an exchange-value that is inseparably connected with, inherent in commodities, seems a contradiction in terms. Let us consider the matter a little more closely.

A given commodity, *e.g.,* a quarter of wheat is exchanged for x blacking, y silk, or z gold, etc.—in short, for other commodities in the most different proportions. Instead of one exchange-value, the wheat has, therefore, a great many. But since x blacking, y silk, or z gold, etc., each represent the exchange-value of one quarter of wheat, x blacking, y silk, z gold, etc., must, as exchange-values, be replaceable by each other, or equal to each other. Therefore, first: the valid exchange-values of a given commodity express something equal; secondly, exchange-value, generally, is only the mode of expression, the phenomenal form, of something contained in it, yet distinguishable from it.

Let us take two commodities, *e.g.,* corn and iron. The proportions in which they are exchangeable, whatever those proportions may be, can always be represented by an equation in which a given quantity of corn is equated to some quantity of iron: *e.g.,* 1 quarter corn = x cwt. iron. What does this equation tell us? It tells us that in two different things—in 1 quarter of corn and x cwt. of iron, there exists in equal quantities something common to both. The two things must therefore be equal to a

---

SOURCE: Marx/Engels Internet Archive.

[i]In bourgeois societies the economic fictio juris prevails, that every one, as a buyer, possesses an encyclopaedic knowledge of commodities. [Marx]

third, which in itself is neither the one nor the other. Each of them, so far as it is exchange-value, must therefore be reducible to this third.

A simple geometrical illustration will make this clear. In order to calculate and compare the areas of rectilinear figures, we decompose them into triangles. But the area of the triangle itself is expressed by something totally different from its visible figure, namely, by half the product of the base multiplied by the altitude. In the same way the exchange-values of commodities must be capable of being expressed in terms of something common to them all, of which thing they represent a greater or less quantity.

This common "something" cannot be either a geometrical, a chemical, or any other natural property of commodities. Such properties claim our attention only in so far as they affect the utility of those commodities, make them use-values. But the exchange of commodities is evidently an act characterised by a total abstraction from use-value. Then one use-value is just as good as another, provided only it be present in sufficient quantity. Or, as old Barbon says, "one sort of wares are as good as another, if the values be equal. There is no difference or distinction in things of equal value. . . . An hundred pounds' worth of lead or iron, is of as great value as one hundred pounds' worth of silver or gold." As use-values, commodities are, above all, of different qualities, but as exchange-values they are merely different quantities, and consequently do not contain an atom of use-value.

If then we leave out of consideration the use-value of commodities, they have only one common property left, that of being products of labour. But even the product of labour itself has undergone a change in our hands. If we make abstraction from its use-value, we make abstraction at the same time from the material elements and shapes that make the product a use-value; we see in it no longer a table, a house, yarn, or any other useful thing. Its existence as a material thing is put out of sight. Neither can it any longer be regarded as the product of the labour of the joiner, the mason, the spinner, or of any other definite kind of productive labour. Along with the useful qualities of the products themselves, we put out of sight both the useful character of the various kinds of labour embodied in them, and the concrete forms of that labour; there is nothing left but what is common to them all: all are reduced to one and the same sort of labour, human labour in the abstract.

Let us now consider the residue of each of these products; it consists of the same unsubstantial reality in each, a mere congelation of homogeneous human labour, of labour-power expended without regard to the mode of its expenditure. All that these things now tell us is, that human labour-power has been expended in their production, that human labour is embodied in them. When looked at as crystals of this social substance, common to them all, they are—Values.

We have seen that when commodities are exchanged, their exchange-value manifests itself as something totally independent of their use-value. But if we abstract from their use-value, there remains their Value as defined above. Therefore, the common substance that manifests itself in the exchange-value of commodities, whenever they are exchanged, is their value. The progress of our investigation will show that exchange-value is the only form in which the value of commodities can manifest itself or be expressed. For the present, however, we have to consider the nature of value independently of this, its form.

A use-value, or useful article, therefore, has value only because human labour in the abstract has been embodied or materialised in it. How, then, is the magnitude of this value to be measured? Plainly, by the quantity of the value-creating substance, the labour, contained in the article. The quantity of labour, however, is measured by its duration, and labour-time in its turn finds its standard in weeks, days, and hours.

Some people might think that if the value of a commodity is determined by the quantity of labour spent on it, the more idle and unskilful the labourer, the more valuable would his commodity be, because more time would be required in its production. The labour, however, that forms the substance of value, is homogeneous human labour, expenditure of one uniform labour-power. The total labour-power of society, which is embodied in the sum total of the values of all commodities produced by that society, counts here as one homogeneous mass of human labour-power, composed though it be of innumerable individual units. Each of these units is the same as any other, so far as it has the character of the average labour-power of society, and takes effect as such; that is, so far as it requires for producing a commodity, no more time than is needed on an average, no more than is socially necessary. The labour-time socially necessary is that required to produce an article under the normal conditions of production, and with the average degree of skill and intensity prevalent at the time. The introduction of power-looms into England probably reduced by one-half the labour required to weave a given quantity of yarn into cloth. The hand-loom weavers, as a matter of fact, continued to require the same time as before; but for all that, the product of one hour of their labour represented after the change only half an hour's social labour, and consequently fell to one-half its former value.

We see then that that which determines the magnitude of the value of any article is the amount of labour socially necessary, or the labour-time socially necessary for its production. Each individual commodity, in this connexion, is to be considered as an average sample of its class. Commodities, therefore, in which equal quantities of labour are embodied, or which can be produced in the same time, have the same value. The value of one commodity is to the value of any other, as the labour-time necessary for the production of the one is to that necessary for the production of the other. "As values, all commodities are only definite masses of congealed labour-time."

The value of a commodity would therefore remain constant, if the labour-time required for its production also remained constant. But the latter changes with every variation in the productiveness of labour. This productiveness is determined by various circumstances, amongst others, by the average amount of skill of the workmen, the state of science, and the degree of its practical application, the social organisation of production, the extent and capabilities of the means of production, and by physical conditions. For example, the same amount of labour in favourable seasons is embodied in 8 bushels of corn, and in unfavourable, only in four. The same labour extracts from rich mines more metal than from poor mines. Diamonds are of very rare occurrence on the earth's surface, and hence their discovery costs, on an average, a great deal of labour-time. Consequently much labour is represented in a small compass. Jacob doubts whether gold has ever been paid for at its full value. This applies still more to diamonds. According to Eschwege, the total produce of the Brazilian diamond mines for the eighty years, ending in 1823, had not realised the price of one-and-a-half years' average produce of the sugar and coffee plantations of the same country, although the diamonds cost much more labour, and therefore represented more value. With richer mines, the same quantity of labour would embody itself in more diamonds, and their value would fall. If we could succeed at a small expenditure of labour, in converting carbon into diamonds, their value might fall below that of bricks. In general, the greater the productiveness of labour, the less is the labour-time required for the production of an article, the less is the amount of labour crystallised in that article, and the less is its value; and *vice versa,* the less the productiveness of labour, the greater is the labour-time required for the

production of an article, and the greater is its value. The value of a commodity, therefore, varies directly as the quantity, and inversely as the productiveness, of the labour incorporated in it.

A thing can be a use-value, without having value. This is the case whenever its utility to man is not due to labour. Such are air, virgin soil, natural meadows, etc. A thing can be useful, and the product of human labour, without being a commodity. Whoever directly satisfies his wants with the produce of his own labour, creates, indeed, use-values, but not commodities. In order to produce the latter, he must not only produce use-values, but use-values for others, social use-values. (And not only for others, without more. The medieval peasant produced quit-rent-corn for his feudal lord and tithe-corn for his parson. But neither the quit-rent-corn nor the tithe-corn became commodities by reason of the fact that they had been produced for others. To become a commodity a product must be transferred to another, whom it will serve as a use-value, by means of an exchange.)[ii] Lastly nothing can have value, without being an object of utility. If the thing is useless, so is the labour contained in it; the labour does not count as labour, and therefore creates no value. . . .

The Fetishism of Commodities and the Secret Thereof

A commodity appears, at first sight, a very trivial thing, and easily understood. Its analysis shows that it is, in reality, a very queer thing, abounding in metaphysical subtleties and theological niceties. So far as it is a value in use, there is nothing mysterious about it, whether we consider it from the point of view that by its properties it is capable of satisfying human wants, or from the point that those properties are the product of human labour. It is as clear as noon-day, that man, by his industry, changes the forms of the materials furnished by Nature, in such a way as to make them useful to him. The form of wood, for instance, is altered, by making a table out of it. Yet, for all that, the table continues to be that common, every-day thing, wood. But, so soon as it steps forth as a commodity, it is changed into something transcendent. It not only stands with its feet on the ground, but, in relation to all other commodities, it stands on its head, and evolves out of its wooden brain grotesque ideas, far more wonderful than "table-turning" ever was.

[ii]I am inserting the parenthesis because its omission has often given rise to the misunderstanding that every product that is consumed by someone other than its producer is considered in Marx a commodity. [Engels, 4th German edition]

The mystical character of commodities does not originate, therefore, in their use-value. Just as little does it proceed from the nature of the determining factors of value. For, in the first place, however varied the useful kinds of labour, or productive activities, may be, it is a physiological fact, that they are functions of the human organism, and that each such function, whatever may be its nature or form, is essentially the expenditure of human brain, nerves, muscles, etc. Secondly, with regard to that which forms the ground-work for the quantitative determination of value, namely, the duration of that expenditure, or the quantity of labour, it is quite clear that there is a palpable difference between its quantity and quality. In all states of society, the labour-time that it costs to produce the means of subsistence, must necessarily be an object of interest to mankind, though not of equal interest in different stages of development. And lastly, from the moment that men in any way work for one another, their labour assumes a social form.

Whence, then, arises the enigmatical character of the product of labour, so soon as it assumes the form of commodities? Clearly from this form itself. The equality of all sorts of human labour is expressed objectively by their products all being equally values; the measure of the expenditure of labour-power by the duration of that expenditure, takes the form of the quantity of value of the products of labour; and finally, the mutual relations of the producers, within which the social character of their labour affirms itself, take the form of a social relation between the products.

A commodity is therefore a mysterious thing, simply because in it the social character of men's labour appears to them as an objective character stamped upon the product of that labour; because the relation of the producers to the sum total of their own labour is presented to them as a social relation, existing not between themselves, but between the products of their labour. This is the reason why the products of labour become commodities, social things whose qualities are at the same time perceptible and imperceptible by the senses. In the same way the light from an object is perceived by us not as the subjective excitation of our optic nerve, but as the objective form of something outside the eye itself. But, in the act of seeing, there is at all events, an actual passage of light from one thing to another, from the external object to the eye. There is a physical relation between physical things. But it is different with commodities. There, the existence of the things *quâ* commodities, and the value-relation between the products of labour which stamps them as

commodities, have absolutely no connexion with their physical properties and with the material relations arising therefrom. There it is a definite social relation between men, that assumes, in their eyes, the fantastic form of a relation between things. In order, therefore, to find an analogy, we must have recourse to the mist-enveloped regions of the religious world. In that world the productions of the human brain appear as independent beings endowed with life, and entering into relation both with one another and the human race. So it is in the world of commodities with the products of men's hands. This I call the Fetishism which attaches itself to the products of labour, so soon as they are produced as commodities, and which is therefore inseparable from the production of commodities.

This Fetishism of commodities has its origin, as the foregoing analysis has already shown, in the peculiar social character of the labour that produces them.

As a general rule, articles of utility become commodities, only because they are products of the labour of private individuals or groups of individuals who carry on their work independently of each other. The sum total of the labour of all these private individuals forms the aggregate labour of society. Since the producers do not come into social contact with each other until they exchange their products, the specific social character of each producer's labour does not show itself except in the act of exchange. In other words, the labour of the individual asserts itself as a part of the labour of society, only by means of the relations which the act of exchange establishes directly between the products, and indirectly, through them, between the producers. To the latter, therefore, the relations connecting the labour of one individual with that of the rest appear, not as direct social relations between individuals at work, but as what they really are, material relations between persons and social relations between things. It is only by being exchanged that the products of labour acquire, as values, one uniform social status, distinct from their varied forms of existence as objects of utility. This division of a product into a useful thing and a value becomes practically important, only when exchange has acquired such an extension that useful articles are produced for the purpose of being exchanged, and their character as values has therefore to be taken into account, beforehand, during production. From this moment the labour of the individual producer acquires socially a two-fold character. On the one hand, it must, as a definite useful kind of labour, satisfy a definite social want, and thus hold its place as part and parcel of the collective labour of all, as a branch of a social division of labour that has sprung up spontaneously. On the other hand, it can satisfy the manifold wants of the individual producer himself, only in so far as the mutual exchangeability of all kinds of useful private labour is an established social fact, and therefore the private useful labour of each producer ranks on an equality with that of all others. The equalisation of the most different kinds of labour can be the result only of an abstraction from their inequalities, or of reducing them to their common denominator, viz., expenditure of human labour-power or human labour in the abstract. The two-fold social character of the labour of the individual appears to him, when reflected in his brain, only under those forms which are impressed upon that labour in every-day practice by the exchange of products. In this way, the character that his own labour possesses of being socially useful takes the form of the condition, that the product must be not only useful, but useful for others, and the social character that his particular labour has of being the equal of all other particular kinds of labour, takes the form that all the physically different articles that are the products of labour, have one common quality, viz., that of having value.

Hence, when we bring the products of our labour into relation with each other as values, it is not because we see in these articles the material receptacles of homogeneous human labour. Quite the contrary: whenever, by an exchange, we equate as values our different products, by that very act, we also equate, as human labour, the different kinds of labour expended upon them. We are not aware of this, nevertheless we do it. Value, therefore, does not stalk about with a label describing what it is. It is value, rather, that converts every product into a social hieroglyphic. Later on, we try to decipher the hieroglyphic, to get behind the secret of our own social products; for to stamp an object of utility as a value, is just as much a social product as language. The recent scientific discovery, that the products of labour, so far as they are values, are but material expressions of the human labour spent in their production, marks, indeed, an epoch in the history of the development of the human race, but, by no means, dissipates the mist through which the social character of labour appears to us to be an objective character of the products themselves. The fact, that in the particular form of production with which we are dealing, viz., the production of commodities, the specific social character of private labour carried on independently, consists in the equality of every kind of that labour, by virtue of its being human labour, which character, therefore, assumes in the product the form of value—this fact appears to the producers, notwithstanding the discovery

above referred to, to be just as real and final, as the fact, that, after the discovery by science of the component gases of air, the atmosphere itself remained unaltered.

What, first of all, practically concerns producers when they make an exchange, is the question, how much of some other product they get for their own? In what proportions the products are exchangeable? When these proportions have, by custom, attained a certain stability, they appear to result from the nature of the products, so that, for instance, one ton of iron and two ounces of gold appear as naturally to be of equal value as a pound of gold and a pound of iron in spite of their different physical and chemical qualities appear to be of equal weight. The character of having value, when once impressed upon products, obtains fixity only by reason of their acting and re-acting upon each other as quantities of value. These quantities vary continually, independently of the will, foresight and action of the producers. To them, their own social action takes the form of the action of objects, which rule the producers instead of being ruled by them. It requires a fully developed production of commodities before, from accumulated experience alone, the scientific conviction springs up, that all the different kinds of private labour, which are carried on independently of each other, and yet as spontaneously developed branches of the social division of labour, are continually being reduced to the quantitative proportions in which society requires them. And why? Because, in the midst of all the accidental and ever fluctuating exchange-relations between the products, the labour-time socially necessary for their production forcibly asserts itself like an over-riding law of Nature. The law of gravity thus asserts itself when a house falls about our ears. The determination of the magnitude of value by labour-time is therefore a secret, hidden under the apparent fluctuations in the relative values of commodities. Its discovery, while removing all appearance of mere accidentality from the determination of the magnitude of the values of products, yet in no way alters the mode in which that determination takes place.

Man's reflections on the forms of social life, and consequently, also, his scientific analysis of those forms, take a course directly opposite to that of their actual historical development. He begins, post festum, with the results of the process of development ready to hand before him. The characters that stamp products as commodities, and whose establishment is a necessary preliminary to the circulation of commodities, have already acquired the stability of natural, self-understood forms of social life, before man seeks to decipher, not their historical character, for in his eyes they are immutable, but their meaning. Consequently it was the analysis of the prices of commodities that alone led to the determination of the magnitude of value, and it was the common expression of all commodities in money that alone led to the establishment of their characters as values. It is, however, just this ultimate money-form of the world of commodities that actually conceals, instead of disclosing, the social character of private labour, and the social relations between the individual producers. When I state that coats or boots stand in a relation to linen, because it is the universal incarnation of abstract human labour, the absurdity of the statement is self-evident. Nevertheless, when the producers of coats and boots compare those articles with linen, or, what is the same thing, with gold or silver, as the universal equivalent, they express the relation between their own private labour and the collective labour of society in the same absurd form.

The categories of bourgeois economy consist of such like forms. They are forms of thought expressing with social validity the conditions and relations of a definite, historically determined mode of production, viz., the production of commodities. The whole mystery of commodities, all the magic and necromancy that surrounds the products of labour as long as they take the form of commodities, vanishes therefore, so soon as we come to other forms of production. . . .

The life-process of society, which is based on the process of material production, does not strip off its mystical veil until it is treated as production by freely associated men, and is consciously regulated by them in accordance with a settled plan. This, however, demands for society a certain material ground-work or set of conditions of existence which in their turn are the spontaneous product of a long and painful process of development.

Political Economy has indeed analysed, however incompletely, value and its magnitude, and has discovered what lies beneath these forms. But it has never once asked the question why labour is represented by the value of its product and labour-time by the magnitude of that value. These formulæ, which bear it stamped upon them in unmistakable letters that they belong to a state of society, in which the process of production has the mastery over man, instead of being controlled by him, such formulæ appear to the bourgeois intellect to be as much a self-evident necessity imposed by Nature as productive labour itself. Hence forms of social production that preceded the bourgeois form, are treated by the bourgeoisie in much the same way as the Fathers of the Church treated pre-Christian religions.

To what extent some economists are misled by the Fetishism inherent in commodities, or by the objective appearance of the social characteristics of labour, is shown, amongst other ways, by the dull and tedious quarrel over the part played by Nature in the formation of exchange-value. Since exchange-value is a definite social manner of expressing the amount of labour bestowed upon an object, Nature has no more to do with it, than it has in fixing the course of exchange.

The mode of production in which the product takes the form of a commodity, or is produced directly for exchange, is the most general and most embryonic form of bourgeois production. It therefore makes its appearance at an early date in history, though not in the same predominating and characteristic manner as now-a-days. Hence its Fetish character is comparatively easy to be seen through. But when we come to more concrete forms, even this appearance of simplicity vanishes. Whence arose the illusions of the monetary system? To it gold and silver, when serving as money, did not represent a social relation between producers but were natural objects with strange social properties. And modern economy, which looks down with such disdain on the monetary system, does not its superstition come out as clear as noon-day, whenever it treats of capital? How long is it since economy discarded the physiocratic illusion, that rents grow out of the soil and not out of society?

But not to anticipate, we will content ourselves with yet another example relating to the commodity-form. Could commodities themselves speak, they would say: Our use-value may be a thing that interests men. It is no part of us as objects. What, however, does belong to us as objects, is our value. Our natural intercourse as commodities proves it. In the eyes of each other we are nothing but exchange-values. Now listen how those commodities speak through the mouth of the economist. "Value"—(*i.e.,* exchange-value) "is a property of things, riches"—(*i.e.,* use-value) "of man. Value, in this sense, necessarily implies exchanges, riches do not." "Riches" (use-value) "are the attribute of men, value is the attribute of commodities. A man or a community is rich, a pearl or a diamond is valuable. . . . A pearl or a diamond is valuable" as a pearl or diamond. So far no chemist has ever discovered exchange-value either in a pearl or a diamond. The economic discoverers of this chemical element, who by-the-by lay special claim to critical acumen, find however that the use-value of objects belongs to them independently of their material properties, while their value, on the other hand, forms a part of them as objects. What confirms them in this view, is the peculiar circumstance that the use-value of objects is realised without exchange, by means of a direct relation between the objects and man, while, on the other hand, their value is realised only by exchange, that is, by means of a social process. Who fails here to call to mind our good friend, Dogberry, who informs neighbour Seacoal, that, "To be a well-favoured man is the gift of fortune; but reading and writing comes by Nature."

THE GENERAL FORMULA FOR CAPITAL

The circulation of commodities is the starting-point of capital. The production of commodities, their circulation, and that more developed form of their circulation called commerce, these form the historical groundwork from which it rises. The modern history of capital dates from the creation in the 16th century of a world-embracing commerce and a world-embracing market.

If we abstract from the material substance of the circulation of commodities, that is, from the exchange of the various use-values, and consider only the economic forms produced by this process of circulation, we find its final result to be money: this final product of the circulation of commodities is the first form in which capital appears.

As a matter of history, capital, as opposed to landed property, invariably takes the form at first of money; it appears as moneyed wealth, as the capital of the merchant and of the usurer. But we have no need to refer to the origin of capital in order to discover that the first form of appearance of capital is money. We can see it daily under our very eyes. All new capital, to commence with, comes on the stage, that is, on the market, whether of commodities, labour, or money, even in our days, in the shape of money that by a definite process has to be transformed into capital.

The first distinction we notice between money that is money only, and money that is capital, is nothing more than a difference in their form of circulation.

The simplest form of the circulation of commodities is C—M—C, the transformation of commodities into money, and the change of the money back again into commodities; or selling in order to buy. But alongside of this form we find another specifically different form: M—C—M, the transformation of money into commodities, and the change of commodities back again into money; or buying in order to sell. Money that circulates in the latter manner is thereby transformed into, becomes capital, and is already potentially capital.

Now let us examine the circuit M—C—M a little closer. It consists, like the other, of two antithetical phases. In the first phase, M—C, or the purchase, the money is changed into a commodity. In the second phase, C—M, or the sale, the commodity is changed back again into money. The combination of these two phases constitutes the single movement whereby money is exchanged for a commodity, and the same commodity is again exchanged for money; whereby a commodity is bought in order to be sold, or, neglecting the distinction in form between buying and selling, whereby a commodity is bought with a commodity. The result, in which the phases of the process vanish, is the exchange of money for money, M—M. If I purchase 2,000 lbs. of cotton for £100, and resell the 2,000 lbs. of cotton for £110, I have, in fact, exchanged £100 for £110, money for money.

Now it is evident that the circuit M—C—M would be absurd and without meaning if the intention were to exchange by this means two equal sums of money, £100 for £100. The miser's plan would be far simpler and surer; he sticks to his £100 instead of exposing it to the dangers of circulation. And yet, whether the merchant who has paid £100 for his cotton sells it for £110, or lets it go for £100, or even £50, his money has, at all events, gone through a characteristic and original movement, quite different in kind from that which it goes through in the hands of the peasant who sells corn, and with the money thus set free buys clothes. We have therefore to examine first the distinguishing characteristics of the forms of the circuits M—C—M and C—M—C, and in doing this the real difference that underlies the mere difference of form will reveal itself.

Let us see, in the first place, what the two forms have in common.

Both circuits are resolvable into the same two antithetical phases, C—M, a sale, and M—C, a purchase. In each of these phases the same material elements—a commodity, and money, and the same economic dramatis personae, a buyer and a seller—confront one another. Each circuit is the unity of the same two antithetical phases, and in each case this unity is brought about by the intervention of three contracting parties, of whom one only sells, another only buys, while the third both buys and sells.

What, however, first and foremost distinguishes the circuit C—M—C from the circuit M—C—M, is the inverted order of succession of the two phases. The simple circulation of commodities begins with a sale and ends with a purchase, while the circulation of money as capital begins with a purchase and ends with a sale. In the one case both the starting-point and the goal are commodities, in the other they are money. In the first form the movement is brought about by the intervention of money, in the second by that of a commodity.

In the circulation C—M—C, the money is in the end converted into a commodity, that serves as a use-value; it is spent once for all. In the inverted form, M—C—M, on the contrary, the buyer lays out money in order that, as a seller, he may recover money. By the purchase of his commodity he throws money into circulation, in order to withdraw it again by the sale of the same commodity. He lets the money go, but only with the sly intention of getting it back again. The money, therefore, is not spent, it is merely advance.

In the circuit C—M—C, the same piece of money changes its place twice. The seller gets it from the buyer and pays it away to another seller. The complete circulation, which begins with the receipt, concludes with the payment, of money for commodities. It is the very contrary in the circuit M—C—M. Here it is not the piece of money that changes its place twice, but the commodity. The buyer takes it from the hands of the seller and passes it into the hands of another buyer. Just as in the simple circulation of commodities the double change of place of the same piece of money effects its passage from one hand into another, so here the double change of place of the same commodity brings about the reflux of the money to its point of departure.

Such reflux is not dependent on the commodity being sold for more than was paid for it. This circumstance influences only the amount of the money that comes back. The reflux itself takes place, so soon as the purchased commodity is resold, in other words, so soon as the circuit M—C—M is completed. We have here, therefore, a palpable difference between the circulation of money as capital, and its circulation as mere money.

The circuit C—M—C comes completely to an end, so soon as the money brought in by the sale of one commodity is abstracted again by the purchase of another.

If, nevertheless, there follow a reflux of money to its starting-point, this can only happen through a renewal or repetition of the operation. If I sell a quarter of corn of £3, and with this £3 buy clothes, the money, so far as I am concerned, is spent and done with. It belongs to the clothes merchant. If I now sell a second quarter of corn, money indeed flows back to me, not however as a sequel to the first transaction, but in consequence of its repetition. The money again leaves me, so soon as I complete this second transaction by a fresh purchase. Therefore, in the circuit C—M—C, the expenditure of money has nothing to do with its reflux. On the other hand, in M—C—M, the reflux of the money is conditioned by the very mode of its expenditure. Without this reflux,

the operation fails, or the process is interrupted and incomplete, owing to the absence of its complementary and final phase, the sale.

The circuit C—M—C starts with one commodity, and finishes with another, which falls out of circulation and into consumption. Consumption, the satisfaction of wants, in one word, use-value, is its end and aim. The circuit M—C—M, on the contrary, commences with money and ends with money. Its leading motive, and the goal that attracts it, is therefore mere exchange-value.

In the simple circulation of commodities, the two extremes of the circuit have the same economic form. They are both commodities, and commodities of equal value. But they are also use-values differing in their qualities, as, for example, corn and clothes. The exchange of products, of the different materials in which the labour of society is embodied, forms here the basis of the movement. It is otherwise in the circulation M—C—M, which at first sight appears purposeless, because tautological. Both extremes have the same economic form. They are both money, and therefore are not qualitatively different use-values; for money is but the converted form of commodities, in which their particular use-values vanish. To exchange £100 for cotton, and then this same cotton again for £110, is merely is roundabout way of exchanging money for money, the same for the same, and appears to be an operation just as purposeless as it is absurd. One sum of money is distinguishable from another only by its amount. The character and tendency of the process M—C—M, is therefore not due to any qualitative difference between its extremes, both being money, but solely to their quantitative difference. More money is withdrawn from circulation at the finish than was thrown into it at the start. The cotton that was bought for £100 is perhaps resold for £100 + £10 or £110. The exact form of this process is therefore M—C—M', where M' + VM = M the original sum advanced, plus an increment. This increment or excess over the original value I call "surplus-value." The value originally advanced, therefore, not only remains intact while in circulation, but adds to itself a surplus-value or expands itself. It is this movement that converts it into capital.

Of course, it is also possible, that in C—M—C, the two extremes C—C, say corn and clothes, may represent different quantities of value. The farmer may sell his corn above its value, or may buy the clothes at less than their value. He may, on the other hand, "be done" by the clothes merchant. Yet, in the form of circulation now under consideration, such differences in value are purely accidental. The fact that the corn and the clothes are equivalents, does not deprive the process of all meaning, as it does in M—C—M. The equivalence of their values is rather a necessary condition to its normal course.

The repetition or renewal of the act of selling in order to buy, is kept within bounds by the very object it aims at, namely, consumption or the satisfaction of definite wants, an aim that lies altogether outside the sphere of circulation. But when we buy in order to sell, we, on the contrary, begin and end with the same thing, money, exchange-value; and thereby the movement becomes interminable. No doubt, M becomes M + VM, £100 become £110. But when viewed in their qualitative aspect alone, £110 are the same as £100, namely money; and considered quantitatively, £110 is, like £100, a sum of definite and limited value. If now, the £110 be spent as money, they cease to play their part. They are no longer capital. Withdrawn from circulation, they become petrified into a hoard, and though they remained in that state till doomsday, not a single farthing would accrue to them. If, then, the expansion of value is once aimed at, there is just the same inducement to augment the value of the £110 as that of the £100; for both are but limited expressions for exchange-value, and therefore both have the same vocation to approach, by quantitative increase, as near as possible to absolute wealth. Momentarily, indeed, the value originally advanced, the £100, is distinguishable from the surplus-value of £10 that is annexed to it during circulation; but the distinction vanishes immediately. At the end of the process, we do not receive with one hand the original £100, and with the other, the surplus-value of £10. We simply get a value of £110, which is in exactly the same condition and fitness for commencing the expanding process, as the original £100 was. Money ends the movement only to begin it again.[iii] Therefore, the final result of every separate circuit, in which a purchase and consequent sale are completed, forms of itself the starting-point of a new circuit. The simple circulation of commodities—selling in order to buy—is a means of carrying out a purpose unconnected with circulation, namely, the appropriation of use-values, the satisfaction of wants. The circulation of money as capital is, on the contrary, an end in itself, for the expansion of value takes place only within this constantly renewed movement. The circulation of capital has therefore no limits.

As the conscious representative of this movement, the possessor of money becomes a capitalist. His person, or rather his pocket, is the point from which the money starts and to which it returns. The expansion of value, which is the objective basis or main-spring of the circulation M—C—M, becomes his subjective aim, and it is only in so far as the appropriation of ever more and more wealth in the abstract becomes the sole motive of his operations, that he functions as a capitalist, that is, as capital personified and endowed with consciousness and a will. Use-values must therefore never be looked upon as the real aim of the capitalist; neither must the profit on any single transaction. The restless never-ending process of profit-making alone is what he aims at. This boundless greed after riches, this passionate chase after exchange-value, is common to the capitalist and the miser; but while the miser is merely a capitalist gone mad, the capitalist is a rational miser. The never-ending augmentation of exchange-value, which the miser strives after, by seeking to save his money from circulation, is attained by the more acute capitalist, by constantly throwing it afresh into circulation.

The independent form, *i.e.,* the money-form, which the value of commodities assumes in the case of simple circulation, serves only one purpose, namely, their exchange, and vanishes in the final result of the movement. On the other hand, in the circulation M—C—M, both the money and the commodity represent only different modes of existence of value itself, the money its general mode, and the commodity its particular, or, so to say, disguised mode. It is constantly changing from one form to the other without thereby becoming lost, and thus assumes an automatically active character. If now we take in turn each of the two different forms which self-expanding value successively assumes in the course of its life, we then arrive at these two propositions: Capital is money: Capital is commodities. In truth, however, value is here the active factor in a process, in which, while constantly assuming the form in turn of money and commodities, it at the same time changes in magnitude, differentiates itself by throwing off surplus-value from itself; the original value, in other words, expands spontaneously. For the movement, in the course of which it adds surplus-value, is its own movement, its expansion, therefore, is automatic expansion. Because it

---

[iii]"Capital is divisible . . . into the original capital and the profit, the increment to the capital . . . although in practice this profit is immediately turned into capital, and set in motion with the original." (F. Engels, "Umrisse zu einer Kritik der Nationalökonomie, in the "Deutsch-Französische Jahrbücher," edited by Arnold Ruge and Karl Marx." Paris, 1844, p. 99.) [Marx]

is value, it has acquired the occult quality of being able to add value to itself. It brings forth living offspring, or, at the least, lays golden eggs.

Value, therefore, being the active factor in such a process, and assuming at one time the form of money, at another that of commodities, but through all these changes preserving itself and expanding, it requires some independent form, by means of which its identity may at any time be established. And this form it possesses only in the shape of money. It is under the form of money that value begins and ends, and begins again, every act of its own spontaneous generation. It began by being £100, it is now £110, and so on. But the money itself is only one of the two forms of value. Unless it takes the form of some commodity, it does not become capital. There is here no antagonism, as in the case of hoarding, between the money and commodities. The capitalist knows that all commodities, however scurvy they may look, or however badly they may smell, are in faith and in truth money, inwardly circumcised Jews, and what is more, a wonderful means whereby out of money to make more money.

In simple circulation, C—M—C, the value of commodities attained at the most a form independent of their use-values, i.e., the form of money; but that same value now in the circulation M—C—M, or the circulation of capital, suddenly presents itself as an independent substance, endowed with a motion of its own, passing through a life-process of its own, in which money and commodities are mere forms which it assumes and casts off in turn. Nay, more: instead of simply representing the relations of commodities, it enters now, so to say, into private relations with itself. It differentiates itself as original value from itself as surplus-value; as the father differentiates himself from himself quâ the son, yet both are one and of one age: for only by the surplus-value of £10 does the £100 originally advanced become capital, and so soon as this takes place, so soon as the son, and by the son, the father, is begotten, so soon does their difference vanish, and they again become one, £110.

Value therefore now becomes value in process, money in process, and, as such, capital. It comes out of circulation, enters into it again, preserves and multiplies itself within its circuit, comes back out of it with expanded bulk, and begins the same round ever afresh. M—M', money which begets money, such is the description of Capital from the mouths of its first interpreters, the Mercantilists.

Buying in order to sell, or, more accurately, buying in order to sell dearer, M—C—M', appears certainly to be a form peculiar to one kind of capital alone, namely merchants' capital. But industrial capital too is money, that is changed into commodities, and by the sale of these commodities, is re-converted into more money.

The events that take place outside the sphere of circulation, in the interval between the buying and selling, do not affect the form of this movement. Lastly, in the case of interest-bearing capital, the circulation M—C—M' appears abridged. We have its result without the intermediate stage, in the form M—M', "en style lapidaire" so to say, money that is worth more money, value that is greater than itself.

M—C—M' is therefore in reality the general formula of capital as it appears prima facie within the sphere of circulation.

Discussion Questions

1. According to Marx's materialist conception of history, what is the relationship between property or the division of labor and consciousness? How might property relations and ideas prevent or promote social change?

2. Do you think that truly communist societies have existed? Can they exist? What are some of the features that such a society must have in order for it to work?

3. What role does private property play in Marx's analysis of the inevitable communist revolution? In his emphasis on class, what factors might Marx have overlooked when accounting for revolutionary change or its absence?

4. Has the proletariat, or working class, sunk deeper and deeper with the advance of industry as Marx suggested? Why or why not? How prevalent is alienation in contemporary capitalist societies? Don't some people like their jobs? If so, have they been "fooled" somehow? Why or why not?

5. Discuss the prevalence of the fetishism of commodities in contemporary capitalist societies. What examples of commodity fetishism do you see in your own life and the lives of your family and friends?

3 ÉMILE DURKHEIM (1858–1917)

Émile Durkheim

Key Concepts

- :: Social facts
- :: Social solidarity
  - ❑ Mechanical solidarity
  - ❑ Organic solidarity
- :: Anomie
- :: Collective conscience
- :: Ritual
- :: Symbol
- :: Collective representations
- :: Sacred and profane

There can be no society which does not feel the need of upholding and reaffirming at regular intervals the collective sentiments and the collective ideas which makes its unity and its personality. Now this moral remaking cannot be achieved except by the means of reunions, assemblies and meetings where the individuals, being closely united to one another, reaffirm in common their common sentiments.

—Durkheim ([1912] 1995:474–75)

Have you ever been to a professional sports event in a stadium full of fans? Or gone to a religious service and taken communion, or to a concert and danced in the aisles (or maybe a mosh pit)? How did these experiences make you feel? What do they have in common? Is it possible to have this same type of experience if/when you are alone? How so or why not?

These are the sorts of issues that intrigued Émile Durkheim. Above all, he sought to explain *what* held societies and social groups together—and *how*. In addressing these twin questions, Durkheim studied a wide variety of phenomena—from suicide and crime, to aboriginal religious totems and symbols. He was especially concerned about how modern, industrial societies can be held together when people don't even know each other and when their experiences and social positions are so varied. In other words, how can social ties, the very basis for society, be maintained in such an increasingly individualistic world?

Yet, Durkheim is an important figure in the history of sociology not only because of his provocative theories about social cohesion but also because he helped found the discipline of sociology. In contrast to some of the other figures whose works you will read in this book, Durkheim sought to delineate, both theoretically and methodologically, how sociology was different from existing schools of philosophy and history (which also examined social issues). Before we discuss his ideas and work, however, let's look at his biography for, like Marx, Durkheim's personal experiences and historical situation deeply influenced his perception and description of the social world.

◼◼ A BIOGRAPHICAL SKETCH

Émile Durkheim was born in a small town in eastern France in 1858. In his youth he followed family tradition, studying Hebrew and the Talmud in order to become a rabbi. However, in his adolescence, Durkheim apparently rejected Judaism. Though he did not disdain traditional religion, as a child of the Enlightenment (see Chapter 1) he came to consider both Christianity and Judaism outmoded in the modern world.

In 1879, Durkheim entered France's most prestigious college, the École Normale Supérieure in Paris, to study philosophy. However, by his third year, Durkheim had become disenchanted with the high-minded, literary humanities curriculum at the Normale. He decided to pursue sociology, which he viewed as eminently more scientific, democratic, and practical. Durkheim still maintained his interest in complex philosophical questions, but he wanted to examine them through a "rational," "scientific" framework. His practical and scientific approach to central social issues would shape his ambition to use sociological methods as a means for reconstituting the moral order of French society, which he saw decaying in the aftermath of the French Revolution (Bellah 1973:xiii–xvi). Durkheim was especially concerned about the abuse of power by political and military leaders, the increasing rates of divorce and suicide, and the rising anti-Semitism. It seemed to Durkheim that social bonds and a sense of community had broken down and social disorder had come to prevail.[1]

Upon graduation from the École Normale, Durkheim began teaching in small lycées (state-run secondary schools) near Paris. In 1887, he married Louise Dreyfus, a woman from the Alsace region of France. In the same year, Durkheim began his career as a professor at the University of Bordeaux, where he quickly gained the reputation for being a committed and exciting teacher. Émile and Louise soon had two children, Marie and Andre.

Durkheim was a serious and productive scholar. His first book, *The Division of Labor in Society,* which was based on his doctoral dissertation, came out in 1893; his second, *The Rules of Sociological Method,* appeared just 2 years later. In 1897, *Suicide,* perhaps his most well-known work, was published. The next year, Durkheim founded the journal *L'Année Sociologique,* which was one of the first sociology journals not only in France, but also in the world. *L'Année Sociologique* was produced annually until the outbreak of World War I in 1914.

---

[1]As indicated in Chapter 1, France has gone through numerous violent changes in government since the French Revolution in 1789. Between 1789 and 1870, there had been three monarchies, two empires, and two republics, culminating in the notorious reign of Napoleon III, who overthrew the democratic government and ruled France for 20 years. Though the French Revolution had brought a brief period of democracy, it also sparked a terrifying persecution of all those who disagreed with the revolutionary leaders. Some 17,000 revolutionaries were executed in the infamous Reign of Terror, led by Maximilien Robespierre. Consequently, political and social divisions in France intensified. French conservatives called for a return to monarchy and a more prominent role for the Catholic Church. In direct contrast, a growing, but still relatively small, class of urban workers demanded political rights and a secular rather than religious education. At the same time, capitalists called for individual rights and free markets, while radical socialists advocated abolishing private property altogether.

In 1902, with his reputation as a leading social philosopher and scientist established, Durkheim was offered a position at the prestigious Sorbonne University in Paris. As he had done previously at Bordeaux, Durkheim quickly gained a large following at the Sorbonne. His education courses were compulsory for all students seeking teaching degrees in philosophy, history, literature, and languages. Durkheim also became an important administrator at the Sorbonne, serving on numerous councils and committees (Lukes 1985:372).

Yet not everyone was enamored with either Durkheim's substantial power or his ideas. Durkheim's notion that *any* social "thing"—including religion—could be studied sociologically (i.e., scientifically) was particularly controversial, as was his adamant insistence on providing students a moral, but secular, education. (These two issues will be discussed further below.) As Steven Lukes (1985:373), noted sociologist and Durkheim scholar, remarked, "To friends he was a prophet and an apostle, but to enemies he was a secular pope."

Moreover, Durkheim identified with some of the goals of socialism, but he was unwilling to commit himself politically. Durkheim believed that sociologists should be committed to education, not political activism—his passion was for dispassionate, scientific research.

This apparent apoliticism, coupled with his focus on the moral constitution of societies (rather than conflict and revolution), has led some analysts to deem Durkheim politically conservative. However, as the eminent sociologist Robert Bellah (1973:xviii) points out, "to try to force Durkheim into the conservative side of some conservative/liberal dichotomy" is inappropriate. It ignores Durkheim's "lifelong preoccupation with orderly, continuous social change toward greater social justice" (ibid.:xvii). In addition, to consider Durkheim politically conservative is also erroneous in light of how he was evaluated in his day. Durkheim was viewed as a radical modernist and liberal, who, though respectful of religion, was most committed to rationality, science, and humanism. Durkheim infuriated religious conservatives, who desired to replace democracy with a monarchy and strengthen the military. He also came under fire because he opposed instituting Catholic education as the basic curriculum.

Moreover, to label Durkheim "conservative" ignores his role in the "Dreyfus affair." Alfred Dreyfus was a Jewish army colonel who was charged and convicted on false charges of spying for Germany. The charges against Dreyfus were rooted in anti-Semitism, which was growing in the 1890s, alongside France's military losses and economic dissatisfaction. Durkheim was very active in the Ligue des droits de l'homme (League of the Rights of Men), which devoted itself to clearing Dreyfus of all charges.

Interestingly, Durkheim's assessment of the Dreyfus affair reflects his lifelong concern for the moral order of society. He saw the Dreyfus affair as symptomatic of a collective moral sickness, rather than merely anti-Semitism at the level of the individual. As Durkheim (1899, as cited by Lukes 1985) states,

> When society undergoes suffering, it feels the need to find someone whom it can hold responsible for its sickness, on whom it can avenge its misfortunes; and those against whom public opinion already discriminates are naturally designated for this role. These are the pariahs who serve as expiatory victims. What confirms me in this interpretation is the way in which the result of Dreyfus's trial was greeted in 1894. There was a surge of joy in the boulevards. People celebrated as a triumph what should have been a cause of public mourning. At least they knew whom to blame for the economic troubles and moral distress in which they lived. The trouble came from the Jews. The charge had been officially proved. By this very fact alone, things already seemed to be getting better and people felt consoled. (p. 345)

In 1912, Durkheim's culminating work, *The Elementary Forms of Religious Life,* was published. Shortly after that World War I broke out, and Durkheim's life was thrown into turmoil. His son, Andre, was killed in battle, spiraling Durkheim into a grief from which he never fully recovered. On October 7, 1916, as he was leaving a committee meeting at the Sorbonne, Durkheim suffered a stroke. He spent the next year resting and seemed to have made much progress toward recovering. But on November 15, 1917, while in Fontainebleau where he had gone for peace and fresh air, Durkheim died. He was 59 years old (ibid.:559).

As indicated previously, Durkheim wrote a number of books and articles on a wide variety of topics. Nevertheless, there are two major themes that transcend all of Durkheim's work. First, Durkheim sought to articulate the nature of society and, hence, his view of sociology as an academic discipline. Durkheim argued that society was a supraindividual force existing independently of the actors who compose it. The task of sociology, then, is to analyze **social facts**—conditions and circumstances external to the individual that, nevertheless, determine one's course of action. Durkheim argued that social facts can be ascertained by using collective data, such as suicide and divorce rates. In other words, through systematic collection of data, the patterns behind and within individual behavior can be uncovered. This emphasis on formal methods and objective data is what distinguished sociology from philosophy and put sociology "on the map" as a viable scientific discipline.

Significant Others

Auguste Comte (1798–1857): The Father of "Social Physics"

Born in southern France during a most turbulent period in French history, Auguste Comte was himself a turbulent figure. Though he excelled as a student, he had little patience for authority. Indeed, his obstinate temperament prevented him from completing his studies at the newly established École Polytechnic, Paris's elite university. Nevertheless, Comte was able to make a name for himself in the intellectual circlesof Paris. In 1817, he began working as a secretary and collaborator to Henri Saint-Simon. Their productive though fractious relationship came to an end seven years later in a dispute over assigning authorship to one of Comte's essays. Comte next set about developing his system of positivist philosophy while working in minor academic positions for meager wages. Beginning in 1926, Comte offered a series of private lectures in an effort to disseminate his views. Though attended by eminent thinkers, the grandiosity of histheoretical system led some to dismiss his ideas. Nevertheless, Comte continued undeterred, and from 1830 to 1842 he worked single-mindedly on his magnum opus, the six-volume *Cours de philosophie positive*. In the series, Comte not only outlines his "Law of Three Stages" discussed earlier, but also delineates the proper methods for his new science of "social physics" as well as its fundamental task—the study of social statics (order) and dynamics (progress). The work was well received in some scientific quarters, and Comte seemed poised to establish himself as a first-rate scholar. Unfortunately, his temperament again proved to be a hindrance to his success, both personal and professional. His troubled marriage ended soon after *Cours* was completed, and his petulance further alienated him from friends and colleagues while costing him a position at the École. Comte's life took a turn for the better, however, when in 1844 he met and fell in love with Colthilde de Vaux. Their affair did not last long; Colthilde developed tuberculosis and died within a year of their first meeting. Comte dedicated the rest of his life to "his angel." In her memory he founded the Religion of Humanity for which he proclaimed himself the High Priest. The new Church was founded on the principle of universal love as Comte abandoned his earlier commitment to science and positivism. Until his death in 1857, Comte sought not supporters for his system of science but also converts to his Positive Church.

NOTE: This account of Comte's biography is based largely on Lewis Coser's (1977) discussion in *Masters of Sociological Thought*.

The significance of Durkheim's position for the development of sociology as a distinct pursuit of knowledge cannot be overstated. As one of the first academics to hold a position in sociology, Durkheim was on the cutting edge of the birth of the new discipline. Nevertheless, his conviction that society is sui generis (an objective reality that is irreducible to the individuals that compose it) and amenable to scientific investigation owes much to the work of **Auguste Comte** (1798–1857). Not only had Comte coined the term *sociology* in 1839; he also contended that the social world could be studied in as rational and scientific a way as physical scientists (chemists, physicists, biologists, etc.) study their respective domains. Moreover, Durkheim's comparative and historical methodology was in large measure a continuation of the approach advocated earlier by Comte.

A second major theme found in Durkheim's work is the issue of **social solidarity**, or the cohesion of social groups. As you will see, all of the selections in this chapter—from *The Division of Labor in Society, Suicide, The Rules of Sociological Method,* and *The Elementary Forms of Religious Life*—explore the nature of the bonds that tie social groups and/or the individual to society. Durkheim was especially concerned about modern societies where people often don't know their neighbors (let alone everyone in the larger community) or worship together and where people often hold jobs in impersonal companies and organizations. Durkheim wondered *how* individuals could feel tied to one another in such an increasingly individualistic world. This issue was of utmost importance, for he maintained that without some semblance of solidarity and moral cohesion, society could not exist.

Significant Others

Herbert Spencer (1820–1903): Survival of the Fittest

Born in the English Midlands, Herbert Spencer's early years were shaped largely by his father and uncle. It was from these two men that Spencer received his education, an education that centered on math, physics, and chemistry. Moreover, it was from them that Spencer was exposed to the radical religious and social doctrines that would inform his staunch individualism. With little formal instruction in history, literature, and languages, Spencer conceded to the limits of his education and at the age of sixteen declined to attend university, opting instead to pursue a "practical" career as an engineer for the London and Birmingham Railway. Nevertheless, he would prove to be an avid student of, and a prolific writer on, a range of social and philosophical topics. With the completion of the railway in 1841, Spencer earned his living by writing essays for a number of radical journals. Of particular note is a series of letters he published through a dissenting paper, *The Nonconformist.* Titled "The Proper Sphere of Government," the letters are an early expression of Spencer's decidedly laissez-faire perspective. In them, Spencer argued that the role of government should be restricted solely to policing, while all other matters, including education, social welfare, and economic activities, should be left to the workings of the private sector. According to Spencer, government regulations interfere with the laws of human evolution which, if left unhampered, ensure the "survival of the fittest." It is not hard to see that Spencer's view of government still resonates with many American politicians and voters. Less sanguine, however, is the racism and sexism that was interjected into Spencer's argument. Following the logic of his view, those who don't survive, that is, succeed, are merely fulfilling their evolutionary destiny. To the extent that women and people of color are less "successful" than white men, their "failure" is deemed a product of their innate inferiority. Spencer ignored that both "success" and "failure" hinge not only on individual aptitude and effort but also on institutional and cultural dynamics that sustain a less than level playing field.

In his emphasis on the nature of solidarity in "traditional" and "modern" societies, Durkheim again drew on Comte's work as well as that of the British sociologist **Herbert Spencer** (1820–1903).[2] Both Comte and Spencer formulated an organic view of society to explain the developmental paths along which societies allegedly evolve. Such a view depicted society as a system of interrelated parts (religious institutions, the economy, government, the family) that work together to form a unitary, stable whole, analogous to how the parts of the human body (lungs, kidneys, brain) function interdependently to sustain its general well-being. Moreover, as the organism (society and the body) grows in size, it becomes increasingly complex, due to the differentiation of its parts.

However, Durkheim was only partly sympathetic to the organic, evolutionary models developed by Comte and Spencer. On one hand, Durkheim's insistence that social solidarity is rooted in shared moral sentiments, and the sense of obligation they evoke, stems from Comte (and Jean-Jacques Rousseau as well). Likewise, his notion that the specialized division of labor characteristic of modern societies leads to greater interdependency and integration owes much to Comte (as well as Saint-Simon). Nevertheless, Durkheim did not embrace Comte's assertion that all societies progress through a series of identifiable evolutionary stages. In particular, he dismissed Comte's "Law of Three Stages" wherein all societies—as well as individual intellectual development—are said to pass from a theological stage characterized by "militaristic" communities led by priests, to a metaphysical stage organized according to "legalistic" principles and controlled by lawyers and clergy, and finally, to a positivist or scientific stage in which "industrial" societies are governed by technocrats and, of course, sociologists.

In terms of Spencer, Durkheim was most influenced by Spencer's theory on the evolution of societies. According to Spencer, just as biological organisms become more differentiated as they grow and mature, so do small-scale, homogeneous communities become increasingly complex and diverse as a result of population growth. The individuals living in simple societies are minimally dependent upon one another for meeting their survival and that of the community as they each carry out similar tasks. As the size of the population increases, however, similarity and likeness is replaced by heterogeneity and a specialized division of labor. Individuals become interdependent upon one another as essential tasks are divided among the society's inhabitants. As a result, an individual's well-being becomes tied more and more to the general welfare of the larger society. Ensuring the functional integration of individuals now becomes the central issue for the survival of the society.

In this regard, Durkheim's perspective is compatible with that of Spencer. As discussed is later, Durkheim hypothesized that a different *kind* of solidarity was prevalent in modern, as opposed to smaller, more traditional, societies. Durkheim's equation of traditional societies with "mechanical" solidarity and modern societies with "organic" solidarity (discussed on pp. 93–94) shares an affinity with Spencer's classification of societies as either "simple" or "compound."

However, the two theorists diverge on the crucial point of integration. Spencer saw society as composed of atomistic individuals each pursuing lines of self-interested conduct. In a classic expression of utilitarian philosophy, Spencer maintained that a stable, well-functioning social whole is the outgrowth of individuals freely seeking to maximize their advantages.

By contrast, Durkheim (and Comte) took a far less utilitarian approach than Spencer. Durkheim emphasized that society is not a result or aftereffect of individual conduct; rather, it exists prior to, and thus shapes, individual action. In other words, individual lines of conduct are the outgrowth of social arrangements, particularly those connected to the developmental stage of the division of labor. Social integration, then, cannot be an unintended consequence of an aggregate of individuals pursuing their self-interest. Instead, it is rooted in a shared moral code, for only it can sustain a harmonious social order. And it is this moral code, along with the feelings of solidarity it generates, that forms the basis of all societies. Without the restraints imposed by a sense of moral obligation to others, the selfish pursuit of interests would destroy the social fabric.

---

[2]Durkheim was influenced by a number of scholars, not only Comte and Spencer. Some of the more important figures in developing his views were the French Enlightenment intellectuals Charles Montesquieu (1689–1775) and Jean-Jacques Rousseau (1712–1778), Henri de Saint-Simon (1760–1825), Charles Renouvier (1815–1903), and the German experimental psychologist Wilhem Wundt (1832–1920).

DURKHEIM'S THEORETICAL ORIENTATION ▪▪

As discussed previously, Durkheim was most concerned with analyzing "social facts": He sought to uncover the preexisting social conditions that shape the parameters for individual behavior. Consequently, Durkheim can be said to take a predominantly collectivist approach to order (see Figures 3.1 and 3.2).

This approach is most readily apparent in *Suicide*. In this study, Durkheim begins with one of the most seemingly individualistic, psychologically motivated acts there is—suicide—in order to illuminate the social and moral parameters behind and within this allegedly "individual" behavior. So, too, Durkheim's emphasis on **collective conscience** and **collective representations** indicates an interest in the collective level of society (see Figure 3.2). By *collective conscience*, Durkheim means the "totality of beliefs and sentiments common to average citizens of the same society" that "forms a determinate system which has its own life" ([1893] 1984:38–39). In later work, Durkheim used the term *collective representations* to refer to much the same thing. In any case, the point is Durkheim's main concern is not with the conscious or psychological state of specific individuals but, rather, with the collective beliefs and sentiments that exist "independent of the particular conditions in which individuals are placed; they pass on and it remains" (ibid.:80).

This leads us to one of the most common criticisms of Durkheim. Because of his preoccupation with social facts and the collective conscience, it is often claimed that he overlooks the role of the individual in producing and reproducing the social order. Durkheim makes it seem as if we're just vessels for society's will. Yet, this criticism ignores two essential points. First, Durkheim not only acknowledged individual autonomy; he also took it for granted as an inevitable condition of modern societies. Durkheim sought to show how, in modern societies, increasing individuation could produce detrimental effects as individuals are often torn between competing normative prescriptions and rules. For instance, in *Suicide,* Durkheim maintains that rather than rest comfortably on all-pervasive norms and values, "a thirst arises for novelties, unfamiliar pleasures, nameless sensations, all of which lose their savor once known . . . [but that] all these new sensations in their infinite quantity cannot form a solid foundation for happiness to support one in days of trial" ([1897] 1951:256). To be sure, the criticism could still be made that Durkheim ignores individual agency in "traditional" societies based on mechanical solidarity. In these societies, Durkheim does in fact posit a lack of individual autonomy, perhaps reflecting the Enlightenment-driven, Eurocentric thinking of his day. (We discuss this issue more fully in the following text.)

Figure 3.1 Durkheim's Basic Theoretical Orientation

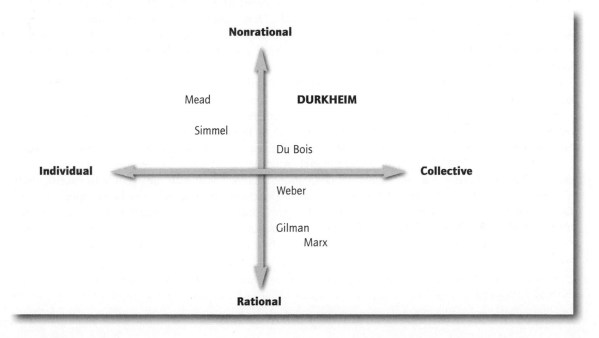

Figure 3.2 Durkheim's Core Concepts

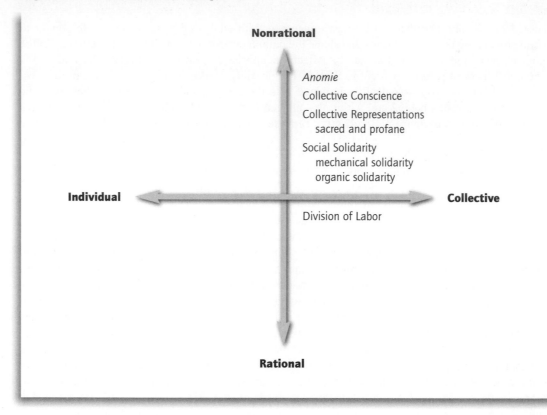

Relatedly, to assert that his orientation was singularly collectivist overlooks Durkheim's assumption that collective life emerges *in* social interaction. For instance, a major part of his analysis of the elementary forms of religious life involved showing how mundane objects, such as lizards and plants, take on the sacredness of the totem (the symbol of the tribe) by virtue of individuals coming together to participate in ritual practices. Similarly, in his study of suicide, Durkheim examined marriage and divorce rates not simply because he was fascinated by abstract, collective dimensions of social life but in order to uncover objective factors that measure the extent to which individuals are bound together in an increasingly individualistic world.

This leads us to the issue of action. In our view, Durkheim is primarily nonrationalist in his orientation (see Figures 3.1 and 3.2); he focuses on how collective representations and moral sentiments are a motivating force, much more so than "rational" or strategic interests connected to economic or political institutions. Yet, it is important to point out that in emphasizing the external nature of social facts, Durkheim also recognized that such facts are not confined to the realm of ideas or feelings but often possess a concrete reality as well. For instance, educational institutions and penal systems are also decisive for shaping the social order and individuals' actions within it. Thus, social facts are capable of exerting both a moral and an institutional force. In the end, however, Durkheim stressed the nonrational aspect of social facts as suggested in his supposition that the penal system (courts, legal codes and their enforcement, etc.) ultimately rests on collective notions of morality, a complex symbolic system as to what is "right" and "wrong." This issue is discussed further in the next section in relationship to the specific selections you will read.

In this section, you will read selections from the four major books that Durkheim published during his lifetime: *The Division of Labor in Society* (1893), *The Rules of Sociological Method* (1895), *Suicide* (1897), and *The Elementary Forms of Religious Life* (1912). We begin with excerpts from *The Rules of Sociological Method* because, as you will see, it is here that Durkheim first laid out his basic conceptualization of sociology as a discipline and delineated his concept of social facts. We then shift to *The Division of Labor in Society,* in which Durkheim sets out the key concepts of mechanical and organic solidarity and collective conscience. This is followed by excerpts from *Suicide,* which is notable, first, in that it exemplifies Durkheim's distinctive approach to the study of the social world and-second, because it further delineates Durkheim's core concept of *anomie.* We conclude this chapter with excerpts from *The Elementary Forms of Religious Life,* which many theorists consider Durkheim's most theoretically significant work. In *The Elementary Forms of Religious Life,* Durkheim takes an explicitly cultural turn, emphasizing the concepts of ritual and symbol, and the sacred and profane, and collective representations.

Introduction to *The Rules of Sociological Method*

In *The Rules of Sociological Method* ([1895] 1966:xiii), Durkheim makes at least three essential points. Durkheim insists that (1) sociology is a distinct field of study and that (2) although the social sciences are distinct from the natural sciences, the methods of the latter can be applied to the former. In addition, Durkheim maintains that (3) the social field is also distinct from the psychological realm. Thus, sociology is the study of social phenomena or "social facts," a very different enterprise from the study of an individual's own ideas or will.

Specifically, Durkheim maintains that there are two different ways that social facts can be identified. First, social facts are "general throughout the extent of a given society" at a given stage in the evolution of that society (ibid.:xv,13). Second, albeit related, a social fact is marked by "any manner of action . . . capable of exercising over the individual exterior constraint" (ibid.). In other words, a "social fact" is recognized by the "coercive power which it exercises or is capable of exercising over individuals" (ibid.:10). This does not mean that there are no "exceptions" to a social fact, but that it is potentially universal in the sense that, given specific conditions, it will be likely to emerge (ibid.:xv).

The "coercive power" of social facts brings us to a critical issue raised in *The Rules of Sociological Method:* crime. Durkheim argues that crime is inevitable or "normal" in all societies because crime defines the moral boundaries of a society and, in doing so, communicates to its inhabitants the range of acceptable behaviors. For Durkheim, crime is "normal" *not* because there will always be "bad" or "wicked" men and/or women in society (i.e., not for individualistic, psychological reasons, though those may well exist too). Rather, Durkheim maintains, "A society exempt from [crime] is utterly impossible" because crime affirms and reaffirms the collective sentiments upon which it is founded and which are necessary for its existence (ibid.:67). In other words, "It is impossible for all to be alike . . . there cannot be a society in which the individuals do not differ more or less from the collective type" (ibid.:69,70). As a result, social mechanisms compelling conformity to existing or new laws inevitably appear. Indeed, Durkheim maintains that even in a hypothetical "society of saints," a "perfect cloister of exemplary individuals," "faults" will appear, which will cause the same "scandal that the ordinary offense does in ordinary consciousnesses" (ibid.:68,69). Crime is "indispensable to the normal evolution of morality and law" because the formation and reformation of the collective conscience is never complete (ibid.).

Simply put, you cannot have a society without "crime" for the same reason that you cannot have a game without rules (i.e., you can do A, but not B) and consequences to rule violations (if you do B, this will happen). Thus, when children make up a new game, they devise both their own rules and consequences for rule infractions (e.g., you have to kick the ball between the tree and the mailbox; if the ball touches your hands, you're out). So, too, one could argue, society is like a game. There are rules (norms and laws), and there are consequences or punishments if you break those norms/rules (whether social ostracism or jail); and, most importantly, it is crime and punishment themselves that help clarify and reaffirm what the rules of the game are and thus the basis of society itself.

The Rules of Sociological Method (1895)

Émile Durkheim

WHAT IS A SOCIAL FACT?

Before inquiring into the method suited to the study of social facts, it is important to know which facts are commonly called "social." This information is all the more necessary since the designation "social" is used with little precision. It is currently employed for practically all phenomena generally diffused within society, however small their social interest. But on that basis, there are, as it were, no human events that may not be called social. Each individual drinks, sleeps, eats, reasons; and it is to society's interest that these functions be exercised in an orderly manner. If, then, all these facts are counted as "social" facts, sociology would have no subject matter exclusively its own, and its domain would be confused with that of biology and psychology.

But in reality there is in every society a certain group of phenomena which may be differentiated from those studied by the other natural sciences. When I fulfill my obligations as brother, husband, or citizen, when I execute my contracts, I perform duties which are defined, externally to myself and my acts, in law and in custom. Even if they conform to my own sentiments and I feel their reality subjectively, such reality is still objective, for I did not create them; I merely inherited them through my education. How many times it happens, moreover, that we are ignorant of the details of the obligations incumbent upon us, and that in order to acquaint ourselves with them we must consult the law and its authorized interpreters! Similarly, the church-member finds the beliefs and practices of his religious life ready-made at birth; their existence prior to his own implies their existence

outside of himself. The system of signs I use to express my thought, the system of currency I employ to pay my debts, the instruments of credit I utilize in my commercial relations, the practices followed in my profession, etc., function independently of my own use of them. And these statements can be repeated for each member of society. Here, then, are ways of acting, thinking, and feeling that present the noteworthy property of existing outside the individual consciousness.

These types of conduct or thought are not only external to the individual but are, moreover, endowed with coercive power, by virtue of which they impose themselves upon him, independent of his individual will. Of course, when I fully consent and conform to them, this constraint is felt only slightly, if at all, and is therefore unnecessary. But it is, nonetheless, an intrinsic characteristic of these facts, the proof thereof being that it asserts itself as soon as I attempt to resist it. If I attempt to violate the law, it reacts against me so as to prevent my act before its accomplishment, or to nullify my violation by restoring the damage, if it is accomplished and reparable, or to make me expiate it if it cannot be compensated for otherwise.

In the case of purely moral maxims; the public conscience exercises a check on every act which offends it by means of the surveillance it exercises over the conduct of citizens, and the appropriate penalties at its disposal. In many cases the constraint is less violent, but nevertheless it always exists. If I do not submit to the conventions of society, if in my dress I do not conform to the customs observed in my country and in my class, the ridicule I provoke, the social isolation in which I am

SOURCE: Reprinted with the permission of The Free Press, a Division of Simon & Schuster Adult Publishing Group from *The Rules of Sociological Method* by Émile Durkheim, translated by Sarah A Solovay and John H. Mueller. Edited by E. G. Catlin. Copyright © 1966 by Sarah A. Solovay, John H. Mueller, George E. G. Catlin. All rights reserved.

kept, produce, although in an attenuated form, the same effects as a punishment in the strict sense of the word. The constraint is nonetheless efficacious for being indirect. I am not obliged to speak French with my fellow-countrymen nor to use the legal currency, but I cannot possibly do otherwise. If I tried to escape this necessity, my attempt would fail miserably. As an industrialist, I am free to apply the technical methods of former centuries; but by doing so, I should invite certain ruin. Even when I free myself from these rules and violate them successfully, I am always compelled to struggle with them. When finally overcome, they make their constraining power sufficiently felt by the resistance they offer. The enterprises of all innovators, including successful ones, come up against resistance of this kind.

Here, then, is a category of facts with very distinctive characteristics: it consists of ways of acting, thinking, and feeling, external to the individual, and endowed with a power of coercion, by reason of which they control him. These ways of thinking could not be confused with biological phenomena, since they consist of representations and of actions; nor with psychological phenomena, which exist only in the individual consciousness and through it. They constitute, thus, a new variety of phenomena; and it is to them exclusively that the term "social" ought to be applied. And this term fits them quite well, for it is clear that, since their source is not in the individual, their substratum can be no other than society, either the political society as a whole or some one of the partial groups it includes, such as religious denominations, political, literary, and occupational associations, etc. On the other hand, this term "social" applies to them exclusively, for it has a distinct meaning only if it designates exclusively the phenomena which are not included in any of the categories of facts that have already been established and classified. These ways of thinking and acting therefore constitute the proper domain of sociology. It is true that, when we define them with this word "constraint," we risk shocking the zealous partisans of absolute individualism. For those who profess the complete autonomy of the individual, man's dignity is diminished whenever he is made to feel that he is not completely self-determinant. It is generally accepted today, however, that most of our ideas and our tendencies are not developed by ourselves but come to us from without. How can they become a part of us except by imposing themselves upon us? This is the whole meaning of our definition. And it is generally accepted, moreover, that

social constraint is not necessarily incompatible with the individual personality.[i]

Since the examples that we have just cited (legal and moral regulations, religious faiths, financial systems, etc.) all consist of established beliefs and practices, one might be led to believe that social facts exist only where there is some social organization. But there are other facts without such crystallized form which have the same objectivity and the same ascendancy over the individual. These are called "social currents." Thus the great movements of enthusiasm, indignation, and pity in a crowd do not originate in any one of the particular individual consciousnesses. They come to each one of us from without and can carry us away in spite of ourselves. Of course, it may happen that, in abandoning myself to them unreservedly, I do not feel the pressure they exert upon me. But it is revealed as soon as I try to resist them. Let an individual attempt to oppose one of these collective manifestations, and the emotions that he denies will turn against him. Now, if this power of external coercion asserts itself so clearly in cases of resistance, it must exist also in the first-mentioned cases, although we are unconscious of it. We are then victims of the illusion of having ourselves created that which actually forced itself from without. If the complacency with which we permit ourselves to be carried along conceals the pressure undergone, nevertheless it does not abolish it. Thus, air is no less heavy because we do not detect its weight. So, even if we ourselves have spontaneously contributed to the production of the common emotion, the impression we have received differs markedly from that which we would have experienced if we had been alone. Also, once the crowd has dispersed, that is, once these social influences have ceased to act upon us and we are alone again, the emotions which have passed through the mind appear strange to us, and we no longer recognize them as ours. We realize that these feelings have been impressed upon us to a much greater extent than they were created by us. It may even happen that they horrify us, so much were they contrary to our nature. Thus, a group of individuals, most of whom are perfectly inoffensive, may, when gathered in a crowd, be drawn into acts of atrocity. And what we say of these transitory outbursts applies similarly to those more permanent currents of opinion on religious, political, literary, or artistic matters which are constantly being formed around us, whether in society as a whole or in more limited circles.

To confirm this definition of the social fact by a characteristic illustration from common experience, one

---

[i]We do not intend to imply, however, that all constraints are normal. We shall return to this point later.

need only observe the manner in which children are brought up. Considering the facts as they are and as they have always been, it becomes immediately evident that all education is a continuous effort to impose on the child ways of seeing, feeling, and acting which he could not have arrived at spontaneously. From the very first hours of his life, we compel him to eat, drink, and sleep at regular hours; we constrain him to cleanliness, calmness, and obedience; later we exert pressure upon him in order that he may learn proper consideration for others, respect for customs and conventions, the need for work, etc. If, in time, this constraint ceases to be felt, it is because it gradually gives rise to habits and to internal tendencies that render constraint unnecessary; but nevertheless it is not abolished, for it is still the source from which these habits were derived. It is true that, according to Spencer, a rational education ought to reject such methods, allowing the child to act in complete liberty; but as this pedagogic theory has never been applied by any known people, it must be accepted only as an expression of personal opinion, not as a fact which can contradict the aforementioned observations. What makes these facts particularly instructive is that the aim of education is, precisely, the socialization of the human being; the process of education, therefore, gives us in a nutshell the historical fashion in which the social being is constituted. This unremitting pressure to which the child is subjected is the very pressure of the social milieu which tends to fashion him in its own image, and of which parents and teachers are merely the representatives and intermediaries.

It follows that sociological phenomena cannot be defined by their universality. A thought which we find in every individual consciousness, a movement repeated by all individuals, is not thereby a social fact. If sociologists have been satisfied with defining them by this characteristic, it is because they confused them with what one might call their reincarnation in the individual. It is, however, the collective aspects of the beliefs, tendencies, and practices of a group that characterize truly social phenomena. As for the forms that the collective states assume when refracted in the individual, these are things of another sort. This duality is clearly demonstrated by the fact that these two orders of phenomena are frequently found dissociated from one another. Indeed, certain of these social manners of acting and thinking acquire, by reason of their repetition, a certain rigidity which on its own account crystallizes

them, so to speak, and isolates them from the particular events which reflect them. They thus acquire a body, a tangible form, and constitute a reality in their own right, quite distinct from the individual facts which produce it. Collective habits are inherent not only in the successive acts which they determine but, by a privilege of which we find no example in the biological realm, they are given permanent expression in a formula which is repeated from mouth to mouth, transmitted by education, and fixed even in writing. Such is the origin and nature of legal and moral rules, popular aphorisms and proverbs, articles of faith wherein religious or political groups condense their beliefs, standards of taste established by literary schools, etc. None of these can be found entirely reproduced in the applications made of them by individuals, since they can exist even without being actually applied.

No doubt, this dissociation does not always manifest itself with equal distinctness, but its obvious existence in the important and numerous cases just cited is sufficient to prove that the social fact is a thing distinct from its individual manifestations. Moreover, even when this dissociation is not immediately apparent, it may often be disclosed by certain devices of method. Such dissociation is indispensable if one wishes to separate social facts from their alloys in order to observe them in a state of purity. Currents of opinion, with an intensity varying according to the time and place, impel certain groups either to more marriages, for example, or to more suicides, or to a higher or lower birthrate, etc. These currents are plainly social facts. At first sight they seem inseparable from the forms they take in individual cases. But statistics furnish us with the means of isolating them. They are, in fact, represented with considerable exactness by the rates of births, marriages, and suicides, that is, by the number obtained by dividing the average annual total of marriages, births, suicides, by the number of persons whose ages lie within the range in which marriages, births, and suicides occur.[ii] Since each of these figures contains all the individual cases indiscriminately, the individual circumstances which may have had a share in the production of the phenomenon are neutralized and, consequently, do not contribute to its determination. The average, then, expresses a certain state of the group mind (*l'âme collective*).

Such are social phenomena, when disentangled from all foreign matter. As for their individual manifestations, these are indeed, to a certain extent, social, since

---

[ii]Suicides do not occur at every age, and they take place with varying intensity at the different ages in which they occur.

they partly reproduce a social model. Each of them also depends, and to a large extent, on the organopsychological constitution of the individual and on the particular circumstances in which he is placed. Thus they are not sociological phenomena in the strict sense of the word. They belong to two realms at once; one could call them sociopsychological. They interest the sociologist without constituting the immediate subject matter of sociology. There exist in the interior of organisms similar phenomena, compound in their nature, which form in their turn the subject matter of the "hybrid sciences," such as physiological chemistry, for example.

The objection may be raised that a phenomenon is collective only if it is common to all members of society, or at least to most of them—in other words, if it is truly general. This may be true; but it is general because it is collective (that is, more or less obligatory), and certainly not collective because general. It is a group condition repeated in the individual because imposed on him. It is to be found in each part because it exists in the whole, rather than in the whole because it exists in the parts. This becomes conspicuously evident in those beliefs and practices which are transmitted to us ready-made by previous generations; we receive and adopt them because, being both collective and ancient, they are invested with a particular authority that education has taught us to recognize and respect. It is, of course, true that a vast portion of our social culture is transmitted to us in this way; but even when the social fact is due in part to our direct collaboration, its nature is not different. A collective emotion which bursts forth suddenly and violently in a crowd does not express merely what all the individual sentiments had in common; it is something entirely different, as we have shown. It results from their being together, a product of the actions and reactions which take place between individual consciousnesses; and if each individual consciousness echoes the collective sentiment, it is by virtue of the special energy resident in its collective origin. If all hearts beat in unison, this is not the result of a spontaneous and pre-established harmony but rather because an identical force propels them in the same direction. Each is carried along by all.

We thus arrive at the point where we can formulate and delimit in a precise way the domain of sociology. It comprises only a limited group of phenomena. A social fact is to be recognized by the power of external coercion which it exercises or is capable of exercising over individuals, and the presence of this power may be recognized in its turn either by the existence of some specific sanction or by the resistance offered against every individual effort that tends to violate it. One can, however,

define it also by its diffusion within the group, provided that, in conformity with our previous remarks, one takes care to add as a second and essential characteristic that its own existence is independent of the individual forms it assumes in its diffusion. This last criterion is perhaps, in certain cases, easier to apply than the preceding one. In fact, the constraint is easy to ascertain when it expresses itself externally by some direct reaction of society, as is the case in law, morals, beliefs, customs, and even fashions. But when it is only indirect, like the constraint which an economic organization exercises, it cannot always be so easily detected. Generality combined with externality may, then, be easier to establish. Moreover, this second definition is but another form of the first; for if a mode of behavior whose existence is external to individual consciousnesses becomes general, this can only be brought about by its being imposed upon them.

But these several phenomena present the same characteristic by which we defined the others. These "ways of existing" are imposed on the individual precisely in the same fashion as the "ways of acting" of which we have spoken. Indeed, when we wish to know how a society is divided politically, of what these divisions themselves are composed, and how complete is the fusion existing between them, we shall not achieve our purpose by physical inspection and by geographical observations; for these phenomena are social, even when they have some basis in physical nature. It is only by a study of public law that a comprehension of this organization is possible, for it is this law that determines the organization, as it equally determines our domestic and civil relations. This political organization is, then, no less obligatory than the social facts mentioned above. If the population crowds into our cities instead of scattering into the country, this is due to a trend of public opinion, a collective drive that imposes this concentration upon the individuals. We can no more choose the style of our houses than of our clothing—at least, both are equally obligatory. The channels of communication prescribe the direction of internal migrations and commerce, etc., and even their extent. Consequently, at the very most, it should be necessary to add to the list of phenomena which we have enumerated as presenting the distinctive criterion of a social fact only one additional category, "ways of existing"; and, as this enumeration was not meant to be rigorously exhaustive, the addition would not be absolutely necessary.

Such an addition is perhaps not necessary, for these "ways of existing" are only crystallized "ways of acting." The political structure of a society is merely the way in which its component segments have become accustomed to live with one another. If their relations are traditionally intimate, the segments tend to fuse with one another, or, in the contrary case, to retain their identity. The type of habitation imposed upon us is merely the way in which our contemporaries and our ancestors have been accustomed to construct their houses. The methods of communication are merely the channels which the regular currents of commerce and migrations have dug, by flowing in the same direction. To be sure, if the phenomena of a structural character alone presented this performance, one might believe that they constituted a distinct species. A legal regulation is an arrangement no less permanent than a type of architecture, and yet the regulation is a "physiological" fact. A simple moral maxim is assuredly somewhat more malleable, but it is much more rigid than a simple professional custom or a fashion. There is thus a whole series of degrees without a break in continuity between the facts of the most articulated structure and those free currents of social life which are not yet definitely molded. The differences between them are, therefore, only differences in the degree of consolidation they present. Both are simply life, more or less crystallized. No doubt, it may be of some advantage to reserve the term "morphological" for those social facts which concern the social substratum, but only on condition of not overlooking the fact that they are of the same nature as the others. Our definition will then include the whole relevant range of facts if we say: *A social fact is every way of acting, fixed or not, capable of exercising on the individual an external constraint; or again, every way of acting which is general throughout a given society, while at the same time existing in its own right independent of its individual manifestations.*[iii] [. . .]

THE NORMAL AND THE PATHOLOGICAL

If there is any fact whose pathological character appears incontestable, that fact is crime. All criminologists are agreed on this point. Although they explain this pathology differently, they are unanimous in recognizing it.

But let us see if this problem does not demand a more extended consideration. . . .

Crime is present not only in the majority of societies of one particular species but in all societies of all types. There is no society that is not confronted with the problem of criminality. Its form changes; the acts thus characterized are not the same everywhere; but, everywhere and always, there have been men who have behaved in such a way as to draw upon themselves penal repression. If, in proportion as societies pass from the lower to the higher types, the rate of criminality, i.e., the relation between the yearly number of crimes and the population, tended to decline, it might be believed that crime, while still normal, is tending to lose this character of normality. But we have no reason to believe that such a regression is substantiated. Many facts would seem rather to indicate a movement in the opposite direction. From the beginning of the [nineteenth] century, statistics enable us to follow the course of criminality. It has everywhere increased. In France the increase is nearly 300 per cent. There is, then, no phenomenon that presents more indisputably all the symptoms of normality, since it appears closely connected with the conditions of all collective life. To make of crime a form of social morbidity would be to admit that morbidity is not something accidental, but, on the contrary, that in certain cases it grows out of the fundamental constitution of the living organism; it would result in wiping out all distinction between the physiological and the pathological. No doubt it is possible that crime itself will have abnormal forms, as, for example, when its rate is unusually high. This excess is, indeed, undoubtedly morbid in nature. What is normal, simply, is the existence of crimi nality, provided that it attains and does not exceed, for each social type, a certain level, which it is perhaps not impossible to fix in conformity with the preceding rules.[iv]

Here we are, then, in the presence of a conclusion in appearance quite paradoxical. Let us make no mistake. To classify crime among the phenomena of normal sociology is not to say merely that it is an inevitable, although regrettable phenomenon, due to the incorrigible wickedness of men; it is to affirm that it is a factor in public health, an

---

[iii]This close connection between life and structure, organ and function, may be easily proved in sociology because between these two extreme terms there exists a whole series of immediately observable intermediate stages which show the bond between them. Biology is not in the same favorable position. But we may well believe that the inductions on this subject made by sociology are applicable to biology and that, in organisms as well as in societies, only differences in degree exist between these two orders of facts.

[iv]From the fact that crime is a phenomenon of normal sociology, it does not follow that the criminal is an individual normally constituted from the biological and psychological points of view. The two questions are independent of each other. This independence will be better understood when we have shown, later on, the difference between psychological and sociological facts.

integral part of all healthy societies. This result is, at first glance, surprising enough to have puzzled even ourselves for a long time. Once this first surprise has been overcome, however, it is not difficult to find reasons explaining this normality and at the same time confirming it.

In the first place crime is normal because a society exempt from it is utterly impossible. Crime, we have shown elsewhere, consists of an act that offends certain very strong collective sentiments. In a society in which criminal acts are no longer committed, the sentiments they offend would have to be found without exception in all individual consciousnesses, and they must be found to exist with the same degree as sentiments contrary to them. Assuming that this condition could actually be realized, crime would not thereby disappear; it would only change its form, for the very cause which would thus dry up the sources of criminality would immediately open up new ones.

Indeed, for the collective sentiments which are protected by the penal law of a people at a specified moment of its history to take possession of the public conscience or for them to acquire a stronger hold where they have an insufficient grip, they must acquire an intensity greater than that which they had hitherto had. The community as a whole must experience them more vividly, for it can acquire from no other source the greater force necessary to control these individuals who formerly were the most refractory. . . .

Imagine a society of saints, a perfect cloister of exemplary individuals. Crimes, properly so called, will there be unknown; but faults which appear venial to the layman will create there the same scandal that the ordinary offense does in ordinary consciousnesses. If, then, this society has the power to judge and punish, it will define these acts as criminal and will treat them as such. For the same reason, the perfect and upright man judges his smallest failings with a severity that the majority reserve for acts more truly in the nature of an offense. Formerly, acts of violence against persons were more frequent than they are today, because respect for individual dignity was less strong. As this has increased, these crimes have become more rare; and also, many acts violating this sentiment have been introduced into the penal law which were not included there in primitive times.[v]

In order to exhaust all the hypotheses logically possible, it will perhaps be asked why this unanimity does not extend to all collective sentiments without exception. Why should not even the most feeble sentiment gather

enough energy to prevent all dissent? The moral consciousness of the society would be present in its entirety in all the individuals, which a vitality sufficient to prevent all acts offending it—the purely conventional faults as well as the crimes. But a uniformity so universal and absolute is utterly impossible; for the immediate physical milieu in which each one of us is placed, the hereditary antecedents, and the social influences vary from one individual to the next, and consequently diversify consciousnesses. It is impossible for all to be alike, if only because each one has his own organism and that these organisms occupy different areas in space. That is why, even among the lower peoples, where individual originality is very little developed, it nevertheless does exist.

Thus, since there cannot be a society in which the individuals do not differ more or less from the collective type, it is also inevitable that, among these divergences, there are some with a criminal character. What confers this character upon them is not the intrinsic quality of a given act but that definition which the collective conscience lends them. If the collective conscience is stronger, if it has enough authority practically to suppress these divergences, it will also be more sensitive, more exacting; and, reacting against the slightest deviations with the energy it otherwise displays only against more considerable infractions, it will attribute to them the same gravity as formerly to crimes. In other words, it will designate them as criminal.

Crime is, then, necessary; it is bound up with the fundamental conditions of all social life, and by that very fact it is useful, because these conditions of which it is a part are themselves indispensable to the normal evolution of morality and law.

Indeed, it is no longer possible today to dispute the fact that law and morality vary from one social type to the next, nor that they change within the same type if the conditions of life are modified. But, in order that these transformations may be possible, the collective sentiments at the basis of morality must not be hostile to change, and consequently must have but moderate energy. If they were too strong, they would no longer be plastic. Every pattern is an obstacle to new patterns, to the extent that the first pattern is inflexible. The better a structure is articulated, the more it offers a healthy resistance to all modification; and this is equally true of functional, as of anatomical, organization. If there were no crimes, this condition could not have been fulfilled; for such a hypothesis presupposes that collective sentiments

---

[v]Calumny, insults, slander, fraud, etc.

have arrived at a degree of intensity unexampled in history. Nothing is good indefinitely and to an unlimited extent. The authority which the moral conscience enjoys must not be excessive; otherwise no one would dare criticize it, and it would too easily congeal into an immutable form. To make progress, individual originality must be able to express itself. In order that the originality of the idealist whose dreams transcend his century may find expression, it is necessary that the originality of the criminal, who is below the level of his time, shall also be possible. One does not occur without the other.

Nor is this all. Aside from this indirect utility, it happens that crime itself plays a useful role in this evolution. Crime implies not only that the way remains open to necessary changes but that in certain cases it directly prepares these changes. Where crime exists, collective sentiments are sufficiently flexible to take on a new form, and crime sometimes helps to determine the form they will take. How many times, indeed, it is only an anticipation of future morality—a step toward what will be! According to Athenian law, Socrates was a criminal, and his condemnation was no more than just. However, his crime, namely, the independence of his thought,

rendered a service not only to humanity but to his country. It served to prepare a new morality and faith which the Athenians needed, since the traditions by which they had lived until then were no longer in harmony with the current conditions of life. Nor is the case of Socrates unique; it is reproduced periodically in history. It would never have been possible to establish the freedom of thought we now enjoy if the regulations prohibiting it had not been violated before being solemnly abrogated. At that time, however, the violation was a crime, since it was an offense against sentiments still very keen in the average conscience. And yet this crime was useful as a prelude to reforms which daily became more necessary. Liberal philosophy had as its precursors the heretics of all kinds who were justly punished by secular authorities during the entire course of the Middle Ages and until the eve of modern times.

From this point of view the fundamental facts of criminality present themselves to us in an entirely new light. Contrary to current ideas, the criminal no longer seems a totally unsociable being, a sort of parasitic element, a strange and inassimilable body, introduced into the midst of society.[vi] On the contrary, he plays a definite role in social life. . . .

▪▪

[vi]We have ourselves committed the error of speaking thus of the criminal, because of a failure to apply our rule (Division du travail social, pp. 395–96).

Introduction to *The Division of Labor in Society*

In Durkheim's first major work, *The Division of Labor in Society* (1893), which was based on his doctoral dissertation, Durkheim explains how the division of labor (or specialization of tasks) characteristic of modern societies affects individuals as well as society as a whole. As you may recall, this issue had been of utmost concern to Marx as well. Marx contended that modern, competitive capitalism, and the specialized division of labor that sustained it, resulted in alienation. In contrast, Durkheim argued that economic specialization was not necessarily "bad" for either the individual or the society as a whole. Instead, he argued that an extensive division of labor could exist without necessarily jeopardizing the moral cohesion of a society or the opportunity for individuals to realize their interests.

How is this possible? Durkheim argued that there were two basic types of solidarity: mechanical and organic.[1] **Mechanical solidarity** is typified by feelings of *likeness*. Mechanical solidarity is rooted in everyone doing/feeling the same thing. Durkheim maintained that this type of solidarity is characteristic of

[1]Durkheim's distinction between mechanical and organic solidarity was developed, in part, as a critical response to the work of the German sociologist Ferdinand Tönnies. In his book, *Gemeinschaft und Gesellschaft,* Tönnies argued that simpler, traditional societies (Gemeinschaft) were more "organic" and beneficial to the formation of social bonds. In contrast to Tönnies's conservative orientation, Durkheim contended that complex, modern societies were, in fact, more "organic" and thus desirable by promoting individual liberties within a context of morally binding, shared social obligations.

small, traditional societies. In these "simple" societies, circumstances compel individuals to be generalists involved in the production and distribution of a variety of goods. Indeed, in small, traditional societies, specialization in one task to the exclusion of others is not possible because the society depends upon each individual providing a host of contributions to the group. For instance, men, women, and children are often all needed to pick crops at harvest time, and all partake in the harvest time celebrations as well.

Durkheim argued that a significant social consequence of the shared work experience characteristic of traditional societies is a shared collective conscience. People in traditional societies tend to feel "one and the same," and it is this feeling of "oneness" that is integral in the maintenance of social order.

Yet, Durkheim saw that in large, complex societies, this type of solidarity was waning. In large, modern societies, labor is specialized; people do not necessarily all engage in the same work or share the same ideas and beliefs. For Durkheim, **organic solidarity** refers to a type of solidarity in which each person is interdependent

Photo 3.1a Durkheim maintained that different types of society exhibit different types of solidarity. Mechanical solidarity, based on likeness, is characteristic of small, traditional societies, such as this village in Namibia (Africa).

Photo 3.1b Organic solidarity, based on specialization, is characteristic of large, modern industrial societies, such as Brasilia (Brazil).

with others, forming a complex web of cooperative associations. In such situations, solidarity (or a feeling of "oneness") comes not from each person believing/doing the same thing, but from cultivating individual differences and knowing that each is doing her part for the good of the whole. Thus, Durkheim argued that the increasing specialization and individuation so readily apparent in modern industrial societies does not necessarily result in a decline in social stability or cohesion. Rather, the growth in a society's density (the number of people living in a community) and consequent increasingly specialized division of labor can result in simply a different *type* of social cohesion.

Significantly, however, Durkheim maintained that organic solidarity does not automatically emerge in modern societies. Rather, it arises only when the division of labor is "spontaneous" or voluntary. States Durkheim, "For the division of labor to produce solidarity, it is not sufficient, then, that each have his task; it is still necessary that this task be fitting to him" ([1893] 1984:375). Moreover, a "normal" division of labor exists only when the specialization of tasks is not exaggerated. If the division of labor is pushed too far, there is a danger for the individual to become "isolated in his special activity." In such cases, the division of labor becomes "a source of disintegration" for both the individual and society (ibid.). The individual "no longer feels the idea of common work being done by those who work side by side with him" (ibid.). Meanwhile, a rigid division of labor can lead to "the institution of classes and castes . . . [which] is often a source of dissension" (ibid.:374). Durkheim used the term **anomie** (a lack of moral regulation) to describe the "pathological" consequences of an overly specialized division of labor. This is an important concept to which we will shortly return.

Most interestingly, then, it is not that Durkheim ignores the potentially harmful aspects of the division of labor in modern societies; on the contrary, Durkheim acknowledges that the division of labor is problematic when it is "forced" and/or pushed to an extreme. This position offers an important similarity as well as difference to that offered by Marx. As we noted previously, Marx saw both alienation and class conflict as inevitable (or "normal") in capitalist societies. By contrast, rather than seeing social conflict as a "normal" condition of capitalism, Durkheim maintained that anomie results only in "abnormal" conditions of *over*specialization, when the rules of capitalism become too rigid and individuals are "forced" into a particular position in the division of labor.

The Division of Labor in Society (1893)

Émile Durkheim

INTRODUCTION: THE PROBLEM

The division of labor is not of recent origin, but it was only at the end of the eighteenth century that social cognizance was taken of the principle, though, until then, unwitting submission had been rendered to it. To be sure, several thinkers from earliest times saw its importance;[i] but Adam Smith was the first to attempt a theory of it. Moreover, he adopted this phrase that social science later lent to biology.

Nowadays, the phenomenon has developed so generally it is obvious to all. We need have no further illusions about the tendencies of modern industry; it advances steadily towards powerful machines, towards great concentrations of forces and capital, and consequently to the extreme division of labor. Occupations

---

[i]Aristotle, Nichomachean Ethics, E, 1133a, 16.

are infinitely separated and specialized, not only inside the factories, but each product is itself a specialty dependent upon others. Adam Smith and John Stuart Mill still hoped that agriculture, at least, would be an exception to the rule, and they saw it as the last resort of small-scale industry. Although one must be careful not to generalize unduly in such matters, nevertheless it is hard to deny today that the principal branches of the agricultural industry are steadily being drawn into the general movement. Finally, business itself is ingeniously following and reflecting in all its shadings the infinite diversity of industrial enterprises; and, while this evolution is realizing itself with unpremeditated spontaneity, the economists, examining its causes and appreciating its results, far from condemning or opposing it, uphold it as necessary. They see in it the supreme law of human societies and the condition of their progress. But the division of labor is not peculiar to the economic world; we can observe its growing influence in the most varied fields of society. The political, administrative, and judicial functions are growing more and more specialized. It is the same with the aesthetic and scientific functions. It is long since philosophy reigned as the science unique; it has been broken into a multitude of special disciplines each of which has its object, method, and though. "Men working in the sciences have become increasingly more specialized."[ii]

MECHANICAL SOLIDARITY

We are now in a position to come to a conclusion.

The totality of beliefs and sentiments common to average citizens of the same society forms a determinate system which has its own life; one may call it the *collective* or *common conscience*. No doubt, it has not a specific organ as a substratum; it is, by definition, diffuse in every reach of society. Nevertheless, it has specific characteristics which make it a distinct reality. It is, in effect, independent of the particular conditions in which individuals are placed; they pass on and it

remains. It is the same in the North and in the South, in great cities and in small, in different professions. Moreover, it does not change with each generation, but, on the contrary, it connects successive generations with one another. It is, thus, an entirely different thing from particular consciences, although it can be realized only through them. It is the psychical type of society, a type which has its properties, its conditions of existence, its mode of development, just as individual types, although in a different way. Thus understood, it has the right to be denoted by a special word. The one which we have just employed is not, it is true, without ambiguity. As the terms, collective and social, are often considered synonymous, one is inclined to believe that the collective conscience is the total social conscience, that is, extend it to include more than the psychic life of society, although, particularly in advanced societies, it is only a very restricted part. Judicial, governmental, scientific, industrial, in short, all special functions are of a psychic nature, since they consist in systems of representations and actions. They, however, are surely outside the common conscience. To avoid the confusion[iii] into which some have fallen, the best way would be to create a technical expression especially to designate the totality of social similitudes. However, since the use of a new word, when not absolutely necessary, is not without inconvenience, we shall employ the well-worn expression, collective or common conscience, but we shall always mean the strict sense in which we have taken it.

We can, then, to resume the preceding analysis, say that an act is criminal when it offends strong and defined states of the collective conscience.[iv]

The statement of this proposition is not generally called into question, but it is ordinarily given a sense very different from that which it ought to convey. We take it as if it expressed, not the essential property of crime, but one of its repercussions. We well know that crime violates very pervasive and intense sentiments, but we believe that this pervasiveness and this intensity derive from the criminal character of the act, which consequently remains to be defined. We do not deny

---

[ii]De Candolle, *Histoire des Sciences et des Savants*, 2nd ed., p. 263.

[iii]The confusion is not without its dangers. Thus, we sometimes ask if the individual conscience varies as the collective conscience. It all depends upon the sense in which the word is taken. If it represents social likenesses, the variation is inverse, as we shall see. If it signifies the total psychic life of society, the relation is direct. It is thus necessary to distinguish them.

[iv]We shall not consider the question whether the collective conscience is a conscience as is that of the individual. By this term, we simply signify the totality of social likenesses, without prejudging the category by which this system of phenomena ought to be defined.

that every delict is universally reproved, but we take as agreed that the reprobation to which it is subjected results from its delictness. But we are hard put to say what this delictness consists of. In immorality which is particularly serious? I wish such were the case, but that is to reply to the question by putting one word in place of another, for it is precisely the problem to understand what this immorality is, and especially this particular immorality which society reproves by means of organized punishment and which constitutes criminality. It can evidently come only from one or several characteristics common to all criminological types. The only one which would satisfy this condition is that opposition between a crime, whatever it is, and certain collective sentiments. It is, accordingly, this opposition which makes crime rather than being a derivative of crime. In other words, we must not say that an action shocks the common conscience because it is criminal, but rather that it is criminal because it shocks the common conscience. We do not reprove it because it is a crime, but it is a crime because we reprove it. As for the intrinsic nature of these sentiments, it is impossible to specify them. They have the most diverse objects and cannot be encompassed in a single formula. We can say that they relate neither to vital interests of society nor to a minimum of justice. All these definitions are inadequate. By this alone can we recognize it: a sentiment, whatever its origin and end, is found in all consciences with a certain degree of force and precision, and every action which violates it is a crime. Contemporary psychology is more and more reverting to the idea of Spinoza, according to which things are good because we like them, as against our liking them because they are good. What is primary is the tendency, the inclination; the pleasure and pain are only derivative facts. It is just so in social life. An act is socially bad because society disproves of it. But, it will be asked, are there not some collective sentiments which result from pleasure and pain which society feels from contact with their ends? No doubt, but they do not all have this origin. A great many, if not the larger part, come from other causes. Everything that leads activity to assume a definite form can give rise to habits, whence result tendencies which must be satisfied. Moreover, it is these latter tendencies which alone are truly fundamental. The others are only special forms and more determinate. Thus, to find charm in such and such an object, collective sensibility must already be constituted so as to be able to enjoy it. If the

corresponding sentiments are abolished, the most harmful act to society will not only be tolerated, but even honored and proposed as an example. Pleasure is incapable of creating an impulse out of whole cloth; it can only link those sentiments which exist to such and such a particular end, provided that the end be in accord with their original nature. . . .

Organic Solidarity

Since negative solidarity does not produce any integration by itself, and since, moreover, there is nothing specific about it, we shall recognize only two kinds of positive solidarity which are distinguishable by the following qualities:

1. The first binds the individual directly to society without any intermediary. In the second, he depends upon society, because he depends upon the parts of which it is composed.

2. Society is not seen in the same aspect in the two cases. In the first, what we call society is a more or less organized totality of beliefs and sentiments common to all the members of the group: this is the collective type. On the other hand, the society in which we are solitary in the second instance is a system of different, special functions which definite relations unite. These two societies really make up only one. They are two aspects of one and the same reality, but none the less they must be distinguished.

3. From this second difference there arises another which helps us to characterize and name the two kinds of solidarity.

The first can be strong only if the ideas and tendencies common to all the members of the society are greater in number and intensity than those which pertain personally to each member. It is as much stronger as the excess is more considerable. But what makes our personality is how much of our own individual qualities we have, what distinguishes us from others. This solidarity can grow only in inverse ratio to personality. There are in each of us, as we have said, two consciences: one which is common to our group in its entirety, which, consequently, is not ourself, but society living and acting within us; the other, on the contrary, represents that in us which is personal and distinct, that which makes us an individual.ᵛ Solidarity which comes from likenesses is at its maximum when the collective conscience completely envelops our whole conscience and coincides in all points with it. But, at that moment, our individuality is nil. It can be born only if the community takes smaller toll of us. There are, here, two contrary forces, one centripetal, the other centrifugal, which cannot flourish at the same time. We cannot, at one and the same time, develop ourselves in two opposite senses. If we have a lively desire to think and act for ourselves, we cannot be strongly inclined to think and act as others do. If our ideal is to present a singular and personal appearance, we do not want to resemble everybody else. Moreover, at the moment when this solidarity exercises its force, our personality vanishes, as our definition permits us to say, for we are no longer ourselves, but the collective life.

The social molecules which can be coherent in this way can act together only in the measure that they have no actions of their own, as the molecules of inorganic bodies. That is why we propose to call this type of solidarity mechanical. The term does not signify that it is produced by mechanical and artificial means. We call it that only by analogy to the cohesion which unites the elements of an inanimate body, as opposed to that which makes a unity out of the elements of a living body. What justifies this term is that the link which thus unites the individual to society is wholly analogous to that which attaches a thing to a person. The individual conscience, considered in this light, is a simple dependent upon the collective type and follows all of its movements, as the possessed object follows those of its owner. In societies where this type of solidarity is highly developed, the individual does not appear, as we shall see later. Individuality is something which the society possesses. Thus, in these social types, personal rights are not yet distinguished from real rights.

It is quite otherwise with the solidarity which the division of labor produces. Whereas the previous type implies that individuals resemble each other, this type presumes their difference. The first is possible only in so far as the individual personality is absorbed into the collective personality; the second is possible only if each one has a sphere of action which is peculiar to him; that is, a personality. It is necessary, then, that the collective conscience leave open a part of the individual conscience in order that special functions may be established there, functions which it cannot regulate. The more this region is extended, the stronger is the cohesion which results from this solidarity. In effect, on the one hand, each one depends as much more strictly on society as labor is more divided; and, on the other, the activity of each is as much more personal as it is more specialized. Doubtless, as circumscribed as it is, it is never completely original. Even in the exercise of our occupation, we conform to usages, to practices which are common to our whole professional brotherhood. But, even in this instance, the yoke that we submit to is much less heavy than when society completely controls us, and it leaves much more place open for the free play of our initiative. Here, then, the individuality of all grows at the same time as that of its parts. Society becomes more capable of collective movement, at the same time that each of its elements has more freedom of movement. This solidarity resembles that which we observe among the higher animals. Each organ, in effect, has its special physiognomy, its autonomy. And, moreover, the unity of the organism is as great as the individuation of the parts is more marked. Because of this analogy, we propose to call the solidarity which is due to the division of labor, organic. . . .

THE CAUSES

We can then formulate the following proposition: The division of labor varies in direct ratio with the volume and density of societies, and, if it progresses in a continuous manner in the course of social development, it is because societies become regularly denser and generally more voluminous.

At all times, it is true, it has been well understood that there was a relation between these two orders of fact, for, in order that functions be more specialized, there must be more co-operators, and they must be related to co-operate. But, ordinarily, this state of societies is seen only as the means by which the division of labor develops, and not as the cause of its development. The latter is made to depend upon individual aspirations toward well-being and happiness, which can be satisfied so much better as societies are more extensive and more condensed. The law we have just established is quite otherwise. We say, not that the growth and

---

ᵛHowever, these two consciences are not in regions geographically distinct from us, but penetrate from all sides.

condensation of societies *permit,* but that they *necessitate* a greater division of labor. It is not an instrument by which the latter is realized; it is its determining cause.[vi]

THE FORCED DIVISION OF LABOR

It is not sufficient that there be rules, however, for sometimes the rules themselves are the cause of evil. This is what occurs in class-wars. The institution of classes and of castes constitutes an organization of the division of labor, and it is a strictly regulated organization, although it often is a source of dissension. The lower classes not being, or no longer being, satisfied with the role which has devolved upon them from custom or by law aspire to functions which are closed to them and seek to dispossess those who are exercising these functions. Thus civil wars arise which are due to the manner in which labor is distributed.

There is nothing similar to this in the organism. No doubt, during periods of crises, the different tissues war against one another and nourish themselves at the expense of others. But never does one cell or organ seek to usurp a role different from the one which it is filling. The reason for this is that each anatomic element automatically executes its purpose. Its constitution, its place in the organism, determines its vocation; its task is a consequence of its nature. It can badly acquit itself, but it cannot assume another's task unless the latter abandons it, as happens in the rare cases of substitution that we have spoken of. It is not so in societies. Here the possibility is greater. There is a greater distance between the hereditary dispositions of the individual and the social function he will fill. The first do not imply the second with such immediate necessity. This space, open to striving and deliberation, is also at the mercy of a multitude of causes which can make individual nature deviate from its normal direction and create a pathological state. Because this organization is more supple, it is also more delicate and

more accessible to change. Doubtless, we are not, from birth, predestined to some special position; but we do have tastes and aptitudes which limit our choice. If no care is taken of them, if they are ceaselessly disturbed by our daily occupations, we shall suffer and seek a way of putting an end to our suffering. But there is no other way out than to change the established order and to set up a new one. For the division of labor to produce solidarity, it is not sufficient, then, that each have his task; it is still necessary that this task be fitting to him. Now, it is this condition which is not realized in the case we are examining. In effect, if the institution of classes or castes sometimes gives rise to anxiety and pain instead of producing solidarity, this is because the distribution of social functions on which it rests does not respond, or rather no longer responds, to the distribution of natural talents. . . .

CONCLUSION

But not only does the division of labor present the character by which we have defined morality; it more and more tends to become the essential condition of social solidarity. As we advance in the evolutionary scale, the ties which bind the individual to his family, to his native soil, to traditions which the past has given to him, to collective group usages, become loose. More mobile, he changes his environment more easily, leaves his people to go elsewhere to live a more autonomous existence, to a greater extent forms his own ideas and sentiments. Of course, the whole common conscience does not, on this account, pass out of existence. At least there will always remain this cult of personality, of individual dignity of which we have just been speaking, and which, today, is the rallying-point of so many people. But how little a thing it is when one contemplates the ever increasing extent of social life, and, consequently, of individual consciences! For, as they become more voluminous, as intelligence becomes richer, activity more varied, in

---

[vi]On this point, we can still rely on Comte as authority. "I must," he said "now indicate the progressive condensation of our species as a last general concurrent element in regulating the effective speed of the social movement. We can first easily recognize that this influence contributes a great deal, especially in origin, in determining a more special division of human labor, necessarily incompatible with a small number of co-operators. Besides, by a most intimate and little known property, although still most important, such a condensation stimulates directly, in a very powerful manner, the most rapid development of social evolution, either in driving individuals to new efforts to assure themselves by more refined means of an existence which otherwise would become more difficult, or by obliging society with more stubborn and better concentrated energy to fight more stiffly against the more powerful effort of particular divergences. With one and the other, we see that it is not a question here of the absolute increase of the number of individuals, but especially of their more intense concourse in a given space." Cours, IV, p. 455.

order for morality to remain constant, that is to say, in order for the individual to remain attached to the group with a force equal to that of yesterday, the ties which bind him to it must become stronger and more numerous. If, then, he formed no others than those which come from resemblances, the effacement of the segmental type would be accompanied by a systematic debasement of morality. Man would no longer be sufficiently obligated; he would no longer feel about and above him this salutary pressure of society which moderates his egoism and makes him a moral being. This is what gives moral value to the division of labor. Through it, the individual becomes cognizant of his dependence upon society; from it come the forces which keep him in check and restrain him. In short, since the division of labor becomes the chief source of social solidarity, it becomes, at the same time, the foundation of the moral order.

We can then say that, in higher societies, our duty is not to spread our activity over a large surface, but to concentrate and specialize it. We must contract our horizon, choose a definite task and immerse ourselves in it completely, instead of trying to make ourselves a sort of creative masterpiece, quite complete, which contains its worth in itself and not in the services that it renders. Finally, this specialization ought to be pushed as far as the elevation of the social type, without assigning any other limit to it.[vii] No doubt, we ought so to work as to realize in ourselves the collective type as it exists. There are common sentiments, common ideas, without which, as has been said, one is not a man. The rule which orders us to specialize remains limited by the contrary rule. Our conclusion is not that it is good to press specialization as far as possible, but as far as necessary. As for the part that is to be played by these two opposing necessities, that is determined by experience and cannot be calculated *a priori*. It is enough for us to have shown that the second is not of a different nature from the first, but that it also is moral, and that, moreover, this duty becomes ever more important and pressing, because the general

qualities which are in question suffice less and less to socialize the individual. . . .

Let us first of all remark that it is difficult to see why it would be more in keeping with the logic of human nature to develop superficially rather than profoundly. Why would a more extensive activity, but more dispersed, be superior to a more concentrated, but circumscribed, activity? Why would there be more dignity in being complete and mediocre, rather than in living a more specialized, but more intense life, particularly if it is thus possible for us to find what we have lost in this specialization, through our association with other beings who have what we lack and who complete us? We take off from the principle that man ought to realize his nature as man, to accomplish his öικ²î^ον ²´ργον, as Aristotle said. But this nature does not remain constant throughout history; it is modified with societies. Among lower peoples, the proper duty of man is to resemble his companions, to realize in himself all the traits of the collective type which are then confounded, much more than today, with the human type. But, in more advanced societies, his nature is, in large part, to be an organ of society, and his proper duty, consequently, is to play his role as an organ.

Moreover, far from being trammelled by the progress of specialization, individual personality develops with the division of labor.

To be a person is to be an autonomous source of action. Man acquires this quality only in so far as there is something in him which is his alone and which individualizes him, as he is something more than a simple incarnation of the generic type of his race and his group. It will be said that he is endowed with free will and that is enough to establish his personality. But although there may be some of this liberty in him, an object of so many discussions, it is not this metaphysical, impersonal, invariable attribute which can serve as the unique basis for concrete personality, which is empirical and variable with individuals. That could not be constituted by the wholly abstract power of choice between two opposites, but it is still necessary for this faculty to be exercised

---

[vii]There is, however, probably another limit which we do not have to speak of since it concerns individual hygiene. It may be held that, in the light of our organico-psychic constitution, the division of labor cannot go beyond a certain limit without disorders resulting. Without entering upon the question, let us straightaway say that the extreme specialization at which biological functions have arrived does not seem favorable to this hypothesis. Moreover, in the very order of psychic and social functions, has not the division of labor, in its historical development, been carried to the last stage in the relations of men and women? Have not there been faculties completely lost by both? Why cannot the same phenomenon occur between individuals of the same sex? Of course, it takes time for the organism to adapt itself to these changes, but we do not see why a day should come when this adaptation would become impossible.

towards ends and aims which are proper to the agent. In other words, the very materials of conscience must have a personal character. But we have seen in the second book of this work that this result is progressively produced as the division of labor progresses. The effacement of the segmental type, at the same time that it necessitates a very great specialization, partially lifts the individual conscience from the organic environment which supports it, as from the social environment which envelops it, and, accordingly, because of this double emancipation, the individual becomes more of an independent factor in his own conduct. The division of labor itself contributes to this enfranchisement, for individual natures, while specializing, become more complex, and by that are in part freed from collective action and hereditary influences which can only enforce themselves upon simple, general things. . . .

▪▪

Introduction to *Suicide*

Suicide (1897) is both a theoretical and methodological exemplar. In this famous study, Durkheim examines a phenomenon that most people think of as an intensely individual act—suicide—and demonstrates its *social* (rather than psychological) roots. His method for doing this is to analyze *rates* of suicide between societies and historical periods and between different social groups within the same society. By linking the different suicide rates of particular societies and social groups to the specific characteristics of that society/social group, Durkheim not only demonstrates that individual pathologies are rooted in social conditions; in addition, Durkheim shows how sociologists can scientifically study social behavior. His innovative examination of suicide rates lent credibility to his conviction that sociology should be considered a viable scientific discipline.

Most importantly, Durkheim argues that the places with the highest rates of alcoholism and mental illness are not the areas with the highest suicide rates (thereby undermining the notion that it is pathological psychological states that are solely determinative of the individual act of suicide). Rather, Durkheim maintains that suicide rates are highest in moments when, and in places where, individuals lack social and moral regulation and/or integration. In addition, as in his first book, *The Division of Labor in Society,* in *Suicide* Durkheim was particularly interested in delineating the fundamental differences between traditional and modern societies. Durkheim sought to explain why suicide is rare in small, simple societies while much more frequent in modern, industrial ones. Parallel to his argument in *The Division of Labor in Society,* Durkheim argues that traditional and modern societies differ not only in their rates of suicide but in the types of suicide that are prevalent as well.

Specifically, Durkheim saw two main characteristics of modern, industrial society: There was (1) a lack of integration of the individual in the social group and (2) a lack of moral regulation. Durkheim used the term *egoism* to refer to the lack of integration of the individual in the social group. He used the term *anomie* to refer to a lack of moral regulation. Durkheim argued that both of these conditions— egoism and anomie—are "chronic" in modern, industrial society; and in extreme, pathological form, both egoism and anomie can result in suicide. Let's look at these two different, albeit intimately interrelated, conditions in turn.

For Durkheim, egoistic suicide results from a pathological weakening of the bonds between the individual and the social group. This lack of integration is evident statistically, in that there are higher rates of suicide among single, divorced, and widowed persons than among married persons and in that there are higher rates of suicide among married persons without children than there are among married persons with children. Additionally, Durkheim argued that egoism helps explain why suicide rates are higher among Protestants than Catholics or Jews: Protestantism emphasizes an individual relationship with God, which means that the individual is less bound to the religious clergy and members of the congregation. Interestingly, then, Durkheim maintains that it is not Catholic doctrine that inhibits the act of suicide; rather, it is Catholics' social and spiritual bonds, their association with the priests, nuns, and other lay members of the congregation, that deters them from this act. Protestant rates of suicide are higher because Protestants are more morally and spiritually isolated than the more communally oriented Jews and Catholics.

Durkheim saw an increase in egoistic suicide as a "natural" outgrowth of the individuation of modern, industrial societies. For instance, today it is quite common—especially in big cities—for people to live alone. By contrast, in many traditional societies, it is virtually unheard of for anyone to live by himself or herself. Children live with parents until they get married; parents move in with children (or vice versa) if a spouse dies; unmarried siblings live with either parents or brothers and sisters. As we noted, Durkheim argued that in its extreme form, the type of social isolation found in modern societies can be literally fatal.

Intertwined with a decrease in social integration in modern, industrial societies is a decrease in moral integration. Durkheim used the term *anomie* to refer to this lack of moral regulation. Anomic suicide is the pathological result of a lack of moral direction, when one feels morally adrift. Durkheim viewed modern societies as "chronically" anomic or characterized by a lack of regulation of the individual by the collective.

Thus, for instance, modern industrial societies are religiously pluralistic, whereby people are more able to freely choose among a variety of religious faiths—or to choose not to "believe" at all. Similarly, today, many people choose to "identify"—or not—with a specific part of their ethnic heritage. That we spend much time and energy searching for "identity"—I'm a punk! I'm Irish!—reflects a lack of moral regulation. To be sure, there are many wonderful benefits from this increasing individuation that contrasts significantly from small, traditional, homogeneous societies in which "who" you are is taken for granted. In small, closed, indigenous societies without so many (or any) options, where there is one religion, one ethnic group, your place in that society is a cultural given—a "place" that may be quite oppressive. Not surprisingly, then, Durkheim asserts that suppressing individuation also can produce pathological consequences (this point is discussed later).

The lack of moral regulation in modern societies is especially prevalent in times of intense social and personal change. During such periods, the authority of the family, the church, and the community may be challenged or questioned, and without moral guidance and authority, individuals may feel like they have no moral anchor. The pursuit of individual desires and goals can overtake moral concerns. However, Durkheim maintains that anomie can result not only from "bad" social change, such as losing one's job, or political crisis, but from "positive" social change as well. Consider, for instance, what happens when someone wins the lottery. Most people think that if they were to win the lottery, they would experience only joy and happiness. Indeed, some people buy lottery tickets thinking, "If I win the 'big one,' all my problems will be solved!" However, Durkheim contends that sudden life-changing events can bring on a battery of social and personal issues that one might not expect.

First, after winning the lottery, one might suddenly find oneself confronted with weighty existential issues. Before the lottery, you may have simply worked—and worked hard—because you needed to earn a living. But now that you've won the lottery, you don't know what to do. By not having to work, you might start thinking about such things as the meaning of life that you had never thought about before. This feeling that you don't know "what to do" and "how to act" is a state of anomie.

In addition, you might start to wonder how much friends and family should get from your winnings. You might begin to feel like everyone just wants your money and that it is hard to tell who likes *you* and who just likes your newfound fame and fortune. You might feel like you can't talk to your friends about your dilemma, that no one in your previous social circle really "understands" you anymore. You may begin to find that you can't relate to the people from your old socioeconomic class, but that you can't relate to anyone in your new class either. Thus, the sudden change brought about by winning the lottery

Photo 3.2 In a modern-day incident of altruistic suicide, a number of South Vietnamese Buddhist monks used self-immolation to protest the persecution of the country's majority Buddhist population at the hands of the Catholic president, Ngo Dinh Diem. Here, Quang Duc burns himself to death on a Saigon street, June 11, 1963.

can lead not only to feeling morally "anchorless" (anomie) but also socially alone (egoism). A most extreme outcome of feeling this moral and social isolation would be suicide.

As we noted previously, Durkheim argued that traditional and modern societies are rooted in different social conditions. Compared to modern societies, social regulation is intensive in traditional societies, thus limiting the development of individuality. In extreme form, such restrictions can lead to altruistic suicide, where an individual gives his life for the social group. According to Durkheim, this is the primary type of suicide that occurs in small, traditional societies where individuation is minimal. The classic type of altruistic suicide was the Aztecs' practice of human sacrifice, in which a person was literally sacrificed for the moral or spiritual benefit of the group.[1]

Today many sociologists find fault with Durkheim's distinction between "modern" and "traditional" societies. This binary opposition seems to be a function of the Eurocentrism of his day: Social scientists tended to imagine that their societies were extremely "complex," while "traditional" societies were just "simple." Indeed, "traditional" and "modern" societies may have more in common than Durkheim let on. The degree of integration of the individual into the collective social group is a complex process rather than a permanent state. For instance, even though Durkheim saw altruistic suicide as more prevalent in "primitive" societies, sadly, it is far from absent in "modern" societies as well. Not unlike the altruistic suicides in primitive societies, modern-day wars and suicide bombings are carried out on the premise that sacrificing one's life is necessary for the fight to preserve or attain a sacred way of life for the group as a whole. Nowhere are the similarities between these expressions of altruistic suicide (soldiers, suicide bombers, and "primitive" human sacrifice) more readily apparent than in the tragic case of the Japanese kamikaze pilots of World War II. Shockingly, kamikaze flights were a principal tactic of Japan in the last year of the war.[2]

---

[1]Durkheim briefly mentioned another type of suicide prevalent in "primitive" societies—"fatalistic suicide." For Durkheim, fatalistic suicide was rooted in hopelessness—the hopelessness of oppressed people, such as slaves, who had not even the slightest chance of changing their personal situation.

[2]In October 1944, some 1,200 kamikaze (which means "god-wind") plunged to their deaths in an attack on a U.S. naval fleet in the Leyte Gulf in the Philippines. Six months later, some 1,900 kamikaze dove to their deaths in the battle of Okinawa, resulting in the death of more than 5,000 American sailors. Most of those involved were men in their teens or early 20s; they were said to have gone to their deaths "joyfully," having followed specific rituals of cleanliness, and equipped with books with uplifting thoughts to "transcend life and death" and "Be always pure-hearted and cheerful" (Daniel Ford, "Review of *Kamikaze: Japan's Suicide Gods,*" by Albert Axell and Hideaki Kase, *Wall Street Journal*, September 10, 2002).

Suicide: A Study in Sociology (1897)

Émile Durkheim

ANOMIC SUICIDE

But society is not only something attracting the sentiments and activities of individuals with unequal force. It is also a power controlling them. There is a relation between the way this regulative action is performed and the social suicide-rate.

I

It is a well-known fact that economic crises have an aggravating effect on the suicidal tendency. . . .

In Vienna, in 1873 a financial crisis occurred which reached its height in 1874; the number of suicides immediately rose. From 141 in 1872, they rose to 153 in 1873 and 216 in 1874. The increase in 1874 is 53 per cent[i] above 1872 and 41 per cent above 1873. What proves this catastrophe to have been the sole cause of the increase is the special prominence of the increase when the crisis was acute, or during the first four months of 1874. From January 1 to April 30 there had been 48 suicides in 1871, 44 in 1872, 43 in 1873; there were 73 in 1874. The increase is 70 per cent.[ii] The same crisis occurring at the same time in Frankfurt-on-Main produced the same effects there. In the years before 1874, 22 suicides were committed annually on the average; in 1874 there were 32, or 45 per cent more. . . .

The famous crash is unforgotten which took place on the Paris Bourse during the winter of 1882. Its consequences were felt not only in Paris but throughout France. From 1874 to 1886 the average annual increase was only 2 per cent; in 1882 it was 7 per cent. Moreover, it was unequally distributed among the different times of year, occurring principally during the first three months or at the very time of the crash. Within these three months alone 59 per cent of the total rise occurred. So

distinctly is the rise the result of unusual circumstances that it not only is not encountered in 1881 but has disappeared in 1883, although on the whole the latter year had a few more suicides than the preceding one:

	1881	1882	1883
Annual Total	6,741	7,213 (plus 7%)	7,267
First Three Months	1,589	1,770 (plus 11%)	1,604

This relation is found not only in some exceptional cases, but is the rule. The number of bankruptcies is a barometer of adequate sensitivity, reflecting the variations of economic life. When they increase abruptly from year to year, some serious disturbance has certainly occurred. From 1845 to 1869 there were sudden rises, symptomatic of crises, on three occasions. While the annual increase in the number of bankruptcies during this period is 3.2 per cent, it is 26 per cent in 1847, 37 per cent in 1854 and 20 per cent in 1861. At these three moments, there is also to be observed an unusually rapid rise in the number of suicides. While the average annual increase during these 24 years was only 2 per cent, it was 17 per cent in 1847, 8 per cent in 1854 and 9 per cent in 1861.

But to what do these crises owe their influence? Is it because they increase poverty by causing public wealth to fluctuate? Is life more readily renounced as it becomes more difficult? The explanation is seductively simple; and it agrees with the popular idea of suicide. But it is contradicted by facts.

Actually, if voluntary deaths increased because life was becoming more difficult, they should diminish perceptibly as comfort increases. Now, although when the price of the most necessary foods rises excessively, suicides generally do the same, they are not found to

SOURCE: Reprinted with the permission of The Free Press, a Division of Simon & Schuster Adult Publishing Group from *Suicide: A Study in Sociology* by Émile Durkheim, translated by John A. Spaulding and George Simpson. Edited by George Simpson. Copyright © 1951 by The Free Press. Copyright renewed © 1979 by The Free Press. All rights reserved.

[i]Durkheim incorrectly gives this figure as 51 per cent.—Ed.

[ii]In 1874 over 1873.—Ed.

fall below the average in the opposite case. In Prussia, in 1850 wheat was quoted at the lowest point it reached during the entire period of 1848–81; it was at 6.91 marks per 50 kilograms; yet at this very time suicides rose from 1,527 where they were in 1849 to 1,736, or an increase of 13 per cent, and continued to increase during the years 1851, 1852 and 1853 although the cheap market held. In 1858–59 a new fall took place; yet suicides rose from 2,038 in 1857 to 2,126 in 1858, and to 2,146 in 1859. From 1863 to 1866 prices which had reached 11.04 marks in 1861 fell progressively to 7.95 marks in 1864 and remained very reasonable for the whole period; suicides during the same time increased 17 per cent (2,112 in 1862, 2,485 in 1866).[iii] Similar facts are observed in Bavaria. According to a curve constructed by Mayr[iv] for the period 1835–61, the price of rye was lowest during the years 1857–58 and 1858–59; now suicides, which in 1857 numbered only 286, rose to 329 in 1858, to 387 in 1859. The same phenomenon had already occurred during the years 1848–50; at that time wheat had been very cheap in Bavaria as well as throughout Europe. Yet, in spite of a slight temporary drop due to political events, which we have mentioned, suicides remained at the same level. There were 217 in 1847, there were still 215 in 1848, and if they dropped for a moment to 189 in 1849, they rose again in 1850 and reached 250.

So far is the increase in poverty from causing the increase in suicide that even fortunate crises, the effect of which is abruptly to enhance a country's prosperity, affect suicide like economic disasters. . . .

The conquest of Rome by Victor-Emmanuel in 1870, by definitely forming the basis of Italian unity, was the starting point for the country of a process of growth which is making it one of the great powers of Europe. Trade and industry received a sharp stimulus from it and surprisingly rapid changes took place. Whereas in 1876, 4,459 steam boilers with a total of 54,000 horse-power were enough for industrial needs, the number of machines in 1887 was 9,983 and their horse-power of 167,000 was threefold more. Of course the amount of production rose proportionately during

the same time.[v] Trade followed the same rising course; not only did the merchant marine, communications and transportation develop, but the number of persons and things transported doubled.[vi] As this generally heightened activity caused an increase in salaries (an increase of 35 per cent is estimated to have taken place from 1873 to 1889), the material comfort of workers rose, especially since the price of bread was falling at the same time.[vii] Finally, according to calculations by Bodio, private wealth rose from 45 and a half billions on the average during the period 1875–80 to 51 billions during the years 1880–85 and 54 billions and a half in 1885–90.[viii]

Now, an unusual increase in the number of suicides is observed parallel with this collective renaissance. From 1866 to 1870 they were roughly stable; from 1871 to 1877 they increased 36 per cent. There were in

1864–70	29 suicides per million
1871	31 suicides per million
1872	33 suicides per million
1873	36 suicides per million
1874	37 suicides per million
1875	34 suicides per million
1876	36.5 suicides per million
1877	40.6 suicides per million

And since then the movement has continued. The total figure, 1,139 in 1877, was 1,463 in 1889, a new increase of 28 per cent.

In Prussia the same phenomenon occurred on two occasions. In 1866 the kingdom received a first enlargement. It annexed several important provinces, while becoming the head of the Confederation of the North. Immediately this growth in glory and power was accompanied by a sudden rise in the number of suicides. There had been 123 suicides per million during the period 1856–60 per average year and only 122 during the years 1861–65. In the five years, 1866–70, in spite of the drop in 1870, the average rose to 133. The

---

[iii]See Starck, *Verbrechen und Vergehen in Preussen,* Berlin, 1884, p. 55.

[iv]*Die Gesetzmässigkeit im Gesellschaftsleben,* p. 345.

[v]See Fornasari di Verce, *La criminalita e le vicende economiche d'Italia,* Turin 1894, pp. 7783.
[vi]Ibid., pp. 108–117.

[vii]Ibid., pp. 86–104.

[viii]The increase is less during the period 1885–90 because of a financial crisis.

year 1867, which immediately followed victory, was that in which suicide achieved the highest point it had reached since 1816 (1 suicide per 5,432 inhabitants, while in 1864 there was only one case per 8,739).

On the morrow of the war of 1870 a new accession of good fortune took place. Germany was unified and placed entirely under Prussian hegemony. An enormous war indemnity added to the public wealth; commerce and industry made great strides. The development of suicide was never so rapid. From 1875 to 1886 it increased 90 per cent, from 3,278 cases to 6,212.

World expositions, when successful, are considered favorable events in the existence of a society. They stimulate business, bring more money into the country and are thought to increase public prosperity, especially in the city where they take place. Yet, quite possibly, they ultimately take their toll in a considerably higher number of suicides. Especially does this seem to have been true of the Exposition of 1878. The rise that year was the highest occurring between 1874 and 1886. It was 8 per cent, that is, higher than the one caused by the crash of 1882. And what almost proves the Exposition to have been the cause of this increase is that 86 per cent of it took place precisely during the six months of the Exposition.

In 1889 things were not identical all over France. But quite possibly the Boulanger crisis neutralized the contrary effects of the Exposition by its depressive influence on the growth of suicides. Certainly at Paris, although the political feeling aroused must have had the same effect as in the rest of the country, things happened as in 1878. For the 7 months of the Exposition, suicides increased almost 10 per cent, 9.66 to be exact, while through the remainder of the year they were below what they had been in 1888 and what they afterwards were in 1890.

It may well be that but for the Boulanger influence the rise would have been greater.

	1888	1889	1890
The seven months of the Exposition	517	567	540
The five other months	319	311	356

What proves still more conclusively that economic distress does not have the aggravating influence often attributed to it, is that it tends rather to produce the opposite effect. There is very little suicide in Ireland, where the peasantry leads so wretched a life. Poverty-stricken Calabria has almost no suicides; Spain has a tenth as many as France. Poverty may even be considered a protection. In the various French departments the more people there are who have independent means, the more numerous are suicides. . . .

If therefore industrial or financial crises increase suicides, this is not because they cause poverty, since crises of prosperity have the same result; it is because they are crises, that is, disturbances of the collective order.[ix] Every disturbance of equilibrium, even though it achieves greater comfort and a heightening of general vitality, is an impulse to voluntary death. Whenever serious readjustments take place in the social order, whether or not due to a sudden growth or to an unexpected catastrophe, men are more inclined to self-destruction. How is this possible? How can something considered generally to improve existence serve to detach men from it?

II

No living being can be happy or even exist unless his needs are sufficiently proportioned to his means. In other words, if his needs require more than can be granted, or even merely something of a different sort, they will be under continual friction and can only function painfully. Movements incapable of production without pain tend not to be reproduced. Unsatisfied tendencies atrophy, and as the impulse to live is merely the result of all the rest, it is bound to weaken as the others relax.

In the animal, at least in a normal condition, this equilibrium is established with automatic spontaneity because the animal depends on purely material conditions. All the organism needs is that the supplies of substance and energy constantly employed in the vital process should be periodically renewed by equivalent quantities; that replacement be equivalent to use. When the void created by existence in its own resources is

---

[ix]To prove that an increase in prosperity diminishes suicides, the attempt has been made to show that they become less when emigration, the escape-valve of poverty, is widely practiced (See Legoyt, pp. 257–259). But cases are numerous where parallelism instead of inverse proportions exist between the two. In Italy from 1876 to 1890 the number of emigrants rose from 76 per 100,000 inhabitants to 335, a figure itself exceeded between 1887 and 1889. At the same time suicides did not cease to grow in nnumbers.

Departments Where Suicides Were Committed (1878–1887; per 100,000 Inhabitants)		Average Number of Persons of Independent Means per 1,000 Inhabitants in Each Group of Department(1886)
Suicides	**Number of Departments**	
From 48 to 43	5	127
From 38 to 31	6	73
From 30 to 24	6	69
From 23 to 18	15	59
From 17 to 13	18	49
From 12 to 8	26	49
From 7 to 3	10	42

filled, the animal, satisfied, asks nothing further. Its power of reflection is not sufficiently developed to imagine other ends than those implicit in its physical nature. On the other hand, as the work demanded of each organ itself depends on the general state of vital energy and the needs of organic equilibrium, use is regulated in turn by replacement and the balance is automatic. The limits of one are those of the other; both are fundamental to the constitution of the existence in question, which cannot exceed them.

This is not the case with man, because most of his needs are not dependent on his body or not to the same degree. Strictly speaking, we may consider that the quantity of material supplies necessary to the physical maintenance of a human life is subject to computation, though this be less exact than in the preceding case and a wider margin left for the free combinations of the will; for beyond the indispensable minimum which satisfies nature when instinctive, a more awakened reflection suggests better conditions, seemingly desirable ends craving fulfillment. Such appetites, however, admittedly sooner or later reach a limit which they cannot pass. But how determine the quantity of well-being, comfort or luxury legitimately to be craved by a human being? Nothing appears in man's organic nor in his psychological constitution which sets a limit to such tendencies. The functioning of individual life does not require them to cease at one point rather than at another; the proof being that they have constantly increased since the beginnings of history, receiving more and more complete satisfaction, yet with no weakening of average health. Above all, how establish their proper variation with different conditions of life, occupations, relative importance of services, etc.? In no society are they equally satisfied in the different stages of the social hierarchy. Yet human nature is substantially the same among all men, in its essential qualities. It is not human nature which can assign the variable limits necessary to our needs. They are thus unlimited so far as they depend on the individual alone. Irrespective of any external regulatory force, our capacity for feeling is in itself an insatiable and bottomless abyss.

But if nothing external can restrain this capacity, it can only be a source of torment to itself. Unlimited desires are insatiable by definition and insatiability is rightly considered a sign of morbidity. Being unlimited, they constantly and infinitely surpass the means at their command; they cannot be quenched. Inextinguishable thirst is constantly renewed torture. It has been claimed, indeed, that human activity naturally aspires beyond assignable limits and sets itself unattainable goals. But how can such an undetermined state be any more reconciled with the conditions of mental life than with the demands of physical life? All man's pleasure in acting, moving and exerting himself implies the sense that his efforts are not in vain and that by walking he has advanced. However, one does not advance when one walks toward no goal, or—which is the same thing—when his goal is infinity. Since the distance between us and it is always the same, whatever road we take, we might as well have made the motions without progress from the spot. Even our glances behind and our feeling of pride at the distance covered can cause only deceptive satisfaction, since the remaining distance is not proportionately reduced. To pursue a goal which is by definition unattainable is to condemn oneself to a state of perpetual unhappiness. Of course, man may hope contrary to all reason, and hope has its pleasures even when unreasonable. It may sustain him for a time; but it cannot survive the repeated disappointments of experience indefinitely.

What more can the future offer him than the past, since he can never reach a tenable condition nor even approach the glimpsed ideal? Thus, the more one has, the more one wants, since satisfactions received only stimulate instead of filling needs. Shall action as such be considered agreeable? First, only on condition of blindness to its uselessness. Secondly, for this pleasure to be felt and to temper and half veil the accompanying painful unrest, such unending motion must at least always be easy and unhampered. If it is interfered with only restlessness is left, with the lack of ease which it, itself, entails. But it would be a miracle if no insurmountable obstacle were never encountered. Our thread of life on these conditions is pretty thin, breakable at any instant.

To achieve any other result, the passions first must be limited. Only then can they be harmonized with the faculties and satisfied. But since the individual has no way of limiting them, this must be done by some force exterior to him. A regulative force must play the same role for moral needs which the organism plays for physical needs. This means that the force can only be moral. The awakening of conscience interrupted the state of equilibrium of the animal's dormant existence; only conscience, therefore, can furnish the means to re-establish it. Physical restraint would be ineffective; hearts cannot be touched by physio-chemical forces. So far as the appetites are not automatically restrained by physiological mechanisms, they can be halted only by a limit that they recognize as just. Men would never consent to restrict their desires if they felt justified in passing the assigned limit. But, for reasons given above, they cannot assign themselves this law of justice. So they must receive it from an authority which they respect, to which they yield spontaneously. Either directly and as a whole or through the agency of one of its organs, society alone can play this moderating role; for it is the only moral power superior to the individual, the authority of which he accepts. It alone has the power necessary to stipulate law and to set the point beyond which the passions must not go. Finally, it alone can estimate the reward to be prospectively offered to every class of human functionary, in the name of the common interest.

As a matter of fact, at every moment of history there is a dim perception, in the moral consciousness of societies, of the respective value of different social services, the relative reward due to each, and the consequent degree of comfort appropriate on the average to workers in each occupation. The different functions are graded in public opinion and a certain coefficient of well-being assigned to each, according to its place in the hierarchy. According to accepted ideas, for example, a certain way of living is considered the upper limit to which a workman may aspire in his efforts to improve his existence, and there is another limit below which he is not willingly permitted to fall unless he has seriously bemeaned himself. Both differ for city and country workers, for the domestic servant and the day-laborer, for the business clerk and the official, etc. Likewise the man of wealth is reproved if he lives the life of a poor man, but also if he seeks the refinements of luxury overmuch. Economists may protest in vain; public feeling will always be scandalized if an individual spends too much wealth for wholly superfluous use, and it even seems that this severity relaxes only in times of moral disturbance.[x] A genuine regimen exists, therefore, although not always legally formulated, which fixes with relative precision the maximum degree of ease of living to which each social class may legitimately aspire. However, there is nothing immutable about such a scale. It changes with the increase or decrease of collective revenue and the changes occurring in the moral ideas of society. Thus what appears luxury to one period no longer does so to another; and the well-being which for long periods was granted to a class only by exception and supererogation, finally appears strictly necessary and equitable.

Under this pressure, each in his sphere vaguely realizes the extreme limit set to his ambitions and aspires to nothing beyond. At least if he respects regulations and is docile to collective authority, that is, has a wholesome moral constitution, he feels that it is not well to ask more. Thus, an end and goal are set to the passions. Truly, there is nothing rigid nor absolute about such determination. The economic ideal assigned each class of citizens is itself confined to certain limits, within which the desires have free range. But it is not infinite. This relative limitation and the moderation it involves, make men contented with their lot while stimulating them moderately to improve it; and this average contentment causes the feeling of calm, active happiness, the pleasure in existing and living which characterizes health for societies as well as for individuals. Each person is then at least, generally speaking, in harmony with

---

[x]Actually, this is a purely moral reprobation and can hardly be judicially implemented. We do not consider any reestablishment of sumptuary laws desirable or even possible.

his condition, and desires only what he may legitimately hope for as the normal reward of his activity. Besides, this does not condemn man to a sort of immobility. He may seek to give beauty to his life; but his attempts in this direction may fail without causing him to despair. For, loving what he has and not fixing his desire solely on what he lacks, his wishes and hopes may fail of what he has happened to aspire to, without his being wholly destitute. He has the essentials. The equilibrium of his happiness is secure because it is defined, and a few mishaps cannot disconcert him.

But it would be of little use for everyone to recognize the justice of the hierarchy of functions established by public opinion, if he did not also consider the distribution of these functions just. The workman is not in harmony with his social position if he is not convinced that he has his deserts. If he feels justified in occupying another, what he has would not satisfy him. So it is not enough for the average level of needs for each social condition to be regulated by public opinion, but another, more precise rule, must fix the way in which these conditions are open to individuals. There is no society in which such regulation does not exist. It varies with times and places. Once it regarded birth as the almost exclusive principle of social classification; today it recognizes no other inherent inequality than hereditary fortune and merit. But in all these various forms its object is unchanged. It is also only possible, everywhere, as a restriction upon individuals imposed by superior authority, that is, by collective authority. For it can be established only by requiring of one or another group of men, usually of all, sacrifices and concessions in the name of the public interest.

Some, to be sure, have thought that this moral pressure would become unnecessary if men's economic circumstances were only no longer determined by heredity. If inheritance were abolished, the argument runs, if everyone began life with equal resources and if the competitive struggle were fought out on a basis of perfect equality, no one could think its results unjust. Each would instinctively feel that things are as they should be.

Truly, the nearer this ideal equality were approached, the less social restraint will be necessary. But it is only a matter of degree. One sort of heredity will always exist, that of natural talent. Intelligence, taste, scientific, artistic, literary or industrial ability, courage and manual dexterity are gifts received by each of us at birth, as the heir to wealth receives his capital or as the nobleman formerly received his title and function. A moral discipline will therefore still be required to make those less favored by nature accept the lesser advantages which they owe to the chance of birth. Shall it be demanded that all have an equal share and that no advantage be given those more useful and deserving? But then there would have to be a discipline far stronger to make these accept a treatment merely equal to that of the mediocre and incapable.

But like the one first mentioned, this discipline can be useful only if considered just by the peoples subject to it. When it is maintained only by custom and force, peace and harmony are illusory; the spirit of unrest and discontent are latent; appetites superficially restrained are ready to revolt. This happened in Rome and Greece when the faiths underlying the old organization of the patricians and plebeians were shaken, and in our modern societies when aristocratic prejudices began to lose their old ascendancy. But this state of upheaval is exceptional; it occurs only when society is passing through some abnormal crisis. In normal conditions the collective order is regarded as just by the great majority of persons. Therefore, when we say that an authority is necessary to impose this order on individuals, we certainly do not mean that violence is the only means of establishing it. Since this regulation is meant to restrain individual passions, it must come from a power which dominates individuals; but this power must also be obeyed through respect, not fear.

It is not true, that human activity can be released from all restraint. Nothing in the world can enjoy such a privilege. All existence being a part of the universe is relative to the remainder; its nature and method of manifestation accordingly depend not only on itself but on other beings, who consequently restrain and regulate it. Here there are only differences of degree and form between the mineral realm and the thinking person. Man's characteristic privilege is that the bond he accepts is not physical but moral; that is, social. He is governed not by a material environment brutally imposed on him, but by a conscience superior to his own, the superiority of which he feels. Because the greater, better part of his existence transcends the body, he escapes the body's yoke, but is subject to that of society.

But when society is disturbed by some painful crisis or by beneficent but abrupt transitions, it is momentarily incapable of exercising this influence; thence come the sudden rises in the curve of suicides which we have pointed out above.

In the case of economic disasters, indeed, something like a declassification occurs which suddenly casts certain individuals into a lower state than their previous one. Then they must reduce their requirements, restrain their needs, learn greater self-control. All the advantages of social influence are lost so far as they are concerned; their moral education has to be recommenced. But society cannot adjust them instantaneously to this new life and teach them to practice the increased self-repression to which they are unaccustomed. So they are not adjusted to the condition forced on them, and its very prospect is intolerable; hence the suffering which detaches them from a reduced existence even before they have made trial of it.

It is the same if the source of the crisis is an abrupt growth of power and wealth. Then, truly, as the conditions of life are changed, the standard according to which needs were regulated can no longer remain the same; for it varies with social resources, since it largely determines the share of each class of producers. The scale is upset; but a new scale cannot be immediately improvised. Time is required for the public conscience to reclassify men and things. So long as the social forces thus freed have not regained equilibrium, their respective values are unknown and so all regulation is lacking for a time. The limits are unknown between the possible and the impossible, what is just and what is unjust, legitimate claims and hopes and those which are immoderate. Consequently, there is no restraint upon aspirations. If the disturbance is profound, it affects even the principles controlling the distribution of men among various occupations. Since the relations between various parts of society are necessarily modified, the ideas expressing these relations must change. Some particular class especially favored by the crisis is no longer resigned to its former lot, and, on the other hand, the example of its greater good fortune arouses all sorts of jealousy below and about it. Appetites, not being controlled by a public opinion become disoriented, no longer recognize the limits proper to them. Besides, they are at the same time seized by a sort of natural erethism simply by the greater intensity of public life. With increased prosperity desires increase. At the very moment when traditional rules have lost their authority, the richer prize offered these appetites stimulates them and makes them more exigent and impatient of control. The state of de-regulation or anomy is thus further heightened by passions being less disciplined, precisely when they need more disciplining.

But then their very demands make fulfillment impossible. Overweening ambition always exceeds the results obtained, great as they may be, since there is no warning to pause here. Nothing gives satisfaction and all this agitation is uninterruptedly maintained without appeasement.

Above all, since this race for an unattainable goal can give no other pleasure but that of the race itself, if it is one, once it is interrupted the participants are left empty-handed. At the same time the struggle grows more violent and painful, both from being less controlled and because competition is greater. All classes contend among themselves because no established classification any longer exists. Effort grows, just when it becomes less productive. How could the desire to live not be weakened under such conditions?

This explanation is confirmed by the remarkable immunity of poor countries. Poverty protects against suicide because it is a restraint in itself. No matter how one acts, desires have to depend upon resources to some extent; actual possessions are partly the criterion of those aspired to. So the less one has the less he is tempted to extend the range of his needs indefinitely. Lack of power, compelling moderation, accustoms men to it, while nothing excites envy if no one has superfluity. Wealth, on the other hand, by the power it bestows, deceives us into believing that we depend on ourselves only. Reducing the resistance we encounter from objects, it suggests the possibility of unlimited success against them. The less limited one feels, the more intolerable all limitation appears. Not without reason, therefore, have so many religions dwelt on the advantages and moral value of poverty. It is actually the best school for teaching self-restraint. Forcing us to constant self-discipline, it prepares us to accept collective discipline with equanimity, while wealth, exalting the individual, may always arouse the spirit of rebellion which is the very source of immorality. This, of course, is no reason why humanity should not improve its material condition. But though the moral danger involved in every growth of prosperity is not irremediable, it should not be forgotten.

III

If anomy never appeared except, as in the above instances, in intermittent spurts and acute crisis, it might cause the social suicide-rate to vary from time to time, but it would not be a regular, constant factor. In one sphere of social life, however—the sphere of trade and industry—it is actually in a chronic state.

For a whole century, economic progress has mainly consisted in freeing industrial relations from all regulation. Until very recently, it was the function of a whole system of moral forces to exert this discipline. First, the influence of religion was felt alike by workers and masters, the poor and the rich. It consoled the former and taught them contentment with their lot by informing them of the providential nature of the social order, that the share of each class was assigned by God himself, and by holding out the hope for just compensation in a world to come in return for the inequalities of this world. It governed the latter, recalling that worldly interests are not man's entire lot, that they must be subordinate to other and higher interests, and that they should therefore not be pursued without rule or measure. Temporal power, in turn, restrained the scope of economic functions by its supremacy over them and by the relatively subordinate role it assigned them. Finally, within the business world proper, the occupational groups by regulating salaries, the price of products and production itself, indirectly fixed the average level of income on which needs are partially based by the very force of circumstances. However, we do not mean to propose this organization as a model. Clearly it would be inadequate to existing societies without great changes. What we stress is its existence, the fact of its useful influence, and that nothing today has come to take its place.

Actually, religion has lost most of its power. And government, instead of regulating economic life, has become its tool and servant. The most opposite schools, orthodox economists and extreme socialists, unite to reduce government to the role of a more or less passive intermediary among the various social functions. The former wish to make it simply the guardian of individual contracts; the latter leave it the task of doing the collective bookkeeping, that is, of recording the demands of consumers, transmitting them to producers, inventorying the total revenue and distributing it according to a fixed formula. But both refuse it any power to subordinate other social organs to itself and to make them converge toward one dominant aim. On both sides nations are declared to have the single or chief purpose of achieving industrial prosperity; such is the implication of the dogma of economic materialism, the basis of both apparently opposed systems. And as these theories merely express the state of opinion, industry, instead of being still regarded as a means to an end transcending itself, has become the supreme end of individuals and societies alike. Thereupon the appetites thus excited have become freed of any limiting authority. By sanctifying them, so to speak, this apotheosis of well-being has placed them above all human law. Their restraint seems like a sort of sacrilege. For this reason, even the purely utilitarian regulation of them exercised by the industrial world itself through the medium of occupational groups has been unable to persist. Ultimately, this liberation of desires has been made worse by the very development of industry and the almost infinite extension of the market. So long as the producer could gain his profits only in his immediate neighborhood, the restricted amount of possible gain could not much overexcite ambition. Now that he may assume to have almost the entire world as his customer, how could passions accept their former confinement in the face of such limitless prospects?

Such is the source of the excitement predominating in this part of society, and which has thence extended to the other parts. There, the state of crisis and anomy is constant and, so to speak, normal. From top to bottom of the ladder, greed is aroused without knowing where to find ultimate foothold. Nothing can calm it, since its goal is far beyond all it can attain. Reality seems valueless by comparison with the dreams of fevered imaginations; reality is therefore abandoned, but so too is possibility abandoned when it in turn becomes reality. A thirst arises for novelties, unfamiliar pleasures, nameless sensations, all of which lose their savor once known. Henceforth one has no strength to endure the least reverse. The whole fever subsides and the sterility of all the tumult is apparent, and it is seen that all these new sensations in their infinite quantity cannot form a solid foundation of happiness to support one during days of trial. The wise man, knowing how to enjoy achieved results without having constantly to replace them with others, finds in them an attachment to life in the hour of difficulty. But the man who has always pinned all his hopes on the future and lived with his eyes fixed upon it, has nothing in the past as a comfort against the present's afflictions, for the past was nothing to him but a series of hastily experienced stages. What blinded him to himself was his expectation always to find further on the happiness he had so far missed. Now he is stopped in his tracks; from now on nothing remains behind or ahead of him to fix his gaze upon. Weariness alone, moreover, is enough to bring disillusionment, for he cannot in the end escape the futility of an endless pursuit.

We may even wonder if this moral state is not principally what makes economic catastrophes of our day so fertile in suicides. In societies where a man is subjected to a healthy discipline, he submits more readily to the blows of chance. The necessary effort for sustaining a little more discomfort costs him relatively little, since he is used to discomfort and constraint. But when every constraint is hateful in itself, how can closer constraint not seem intolerable? There is no tendency to resignation in the feverish impatience of men's lives. When there is no other aim but to outstrip constantly the point arrived at, how painful to be thrown back! Now this very lack of organization characterizing our economic condition throws the door wide to every sort of adventure. Since imagination is hungry for novelty, and ungoverned, it gropes at random. Setbacks necessarily increase with risks and thus crises multiply, just when they are becoming more destructive.

Yet these dispositions are so inbred that society has grown to accept them and is accustomed to think them normal. It is everlastingly repeated that it is man's nature to be eternally dissatisfied, constantly to advance, without relief or rest, toward an indefinite goal. The longing for infinity is daily represented as a mark of moral distinction, whereas it can only appear within unregulated consciences which elevate to a rule the lack of rule from which they suffer. The doctrine of the most ruthless and swift progress has become an article of faith. But other theories appear parallel with those praising the advantages of instability, which, generalizing the situation that gives them birth, declare life evil, claim that it is richer in grief than in pleasure and that it attracts men only by false claims. Since this disorder is greatest in the economic world, it has most victims there.

Industrial and commercial functions are really among the occupations which furnish the greatest number of suicides (see Table XXIV). Almost on a level with the liberal professions, they sometimes surpass them; they are especially more afflicted than agriculture, where the old regulative forces still make their appearance felt most and where the fever of business has least penetrated. Here is best called what was once the general constitution of the economic order. And the divergence would be yet greater if, among the suicides of industry, employers were distinguished from workmen, for the former are probably most stricken by the state of anomy. The enormous rate of those with independent means (720 per million) sufficiently shows that the possessors of most comfort suffer most. Everything that enforces subordination attenuates the effects of this state. At least the horizon of the lower classes is limited by those above them, and for this same reason their desires are more modest. Those who have only empty space above them are almost inevitably lost in it, if no force restrains them.

Anomy, therefore, is a regular and specific factor in suicide in our modern societies; one of the springs from which the annual contingent feeds. So we have here a new type to distinguish from the others. It differs from them in its dependence, not on the way in which individuals are attached to society, but on how it regulates them. Egoistic suicide results from man's no longer finding a basis for existence in life; altruistic suicide, because this basis for existence appears to man situated beyond life itself. The third sort of suicide, the existence of which has just been shown, results from man's activity's lacking regulation and his consequent sufferings. By virtue of its origin we shall assign this last variety the name of *anomic suicide.*

Certainly, this and egoistic suicide have kindred ties. Both spring from society's insufficient presence in individuals. But the sphere of its absence is not the same in both cases. In egoistic suicide it is deficient in truly collective activity, thus depriving the latter of object and meaning. In anomic suicide, society's influence is lacking in the basically individual passions, thus leaving them without a check-rein. In spite of their relationship, therefore, the two types are independent of each other. We may offer society everything social in us, and still be unable to control our desires; one may live in an anomic state without being egoistic, and vice versa. These two sorts of suicide therefore do not draw their chief recruits from the same social environments; one has its principal field among intellectual careers, the world of thought—the other, the industrial or commercial world.

IV

But economic anomy is not the only anomy which may give rise to suicide.

The suicides occurring at the crisis of widowhood, of which we have already spoken[xi] are really due to domestic anomy resulting from the death of husband or wife. A family catastrophe occurs which affects the survivor. He is not adapted to the new situation in which he finds himself and accordingly offers less resistance to suicide.

But another variety of anomic suicide should draw greater attention, both because it is more chronic and because it will serve to illustrate the nature and functions of marriage.

In the *Annales de demographie internationale* (September 1882), Bertillon published a remarkable study of divorce, in which he proved the following proposition: throughout Europe the number of suicides varies with that of divorces and separations [Table XXV illustrates such variations]. . . .

[xi]See above, Book II, Ch. 3.

Table XXIV   Suicides per Million Persons of Different Occupations

	Trade	Transportation	Industry	Agriculture	Liberal* Professions
France (1878–87)†	440	–	340	240	300
Switzerland (1876)	664	1,514	577	304	558
Italy (1866–76)	277	152.6	80.4	26.7	618‡
Prussia (1883–90)	754	–	456	315	832
Bavaria (1884–91)	465	–	369	153	454
Belgium (1886–90)	421	–	160	160	100
Wurttemberg (1873–78)	273	–	190	206	–
Saxony (1878)		341.59§		71.17	–

*When statistics distinguish several different sorts of liberal occupation, we show as a specimen the one in which the suicide-rate is highest.

†From 1826 to 1880 economic functions seem less affected (see Compte-rendu of 1880); but were occupational statistics very accurate?

‡This figure is reached only by men of letters.

§Figure represents Trade, Transportation, and Industry combined for Saxony. Ed.

INDIVIDUAL FORMS OF THE DIFFERENT TYPES OF SUICIDE

One result now stands out prominently from our investigation: namely, that there are not one but various forms of suicide. Of course, suicide is always the act of a man who prefers death to life. But the causes determining him are not of the same sort in all cases: they are even sometimes mutually opposed. Now, such difference in causes must reappear in their effects. We may therefore be sure that there are several sorts of suicide which are distinct in quality from one another. But the certainty that these differences exist is not enough; we need to observe them directly and know of what they consist. We need to see the characteristics of special suicides grouped in distinct classes corresponding to the types just distinguished. Thus we would follow the various currents which generate suicide from their social origins to their individual manifestations.

This morphological classification, which was hardly possible at the commencement of this study, may be undertaken now that an aetiological classification forms its basis. Indeed, we only need to start with the three kinds of factors which we have just assigned to suicide and discover whether the distinctive properties it assumes in manifesting itself among individual persons may be derived from them, and if so, how. Of course, not all the peculiarities which suicide may present can be deduced in this fashion; for some may exist which depend solely on the person's own nature. Each victim of suicide gives his act a personal stamp which expresses his temperament, the special conditions in which he is involved, and which, consequently, cannot be explained by the social and general causes of the phenomenon. But these causes in turn must stamp the suicides they determine with a shade all their own, a special mark expressive of them. This collective mark we must find.

To be sure, this can be done only approximately. We are not in a position to describe methodically all the suicides daily committed by men or committed in the course of history. We can only emphasize the most general and striking characteristics without even having an objective criterion for making the selection. Moreover, we can only proceed deductively in relating them to the respective causes from which they seem to spring. All that we can do is to show their logical implication, though the reasoning may not always be able to receive experimental confirmation. We do not forget that a deduction uncontrolled by

Table XXV   Comparison of European States from the Point of View of Both Divorce and Suicide

	Annual Divorces per 1,000 Marriages		Suicides per Million Inhabitants
I. Countries Where Divorce and Separation Are Rare			
Norway	0.54	(1875–80)	73
Russia	1.6	(1871–77)	30
England and Wales	1.3	(1871–79)	68
Scotland	2.1	(1871–81)	–
Italy	3.05	(1871–73)	31
Finland	3.9	(1875–79)	30.8
Averages	2.07		46.5
II. Countries Where Divorce and Separation Are of Average Frequency			
Bavaria	5.0	(1881)	90.5
Belgium	5.1	(1871–80)	68.5
Holland	6.0	(1871–80)	35.5
Sweden	6.4	(1871–80)	81
Baden	6.5	(1874–79)	156.6
France	7.5	(1871–79)	150
Wurttemberg	8.4	(1876–78)	162.4
Prussia	–		133
Averages	6.4		109.6
III. Countries Where Divorce and Separation Are Frequent			
Kingdom of Saxony	26.9	(1876–80)	299
Denmark	38	(1871–80)	258
Switzerland	47	(1876–80)	216
Averages	37.3		257

experiment is always questionable. Yet this research is far from being useless, even with these reservations. Even though it may be considered only a method of illustrating the preceding results by examples, it would still have the worth of giving them a more concrete character by connecting them more closely with the data of sense-perception and with the details of daily experience. It will also introduce some little distinctiveness into this mass of facts usually lumped together as though varying only by shades, though there are striking differences among them. Suicide is like mental alienation. For the popular mind the latter consists in a single state, always identical, capable only of superficial differentiation according to circumstances. For the alienist, on the contrary, the word denotes many nosological types. Every suicide is, likewise, ordinarily considered a victim of melancholy whose life has become a burden to him. Actually, the acts by which a man renounces life belong to different species, of wholly different moral and social significance.

Introduction to *The Elementary Forms of Religious Life*

In his final and most theoretically acclaimed book, *The Elementary Forms of Religious Life* (1912), Durkheim sought to explain the way the moral realm worked by focusing on religion. Durkheim saw religious ceremonies not merely as a celebration of supernatural deities, but as a worshipping of social life itself, such that as long as there are societies, there will be religion (Robertson 1970:13).

In other words, for Durkheim, social life—whether in traditional or modern society—is inherently religious, for "religious force is nothing other than the collective and anonymous force" of society ([1912] 1995:210). The worship of transcendent gods or spirits and the respect and awe accorded to their power is in actuality the worship of the social group and the force it exerts over the individual. No matter how "simple" or "complex" the society, religion is thus a "system of ideas with which the individuals represent to themselves the society of which they are members, and the obscure but intimate relations which they have with it . . . for it is an eternal truth that outside of us there exists something greater than us, with which we enter into communion" (ibid. 257). For Durkheim, this outside power, this "something greater" is society.

In saying that social life is inherently religious, Durkheim defined religion in a very broad way. For Durkheim, "religion" does not mean solely "churchly" or institutional things; rather, religion is a system of symbols and rituals about the sacred that is practiced by a community of believers. This definition of religion is often called "functionalist" rather than "substantive" because it emphasizes not the substantive content of religion, such as particular rituals or doctrines (e.g., baptisms or bar mitzvahs, or belief in an afterlife, higher beings, etc.), but the social *function* of religion.

For Durkheim, the primary function of religion is to encode the system of relations of the group (Eliade and Couliano 1991:2). It focuses and reaffirms the collective sentiments and ideas that hold the group together. Religious practices, accordingly, serve to bind participants together in celebration of the society (Robertson 1970:15). As Durkheim ([1912] 1995) states,

> There can be no society which does not feel the need of upholding and reaffirming at regular intervals the collective sentiments and the collective ideas which makes its unity and its personality. Now this moral remaking cannot be achieved except by the means of reunions, assemblies and meetings where the individuals, being closely united to one another, reaffirm in common their common sentiments. (p. 429)

This communal function of religion is carried out through the dual processes of ritualization and symbolization. A **ritual** is a highly routinized act, such as taking communion. As the name reveals, the Christian ritual of communion not only commemorates an historical event in the life of Jesus; it also represents participation in the unity ("communion") of believers (McGuire 1997:187). Most interestingly, because they are practices (not beliefs or values), rituals can unite a social group regardless of individual differences in beliefs or strength of convictions. It is the common *experience* and *focus* that binds the participants together (see Photos 3.3a and 3.3b).

This is why, for Durkheim, there is no essential difference between "religious" and "secular" ritual acts. "Let us pray" (an opening moment in a religious service) and "Let us stand for the national anthem" (an opening moment of a baseball game) are both ritual acts that bond the individual to a community. In exactly the same way, Durkheim suggested that there is no essential difference between religious holidays, such as Passover or Christmas, and secular holidays, such as Independence Day or Thanksgiving. Both are collective celebrations of identity and community (see Edles 2002:27–30).

As noted above, in addition to ritual practices, there is another important means through which the communal function of religion is achieved: symbolization. A **symbol** is something that stands for something else. It is a representation that calls up collective ideas and meanings. Thus, for instance, the "cross" is a marker that symbolizes Christian spirituality and/or tradition. Wearing a cross on a necklace often *means* that one is a Christian. It identifies the wearer as a member of a specific religious community and/or specific shared ideas (e.g., a religious tradition in which Jesus Christ is understood as the son of God). Most importantly, symbols such as the cross are capable of calling up and reaffirming shared meaning and the feeling of community in between periodic ritual acts (such as religious celebrations and weekly church services). As Durkheim ([1912] 1995:232) states, "Without symbols, social sentiments could have only a precarious existence."

In *The Elementary Forms of Religious Life,* Durkheim explains that symbols are classified as fundamentally sacred or profane. The **sacred** refers to the extraordinary, that which is set apart from and "above and beyond" the everyday world. In direct contrast to the sacred realm, is the realm of the everyday world of the mundane or routine, or the **profane**. Most importantly, objects are intrinsically neither sacred nor profane; rather, their meaning or classification is continually produced and reproduced (and/or altered) in collective processes of ritualization and symbolization. Thus, for instance, lighting a candle can either be a relatively mundane task to enhance one's dinner table or it can be a sacred act, as in the case of the Jewish ritual of lighting a candle to commemorate the Sabbath (McGuire 1997:17). In the latter context, this act denotes a sacred *moment* as well as celebration. This points to the central function of the distinction between the sacred and the profane. It imposes an orderly system on the inherently untidy experience of living (Gamson 1998:141). Thus, for instance, ritual practices (e.g., standing for the national anthem or lighting a candle to commemorate the Sabbath) transform a profane moment into a sacred moment; while sacred sites (churches, mosques, synagogues) differentiate "routine" places from those that compel attitudes of awe and inspiration. The symbolic plasticity of time and space

Photo 3.3a Congregation Taking Communion at a Catholic Church

Photo 3.3b Fans at Sporting Event Doing "the Wave"

Both church goers and sports fans engage in communal ritual acts. As Durkheim ([1912] 1995:262) states, "It is by uttering the same cry, pronouncing the same word, or performing the same gesture in regard to some object that they become and feel themselves to be in unison."

is especially apparent in the way devout Muslims (who often must pray in everyday, mundane settings in order to fulfill their religious duties) carry out the frequent prayers required by their religion. They lay down a (sacred) prayer carpet in their office or living room, thereby enabling them to convert a profane time and space into a sacred time and space. This temporal and spatial reordering transforms the profane realm of work or home into a spiritual, sacred domain. Such acts, and countless others, help order and organize our experience of the world by carving it into that which is extraordinary or sacred and that which is unremarkable or profane.

The Elementary Forms of Religious Life (1912)

Émile Durkheim

ORIGINS OF THESE BELIEFS

It is obviously not out of the sensations which the things serving as totems are able to arouse in the mind; we have shown that these things are frequently insignificant. The lizard, the caterpillar, the rat, the ant, the frog, the turkey, the bream-fish, the plum-tree, the cockatoo, etc., to cite only those names which appear frequently in the lists of Australian totems, are not of a nature to produce upon men these great and strong impressions which in a way resemble religious emotions and which impress a sacred character upon the objects they create. It is true that this is not the case with the stars and the great atmospheric phenomena, which have, on the contrary, all that is necessary to strike the imagination forcibly; but as a matter of fact, these serve only very exceptionally as totems. It is even probable that they were very slow in taking this office. So it is not the intrinsic nature of the thing whose name the clan bears that marked it out to become the object of a cult. Also, if the sentiments which it inspired were really the determining cause of the totemic rites and beliefs, it would be the pre-eminently sacred thing; the animals or plants employed as totems would play an eminent part in the religious life. But we know that the centre of the cult is actually elsewhere. It is the figurative representations of this plant or animal and the totemic emblems and symbols of every sort, which have the greatest sanctity; so it is in them that is found the source of that religious nature, of which the real objects represented by these emblems receive only a reflection.

Thus the totem is before all a symbol, a material expression of something else. But of what?

From the analysis to which we have been giving our attention, it is evident that it expresses and symbolizes two different sorts of things. In the first place, it is the outward and visible form of what we have called the totemic principle or god. But it is also the symbol of the determined society called the clan. It is its flag; it is the sign by which each clan distinguishes itself from the others, the visible mark of its personality, a mark borne by everything which is a part of the clan under any title whatsoever, men, beasts or things. So if it is at once the symbol of the god and of the society, is that not because the god and the society are only one? How could the emblem of the group have been able to become the figure of this quasi-divinity, if the group and the divinity were two distinct realities? The god of the clan, the totemic principle, can therefore be nothing else than the clan itself, personified and represented to the imagination under the visible form of the animal or vegetable which serves as totem.

But how has this apotheosis been possible, and how did it happen to take place in this fashion?

II

In a general way, it is unquestionable that a society has all that is necessary to arouse the sensation of the divine in minds, merely by the power that it has over them; for to its members it is what a god is to his worshippers. In fact, a god is, first of all, a being whom men think of as superior to themselves, and upon whom they feel that they depend. Whether it be a conscious personality, such as Zeus or Jahveh, or merely abstract forces such as those in play in totemism, the worshipper, in the one case as in the other, believes himself held to certain manners of acting which are imposed upon him by the nature of the sacred principle with which he feels that he is in communion. Now society also gives us the sensation of a perpetual dependence. Since it has a nature which is peculiar to itself and different from our individual nature, it pursues ends which are likewise special to it; but, as it cannot attain them except through our intermediacy, it imperiously demands our aid. It requires that, forgetful of our own interest, we make ourselves its servitors, and it submits us to every sort of inconvenience, privation and sacrifice, without which social life would be impossible. It is because of this that at every instant we are obliged to submit ourselves to rules of conduct and of thought which we have neither made nor desired, and which are sometimes even contrary to our most fundamental inclinations and instincts.

Even if society were unable to obtain these concessions and sacrifices from us except by a material constraint, it might awaken in us only the idea of a physical force to which we must give way of necessity, instead of that of a moral power such as religious adore. But as a matter of fact, the empire which it holds over consciences is due much less to the physical supremacy of which it has the privilege than to the moral authority with which it is invested. If we yield to its orders, it is not merely because it is strong enough to triumph over our resistance; it is primarily because it is the object of a venerable respect.

We say that an object, whether individual or collective, inspires respect when the representation expressing it in the mind is gifted with such a force that it automatically causes or inhibits actions, *without regard for any consideration relative to their useful or injurious effects.* When we obey somebody because of the moral authority which we recognize in him, we follow out his opinions, not because they seem wise, but because a certain sort of physical energy is imminent in the idea that we form of this person, which conquers our will and inclines it in the indicated direction. Respect is the emotion which we experience when we feel this interior and wholly spiritual pressure operating upon us. Then we are not determined by the advantages or inconveniences of the attitude which is prescribed or recommended to us; it is by the way in which we represent to ourselves the person recommending or prescribing it. This is why commands generally take a short, peremptory form leaving no place for hesitation; it is because, in so far as it is a command and goes by its own force, it excludes all idea of deliberation or calculation; it gets its efficacy from the intensity of the mental state in which it is placed. It is this intensity which creates what is called a moral ascendancy.

Now the ways of action to which society is strongly enough attached to impose them upon its members, are, by that very fact, marked with a distinctive sign provocative of respect. Since they are elaborated in common, the vigour with which they have been thought of by each particular mind is retained in all the other minds, and reciprocally. The representations which express them within each of us have an intensity which no purely private states of consciousness could ever attain; for they have the strength of the innumerable individual representations which have served to form each of them. It is society who speaks through the mouths of those who affirm them in our presence; it is society whom we hear in hearing them; and the voice of all has an accent which that of one alone could never have.[i] The very violence with which society reacts, by way of blame or material suppression, against every attempted dissidence, contributes to strengthening its empire by manifesting the common conviction through this burst of ardour[ii] In a word, when something is the object of such a state of opinion, the representation which each individual has of it gains a power of action from its origins and the conditions in which it was born, which even those feel who do not submit themselves to it. It tends to repel the representations which contradict it, and it keeps them at a distance; on the other hand, it commands those acts which will realize it, and it does so, not by a material coercion or by the perspective of something of this sort, but by the simple radiation of the mental energy which it contains. It has an efficacy coming solely from its psychical properties, and it is by just this sign that moral authority is recognized. So opinion, primarily a social thing, is a source of authority, and it might even be asked whether all authority is not the daughter of opinion.[iii] It may be objected that science is often the antagonist of opinion, whose errors it combats and rectifies. But it cannot succeed in this task if it does not have sufficient authority, and it can obtain this authority only from opinion itself. If a people did not have faith in science, all the scientific demonstrations in the world would be without any influence whatsoever over their minds. Even to-day, if science happened to resist a very

---

[i]See our *Division du travail social,* 3rd ed., pp. 64 ff.

[ii]Ibid., p. 76.

[iii]This is the case at least with all moral authority recognized as such by the group as a whole.

strong current of public opinion, it would risk losing its credit there.[iv]

Since it is in spiritual ways that social pressure exercises itself, it could not fail to give men the idea that outside themselves there exist one or several powers, both moral and, at the same time, efficacious, upon which they depend. They must think of these powers, at least in part, as outside themselves, for these address them in a tone of command and sometimes even order them to do violence to their most natural inclinations. It is undoubtedly true that if they were able to see that these influences which they feel emanate from society, then the mythological system of interpretations would never be born. But social action follows ways that are too circuitous and obscure, and employs psychical mechanisms that are too complex to allow the ordinary observer to see when it comes. As long as scientific analysis does not come to teach it to them, men know well that they are acted upon, but they do not know by whom. So they must invent by themselves the idea of these powers with which they feel themselves in connection, and from that, we are able to catch a glimpse of the way by which they were led to represent them under forms that are really foreign to their nature and to transfigure them by thought.

But a god is not merely an authority upon whom we depend; it is a force upon which our strength relies. The man who has obeyed his god and who for this reason, believes the god is with him, approaches the world with confidence and with the feeling of an increased energy. Likewise, social action does not confine itself to demanding sacrifices, privations and efforts from us. For the collective force is not entirely outside of us; it does not act upon us wholly from without; but rather, since society cannot exist except in and through individual consciousness,[v] this force must also penetrate us and organize itself within us; it thus becomes an integral part of our being and by that very fact this is elevated and magnified.

There are occasions when this strengthening and vivifying action of society is especially apparent. In the midst of an assembly animated by a common passion, we become susceptible of acts and sentiments of which we are incapable when reduced to our own forces; and when the assembly is dissolved and when, finding ourselves alone again, we fall back to our ordinary level, we are then able to measure the height to which we have been raised above ourselves. History abounds in examples of this sort. It is enough to think of the night of the Fourth of August, 1789, when an assembly was suddenly led to an act of sacrifice and abnegation which each of its members had refused the day before, and at which they were all surprised the day after.[vi] This is why all parties political, economic or confessional, are careful to have periodical reunions where their members may revivify their common faith by manifesting it in common. To strengthen those sentiments which, if left to themselves, would soon weaken, it is sufficient to bring those who hold them together and to put them into closer and more active relations with one another. This is the explanation of the particular attitude of a man speaking to a crowd, at least if he has succeeded in entering into communion with it. His language has a grandiloquence that would be ridiculous in ordinary circumstances; his gestures show a certain domination; his very thought is impatient of all rules, and easily falls into all sorts of excesses. It is because he feels within him an abnormal over-supply of force which overflows and tries to burst out from him; sometimes he even has the feeling that he is dominated by a moral force which

---

[iv]We hope that this analysis and those which follow will put an end to an inexact interpretation of our thought, from which more than one misunderstanding has resulted. Since we have made constraint the *outward sign* by which social facts can be the most easily recognized and distinguished from the facts of individual psychology, it has been assumed that according to our opinion, physical constraint is the essential thing for social life. As a matter of fact, we have never considered it more than the material and apparent expression of an interior and profound fact which is wholly ideal: this is *moral authority*. The problem of sociology—if we can speak of a sociological problem—consists in seeking, among the different forms of external constraint, the different sorts of moral authority corresponding to them and in discovering the causes which have determined these latter. The particular question which we are treating in this present work has as its principal object, the discovery of the form under which that particular variety of moral authority which is inherent in all that is religious has been born, and out of what elements it is made. It will be seen presently that even if we do make social pressure one of the distinctive characteristics of sociological phenomena, we do not mean to say that it is the only one. We shall show another aspect of the collective life, nearly opposite to the preceding one, but none the less real.

[v]Of course this does not mean to say that the collective [conscience] does not have distinctive characteristics of its own (on this point, see *Représentations individuelles et représentations collectives,* in *Revue de Métaphysique et de Morale,* 1898, pp. 273 ff.).

[vi]This is proved by the length and passionate character of the debates where a legal form was given to the resolutions made in a moment of collective enthusiasm. In the clergy as in the nobility, more than one person called this celebrated night the dupe's night, or, with Rivarol, the St. Bartholomew of the estates (see Stoll, *Suggestion und Hypnotismus in de Völkerpsychologie,* 2nd ed., p. 618, n. 2).

is greater than he and of which he is only the interpreter. It is by this trait that we are able to recognize what has often been called the demon of oratorical inspiration. Now this exceptional increase of force is something very real; it comes to him from the very group which he addresses. The sentiments provoked by his words come back to him, but enlarged and amplified, and to this degree they strengthen his own sentiment. The passionate energies he arouses re-echo within him and quicken his vital tone. It is no longer a simple individual who speaks; it is a group incarnate and personified.

Besides these passing and intermittent states, there are other more durable ones, where this strengthening influence of society makes itself felt with greater consequences and frequently even with greater brilliancy. There are periods in history when, under the influence of some great collective shock, social interactions have become much more frequent and active. Men look for each other and assemble together more than ever. That general effervescence results which is characteristic of revolutionary or creative epochs. Now this greater activity results in a general stimulation of individual forces. Men see more and differently now than in normal times. Changes are not merely of shades and degrees; men become different. The passions moving them are of such an intensity that they cannot be satisfied except by violent and unrestrained actions, actions of superhuman heroism or of bloody barbarism. This is what explains the Crusades,[vii] for example, or many of the scenes, either sublime or savage, of the French Revolution.[viii] Under the influence of the general exaltation, we see the most mediocre and inoffensive bourgeois become either a hero or a butcher.[ix] And so clearly are all these mental processes the ones that are also at the root of religion that the individuals themselves have often pictured the pressure before which they thus gave way in a distinctly religious form. The Crusaders believed that they felt God present in the midst of them, enjoining them to go to the conquest of the Holy Land; Joan of Arc believed that she obeyed celestial voices.[x]

But it is not only in exceptional circumstances that this stimulating action of society makes itself felt; there is not,

so to speak, a moment in our lives when some current of energy does not come to us from without. The man who has done his duty finds, in the manifestations of every sort expressing the sympathy, esteem or affection which his fellows have for him, a feeling of comfort, of which he does not ordinarily take account, but which sustains him, none the less. The sentiments which society has for him raise the sentiments which he has for himself. Because he is in moral harmony with his comrades, he has more confidence, courage and boldness in action, just like the believer who thinks that he feels the regard of his god turned graciously towards him. It thus produces, as it were, a perpetual sustenance of our moral nature. Since this varies with a multitude of external circumstances, as our relations with the groups about us are more or less active and as these groups themselves vary, we cannot fail to feel that this moral support depends upon an external cause; but we do not perceive where this cause is nor what it is. So we ordinarily think of it under the form of a moral power which, though immanent in us, represents within us something not ourselves: this is the moral conscience, of which, by the way, men have never made even a slightly distinct representation except by the aid of religious symbols.

In addition to these free forces which are constantly coming to renew our own, there are others which are fixed in the methods and traditions which we employ. We speak a language that we did not make; we use instruments that we did not invent; we invoke rights that we did not found; a treasury of knowledge is transmitted to each generation that it did not gather itself, etc. It is to society that we owe these varied benefits of civilization, and if we do not ordinarily see the source from which we get them, we at least know that they are not our own work. Now it is these things that give man his own place among things; a man is a man only because he is civilized. So he could not escape the feeling that outside of him there are active causes from which he gets the characteristic attributes of his nature and which, as benevolent powers, assist him, protect him and assure him of a privileged fate. And of course he must attribute to these powers a dignity corresponding to the great value of the good things he attributes to them.[xi]

[vii]See Stoll, *op. cit.,* pp. 353 ff.

[viii]Ibid., pp. 619, 635.

[ix]Ibid., pp. 622 ff.

[x]The emotions of fear and sorrow are able to develop similarly and to become intensified under these same conditions. As we shall see, they correspond to quite another aspect of the religious life (Bk. III, ch. v).

[xi]This is the other aspect of society which, while being imperative, appears at the same time to be good and gracious. It dominates us and assists us. If we have defined the social fact by the first of these characteristics rather than the second, it is because it is more readily observable, for it is translated into outward and visible signs; but we have never thought of denying the second (see our *Règles de la Méthode Sociologique,* preface to the second edition, p. xx, n. 1).

Thus the environment in which we live seems to us to be peopled with forces that are at once imperious and helpful, august and gracious, and with which we have relations. Since they exercise over us a pressure of which we are conscious, we are forced to localize them outside ourselves, just as we do for the objective causes of our sensations. But the sentiments which they inspire in us differ in nature from those which we have for simple visible objects. As long as these latter are reduced to their empirical characteristics as shown in ordinary experience, and as long as the religious imagination has not metamorphosed them, we entertain for them no feeling which resembles respect, and they contain within them nothing that is able to raise us outside ourselves. Therefore, the representations which express them appear to us to be very different from those aroused in us by collective influences. The two form two distinct and separate mental states in our consciousness, just as do the two forms of life to which they correspond. Consequently, we get the impression that we are in relations with two distinct sorts of reality and that a sharply drawn line of demarcation separates them from each other: on the one hand is the world of profane things, on the other, that of sacred things.

Also, in the present day just as much as in the past, we see society constantly creating sacred things out of ordinary ones. If it happens to fall in love with a man and if it thinks it has found in him the principal aspirations that move it, as well as the means of satisfying them, this man will be raised above the others and, as it were, deified. Opinion will invest him with a majesty exactly analogous to that protecting the gods. This is what has happened to so many sovereigns in whom their age had faith: if they were not made gods, they were at least regarded as direct representatives of the deity. And the fact that it is society alone which is the author of these varieties of apotheosis, is evident since it frequently chances to consecrate men thus who have no right to it from their own merit. The simple deference inspired by men invested with high social functions is not different in nature from religious respect. It is expressed by the same movements: a man keeps at a distance from a high personage; he approaches him only with precautions; in conversing with him, he uses other gestures and language than those used with ordinary mortals. The sentiment felt on these occasions is so closely related to the religious sentiment that many peoples have confounded the two. In order to explain the consideration accorded to princes, nobles and political chiefs, a sacred character has been attributed to them. In Melanesia and Polynesia, for example, it is said that an influential man has *mana,* and that his influence is due to this *mana.*[xii] However, it is evident that his situation is due solely to the importance attributed to him by public opinion. Thus the moral power conferred by opinion and that with which sacred beings are invested are at bottom of a single origin and made up of the same elements. That is why a single word is able to designate the two.

In addition to men, society also consecrates things, especially ideas. If a belief is unanimously shared by a people, then, for the reason which we pointed out above, it is forbidden to touch it, that is to say, to deny it or to contest it. Now the prohibition of criticism is an interdiction like the others and proves the presence of something sacred. Even to-day, howsoever great may be the liberty which we accord to others, a man who should totally deny progress or ridicule the human ideal to which modern societies are attached, would produce the effect of a sacrilege. There is at least one principle which those the most devoted to the free examination of everything tend to place above discussion and to regard as untouchable, that is to say, as sacred: this is the very principle of free examination.

This aptitude of society for setting itself up as a god or for creating gods was never more apparent than during the first years of the French Revolution. At this time, in fact, under the influence of the general enthusiasm, things purely laïcal by nature were transformed by public opinion into sacred things: these were the Fatherland, Liberty, Reason.[xiii] A religion tended to become established which had its dogmas,[xiv] symbols,[xv] altars[xvi] and feasts.[xvii] It was to these spontaneous aspirations that the cult of Reason and the Supreme Being attempted to give a sort of official satisfaction. It is true that this religious renovation had only an ephemeral duration. But that was because the patriotic enthusiasm which at first transported the masses

---

[xii]Codrington, *The Melanesians,* pp. 50, 103, 120. It is also generally thought that in the Polynesian languages, the word *mana* primitively had the sense of authority (see Tregear, *Maori Comparative Dictionary, s.v.*).

[iii]See Albert Mathiez, *Les origines des cultes révolutionnaires* (1789–1792).

[xiv]Ibid., p. 24.

[xv]Ibid., pp. 29, 32.

[xvi]Ibid., p. 30.

[xvii]Ibid., p. 46.

soon relaxed.[xviii] The cause being gone, the effect could not remain. But this experiment, though short-lived, keeps all its sociological interest. It remains true that in one determined case we have seen society and its essential ideas become, directly and with no transfiguration of any sort, the object of a veritable cult.

All these facts allow us to catch glimpses of how the clan was able to awaken within its members the idea that outside of them there exist forces which dominate them and at the same time sustain them, that is to say in fine, religious forces: it is because there is no society with which the primitive is more directly and closely connected. The bonds uniting him to the tribe are much more lax and more feebly felt. Although this is not at all strange or foreign to him, it is with the people of his own clan that he has the greatest number of things in common; it is the action of this group that he feels the most directly; so it is this also which, in preference to all others, should express itself in religious symbols. . . .

III

One can readily conceive how, when arrived at this state of exaltation, a man does not recognize himself any longer. Feeling himself dominated and carried away by some sort of an external power which makes him think and act differently than in normal times, he naturally has the impression of being himself no longer. It seems to him that he has become a new being: the decorations he puts on and the masks that cover his face and figure materially in this interior transformation, and to a still greater extent, they aid in determining its nature. And as at the same time all his companions feel themselves transformed in the same way and express this sentiment by their cries, their gestures and their general attitude, everything is just as though he really were transported into a special world, entirely different from the one where he ordinarily lives, and into an environment filled with exceptionally intense forces that take hold of him and metamorphose him. How could such experiences as these, especially when they are repeated every day for weeks, fail to leave in him the conviction that there really exist two heterogeneous and mutually incomparable worlds? One is that where his daily life drags wearily along; but he cannot penetrate into the other without at once entering into relations with extraordinary powers that excite him to the point of frenzy. The first is the profane world, the second, that of sacred things.

So it is in the midst of these effervescent social environments and out of this effervescence itself that the religious idea seems to be born. The theory that this is really its origin is confirmed by the fact that in Australia the really religious activity is almost entirely confined to the moments when these assemblies are held. To be sure, there is no people among whom the great solemnities of the cult are not more or less periodic; but in the more advanced societies, there is not, so to speak, a day when some prayer or offering is not addressed to the gods and some ritual act is not performed. But in Australia, on the contrary, apart from the celebrations of the clan and tribe, the time is nearly all filled with lay and profane occupations. Of course there are prohibitions that should be and are preserved even during these periods of temporal activity; it is never permissible to kill or eat freely of the totemic animal, at least in those parts where the interdiction has retained its original vigour; but almost no positive rites are then celebrated, and there are no ceremonies of any importance. These take place only in the midst of assembled groups. The religious life of the Australian passes through successive phases of complete lull and of superexcitation, and social life oscillates in the same rhythm. This puts clearly into evidence the bond uniting them to one another, but among the peoples called civilized, the relative continuity of the two blurs their relations. It might even be asked whether the violence of this contrast was not necessary to disengage the feeling of sacredness in its first form. By concentrating itself almost entirely in certain determined moments, the collective life has been able to attain its greatest intensity and efficacy, and consequently to give men a more active sentiment of the double existence they lead and of the double nature in which they participate. . . .

Now the totem is the flag of the clan. It is therefore natural that the impressions aroused by the clan in individual minds—impressions of dependence and of increased vitality—should fix themselves to the idea of the totem rather than that of the clan: for the clan is too complex a reality to be represented clearly in all its complex unity by such rudimentary intelligences. More than that, the primitive does not even see that these impressions come to him from the group. He does not know that the coming together of a number of men associated in the same life results in disengaging new energies,

---

[xviii]See Mathiez, La Théophilanthropie et la Culte décadaire, p. 36.

which transform each of them. All that he knows is that he is raised above himself and that he sees a different life from the one he ordinarily leads. However, he must connect these sensations to some external object as their cause. Now what does he see about him? On every side those things which appeal to his senses and strike his imagination are the numerous images of the totem. They are the waninga and the nurtunja, which are symbols of the sacred being. They are churinga and bull-roarers, upon which are generally carved combinations of lines having the same significance. They are the decorations covering the different parts of his body, which are totemic marks. How could this image, repeated everywhere and in all sorts of forms, fail to stand out with exceptional relief in his mind? Placed thus in the centre of the scene, it becomes representative. The sentiments experienced fix themselves upon it, for it is the only concrete object upon which they can fix themselves. It continues to bring them to mind and to evoke them even after the assembly has dissolved, for it survives the assembly, being carved upon the instruments of the cult, upon the sides of rocks, upon bucklers, etc. By it, the emotions experienced are perpetually sustained and revived. Everything happens just as if they inspired them directly. It is still more natural to attribute them to it for, since they are common to the group, they can be associated only with something that is equally common to all. Now the totemic emblem is the only thing satisfying this condition. By definition, it is common to all. During the ceremony, it is the centre of all regards. While generations change, it remains the same; it is the permanent element of the social life. So it is from it that those mysterious forces seem to emanate with which men feel that they are related, and thus they have been led to represent these forces under the form of the animate or inanimate being whose name the clan bears.

When this point is once established, we are in a position to understand all that is essential in the totemic beliefs.

Since religious force is nothing other than the collective and anonymous force of the clan, and since this can be represented in the mind only in the form of the totem, the totemic emblem is like the visible body of the god. Therefore, it is from it that those kindly and dreadful actions seem to emanate, which the cult seeks to provoke or prevent; consequently, it is to it that the cult is addressed. This is the explanation of why it holds the first place in the series of sacred things.

But the clan, like every other sort of society, can live only in and through the individual consciousnesses that compose it. So if religious force, in so far as it is conceived as incorporated in the totemic emblem, appears to be outside of the individuals and to be endowed with a sort of transcendence over them, it, like the clan of which it is the symbol, can be realized only in and through them; in this sense, it is imminent in them and they necessarily represent it as such. They feel it present and active within them, for it is this which raises them to a superior life. This is why men have believed that they contain within them a principle comparable to the one residing in the totem, and consequently, why they have attributed a sacred character to themselves, but one less marked than that of the emblem. It is because the emblem is the pre-eminent source of the religious life; the man participates in it only indirectly, as he is well aware; he takes into account the fact that the force that transports him into the world of sacred things is not inherent in him, but comes to him from the outside. . . .

But if this theory of totemism has enabled us to explain the most characteristic beliefs of this religion, it rests upon a fact not yet explained. When the idea of the totem, the emblem of the clan, is given, all the rest follows; but we must still investigate how this idea has been formed. This is a double question and may be subdivided as follows: What has led the clan to choose an emblem? and why have these emblems been borrowed from the animal and vegetable worlds, and particularly from the former?

That an emblem is useful as a rallying-centre for any sort of a group it is superfluous to point out. By expressing the social unity in a material form, it makes this more obvious to all, and for that very reason the use of emblematic symbols must have spread quickly when once thought of. But more than that, this idea should spontaneously arise out of the conditions of common life; for the emblem is not merely a convenient process for clarifying the sentiment society has of itself: it also serves to create this sentiment; it is one of its constituent elements.

In fact, if left to themselves, individual consciousnesses are closed to each other; they can communicate only by means of signs which express their internal states. If the communication established between them is to become a real communion, that is to say, a fusion of all particular sentiments into one common sentiment, the signs expressing them must themselves be fused into one single and unique resultant. It is the appearance of this that informs individuals that they are in harmony and makes them conscious of their moral unity. It is by uttering the same cry, pronouncing the same word, or performing the same gesture in regard to some object that they become and feel themselves to be in unison. It is true that individual representations also cause reactions in the organism that are not without importance; however, they can be thought of apart from these physical reactions which accompany them or follow them, but which do not constitute them. But it is quite another matter with collective representations. They presuppose that minds act and react upon one another; they are the product of these actions and reactions which are themselves possible only through material intermediaries. These latter do not confine themselves to revealing the mental state with which they are associated; they aid in creating it. Individual minds cannot come in contact and communicate with each other except by coming out of themselves; but they cannot do this except by movements. So it is the homogeneity of these movements that gives the group consciousness of itself and consequently makes it exist. When this homogeneity is once established and these movements have once taken a stereotyped form, they serve to symbolize the corresponding representations. But they symbolize them only because they have aided in forming them.

Moreover, without symbols, social sentiments could have only a precarious existence. Though very strong as long as men are together and influence each other reciprocally, they exist only in the form of recollections after the assembly has ended, and when left to themselves, these become feebler and feebler; for since the group is now no longer present and active, individual temperaments easily regain the upper hand. The violent passions which may have been released in the heart of a crowd fall away and are extinguished when this is dissolved, and men ask themselves with astonishment how they could ever have been so carried away from their normal character. But if the movements by which these sentiments are expressed are connected with something that endures, the sentiments themselves become more durable. These other things are constantly bringing them to mind and arousing them; it is as though the cause which excited them in the first place continued to act. Thus these systems of emblems, which are necessary if society is to become conscious of itself, are no less indispensable for assuring the continuation of this consciousness.

So we must refrain from regarding these symbols as simple artifices, as sorts of labels attached to representations already made, in order to make them more manageable: they are an integral part of them. Even the fact that collective sentiments are thus attached to things completely foreign to them is not purely conventional: it illustrates under a conventional form a real characteristic of social facts, that is, their transcendence over individual minds. In fact, it is known that social phenomena are born, not in individuals, but in the group. Whatever part we may take in their origin, each of us receives them from without.[xix]

So when we represent them to ourselves as emanating from a material object, we do not completely misunderstand their nature. Of course they do not come from the specific thing to which we connect them, but nevertheless, it is true that their origin is outside of us. If the moral force sustaining the believer does not come from the idol he adores or the emblem he venerates, still it is from outside of him, as he is well aware. The objectivity of its symbol only translates its eternalness.

Thus social life, in all its aspects and in every period of its history, is made possible only by a vast symbolism. The material emblems and figurative representations with which we are more especially concerned in our present study, are one form of this; but there are many others. Collective sentiments can just as well become incarnate in persons or formulæ: some formulæ are flags, while there are persons, either real or mythical, who are symbols. . . .

Conclusion

As we have progressed, we have established the fact that the fundamental categories of thought, and consequently of science, are of religious origin. We have seen that the same is true for magic and consequently for the different processes which have issued from it. On the other hand, it has long been known that up until a relatively advanced moment of evolution, moral and legal rules have been indistinguishable from ritual prescriptions. In summing up, then, it may be said that nearly all the great social institutions have been born in religion.[xx] Now in order that these principal aspects of the collective life may have commenced by being only varied aspects of the religious life, it is obviously necessary that the religious life be the eminent form and, as it were, the concentrated expression of the whole collective life. If religion has given birth to all that is essential in society, it is because the idea of society is the soul of religion.

Religious forces are therefore human forces, moral forces. It is true that since collective sentiments can become conscious of themselves only by fixing themselves upon external objects, they have not been able to take form without adopting some of their characteristics from other things: they have thus acquired a sort of physical nature; in this way they have come to mix themselves with the life of the material world, and then have considered themselves capable of explaining what passes there. But when they are considered only from this point of view and in this role, only their most superficial aspect is seen. In reality, the essential elements of which these collective sentiments are

---

[xix]On this point see *Régles de la méthode sociologique,* pp. 5 ff.

made have been borrowed by the understanding. It ordinarily seems that they should have a human character only when they are conceived under human forms;[xxi] but even the most impersonal and the most anonymous are nothing else than objectified sentiments.

It is only by regarding religion from this angle that it is possible to see its real significance. If we stick closely to appearances, rites often give the effect of purely manual operations: they are anointings, washings, meals. To consecrate something, it is put in contact with a source of religious energy, just as to-day a body is put in contact with a source of heat or electricity to warm or electrize it; the two processes employed are not essentially different. Thus understood, religious technique seems to be a sort of mystic mechanics. But these material manoeuvres are only the external envelope under which the mental operations are hidden. Finally, there is no question of exercising a physical constraint upon blind and, incidentally, imaginary forces, but rather of reaching individual consciousnesses of giving them a direction and of disciplining them. It is sometimes said that inferior religions are materialistic. Such an expression is inexact. All religions, even the crudest, are in a sense spiritualistic: for the powers they put in play are before all spiritual, and also their principal object is to act upon the moral life. Thus it is seen that whatever has been done in the name of religion cannot have been done in vain: for it is necessarily the society that did it, and it is humanity that has reaped the fruits. . . .

II

Thus there is something eternal in religion which is destined to survive all the particular symbols in which religious thought has successively enveloped itself. There can be no society which does not feel the need of upholding and reaffirming at regular intervals the collective sentiments and the collective ideas which make its unity and its personality. Now this moral remaking cannot be achieved except by the means of reunions, assemblies and meetings where the individuals, being closely united to one another, reaffirm in common their common sentiments; hence come ceremonies which do not differ from regular religious ceremonies, either in their object, the results which they produce, or the processes employed to attain these results. What essential difference is there between an assembly of Christians celebrating the principal dates of the life of Christ, or of Jews remembering the exodus from Egypt or the promulgation of the decalogue, and a reunion of citizens commemorating the promulgation of a new moral or legal system or some great event in the national life?. . .

In summing up, then, we must say that society is not at all the illogical or a-logical, incoherent and fantastic being which it has too often been considered. Quite on the contrary, the collective consciousness is the highest form of the psychic life, since it is the consciousness of the consciousnesses. Being placed outside of and above individual and local contingencies, it sees things only in their permanent and essential aspects, which it crystallizes into communicable ideas. At the same time that it sees from above, it sees farther; at every moment of time, it embraces all known reality; that is why it alone can furnish the mind with the moulds which are applicable to the totality of things and which make it possible to think of them. It does not create these moulds artificially; it finds them within itself; it does nothing but become conscious of them . . .

Discussion Questions

1. Define mechanical and organic solidarity. Do these concepts help explain the division of labor in your family of origin? In your current (or most recent) place of employment? How so or why not? Be specific.

2. Discuss the various types of suicide that Durkheim delineates using specific examples. To what extent do you agree or disagree with the notion that different *types* of suicide prevail in "modern" as opposed to "traditional" societies? Give concrete examples.

3. Define and compare and contrast Marx's concept of alienation and Durkheim's concept of anomie. How exactly do these concepts overlap? How are they different?

4. Discuss Durkheim's notion of collective conscience. Why is it, that is, how can it be, that the collective conscience is *not* just a "sum" of individual consciousnesses? Use concrete examples to explain.

5. Discuss specific moments of collective effervescence that you have experienced (e.g., concerts, church, etc.). What particular symbols and rituals were called up and used to arouse this social state?

4 MAX WEBER (1864–1920)

Max Weber

Key Concepts

- *Verstehen*
- Ideal types
- Protestant ethic
- Calling
- Iron cage
- Rationalization
- Bureaucracy
- Authority
- Charisma
- Class, status, and party

No one knows who will live in this cage in the future, or whether at the end of this tremendous development entirely new prophets will arise, or there will be a great rebirth of old ideas and ideals, or, if neither, mechanized petrification embellished with a sort of convulsive self-importance. For of the last stage of this cultural development it might well be truly said: "Specialists without spirit, sensualists without heart; this nullity imagines that it has attained a level of civilization never before achieved."

—Weber ([1904–05] 1958:182)

From the course requirements necessary to earn your degree, to the paperwork and tests you must complete in order to receive your driver's license, to the record keeping and mass of files that organize most every business enterprise, our everyday life is channeled in large measure through formalized, codified procedures. Indeed, in Western cultures few aspects of life have been untouched by the general tendency toward rationalization and the adoption of methodical practices. So, whether it's developing a long-term financial plan for one's business, following the advice written in sex manuals, or even planning for one's own death, little in modern life is left to chance. It was toward an examination of the causes and consequences of this "disenchantment" of everyday life that Max Weber's wide-ranging work crystallized. In this chapter, we explore Weber's study of this general trend in modern society as well as other aspects of his writings. But while Weber did not self-consciously set out to develop a unified theoretical model, making his intellectual path unlike that followed by both Marx and Durkheim, it is this characteristic of his work that has made it a continual wellspring of inspiration for other scholars. Perhaps

125

the magnitude of Weber's impact on the development of sociology is captured best by the prominent social theorist, Raymond Aron, who described Weber as "the greatest of the sociologists" (Aron [1965] 1970:294).

❚❚ A BIOGRAPHICAL SKETCH

Max Weber, Jr., was born in Erfurt, Germany, in 1864. He was the eldest of eight children born to Max Weber, Sr., and Helene Fallenstein Weber, although only six survived to adulthood. Max Jr. was a sickly child. When he was four years old, he became seriously ill with meningitis. Though he eventually recovered, throughout the rest of his life he suffered the physical and emotional aftereffects of the disease, most apparently anxiety and nervous tension. From an early age, books were central in Weber's life. He read whatever he could get his hands on, including Kant, Machiavelli, Spinoza, Goethe, and Schopenhauer, and he wrote two historical essays before his 14th birthday. But Weber paid little attention in class and did almost no work for school. According to his widow Marianne, although "he was not uncivil to his teachers, he did not respect them. . . . If there was a gap in his knowledge, he went to the root of the matter and then gladly shared what he knew" (Marianne Weber [1926] 1975:48).

In 1882, at 18 years old, Weber took his final high school examinations. His teachers acknowledged his outstanding intellectual accomplishments and thirst for knowledge, but expressed doubts about his "moral maturity." Weber went to the University of Heidelberg for three semesters and then completed one year of military service in Strasbourg. When his service ended, he enrolled at the University of Berlin and, for the next eight years, lived at his parents' home. Upon passing his first examination in law in 1886, Weber began work as a full-time legal apprentice. While working as a junior barrister, he earned a PhD in economic and legal history in 1889. He then took a position as lecturer at the University of Berlin.

Throughout his life, Weber was torn by the personal struggles between his mother and his father. Weber admired his mother's extraordinary religious piety and devotion to her family, and loathed his father's abusive treatment of her. At the same time, Weber admired his father's intellectual prowess and achievements and reviled his mother's passivity. Weber followed in his father's footsteps by becoming a lawyer and joining the same organizations as his father had at the University of Heidelberg. Like his father, he was active in government affairs as well. As a member of the National Liberal Party, Max Sr. was elected to the Reichstag (national legislature) and later appointed by Chancellor Bismarck to the Prussian House of Deputies. For his part, Max Jr. was a committed nationalist and served the government in numerous capacities, including as a delegate to the German Armistice Commission in Versailles following Germany's defeat in World War I. But he was also imbued with a sense of moral duty quite similar to that of his mother. Weber's feverish work ethic—he drove himself mercilessly, denying himself all leisure—can be understood as an inimitable combination of his father's intellectual accomplishments and his mother's moral resolve.

In 1893, at the age of 29, Weber married Marianne Schnitger, a distant cousin, and finally left his childhood home. Today, Marianne Weber is recognized as an important feminist, intellectual, and sociologist in her own right. She was a popular public speaker on social and sexual ethics and wrote many books and articles. Her most influential works, *Marriage and Motherhood in the Development of Law* (1907) and *Women and Love* (1935), examined feminist issues and the reform of marriage. However, Marianne is known best as the intellectual partner of her husband. She and Max made a conscious effort to establish an egalitarian relationship, and worked together on intellectual projects. Interestingly, Marianne referred to Max as her "companion" and implied that theirs was an unconsummated marriage. (It is rumored that Max had a long-lasting affair with a woman of Swiss nobility who was a member of the Tobleron family.) Despite her own intellectual accomplishments, Marianne's 700-page treatise, *Max Weber: A Biography,* first published in 1926, has received the most attention, serving as the central source of biographical information on her husband (and vital to this introduction as well).

In 1894, Max Weber joined the faculty at Freiburg University as a full professor of economics. Shortly thereafter, in 1896, Weber accepted a position as chair of economics at the University of Heidelberg, where he first began his academic career. But in 1897, he suffered a serious nervous breakdown.

According to Marianne, the breakdown was triggered by the inexorable guilt Weber experienced after his father's sudden death. Just seven weeks before he died, Weber had rebuked his father over his tyrannical treatment of his mother. The senior Weber had prohibited his wife Helene from visiting Max and Marianne at their home in Heidelberg without him. When he and Helene showed up together for the visit, his son forced him to leave. Unfortunately, that was the last time father and son ever spoke.

Weber experienced debilitating anxiety and insomnia throughout the rest of his life. He often resorted to taking opium in order to sleep. Despite resigning his academic posts, traveling, and resting, the anxiety could not be dispelled. Nevertheless, he had spurts of manic intellectual activity and continued to write as an independent scholar. In 1904, Weber traveled to the United States and began to formulate the argument of what would be his most celebrated work, *The Protestant Ethic and the Spirit of Capitalism* (Weber [1904–05] 1958).

After returning to Europe, Weber resumed his intellectual activity. He met with the brilliant thinkers of his day, including Werner Sombart, Paul Hensel, Ferdinand Tönnies, Ernst Troeltsch, and Georg Simmel (see Chapter 6). He helped establish the Heidelberg Academy of the Sciences in 1909 and the Sociological Society in 1910 (Marianne Weber [1926] 1975:425). However, Weber was still plagued by compulsive anxiety. In 1918, he helped draft the constitution of the Weimar Republic while giving his first university lectures in 19 years at the University of Vienna. He suffered tremendously, however, and turned down an offer for a permanent post (Weber 1958:23). In 1920, at the age of 56, Max Weber died of pneumonia. Marianne lived for another 34 years and completed several important manuscripts left unfinished at her husband's death.

INTELLECTUAL INFLUENCES AND CORE IDEAS ▓▓

Weber's work encompasses a wide scope of substantive interests. Most, if not all, of his writing has had a profound impact on sociology. As such, an attempt to fully capture the breadth and significance of his scholarship exceeds the limitations of a single chapter. Nevertheless, we can isolate several aspects of his work that, taken together, serve as a foundation for understanding the impetus behind much of his writing. To this end, we divide our discussion in this section into two major parts: (1) Weber's view of the science of sociology and (2) his engagement with the work of Friedrich Nietzsche and Karl Marx.

Sociology

Weber defined sociology as "a science which attempts the interpretive understanding of social action in order thereby to arrive at a causal explanation of its course and effects" (Weber 1947:88). In casting "interpretive understanding," or ***Verstehen***, as the principal objective, Weber's vision of sociology offers a distinctive counter to those who sought to base the young discipline on the effort to uncover universal laws applicable to all societies. Thus, unlike Durkheim, who analyzed objective, sui generis "social facts" that operated independently of the individuals making up a society, Weber turned his attention to the subjective dimension of social life, seeking to understand the states of mind or motivations that guide individuals' behavior.

In delimiting the subject matter of sociology, Weber further specified "social action" to mean that which, "by virtue of the subjective meaning attached to it by the acting individual (or individuals), it takes account of the behaviour of others and is thereby oriented in its course" (Weber 1947:88). Such action can be either observable or internal to the actor's imagination, and it can involve a deliberate intervening in a given situation, an abstaining of involvement, or acquiescence. The task for the sociologist is to understand the meanings individuals assign to the contexts in which they are acting and the consequences that such meanings have for their conduct.

To systematize interpretive analyses of meaning, Weber distinguished four types of social action. In doing so, he clearly demonstrates his multidimensional approach to the problem of action (see Figure 4.1). First is *instrumental-rational action*. Such action is geared toward the efficient pursuit of goals through

Figure 4.1 Weber's Four Types of Social Action

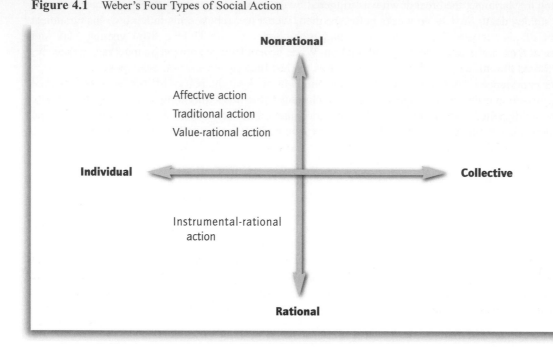

calculating the advantages and disadvantages associated with the possible means for realizing them. Under this category would fall the decision of a labor union to strike in order to bargain for greater employment benefits. Rehearsing one's performance for an upcoming job interview is another example of instrumental-rational action.

Like instrumental-rational action, *value-rational action* involves the strategic selection of means capable of effectively achieving one's goals. However, value-rational action is pursued as an end in itself, not because it serves as a means for achieving an ulterior goal. As such, it "always involves 'commands' or 'demands'" that compel the individual to follow a line of conduct for its own sake—because it is the "right" thing to do (Weber 1947:116). Examples of this type of action include risking arrest to further an environmental cause, or refraining from cheating on exams.

The third type of social action outlined by Weber is *traditional action,* where behaviors are determined by habit or long-standing custom. Here, an individual's conduct is shaped not by a concern with maximizing efficiency or commitment to an ethical principle, but rather by an unreflective adherence to established routines. This category includes religious rites of passage such as confirmations and bar mitzvahs, singing the national anthem at the start of sporting events, and eating turkey at Thanksgiving with one's family.

The fourth type is *affective action,* which is marked by impulsiveness or a display of unchecked emotions. Absent from this behavior is the calculated weighing of means for a given end. Examples of affective action are a baseball player arguing an umpire's called strike or parents crying at their child's wedding ceremony.

It is important to point out that in everyday life a given behavior or course of conduct is likely to exhibit characteristics of more than one type of social action. Thus, a person may pursue a career in social work not only because it is a means for earning a salary, but also because he or she is committed to the goal of helping others as a value in its own right. Weber's categories of social action, then, serve as **ideal types** or analytical constructs against which real-life cases can be compared. Such "pure" categories are not realized in concrete cases but instead are a conceptual yardstick for examining differences and similarities, as well as causal connections, between the social processes under investigation. Thus, "ideal" refers to an emphasis on particular aspects of social life specified by the researcher, not to a value judgment as to whether something is "good" or "bad." As you will read in the selections that follow, Weber's work is guided in large measure by constructing ideal types. For instance, his essay on bureaucracy consists in the main of a discussion of the ideal characteristics of such an organization. Similarly, his essay on the three forms of domination involves isolating the features specific to each ideal type, none of which actually exists in pure form.

Weber's notion of sociology as an interpretive science based on *Verstehen* (understanding) and his focus on constructing ideal types marks his ties to important intellectual debates that were taking shape in German universities (Bendix 1977). At the heart of the debates was the distinction drawn between the natural and social sciences, and the methodologies appropriate to each. The boundary separating biology, chemistry, and physics from history, economics, psychology, and sociology was an outgrowth of German Idealism and the philosophy of **Immanuel Kant** (1724–1804). Kant argued that the realm of mind and "spirit" was radically different from the external, physical world of objects. According to Kant, because individuals create meaning and ultimately are free to choose their course of action, it is not possible to construct universal laws regarding human behavior. As a result, social life is not amenable to scientific investigation. On the other hand, absent of consciousness, objects and processes occurring in the natural world are open to scientific analysis and the development of general laws regarding their actions.

Among the scholars grappling with the implications of the Kantian division were the historical economists **Wilhelm Dilthey** (1833–1911) and **Heinrich Rickert** (1863–1936), whose work would have a profound impact on Weber. It was Dilthey who articulated the view that historical studies, and the social sciences more generally, should seek to *understand* particular events and their relationship to the specific contexts in which they occur. The task of history, then, is to interpret the subjective meanings actors assign to their conduct, not to search for causal explanations couched in terms of universal laws. According to Dilthey, any attempt to produce general causal laws regarding human behavior would not capture the unique historical conditions that shaped the events in question or a society's development. Moreover, such efforts would fail to study the very things that separate social life from the physical world of objects—human intent and motivation. Unlike the natural sciences and their analyses of the regularities governing observable objects and events, the social sciences aim to understand the internal states of actors and their relationship to behaviors.

In Weber's own definition of sociology, quoted above, we clearly see his indebtedness to Dilthey's work. Following Dilthey, Weber cast the social sciences as a branch of knowledge dedicated to developing an interpretive understanding of the subjective meanings actors attach to their conduct. However, Weber maintained a view not shared by Dilthey—that the social sciences, like the natural sciences, are conducted by making use of abstract and generalizing concepts. Here lies the impetus behind Weber's development of ideal types as a method for producing generalizable findings based on the study of historically specific events. For Weber, scientific knowledge is distinguished from nonscientific analyses not on the basis of the subject matter under consideration but rather on how such studies are carried out. Thus, in constructing ideal types of action, Weber argued that analyses of the social world were not inherently less scientific or generalizable than investigations of the physical world. Nevertheless, Weber's Verstehen approach led him to contend that the search for universal laws of human action would lose sight of what is human—the production of meaningful behavior as it is grounded within a specific historical context.

It is in his notion of ideal types that we find Weber's links to the work of Heinrich Rickert. As a neo-Kantian thinker, Rickert accepted the distinction between the natural and social sciences as self-evident. However, he saw the differences between the two branches of knowledge as tied to the method of inquiry appropriate to each, not to any inherent differences in subject matter, as did Dilthey. According to Rickert, regardless of whether an investigator is trying to understand the meanings that motivate actors or attempting to uncover universal laws that govern the world of physical objects, they would study both subjects by way of concepts. Moreover, it is through the use of concepts that the investigator is able to select the aspects of the social or natural world most relevant to the purpose of her inquiry. The difference between the sciences lies, then, in *how* concepts are used to generate knowledge.

While the natural sciences used concepts as a way to generate abstract principles that explain the uniformities that shape the physical world, Rickert maintained that concepts used in the social sciences are best directed toward detailing the particular features that account for the uniqueness of an event or a society's development. In short, for Rickert the natural sciences were driven by the deductive search for universal laws. On the other hand, the social sciences were committed to producing inductive descriptions of historically specific phenomena.

For example, in subjecting molecules to changes in temperature and pressure, a physicist is interested in explaining the molecules' reactions in terms of causal laws whose validity is not restricted to any specific time or setting. Conversely, social scientists studying episodes of protests, for instance, should seek to understand why individuals chose to act and how the cultural and institutional contexts shaped their behaviors. But because the contexts in which, for instance, the French Revolution, the Boston Tea Party, and the women's suffrage movement occurred were historically unique, it is not possible to formulate generalized explanations of protests on the basis of such specific, unreplicable events. Attempts to do so would require a level of conceptual abstraction that would necessarily lose sight of the particulars that made the events historically meaningful.

Weber's use of ideal types as a method for framing his analyses stems in important respects from Rickert's discussions on the role of concepts in the sciences. However, he did not share Rickert's view that the social sciences are unable to construct general causal explanations of historical events or societal development. Here, Weber sought to forge a middle ground between the generating of abstract laws characteristic of the natural sciences and the accumulation of historically specific facts that some contended must guide the social sciences. To this end, he cast the determination of causality as an attempt to establish the *probability* that a series of actions or events are related or have an *elective affinity*. Hence, Weber's notion of causality is fundamentally different from the conventional scientific usage, which sees it as the positing of invariant and necessary relationships between variables. According to Weber, the complexities of social life make it unamenable to formulating strict causal arguments such as those found in the natural sciences. While it can be stated that temperatures above 32 degrees Fahrenheit (x) will cause ice to melt (y), such straightforward, universal relationships between variables cannot be isolated when analyzing social processes; individual conduct and societal developments are not carried out with the constancy and singular causal "elegance" that characterizes the physical world. Thus, a sociologist cannot say with the same degree of certainty that an increase in educational attainment (x) will cause a rise in income (y), because while this relationship between the two variables may be probable; it is not inevitable. One need only keep in mind that a university professor with a PhD typically makes far less money than a corporate executive with a bachelor's degree. As a result, sociologists should set out to determine the set of factors that, when taken together, have an elective affinity with a particular outcome. Armed with ideal types, the sociologist can then develop general arguments that establish the probable relationship between a combination of causes and a particular consequence.

Of Nietzsche and Marx

> The honesty of a contemporary scholar . . . can be measured by the position he takes vis-à-vis Nietzsche and Marx. Whoever fails to acknowledge that he could not carry out the most important part of his own work without the work done by both . . . deceives himself and others. The intellectual world in which we live is a world which to a large extent bears the imprint of Marx and Nietzsche.[1]

Such were the words spoken by Max Weber to his students shortly before his death. While his vision of sociology as a discipline was shaped in large measure by his links to German Idealism and the controversies surrounding historical studies, his substantive interests bear important connections to the work of Friedrich Nietzsche (1844–1900) and Karl Marx (1818–1883).

Evidencing his connection to Nietzsche, a major theme running throughout the whole of Weber's work is **rationalization**. By rationalization, Weber was referring to an ongoing process in which social interaction

---

[1]Quoted in Antonio (1995:3).

and institutions become increasingly governed by methodical procedures and calculable rules. Thus, in steering the course of societal development, values, traditions, and emotions were being displaced in favor of formal and impersonal bureaucratic practices. While such practices may breed greater efficiency in obtaining designated ends, they also lead to the "disenchantment of the world" where "there are no mysterious incalculable forces that come into play, but rather that one can, in principle, master all things by calculation" (Weber [1919] 1958:139).

Few domains within modern Western societies have escaped from the trend toward rationalization. For instance, music became thoroughly codified by the 1500s with the development of scales derived from mathematical formulas and tonal and rhythmic notation. While musical improvisation by no means disappeared, it henceforth was based on underlying systematized principles of melody and harmony. The visual arts likewise became codified according to principles of perspective, composition, and color against which the avant-garde purposively rebels (and is thus no less subject to). Sex as an "irrational" bodily pleasure or as a rite tied to orgiastic rituals has been replaced by sex as a rational practice necessary for procreation. And procreation has itself come under increasing scientific control as advances in birth control and in vitro fertilization make it possible to plan when a birth will occur, to circumvent a person's natural infertility, and even to prenatally select specific traits. The transformation of sex was itself part of the broader displacement of magical belief systems by doctrinal religions, which were themselves later marginalized by an instrumental, scientific worldview. With each step, the work of fortune and fate, and mysterious and unknown powers were further removed from everyday life. The pantheon of gods and spirits that once ruled the universe would be distilled and simplified into the one all-knowing, omnipotent God, who would eventually lose his throne to the all-seeing telescope. And finally, Weber places special emphasis on the changes to social life brought on by the rationalization of capitalistic economic activity, as you will read later.

The ambivalence with which Weber viewed the process of rationalization stems from the loss of ultimate meaning that accompanied the growing dominance of an instrumental and scientific orientation to life. While science can provide technological advances that enable us to address more efficiently *how* to do things, it cannot provide us with a set of meanings and values that answer the more fundamental question: *Why?* Unlike those who saw in the Enlightenment's debunking of magical superstitions and religious beliefs the road to progress, Weber maintained that rationalization—and the scientific, calculative outlook in which it is rooted—does not generate "an increased and general knowledge of the conditions under which one lives" (Weber [1919] 1958:139). They offer, instead, techniques empty of ultimate meaning.

Weber's reluctance to champion the progress brought by science and technological advances was influenced by Nietzsche's own nihilistic view of modernity expressed most boldly in his assertion "God is dead" (Nietzsche). Nietzsche's claim reflected his conviction that the eclipse of religious and philosophical absolutes brought on by the rise of science and instrumental reasoning had created an era of nihilism or meaninglessness. Without religious or philosophical doctrines to provide a foundation for moral direction, life itself would cease to have an ultimate purpose. No longer could ethical distinctions be made between what one ought to do and what one can do (Nietzsche [1919] 1958).

Weber was unwilling to assign a determinative end to history, however. Whether or not the spiritual void created by the disenchantment of the modern world would continue was, for him, an open question. The search for meaning—which Weber saw as the essence of the human condition—carried out in a meaningless world sparked the rise of charismatic leaders who were capable of offering their followers purpose and direction in their lives. (See "The Types of Legitimate Domination" in this chapter's Readings.) Ruling over others by virtue of their professed "state of grace," such figures were capable of radically transforming the existing social order. Weber's depiction of the power of charismatic leaders, with their ability to transcend the conventions and expectations imposed by the social order, bears important similarities to Nietzsche's notion of the *Übermensch*, or "superman." For Nietzsche, the fate of humanity and what is truly human lay in the hands of the *Übermenschen*, who alone are capable of overcoming the moral and spiritual bankruptcy that he believed corrupted the modern age (Nietzsche [1883] 1978).

Friedrich Nietzsche (1844–1900): Is God Dead?

It is difficult to overstate the influence that the work of German philosopher and social critic Friedrich Nietzsche has had on twentieth-century thought. From theologians and psychologists, to philosophers and sociologists, to poets and playwrights, Nietzsche's ideas have penetrated virtually every domain of modern intellectual culture. It was not until after his death, however, that he would earn such acclaim, for during his life his writings attracted but the smallest of audiences.

Beset with a host of physical ailments, and stricken by a complete mental breakdown at the age of 45, Nietzsche, nevertheless, managed to develop a number of themes that would usher in a thoroughgoing critique of seemingly unassailable truths. Rejecting the Enlightenment notion that reason offers the pathway to human emancipation, Nietzsche believed that the essence of humanity lies in emotional and physical experiences. Moreover, he repudiated Christianity's ascetic ethic as a renunciation or avoidance of life, and championed, instead, the embracing of all that life offers, even the most tragic of sufferings, as the ultimate expression of greatness.

The man who declared, "God is dead" and who argued that truth, values, and morals are not based on some intrinsic, ahistorical criteria, but, instead, are established by the victors in the unending struggle for power, did not enter the canon of liberal academia without controversy. Owing to the intentional distortions and forgeries of some of his writing by his sister, Elisabeth, Nietzsche was often interpreted as an anti-Semitic fascist. Though he abhorred such hatred as "slavish" and "herd-like," Hitler's Third Reich reinvented Nietzsche's notion of the "will to power" and the Übermensch or "superman" as a justification for its military aggression and genocidal practices. Fortunately, contemporary scholars of Nietzsche's work have corrected many of Elisabeth's falsities, allowing the true intention of his piercing, original insights into modern culture to be realized.

In addition to drawing inspiration from Nietzsche's work, much of Weber's writing reflects a critical engagement with and extension of Marx's theory of historical materialism.[2] As we noted in Chapter 2, Marx saw class struggles as the decisive force in the evolution of history. Class struggles were, in turn, the inevitable outcome of the inherent contradictions found in all precommunist economic systems. While finding much convincing in Marx's argument, Weber nevertheless did not embrace it in its entirety. In constructing his own theoretical framework, Weber departed from Marx in a number of respects, three of which we outline here.

First, Weber maintained that social life did not evolve according to some immanent or necessary law. Thus, unlike Marx, Weber did not foresee a definitive "end of prehistory" toward which social evolution progressed. Instead, he saw the future of modern society as an open question, the answer to which it is impossible to foretell. This position, coupled with his view that rationalizing processes had transformed modern society into an "iron cage" (p. 137), accounts for Weber's unwillingness to accept a utopian vision of humanity's future.

---

[2]It is important to point out that Weber's critique of Marx was based more on secondary interpretations of Marx's work than on a thorough, firsthand encounter with his writings, since much of it was unavailable. In Weber's time, and continuing today, Marx was (is) often miscast by his followers and critics alike as an economic determinist. Perhaps more accurately, then, Weber was responding to a "crude," reductionist version of Marxism.

Second, he contended that the development of societies could not be adequately explained on the basis of a single or primary causal mechanism. The analysis of economic conditions and class dynamics alone could not capture the complex social and cultural processes responsible for shaping a society's trajectory. In particular, Weber maintained that Marx, in emphasizing economic factors and class-based interests, underestimated the role that *ideas* play in determining a society's course of development. On this point, Weber sought to incorporate Marx's argument into his own work while offering what he saw as a necessary corrective, remarking, "Not ideas, but material and ideal interests, directly govern men's conduct. Yet very frequently the 'world images' that have been created by 'ideas' have, like switchmen, determined the tracks along which action has been pushed by the dynamic of interest" (Weber [1915] 1958:280).

Acknowledging the powerful sway that "interests" hold over individuals as they chart their course of action, Weber nevertheless argued that ideas play a central role in shaping the paths along which interests are realized. He saw ideas as independent cultural forces and not as a reflections of material conditions or the existing mode of production. As the source for constructing meaning and purposeful lines of action, ideas are not simply one element among others confined to the "superstructure." Instead, they serve as the bases on which individuals carve out possible avenues of action, and, more dramatically, when advanced by a charismatic leader, ideas can inspire revolution.

A third difference lies in where the two theorists located the fundamental problems facing modern industrial society. As you read previously, Marx identified capitalism as the primary source of humanity's inhumanity. The logic of capitalism necessarily led to the exploitation of the working class as well as to the alienation of the individual from his work, himself, and others. For Weber, however, it was not capitalism but the process of rationalization and the increasing dominance of bureaucracies that threatened to destroy creativity and individuality. By design, bureaucratic organizations—and the rational procedures that govern them—routinize and standardize people and products. Though making for greater efficiency and predictability in the spheres of life they have touched, the impersonality of bureaucracies, their indifference to difference, has created a "cold" and empty world. (See Weber's essay "Bureaucracy," excerpted in this chapter's Readings.)

Not surprisingly, then, Weber, unlike many of his contemporaries, did not see in socialism the cure for society's ills. In taking control of a society's productive forces, socialist forms of government would only further bureaucratize the social order, offering a poor alternative to capitalism. Indeed, Weber believed capitalism was a "better" economic system to the extent that its competitiveness allowed more opportunities to express one's individuality and creative impulses. Clearly, Weber did not embrace Marx's or his followers' calls for a communist revolution, because such a movement, to the extent that it led to an expansion of the scope of bureaucracies, would accelerate the hollowing out of human life.

Significant Others

Robert Michels (1876–1936): The Iron Law of Oligarchy

Political activist and sociologist, Robert Michels is best known for his studies on the organization of political parties. Influenced by the ideas of his teacher and mentor, Max Weber, Michels argued that all large-scale organizations have a tendency to evolve into hierarchical bureaucracies regardless of their original formation and ultimate goals. Even organizations that adopt an avowedly democratic agenda are inevitably subject to this "iron law of oligarchy" because leadership is necessarily transferred to an elite decision-making body.

Michels developed his argument in *Political Parties* (1911) in which he examined the organizational structures of western European socialist trade unions and political parties. During the late 1800s, the democratic ethos was particularly strong within these revolutionary socialist parties whose

(Continued)

(Continued)

principal aim was to overthrow aristocratic or oligarchic regimes and replace them with governing bodies controlled directly by the people. Despite their intent on destroying elite rule—the rule of the many by the few—these parties were themselves unable to escape the tendency toward oligarchy.

Michels advanced his ideas in part as a response to his disillusionment with the German Social Democratic Party. An active member of the party, he witnessed firsthand its growing political conservatism. (Michels was censured by German government authorities for his political radicalism, compelling him to take positions at universities in Italy and Switzerland.) Established in the 1870s as an advocate for the working class, the Marxist-inspired party abandoned its revolutionary program soon after its formation, as its ambitions to wrest control of the means of production into the hands of the people was replaced by the conservative goals of increasing its membership, amassing funds for its war chest, and winning electoral seats in the German legislature through which piecemeal reform might be gained. Considered a vanguard of the proletariat revolution, this dramatic shift in party tactics signaled a rejection of Marxist principles and the abandonment of the struggle for realizing an ideal democracy where workers controlled their labor and freedom from want existed throughout society.

What led to the cooptation of this and similarly driven parties' ideals? The answer lies in the working classes' lack of economic and political power. In order to effect democratic change, the otherwise powerless working-class individuals must first organize; their strength as a movement is directly related to their strength in numbers. Numbers, however, require representation through individual delegates who are entrusted by the mass to act on its behalf. Despite Marx's utopian promise, the growth in numbers necessary to achieve power makes it impossible for the people to exercise direct control over their destinies. Instead, the success of working-class parties hinges on creating an organization committed to representing its interests: "Organization is . . . the source from which the conservative currents flow over the plain of democracy, occasioning there disastrous floods and rendering the plain unrecognizable" (Michels [1911] 1958:26). The inevitable rise of an organization brings with it the equally inevitable need for technical expertise, centralized authority, and a professional staff to ensure its efficient functioning. Bureaucratization transforms the party from a means to an end, to an end in itself. The preservation of the organization itself becomes the essential aim, and its original democratic ambitions are preserved only in talk, because aggressive action against the state would surely threaten its continued existence.

Thus, while "[d]emocracy is inconceivable without organization" (Michels [1911] 1958:25), the inherently oligarchic and bureaucratic nature of party organizations saps its revolutionary zeal and replaces it with the pursuit of disciplined, cautious policies intended to defend its own long-term interests, which do not necessarily coincide with the interests of the class it represents. As Michels notes, it would seem

society cannot exist without a "dominant" or "political" class [that] . . . constitutes the only factor of sufficiently durable efficacy in the history of human development. . . . [T]he state, cannot be anything other than the organization of a minority. It is the aim of this minority to impose upon the rest of society a "legal order," which is the outcome of the exigencies of dominion and of the exploitation of the mass of helots effected by the ruling minority, and can never be truly representative of the majority. The majority is thus permanently incapable of self-government. Even when the discontent of the masses culminates in a successful attempt to deprive the bourgeoisie of power, this is . . . effected only in appearance; always and necessarily there springs from the masses a new organized minority which raises itself to the rank of a governing class. Thus the majority of human beings, in a condition of eternal tutelage, are predestined by tragic necessity to submit to the dominion of a small minority, and must be content to constitute the pedestal of an oligarchy. (Michels [1911] 1958:406, 407)

Weber's work is avowedly multidimensional. This is depicted in Figure 4.2 by his positioning relative to the other theorists discussed in this text. He explicitly recognized that individual action is channeled through a variety of motivations that encompass both rationalist and nonrationalist dimensions. Moreover, his definition of sociology as a science aimed at the interpretive understanding of social action squarely places the individual and his or her conduct at the center of analysis. Complementing this position are Weber's substantive interests that led him to study religious idea systems, institutional arrangements, class and status structures, forms of domination, and broad historical trends; in short, elements aligned with the collective dimension of social life.

Figure 4.2   Weber's Basic Theoretical Orientation

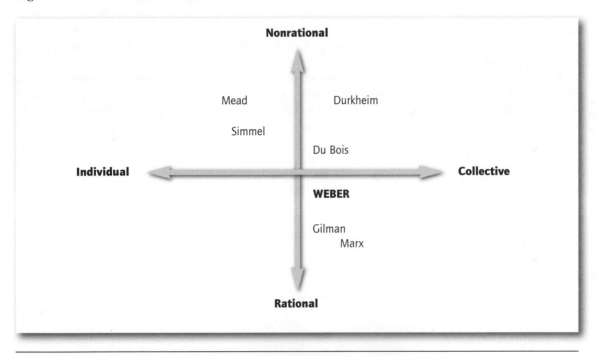

SOURCE: Courtesy of Activision, Inc. Copyright © 1993. Used by permission.

Of course, not every essay incorporates elements from each of the four dimensions. For instance, Weber's discussion of bureaucracy (excerpted in this chapter's Readings) focuses on the administrative functions and rules that account for the efficiency and impersonality that mark this organizational form. As a result, he emphasizes the structural or collectivist aspects of bureaucracies and how they work down to shape a given individual's behaviors and attitudes within them. Thus, you will find Weber remarking, "The individual bureaucrat cannot squirm out of the apparatus into which he has been harnessed. . . . [H]e is only a small cog in a ceaselessly moving mechanism which prescribes to him an essentially fixed route of march" (Weber [1925d] 1978:988). Weber's interest, then, lies here in describing the bureaucratic apparatus replete with its institutionalized demands for technical expertise and leveling of social differences.[3]

---

[3]While Weber's approach is clearly multidimensional, it is due to arguments like the one expressed in his essay on bureaucracy that we position the body of his work "off-center," ultimately in the collectivist/rationalist quadrant of our diagram. In the end, his emphasis lies in examining the rationalizing (i.e., rationalist) processes that have shaped the development of modern Western institutions (i.e., collectivist).

In Figure 4.3, we have highlighted a number of key concepts found in our preceding remarks or in the primary selections that follow. From the chart, it is readily apparent that Weber's theoretical orientation spans each of the four dimensions. Because some of these concepts were discussed previously (for instance, those regarding the types of action) and others will be addressed later in our introductions to the selections, we will restrict our comments in this section to a single example that underscores Weber's multidimensional approach.

In *The Protestant Ethic and the Spirit of Capitalism* (1904–05), Weber discusses the importance of the **calling** in motivating individuals to pursue worldly success. A doctrine first espoused by the Protestant reformer **Martin Luther** (1483–1546), the idea that each individual has a calling or "life-task" has its roots in a religious quest for salvation. In terms of our theoretical map, then, the calling reflects a nonrationalist orientation to action. The actions of the religiously faithful were motivated by the *moral* obligation to perform the duties of his labor to the best of his abilities. Here, the individual's actions are inspired by his desire to glorify God and thus gain confidence in the certainty of His grace, not by a desire to accumulate wealth as a means for purchasing material goods. Moreover, the calling is an individualist concept. It serves as the basis on which individuals make sense of their life circumstances as they chart their chances for attaining worldly success and eternal salvation.

Weber's analysis of the calling, however, was not tied solely to an examination of how religious ideas motivate individual conduct. For Weber, the significance of the calling also lies in its fueling a dramatic social transformation: the growth and eventual dominance of capitalism and the accompanying rationalization of much of social life. While oversimplifying his argument, Weber contended that the development of modern forms of capitalism was tied to the ascetic lifestyle demanded by the pursuit of one's calling. Originally a religious injunction to lead a life freed from the "temptations of the flesh," the secularization of the calling was a major force contributing to the explosive growth of capitalism in the West as businesses were increasingly organized on the basis of impersonal, methodical practices aimed at the efficient

Figure 4.3 Weber's Core Concepts

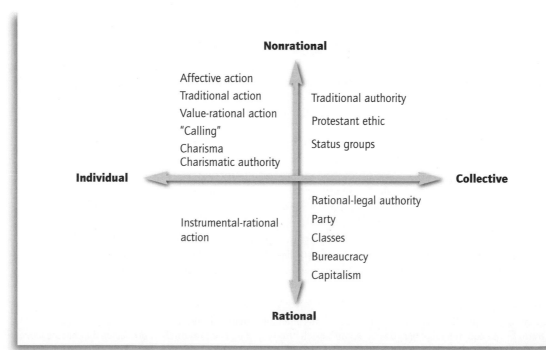

production of goods and services. Profit was now sought not to ensure one's state of grace, but because it was in one's self-interest to do so. Stripped of its religious impulses and spiritual moorings, the calling was further transformed into an overarching rationalist orientation to action that, as we remarked earlier, introduced methodical and calculative procedures into not only economic practices, but also into numerous spheres of life including politics, art, and sex, to name only a few.

Last, Weber's argument reveals a decidedly collectivist element as well. The ascetic ideals lying at the heart of the **Protestant ethic** were carried into the practical affairs of economic activity and social life more generally. This unleashed the process of rationalization, disenchanting Western society and creating an **iron cage** from which the individual is left with little power to escape. The dominance of capitalism and impersonal, bureaucratic forms of organization was a collective force that determined the life-chances of the individual. This dynamic is illustrated in the following passage taken from *The Protestant Ethic* and with which we end this section:

> The Puritan wanted to work in a calling; we are forced to do so. For when asceticism was carried out of monastic cells into everyday life, and began to dominate worldly morality, it did its part in building the tremendous cosmos of the modern economic order. This order is now bound to the technical and economic conditions of machine production which to-day [sic] determine the lives of all the individuals who are born into this mechanism, not only those directly concerned with economic acquisition, with irresistible force. ([1904–05] 1958:181)

Readings

In the selections that follow you will be introduced to five of Weber's most influential writings. In the first reading, excerpts from *The Protestant Ethic and the Spirit of Capitalism* (Weber 1904–05) offer Weber's analysis of the relationship between Protestantism and the economic and cultural life of modern Western society. In the second reading, from "The Social Psychology of the World Religions," Weber expands this theme in an examination of the psychological motivations underlying the "world religions." In the third reading, Weber investigates the crosscutting sources of power: class, status, and party. A parallel theme is addressed in the fourth selection, "The Types of Legitimate Domination," in which Weber outlines three distinct types of domination or authority. Finally, in "Bureaucracy," we end with Weber's description of bureaucracy, the predominant form of modern social organizations.

Introduction to *The Protestant Ethic and the Spirit of Capitalism*

Beyond doubt, one of the most influential sociology books ever written, *The Protestant Ethic* masterfully captures the two subjects that preoccupied Weber's intellectual activities: (1) the rationalizing tendencies so prevalent in Western society and (2) the role of ideas in shaping them. In addressing these twin issues, Weber argues that a religious belief system, intended to explain the path to a transcendent eternal salvation, paradoxically fueled the creation of a secular world in which "material goods have gained an increasing and finally an inexorable power over the lives of men as at no previous period in history" (Weber [1904–05] 1958:181).

Unlike Marx, who viewed religion as "the opiate of masses," or as an ideology that served the economic interests of the ruling class, and unlike Durkheim, who saw in religion humanity's worship of itself, Weber saw in religious beliefs a system of meaning aimed at explaining the existence of suffering and evil in the world. For Weber, such explanations have a profound impact on individuals' actions, and consequently on the broader social order. Of particular import is whether in addressing these ultimate issues, a belief system orients its adherents toward a "mastery" of the world or a mystical or contemplative escape from it. Thus, Protestantism, and Calvinism in particular, demanded that its followers serve as the "instruments" of God in order to fashion the world in His image. Conversely, Eastern religions such as Buddhism and Hinduism required their faithful to become "vessels" for the divine spirit in order to commune with otherworldly cosmic powers. The active engagement with the external, secular world called for by the Protestant belief system functioned as a potent impetus for social change, while the inward search for spiritual awakening characteristic of the major Eastern religions proved to be a socially conservative force.

In developing a scientifically based account of the independent role religious ideas can play in shaping the social order and, in particular, economic systems, Weber offered a powerful critique of Marxist theories of capitalism. As we discussed previously, he saw in historical materialism a one-sided causal interpretation, and, in several passages of *The Protestant Ethic,* you will read Weber clearly setting his sights on piercing this doctrine. As a counter to Marx's emphasis on property relations and class struggle, Weber maintained that the extraordinarily methodical *attitude* that characterized Protestant asceticism was integral to the rise and eventual dominance of Western capitalism.[1] Thus, Weber sought to demonstrate that not only "material" factors but also "ideal" factors can be instrumental in producing social change. In doing so, he sparked one of the most important and enduring debates in the history of sociology.

Having already highlighted several key elements of *The Protestant Ethic* when we outlined Weber's theoretical orientation, we briefly call attention to the book's main ideas. Weber traced the rise of individualism to the late sixteenth century and the Protestant Reformation, which, among other things, redefined the nature of the relationship between man and God. Led by **Martin Luther** (1483–1546), the Protestant Reformers insisted that each individual must methodically strive to realize a moral and righteous life each and every day in all their practical activities, as a constant expression of their devotion to the glorification of God. This methodical individualism challenged the previously dominant religious practice in which a handful of religious professionals (clergy) performed rituals in order to appease the gods either on behalf of the whole society or on behalf of those who paid them for their services. But Luther maintained that these token, periodic rituals (for instance, the Catholic confessional) or occasional "good works" could never placate or gain the favor of a great and all-powerful God. Instead, it was the duty of each to *submit* to the will of God through faithful dedication to his calling. It was demanded of rich and poor alike to be content with their lot, for it was God's unfathomable will that had assigned to each his station in life.

With its emphasis on submission and faith, Luther's view of the calling, like the Catholicism it rebelled against, promoted a traditional economic ethic that discouraged both laboring and profit seeking beyond what had long been established through custom. Workers and merchants sought simply to maintain the level of productivity and standard of living associated with the vocation in which they were engaged. However, in the hands of later Puritans leaders, the meaning of the calling was transformed. Under **John Calvin** (1509– 1564) and **Richard Baxter** (1615– 1691), the calling was interpreted as God's commandment to *work* for His divine glory. With submission and faith no longer sufficient for gaining confidence in one's salvation, how could the believer know that he was fulfilling his calling and thus might be one of His elect? Existing beyond the influence of mortals, only God knows who will be saved; there could be no certainty of proof of one's state of grace. The best one could hope for was a divinely granted sign.

---

[1]Significantly, Weber's central point was not that the Protestant ethic caused the emergence and growth of Western capitalism. Protestantism alone was not sufficient for creating this profound economic change. Rather, he argued that Protestant asceticism combined with a number of other important structural and social factors to produce the dominance of Western capitalism. In particular, Weber pointed to the separation of business pursuits from the home; the development of rational bookkeeping methods; technological advances in methods of production, distribution, and communication; the development of a rational legal system based on impersonal, formal rules; and, most importantly, the rational organization of free labor.

And that sign?: success and profit in worldly affairs, the pursuit of which was now religiously enjoined. Baxter stated the injunction thusly, "If God show you a way in which you may lawfully get more than in another way . . . If you refuse this, and choose the less gainful way, you cross one of the ends of your calling, and you refuse to be God's steward, and to accept His gifts and use them for Him when He requireth it" (Weber [1904–05] 1958:162). Profit was now understood to be a visible blessing from God that allowed the faithful to answer the most burning of all questions: Am I saved? Possessed by this "new spirit," one's predestined, eternal fate was now tied to the success of his conduct in work, a sphere of activity that was catapulted to the center of the believer's existence.

It was not success itself that offered proof, however. Rather, it was *how* success was achieved that marked a person as one of God's elect. Baxter cautioned his followers that "You may labour to be rich for God, though not for the flesh and sin" (Weber [1904–05] 1958:162). In this proscription lies the seeds for the subjective disposition that would ignite the growth of capitalism. Wealth served as confirmation of one's salvation only if it did not lead to idleness or the enjoyment of luxuries. Profitableness, moreover, was best guaranteed when economic pursuits were carried out on the basis of methodical and rational planning. Thus, ascetic restrictions on consumption were combined with the religiously derived compulsion to increase one's wealth. The ethical imperative to save and invest one's wealth would become the spiritual foundation for the spread of capitalism.

It would not be long, however, before the rational pursuit of wealth and bureaucratic structures necessary to modern capitalism would render obsolete the religious ethic that first had imbued work with a sense of meaning and purpose.[2] Chained by unquenchable consumption, modern humanity is now left to live in a disenchanted world where "material goods have gained an increasing and finally inexorable power over the lives of men" (Weber [1904–05] 1958:180). "In Baxter's view the care for external goods should only lie on the shoulders of the 'saint like a light cloak, which can be thrown aside at any moment.' But fate decreed that the cloak should become an iron cage" (ibid.:181).

And what of the iron cage today? Consider some statistics from the U.S. Commerce Department and the Federal Reserve Board: The average household is saddled with a credit card debt of $8,000, while the nation's credit card debt currently stands at $880 million. Not including home mortgages, in 2003, the average household was faced with more than $18,000 in total debt. As a nation, consumer debt soared to nearly 2 trillion dollars, an increase of 40 percent from 1998's total. Not surprisingly, personal savings rates have declined. After essential expenditures, Americans saved 9 percent of their disposable income during the 1980s. This rate fell to 5 percent during the 1990s, and in 2006 Americans registered a negative savings rate (–1 percent) for the first time since the Great Depression. Currently, 40 percent of Americans spend more than they earn. Far from being a "light cloak," our "care for external goods" has become central to our personal identity and sense of self. We define ourselves through the cars we drive, the clothes we wear, the places we vacation, and the neighborhoods we live in rather than through a sense of ultimate purpose or meaning to life. Whether it's trying to keep up with the Joneses or to distinguish ourselves from the herd, we are in continual "need" of new and better products, the purchasing of which requires ever-longer working hours in order to earn more money, so we can spend more money. To keep pace with the growing accumulation of products, over the last 50 years the average home size has doubled. Still, we can't seem to fit everything in so we hire companies to organize our closets and garages, or, when that fails, we pay to pack our "unessential" belongings into one of the thousands of self-storage spaces that dot the landscape. Like Marx's views on the fetishism of commodities and Veblen's notion of conspicuous consumption, the iron cage has imprisoned us in the pursuit of the "lifestyles of the rich and famous" whether or not we can afford to live like the affluent.

---

[2]One need merely note the spread of capitalism to countries and regions of the world that have not been exposed in any significant degree to Protestantism.

The Protestant Ethic and the Spirit of Capitalism (1904)

Max Weber

THE SPIRIT OF CAPITALISM

In the title of this study is used the somewhat pretentious phrase, the spirit of capitalism. What is to be understood by it? The attempt to give anything like a definition of it brings out certain difficulties which are in the very nature of this type of investigation. . . .

Thus, if we try to determine the object, the analysis and historical explanation of which we are attempting, it cannot be in the form of a conceptual definition, but at least in the beginning only a provisional description of what is here meant by the spirit of capitalism. Such a description is, however, indispensable in order clearly to understand the object of the investigation. For this purpose we turn to a document of that spirit which contains what we are looking for in almost classical purity, and at the same time has the advantage of being free from all direct relationship to religion, being thus, for our purposes, free of preconceptions.

"Remember, that time is money. He that can earn ten shillings a day by his labour, and goes abroad, or sits idle, one half of that day, though he spends but sixpence during his diversion or idleness, ought not to reckon that the only expense; he has really spent, or rather thrown away, five shillings besides.

"Remember, that *credit* is money. If a man lets his money lie in my hands after it is due, he gives me the interest, or so much as I can make of it during that time. This amounts to a considerable sum where a man has good and large credit, and makes good use of it.

"Remember, that money is of the prolific, generating nature. Money can beget money, and its offspring can beget more, and so on. Five shillings turned is six, turned again it is seven and threepence, and so on, till it becomes a hundred pounds. The more there is of it, the more it produces every turning, so that the profits rise quicker and quicker. He that kills a breeding-sow, destroys all her offspring to the thousandth generation. He that murders a crown, destroys all that it might have produced, even scores of pounds."

"Remember this saying, *The good paymaster is lord of another man's purse.* He that is known to pay punctually and exactly to the time he promises, may at any time, and on any occasion, raise all the money his friends can spare. This is sometimes of great use. After industry and frugality, nothing contributes more to the raising of a young man in the world than punctuality and justice in all his dealings; therefore never keep borrowed money an hour beyond the time you promised, lest a disappointment shut up your friend's purse for ever.

"The most trifling actions that affect a man's credit are to be regarded. The sound of your hammer at five in the morning, or eight at night, heard by a creditor, makes him easy six months longer; but if he sees you at a billiard-table, or hears your voice at a tavern, when you should be at work, he sends for his money the next day; demands it, before he can receive it, in a lump.

"It shows, besides, that you are mindful of what you owe; it makes you appear a careful as well as an honest man, and that still increases your credit.

"Beware of thinking all your own that you possess, and of living accordingly. It is a mistake that many people who have credit fall into. To prevent this, keep an exact account for some time both of your expenses and your income. If you take the pains at first to mention particulars, it will have this good effect: you will discover how wonderfully small, trifling expenses mount up to large sums, and will discern what might have been, and may for the future be saved, without occasioning any great inconvenience.

"For six pounds a year you may have the use of one hundred pounds, provided you are a man of known prudence and honesty.

"He that spends a groat a day idly, spends idly above six pounds a year, which is the price for the use of one hundred pounds.

"He that wastes idly a groat's worth of his time per day, one day with another, wastes the privilege of using one hundred pounds each day.

"He that idly loses five shillings' worth of time, loses five shillings, and might as prudently throw five shillings into the sea.

"He that loses five shillings, not only loses that sum, but all the advantage that might be made by turning it in dealing, which by the time that a young man becomes old, will amount to a considerable sum of money."

It is Benjamin Franklin who preaches to us in these sentences, the same which Ferdinand Kürnberger satirizes in his clever and malicious *Picture of American Culture* as the supposed confession of faith of the Yankee. That it is the spirit of capitalism which here speaks in characteristic fashion, no one will doubt, however little we may wish to claim that everything which could be understood as pertaining to that spirit is contained in it. Let us pause a moment to consider this passage, the philosophy of which Kürnberger sums up in the words, "They make tallow out of cattle and money out of men." The peculiarity of this philosophy of avarice appears to be the ideal of the honest man of recognized credit, and above all the idea of a duty of the individual toward the increase of his capital, which is assumed as an end in itself. Truly what is here preached is not simply a means of making one's way in the world, but a peculiar ethic. The infraction of its rules is treated not as foolishness but as forgetfulness of duty. That is the essence of the matter. It is not mere business astuteness, that sort of thing is common enough, it is an ethos. *This* is the quality which interests us. . . .

Now, all Franklin's moral attitudes are coloured with utilitarianism. Honesty is useful, because it assures credit; so are punctuality, industry, frugality, and that is the reason they are virtues. A logical deduction from this would be that where, for instance, the appearance of honesty serves the same purpose, that would suffice, and an unnecessary surplus of this virtue would evidently appear to Franklin's eyes as unproductive waste. And as a matter of fact, the story in his autobiography of his conversion to those virtues, or the discussion of the value of a strict maintenance of the appearance of modesty, the assiduous belittlement of one's own deserts in order to gain general recognition later, confirms this impression. According to Franklin, those virtues, like all others, are only in so far virtues as they are actually useful to the individual, and the surrogate of mere appearance is always sufficient when it accomplishes the end in view. It is a conclusion which is inevitable for strict utilitarianism. The impression of many Germans that the virtues professed by Americanism are pure hypocrisy seems to have been confirmed by this striking case. But

in fact the matter is not by any means so simple. Benjamin Franklin's own character, as it appears in the really unusual candidness of his autobiography, belies that suspicion. The circumstance that he ascribes his recognition of the utility of virtue to a divine revelation which was intended to lead him in the path of righteousness, shows that something more than mere garnishing for purely egocentric motives is involved.

In fact, the *summum bonum* of this ethic, the earning of more and more money, combined with the strict avoidance of all spontaneous enjoyment of life, is above all completely devoid of any eudæmonistic, not to say hedonistic, admixture. It is thought of so purely as an end in itself, that from the point of view of the happiness of, or utility to, the single individual, it appears entirely transcendental and absolutely irrational. Man is dominated by the making of money, by acquisition as the ultimate purpose of his life. Economic acquisition is no longer subordinated to man as the means for the satisfaction of his material needs. This reversal of what we should call the natural relationship, so irrational from a naïve point of view, is evidently as definitely a leading principle of capitalism as it is foreign to all peoples not under capitalistic influence. At the same time it expresses a type of feeling which is closely connected with certain religious ideas. If we thus ask, *why* should "money be made out of men," Benjamin Franklin himself, although he was a colourless deist, answers in his autobiography with a quotation from the Bible, which his strict Calvinistic father drummed into him again and again in his youth: "Seest thou a man diligent in his business? He shall stand before kings" (Prov. xxii. 29). The earning of money within the modern economic order is, so long as it is done legally, the result and the expression of virtue and proficiency in a calling; and this virtue and proficiency are, as it is now not difficult to see, the real Alpha and Omega of Franklin's ethic, as expressed in the passages we have quoted, as well as in all his works without exception.

And in truth this peculiar idea, so familiar to us to-day, but in reality so little a matter of course, of one's duty in a calling, is what is most characteristic of the social ethic of capitalistic culture, and is in a sense the fundamental basis of it. It is an obligation which the individual is supposed to feel and does feel towards the content of his professional activity, no matter in what it consists, in particular no matter whether it appears on the surface as a utilization of his personal powers, or only of his material possessions (as capital).

Of course, this conception has not appeared only under capitalistic conditions. On the contrary, we shall later trace its origins back to a time previous to the advent of capitalism. Still less, naturally, do we maintain that a conscious acceptance of these ethical maxims on the part of the individuals, entrepreneurs or labourers, in modern capitalistic enterprises, is a condition of the further existence of present-day capitalism. The capitalistic economy of the present day is an immense cosmos into which the individual is born, and which presents itself to him, at least as an individual, as an unalterable order of things in which he must live. It forces the individual, in so far as he is involved in the system of market relationships, to conform to capitalistic rules of action. The manufacturer who in the long run acts counter to these norms, will just as inevitably be eliminated from the economic scene as the worker who cannot or will not adapt himself to them will be thrown into the streets without a job.

Thus the capitalism of to-day, which has come to dominate economic life, educates and selects the economic subjects which it needs through a process of economic survival of the fittest. But here one can easily see the limits of the concept of selection as a means of historical explanation. In order that a manner of life so well adapted to the peculiarities of capitalism could be selected at all, i.e. should come to dominate others, it had to originate somewhere, and not in isolated individuals alone, but as a way of life common to whole groups of men. This origin is what really needs explanation. Concerning the doctrine of the more naïve historical materialism, that such ideas originate as a reflection or superstructure of economic situations, we shall speak more in detail below. At this point it will suffice for our purpose to call attention to the fact that without doubt, in the country of Benjamin Franklin's birth (Massachusetts), the spirit of capitalism (in the sense we have attached to it) was present before the capitalistic order. . . . It is further undoubted that capitalism remained far less developed in some of the neighbouring colonies, the later Southern States of the United States of America, in spite of the fact that these latter were founded by large capitalists for business motives, while the New England colonies were founded by preachers and seminary graduates with the help of small bourgeois, craftsmen and yoemen, for religious reasons. In this case the causal relation is certainly the reverse of that suggested by the materialistic standpoint.

But the origin and history of such ideas is much more complex than the theorists of the superstructure suppose. The spirit of capitalism, in the sense in which we are using the term, had to fight its way to supremacy against a whole world of hostile forces. A state of mind such as that expressed in the passages we have quoted from Franklin, and which called forth the applause of a whole people, would both in ancient times and in the Middle Ages have been proscribed as the lowest sort of avarice and as an attitude entirely lacking in self-respect. It is, in fact, still regularly thus looked upon by all those social groups which are least involved in or adapted to modern capitalistic conditions. This is not wholly because the instinct of acquisition was in those times unknown or undeveloped, as has often been said. Nor because the *auri sacra fames,* the greed for gold, was then, or now, less powerful outside of bourgeois capitalism than within its peculiar sphere, as the illusions of modern romanticists are wont to believe. The difference between the capitalistic and pre-capitalistic spirits is not to be found at this point. The greed of the Chinese Mandarin, the old Roman aristocrat, or the modern peasant, can stand up to any comparison. And the *auri sacra fames* of a Neapolitan cab-driver or *barcaiuolo,* and certainly of Asiatic representatives of similar trades, as well as of the craftsmen of southern European or Asiatic countries, is, as anyone can find out for himself, very much more intense, and especially more unscrupulous than that of, say, an Englishman in similar circumstances. . . .

The most important opponent with which the spirit of capitalism, in the sense of a definite standard of life claiming ethical sanction, has had to struggle, was that type of attitude and reaction to new situations which we may designate as traditionalism. . . .

One of the technical means which the modern employer uses in order to secure the greatest possible amount of work from his men is the device of piece-rates. In agriculture, for instance, the gathering of the harvest is a case where the greatest possible intensity of labour is called for, since, the weather being uncertain, the difference between high profit and heavy loss may depend on the speed with which the harvesting can be done. Hence a system of piece-rates is almost universal in this case. And since the interest of the employer in a speeding-up of harvesting increases with the increase of the results and the intensity of the work, the attempt has again and again been made, by increasing the piece-rates of the workmen, thereby giving them an opportunity to earn what is for them a very high wage, to interest them in increasing their own efficiency. But a peculiar difficulty has been met with surprising frequency: raising the piece-rates has often had the result that not more but less has been accomplished in the same time, because the worker reacted to the increase not by increasing but by decreasing the amount of his work. A man, for instance, who at the rate of 1 mark per acre mowed 2½ acres per day and earned 2½ marks, when the rate was raised to 1.25 marks per acre mowed, not 3 acres, as he might easily have done, thus earning 3.75 marks, but only 2 acres, so that he could still earn the 2½ marks to which he was accustomed. The opportunity of earning more was less attractive than that of working less. He did not ask: how much can I earn in a day if I do as much work as possible? but: how much must I work in order to earn the wage, 2½ marks, which I earned before and which takes care of my traditional needs? This is an example of what is here meant by traditionalism. A man does not "by nature" wish to earn more and more money, but simply to live as he is accustomed to live and to earn as much as is necessary for that purpose. Wherever modern capitalism has begun its work of increasing the productivity of human labour by increasing its intensity, it has encountered the immensely stubborn resistance of this leading trait of pre-capitalistic labour. And to-day it encounters it the more, the more backward (from a capitalistic point of view) the labouring forces are with which it has to deal.

Another obvious possibility, to return to our example, since the appeal to the acquisitive instinct through higher wage-rates failed, would have been to try the opposite policy, to force the worker by reduction of his wage-rates to work harder to earn the same amount than he did before. Low wages and high profits seem even to-day to a superficial observer to stand in correlation; everything which is paid out in wages seems to involve a corresponding reduction of profits. That road capitalism has taken again and again since its beginning. For centuries it was an article of faith, that low wages were productive, i.e. that they increased the material results of labour so that, as Pieter de la Cour, on this point, as we shall see, quite in the spirit of the old Calvinism, said long ago, the people only work because and so long as they are poor.

But the effectiveness of this apparently so efficient method has its limits. Of course the presence of a surplus population which it can hire cheaply in the labour market is a necessity for the development of capitalism. But though too large a reserve army may in certain cases favour its quantitative expansion, it checks its qualitative development, especially the transition to types of enterprise which make more intensive use of labour. Low wages are by no means identical with cheap labour. From a purely quantitative point of view the efficiency of labour decreases with a wage which is physiologically insufficient, which may in the long run even mean a survival of the unfit. . . . Low wages fail even from a purely business point of view wherever it is a question of producing goods which require any sort of skilled labour, or the use of expensive machinery which is easily damaged, or in general wherever any great amount of sharp attention or of initiative is required. Here low wages do not pay, and their effect is the opposite of what was intended. For not only is a developed sense of responsibility absolutely indispensable, but in general also an attitude which, at least during working hours, is freed from continual calculations of how the customary wage may be earned with a maximum of comfort and a minimum of exertion. Labour must, on the contrary, be performed as if it were an absolute end in itself, a calling. But such an attitude is by no means a product of nature. It cannot be evoked by low wages or high ones alone, but can only be the product of a long and arduous process of education. To-day, capitalism, once in the saddle, can recruit its labouring force in all industrial countries with comparative ease. In the past this was in every case an extremely difficult problem. And even to-day it could probably not get along without the support of a powerful ally along the way, which, as we shall see below, was at hand at the time of its development. . . .

Now, how could activity, which was at best ethically tolerated, turn into a calling in the sense of Benjamin Franklin? The fact to be explained historically is that in the most highly capitalistic centre of that time, in Florence of the fourteenth and fifteenth centuries, the money and capital market of all the great political Powers, this attitude was considered ethically unjustifiable, or at best to be tolerated. But in the backwoods small bourgeois circumstances of Pennsylvania in the eighteenth century, where business threatened for simple lack of money to fall back into barter, where there was hardly a sign of large enterprise, where only the earliest beginnings of banking were to be found, the same thing was considered the essence of moral conduct, even commanded in the name of duty. To speak here of a reflection of material conditions in the ideal superstructure would be patent nonsense. What was the background of ideas which could account for the sort of activity apparently directed toward profit alone as a calling toward which the individual feels himself to have an ethical obligation? For it was this idea which gave the way of life of the new entrepreneur its ethical foundation and justification. . . .

ASCETICISM AND THE SPIRIT OF CAPITALISM

In order to understand the connection between the fundamental religious ideas of ascetic Protestantism and its maxims for everyday economic conduct, it is necessary to examine with especial care such writings as have evidently been derived from ministerial practice. For in a time in which the beyond meant everything, when the social position of the Christian depended upon his admission to the communion, the clergyman, through his ministry, Church discipline, and preaching, exercised and influence (as a glance at collections of *consilia, casus conscientiæ,* etc., shows) which we modern men are entirely unable to picture. In such a time the religious forces which express themselves through such channels are the decisive influences in the formation of national character.

For the purposes of this chapter, though by no means for all purposes, we can treat ascetic Protestantism as a single whole. But since that side of English Puritanism which was derived from Calvinism gives the most consistent religious basis for the idea of the calling, we shall, following our previous method, place one of its representatives at the centre of the discussion. Richard Baxter stands out above many other writers on Puritan ethics, both because of his eminently practical and realistic attitude, and, at the same time, because of the universal recognition accorded to his works, which have gone through many new editions and translations. He was a Presbyterian and an apologist of the Westminster Synod, but at the same time, like so many of the best spirits of his time, gradually grew away from the dogmas of pure Calvinism. . . . His *Christian Directory* is the most complete compendium of Puritan ethics, and is continually adjusted to the practical experiences of his own ministerial activity. In comparison we shall make use of Spener's *Theologische Bedenken,* as representative of German Pietism, Barclay's *Apology* for the Quakers, and some other representatives of ascetic ethics, which, however, in the interest of space, will be limited as far as possible.

Now, in glancing at Baxter's *Saints' Everlasting Rest,* or his *Christian Directory,* or similar works of others, one is struck at first glance by the emphasis placed, in the discussion of wealth and its acquisition, on the ebionitic elements of the New Testament. Wealth as such is a great danger; its temptations never end, and its pursuit is not only senseless as compared with the dominating importance of the Kingdom of God, but it is morally suspect. Here asceticism seems to have turned much more sharply against the acquisition of earthly goods than it did in Calvin, who saw no hindrance to the effectiveness of the clergy in their wealth, but rather a thoroughly desirable enhancement of their prestige. Hence he permitted them to employ their means profitably. Examples of the condemnation of the pursuit of money and goods may be gathered without end from Puritan writings, and may be contrasted with the late mediæval ethical literature, which was much more open-minded on this point.

Moreover, these doubts were meant with perfect seriousness; only it is necessary to examine them somewhat more closely in order to understand their true ethical significance and implications. The real moral objection is to relaxation in the security of possession, the enjoyment of wealth with the consequence of idleness and the temptations of the flesh, above all of distraction from the pursuit of a righteous life. In fact, it is only because possession involves this danger of relaxation that it is objectionable at all. For the saints' everlasting rest in the next world; on earth man must, to be certain of his state of grace, "do the works of him who sent him, as long as it is yet day." Not leisure and enjoyment, but only activity serves to increase the glory of God, according to the definite manifestations of His will.

Waste of time is thus the first and in principle the deadliest of sins. The span of human life is infinitely short and precious to make sure of one's own election. Loss of time through sociability, idle talk, luxury, even more sleep than is necessary for health, six to at most eight hours, is worthy of absolute moral condemnation. It does not yet hold, with Franklin, that time is money, but the proposition is true in a certain spiritual sense. It is infinitely valuable because every hour lost is lost to labour for the glory of God. Thus inactive contemplation is also valueless, or even directly reprehensible if it is at the expense of one's daily work. For it is less pleasing to God than the active performance of His will in a calling. Besides, Sunday is provided for that, and, according to Baxter, it is always those who are not diligent in their callings who have no time for God when the occasion demands it.

Accordingly, Baxter's principal work is dominated by the continually repeated, often almost passionate preaching of hard, continuous bodily or mental labour. It is due to a combination of two different motives. Labour is, on the one hand, an approved ascetic technique, as it always has been in the Western Church, in sharp contrast not only to the Orient but to almost all monastic rules the world over. It is in particular the specific defence against all those temptations which Puritanism united under the name of the unclean life, whose rôle for it was by no means small. The sexual asceticism of Puritanism differs only in degree, not in fundamental principle, from that of monasticism; and on account of the Puritan conception of marriage, its practical influence is more far-reaching than that of the latter. For sexual intercourse is permitted, even within marriage, only as the means willed by God for the increase of His glory according to the commandment, "Be fruitful and multiply." Along with a moderate vegetable diet and cold baths, the same prescription is given for all sexual temptations as is used against religious doubts and a sense of moral unworthiness: "Work hard in your calling." But the most important thing was that even beyond that labour came to be considered in itself the end of life, ordained as such by God. St. Paul's "He who will not work shall not eat" holds unconditionally for everyone. Unwillingness to work is symptomatic of the lack of grace. . . .

[Not] only do these exceptions to the duty to labour naturally no longer hold for Baxter, but he holds most emphatically that wealth does not exempt anyone from the unconditional command. Even the wealthy shall not eat without working, for even though they do not need to labour to support their own needs, there is God's commandment which they, like the poor, must obey. For everyone without exception God's Providence has prepared a calling, which he should profess and in which he should labour. And this calling is not, as it was for the Lutheran, a fate to which he must submit and which he must make the best of, but God's commandment to the individual to work for the divine glory. This seemingly subtle difference had far-reaching psychological consequences, and became connected with a further development of the providential interpretation of the economic order which had begun in scholasticism.

The phenomenon of the division of labour and occupations in society had, among others, been interpreted by Thomas Aquinas, to whom we may most conveniently refer, as a direct consequence of the divine scheme of things. But the places assigned to each man in this cosmos follow *ex causis naturalibus* and are fortuitous (contingent in the Scholastic terminology). The differentiation of men into the classes and occupations established through historical development became for Luther, as we have seen, a direct result of the divine will. The perseverance of the individual in the place and within the limits which God had assigned to him was a religious duty. . . .

But in the Puritan view, the providential character of the play of private economic interests takes on a somewhat different emphasis. True to the Puritan tendency to pragmatic interpretations, the providential purpose of the division of labour is to be known by its fruits. . . .

But the characteristic Puritan element appears when Baxter sets at the head of his discussion the statement that "outside of a well-marked calling the accomplishments of a man are only casual and irregular, and he spends more time in idleness than at work," and when he concludes it as follows: "and he [the specialized worker] will carry out his work in order while another remains in constant confusion, and his business knows neither time nor place . . . therefore is a certain calling the best for everyone." Irregular work, which the ordinary labourer is often forced to accept, is often unavoidable, but always an unwelcome state of transition. A man without a calling thus lacks the systematic, methodical character which is, as we have seen, demanded by worldly asceticism.

The Quaker ethic also holds that a man's life in his calling is an exercise in ascetic virtue, a proof of his state of grace through his conscientiousness, which is expressed in the care and method with which he pursues his calling. What God demands is not labour in itself, but rational labour in a calling. In the Puritan concept of the calling the emphasis is always placed on this methodical character of worldly asceticism, not, as with Luther, on the acceptance of the lot which God has irretrievably assigned to man.

Hence the question whether anyone may combine several callings is answered in the affirmative, if it is useful for the common good or one's own, and not injurious to anyone, and if it does not lead to unfaithfulness in one of the callings. Even a change of calling is by no means regarded as objectionable, if it is not thoughtless and is made for the purpose of pursuing a calling more pleasing to God, which means, on general principles, one more useful.

It is true that the usefulness of a calling, and thus its favour in the sight of God, is measured primarily in moral terms, and thus in terms of the importance of the goods produced in it for the community. But a further, and, above all, in practice the most important, criterion is found in private profitableness. For if that God, whose hand the Puritan sees in all the occurrences of life, shows one of His elect a chance of profit, he must do it with a purpose. Hence the faithful Christian must follow the call by taking advantage of the opportunity. "If God show you a way in which you may lawfully get more than in another way (without wrong to your soul or to any other), if you refuse this, and choose the less gainful way, you cross one of the ends of your calling, and you refuse to be God's steward, and to accept His gifts and use them for Him when He requireth it: you may labour to be rich for God, though not for the flesh and sin."

Wealth is thus bad ethically only in so far as it is a temptation to idleness and sinful enjoyment of life, and its acquisition is bad only when it is with the purpose of later living merrily and without care. But as a performance of duty in a calling it is not only morally permissible, but actually enjoined. The parable of the servant who was rejected because he did not increase the talent which was entrusted to him seemed to say so directly. To wish to be poor was, it was often argued, the same as wishing to be unhealthy; it is objectionable as a glorification of works and derogatory to the glory of God. Especially begging, on the part of one able to work, is not only the sin of slothfulness, but a violation of the duty of brotherly love according to the Apostle's own word.

The emphasis on the ascetic importance of a fixed calling provided an ethical justification of the modern specialized division of labour. In a similar way the providential interpretation of profit-making justified the activities of the business man. The superior indulgence of the seigneur and the parvenu ostentation of the nouveau riche are equally detestable to asceticism. But, on the other hand, it has the highest ethical appreciation of the sober, middle-class, self-made man. "God blesseth His trade" is a stock remark about those good men who had successfully followed the divine hints. The whole power of the God of the Old Testament, who rewards His people for their obedience in this life, necessarily exercised a similar influence on the Puritan who, following Baxter's advice, compared his own state of grace with that of the heroes of the Bible, and in the process interpreted the statements of the Scriptures as the articles of a book of statutes. . . .

Let us now try to clarify the points in which the Puritan idea of the calling and the premium it placed upon ascetic conduct was bound directly to influence the development of a capitalistic way of life. As we have seen, this asceticism turned with all its force against one thing: the spontaneous enjoyment of life and all it had to offer. . . .

As against this the Puritans upheld their decisive characteristic, the principle of ascetic conduct. For otherwise the Puritan aversion to sport, even for the Quakers, was by no means simply one of principle. Sport was accepted if it served a rational purpose, that of recreation necessary for physical efficiency. But as a means for the spontaneous expression of undisciplined impulses, it was under suspicion; and in so far as it became purely a means of enjoyment, or awakened pride, raw instincts or the irrational gambling instinct, it was of course strictly condemned. Impulsive enjoyment of life, which leads away both from work in a calling and from religion, was as such the enemy of rational asceticism, whether in the form of seigneurial sports, or the enjoyment of the dance-hall or the public-house of the common man. . . .

The theatre was obnoxious to the Puritans, and with the strict exclusion of the erotic and of nudity from the realm of toleration, a radical view of either literature or art could not exist. The conceptions of idle talk, of superfluities, and of vain ostentation, all designations of an irrational attitude without objective purpose, thus not ascetic, and especially not serving the glory of God, but of man, were always at hand to serve in deciding in favour of sober utility as against any artistic tendencies. This was especially true in the case of decoration of the person, for instance clothing. That powerful tendency toward uniformity of life, which to-day so immensely aids the capitalistic interest in the standardization of production, had its ideal foundations in the repudiation of all idolatry of the flesh. . . .

Although we cannot here enter upon a discussion of the influence of Puritanism in all these directions, we should call attention to the fact that the toleration of pleasure in cultural goods, which contributed to purely aesthetic or athletic enjoyment, certainly always ran up against one characteristic limitation: they must not cost anything. Man is only a trustee of the goods which have come to him through God's grace. He must, like the servant in the parable, give an account of every penny entrusted to him, and it is at least hazardous to spend any of it for a purpose which does not serve the glory of God but only one's own enjoyment. What person, who keeps his eyes open, has not met representatives of this view-point even in the present? The idea of a man's duty to his possessions, to which he subordinates himself as an obedient steward, or even as an acquisitive machine, bears with chilling weight on his life. The greater the possessions the heavier, if the ascetic attitude toward life stands the test, the feeling of responsibility for them, for holding them undiminished for the glory of God and increasing them by restless effort. The origin of this type of life also extends in certain roots, like so many aspects of the spirit of capitalism, back into the Middle Ages. But it was in the ethic of ascetic Protestantism that it first found a consistent ethical foundation. Its significance for the development of capitalism is obvious.

This worldly Protestant asceticism, as we may recapitulate up to this point, acted powerfully against the spontaneous enjoyment of possessions; it restricted consumption, especially of luxuries. On the other hand, it had the psychological effect of freeing the acquisition of goods from the inhibitions of traditionalistic ethics. It broke the bonds of the impulse of acquisition in that it not only legalized it, but (in the sense discussed) looked upon it as directly willed by God. The campaign against the temptations of the flesh, and the dependence on external things, was, as besides the Puritans the great Quaker apologist Barclay expressly says, not a struggle against the rational acquisition, but against the irrational use of wealth.

But this irrational use was exemplified in the outward forms of luxury which their code condemned as idolatry of the flesh, however natural they had appeared to the feudal mind. On the other hand, they approved the rational and utilitarian uses of wealth which were willed by God for the needs of the individual and the community. They did not wish to impose mortification on the man of wealth, but the use of his means for necessary and practical things. The idea of comfort characteristically limits the extent of ethically permissible expenditures. It is naturally no accident that the development of a manner of living consistent with that idea may be observed earliest and most clearly among the most consistent representatives of this whole attitude toward life. Over against the glitter and ostentation of feudal magnificence which, resting on an unsound economic basis, prefers a sordid elegance to a sober simplicity, they set the clean and solid comfort of the middle-class home as an ideal.

On the side of the production of private wealth, asceticism condemned both dishonesty and impulsive avarice. What was condemned as covetousness, Mammonism, etc., was the pursuit of riches for their own sake. For wealth in itself was a temptation. But here asceticism was the power "which ever seeks the good but ever creates evil"; what was evil in its sense was possession and its temptations. For, in conformity with the Old Testament and in analogy to the ethical valuation of good works, asceticism looked upon the pursuit of wealth as an end in itself as highly reprehensible; but the attainment of it as a fruit of labour in a calling was a sign of God's blessing. And even more important: the religious valuation of restless, continuous, systematic work in a worldly calling, as the highest means to asceticism, and at the same time the surest and most evident proof of rebirth and genuine faith, must have been the most powerful conceivable lever for the expansion of that attitude toward life which we have here called the spirit of capitalism.

When the limitation of consumption is combined with this release of acquisitive activity, the inevitable practical result is obvious: accumulation of capital through ascetic compulsion to save. The restraints which were imposed upon the consumption of wealth naturally served to increase it by making possible the productive investment of capital. . . .

As far as the influence of the Puritan outlook extended, under all circumstances—and this is, of course, much more important than the mere encouragement of capital accumulation—it favoured the development of a rational bourgeois economic life; it was the most important, and above all the only consistent influence in the development of that life. It stood at the cradle of the modern economic man.

To be sure, these Puritanical ideals tended to give way under excessive pressure from the temptations of wealth, as the Puritans themselves knew very well. With great regularity we find the most genuine adherents of Puritanism among the classes which were rising from a lowly status, the small bourgeois and farmers, while the *beati possidentes,* even among Quakers, are often found tending to repudiate the old ideals. It was the same fate which again and again befell the predecessor of this worldly asceticism, the monastic asceticism of the Middle Ages. In the latter case, when rational economic activity had worked out its full effects by strict regulation of conduct and limitation of consumption, the wealth accumulated either succumbed directly to the nobility, as in the time before the Reformation, or monastic discipline threatened to break down, and one of the numerous reformations became necessary.

In fact the whole history of monasticism is in a certain sense the history of a continual struggle with the problem of the secularizing influence of wealth. The same is true on a grand scale of the worldly asceticism of Puritanism. The great revival of Methodism, which preceded the expansion of English industry toward the end of the eighteenth century, may well be compared with such a monastic reform. We may hence quote here a passage from John Wesley himself which might well serve as a motto for everything which has been said above. For it shows that the leaders of these ascetic movements understood the seemingly paradoxical relationships which we have here analysed perfectly well, and in the same sense that we have given them. He wrote:

"I fear, wherever riches have increased, the essence of religion has decreased in the same proportion. Therefore I do not see how it is possible, in the nature of things, for any revival of true religion to continue long. For religion must necessarily produce both industry and frugality, and these cannot but produce riches. But as riches increase, so will pride, anger, and love of the world in all its branches. How then is it possible that Methodism, that is, a religion of the heart, though it flourishes now as a green bay tree, should continue in this state? For the Methodists in every place grow diligent and frugal; consequently they increase in goods. Hence they proportionately increase in pride, in anger, in the desire of the flesh, the desire of the eyes, and the pride of life. So, although the form of religion remains, the spirit is swiftly vanishing away. Is there no way to prevent this—this continual decay of pure religion? We ought not to prevent people from being diligent and frugal; *we must exhort all Christians to gain all they can, and to save all they can; that is, in effect, to grow rich.*"

There follows the advice that those who gain all they can and save all they can should also give all they can, so that they will grow in grace and lay up a treasure in heaven. It is clear that Wesley here expresses, even in detail, just what we have been trying to point out.

As Wesley here says, the full economic effect of those great religious movements, whose significance for economic development lay above all in their ascetic educative influence, generally came only after the peak of the purely religious enthusiasm was past. Then the intensity of the search for the Kingdom of God commenced gradually to pass over into sober economic virtue; the religious roots died out slowly, giving way to utilitarian worldliness. . . .

A specifically bourgeois economic ethic had grown up. With the consciousness of standing in the fullness of God's grace and being visibly blessed by Him, the bourgeois business man, as long as he remained within the bounds of formal correctness, as long as his moral conduct was spotless and the use to which he put his wealth was not objectionable, could follow his pecuniary interests as he would and feel that he was fulfilling a duty in doing so. The power of religious asceticism provided him in addition with sober, conscientious, and unusually industrious workmen, who clung to their work as to a life purpose willed by God.

Finally, it gave him the comforting assurance that the unequal distribution of the goods of this world was a special dispensation of Divine Providence, which in these differences, as in particular grace, pursued secret ends unknown to men. Calvin himself had made the much-quoted statement that only when the people, i.e. the mass of labourers and craftsmen, were poor did they remain obedient to God. In the Netherlands (Pieter de la Court and others), that had been secularized to the effect that the mass of men only labour when necessity forces them to do so. This formulation of a leading idea of capitalistic economy later entered into the current theories of the productivity of low wages. Here also, with the dying out of the religious root, the utilitarian interpretation crept in unnoticed, in the line of development which we have again and again observed . . .

Now naturally the whole ascetic literature of almost all denominations is saturated with the idea that faithful labour, even at low wages, on the part of those whom life offers no other opportunities, is highly pleasing to God. In this respect Protestant Asceticism added in itself nothing new. But it not only deepened this idea most powerfully, it also created the force which was alone decisive for its effectiveness: the psychological sanction of it through the conception of this labour as a calling, as the best, often in the last analysis the only means of attaining certainty of grace. And on the other hand it legalized the exploitation of this specific willingness to work, in that it also interpreted the employer's business activity as a calling. It is obvious how powerfully the exclusive search for the Kingdom of God only through the fulfilment of duty in the calling, and the strict asceticism which Church discipline naturally imposed, especially on the propertyless classes, was bound to affect the productivity of labour in the capitalistic sense of the word. The treatment of labour as a calling became as characteristic of the modern worker as the corresponding attitude toward acquisition of the business man. It was a perception of this situation, new at his time, which caused so able an observer as Sir William Petty to attribute the economic power of Holland in the seventeenth century to the fact that the very numerous dissenters in that country (Calvinists and Baptists) "are for the most part thinking, sober men, and such as believe that Labour and Industry is their duty towards God." . . .

One of the fundamental elements of the spirit of modern capitalism, and not only of that but of all modern culture: rational conduct on the basis of the idea of the calling, was born—that is what this discussion has sought to demonstrate—from the spirit of Christian asceticism. One has only to re-read the passage from Franklin, quoted at the beginning of this essay, in order to see that the essential elements of the attitude which was there called the spirit of capitalism are the same as what we have just shown to be the content of the Puritan worldly asceticism, only without the religious basis, which by Franklin's time had died away. The idea that modern labour has an ascetic character is of course not new. Limitation to specialized work, with a renunciation of the Faustian universality of man which it involves, is a condition of any valuable work in the modern world; hence deeds and renunciation inevitably condition each other today. This fundamentally ascetic trait of middle-class life, if it attempts to be a way of life at all, and not simply the absence of any, was what Goethe wanted to teach, at the height of his wisdom, in the *Wanderjahren,* and in the end which he gave to the life of his *Faust.* For him the realization meant a renunciation, a departure from an age of full and beautiful humanity, which can no more be repeated in the course of our cultural development than can the flower of the Athenian culture of antiquity.

The Puritan wanted to work in a calling; we are forced to do so. For when asceticism was carried out of monastic cells into everyday life, and began to dominate worldly morality, it did its part in building the tremendous cosmos of the modern economic order. This order is now bound to the technical and economic conditions of machine production which to-day determine the lives of all the individuals who are born into this mechanism, not only those directly concerned with economic acquisition, with irresistible force. Perhaps it will so determine them until the last ton of fossilized coal is burnt. In Baxter's view the care for external goods should only lie on the shoulders of the "saint like a light cloak, which can be thrown aside at any moment." But fate decreed that the cloak should become an iron cage.

Since asceticism undertook to remodel the world and to work out its ideals in the world, material goods have gained an increasing and finally an inexorable power over the lives of men as at no previous period in history. To-day the spirit of religious asceticism—whether finally, who knows?—has escaped from the cage. But victorious capitalism, since it rests on mechanical foundations, needs its support no longer. The rosy blush of its laughing heir, the Enlightenment, seems also to be irretrievably fading, and the idea of duty in one's calling prowls about in our lives like the ghost of dead religious beliefs. Where the fulfilment of the calling cannot directly be related to the highest spiritual and cultural values, or when, on the other hand, it need not be felt simply as economic compulsion, the individual generally abandons the attempt to justify it at all. In the field of its highest development, in the United States, the pursuit of wealth, stripped of its religious and ethical meaning, tends to become associated with purely mundane passions, which often actually give it the character of sport.

No one knows who will live in this cage in the future, or whether at the end of this tremendous development entirely new prophets will arise, or there will be a great rebirth of old ideas and ideals, or, if neither, mechanized petrification, embellished with a sort of convulsive self-importance. For of the last stage of this cultural development, it might well be truly said: "Specialists without spirit, sensualists without heart; this nullity imagines that it has attained a level of civilization never before achieved."

But this brings us to the world of judgments of value and of faith, with which this purely historical discussion need not be burdened. The next task would be rather to show the significance of ascetic rationalism, which has only been touched in the foregoing sketch, for the content of practical social ethics, thus for the types of organization and the functions of social groups from the conventicle to the State. Then its relations to humanistic rationalism, its ideals of life and cultural influence; further to the development of philosophical and scientific empiricism, to technical development and to spiritual ideals would have to be analysed. Then its historical development from the mediæval beginnings of worldly asceticism to its dissolution into pure utilitarianism would have to be traced out through all the areas of ascetic religion. Only then could the quantitative cultural significance of ascetic Protestantism in its relation to the other plastic elements of modern culture be estimated.

Here we have only attempted to trace the fact and the direction of its influence to their motives in one, though a very important point. But it would also further be necessary to investigate how Protestant Asceticism was in turn influenced in its development and its character by the totality of social conditions, especially economic. The modern man is in general, even with the best will, unable to give religious ideas a significance for culture and national character which they deserve. But it is, of course, not my aim to substitute for a one-sided materialistic an equally one-sided spiritualistic causal interpretation of culture and of history. Each is equally possible, but each, if it does not serve as the preparation, but as the conclusion of an investigation, accomplishes equally little in the interest of historical truth.

▪▪

Introduction to "The Social Psychology of the World Religions"

In this essay, Weber extends his analysis developed in *The Protestant Ethic* by taking up five major world religions—Confucianism, Hinduism, Buddhism, Islam, and Christianity—to address more generally the relationship between religion and "economic ethics." (He was completing his studies on Judaism when he died.) In doing so, he again provides an account of religious experience that diverges from those offered by Marx and Durkheim. Drawing a contrast with Marxist views, Weber asserts that religion is not a "simple 'function' of the social situation of the stratum which appears as its characteristic bearer" nor does it represent "the stratum's 'ideology' [nor is it] a 'reflection' of a stratum's material or ideal interest-situation" (Weber 1958:269–70). Religion, instead, shapes economic, practical behavior just as much as such behavior shapes religious doctrines. Most importantly, religions address the psychological need of the fortunate to legitimate their good fortune, while for the less fortunate they offer the promise of a future salvation. While this "religious need" may be universal, the form in which it is met varies across different social strata (warriors, peasants, political officials, intellectuals, "civic") that exhibit an affinity for particular religious worldviews. Nevertheless, these worldviews have their own impact on behavior that cannot be understood simply as a reflection of its bearer's material position. This is particularly the case for religious virtuosos whose quest for salvation is guided by authentically spiritual motives. For the devout, actively proving oneself as an instrument or tool of God's will, communing contemplatively with the cosmic love of Nirvana, or striving for orgiastic ecstasy, represents genuine religious aims that cannot be reduced to some sort of underlying "distorted" class interest. Nor can the motives of the devout be understood as misguided intentions to deify society or as expressions of the collective conscience, as Durkheim would contend.

Weber also notes how religions have fostered the "rationalization of reality." Offering a promise of redemption, whether it be from social oppression, evil spirits, the cycle of rebirths, human imperfections, or any number of other forces, all religions counter a "senseless" world with the belief that "the world in its totality is, could, and should somehow be a meaningful 'cosmos'" (Weber 1958:281). The specific religious form of meaning is derived from a "systematic and rationalized 'image of the world'" that determines "'[f]rom what' and 'for what' one wished to be redeemed and . . . 'could be' redeemed" (ibid.:280). Religion declares that the world is not a playground for chance; instead, it is ruled by reasons and fates that can be "known." Knowing how to redeem oneself and how to obtain salvation requires that one knows how the world "works." In devising answers for such concerns, religions have developed along two primary paths: "exemplary" prophecy and "emissary" prophecy.

Exemplary prophecy is rooted in the conception of a supreme, impersonal being accessible only through contemplation, while emissary prophecy conceives of a personal God who is vengeful and loving, forgiving and punishing, and who demands of the faithful active, ethical conduct in order to serve

His commandments. Though the masses may be religiously "unmusical," the religiosity of the devout (monks, prophets, shamans, ascetics) nevertheless "has been of decisive importance for the development of the way of life of the masses," particularly with regard to regulating practical, economic activity (Weber 1958:289). Thus, religions grounded in an exemplary prophecy (e.g., Buddhism, Hinduism) lead adherents away from workaday life by seeking salvation through extraordinary psychic states attained through mystical, orgiastic, or ecstatic experiences. The virtuoso's hostility toward economic activity discourages this-worldly practical conduct by viewing it as "religiously inferior," a distraction from communing with the divine. Absent from the contemplative, mystical "flight from the world" is any psychological motivation to engage in worldly action as a path for redemption. As a result, a rationalized economic ethic remains underdeveloped.

Conversely, religions based on an emissary prophecy (e.g., Judaism, Christianity, Islam) require the devout to actively fashion the world according to the will of their god. Not contemplative "flight from," but, rather, ascetic "work in" this world is the path for redemption according this prophecy. Seeking mystical union with the cosmos is understood here as an irrational act of hedonism that devalues the God-created world. The virtuoso is instead compelled to "prove" himself as a worthy instrument of God through the ethical quality of his everyday activity. This psychological imperative leads to the development of rational, economic ethic that transforms work into a "holy," worldly calling. Everyday life is here the setting for the "methodical and rationalized routine-activities of workaday life in the service of the Lord" (Weber 1958:289). Yet, as Weber argued in *The Protestant Ethic,* this worldview, while faithful to God's commandments and devoted to creating His Kingdom on earth, leads to a thoroughgoing "disenchantment of the world."

"The Social Psychology of the World Religions" (1915)

Max Weber

By "world religions," we understand the five religions or religiously determined systems of life-regulation which have known how to gather multitudes of confessors around them. The term is used here in a completely value-neutral sense. The Confucian, Hinduist, Buddhist, Christian, and Islamist religious ethics all belong to the category of world religion. A sixth religion, Judaism, will also be dealt with. It is included because it contains historical preconditions decisive for understanding Christianity and Islamism, and because of its historic and autonomous significance for the development of the modern economic ethic of the Occident—a significance, partly real and partly alleged, which has been discussed several times recently. . . .

What is meant by the "economic ethic" of a religion will become increasingly clear during the course of our presentation. . . . The term "economic ethic" points to the practical impulses for action which are founded in the psychological and pragmatic contexts of religions.

The following presentation may be sketchy, but it will make obvious how complicated the structures and how many-sided the conditions of a concrete economic ethic usually are. Furthermore, it will show that externally similar forms of economic organization may agree with very different economic ethics and, according to the unique character of their economic ethics, how such forms of economic organization may produce very different historical results. An economic ethic is not a simple "function" of a form of economic organization; and just as little does the reverse hold, namely, that economic ethics unambiguously stamp the form of the economic organization.

No economic ethic has ever been determined solely by religion. In the face of man's attitudes towards the world—as determined by religious or other (in our sense) "inner" factors—an economic ethic has, of course, a high measure of autonomy. Given factors of economic geography and history determine this measure of autonomy in

SOURCE: Translation of the Introduction to The Economic Ethic of the World Religions by Max Weber, 1915.

the highest degree. The religious determination of life-conduct, however, is also one—note this—only one, of the determinants of the economic ethic. Of course, the religiously determined way of life is itself profoundly influenced by economic and political factors operating within given geographical, political, social, and national boundaries. We should lose ourselves in these discussions if we tried to demonstrate these dependencies in all their singularities. Here we can only attempt to peel off the directive elements in the life-conduct of those social *strata* which have most strongly influenced the practical ethic of their respective religions. These elements have stamped the most characteristic features upon practical ethics, the features that distinguish one ethic from others; *and,* at the same time, they have been important for the respective economic ethics. . . .

It is not our thesis that the specific nature of a religion is a simple "function" of the social situation of the stratum which appears as its characteristic bearer, or that it represents the stratum's "ideology," or that it is a "reflection" of a stratum's material or ideal interest-situation. On the contrary, a more basic misunderstanding of the standpoint of these discussions would hardly be possible.

However incisive the social influences, economically and politically determined, may have been upon a religious ethic in a particular case, it receives its stamp primarily from religious sources, and, first of all, from the content of its annunciation and its promise. Frequently the very next generation reinterprets these annunciations and promises in a fundamental fashion. Such reinterpretations adjust the revelations to the needs of the religious community. If this occurs, then it is at least usual that religious doctrines are adjusted to *religious needs*. Other spheres of interest could have only a secondary influence; often, however, such influence is very obvious and sometimes it is decisive.

For every religion we shall find that a change in the socially decisive strata has usually been of profound importance. On the other hand, the type of a religion, once stamped, has usually exerted a rather far-reaching influence upon the life-conduct of very heterogeneous strata. In various ways people have sought to interpret the connection between religious ethics and interest-situations in such a way that the former appear as mere "functions" of the latter. Such interpretation occurs in so-called historical materialism—which we shall not here discuss—as well as in a purely psychological sense. . . .

In treating suffering as a symptom of odiousness in the eyes of the gods and as a sign of secret guilt, religion has psychologically met a very general need. The

fortunate is seldom satisfied with the fact of being fortunate. Beyond this, he needs to know that he has a *right* to his good fortune. He wants to be convinced that he "deserves" it, and above all, that he deserves it in comparison with others. He wishes to be allowed the belief that the less fortunate also merely experience his due. Good fortune thus wants to be "legitimate" fortune.

If the general term "fortune" covers all the "good" of honor, power; possession, and pleasure, it is the most general formula for the service of legitimation, which religion has had to accomplish for the external and the inner interests of all ruling men, the propertied, the victorious, and the healthy. In short, religion provides the theodicy of good fortune for those who are fortunate. This theodicy is anchored in highly robust ("pharisaical") needs of man and is therefore easily understood, even if sufficient attention is often not paid to its effects. . . .

The annunciation and the promise of religion have naturally been addressed to the masses of those who were in need of salvation. They and their interests have moved into the center of the professional organization for the "cure of the soul," which, indeed, only therewith originated. The typical service of magicians and priests becomes the determination of the factors to be blamed for suffering, that is, the confession of "sins." At first, these sins were offenses against ritual commandments. The magician and priest also give counsel for behavior fit to remove the suffering. The material and ideal interests of magicians and priests could thereby actually and increasingly enter the service of specifically *plebeian* motives. A further step along this course was signified when, under the pressure of typical and ever-recurrent distress, the religiosity of a "redeemer" evolved. This religiosity presupposed the myth of a savior, hence (at least relatively) of a *rational* view of the world. Again, suffering became the most important topic. The primitive mythology of nature frequently offered a point of departure for this religiosity. The spirits who governed the coming and going of vegetation and the paths of celestial bodies important for the seasons of the year became the preferred carriers of the myths of the suffering, dying, and resurrecting god to needful men. The resurrected god guaranteed the return of good fortune in this world or the security of happiness in the world beyond. . . .

The need for an ethical interpretation of the "meaning" of the distribution of fortunes among men increased with the growing rationality of conceptions of the world. As the religious and ethical reflections upon the world were increasingly rationalized and primitive, and magical notions were eliminated, the theodicy of suffering encountered increasing difficulties. Individually "undeserved" woe was all too frequent; not "good" but "bad" men succeeded—even when "good" and "bad" were measured by the yardstick of the master stratum and not by that of a "slave morality."

One can explain suffering and injustice by referring to individual sin committed in a former life (the migration of souls), to the guilt of ancestors, which is avenged down to the third and fourth generation, or—the most principled—to the wickedness of all creatures *per se.* As compensatory promises, one can refer to hopes of the individual for a better life in the future in this world (transmigration of souls) or to hopes for the successors (Messianic realm), or to a better life in the hereafter (paradise). . . .

The distrust of wealth and power, which as a rule exists in genuine religions of salvation, has had its natural basis primarily in the experience of redeemers, prophets, and priests. They understood that those strata which were "satiated" and favored in this world had only a small urge to be saved, regardless of the kind of salvation offered. Hence, these master strata have been less "devout" in the sense of salvation religions. The development of a rational religious ethic has had positive and primary roots in the inner conditions of those social strata which were less socially valued.

Strata in solid possession of social honor and power usually tend to fashion their status-legend in such a way as to claim a special and intrinsic quality of their own, usually a quality of blood; their sense of dignity feeds on their actual or alleged being. The sense of dignity of socially repressed strata or of strata whose status is negatively (or at least not positively) valued is nourished most easily on the belief that a special "mission" is entrusted to them; their worth is guaranteed or constituted by an *ethical imperative,* or by their own functional *achievement.* Their value is thus moved into something beyond themselves, into a "task" placed before them by God. One source of the ideal power of ethical prophecies among socially disadvantaged strata lies in this fact. . . .

Psychologically considered, man in quest of salvation has been primarily preoccupied by attitudes of the here and now. The puritan *certitudo salutis,* the permanent state of grace that rests in the feeling of "having proved oneself," was psychologically the only concrete object among the sacred values of this ascetic religion. The Buddhist monk, certain to enter Nirvana, seeks the sentiment of a cosmic love; the devout Hindu seeks

either Bhakti (fervent love in the possession of God) or apathetic ecstasy. The Chlyst with his radjeny, as well as the dancing Dervish, strives for orgiastic ecstasy. Others seek to be possessed by God and to possess God, to be a bridegroom of the Virgin Mary, or to be the bride of the Savior. The Jesuit's cult of the heart of Jesus, quietistic edification, the pietists' tender love for the child Jesus and its "running sore," the sexual and semi-sexual orgies at the wooing of Krishna, the sophisticated cultic dinners of the Vallabhacharis, the gnostic onanist cult activities, the various forms of the *unio mystica,* and the contemplative submersion in the All-one—these states undoubtedly have been sought, first of all, for the sake of such emotional value as they directly offered the devout. In this respect, they have in fact been absolutely equal to the religious and alcoholic intoxication of the Dionysian or the soma cult; to totemic meat-orgies, the cannibalistic feasts, the ancient and religiously consecrated use of hashish, opium, and nicotine; and, in general, to all sorts of magical intoxication. They have been considered specifically consecrated and divine because of their psychic extraordinariness and because of the intrinsic value of the respective states conditioned by them. . . .

The two highest conceptions of sublimated religious doctrines of salvation are "rebirth" and "redemption." Rebirth, a primeval magical value, has meant the acquisition of a new soul by means of an orgiastic act or through methodically planned asceticism. Man transitorily acquired a new soul in ecstasy; but by means of magical asceticism, he could seek to gain it permanently. The youth who wished to enter the community of warriors as a hero, or to participate in its magical dances or orgies, or who wished to commune with the divinities in cultic feasts, had to have a new soul The heroic and magical asceticism, the initiation rites of youths, and the sacramental customs of rebirth at important phases of private and collective life are thus quite ancient. The means used in these activities varied, as did their ends: that is, the answers to the question, "For what should I be reborn?" . . .

The kind of empirical state of bliss or experience of rebirth that is sought after as the supreme value by a religion has obviously and necessarily varied according to the character of the stratum which was foremost in adopting it. The chivalrous, warrior class, peasants, business classes, and intellectuals with literary education have naturally pursued different religious tendencies. As will become evident, these tendencies have not by themselves determined the psychological character of religion; they have, however, exerted a very lasting influence upon it. The contrast between warrior and peasant classes, and intellectual and business classes, is of special importance. Of these groups, the intellectuals have always been the exponents of a rationalism which in their case has been relatively theoretical. The business classes (merchants and artisans) have been at least possible exponents of rationalism of a more practical sort. Rationalism of either kind has borne very different stamps, but has always exerted a great influence upon the religious attitude.

Above all, the peculiarity of the intellectual strata in this matter has been in the past of the greatest importance for religion. At the present time, it matters little in the development of a religion whether or not modern intellectuals feel the need of enjoying a "religious" state as an "experience," in addition to all sorts of other sensations, in order to decorate their internal and stylish furnishings with paraphernalia guaranteed to be genuine and old. A religious revival has never sprung from such a source. In the past, it was the work of the intellectuals to sublimate the possession of sacred values into a belief in "redemption." The conception of the idea of redemption, as such, is very old, if one understands by it a liberation from distress, hunger, drought, sickness, and ultimately from suffering and death. Yet redemption attained a specific significance only where it expressed a systematic and rationalized "image of the world" and represented a stand in the face of the world. For the meaning as well as the intended and actual psychological quality of redemption has depended upon such a world image and such a stand. Not ideas, but material and ideal interests, directly govern men's conduct. Yet very frequently the "world images" that have been created by "ideas" have, like switchmen, determined the tracks along which action has been pushed by the dynamic of interest. "From what" and "for what" one wished to be redeemed and, let us not forget, "could be" redeemed, depended upon one's image of the world.

There have been very different possibilities in this connection: One could wish to be saved from political and social servitude and lifted into a Messianic realm in the future of this world; or one could wish to be saved from being defiled by ritual impurity and hope for the pure beauty of psychic and bodily existence. One could wish to escape being incarcerated in an impure body and hope for a purely spiritual existence. One could wish to be saved from the eternal and senseless play of human passions and desires and hope for the quietude of the pure beholding of the divine. One could wish to be saved from radical evil and the servitude of sin and hope for the eternal and free benevolence in the lap of a fatherly god. One could wish to be saved from peonage under the astrologically conceived determination of stellar constellations and long for the dignity of freedom and partaking of the substance of the hidden deity. One could wish to be redeemed from the barriers to the finite, which express themselves in suffering, misery and death, and the threatening punishment of hell, and hope for an eternal bliss in an earthly or paradisical future existence. One could wish to be saved from the cycle of rebirths with their inexorable compensations for the deeds of the times past and hope for eternal rest. One could wish to be saved from senseless brooding and events and long for the dreamless sleep. Many more varieties of belief have, of course, existed. Behind them always lies a stand towards something in the actual world which is experienced as specifically "senseless." Thus, the demand has been implied: that the world order in its totality is, could, and should somehow be a meaningful "cosmos." This quest, the core of genuine religious rationalism, has been borne precisely by strata of intellectuals. The avenues, the results, and the efficacy of this metaphysical need for a meaningful cosmos have varied widely. Nevertheless, some general comments may be made.

The general result of the modern form of thoroughly rationalizing the conception of the world and of the way of life, theoretically and practically, in a purposive manner, has been that religion has been shifted into the realm of the irrational. This has been the more the case the further the purposive type of rationalization has progressed, if one takes the standpoint of an intellectual articulation of an image of the world. This shift of religion into the irrational realm has occurred for several reasons. On the one hand, the calculation of consistent rationalism has not easily come out even with nothing left over. In music, the Pythagorean "comma" resisted complete rationalization oriented to tonal physics. The various great systems of music of all peoples and ages have differed in the manner in which they have either covered up or bypassed this inescapable irrationality or, on the other hand, put irrationality into the service of the richness of tonalities. The same has seemed to happen to the theoretical conception of the world, only far more so; and above all, it has seemed to happen to the rationalization of practical life. The various great ways of leading a rational and methodical life have been characterized by irrational presuppositions, which have been accepted simply as "given" and which have been incorporated into such ways of life. What these presuppositions have been is historically and socially determined, at least to a very large extent, through the peculiarity of those strata that have been the carriers of the ways of life during its formative and decisive period. The *interest* situation of these strata, as determined socially and psychologically, has made for their peculiarity, as we here understand it.

Furthermore, the irrational elements in the rationalization of reality have been the *loci* to which the irrepressible quest of intellectualism for the possession of supernatural values has been compelled to retreat. That is the more so the more denuded of irrationality the world appears to be. The unity of the primitive image of the world, in which everything was concrete magic, has tended to split into rational cognition and mastery of nature, on the one hand, and into "mystic" experiences, on the other. The inexpressible contents of such experiences remain the only possible "beyond," added to the mechanism of a world robbed of gods. In fact, the beyond remains an incorporeal and metaphysical realm in which individuals intimately possess the holy. Where this conclusion has been drawn without any residue, the individual can pursue his quest for salvation only as an individual. This phenomenon appears in some form, with progressive intellectualist rationalism, wherever men have ventured to rationalize the image of the world as being a cosmos governed by impersonal rules. Naturally it has occurred most strongly among religions and religious ethics which have been quite strongly determined by genteel strata of intellectuals devoted to the purely cognitive comprehension of the world and of its "meaning." This was the case with Asiatic and, above all, Indian world religions. For all of them, contemplation became the supreme and ultimate religious value accessible to man. Contemplation offered them entrance into the profound and blissful tranquillity [sic] and immobility of the All-one. All other forms of religious states, however, have been at best considered a relatively valuable *Ersatz* for contemplation. This has had far-reaching consequences for the relation of religion to life, including economic life, as we shall repeatedly see. Such consequences flow from the general character of "mystic" experiences, in the contemplative sense, and from the psychological preconditions of the search for them.

The situation in which strata decisive for the development of a religion were active in practical life has been entirely different. Where they were chivalrous warrior heroes, political officials, economically acquisitive classes, or, finally, where an organized hierocracy dominated religion, the results were different than where genteel intellectuals were decisive.

The rationalism of hierocracy grew out of the professional preoccupation with cult and myth or—to a far higher degree—out of the cure of souls, that is, the confession of sin and counsel to sinners. Everywhere hierocracy has sought to monopolize the administration of religious values. They have also sought to bring and to temper the bestowal of religious goods into the form of "sacramental" or "corporate grace," which could be ritually bestowed only by the priesthood and could not be attained by the individual. The individual's quest for salvation or the quest of free communities by means of contemplation, orgies, or asceticism, has been considered highly suspect and has had to be regulated ritually and, above all, controlled hierocratically. From the standpoint of the interests of the priesthood in power, this is only natural.

Every body of *political* officials, on the other hand, has been suspicious of all sorts of individual pursuits of salvation and of the free formation of communities as sources of emancipation from domestication at the hands of the institution of the state. Political officials have distrusted the competing priestly corporation of grace and, above all, at bottom they have despised the very quest for these impractical values lying beyond utilitarian and worldly ends. For all political bureaucracies, religious duties have ultimately been simply official or social obligations of the citizenry and of status groups. . . .

It is also usual for a stratum of *chivalrous* warriors to pursue absolutely worldly interests and to be remote from all "mysticism." Such strata, however, have lacked—and this is characteristic of heroism in general—the desire as well as the capacity for a rational mastery of reality. The irrationality of "fate" and, under certain conditions, the idea of a vague and deterministically conceived "destiny" (the Homeric *Moira*) has stood above and behind the divinities and demons who were conceived of as passionate and strong heroes, measuring out assistance and hostility, glory and booty, or death to the human heroes.

Peasants have been inclined towards magic. Their whole economic existence has been specifically bound to nature and has made them dependent upon elemental forces. They readily believe in a compelling sorcery directed against spirits who rule over or through natural forces, or they believe in simply buying divine benevolence. Only tremendous transformations of life-orientation have succeeded in tearing them away from this universal and primeval form of religiosity. Such transformations have been derived either from other strata or from mighty prophets, who, through the power of miracles, legitimize themselves as sorcerers. Orgiastic and ecstatic states of "possession," produced by means of toxics or by the dance, are strange to the status honor of knights because they are considered undignified. Among the peasants, however, such states have taken the place that "mysticism" holds among the intellectuals.

Finally, we may consider the strata that in the western European sense are called "civic," as well as those which elsewhere correspond to them: artisans, traders, enterprisers engaged in cottage industry, and their derivatives existing only in the modern Occident. Apparently these strata have been the most ambiguous with regard to the religious stands open to them. And this is especially important to us. . . .

Of course, the religions of all strata are certainly far from being unambiguously dependent upon the character of the strata we have presented as having special affinities with them. Yet, at first sight, civic strata appear, in this respect and on the whole, to lend themselves to a more varied determination. Yet it is precisely among these strata that elective affinities for special types of religion stand out. The tendency towards a *practical* rationalism in conduct is common to all civic strata; it is conditioned by the nature of their way of life, which is greatly detached from economic bonds to nature. Their whole existence has been based upon technological or economic calculations and upon the mastery of nature and of man, however primitive the means at their disposal. The technique of living handed down among them may, of course, be frozen in traditionalism, as has occurred repeatedly and everywhere. But precisely for these, there has always existed the possibility—even though in greatly varying measure—of letting an *ethical* and rational regulation of life arise. This may occur by the linkage of such an ethic to the tendency of technological and economic rationalism. Such regulation has not always been able to make headway against traditions which, in the main, were magically stereotyped. But where prophecy has provided a religious basis, this basis could be one of two fundamental types of prophecy which we shall repeatedly discuss: "exemplary" prophecy, and "emissary" prophecy.

Exemplary prophecy points out the path to salvation by exemplary living, usually by a contemplative and apathetic-ecstatic life. The emissary type of prophecy addresses its *demands* to the world in the name of a god. Naturally these demands are ethical; and they are often of an active ascetic character.

It is quite understandable that the more weighty the civic strata as such have been, and the more they have been torn from bonds of taboo and from divisions into sibs and castes, the more favorable has been the soil for religions that call for action in this world. Under these conditions, the preferred religious attitude could become the attitude of active asceticism, of God-willed *action* nourished by the sentiment of being God's "tool," rather than the possession of the deity or the inward and contemplative surrender to God, which has appeared as the supreme value to religions influenced by strata of genteel intellectuals. In the Occident the attitude of active asceticism has repeatedly retained supremacy over contemplative mysticism and orgiastic or apathetic ecstasy, even though these latter types have been well known in the Occident. . . .

In the missionary prophecy the devout have not experienced themselves as vessels of the divine but rather as instruments of a god. This emissary prophecy has had a profound elective affinity to a special conception of God: the conception of a supra-mundane, personal, wrathful, forgiving, loving, demanding, punishing Lord of Creation. Such a conception stands in contrast to the supreme being of exemplary prophecy. As a rule, though by no means without exception, the supreme being of an exemplary prophecy is an impersonal being because, as a static state, he is accessible only by means of contemplation. The conception of an active God, held by emissary prophecy, has dominated the Iranian and Mid-Eastern religions and those Occidental religions which are derived from them. The conception of a supreme and static being, held by exemplary prophecy, has come to dominate Indian and Chinese religiosity.

These differences are not primitive in nature. On the contrary, they have come into existence only by means of a far-reaching sublimation of primitive conceptions of animist spirits and of heroic deities which are everywhere similar in nature. Certainly the connection of conceptions of God with religious states, which are evaluated and desired as sacred values, have also been strongly influential in this process of sublimation. These religious states have simply been interpreted in the direction of a different conception of God, according to whether the holy states, evaluated as supreme, were contemplative mystic experiences or apathetic ecstasy, or whether they were the orgiastic possession of god, or visionary inspirations and "commands." . . .

The rational elements of a religion, its "doctrine," also have an autonomy: for instance, the Indian doctrine of Kharma, the Calvinist belief in predestination, the Lutheran justification through faith, and the Catholic doctrine of sacrament. The rational religious pragmatism of salvation, flowing from the nature of the images of God and of the world, have under certain conditions had far-reaching results for the fashioning of a practical way of life.

These comments presuppose that the nature of the desired sacred values has been strongly influenced by the nature of the external interest-situation and the corresponding way of life of the ruling strata and thus by the social stratification itself. But the reverse also holds: wherever the direction of the whole way of life has been methodically rationalized, it has been profoundly determined by the ultimate values toward which this rationalization has been directed. These values and positions were thus *religiously* determined. Certainly they have not always, or exclusively, been decisive; however, they have been decisive in so far as an *ethical* rationalization held sway, at least so far as its influence reached. As a rule, these religious values have been also, and frequently absolutely, decisive. . . .

The empirical fact, important for us, that men are *differently qualified* in a religious way stands at the beginning of the history of religion. This fact had been dogmatized in the sharpest rationalist form in the "particularism of grace," embodied in the doctrine of predestination by the Calvinists. The sacred values that have been most cherished, the ecstatic and visionary capacities of shamans, sorcerers, ascetics, and pneumatics of all sorts, could not be attained by everyone. The possession of such faculties is a "charisma," which, to be sure, might be awakened in some but not in all. It follows from this that all intensive religiosity has a tendency toward a sort of *status stratification,* in accordance with differences in the charismatic qualifications. "Heroic" or "virtuoso" religiosity is opposed to mass religiosity. By "mass" we understand those who are religiously "unmusical"; we do not, of course, mean those who occupy an inferior position in the secular status order. . . .

Now, every hierocratic and official authority of a "church"—that is, a community organized by officials into an institution which bestows gifts of grace—fights principally against all virtuoso-religion and against its autonomous development. For the church, being the holder of institutionalized grace, seeks to organize the religiosity of the masses and to put its own officially monopolized and mediated sacred values in the place of the autonomous and religious status qualifications of

the religious virtuosos. By its nature, that is, according to the interest-situation of its officeholders, the church must be "democratic" in the sense of making the sacred values generally accessible. This means that the church stands for a universalism of grace and for the ethical sufficiency of all those who are enrolled under its institutional authority. Sociologically, the process of leveling constitutes a complete parallel with the political struggles of the bureaucracy against the political privileges of the aristocratic estates. As with hierocracy, every full-grown political bureaucracy is necessarily and in a quite similar sense "democratic"—namely, in the sense of leveling and of fighting against status privileges that compete with its power. . . .

The religious virtuosos saw themselves compelled to adjust their demands to the possibilities of the religiosity of everyday life in order to gain and to maintain ideal and material mass-patronage. The nature of their concessions have naturally been of primary significance for the way in which they have religiously influenced everyday life. In almost all Oriental religions, the virtuosos allowed the masses to remain stuck in magical tradition. Thus, the influence of religious virtuosos has been infinitely smaller than was the case where religion has undertaken ethically and generally to rationalize everyday life. This has been the case even when religion has aimed precisely at the masses and has cancelled however many of its ideal demands. Besides the relations between the religiosity of the virtuosos and the religion of the masses, which finally resulted from this struggle, the peculiar nature of the concrete religiosity of the virtuosos has been of decisive importance for the development of the way of life of the masses. This virtuoso religiosity has therefore also been important for the economic ethic of the respective religion. The religion of the virtuoso has been the genuinely "exemplary" and practical religion. According to the way of life his religion prescribed to the virtuoso, there have been various possibilities of establishing a rational ethic of everyday life. The relation of virtuoso religion to *workaday life* in the locus of the economy has varied, especially according to the peculiarity of the sacred values desired by such religions.

Wherever the sacred values and the redemptory means of a virtuoso religion bore a contemplative or orgiastic-ecstatic character, there has been no bridge between religion and the practical action of the workaday world. In such cases, the economy and all other action in the world has been considered religiously inferior, and no psychological motives for worldly action could be derived from the attitude cherished as the supreme value. In their innermost beings, contemplative and ecstatic religions have been rather specifically hostile to economic life. Mystic, orgiastic, and ecstatic experiences are extraordinary psychic states; they lead away from everyday life and from all expedient conduct. Such experiences are, therefore, deemed to be "holy." With such religions, a deep abyss separates the way of life of the laymen from that of the community of virtuosos. The rule of the status groups of religious virtuosos over the religious community readily shifts into a magical anthropolatry; the virtuoso is directly worshipped as a Saint, or at least laymen buy his blessing and his magical powers as a means of promoting mundane success or religious salvation. As the peasant was to the landlord, so the layman was to the Buddhist and Jainist bhikshu: ultimately, mere sources of tribute. Such tribute allowed the virtuosos to live entirely for religious salvation without themselves performing profane work, which always would endanger their salvation. Yet the conduct of the layman could still undergo a certain ethical regulation, for the virtuoso was the layman's spiritual adviser, his father confessor and *directeur de l'âme*. Hence, the virtuoso frequently exercises a powerful influence over the religiously "unmusical" laymen; this influence might not be in the direction of his (the virtuoso's) own religious way of life; it might be an influence in merely ceremonious, ritualist, and conventional particulars. For action in this world remained in principle religiously insignificant; and compared with the desire for the religious end, action lay in the very opposite direction.

In the end, the charisma of the pure "mystic" serves only himself. The charisma of the genuine magician serves others.

Things have been quite different where the religiously qualified virtuosos have combined into an ascetic sect, striving to mould life in this world according to the will of a god. To be sure, two things were necessary before this could happen in a genuine way. First, the supreme and sacred value must not be of a contemplative nature; it must not consist of a union with a supra-mundane being who, in contrast to the world, lasts forever; nor in a *unia mystica* to be grasped orgiastically or apathetic-ecstatically. For these ways always lie apart from everyday life and beyond the real world and lead away from it. Second, such a religion must, so far as possible, have given up the purely magical or sacramental character of the *means* of grace. For these means always devalue action in this world as, at best,

merely relative in their religious significance, and they link the decision about salvation to the success of processes which are *not* of a rational everyday nature.

When religious virtuosos have combined into an active asceticist sect, two aims are completely attained: the disenchantment of the world and the blockage of the path to salvation by a flight from the world. The path to salvation is turned away from a contemplative "flight from the world" and towards an active ascetic "work in this world." If one disregards the small rationalist sects, such as are found all over the world, this has been attained only in the great church and sect organizations of Occidental and asceticist Protestantism. The quite distinct and the purely historically determined destinies of Occidental religions have co-operated in this matter. Partly the social environment exerted an influence, above all, the environment of the stratum that was decisive for the development of such religion. Partly, however—and just as strongly—the intrinsic character of Christianity exerted an influence: the supra-mundane God and the specificity of the means and paths of salvation as determined historically, first by Israelite prophecy and the Thora doctrine.

The religious virtuoso can be placed in the world as the instrument of a God and cut off from all magical means of salvation. At the same time, it is imperative for the virtuoso that he "prove" himself before God, as being called *solely* through the ethical quality of his conduct in this world. This actually means that he "prove" himself to himself as well. No matter how much the "world" as such is religiously devalued and rejected as being creatural and a vessel of sin, yet psychologically the world is all the more affirmed as the theatre of God-willed activity in one's worldly "calling." For this inner-worldly asceticism rejects the world in the sense that it despises and taboos the values of dignity and beauty, of the beautiful frenzy and the dream, purely secular power, and the purely worldly pride of the hero. Asceticism outlawed these values as competitors of the kingdom of God. Yet precisely because of this rejection, asceticism did not fly from the world, as did contemplation. Instead, asceticism has wished to rationalize the world ethically in accordance with God's commandments. It has therefore remained oriented towards the world in a more specific and thoroughgoing sense than did the naive "affirmation of the world" of unbroken humanity, for instance, in Antiquity and in lay-Catholicism. In inner-worldly asceticism, the grace and the chosen state of the religiously qualified man prove themselves in everyday life. To be sure, they do so not in the everyday life as it is given, but in methodical and rationalized routine-activities of work-aday life in the service of the Lord. Rationally raised into a vocation, everyday conduct becomes the locus for proving one's state of grace. The Occidental sects of the religious virtuosos have fermented the methodical rationalization of conduct, including economic conduct. These sects have not constituted valves for the longing to escape from the senselessness of work in this world, as did the Asiatic communities of the ecstatics: contemplative, orgiastic, or apathetic. . . .

We have to remind ourselves in advance that "rationalism" may mean very different things. It means one thing if we think of the kind of rationalization the systematic thinker performs on the image of the world: an increasing theoretical mastery of reality by means of increasingly precise and abstract concepts. Rationalism means another thing if we think of the methodical attainment of a definitely given and practical end by means of an increasingly precise calculation of adequate means. These types of rationalism are very different, in spite of the fact that ultimately they belong inseparately together. . . .

"Rational" may also mean a "systematic arrangement." In this sense, the following methods are rational: methods of mortificatory or of magical asceticism, of contemplation in its most consistent forms—for instance, in *yoga*—or in the manipulations of the prayer machines of later Buddhism.

In general, all kinds of practical ethics that are systematically and unambiguously oriented to fixed goals of salvation are "rational," partly in the same sense as formal method is rational, and partly in the sense that they distinguish between "valid" norms and what is empirically given.

■■

Introduction to "The Distribution of Power Within the Political Community: Class, Status, Party"

In "Class, Status, Party," we again find Weber engaged in an implicit debate with Marx. While Marx saw interests, and the power to realize them, tied solely to class position, Weber saw the two as flowing from several sources. In fact, he argued that distinct interests and forms of power were connected to economic **classes, status** groups, and political **parties.** (See Table 4.1.) The result is a discarding of Marx's model in favor of a more complex view of how interests shape individuals' actions and the organization of societies.

Weber begins this essay with a definition of power, a definition that to this day guides work in political sociology. He defines it as "the chance of a man or of a number of men to realize their own will in a social action even against the resistance of others" (Weber [1925b] 1978:926). Such chances, however, are not derived from a single source, nor is power valued for any one particular reason. Power may be exercised for economic gain, to increase one's "social honor" (or **status),** or for its own sake. Moreover, power stemming from one source, for instance economic power, may not translate into other domains. Thus, a person who has achieved substantial economic wealth through criminal activity will not have a high degree of status in the general society. Conversely, academics have a relatively high degree of status, but little economic power. Whatever power intellectuals have stems from their social honor, not from their ability to "realize their own will" through financial influence.

This essay is significant not only for its picture of the crosscutting sources of interests and power. Weber also offers here a distinct definition of class as well as his conception of status groups and parties. Recall that for Marx classes are based on a group's more or less stable relationship to the means of production (owners of capital versus owners of labor power). For Weber, however, classes are not stable groups or "communities" produced by existing property relations. Instead, they are people who share "life chances" or possibilities that are determined by "economic interests in the possession of goods and opportunities for income" within the commodity and labor markets (Weber [1925b] 1978:927). While recognizing with Marx that "property" and "lack of property" form the basic distinction between classes, Weber

Table 4.1   Weber's Notion of Class, Status, and Party

		ORDER	
		Individual	**Collective**
ACTION	**Nonrational**	Status	*Status Groups:* "A specific, positive or negative, social estimation of *honor.*"
	Rational	Interests	*Class:* People who share "life chances" or possibilities that are "determined by economic interests in the possession of goods and opportunities for income."
			Party: Aimed at "influencing a communal action no matter what its content may be."

nevertheless argued that classes are themselves the product of a shared "class situation"—a situation that reflects the type and amount of exchanges one can pursue in the market.

Status groups, on the other hand, are communities. The fate of such communities is determined not by their chances on the commodity or labor markets, however, but by "a specific, positive or negative, social estimation of *honor*" (Weber [1925b] 1978:932, emphasis in the original). Such "honor" is expressed through "styles of life" or "conventions" that identify individuals with specific social circles. Race, ethnicity, religion, taste in fashion and the arts, and occupation have often formed a basis for making status distinctions. More than anything, membership in status groups serves to restrict an individual's chances for social interaction. For instance, the selection of marriage partners has frequently depended on a potential mate's religion or ethnicity. Even in modern, "egalitarian" societies like the United States, interracial marriages are relatively uncommon.

Additionally, regardless of possessing significant economic power or material wealth, one's race or religion can either close or open a person to educational and professional opportunities, as well as to membership in various clubs or associations.[1] Indeed, once membership into a style of life or institution can be bought, its ability to function as an expression of social honor or sign of exclusivity is threatened. This dynamic can be seen in shifting fashions in clothes and tastes in music, as well as in the democratization of education whereby proper "breeding" is no longer a prerequisite for getting a college diploma.

The third domain from which distinct interests are generated and power is exercised is the "legal order." Here, "'parties' reside in the sphere of power" (Weber [1925b] 1978:938). They include not only explicitly political groups, but also rationally organized groups more generally. As such, parties are characterized by the strategic pursuit of goals and the maintenance of a staff capable of implementing their objectives. Moreover, they are not necessarily tied to either class or status group interests, but are aimed instead at "influencing a communal action no matter what its content may be" (ibid.). Examples of parties include labor unions, which, through bureaucratic channels and the election of officers, seek to win economic

---

[1]During the early years of unionizing in the United States, trade unions were segregated racially, and at times ethnically. Thus, while sharing a common "class situation," workers, nevertheless, were divided by status group memberships. Some sociologists and labor historians have argued that the overriding salience of racial (i.e., status group) divisions fractured the working class, preventing workers from achieving more fully their class-based interests. Similar arguments have been made with regard to the feminist movement. In this case, white, middle-class women are charged with forsaking the plight of nonwhite and lower-class women in favor of pursuing goals that derive from their unique class situation.

benefits on behalf of workers, and, of course, the Republican and Democratic parties, which pursue legislative action that alternates between serving the class interests of their constituents (e.g., tax policy, trade regulations) and the interests of varying status groups (e.g., affirmative action, abortion rights, and gun control).

"The Distribution of Power Within the Political Community: Class, Status, Party" (1925)

Max Weber

A. *Economically determined power and the status order*.

The structure of every legal order directly influences the distribution of power, economic or otherwise, within its respective community. This is true of all legal orders and not only that of the state. In general, we understand by "power" the chance of a man or a number of men to realize their own will in a social action even against the resistance of others who are participating in the action.

"Economically conditioned" power is not, of course, identical with "power" as such. On the contrary, the emergence of economic power may be the consequence of power existing on other grounds. Man does not strive for power only in order to enrich economically. Power, including economic power, may be valued for its own sake. Very frequently the striving for power is also conditioned by the social honor it entails. Not all power, however, entails social honor: The typical American Boss, as well as the typical big speculator, deliberately relinquishes social honor. Quite generally, "mere economic" power, and especially "naked" money power, is by no means a recognized basis of social honor. Nor is power the only basis of social honor. Indeed, social honor, or prestige, may even be the basis of economic power, and very frequently has been. Power, as well as honor, may be guaranteed by the legal order, but, at least normally, it is not their primary source. The legal order is rather an additional factor that enhances the chance to hold power or honor; but it cannot always secure them.

The way in which social honor is distributed in a community between typical groups participating in this distribution we call the "status order." The social order and the economic order are related in a similar manner to the legal order. However, the economic order merely defines the way in which economic goods and services are distributed and used. Of course, the status order is strongly influenced by it, and in turn reacts upon it.

Now: "classes," "status groups," and "parties" are phenomena of the distribution of power within a community.

B. *Determination of class situation by market situation*.

In our terminology, "classes" are not communities; they merely represent possible, and frequent, bases for social action. We may speak of a "class" when (1) a number of people have in common a specific causal component of their life chances, insofar as (2) this component is represented exclusively by economic interests in the possession of goods and opportunities for income, and (3) is represented under the conditions of the commodity or labor markets. This is "class situation."

It is the most elemental economic fact that the way in which the disposition over material property is distributed among a plurality of people, meeting competitively in the market for the purpose of exchange, in itself creates specific life chances. The mode of distribution, in accord with the law of marginal utility, excludes the non-wealthy from competing for highly valued goods; it favors the owners and, in fact, gives to them a monopoly to acquire such goods. Other things being equal, the mode of distribution monopolizes the opportunities for profitable deals for all those who, provided with goods, do not necessarily have to exchange them. It increases, at least generally, their power in the price struggle with those who, being propertyless, have nothing to offer but their labor or the resulting products, and who are compelled to get rid of these products in order to subsist at all. The mode of distribution gives to the propertied a monopoly on the possibility of transferring property from the sphere of use as "wealth" to the sphere of "capital," that is, it gives them the entrepreneurial function and all chances to share directly or indirectly in returns on capital. All this holds true within the area in which pure market conditions prevail. "Property" and "lack of property" are, therefore, the basic categories of all class situations. It does not matter whether these two categories become effective in the competitive struggles of the consumers or of the producers.

Within these categories, however, class situations are further differentiated: on the one hand, according to the kind of property that is usable for returns; and, on the other hand, according to the kind of services that can be offered in the market. Ownership of dwellings; workshops; warehouses; stores; agriculturally usable land in large or small holdings—a quantitative difference with possibly qualitative consequences; ownership of mines; cattle; men (slaves); disposition over mobile instruments of production, or capital goods of all sorts, especially money or objects that can easily be exchanged for money; disposition over products of one's own labor or of others' labor differing according to their various distances from consumability; disposition over transferable monopolies of any kind—all these distinctions differentiate the class situations of the propertied just as does the "meaning" which they can give to the use of property, especially to property which has money equivalence. Accordingly, the propertied, for instance, may belong to the class of rentiers or to the class of entrepreneurs.

Those who have no property but who offer services are differentiated just as much according to their kinds of services as according to the way in which they make use of these services, in a continuous or discontinuous relation to a recipient. But always this is the generic connotation of the concept of class: that the kind of chance in the *market* is the decisive moment which presents a common condition for the individual's fate. Class situation is, in this sense, ultimately market situation. The effect of naked possession *per se,* which among cattle breeders gives the non-owning slave or serf into the power of the cattle owner, is only a fore-runner of real "class" formation. However, in the cattle loan and in the naked severity of the law of debts in such communities for the first time mere "possession" as such emerges as decisive for the fate of the individual; this is much in contrast to crop-raising communities, which are based on labor. The creditor-debtor relation becomes the basis of "class situations" first in the cities, where a "credit market," however primitive, with rates of interest increasing according to the extent of dearth and factual monopolization of lending in the hands of a plutocracy could develop. Therewith "class struggles" begin.

Those men whose fate is not determined by the chance of using goods or services for themselves on the market, e.g., slaves, are not, however, a class in the technical sense of the term. They are, rather, a status group.

C. *Social action flowing from class interest*

According to our terminology, the factor that creates "class" is unambiguously economic interest, and indeed, only those interests involved in the existence of the market. Nevertheless, the concept of class-interest is an ambiguous one: even as an empirical concept it is ambiguous as soon as one understands by it something other than the factual direction of interests following with a certain probability from the class situation for a certain average of those people subjected to the class situation. The class situation and other circumstances remaining the same, the direction in which the individual worker, for instance, is likely to pursue his interests may vary widely, according to whether he is constitutionally qualified for the task at hand to a high, to an average, or to a low degree. In the same way, the direction of interests may vary according to whether or not social action of a larger or smaller portion of those commonly affected by the class situation, or even an association among them, e.g., a trade union, has grown out of the class situation, from which the individual may expect promising results for himself. The emergence of an association or even of mere social action from a common class situation is by no means a universal phenomenon.

The class situation may be restricted in its efforts to the generation of essentially *similar* reactions, that is to say, within our terminology, of "mass behavior." However, it may not even have this result. Furthermore, often merely amorphous social action emerges. For example, the "grumbling" of workers known in ancient Oriental ethics: The moral disapproval of the workmaster's conduct, which in its practical significance was probably equivalent to an increasingly typical phenomenon of precisely the latest industrial development, namely, the slowdown of laborers by virtue of tacit agreement. The degree in which "social action" and possibly associations emerge from the mass behavior of the members of a class is linked to general cultural conditions, especially to those of an intellectual sort. It is also linked to the extent of the contrasts that have already evolved, and is especially linked to the transparency of the connections between the causes and the consequences of the class situation. For however different life chances may be, this fact in itself, according to all experience, by no means gives birth to "class action" (social action by the members of a class). For that, the real conditions and the results of the class situation must be distinctly recognizable. For only then the contrast of life chances can be felt not as an absolutely given fact to be accepted, but as a resultant from either (1) the given distribution of property, or (2) the structure of the concrete economic order. It is only then that people may react against the class structure not only through acts of intermittent and irrational protest, but in the form of rational association. There have been "class situations" of the first category (1), of a specifically naked and transparent sort, in the urban centers of Antiquity and during the Middle Ages; especially then when great fortunes were accumulated by factually monopolized trading in local industrial products or in foodstuffs; furthermore, under certain conditions, in the rural economy of the most diverse periods, when agriculture was increasingly exploited in a profit-making manner. The most important historical example of the second category (2) is the class situation of the modern proletariat.

D. *Types of class struggle*

Thus every class may be the carrier of any one of the innumerable possible forms of class action, but this is not necessarily so. In any case, a class does not in itself constitute a group (Gemeinschaft). To treat "class" conceptually as being equivalent to "group" leads to distortion. That men in the same class situation regularly react in mass actions to such tangible situations as economic ones in the direction of those interests that are most adequate to their average number is an important and after all simple fact for the understanding of historical events. However, this fact must not lead to that kind of pseudo-scientific operation with the concepts of class and class interests which is so frequent these days and which has found its most classic expression in the statement of a talented author, that the individual may be in error concerning his interests but that the class is infallible about its interests.

If classes as such are not groups, nevertheless class situations emerge only on the basis of social action. However, social action that brings forth class situations is not basically action among members of the identical class; it is an action among members of different classes. Social actions that directly determine the class situation of the worker and the entrepreneur are: the labor market, the commodities market, and the capitalistic enterprise. But, in its turn, the existence of a capitalistic enterprise presupposes that a very specific kind of social action exists to protect the possession of goods *per se,* and especially the power of individuals to dispose, in principle freely, over the means of production: a certain kind of legal order. Each kind of class situation, and above all when it rests upon the power of property *per se,* will become most clearly efficacious when all other determinants of reciprocal relations are, as far as possible, eliminated in their significance. It is in this way that the use of the power of property in the market obtains its most sovereign importance.

Now status groups hinder the strict carrying through of the sheer market principle. In the present context they are of interest only from this one point of view. Before we briefly consider them, note that not much of a general nature can be said about the more specific kinds of antagonism between classes (in our meaning of the term). The great shift, which has been going on continuously in the past, and up to our times, may be summarized, although at a cost of some precision: the struggle in which class situations are effective has progressively shifted from consumption credit toward, first, competitive struggles in the commodity market and then toward wage disputes on the labor market. The class struggles of Antiquity—to the extent that they were genuine class struggles and not struggles between status groups—were initially carried on by peasants and perhaps also artisans threatened by debt bondage and struggling against urban creditors. . . .

The propertyless of Antiquity and of the Middle Ages protested against monopolies, pre-emption, forestalling, and the withholding of goods from the market in order to raise prices. Today the central issue is the determination of the price of labor. The transition is represented by the fight for access to the market and for the determination of the price of products. Such fights went on between merchants and workers in the putting-out system of domestic handicraft during the transition to modern times. Since it is quite a general phenomenon we must mention here that the class antagonisms that are conditioned through the market situations are usually most bitter between those who actually and directly participate as opponents in price wars. It is not the rentier, the share-holder, and the banker who suffer the ill will of the worker, but almost exclusively the manufacturer and the business executives who are the direct opponents of workers in wage conflicts. This is so in spite of the fact that it is precisely the cash boxes of the rentier, the shareholder, and the banker into which the more or less unearned gains flow, rather than into the pockets of the manufacturers or of the business executives. This simple state of affairs has very frequently been decisive for the role the class situation has played in the formation of political parties. For example, it has made possible the varieties of patriarchal socialism and the frequent attempts—formerly, at least—of threatened status groups to form alliances with the proletariat against the bourgeoisie.

E. *Status honor*

In contrast to classes, *Stände* (status groups) are normally groups. They are, however, often of an amorphous kind. In contrast to the purely economically determined "class situation," we wish to designate as *status situation* every typical component of the life of men that is determined by a specific, positive or negative, social estimation of *honor*. This honor may be connected with any quality shared by a plurality, and, of course, it can be knit to a class situation: class distinctions are linked in the most varied ways with status distinctions. Property as such is not always recognized as a status qualification, but in the long run it is, and with extraordinary regularity. In the subsistence economy of neighborhood associations, it is often simply the richest who is the "chieftain." However, this often is only an honorific preference. For example, in the so-called pure modern democracy, that is, one devoid of any expressly ordered status privileges for individuals, it may be that only the families coming under approximately the same tax class dance with one another. This example is reported of certain smaller Swiss cities. But status honor need not necessarily be linked with a class situation. On the contrary, it normally stands in sharp opposition to the pretensions of sheer property.

Both propertied and propertyless people can belong to the same status group, and frequently they do with very tangible consequences. This equality of social esteem may, however, in the long run become quite precarious. The equality of status among American gentlemen, for instance, is expressed by the fact that outside the subordination determined by the different functions of business, it would be considered strictly repugnant—wherever the old tradition still prevails—if even the richest boss, while playing billiards or cards in his club would not treat his clerk as in every sense fully his equal in birthright, but would bestow upon him the condescending status-conscious "benevolence" which the German boss can never dissever from his attitude. This is one of the most important reasons why in America the German clubs have never been able to attain the attraction that the American clubs have.

In content, status honor is normally expressed by the fact that above all else a specific *style of life* is expected from all those who wish to belong to the circle. Linked with this expectation are restrictions on social intercourse (that is, intercourse which is not subservient to economic or any other purposes). These restrictions may confine normal marriages to within the status circle and may lead to complete endogamous closure. Whenever this is not a mere individual and socially irrelevant imitation of another style of life, but consensual action of this closing character, the status development is under way.

In its characteristic form, stratification by status groups on the basis of conventional styles of life evolves at the present time in the United States out of the traditional democracy. For example, only the resident of a certain street ("the Street") is considered as belonging to "society," is qualified for social intercourse, and is visited and invited. Above all, this differentiation evolves in such a way as to make for strict submission to the fashion that is dominant at a given time in society. This submission to fashion also exists among men in America to a degree unknown in Germany; it appears as an indication of the fact that a given man puts forward a *claim* to qualify as a gentleman. This submission decides, at least *prima facie,* that he will be treated as such. And this recognition becomes just as important for his employment chances in swank establishments, and above all, for social intercourse and marriage with "esteemed" families, as the qualification for dueling among Germans. As for the rest, status honor is usurped by certain families resident for a long time, and, of course, correspondingly wealthy (e.g., F.F.V., the First Families of Virginia), or by the actual or alleged descendants of the "Indian Princess" Pocahontas, of the Pilgrim fathers, or of the Knickerbockers, the members of almost inaccessible sects and all sorts of circles setting themselves apart by means of any other characteristics and badges. In this case stratification is purely conventional and rests largely on usurpation (as does almost all status honor in its beginning). But the road to legal privilege, positive or negative, is easily traveled as soon as a certain stratification of the social order has in fact been "lived in" and has achieved stability by virtue of a stable distribution of economic power.

F. *Ethnic segregation and caste*

Where the consequences have been realized to their full extent, the status group evolves into a closed caste. Status distinctions are then guaranteed not merely by conventions and laws, but also by religious sanctions. This occurs in such a way that every physical contact with a member of any caste that is considered to be lower by the members of a higher caste is considered as making for a ritualistic impurity and a stigma which must be expiated by a religious act. In addition, individual castes develop quite distinct cults and gods.

In general, however, the status structure reaches such extreme consequences only where there are underlying differences which are held to be "ethnic." The caste is, indeed, the normal form in which ethnic communities that believe in blood relationship and exclude exogamous marriage and social intercourse usually associate with one another. Such a caste situation is part of the phenomenon of pariah peoples and is found all over the world. These people form communities, acquire specific occupational traditions of handicrafts or of other arts, and cultivate a belief in their ethnic community. They live in a diaspora strictly segregated from all personal intercourse, except that of an unavoidable sort, and their situation is legally precarious. Yet, by virtue of their economic indispensability, they are tolerated, indeed frequently privileged, and they live interspersed in the political communities. The Jews are the most impressive historical example.

A status segregation grown into a caste differs in its structure from a mere ethnic segregation: the caste structure transforms the horizontal and unconnected coexistences of ethnically segregated groups into a vertical social system of super- and subordination. Correctly formulated: a comprehensive association integrates the ethnically divided communities into one political unit. They differ precisely in this way: ethnic coexistence, based on mutual repulsion and disdain, allows each ethnic community to consider its own honor as the highest one; the caste structure brings about a social subordination and an acknowledgement of "more honor" in favor of the privileged caste and status groups. This is due to the fact that in the caste structure ethnic distinctions as such have become "functional" distinctions within the political association (warriors, priests, artisans that are politically important for war and for building, and so on). But even pariah peoples who are most despised (for example, the Jews) are usually apt to continue cultivating the belief in their own specific "honor," a belief that is equally peculiar to ethnic and to status groups.

However, with the negatively privileged status groups the sense of dignity takes a specific deviation. A sense of dignity is the precipitation in individuals of social honor and of conventional demands which a positively privileged status group raises for the deportment of its members. The sense of dignity that characterizes positively privileged status groups is naturally related to their "being" which does not transcend itself, that is, it is related to their "beauty and excellence" (ὅικεῖον ἔργον). Their kingdom is "of this world." They live for the present and by exploiting their great past. The sense of dignity of the negatively privileged strata naturally refers to a future lying beyond the present, whether it is of this life or of another. In other words, it must be nurtured by the belief in a providential mission and by a belief in a specific honor before God. The chosen people's dignity is nurtured by a belief either that in the beyond "the last will be the first," or that in this life a Messiah will appear to bring forth into the light of the world which has cast them out the hidden honor of the pariah people. This simple state of affairs, and not the resentment which is so strongly emphasized in Nietzsche's much-admired construction in the *Genealogy of Morals,* is the source of the religiosity cultivated by pariah status groups. . . .

For the rest, the development of status groups from ethnic segregations is by no means the normal phenomenon. On the contrary. Since objective "racial differences" are by no means behind every subjective sentiment of an ethnic community, the question of an ultimately racial foundation of status structure is rightly a question of the concrete individual case. Very frequently a status group is instrumental in the production of a thoroughbred anthropological type. Certainly status groups are to a high degree effective in producing extreme types, for they select personally qualified individuals (e.g., the knighthood selects those who are fit for warfare, physically and psychically). But individual selection is far from being the only, or the predominant, way in which status groups are formed: political membership or class situation has at all times been at least as frequently decisive. And today the class situation is by far the predominant factor. After all, the possibility of a style of life expected for members of a status group is usually conditioned economically.

G. *Status privileges*

For all practical purposes, stratification by status goes hand in hand with a monopolization of ideal and material goods or opportunities, in a manner we have come to know as typical. Besides the specific status honor, which always rests upon distance and exclusiveness, honorific preferences may consist of the privilege of wearing special costumes, of eating special dishes taboo to others, of carrying arms—which is most obvious in its consequences—, the right to be a dilettante, for example, to play certain musical instruments. However, material monopolies provide the most effective motives for the exclusiveness of a status group; although, in themselves, they are rarely sufficient, almost always they come into play to some extent. Within a status circle there is the question of intermarriage: the interest of the families in the monopolization of potential bridegrooms is at least of equal importance and is parallel to the interest in the monopolization of daughters. The daughters of the members must be provided for. With an increased closure of the status group, the conventional preferential opportunities for special employment grow into a legal monopoly of special offices for the members. Certain goods become objects for monopolization by status groups, typically, entailed estates, and frequently also the possession of serfs or bondsmen and, finally, special trades. This monopolization occurs positively when the status group is exclusively entitled to own and to manage them; and negatively when, in order to maintain its specific way of life, the status group must not own and manage them. For the decisive role of a style of life in status honor means that status groups are the specific bearers of all conventions. In whatever way it may be manifest, all stylization of life either originates in status groups or is at least conserved by them. Even if the principles of status conventions differ greatly, they reveal certain typical traits, especially among the most privileged strata. Quite generally, among privileged status groups there is a status disqualification that operates against the performance of common physical labor. This disqualification is now "setting in" in America against the old tradition of esteem for labor. Very frequently every rational economic pursuit, and especially entrepreneurial activity, is looked upon as a disqualification of status. Artistic and literary activity is also considered degrading work as soon as it is exploited for income, or at least when it is connected with hard physical exertion.

An example is the sculptor working like a mason in his dusty smock as over against the painter in his salon-like studio and those forms of musical practice that are acceptable to the status group.

H. *Economic conditions and effects of status stratification.*

The frequent disqualification of the gainfully employed as such is a direct result of the principle of status stratification, and of course, of this principle's opposition to a distribution of power which is regulated exclusively through the market. These two factors operate along with various individual ones, which will be touched upon below.

We have seen above that the market and its processes knows no personal distinctions: "functional" interests dominate it. It knows nothing of honor. The status order means precisely the reverse: stratification in terms of honor and styles of life peculiar to status groups as such. The status order would be threatened at its very root if mere economic acquisition and naked economic power still bearing the stigma of its extra-status origin could bestow upon anyone who has won them the same or even greater honor as the vested interests claim for themselves. After all, given equality of status honor, property *per se* represents an addition even if it is not overtly acknowledged to be such. Therefore all groups having interest in the status order react with special sharpness precisely against the pretensions of purely economic acquisition. In most cases they react the more vigorously the more they feel themselves threatened. . . . Precisely because of the rigorous reactions against the claims of property *per se*, the "parvenu" is never accepted, personally and without reservation, by the privileged status groups, no matter how completely his style of life has been adjusted to theirs. They will only accept his descendants who have been educated in the conventions of their status group and who have never besmirched its honor by their own economic labor.

As to the general *effect* of the status order, only one consequence can be stated, but it is a very important one: the hindrance of the free development of the market. This occurs first for those goods that status groups directly withhold from free exchange by monopolization, which may be effected either legally or conventionally. For example, in many Hellenic cities during the "status era" and also originally in Rome, the inherited estate (as shown by the old formula for placing spendthrifts under a guardian) was monopolized, as were the estates of knights, peasants, priests, and especially the clientele of the craft and merchant guilds. The market is restricted, and the power of naked property *per se,* which gives its stamp to class formation, is pushed into the background. The results of this process can be most varied. Of course, they do not necessarily weaken the contrasts in the economic situation. Frequently they strengthen these contrasts, and in any case, where stratification by status permeates a community as strongly as was the case in all political communities of Antiquity and of the Middle Ages, one can never speak of a genuinely free market competition as we understand it today. There are wider effects than this direct exclusion of special goods from the market. From the conflict between the status order and the purely economic order mentioned above, it follows that in most instances the notion of honor peculiar to status absolutely abhors that which is essential to the market: hard bargaining. Honor abhors hard bargaining among peers and occasionally it taboos it for the members of a status group in general. Therefore, everywhere some status groups, and usually the most influential, consider almost any kind of overt participation in economic acquisition as absolutely stigmatizing.

With some over-simplification, one might thus say that classes are stratified according to their relations to the production and acquisition of goods; whereas status groups are stratified according to the principles of their *consumption* of goods as represented by special styles of life.

An "occupational status group," too, is a status group proper. For normally, it successfully claims social honor only by virtue of the special style of life which may be determined by it. The differences between classes and status groups frequently overlap. It is precisely those status communities most strictly segregated in terms of honor (viz., the Indian castes) who today show, although within very rigid limits, a relatively high degree of indifference to pecuniary income. However, the Brahmins seek such income in many different ways.

As to the general economic conditions making for the predominance of stratification by status, only the following can be said. When the bases of the acquisition and distribution of goods are relatively stable, stratification by status is favored. Every technological repercussion and economic transformation threatens stratification by status and pushes the class situation into the foreground. Epochs and countries in which the naked class situation is of predominant significance are regularly the periods of technical and economic transformations. And every slowing down of the change in economic stratification leads, in due course, to the growth of status structures and makes for a resuscitation of the important role of social honor.

I. *Parties*

Whereas the genuine place of classes is within the economic order, the place of status groups is within the social order, that is, within the sphere of the distribution of honor. From within these spheres, classes and status groups influence one another and the legal order and are in turn influenced by it. *"Parties"* reside in the sphere of power. Their action is oriented toward the acquisition of social power, that is to say, toward influencing social action no matter what its content may be. In principle, parties may exist in a social club as well as in a state. As over against the actions of classes and status groups, for which this is not necessarily the case, party-oriented social action always involves association. For it is always directed toward a goal which is striven for in a planned manner. This goal may be a cause (the party may aim at realizing a program for ideal or material purposes), or the goal may be personal (sinecures, power, and from these, honor for the leader and the followers of the party). Usually the party aims at all these simultaneously. Parties are, therefore, only possible within groups that have an associational character, that is, some rational order and a staff of persons available who are ready to enforce it. For parties aim precisely at influencing this staff, and if possible, to recruit from it party members.

In any individual case, parties may represent interests determined through class situation or status situation, and they may recruit their following respectively from one or the other. But they need be neither purely class nor purely status parties; in fact, they are more likely to be mixed types, and sometimes they are neither. They may represent ephemeral or enduring structures. Their means of attaining power may be quite varied, ranging from naked violence of any sort to canvassing for votes with coarse or subtle means: money, social influence, the force of speech, suggestion, clumsy hoax, and so on to the rougher or more artful tactics of obstruction in parliamentary bodies.

The sociological structure of parties differs in a basic way according to the kind of social action which they struggle to influence; that means, they differ according to whether or not the community is stratified by status or by classes. Above all else, they vary according to the structure of domination. For their leaders normally deal with its conquest. In our general terminology, parties are not only products of modern forms of domination. We shall also designate as parties the ancient and medieval ones, despite the fact that they differ basically from modern parties. Since a party always struggles for political control (*Herrschaft*), its organization too is frequently strict and "authoritarian." Because of these variations between the forms of domination, it is impossible to say anything about the structure of parties without discussing them first. Therefore, we shall now turn to this central phenomenon of all social organization.

Before we do this, we should add one more general observation about classes, status groups and parties: The fact that they presuppose a larger association, especially the framework of a polity, does not mean that they are confined to it. On the contrary, at all times it has been the order of the day that such association (even when it aims at the use of military force in common) reaches beyond the state boundaries. This can be seen in the [interlocal] solidarity of interests of oligarchs and democrats in Hellas, of Guelphs and Ghibellines in the Middle Ages, and within the Calvinist party during the age of religious struggles; and all the way up to the solidarity of landlords (International Congresses of Agriculture), princes (Holy Alliance, Karlsbad Decrees [of 1819]), socialist workers, conservatives (the longing of Prussian conservatives for Russian intervention in 1850). But their aim is not necessarily the establishment of a new territorial dominion. In the main they aim to influence the existing polity.

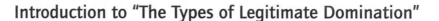

Introduction to "The Types of Legitimate Domination"

In this selection, Weber defines three "ideal types" of legitimate domination: **rational** or **legal authority**, **traditional authority**, and **charismatic authority**. (See Table 4.2.) As abstract constructs, none of the ideal types actually exists in pure form. Instead, public authority is based on some mixture of the three types. Nevertheless, social systems generally exhibit a predominance of one form or another of domination.

Before briefly describing the forms of legitimate authority, we first need to clarify Weber's definition of legitimacy. By "legitimacy," Weber was referring to the belief systems on which valid commands issuing from authority figures are based. Such belief systems supply the justifications and motives for demanding obedience and allow those in authority to rightfully exercise domination over others. It is to these justifications that authority figures turn when seeking to legitimate their actions and the actions of those subjected to their commands.

Modern states are ruled through rational-legal authority. This form of domination is based on the rule of rationally established laws. Legitimacy thus rests "on a belief in the legality of enacted rules and the right of those elevated to authority under such rules to issue commands" (Weber [1925c] 1978:215). Obedience is owed not to the person who exercises authority, but to the office or position in which authority is vested. It is the impersonal, legal order that vests the superior with the authority to demand compliance, a right that is ceded on vacating the office. Once retired, a police officer or judge is but another civilian and as such no longer has the power to enforce the law.

Photo 4.1a William Jefferson Clinton, the 42nd President of the United States.

Traditional authority is the authority of "eternal yesterday." It rests on an "established belief in the sanctity of immemorial traditions" (ibid.:215). This is the rule of kings and tribal chieftains. Leadership is attained not on the basis of impersonally measured merit, but on lines of heredity or rites of passage. Subjects owe their allegiance not to bureaucratically imposed rules and laws that are open to change, but to their personal "master" whose demands for compliance and loyalty are legitimated by sacred, inviolable traditions.

Weber's third type of authority derives from the **charisma** possessed by the leader. Demands for obedience are legitimated by the leader's "gift of grace," which is demonstrated through extraordinary feats, acts of heroism, or revelations—in short, the miracles of heroes and prophets. Like traditional authority, loyalty is owed to the person and not to an office defined through impersonal rules. But unlike traditional authority, legitimacy is not based on appeals to sacred traditions or on the exalting of "what has always been." Instead, compliance from "disciples" is demanded on the basis of the "conception that it is the duty of those subject to charismatic authority to recognize its genuineness and to act accordingly" ([1925c] 1978:242).

History is replete with charismatic leaders who have inspired intense personal devotion to themselves and their cause. From Jesus and Muhammad, Joan of Arc and Gandhi, to Napoleon and Hitler, such leaders have proved to be a powerful force for social change, both good and bad. Indeed, in its rejection of both tradition and rational, formal rules, charismatic authority, by its very nature, poses a challenge to existing political order. In breaking from history as well as objective laws, charisma is a creative force that carries the commandment: "It is written, but I say unto you."

However, the revolutionary potential of charismatic authority makes it inherently unstable. Charisma lasts only as long as its possessor is able to provide benefits to his followers. If the leader's prophecies are proved wrong, if enemies are not defeated, if miraculous deeds begin to "dry up," then his or her legitimacy will be called into question. On the other hand, even if such deeds or benefits provide a continued source of legitimacy, the leader at some point will die. With authority resting solely in the charismatic individual, the movement he inspired will collapse along with his rule, unless designs for a successor are developed. Often, the transferring of authority eventually leads to the "routinization of charisma" and the transformation of legitimacy into either a rational-legal or traditional type—witness the Catholic Church.

Photo 4.1b England's Queen Victoria (1819–1901)

Embodiments of legitimate domination: President Clinton exercised rational-legal authority; Queen Victoria ruled on the basis of traditional authority; Mahatma Gandhi possessed charismatic authority.

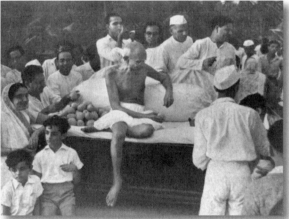

Photo 4.1c Mahatma Gandhi (1869–1948), India's Past Spiritual and Political Leader.

Table 4.2 Weber's Types of Legitimate Domination

		ORDER	
		Individual	**Collective**
ACTION	**Nonrational**	*Charismatic:* "Gift of grace" of leader	*Traditional:* "Established belief in the sanctity of immemorial traditions"
	Rational		*Rational-Legal:* "Belief in the legality of enacted rules and the right of those elevated to authority under such rules to issue commands"

"The Types of Legitimate Domination" (1925)

Max Weber

DOMINATION AND LEGITIMACY

Domination was defined as the probability that certain specific commands (or all commands) will be obeyed by a given group of persons. It thus does not include every mode of exercising "power" or "influence" over other persons. Domination ("authority") in this sense may be based on the most diverse motives of compliance: all the way from simple habituation to the most purely rational calculation of advantage. Hence every genuine form of domination implies a minimum of voluntary compliance, that is, an *interest* (based on ulterior motives or genuine acceptance) in obedience.

Not every case of domination makes use of economic means; still less does it always have economic objectives. However, normally the rule over a considerable number of persons requires a staff, that is, a *special* group which can normally be trusted to execute the general policy as well as the specific commands. The members of the administrative staff may be bound to obedience to their superior (or superiors) by custom, by affectual ties, by a purely material complex of interests, or by ideal (*wertrationale*) motives. The quality of these motives largely determines the type of domination. *Purely* material interests and calculations of advantages as the basis of solidarity between the chief and his administrative staff result, in this as in other connexions, in a relatively unstable situation. Normally other elements, affectual and ideal, supplement such interests. In certain exceptional cases the former alone may be decisive. In everyday life these relationships, like others, are governed by custom and material calculation of advantage. But custom, personal advantage, purely affectual or ideal motives of solidarity, do not form a sufficiently reliable basis for a given domination. In addition there is normally a further element, the belief in *legitimacy*.

Experience shows that in no instance does domination voluntarily limit itself to the appeal to material or affectual or ideal motives as a basis for its continuance. In addition every such system attempts to establish and to cultivate the belief in its legitimacy. But according to the kind of legitimacy which is claimed, the type of obedience, the kind of administrative staff developed to guarantee it, and the mode of exercising authority, will all differ fundamentally. Equally fundamental is the variation in effect. Hence, it is useful to classify the types of domination according to the kind of claim to legitimacy typically made by each. In doing this, it is best to start from modern and therefore more familiar examples. . . .

The Three Pure Types of Authority

There are three pure types of legitimate domination. The validity of the claims to legitimacy may be based on:

1. Rational grounds—resting on a belief in the legality of enacted rules and the right of those elevated to authority under such rules to issue commands (legal authority).

2. Traditional grounds—resting on an established belief in the sanctity of immemorial traditions and the legitimacy of those exercising authority under them (traditional authority); or finally,

3. Charismatic grounds—resting on devotion to the exceptional sanctity, heroism or exemplary character of an individual person, and of the normative patterns or order revealed or ordained by him (charismatic authority).

In the case of legal authority, obedience is owed to the legally established impersonal order. It extends to the persons exercising the authority of office under it by virtue of the formal legality of their commands and only within the scope of authority of the office. In the case of traditional authority, obedience is owed to the *person* of the chief who occupies the traditionally sanctioned position of authority and who is (within its sphere) bound by tradition. But here the obligation of obedience is a matter of personal loyalty within the

SOURCE: Originally published in *The Theory of Social and Economic Organization* by Max Weber. Translated by A. M. Henderson and Talcott Parsons, edited with an introduction by Talcott Parsons. Copyright © 1947 by The Free Press, Copyright renewed © 1975 by Talcott Parons. As appears, with revisions, in *Economy and Society,* Volume 1 by Max Weber. Edited by Guenther Roth and Claus Wittich. Copyright © 1978 The Regents of the University of California. Reprinted with permission.

area of accustomed obligations. In the case of charismatic authority, it is the charismatically qualified leader as such who is obeyed by virtue of personal trust in his revelation, his heroism or his exemplary qualities so far as they fall within the scope of the individual's belief in his charisma. . . .

LEGAL AUTHORITY WITH A BUREAUCRATIC STAFF

Legal Authority: The Pure Type

Legal authority rests on the acceptance of the validity of the following mutually inter-dependent ideas.

1. That any given legal norm may be established by agreement or by imposition, on grounds of expediency or value-rationality or both, with a claim to obedience at least on the part of the members of the organization. This is, however, usually extended to include all persons within the sphere of power in question—which in the case of territorial bodies is the territorial area—who stand in certain social relationships or carry out forms of social action which in the order governing the organization have been declared to be relevant.

2. That every body of law consists essentially in a consistent system of abstract rules which have normally been intentionally established. Furthermore, administration of law is held to consist in the application of these rules to particular cases; the administrative process in the rational pursuit of the interests which are specified in the order governing the organization within the limits laid down by legal precepts and following principles which are capable of generalized formulation and are approved in the order governing the group, or at least not disapproved in it.

3. That thus the typical person in authority, the "superior," is himself subject to an impersonal order by orienting his actions to it in his own dispositions and commands. (This is true not only for persons exercising legal authority who are in the usual sense "officials," but, for instance, for the elected president of a state.)

4. That the person who obeys authority does so, as it is usually stated, only in his capacity as a "member" of the organization and what he obeys is only "the

law." (He may in this connection be the member of an association, of a community, of a church, or a citizen of a state.)

5. In conformity with point 3, it is held that the members of the organization, insofar as they obey a person in authority, do not owe this obedience to him as an individual, but to the impersonal order. Hence, it follows that there is an obligation to obedience only within the sphere of the rationally delimited jurisdiction which, in terms of the order, has been given to him. . . .

The purest type of exercise of legal authority is that which employs a bureaucratic administrative staff. Only the supreme chief of the organization occupies his position of dominance (*Herrenstellung*) by virtue of appropriation, of election, or of having been designated for the succession. But even *his* authority consists in a sphere of legal "competence." The whole administrative staff under the supreme authority then consist, in the purest type, of individual officials (constituting a "monocracy" as opposed to the "collegial" type, which will be discussed below) who are appointed and function according to the following criteria:

1. They are personally free and subject to authority only with respect to their impersonal official obligations.

2. They are organized in a clearly defined hierarchy of offices.

3. Each office has a clearly defined sphere of competence in the legal sense.

4. The office is filled by a free contractual relationship. Thus, in principle, there is free selection.

5. Candidates are selected on the basis of technical qualifications. In the most rational case, this is tested by examination or guaranteed by diplomas certifying technical training, or both. They are *appointed*, not elected.

6. They are remunerated by fixed salaries in money, for the most part with a right to pensions. Only under certain circumstances does the employing authority, especially in private organizations, have a right to terminate the appointment, but the official is always free to resign. The salary scale is graded according to rank in the hierarchy; but in addition to this criterion, the responsibility of

the position and the requirements of the incumbent's social status may be taken into account.

7. The office is treated as the sole, or at least the primary, occupation of the incumbent.

8. It constitutes a career. There is a system of "promotion" according to seniority or to achievement, or both. Promotion is dependent on the judgment of superiors.

9. The official works entirely separated from ownership of the means of administration and without appropriation of his position.

10. He is subject to strict and systematic discipline and control in the conduct of the office.

This type of organization is in principle applicable with equal facility to a wide variety of different fields. It may be applied in profit-making business or in charitable organizations, or in any number of other types of private enterprises serving ideal or material ends. It is equally applicable to political and to hierocratic organizations. With the varying degrees of approximation to a pure type, its historical existence can be demonstrated in all these fields. . . .

TRADITIONAL AUTHORITY: THE PURE TYPE

Authority will be called traditional if legitimacy is claimed for it and believed in by virtue of the sanctity of age-old rules and powers. The masters are designated according to traditional rules and are obeyed because of their traditional status (*Eigenwürde*). This type of organized rule is, in the simplest case, primarily based on personal loyalty which results from common upbringing. The person exercising authority is not a "superior," but a personal master, his administrative staff does not consist mainly of officials but of personal retainers, and the ruled are not "members" of an association but are either his traditional "comrades" or his "subjects." Personal loyalty, not the official's impersonal duty, determines the relations of the administrative staff to the master.

Obedience is owed not to enacted rules but to the person who occupies a position of authority by tradition or who has been chosen for it by the traditional master. The commands of such a person are legitimized in one of two ways:

A. partly in terms of traditions which themselves directly determine the content of the command and are believed to be valid within certain limits that cannot be overstepped without endangering the master's traditional status;

B. partly in terms of the master's discretion in that sphere which tradition leaves open to him; this traditional prerogative rests primarily on the fact that the obligations of personal obedience tend to be essentially unlimited.

Thus there is a double sphere:

A. that of action which is bound to specific traditions;

B. that of action which is free of specific rules.

In the latter sphere, the master is free to do good turns on the basis of his personal pleasure and likes, particularly in return for gifts—the historical sources of dues (*Gebühren*). So far as his action follows principles at all, these are governed by considerations of ethical common sense, of equity or of utilitarian expediency. They are not formal principles, as in the case of legal authority. The exercise of power is oriented toward the consideration of how far master and staff can go in view of the subjects' traditional compliance without arousing their resistance. When resistance occurs, it is directed against the master or his servant personally, the accusation being that he failed to observe the traditional limits of his power. Opposition is not directed against the system as such—it is a case of "traditionalist revolution."

In the pure type of traditional authority it is impossible for law or administrative rule to be deliberately created by legislation. Rules which in fact are innovations can be legitimized only by the claim that they have been "valid of yore," but have only now been recognized by means of "Wisdom" [the *Weistum* of ancient Germanic law]. Legal decisions as "finding of the law" (*Rechtsfindung*) can refer only to documents of tradition, namely to precedents and earlier decisions. . . .

In the pure type of traditional rule, the following features of a bureaucratic administrative staff are absent:

A. a clearly defined sphere of competence subject to impersonal rules,

B. a rationally established hierarchy,

C. a regular system of appointment on the basis of free contract, and orderly promotion,

D. technical training as a regular requirement,

E. (frequently) fixed salaries, in the type case paid in money. . . .

CHARISMATIC AUTHORITY

The term "charisma" will be applied to a certain quality of an individual personality by virtue of which he is considered extraordinary and treated as endowed with supernatural, superhuman, or at least specifically exceptional powers or qualities. These are such as are not accessible to the ordinary person, but are regarded as of divine origin or as exemplary, and on the basis of them the individual concerned is treated as a "leader." In primitive circumstances this peculiar kind of quality is thought of as resting on magical powers, whether of prophets, persons with a reputation for therapeutic or legal wisdom, leaders in the hunt, or heroes in war. How the quality in question would be ultimately judged from any ethical, aesthetic, or other such point of view is naturally entirely indifferent for purposes of definition. What is alone important is how the individual is actually regarded by those subject to charismatic authority, by his "followers" or "disciples." . . .

I. It is recognition on the part of those subject to authority which is decisive for the validity of charisma. This recognition is freely given and guaranteed by what is held to be a proof, originally always a miracle, and consists in devotion to the corresponding revelation, hero worship, or absolute trust in the leader. But where charisma is genuine, it is not this which is the basis of the claim to legitimacy. This basis lies rather in the conception that it is the duty of those subject to charismatic authority to recognize its genuineness and to act accordingly. Psychologically this recognition is a matter of complete personal devotion to the possessor of the quality, arising out of enthusiasm, or of despair and hope. . . .

II. If proof and success elude the leader for long, if he appears deserted by his god or his magical or heroic powers, above all, if his leadership fails to benefit his followers, it is likely that his charismatic authority will disappear. This is the genuine meaning of the divine right of kings (*Gottesgnadentum*). . . .

III. An organized group subject to charismatic authority will be called a charismatic community (*Gemeinde*). It is based on an emotional form of communal relationship (*Vergemeinschaftung*). The administrative staff of a charismatic leader does not consist of "officials"; least of all are its members technically trained. It is not chosen on the basis of social privilege nor from the point of view of domestic or personal dependency. It is rather chosen in terms of the charismatic qualities of its members. The prophet has his disciples; the warlord his bodyguard; the leader, generally, his agents (*Vertrauensmänner*). There is no such thing as appointment or dismissal, no career, no promotion. There is only a call at the instance of the leader on the basis of the charismatic qualification of those he summons. There is no hierarchy; the leader merely intervenes in general or in individual cases when he considers the members of his staff lacking in charismatic qualification for a given task. There is no such thing as a bailiwick or definite sphere of competence, and no appropriation of official powers on the basis of social privileges. There may, however, be territorial or functional limits to charismatic powers and to the individual's mission. There is no such thing as a salary or a benefice.

Disciples or followers tend to live primarily in a communistic relationship with their leader on means which have been provided by voluntary gift. There are no established administrative organs. In their place are agents who have been provided with charismatic authority by their chief or who possess charisma of their own. There is no system of formal rules, of abstract legal principles, and hence no process of rational judicial decision oriented to them. But equally there is no legal wisdom oriented to judicial precedent. Formally concrete judgments are newly created from case to case and are originally regarded as divine judgments and revelations. From a substantive point of view, every charismatic authority would have to subscribe to the proposition, "It is written . . . but I say unto you . . ." The genuine prophet, like the genuine military leader and every true leader in this sense, preaches, creates, or demands *new* obligations—most typically, by virtue of revelation, oracle, inspiration, or of his own will, which are recognized by the members of the religious, military, or party group because they come from such a source. Recognition is a duty. When such an authority comes into conflict with the competing authority of another who also claims charismatic sanction, the only recourse is to some kind of a contest, by magical means or an actual physical battle of the leaders. In principle, only one side can be right in such a conflict; the other must be guilty of a wrong which has to be expiated.

Since it is "extra-ordinary," charismatic authority is sharply opposed to rational, and particularly bureaucratic, authority, and to traditional authority, whether in its patriarchal, patrimonial, or estate variants, all of which are everyday forms of domination; while the charismatic type is the direct antithesis of this. Bureaucratic authority is specifically rational in the sense of being bound to intellectually analysable rules; while charismatic authority is specifically irrational in the sense of being foreign to all rules. Traditional authority is bound to the precedents handed down from the past and to this extent is also oriented to rules. Within the sphere of its claims, charismatic authority repudiates the past, and is in this sense a specifically revolutionary force. It recognizes no appropriation of positions of power by virtue of the possession of property, either on the part of a chief or of socially privileged groups. The only basis of legitimacy for it is personal charisma so long as it is proved; that is, as long as it receives recognition and as long as the followers and disciples prove their usefulness charismatically. . . .

IV. Pure charisma is specifically foreign to economic considerations. Wherever it appears, it constitutes a "call" in the most emphatic sense of the word, a "mission" or a "spiritual duty." In the pure type, it disdains and repudiates economic exploitation of the gifts of grace as a source of income, though, to be sure, this often remains more an ideal than a fact. It is not that charisma always demands a renunciation of property or even of acquisition, as under certain circumstances prophets and their disciples do. The heroic warrior and his followers actively seek booty; the elective ruler or the charismatic party leader requires the material means of power. The former in addition requires a brilliant display of his authority to bolster his prestige. What is despised, so long as the genuinely charismatic type is adhered to, is traditional or rational everyday economizing, the attainment of a regular income by continuous economic activity devoted to this end. Support by gifts, either on a grand scale involving donation, endowment, bribery and honoraria, or by begging, constitute the voluntary type of support. On the other hand, "booty" and extortion, whether by force or by other means, is the typical form of charismatic provision for needs. From the point of view of rational economic activity, charismatic want satisfaction is a typical anti-economic force. It repudiates any sort of involvement in the everyday routine world. It can only tolerate, with an attitude of complete emotional indifference, irregular, unsystematic acquisitive acts. In that it relieves the recipient of economic concerns, dependence on property income can be the economic basis of a charismatic mode of life for some groups; but that is unusual for the normal charismatic "revolutionary." . . .

V. In traditionalist periods, charisma is *the* great revolutionary force. The likewise revolutionary force of "reason" works from *without:* by altering the situations of life and hence its problems, finally in this way changing men's attitudes toward them; or it intellectualizes the individual. Charisma, on the other hand, *may* effect a subjective or *internal* reorientation born out of suffering, conflicts, or enthusiasm. It may then result in a radical alteration of the central attitudes and directions of action with a completely new orientation of all attitudes toward the different problems of the "world." In prerationalistic periods, tradition and charisma between them have almost exhausted the whole of the orientation of action.

The Routinization of Charisma

In its pure form charismatic authority has a character specifically foreign to everyday routine structures. The social relationships directly involved are strictly personal, based on the validity and practice of charismatic personal qualities. If this is not to remain a purely transitory phenomenon, but to take on the character of a permanent relationship, a "community" of disciples or followers or a party organization or any sort of political or hierocratic organization, it is necessary for the character of charismatic authority to become radically changed. Indeed, in its pure form charismatic authority may be said to exist only *in statu nascendi.* It cannot remain stable, but becomes either traditionalized or rationalized, or a combination of both.

The following are the principal motives underlying this transformation: (a) The ideal and also the material interests of the followers in the continuation and the continual reactivation of the community, (b) the still stronger ideal and also stronger material interests of the members of the administrative staff, the disciples, the party workers, or others in continuing their relationship. Not only this, but they have an interest in continuing it in such a way that both from an ideal and a material point of view, their own position is put on a stable everyday basis. This means, above all, making it possible to participate in normal family relationships or at least to enjoy a secure social position in place of the kind of discipleship which is cut off from ordinary worldly connections, notably in the family and in economic relationships.

These interests generally become conspicuously evident with the disappearance of the personal charismatic leader and with the problem of *succession*. The way in which this problem is met—if it is met at all and the charismatic community continues to exist or now begins to emerge—is of crucial importance for the character of the subsequent social relationships. . . .

Concomitant with the routinization of charisma with a view to insuring adequate succession, go the interests in its routinization on the part of the administrative staff. It is only in the initial stages and so long as the charismatic leader acts in a way which is completely outside everyday social organization, that it is possible for his followers to live communistically in a community of faith and enthusiasm, on gifts, booty, or sporadic acquisition. Only the members of the small group of enthusiastic disciples and followers are prepared to devote their lives purely idealistically to their call. The great majority of disciples and followers will in the long run "make their living" out of their "calling" in a material sense as well. Indeed, this must be the case if the movement is not to disintegrate.

Hence, the routinization of charisma also takes the form of the appropriation of powers and of economic advantages by the followers or disciples, and of regulating recruitment. This process of traditionalization or of legalization, according to whether rational legislation is involved or not, may take any one of a number of typical forms. . . .

For charisma to be transformed into an everyday phenomenon, it is necessary that its anti- economic character should be altered. It must be adapted to some form of fiscal organization to provide for the needs of the group and hence to the economic conditions necessary for raising taxes and contributions. When a charismatic movement develops in the direction of prebendal provision, the "laity" becomes differentiated from the "clergy"—derived from κΠρος, meaning a "share"—, that is, the participating members of the charismatic administrative staff which has now become routinized. These are the priests of the developing "church." Correspondingly, in a developing political body—the "state" in the rational case—vassals, benefice-holders, officials or appointed party officials (instead of voluntary party workers and functionaries) are differentiated from the "tax payers." . . .

It follows that, in the course of routinization, the charismatically ruled organization is largely transformed into one of the everyday authorities, the patrimonial form, especially in its estate-type or bureaucratic variant. Its original peculiarities are apt to be retained in the charismatic status honor acquired by heredity or office- holding. This applies to all who participate in the appropriation, the chief himself and the members of his staff. It is thus a matter of the type of prestige enjoyed by ruling groups. A hereditary monarch by "divine right" is not a simple patrimonial chief, patriarch, or sheik; a vassal is not a mere household retainer or official. Further details must be deferred to the analysis of status groups.

As a rule, routinization is not free of conflict. In the early stages personal claims on the charisma of the chief are not easily forgotten and the conflict between the charisma of the office or of hereditary status with personal charisma is a typical process in many historical situations.

Introduction to "Bureaucracy"

In this essay, Weber defines the "ideal type" of bureaucracy, outlining its unique and most significant features. The salience of Weber's description lies in the fact that bureaucracies have become the dominant form of social organization in modern society. Indeed, bureaucracies are indispensable to modern life. Without them, a multitude of necessary tasks could not be performed with the degree of efficiency required for serving large numbers of individuals. For instance, strong and effective armies could not be maintained, the mass production of goods and their sale would slow to a trickle, the thousands of miles of public roadways could not be paved, hospitals could not treat the millions of patients in need of care, and establishing a university capable of educating 20,000 students would be impossible. Of course, all of these tasks and countless others are themselves dependent on a bureaucratic organization capable of collecting tax dollars from millions of people.

Despite whatever failings particular bureaucracies may exhibit, the form of organization is as essential to modern life as the air we breathe. In accounting for the ascendancy of bureaucracies, Weber is clear:

> The decisive reason for the advance of bureaucratic organization has also been its purely *technical* superiority over any other form of organization. . . . Precision, speed, unambiguity, knowledge of the files, continuity, discretion, unity, strict subordination, reduction of friction and of material and personal costs—these are raised to the optimum point in the strictly bureaucratic administration. . . . As compared with all [other] forms of administration, trained bureaucracy is superior on all these points. (Weber [1925d] 1978:973, emphasis in the original)

A number of features ensure the technical superiority of bureaucracies. First, authority is hierarchically structured, making for a clear chain of command. Second, selection of personnel is competitive and based on demonstrated merit. This reduces the likelihood of incompetence that can result from appointing officials through nepotism or by virtue of tradition. Third, a specialized division of labor allows for the more efficient completion of assigned tasks. Fourth, bureaucracies are governed by formal, impersonal rules that regulate all facets of the organization. As a result, predictability of action and the strategic planning that it makes possible are better guaranteed.

As the epitome of the process of rationalization, however, Weber by no means embraced unequivocally the administrative benefits provided by bureaucracies. While in important respects, bureaucracies are dependent on the development of mass democracy for their fullest expression, they nevertheless create new elite groups of experts and technocrats. Moreover, he contended that their formal rules and procedures led to the loss of individual freedom.[1] For those working in bureaucracies (and countless do), Weber saw the individual "chained to his activity in his entire economic and ideological existence" (Weber [1925d] 1978:988). The bureaucrat adopts as his own the detached, objective attitudes on which the efficiency and predictability of bureaucracies depend. Operating "'[w]ithout regard for persons . . . [b]ureaucracy develops the more perfectly, the more it is 'dehumanized,' the more completely it succeeds in eliminating from official business love, hatred, and all purely personal, irrational, and emotional elements which escape calculation" (ibid.:975). Whether as an employee or as a client, who among us has not been confronted with the faceless impersonality of a bureaucracy immune to the "special circumstances" that, after all, make up the very essence of our individuality?

---

[1]As we noted earlier, Weber's analysis of bureaucratic organizations offers an important critique of Marx's perspective. While Marx argued that capitalism is the source of alienation in modern society, Weber saw the source lying in bureaucracies and the rational procedures they embody. Additionally, in recognizing that bureaucracies create elite groups of technocrats who pursue their own professional interests, Weber also suggested that such organizational leaders (i.e., state officials) do not necessarily advance the interests of a ruling capitalist class. A related theme can likewise be found in "Class, Status, Party."

"Bureaucracy" (1925)

Max Weber

CHARACTERISTICS OF MODERN BUREAUCRACY

Modern officialdom functions in the following manner:

I. There is the principle of official *jurisdictional areas,* which are generally ordered by rules, that is, by laws or administrative regulations. This means:

1. The regular activities required for the purposes of the bureaucratically governed structure are assigned as official duties.

2. The authority to give the commands required for the discharge of these duties is distributed in a stable way and is strictly delimited by rules concerning the coercive means, physical, sacerdotal, or otherwise, which may be placed at the disposal of officials.

3. Methodical provision is made for the regular and continuous fulfillment of these duties and for the exercise of the corresponding rights; only persons who qualify under general rules are employed.

In the sphere of the state these three elements constitute a bureaucratic *agency,* in the sphere of the private economy they constitute a bureaucratic *enterprise.* Bureaucracy, thus understood, is fully developed in political and ecclesiastical communities only in the modern state, and in the private economy only in the most advanced institutions of capitalism. Permanent agencies, with fixed jurisdiction, are not the historical rule but rather the exception. This is even true of large political structures such as those of the ancient Orient, the Germanic and Mongolian empires of conquest, and of many feudal states. In all these cases, the ruler executes the most important measures through personal trustees, table-companions, or court-servants. Their commissions and powers are not precisely delimited and are temporarily called into being for each case.

II. The principles of *office hierarchy* and of channels of appeal (*Instanzenzug*) stipulate a clearly established system of super- and subordination in which there is a supervision of the lower offices by the higher ones. Such a system offers the governed the possibility of appealing, in a precisely regulated manner, the decision of a lower office to the corresponding superior authority. With the full development of the bureaucratic type, the office hierarchy is *monocratically* organized. The principle of hierarchical office authority is found in all bureaucratic structures: in state and ecclesiastical structures as well as in large party organizations and private enterprises. It does not matter for the character of bureaucracy whether its authority is called "private" or "public."

When the principle of jurisdictional "competency" is fully carried through, hierarchical subordination—at least in public office—does not mean that the "higher" authority is authorized simply to take over the business of the "lower." Indeed, the opposite is the rule; once an office has been set up, a new incumbent will always be appointed if a vacancy occurs.

III. The management of the modern office is based upon written documents (the "files"), which are preserved in their original or draft form, and upon a staff of subaltern officials and scribes of all sorts. The body of officials working in an agency along with the respective apparatus of material implements and the files makes up a *bureau* (in private enterprises often called the "counting house," *Kontor*).

In principle, the modern organization of the civil service separates the bureau from the private domicile of the official and, in general, segregates official activity from the sphere of private life. Public monies and equipment are divorced from the private property of the official. This condition is everywhere the product of a long development. Nowadays, it is found in public as well as in private enterprises; in the latter, the principle extends even to the entrepreneur at the top. In principle, the *Kontor* (office) is separated from the household, business from private correspondence, and business assets from private wealth. The more consistently the modern type of business management has been carried through, the more are these separations the case. The beginnings of this process are to be found as early as the Middle Ages.

It is the peculiarity of the modern entrepreneur that he conducts himself as the "first official" of his enterprise, in the very same way in which the ruler of a specifically modern bureaucratic state [Frederick II of Prussia] spoke of himself as "the first servant" of the state. The idea that the bureau activities of the state are intrinsically different in character from the management of private offices is a continental European notion and, by way of contrast, is totally foreign to the American way.

IV. Office management, at least all specialized office management—and such management is distinctly modern—usually presupposes thorough training in a field of specialization. This, too, holds increasingly for the modern executive and employee of a private enterprise, just as it does for the state officials.

V. When the office is fully developed, official activity demands the *full working capacity* of the official, irrespective of the fact that the length of his obligatory working hours in the bureau may be limited. In the normal case, this too is only the product of a long development, in the public as well as in the private office. Formerly the normal state of affairs was the reverse: Official business was discharged as a secondary activity.

VI. The management of the office follows *general rules,* which are more or less stable, more or less exhaustive, and which can be learned. Knowledge of these rules represents a special technical expertise which the officials possess. It involves jurisprudence, administrative or business management.

The reduction of modern office management to rules is deeply embedded in its very nature. The theory of modern public administration, for instance, assumes that the authority to order certain matters by decree—which has been legally granted to an agency—does not entitle the agency to regulate the matter by individual commands given for each case, but only to regulate the matter abstractly. This stands in extreme contrast to the regulation of all relationships through individual privileges and bestowals of favor, which, as we shall see, is absolutely dominant in patrimonialism, at least in so far as such relationships are not fixed by sacred tradition.

The Position of the Official Within and Outside of Bureaucracy

All this results in the following for the internal and external position of the official:

I. Office Holding as a Vocation

That the office is a "vocation" (*Beruf*) finds expression, first, in the requirement of a prescribed course of training, which demands the entire working capacity for a long period of time, and in generally prescribed special examinations as prerequisites of employment. Furthermore, it finds expression in that the position of the official is in the nature of a "duty" (*Pflicht*). This determines the character of his relations in the following manner: Legally and actually, office holding is not considered ownership of a source of income, to be exploited for rents or emoluments in exchange for the rendering of certain services, as was normally the case during the Middle Ages and frequently up to the threshold of recent times, nor is office holding considered a common exchange of services, as in the case of free employment contracts. Rather, entrance into an office, including one in the private economy, is considered an acceptance of a specific duty of fealty to the purpose of the office (*Amtstreue*) in return for the grant of a secure existence. It is decisive for the modern loyalty to an office that, in the pure type, it does not establish a relationship to a *person,* like the vassal's or disciple's faith under feudal or patrimonial authority, but rather is devoted to *impersonal* and *functional* purposes. These purposes, of course, frequently gain an ideological halo from cultural values, such as state, church, community, party or enterprise, which appear as surrogates for a this-worldly or other-worldly personal master and which are embodied by a given group.

The political official—at least in the fully developed modern state—is not considered the personal servant of a ruler. Likewise, the bishop, the priest and the preacher are in fact no longer, as in early Christian times, carriers of a purely personal charisma, which offers other-worldly sacred values under the personal mandate of a master, and in principle responsible only to him, to everybody who appears worthy of them and asks for them. In spite of the partial survival of the old theory, they have become officials in the service of a functional purpose, a purpose which in the present-day "church" appears at once impersonalized and ideologically sanctified.

II. The Social Position of the Official

A. Social esteem and status convention. Whether he is in a private office or a public bureau, the modern official, too, always strives for and usually attains a distinctly elevated *social esteem* vis-à-vis the governed. His social position is protected by prescription about rank order and, for the political official, by special prohibitions of the criminal code against "insults to the office" and "contempt" of state and church authorities.

The social position of the official is normally highest where, as in old civilized countries, the following conditions prevail: a strong demand for administration by trained experts; a strong and stable social differentiation, where the official predominantly comes from socially and economically privileged strata because of the social distribution of power or the costliness of the required training and of status conventions. The possession of educational certificates or patents . . . is usually linked with qualification for office; naturally, this enhances the "status element" in the social position of the official. Sometimes the status factor is explicitly acknowledged; for example, in the prescription that the acceptance of an aspirant to an office career depends upon the consent ("election") by the members of the official body. . . .

Usually the social esteem of the officials is especially low where the demand for expert administration and the hold of status conventions are weak. This is often the case in new settlements by virtue of the great economic opportunities and the great instability of their social stratification: witness the United States.

B. Appointment versus election: Consequences for expertise. Typically, the bureaucratic official is appointed by a superior authority. An official elected by the governed is no longer a purely bureaucratic figure. Of course, a formal election may hide an appointment—in politics especially by party bosses. This does not depend upon legal statutes, but upon the way in which the party mechanism functions. Once firmly organized, the parties can turn a formally free election into the mere acclamation of a candidate designated by the party chief, or at least into a contest, conducted according to certain rules, for the election of one of two designated candidates.

In all circumstances, the designation of officials by means of an election modifies the rigidity of hierarchical subordination. In principle, an official who is elected has an autonomous position vis-à-vis his superiors, for he does not derive his position "from above" but "from below," or at least not from a superior authority of the official hierarchy but from powerful party men ("bosses"), who also determine his further career. The career of the elected official is not primarily dependent upon his chief in the administration. The official who is not elected, but appointed by a master, normally functions, from a technical point of view, more accurately because it is more likely that purely functional points of consideration and qualities will determine his selection and career. As laymen, the governed can evaluate the expert qualifications of a candidate for office only in terms of experience, and hence only after his service. Moreover, if political parties are involved in any sort of selection of officials by election, they quite naturally tend to give decisive weight not to technical competence but to the services a follower renders to the party boss. This holds for the designation of otherwise freely elected officials by party bosses when they determine the slate of candidates as well as for the free appointment of officials by a chief who has himself been elected. The contrast, however, is relative: substantially similar conditions hold where legitimate monarchs and their subordinates appoint officials, except that partisan influences are then less controllable.

Where the demand for administration by trained experts is considerable, and the party faithful have to take into account an intellectually developed, educated, and free "public opinion," the use of unqualified officials redounds upon the party in power at the next election. Naturally, this is more likely to happen when the officials are appointed by the chief. The demand for a trained administration now exists in the United States, but wherever, as in the large cities, immigrant votes are "corralled," there is, of course, no effective public opinion. Therefore, popular election not only of the administrative chief but also of his subordinate officials usually endangers, at least in very large administrative bodies which are difficult to supervise, the expert qualification of the officials as well as the precise functioning of the bureaucratic mechanism, besides weakening the dependence of the officials upon the hierarchy. The superior qualification and integrity of Federal judges appointed by the president, as over and against elected judges, in the United States is well known, although both types of officials are selected primarily in terms of party considerations. The great changes in American metropolitan administrations demanded by reformers have been effected essentially by elected mayors working with an apparatus of officials who were appointed by them. These reforms have thus come about in a "caesarist" fashion. Viewed technically, as an organized form of domination, the efficiency of "caesarism," which often grows out of democracy, rests in general upon the position of the "caesar" as a free trustee of the masses (of the army or of the citizenry), who is unfettered by tradition. The "caesar" is thus the unrestrained master of a body of highly qualified military officers and officials whom he selects freely and personally without regard to tradition or to any other impediments. Such "rule of the personal genius," however, stands in conflict with the formally "democratic" principle of a generally elected officialdom.

C. Tenure and the inverse relationship between judicial independence and social prestige. Normally, the position of the official is held for life, at least in public bureaucracies, and this is increasingly the case for all similar structures. As a factual rule, *tenure for life* is presupposed even where notice can be given or periodic reappointment occurs. In a private enterprise, the fact of such tenure normally differentiates the official from the worker. Such legal or actual life-tenure, however, is not viewed as a proprietary right of the official to the possession of office as was the case in many structures of authority of the past. Wherever legal guarantees against discretionary dismissal or transfer are developed, as in Germany for all judicial and increasingly also for administrative officials, they merely serve the purpose of guaranteeing a strictly impersonal discharge of specific office duties. . . .

D. Rank as the basis of regular salary. The official as a rule receives a *monetary* compensation in the form of a *salary,* normally fixed, and the old age security provided by a pension. The salary is not measured like a wage in terms of work done, but according to "status," that is, according to the kind of function (the "rank") and, possibly, according to the length of service. The relatively great security of the official's income, as well as the rewards of social esteem, make the office a sought-after position, especially in countries which no longer provide opportunities for colonial profits. In such countries, this situation permits relatively low salaries for officials.

E. Fixed career lines and status rigidity. The official is set for a "career" within the hierarchical order of the public service. He expects to move from the lower, less

important and less well paid, to the higher positions. The average official naturally desires a mechanical fixing of the conditions of promotion: if not of the offices, at least of the salary levels. He wants these conditions fixed in terms of "seniority," or possibly according to grades achieved in a system of examinations. Here and there, such grades actually form a *character indelebilis* of the official and have lifelong effects on his career. To this is joined the desire to reinforce the right to office and to increase status group closure and economic security. All of this makes for a tendency to consider the offices as "prebends" of those qualified by educational certificates. The necessity of weighing general personal and intellectual qualifications without concern for the often subaltern character of such patents of specialized education, has brought it about that the highest political offices, especially the "ministerial" positions, are as a rule filled without reference to such certificates. . . .

THE TECHNICAL SUPERIORITY OF BUREAUCRATIC ORGANIZATION OVER ADMINISTRATION BY NOTABLES

The decisive reason for the advance of bureaucratic organization has always been its purely *technical* superiority over any other form of organization. The fully developed bureaucratic apparatus compares with other organizations exactly as does the machine with the non-mechanical modes of production. Precision, speed, unambiguity, knowledge of the files, continuity, discretion, unity, strict subordination, reduction of friction and of material and personal costs—these are raised to the optimum point in the strictly bureaucratic administration, and especially in its monocratic form. As compared with all collegiate, honorific, and avocational forms of administration, trained bureaucracy is superior on all these points. And as far as complicated tasks are concerned, paid bureaucratic work is not only more precise but, in the last analysis, it is often cheaper than even formally unremunerated honorific service. . . .

Today, it is primarily the capitalist market economy which demands that the official business of public administration be discharged precisely, unambiguously, continuously, and with as much speed as possible. Normally, the very large modern capitalist enterprises are themselves unequalled models of strict bureaucratic organization.

Business management throughout rests on increasing precision, steadiness, and, above all, speed of operations. This, in turn, is determined by the peculiar nature of the modern means of communication, including, among other things, the news service of the press. The extraordinary increase in the speed by which public announcements, as well as economic and political facts, are transmitted exerts a steady and sharp pressure in the direction of speeding up the tempo of administrative reaction towards various situations. The optimum of such reaction time is normally attained only by a strictly bureaucratic organization. (The fact that the bureaucratic apparatus also can, and indeed does, create certain definite impediments for the discharge of business in a manner best adapted to the individuality of each case does not belong in the present context.)

Bureaucratization offers above all the optimum possibility for carrying through the principle of specializing administrative functions according to purely objective considerations. Individual performances are allocated to functionaries who have specialized training and who by constant practice increase their expertise. "Objective" discharge of business primarily means a discharge of business according to *calculable rules* and "without regard for persons."

"Without regard for persons," however, is also the watchword of the market and, in general, of all pursuits of naked economic interests. Consistent bureaucratic domination means the leveling of "status honor." Hence, if the principle of the free market is not at the same time restricted, it means the universal domination of the "class situation." That this consequence of bureaucratic domination has not set in everywhere proportional to the extent of bureaucratization is due to the differences between possible principles by which polities may supply their requirements. However, the second element mentioned, calculable rules, is the most important one for modern bureaucracy. The peculiarity of modern culture, and specifically of its technical and economic basis, demands this very "calculability" of results. When fully developed, bureaucracy also stands, in a specific sense, under the principle of *sine ira ac studio.* Bureaucracy develops the more perfectly, the more it is "dehumanized," the more completely it succeeds in eliminating from official business love, hatred, and all purely personal, irrational, and emotional elements which escape calculation. This is appraised as its special virtue by capitalism.

The more complicated and specialized modern culture becomes, the more its external supporting apparatus demands the personally detached and strictly objective *expert,* in lieu of the lord of older social structures who was moved by personal sympathy and favor, by grace and gratitude. Bureaucracy offers the attitudes demanded by the external apparatus of modern culture in the most favorable combination. In particular, only bureaucracy has established the foundation for the administration of a rational law conceptually systematized on the basis of "statutes," such as the later Roman Empire first created with a high degree of technical perfection. During the Middle Ages, the reception of this [Roman] law coincided with the bureaucratization of legal administration: The advance of the rationally trained expert displaced the old trial procedure which was bound to tradition or to irrational presuppositions. . . .

THE LEVELING OF SOCIAL DIFFERENCES

In spite of its indubitable technical superiority, bureaucracy has everywhere been a relatively late development. A number of obstacles have contributed to this, and only under certain social and political conditions have they definitely receded into the background.

A. Administrative Democratization

Bureaucratic organization has usually come into power on the basis of a leveling of economic and social differences. This leveling has been at least relative, and has concerned the significance of social and economic differences for the assumption of administrative functions.

Bureaucracy inevitably accompanies modern *mass democracy,* in contrast to the democratic self-government of small homogeneous units. This results from its characteristic principle: the abstract regularity of the exercise of authority, which is a result of the demand for "equality before the law" in the personal and functional sense—hence, of the horror of "privilege," and the principled rejection of doing business "from case to case." Such regularity also follows from the social preconditions of its origin. Any non-bureaucratic administration of a large social structure rests in some way upon the fact that existing social, material, or honorific preferences and ranks are connected with administrative functions and duties. This usually means that an economic or a social exploitation of position, which every sort of administrative activity provides to its bearers, is the compensation for the assumption of administrative functions.

Bureaucratization and democratization within the administration of the state therefore signify an increase of the cash expenditures of the public treasury, in spite of the fact that bureaucratic administration is usually more "economical" in character than other forms. Until recent times—at least from the point of view of the treasury—the cheapest way of satisfying the need for administration was to leave almost the entire local administration and lower judicature to the landlords of Eastern Prussia. The same is true of the administration by justices of the peace in England. Mass democracy which makes a clean sweep of the feudal, patrimonial, and—at least in intent—the plutocratic privileges in administration unavoidably has to put paid professional labor in place of the historically inherited "avocational" administration by notables.

B. Mass Parties and the Bureaucratic Consequences of Democratization

This applies not only to the state. For it is no accident that in their own organizations the democratic mass parties have completely broken with traditional rule by notables based upon personal relationships and personal esteem. Such personal structures still persist among many old conservative as well as old liberal parties, but democratic mass parties are bureaucratically organized under the leadership of party officials, professional party and trade union secretaries, etc. In Germany, for instance, this has happened in the Social Democratic party and in the agrarian mass-movement; in England earliest in the caucus democracy of Gladstone and Chamberlain which spread from Birmingham in the 1870's. In the United States, both parties since Jackson's administration have developed bureaucratically. In France, however, attempts to organize disciplined political parties on the basis of an election system that would compel bureaucratic organization have repeatedly failed. The resistance of local circles of notables against the otherwise unavoidable bureaucratization of the parties, which would encompass the entire country and break their influence, could not be overcome. Every advance of simple election techniques based on numbers alone as, for instance, the system of proportional representation, means a strict and inter-local bureaucratic organization of the parties and therewith an increasing domination of party bureaucracy and discipline, as well as the elimination of the local circles of notables—at least this holds for large states.

The progress of bureaucratization within the state administration itself is a phenomenon paralleling the development of democracy, as is quite obvious in France, North America, and now in England. Of course, one must always remember that the term "democratization" can be misleading. The *demos,* itself, in the sense of a shapeless mass, never "governs" larger associations, but rather is governed. What changes is only the way in which the executive leaders are selected and the measure of influence which the *demos,* or better, which social circles from its midst are able to exert upon the content and the direction of administrative activities by means of "public opinion." "Democratization," in the sense here intended, does not necessarily mean an increasingly active share of the subjects in government. This may be a result of democratization, but it is not necessarily the case.

We must expressly recall at this point that the political concept of democracy, deduced from the "equal rights" of the governed, includes these further postulates: (1) prevention of the development of a closed status group of officials in the interest of a universal accessibility of office, and (2) minimization of the authority of officialdom in the interest of expanding the sphere of influence of "public opinion" as far as practicable. Hence, wherever possible, political democracy strives to shorten the term of office through election and recall, and to be relieved from a limitation to candidates with special expert qualifications. Thereby democracy inevitably comes into conflict with the bureaucratic tendencies which have been produced by its very fight against the notables. The loose term "democratization" cannot be used here, in so far as it is understood to mean the minimization of the civil servants' power in favor of the greatest possible "direct" rule of the *demos,* which in practice means the respective party leaders of the *demos.* The decisive aspect here—indeed it is rather exclusively so—is *the leveling of the governed* in face of the governing and bureaucratically articulated group, which in its turn may occupy a quite autocratic position, both in fact and in form. . . .

THE OBJECTIVE AND SUBJECTIVE BASES OF BUREAUCRATIC PERPETUITY

Once fully established, bureaucracy is among those social structures which are the hardest to destroy. Bureaucracy is *the* means of transforming social action into rationally organized action. Therefore, as an instrument of rationally organizing authority relations, bureaucracy was and is a power instrument of the first order for one who controls the bureaucratic apparatus. Under otherwise equal conditions, rationally organized and directed action (*Gesellschaftshandeln*) is superior to every kind of collective behavior (*Massenhandeln*) and also social action (*Gemeinschaftshandeln*) opposing it. Where administration has been completely bureaucratized, the resulting system of domination is practically indestructible.

The individual bureaucrat cannot squirm out of the apparatus into which he has been harnessed. In contrast to the "notable" performing administrative tasks as a honorific duty or as a subsidiary occupation (avocation), the professional bureaucrat is chained to his activity in his entire economic and ideological existence. In the great majority of cases he is only a small cog in a ceaselessly moving mechanism which prescribes to him an essentially fixed route of march. The official is entrusted with specialized tasks, and normally the mechanism cannot be put into motion or arrested by him, but only from the very top. The individual bureaucrat is, above all, forged to the common interest of all the functionaries in the perpetuation of the apparatus and the persistence of its rationally organized domination.

The ruled, for their part, cannot dispense with or replace the bureaucratic apparatus once it exists, for it rests upon expert training, a functional specialization of work, and an attitude set on habitual virtuosity in the mastery of single yet methodically integrated functions. If the apparatus stops working, or if its work is interrupted by force, chaos results, which it is difficult to master by improvised replacements from among the governed. This holds for public administration as well as for private economic management. Increasingly the material fate of the masses depends upon the continuous and correct functioning of the ever more bureaucratic organizations of private capitalism, and the idea of eliminating them becomes more and more utopian.

Increasingly, all order in public and private organizations is dependent on the system of files and the discipline of officialdom, that means, its habit of painstaking obedience within its wonted sphere of action. The latter is the more decisive element, however important in practice the files are. The naive idea of Bakuninism of destroying the basis of "acquired rights" together with "domination" by destroying the public documents overlooks that the settled orientation of *man* for observing the accustomed rules and regulations will survive independently of the documents. Every reorganization of defeated or scattered army units, as well as every restoration of an administrative order destroyed by revolts, panics, or other catastrophes, is effected by an appeal to this conditioned orientation, bred both in the officials and in the subjects, of obedient adjustment to such [social and political] orders. If the appeal is successful it brings, as it were, the disturbed mechanism to "snap into gear" again.

The objective indispensability of the once-existing apparatus, in connection with its peculiarly "impersonal" character, means that the mechanism—in contrast to the feudal order based upon personal loyalty—is easily made to work for anybody who knows how to gain control over it. A rationally ordered officialdom .continues to function smoothly after the enemy has occupied the territory; he merely needs to change the top officials. It continues to operate because it is to the vital interest of everyone concerned, including above all the enemy. After Bismarck had, during the long course of his years in power, brought his ministerial colleagues into unconditional bureaucratic dependence by eliminating all independent statesmen, he saw to his surprise that upon his resignation they continued to administer their offices unconcernedly and undismayedly, as if it had not been the ingenious lord and very creator of these tools who had left, but merely some individual figure in the bureaucratic machine which had been exchanged for some other figure. In spite of all the changes of masters in France since the time of the First Empire, the power apparatus remained essentially the same.

Such an apparatus makes "revolution," in the sense of the forceful creation of entirely new formations of authority, more and more impossible—technically, because of its control over the modern means of communication (telegraph etc.), and also because of its increasingly rationalized inner structure. The place of "revolutions" is under this process taken by *coups d'état,* as again France demonstrates in the classical manner since all successful transformations there have been of this nature. . . .

Discussion Questions

1. How can the rise of "new age" movements, extreme sports, religious fundamentalisms, and spiritual healers be explained in light of Weber's discussion of rationalization and the "disenchantment of the world"?

2. What are some of the essential differences between Weber's view of religion and Durkheim's? Which view better explains the role of religion in contemporary life?

3. In developing his ideal type of bureaucracy, Weber highlights the rational aspects of this organizational form. In what ways might bureaucracies exhibit "irrational" or inefficient features? How have bureaucracies "dehumanized" social life, transforming modernity into an iron cage?

4. Given Weber's three types of legitimate domination, the political system in the United States is best characterized as based on legal authority. What elements of the other types of authority can, nevertheless, still be found? How might political controversies result from the "illegitimate mixing" of different types of authority? Provide examples.

5. Following Weber's arguments in *The Protestant Ethic* and *Spirit of Capitalism*, what role did the "calling" and outward signs of grace play in the development of capitalism? When capitalism was firmly established, what effect did it have on the religiously based ideas?

6. In what way(s) is Weber's analysis of class, status, and party different from Marx's understanding of class and social stratification? What are the implications of the difference(s) for designating the proletariat a revolutionary force for social change and for understanding the exercise of power in the United States?

Part II

CLASSICAL SOCIOLOGICAL THEORY

Expanding the Foundation

5 CHARLOTTE PERKINS GILMAN (1860–1935)

Charlotte Perkins Gilman

Key Concepts

:: Gender inequality

:: Women's economic independence

The labor of women in the house, certainly, enables men to produce more than they otherwise could; and in this way women are economic factors in society. But so are horses.

—Gilman (1898/1998:7)

In 1980, the United Nations summed up the burden of **gender inequality:** Women comprised half the world's population, did two-thirds of the world's work, earned one tenth of the world's income, and owned one hundredth of the world's property. In the past 30 years, not that much has changed. Even in one of the most developed nations in the world, the United States, significant economic gender inequities continue: According to the United States Department of Labor, in 2009 American women earned approximately 77 percent of what men with similar educational and other qualifications earned. Imagine, then, what life was like for women in the *nineteenth* century, when, legally speaking, women were analogous to children, without the right either to own property or to vote.

In the nineteenth and early twentieth centuries, American and European women were not only legally prevented from owning property or voting; they also were denied access to higher education. Yet, despite this absence of equality, since at least the 1800s a number of self-educated women have done remarkable scholarly work.[1] In this chapter, we consider one of these extraordinary women: Charlotte Perkins Gilman (1860–1935). Gilman was an accomplished writer, feminist, and sociologist—though in her own day she was not widely recognized as a sociologist. However, despite her lack of institutional credentials, there are a number of reasons for regarding Gilman as a "sociologist." As Deegan (1997:11) points out, Gilman

1. identified herself as a sociologist;

2. was identified by others as a sociologist;

3. taught sociology courses (though she declined an academic appointment in sociology, she was a frequent freelance lecturer on college campuses);

4. wrote sociological books and articles, some of which were published in the influential *American Journal of Sociology;*

5. was a charter member of the American Sociological Society and remained a member for 25 years.

In addition, Gilman is a pivotal feminist theorist, for she was one of the first to seek to explain *how* women and men came to have their respective societal roles and *why* societies developed gender inequalities. Significantly, the three main theoretical traditions from which Gilman drew—social Darwinism, symbolic interactionism, and Marxist theory (and their offshoots)—are the main traditions around which feminist social theories still revolve today. It is for these reasons, then, that we consider Gilman a core classical theorist.

A BIOGRAPHICAL SKETCH ▦

Charlotte Perkins was born on July 3, 1860, to Mary Wescott Perkins and Frederic Beecher Perkins. She was the third of four children, although the Perkins's first child died at birth, and their fourth child died in infancy. Charlotte's mother, Mary, was said to be an attractive woman who had had many suitors. Charlotte's father, Frederic, was part of the distinguished New England Beecher clan. His grandfather was the influential theologian Lyman Beecher; his uncle was the famous clergyman and abolitionist, Henry Ward Beecher; and his aunt was the famous abolitionist and author of *Uncle Tom's Cabin,* Harriet Beecher Stowe. Frederic Beecher Perkins himself was a librarian and writer of some distinction, but he abandoned the family soon after Charlotte was born. Although Frederic provided some financial support to his family, Charlotte and her brother and mother suffered financially. They moved 19 times in 18 years; 14 of the moves were from city to city. Although a lack of money prevented Charlotte from receiving much in the way of a formal education, she attended the Rhode Island School of Design in 1878–1879. The training she received there enabled her to earn money as a commercial artist. But most of Charlotte's education was gained through her voracious reading, some of which was overseen by her father. Although he was only an "occasional visitor," he sent Charlotte books and catalogues of books, as well as lists of books to read (Degler 1966:viii–ix; Gilman [1935] 1972:5).

---

[1]These women include Harriet Martineau (1802–1876, see Significant Others box below), Jane Addams (1860–1953), Anna Julia Cooper (1858–1964, see Significant Others box, pages 328–329, Chapter7), Marianne Weber (1870–1954), and Beatrice Potter Webb (1858–1943). In addition to this brief list of "first wave" (i.e., nineteenth and early twentieth centuries) feminists who wrote sociologically about the origins and dynamics of gender in equality, are the many feminist activists devoted to remedying gender in equality, e.g., Sojourner Truth (1797–1883), Elizabeth Cady Stanton (1815–1902), SusanB. Anthony (1820–1906), and Ida B.Wells-Barnett (1862–1931).

Harriet Martineau (1802–1876): The First Woman Sociologist

Harriet Martineau was born the sixth of eight children in a well-to-do English family. Her father was a successful textile manufacturer and a devout Unitarian whose relatively progressive views allowed Harriet to pursue academic subjects normally reserved for men. Despite the comforts her father's career afforded the family, Martineau's life was far from idyllic. By the age of 12, she was deaf, and she did not possess the senses of smell or taste. Her father died in 1826, and shortly thereafter the family business closed. During this period, her fiancé died as well, and Martineau was left without financial support.

As it would turn out, Martineau's misfortunes thrust her into the world of professional writing as a means of earning a living. In keeping with her upbringing, she began her career by publishing articles for the Unitarian journal, the *Monthly Repository,* as well as writing religious books. However, she quickly expanded her range by publishing works on political economy, sociology, and English history; novels; travel books; and children's stories. Moreover, she worked as a journalist for the *Daily News* where over the course of some 15 years she contributed more than 1,500 articles.

In much of her writing and public life, she proved to be a staunch advocate for women's rights and the abolition of slavery. As for her more sociological writings, she examined many of the issues that would later occupy the attention of Marx, Durkheim, and Weber, including class relations, suicide, religion, and social science methodology. Her analyses of these subjects are particularly noteworthy, given that the discipline of sociology had not yet been born at the time of her writing. Of her works, two stand out for their impact on the field: *The Positive Philosophy of Auguste Comte* (1853), a translation and condensing of Comte's six-volume opus, and *Society in America* (1837). Considered by many to be a masterpiece study of American life, Society in America was based on Martineau's 2-year stay in the United States.

In it, she compared the democratic principles of equality and freedom on which the young nation was founded with its actual institutionalized practices. Certainly, were it not for the sexism of the Victorian era and its lingering effects in academia, Martineau's position in the discipline would be more secure.

NOTE: This account is based largely on Valerie K. Pichanick's *Harriet Martineau, The Woman and Her Work, 1802–76* (Ann Arbor: University of Michigan Press, 1980).

Charlotte was known as a very "willful" child. She had little interest in the pursuits deemed proper for a girl, and as she matured, Charlotte spurned the traditional roles assigned to women. She refused to play the part of the precious and frail coquette, choosing instead to exercise vigorously and develop her physical strength (Degler 1966:x).

Charlotte cherished her independence and vowed never to marry. However, a young artist named Charles Walter Stetson fell in love with Charlotte and proposed to her, and despite some initial reluctance, Charlotte agreed to marry him in 1884. But her reservations about marriage proved to be well founded. In the course of the first year of their marriage, she became increasingly and inexplicably despondent. The birth of a daughter, Katharine, just 10 and a half months after her wedding, seemed to make her depression even more severe. At the time, the prescribed remedy for women's melancholia was rest and no intellectual activity, which nearly drove the young, spirited, intellectually curious Perkins into madness. In a now-famous passage in her autobiography, Perkins described the "rest cure" that nearly did her in: "Live as domestic a life as possible. Have your child with you all the time. . . . Lie down an hour after each meal. Have but two hours' intellectual life a day. And never touch pen, brush or pencil again" ([1935] 1972:96).

A trip away from her daughter and her husband made Perkins recognize that it was not rest but activity she needed. In 1890, Charlotte decided to separate from Stetson and move to California: "Better for that dear child [Katharine] to have separated parents than a lunatic mother," she wrote in her autobiography (ibid.:97).

Life in California was very difficult for Perkins. She undertook what work she could find—lecturing on women's rights, writing for small periodicals, and even managing a boarding house after her destitute mother came to live with her in Oakland (Degler 1966:xii). Yet, during this time, Perkins published several important works. In 1892, "The Yellow Wallpaper" was published. This poignant, semiautobiographical account of her experience with depression would become one of her most well-known and highly acclaimed stories. The following year, Perkins's *In This Our World* was published. This sensational book of poems "enjoyed a near cult following in the United States and England" (Golden and Zangrando 2000:11).

In 1894, Charlotte Perkins and Charles Walter Stetson were divorced, and Stetson married one of Charlotte's closest friends, Grace Channing. Charlotte freely gave the couple her blessing, and she permitted her daughter, Katharine, to live with Stetson and his new wife. This was a flagrant departure from the accepted attitude of the day, and Perkins was pilloried in the newspapers for giving away both her husband and her child to another woman. Nevertheless, she remained on close terms with Stetson and Channing (Degler 1966:xi–xii).

With her daughter no longer her direct responsibility, Perkins expanded her lecturing tours from California to the rest of the nation. She became well known in women's suffrage circles for her provocative ideas about women's rights as well as other social issues. After the publication of *Women and Economics* in 1898, her reputation as a feminist was secured. In this book, which was translated into seven languages, Perkins denounced women's economic dependence on men and advocated for public day care and cooperative kitchens.

Not surprisingly, Perkins was allied with progressive political movements. Most noteworthy was her commitment to Fabian socialism, a brand of socialism that called for the collective ownership and democratic control of resources. While advocating equality for all, Fabian socialists, nevertheless, rejected the classical Marxist theory of revolutionary class struggle in favor of peaceful and gradual social change.

In conjunction with Fabian socialism, Perkins emphasized how the insular, nuclear family was dysfunctional for women. Gilman believed in communal kitchens and child care, rather than individual mothers "doing it all"—cooking, cleaning, and childrearing—while isolated in their own homes. Incidentally, this same inefficient individualism is readily apparent today, for instance, in the gridlock of SUVs in suburban school parking lots as mothers (and some fathers) arrive en masse—each in her (or his) own car—to drop off and pick up children.

In 1900, at the age of 40, Perkins married George Houghton Gilman, a man 7 years her junior. Between her second marriage and 1914, Gilman reached the peak of her public activity and fame. She published nine books during this period, including *Concerning Children* (1900), *The Home* (1903), *Human Work* (1904), and *The Man-Made World* (1911). She also founded the journal *Forerunner* (1909–16), in which she published feminist stories and articles. In 1932, Gilman was diagnosed with breast cancer. She continued to write and lecture for a few more years, but, unfortunately, the cancer could not be arrested. On August 17, 1935, Charlotte Perkins Gilman took her own life, writing in her suicide note that she "preferred chloroform to cancer" (Degler 1966:xvii). Gilman's autobiography, *The Living of Charlotte Perkins Gilman* ([1935] 1972), was published shortly after her death.

INTELLECTUAL INFLUENCES AND CORE IDEAS ▪▪

As noted above, Gilman's approach to gender reflects the basic building blocks that inform feminist theory to this day: She drew from a variety of theoretical wells including Marxism, symbolic interactionism, and social Darwinism. In this section, we outline her indebtedness to each of these frameworks. As you will see, her multidimensional theory of gender inequality combines (1) a Marxist emphasis on the economic and political basis for gender inequality, (2) a symbolic interactionist emphasis on how these gender differences are reinforced and institutionalized through the process of socialization, and (3) a sociobiological emphasis on the evolutionary advantages or roots of gender differences.

First, following the Marxist tradition, Gilman analyzed the political and economic factors that produce and reproduce gender inequality. Gilman sought to show that the division of labor of the traditional family (breadwinner husband/stay-at-home wife) was inherently problematic because it makes women economically dependent on men. To be sure, Gilman did not focus on the evils of capitalism as did Marx. However, just as Marx considered the system of capitalism inherently exploitative because workers do not own the means of production, Gilman considered the traditional family structure inherently exploitative because the economic compensation of women bears absolutely no relation to her labor. Regardless of how much work she actually does (or doesn't do) in the home, the housewife's social and economic standing comes from her husband; thus, her labor belongs to her husband, not to her. This is the major point of the selection, *Women and Economics,* that you will read below.

Second, in conjunction with the Chicago School and the symbolic interactionist tradition (discussed further in Chapters 8 and 12), Gilman emphasized how differential socialization leads to and sustains gender inequality. In doing so, she challenged the long-standing assumption that inherent biological differences precluded men and women from effectively pursuing overlapping social activities (Degler 1966:xii–xxiii). Instead, Gilman maintained that from the earliest age, young girls were encouraged, if not forced, to act, think, look, and talk differently from boys, though their interests and capabilities at that age might be identical. For instance, Gilman states,

> One of the first things we force upon the child's dawning consciousness is the fact that he is a boy or that she is a girl, and that, therefore, each must regard everything from a different point of view. They must be dressed differently, not on account of their personal needs, which are exactly similar at this period, but so that neither they, nor any one beholding them, may for a moment forget the distinction of sex. ([1898] 1998:28)

Ironically, differential gender socialization is not only evident today as in Gilman's time—in some ways, it is even *more* apparent. Because of consumerism and marketing, as well as the fact that parents often find out the sex of their child even before he or she is born, there is a huge array of gendered baby paraphernalia (up to and including pink and blue diapers). Today, children are subjected to extensive marketing campaigns, most notably by fast-food restaurants and toy manufacturers, which from a very early age "force upon the child" a sense of being a "boy" or a "girl." All one need do is stroll down the toy aisle at a local department store to see how children are not taught to view themselves (and each other) as *human beings;* rather, they are taught to view themselves (and view each other) as "boys" and "girls."

Photo 5.1   One of the most gendered sites today: The toy store

However, despite her emphasis on differential socialization, Gilman did not deny that biological differences exist between men and women. On the contrary, Gilman borrowed theoretically not only from Fabian socialism (with its origins in Marxism) and symbolic interactionism but also from social Darwinism. Social Darwinists, such as Herbert Spencer in England and William Graham Sumner in America, applied Darwin's theory of evolution to human societies and maintained that human existence was based on "survival of the fittest." Like many social Darwinists, Gilman was fascinated by the animal world, and she used animal analogies to explain the human condition as well as biological and behavioral differences between the sexes. Specifically, Gilman contended that women and men, in general, have different biological "principles" to which they adhere. She maintained that women's unique capabilities—particularly their love and concern for others—have tremendous social value, though they are grossly underappreciated.

Indeed, Gilman went so far as to assert the natural superiority of the female sex. She enthusiastically endorsed the "scientifically" based gynocentric theory promoted by the Harvard sociologist Lester Ward, suggesting that that the civilizing capacities of women could compensate for the destructive combativeness of men. As Gilman notes,

> The innate underlying difference [between the sexes] is one of principle. On the one hand, the principle of struggle, conflict, and competition. . . . On the other, the principle of growth, of culture, of applying services and nourishment in order to produce improvement. (as cited in Hill 1989:45)

Thus, Gilman maintained that, in contrast to men, women did not want to fight, to take, to oppress, but rather to love. Women exhibited "the growing altruism of work, founded in mother love, in the antiselfish instinct of reproduction" (ibid.).

However, as the acclaimed Gilman historian Mary Hill (1989:45) notes, Gilman's view of women as "saintly givers" and men as "warring beasts" is problematic. For when she glorifies the female "instincts" of love and service, her radical feminist theory dissolves into a "sentimental worship of the status quo." Her insistence on the "giving" nature of women as compared to the "combative" nature of men seems to indicate that women must be and should be the primary caretaker of children and that there is only so far a man can go in his role as nurturer. In short, Gilman's biological determinism seems quite antiquated today. Many contemporary fathers are far more nurturing toward their children than were their fathers, though they have the same basic biological constitution.

Despite significant shortcomings in her biological arguments, however, Gilman raised interesting issues that today are being explored by brain researchers. Contemporary neuroscientists, geneticists, evolutionary psychologists, and others are breathing new life into some of the same gender differences that Gilman noted more than a century ago. For instance, researchers today find that females tend to be more highly sensitive to touch, sound, and smell than males. In one study of day-old infants, researchers found that though both male and female infants reacted most intensely to the sound of another's trouble (as opposed to other sounds), infant females reacted more strongly than males. The researchers suggested that the infant girls were more finely attuned to an empathetic response, and that this sensitivity would run like an "underground stream" throughout their entire lives (Blum 1997:66–67).

Most importantly, in contrast to the static "nature-versus-nurture" debates of the past, contemporary neurological research illuminates the interconnectedness of nature and nurture. Researchers now are investigating the complex ways in which the environment sparks significant neurological or chemical changes and developments, and vice versa. Nature is understood to be more malleable and more of a process than social Darwinists ever imagined. For example, even genetic instructions that are often thought to be determinate, such as height, are known to be influenced by the environment. A baby whose genes blueprint him to grow six feet tall, but who does not receive adequate nutrition, will fall short by an inch or more. Some researchers even suggest that stress in childhood can interfere with height by suppressing growth hormones (Blum 1997:21–22). To be sure, it is outside the scope of this chapter to delve into these highly complex and contentious neurological arguments here, but the point is that the questions about nature versus nurture that Gilman raised continue to be at the heart of gender theory and research today.

Far more problematic than her biological arguments about gender were Gilman's biological arguments about race. Drawing not only on social Darwinist theories of "survival of the fittest" but also on the "commonsense" notions of "manifest destiny" and the "white man's burden" dominant in her day (see Chapter 7; and Edles 2002:9), Gilman made patently racist remarks. For instance, she maintained that some races

combine well, making a good blend, [but] some do not. We are perfectly familiar in this country with the various blends of black and white, and the wisest of both races prefer the pure stock. The Eurasian mixture is generally considered unfortunate by most observers. (1923; see also Scharnhorst 2000:69)

Less insidiously, but no less problematic, Gilman also implicitly assumed she was speaking about all women when she was really referring to white women. Certainly, Gilman did not discuss or consider the resources or situations of nonwhite women (though, to be sure, this implicit privileging of "white lives" was typical of *all* of sociology's white classical figures).

For instance, in her discussion of the isolation of traditional housewives, Gilman ignored the fact that African American women may have had traditions of support and community that white women did not (see Stack 1974). And, of course, the entire notion of the isolated stay-at-home mom is out of sync with the fact that paid work has long been "an important and valued dimension of Afrocentric definitions of Black motherhood" (Collins 1987:5). Yet, despite her racist assumptions, Gilman often made pointedly antiracist comments too. She decried slavery and the continued oppression of African Americans, and she spoke out against the genocide and ill treatment of Native Americans and Native Hawaiians.

▪▪ GILMAN'S THEORETICAL ORIENTATION

Given that Gilman melds the distinct traditions of neo-Marxism, symbolic interactionism, and social Darwinism, it should come as no surprise that Gilman's approach to gender is theoretically multidimensional. Specifically, as shown in Figure 5.2, Gilman highlights differential gender socialization (i.e., how boys and girls are taught to *behave* differently) as well as the distinct sex "principles" with which they are born. These concerns reflect a nonrational orientation to action at the individual level. At the collective level, she highlights both (rationalist) political and economic structures—most importantly, those that prohibit **women's economic independence**—and the (nonrationalist) normative, symbolic structures or codes that ensure differential socialization.

Nevertheless, in her sociological work, especially *Women and Economics,* it is the structural, institutional basis of inequality with which Gilman is most concerned. Hence, as shown in Figure 5.1, we situate her at the more rational/collective end of the theoretical continuum. As indicated previously, Figure 5.1 and the related figures in other chapters reflect our perception of each major theorist's *overall* or most basic theoretical orientation. The point is not that each theorist is situated in a particular "box." Rather, each of these figures is a heuristic device with which to compare and contrast the theoretical orientation of each sociologist discussed in this book (and it can be applied to any other theorist). These positions can and should be discussed and contested; that is why there are no fixed points in these figures.

In any case, Gilman's particular theoretical approach (as well as her similarities with Marx) is readily apparent in the following passage on women's corsets, which can be considered a metaphor for the general constraints placed on women:

Put a corset, even a loose one, on a vigorous man or woman who never wore one, and there is intense discomfort, and a vivid consciousness thereof. The healthy muscles of the trunk resent the pressure, the action of the whole body is checked in the middle, the stomach is choked, the process of digestion is interfered with; and the victim says, "how can you bear such a thing?" ([1898] 1998:40)

Just as the corset "chokes" the stomach, so, too, do the traditional institutional features of the family "choke" women. Just as "healthy muscles" resent the "pressure," so, too, do healthy women. In short, the

Figure 5.1   Gilman's Basic Theoretical Orientation

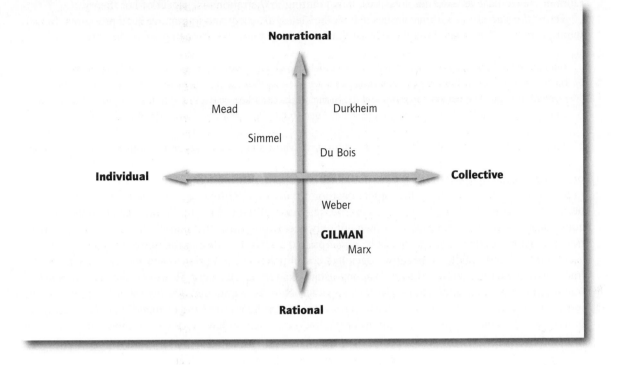

Figure 5.2   Gilman's Multidimensional Explanation of Gender Inequality

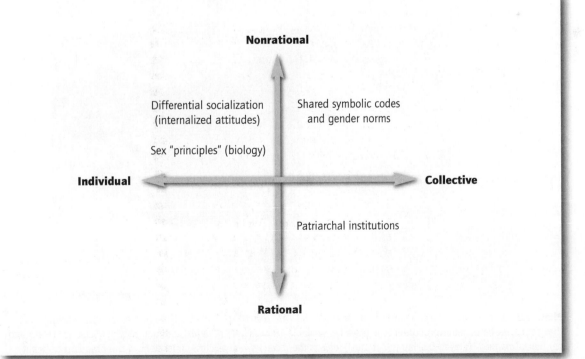

metaphor of the corset reflects that the constraints placed on women originate outside her; as such, Gilman views these as external pressures, thus pointing to the rationalist aspect of her theory.

Yet, at the same time, Gilman argues that women learn to accept and internalize such pressures. In fact, women are so indoctrinated that they resist their own "freedom." As Gilman goes on to note,

> But the person habitually wearing a corset does not feel these evils. They exist, assuredly, the facts are there, the body is not deceived; but the nerves have become accustomed to these disagreeable sensations, and no longer respond to them. The person "does not feel it." In fact, the wearer becomes so used to the sensations that when they are removed,—with the corset,—there is a distinct sense of loss and discomfort. ([1898] 1998:40)

Thus, Gilman's metaphor of the corset is similar to Marx's notion of false consciousness. In both cases, "the facts are there"—the inequality is there—but the person "does not feel it"; he does not see or know of it. She has internalized the pressures and constraints as her own. This view reflects Gilman's incorporation of a more nonrationalist theoretical position. In addition, Gilman acknowledges that women may be resistant to developing a "true" consciousness because of the safety and familiarity provided by false consciousness. Gilman argues that women not only accept but also believe in the legitimacy of the traditional division of labor; thus, it should be no surprise that they feel discomfited by any other reality.

In terms of order, Gilman acknowledges that social patterns exist prior to any individual; that is, the traditional gendered division of labor is a long-established social structure. However, she also emphasizes that it is in individual interactions that this social structure is maintained. On the other hand, it is at the individual level of "free will" that such social structures can be resisted and changed. Yet, the exertion of individual will against existing conditions or the forces of "natural law" is not a common trait. Instead, it

Photo 5.2   In this famous scene from the 1939 film *Gone With the Wind,* Scarlett O'Hara (played by Vivien Leigh) insists that her corset be pulled even tighter to achieve her enviable 18-inch waist. The juxtaposition of white Scarlett O'Hara and her black "mammy" (played by Hattie McDaniel) reflects the racial as well as class dimensions of this standard of beauty.

takes an "advanced" individual to see through the taken-for-granted symbolic codes that legitimate social hierarchies and oppression. As Gilman remarks,

> In the course of social evolution there are developed individuals so constituted as not to fit existing conditions, but to be organically adapted to more advanced conditions. These advanced individuals respond in sharp and painful consciousness to existing conditions, and cry out against them according to their lights. The history of religion, of political and social reform, is full of familiar instances of this. The heretic, the reformer, the agitator, these feel what their compeers do not, see what they do not, and naturally, say what they do not. The mass of the people are invariably displeased by the outcry of these uneasy spirits. In simple primitive periods they were promptly put to death. ([1898] 1998:41)

The autobiographical bent of this passage is readily apparent. Much more so than other social scientists who challenged existing dogma, feminists such as Gilman challenged and threatened core, sacred "family values." As indicated previously, Gilman was perceived as "giving away" both her husband and her daughter, and Gilman was publicly skewered for her actions—not only for her beliefs. No doubt Gilman felt very much like the isolated heretic.

Nevertheless, theoretically, Gilman's assumption that social change rests entirely at the level of the individual is problematic. Gilman did not recognize that it was not solely individual "free will" or social "advancement" but the examination and study of alternative social and cultural schemas (e.g., Fabian socialism) that allowed her to see through and cast off the "corsets" of her day. Moreover, because intellectuals are no different from anyone else, except that they may have more intellectual "wells" from which to draw, Gilman fails to acknowledge that there might be a wealth of other social "corsets" that the intellectual herself never perceives and consequently never rejects. For instance, as we have seen, in some of her essays that draw on social Darwinism, Gilman reinforces some of the racist assumptions common in her day. In sum, *all* individuals are necessarily affected by their environment, not just those who are not "advanced." But social change can arise when people—intellectuals or not—begin to see their lives in a new way. This process occurs not just because of revelations at the level of the individual, but because of new social and cultural conditions. The virtue of social theorists, and intellectuals and social activists more generally, is that they can produce and disseminate new ideas and symbolic schemes that potentially can lead to positive social change.

Reading

Introduction to *Women and Economics*

Women and Economics is Charlotte Perkins Gilman's most sociological, as well as most theoretical, work of nonfiction. In this highly acclaimed book, Gilman seeks to show that the traditional division of labor (breadwinner husband/stay-at-home wife) is inherently problematic. Contrary to the (still prevalent) commonsense rhetoric, which holds that in the traditional division of labor, women are equal partners to men (or that both the man and the woman are dependent on each other), Gilman maintains that the traditional division of labor renders women economically dependent on men and, hence, necessarily strips women of their freedom. In this arrangement, the woman receives both her social status and her economic viability not through her own labor, but through that of her husband. This makes her labor not her "own," but a property of the male. Indeed, as the opening quote in this chapter reflects, rather than viewing the woman in the traditional family as an "equal partner" to her husband, Gilman compares the traditional position of the woman to the domesticated horse: Neither the horse nor the woman is "free."

Specifically, Gilman argues that if women were actually compensated for their work in the home (and not "given" the status of their husband), poor women with lots of children would get the most money (for they are doing the most work), while women with no children and those who do no work in the home (i.e., those who have nannies, maids, etc.) would get no compensation. But, of course, the fact is that poor women (i.e., women married to unemployed or working-class men) do the most amount of work and receive the least amount of money. They work long and hard, cleaning, cooking, and raising children. Meanwhile, rich women (women married to wealthy men) do the least amount of work and have the most money to spend—for these women have domestic help, servants, and nannies who perform the household and childrearing labor for them. As Gilman ([1898] 1998:8) states,

> Whatever the economic value of the domestic industry of women is, they do not get it. The women who do the most work get the least money, and the women who have the most money do the least work. Their labor is neither given nor taken as a factor in economic exchange. It is held to be their duty as women to do this work; and their economic status bears no relation to their domestic labors, unless an inverse one.

For those who argue that "a woman's place is in the home" because of her *childbearing* responsibilities, Gilman argues that "women's work" is actually mostly *house* service (cooking, cleaning, mending, etc.), not *child* service (bearing children, breastfeeding, etc.). Thus, Gilman contends that the traditional division of labor is not biologically driven. On this point, Gilman asserts,

> The poor man's wife has far too much of other work to do to spend all her time in waiting on her children. The rich man's wife could do it, but does not, partly because she hires some one to do it for her, and partly because she, too, has other duties to occupy her time. (ibid.:94)

Most provocatively, however, Gilman maintains that her economic dependency makes the woman more akin to a horse than an equal partner in traditional marriage. As Gilman states,

> The horse, in his free natural condition, is economically independent. He gets his living by his own exertions irrespective of any other creature. The horse, in his present condition of slavery, is economically dependent. He gets his living at the hands of his master; and his exertions, though strenuous, bear no direct relation to his living. . . . The horse works, it is true; but what he gets to eat depends on the power and will of his master. His living comes through another. He is economically dependent. (ibid.:4)

Translated into the human condition, Gilman remarks,

> From the day laborer to the millionaire, the wife's worn dress or flashing jewels, her low roof or her lordly one, her weary feet or her rich equipage,—these speak of the economic ability of the husband. The comfort, the luxury, the necessities of life itself, which the woman receives, are obtained by the husband and given her by him. And, when the woman, left alone with no man to "support" her, tries to meet her own economic necessities, the difficulties which confront her prove conclusively what the general economic status of the woman is. (ibid.:5)

In short, like a horse, women are subject to the "power and will of another" because their domestic labor, for which no wages are received in return, belongs not to themselves but to their husbands. Women are thus rendered economically dependent.

Consequently, Gilman argues, rather than develop her own capabilities, women reduce themselves to attracting a viable life partner. Economically, this makes sense for women, because "their profit comes through the power of sex-attraction," not through their own talents (ibid.:33). As evidence for this state of affairs, Gilman remarks that

> when we honestly care as much for motherhood as we pretend, we shall train the woman for her duty, not the girl for her guileless maneuvers to secure a husband. We talk about the noble duties of the mother, but our maidens are educated for economically successful marriage. (ibid.:100)

Thus, the problem with women's economic dependence on men is that their energies are focused on "catching" a man rather than on being productive citizens. Gilman saw it as a tragic waste that women were forced to spend their time and energy on grooming and "finding a man" rather than on intellectual concerns. Moreover, in denying her capabilities, she reduces herself to being, literally, the "weaker sex." As Gilman states,

> The degree of feebleness and clumsiness common to women, the comparative inability to stand, walk, run, jump, climb, and perform other race-functions common to both sexes, is an excessive sex-distinction; and the ensuing transmission of this relative feebleness to their children, boys and girls alike, retards human development. . . . The relative weakness of women is a sex-distinction. It is apparent in her to a degree that injures motherhood, that injures wifehood, that injures the individual. ([1898] 1998:24)

This brings us back to the issue of socialization with which we began this section. Women (especially middle- and upper-class women) are encouraged not to use, but to *deny,* their talents and capabilities:

> The daughters and wives of the rich fail to perform even the domestic service expected of the women of poorer families. They are from birth to death absolutely nonproductive in goods of labor of economic value, and consumers of such goods and labor to an extent limited only by the purchasing power of their male relatives. (ibid.:85)

This is the *sociobiological* tragedy that Gilman perceives: She contends that women are not "underdeveloped men, but the feminine half of humanity in undeveloped form." Women are "oversexed," there is too much emphasis on their sex distinction. Rather than a healthy "survival of the fittest" in which women would be taught to be strong and productive, bourgeois women are mandated to be soft and weak, dependent, emotional, and frail.

It is this emphasis on economic dependency that distinguishes Gilman's perspective from that of Marx. While Marx implies that bourgeois women are privileged because of their economic status, Gilman sees bourgeois women as economically dependent and, therefore, also oppressed (though she fully recognizes that bourgeois women are economically privileged in comparison to poor women). To be sure, Marxist theorists would be quick to point out that in *The Origin of Family, Private Property, and the State* (1884), Marx's coauthor, Friedrich Engels, maintains that in traditional marriage, the man "is the bourgeois; the wife represents the proletariat" (in Tucker 1978:744); in other words, just as the proletariat are exploited by the bourgeoisie, so, too, women are exploited by men. Nevertheless, whether it was because of their own division of labor or his intellectual predisposition, it was Engels and not Marx who developed an explicit theory of gender oppression.

Despite significant institutional advances in educational and professional opportunities, legal rights, and other spheres, some of the social and cultural gender inequities that Gilman discussed are still readily apparent today. Women are still encouraged to "catch" a man by using their bodies rather than their minds, to focus on their looks at the expense of what Gilman considered "matters of real importance." Indeed, one might argue that this male-oriented preoccupation is even more prevalent in the media today than in previous eras (although, confusingly, women also are encouraged to pursue a career and to be athletic). Of course, sadly, there also are many contemporary societies where women still lack basic legal and civil rights, in addition to enduring major cultural and social inequality. For instance, M. Steven Fish (2003) finds that while the average literacy gap between the sexes in most non-Muslim countries is 7 percentage points; it is 20 percentage points in Iran, 23 in Egypt, and 28 in Syria. Many women in these countries also are confronted with significantly inferior health care and restricted educational opportunities, while being denied basic legal rights.

Women and Economics (1898)

Charlotte Perkins Gilman

PREFACE

This book is written to offer a simple and natural explanation of one of the most common and most perplexing problems of human life,—a problem which presents itself to almost every individual for practical solution, and which demands the most serious attention of the moralist, the physician, and the sociologist—

To show how some of the worst evils under which we suffer, evils long supposed to be inherent and ineradicable in our natures, are but the result of certain arbitrary conditions of our own adoption, and how, by removing those conditions, we may remove the evil resultant—

To point out how far we have already gone in the path of improvement, and how irresistibly the social forces of to-day are compelling us further, even without our knowledge and against our violent opposition,—an advance which may be greatly quickened by our recognition and assistance—

To reach in especial the thinking women of to-day, and urge upon them a new sense, not only of their social responsibility as individuals, but of their measureless racial importance as makers of men.

It is hoped also that the theory advanced will prove sufficiently suggestive to give rise to such further study and discussion as shall prove its error or establish its truth.

—Charlotte Perkins Stetson

I

Since we have learned to study the development of human life as we study the evolution of species throughout the animal kingdom, some peculiar phenomena which have puzzled the philosopher and moralist for so long, begin to show themselves in a new light. We begin to see that, so far from being inscrutable problems, requiring another life to explain, these sorrows and perplexities of our lives are but the natural results of natural causes, and that, as soon as we ascertain the causes, we can do much to remove them.

In spite of the power of the individual will to struggle against conditions, to resist them for a while, and sometimes to overcome them, it remains true that the human creature is affected by his environment, as is every other living thing. The power of the individual will to resist natural law is well proven by the life and death of the ascetic. In any one of those suicidal martyrs may be seen the will, misdirected by the ill-informed intelligence, forcing the body to defy every natural impulse,—even to the door of death, and through it.

But, while these exceptions show what the human will can do, the general course of life shows the inexorable effect of conditions upon humanity. Of these conditions we share with other living things the environment of the material universe. We are affected by climate and locality, by physical, chemical, electrical forces, as are all animals and plants. With the animals, we farther share the effect of our own activity, the reactionary force of exercise. What we do, as well as what is done to us, makes us what we are. But, beyond these forces, we come under the effect of a third set of conditions peculiar to our human status; namely, social conditions. In the organic interchanges which constitute social life, we are affected by each other to a degree beyond what is found even among the most gregarious of animals. This third factor, the social environment, is of enormous force as a modifier of human life. Throughout all these environing conditions, those which affect us through our economic necessities are most marked in their influence.

Without touching yet upon the influence of the social factors, treating the human being merely as an individual animal, we see that he is modified most by his economic conditions, as is every other animal. Differ as they may in color and size, in strength and speed, in minor adaptation to minor conditions, all animals that live on grass have distinctive traits in common, and all animals that eat flesh have distinctive traits in common,—so distinctive

SOURCE: Charlotte Perkins Gilman. [1898] 1998. *Women and Economics.* New York: Dover.

and so common that it is by teeth, by nutritive apparatus in general, that they are classified, rather than by means of defence or locomotion. The food supply of the animal is the largest passive factor in his development; the processes by which he obtains his food supply, the largest active factor in his development. It is these activities, the incessant repetition of his exertions by which he is fed, which most modify his structure and develope his functions. The sheep, the cow, the deer, differ in their adaptation to the weather, their locomotive ability, their means of defence; but they agree in main characteristics, because of their common method of nutrition.

The human animal is no exception to this rule. Climate affects him, weather affects him, enemies affect him; but most of all he is affected, like every other living creature, by what he does for his living. Under all the influence of his later and wider life, all the reactive effect of social institutions, the individual is still inexorably modified by his means of livelihood: "the hand of the dyer is subdued to what he works in." As one clear, world-known instance of the effect of economic conditions upon the human creature, note the marked race-modification of the Hebrew people under the enforced restrictions of the last two thousand years. Here is a people rising to national prominence, first as a pastoral, and then as an agricultural nation; only partially commercial through race affinity with the Phoenicians, the pioneer traders of the world. Under the social power of a united Christendom—united at least in this most unchristian deed—the Jew was forced to get his livelihood by commercial methods solely. Many effects can be traced in him to the fierce pressure of the social conditions to which he was subjected: the intense family devotion of a people who had no country, no king, no room for joy and pride except the family; the reduced size and tremendous vitality and endurance of the pitilessly selected survivors of the Ghetto; the repeated bursts of erratic genius from the human spirit so inhumanly restrained. But more patent still is the effect of the economic conditions,—the artificial development of a race of traders and dealers in money, from the lowest pawnbroker to the house of Rothschild; a special kind of people, bred of the economic environment in which they were compelled to live.

One rough but familiar instance of the same effect, from the same cause, we can all see in the marked distinction between the pastoral, the agricultural, and the manufacturing classes in any nation, though their other conditions be the same. On the clear line of argument that functions and organs are developed by use, that what we use most is developed most, and that the daily processes of supplying economic needs are the processes that we most use, it follows that, when we find special economic conditions affecting any special class of people, we may look for special results, and find them.

In view of these facts, attention is now called to a certain marked and peculiar economic condition affecting the human race, and unparalleled in the organic world. We are the only animal species in which the female depends on the male for food, the only animal species in which the sex-relation is also an economic relation. With us an entire sex lives in a relation of economic dependence upon the other sex, and the economic relation is combined with the sex-relation. The economic status of the human female is relative to the sex-relation.

It is commonly assumed that this condition also obtains among other animals, but such is not the case. There are many birds among which, during the nesting season, the male helps the female feed the young, and partially feeds her; and, with certain of the higher carnivora, the male helps the female feed the young, and partially feeds her. In no case does she depend on him absolutely, even during this season, save in that of the hornbill, where the female, sitting on her nest in a hollow tree, is walled in with clay by the male, so that only her beak projects; and then he feeds her while the eggs are developing. But even the female hornbill does not expect to be fed at any other time. The female bee and ant are economically dependent, but not on the male. The workers are females, too, specialized to economic functions solely. And with the carnivora, if the young are to lose one parent, it might far better be the father: the mother is quite competent to take care of them herself. With many species, as in the case of the common cat, she not only feeds herself and her young, but has to defend the young against the male as well. In no case is the female throughout her life supported by the male.

In the human species the condition is permanent and general, though there are exceptions, and though the present century is witnessing the beginnings of a great change in this respect. We have not been accustomed to face this fact beyond our loose generalization that it was "natural," and that other animals did so, too.

To many this view will not seem clear at first; and the case of working peasant women or females of savage tribes, and the general household industry of women, will be instanced against it. Some careful and honest discrimination is needed to make plain to ourselves the essential facts of the relation, even in these cases. The horse, in his free natural condition, is economically independent. He gets his living by his own exertions, irrespective of any other creature. The horse, in his present condition of slavery, is economically dependent. He gets his living at the hands of his master; and his exertions, though strenuous, bear no direct relation to his living. In fact, the horses who are the best fed and cared for and the horses who are the hardest worked are quite different animals. The horse works, it is true; but what he gets to eat depends on the power and will of his master. His living comes through another. He is economically dependent. So with the hard-worked savage or peasant women. Their labor is the property of another: they work under another will; and what they receive depends not on their labor, but on the power and will of another. They are economically dependent. This is true of the human female both individually and collectively.

In studying the economic position of the sexes collectively, the difference is most marked. As a social animal, the economic status of man rests on the combined and exchanged services of vast numbers of progressively specialized individuals. The economic progress of the race, its maintenance at any period, its continued advance, involve the collective activities of all the trades, crafts, arts, manufactures, inventions, discoveries, and all the civil and military institutions that go to maintain them. The economic status of any race at any time, with its involved effect on all the constituent individuals, depends on their world-wide labors and their free exchange. Economic progress, however, is almost exclusively masculine. Such economic processes as women have been allowed to exercise are of the earliest and most primitive kind. Were men to perform no economic services save such as are still performed by women, our racial status in economics would be reduced to most painful limitations.

To take from any community its male workers would paralyze it economically to a far greater degree than to remove its female workers. The labor now performed by the women could be performed by the men, requiring only the setting back of many advanced workers into earlier forms of industry; but the labor now performed by the men could not be performed by the women without generations of effort and adaptation. Men can cook, clean, and sew as well as women; but the making and managing of the great engines of modern industry, the threading of earth and sea in our vast systems of transportation, the handling of our elaborate machinery of trade, commerce, government,—these things could not be done so well by women in their present degree of economic development.

This is not owing to lack of the essential human faculties necessary to such achievements, nor to any inherent disability of sex, but to the present condition of woman, forbidding the development of this degree of economic ability. The male human being is thousands of years in advance of the female in economic status. Speaking collectively, men produce and distribute wealth; and women receive it at their hands. As men hunt, fish, keep cattle, or raise corn, so do women eat game, fish, beef, or corn. As men go down to the sea in ships, and bring coffee and spices and silks and gems from far away, so do women partake of the coffee and spices and silks and gems the men bring.

The economic status of the human race in any nation, at any time, is governed mainly by the activities of the male: the female obtains her share in the racial advance only through him.

Studied individually, the facts are even more plainly visible, more open and familiar. From the day laborer to the millionaire, the wife's worn dress or flashing jewels, her low roof or her lordly one, her weary feet or her rich equipage,—these speak of the economic ability of the husband. The comfort, the luxury, the necessities of life itself, which the woman receives, are obtained by the husband, and given her by him. And, when the woman, left alone with no man to "support" her, tries to meet her own economic necessities, the difficulties which confront her prove conclusively what the general economic status of the woman is. None can deny these patent facts,—that the economic status of women generally depends upon that of men generally, and that the economic status of women individually depends upon that of men individually, those men to whom they are related. But we are instantly confronted by the commonly received opinion that, although it must be admitted that men make and distribute the wealth of the world, yet women earn their share of it as wives. This assumes either that the husband is in the position of employer and the wife as employee, or that marriage is a "partnership," and the wife an equal factor with the husband in producing wealth.

Economic independence is a relative condition at best. In the broadest sense, all living things are economically dependent upon others,—the animals upon the vegetables, and man upon both. In a narrower sense, all social life is economically interdependent, man producing collectively what he could by no possibility produce separately. But, in the closest interpretation, individual economic independence among human beings means that the individual pays for what he gets, works for what he gets, gives to the other an equivalent for what the other gives him. I depend on the shoemaker for shoes, and the tailor for coats; but, if I give the shoemaker and the tailor enough of my own labor as a house-builder to pay for the shoes and coats they give me, I retain my personal independence. I have not taken of their product, and given nothing of mine. As long as what I get is obtained by what I give, I am economically independent.

Women consume economic goods. What economic product do they give in exchange for what they consume? The claim that marriage is a partnership, in which the two persons married produce wealth which neither of them, separately, could produce, will not bear examination. A man happy and comfortable can produce more than one unhappy and uncomfortable, but this is as true of a father or son as of a husband. To take from a man any of the conditions which make him happy and strong is to cripple his industry, generally speaking. But those relatives who make him happy are not therefore his business partners, and entitled to share his income.

Grateful return for happiness conferred is not the method of exchange in a partnership. The comfort a man takes with his wife is not in the nature of a business partnership, nor are her frugality and industry. A housekeeper, in her place, might be as frugal, as industrious, but would not therefore be a partner. Man and wife are partners truly in their mutual obligation to their children,—their common love, duty, and service. But a manufacturer who marries, or a doctor, or a lawyer, does not take a partner in his business, when he takes a partner in parenthood, unless his wife is also a manufacturer, a doctor, or a lawyer. In his business, she cannot even advise wisely without training and experience. To love her husband, the composer, does not enable her to compose; and the loss of a man's wife, though it may break his heart, does not cripple his business, unless his mind is affected by grief. She is in no sense a business partner, unless she contributes capital or experience or labor, as a man would in like relation. Most men would hesitate very seriously before entering a business partnership with any woman, wife or not.

If the wife is not, then, truly a business partner, in what way does she earn from her husband the food, clothing, and shelter she receives at his hands? By house service, it will be instantly replied. This is the general misty idea upon the subject,—that women earn all they get, and more, by house service. Here we come to a very practical and definite economic ground. Although not producers of wealth, women serve in the final processes of preparation and distribution. Their labor in the household has a genuine economic value.

For a certain percentage of persons to serve other persons, in order that the ones so served may produce more, is a contribution not to be overlooked. The labor of women in the house, certainly, enables men to produce more wealth than they otherwise could; and in this way women are economic factors in society. But so are horses. The labor of horses enables men to produce more wealth than they otherwise could. The horse is an economic factor in society. But the horse is not economically independent, nor is the woman. If a man plus a valet can perform more useful service than he could minus a valet, then the valet is performing useful service. But, if the valet is the property of the man, is obliged to perform this service, and is not paid for it, he is not economically independent.

The labor which the wife performs in the household is given as part of her functional duty, not as employment. The wife of the poor man, who works hard in a small house, doing all the work for the family, or the wife of the rich man, who wisely and gracefully manages a large house and administers its functions, each is entitled to fair pay for services rendered.

To take this ground and hold it honestly, wives, as earners through domestic service, are entitled to the wages of cooks, housemaids, nursemaids, seamstresses, or housekeepers, and to no more. This would of course reduce the spending money of the wives of the rich, and put it out of the power of the poor man to "support" a wife at all, unless, indeed, the poor man faced the situation fully, paid his wife her wages as house servant, and then she and he combined their funds in the support of their children. He would be keeping a servant: she would be helping keep the family. But nowhere on earth would there be "a rich woman" by these means. Even the highest class of private housekeeper, useful as her services are, does not accumulate a fortune. She does not buy diamonds and sables and keep a carriage. Things like these are not earned by house service.

But the salient fact in this discussion is that, whatever the economic value of the domestic industry of women is, they do not get it. The women who do the most work get the least money, and the women who have the most money do the least work. Their labor is neither given nor taken as a factor in economic exchange. It is held to be their duty as women to do this work; and their economic status bears no relation to their domestic labors, unless an inverse one. Moreover, if they were thus fairly paid,—given what they earned, and no more,—all women working in this way would be reduced to the economic status of the house servant.

Few women—or men either—care to face this condition. The ground that women earn their living by domestic labor is instantly forsaken, and we are told that they obtain their livelihood as mothers. This is a peculiar position. We speak of it commonly enough, and often with deep feeling, but without due analysis.

In treating of an economic exchange, asking what return in goods or labor women make for the goods and labor given them,—either to the race collectively or to their husbands individually,—what payment women make for their clothes and shoes and furniture and food and shelter, we are told that the duties and services of the mother entitle her to support.

If this is so, if motherhood is an exchangeable commodity given by women in payment for clothes and food, then we must of course find some relation between the quantity or quality of the motherhood and the quantity and quality of the pay. This being true, then the women who are not mothers have no economic status at all; and the economic status of those who are must be shown to be relative to their motherhood. This is obviously absurd. The childless wife has as much money as the mother of many,—more; for the children of the latter consume what would otherwise be hers; and the inefficient mother is no less provided for than the efficient one. Visibly, and upon the face of it, women are not maintained in economic prosperity proportioned to their motherhood. Motherhood bears no relation to their economic status. Among primitive races, it is true,—in the patriarchal period, for instance,—there was some truth in this position. Women being of no value whatever save as bearers of children, their favor and indulgence did bear direct relation to maternity; and they had reason to exult on more grounds than one when they could boast a son. To-day, however, the maintenance of the woman is not conditioned upon this. A man is not allowed to discard his wife because she is barren. The claim of motherhood as a factor in economic exchange is false to-day. But suppose it were true. Are we willing to hold this ground, even in theory? Are we willing to consider motherhood as a business, a form of commercial exchange? Are the cares and duties of the mother, her travail and her love, commodities to be exchanged for bread?

It is revolting so to consider them; and, if we dare face our own thoughts, and force them to their logical conclusion, we shall see that nothing could be more repugnant to human feeling, or more socially and individually injurious, than to make motherhood a trade. Driven off these alleged grounds of women's economic independence; shown that women, as a class, neither produce nor distribute wealth; that women, as individuals, labor mainly as house servants, are not paid as such, and would not be satisfied with such an economic status if they were so paid; that wives are not business partners or co-producers of wealth with their husbands, unless they actually practise the same profession; that they are not salaried as mothers, and that it would be unspeakably degrading if they were,—what remains to those who deny that women are supported by men? This (and a most amusing position it is),—that the function of maternity unfits a woman for economic production, and, therefore, it is right that she should be supported by her husband.

The ground is taken that the human female is not economically independent, that she is fed by the male of her species. In denial of this, it is first alleged that she is economically independent,—that she does support herself by her own industry in the house. It being shown that there is no relation between the economic status of woman and the labor she performs in the home, it is then alleged that not as house servant, but as mother, does woman earn her living. It being shown that the economic status of woman bears no relation to her motherhood, either in quantity or quality, it is then alleged that motherhood renders a woman unfit for economic production, and that, therefore, it is right that she be supported by her husband. Before going farther, let us seize upon this admission,—that she *is* supported by her husband.

Without going into either the ethics or the necessities of the case, we have reached so much common ground: the female of genus homo is supported by the male. Whereas, in other species of animals, male and female alike graze and browse, hunt and kill, climb, swim, dig, run, and fly for their livings, in our species the female does not seek her own living in the specific activities of our race, but is fed by the male.

Now as to the alleged necessity. Because of her maternal duties, the human female is said to be unable to get her own living. As the maternal duties of other females do not unfit them for getting their own living and also the livings of their young, it would seem that the human maternal duties require the segregation of the entire energies of the mother to the service of the child during her entire adult life, or so large a proportion of them that not enough remains to devote to the individual interests of the mother.

Such a condition, did it exist, would of course excuse and justify the pitiful dependence of the human female, and her support by the male. As the queen bee, modified entirely to maternity, is supported, not by the male, to be sure, but by her co-workers, the "old maids," the barren working bees, who labor so patiently and lovingly in their branch of the maternal duties of the hive, so would the human female, modified entirely to maternity, become unfit for any other exertion, and a helpless dependant.

Is this the condition of human motherhood? Does the human mother, by her motherhood, thereby lose control of brain and body, lose power and skill and desire for any other work? Do we see before us the human race, with all its females segregated entirely to the uses of motherhood, consecrated, set apart, specially developed, spending every power of their nature on the service of their children?

We do not. We see the human mother worked far harder than a mare, laboring her life long in the service, not of her children only, but of men; husbands, brothers, fathers, whatever male relatives she has; for mother and sister also; for the church a little, if she is allowed; for society, if she is able; for charity and education and reform,—working in many ways that are not the ways of motherhood.

It is not motherhood that keeps the housewife on her feet from dawn till dark; it is house service, not child service. Women work longer and harder than most men, and not solely in maternal duties. The savage mother carries the burdens, and does all menial service for the family tribe. The peasant mother toils in the fields, and the workingman's wife in the home. Many mothers, even now, are wage-earners for the family, as well as bearers and rearers of it. And the women who are not so occupied, the women who belong to rich men,—here perhaps is the exhaustive devotion to maternity which is supposed to justify an admitted economic dependence. But we do not find it even among these. Women of ease and wealth provide for their children better care than the poor woman can; but they do not spend more time upon it themselves, nor more care and effort. They have other occupation.

In spite of her supposed segregation to maternal duties, the human female, the world over, works at extra-maternal duties for hours enough to provide her with an independent living, and then is denied independence on the ground that motherhood prevents her working!

If this ground were tenable, we should find a world full of women who never lifted a finger save in the service of their children, and of men who did *all* the work besides, and waited on the women whom motherhood prevented from waiting on themselves. The ground is not tenable. A human female, healthy, sound, has twenty-five years of life before she is a mother, and should have twenty-five years more after the period of such maternal service as is expected of her has been given. The duties of grandmotherhood are surely not alleged as preventing economic independence.

The working power of the mother has always been a prominent factor in human life. She is the worker *par excellence,* but her work is not such as to affect her economic status. Her living, all that she gets,—food, clothing, ornaments, amusements, luxuries,—these bear no relation to her power to produce wealth, to her services in the house, or to her motherhood. These things bear relation only to the man she marries, the man she depends on,—to how much he has and how much he is willing to give her. The women whose splendid extravagance dazzles the world, whose economic goods are the greatest, are often neither houseworkers nor mothers, but simply the women who hold most power over the men who have the most money. The female of genus homo is economically dependent on the male. He is her food supply.

II

Knowing how important a factor in the evolution of species is the economic relation, and finding in the human species an economic relation so peculiar, we may naturally look to find effects peculiar to our race. We may expect to find phenomena in the sex-relation and in the economic relation of humanity of a unique character,— phenomena not traceable to human superiority, but singularly derogatory to that superiority; phenomena so marked, so morbid, as to give rise to much speculation as to their cause. Are these natural inferences fulfilled? Are these peculiarities in the sex-relation and in the economic relation manifested in human life? Indisputably these are,—so plain, so prominent, so imperiously demanding attention, that human thought has been occupied from its first consciousness in trying some way to account for them. To explain and relate these phenomena, separating what is due to normal race-development from what is due to this abnormal sexuo-economic relation, is the purpose of the line of study here suggested.

As the racial distinction of humanity lies in its social relation, so we find the distinctive gains and losses of humanity to lie also in its social relation. We are more affected by our relation to each other than by our physical environment.

Disadvantages of climate, deficiencies in food supply, competition from other species,—all these conditions society, in its organic strength, is easily able to overcome or to adjust. But in our inter-human relations we are not so successful. The serious dangers and troubles of human life arise from difficulties of adjustment with our social environment, and not with our physical environment. These difficulties, so far, have acted as a continual check to social progress. The more absolutely a nation has triumphed over physical conditions, the more successful it has become in its conquest of physical enemies and obstacles, the more it has given rein to the action of social forces which have ultimately destroyed the nation, and left the long ascent to be begun again by others.

> There is the moral of all human tales:
> 'Tis but the same rehearsal of the past,—
> First Freedom, and then Glory; when that fails,
> Wealth, Vice, Corruption,—barbarism at last.
> And History, with all her volumes vast,
> Hath but *one* page.[i]

The path of history is strewn with fossils and faint relics of extinct races,—races which died of what the sociologist would call internal diseases rather than natural causes. This, too, has been clear to the observer in all ages. It has been easily seen that there was something in our own behavior which did us more harm than any external difficulty; but what we have not seen is the natural cause of our unnatural conduct, and how most easily to alter it.

Rudely classifying the principal fields of human difficulty, we find one large proportion lies in the sex-relation, and another in the economic relation, between the individual constituents of society. To speak broadly, the troubles of life as we find them are mainly traceable to the heart or the purse. The other horror of our lives—disease—comes back often to these causes,—to something wrong either in economic relation or in sex-relation. To be ill-fed or ill-bred, or both, is largely what makes us the sickly race we are. In this wrong breeding, this maladjustment of the sex-relation in humanity, what are the principal features? We see in social evolution two main lines of action in this department of life. One is a gradual orderly development of monogamous marriage, as the form of sex-union best calculated to advance the interests of the individual and of society. It should be clearly understood that this is a natural development,

inevitable in the course of social progress; not an artificial condition, enforced by laws of our making. Monogamy is found among birds and mammals: it is just as natural a condition as polygamy or promiscuity or any other form of sex-union; and its permanence and integrity are introduced and increased by the needs of the young and the advantage to the race, just as any other form of reproduction was introduced. Our moral concepts rest primarily on facts. The moral quality of monogamous marriage depends on its true advantage to the individual and to society. If it were not the best form of marriage for our racial good, it would not be right. All the way up, from the promiscuous horde of savages, with their miscellaneous matings, to the lifelong devotion of romantic love, social life has been evolving a type of sex-union best suited to develope and improve the individual and the race. This is an orderly process, and a pleasant one, involving only such comparative pain and difficulty as always attend the assumption of new processes and the extinction of the old; but accompanied by far more joy than pain.

But with the natural process of social advancement has gone an unnatural process,—an erratic and morbid action, making the sex-relation of humanity a frightful source of evil. So prominent have been these morbid actions and evil results that hasty thinkers of all ages have assumed that the whole thing was wrong, and that celibacy was the highest virtue. Without the power of complete analysis, without knowledge of the sociological data essential to such analysis, we have sweepingly condemned as a whole what we could easily see was so allied with pain and loss. But, like all natural phenomena, the phenomena of sex may be studied, both the normal and the abnormal, the physiological and the pathological; and we are quite capable of understanding why we are in such evil case, and how we may attain more healthful conditions.

So far, the study of this subject has rested on the assumption that man must be just as we find him, that man behaves just as he chooses, and that, if he does not choose to behave as he does, he can stop. Therefore, when we discovered that human behavior in the sex-relation was productive of evil, we exhorted the human creature to stop so behaving, and have continued so to exhort for many centuries. By law and religion, by education and custom, we have sought to enforce upon the human individual the kind of behavior which our social sense so clearly showed was right.

---

[i]Childe Harold's Pilgrimage, Canto IV., cviii.

But always there has remained the morbid action. Whatever the external form of sex-union to which we have given social sanction, however Bible and Koran and Vedas have offered instruction, some hidden cause has operated continuously against the true course of social evolution, to pervert the natural trend toward a higher and more advantageous sex-relation; and to maintain lower forms, and erratic phases, of a most disadvantageous character.

Every other animal works out the kind of sex-union best adapted to the reproduction of his species, and peacefully practises it. We have worked out the kind that is best for us,—best for the individuals concerned, for the young resultant, and for society as a whole; but we do not peacefully practise it. So palpable is this fact that we have commonly accepted it, and taken it for granted that this relation must be a continuous source of trouble to humanity. "Marriage is a lottery," is a common saying among us. "The course of true love never did run smooth." And we quote with unction Punch's advice to those about to marry,—"Don't!" That peculiar sub-relation which has dragged along with us all the time that monogamous marriage has been growing to be the accepted form of sex-union—prostitution—we have accepted, and called a "social necessity." We also call it "the social evil." We have tacitly admitted that this relation in the human race must be more or less uncomfortable and wrong, that it is part of our nature to have it so.

Now let use examine the case fairly and calmly, and see whether it is as inscrutable and immutable as hitherto believed. What are the conditions? What are the natural and what the unnatural features of the case? To distinguish these involves a little study of the evolution of the processes of reproduction.

Very early in the development of species it was ascertained by nature's slow but sure experiments that the establishment of two sexes in separate organisms, and their differentiation, was to the advantage of the species. Therefore, out of the mere protoplasmic masses, the floating cells, the amorphous early forms of life, grew into use the distinction of the sexes,—the gradual development of masculine and feminine organs and functions in two distinct organisms. Developed and increased by use, the distinction of sex increased in the evolution of species. As the distinction increased, the attraction increased, until we have in all the higher races two markedly different sexes, strongly drawn together by the attraction of sex, and fulfilling their use in the reproduction of species. These are the natural features of sex-distinction and sex-union, and they are found in the human

species as in others. The unnatural feature by which our race holds an unenviable distinction consists mainly in this,—a morbid excess in the exercise of this function.

It is this excess, whether in marriage or out, which makes the health and happiness of humanity in this relation so precarious. It is this excess, always easily seen, which law and religion have mainly striven to check. Excessive sex-indulgence is the distinctive feature of humanity in this relation.

To define "excess" in this connection is not difficult. All natural functions that require our conscious co-operation for their fulfilment are urged upon our notice by an imperative desire. We do not have to desire to breathe or to digest or to circulate the blood, because that is done without our volition; but we do have to desire to eat and drink, because the stomach cannot obtain its supplies without in some way spurring the whole organism to secure them. So hunger is given us as an essential factor in our process of nutrition. In the same manner sex-attraction is an essential factor in the fulfilment of our processes of reproduction. In a normal condition the amount of hunger we feel is exactly proportioned to the amount of food we need. It tells us when to eat and when to stop. In some diseased conditions "an unnatural appetite" sets in; and we are impelled to eat far beyond the capacity of the stomach to digest, of the body to assimilate. This is an excessive hunger.

We, as a race, manifest an excessive sex-attraction, followed by its excessive indulgence, and the inevitable evil consequence. It urges us to a degree of indulgence which bears no relation to the original needs of the organism, and which is even so absurdly exaggerated as to react unfavorably on the incidental gratification involved; an excess which tends to pervert and exhaust desire as well as to injure reproduction.

The human animal manifests an excess in sex-attraction which not only injures the race through its morbid action on the natural processes of reproduction, but which injures the happiness of the individual through its morbid reaction on his own desires.

What is the cause of this excessive sex- attraction in the human species? The immediately acting cause of sex-attraction is sex-distinction. The more widely the sexes are differentiated, the more forcibly they are attracted to each other. The more highly developed becomes the distinction of sex in either organism, the more intense is its attraction for the other. In the human species we find sex-distinction carried to an excessive degree. Sex-distinction in humanity is so marked as to retard and confuse race-distinction, to check individual distinction, seriously to injure the race. Accustomed as we are simply to accept the facts of life as we find them, to consider people as permanent types instead of seeing them and the whole race in continual change according to the action of many forces, it seems strange at first to differentiate between familiar manifestations of sex-distinction, and to say: "This is normal, and should not be disturbed. This is abnormal, and should be removed." But that is precisely what must be done.

Normal sex-distinction manifests itself in all species in what are called primary and secondary sex-characteristics. The primary are those organs and functions essential to reproduction; the secondary, those modifications of structure and function which subserve the uses of reproduction ultimately, but are not directly essential,—such as the horns of the stag, of use in sex-combat; the plumage of the peacock, of use in sex- competition. All the minor characteristics of beard or mane, comb, wattles, spurs, gorgeous color or superior size, which distinguish the male from the female,—these are distinctions of sex. These distinctions are of use to the species through reproduction only, the processes of race-preservation. They are not of use in self-preservation. The creature is not profited personally by his mane or crest or tail-feathers: they do not help him get his dinner or kill his enemies.

On the contrary, they react unfavorably upon his personal gains, if, through too great development, they interfere with his activity or render him a conspicuous mark for enemies. Such development would constitute excessive sex-distinction, and this is precisely the condition of the human race. Our distinctions of sex are carried to such a degree as to be disadvantageous to our progress as individuals and as a race. The sexes in our species are differentiated not only enough to perform their primal functions; not only enough to manifest all sufficient secondary sexual characteristics and fulfil their use in giving rise to sufficient sex-attraction; but so much as seriously to interfere with the processes of self-preservation on the one hand; and, more conspicuous still, so much as to react unfavorably upon the very processes of race-preservation which they are meant to serve. Our excessive sex-distinction, manifesting the characteristics of sex to an abnormal degree, has given rise to a degree of attraction which demands a degree of indulgence that directly injures motherhood and fatherhood. We are not better as parents, nor better as people, for our existing degree of sex-distinction, but visibly worse. To what conditions are we to look for the developing cause of these phenomena?

Let us first examine the balance of forces by which these two great processes, self-preservation and race-preservation, are conducted in the world. Self-preservation involves the expenditure of energy in those acts, and their ensuing modifications of structure and function, which tend to the maintenance of the individual life. Race-preservation involves the expenditure of energy in those acts, and their ensuing modifications of structure and function, which tend to the maintenance of the racial life, even to the complete sacrifice of the individual. This primal distinction should be clearly held in mind. Self-preservation and race-preservation are in no way identical processes, and are often directly opposed. In the line of self-preservation, natural selection, acting on the individual, developes those characteristics which enable it to succeed in "the struggle for existence," increasing by use those organs and functions by which it directly profits. In the line of race-preservation, sexual selection, acting on the individual, developes those characteristics which enable it to succeed in what Drummond has called "the struggle for the existence of others," increasing by use those organs and functions by which its young are to profit, directly or indirectly. The individual has been not only modified to its environment, under natural selection, but modified to its mate, under sexual selection, each sex developing the qualities desired by the other by the simple process of choice, those best sexed being first chosen, and transmitting their sex-development as well as their racial development.

The order mammalia is the resultant of a primary sex-distinction developed by natural selection; but the gorgeous plumage of the peacock's tail is a secondary sex-distinction developed by sexual selection. If the peacock's tail were to increase in size and splendor till it shone like the sun and covered an acre,—if it tended so to increase, we will say,—such excessive sex-distinction would be so inimical to the personal prosperity of that peacock that he would die, and his tail-tendency would perish with him. If the pea-hen, conversely, whose sex-distinction attracts in the opposite direction, not by being large and splendid, but small and dull,—if she should grow so small and dull as to fail to keep herself and her young fed and defended, then she would die; and there would be another check to excessive sex-distinction. In herds of deer and cattle the male is larger and stronger, the female smaller and weaker; but, unless the latter is large and strong enough to keep up with the male in the search for food or the flight from foes, one is taken and the other left, and there is no more of that kind of animal. Differ as they may in sex, they must remain alike in species, equal in race-development, else destruction overtakes them. The force of natural selection, demanding and producing identical race-qualities, acts as a check on sexual selection, with its production of different sex-qualities. As sexes, they perform different functions, and therefore tend to develope differently. As species, they perform the same functions, and therefore tend to develope equally.

And as sex-functions are only used occasionally, and race-functions are used all the time,—as they mate but yearly or tri-monthly, but eat daily and hourly,—the processes of obtaining food or of opposing constant enemies act more steadily than the processes of reproduction, and produce greater effect.

We find the order mammalia accordingly producing and suckling its young in the same manner through a wide variety of species which obtain their living in a different manner. The calf and colt and cub and kitten are produced by the same process; but the cow and horse, the bear and cat, are produced by different processes. And, though cow and bull, mare and stallion, differ as to sex, they are alike in species; and the likeness in species is greater than the difference in sex. Cow, mare, and cat are all females of the order mammalia, and so far alike; but how much more different they are than similar!

Natural selection developes race. Sexual selection developes sex. Sex-development is one throughout its varied forms, tending only to reproduce what is. But race-development rises ever in higher and higher manifestation of energy. As sexes, we share our distinction with the animal kingdom almost to the beginning of life, and with the vegetable world as well. As races, we differ in ascending degree; and the human race stands highest in the scale of life so far.

When, then, it can be shown that sex- distinction in the human race is so excessive as not only to affect injuriously its own purposes, but to check and pervert the progress of the race, it becomes a matter for most serious consideration. Nothing could be more inevitable, however, under our sexuo-economic relation. By the economic dependence of the human female upon the male, the balance of forces is altered. Natural selection no longer checks the action of sexual selection, but co-operates with it. Where both sexes obtain their food through the same exertions, from the same sources, under the same conditions, both sexes are acted upon alike, and developed alike by their environment. Where the two sexes obtain their food under different conditions, and where that difference consists in one of them being fed by the other, then the feeding sex becomes the environment of the fed. Man, in supporting woman, has

become her economic environment. Under natural selection, every creature is modified to its environment, developing perforce the qualities needed to obtain its livelihood under that environment. Man, as the feeder of woman, becomes the strongest modifying force in her economic condition. Under sexual selection the human creature is of course modified to its mate, as with all creatures. When the mate becomes also the master, when economic necessity is added to sex-attraction, we have the two great evolutionary forces acting together to the same end; namely, to develope sex-distinction in the human female. For, in her position of economic dependence in the sex-relation, sex-distinction is with her not only a means of attracting a mate, as with all creatures, but a means of getting her livelihood, as is the case with no other creature under heaven. Because of the economic dependence of the human female on her mate, she is modified to sex to an excessive degree. This excessive modification she transmits to her children; and so is steadily implanted in the human constitution the morbid tendency to excess in this relation, which has acted so universally upon us in all ages, in spite of our best efforts to restrain it. It is not the normal sex-tendency, common to all creatures, but an abnormal sex-tendency, produced and maintained by the abnormal economic relation which makes one sex get its living from the other by the exercise of sex-functions. This is the immediate effect upon individuals of the peculiar sexuo-economic relation which obtains among us.

III

In establishing the claim of excessive sex- distinction in the human race, much needs to be said to make clear to the general reader what is meant by the term. To the popular mind, both the coarsely familiar and the over-refined, "sexual" is thought to mean "sensual"; and the charge of excessive sex-distinction seems to be a reproach. This should be at once dismissed, as merely showing ignorance of the terms used. A man does not object to being called "masculine," nor a woman to being called "feminine." Yet whatever is masculine or feminine is sexual. To be distinguished by femininity is to be distinguished by sex. To be over-feminine is to be over-sexed. To manifest in excess any of the distinctions of sex, primary or secondary, is to be over-sexed. Our hypothetical peacock, with his too large and splendid tail, would be over-sexed, and no offence to his moral character!

The primary sex-distinctions in our race as in others consist merely in the essential organs and functions of reproduction. The secondary distinctions, and this is where we are to look for our largest excess—consist in all those differences in organ and function, in look and action, in habit, manner, method, occupation, behavior, which distinguish men from women. In a troop of horses, seen at a distance, the sexes are indistinguishable. In a herd of deer the males are distinguishable because of their antlers. The male lion is distinguished by his mane, the male cat only by a somewhat heavier build. In certain species of insects the male and female differ so widely in appearance that even naturalists have supposed them to belong to separate species. Beyond these distinctions lies that of conduct. Certain psychic attributes are manifested by either sex. The intensity of the maternal passion is a sex-distinction as much as the lion's mane or the stag's horns. The belligerence and dominance of the male is a sex-distinction: the modesty and timidity of the female is a sex-distinction. The tendency to "sit" is a sex-distinction of the hen: the tendency to strut is a sex-distinction of the cock. The tendency to fight is a sex-distinction of males in general: the tendency to protect and provide for is a sex- distinction of females in general.

With the human race, whose chief activities are social, the initial tendency to sex-distinction is carried out in many varied functions. We have differentiated our industries, our responsibilities, our very virtues, along sex lines. It will therefore be clear that the claim of excessive sex-distinction in humanity, and especially in woman, does not carry with it any specific "moral" reproach, though it does in the larger sense prove a decided evil in its effect on human progress.

In primary distinctions our excess is not so marked as in the farther and subtler development; yet, even here, we have plain proof of it. Sex-energy in its primal manifestation is exhibited in the male of the human species to a degree far greater than is necessary for the processes of reproduction,—enough, indeed, to subvert and injure those processes. The direct injury to reproduction from the excessive indulgence of the male, and the indirect injury through its debilitating effect upon the female, together with the enormous evil to society produced by extra-marital indulgence,—these are facts quite generally known. We have recognized them for centuries; and sought to check the evil action by law, civil, social, moral. But we have treated it always as a field of voluntary action, not as a condition of morbid development. We have held it as right that man should be so, but wrong that man should do so. Nature does not work in that way.

What it is right to be, it is right to do. What it is wrong to do, it is wrong to be. This inordinate demand in the human male is an excessive sex-distinction. In this, in a certain over-coarseness and hardness, a too great belligerence and pride, a too great subservience to the power of sex-attraction, we find the main marks of excessive sex-distinction in men. It has been always checked and offset in them by the healthful activities of racial life. Their energies have been called out and their faculties developed along all the lines of human progress. In the growth of industry, commerce, science, manufacture, government, art, religion, the male of our species has become human, far more than male. Strong as this passion is in him, inordinate as is his indulgence, he is a far more normal animal than the female of his species,—far less over-sexed. To him this field of special activity is but part of life,—an incident. The whole world remains besides. To her it is the world. This has been well stated in the familiar epigram of Madame de Staël,—"Love with man is an episode, with woman a history." It is in woman that we find most fully expressed the excessive sex-distinction of the human species,—physical, psychical, social. See first the physical manifestation.

To make clear by an instance the difference between normal and abnormal sex-distinction, look at the relative condition of a wild cow and a "milch cow," such as we have made. The wild cow is a female. She has healthy calves, and milk enough for them; and that is all the femininity she needs. Otherwise than that she is bovine rather than feminine. She is a light, strong, swift, sinewy creature, able to run, jump, and fight, if necessary. We, for economic uses, have artificially developed the cow's capacity for producing milk. She has become a walking milk-machine, bred and tended to that express end, her value measured in quarts. The secretion of milk is a maternal function,—a sex-function. The cow is over-sexed. Turn her loose in natural conditions, and, if she survive the change, she would revert in a very few generations to the plain cow, with her energies used in the general activities of her race, and not all running to milk.

Physically, woman belongs to a tall, vigorous, beautiful animal species, capable of great and varied exertion. In every race and time when she has opportunity for racial activity, she developes accordingly, and is no less a woman for being a healthy human creature. In every race and time where she is denied this opportunity,—and few, indeed, have been her years of freedom,—she has developed in the lines of action to which she was confined; and those were always lines of sex-activity. In consequence the body of woman, speaking in the largest generalization, manifests sex-distinction predominantly.

Woman's femininity—and "the eternal feminine" means simply the eternal sexual—is more apparent in proportion to her humanity than the femininity of other animals in proportion to their caninity or felinity or equinity. "A feminine hand" or "a feminine foot" is distinguishable anywhere. We do not hear of "a feminine paw" or "a feminine hoof." A hand is an organ of prehension, a foot an organ of locomotion: they are not secondary sexual characteristics. The comparative smallness and feebleness of woman is a sex-distinction. We have carried it to such an excess that women are commonly known as "the weaker sex." There is no such glaring difference between male and female in other advanced species. In the long migrations of birds, in the ceaseless motion of the grazing herds that used to swing up and down over the continent each year, in the wild, steep journeys of the breeding salmon, nothing is heard of the weaker sex. And among the higher carnivora, where longer maintenance of the young brings their condition nearer ours, the hunter dreads the attack of the female more than that of the male. The disproportionate weakness is an excessive sex-distinction. Its injurious effect may be broadly shown in the Oriental nations, where the female in curtained harems is confined most exclusively to sex-functions and denied most fully the exercise of race-functions. In such peoples the weakness, the tendency to small bones and adipose tissue of the over-sexed female, is transmitted to the male, with a retarding effect on the development of the race. Conversely, in early Germanic tribes the comparatively free and humanly developed women—tall, strong, and brave—transmitted to their sons a greater proportion of human power and much less of morbid sex-tendency.

The degree of feebleness and clumsiness common to women, the comparative inability to stand, walk, run, jump, climb, and perform other race-functions common to both sexes, is an excessive sex-distinction; and the ensuing transmission of this relative feebleness to their children, boys and girls alike, retards human development. Strong, free, active women, the sturdy, field-working peasant, the burden-bearing savage, are no less good mothers for their human strength. But our civilized "feminine delicacy," which appears somewhat less delicate when recognized as an expression of sexuality in excess,—makes us no better mothers, but worse. The relative weakness of women is a sex-distinction. It is apparent in her to a degree that injures motherhood, that injures wifehood, that injures the individual. The sex-usefulness and the human usefulness of women, their general duty to their kind, are greatly injured by this degree of distinction. In every way the over-sexed condition of the human female reacts unfavorably upon herself, her husband, her children, and the race.

In its psychic manifestation this intense sex-distinction is equally apparent. The primal instinct of sex-attraction has developed under social forces into a conscious passion of enormous power, a deep and lifelong devotion, overwhelming in its force. This is excessive in both sexes, but more so in women than in men,—not so commonly in its simple physical form, but in the unreasoning intensity of emotion that refuses all guidance, and drives those possessed by it to risk every other good for this one end. It is not at first sight easy, and it may seem an irreverent and thankless task, to discriminate here between what is good in the "master passion" and what is evil, and especially to claim for one sex more of this feeling than for the other; but such discrimination can be made.

It is good for the individual and for the race to have developed such a degree of passionate and permanent love as shall best promote the happiness of individuals and the reproduction of species. It is not good for the race or for the individual that his feeling should have become so intense as to override all other human faculties, to make a mock of the accumulated wisdom of the ages, the stored power of the will; to drive the individual—against his own plain conviction—into a union sure to result in evil, or to hold the individual helpless in such an evil union, when made.

Such is the condition of humanity, involving most evil results to its offspring and to its own happiness. And, while in men the immediate dominating force of the passion may be more conspicuous, it is in women that it holds more universal sway. For the man has other powers and faculties in full use, whereby to break loose from the force of this; and the woman, specially modified to sex and denied racial activity, pours her whole life into her love, and, if injured here, she is injured irretrievably. With him it is frequently light and transient, and, when most intense, often most transient. With her it is a deep, all-absorbing force, under the action of which she will renounce all that life offers, take any risk, face any hardships, bear any pain. It is maintained in her in the face of a lifetime of neglect and abuse. The common instance of the police court trials—the woman cruelly abused who will not testify against her husband—shows this. This devotion, carried to such a degree as to lead to the mismating of individuals with its personal and social injury, is an excessive sex-distinction.

But it is in our common social relations that the predominance of sex-distinction in women is made most manifest. The fact that, speaking broadly, women have, from the very beginning, been spoken of expressively enough as "the sex," demonstrates clearly that this is the main impression which they have made upon observers and recorders. Here one need attempt no farther proof than to turn the mind of the reader to an unbroken record of facts and feelings perfectly patent to every one, but not hitherto looked at as other than perfectly natural and right. So utterly has the status of woman been accepted as a sexual one that it has remained for the woman's movement of the nineteenth century to devote much contention to the claim that women are persons! That women are persons as well as females,—an unheard of proposition!

In a "Handbook of Proverbs of All Nations," a collection comprising many thousands, these facts are to be observed: first, that the proverbs concerning women are an insignificant minority compared to those concerning men; second, that the proverbs concerning women almost invariably apply to them in general,—to the sex. Those concerning men qualify, limit, describe, specialize. It is "a lazy man," "a violent man," "a man in his cups." Qualities and actions are predicated of man individually, and not as a sex, unless he is flatly contrasted with woman, as in "A man of straw is worth a woman of gold," "Men are deeds, women are words," or "Man, woman, and the devil are the three degrees of comparison." But of woman it is always and only "a woman," meaning simply a female, and recognizing no personal distinction: "As much pity to see a woman weep as to see a goose go barefoot." "He that hath an eel by the tail and a woman by her word hath a slippery handle." "A woman, a spaniel, and a walnut-tree,—the more you beat 'em, the better they be." Occasionally a distinction is made between "a fair woman" and "a black woman"; and Solomon's "virtuous woman," who commanded such a high price, is familiar to us all. But in common thought it is simply "a woman" always. The boast of the profligate that he knows "the sex," so recently expressed by a new poet,—"The things you will learn from the Yellow and Brown, they'll 'elp you an' 'eap with the White"; the complaint of the angry rejected that "all women are just alike!"—the consensus of public opinion of all time goes to show that the characteristics common to the sex have predominated over the characteristics distinctive of the individual,—a marked excess in sex-distinction.

From the time our children are born, we use every means known to accentuate sex-distinction in both boy and girl; and the reason that the boy is not so hopelessly marked by it as the girl is that he has the whole field of human expression open to him besides. In our steady insistence on proclaiming sex-distinction we have grown to consider most human attributes as masculine attributes, for the simple reason that they were allowed to men and forbidden to women.

A clear and definite understanding of the difference between race-attributes and sex-attributes should be established. Life consists of action. The action of a living thing is along two main lines,—self-preservation and race-preservation. The processes that keep the individual alive, from the involuntary action of his internal organs to the voluntary action of his external organs,—every act, from breathing to hunting his food, which contributes to the maintenance of the individual life,—these are the processes of self-preservation. Whatever activities tend to keep the race alive, to reproduce the individual, from the involuntary action of the internal organs to the voluntary action of the external organs; every act from the development of germ-cells to the taking care of children, which contributes to the maintenance of the racial life,—these are the processes of race-preservation. In race-preservation, male and female have distinctive organs, distinctive functions, distinctive lines of action. In self-preservation, male and female have the same organs, the same functions, the same lines of action. In the human species our processes of race-preservation have reached a certain degree of elaboration; but our processes of self-preservation have gone farther, much farther.

All the varied activities of economic production and distribution, all our arts and industries, crafts and trades, all our growth in science, discovery, government, religion,—these are along the line of self-preservation: these are, or should be, common to both sexes. To teach, to rule, to make, to decorate, to distribute,—these are not sex-functions: they are race-functions. Yet so inordinate is the sex-distinction of the human race that the whole field of human progress has been considered a masculine prerogative. What could more absolutely prove the excessive sex-distinction of the human race? That this difference should surge over all its natural boundaries and blazon itself across every act of life, so that every step of the human creature is marked "male" or "female,"—surely, this is enough to show our over-sexed condition.

Little by little, very slowly, and with most unjust and cruel opposition, at cost of all life holds most dear, it is being gradually established by many martyrdoms that human work is woman's as well as man's. Harriet Martineau must conceal her writing under her sewing when callers came, because "to sew" was a feminine verb, and "to write" a masculine one. Mary Somerville must struggle to hide her work from even relatives, because mathematics was a "masculine" pursuit. Sex has been made to dominate the whole human world,— all the main avenues of life marked "male," and the female left to be a female, and nothing else.

But while with the male the things he fondly imagined to be "masculine" were merely human, and very good for him, with the female the few things marked "feminine" were feminine, indeed; and her ceaseless reiterance of one short song, however sweet, has given it a conspicuous monotony. In garments whose main purpose is unmistakably to announce her sex; with a tendency to ornament which marks exuberance of sex-energy, with a body so modified to sex as to be grievously deprived of its natural activities; with a manner and behavior wholly attuned to sex-advantage, and frequently most disadvantageous to any human gain; with a field of action most rigidly confined to sex-relations; with her overcharged sensibility, her prominent modesty, her "eternal femininity,"—the female of genus homo is undeniably over-sexed.

This excessive distinction shows itself again in a marked precocity of development. Our little children, our very babies, show signs of it when the young of other creatures are serenely asexual in general appearance and habit. We eagerly note this precocity. We are proud of it. We carefully encourage it by precept and example, taking pains to develope the sex-instinct in little children, and think no harm. One of the first things we force upon the child's dawning consciousness is the fact that he is a boy or that she is a girl, and that, therefore, each must regard everything from a different point of view. They must be dressed differently, not on account of their personal needs, which are exactly similar at this period, but so that neither they, nor any one beholding them, may for a moment forget the distinction of sex.

Our peculiar inversion of the usual habit of species, in which the male carries ornament and the female is dark and plain, is not so much a proof of excess indeed, as a proof of the peculiar reversal of our position in the matter of sex-selection. With the other species the males compete in ornament, and the females select. With us the females compete in ornament, and the males select. If this theory of sex-ornament is disregarded, and we prefer rather to see in masculine decoration merely a form of exuberant sex-energy, expending itself in non-productive excess, then, indeed, the fact that with us the females manifest such a display of gorgeous adornment is another sign of excessive sex-distinction. In either case the forcing upon girl-children of an elaborate ornamentation which interferes with their physical activity and unconscious

freedom, and fosters a premature sex-consciousness, is as clear and menacing a proof of our condition as could be mentioned. That the girl-child should be so dressed as to require a difference in care and behavior, resting wholly on the fact that she is a girl,—a fact not otherwise present to her thought at that age,—is a precocious insistence upon sex-distinction, most unwholesome in its results. Boys and girls are expected, also, to behave differently to each other, and to people in general,—a behavior to be briefly described in two words. To the boy we say, "Do"; to the girl, "Don't." The little boy must "take care" of the little girl, even if she is larger than he is. "Why?" he asks. Because he is a boy. Because of sex. Surely, if she is the stronger, she ought to take care of him, especially as the protective instinct is purely feminine in a normal race. It is not long before the boy learns his lesson. He is a boy, going to be a man; and that means all. "I thank the Lord that I was not born a woman," runs the Hebrew prayer. She is a girl, "only a girl," "nothing but a girl," and going to be a woman,—only a woman. Boys are encouraged from the beginning to show the feelings supposed to be proper to their sex. When our infant son bangs about, roars, and smashes things, we say proudly that he is "a regular boy!" When our infant daughter coquettes with visitors, or wails in maternal agony because her brother has broken her doll, whose sawdust remains she nurses with piteous care, we say proudly that "she is a perfect little mother already!" What business has a little girl with the instincts of maternity? No more than the little boy should have with the instincts of paternity. They are sex-instincts, and should not appear till the period of adolescence. The most normal girl is the "tom-boy,"—whose numbers increase among us in these wiser days,—a healthy young creature, who is human through and through, not feminine till it is time to be. The most normal boy has calmness and gentleness as well as vigor and courage. He is a human creature as well as a male creature, and not aggressively masculine till it is time to be. Childhood is not the period for these marked manifestations of sex. That we exhibit them, that we admire and encourage them, shows our over-sexed condition.

IV

Having seen the disproportionate degree of sex-distinction in humanity and its greater manifestation in the female than in the male, and having seen also the unique position of the human female as an economic dependant on the male of her species, it is not difficult to establish a relation between these two facts. The general law acting to produce this condition of exaggerated sex-development was briefly referred to in the second chapter. It is as follows: the natural tendency of any function to increase in power by use causes sex-activity to increase under the action of sexual selection. This tendency is checked in most species by the force of natural selection, which diverts the energies into other channels and developes race-activities. Where the female finds her economic environment in the male, and her economic advantage is directly conditioned upon the sex-relation, the force of natural selection is added to the force of sexual selection, and both together operate to develope sex-activity. In any animal species, free from any other condition, such a relation would have inevitably developed sex to an inordinate degree, as may be readily seen in the comparatively similar cases of those insects where the female, losing economic activity and modified entirely to sex, becomes a mere egg-sac, an organism with no powers of self-preservation, only those of race-preservation. With these insects the only race-problem is to maintain and reproduce the species, and such a condition is not necessarily evil; but with a race like ours, whose development as human creatures is but comparatively begun, it is evil because of its check to individual and racial progress. There are other purposes before us besides mere maintenance and reproduction.

It should be clear to any one accustomed to the working of biological laws that all the tendencies of a living organism are progressive in their development, and are held in check by the interaction of their several forces. Each living form, with its dominant characteristics, represents a balance of power, a sort of compromise. The size of earth's primeval monsters was limited by the tensile strength of their material. Sea monsters can be bigger, because the medium in which they move offers more support. Birds must be smaller for the opposite reason. The cow requires many stomachs of a liberal size, because her food is of low nutritive value; and she must eat large quantities to keep her machine going. The size of arboreal animals, such as monkeys or squirrels, is limited by the nature of their habitat: creatures that live in trees cannot be so big as creatures that live on the ground. Every quality of every creature is relative to its condition, and tends to increase or decrease accordingly; and each quality tends to increase in proportion to its use, and to decrease in proportion to its disuse. Primitive man and his female were animals, like other animals. They were strong, fierce, lively beasts; and she was as nimble and ferocious as he, save for the added belligerence of the males in their sex-competition. In this competition, he, like the other male creatures, fought savagely with his hairy rivals; and she, like the other female creatures, complacently viewed their struggles, and mated with the victor. At other times she ran about in the forest, and helped herself to what there was to eat as freely as he did.

There seems to have come a time when it occurred to the dawning intelligence of this amiable savage that it was cheaper and easier to fight a little female, and have it done with, than to fight a big male every time. So he instituted the custom of enslaving the female; and she, losing freedom, could no longer get her own food nor that of her young. The mother ape, with her maternal function well fulfilled, flees leaping through the forest,—plucks her fruit and nuts, keeps up with the movement of the tribe, her young one on her back or held in one strong arm. But the mother woman, enslaved, could not do this. Then man, the father, found that slavery had its obligations: he must care for what he forbade to care for itself, else it died on his hands. So he slowly and reluctantly shouldered the duties of his new position. He began to feed her, and not only that, but to express in his own person the thwarted uses of maternity: he had to feed the children, too. It seems a simple arrangement. When we have thought of it at all, we have thought of it with admiration. The naturalist defends it on the ground of advantage to the species through the freeing of the mother from all other cares and confining her unreservedly to the duties of maternity. The poet and novelist, the painter and sculptor, the priest and teacher, have all extolled this lovely relation. It remains for the sociologist, from a biological point of view, to note its effects on the constitution of the human race, both in the individual and in society.

When man began to feed and defend women, she ceased proportionately to feed and defend herself. When he stood between her and her physical environment, she ceased proportionately to feel the influence of that environment and respond to it. When he became her immediate and all-important environment, she began proportionately to respond to this new influence, and to be modified accordingly. In a free state, speed was of as great advantage to the female as to the male, both in enabling her to catch prey and in preventing her from being caught by enemies; but, in her new condition, speed was a disadvantage. She was not allowed to do the catching, and it profited her to be caught by her new master. Free creatures, getting their own food and maintaining their own lives, develope an active capacity for attaining their ends. Parasitic creatures, whose living is obtained by the exertions of others, develope powers of absorption and of tenacity,—the powers by which they profit most. The human female was cut off from the direct action of natural selection, that mighty force which heretofore had acted on male and female alike with inexorable and beneficial effect, developing strength, developing skill, developing endurance, developing courage,—in a word, developing species. She now met the influence of natural selection acting indirectly through the male, and developing, of course, the faculties required to secure and obtain a hold on him. Needless to state that these faculties were those of sex-attraction, the one power that has made him cheerfully maintain, in what luxury he could, the being in whom he delighted. For many, many centuries she had no other hold, no other assurance of being fed. The young girl had a prospective value, and was maintained for what should follow; but the old woman, in more primitive times, had but a poor hold on life. She who could best please her lord was the favorite slave or favorite wife, and she obtained the best economic conditions.

With the growth of civilization, we have gradually crystallized into law the visible necessity for feeding the helpless female; and even old women are maintained by their male relatives with a comfortable assurance. But to this day—save, indeed, for the increasing army of women wage-earners, who are changing the face of the world by their steady advance toward economic independence—the personal profit of women bears but too close a relation to their power to win and hold the other sex. From the odalisque with the most bracelets to the débutante with the most bouquets, the relation still holds good,—woman's economic profit comes through the power of sex-attraction.

When we confront this fact boldly and plainly in the open market of vice, we are sick with horror. When we see the same economic relation made permanent, established by law, sanctioned and sanctified by religion, covered with flowers and incense and all accumulated sentiment, we think it innocent, lovely, and right. The transient trade we think evil. The bargain for life we think good. But the biological effect remains the same. In both cases the female gets her food from the male by virtue of her sex-relationship to him. In both cases, perhaps even more in marriage because of its perfect acceptance of the situation, the female of genus homo, still living under natural law, is inexorably modified to sex in an increasing degree.

Followed in specific detail, the action of the changed environment upon women has been in given instances as follows: In the matter of mere passive surroundings she has been immediately restricted in her range. This one factor has an immense effect on man and animal alike. An absolutely uniform environment, one shape, one size, one color, one sound, would render life, if any life could be, one helpless, changeless thing. As the environment increases and varies, the development of the creature must increase and vary with it; for he acquires knowledge and power, as the material for knowledge and the need for power appear. In migratory species the female is free to acquire the same knowledge as the male by the same means, the same development by the same experiences. The human female has been restricted in range from the earliest beginning. Even among savages, she has a much more restricted knowledge of the land she lives in. She moves with the camp, of course, and follows her primitive industries in its vicinity; but the war-path and the hunt are the man's. He has a far larger habitat. The life of the female savage is freedom itself, however, compared with the increasing constriction of custom closing in upon the woman, as civilization advanced, like the iron torture chamber of romance. Its culmination is expressed in the proverb: "A woman should leave her home but three times,—when she is christened, when she is married, and when she is buried." Or this: "The woman, the cat, and the chimney should never leave the house." The absolutely stationary female and the wide-ranging male are distinctly human institutions, after we leave behind us such low forms of life as the gypsy moth, whose female seldom moves more than a few feet from the pupa moth. She has aborted wings, and cannot fly. She waits humbly for the winged male, lays her myriad eggs, and dies,—a fine instance of modification to sex.

To reduce so largely the mere area of environment is a great check to race-development; but it is not to be compared in its effects with the reduction in voluntary activity to which the human female has been subjected. Her restricted impression, her confinement to the four walls of the home, have done great execution, of course, in limiting her ideas, her information, her thought-processes, and power of judgment; and in giving a disproportionate prominence and intensity to the few things she knows about; but this is innocent in action compared with her restricted expression, the denial of freedom to act. A living organism is modified far less through the action of external circumstances upon it and its reaction thereto, than through the effect of its own exertions. Skin may be thickened gradually by exposure to the weather; but it is thickened far more quickly by being rubbed against something, as the handle of an oar or of a broom. To be surrounded by beautiful things has much influence upon the human creature: to make beautiful things has more. To live among beautiful surroundings and make ugly things is more directly lowering than to live among ugly surroundings and make beautiful things. What we do modifies us more than what is done to us. The freedom of expression has been more restricted in women than the freedom of impression, if that be possible. Something of the world she lived in she has seen from her barred windows. Some air has come through the purdah's folds, some knowledge has filtered to her eager ears from the talk of men. Desdemona learned somewhat of Othello. Had she known more, she might have lived longer. But in the ever-growing human impulse to create, the power and will to make, to do, to express one's new spirit in new forms,—here she has been utterly debarred. She might work as she had worked from the beginning,—at the primitive labors of the household; but in the inevitable expansion of even those industries to professional levels we have striven to hold her back. To work with her own hands, for nothing, in direct body-service to her own family,—this has been permitted,—yes, compelled. But to be and do anything further from this she has been forbidden. Her labor has not only been limited in kinds, but in degree. Whatever she has been allowed to do must be done in private and alone, the first-hand industries of savage times.

Our growth in industry has been not only in kind, but in class. The baker is not in the same industrial grade with the house-cook, though both make bread. To specialize any form of labor is a step up: to organize it is another step. Specialization and organization are the basis of human progress, the organic methods of social life. They have been forbidden to women almost absolutely. The greatest and most beneficent change of this century is the progress of women in these two lines of advance. The effect of this check in industrial development, accompanied as it was by the constant inheritance of increased racial power, has been to intensify the sensations and emotions of women, and to develop great activity in the lines allowed. The nervous energy that up to present memory has impelled women to labor incessantly at something, be it the veriest folly of fancy work, is one mark of this effect.

In religious development the same dead-line has held back the growth of women through all the races and ages. In dim early times she was sharer in the mysteries and rites; but, as religion developed, her place receded, until Paul commanded her to be silent in the churches. And she has been silent until to-day. Even now, with all the ground gained, we have but the beginnings—the slowly forced and disapproved beginnings—of religious equality for the sexes. In some nations, religion is held to be a masculine attribute exclusively, it being even questioned whether women have souls. An early Christian council settled that important question by vote, fortunately deciding that they had. In a church whose main strength has always been derived from the adherence of women, it would have been an uncomfortable reflection not to have allowed them souls. Ancient family worship ran in the male line. It was the son who kept the sacred grandfathers in due respect, and poured libations to their shades. When the woman married, she changed her ancestors, and had to worship her husband's progenitors instead of her own. This is why the Hindu and the Chinaman and many others of like stamp must have a son to keep them in countenance,—a deep-seated sex-prejudice, coming to slow extinction as women rise in economic importance.

It is painfully interesting to trace the gradual cumulative effect of these conditions upon women: first, the action of large natural laws, acting on her as they would act on any other animal; then the evolution of social customs and laws (with her position as the active cause), following the direction of mere physical forces, and adding heavily to them; then, with increasing civilization, the unbroken accumulation of precedent, burnt into each generation by the growing force of education, made lovely by art, holy by religion, desirable by habit; and, steadily acting from beneath, the unswerving pressure of economic necessity upon which the whole structure rested. These are strong modifying conditions, indeed.

The process would have been even more effective and far less painful but for one important circumstance. Heredity has no Salic law. Each girl-child inherits from her father a certain increasing percentage of human development, human power, human tendency; and each boy as well inherits from his mother the increasing percentage of sex-development, sex-power, sex-tendency. The action of heredity has been to equalize what every tendency of environment and education made to differ. This has saved us from such a female as the gypsy moth. It has held up the woman, and held down the man. It has set iron bounds to our absurd effort to make a race with one sex a million years behind the other. But it has added terribly to the pain and difficulty of human life,— a difficulty and a pain that should have taught us long since that we were living on wrong lines. Each woman born, re-humanized by the current of race activity carried on by her father and re-womanized by her traditional position, has had to live over again in her own person the same process of restriction, repression, denial; the smothering "no" which crushed down all her human desires to create, to discover, to learn, to express, to advance. Each woman has had, on the other hand, the same single avenue of expression and attainment; the same one way in which alone she might do what she could, get what she might. All other doors were shut, and this one always open; and the whole pressure of advancing humanity was upon her. No wonder that young Daniel in the apocryphal tale proclaimed: "The king is strong! Wine is strong! But women are stronger!"

To the young man confronting life the world lies wide. Such powers as he has he may use, must use. If he chooses wrong at first, he may choose again, and yet again. Not effective or successful in one channel, he may do better in another. The growing, varied needs of all mankind call on him for the varied service in which he finds his growth. What he wants to be, he may strive to be. What he wants to get, he may strive to get. Wealth, power, social distinction, fame,—what he wants he can try for.

To the young woman confronting life there is the same world beyond, there are the same human energies and human desires and ambition within. But all that she may wish to have, all that she may wish to do, must come through a single channel and a single choice. Wealth, power, social distinction, frame,—not only these, but home and happiness, reputation, ease and pleasure, her bread and butter,—all must come to her through a small gold ring. This is a heavy pressure. It has accumulated behind her through heredity, and continued about her through environment. It has been subtly trained into her through education, till she herself has come to think it a right condition, and pours its influence upon her daughter with increasing impetus. Is it any wonder that women are over-sexed? But for the constant inheritance from the more human male, we should have been queen bees, indeed, long before this. But the daughter of the soldier and the sailor, of the artist, the inventor, the great merchant, has inherited in body and brain her share of his development in each generation, and so stayed somewhat human for all her femininity.

All morbid conditions tend to extinction. One check has always existed to our inordinate sex-development,—nature's ready relief, death. Carried to its furthest excess, the individual has died, the family has become extinct, the nation itself has perished, like Sodom and Gomorrah. Where one function is carried to unnatural excess, others are weakened, and the organism perishes. We are familiar with this in individual cases,—at least, the physician is. We can see it somewhat in the history of nations. From younger races, nearer savagery, nearer the healthful equality of pre-human creatures, has come each new start in history. Persia was older than Greece, and its highly differentiated sexuality had produced the inevitable result of enfeebling the racial qualities. The Greek commander stripped the rich robes and jewels from his Persian captives, and showed their unmanly feebleness to his men. "You have such bodies as these to fight for such plunder as this," he said. In the country, among peasant classes, there is much less sex-distinction than in cities, where wealth enables the women to live in absolute idleness; and even the men manifest the same characteristics. It is from the country and the lower classes that the fresh blood pours into the cities, to be weakened in its turn by the influence of this unnatural distinction until there is none left to replenish the nation.

The inevitable trend of human life is toward higher civilization; but, while that civilization is confined to one sex, it inevitably exaggerates sex-distinction, until the increasing evil of this condition is stronger than all the good of the civilization attained, and the nation falls. Civilization, be it understood, does not consist in the acquisition of luxuries. Social development is an organic development. A civilized State is one in which the citizens live in organic industrial relation. The more full, free, subtle, and easy that relation; the more perfect the differentiation of labor and exchange of product, with their correlative institutions,—the more perfect is that civilization. To eat, drink, sleep, and keep warm,—these are common to all animals, whether the animal couches in a bed of leaves or one of eiderdown, sleeps in the sun to avoid the wind or builds a furnace-heated house, lies in wait for game or orders a dinner at a hotel. These are but individual animal processes. Whether one lays an egg or a million eggs, whether one bears a cub, a kitten, or a baby, whether one broods its chickens, guards its litter, or tends a nursery full of children, these are but individual animal processes. But to serve each other more and more widely; to live only by such service; to develope special functions, so that we depend for our living on society's return for services that can be of no direct use to ourselves,—this is civilization, our human glory and race-distinction.

All this human progress has been accomplished by men. Women have been left behind, outside, below, having no social relation whatever, merely the sex-relation, whereby they lived. Let us bear in mind that all the tender ties of family are ties of blood, of sex-relationship. A friend, a comrade, a partner,—this is a human relative. Father, mother, son, daughter, sister, brother, husband, wife,—these are sex-relatives. Blood is thicker than water, we say. True. But ties of blood are not those that ring the world with the succeeding waves of progressive religion, art, science, commerce, education, all that makes us human. Man is the human creature. Woman has been checked, starved, aborted in human growth; and the swelling forces of race-development have been driven back in each generation to work in her through sex-functions alone.

This is the way in which the sexuo-economic relation has operated in our species, checking race-development in half of us, and stimulating sex-development in both.

V

The facts stated in the foregoing chapters are familiar and undeniable, the argument seems clear; yet the mind reacts violently from the conclusions it is forced to admit, and tries to find relief in the commonplace conditions of every-day life. From this looming phantom of the over-sexed female of genus homo we fly back in satisfaction to familiar acquaintances and relatives,—to Mrs. John Smith and Miss Imogene Jones, to mothers and sisters and daughters and sweethearts and wives. We feel that such a dreadful state of things cannot be true, or we should surely have noticed it. We may even perform that acrobatic feat so easy to most minds,—admit that the statement may be theoretically true, but practically false!

Two simple laws of brain action are responsible for the difficulty of convincing the human race of any large general truths concerning itself. One is common to all brains, to all nerve sensations indeed, and is cheerfully admitted to have nothing to do with the sexuo-economic relation. It is this simple fact, in popular phrase,—that what we are used to we do not notice. This rests on the law of adaptation, the steady, ceaseless pressure that tends to fit the organism to the environment. A nerve touched for the first time with a certain impression feels this first impression far more than the hundredth or thousandth, though the thousandth be far more violent than the first. If an impression be constant and regular, we become utterly insensitive to it, and only respond under some special condition, as the ticking of a clock, the noise of running water or waves on the beach, even the clatter of railroad trains, grows imperceptible to those who hear it constantly. It is perfectly possible for an individual to become accustomed to the most disadvantageous conditions, and fail to notice them.

It is equally possible for a race, a nation, a class, to become accustomed to most disadvantageous conditions, and fail to notice them. Take, as an individual instance, the wearing of corsets by women. Put a corset, even a loose one, on a vigorous man or woman who never wore one, and there is intense discomfort, and a vivid consciousness thereof. The healthy muscles of the trunk resent the pressure, the action of the whole body is checked in the middle, the stomach is choked, the process of digestion interfered with; and the victim says, "How can you bear such a thing?"

But the person habitually wearing a corset does not feel these evils. They exist, assuredly, the facts are there, the body is not deceived; but the nerves have become accustomed to these disagreeable sensations, and no longer respond to them. The person "does not feel it." In fact, the wearer becomes so used to the sensations that, when they are removed,—with the corset,—there is a distinct sense of loss and discomfort. The heavy folds of the cravat, stock, and neckcloth of earlier men's fashions, the heavy horse-hair peruke, the stiff high collar of to-day, the kind of shoes we wear,—these are perfectly familiar instances of the force of habit in the individual.

This is equally true of racial habits. That a king should rule because he was born, passed unquestioned for thousands of years. That the eldest son should inherit the titles and estates was a similar phenomenon as little questioned. That a debtor should be imprisoned, and so entirely prevented from paying his debts, was common law. So glaring an evil as chattel slavery was an unchallenged social institution from earliest history to our own day among the most civilized nations of the earth. Christ himself let it pass unnoticed. The hideous injustice of Christianity to the Jew attracted no attention through many centuries. That the serf went with the soil, and was owned by the lord thereof, was one of the foundations of society in the Middle Ages.

Social conditions, like individual conditions, become familiar by use, and cease to be observed. This is the reason why it is so much easier to criticise the customs of other persons or other nations than our own. It is also the reason why we so naturally deny and resent the charges of the critic. It is not necessarily because of any injustice on the one side or dishonesty on the other, but because of a simple and useful law of nature. The Englishman coming to America is much struck by America's political corruption; and, in the earnest desire to serve his brother, he tells us all about it. That which he has at home he does not observe, because he is used to it. The American in England finds also something to object to, and omits to balance his criticism by memories of home.

When a condition exists among us which began in those unrecorded ages back of tradition even, which obtains in varying degree among every people on earth, and which begins to act upon the individual at birth, it would be a miracle past all belief if people should notice it. The sexuo-economic relation is such a condition. It began in primeval savagery. It exists in all nations. Each boy and girl is born into it, trained into it, and has to live in it. The world's progress in matters like these is attained by a slow and painful process, but one which works to good ends.

In the course of social evolution there are developed individuals so constituted as not to fit existing conditions, but to be organically adapted to more advanced conditions. These advanced individuals respond in sharp and painful consciousness to existing conditions, and cry out against them according to their lights. The history of religion, of political and social reform, is full of familiar instances of this. The heretic, the reformer, the agitator, these feel what their compeers do not, see what they do not, and, naturally, say what they do not. The mass of the people are invariably displeased by the outcry of these uneasy spirits. In simple primitive periods they were promptly put to death. Progress was slow and difficult in those days. But this severe process of elimination developed the kind of progressive person known as a martyr; and this remarkable sociological law was manifested: that the strength of a current of social force is increased by the sacrifice of individuals who are willing to die in the effort to promote it. "The blood of the martyrs is the seed of the church." This is so commonly known to-day, though not formulated, that power hesitates to persecute, lest it intensify the undesirable heresy. A policy of "free speech" is found to let pass most of the uneasy pushes and spurts of these stirring forces, and lead to more orderly action. Our great anti-slavery agitation, the heroic efforts of the "women's rights" supporters, are fresh and recent proofs of these plain facts: that the mass of the people do not notice existing conditions, and that they are not pleased with those who do. This is one strong reason why the sexuo-economic relation passes unobserved among us, and why any statement of it will be so offensive to many. . . .

Discussion Questions

1. One of Gilman's main points is that women cannot be equal unless they are economically independent. Do you agree or disagree? Why or why not? Use concrete examples to explain and support your point of view.

2. For Gilman, "women's work" was lonely and demeaning. Do you think "parentwork" and housework has to be this way? Why or why not? What specific measures, institutions, and/or practices might help prevent or combat these tendencies?

3. Discuss the specific advances in gender equality that have occurred since Gilman's day. What specific issues highlighted by Gilman do you consider still problematic, and which problems do you consider "eradicated" (at least in the United States)?

4. Compare and contrast Gilman's theory as to the oppression of women in patriarchy with Marx's theory as to how and why workers are oppressed under capitalism. What similarities do you see in their arguments? What are the differences in these two theories of oppression?

5. Compare and contrast Gilman's discussion of social bonds and mental health with that of Durkheim. Does each theorist conceive of social bonds as working in the same way? How so or why not? Discuss how each theorist would construe the social bonds between (1) mothers and their children and (2) husbands and wives.

6 GEORG SIMMEL (1858–1918)

Georg Simmel

Key Concepts

- Duality
- Form
- Content
- Types
- Tragedy of culture
- Blasé attitude

Just as the universe needs "love and hate," that is, attractive and repulsive forces, in order to have any form at all, so society, too, in order to attain a determinate shape, needs some quantitative ratio of harmony and disharmony, of association and competition, of favorable and unfavorable tendencies.

—Simmel (1908b/1971:72)

As the quotation above suggests, Georg Simmel's work is informed by a combining of opposites into a whole, where social life is based on seeming contradictions. Consider, for instance, one example from Simmel's work that expresses this dynamic: fashion. Sociologically, the fascinating thing about fashion is that whether you wear baggy jeans, toe rings and tattoos, or Armani suits, fashion signals both your individuality *and* your attachment to specific social groups. Whether age related (e.g., youth), hobby related (e.g., skaters), or attitude related (e.g., hip, urban), fashion is at once a process of conforming to some groups, while distancing yourself from others. Moreover, even within a specific attachment, you strive for your "own" look; that is, you don't want to look identical to your friends. Fashion, then, is not a singular, "pure" expression, but is built on two opposing forces: differentiation/individuality and conformity/imitation.

Simmel's analyses of social life have produced a unique legacy among the classical theorists. Unlike Marx, Durkheim, Weber, and Mead (see Chapter 8), no definable theoretical school has coalesced around his work. Nevertheless, it should not be inferred from this that Simmel's work has proved less fertile for sociologists. On the contrary, his impact on the discipline has been far reaching. The influence of

Simmel's insights is found in the work of his contemporaries (most notably Weber and Mead) as well as in that of succeeding generations of sociologists. Indeed, the names of those who have drawn inspiration from Simmel reads as a veritable list of "who's who" in sociology. Theorists have drawn on his work in their explorations of topics ranging from small group dynamics, networks of interpersonal relationships, processes of exchange behavior, and the nature of social conflict, to the character of the urban environment and its effects on the individual and social life, the consequences of modernization for culture and individual personality, and the patterns of apprehension through which we experience our world.

In a sense, Simmel himself predicted this outcome:

> I know that I shall die without intellectual heirs, and that is as it should be. My legacy will be, as it were, in cash, distributed to many heirs, each transforming his part into use conformed to *his* nature: a use which will reveal no longer its indebtedness to this heritage. (Levine 1971:xii)

A BIOGRAPHICAL SKETCH ▪▪

Georg Simmel was born on March 1, 1858, in Berlin, the urban center of Germany. His father, a successful businessman and part owner of a Berlin chocolate factory, died when Georg was an infant. However, Georg's financial future was secured when a friend of the family, the owner of a profitable music publishing house, was appointed his guardian. The guardianship provided Georg with a comfortable childhood and, upon adulthood, allowed him to pursue his intellectual interests with relatively few monetary worries.

As fate would have it, Simmel's inheritance proved more beneficial than might appear at first sight. Though his reputation as a scholar drew international attention, and his skills as a lecturer earned him a degree of popularity among students and Berlin's cultural elite, Simmel was unable to obtain a permanent academic position throughout most of his career. This was due, in part, to the seemingly fragmentary nature of his work. Simmel wrote on a wide range of subjects that crossed disciplinary boundaries. For instance, he wrote books on the artists Rembrandt, Goethe, and Rodin, and published essays, many of which appeared in nonacademic journals, on unusual topics such as the sociology of smell and the sociology of secrecy. Such eclecticism was frowned on in the conservative climate of German universities. Instead, "real" scholars were those who committed themselves to sustained analyses of a limited set of questions. Moreover, the discipline of sociology, itself only recently born, was met with significant resistance within the university establishment. Thus, appointments to full-time positions in sociology departments were far less available relative to the other disciplines.

However, in addition to the more academic obstacles, there is another reason for Simmel's lack of professional success. The German university establishment—and, for that matter, much of Europe more generally—was tainted by anti-Semitism. As a Jew, Simmel encountered discrimination firsthand. In 1908, he was recommended for the chair of philosophy at the University of Heidelberg. His candidacy was rejected, however, when the minister of education in Baden requested an evaluation of Simmel's qualifications from Professor Dietrich Schaefer, a prominent historian at the University of Berlin. Schaefer wrote of Simmel,

> [h]e is . . . a dyed-in-the-wool Israelite, in his outward appearance, in his bearing, and in his manner of thinking. . . . He spices his words with clever sayings. And the audience he recruits is composed accordingly. The ladies constitute a very large portion. . . . For the rest, there [appears at his lectures] an extraordinarily numerous contingent of the oriental world. (Quoted in Coser 1977:209)

Simmel was thus "guilty" on three counts: (1) he was Jewish, (2) he was "clever" (i.e., his analyses were superficial), and (3) his lectures, though well attended, attracted intellectually inferior "Orientals" (i.e., foreigners) and women.

Anti-Semitism together with his own refusal to pursue a conventional intellectual track left Simmel confined to the margins of academia for most of his career. In 1885, he was appointed a *Privatdozent,* or unpaid lecturer, at the University of Berlin. After 16 years, he was granted the purely honorary title of *Außerordentlicher Professor,* or Extraordinary Professor, a position that still precluded him from fully participating in departmental and university affairs. Finally, in 1914, at the age of 56, Simmel was awarded a full professorship at the University of Strasbourg. His time at Strasbourg was far from ideal, however. Shortly after his arrival, the dormitories and lecture halls were turned over to the military for use as barracks and hospitals; World War I had just erupted. Simmel would die four years later of liver cancer.

Simmel's marginal status within the university establishment did not prevent him from playing a pivotal role in the development of sociology, however. What Durkheim was to sociology in France, Simmel was in Germany. He taught some of the first sociology courses in Germany (Ashley and Orenstein 1998:309) and, together with Max Weber and Ferdinand Tönnies (1855–1936), founded the German Society for Sociology. Indeed, it was through the efforts of these three that the emerging discipline gained a foothold within the German university system. In addition, Simmel was an active figure in a number of Berlin's intellectual circles. Through the salons, he developed friendships with writers, poets, journalists, and artists. He was also friends with a number of leading academics including Weber, the philosopher and founder of phenomenology Edmund Husserl, and the philosopher Heinrich Rickert. Each of these men lent their continued support to and endorsement of Simmel as he sought to win a full-time university appointment. Finally, lack of recognition from the university establishment did not adversely affect Simmel's productivity. During his life, he published more than 200 articles and 30 books, some of which were translated into five languages (Ashley and Orenstein 1998:311; Coser 1977:195).

◫ Intellectual Influences and Core Ideas

Outlining the core ideas that make up Simmel's work is a task made particularly difficult for a number of reasons. First, while he saw himself first and foremost as a philosopher (Frisby 1984:25), Simmel's intellectual interests spanned three disciplines: philosophy, history, and sociology (Levine 1971:xxi). Second, even within these three fields, his intellectual pursuits led him into a number of directions that were not necessarily related. In addition to his more sociological writings, varied as they are, Simmel published works on aesthetics, ethics, religion, the philosophy of history, the philosophies of Nietzsche and Schopenhauer, and the metaphysics of individuality. Third, Simmel, unlike, say, Marx or Mead, did not set out to construct a coherent theoretical scheme, nor did he explicitly aim to develop a systematic critique of or to build on a specific theoretical paradigm. As a result, his work perhaps is best seen as a "collection of insights" (Collins and Makowsky 1998:160) that provides for the sociologist a unique and subtle understanding of social interaction.[1]

In this section, we restrict ourselves to presenting an overview of Simmel's central sociological ideas. To this end, we touch on the following issues: (1) Simmel's image of society, (2) his view of sociology as a discipline, and (3) the plight of the individual in modern society.

Society

Simmel's conception of society stood in contrast both to the organic view developed by Comte, Spencer, and Durkheim, and to the purely abstract view articulated within German idealist philosophy. From the former vantage point, society is seen as having a reality outside or independent of the

---

[1]There is yet another reason why tracing the connections between Simmel's ideas and those articulated by other scholars is a complicated endeavor: He refused to put footnotes or citations in his publications. His avoidance of the professionally expected practice was one more reason for the disdain he evoked from many within the academic establishment.

existence of the interacting individuals who compose it. Thus, society is seen as working down on individuals as it shapes, or even determines, their behavior, attitudes, and interests. On the other hand, idealists reject the notion of society as a reality sui generis, arguing instead that "society" is only an abstract label. Despite whatever practical uses the label may have, society is not a real object or thing. From this latter perspective, then, what "really" exist are unique individuals and the ideas that motivate their actions.

Simmel attempted to carve out a middle ground with respect to these competing visions. For him, the essence of society lies in the *interactions* that take place between individuals and groups. Society, then, is not a system of overarching institutions or symbolic codes nor is it merely an abstract idea used to describe a collection of individuals atomistically pursuing lines of conduct. Thus, Simmel defined society as a "number of individuals connected by interaction. . . . It is not a 'substance,' nothing concrete, but an *event:* It is the function of receiving and affecting the fate and development of one individual by the other" (Simmel [1917] 1950:10, 11; emphasis in original). The centrality of interaction for Simmel's view of society is expressed further when he states, "the significance of interactions among men [*sic*] lies in the fact that it is because of them that the individuals . . . form a unity, that is, a society" (Simmel [1908d] 1971:23). In short, society is the array of interactions engaged in by individuals.

But what of those large-scale institutions and organizations—corporations, governments, schools, advocacy groups—that often come to mind when talking about society? Indeed, are not such supra-individual systems often equated with society itself? For Simmel, they "are nothing but immediate interactions that . . . have become crystallized as permanent fields, as autonomous phenomena" (Simmel [1917] 1950:10). Here, he acknowledges that while such institutions develop under their own laws and logic and, thus, confront the individual as a seeming outside force, society is a process "constantly being realized" (ibid.). Society is something individuals *do* as they influence and are influenced by each other.

Thus, according to Simmel, society and the individuals that compose it constitute an interdependent **duality**. In other words, the existence of one presupposes the existence of the other. Moreover, this duality has a profound effect on the nature of individuality. For while who you are as an individual is in an important sense defined and made possible by the groups to which you belong, preserving your individuality demands that your identity not be completely submerged into or engulfed by group membership. Otherwise, you have no self that you can call your own.[2] As Simmel remarks, society is

> a structure which consists of beings who stand inside and outside of it at the same time. This fact forms the basis for one of the most important sociological phenomena, namely, that between a society and its component individuals a relation may exist as if between two parties. . . . [T]he individual can never stay within a unit which he does not at the same time stay outside of, that he is not incorporated into any order without also confronting it. (Simmel [1908e] 1971:14, 15)

In other words, your self-directed efforts to express and satisfy your interests and desires require that you engage in interaction with others. Yet interaction, in turn, shapes which aspects of your self can be expressed and how they are expressed. In an important sense, then, your "individuality" is created out of a synthesis of two seemingly contradictory forces: You are at the same time both an autonomous being with a unique disposition and history, and a product of society.

---

[2]Simmel's emphasis on the dualistic relationship between the individual and society is shaped in large measure by the philosophy of Immanuel Kant (1724–1804). In asking, "How is nature possible?" Kant argued that a priori (pre-existing) "categories" or concepts that exist within the human mind structure our experience of the external world. The never-ending streams of sensory impressions that confront the individual have no underlying unity or pattern. It is only through the mind's a priori categories (particularly our conceptions of time, space, and causality) that the raw data that enter our consciousness are given a form, and hence meaning. However, without sensory impressions, the categories have no effect on human existence.

Sociology

Simmel's view of society provides, not surprisingly, an important glimpse into his understanding of sociology as a discipline.[3] His emphasis on the duality existing between society and the individual led him to define sociology as the study of social interaction or, as he often called it, "sociation." But it was not interaction per se that interested Simmel. Rather, he sought to analyze the **forms** in which interaction takes place. For instance, understanding the specific **content** of interactions that take place between an employer and an employee—what they talk about and why—is not of central concern to sociologists. What is of sociological significance, however, is determining the uniformities or commonalities that such interactions have with those between, say, a husband and a wife in a patriarchal society or between a ruler and his subjects. Recognizing the commonalities leads to the realization that each is based on a reciprocal relationship of domination and subordination. The main task of sociology, then, is to uncover the basic forms of interaction through which individuals pursue their interests or satisfy their desires, for it is only in relating to others, acting both with them and against them, that we are able to satisfy our ambitions.

Let's further clarify Simmel's notion of content and form. The content of interaction refers to the drives, purposes, interests, or inclinations that individuals have for interacting with another. In themselves, such motivations are not social, but rather are simply isolated psychological or biological impulses. What is social, however, are the actions that we take in concert with others in order to fulfill our drives or realize our interests. Moreover, when joined with others, our actions take on an identifiable, though not necessarily stable, form. Thus, our interactions may take on the form of conflict or cooperation, for example, or perhaps domination or equality (Simmel [1908d] 1971:24).

Simmel also noted that while, on one hand, the most dissimilar of contents (individual motivations) may be realized in an identical form of sociation, on the other hand, the same motivations or interests can be expressed in a range of forms of interaction. For instance, interaction within and between families, gangs, corporations, political organizations, and governments may all take the form of conflict. Thus, despite the varied interests or purposes that led to interaction in each of these cases, the individuals involved all may find themselves facing an opposing party that hinders the realization of their impulses or desires. Conversely, the same drive or interest can be expressed through a number of forms. Attempts to gain economic advantage, for instance, can be asserted through cooperative agreements among parties as well as through forcing submission from others.

Simmel's distinction between the forms and contents of interaction led him to draw a parallel between the subject matter of sociology and that of geometry. Both sciences develop general principles from determining the regularities that exist among diverse materials. Simmel thus saw himself as having devised a geometry of social life:

> Geometric abstraction investigates only the spatial forms of bodies, although empirically, these forms are given merely as the forms of some material content. Similarly, if society is conceived as interaction among individuals, the description of the forms of this interaction is the task of the science of society in its strictest and most essential sense. (Simmel [1917] 1950:21, 22)

To take a simple example, while a car tire, a clock face, and a nickel are different things or "contents" that serve different purposes (you wouldn't look at your tire to find out what time it is), to the mathematician they all take the form of a three-dimensional circle and thus share spatial properties. On the other hand, although giving a wedding present, paying for music lessons, and volunteering at the food co-op are motivated by different intentions or "contents," to the sociologist they all take the form of an exchange relation and thus share interaction properties.

Later in the chapter we provide excerpts from Simmel's essay on the unique properties of exchange as a form of interaction. Here, we discuss two other forms of interaction: conflict and sociability. Simmel maintains that conflict, as a form of interaction, is an inevitable, and in many ways beneficial, feature

---

[3]Bear in mind that at this time sociology was confronted with the task of justifying its existence as an academic discipline. This required, among other things, carving out a unique subject matter and methodology to distinguish it from other disciplines.

of social life. Unlike indifference towards another, which expresses an absence of a relationship and is thus purely "negative," conflict "resolves the tension between contrasts: the synthesis of elements that work both against and for one another" ([1908b] 1971:71). Conflict is not in itself a destructive force; rather it is based on an integration of unity and discord, that is, "positive" and "negative" aspects found within any social relation. Indeed, as the opening quote to this chapter observes, society necessarily is formed out of "some quantitative ratio of harmony and disharmony, of association and competition, of favorable and unfavorable tendencies" ([1908b] 1971:72). In other words, while a superficial view may suggest that it is something to be avoided, no society could exist without some measure of conflict.

Conflict serves a positive role in developing one's sense of self as well as in creating group unity—both are unimaginable without confronting antagonisms of one kind or another. It is often the case that conflict or competition with others spurs us to increase our efforts in the pursuit of goals, whether it be an athlete honing her physical skills, or an entrepreneur devising a more efficient business enterprise to outpace competitors. And in intimate relations, such as a marriage, conflict can be evidence of a deep and unreserved commitment whose foundation is not easily shaken. Conversely, intimate relations in which conflict is avoided often point to an absence of unconditional emotional devotion. For "the strongest love can stand a blow most easily, and hence it does not even occur to it, as is characteristic of a weaker one, to fear that the consequences of such a blow cannot be faced, and it must therefore be avoided by all means" ([1908b] 1971:94). Moreover, just as our individuality is forged in opposition to (i.e., in conflict with) the demands placed on us by other individuals and groups, so too we require someone or something to stand against in order to realize our essence. Voicing our opposition or unwillingness to capitulate to those we find unreasonable or oppressive often makes an otherwise unbearable relationship sufferable. As Simmel remarks, "Our opposition makes us feel that we are not completely victims of circumstances. It allows us to prove our strength consciously and only thus gives vitality and reciprocity to conditions from which, without such corrective, we would withdraw at any cost" ([1908b] 1971:75). Who hasn't at the very least muttered under his or her breath disapproval of a boss or a parent? Though such behaviors may not lead to changes in the relationship, by producing a sense of "virtual power" they perform the positive function of preserving the relationship as well as one's individuality.

With regard to group life, conflict generates a clearer distinction between those who belong to a group and those who do not, thereby breeding an intensified sense of group membership. For instance, nothing stokes the flames of school unity like the existence of a rival school. Antagonism directed toward the "rival" school promotes cohesion or integration within the group. So, too, is patriotism greater in times of war when there is a clear enemy toward which to direct our antagonisms. Group cohesion can also be sparked when a community rallies together in the wake of a natural disaster. In short, Simmel maintains that without having to overcome a common crisis or attain a common goal in the face of obstacles (in short, without some measure of conflict), there would be no basis for cooperation, group feelings, or "harmony of interest"—the unifying forces that make society possible.

Of course, it is not always the case that people engage in interactions for strategic purposes or because they have a specific goal they seek to realize. Sometimes we find ourselves interacting with others simply for the sake of the connection itself. Simmel called this form of interaction "sociability" or the "play-form of association." During sociability, the content of conversations has no significance outside of the encounter; instead, "talking is an end in itself" (Simmel [1910] 1971:136). Here, we talk about movies or concerts we've seen, classes we've taken, or the latest news. The impetus for such conversations lies not in resolving the topics discussed or in the attempt to gain personal advantage, but in the pleasure of conversing and satisfying the impulse to associate with others. As soon as the truthfulness of the conversation's content or the striving for personal gain of any type is made the focus, the encounter loses its playfulness. Simmel's point is not that in order to be sociable, conversations must avoid discussions or arguments over serious issues. Rather, it is that regardless of the seriousness of subject matter, sociability finds "its justification, its place, and its purpose only in the functional play of conversation as such" and not in the substance of the conversation (ibid.).

Consider, for example, a conversation you strike up with someone at a party. So long as personal opinions are not expressed earnestly and personal problems or triumphs remain unspoken, sociability will reign. However, should the person dismissively reply, "You actually *liked* that movie?! I thought it was absolutely *terrible*!" or pompously pronounce, "In high school I was the coolest kid in my class. *Everybody* wanted to hang out with me," then the focus of the conversation will quickly turn to its contents, the defense of which both of you will have something to gain or lose.

An important element of sociability that contributes to its playfulness is its "democratic" nature. To the extent that it is a form of interaction freed from the conflicts and pressures that make up "real" life, sociability establishes an "artificial" world, a world without friction. In this self-contained, temporary escape, actors leave their personal concerns and ambitions tied to real life behind, thus encouraging the treatment of others as equals. A person's status in the wider society or personal merits or values must not encroach upon the content of the interaction. Simmel speaks of it thusly:

> Inasmuch as sociability is the abstraction of association—an abstraction of the character of art or of play—it demands the purest, most transparent, most engaging kind of interaction—that among *equals*. It must, because of its very nature, posit beings who give up so much of their objective content, who are so modified in both their outward and their inner significance, that they are sociably equal, and every one of them can win sociability values for himself only under the condition that the others, interacting with him, can also win them. It is a game in which one "acts" as though all were equal, as though he especially esteemed everyone. ([1910] 1971:133, emphasis in the original)

Photo 6.1 It's all in the look: Maybe yes . . . or maybe no.

A particular kind of sociability that epitomizes the duality of social life discussed previously is flirtation or coquetry. As depicted in Figure 6.1, flirtation is a type of erotic playfulness in which an actor continuously alters between consent and denial. The art of flirtation lies in playing "hinted consent and hinted denial against each other to draw the man on without letting matters come to a decision, to rebuff him without making him lose all hope" (ibid.:134). At the same time, the man must likewise play along, substituting the reality of erotic desire with its "remote symbol." Underlying flirtation is the tension arising from a playful unwillingness to surrender oneself to another that could lead to submission and the threat of the other's refusal or withdrawal. However, should a final decision be revealed, resolving the tension between consent and denial, the "play" is over. Either the other's desire to continue the performance is denied or one's attraction for the other has been made real. In both instances, sociability comes to an end as serious positions have been staked.

Figure 6.1 Duality of Flirtation

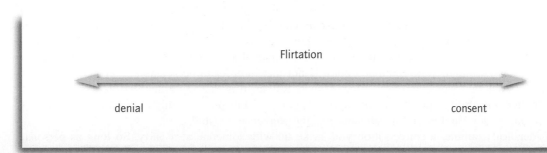

Flirtation

denial   consent

The Individual in Modern Society

Nowhere is the duality between individual identity and the "web of association" expressed more vividly than in Simmel's discussion of the nature of modern society. For Simmel, modern, urban societies not only allow individuals to cultivate their unique talents and interests, but at the same time also lead to a "tragic" leveling of the human spirit.

Underlying this process is the changing nature of group life in modern societies. In contrast to premodern societies, modern societies are characterized by an increasingly specialized division of labor and expanding circles of affiliations. No longer is an individual's total personality absorbed into a particular group or controlled by a particular leader or landlord as tribal life or feudal relations once demanded. Unlike the peasant of feudal times who was economically, socially, and legally dependent on the landlord, the modern individual is not bound so extensively to any specific person. Thus, for example, employers today do not establish or adjudicate your legal status, nor do they determine whom you can associate with or marry. The demands of group membership in modern society are not totalizing. Modern, functionally specific organizations require only a "part" of the self in the service of its aims.

Echoing Durkheim's earlier argument, Simmel notes how occupational specialization and membership in an array of social circles has allowed individuals to differentiate themselves from others while developing their unique personalities. The modern individual has separate relationships that tie her to work, family, community, creative pursuits, or hobbies, as well as to her religious and ethnic identity (see Figure 6.1). Moreover, in extending the network of their affiliations, modern individuals become less dependent on any one group for meeting their needs. Because individuals are not completely immersed in any specific group, they are freed from the dominating control of group life that characterized premodern societies. Expanding group memberships thus brings with it liberation and the potential to develop one's individuality.

Figure 6.2   Simmel's Web of Group Affiliations

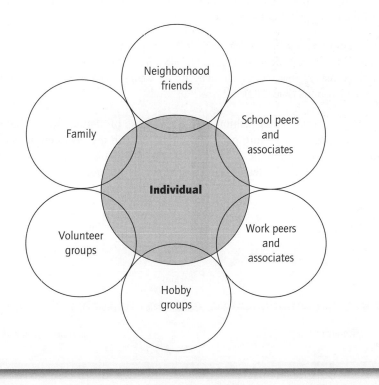

This freedom not only prevents the individual's entire identity from being absorbed by a single set of relations, but also enables a person to refine his skills, talents, and personal preferences. Individuality is enhanced because as the number of affiliations increases, the less likely it becomes that the whole of an individual's multiple memberships is identical to another person's. Because individual identity is forged through group affiliations, one's occupying a distinct position in the space of possible affiliations allows for the development of a unique personality (Simmel [1908a] 1955).

As the example at the beginning of this chapter indicated, Simmel explored the world of fashion as yet another aspect of modern social life built upon the duality of individuation and group membership. Whether it be taste in music or in cars, the design of clothes or furniture, Simmel saw fashion as an expression of individualization and differentiation on the one hand, while on the other an expression of imitation and conformity. (See Figure 6.3 below.) As Simmel contends,

> From the fact that fashion as such can never be generally in vogue, the individual derives the satisfaction of knowing that as adopted by him it still represents something special and striking, while at the same time he feels inwardly supported by a set of persons who are striving for the same thing, not as in the case of other social satisfactions, by a set actually doing the same thing. ([1904] 1971:304)

As a result, adopting a particular fashion (or rejecting fashion which can itself be fashionable) allows the individual to cultivate his or her uniqueness and sense of self-identity with the security of knowing that should the trend be met with reproach, he is not responsible for creating it. After all, in expressing his union with others who have adopted the trend, he is merely following the latest fashion. Moreover, that fashion is primarily intended to serve such social purposes is demonstrated by the fact that there is no functional or aesthetic reason why jacket lapels should be wide one season and narrow the next, why pastels should "replace" earth tones, or why bell-bottom pants should reappear 30 years after becoming passé.

The passage quoted above suggests an interesting question: Why can't a fashion ever be generally in vogue? To this, Simmel offers an answer that, not surprisingly, speaks to the paradoxical nature of social life. Simply put, fashions remain fashionable only to the extent that the general population does not adopt them. For once fashions become widely disseminated and take on an air of permanence, they become a common fact of life. Turning again to Simmel,

> The very character of fashion demands that it should be exercised at one time only by a portion of the given group, the great majority merely being on the road to adopting it. As soon as . . . anything that was originally done only by few has really come to be practiced by all . . . we no longer speak of fashion. As fashion spreads, it gradually goes to its doom. (ibid.:302)

Thus, the capacity for a particular fashion to create a sense of distinction for the individuals who first adopt it is destroyed as more and more people practice it. With the destruction of its very purpose—to cultivate individuality—the fashion dies, to be replaced by a new trend that, through its inevitable spread, will also face its equally inevitable death. And so the cycle continues.

Figure 6.3   Duality of Fashion

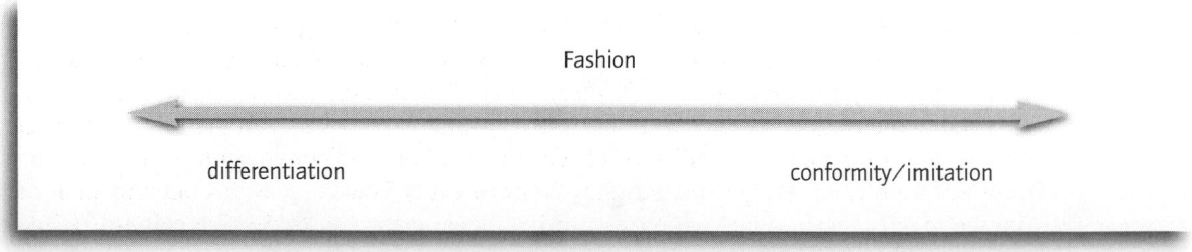

Fashion

differentiation   conformity/imitation

Photo 6.2 "Punk" fashion: Looking the same only different

The ebb and flow of fashion trends raises an important issue: Aside from an individual's quest to be fashionable, what group forces shape the rise and demise of trends? In addressing this question, Simmel looks to the dynamics of class relations. Indeed, the cycle described above is in large measure a consequence of the upper classes within a society attempting to distance themselves from the lower classes. Fashion is a visible and easily identifiable sign of class position, making it a domain well suited for publicly demonstrating one's place in the class hierarchy. (The overt connection between cars and class position clearly indicates this.) However, as the lower classes set out to imitate those above them in the "externals of life," the upper classes necessarily must seek out an alternative form of fashion in order to retain and express their distinctiveness.

The pace at which styles change is quickened in modern societies. To the extent that the lower classes in advanced societies possess greater wealth relative to those in less-developed or premodern societies, they have an advantage in chasing the fashion trends established by the upper classes. Likewise with the mass production of goods, the costs for manufacturing them decreases, in turn making them more affordable to the lower classes. More purchasing power coupled with cheaper products and increased supply shortens the life span of fashions first adopted by the upper classes in their attempt to differentiate themselves from the masses. Finally, technological development has produced yet another Simmelian irony: Insofar as we try to express our uniqueness or individuality through fashion, we often turn to buying mass-produced, standardized goods.

While the evolution from homogeneous, premodern ways of life to heterogeneous, modern societies has created an increased potential for freedom and the cultivation of individuality, it has also produced new types of constraints. Reminiscent of Marx's discussion of alienation and commodity fetishism, Simmel speaks of the **tragedy of culture** where "objective culture"—the ideas and products of human creativity—comes to dominate individual will and self-development or "subjective culture." In short, the cultivation of individual potential (subjective culture) is inextricably bound up with the cultivation of products that are the tangible, objectified expression of individual creativity (objective culture). While

Marx centered his critique on the oppression inherent in the capitalist mode of production, Simmel, like Weber, saw advances in science, technology, and the arts, as well as in the economy, bringing in their wake an increasing rationalization and routinization of many domains of life.

The alienating effects are an inevitable consequence of modernization. As the complexities of modern societies and its extensive division of labor create a need for the specialization of skills, they also sow the seeds for individual creativity and innovation. The freedom to cultivate one's desires and talents that accompanies individualization leads to an ever-increasing number of cultural objects—things such as new goods to consume, new scientific discoveries, and new forms of technology in production, transportation, and communication. However, these products of individual invention become reified; that is, they take on a life independent of their creators. As a result, what was initially the expression of individual growth and creativity (subjective culture) later confronts individuals as an autonomous force that requires submission to its own internal logic, thus compromising the self-development that these creations were intended to foster. This is one reason, for instance, that politicians and military specialists feel compelled to produce ever-more-lethal weapons as a way to prevent war. Mutual annihilation made possible by scientific discoveries has promoted a logic that guarantees peace through the fear of destruction. This is also why objects that did not exist 20 years ago (e.g., the cell phone) seem indispensable today. Indeed, our homes have become bigger and bigger in order to contain all the stuff we think we cannot live without. In the past century, the average square footage of the American home has almost tripled, going from 800 square feet in 1900 to 983 square feet in 1950 to 2,265 square feet in 2003 (National Association of Home Builders, as cited in the *Los Angeles Times,* September 14, 2003). And that home is itself part of the vast suburbanization of the country that, while offering the "American dream" to many, has also led to homogeneous neighborhoods, insular families, and the loss of a sense of community (Putnam 2001). Meanwhile, the network of highways that has made suburban sprawl possible is converted daily into a veritable parking lot with average speeds of less than 30 miles per hour commonplace during rush hour commutes in cities such as Los Angeles.

Yet, the issue here is not whether such objects and developments represent real advances, but rather that they are fetishized. For example, the car is not only a means of transportation that, among other things, has allowed individuals to be liberated from the once-confining aspects of distance and thereby experience new people and places: it is also an object that is infused with expressive meaning. Seen as an extension of one's self, cars are invested with a "personality" though they are in reality an assemblage of metal and wires. Much like Weber's iron cage, such is the tragedy of culture in which the very cultural objects (developments in science, technology, politics, religion, the production of goods, and the arts) that were once the expression of individual ingenuity have imprisoned us. We are dependent on them not only for meeting our everyday needs, but also because they are, paradoxically, the alienating mediums through which we continue to express our individual creativity and contribute to subjective culture. As Simmel notes, in modern society,

> [t]hings become more perfected, more intellectual, and to some degree more controlled by an internal, objective logic tied to their instrumentality; but the supreme cultivation, that of subjects, does not increase proportionately. . . . The dissonance in modern life—in particular manifested in the improvement of technique in every area and the simultaneous deep dissatisfaction with technical progress—is caused in large part by the fact that things are becoming more and more cultivated, while men are less able to gain from the perfection of objects a perfection of subjective life. (Simmel [1908f] 1971:234)

Significant Others

Ferdinand Tönnies (1855–1936): Gemeinschaft and Gesellschaft

A contemporary of both Georg Simmel and Max Weber, Tönnies crafted a sociological framework that shared much with the theories of his compatriots as well as with the evolutionary social theories offered by Herbert Spencer and Émile Durkheim. In his most influential work, *Gemeinschaft und Gesellschaft* (Community and Society), published in 1887, he set out a basic dichotomy describing the foundation of both individual and social action, as well as the nature of the broader social order. The ideal types that served as the polar ends of this dichotomy are *Gemeinschaft* ("community") and *Gesellschaft* ("society"), and each is rooted in a particular "will" that possesses "binding force within or determine[s] the individual wills" (Tönnies [1887] 1957:205). According to Tönnies, Gemeinschaft is the principle that structures agricultural and handicraft life within families, villages, and towns. It gives rise to a "natural will" or mentality that inclines individuals and groups to associate with one another on the basis of concord (harmony), folkways, and religious beliefs. These likewise form the core of the normative order as it is manifested in both morality and laws.

In contrast to the warmth and emotional depth nurtured by natural will, Gesellschaft is rooted in a "rational will" that promotes "cold" instrumental relationships in which individuals are viewed as "tools" or as means to further one's ends. Rational self-interest, not feelings of sympathy, ties individuals to one another. This outlook predominates in the affairs of industry and commerce that make up city, national, and cosmopolitan (international) life. It is not concord, folkways, and religion that structure actions and attitudes, morality and law, but rather conventions, state-sponsored legislation, and scientifically informed public opinion. Social life is regulated according to legally sanctioned contracts, not long-standing customs, while "the intellectual attitude of the individual becomes gradually less and less influenced by religion and more and more influenced by science" (ibid.:226).

Like Durkheim, Tönnies viewed societies transitioning from a form of solidarity based on a relatively simple, common way of life and a widely shared set of normative and moral codes, to a more complex social bond based on individualism and interdependence. Interestingly, however, unlike Durkheim, Tönnies described the evolution from Gemeinschaft to Gesellschaft as a shift from an organic to a mechanical form of social life. As he characterized the two forms, "Gemeinschaft . . . is the lasting and genuine form of living together. In contrast to Gemeinschaft, Gesellschaft is transitory and superficial. Accordingly, Gemeinschaft should be understood as a living organism, Gesellschaft as a mechanical aggregate and artifact" (ibid.:35). Gemeinschaft is "intimate," "real and organic life" as opposed to the "imaginary and mechanical structure" of the public Gesellschaft based on the "mere coexistence of people independent of each other" (ibid.:33, 34).

From these remarks, it is apparent that Tönnies did not hail the transformation to Gesellschaft as an indisputable step toward progress. In fact, even more than Weber and Simmel he harbored a pronounced pessimism with regard to modernity where "money and capital are unlimited and almighty" (ibid.:228). However, he was troubled not only by the increasing rationalization of modern life and the alienating effects of the money economy: like Marx, he saw in capitalism a source of conflict and exploitation that could only be overcome through revolutionary class struggle and the destruction of Gesellschaft. As a social order in which "only the upper strata, the rich and the cultured, are really active and alive . . . city life and Gesellschaft down the common people to decay and death" (ibid.:227, 230). Only a return to Gemeinschaft can rescue humanity from this "decaying" society where "individuals remain in isolation and veiled hostility toward each other so that only fear of clever retaliation restrains them from attacking one another, and, therefore, even peaceful and neighborly relations are in reality based upon a warlike situation . . . and underlying mutual fear" (ibid.:224).

The Individual and Money

Perhaps the most profound expression of the "tragedy" inherent in modern society is money. While money, as a medium of exchange, allows us to enter into a wide range of relationships as we seek to satisfy our individual needs, it does so at the cost of transforming the nature of our ties to others. It expands the freedom to pursue self-fulfillment and self-expression as it trivializes the personal qualities we share with others and ushers in superficial, impersonal, and fragmented social relationships. A money economy, then, represents in particularly clear fashion the alienating duality between objective and subjective culture. Just as clearly revealed in his discussion of money is Simmel's own ambivalence regarding the "advances" offered by modern society.

What are the properties of money that produce these effects on the individual and social relationships more generally? First, money is an abstract, general standard for measuring the value of goods and services. As such, it can be used as a medium of exchange in a vast number of interactions. Few things cannot be bought or sold for money. It can pay for food, entertainment, education, shelter, sex, transportation—the list is virtually endless. The flexibility of money opens up greater opportunity for individuals to satisfy an expanded range of desires and needs (Simmel [1900a] 1978:212, 213).

Second, to the extent that we are dependent on money to meet our needs, we become less dependent on others. As the locus of dependency shifts from people to money, the possibilities for freedom of expression are enhanced. In modern societies, the pursuit of interests and fulfillment of desires are limited more by the amount of money you possess than by the demands or whims of others or the constraints imposed by tradition or custom. Consider, for instance, the teenager's summer job that often makes possible her first foray into independence. More pressingly, the fight for women's equality has always hinged on the ability not only to enter into a full range of occupations but also to earn equal pay for equal work.

Third, because money retains the same value regardless of who possesses it, it can cancel out social inequalities. Thus, one's position in society is related more to the objective, impersonal issue of how much money one has than to tradition, personal qualities, or even ethnicity and gender. In a money economy, group membership, whether it is membership in a neighborhood, a college, a profession, or a gardening club, is tied more to the *quantity* of money an individual has than to his personal *qualities.* For instance, African American "bad boy" Dennis Rodman owns both a home and a restaurant in Newport Beach, California, a wealthy, predominantly white community. Though his personal qualities and antics (wild parties, public drunkenness, etc.) continually infuriate his neighbors, Rodman's money prevents his neighbors from keeping him out.

Fourth, the impersonal quality of money allows for a qualitative and quantitative expansion of one's social network. As more needs and desires are satisfied through monetary transactions with others, money promotes contacts with a greater range of people. This leads to an expansion of avenues for the cultivation of personal freedom, because individuality can be expressed only in relationships with others. Money buys us membership in groups (fraternities, sororities, professional associations) or access to activities (education, travel, athletics, pottery classes). With the widening of social circles that money makes possible comes the ability to pursue self-expressive interests and thus further individuation (Simmel [1900a] 1978:344, 345).

However, while money offers a number of advantages for individuals, Simmel also notes the far less beneficial consequences it brings about. Most important, in modern societies in which money serves as a nexus for social integration, relations with others become less personal and intimate. Without significant emotional attachment or investment, relations become more and more standardized, like money itself. The impersonality and generalizability of money as a medium of exchange transform the nature of social interaction. Money breeds relationships based on rational calculation and instrumental purposes as opposed to personal ties of family or attraction to another's individual, subjective qualities. Coworkers come into and go out of our lives with little notice and only passing concern. We pick doctors based on the convenience of their office location and hours. Dating services, for a fee, will "match" individuals by way of quantitative surveys. Sharing an affinity with Weber's analysis of Western societies, Simmel argues that money plays an integral role in the trend toward the rationalization of social life. As money becomes the standard by which the value of all objects and relations are quantified, life is emptied of emotional connections and the enchantment of traditions. Personal loyalties and emotional attachment have been replaced with indifference ([1900a] 1978:297–303). Moreover, the expanding social relations and involvements made possible through a money economy leave individuals able only to partly invest their self in any one activity. Thus, the self becomes atomized or fragmented, while others with whom the individual interacts "know" but a small part of his self (ibid.:342–345).

Turning to our metatheoretical framework, Simmel's work is predominantly individualistic and non-rationalist in orientation (Figure 6.4). Let's consider first the individualistic dimension of his theory. Simmel's view of the social order—the routines and patterns of behavior that comprise everyday life—emphasizes the importance of *interaction*. As we noted above, Simmel did not see society as a "thing" or reality that existed independently of the individuals who compose it. Instead, Simmel saw society as an "event"—that is, the ongoing interactions that individuals engage in as they move through their daily lives seeking to satisfy their needs and realize their desires. The regularities or unity that characterize social life stems from the basic *forms* of interaction or structure of relationships, not from some overarching system of institutional arrangements that determines one's interests, behavior, or consciousness.

To be sure, although Simmel viewed society primarily as an ongoing process of interaction in which individuals mutually influence one another, his perspective incorporates collectivist elements as well. In particular, as discussed previously, Simmel saw "objective culture" taking on a seemingly autonomous presence in modern, urban societies. Here, the increasingly complex division of labor promotes the specialization of individual skills and, with it, a greater cultivation of individuality and personal expression. However, the sense of liberation is countered by external objects that confront individuals as dominating forces. Such alien objects are in reality the products of individual innovation that have become reified. Advances in science, industry, technology, and the political and legal administration of the populace are often experienced as oppressive systems that compel conformity. Individuals' lives move to the impersonal rhythms of a money economy and the institutional demands around which urban society is organized. This duality between "objective" and "subjective" culture—and the individualist/nonrationalist versus collectivist/rationalist presuppositions—is illustrated in Table 6.1.

Figure 6.4   Simmel's Basic Theoretical Orientation

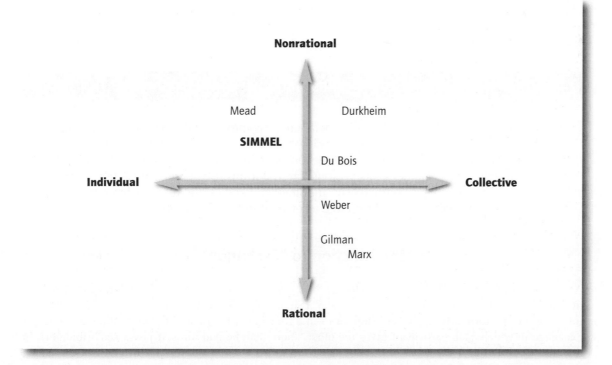

Table 6.1 Simmel's Duality Regarding Subjective/Objective Culture

		ORDER	
		Individual	**Collective**
ACTION	**Nonrational**	*Subjective culture:* Ongoing interactions that individuals engage in as they move through their daily lives seeking to satisfy their needs and realize their desires.	
	Rational		*Objective culture:* Reification; external objects confront individuals as oppressive systems that compel conformity.

With regard to his view on the nature of action, Simmel adopts a predominantly (though not exclusively) nonrationalist approach. A variety of conscious and unconscious motivations can impel individuals to interact with one another. As Simmel states,

> interaction always arises on the basis of certain drives or for the sake of certain purposes. Erotic, religious, or merely associative impulses; and purposes of defense, attack, play, gain, aid, or instruction—these and countless others cause man to live with other men, to act for them, with them, against them, and thus to correlate his condition with theirs. ([1908e] 1971:23)

Thus, while Simmel acknowledges that action can be based on calculative attempts to minimize costs and maximize gains, by no means does he restrict motivations to such a rationalist basis. Indeed, Simmel points out that we often are led to interact simply for the pleasure we derive from being in the company of others.

Readings

As we noted previously, Simmel saw in social life a synthesis of contradictions. This emphasis on the dualities underlying modern society and interaction is found in all his writings. Within this general tendency, however, three topics are most relevant. In the following selections, you will read Simmel's discussions on (1) forms of social interaction, namely, "exchange"; (2) social types, particularly "the stranger"; and (3) the relationship between the individual and the broader social structure as evident in the metropolis.

Introduction to "Exchange"

As we noted above, Simmel saw in exchange "the purest and most concentrated form of all human interactions in which serious interests are at stake" (Simmel [1900b] 1971:43). Indeed, for Simmel every interaction (a performance, a conversation, or even a romantic affair) could be understood as a form of exchange in which each participant gives the other "more than he had himself possessed" (ibid.:44). In the reading that follows, you will find Simmel focusing his attention on a specific form of exchange, *economic* exchange, particularly as it relates to the creation of value. Here, you also will see Simmel engaged in an important debate with Marx's view on the source of economic value.

What separates specifically economic forms of exchange from more general exchange interactions is *sacrifice*. For Simmel, the measure of sacrifice necessary to attain goods or goals is the source of their economic value. Unlike Marx, who argued that the value of goods is equal to the amount of labor power invested in their production, Simmel found in sacrifice—the giving up of one's money, time, services, possessions—"the condition of all value" ([1900b] 1971:49). Hence, there can be no universal, objective standard by which value can be established (see Table 6.2). For instance, while an avid collector of comic books may be willing to buy a rare comic "valued" at $200, most of us would not, instead preferring to sacrifice our money on other goods such as clothes or CDs. As a result, the value of the comic book, the ratio of desire to sacrifice, would be far less than $200 for most of us.

Table 6.2 Comparison of Simmel and Marx on the Issue of Economic Value

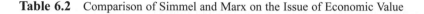

		ORDER	
		Individual	**Collective**
ACTION	**Nonrational**	*Simmel:* Value of goods is determined in *interaction*, as actors weigh their *desire* for goods against *sacrifice* required to attain them.	
	Rational		*Marx:* Value of goods is equal to the amount of labor invested in *production*.

Value, then, is always subjective and relative. It is determined by the interaction at hand in which actors weigh their desire for the goods in question against the amount of sacrifice required to attain them. Moreover, without having to endure obstacles or some form of self-denial, not even the most intensely felt desire for an object will make it valuable. This is because value is created out of the "distance" that separates desire from its satisfaction and the willingness to sacrifice something in order to overcome that distance. This is precisely the lesson that some parents try to teach their children when they assign chores. By having children earn money in exchange for sacrificing their time and energies, the toys they purchase for themselves will acquire value and thus won't be so freely lost or discarded—at least in theory!

"Exchange" (1907)

Georg Simmel

Most relationships among men can be considered under the category of exchange. Exchange is the purest and most concentrated form of all human interactions in which serious interests are at stake.

Many actions which at first glance appear to consist of mere unilateral process in fact involve reciprocal effects. The speaker before an audience, the teacher before a class, the journalist writing to his public—each appears to be the sole source of influence in such situations, whereas each of them is really acting in response to demands and directions that emanate from apparently passive, ineffectual groups. The saying "I am their leader, therefore I must follow them" holds good for politicians the world over. Even in hypnosis, which is manifestly the most clear-cut case where one person exercises influence and the other shows total passivity, reciprocity still obtains. As an outstanding hypnotist has recently stressed, the hypnotic effect would not be realized were it not for a certain ineffable reaction of the person hypnotized back on the hypnotist himself.

INTERACTIONS AS EXCHANGE

Now every interaction is properly viewed as a kind of exchange. This is true of every conversation, every love (even when requited unfavorably), every game, every act of looking one another over. It might seem that the two categories are dissimilar, in that in interaction one gives something one does not have, whereas in exchange one gives only what one does have, but this distinction does not really hold. What one expends in interaction can only be one's own energy, the transmission of one's own substance. Conversely, exchange takes place not for the sake of an object previously possessed by another person, but rather for the sake of one's own feeling about an object, a feeling which the other previously did not possess. The meaning of exchange, moreover, is that the sum of values is greater afterward than it was before, and this implies that each party gives the other more than he had himself possessed.

Interaction is, to be sure, the broader concept, exchange the narrower one. In human relations, however, interaction generally appears in forms which lend themselves to being viewed as exchange. The ordinary vicissitudes of daily life produce a continuous alternation of profit and loss, an ebbing and flowing of the contents of life. Exchange has the effect of rationalizing these vicissitudes, through the conscious act of setting the one *for* the other. The same synthetic process of mind that from the mere juxtaposition of things creates a with-another and for-another—the same ego which, permeated by sense data, informs them with its own unified character—has through the category of exchange seized that naturally given rhythm of our existence and organized its elements into a meaningful nexus.

THE NATURE OF ECONOMIC EXCHANGE

Of all kinds of exchange, the exchange of economic values is the least free of some tinge of sacrifice. When we exchange love for love, we release an inner energy we would otherwise not know what to do with. Insofar as we surrender it, we sacrifice no real utility (apart from what may be the external consequences of involvement). When we communicate intellectual matters in conversation, these are not thereby diminished. When we reveal a picture of our personality in the course of taking in that of others, this exchange in no way decreases our possession of ourselves. In all these exchanges the increase of value does not occur through the calculation of profit and loss. Either the contribution of each party stands beyond such a consideration, or else simply to be allowed to contribute is itself a gain—in which case we perceive the response of the other, despite our own offering, as an unearned gift. In contrast, economic exchange—whether it involves substances, labor, or labor power invested in substances—always entails the sacrifice of some good that has other potential uses, even though utilitarian gain may prevail in the final analysis.

The idea that all economic action is interaction, in the specific sense of exchange that involves sacrifice, may be met with the same objection which has been raised against the doctrine that equates all economic value with exchange value. The point has been made that the totally isolated economic man, who neither buys nor sells, would still have to evaluate his products and means of production—would therefore have to construct a concept of value independent of all exchange—if his expenditures and results were to stand in proper relation to one another. This fact, however, proves exactly what it is supposed to disprove, for all consideration whether a certain product is worth enough to justify a certain expenditure of labor or other goods is, for the economic agent, precisely the same as the appraisal which takes place in connection with exchange.

In dealing with the concept of exchange there is frequently a confusion of thought which leads one to speak of a relationship as though it were something external to the elements between which it occurs. Exchange means, however, only a condition of or a change within each of these elements, nothing that is *between* them in the sense of an object separated in space between the two other objects. When we subsume the two acts or changes of condition which occur in reality under the concept "exchange," it is tempting to think that with the exchange something has happened in addition to or beyond that which took place in each of the contracting parties.

This is just like being misled by the substantive concept of "the kiss" (which to be sure is also "exchanged") into thinking that a kiss is something that lies outside of the two pairs of lips, outside of their movements and sensations. Considered with reference to its immediate content, exchange is nothing more than the causally connected repetition of the fact that an actor now has something which he previously did not have, and for that has lost something which he previously did have.

That being the case, the isolated economic man, who surely must make certain sacrifices in order to gain certain fruits, behaves exactly like the one who makes exchanges. The only difference is that the party with whom he contracts is not a second free agent, but the natural order and regularity of things, which no more satisfy our desires without a sacrifice on our part than would another person. His calculations of value, in accordance with which he governs his actions, are generally the same as in exchange. For the economic actor as such it is surely quite immaterial whether the substances or labor capacities which he possesses are sunk into the ground or given to another man, if what he gains from the sacrifice is exactly the same in both cases.

This subjective process of sacrifice and gain within the individual psyche is by no means something secondary or imitative in relation to interindividual exchange. On the contrary, the give-and-take between sacrifice and attainment within the individual is the fundamental presupposition and, as it were, the essence of every two-sided exchange. The latter is only a subspecies of the former; that is, it is the sort in which the sacrifice is occasioned by the demand of another individual, whereas the sacrifice can be occasioned by things and their natural properties with the same sort of consequences for the actor.

It is extremely important to carry through this reduction of the economic process to that which takes place in *actuality,* that is, within the psyche of every economic actor. We should not let ourselves be misled because in exchange this process is reciprocal, conditioned by a similar process within another party. The natural and "solipsistic" economic transaction goes back to the same fundamental form as the two-sided exchange: to the process of balancing two subjective events within an individual. This is basically unaffected by the secondary question whether the process is instigated by the nature of things or the nature of man, whether it is a matter of purely natural economy or exchange economy. All feelings of value, in other words, which are set free by producible objects are in general to be gained only by foregoing other values. Such self-denial consists not only in that indirect labor for ourselves which appears as labor for others, but

frequently enough in direct labor on behalf of our own personal ends. . . .

THE SIGNIFICANCE OF SACRIFICE

The fact that value is the issue of a process of sacrifice discloses the infinity of riches for which our life is indebted to this basic form. Because we strive to minimize sacrifice and perceive it as painful, we tend to suppose that only with its complete disappearance would life attain its highest level of value. But this notion overlooks the fact that sacrifice is by no means always an external barrier to our goals. It is rather the *inner* condition of the goal and of the way to it. Because we dissect the problematic unity of our practical relations to things into the categories of sacrifice and profit, of obstacle and attainment, and because these categories are frequently separated into differentiated temporal stages, we forget that if a goal were granted to us without the interposition of obstacles it would no longer be the same goal.

The resistance which has to be eliminated is what gives our powers the possibility of proving themselves. Sin, after whose conquest the soul ascends to salvation, is what assures that special "joy in heaven" which those who were upright from the outset do not possess there. Every synthesis requires at the same time an effective analytic principle, which actually negates it (for without this it would be an absolute unity rather than a synthesis of several elements). By the same token every analysis requires a synthesis, in the dissolution of which it consists (for analysis demands always a certain coherence of elements if it is not to amount to a mere congeries without relations). The most bitter enmity is still more of a connection than simple indifference, indifference still more than not even knowing of one another. In short: the inhibiting countermovement, the diversion of which signifies sacrifice, is often—perhaps, seen from the point of view of elementary processes, even always—the positive presupposition of the goal itself. Sacrifice by no means belongs in the category of the undesirable, though superficiality and greed might portray it as such. It is not only the condition of individual values but, in what concerns us here, the economic realm, sacrifice is the condition of all value; not only the price to be paid for individual values that are already established, but that through which alone values can come into being.

Exchange occurs in two forms, which I shall discuss here in connection with the value of labor. All labor is indisputably a sacrifice if it is accompanied by a desire for leisure, for the mere self-satisfying play of skills, or for the avoidance of strenuous exertion. In addition to such desires, however, there exists a quantum of latent work energy which either we do not know what to do with or which presents itself as a drive to carry out voluntary labor, labor called forth neither by necessity nor by ethical motives. The expenditure of this energy is in itself no sacrifice, yet for this quantum of energy there compete a number of demands all of which it cannot satisfy. For every expenditure of the energy in question one or more possible and desirable alternative uses of it must be sacrificed. Could we not usefully spend the energy with which we accomplish task A also on task B, then the first would not entail any sacrifice; the same would hold for B in the event we chose it rather than A. In this utilitarian loss what is sacrificed is not labor, but *non-labor.* What we pay for A is not the sacrifice of labor—for our assumption here is that the latter in itself poses not the slightest hardship on us—but the giving up of task B.

The sacrifice which we make of labor in exchange is therefore of two sorts, of an absolute and a relative sort. The discomfort we accept is in the one case directly bound up with the labor itself, because the labor is annoying and troublesome. In the case where the labor itself is of eudaemonistic irrelevance or even of positive value, and when we can attain one object only at the cost of denying ourselves another, the frustration is indirect. The instances of happily done labor are thereby reduced to the form of exchange entailing renunciation, the form which characterizes all aspects of economic life. . . .

The Source of Value

If we regard economic activity as a special case of the universal life-form of exchange, as a sacrifice in return for a gain, we shall from the very beginning intuit something of what takes place within this form, namely, that the value of the gain is not, so to speak, brought with it, ready-made, but accrues to the desired object, in part or even entirely through the measure of the sacrifice demanded in acquiring it. These cases, which are as frequent as they are important for the theory of value, seem, to be sure, to harbor an inner contradiction: they have us making a sacrifice of a value for things which in themselves are worthless.

No one in his right mind would forego value without receiving for it at least an equal value; that, on the contrary, an end should receive its value only through the price that we must give for it could be the case

only in an absurd world. Yet common sense can readily see why this is so.

The value which an actor surrenders for another value can never be greater, for the subject himself under the actual circumstances of the moment, than that for which it is given. All contrary appearances rest on the confusion of the value actually estimated by the actor with the value which the object of exchange in question usually has or has by virtue of some apparently objective assessment. Thus if someone at the point of death from hunger gives away a jewel for a piece of bread, he does so because the latter is worth more to him under the circumstances than the former. Some particular circumstances, however, are *always* involved when one attaches a feeling of value to an object. Every such feeling of value is lodged in a whole complex system of our feelings which is in constant flux, adaptation, and reconstruction. Whether these circumstances are exceptional or relatively constant is obviously in principle immaterial. Through the fact that the starving man gives away his jewel he shows unambiguously that the bread is worth more to him.

There can thus be no doubt that in the moment of the exchange, of the making of the sacrifice, the value of the exchanged object forms the limit which is the highest point to which the value of the object being given away can rise. Quite independent of this is the question whence that former object derives its exigent value, and whether it may not come from the sacrifices to be offered for it, such that the equivalence between gain and cost would be established a posteriori, so to speak, and by virtue of the latter. We will presently see how frequently value comes into being psychologically in this apparently illogical manner.

Given the existence of the value, however, it is psychologically necessary to regard it, no less than values constituted in every other way, as a positive good at least as great as the negative of what has been sacrificed for it. There is in fact a whole range of cases known to the untrained psychological observer in which sacrifice not only heightens the value of the goal, but even generates it by itself. What comes to expression in this process is the desire to prove one's strength, to overcome difficulties, indeed often to oppose for the sheer joy of opposition. The detour required to attain certain things is often the occasion, often the cause as well, of perceiving them as values. In human relationships, most frequently and clearly in erotic relations, we notice how reserve, indifference, or rejection inflames the most passionate desire to prevail over these obstacles, and spurs us to efforts and sacrifices which, without these obstacles, would surely seem to us excessive. For many people the aesthetic gain from climbing the high Alps would not be considered worth further notice if it did not demand the price of extraordinary exertion and dangers and thereby acquire character, appeal, and consecration.

The charm of antiques and curios is frequently of the same sort. Even if antiques possess no intrinsic aesthetic or historical interest, a substitute for this is furnished by the mere difficulty of acquiring them: they are worth as much as they cost. It then comes to appear that they cost what they are worth. Furthermore, all ethical merit signifies that for the sake of the morally desirable deed contrary drives and wishes must be combatted and given up. If the act occurs without any conquest, as the direct issue of uninhibited impulses, its content may be objectively desirable, but it is not accorded a subjective moral value in the same sense. Only through the sacrifice of the lower and yet so seductive goods does one reach the height of ethical merit; and the more tempting the seductions and the more profound their sacrifice, the loftier the height. If we observe which human achievements attain to the highest honors and evaluations, we find them always to be those which manifest, or at least appear to manifest, the most depth, the most exertion, the most persistent concentration of the whole being—which is to say the most self-denial, sacrifice of all that is subsidiary, and devotion of the subjective to the objective ideal.

And if, in contrast with all this, aesthetic production and everything sweet and light, flowing from the naturalness of impulse, unfolds an incomparable charm, this charm derives its special quality from feelings associated with the burdens and sacrifices which are ordinarily required to gain such things. The liability and inexhaustible richness of combination of the contents of our minds frequently transform the significance of a connection into its exact converse, somewhat as the association between two ideas follows equally whether they are asserted or denied of each other. We perceive the specific value of something obtained without difficulty as a gift of fortune only on the grounds of the significance which things have for us that are hard to come by and measured by sacrifice. It is the same value, but with the negative sign; and the latter is the primary from which the former may be derived—but not vice versa.

We may be speaking here of course of exaggerated or exceptional cases. To find their counterpart in the whole realm of the economy it seems necessary, first of all, to make an analytic distinction between the universal substance of value, and economic activity as a differentiated form thereof. If for the moment we take

value as something given, then in accord with our foregoing discussion the following proposition is established beyond doubt: *Economic value as such does not inhere in an object in its isolated self-existence, but comes to an object only through the expenditure of another object which is given for it.* Wild fruit picked without effort, and not given in exchange, but immediately consumed, is no economic good. It can at most count as such only when its consumption saves some other economic expense. If, however, all of life's requirements were to be satisfied in this manner, so that at no point was sacrifice involved, men would simply not have *economic* activity, any more than do birds or fish or the denizens of fairyland. Whatever the way two objects, A and B, became values, A becomes an *economic* value only because I must give B for it, B only because I can obtain A for it. As mentioned above, it is in principle immaterial here whether the sacrifice takes place by transferring a value to another person, that is, through interindividual exchange, or within the circle of the individual's own interests, through a balancing of efforts and results. In articles of commerce there is simply nothing else to be found other than the meaning each one directly or indirectly has for our consumption needs and the exchange which takes place between them. Since, as we have seen, the former does not of itself suffice to make a given object an object of economic activity, it follows that the latter alone can supply to it the specific difference which we call economic.

This distinction between value and its economic form, is, however, an artificial one. If at first economy appears to be a mere form, in the sense that it presumes values as its contents, in order to be able to draw them into the process of balancing between sacrifice and profit, in reality the same process which forms the presumed values into an economy can be shown to be the creator of the economic *values themselves.* This will now be demonstrated.

The economic form of the value stands between two boundaries: on the one hand, *desire* for the object, connected with the anticipated feeling of satisfaction from its possession and enjoyment; on the other hand, this *enjoyment* itself which, strictly speaking, is not an economic act. That is, as soon as one concedes, as was shown above, that the immediate consumption of wild fruit is not an economic act and therefore the fruit itself is not an economic value (except insofar as it saves the production of economic values), then the consumption of real economic values is no longer economic, for the act of consumption in the latter case is not distinguishable from that in the former. Whether the fruit someone eats has been accidentally found, stolen, home-grown, or bought makes not the slightest difference in the act of eating and its direct consequences for the eater. . . .

The qualities of objects which account for their subjective desirability cannot, consequently, be credited with producing an absolute amount of value. It is always the relation of desires to one another, realized in exchange, which turns their objects into economic values. With respect to scarcity, the other element supposed to constitute value, this consideration is more directly apparent. Exchange is, indeed, nothing other than the interindividual attempt to improve an unfavorable situation arising out of a shortage of goods; that is, to reduce as much as possible the amount of subjective abstinence by the mode of distributing the available supply. Thereupon follows immediately a universal correlation between what is called scarcity-value (a term justly criticized) and what is called exchange-value.

For us, however, the connection is more important in the reverse direction. As I have already emphasized, the fact that goods are scarce would not lead us to value them unless we could not somehow modify that scarcity. It is modifiable in only two ways: by expending labor to increase the supply of goods, or by giving up objects already possessed in order to make whatever items an individual most desires less scarce for him. One can accordingly say that the scarcity of goods in relation to the desires directed to them objectively conditions exchange, but that it is exchange alone that makes scarcity a factor in value. It is a mistake of many theories of value to assume that, when utility and scarcity are given, economic value—that is, the exchange process—is something to be taken for granted, a conceptually necessary consequence of those premises. In this they are by no means correct. If, for instance, those conditions are accompanied by ascetic resignation, or if they instigated only combat or robbery—which, to be sure, is indeed true often enough—no economic value and no economic life would emerge. . . .

The Process of Value Formation: Creating Objects Through Exchange

Now an object is not a value so long as it remains a mere emotional stimulus enmeshed in the subjective process—a natural part of our sensibility, as it were. It must first be separated from this subjective sensibility for it to attain the peculiar significance which we call value. For not only is it certain that desire in and of itself could not establish any value if it did not encounter obstacles—trade in economic values could never

have arisen if every desire was satisfied without struggle or exertion—but even desire itself would never have ascended to such a considerable height if it could be satisfied without further ado. It is only the postponement of satisfaction through impediment, the anxiety that the object may escape, the tension of struggle for it, that brings about the cumulation of desires to a point of intensified volition and continuous striving.

If, however, even the highest pitch of desire were generated wholly from within, we still would not confer value on the object which satisfies it if the object were available to us in unlimited abundance. The important thing in that case would be the total enjoyment, the existence of which guarantees to us the satisfaction of our wishes, but not that particular quantum which we actually take possession of, since this could be replaced quite as easily by another. Even that totality would acquire some sense of value only by virtue of the thought of its possible shortage. Our consciousness would in this case be filled simply with the rhythm of subjective desires and satisfactions, without attaching any attention to the mediating object. Neither need nor enjoyment contains in itself value or economic process. These are actualized simultaneously through exchange between two subjects, each of whom requires some self-denial by the other as a condition of feeling satisfied, or through the counterpart of this process in the solipsistic economy. Through exchange, economic process and economic values emerge simultaneously, because exchange is what sustains or produces the distance between subject and object which transmutes the subjective state of feeling into objective valuation.

Kant once summarized his theory of knowledge in the proposition: "The conditions of experience are at the same time the conditions of the objects of experience." By this he meant that the process we call experience and the concepts which constitute its contents or objects are subject to the same laws of reason. The objects can enter into our experience, that is, can be experienced by us, because they exist as concepts within us, and the same energy which forms and defines the experience manifests itself in the formation of those concepts. In the same spirit we may say here that the possibility of economy is at the same time the possibility of the objects of economy. The very transaction between two possessors of objects (substances, labor energies, rights) which brings them into the so-called economic relation, namely, reciprocal sacrifice, at the same time elevates each of these objects into the category of value. The logical difficulty raised by the argument that values must first exist, and exist as values, in order to enter into the form and process of economic action, is now removed. It is removed thanks to the significance we have perceived in that psychic relationship which we designated as the distance between us and things. This distance differentiates the original subjective state of feeling into (1) a desiring subject, anticipating feelings, and (2) counterposed to him, an object that is now imbued with value; while the distance, on its side, is produced in the economic realm by exchange, that is, by the two-sided operation of barriers, restraint, and self-denial. Economic *values* thus emerge through the same reciprocity and relativity in which the *economic condition* of values consists.

Exchange is not merely the addition of the two processes of giving and receiving. It is, rather, something new. Exchange constitutes a third process, something that emerges when each of those two processes is simultaneously the cause and the effect of the other. Through this process, the value which the necessity of self-denial for an object imparts to it becomes an economic value. If it is true that value arises in general in the interval which obstacles, renunciations, and sacrifices interpose between desire and its satisfaction, and if the process of exchange consists in that reciprocally conditioned taking and giving, there is no need to invoke a prior process of valuation which makes a value of an isolated object for an isolated subject. What is required for this valuation takes place in the very act of exchange itself. . . .

VALUE AND PRICE

If value is, as it were, the offspring of price, then it seems logical to assert that their amounts must be the same. I refer now to what has been established above, that in each individual case no contrasting party pays a price which to him under the given circumstances is too high for the thing obtained. If, in the poem of Chamisso, the highwayman with pistol drawn compels the victim to sell his watch and ring for three coppers, the fact is that under the circumstances, since the victim could not otherwise save his life, the thing obtained in exchange is actually worth the price. No one would work for starvation wages if, in the situation in which he actually found himself, he did not prefer this wage to not working. The appearance of paradox in the assertion that value and price are equivalent in every individual case arises from the fact that certain conceptions of other kinds of equivalence of value and price are brought into our estimate.

Two kinds of considerations bring this about: (1) the relative stability of the relations which determine the majority of exchange transactions, and (2) the analogies which set still uncertain value-relations according to the norms of those that already exist. Together these produce the notion that if for a certain object this and that other object were exchange equivalents, then these two objects, or the circle of objects which they define, would have the same position in the scale of values. They also give rise to the related notion that if abnormal circumstances caused us to exchange this object for values that lie higher or lower in the scale, price and value would become discrepant—although in each individual case, considering *its* circumstances, we would find them actually to coincide. We should not forget that the objective and just equivalence of value and price, which we make the norm of the actual and individual case, holds good only under very specific historical and technical conditions; and that, with the change of these conditions, the equivalence vanishes at once. Between the norm itself and the cases which it defines as either exceptional or standard there is no difference of kind: there is, so to speak, only a quantitative difference. This is somewhat like when we say of an extraordinarily elevated or degraded individual, "He is really no longer a man." The fact is that this idea of man is only an average. It would lose its normative character at the moment a majority of men ascended or descended to that level of character, which would then pass for the generically "human."

To perceive this requires an energetic effort to disentangle two deeply rooted conceptions of value which have substantial practical justification. In relations that are somewhat evolved these conceptions are lodged in two superimposed levels. One kind of standard is formed from the traditions of society, from the majority of experiences, from demands that seem to be purely logical; the other, from individual constellations, from demands of the moment, from the constraints of a capricious environment. Looking at the rapid changes which take place within the latter sphere, we lose sight of the slow evolution of the former and its development out of the sublimation of the latter; and the former seems suitably justified as the expression of an objective proportion. In an exchange that takes place under such circumstances, when the feelings of loss and gain at least balance each other (for otherwise no actor who made any comparisons at all would consummate the exchange) yet when these same feelings of value are discrepant when measured by those general standards, one speaks of a divergence between value and price. This occurs most conspicuously under two conditions, which almost always go together: (1) when a single value-quality is counted as the economic value and two objects consequently are adjudged equal in value only insofar as the same quantum of that fundamental value is present in them, and (2) when a certain proportion between two values is expected not only in an objective sense but also as a moral imperative.

The conception, for example, that the real value-element in all values is the socially required labor time objectified in them has been applied in both of these ways, and provides a standard, directly or indirectly applicable, which makes value fluctuate positively and negatively with respect to price. The fact of that single *standard of value* in no way establishes how labor power comes to be a value in the first place. It could hardly have done so if the labor power had not, by acting on various materials and fashioning various products, created the possibility of exchange, or if the use of the labor power were not perceived as a sacrifice which one makes for the sake of its fruits. Labor energy also, then, is aligned with the category of value only through the possibility and reality of exchange, irrespective of the fact that subsequently *within* this category of value labor may itself provide a standard for the remaining contents. If the labor power therefore is also the content of every value, it receives its form as value in the first place only because it enters into the relations between sacrifice and gain, or profit and value (here in the narrower sense).

In the cases of discrepancy between price and value the one contracting party would, according to this theory, give a certain amount of immediately objectified labor power for a lesser amount of the same. Other factors, not involving labor power, would then lead the party to complete the exchange, factors such as the satisfaction of a terribly urgent need, amateurish fancy, fraud, monopoly, and so on. In the wider and subjective sense, therefore, the equivalence of value and countervalue holds fast in these cases, and the single norm, labor power, which makes the discrepancy possible, does not cease to derive the genesis of its character as a value from exchange. . . .

From all the foregoing it appears that exchange is a sociological structure sui generis, a primary form and function of interindividual life. By no means does it follow logically from those qualitative and quantitative properties of things which we call utility and scarcity. On the contrary, both these properties derive their significance as generators of value only under the presupposition of exchange. Where exchange, offering a sacrifice for the sake of a gain, is impossible for any reason, no degree of scarcity of a desired object can convert it to an economic value until the possibility of that relation reappears.

The meaning that an object has for an individual always rests solely in its desirability. For whatever an object is to accomplish for us, its qualitative character is decisive. When we possess it, it is a matter of indifference whether in addition there exist many, few, or no other specimens of its kind. (I do not distinguish here those cases in which scarcity itself is a kind of qualitative property which makes the object desirable to us, such as old postage stamps, curiosities, antiques without aesthetic or historical value, etc.) The sense of difference, incidentally, important for enjoyment in the narrower sense of the word, may be everywhere conditioned by a scarcity of the object, that is, by the fact that it is not enjoyed everywhere and at all times. This inner psychological condition of enjoyment, however, is not a practical factor, because it would have to lead not to the overcoming of scarcity but to its conservation, its increase even—which is patently not the case. The only relevant question apart from the direct enjoyment of things for their qualities is the question of the way to it. As soon as this way is a long and difficult one, involving sacrifices in patience, disappointment, toil, inconvenience, feats of self-denial, and so on, we call the object "scarce." One can express this directly: things are not difficult to obtain because they are scarce, but they are scarce because they are difficult to obtain. The inflexible external fact that the supply of certain goods is too small to satisfy all our desires for them would be in itself insignificant. There are many objectively scarce things which are not scarce in the economic sense of the term. Whether they are scarce in this sense depends entirely upon what measure of energy, patience, and devotion is necessary for their acquisition—sacrifices which naturally presume the desirability of the object.

The difficulty of attainment, that is, the magnitude of the sacrifice involved in exchange, is thus the element that peculiarly constitutes value. Scarcity constitutes only the outer appearance of this element, only its objectification in the form of quantity. One often fails to observe that scarcity, purely as such, is only a negative property, an existence characterized by nonexistence. The nonexistent, however, cannot be operative. Every positive consequence must be the issue of a positive property and force, of which that negative property is only the shadow. These concrete forces are, however, manifestly the only ingredients of exchange. The aspect of concreteness is in no wise reduced because we are not dealing here with individuals as such. Relativity among things has a peculiar property: it involves reaching out beyond the individual, it subsists only within a plurality, and yet it does not constitute a mere conceptual generalization and abstraction.

Herewith is expressed the profound relation between relativity and society, which is the most immediate demonstration of relativity in regard to the material of humanity: society is the supra-singular structure which is nonetheless not abstract. Through this concept historical life is spared the alternatives of having to run either in mere individuals or in abstract generalities. Society is the generality that has, simultaneously, concrete vitality. From this can be seen the unique meaning which exchange, as the economic realization of the relativity of things, has for society. It lifts the individual thing and its significance for the individual man out of their singularity, not into the sphere of the abstract but into the liveliness of interaction, which is, so to speak, the body of economic value. We may examine an object ever so closely with respect to its self-sufficient properties, but we shall not find its economic value. For this consists exclusively in the *reciprocal relationship* which comes into being among several objects on the basis of these properties, each determining the other and each returning to the other the significance it has received therefrom.

▓

Introduction to "The Stranger"

Simmel juxtaposed his analysis of the forms of social interaction with a discussion of social **types**. Social types derive not from qualities intrinsic to the individual in question, nor from an individual's choice to be one "type" or another. Rather, being assigned or identified as a type of individual is a product of one's *relationship* to others. Simmel identified a number of social types, including "the poor," "the nobility," "the miser," "the spendthrift," and "the adventurer," but it is his analysis of "the stranger" that has become most well known.

The relationship of the stranger to the group is rooted in a unique synthesis of opposites: "nearness" and "remoteness" (see Figure 6.5). As a distinct social type, the stranger is near or close to us insofar as we share with him general, impersonal qualities, such as nationality, gender, or race. But because such similarities connect us with so many others, the stranger is indistinct or "remote." No unique or specific qualities are shared with him that could in turn form the basis of a personal relationship. As a result, the stranger is seen not as an individual, but, rather, as a "type" of person whose *particular* characteristics make him fundamentally different from the group. Yet, this unique position of the stranger relative to the group allows him to provide services that are otherwise unattainable or "unfit" for the in-group to perform. Often this attachment makes the stranger an indispensable element of the group, though his positive contributions are dependent on his outsider status. Though he is part of the group, the stranger, then, exists outside of and is thus confronted by the group.

For Simmel, the classical example of the stranger was the European Jew who served as a trader. The trader is a "middleman" who makes possible the exchange of goods with people who live beyond the boundaries of the group. Jews became traders because they were denied many of the legal, political, and property rights granted to ordinary citizens.[1] As a result, European Jews were often restricted in their professional activities to "mobile" occupations such as trading and finance. Simmel argued that it is the mobility of the stranger within a group—he is "no landowner"—that makes the position a "synthesis of nearness and remoteness." As Simmel remarked, "The purely mobile person comes incidentally into contact with *every* single element [of a group] but is not bound up organically, through established ties of kinship, locality, or occupation, with any single one" ([1908c] 1971:145, emphasis in the original).

---

[1]As you will read in the next chapter, there are many interesting parallels between Simmel's notion of the "stranger" and Du Bois's notion of "double consciousness" and the "place" of African Americans in the United States.

Figure 6.5 Duality of the "Stranger"

Stranger

remoteness/wandering   nearness/attachment

In addition to these occupational consequences, the unique relation of the stranger to the larger group allows the stranger to adopt an objective attitude toward internal conflicts. Nonpartisanship grants the stranger a position of objectivity in efforts to resolve disputes. In other words, the stranger is not likely to be committed in advance to any one party should disagreements arise between individuals who possess full group membership. Furthermore, the remoteness and freedom from prejudiced understanding that objectivity entails can also make the stranger a valued confidant. Certainly, in modern society it is not uncommon to be willing to share otherwise unaired, intimate details about one's life and relationships with a complete stranger. Indeed, now such strangers—professional therapists and counselors—are paid.

Significantly, then, the stranger is not the same as the complete "outcast." The stranger has elements of nearness *and* remoteness—he is attached, but not completely—while the social outcast is only remote. However, despite the services that strangers are able to provide to a community, nonetheless we should be careful not to romanticize the position of this social type. Strangers often are exceptionally vulnerable to discrimination, if not violence.[2]

"The Stranger" (1908)

Georg Simmel

If Wandering, considered as a state of detachment from every given point in space, is the conceptual opposite of attachment to any point, then the sociological form of "the stranger" presents the synthesis, as it were, of both of these properties. (This is another indication that spatial relations not only are determining conditions of relationships among men, but are also symbolic of those relationships.) The stranger will thus not be considered here in the usual sense of the term, as the wanderer who comes today and goes tomorrow, but rather as the man who comes today and stays tomorrow—the potential wanderer, so to speak, who, although he has gone no further, has not quite got over the freedom of coming and going. He is fixed within a certain spatial circle—or within a group whose boundaries are analogous to spatial boundaries—but his position within it is fundamentally

[2]While Jews historically have been strangers throughout Europe, America has its own history of "strangers." African Americans first were enslaved, then, after abolition, were kept "second-class" citizens with the establishment of Jim Crow laws. During the late nineteenth and early twentieth centuries, immigrants from southern and eastern Europe faced intense hatred and discrimination. Most ironic, however, is the plight suffered by Native Americans and Mexican Americans, who have always been treated as strangers in their own land as a result of colonization. All the while, they have played important roles by risking their lives in the armed forces, building the nation's infrastructure, and performing jobs that are considered "beneath" the dominant groups.

affected by the fact that he does not belong in it initially and that he brings qualities into it that are not, and cannot be, indigenous to it.

In the case of the stranger, the union of closeness and remoteness involved in every human relationship is patterned in a way that may be succinctly formulated as follows: the distance within this relation indicates that one who is close by is remote, but his strangeness indicates that one who is remote is near. The state of being a stranger is of course a completely positive relation; it is a specific form of interaction. The inhabitants of Sirius are not exactly strangers to us, at least not in the sociological sense of the word as we are considering it. In that sense they do not exist for us at all; they are beyond being far and near. The stranger is an element of the group itself, not unlike the poor and sundry "inner enemies"—an element whose membership within the group involves both being outside it and confronting it.

The following statements about the stranger are intended to suggest how factors of repulsion and distance work to create a form of being together, a form of union based on interaction.

In the whole history of economic activity the stranger makes his appearance everywhere as a trader, and the trader makes his as a stranger. As long as production for one's own needs is the general rule, or products are exchanged within a relatively small circle, there is no need for a middleman within the group. A trader is required only for goods produced outside the group. Unless there are people who wander out into foreign lands to buy these necessities, in which case they are themselves "strange" merchants in this other region, the trader *must* be a stranger; there is no opportunity for anyone else to make a living at it.

This position of the stranger stands out more sharply if, instead of leaving the place of his activity, he settles down there. In innumerable cases even this is possible only if he can live by trade as a middleman. Any closed economic group where land and handicrafts have been apportioned in a way that satisfies local demands will still support a livelihood for the trader. For trade alone makes possible unlimited combinations, and through it intelligence is constantly extended and applied in new areas, something that is much harder for the primary producer with his more limited mobility and his dependence on a circle of customers that can be expanded only very slowly. Trade can always absorb more men than can primary production. It is therefore the most suitable activity for the stranger, who intrudes as a supernumerary, so to speak, into a group in which all the economic

positions are already occupied. The classic example of this is the history of European Jews. The stranger is by his very nature no owner of land—land not only in the physical sense but also metaphorically as a vital substance which is fixed, if not in space, then at least in an ideal position within the social environment.

Although in the sphere of intimate personal relations the stranger may be attractive and meaningful in many ways, so long as he is regarded as a stranger he is no "landowner" in the eye of the other. Restriction to intermediary trade and often (although sublimated from it) to pure finance gives the stranger the specific character of *mobility*. The appearance of this mobility within a bounded group occasions that synthesis of nearness and remoteness which constitutes the formal position of the stranger. The purely mobile person comes incidentally into contact with *every* single element but is not bound up organically, through established ties of kinship, locality, or occupation, with any single one.

Another expression of this constellation is to be found in the objectivity of the stranger. Because he is not bound by roots to the particular constituents and partisan dispositions of the group, he confronts all of these with a distinctly "objective" attitude, an attitude that does not signify mere detachment and nonparticipation but is a distinct structure composed of remoteness and nearness, indifference and involvement. I refer to my analysis of the dominating positions gained by aliens, in the discussion of superordination and subordination, typified by the practice in certain Italian cities of recruiting their judges from outside, because no native was free from entanglement in family interest and factionalism.

Connected with the characteristic of objectivity is a phenomenon that is found chiefly, though not exclusively, in the stranger who moves on. This is that he often receives the most surprising revelations and confidences, at times reminiscent of a confessional about matters which are kept carefully hidden from everybody with whom one is close. Objectivity is by no means nonparticipation, a condition that is altogether outside the distinction between subjective and objective orientations. It is rather a positive and definite kind of participation, in the same way that the objectivity of theoretical observation clearly does not mean that the mind is a passive tabula rasa on which things inscribe their qualities, but rather signifies the full activity of a mind working according to its own laws, under conditions that exclude accidental distortions and emphases whose individual and subjective differences would produce quite different pictures of the same object.

Objectivity can also be defined as freedom. The objective man is not bound by ties which could prejudice his perception, his understanding, and his assessment of data. This freedom, which permits the stranger to experience and treat even his close relationships as though from a bird's-eye view, contains many dangerous possibilities. From earliest times, in uprisings of all sorts the attacked party has claimed that there has been incitement from the outside, by foreign emissaries and agitators. Insofar as this has happened, it represents an exaggeration of the specific role of the stranger: he is the freer man, practically and theoretically; he examines conditions with less prejudice; he assesses them against standards that are more general and more objective; and his actions are not confined by custom, piety, or precedent.

Finally, the proportion of nearness and remoteness which gives the stranger the character of objectivity also finds practical expression in the more *abstract* nature of the relation to him. That is, with the stranger one has only certain *more general* qualities in common, whereas the relation with organically connected persons is based on the similarity of just those specific traits which differentiate them from the merely universal. In fact, all personal relations whatsoever can be analyzed in terms of this scheme. They are not determined only by the existence of certain common characteristics which the individuals share in addition to their individual differences, which either influence the relationship or remain outside of it. Rather, the kind of effect which that commonality has on the relation essentially depends on whether it exists only among the participants themselves, and thus, although general within the relation, is specific and incomparable with respect to all those on the outside, or whether the participants feel that what they have in common is so only because it is common to a group, a type, or mankind in general. In the latter case, the effect of the common features becomes attenuated in proportion to the size of the group bearing the same characteristics. The commonality provides a basis for unifying the members, to be sure; but it does not specifically direct *these* particular persons to one another. A similarity so widely shared could just as easily unite each person with every possible other. This, too, is evidently a way in which a relationship includes both nearness and remoteness simultaneously. To the extent to which the similarities assume a universal nature, the warmth of the connection based on them will acquire an element of coolness, a sense of the contingent nature of precisely *this* relation—the connecting forces have lost their specific, centripetal character.

In relation to the stranger, it seems to me, this constellation assumes an extraordinary preponderance in principle over the individual elements peculiar to the relation in question. The stranger is close to us insofar as we feel between him and ourselves similarities of nationality or social position, of occupation or of general human nature. He is far from us insofar as these similarities extend beyond him and us, and connect us only because they connect a great many people.

A trace of strangeness in this sense easily enters even the most intimate relationships. In the stage of first passion, erotic relations strongly reject any thought of generalization. A love such as this has never existed before; there is nothing to compare either with the person one loves or with our feelings for that person. An estrangement is wont to set in (whether as cause or effect is hard to decide) at the moment when this feeling of uniqueness disappears from the relationship. A skepticism regarding the intrinsic value of the relationship and its value for us adheres to the very thought that in this relation, after all, one is only fulfilling a general human destiny, that one has had an experience that has occurred a thousand times before, and that, if one had not accidentally met this precise person, someone else would have acquired the same meaning for us.

Something of this feeling is probably not absent in any relation, be it ever so close, because that which is common to two is perhaps never common *only* to them but belongs to a general conception which includes much else besides, many *possibilities* of similarities. No matter how few of these possibilities are realized and how often we may forget about them, here and there, nevertheless, they crowd in like shadows between men, like a mist eluding every designation, which must congeal into solid corporeality for it to be called jealousy. Perhaps this is in many cases a more general, at least more insurmountable, strangeness than that due to differences and obscurities. It is strangeness caused by the fact that similarity, harmony, and closeness are accompanied by the feeling that they are actually not the exclusive property of this particular relation, but stem from a more general one—a relation that potentially includes us and an indeterminate number of others, and therefore prevents that relation which alone was experienced from having an inner and exclusive necessity.

On the other hand, there is a sort of "strangeness" in which this very connection on the basis of a general quality embracing the parties is precluded. The relation of the Greeks to the barbarians is a typical example; so are all the cases in which the general characteristics one takes as peculiarly and merely human are disallowed to the other. But here the expression "the stranger" no longer has any positive meaning. The relation with him is a non-relation; he is not what we have been discussing here: the stranger as a member of the group itself.

As such, the stranger is near and far *at the same time,* as in any relationship based on merely universal human similarities. Between these two factors of nearness and distance, however, a peculiar tension arises, since the consciousness of having only the absolutely general in common has exactly the effect of putting a special emphasis on that which is not common. For a stranger to the country, the city, the race, and so on, what is stressed is again nothing individual, but alien origin, a quality which he has, or could have, in common with many other strangers. For this reason strangers are not really perceived as individuals, but as strangers of a certain type. Their remoteness is no less general than their nearness.

This form appears, for example, in so special a case as the tax levied on Jews in Frankfurt and elsewhere during the Middle Ages. Whereas the tax paid by Christian citizens varied according to their wealth at any given time, for every single Jew the tax was fixed once and for all. This amount was fixed because the Jew had his social position as a *Jew,* not as the bearer of certain objective contents. With respect to taxes every other citizen was regarded as possessor of a certain amount of wealth, and his tax could follow the fluctuations of his fortune. But the Jew as taxpayer was first of all a Jew, and thus his fiscal position contained an invariable element. This appears most forcefully, of course, once the differing circumstances of individual Jews are no longer considered, limited though this consideration is by fixed assessments, and all strangers pay exactly the same head tax.

Despite his being inorganically appended to it, the stranger is still an organic member of the group. Its unified life includes the specific conditioning of this element. Only we do not know how to designate the characteristic unity of this position otherwise than by saying that it is put together of certain amounts of nearness and of remoteness. Although both these qualities are found to some extent in all relationships, a special proportion and reciprocal tension between them produce the specific form of the relation to the "stranger."

Introduction to "The Metropolis and Mental Life"

"The Metropolis and Mental Life" addresses several key themes that we have already discussed. For instance, you will find Simmel examining the nature of the struggle for individuality in modern societies as well as the relationship between objective culture and subjective or "individual" culture. You will also read Simmel's views on money and its psychological effects on the individual and her relationships with others.

This essay also contains an important theme not discussed previously: the intensity of stimuli created by the urban environment and its consequences for the psychology of the city dweller. Unlike the slower tempo and rhythms of small-town life, city life is characterized by a "swift and continuous shift of external and internal stimuli . . . the rapid telescoping of changing images, pronounced differences within what is grasped at a single glance" (Simmel [1903] 1971:325). While the slower tempo and limited social contacts within small towns foster the development of emotional bonds that tie its inhabitants together, the metropolis, with its unceasing fluctuations of stimuli and expansiveness of interpersonal contacts, is antithetical to nurturing a rich emotional life. Indeed, it is impossible for the city dweller to absorb or become emotionally invested in all the happenings and encounters that make up his daily life. Attempting to do so would lead to an overstimulation of the senses that would in turn produce a virtual psychological and emotional paralysis. This leaves the urbanite to react to her world "primarily in a rational manner. . . . Thus the reaction of the metropolitan person to those events is moved to a sphere of mental activity which is least sensitive and which is furthest removed from the depths of personality" (ibid.:326).

Photo 6.3   Simmel's Metropolis: Berlin, circa 1900

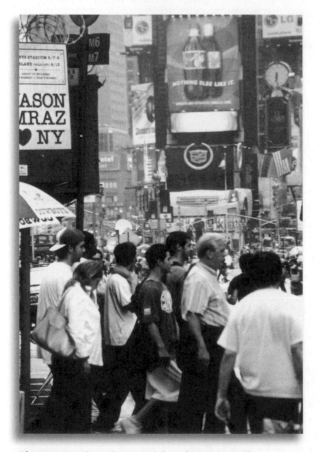

Photo 6.4   The quintessential modern metropolis: New York City, circa 2000

Bombarded with sensory impressions, the metropolitan person, in order to shield himself from the onslaught of stimuli and disruptions, adopts out of necessity an intellectualized approach to life. This psychological disposition, or **blasé attitude**, produces a dulling of differences, an emotional "graying" of reactions, which protects the individual from becoming overwhelmed by the sensory intensity of city life. As a result, the metropolitan personality experiences "quality" and differences as meaningless: "They appear to the blasé person in a homogenous, flat and gray color with no one of them worthy of being preferred to another" ([1903] 1971:330). The blasé attitude, while an adaptive outlook, is coupled with a money economy that further hinders the development of an emotionally meaningful life. As we discussed previously, money is a standardized, impersonal measure of value. It levels all subjective qualitative distinctions into objective differences of quantity. The emphasis on exactness and calculability required by the urban, capitalist economy finds its expression in the life of the individual to the extent that he likewise becomes indifferent to the qualitative distinctions in his surroundings and in his relationships. The more money mediates our relationships and serves as the medium for self-expression, the more life itself takes on a quantitative quality.

Like Marx and Weber, Simmel's analysis of urban life, in part, was intended as a critique of modernity and its corruption of individuality and the human spirit. In the end, however, his project was more an intellectual journey whose goal was to further our insight into the duality that exists between the individual and the "sovereign powers of society." "To the extent that such forces have been integrated . . . into the root as well as the crown of the totality of historical life to which we belong—it is our task not to complain or condone but only to understand" ([1903] 1971:339).

Photo 6.5a   During the late nineteenth and early twentieth centuries, Impressionist painters often depicted the distinctly metropolitan blasé attitude in their art, as here in Edouard Manet's *Bar at Folies Bergère.*

Photo 6.5b   Today, this metropolitan attitude often finds its expression in our impersonal indifference to the plight of others— particularly those with whom we would prefer not to come into contact.

"The Metropolis and Mental Life" (1903)

Georg Simmel

The deepest problems of modern life flow from the attempt of the individual to maintain the independence and individuality of his existence against the sovereign powers of society, against the weight of the historical heritage and the external culture and technique of life. This antagonism represents the most modern form of the conflict which primitive man must carry on with nature for his own bodily existence. The eighteenth century may have called for liberation from all the ties which grew up historically in politics, in religion, in morality and in economics in order to permit the original natural virtue of man, which is equal in everyone, to develop without inhibition; the nineteenth century may have sought to promote, in addition to man's freedom, his individuality (which is connected with the division of labor) and his achievements which make him unique and indispensable but which at the same time make him so much the more dependent on the complementary activity of others; Nietzsche may have seen the relentless struggle of the individual as the prerequisite for his full development, while Socialism found the same thing in the suppression of all competition—but in each of these the same fundamental motive was at work, namely the resistance of the individual to being levelled, swallowed up in the social-technological mechanism. When one inquires about the products of the specifically modern aspects of contemporary life with reference to their inner meaning—when, so to speak, one examines the body of culture with reference to the soul, as I am to do concerning the metropolis today—the answer will require the investigation of the relationship which such a social structure promotes between the individual aspects of life and those which transcend the existence of single individuals. It will require the investigation of the adaptations made by the personality in its adjustment to the forces that lie outside of it.

The psychological foundation, upon which the metropolitan individuality is erected, is the intensification of emotional life due to the swift and continuous shift of external and internal stimuli. Man is a creature whose existence is dependent on differences, i.e., his mind is stimulated by the difference between present impressions and those which have preceded. Lasting impressions, the slightness in their differences, the habituated regularity of their course and contrasts between them, consume, so to speak, less mental energy than the rapid telescoping of changing images, pronounced differences within what is grasped at a single glance, and the unexpectedness of violent stimuli. To the extent that the metropolis creates these psychological conditions—with every crossing of the street, with the tempo and multiplicity of economic, occupational and social life—it creates in the sensory foundations of mental life, and in the degree of awareness necessitated by our organization as creatures dependent on differences, a deep contrast with the slower, more habitual, more smoothly flowing rhythm of the sensory-mental phase of small town and rural existence. Thereby the essentially intellectualistic character of the mental life of the metropolis becomes intelligible as over against that of the small town which rests more on feelings and emotional relationships. These latter are rooted in the unconscious levels of the mind and develop most readily in the steady equilibrium of unbroken customs. The locus of reason, on the other hand, is in the lucid, conscious upper strata of the mind and it is the most adaptable of our inner forces. In order to adjust itself to the shifts and contradictions in events, it does not require the disturbances and inner upheavals which are the only means whereby more conservative personalities are able to adapt themselves to the same rhythm of events. Thus the metropolitan type—which naturally takes on a thousand individual modifications—creates a protective organ for itself against the profound disruption with which the fluctuations and discontinuities of the external milieu threaten it. Instead of reacting emotionally, the metropolitan type reacts primarily in a rational manner, thus creating a mental predominance through the intensification of consciousness, which in turn is caused by it. Thus the reaction of the metropolitan person to those events is moved to a sphere of mental activity which is least

SOURCE: From *On Individuality and Social Forms,* edited by Donald Levine (1971). Copyright © 1971 by The University of Chicago. All rights reserved. Used by permission of The University of Chicago Press.

sensitive and which is furthest removed from the depths of the personality.

This intellectualistic quality which is thus recognized as a protection of the inner life against the domination of the metropolis, becomes ramified into numerous specific phenomena. The metropolis has always been the seat of money economy because the many-sidedness and concentration of commercial activity have given the medium of exchange an importance which it could not have acquired in the commercial aspects of rural life. But money economy and the domination of the intellect stand in the closest relationship to one another. They have in common a purely matter-of-fact attitude in the treatment of persons and things in which a formal justice is often combined with an unrelenting hardness. The purely intellectualistic person is indifferent to all things personal because, out of them, relationships and reactions develop which are not to be completely understood by purely rational methods—just as the unique element in events never enters into the principle of money. Money is concerned only with what is common to all, i.e., with the exchange value which reduces all quality and individuality to a purely quantitative level. All emotional relationships between persons rest on their individuality, whereas intellectual relationships deal with persons as with numbers, that is, as with elements which, in themselves, are indifferent, but which are of interest only insofar as they offer something objectively perceivable. It is in this very manner that the inhabitant of the metropolis reckons with his merchant, his customer, and with his servant, and frequently with the persons with whom he is thrown into obligatory association. These relationships stand in distinct contrast with the nature of the smaller circle in which the inevitable knowledge of individual characteristics produces, with an equal inevitability, an emotional tone in conduct, a sphere which is beyond the mere objective weighting of tasks performed and payments made. What is essential here as regards the economic-psychological aspect of the problem is that in less advanced cultures production was for the customer who ordered the product so that the producer and the purchaser knew one another. The modern city, however, is supplied almost exclusively by production for the market, that is, for entirely unknown purchasers who never appear in the actual field of vision of the producers themselves. Thereby, the interests of each party acquire a relentless matter-of- factness, and its rationally calculated economic egoism need not fear any divergence from its set path because of the imponderability of personal relationships. This is all the more

the case in the money economy which dominates the metropolis in which the last remnants of domestic production and direct barter of goods have been eradicated and in which the amount of production on direct personal order is reduced daily. Furthermore, this psychological intellectualistic attitude and the money economy are in such close integration that no one is able to say whether it was the former that effected the latter or *vice versa*. What is certain is only that the form of life in the metropolis is the soil which nourishes this interaction most fruitfully, a point which I shall attempt to demonstrate only with the statement of the most outstanding English constitutional historian to the effect that through the entire course of English history London has never acted as the heart of England but often as its intellect and always as its money bag.

In certain apparently insignificant characters or traits of the most external aspects of life are to be found a number of characteristic mental tendencies. The modern mind has become more and more a calculating one. The calculating exactness of practical life which has resulted from a money economy corresponds to the ideal of natural science, namely that of transforming the world into an arithmetical problem and of fixing every one of its parts in a mathematical formula. It has been money economy which has thus filled the daily life of so many people with weighing, calculating, enumerating and the reduction of qualitative values to quantitative terms. Because of the character of calculability which money has there has come into the relationships of the elements of life a precision and a degree of certainty in the definition of the equalities and inequalities and an unambiguousness in agreements and arrangements, just as externally this precision has been brought about through the general diffusion of pocket watches. It is, however, the conditions of the metropolis which are cause as well as effect for this essential characteristic. The relationships and concerns of the typical metropolitan resident are so manifold and complex that, especially as a result of the agglomeration of so many persons with such differentiated interests, their relationships and activities intertwine with one another into a many-membered organism. In view of this fact, the lack of the most exact punctuality in promises and performances would cause the whole to break down into an inextricable chaos. If all the watches in Berlin suddenly went wrong in different ways even only as much as an hour, its entire economic and commercial life would be derailed for some time. Even though this may seem more superficial in its significance, it transpires that the magnitude of distances results in making all waiting and the breaking of appointments an ill-afforded waste of time. For this reason the technique of metropolitan life in general is not conceivable without all of its activities and reciprocal relationships being organized and coordinated in the most punctual way into a firmly fixed framework of time which transcends all subjective elements. But here too there emerge those conclusions which are in general the whole task of this discussion, namely, that every event, however restricted to this superficial level it may appear, comes immediately into contact with the depths of the soul, and that the most banal externalities are, in the last analysis, bound up with the final decisions concerning the meaning and the style of life. Punctuality, calculability, and exactness, which are required by the complications and extensiveness of metropolitan life are not only most intimately connected with its capitalistic and intellectualistic character but also color the content of life and are conductive to the exclusion of those irrational, instinctive, sovereign human traits and impulses which originally seek to determine the form of life from within instead of receiving it from the outside in a general, schematically precise form. Even though those lives which are autonomous and characterised by these vital impulses are not entirely impossible in the city, they are, none the less, opposed to it *in abstracto*. It is in the light of this that we can explain the passionate hatred of personalities like Ruskin and Nietzsche for the metropolis—personalities who found the value of life only in unschematized individual expressions which cannot be reduced to exact equivalents and in whom, on that account, there flowed from the same source as did that hatred, the hatred of the money economy and of the intellectualism of existence.

The same factors which, in the exactness and the minute precision of the form of life, have coalesced into a structure of the highest impersonality, have, on the other hand, an influence in a highly personal direction. There is perhaps no psychic phenomenon which is so unconditionally reserved to the city as the blasé outlook. It is at first the consequence of those rapidly shifting stimulations of the nerves which are thrown together in all their contrasts and from which it seems to us the intensification of metropolitan intellectuality seems to be derived. On that account it is not likely that stupid persons who have been hitherto intellectually dead will be blasé. Just as an immoderately sensuous life makes one blasé because it stimulates the nerves to their utmost reactivity until they finally can no longer produce any reaction at all, so, less harmful stimuli, through the rapidity and the contradictoriness of their shifts, force the nerves to make such violent responses, tear them about so brutally that they exhaust their last reserves of strength and, remaining in the same milieu, do not have time for new reserves to form. This incapacity to react to new stimulations with the required amount of energy constitutes in fact that blasé attitude which every child of a large city evinces when compared with the products of the more peaceful and more stable milieu.

Combined with this physiological source of the blasé metropolitan attitude there is another which derives from a money economy. The essence of the blasé attitude is an indifference toward the distinctions between things. Not in the sense that they are not perceived, as is the case of mental dullness, but rather that the meaning and the value of the distinctions between things, and therewith of the things themselves, are experienced as meaningless. They appear to the blasé person in a homogeneous, flat and gray color with no one of them worthy of being preferred to another. This psychic mood is the correct subjective reflection of a complete money economy to the extent that money takes the place of all the manifoldness of things and expresses all qualitative distinctions between them in the distinction of "how much." To the extent that money, with its colorlessness and its indifferent quality, can become a common denominator of all values it becomes the frightful leveler—it hollows out the core of things, their peculiarities, their specific values and their uniqueness and incomparability in a way which is beyond repair. They all float with the same specific gravity in the constantly moving stream of money. They all rest on the same level and are distinguished only by their amounts. In individual cases this coloring, or rather this de-coloring of things, through their equation with money, may be imperceptibly small. In the relationship, however, which the wealthy person has to objects which can be bought for money, perhaps indeed in the total character which, for this reason, public opinion now recognizes in these objects, it takes on very considerable proportions. This is why the metropolis is the seat of commerce and it is in it that the purchasability of things appears in quite a different aspect than in simpler economies. It is also the peculiar seat of the blasé attitude. In it is brought to a peak, in a certain way, that achievement in the concentration of purchasable things which stimulates the individual to the highest degree of nervous energy. Through the mere quantitative intensification of the same conditions this achievement is transformed into its opposite, into this peculiar adaptive phenomenon—the blasé attitude—in which the nerves reveal their final possibility of adjusting themselves to the content and the form of metropolitan life by renouncing the response to them. We see that the self-preservation of certain types of personalities is obtained at the cost of devaluing the entire objective world, ending inevitably in dragging the personality downward into a feeling of its own valuelessness.

Whereas the subject of this form of existence must come to terms with it for himself, his self-preservation in the face of the great city requires of him a no less negative type of social conduct. The mental attitude of the people of the metropolis to one another may be designated formally as one of reserve. If the unceasing external contact of numbers of persons in the city should be met by the same number of inner reactions as in the small town, in which one knows almost every person he meets and to each of whom he has a positive relationship, one would be completely atomized internally and would fall into an unthinkable mental condition. Partly this psychological circumstance and partly the privilege of suspicion which we have in the face of the elements of metropolitan life (which are constantly touching one another in fleeting contact) necessitates in us that reserve, in consequence of which we do not know by sight neighbors of years standing and which permits us to appear to small-town folk so often as cold and uncongenial. Indeed, if I am not mistaken, the inner side of this external reserve is not only indifference but more frequently than we believe, it is a slight aversion, a mutual strangeness and repulsion which, in a close contact which has arisen any way whatever, can break out into hatred and conflict. The entire inner organization of such a type of extended commercial life rests on an extremely varied structure of sympathies, indifferences and aversions of the briefest as well as of the most enduring sort. This sphere of indifference is, for this reason, not as great as it seems superficially. Our minds respond, with some definite feeling, to almost every impression emanating from another person. The unconsciousness, the transitoriness and the shift of these feelings seem to raise them only into indifference. Actually this latter would be as unnatural to us as immersion into a chaos of unwished-for suggestions would be unbearable. From these two typical dangers of metropolitan life we are saved by antipathy which is the latent adumbration of actual antagonism since it brings about the sort of distanciation and

deflection without which this type of life could not be carried on at all. Its extent and its mixture, the rhythm of its emergence and disappearance, the forms in which it is adequate—these constitute, with the simplified motives (in the narrower sense) an inseparable totality of the form of metropolitan life. What appears here directly as dissociation is in reality only one of the elementary forms of socialization.

This reserve with its overtone of concealed aversion appears once more, however, as the form or the wrappings of a much more general psychic trait of the metropolis. It assures the individual of a type and degree of personal freedom to which there is no analogy in other circumstances. It has its roots in one of the great developmental tendencies of social life as a whole; in one of the few for which an approximately exhaustive formula can be discovered. The most elementary stage of social organization which is to be found historically, as well as in the present, is this: a relatively small circle almost entirely closed against neighboring foreign or otherwise antagonistic groups but which has however within itself such a narrow cohesion that the individual member has only a very slight area for the development of his own qualities and for free activity for which he himself is responsible. Political and familial groups began in this way as do political and religious communities; the self-preservation of very young associations requires a rigorous setting of boundaries and a centripetal unity and for that reason it cannot give room to freedom and the peculiarities of inner and external development of the individual. From this stage social evolution proceeds simultaneously in two divergent but none the less corresponding directions. In the measure that the group grows numerically, spatially, and in the meaningful content of life, its immediate inner unity and the definiteness of its original demarcation against others are weakened and rendered mild by reciprocal interactions and interconnections. And at the same time the individual gains a freedom of movement far beyond the first jealous delimitation, and gains also a peculiarity and individuality to which the division of labor in groups, which have become larger, gives both occasion and necessity. However much the particular conditions and forces of the individual situation might modify the general scheme, the state and Christianity, guilds and political parties and innumerable other groups have developed in accord with this formula. This tendency seems, to me, however to be quite clearly recognizable also in the development of individuality within the framework of city life. Small town life in antiquity as well as in the Middle Ages imposed such limits upon the movements of the individual in his relationships with the outside world and on his inner independence and differentiation that the modern person could not even breathe under such conditions. Even today the city dweller who is placed in a small town feels a type of narrowness which is very similar. The smaller the circle which forms our environment and the more limited the relationships which have the possibility of transcending the boundaries, the more anxiously the narrow community watches over the deeds, the conduct of life and the attitudes of the individual and the more will a quantitative and qualitative individuality tend to pass beyond the boundaries of such a community.

The ancient *polis* seems in this regard to have had a character of a small town. The incessant threat against its existence by enemies from near and far brought about that stern cohesion in political and military matters, that supervision of the citizen by other citizens, and that jealousy of the whole toward the individual whose own private life was repressed to such an extent that he could compensate himself only by acting as a despot in his own household. The tremendous agitation and excitement, and the unique colorfulness of Athenian life is perhaps explained by the fact that a people of incomparably individualized personalities were in constant struggle against the incessant inner and external oppression of a de-individualizing small town. This created an atmosphere of tension in which the weaker were held down and the stronger were impelled to the most passionate type of self- protection. And with this there blossomed in Athens, what, without being able to define it exactly, must be designated as "the general human character" in the intellectual development of our species. For the correlation, the factual as well as the historical validity of which we are here maintaining, is that the broadest and the most general contents and forms of life are intimately bound up with the most individual ones. Both have a common prehistory and also common enemies in the narrow formations and groupings, whose striving for self-preservation set them in conflict with the broad and general on the outside, as well as the freely mobile and individual on the inside. Just as in feudal times the "free" man was he who stood under the law of the land, that is, under the law of the largest social unit, but he was unfree who derived his legal rights only from the narrow circle of a feudal community—so today in an intellectualized and refined sense the citizen of the metropolis is "free" in contrast with the trivialities and prejudices which bind the small town person. The mutual reserve and indifference, and the intellectual conditions of life in large social units are never more sharply appreciated in their significance for the independence of the individual than in the dense crowds of the metropolis because the bodily closeness and lack of space make intellectual distance really perceivable for the first time. It is obviously only the obverse of this freedom that, under certain circumstances, one never feels as lonely and as deserted as in this metropolitan crush of persons. For here, as elsewhere, it is by no means necessary that the freedom of man reflect itself in his emotional life only as a pleasant experience.

It is not only the immediate size of the area and population which, on the basis of world-historical correlation between the increase in the size of the social unit and the degree of personal inner and outer freedom, makes the metropolis the locus of this condition. It is rather in transcending this purely tangible extensiveness that the metropolis also becomes the seat of cosmopolitanism. Comparable with the form of the development of wealth—(beyond a certain point property increases in ever more rapid progression as out of its own inner being)—the individual's horizon is enlarged. In the same way, economic, personal and intellectual relations in the city (which are its ideal reflection), grow in a geometrical progression as soon as, for the first time, a certain limit has been passed. Every dynamic extension becomes a preparation not only for a similar extension but rather for a larger one and from every thread which is spun out of it there continue, growing as out of themselves, an endless number of others. This may be illustrated by the fact that within the city the "unearned increment" of ground rent, through a mere increase in traffic, brings to the owner profits which are self-generating. At this point the quantitative aspects of life are transformed qualitatively. The sphere of life of the small town is, in the main, enclosed within itself. For the metropolis it is decisive that its inner life is extended in a wave-like motion over a broader national or international area. Weimar was no exception because its significance was dependent upon individual personalities and died with them, whereas the metropolis is characterised by its essential independence even of the most significant individual personalities; this is rather its antithesis and it is the price of independence which the individual living in it enjoys. The most significant aspect of the metropolis lies in this functional magnitude beyond its actual physical boundaries and this effectiveness reacts upon the latter and gives to it life, weight, importance and responsibility. A person does not end with limits of his physical body or with the area to which his physical activity is immediately confined but embraces, rather, the totality of meaningful effects which emanates from him temporally and spatially. In the same way the city exists only in the totality of the effects which transcend their immediate sphere. These really are the actual extent in which their existence is expressed. This is already expressed in the fact that individual freedom, which is the logical historical complement of such extension, is not only to be understood in the negative sense as mere freedom of movement and emancipation from prejudices and philistinism. Its essential characteristic is rather to be found in the fact that the particularity and incomparability which ultimately every person possesses in some way is actually expressed, giving form to life. That we follow the laws of our inner nature—and this is what freedom is—becomes perceptible and convincing to us and to others only when the expressions of this nature distinguish themselves from others; it is our irreplaceability by others which shows that our mode of existence is not imposed upon us from the outside.

Cities are above all the seat of the most advanced economic division of labor. They produce such extreme phenomena as the lucrative vocation of the *quatorzieme* in Paris. These are persons who may be recognized by shields on their houses and who hold themselves ready at the dinner hour in appropriate costumes so they can be called upon on short notice in case thirteen persons find themselves at the table. Exactly in the measure of its extension the city offers to an increasing degree the determining conditions for the division of labor. It is a unit which, because of its large size, is receptive to a highly diversified plurality of achievements while at the same time the agglomeration of individuals and their struggle for the customer forces the individual to a type of specialized accomplishment in which he cannot be so easily exterminated by the other. The decisive fact here is that in the life of a city, struggle with nature for the means of life is transformed into a conflict with human beings and the gain which is fought for is granted, not by nature, but by man. For here we find not only the previously mentioned source of specialization but rather the deeper one in which the seller must seek to produce in the person to whom he wishes to sell ever new and unique needs. The necessity to specialize one's product in order to find a source of income which is not yet exhausted and also to specialize a function which cannot be easily supplanted is conducive to differentiation, refinement and enrichment of the needs of the public which obviously must lead to increasing personal variation within this public.

All this leads to the narrower type of intellectual individuation of mental qualities to which the city gives rise in proportion to its size. There is a whole series of causes for this. First of all there is the difficulty of giving one's own personality a certain status within the framework of metropolitan life. Where quantitative increase of value and energy has reached its limits, one seizes on qualitative distinctions, so that, through taking advantage of the existing sensitivity to differences, the attention of the social world can, in some way, be won for oneself. This leads ultimately to the strangest eccentricities, to specifically metropolitan extravagances of self-distanciation, of caprice, of fastidiousness, the meaning of which is no longer to be found in the content of such activity itself but rather in its being a form of "being different"—of making oneself noticeable. For many types of persons these are still the only means of saving for oneself, through the attention gained from others, some sort of self-esteem and the sense of filling a position. In the same sense there operates an apparently insignificant factor which in its effects however is perceptibly cumulative, namely, the brevity and rarity of meetings which are allotted to each individual as compared with social intercourse in a small city. For here we find the attempt to appear to-the-point, clear-cut and individual with extraordinarily greater frequency than where frequent and long association assures to each person an unambiguous conception of the other's personality.

This appears to me to be the most profound cause of the fact that the metropolis places emphasis on striving for the most individual forms of personal existence—regardless of whether it is always correct or always successful. The development of modern culture is characterised by the predominance of what one can call the objective spirit over the subjective; that is, in language as well as in law, in the technique of production as well as in art, in science as well as in the objects of domestic environment, there is embodied a sort of spirit [*Geist*], the daily growth of which is followed only imperfectly and with an even greater lag by the intellectual development of the individual. If we survey for instance the vast culture which during the last century has been embodied in things and in knowledge, in institutions and comforts, and if we compare them with the cultural progress of the individual during the same period—at least in the upper classes—we would see a frightful difference in rate of growth between the two which represents, in many points, rather a regression of the culture of the individual with reference to spirituality, delicacy and idealism. This discrepancy is in essence the result of the success of the growing division of labor. For it is this which requires from the individual an ever more one-sided type of achievement which, at its highest point, often permits his personality as a whole to fall into neglect. In any case this overgrowth of objective culture has been less and less satisfactory for the individual. Perhaps less conscious than in practical activity and in the obscure complex of feelings which flow from him, he is reduced to a negligible quantity. He becomes a single cog as over against the vast overwhelming organization of things and forces which gradually take out of his hands everything connected with progress, spirituality and value. The operation of these forces results in the transformation of the latter from a subjective form into one of purely objective existence. It need only be pointed out that the metropolis is the proper arena for this type of culture which has outgrown every personal element. Here in buildings and in educational institutions, in the wonders and comforts of space-conquering technique, in the formations of social life and in the concrete institutions of the State is to be found such a tremendous richness of crystallizing, depersonalized cultural accomplishments that the personality can, so to speak, scarcely maintain itself in the face of it. From one angle life is made infinitely more easy in the sense that stimulations, interests, and the taking up of time and attention, present themselves from all sides and carry it in a stream which scarcely requires any individual efforts for its ongoing. But from another angle, life is composed more and more of these impersonal cultural elements and existing goods and values which seek to suppress peculiar personal interests and incomparabilities. As a result, in order that this most personal element be saved, extremities and peculiaritics and individualizations must be produced and they must be over-exaggerated merely to be brought into the awareness even of the individual himself. The atrophy of individual culture through the hypertrophy of objective culture lies at the root of the bitter hatred which the preachers of the most extreme individualism, in the footsteps of Nietzsche, directed against the metropolis. But it is also the explanation of why indeed they are so passionately loved in the metropolis and indeed appear to its residents as the saviors of their unsatisfied yearnings.

When both of these forms of individualism which are nourished by the quantitative relationships of the metropolis, i.e., individual independence and the elaboration of personal peculiarities, are examined with reference to their historical position, the metropolis attains an entirely new value and meaning in the world history of the spirit. The eighteenth century found the individual in the grip of powerful bonds which had become meaningless—bonds of a political, agrarian, guild and religious nature—delimitations which imposed upon the human being at the same time an unnatural form and for a long time an unjust inequality. In this situation arose the cry for freedom and equality—the belief in the full freedom of movement of the individual in all his social and intellectual relationships which would then permit the same noble essence to emerge equally from all individuals as Nature had placed it in them and as it had been distorted by social life and historical development. Alongside of this liberalistic ideal there grew up in the nineteenth century from Goethe and the Romantics, on the one hand, and from the economic division of labor on the other, the further tendency, namely, that individuals who had been liberated from their historical bonds sought now to distinguish themselves from one another. No longer was it the "general human quality" in every individual but rather his qualitative uniqueness and irreplaceability that now became the criteria of his value. In the conflict and shifting interpretations of these two ways of defining the position of the individual within the totality is to be found the external as well as the internal history of our time. It is the function of the metropolis to make a place for the conflict and for the attempts at unification of both of these in the sense that its own peculiar conditions have been revealed to us as the occasion and the stimulus for the development of both. Thereby they attain a quite unique place, fruitful with an inexhaustible richness of meaning in the development of the mental life. They reveal themselves as one of those great historical structures in which conflicting life-embracing currents find themselves with equal legitimacy. Because of this, however, regardless of whether we are sympathetic or antipathetic with their individual expressions, they transcend the sphere in which a judge-like attitude on our part is appropriate. To the extent that such forces have been integrated, with the fleeting existence of a single cell, into the root as well as the crown of the totality of historical life to which we belong—it is our task not to complain or to condone but only to understand.

Discussion Questions

1. With his concept of the tragedy of culture, Simmel explored how cultural objects, once they confront their users as external forces, often constrain our attempts to cultivate freedom and individuality. How do cars and computers allow us to fulfill some of our needs, while at the same time constraining our pursuit of independence and individual goals? What other cultural objects possess these opposing elements? How do Simmel's views on such issues compare with those offered by Marx's understanding of alienation and commodity fetishism?

2. In his essay "Exchange," Simmel explores the origins of economic value. According to his argument, what role does interaction play in establishing the value of goods or objects? What roles does the quality and scarcity of goods and objects play in establishing their value? How does Simmel's view on the source of value compare with that offered by Marx?

3. Simmel describes the stranger as someone who is near and far, or close and remote, at the same time. How might his analysis of the stranger shed light on current debates regarding immigration and the influx of "illegal aliens" into the United States?

4. According to Simmel, what effects does the metropolis have on the emotions and intellect of the individual? What are the causes responsible for producing these effects? In what ways is the metropolis both "freeing" and "dehumanizing" with regard to the cultivation of individuality?

5. How does Simmel's analysis of modern society as outlined in "The Metropolis and Mental Life" compare to the views offered by Marx, Durkheim, and Weber? According to their respective arguments, where does the locus for the modern condition lie? What prospects, if any, does each hold out for change or further societal development?

7 W. E. B. Du Bois (1868–1963)

W. E. B. Du Bois

Key Concepts

■ The color line

■ Double consciousness

■ The veil

The problem of the 20th century is the problem of the color line—the relation of the darker to the lighter races of men in Asia and Africa, in America and the islands of the sea.

—Du Bois (1903/1989:13)

From Jim Crow, to the civil rights movement, to affirmative action; from Willie Horton to Rodney King and O. J. Simpson to the "Million Man March," it seems clear that W. E. B. Du Bois's famous prophecy as to the salience of the "color line" in the twentieth century has proven true. Yet, in the above passage, Du Bois was not talking only about the United States. Du Bois saw that the forces of white oppression were "all but world-wide" (1920/2003d:66–67). He was perhaps the first social theorist to critically remark on the global racial order, to understand not only the economic but also the *racial* dimensions of the European colonization of Africa, Asia, India, and Latin America.

Ironically, Du Bois's own life is further testament to the existence and pervasiveness of the color line. Despite his exceptional work and provocative sociological insights, Du Bois was virtually ignored by fellow American sociologists in his day. This, despite the fact that he was the first African American to receive a PhD from Harvard University, and despite the fact that his most famous book of essays,

The Souls of Black Folk, which has been republished in no fewer than 119 editions since 1903, has been called one the most important books ever on the subject of "race."[1] Yet, at the turn of the twentieth century, Du Bois's work on African American communities was neither well known nor highly respected. Most white reviewers could not see past the "bitterness" and "hateful" stance that was said to characterize his works (Feagin 2003:22). What this demonstrates, of course, is that the canon of sociological theory is a product not only of intellectual accomplishments but also of social, cultural, and historical dynamics. Many people wrote insightful sociological treatises at the same time as our founding figures but were not widely recognized by their peers. One of the most important of these, however, is W. E. B. Du Bois.

At the turn of the twentieth century, racism precluded most social scientists from a cogent understanding of race. Indeed, during the nineteenth and early twentieth centuries, most scientists and essayists writing about "Negroes" denied the very humanity of African Americans. It was widely believed that Negro brains weighed less than those of whites and that "prefrontal deficiencies" caused inferior "psychic faculties, especially reason, judgment, self control or voluntary inhibitions" (Rudwick [1960] 1982:123). Moreover, between the end of the Civil War and the dawn of the twentieth century, racial violence and tension, segregation, and vigilantism were not only widespread—they were increasing. In 1891, the Georgia legislature passed legislation segregating all public streetcars, and by 1900, most southern states had similar codes. At the same time, between 1882 and 1901 there were some 150 lynchings each year. Most appallingly, in the late nineteenth and early twentieth centuries, lynchings were not the domain of deviants and outlaws; rather, they were communal events, with the tacit, if not explicit, support of many "upstanding" white citizens, including religious clergy (Patterson 1998:173). As Du Bois ([1935] 1962) maintained, it was the "nucleus of ordinary men that continually [gave] the mob its initial and awful impetus" (p. 678).

Within this context of social, as well as scientific, ignorance and injustice, Du Bois stood virtually alone in his quest to sociologically explain the complex, intertwined dimensions of race and class. This is not to say that Du Bois was the only turn-of-the-century intellectual to astutely explore the race question. On the contrary, many activists and essayists—such as Sojourner Truth (1797–1883), Frederick Douglass (1817–1895), Booker T. Washington (1856–1915), Ida B. Wells-Barnett (1862–1931), and Anna Julia Cooper (1858–1964)—spoke and/or wrote eloquently about the issue of race. The point here, however, is that Du Bois sought to investigate and explain racial issues *sociologically.* Du Bois viewed sociological investigation as the solution to racism. As he maintained,

> The Negro problem was in my mind a matter of systematic investigation and intelligent understanding. The world was thinking wrong about race because it did not know. The ultimate evil was stupidity. The cure for it was knowledge based on scientific investigation. ([1940] 1984:58)

Consequently, Du Bois set the theoretical as well as empirical parameters in the field of race relations, and his recognition of the interconnections between race and class continues to steer sociological inquiry to this day.

While few sociologists deny the importance of Du Bois in the field of race relations, not all sociologists consider him to be a significant social theorist. This perception stems from two somewhat contradictory beliefs: (1) Du Bois's work is perceived as being too empirical and theoretically narrow to qualify him as a theorist, and (2) his work is deemed too subjective to merit the label of "theory." In fact, Du Bois did not write explicitly about the "big" theoretical questions such as the nature of action or social order. Instead, much of his work centered on describing and explaining the actual social conditions in which African Americans lived. Additionally, in contrast to the other founding figures of sociology, Du Bois not only wrote in the subjective form of essays; he sometimes wrote in the first person. As you will see, Du Bois uses poetry, slave songs, and his own autobiographical experience in his most widely read book of essays, *The Souls of Black Folk* (1903).

---

[1]We use quotation marks around the concept of "race" to reflect that we are talking about the social perception and experience of race—not a biological concept of race. (See Edles 2002.) This is the meaning to be taken throughout the remainder of the chapter.

Photo 7.1 A crowd mills around the bodies of Abram South, 19, and Thomas Shipp, 18, who were taken from the Grand County Jail, in Marion, Indiana, and lynched in the public square in 1930.

But interestingly, Du Bois himself made no apologies for his style. He maintained that European thought is but one kind of thought; it does not represent a "universal," let alone superior, approach to the world. As Du Bois states,

> The style is tropical—African. This needs no apology. The blood of my fathers spoke through me and cast off the English restraint of my training and surroundings. . . . One who is born with a cause is predestined to certain narrowness of view, and at the same time to some clearness of vision within his limits with which the world often finds it well to reckon. (Du Bois, 1904, as quoted in E. Griffin 2003:32)

Given his standing among his contemporaries and succeeding generations of sociologists, the question remains: Why include Du Bois's work in a theory reader such as this? Our answer to this question is two-fold: First, we include Du Bois because his work brings to the canon a "clearness of vision" regarding specific phenomena—most obviously race, but also epistemology, or "how we know what we know." Anyone writing on these issues today is in some fashion indebted to Du Bois's groundbreaking insights. Secondly, akin to the case of Charlotte Perkins Gilman (Chapter 5), we include Du Bois for historical reasons. In the nineteenth and early twentieth centuries, African Americans were marginalized, both explicitly and implicitly, from formal sociological theorizing. As a result, we must broaden our notion of "theory" if we want to uncover the ideas of African Americans such as Du Bois who lived during this patently racist period. Indeed, as the acclaimed contemporary sociologist Lawrence Bobo (2000) maintains,

> Were it not for the deeply entrenched racism in the United States during his early professional years, Du Bois would be recognized alongside the likes of Albion Small, Edward A. Ross, Robert E. Part, Lewis With, and W. I. Thomas as one of the fountainheads of American sociology. Had not racism so thoroughly excluded him from placement in the center of the academy, he might arguably have come to rank with Max Weber or Émile Durkheim in stature. (p. 187)

Because his work is so intimately tied to his personal life, in the next section we provide a rather lengthy biographical sketch of Du Bois; then we turn to more theoretical concerns.

Anna Julia Cooper (1858–1964): *A Voice from the South*

Born in 1858 to a slave mother, Hannah Stanley Haywood, and her white master, George Washington Haywood, Anna Julia Cooper is one of the earliest individuals to offer what would today be recognized as a social theory of women of color. Cooper articulated her perspective in her major work, *A Voice from the South by a Black Woman of the South,* first published in 1892. This provocative and compelling book turned Cooper into a leading spokeswoman of her time. However, Cooper dedicated most of her long and extraordinary life (she died at 105 years old) to the education of others, and in particular "to the education of neglected people" (Washington 1988:xxviii). In addition to being a high school teacher and principal, she started a night school for working people who could not attend college during the day. Cooper maintained that the ideals and integrity of any group can be measured "by its treatment of those who suffer the greatest oppression."

Cooper's *A Voice From the South* is lauded as "the first *systematic* working out of the insistence that no one social category can capture the reality of black women" (Lemert 1998:15, emphasis in original). In this inimitable book, Cooper points out that categories such as race, gender, and class do not capture, by themselves, the situation of black women. Instead, the black woman "is confronted by both a woman question and a race problem," as Cooper bluntly states (Cooper [1892] 1988:112, as cited by Lemert 1998:15). In addition, Cooper sought to counteract the prevailing assumptions about black women as immoral and ignorant. She criticizes black men for securing higher education for themselves while erecting roadblocks to deny women access to those same opportunities. Cooper herself had to fight against both black and white men in order to attain her own education. Nevertheless, she attained bachelors and masters degrees from Oberlin College in 1984 and 1987, respectively, and a PhD from the Sorbonne in Paris in 1925, at 66 years of age. If black men are a "muffled chord," she maintained, then black women are "the mute and voiceless note" of the race, "with no language but a cry" (Washington 1988:xxix). Cooper was equally critical of the white women's movement for its elitism, and she challenged white women to link their cause with all of the "undefended" (ibid.).

Cooper was passionately committed to higher education as the key to ending women's physical, emotional, and economic dependence on men. Indeed, she was forced to resign from her job as principal of the only black high school in Washington, D.C., when, in defiance of her white supervisor, who told her that black children should be taught trades, Cooper continued to prepare them for college (Washington 1988:xxxiv). (Several of her students attended prestigious universities such as Harvard, Brown, Oberlin, Yale, Amherst, Dartmouth, and Radcliffe.) Like Du Bois, Cooper opposed Booker T. Washington's prescriptions for advancing the cause of African Americans. Cooper and Du Bois also shared an emphasis on "voice": both sought to expose the standpoint of oppressed folks, to present submerged points of view. As Cooper (1892:60) states, "there is a feminine as well as a masculine side to truth; that these are related not as inferior and superior, not as better and worse, not as weaker and stronger, but as complements—complements in one necessary and symmetric whole" (ibid:60). Ironically, however, Cooper was largely ignored by the leading black male intellectuals of her day, including Du Bois, though she did not blame him but retained hope:

> It is no fault of man's that he has not been able to see truth from her standpoint. It does credit both to his head and heart that no greater mistakes have been committed or even wrongs perpetrated while she sat making tatting and snipping paper flowers. Man's own innate chivalry and the mutual interdependence of their interests have insured his treating her causes, in the main at least, as his own. And his is pardonably surprised and even a little chagrined, perhaps to find his legislation not considered "perfectly lovely" in every respect." But in any case his work is only impoverished by her remaining dumb. The world has had to limp along with the wobbling gait and one-sided hesitancy of a man with one eye. Suddenly the bandage is removed from the other eye and the whole body is filled with light. It sees a circle where before it saw a segment. The darkened eye restored, every member rejoices with it. (ibid.:122)

▪ A BIOGRAPHICAL SKETCH

William Edward Burghardt (W. E. B.) Du Bois was born in Great Barrington, Massachusetts, in 1868.[2] Du Bois's father, Alfred, was a restless, unhappy, "just visibly colored" man, who tried a wide variety of occupations, including barbering and even preaching, before he abandoned his wife and child and drifted away, never to return. Du Bois's mother, Mary Burghardt, was a "silent, repressed," "dark shining bronze" woman who struggled to make ends meet by working as a domestic and boarding a relative.[3]

While Du Bois was still a child, his mother suffered a stroke and became partly paralyzed. Though she was able to continue working as a maid, William began to shoulder a share of the burden for maintaining their small family, seeking out odd jobs wherever he could. Despite his new obligations, William did exceptionally well in school.

Yet, Du Bois soon came to recognize that his race set him apart from his schoolmates. As he stated later, "In early youth a great bitterness entered my life and kindled a great ambition" (Marable 1986:6). He began to recognize not only the existence of the color line but also the ways in which it was distinct from class-based inequality. Indeed, by age 16, Du Bois had formulated the basic premise of the social and political philosophy that would guide him the rest of his years: Black Americans must organize themselves as a race-conscious bloc in order to win and exercise their freedom. They must "act in concert" if they want to become a "power not to be despised" (ibid.). Although such race consciousness would seem to run counter to the individualism and egalitarianism supposedly underlying American society, Du Bois maintained that in a racist society, democracy could not be colorblind.

Du Bois was the first black graduate in his high school's history. Shortly after his graduation, his mother suddenly died. Du Bois's formal education might have ended right then were it not for the benevolence of four liberal white men who had taken great interest in Du Bois and had agreed to pay for his college education. These community leaders (who included Du Bois's high school principal and several prominent pastors) had decided that Du Bois should attend Fisk University—a historically black school in Nashville, Tennessee—rather than Harvard, as Du Bois initially desired.

In fall 1885, Du Bois arrived at Fisk University. In his autobiography, he described himself as

> thrilled to be for the first time among so many people of my own color or rather of such various and such extraordinary colors, which I had only glimpsed before, but who it seemed were bound to me by new and exciting and eternal ties. (Du Bois 1968:107)

For Du Bois, Fisk represented the tremendous yet unfulfilled potential of the entire black world that had been "held back by race prejudice and legal bonds, as well as by deep ignorance and dire poverty" (Marable 1986:9).

Du Bois witnessed "the real seat of slavery" not at Fisk, but in the east Tennessee countryside where he went to teach summer school in 1886 and 1887. There he taught the poorest of the black poor in a schoolhouse he described as a "log hut" with "neither door or furniture" ([1903] 1989:55–56). Du Bois eloquently describes this experience in an essay that would later appear in *The Souls of Black Folk* (1903).

In 1888, Du Bois transferred to Harvard, and in 1892 he received his master's degree in philosophy (Harvard would not offer a degree in sociology for another 20 years). Du Bois then set off to study further

---

[2]Great Barrington was a small town of around 5,000 mostly middle-class families of European descent. The wealthier farm and factory owners tended primarily to be of English and Dutch stock, while the small farmers, merchants, and skilled laborers tended to be mostly of German and Irish ancestry. There were a handful of black families in Great Barrington, and they (including the Burghardts) were among the town's oldest residents. They worked primarily as laborers and servants. According to Du Bois, "The color line was manifest and yet not absolutely drawn" in Great Barrington, allowing blacks to organize their own social life to an extent (Rudwick [1960] 1982:16; Du Bois [1940] 1984:10.

[3]These descriptions of his parents' skin color come from Du Bois's autobiography, *The Autobiography of W. E. B. Du Bois* (1968). Du Bois described his own racial background as "a flood of Negro blood, a strain of French, a bit of Dutch but thank God! no 'Anglo-Saxon'" (Rudwick [1960] 1982:15–16).

at the University of Berlin. Germany was at the heart of the fledgling new discipline of sociology, and Du Bois attended lectures and communicated with some of the great sociologists of the day, including Max Weber. In addition, Du Bois attended local meetings of the German Social Democratic party, at that time the largest socialist party in the world. Enthusiastic about this new intersection of philosophy and social reform, Du Bois sought to apply "philosophy to an historical interpretation of race relations," that is, "to take my first steps toward sociology as the science of human action" (1968:148).

In Europe, Du Bois's racial consciousness was transformed just as it had been seven years earlier when he moved to the American South. In Europe, Du Bois encountered white folks who exhibited little or no racial prejudice. As Du Bois maintains,

> In Germany in 1892, I found myself on the outside of the American world, looking in. With me were white folk—students, acquaintances, teachers—who viewed the scene with me. They did not always pause to regard me as a curiosity, or something sub-human; I was just a man of the somewhat privileged student rank, with whom they were glad to meet and talk over the world; particularly, the part of the world whence I came. (1968:157)

Correspondingly, Du Bois's view of white folks was itself transformed. White Europeans became "not white folks, but folks. The unity beneath all life clutched me" (Marable 1986:20). Nevertheless, Du Bois fully recognized that racism in the United States was only one virulent example of racial/ethnic/national subordination and that German anti-Semitism had "much in common with our own race question" (ibid.:17, 18).

For 2 years, Du Bois "dreamed and loved and wandered and sang" in Europe, but in 1894, low on funds, he was forced to return to the United States. "Dropped suddenly back into 'nigger'-hating America" (1968:183), Du Bois began teaching at Wilberforce University, a small African Methodist Episcopal school in central Ohio. While there, Du Bois also finished his doctoral thesis (which he had begun at Harvard before he left for Europe) and turned it into his first book, *The Suppression of the African Slave-Trade to the United States of America, 1638–1870* (1896). This was the first major research study on the U.S. slave trade (Feagin 2003:23).

Only 2 years would pass before he published his second book, *The Philadelphia Negro* (1899), which was the first sociological text on an African American community published in the United States. In this ground-breaking book, Du Bois presents a rich array of empirical information about the lives of Philadelphia African Americans, focusing on the "social condition of the Colored People of the Seventh Ward of Philadelphia," an impoverished area of about 10,000 African Americans (nearly one-fourth of the entire Philadelphia black population). Du Bois carried out some 5,000 surveys and interviews with African Americans in the Seventh Ward. Du Bois and his wife, Nina Gomer, whom he married in 1896, also directly experienced many of the conditions he described so vividly, living as they did in the Settlement House, in the heart of the seventh ward. In addition to his careful statistical and ethnographic research, Du Bois also paid particular attention to historical information because he felt that "one cannot study the Negro in freedom . . . without knowing his history in slavery" (Marable 1986:24, 25). Du Bois adamantly believed that careful sociological documentation combined with social and cultural understanding of a social group (reminiscent of Weber's *Verstehen;* see Chapter 4) would make possible a cogent agenda of social reform.

Less than 2 years after Du Bois and Nina Gomer were married, their first child was born. Sadly, the child, named Burghardt, died 18 months later, allegedly a victim of sewage pollution in the city's water system (Marable 1986:30). Du Bois scholars such as Manning Marable suggest that it was perhaps Du Bois's deep sorrow that prompted the softer, more sympathetic tone apparent in his masterpiece, *The Souls of Black Folk,* published in 1903. This work would become not only Du Bois's most famous book, but also arguably one of the most important books on race and class ever written in the United States. (The significance of this book is discussed further in this chapter.)

In 1897, Du Bois began teaching at Atlanta University in Georgia, where he would remain until 1910. He wrote prolifically during this time, while receiving the recognition of a few leading sociologists, such as Max Weber, who hailed Du Bois's *The Souls of Black Folk* as a "splendid work" that "ought to be translated in German" (Marable:63). Indeed, Europeans appreciated Du Bois's sociological research and

Photo 7.2 The seventh ward of Philadelphia, circa 1900

writing much more than his fellow (white) American intellectuals.

Soon after arriving in Atlanta, Du Bois turned his attention to political activism. In 1905, he helped found a militant civil rights organization, the Niagara Movement, devoted to "freedom of speech and criticism," "suffrage," "the abolition of all caste distinctions based simply on race or color," and "the principle of human brotherhood as a practical present creed" (Marable:55). In 1910, Du Bois left his academic post at Atlanta University in order to work full-time for the National Association for the Advancement of Colored People (NAACP) in New York City. Against the advice of some NAACP board members (several of whom were white), Du Bois founded the journal *Crisis,* which from the beginning had a distinctly militant tone. The journal became a forum for Du Bois, who wrote dozens of articles and editorials on the profound discrepancy between American democratic principles and racism, beseeching the country to expand political and economic opportunities for blacks (Rudwick [1960] 1982:151). He also challenged the laws against miscegenation, asserting that a black man should be able "to marry any sane grown person who wants to marry him" (Marable 1986:79).

During his tenure as the editor of *Crisis,* Du Bois managed to write several more books, including the highly acclaimed *Darkwater: Voices From Within the Veil* (1920), *The Gift of Black Folk* (1924), and *Dark Princess* (1928), as well as books on Africa: *The Negro* (1915), *Africa—Its Place in Modern History* (1930) and *Africa—Its Geography, People, and Products* (1930). Du Bois, moreover, was sensitive to the dual oppression faced by black women, as is evident in his poem, "The Burden of Black Women" (first published as "The Riddle of the Sphinx" in 1907), and his essays "The Black Mother" (1912), "Hail Columbia" (1913), and "The Damnation of Women" (1920).[4]

Historians and biographers have held that Du Bois was not an easy man to work with. He was known to be stubborn, snobbish, arrogant, and demanding. His sharp tongue and harsh editorials created a series of problems for the NAACP board, and over the years there were significant power struggles between Du Bois and the board members of the NAACP. In 1934, these struggles came to a head, and Du Bois took a leave of absence from the NAACP. He then returned to Atlanta University where he became professor and chair of the department of sociology. Meanwhile, Du Bois's political position shifted farther and farther to the left. Drawing on Marxist-Leninism and the program of the Communist Party, he portrayed slaves who abandoned plantations as proletariat engaged in "general strikes" and romantically deemed the Reconstruction governments a Marxist experiment (Rudwick [1960] 1982:287).

---

[4]As Farah Jasmine Griffin (2000:30, 31) aptly notes, this does not mean that Du Bois completely escaped the sexism of his day. Rather, Du Bois can be considered sexist, elitist, and "color-struck," in that his image of the "Talented Tenth," or African American intellectual leadership, privileged "biracial" or light-skinned, "cultured" males.

From the 1940s until his death in 1963, Du Bois wrote eloquently and provocatively about the twin evils of colonialism and imperialism. He considered colonies "the slums of the world . . . centers of helplessness, of discouragement of initiative, or forced labor, and of legal suppression of all activities or thoughts which the master country fears or dislikes" (1945:17), and he assailed capitalist nations and corporations for profiting from war. Maintaining that class divisions within America's Negro communities had undermined the prospects for black unity and social advance, he advocated a racially segregated, socialized economy for the United States. As Du Bois became increasingly interested in socialist political alternatives, he called for the creation of a unified, socialist African state and began to devote himself almost exclusively to the Pan-African movement.

Given his outspoken views on racial injustice and stated sympathies with the Soviet Union, it is not surprising that in the early 1950s, Du Bois became a target in the cold war. The Justice Department ordered Du Bois to register as an "agent of a foreign principal." He refused and faced a possible fine of $10,000 and 5 years in federal prison (Marable 1986:182). At his arraignment, Du Bois coolly read the following statement:

Photo 7.3 Du Bois with Mao Tse-tung at Mao's villa in South Central China, May 1959

> It is a sad commentary that we must enter a courtroom today to plead Not Guilty to something that cannot be a crime—advocating peace and friendship between the American people and the peoples of the world. . . . In a world which has barely emerged from the horrors of the Second World War and which trembles on the brink of atomic catastrophe, can it be criminal to hope and work for peace? (ibid.)

After a long and costly trial, Du Bois was acquitted, though his passport was retained by the Justice Department and he was refused the right to travel abroad. Du Bois was not intimidated by McCarthyism and the Red Scare, however. The following year, when the State Department asked Du Bois to make a sworn statement "as to whether you are now or ever have been [a] communist," Du Bois indignantly refused, stating to the press that his political beliefs "are none of your business" (Marable:203).

Meanwhile, the topic of Africa continued to dominate much of Du Bois's intellectual thought. In 1960, the president of Ghana, Kwame Nkrumah, invited Du Bois to direct a major scholarly project, an *Encyclopedia Africana*. Du Bois was reluctant to accept the offer, as he was already 92 years old. But President Kwame Nkrumah maintained that as the father of Pan-Africanism and the leading advocate of socialism in the black world, Du Bois was the best possible person for the project, and in the end the president's pleas won out.

In Ghana, Du Bois officially joined the Communist Party of the United States and made clear his affinities with Marxist theory. Remarking on his commitment to communism, he stated,

> Capitalism cannot reform itself; it is doomed to self-destruction. No universal selfishness can bring social good to all. Communism—the effort to give all men what they need and to ask of each the best they can contribute—this is the only way of human life. . . . In the end communism will triumph. I want to help bring that day. The path of the American Communist Party is clear: It will provide the United States with a real Third party and thus restore democracy to this land. (Marable:212)

Significant costs came with joining the Communist Party. Du Bois was prevented from returning to the United States, as the American Consulate refused to renew his passport. For that matter, it was illegal for members of the Communist party to possess an American passport, leaving Du Bois subject to 10 years of imprisonment for his crime (Marable:213). Du Bois had no choice but to renounce his American citizenship. In March 1963, he became a citizen of Ghana. Less than 6 months later, at the age of 95, Du Bois died.

As indicated previously, Du Bois was primarily concerned with the nature and inter-section of race and class—an issue virtually ignored by the discipline's "founding figures." Within this program, Du Bois conducted three types of research: (1) empirical studies illuminating the actual social conditions of African Americans (e.g., *The Philadelphia Negro* [1899]); (2) interpretive essays informed by careful historical research and personal experience, as well as keen observation (e.g., *The Souls of Black Folk* [1903]) that emphasized the subjective experience and sources of inequality; and (3) explicitly political essays focusing on Pan-Africanist and socialist solutions to inequality and racism (e.g., *Color and Democracy: Colonies and Peace* [1945]). To be sure, the three styles overlap, as elements of each are apparent in most of Du Bois's work.

On the surface, Du Bois's first book, *The Suppression of the African Slave-Trade to the United States of America, 1638–1870,* appears to be a formal, legalistic study, but underneath, it is "a sharp moral and ethical repudiation of chattel slavery" (Marable 1986:22). In it, Du Bois harshly condemns the United States for its moral and political hypocrisy and the inability of its leaders to check "this real, existent, growing evil" that led "straight to the Civil War" (ibid.:23). This same detachment coupled with an implicit moralistic tone is also evident in *The Philadelphia Negro.*

At the same time, Du Bois did not hesitate to criticize the black community. He censured black parents for not sufficiently reinforcing the value of formal education, assailed the black Church for not adequately combating social corruption and moral decay, and lamented the existence of "the usual substratum of loafers and semi-criminals who will not work" (Marable:28). Nevertheless, he insisted that the vast majority of African Americans suffered not from moral laxity but from "irregular employment" (ibid.).

Du Bois studied less impoverished and oppressed African American communities too. In *The College-Bred Negro,* he identified and contacted thousands of African American college graduates. He also wrote on "the Negro Landholders of Georgia" (1901) and African American artisans and skilled workers (1902). Du Bois maintained that "the better classes of Negroes should recognize their duty toward the masses," an idea that would become one of Du Bois's most famous. Du Bois believed that the burden for winning freedom and justice for all African Americans rested on the shoulders of those who were best prepared, educationally and economically. It would thus fall to the so-called Talented Tenth to lead the fight against racial discrimination (Marable:51; see also Du Bois 1968:236).

In 1935, Du Bois published *Black Reconstruction in America: An Essay Toward a History of the Part Which Black Folk Played in the Attempt to Reconstruct Democracy in America 1860–1880.* This book, based on ignored data about blacks' role in Reconstruction, is most notable in that it challenged the elitism that pervaded historical studies. In contrast to an elitist emphasis on a few "great men" and magnificent military battles, Du Bois sought to write about "ordinary" people—"ordinary" black people—though it would be a full generation later before a "people's" social history would find its place in academia. Indeed, *Black Reconstruction* uncovered the latent elitism and ethnocentrism of the entire historical profession, which under the guise of scientific objectivity was actually, according to Du Bois, a font of racist oppression, as historians' monographs reflected their prejudices (Rudwick [1960] 1982:302).

In 1939, Du Bois founded the journal *Phylon: A Review of Race and Culture* as a forum for passionate and scientific works dedicated to promoting black freedom and justice. He explicitly rejected the "sterility of a cloistered research approach" in favor of an interventionist social science, even though this kind of sociology would not be accepted (and in some circles, has yet to be accepted) for another 50 years. In both his history and his sociology, Du Bois refused to write from a position of value neutrality even as he strove for the truth—a stance that makes perfect sense in the contemporary context of humanistic and interpretive social science, but that was quite unheard of in his day.

It is no wonder then that Du Bois has been proclaimed "an authentic American genius" not only ahead of *his* time, but "*ahead of our time*" (Feagin 2003:22, 23, emphasis in original). No wonder Du Bois was nominated for *Time* magazine's "Person of the Twentieth Century" (Zuckerman 2000:1). Du Bois's sociological legacy is based not only on his intrepid exploration of the workings of race and class but also

Photo 7.4 Du Bois believed that higher education was essential to developing the leadership capacity of African Americans and achieving political and civil equality. These graduates of Fisk University, circa 1906, would be considered part of "the Talented Tenth," that is, among the most able 10 percent of African Americans, those who could dedicate themselves to "leavening the lump" and "inspiring the masses" (Du Bois [1903] 2004).

on his interpretive commitment to truth and justice as well. Indeed, as Reverend Martin Luther King, Jr., stated in a tribute to him in 1968,

> Dr. Du Bois was a tireless explorer and a gifted discoverer of social truths. His singular greatness lay in his quest for truth about his own people. . . . [His] greatest virtue was his committed empathy with all the oppressed and his divine dissatisfaction with all forms of injustice. (1970:181–83)

DU BOIS'S THEORETICAL ORIENTATION ██

Du Bois's work is exemplary in that it illuminates the intertwined structural and subjective causes and consequences of class, race, and racism. This multidimensionality is evident in the multiple approaches that Du Bois used. As discussed previously, Du Bois incorporated rich empirical data and rigorous historical information as well as autobiographical and literary styles into his work. In so doing, Du Bois addressed both the distinct subjectivities that motivate action and shape our perception and experience of the world; and, at the same time, he underscored the economic, and more specifically capitalistic, social structures behind such social and cultural patterns.

As shown in Figure 7.1, in terms of order, Du Bois tended to be collectivist in orientation; he was primarily interested in examining how broad social and cultural patterns—most important, class and racial stratification systems—shaped individual behavior and perceptions. In terms of action, Du Bois continually emphasized both that race and racism are formed by social structural forces, and that they are lived and felt experiences (see Figure 7.2). However, here we consider Du Bois primarily nonrationalist in terms of action. Although Du Bois was, as you will see, quite theoretically multidimensional, it is for his groundbreaking work on racial *consciousness* (as motivating particular lines of conduct) that he is best known. Du Bois powerfully demonstrated how race and racism are rooted in taken- for-granted symbolic structures, thereby reflecting the nonrational realm.

Figure 7.1   Du Bois's Basic Theoretical Orientation

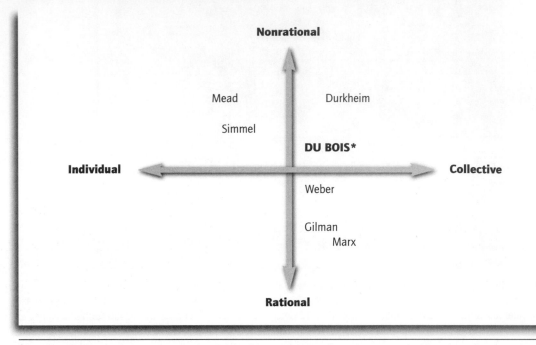

*Our placement of Du Bois on the nonrational side of the continuum reflects the excerpts in this volume that were chosen because of their *theoretical* significance. In our view, it is his understanding of racial *consciousness* that constitutes his single most important theoretical contribution. However, he continually underscored the intertwined, structural underpinnings of race and class which, in the latter part of his life, led him to adopt a predominantly rationalist, Marxist-inspired orientation. In our view, however, Du Bois's later work has more empirical than theoretical significance.

Figure 7.2   Du Bois's Multidimensional Approach to Race and Class

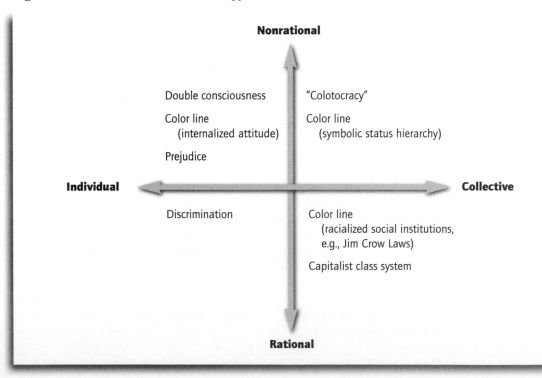

Du Bois's multidimensional theoretical orientation is clearly evident in *The Philadelphia Negro*. On one hand, Du Bois develops an empirical typology of four economic classes within the Philadelphia black community: the well-to-do; the decent hard workers, who were doing well; the "worthy poor," who were barely making ends meet; and the "submerged tenth," who were beneath the level of socioeconomic viability. Du Bois emphasizes that this stratification system resulted from the increasing industrialization of the time, reflecting a collectivist and rationalist approach to order and action (see Figure 7.2). In short, he contends that the economic system of capitalism led to industrialization and, hence, the decay of the black community within Philadelphia.

Nevertheless, in *The Philadelphia Negro,* Du Bois recognizes the power of nonrational factors too. For instance, Du Bois's typology depicts an upper-class "colortocracy" within the black community: Light-skinned blacks were a caste apart and formed the bulk of Du Bois's so-called Talented Tenth (Anderson 2000:56). This "colortocracy" reflects the nonrational realm (at the collective level) in the sense that it represents a symbolic code or scheme in Durkheimian terms, a status system in Weber's terms, and an ideology in Marxist terms (see Figure 7.2). As Du Bois states,

> Individual members of the colortocracy at times developed a notorious but distinctive racial complex involving an ideology that set them apart from those they viewed as their inferiors. They would take excessive pride in their "white" features, including light skin, thin noses and lips, and "good" hair. Often "colorstruck," they mimicked and voiced the anti-black prejudices of whites, whose fears, concerns, and values they understood and partly shared. (ibid.:57)

Photo 7.5 Segregated drinking fountains labeled "white" and "colored" in the Dougherty County Courthouse, Albany, Georgia, circa 1963. Note the unequal size of the fountains.

In addition, while Du Bois emphasized the structural or collectivist dimension of class and race, he also recognized that their institutional and symbolic features were perpetuated and disseminated in everyday interaction, at the level of the individual. Consider, for instance, the following passage about his experiences as a waiter at a summer resort hotel:

> I did not mind the actual work or the kind of work, but it was the dishonesty and deception, the flattery and cajolery, the unnatural assumption that worker and diner had no common humanity. It was uncanny. It was inherently and fundamentally wrong. I stood staring and thinking, while the other boys hustled about. Then I noticed one fat hog, feeding at a heavily gilded trough, who could not find his waiter. He beckoned me. It was not his voice, for his mouth was too full. It was his way, his air, his assumption. Thus Caesar ordered his legionnaires or Cleopatra her slaves. Dogs recognized the gesture. I did not. He may be beckoning yet for all I know, for something froze within me. I did not look his way again. Then and there I disowned menial service for me and my people. . . . When I finally walked out of that hotel and out of menial service forever, I felt as though, in a field of flowers, my nose had been unpleasantly long to the worms and manure at their roots. ([1920] 2003c:128–29)

On one hand, in this passage, Du Bois describes how it *felt* to be treated as a second-class citizen, a "dog," a "slave." Speaking to the nonrational realm, he illuminates the profound dehumanization that is at the heart of racism (at the level of the individual). On the other hand, Du Bois also emphasizes here that the font of such dehumanization is a hierarchical racial and class *system* that enables and reaffirms both this symbolic code and the hierarchical social roles. The point is (and this is made more explicit in his later, more Marxist-oriented works) that it is the menial service role that makes for such degradation or alienation. This is why Du Bois followed Marx in suggesting that capitalism itself precluded equality and democracy, resting as it did on exploitation of labor. So too, this passage suggests that the sordid "assumptions" and "airs" of white elites stem from their privileged position at the top of the racial and class hierarchies. Du Bois's multidimensional approach to race is illustrated in Figure 7.2.

Readings

We begin with Du Bois's most well-known and highly acclaimed *The Souls of Black Folk* (1903), a book of essays that relies not only on "objective" "social facts" but on subjective, interpretative experience as well. This selection is followed by "The Souls of White Folk," a provocative, albeit less-well-known, essay that was first published in 1910 but was revised with references to World War I for republication in *Darkwater* in 1920.

Introduction to *The Souls of Black Folk*

Du Bois's most famous work, *The Souls of Black Folk* (1903), is a compilation of 14 essays that Du Bois published previously in the *Atlantic Monthly,* the *New World,* and other journals. As Marable (1986:47) quite rightly observes, "Its grace and power is still overwhelming." Yet, *The Souls of Black Folk* is notable not only for its grace, but also for at least three distinct, albeit interrelated, reasons.

First, historically, *The Souls of Black Folk* is important because it explicitly exposed an important intellectual and political schism in the black community, between the more moderate Booker T. Washington and more radical political activists, such as Du Bois (and also journalist Ida B. Wells-Barnett, a feminist and anti-lynching activist). In his essay "Of Mr. Booker T. Washington and Others" (see p. 276), Du Bois pointedly attacks Washington for his "capitulating" agenda. Du Bois was enraged by Washington's "Atlanta Compromise" (favored by many whites) in which blacks would disavow open agitation in exchange for developing their own segregated institutions—which Du Bois interpreted as nothing more than "the old attitude of adjustment and submission" ([1903] 1989). Du Bois also abhorred Washington's "gradualist" solution to the issue of suffrage and his notion that Negro education should primarily be in the form of trade schools (rather than "abstract" knowledge).

Second, from a social science standpoint, *The Souls of Black Folk* is significant because Du Bois writes in a new "voice." Disgusted by the failure of sound empirical research to lead to desperately needed social change for the African American community and having done the empirical work himself, Du Bois became convinced that empirical data alone would never convince white Americans of the true workings of racial discrimination and prejudice. As a result, he turned away from the more empirical, strictly scientific accounting of race after writing *The Philadelphia Negro,* convinced that facts alone did nothing to influence people toward improving conditions for blacks. He began to write with a more "soulful" voice, because, quite rightly (in our view), he recognized that race does not work or exist solely at the "rational" level.[1]

As Du Bois states,

> For many years it was the theory of most Negro leaders that . . . white America did not know or realize the continuing plight of the Negro. Accordingly, for the last two decades, we have striven by book and periodical,

---

[1]Nevertheless, Du Bois did not completely abandon a positivist stance; he continued to seek "hard data," be they from official censuses, government documents, or field notes.

by speech and appeal, by various dramatic methods of agitation to put the essential facts before the American people. Today there can be no doubt that Americans know the facts; and yet they remain for the most part indifferent and unmoved. (Du Bois, 1898, as quoted in Berry, 2000:106)

Thus, Du Bois himself "stepped within the Veil, raising it that you [the presumably white reader] may view faintly its deeper recesses,—the meaning of its religion, the passion of its human sorrow, and the struggle of its greater souls" ([1903] 1989:1, 2). Adding, "Need I add that I who speak here am bone of the bone and flesh of the flesh of them that live within the Veil?," Du Bois pointedly used his own biographical experience to illuminate the reality of race in the United States (ibid.:2).

Relatedly, although Du Bois was intensely critical of the black church for failing to address the real economic needs of the black community, in *The Souls of Black Folk,* Du Bois turns his attention to the essence and power of black spirituality. He begins each chapter with "a bar of the Sorrow Songs,—some echo of haunting melody from the only American music which welled up from black souls in the dark past" to convey the essence of blackness and "our spiritual strivings" (ibid.:2–3). As Du Bois eloquently points out in "Of the Faith of the Fathers" (see p. 278), "Under the stress of law and whip, it [the Music of the Negro] became the one true expression of a people's sorrow, despair, and hope" (ibid.:156). In addition, Du Bois underscores the tremendous social significance of the Negro church. He calls the church the "social center of Negro life in the United States" (ibid.:157) and notes that "in the South, at least, practically every American Negro is a church member" (ibid.:158). Of course, this is precisely why Du Bois was critical of the Negro church as well; given its social and institutional vitality, Du Bois felt that the Black church could and should do more for its people.

Social scientifically, then, *Souls of Black Folk* contains a crucial methodological lesson: The workings of such complex phenomena as race and class cannot be fully understood using only "scientific" means. Du Bois explored subjectivity because he believed that race and racism did not work at a strictly rational level. Interestingly, today this position is not only fully embraced by postmodernists—it is taken to a radical extreme. Whereas Du Bois sought to combine empirical, scientific, and historical data with more subjective, intuitive understandings, postmodernists deny the existence of "social scientific" ways of knowing altogether, saying that all knowledge, in the end, is subjective and idiosyncratic.

Put in another way, Du Bois's approach was a precursor to the interpretive shift that began to emerge in sociology (and the social sciences in general) in the 1980s. Interpretive social scientists emphasize that complete objectivity is neither possible nor desirable. So-called value-free, positivist sociology is disingenuous, if not impossible to practice; a sociologist who does not acknowledge his or her own subjectivity runs the risk of reifying existing prejudices and beliefs. From this point of view, the notion that truth is arrived at through objectivity is unfounded. We cannot adequately capture the richness and complexity of reality using only objectivist tools (e.g., historical documents, surveys, statistics); we can and must use our own subjective experience and humanistic understanding in order to explore and explain the complexities of reality.

Third, *The Souls of Black Folks* is important theoretically because it contains three interrelated concepts for which Du Bois is now famous: the **color line**, **double consciousness**, and the **veil**. All of these concepts reflect the intertwined dimensions of race discussed previously (see Figure 7.2). Consider, first, one of Du Bois's most oft-cited sentences, which appears in this work: "The problem of the twentieth century is the problem of the color line—the relation of the darker to the lighter races of men in Asia and Africa, in America and the islands of the sea" (ibid.:13). Du Bois's concept of the "color line" is most intriguing, in part, because it speaks to the collective/rational, collective/nonrational, and individual/nonrational realms; as shown in Figure 7.2, the color line is both a preexisting social and cultural structure and an internalized attitude.

Specifically, on one hand, the color line addresses the historical and institutional (particularly colonial) dimensions of race. For instance, Du Bois maintains that once "the 'Color Line' began to pay dividends" through the colonization and exploitation of Africa and Africans beginning in the fifteenth century, race became central to world history (Monteiro 2000:229). Africa's poverty is inexorably linked to colonialism and imperial domination; the wealth of the colonial empires of England, France, Germany, and the United States "comes directly from the darker races of the world" (Du Bois [1947] 1995:645).

At the same time, the color line has important subjective or nonrational dimensions (see Figure 7.2). Du Bois not only examines race in its objective, demographic, and historical aspects but also addresses race as a symbolic and experiential reality. This emphasis on the nonrational workings of the color line is highlighted in his question, "How does it feel to be a problem?" found at the very beginning of *The Souls of Black Folk* ([1903] 1989:3).

This focus on the subjectivity integral to race and racism is also readily apparent in Du Bois's concepts of the veil and double consciousness:

> The Negro is a sort of seventh son, born with a veil, and gifted with second sight in this American world—a world which yields him no self-consciousness, but only lets him see himself through the revelation of the other world. It is a peculiar sensation, this double-consciousness, this sense of always looking at one's self through the eyes of others, of measuring one's soul by the tape of a world that looks on in amused contempt and pity. One ever feels his two-ness—an American, a Negro; two souls, two thoughts, two unreconciled strivings; two warring ideals in one dark body, whose dogged strength alone keeps it from being torn asunder. ([1903] 1989:5)

Here Du Bois reflects not on the structural causes, but rather on the subjective (or nonrational) consequences of being black in America. He explores the "peculiar *sensation*" (emphasis added) of being black in America; he wonders, "Why did God make me an outcast and a stranger in mine own house?" (ibid.:5). Nevertheless, as discussed previously, in exploring the subjective dimensions of racism, Du Bois never lost sight of its "objective," historical origins; the subjective experience of race—"double consciousness"—is rooted ultimately in the marginal structural position of "blacks." In this way, Du Bois's argument can be said to parallel that of Marx. As you will recall (see Chapter 2), Marx always tied the subjective experience of "alienation" and "estrangement" to workers' relationship to the means of production, that is, to class structure.

Interestingly, in some ways, Du Bois's notion of double consciousness parallels Simmel's discussion of the stranger discussed in the previous chapter. Both theorists emphasized the sense of otherness that not only inhibits social solidarity, but also prevents the formation of a unified sense of self (Calhoun et al. 2002:237, 238). Moreover, both Simmel and Du Bois used the concept of a veil to describe the social distance between people. Of course, the theoretical power of these terms is that they can be used to discuss all sorts of social groups—from the disabled, to the transgendered, to the "undocumented" populations. Today these concepts are especially familiar to feminists. In the 1970s, Dorothy Smith coined the term "bifurcation of consciousness" to refer to how women in a man's world were forced to live a double existence as they sought to reconcile the male-dominated, abstract "governing consciousness" with their own practical consciousness rooted in the everyday world.

Put in another way, while until the 1970s, the reality of double consciousness was submerged in the context of white male hegemony, today it is more transparent, and the advantages of it are more clear. The same double consciousness that enabled Du Bois to banter with both Harvard intellectuals and Philadelphians from the seventh ward—while being completely oblivious to whites—is overt in the person of Barack Obama. Obama's dual Kansan and Kenyan background is considered an asset rather than scandalous or exotic. From this point of view, Obama's phenomenal success reflects not only his personal charisma and intelligence, but also his immersion in and ability to capitalize on multiple symbolic worlds. Working-class tropes are prevalent today as well. White, upper-class intellectual elites and "bluebloods" are considered suspect and "out of touch" (this was the symbolic problem that plagued both Al Gore and John Kerry in their presidential bids). Instead, it is the "realness" of "Joe the Plumber" that commands attention and offers credibility. To be sure, such shifts in meanings do not necessarily directly reflect or impact institutionalized systems of social stratification. The point is that multiculturalism is our new reality, a critical component of our globalized and hypermediated world.

Indeed, given our globalized multicultural world, and the recent election of Barack Obama to presidency of the United States, some might maintain that DuBois' notion of "the color line" is outdated and that the "problem of the color line" has been resolved. From this point of view, since African Americans make up only some 13 percent of the American population, his electoral victory *must* be evidence of a fundamental reduction in prejudice and racism in the United States. One might argue that as holder of the most powerful position in the nation (if not the world), with his picture adorning the walls of the White House as well as public schools and government offices across the land (and around the globe), it seems absurd that President Obama might consider himself "an outcast and a stranger in mine own house" as did Du Bois and African Americans of his generation.

Certainly, the election of Barack Obama reflects a tremendous sea change in the history of the United States. For instance, children today (both white and nonwhite) will be less likely to take white supremacy for granted, simply because their biographical experience will include living in a moment in which educated, highly esteemed African Americans are president and first lady. Yet, despite this phenomenal turn of events, huge reservoirs of inequality in the United States persist and important issues regarding race and class remain. Based on current rates of first incarceration, an estimated 32% of black males will enter State or Federal prison during their lifetime, compared to 17% of Hispanic males and 5.9% of white males.[2] So, too, white households continue to be far wealthier than black or Hispanic households, and educational differences extend directly from those differentials in class and wealth. For instance, students with family incomes of more than $200,000 had an average SAT math score of 570, those in the $80,000–$100,000 cohort had an average of 525, and those with family income up to $20,000 had an average of 456.

Yet, from a Duboisian point of view, what the election of Obama makes clear is that the "place" of African Americans in the United States' system of stratification is complex. As Du Bois readily recognized when he coined the term "Talented Tenth," even before legal emancipation in 1865, when "race" was understood as literally an issue of "blood," the binary symbolic system of "black" and "white" oversimplified a far more nuanced scheme of racial classification. In the past two centuries, we have witnessed a proliferation of even more multifarious and contradictory systems of meaning. On one hand, Obama's remarkable journey parallels that of Du Bois himself; both are the crème de la crème of the "Talented Tenth" of their respective generations. In Du Bois's day, the divide was between "good" (subservient, often mulatto) "house" blacks who "knew their place" and ("uppity") "city blacks" or (unrefined) "field" blacks who were arrogant or ignorant, respectively. Today, the reality is that though the two images of blacks, that is, the divide between "good" (upper-middle class, upstanding) citizens who are black—for example, Obama, as well as Oprah Winfrey, Colin Powell, and Condoleezza Rice—and "bad" blacks (inner-city thugs and gangsters) remain, African Americans are vulnerable to either symbolic classificatory scheme. (Thus, for instance, Obama was charged with being "uppity" and "white-washed" because of his degree from Harvard.) The media continuously juxtaposes regal images of Barack and Michelle Obama with threatening images of inner-city African American thugs and gangsters (e.g., Rodney King and Willie Horton). But even this dual imagery is more complex than this narrative lets on: "gangstas" might be glorified or vilified, epitomizing celebrity idols or abject poverty and bad behavior. And of course even the most esteemed blacks can instantaneously spiral downward (like O. J. Simpson—see Edles 2002:121–27).

---

[2]http://www.ojp.usdoj.gov/bjs/crimoff.htm#inmates

The Souls of Black Folk (1903)

W. E. B. Du Bois

THE FORETHOUGHT

Herein lie buried many things which if read with patience may show the strange meaning of being black here in the dawning of the Twentieth Century. This meaning is not without interest to you, Gentle Reader; for the problem of the Twentieth Century is the problem of the color-line.

I pray you, then, receive my little book in all charity, studying my words with me, forgiving mistake and

foible for sake of the faith and passion that is in me, and seeking the grain of truth hidden there.

I have sought here to sketch, in vague, uncertain outline, the spiritual world in which ten thousand thousand Americans live and strive. First, in two chapters I have tried to show what Emancipation meant to them, and what was its aftermath. In a third chapter I have pointed out the slow rise of personal leadership, and criticised candidly the leader who bears the chief burden of his race to-day. Then, in two other chapters I

SOURCE: From *The Souls of Black* Folk by W. E. B. Du Bois, 1903.

have sketched in swift outline the two worlds within and without the Veil, and thus have come to the central problem of training men for life. Venturing now into deeper detail, I have in two chapters studied the struggles of the massed millions of the black peasantry, and in another have sought to make clear the present relations of the sons of master and man.

Leaving, then, the world of the white man, I have stepped within the Veil, raising it that you may view faintly its deeper recesses,—the meaning of its religion, the passion of its human sorrow, and the struggle of its greater souls. All this I have ended with a tale twice told but seldom written. . . .

Before each chapter, as now printed, stands a bar of the Sorrow Songs,—some echo of haunting melody from the only American music which welled up from black souls in the dark past. And, finally, need I add that I who speak here am bone of the bone and flesh of the flesh of them that live within the Veil?

OF OUR SPIRITUAL STRIVINGS

Between me and the other world there is ever an unasked question: unasked by some through feelings of delicacy; by others through the difficulty of rightly framing it. All, nevertheless, flutter round it. They approach me in a half- hesitant sort of way, eye me curiously or compassionately, and then, instead of saying directly, How does it feel to be a problem? they say, I know an excellent colored man in my town; or, I fought at Mechanicsville; or, Do not these Southern outrages make your blood boil? At these I smile, or am interested, or reduce the boiling to a simmer, as the occasion may require. To the real question, How does it feel to be a problem? I answer seldom a word.

And yet, being a problem is a strange experience,—peculiar even for one who has never been anything else, save perhaps in babyhood and in Europe. It is in the early days of rollicking boyhood that the revelation first bursts upon one, all in a day, as it were. I remember well when the shadow swept across me. I was a little thing, away up in the hills of New England, where the dark Housatonic winds between Hoosac and Taghkanic to the sea. In a wee wooden schoolhouse, something put it into the boys' and girls' heads to buy gorgeous visiting-cards—ten cents a package—and exchange. The exchange was merry, till one girl, a tall newcomer, refused my card,—refused it peremptorily, with a glance. Then it dawned upon me with a certain suddenness that I was different

from the others; or like, mayhap, in heart and life and longing, but shut out from their world by a vast veil. I had thereafter no desire to tear down that veil, to creep through; I held all beyond it in common contempt, and lived above it in a region of blue sky and great wandering shadows. That sky was bluest when I could beat my mates at examination-time, or beat them at a foot-race, or even beat their stringy heads. Alas, with the years all this fine contempt began to fade; for the worlds I longed for, and all their dazzling opportunities, were theirs, not mine. But they should not keep these prizes, I said; some, all, I would wrest from them. Just how I would do it I could never decide: by reading law, by healing the sick, by telling the wonderful tales that swam in my head,—some way. With other black boys the strife was not so fiercely sunny: their youth shrunk into tasteless sycophancy, or into silent hatred of the pale world about them and mocking distrust of everything white; or wasted itself in a bitter cry, Why did God make me an outcast and a stranger in mine own house? The shades of the prison-house closed round about us all: walls strait and stubborn to the whitest, but relentlessly narrow, tall, and unscalable to sons of night who must plod darkly on in resignation, or beat unavailing palms against the stone, or steadily, half hopelessly, watch the streak of blue above.

After the Egyptian and Indian, the Greek and Roman, the Teuton and Mongolian, the Negro is a sort of seventh son, born with a veil, and gifted with second-sight in this American world,—a world which yields him no true self-consciousness, but only lets him see himself through the revelation of the other world. It is a peculiar sensation, this double-consciousness, this sense of always looking at one's self through the eyes of others, of measuring one's soul by the tape of a world that looks on in amused contempt and pity. One ever feels his two-ness,—an American, a Negro; two souls, two thoughts, two unreconciled strivings; two warring ideals in one dark body, whose dogged strength alone keeps it from being torn asunder.

The history of the American Negro is the history of this strife—this longing to attain self-conscious manhood, to merge his double self into a better and truer self. In this merging he wishes neither of the older selves to be lost. He would not Africanize America, for America has too much to teach the world and Africa. He would not bleach his Negro soul in a flood of white Americanism, for he knows that Negro blood has a message for the world. He simply wishes to make it possible for a man to be both a Negro and an American, without being cursed and spit upon by his fellows, without having the doors of Opportunity closed roughly in his face.

This, then, is the end of his striving: to be a co-worker in the kingdom of culture, to escape both death and isolation, to husband and use his best powers and his latent genius. These powers of body and mind have in the past been strangely wasted, dispersed, or forgotten. The shadow of a mighty Negro past flits through the tale of Ethiopia the Shadowy and of Egypt the Sphinx. Throughout history, the powers of single black men flash here and there like falling stars, and die sometimes before the world has rightly gauged their brightness. Here in America, in the few days since Emancipation, the black man's turning hither and thither in hesitant and doubtful striving has often made his very strength to lose effectiveness, to seem like absence of power, like weakness. And yet it is not weakness,—it is the contradiction of double aims. The double-aimed struggle of the black artisan—on the one hand to escape white contempt for a nation of mere hewers of wood and drawers of water, and on the other hand to plough and nail and dig for a poverty-stricken horde—could only result in making him a poor craftsman, for he had but half a heart in either cause. By the poverty and ignorance of his people, the Negro minister or doctor was tempted toward quackery and demagogy; and by the criticism of the other world, toward ideals that made him ashamed of his lowly tasks. The would-be black *savant* was confronted by the paradox that the knowledge his people needed was a twice-told tale to his white neighbors, while the knowledge which would teach the white world was Greek to his own flesh and blood. The innate love of harmony and beauty that set the ruder souls of his people a-dancing and a-singing raised but confusion and doubt in the soul of the black artist; for the beauty revealed to him was the soul-beauty of a race which his larger audience despised, and he could not articulate the message of another people. This waste of double aims, this seeking to satisfy two unreconciled ideals, has wrought sad havoc with the courage and faith and deeds of ten thousand thousand people,—has sent them often wooing false gods and invoking false means of salvation, and at times has even seemed about to make them ashamed of themselves.

Away back in the days of bondage they thought to see in one divine event the end of all doubt and disappointment; few men ever worshipped Freedom with half such unquestioning faith as did the American Negro for two centuries. To him, so far as he thought and dreamed, slavery was indeed the sum of all villainies, the cause of all sorrow, the root of all prejudice; Emancipation was the key to a promised land of sweeter beauty than ever stretched before the eyes of wearied Israelites. In song and exhortation swelled one refrain—Liberty; in his tears and curses the God he implored had Freedom in his right hand. At last it came,—suddenly, fearfully, like a dream. With one wild carnival of blood and passion came the message in his own plaintive cadences:—

"Shout, O children!
Shout, you're free!
For God has bought your liberty!"

Years have passed away since then,—ten, twenty, forty; forty years of national life, forty years of renewal and development, and yet the swarthy spectre sits in its accustomed seat at the Nation's feast. In vain do we cry to this our vastest social problem:—

"Take any shape but that, and my firm nerves
Shall never tremble!"

The Nation has not yet found peace from its sins; the freedman has not yet found in freedom his promised land. Whatever of good may have come in these years of change, the shadow of a deep disappointment rests upon the Negro people,—a disappointment all the more bitter because the unattained ideal was unbounded save by the simple ignorance of a lowly people.

The first decade was merely a prolongation of the vain search for freedom, the boon that seemed ever barely to elude their grasp,—like a tantalizing will-o'-the-wisp, maddening and misleading the headless host. The holocaust of war, the terrors of the Ku-Klux Klan, the lies of carpet-baggers, the disorganization of industry, and the contradictory advice of friends and foes, left the bewildered serf with no new watch-word beyond the old cry for freedom. As the time flew, however, he began to grasp a new idea. The ideal of liberty demanded for its attainment powerful means, and these the Fifteenth Amendment gave him. The ballot, which before he had looked upon as a visible sign of freedom, he now regarded as the chief means of gaining and perfecting the liberty with which war had partially endowed him. And why not? Had not votes made war and emancipated millions? Had not votes enfranchised the freedmen? Was anything impossible to a power that had done all this? A million black men started with renewed zeal to vote themselves into the kingdom. So the decade flew away, the revolution of 1876 came, and left the half-free serf weary, wondering, but still inspired. Slowly but steadily, in the following years, a new vision began gradually to replace the dream of political power,—a powerful movement, the rise of another ideal to guide the unguided, another pillar of fire by night after a clouded day. It was the ideal of "book-learning"; the curiosity, born of compulsory ignorance, to know and test the power of the cabalistic letters of the white man, the longing to know. Here at last seemed to have been discovered the mountain path to Canaan; longer than the highway of Emancipation and law, steep and rugged, but straight, leading to heights high enough to overlook life.

Up the new path the advance guard toiled, slowly, heavily, doggedly; only those who have watched and guided the faltering feet, the misty minds, the dull understandings, of the dark pupils of these schools know how faithfully, how piteously, this people strove to learn. It was weary work. The cold statistician wrote down the inches of progress here and there, noted also where here and there a foot had slipped or some one had fallen. To the tired climbers, the horizon was ever dark, the mists were often cold, the Canaan was always dim and far away. If, however, the vistas disclosed as yet no goal, no resting-place, little but flattery and criticism, the journey at least gave leisure for reflection and self-examination; it changed the child of Emancipation to the youth with dawning self-consciousness, self-realization, self-respect. In those sombre forests of his striving his own soul rose before him, and he saw himself,—darkly as through a veil; and yet he saw in himself some faint revelation of his power, of his mission. He began to have a dim feeling that, to attain his place in the world, he must be himself, and not another. For the first time he sought to analyze the burden he bore upon his back, that deadweight of social degradation partially masked behind a half-named Negro problem. He felt his poverty; without a cent, without a home, without land, tools, or savings, he had entered into competition with rich, landed, skilled neighbors. To be a poor man is hard, but to be a poor race in a land of dollars is the very bottom of hardships. He felt the weight of his ignorance,—not simply of letters, but of life, of business, of the humanities; the accumulated sloth and shirking and awkwardness of decades and centuries shackled his hands and feet. Nor was his burden all poverty and ignorance. The red stain of bastardy, which two centuries of systematic legal

defilement of Negro women had stamped upon his race, meant not only the loss of ancient African chastity, but also the hereditary weight of a mass of corruption from white adulterers, threatening almost the obliteration of the Negro home.

A people thus handicapped ought not to be asked to race with the world, but rather allowed to give all its time and thought to its own social problems. But alas! while sociologists gleefully count his bastards and his prostitutes, the very soul of the toiling, sweating black man is darkened by the shadow of a vast despair. Men call the shadow prejudice, and learnedly explain it as the natural defence of culture against barbarism, learning against ignorance, purity against crime, the "higher" against the "lower" races. To which the Negro cries Amen! and swears that to so much of this strange prejudice as is founded on just homage to civilization, culture, righteousness, and progress, he humbly bows and meekly does obeisance. But before that nameless prejudice that leaps beyond all this he stands helpless, dismayed, and well-nigh speechless; before that personal disrespect and mockery, the ridicule and systematic humiliation, the distortion of fact and wanton license of fancy, the cynical ignoring of the better and the boisterous welcoming of the worse, the all-pervading desire to inculcate disdain for everything black, from Toussaint to the devil,—before this there rises a sickening despair that would disarm and discourage any nation save that black host to whom "discouragement" is an unwritten word.

But the facing of so vast a prejudice could not but bring the inevitable self-questioning, self-disparagement, and lowering of ideals which ever accompany repression and breed in an atmosphere of contempt and hate. Whisperings and portents came borne upon the four winds: Lo! we are diseased and dying, cried the dark hosts; we cannot write, our voting is vain; what need of education, since we must always cook and serve? And the Nation echoed and enforced this self-criticism, saying: Be content to be servants, and nothing more; what need of higher culture for half-men? Away with the black man's ballot, by force or fraud,—and behold the suicide of a race! Nevertheless, out of the evil came something of good,—the more careful adjustment of education to real life, the clearer perception of the Negroes' social responsibilities, and the sobering realization of the meaning of progress.

So dawned the time of *Sturm und Drang:* storm and stress today rocks our little boat on the mad waters of the world-sea; there is within and without the sound of conflict, the burning of body and rending of soul; inspiration strives with doubt, and faith with vain questionings. The bright ideals of the past,—physical freedom, political power, the training of brains and the training of hands,—all these in turn have waxed and waned, until even the last grows dim and overcast. Are they all wrong,—all false? No, not that, but each alone was oversimple and incomplete,—the dreams of a credulous race-childhood, or the fond imaginings of the other world which does not know and does not want to know our power. To be really true, all these ideals must be melted and welded into one. The training of the schools we need to-day more than ever,—the training of deft hands, quick eyes and ears, and above all the broader, deeper, higher culture of gifted minds and pure hearts. The power of the ballot we need in sheer self-defence,—else what shall save us from a second slavery? Freedom, too, the long-sought, we still seek,—the freedom of life and limb, the freedom to work and think, the freedom to love and aspire. Work, culture, liberty,—all these we need, not singly but together, not successively but together, each growing and aiding each, and all striving toward that vaster ideal that swims before the Negro people, the ideal of human brotherhood, gained through the unifying ideal of Race; the ideal of fostering and developing the traits and talents of the Negro, not in opposition to or contempt for other races, but rather in large conformity to the greater ideals of the American Republic, in order that some day on American soil two world-races may give each to each those characteristics both so sadly lack. We the darker ones come even now not altogether empty-handed: there are to-day no truer exponents of the pure human spirit of the Declaration of Independence than the American Negroes; there is no true American music but the wild sweet melodies of the Negro slave; the American fairy tales and folklore are Indian and African; and, all in all, we black men seem the sole oasis of simple faith and reverence in a dusty desert of dollars and smartness. Will America be poorer if she replace her brutal dyspeptic blundering with light-hearted but determined Negro humility? or her coarse and cruel wit with loving jovial good-humor? or her vulgar music with the soul of the Sorrow Songs?

Merely a concrete test of the underlying principles of the great republic is the Negro Problem, and the spiritual striving of the freedmen's sons is the travail of souls whose burden is almost beyond the measure of their strength, but who bear it in the name of an historic race, in the name of this the land of their fathers' fathers, and in the name of human opportunity.

And now what I have briefly sketched in large outline let me on coming pages tell again in many ways, with loving emphasis and deeper detail, that men may listen to the striving in the souls of black folk.

OF THE DAWN OF FREEDOM

The problem of the twentieth century is the problem of the color-line,—the relation of the darker to the lighter races of men in Asia and Africa, in America and the islands of the sea. It was a phase of this problem that caused the Civil War; and however much they who marched South and North in 1861 may have fixed on the technical points of union and local autonomy as a shibboleth, all nevertheless knew, as we know, that the question of Negro slavery was the real cause of the conflict. Curious it was, too, how this deeper question ever forced itself to the surface despite effort and disclaimer. No sooner had Northern armies touched Southern soil than this old question, newly guised, sprang from the earth,—What shall be done with Negroes? Peremptory military commands, this way and that, could not answer the query; the Emancipation Proclamation seemed but to broaden and intensify the difficulties; and the War Amendments made the Negro problems of to-day. . . .

The passing of a great human institution before its work is done, like the untimely passing of a single soul, but leaves a legacy of striving for other men. The legacy of the Freedmen's Bureau is the heavy heritage of this generation. To-day, when new and vaster problems are destined to strain every fibre of the national mind and soul, would it not be well to count this legacy honestly and carefully? For this much all men know: despite compromise, war, and struggle, the Negro is not free. In the backwoods of the Gulf States, for miles and miles, he may not leave the plantation of his birth; in well-nigh the whole rural South the black farmers are peons, bound by law and custom to an economic slavery, from which the only escape is death or the penitentiary. In the most cultured sections and cities of the South the Negroes are a segregated servile caste, with restricted rights and privileges. Before the courts, both in law and custom, they stand on a different and peculiar basis. Taxation without representation is the rule of their political life. And the result of all this is, and in nature must have been, lawlessness and crime. That is the large legacy of the Freedmen's Bureau, the work it did not do because it could not.

I have seen a land right merry with the sun, where children sing, and rolling hills lie like passioned women wanton with harvest. And there in the King's Highway sat and sits a figure veiled and bowed, by which the traveller's footsteps hasten as they go. On the tainted air broods fear. Three centuries' thought has been the raising and unveiling of that bowed human heart, and now behold a century new for the duty and the deed. The problem of the Twentieth Century is the problem of the color-line.

OF MR. BOOKER T. WASHINGTON AND OTHERS

Easily the most striking thing in the history of the American Negro since 1876 is the ascendancy of Mr. Booker T. Washington. It began at the time when war memories and ideals were rapidly passing; a day of astonishing commercial development was dawning; a sense of doubt and hesitation overtook the freedmen's sons,—then it was that his leading began. Mr. Washington came, with a simple definite programme, at the psychological moment when the nation was a little ashamed of having bestowed so much sentiment on Negroes, and was concentrating its energies on Dollars. His programme of industrial education, conciliation of the South, and submission and silence as to civil and political rights, was not wholly original; the Free Negroes from 1830 up to wartime had striven to build industrial schools, and the American Missionary Association had from the first taught various trades; and Price and others had sought a way of honorable alliance with the best of the Southerners. But Mr. Washington first indissolubly linked these things; he put enthusiasm, unlimited energy, and perfect faith intro this programme, and changed it from a by-path into a veritable Way of Life. And the ale of the methods by which he did this is a fascinating study of human life.

It startled the nation to hear a Negro advocating such a programme after many decades of bitter complaint; it startled and won the applause of the South, it interested and won the admiration of the North; and after a confused murmur of protest, it silenced if it did not convert the Negroes themselves.

To gain the sympathy and cooperation of the various elements comprising the white South was Mr. Washington's first task; and this, at the time Tuskegee was founded, seemed, for a black man, well-nigh impossible. And yet ten years later it was done in the word spoken at Atlanta: "In all things purely social we can be as separate as the five fingers, and yet one as the hand in all things essential to mutual progress." This "Atlanta Compromise" is by all odds the most notable thing in Mr. Washington's career. The South interpreted it in different ways: The radicals received it as a complete surrender of the demand for civil and political equality; the conservatives, as a generously conceived working basis for mutual understanding. So both approved it, and to-day its author is certainly the most distinguished Southerner since Jefferson Davis, and the one with the largest personal following.

Next to this achievement comes Mr. Washington's work in gaining place and consideration in the North. Others less shrewd and tactful had formerly essayed to sit on these two stools and had fallen between them; but as Mr. Washington knew the heart of the South from birth and training, so by singular insight he intuitively grasped the spirit of the age which was dominating the North. And so thoroughly did he learn the speech and thought of triumphant commercialism, and the ideals of material prosperity, that the picture of a lone black boy poring over a French grammar amid the weeds and dirt of a neglected home soon seemed to him the acme of absurdities. One wonders what Socrates and St. Francis of Assisi would say to this.

And yet this very singleness of vision and thorough oneness with his age is a mark of the successful man. It is as though Nature must needs make men narrow in order to give them force. So Mr. Washington's cult has gained unquestioning followers, his work has wonderfully prospered, his friends are legion, and his enemies are confounded. To-day he stands as the one recognized spokesman of his ten million fellows, and one of the most notable figures in a nation of seventy millions. One hesitates, therefore, to criticise a life which, beginning with so little, has done so much. And yet the time is come when one may speak in all sincerity and utter courtesy of the mistakes and shortcomings of Mr. Washington's career, as well as of his triumphs, without being thought captious or envious, and without forgetting that it is easier to do ill than well in the world.

The criticism that has hitherto met Mr. Washington has not always been of this broad character. In the South especially has he had to walk warily to avoid the harshest judgments,—and naturally so, for he is dealing with the one subject of deepest sensitiveness to that section. Twice—once when at the Chicago celebration of the Spanish-American War he alluded to the color-prejudice that is "eating away the vitals of the South," and once when he dined with President Roosevelt—has the resulting Southern criticism been violent enough to threaten seriously his popularity. In the North the feeling has several times forced itself into words, that Mr. Washington's counsels of submission overlooked certain elements of true manhood, and that his educational programme was unnecessarily narrow. Usually, however, such criticism has not found open expression, although, too, the spiritual sons of the Abolitionists have not been prepared to acknowledge that the schools founded before Tuskegee, by men of broad ideals and self-sacrificing spirit, were wholly failures or worthy of ridicule. While, then, criticism has not failed to follow Mr. Washington, yet the prevailing public opinion of the land has been but too willing to deliver the solution of a wearisome problem into his hands, and say, "If that is all you and your race ask, take it."

Among his own people, however, Mr. Washington has encountered the strongest and most lasting opposition, amounting at times to bitterness, and even to-day continuing strong and insistent even though largely silenced in outward expression by the public opinion of the nation. Some of this opposition is, of course, mere envy; the disappointment of displaced demagogues and the spite of narrow minds. But aside from this, there is among educated and thoughtful colored men in all parts of the land a feeling of deep regret, sorrow, and apprehension at the wide currency and ascendancy which some of Mr. Washington's theories have gained. These same men admire his sincerity of purpose, and are willing to forgive much to honest endeavor which is doing something worth the doing. They coöperate with Mr. Washington as far as they conscientiously can; and, indeed, it is no ordinary tribute to this man's tact and power that, steering as he must between so many diverse interests and opinions, he so largely retains the respect of all. . . .

Mr. Washington represents in Negro thought the old attitude of adjustment and submission; but adjustment at such a peculiar time as to make his programme unique. This is an age of unusual economic development, and Mr. Washington's programme naturally takes an economic cast, becoming a gospel of Work and Money to such an extent as apparently almost completely to overshadow the higher aims of life. Moreover, this is an age when the more advanced races are coming in closer contact with the less developed races, and the race-feeling is therefore intensified; and Mr. Washington's programme practically accepts the alleged inferiority of the Negro races. Again, in our own land, the reaction from the sentiment of war time has

given impetus to race-prejudice against Negroes, and Mr. Washington withdraws many of the high demands of Negroes as men and American citizens. In other periods of intensified prejudice all the Negro's tendency to self-assertion has been called forth; at this period a policy of submission is advocated. In the history of nearly all other races and peoples the doctrine preached at such crises has been that manly self-respect is worth more than lands and houses, and that a people who voluntarily surrender such respect, or cease striving for it, are not worth civilizing.

In answer to this, it has been claimed that the Negro can survive only through submission. Mr. Washington distinctly asks that black people give up, at least for the present, three things,—First, political power,

Second, insistence on civil rights,

Third, higher education of Negro youth,—

And concentrate all their energies on industrial education, the accumulation of wealth, and the conciliation of the South. This policy has been courageously and insistently advocated for over fifteen years, and has been triumphant for perhaps ten years. As a result of this tender of the palm-branch, what has been the return? In these years there have occurred:

1. The disfranchisement of the Negro.

2. The legal creation of a distinct status of civil inferiority for the Negro.

3. The steady withdrawal of aid from institutions for the higher training of the Negro.

These movements are not, to be sure, direct results of Mr. Washington's teachings; but his propaganda has, without a shadow of doubt, helped their speedier accomplishment. The question then comes: Is it possible, and probable, that nine millions of men can make effective progress in economic lines if they are deprived of political rights, made a servile caste, and allowed only the most meagre chance for developing their exceptional men? If history and reason give any distinct answer to these questions, it is an emphatic *No.* And Mr. Washington thus faces the triple paradox of his career:

1. He is striving nobly to make Negro artisans business men and property-owners; but it is utterly impossible, under modern competitive methods, for workingmen and property-owners to defend their rights and exist without the right of suffrage.

2. He insists on thrift and self-respect, but at the same time counsels a silent submission to civic inferiority such as is bound to sap the manhood of any race in the long run.

3. He advocates common-school and industrial training, and depreciates institutions of higher learning; but neither the Negro common-schools, nor Tuskegee itself, could remain open a day were it not for teachers trained in Negro colleges, or trained by their graduates . . .

To-day even the attitude of the Southern whites toward the blacks is not, as so many assume, in all cases the same; the ignorant Southerner hates the Negro, the workingmen fear his competition, the money-makers wish to use him as a laborer, some of the educated see a menace in his upward development, while others—usually the sons of the masters—wish to help him to rise. National opinion has enabled this last class to maintain the Negro common schools, and to protect the Negro partially in property, life, and limb. Through the pressure of the money-makers, the Negro is in danger of being reduced to semi-slavery, especially in the country districts; the workingmen, and those of the educated who fear the Negro, have united to disfranchise him, and some have urged his deportation; while the passions of the ignorant are easily aroused to lynch and abuse any black man . . .

The South ought to be led, by candid and honest criticism, to assert her better self and do her full duty to the race she has cruelly wronged and is still wronging. The North—her co- partner in guilt—cannot salve her conscience by plastering it with gold. We cannot settle this problem by diplomacy and suaveness, by "policy" alone. If worse comes to worst, can the moral fibre of this country survive the slow throttling and murder of nine millions of men?

The black men of America have a duty to perform, a duty stern and delicate,—a forward movement to oppose a part of the work of their greatest leader. So far as Mr. Washington preaches Thrift, Patience, and Industrial Training for the masses, we must hold up his hands and strive with him, rejoicing in his honors and glorying in the strength of this Joshua called of God and of man to lead the headless host. But so far as Mr. Washington apologizes for injustice, North or South, does not rightly value the prvilege and duty of voting, belittles the emasculating effects of caste distinctions, and opposes the higher training and ambition

of our brighter minds,—so far as he, the South, or the Nation, does this,—we must unceasingly and firmly oppose them. By every civilized and peaceful method we must strive for the rights which the world accords to men, clinging unwaveringly to those great words which the sons of the Fathers would fain forget: "We hold these truths to be self-evident: That all men are created equal; that they are endowed by their Creator with certain unalienable rights; that among these are life, liberty, and the pursuit of happiness."

OF THE FAITH OF THE FATHERS

It was out in the country, far from home, far from my foster home, on a dark Sunday night. The road wandered from our rambling log-house up the stony bed of a creek, past wheat and corn, until we could hear dimly across the fields a rhythmic cadence of song,—soft, thrilling, powerful, that swelled and died sorrowfully in our ears. I was a country schoolteacher then, fresh from the East, and had never seen a Southern Negro revival. To be sure, we in Berkshire were not perhaps as stiff and formal as they in Suffolk of olden time; yet we were very quiet and subdued, and I know not what would have happened those clear Sabbath mornings had someone punctuated the sermon with a wild scream, or interrupted the long prayer with a loud Amen! And so most striking to me, as I approached the village and the little plain church perched aloft, was the air of intense excitement that possessed that mass of black folk. A sort of suppressed terror hung in the air and seemed to seize us,—a pythian madness, a demoniac possession, that lent terrible reality to song and word. The black and massive form of the preacher swayed and quivered as the words crowded to his lips and flew at us in singular eloquence. The people moaned and fluttered, and then the gaunt-cheeked brown woman beside me suddenly leaped straight into the air and shrieked like a lost soul, while round about came wail and groan and outcry, and a scene of human passion such as I had never conceived before.

Those who have not thus witnessed the frenzy of a Negro revival in the untouched backwoods of the South can but dimly realize the religious feeling of the slave; as described, such scenes appear grotesque and funny, but as seen they are awful. Three things characterized this religion of the slave,—the Preacher, the Music, and the Frenzy. The preacher is the most unique personality developed by the Negro on American soil. A leader, a politician, an orator, a "boss," an intriguer, an idealist,—all these he is, and ever, too, the centre of a group of men, now twenty, now a thousand in number. The combination of a certain adroitness with deep-seated earnestness, of tact with consummate ability, gave him his preeminence, and helps him maintain it. The type, of course, varies according to time and place, from the West Indies in the sixteenth century to New England in the nineteenth, and from the Mississippi bottoms to cities like New Orleans or New York.

The Music of Negro religion is that plaintive rhythmic melody, with its touching minor cadences, which, despite caricature and defilement, still remains the most original and beautiful expression of human life and longing yet born on American soil. Sprung from the African forests, where its counterpart can still be heard, it was adapted, changed, and intensified by the tragic soul-life of the slave, until, under the stress of law and whip, it became the one true expression of a people's sorrow, despair, and hope.

Finally the Frenzy of "Shouting," when the Spirit of the Lord passed by, and, seizing the devotee, made him mad with supernatural joy, was the last essential of Negro religion and the one more devoutly believed in than all the rest. It varied in expression from the silent rapt countenance or the low murmur and moan to the mad abandon of physical fervor,—the stamping, shrieking, and shouting, the rushing to and fro and wild waving of arms, the weeping and laughing, the vision and the trance. All this is nothing new in the world, but old as religion, as Delphi and Endor. And so firm a hold did it have on the Negro, that many generations firmly believed that without this visible manifestation of the God there could be no true communion with the Invisible.

These were the characteristics of Negro religious life as developed up to the time of Emancipation. Since under the peculiar circumstances of the black man's environment they were the one expression of his higher life, they are of deep interest to the student of his development, both socially and psychologically. Numerous are the attractive lines of inquiry that here group themselves. What did slavery mean to the African savage? What was his attitude toward the World and Life? What seemed to him good and evil,—God and Devil? Whither went his longings and strivings, and wherefore were his heart-burnings and disappointments? Answers to such questions can come only from a study of Negro religion as a development, through its gradual changes from the heathenism of the Gold Coast to the institutional Negro church of Chicago.

Moreover, the religious growth of millions of men, even though they be slaves, cannot be without potent influence upon their contemporaries. The Methodists and Baptists of America owe much of their condition to the silent but potent influence of their millions of Negro converts. Especially is this noticeable in the South, where theology and religious philosophy are on this account a long way behind the North, and where the religion of the poor whites is a plain copy of Negro thought and methods. The mass of "gospel" hymns which has swept through American churches and well-nigh ruined our sense of song consists largely of debased imitations of Negro melodies made by ears that caught the jingle but not the music, the body but not the soul, of the Jubilee songs. It is thus clear that the study of Negro religion is not only a vital part of the history of the Negro in America, but no uninteresting part of American history.

The Negro church of to-day is the social centre of Negro life in the United States, and the most characteristic expression of African character. Take a typical church in a small Virginia town: it is the "First Baptist"—a roomy brick edifice seating five hundred or more persons, tastefully finished in Georgia pine, with a carpet, a small organ, and stained-glass windows. Underneath is a large assembly room with benches. This building is the central club-house of a community of a thousand or more Negroes. Various organizations meet here,—the church proper, the Sunday-school, two or three insurance societies, women's societies, secret societies, and mass meetings of various kinds. Entertainments, suppers, and lectures are held beside the five or six regular weekly religious services. Considerable sums of money are collected and expended here, employment is found for the idle, strangers are introduced, news is disseminated and charity distributed. At the same time this social, intellectual, and economic centre is a religious centre of great power. Depravity, Sin, Redemption, Heaven, Hell, and Damnation are preached twice a Sunday after the crops are laid by; and few indeed of the community have the hardihood to withstand conversion. Back of this more formal religion, the Church often stands as a real conserver of morals, a strengthener of family life, and the final authority on what is Good and Right.

Thus one can see in the Negro church to-day, reproduced in microcosm, all the great world from which the Negro is cut off by color-prejudice and social condition. In the great city churches the same tendency is noticeable and in many respects emphasized. A great church like the Bethel of Philadelphia has over eleven hundred members, an edifice seating fifteen hundred persons and valued at one hundred thousand dollars, an annual budget of five thousand dollars, and a government consisting of a pastor with several assisting local preachers, an executive and legislative board, financial boards and tax collectors; general church meetings for making laws; sub-divided groups led by class leaders, a company of militia, and twenty-four auxiliary societies. The activity of a church like this is immense and far-reaching, and the bishops who preside over these organizations throughout the land are among the most powerful Negro rulers in the world.

Such churches are really governments of men, and consequently a little investigation reveals the curious fact that, in the South, at least, practically every American Negro is a church member. Some, to be sure, are not regularly enrolled, and a few do not habitually attend services; but, practically, a proscribed people must have a social centre and that centre for this people is the Negro church. The census of 1890 showed nearly twenty-four thousand Negro churches in the country, with a total enrolled membership of over two and a half millions, or ten actual church members to every twenty eight persons, and in some Southern states one in every two persons. Besides these there is the large number who, while not enrolled as members, attend and take part in many of the activities of the church. There is an organized Negro church for every sixty black families in the nation, and in some States for every forty families, owning, on an average, a thousand dollars' worth of property each, or nearly twenty-six million dollars in all.

Such, then, is the large development of the Negro church since Emancipation. The question now is, What have been the successive steps of this social history and what are the present tendencies? First, we must realize that no such institution as the Negro church could rear itself without definite historical foundations. These foundations we can find if we remember that the social history of the Negro did not start in America. He was from a definite social environment—the polygamous clan life under the headship of the chief and the potent influence of the priest. His religion was nature-worship, with profound belief in invisible surrounding influences, good and bad, and his worship was through incantation and sacrifice. The first rude change in this life was the slave ship and the West Indian sugar-fields. The plantation organization replaced the clan and tribe, and the white master replaced the chief with far greater and more despotic powers. Forced and long- continued

toil became the rule of life, the old ties of blood relationship and kinship disappeared, and instead of the family appeared a new polygamy and polyandry, which, in some cases, almost reached promiscuity. It was a terrific social revolution, and yet some traces were retained of the former group life, and the chief remaining institution was the Priest or Medicine-man. He early appeared on the plantation and found his function as the healer of the sick, the interpreter of the Unknown, the comforter of the sorrowing, the supernatural avenger of wrong, and the one who rudely but picturesquely expressed the longing, disappointment, and resentment of a stolen and oppressed people. Thus, as bard, physician, judge, and priest, within the narrow limits allowed by the slave system, rose the Negro preacher, and under him the first church was not at first by any means Christian nor definitely organized; rather it was an adaptation and mingling of heathen rites among the members of each plantation, and roughly designated as Voodooism. Association with the masters, missionary effort and motive of expediency gave these rites an early veneer of Christianity, and after the lapse of many generations the Negro church became Christian.

Two characteristic things must be noticed in regard to the church. First, it became almost entirely Baptist and Methodist in faith; secondly, as a social institution it antedated by many decades the monogamic Negro home. From the very circumstances of its beginning, the church was confined to the plantation, and consisted primarily of a series of disconnected units; although, later on, some freedom of movement was allowed, still this geographical limitation was always important and was one cause of the spread of the decentralized and democratic Baptist faith among the slaves. At the same time, the visible rite of baptism appealed strongly to their mystic temperament. To-day the Baptist Church is still largest in membership among Negroes, and has a million and a half communicants. Next in popularity came the churches organized in connection with the white neighboring churches, chiefly Baptist and Methodist, with a few Episcopalian and others. The Methodists still form the second greatest denomination, with nearly a million members. The faith of these two leading denominations was more suited to the slave church from the prominence they gave to religious feeling and fervor. The Negro membership in other denominations has always been small and relatively unimportant, although the Episcopalian and Presbyterians are gaining among the more intelligent classes to-day, and the Catholic Church is making headway in certain sections. After Emancipation, and still earlier in the North, the Negro churches largely severed such affiliations as they had had with the white churches, either by choice or by compulsion. The Baptist churches became independent, but the Methodists were compelled early to unite for purposes of episcopal government. This gave rise to the great African Methodist Church, the greatest Negro organization in the world, to the Zion Church and the Colored Methodist, and to the black conferences and churches in this and other denominations.

The second fact noted, namely, that the Negro church antedates the Negro home, leads to an explanation of much that is paradoxical in this communistic institution and in the morals of its members. But especially it leads us to regard this institution as peculiarly the expression of the inner ethical life of a people in a sense seldom true elsewhere. Let us turn, then, from the outer physical development of the church to the more important inner ethical life of the people who compose it. The Negro has already been pointed out many times as a religious animal—a being of that deep emotional nature which turns instinctively toward the supernatural. Endowed with a rich tropical imagination and a keen, delicate appreciation of Nature, the transplanted African lived in a world animate with gods and devils, elves and witches; full of strange influences,—of Good to be implored, of Evil to be propitiated. Slavery, then, was to him the dark triumph of Evil over him. All the hateful powers of the Underworld were striving against him, and a spirit of revolt and revenge filled his heart. He called up all the resources of heathenism to aid,—exorcism and witchcraft, the mysterious Obi worship with its barbarious rites, spells, and blood-sacrifice even, now and then, of human victims. Weird midnight orgies and mystic conjurations were invoked, the witch-woman and the voodoo-priest became the centre of Negro group life, and that vein of vague superstition which characterizes the unlettered Negro even to-day was deepened and strengthened.

In spite, however, of such success as that of the fierce Maroons, the Danish blacks, and others, the spirit of revolt gradually died away under the untiring energy and superior strength of the slave masters. By the middle of the eighteenth century the black slave had sunk, with hushed murmurs, to his place at the bottom of a new economic system, and was unconsciously ripe for a new philosophy of life. Nothing suited his condition then better than the doctrines of passive submission embodied in the new newly learned Christianity. Slave masters early

realized this, and cheerfully aided religious propaganda within certain bounds. The long system of repression and degradation of the Negro tended to emphasize the elements of his character which made him a valuable chattel: courtesy became humility, moral strength degenerated into submission, and the exquisite native appreciation of the beautiful became an infinite capacity for dumb suffering. The Negro, losing the joy of this world, eagerly seized upon the offered conceptions of the next; the avenging Spirit of the Lord enjoining patience in this world, under sorrow and tribulation until the Great Day when He should lead His dark children home,—this became his comforting dream. His preacher repeated the prophecy, and his bards sang,—

"Children, we all shall be free
When the Lord shall appear!"

This deep religious fatalism, painted so beautifully in "Uncle Tom," came soon to breed, as all fatalistic faiths will, the sensualist side by side with the martyr. Under the lax moral life of the plantation, where marriage was a farce, laziness a virtue, and property a theft, a religion of resignation and submission degenerated easily, in less strenuous minds, into a philosophy of indulgence and crime. Many of the worst characteristics of the Negro masses of to-day had their seed in this period of the slave's ethical growth. Here it was that the Home was ruined under the very shadow of the Church, white and black; here habits of shiftlessness took root, and sullen hopelessness replaced hopeful strife.

With the beginning of the abolition movement and the gradual growth of a class of free Negroes came a change. We often neglect the influence of the freedman before the war, because of the paucity of his numbers and the small weight he had in the history of the nation. But we must not forget that his chief influence was internal,—was exerted on the black world; and that there he was the ethical and social leader. Huddled as he was in a few centres like Philadelphia, New York, and New Orleans, the masses of the freedmen sank into poverty and listlessness; but not all of them. The free Negro leader early arose and his chief characteristic was intense earnestness and deep feeling on the slavery question. Freedom became to him a real thing and not a dream. His religion became darker and more intense, and into his ethics crept a note of revenge, into his songs a day of reckoning close at hand. The "Coming of the Lord" swept this side of Death, and came to be a thing to be hoped for in this day. Through fugitive

slaves and irrepressible discussion this desire for freedom seized the black millions still in bondage, and
became their one ideal of life. The black bards caught
new notes, and sometimes even dared to sing,—

"O Freedom, O Freedom,
O Freedom over me!
Before I'll be a slave
I'll be buried in my grave,
And go home to my Lord
And be free."

For fifty years Negro religion thus transformed itself
and identified itself with the dream of Abolition, until
that which was a radical fad in the white North and an
anarchistic plot in the white South had become a
religion to the black world.

Thus, when Emancipation finally came, it seemed to
the freedman a literal Coming of the Lord. His fervid
imagination was stirred as never before, by the tramp of
armies, the blood and dust of battle, and the wail and
whirl of social upheaval. He stood dumb and motion-
less before the whirlwind: what had he to do with it?
Was it not the Lord's doing, and marvellous in his eyes?
Joyed and bewildered with what came, he stood awaiting new wonders till the inevitable Age of Reaction
swept over the nation and brought the crisis of to-day.

It is difficult to explain clearly the present critical
stage of Negro religion. First, we must remember that
living as the blacks do in close contact with a great
modern nation, and sharing, although imperfectly, the
soul-life of that nation, they must necessarily be
affected more or less directly by all the religious and
ethical forces that are to-day moving the United States.
These questions and movements are, however, over-
shadowed and dwarfed by the (to them) all-important
question of their civil, political, and economic status.
They must perpetually discuss the "Negro Problem,"—
must live, move, and have their being in it, and interpret
all else in its light or darkness. With this come, too,
peculiar problems of their inner life,—of the status of
women, the maintenance of Home, the training of children, the accumulation of wealth, and the prevention of
crime. All this must mean a time of intense ethical fer-
ment, of religious heart-searching and intellectual
unrest. From the double life every American Negro
must live, as a Negro and as an American, as swept on
by the current of the nineteenth while yet struggling in
the eddies of the fifteenth century,—from this must

arise a painful self-consciousness, an almost morbid
sense of personality and a moral hesitancy which is
fatal to self-confidence. The worlds within and without
the Veil of Color are changing, and changing rapidly,
but not at the same rate, not in the same way; and this
must produce a peculiar wrenching of the soul, a peculiar sense of doubt and bewilderment. Such a double
life, with double thoughts, double duties, and double
social classes, must give rise to double words and double ideals, and tempt the mind to pretence or revolt, to
hypocrisy or radicalism.

In some such doubtful words and phrases can one
perhaps most clearly picture the peculiar ethical paradox that faces the Negro of to-day and is tingeing and
changing his religious life. Feeling that his rights and
his dearest ideals are being trampled upon, that the
public conscience is ever more deaf to his righteous
appeal, and that all the reactionary forces of prejudice,
greed, and revenge are daily gaining new strength and
fresh allies, the Negro faces no enviable dilemma.
Conscious of his impotence, and pessimistic, he often
becomes bitter and vindictive; and his religion, instead
of a worship, is a complaint and a curse, a wail rather
than a hope, a sneer rather than a faith. On the other
hand, another type of mind, shrewder and keener and
more tortuous too, sees in the very strength of the anti-
Negro movement its patent weaknesses, and with
Jesuitic casuistry is deterred by no ethical considerations in the endeavor to turn this weakness to the
black man's strength. Thus we have two great and
hardly reconcilable streams of thought and ethical
strivings; the danger of the one lies in anarchy, that of
the other in hypocrisy. The one type of Negro stands
almost ready to curse God and die, and the other is too
often found a traitor to right and a coward before
force; the one is wedded to ideals remote, whimsical,
perhaps impossible of realization; the other forgets
that life is more than meat and the body more than rai-
ment. But, after all, is not this simply the writhing of
the age translated into black, the triumph of the Lie
which today, with its false culture, faces the hideous-
ness of the anarchist assassin?

To-day the two groups of Negroes, the one in the
North, the other in the South, represent these divergent
ethical tendencies, the first tending toward radicalism,
the other toward hypocritical compromise. It is no idle
regret with which the white South mourns the loss of the
old-time Negro,—the frank, honest, simple old servant
who stood for the earlier religious age of submission and

humility. With all his laziness and lack of many elements of true manhood, he was at least open-hearted, faithful, and sincere. To-day he is gone, but who is to blame for his going? Is it not those very persons who mourn for him? Is it not the tendency, born of Reconstruction and Reaction, to found a society on lawlessness and deception, to tamper with the moral fibre of a naturally honest and straight-forward people until the whites threaten to become ungovernable tyrants and the blacks criminals and hypocrites? Deception is the natural defence of the weak against the strong, and the South used it for many years against its conquerors; to-day it must be prepared to see its black proletariat turn that same two-edged weapon against itself. And how natural this is! The death of Denmark Vesey and Nat Turner proved long since to the Negro the present hopelessness of physical defence. Political defence is becoming less and less available, and economic defence is still only partially effective. But there is a patent defence at hand,—the defence of deception and flattery, of cajoling and lying. It is the same defence which peasants of the Middle Age used and which left its stamp on their character for centuries. To-day the young Negro of the South who would succeed cannot be frank and outspoken, honest and self-assertive, but rather he is daily tempted to be silent and wary, politic and sly; he must flatter and be pleasant, endure petty insults with a smile, shut his eyes to wrong; in too many cases he sees positive personal advantage in deception and lying. His real thoughts, his real aspirations, must be guarded in whispers; he must not criticise, he must not complain. Patience, humility, and adroitness must, in these growing black youth, replace impulse, manliness, and courage. With this sacrifice there is an economic opening, and perhaps peace and some prosperity. Without this there is riot, migration, or crime. Nor is this situation peculiar to the Southern United States, is it not rather the only method by which undeveloped races have gained the right to share modern culture? The price of culture is a Lie.

On the other hand, in the North the tendency is to emphasize the radicalism of the Negro. Driven from his birthright in the South by a situation at which every fibre of his more outspoken and assertive nature revolts, he finds himself in a land where he can scarcely earn a decent living amid the harsh competition and the color discrimination. At the same time, through schools and periodicals, discussions and lectures, he is intellectually quickened and awakened. The soul, long pent up and dwarfed, suddenly expands in new-found freedom. What wonder that every tendency is to excess, radical complaint, radical remedies, bitter denunciation or angry silence. Some sink, some rise. The criminal and the sensualist leave the church for the gambling-hell and the brothel, and fill the slums of Chicago and Baltimore; the better classes segregate themselves from the group-life of both white and black, and form an aristocracy, cultured but pessimistic, whose bitter criticism stings while it points out no way of escape. They despise the submission and subserviency of the Southern Negroes, but offer no other means by which a poor and oppressed minority can exist side by side with its masters. Feeling deeply and keenly the tendencies and opportunities of the age in which they live, their souls are bitter at the fate which drops the Veil between; and the very fact that this bitterness is natural and justifiable only serves to intensify it and make it more maddening.

Between the two extreme types of ethical attitude which I have thus sought to make clear wavers the mass of the millions of Negroes, North and South; and their religious life and activity partake of this social conflict within their ranks. Their churches are differentiating,— now into groups of cold, fashionable devotees, in no way distinguishable from similar white groups save in color of skin; now into large social and business institutions catering to the desire for information and amusement of their members, warily avoiding unpleasant questions both within and without the black world, and preaching in effect if not in word: *Dum vivimus, vivamus.*

But back of this still broods silently the deep religious feeling of the real Negro heart, the stirring, unguided might of powerful human souls who have lost the guiding star of the past and seek in the great night a new religious ideal. Some day the Awakening will come, when the pent-up vigor of ten million souls shall sweep irresistibly toward the Goal, out of the Valley of the Shadow of Death, where all that makes life worth living—Liberty, Justice, and Right—is marked "For White People Only."

Introduction to "The Souls of White Folk"

Throughout his long life, Du Bois's basic argument remained unchanged. However, after World War I, Du Bois stepped up his demands for emancipation, and his work acquired a more aggressive tone. African Americans coming home from battle in Europe were incensed to find that despite their wartime contributions, they were treated, at best, as second-class citizens. In addition, between 1910 and 1920, black literacy and militancy had been increasing, and a "New Negro" movement was taking hold. Membership in the NAACP nearly doubled between 1918 and 1919, and there were dozens of riots in scores of American cities. In a single riot in Chicago, 15 whites and 23 blacks were killed, and more than 500 were injured (Lewis 2000:8; Rudwick [1960] 1982:237).

"The Souls of White Folk" was originally published in 1910, but it was revised with references to World War I for re-publication in *Darkwater* in 1920. *Darkwater* set off a tremendous public storm, in large part because Du Bois asserted that "the dark was preparing to meet the white world in battle" (Rudwick [1960] 1982:242). Of course, Du Bois had already been stirring up controversy with his provocative editorials in the *Crisis*. For instance, in the May 1919 issue of that journal (which sold approximately 100,000 copies), Du Bois maintained,

> By the God of Heaven, we are cowards and jackasses if now that the war is over, we do not marshal every ounce of our brain and brawn to fight a sterner, longer, more unbending battle against the forces of hell in our own land. (ibid.:238)

Sadly (albeit predictably), many whites reacted hysterically to the "New Negro" movement. In 1919, 77 African Americans were lynched in various parts of the country (Rudwick [1960] 1982:237). Racial tensions completely exploded in rural Arkansas in October 1919 after gunfire from black sharecroppers who had been meeting in a church left a white deputy sheriff dead and several white citizens wounded. In response, white planters and farmers instigated a 7-day frenzy of violence, during which time nearly 200 people were killed. Within 2 days of the bloodbath, more than 1,000 African Americans were rounded up; 12 were hanged for "conspiracy to seize control of the county," and 67 received prison terms of 5 to 25 years. Meanwhile, because of the five white deaths, the whites' actions were deemed legal (Lewis 2000:8).

Thus, although you will see similarities between "The Souls of White Folk" and *The Souls of Black Folk* (most obviously, both explore the dual workings of race and class), "The Souls of White Folk" reads far more militantly. Another notable feature of this essay is that Du Bois reverses the gaze of racial domination. Here it is *white* consciousness (or the lack thereof) that is explored. Indeed, "The Souls of White Folk" has been called "the first major analysis in Western intellectual history to probe deeply White identity and the meaning of Whiteness" (Feagin 2003:11).

There are several important points in this essay worth highlighting. First, Du Bois suggests that white privilege, ironically, is invisible to whites. Indeed, Du Bois contends that while African Americans have a "double consciousness," whites have no racial consciousness at all. In other words, just as the fish in the fishbowl is the last to know that it lives in water, the white man does not realize what it is, and what it entails, to be white. However, Du Bois suggests that, unbeknownst to whites, African Americans *can* see what it means to be white. Blacks' "clairvoyance" comes from their servile position. As servants in one form or another, blacks are exposed to the intimate details of whites' lives; hence they see whites as they really are.

Most important, what blacks see is that whites typically practice the very opposite of what they preach. Du Bois condemns whites not only for their hypocrisy, but also for their delusion. Thus, Du Bois declares white Christianity "a miserable failure" because the number of whites who actually practice "the democracy and unselfishness of Jesus Christ" is so small as to be farcical. So too, Du Bois finds the United States' call to make "the World Safe for Democracy" ludicrous. Asks Du Bois, "How could America condemn in Germany that which she commits, just as brutally, within her own borders?" ([1920] 2003d:59).

Of course, this is the very same message preached by Dr. Martin Luther King, Jr.—albeit more positively—30 years later. Dr. King challenged America to live up to its own moral and democratic code. He shamed America by showing that it did not live up to its own sacred ideals—most important, "freedom and justice for all." Indeed, in his final tribute to Du Bois, Dr. King still challenged his audience "to be dissatisfied":

Let us be dissatisfied until every man can have food and material necessities for his body, culture, and education for his mind, freedom and human dignity for his spirit. Let us be dissatisfied until rat-infested, vermin-filled slums will be a thing of a dark past and every family will have a decent sanitary house in which to live. Let us be dissatisfied until the empty stomachs of Mississippi are filled and the idle industries of Appalachia are revitalized. Let us be dissatisfied until brotherhood is no longer a meaningless word at the end of a prayer but the first order of business on every legislative agenda. Let us be dissatisfied until our brother of the Third World—Asia, Africa, and Latin America—will no longer be the victim of imperialist exploitation but will be lifted from the long night of poverty, illiteracy, and disease. Let us be dissatisfied until this pending cosmic elegy will be transformed into a creative psalm of peace and "justice will roll down like waters from a mighty stream." (1970:183)

"The Souls of White Folk" From *Darkwater: Voice From Within the Veil* (1920)

W. E. B. Du Bois

High in the tower, where I sit above the loud complaining of the human sea, I know many souls that toss and whirl and pass, but none there are that intrigue me more than the Souls of White Folk.

Of them I am singularly clairvoyant. I see in and through them. I view them from unusual points of vantage. Not as a foreigner do I come, for I am native, not foreign, bone of their thought and flesh of their language. Mine is not the knowledge of the traveler or the colonial composite of dear memories, words and wonder. Nor yet is my knowledge that which servants have of masters, or mass of class, or capitalist of artisan. Rather I see these souls undressed and from the back and side. I see the working of their entrails. I know their thoughts and they know that I know. This knowledge makes them now embarrassed, now furious! They deny my right to live and be and call me misbirth! My word is to them mere bitterness and my soul, pessimism. And yet as they preach and strut and shout and threaten, crouching as they clutch at rags of facts and fancies to hide their nakedness, they go twisting, flying by my tired eyes and I see them ever stripped,—ugly, human.

The discovery of personal whiteness among the world's peoples is a very modern thing,—a nineteenth and twentieth century matter, indeed. The ancient world would have laughed at such a distinction. The Middle Age regarded skin color with mild curiosity; and even up into the eighteenth century we were hammering our national manikins into one, great, Universal Man, with fine frenzy which ignored color and race even more than birth. Today we have changed all that, and the world in a sudden, emotional conversion has discovered that it is white and by that token, wonderful!

This assumption that of all the hues of God whiteness alone is inherently and obviously better than brownness or tan leads to curious acts; even the sweeter souls of the dominant world as they discourse with me on weather, weal, and woe are continually playing above their actual words an obligato of tune and tone, saying:

"My poor, un-white thing! Weep not nor rage. I know, too well, that the curse of God lies heavy on you. Why? That is not for me to say, but be brave! Do your work in your lowly sphere, praying the good Lord that into heaven above, where all is love, you may, one day, be born—white!"

I do not laugh. I am quite straight-faced as I ask soberly:

"But what on earth is whiteness that one should so desire it?" Then always, somehow, some way, silently but clearly, I am given to understand that whiteness is the ownership of the earth forever and ever, Amen!

Now what is the effect on a man or a nation when it comes passionately to believe such an extraordinary dictum as this? That nations are coming to believe it is manifest daily. Wave on wave, each with increasing virulence, is dashing this new religion of whiteness on the shores of our time. Its first effects are funny: the strut of the Southerner, the arrogance of the Englishman amuck, the whoop of the hoodlum who vicariously leads your mob. Next it appears dampening generous enthusiasm in what we once counted glorious; to free the slave is discovered to be tolerable only in so far as

SOURCE: From *Dark Water: Voices from Within the Veil* by W. E. B. DuBois, pp. 55–60 and 65–70 (Amherst, NY: Prometheus Books). Published in 2003 (Classics in Black Studies Series).

it freed his master! Do we sense somnolent writhings in black Africa or angry groans in India or triumphant banzais in Japan? "To your tents, O Israel!" These nations are not white!

After the more comic manifestations and the chilling of generous enthusiasm come subtler, darker deeds. Everything considered, the title to the universe claimed by White Folk is faulty. It ought, at least, to look plausible. How easy, then, by emphasis and omission to make children believe that every great soul the world ever saw was a white man's soul; that every great thought the world ever knew was a white man's thought; that every great deed the world ever did was a white man's deed; that every great dream the world ever sang was a white man's dream. In fine, that if from the world were dropped everything that could not fairly be attributed to White Folk, the world would, if anything, be even greater, truer, better than now. And if all this be a lie, is it not a lie in a great cause?

Here it is that the comedy verges to tragedy. The first minor note is struck, all unconsciously, by those worthy souls in whom consciousness of high descent brings burning desire to spread the gift abroad,—the obligation of nobility to the ignoble. Such sense of duty assumes two things: a real possession of the heritage and its frank appreciation by the humble-born. So long, then, as humble black folk, voluble with thanks, receive barrels of old clothes from lordly and generous whites, there is much mental peace and moral satisfaction. But when the black man begins to dispute the white man's title to certain alleged bequests of the Fathers in wage and position, authority and training; and when his attitude toward charity is sullen anger rather than humble jollity; when he insists on his human right to swagger and swear and waste,—then the spell is suddenly broken and the philanthropist is ready to believe that Negroes are impudent, that the South is right, and that Japan wants to fight America.

After this the descent to Hell is easy. On the pale, white faces which the great billows whirl upward to my tower I see again and again, often and still more often, a writing of human hatred, a deep and passionate hatred, vast by the very vagueness of its expressions. Down through the green waters, on the bottom of the world, where men move to and fro, I have seen a man— an educated gentleman—grow livid with anger because a little, silent, black woman was sitting by herself in a Pullman car. He was a white man. I have seen a great, grown man curse a little child, who had wandered into the wrong waiting-room, searching for its mother: "Here, you damned black—" He was white. In Central Park I have seen the upper lip of a quiet, peaceful man curl back in a tigerish snarl of rage because black folk rode by in a motor car. He was a white man. We have seen, you and I, city after city drunk and furious with ungovernable lust of blood; mad with murder, destroying, killing, and cursing; torturing human victims because somebody accused of crime happened to be of the same color as the mob's innocent victims and because that color was not white! We have seen,—Merciful God! in these wild days and in the name of Civilization, Justice, and Motherhood,— what have we not seen, right here in America, of orgy, cruelty, barbarism, and murder done to men and women of Negro descent.

Up through the foam of green and weltering waters wells this great mass of hatred, in wilder, fiercer violence, until I look down and know that today to the millions of my people no misfortune could happen,—of death and pestilence, failure and defeat—that would not make the hearts of millions of their fellows beat with fierce, vindictive joy! Do you doubt it? Ask your own soul what it would say if the next census were to report that half of black America was dead and the other half dying.

Unfortunate? Unfortunate. But where is the misfortune? Mine? Am I, in my blackness, the sole sufferer? I suffer. And yet, somehow, above the suffering, above the shackled anger that beats the bars, above the hurt that crazes there surges in me a vast pity,—pity for a people imprisoned and enthralled, hampered and made miserable for such a cause, for such a phantasy!

Conceive this nation, of all human peoples, engaged in a crusade to make the "World Safe for Democracy"! Can you imagine the United States protesting against Turkish atrocities in Armenia, while the Turks are silent about mobs in Chicago and St. Louis; what is Louvain compared with Memphis, Waco, Washington, Dyersburg, and Estill Springs? In short, what is the black man but America's Belgium, and how could America condemn in Germany that which she commits, just as brutally, within her own borders?

A true and worthy ideal frees and uplifts a people; a false ideal imprisons and lowers. Say to men, earnestly and repeatedly: "Honesty is best, knowledge is power; do unto others as you would be done by." Say this and act it and the nation must move toward it, if not to it. But say to a people: "The one virtue is to be white," and the people rush to the inevitable conclusion, "Kill the 'nigger'!"

Is not this the record of present America? Is not this its headlong progress? Are we not coming more and more, day by day, to making the statement "I am white," the one fundamental tenet of our practical morality? Only when this basic, iron rule is involved is our defense of right nationwide and prompt. Murder may swagger, theft may rule and prostitution may flourish and the nation gives but spasmodic, intermittent and lukewarm attention. But let the murderer be black or the thief brown or the violator of womanhood have a drop of Negro blood, and the righteousness of the indignation sweeps the world. Nor would this fact make the indignation less justifiable did not we all know that it was blackness that was condemned and not crime.

In the awful cataclysm of World War, where from beating, slandering, and murdering us the white world turned temporarily aside to kill each other, we of the Darker Peoples looked on in mild amaze. . . .

The European world is using black and brown men for all the uses which men know. Slowly but surely white culture is evolving the theory that "darkies" are born beasts of burden for white folk. It were silly to think otherwise, cries the cultured world, with stronger and shriller accord. The supporting arguments grow and twist themselves in the mouths of merchant, scientist, soldier, traveler, writer, and missionary: Darker peoples are dark in mind as well as in body; of dark, uncertain, and imperfect descent; of frailer, cheaper stuff; they are cowards in the face of mausers and maxims; they have no feelings, aspirations, and loves; they are fools, illogical idiots,—"half-devil and half-child."

Such as they are civilization must, naturally, raise them, but soberly and in limited ways. They are not simply dark white men. They are not "men" in the sense that Europeans are men. To the very limited extent of their shallow capacities lift them to be useful to whites, to raise cotton, gather rubber, fetch ivory, dig diamonds,—and let them be paid what men think they are worth—white men who know them to be well-nigh worthless.

Such degrading of men by men is as old as mankind and the invention of no one race or people. Ever have men striven to conceive of their victims as different from the victors, endlessly different, in soul and blood, strength and cunning, race and lineage. It has been left, however, to Europe and to modern days to discover the eternal worldwide mark of meanness,—color!

Such is the silent revolution that has gripped modern European culture in the later nineteenth and twentieth centuries. Its zenith came in Boxer times: White supremacy was all but world-wide, Africa was dead, India conquered, Japan isolated, and China prostrate, while white America whetted her sword for mongrel Mexico and mulatto South America, lynching her own Negroes the while. Temporary halt in this program was made by little Japan and the white world immediately sensed the peril of such "yellow" presumption! What sort of a world would this be if yellow men must be treated "white"? Immediately the eventual overthrow of Japan became a subject of deep thought and intrigue, from St. Petersburg to San Francisco, from the Key of Heaven to the Little Brother of the Poor.

The using of men for the benefit of masters is no new invention of modern Europe. It is quite as old as the world. But Europe proposed to apply it on a scale and with an elaborateness of detail of which no former world ever dreamed. The imperial width of the thing,—the heaven-defying audacity—makes its modern newness.

The scheme of Europe was no sudden invention, but a way out of long-pressing difficulties. It is plain to modern white civilization that the subjection of the white working classes cannot much longer be maintained. Education, political power, and increased knowledge of the technique and meaning of the industrial process are destined to make a more and more equitable distribution of wealth in the near future. The day of the very rich is drawing to a close, so far as individual white nations are concerned. But there is a loophole. There is a chance for exploitation on an immense scale for inordinate profit, not simply to the very rich, but to the middle class and to the laborers. This chance lies in the exploitation of darker peoples. It is here that the golden hand beckons. Here are no labor unions or votes or questioning onlookers or inconvenient consciences. These men may be used down to the very bone, and shot and maimed in "punitive" expeditions when they revolt. In these dark lands "industrial development" may repeat in exaggerated form every horror of the industrial history of Europe, from slavery and rape to disease and maiming, with only one test of success,—dividends!

This theory of human culture and its aims has worked itself through warp and woof of our daily thought with a thoroughness that few realize. Everything great, good, efficient, fair, and honorable is "white"; everything mean, bad, blundering, cheating, and dishonorable is "yellow"; a bad taste is "brown"; and the devil is "black." The changes of this theme are continually rung in picture and story, in newspaper heading and moving-picture, in sermon and school book, until, of course, the King can do no wrong,—a White Man is always right and a Black Man has no rights which a white man is bound to respect.

There must come the necessary despisings and hatreds of these savage half-men, this unclean *canaille* of the world—these dogs of men. All through the world this gospel is preaching. It has its literature, it has its priests, it has its secret propaganda and above all—it pays!

There's the rub,—it pays. Rubber, ivory, and palm-oil; tea, coffee, and cocoa; bananas, oranges, and other fruit; cotton, gold, and copper—they, and a hundred other things which dark and sweating bodies hand up to the white world from their pits of slime, pay and pay well, but of all that the world gets the black world gets only the pittance that the white world throws it disdainfully.

Small wonder, then, that in the practical world of things-that-be there is jealousy and strife for the possession of the labor of dark millions, for the right to bleed and exploit the colonies of the world where this golden stream may be had, not always for the asking, but surely for the whipping and shooting. It was this competition for the labor of yellow, brown, and black folks that was the cause of the World War. Other causes have been glibly given and other contributing causes there doubtless were, but they were subsidiary and subordinate to this vast quest of the dark world's wealth and toil.

Colonies, we call them, these places where "niggers" are cheap and the earth is rich; they are those outlands where like a swarm of hungry locusts white masters may settle to be served as kings, wield the lash of slave-drivers, rape girls and wives, grow as rich as Croesus and send homeward a golden stream. They belt the earth, these places, but they cluster in the tropics, with its darkened peoples: in Hong Kong and Anam, in Borneo and Rhodesia, in Sierra Leone and Nigeria, in Panama and Havana—these are the El Dorados toward which the world powers stretch itching palms.

Germany, at last one and united and secure on land, looked across the seas and seeing England with sources of wealth insuring a luxury and power which Germany could not hope to rival by the slower processes of exploiting her own peasants and workingmen, especially with these workers half in revolt, immediately built her navy and entered into a desperate competition for possession of colonies of darker peoples. To South America, to China, to Africa, to Asia Minor, she turned like a hound quivering on the leash, impatient, suspicious, irritable, with blood-shot eyes and dripping fangs, ready for the awful word. England and France crouched watchfully over their bones, growling and wary, but gnawing industriously, while the blood of the dark world whetted their greedy appetites. In the background, shut out from the highway to the seven seas, sat Russia and Austria, snarling and snapping at each other and at the last Mediterranean gate to the El Dorado, where the Sick Man enjoyed bad health, and where millions of serfs in the Balkans, Russia, and Asia offered a feast to greed wellnigh as great as Africa.

The fateful day came. It had to come. The cause of war is preparation for war; and of all that Europe has done in a century there is nothing that has equaled in energy, thought, and time her preparation for wholesale murder. The only adequate cause of this preparation was conquest and conquest, not in Europe, but primarily among the darker peoples of Asia and Africa; conquest, not for assimilation and uplift, but for commerce and degradation. For this, and this mainly, did Europe gird herself at frightful cost for war.

Discussion Questions

1. Explain Du Bois's concept of the color line. Using specific examples from Du Bois's work as well as real life, discuss how this concept encompasses both nonrational and rational motivation for action, at the individual as well as collective level.

2. Compare and contrast Simmel's notion of the stranger and Du Bois's notion of double consciousness. What similarities do these concepts share? How are they different?

3. Discuss Weber's distinction between status and class in the context of Du Bois's work. Is race *merely* a "status"? How so or why not?

4. Compare and contrast Du Bois's theory as to the oppression of African Americans with Gilman's theory as to the oppression of women. What similarities do you see in their arguments? What are the differences in these two theories of oppression?

5. Compare and contrast the Marxist notion of false consciousness with the delusion of whites, as described by Du Bois. Who are affected by each type of delusion, and how? Do you see any similarities in these two types of delusion? How so or why not? Discuss the Marxist dimensions of Du Bois's work in general.

8 George Herbert Mead (1863–1931)

George Herbert Mead

Key Concepts

- "I"
- "Me"
- Generalized other
- Meaning
- Significant symbols
- Taking the attitude of the other
- Play stage
- Game stage

The individual experiences himself as such, not directly, but only indirectly, from the particular standpoints of other individual members of the same social group, or from the generalized standpoint of the social group as a whole to which he belongs... and he becomes an object to himself only by taking the attitudes of other individuals toward himself.

—Mead ([1934] 1962:138)

It's time to get ready for tonight's date and you're trying on different clothes. As you put them on, what are you doing? More than likely, you're looking in a mirror and *thinking*—"How do I look in this? What impression about myself am I giving?" That is, in deciding what to wear, you're talking to yourself. But this conversation involves someone other than you. It's a dialogue between you and an *imagined* other, your date. Through the use of language, you are able to answer your own questions as if you were another person looking at yourself. Through language, you are able to see yourself as an object, just as your date does, and react to your appearance as she or he does. In an important sense, then, you experience your "self" socially, through taking the presumed attitudes of others toward your own behaviors, and on the basis of these attitudes, you control your own conduct.

What is the link between thinking and behavior? When you are thinking, what are you doing? What role does an individual's thinking play in the evolution of society? These are the questions that fascinated George Herbert Mead, the pragmatist philosopher whose views laid the foundation for the theory of symbolic interactionism, one of the major theoretical paradigms within sociology. In contrast to theorists who define society as a system of interrelated parts that promote consensus and stability or conflict and domination, Mead defined society as "generalized social attitudes" that continually *emerge* through coordinated interaction between individuals and groups. In this chapter, we examine Mead's perspective on the relationship between the individual and society through selections from his posthumously published book, *Mind, Self, and Society* (Mead [1934] 1962). Before turning to the primary texts, however, we first provide a brief sketch of Mead's biography and then an overview of the intellectual currents that informed his philosophy.

▦ A Biographical Sketch

Mead was born in 1863 in South Hadley, Massachusetts. At the age of seven, Mead moved to Ohio when his father, a Congregationalist minister, had accepted a position at Oberlin Theological Seminary. After graduating from Oberlin College in 1883, Mead left for the Northwest, where he worked for three years as a tutor and as a railroad surveyor. Undecided on his career options, Mead's next move took him to Harvard University, where he pursued graduate studies in philosophy and psychology. During this period, Mead encountered some of the thinkers who would prove pivotal in formulating his own philosophy of the mind and self—perhaps most important, the American pragmatist William James and the German philosopher Wilhelm Wundt.

Upon completing his master's degree at Harvard in 1888, Mead was hired as a philosophy instructor at the University of Michigan. Here, Mead kindled important intellectual relationships with eminent American scholars John Dewey and Charles Horton Cooley. Mead's stay at Michigan was short-lived, however. Two years after arriving, Mead left for the recently founded University of Chicago, where Dewey had been appointed to head the Department of Philosophy. Mead remained at Chicago until his death in 1931.

Mead's influence on sociology stemmed largely from his graduate course in social psychology. A dynamic speaker, Mead's lectures attracted students from an array of disciplines including, of course, sociology. Recognizing the significance of the ideas Mead presented in the classroom, some of his students had his lectures transcribed. Shortly after Mead's death, the transcriptions were edited and published in *Mind, Self, and Society* ([1934] 1962), a text that has since earned a central position in social theory literature.

Mead's academic career was complemented by his ongoing commitment to progressive social causes. Not unlike Marx, Mead envisioned the creation of a social utopia. Mead's adherence to liberal ideals (e.g., individualism, volunteerism), however, led him to advocate reform over revolution as the means for realizing a more just and democratic society. For Mead, such a society would be based on expanding universal rights that enabled individuals to pursue their own interests, while at the same time creating a more cooperative, united democratic order. At the root of this evolutionary process stood individuals coordinating their actions with one another through sympathetic understanding and the self-inhibition of behaviors detrimental to social cohesion. Rejecting a social gospel based on religious doctrine, Mead nevertheless was active in the reform movement of the early twentieth century. His optimistic faith in science and reason as means of creating a full-fledged, participatory democracy spurred his efforts in educational reform and feminist causes as well as his work with Jane Addams (1860–1935), the pioneer activist, reformer, and architect of professional social work.[1]

INTELLECTUAL INFLUENCES AND CORE IDEAS ▪▪

Mead's philosophy was informed by a number of intellectual currents. In this section, we briefly discuss three of the most significant theoretical influences on his work: pragmatism, behaviorism, and evolutionism.

Pragmatism

Pragmatism is a uniquely American philosophical doctrine developed primarily through the work of Charles S. Peirce (1839–1914), William James (1842–1910), and John Dewey (1859–1952). Unlike most European philosophical schools, pragmatism was not oriented toward uncovering general "Truths" or formal principles of human behaviors or desires. Instead, it was offered as a method or instrument for studying the meaning of behavior. More specifically, the early pragmatists argued that the meaning of objects and actions lie in their practical aspects, that is, how they allow individuals to adapt to and solve the problems they confront. "Truth," or meaning, is thus found in what "works." It is a result, not a fixed ideal or an intrinsic feature of a given object, event, or situation.

For instance, consider something as ordinary as water. What *is* water? Is it fundamentally only a combination of hydrogen and oxygen? From the point of view of pragmatism, what water is—its meaning—depends on the situation in which it is encountered and thus the use to which it will be put. For the chemist, water *is* a solvent used to dissolve substances. For the athlete, water *is* a liquid used for hydration or as a means for exercise or sport. For the firefighter, water *is* a tool used for extinguishing fires, or even perhaps a potential danger, depending on the type of fire. For the gardener, water *is* a necessary ingredient that must be used in order to grow healthy plants. The point is that the meaning of objects or social interaction is rooted in action, in everyday practical conduct, that is, the uses that are made of them as individuals go about the business of constructing their behavior.

Although such a position may not appear to be particularly controversial, it is based on a host of assumptions regarding the nature of the individual and his relationship to the external world. Particularly important in this regard is the indebtedness of Mead (and pragmatists more generally) to German idealism and the philosophy of Immanuel Kant (1724–1804) and Georg Hegel (1770–1831). It was Kant who proposed the existence of a "twofold self or "double I": in being conscious of one's self, the individual "splits" into "the I as subject and the I as object" (Wrong 1994:61, 62). Indeed, for Kant and many other philosophers, it is this capacity for "reflexivity," or the ability to experience one's self simultaneously as a thinking, perceiving subject, *and* as a perceived object, that separates humans from all other animal forms. The influence of these ideas on Mead is most apparent in his conversion of Kant's twofold self into a distinction between the "I" and the "me" (see a discussion of Mead's distinction below).

Equally important, from a Kantian view, there is no reality separable from the perceiving subject. Instead, the external world of objects and events exists only through the conscious apprehension of them. The world is thus not something "out there" to be experienced by the subject, but rather is "a task to be

---

[1]Lewis Coser's (1977) *Masters of Sociological Thought* and David Ashley and Michael Orenstein's (1998) *Sociological Theory: Classical Statements* were particularly useful for preparing this brief biography.

accomplished" (Coser 1977:350). Moreover, consciousness itself cannot be said to exist unless there is some object in the external world that one notes or becomes aware of. The act of knowing (subject) and the known (object) are thus intimately connected.[2]

Behaviorism

The pragmatist view of consciousness and its importance for guiding our actions is not without its detractors. Of particular relevance here is the behaviorist branch of psychology, in response to which Mead fashioned his own theoretical framework that he labeled "social behaviorism." Psychological behaviorism is a resolutely empirical branch of psychology that focuses solely on observable actions. Indeed, its proponents argue that only overt behaviors are open to scientific investigation. From this it follows that because states of mind, feelings, desires, and thinking cannot be observed, they cannot be studied scientifically. As a result, behaviorism is confined to studying the links between visible stimuli and the learned responses that are associated with them. As you will read, some of Mead's most significant contributions stem from his counterargument that thinking is indeed a behavior and thus available for analysis.

During Mead's time, the leading figure in behaviorism was the psychologist **John B. Watson** (1878–1958). Watson argued that human behavior differed little in principle from animal behavior—both could be explained and predicted on the basis of laws that govern the association of behavioral responses to external stimuli. Thus, as a *learned* response to a specific stimulus, a rabbit's retreat from the path of a snake is no different in kind from a person's efforts to perform well in her job. Watson's conviction regarding the power of behaviorism to unlock the secrets of human development and action is perhaps nowhere more plainly asserted than in his claim:

> Give me a dozen healthy infants, well-formed, and my own specified world to bring them up in and I'll guarantee to take any one at random and train him to become any type of specialist I might select—doctor, lawyer, artist, merchant-chief, and yes, even beggarman and thief, regardless of his talents, penchants, tendencies, abilities, vocations, and race of his ancestors. (Watson [1924] 1966:104)

It is against this picture of passive, nonreflexive individuals that Mead fashioned his social behaviorism. For Mead, the mind is not an ephemeral "black box" that is inaccessible to investigation. Instead, Mead viewed the mind as a *behavioral process* that entails a "conversation of significant gestures." In this internal conversation, an individual **takes the attitude of the other**, arousing in his own mind the same responses to his potential action that are aroused in the other person's.[3] Individuals, then, shape their actions on the basis of the imagined responses they attribute to others. Self-control of what we say and do is thus in actuality a form of social control as we check our behaviors—discarding some options while pursuing others—against the responses that we anticipate will be elicited from others.

---

[2]Although Mead borrowed from Kant's philosophy of knowledge, he did not adopt some of its fundamental principles. Perhaps most important is that Mead, unlike Kant, did not maintain that the mental categories (e.g., our experience of time, space, or beauty) through which our perceptions are organized and given meaning exist independently of the knowing subject. Instead, Mead argued that such categories are learned and emerge through the process of interaction with others.

[3]Depending on the situation, "others" might be a specific person (a sibling or neighbor), an identifiable group (classmates or coworkers), or the community at large, which Mead referred to as the "generalized other" (e.g., Southerners, Americans).

Significant Others

Charles Horton Cooley (1864–1929): The "Looking-Glass Self"

Born in Ann Arbor, Michigan, Charles Cooley spent most of his life revolved around that city's hub, the University of Michigan. It was here that his father served on the faculty of the Michigan Law School and where he would attend school, earning a degree in engineering and later a PhD in political economy and sociology. While pursuing his graduate degree, Cooley began teaching at the University of Michigan, where he remained on the faculty for his entire career. Mead's notion of "taking the attitude of the other" and the internal conversation that it entails shares much with Cooley's discussion of the "looking-glass self." Cooley argues that an individual's sense of self is developed through interaction with others. More specifically, there are three facets to the looking-glass self: "the imagination of our appearance to the other person, the imagination of his judgment of that appearance, and some sort of self-feeling, such as pride or mortification" (Cooley [1902] 1964:184). His underscoring of the role of imaginations in the construction of self-identity likewise informed his general perspective on society and sociology. Indeed, Cooley maintained that the "imaginations which people have of one another are the *solid* facts of society, and . . . to observe and interpret these must be the chief aim of sociology" (ibid.:121, emphasis in the original). It is this overly "mentalistic" understanding of society and of the task of sociology that has been the primary cause of the marginalization of Cooley's ideas in the discipline. Perhaps, in some ways, his focus on mental processes stemmed from his personal disposition. Cooley was a shy, semi-invalid whose list of ailments included a speech impediment. As a result, he led a quiet social life, preferring books and his own imaginative introspections to the company of others.

SOURCE: Coser (1977:314–18).

It is important to note that Mead's view of the relationship between self-control and social control is not based on a deterministic understanding of individual action. A deterministic view contends that social structures or divisions within society determine behaviors or attitudes, leaving little possibility for individuals to shape their life courses and counteract the destinies supposedly derived from broad social forces. Instead, Mead emphasized that the individual and the society to which she belongs are mutually dependent; each requires the other for its progressive evolution. Thus, for Mead, "social control, so far from tending to crush out the human individual or to obliterate his self-conscious individuality, is, on the contrary, actually constitutive of and inextricably associated with that individuality" (Mead [1934] 1962:255).

This view of the mind as a process of thinking that entails both self- and social control can be illustrated through any number of commonplace examples, for it is something we all continually experience. Consider, for instance, the internal conversation you engage in before asking someone out for a date, going on a job interview, determining how you will resolve an argument with a friend, or deciding whether to ask a question in class. In each case, you take the attitude of the other, viewing yourself as an object as other individuals do during interaction. So, you may ask yourself, "What type of movie should I take him or her to?" or "What music should we listen to when we're driving in the car?" Similarly, before raising your hand, you may think, "Will I seem stupid if I ask the professor this question? Will it will seem like I'm trying to score 'brownie' points?" The answers to such questions, and thus the behavior you intend to undertake, are shaped by the responses evoked in your mind. The responses are not entirely your own, however; instead, they reflect the assumed attitude that others take toward your behavior.

Evolutionism

A third major stream in Mead's framework is found in Darwinian evolutionism. The impact of evolutionism on Mead's thought comes in two forms. First, Mead, like Darwin, saw humans as the most advanced species. Humankind's superiority lies in the capacity to communicate symbolically. Through the use of language, we are able to take the attitude of the other and respond to our behavior in the same way as those to whom it is directed. Language thus allows us to see ourselves as an object. Symbolically casting off one's self as an object is the essence of self-consciousness, something other animals do not possess. Instead, other animals react to each other and their environment on the basis of instinct without anticipating the possible effects of their behavior.

The human capacity for self-consciousness, then, allows us to temporarily suspend our behaviors as we symbolically test different solutions to the situations we face. The ability to interact symbolically with others, ourselves, and the environment not only provides for an individual's adjustment to problems that may arise; it also creates the mechanism for social evolution. The second evolutionary theme in Mead's work is thus found in his view of social progress. Mead outlined both a practical and a utopian version of progress. His practical view highlighted the role of science as an instrument for addressing the problems that we confront in the natural and social environments. For Mead, science is "the evolutionary process grown self-conscious" (Mead [1936] 1964:23). While the process of natural selection may take centuries to enable a species of plant or animal to adapt better to its environment, science is a conscious endeavor to identify and solve problems. Whether it is the problem of crime, pollution, or illness, the scientific method generates alternative solutions that can then be tested. When the problem has been fixed (a vaccine is invented), evolution is furthered. Science thus allows for an orderly adaptation to and increasing advantage over our environment.

Mead's utopian view of social progress rests with his emphasis on language and the eventual creation of a democratic "universal society." For Mead (and other early social theorists), the problem to be solved in modern, differentiated societies lies in fashioning an orderly coordination of social activities. Yet, how can such coordination be achieved without stifling innovation and adjustment to a changing environment? How can social stability be maintained without diminishing individual expression and freedom? In short, how can a "true" democracy be attained in light of the opposing interests and unequal access to resources that make up a highly diverse society?

Social progress thus requires accommodating the need for both social control and individual creativity. The key to creating the ideal society lies in how these twin processes evolve. In modern, heterogeneous societies characterized by functional interdependence, individuals rely on expanding networks of increasingly distant others who perform specialized or differentiated tasks. The proper balance between the needs of the individual and the needs of society emerges when an individual or group asserts its "functional superiority" without seeking to dominate others. In doing so, the individual or group is able to exercise its superior skills and develop its distinctive sense of self while recognizing that one's unique contribution is made possible only in relation with others.

The democratic expression of functional superiority is effected through taking the attitude of the other and the self-conscious control of action it fosters. The ever-widening circle of social relations characteristic of modern societies expands the vocabulary of significant symbols allowing individuals and groups to take the attitude of a broader range of others. A more highly evolved empathetic understanding would then breed a "universal discourse" of shared meanings and with it an ideal democracy based on cooperation and a full recognition of the rights and duties to be freely and equally exercised by all. Mead notes the connection between universal discourse and the ideal society thusly:

> The ideal of human society is one which brings people so closely together in their interrelationships, so fully develops the necessary system of communication [i.e., significant symbols], that the individuals who exercise

their peculiar functions can take the attitude of those whom they affect. . . . If that system of communication could be made theoretically perfect, the individual would affect himself as he affects others in every way. That would be the ideal of communication. . . . The meaning of that which is said is here the same to one as it is to everybody else. Universal discourse is then the formal ideal of communication. If communication can be carried through and made perfect, then there would exist the kind of democracy . . . in which each individual would carry just the response in himself that he knows he calls out in the community. (Mead [1934] 1962:327)

Such a process can occur not only on an individual level within societies, but on a collective level across societies as whole societies become increasingly interdependent on each other for their continued development. Indeed, this more highly developed ability to take the attitude of the other found expression in the creation of the League of Nations shortly after World War I as countries realized their growing mutual dependence on one another for their own continued development. For Mead, the establishment of this international organization, "where every community recognizes every other community in the very process of asserting itself" ([1934] 1962:287), marked the beginning stages of a truly democratic, universal society. Unfortunately, modern history has rendered Mead's vision of a utopian democracy and its potential realization in organizations such as the League of Nations more and more a fantasy. The League dissolved in 1946, in no small measure due to the United States' unwillingness to join. While the establishment of the United Nations in 1945 perhaps has allowed for a lingering hope for the creation of a universal democracy, it, too, has been beset by internal conflicts. Sadly, many of the conflicts have been sparked by the United States' blocking the passage of international policy or refusing to abide by existing regulations. Nowhere is this "might makes right" expression of functional superiority more glaring than in the Iraq war and the avowal of American political leaders to "go it alone" without passage of a UN-backed resolution.

Photo 8.1   Today's democratic, universal community? Delegates to the United Nations participate in international policy making.

Significant Others

William James (1842–1910): Consciousness and the Self

William James was one of America's leading scholars at the turn of the twentieth century. Born into one of the most illustrious families in American intellectual life, his father was an independently wealthy eccentric and amateur theologian, and his younger brother, Henry James, Jr., is one of the great names in American literature. Though James's formal academic degree was in medicine (Harvard, 1869), his work has been most influential in the fields of psychology and philosophy, two disciplines that better fit his quest to resolve his own emotional, spiritual, and intellectual dilemmas. Often house-bound due to physical illnesses such as smallpox, as well as serious bouts of depression and a mental condition he described as "soul sickness," James nevertheless published a number of major works, most importantly his two-volume, 1,200-page opus, The *Principles of Psychology* (1890), and *The Varieties of Religious Experience* (1902). Through his books, essays, and lectures, he tackled such topics as the nature of truth (it is "good" because it is "useful"), the connection between physiological and emotional responses to stimuli (the James-Lange theory), the constitution and experience of the self, religious experiences of the "divine," and the nature of consciousness and its relationship to reality.

For James, consciousness does not exist as a sort of substance or entity that is brought to experiences; rather, it is a continuous series of changing relations within experiences that perform the "function of knowing." Life flows along an ongoing "stream of consciousness" that selectively attends to and emphasizes some features of our outer world while ignoring most of what transpires around us. "Now we are seeing, now hearing; now reasoning, now willing; now recollecting, now expecting; now loving, now hating. . . . Like a bird's life, [consciousness] seems to be an alternation of flights and perchings" in which our every thought, our every state of mind, is always unique (James [1892] 2001:27). Out of the indistinguishable, chaotic swirl of movements that make up our environment, our attention is focused only on those things that happen to be of interest to us and, which, having singled out its independent existence, we thus name.

Of all the things that flow through our stream of consciousness, one holds our interest unlike any other: the "Empirical Self" or "me"—that "empirical aggregate of things objectively known" (James [1890] 2007:215) by the individual. James divides the self or "me" into three parts: its "constituents" (which he further divides into the "material Self," the "social Self," the "spiritual Self," and the "pure Ego"), the feelings and emotions aroused by the various constituents of the self ("Self-feelings"), and the actions prompted by the self ("Self-seeking" and "Self-preservation"). Taken together, "*a man's Self is the sum total of all that he* CAN *call his*" (ibid.:291; emphasis in original). Thus, the self is composed of one's physical body and mental powers, material possessions and social relationships, reputation and deeds, and all the feelings and actions to which they give rise. These "MUST *be the supremely interesting* OBJECTS *for each human mind*" (ibid.:324; emphasis in original).

▪▪ MEAD'S THEORETICAL ORIENTATION

With regard to our metatheoretical framework, Mead's work is predominantly individualistic and nonrationalist in orientation. (See Figure 8.1.) His vision is one where the social order is continually emerging through the ongoing activities of individuals (individualistic) who are attempting to navigate or make sense of the situations in which they find themselves (nonrationalist).

To explicate this understanding of Mead's perspective, consider first the individualistic dimension of his theory. As we discussed, Mead saw society as the product of coordinated activities undertaken by *individuals* reflexively taking the attitude of the other(s) with whom they are interacting. Society is neither a "thing," an overarching system of preexisting structures nor a form of relationship that connects

Figure 8.1 Mead's Basic Theoretical Orientation

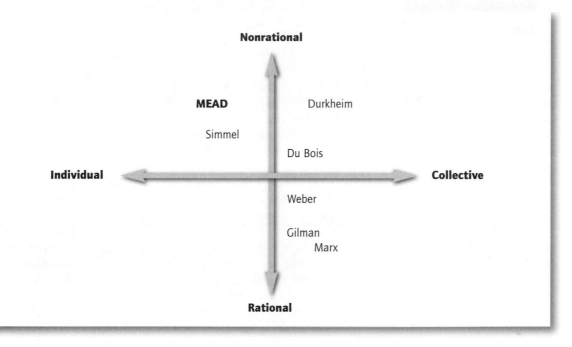

individuals with one another. Instead, Mead envisioned society as shared attitudes that consciously shape individuals' behaviors. Even institutions, such as legal, educational, or kinship systems, were defined by Mead as "social habits" or a "common response on the part of all members of the community to a particular situation" (Mead [1934] 1962:261). Hence, they affect our behavior insofar as we adopt the attitude of others with whom we interact in recurring situations. Institutions (e.g., courts, schools, marriage) continually *emerge* as patterns of action are reenacted by participants in the encounter.

Unlike Marx or Weber, then, Mead granted little attention to *categories* such as institutions, class, or status, because he did not believe that these collectivist or structural forces determined one's consciousness or behavior. Certainly, he did not deny society a role in shaping an individual's attitudes and behaviors. Indeed, our self is essentially a social construct, given its basis in perceptions of how *others* will respond to our behaviors. In important aspects, then, Mead viewed the individual and society as dialectically intertwined—each one shapes and cannot exist without the other. Nevertheless, for Mead, the central elements of social interaction take place in one's imagination. In turn, he emphasized the importance of the "I," "me," "generalized other," and the role of thinking and language in social life. Importantly, Mead viewed language as a neutral means for symbolically communicating with others. Language, in other words, was not seen as reflecting relations of domination and subordination based, for example, on class or gender.[4] One's mind or thinking, which for Mead is intimately tied to language, is therefore not a product of one's station in society. Compare this view with that expressed by Marx, who argued that an individual's consciousness and view of the world are determined by her class position.

But what of the nonrationalist presuppositions underlying Mead's theory? Here, his position is perhaps more difficult to glean. Dennis Wrong points to the source of this difficulty when he remarks that Mead was "only casually and tangentially interested in the self as a source of motivational energy, or as an

---

[4] Mead's assumptions regarding language are by no means unchallenged. For instance, many observers argue that the use of "he" as a universal pronoun renders women invisible. In nullifying women's experiences, the reality of patriarchal relations is obscured and thus further reinforced. Similarly, some argue that the wording of SATs reflects class, race, and gender biases. To the extent that occupation and wage are tied to educational attainment, the ostensibly "objective" test therefore perpetuates structurally based patterns of economic inequality.

object of affective attachment" (1994:65). Because our distinction between rationalist and nonrationalist refers precisely to a theorist's view of the motivational foundations of action, Mead's "casual interest" on this issue leaves us with less to work with than we may like. Nevertheless, we can point out the general tendencies in Mead's perspective.

First, Mead's account of the self and interaction posits individuals as essentially cognitive. In other words, he offers much insight into the mental behaviors that we undertake in preparing to act with others and in assigning meaning to conduct and events. Hence, Mead details the role of language and self-objectification in *thinking*. This focus should not, however, be confused with a rationalist orientation to motivation. As we discussed in Chapter 1, rationalist presuppositions portray actors and groups as *motivated* by their self-interested maximization of rewards or pleasures and minimizing of costs. Nowhere does Mead make such claims regarding the motivational source of behavior.

Based on our definition of rationalism, we are left with Mead as a theorist given to a nonrationalist orientation. However, such a description is based on more than a simple process of elimination. For instance, in describing institutions as "social habits," Mead suggests that actions are motivated out of just that—habit. Also, in arguing that we approach situations pragmatically, Mead suggests that our actions are guided by an attempt to solve problems. Thus, we pursue behaviors that work and are less preoccupied with fashioning the optimal line of action with regard to any particular payoff. Last, Mead occasionally notes the role of disapproval in shaping our behaviors. Here, he argues that in taking the attitude of the other and responding to our gestures as the other might, we suppress actions that would elicit disapproval from others. To the extent that experiencing the disapproval of others arouses negative emotions or affects, we can say it is a nonrationalist motivating force.[5] As central points of debate, it is, of course, up to the reader to determine where Mead's work falls in the metatheoretical framework.

Figure 8.2 Mead's Core Concepts

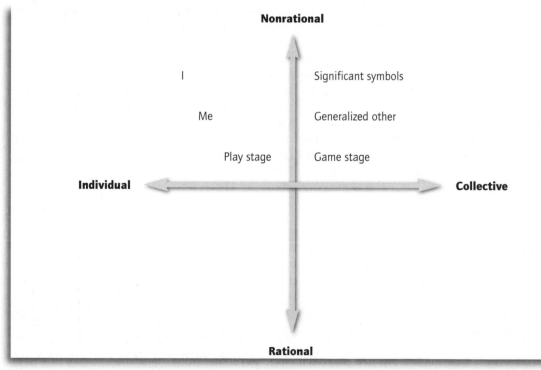

---

[5]One could argue that it is rational to avoid disapproval because failing to do so might compromise our attempts to gain rewards. However, this leads to tautological or circular reasoning, where it is impossible for motivations of any kind to fall outside rational, cost-benefit calculations. In turn, this logic would suggest that it is rational to act on the basis of tradition or habit, rational to be in love or desire companionship, or rational to be fearful in certain situations.

Readings

In what follows, you will read selections from Mead's *Mind, Self, and Society*. The first excerpts are from the portion titled "Mind" in which Mead discusses the social and symbolic bases of thinking and of the construction of meaning. We then move to excerpts taken from the second portion of the book, titled "Self." Here, Mead explores, among other issues, the "phases" of the self and their development through social experiences.

Introduction to "Mind"

In these first selections, drawn from *Mind, Self, and Society,* you will encounter three central themes in Mead's work: (1) mind, (2) symbols and language, and (3) the essence of meaning. Here, you will read that, for Mead, "mind" is a process or behavior that allows for the conscious control of one's actions. More specifically, the mind involves an internal conversation of gestures that makes possible the imagined testing of alternative lines of conduct. In the delay of responses produced by such testing lies the crux of intelligent behavior: controlling one's present action with reference to *ideas* about possible future consequences.

It is through symbols, or language, that we point out objects to ourselves and orient our behavior. For instance, when looking for a place to sit at the movies, you probably will have an image of an empty seat in your mind. You know what it is you are looking for even if you do not actually see an empty seat because thinking is carried out through symbols. "Symbols stand for the meanings of those things or objects which have meanings" (Mead [1934] 1962:122). Moreover, it is our capacity to think symbolically—to hold within our minds the meanings of things—that allows us to mentally rehearse lines of action without actually performing them. Herein lies the locus of behavioral control, yet controlling one's behavior through thinking is a social process. This is because the mind "emerges" as we point out to others and to ourselves the meaning of things.

Our awareness of the meaning of an object, gesture, or event makes it possible for us to form meaningful responses—that is, responses that indicate to others and ourselves how we are going to act in reference to the situation at hand. And, in indicating our forthcoming responses, we are at the same time indicating the meaning of the object, gesture, or event. This viewpoint, however, does not indicate *where* meaning is produced. Again, we find Mead emphasizing the social nature of individual experience.

Mead defines **meaning** as a "threefold relationship" between (1) an individual's gesture, (2) the adjustive response by another to that gesture, and (3) the completion of the social act initiated by the gesture of the first individual. A gesture thus signifies (1) what the individual making it intends to do, (2) what the individual to whom it is directed is expected to do, and (3) the coordinated activity that is to take place. Meaning, in turn, is not an *idea,* but a *response* to a gesture developed within a social act. In other words, meaning is not intrinsic to a given object or action, nor does it exist within one's own consciousness independently of interaction, as the earlier German idealist philosophers argued. One's gesture to another (for example, asking someone, "May I borrow a pencil?") refers to a desired result (getting a pencil) of the interaction. But an individual's gesture is socially meaningful, or significant, *only* if it elicits the desired response from the person to whom it is directed. That is, your gesture is understood only if the other person responds to your words or actions as you responded in your own mind. If a person hands you a book instead of the pencil you asked for, your gesture lacked meaning, as the other's behavior did not lead to the successful completion of the act. (This point underscores Mead's roots in pragmatic philosophy.) How many times have you said to someone, "You don't understand what I'm saying," when he or she did not respond as you had imagined they would? Such instances illustrate how meaning develops through a social process.

If the meaning of one's gesture is the response to that gesture by another, how then are you able to ensure the proper interpretation of your actions? How do you make your intentions known such that you are able to bring out the desired responses from others? For Mead, such coordinated activity becomes possible only with the development of language in the form of significant symbols. **Significant symbols** are words and gestures that have the same meaning for all those involved in a social act. They call out the same response in the person who initiates the gesture as they do in those to whom the gesture is directed. Shared meanings provide the basis from which actions can be planned and carried out. Without such a consensus, whether it is preexisting or created by the participants, social interaction would proceed haltingly, as we would be unable to indicate our intentions to others in light of their presumed interpretation of our actions.

"Mind" (1934)

George Herbert Mead

MIND AND THE SYMBOL

The mechanism of the central nervous system enables us to have now present, in terms of attitudes or implicit responses, the alternative possible overt completions of any given act in which we are involved; and this fact must be realized and recognized, in virtue of the obvious control which later phases of any given act exert over its earlier phases. More specifically, the central nervous system provides a mechanism of implicit response which enables the individual to test out implicitly the various possible completions of an already initiated act in advance of the actual completion of the act—and thus to choose for himself, on the basis of this testing, the one which it is most desirable to perform explicitly or carry into overt effect. The central nervous system, in short, enables the individual to exercise conscious control over his behavior. It is the possibility of delayed response which principally differentiates reflective conduct from non-reflective conduct in which the response is always immediate. The higher centers of the central nervous system are involved in the former type of behavior by making possible the interposition, between stimulus and

response in the simple stimulus-response arc, of a process of selecting one or another of a whole set of possible responses and combinations of responses to the given stimulus.

Mental processes take place in this field of attitudes as expressed by the central nervous system; and this field is hence the field of ideas: the field of the control of present behavior in terms of its future consequences, or in terms of future behavior; the field of that type of intelligent conduct which is peculiarly characteristic of the higher forms of life, and especially of human beings. The various attitudes expressible through the central nervous system can be organized into different types of subsequent acts; and the delayed reactions or responses thus made possible by the central nervous system are the distinctive feature of mentally controlled or intelligent behavior.[i]

What is the mind as such, if we are to think in behavioristic terms? Mind, of course, is a very ambiguous term, and I want to avoid ambiguities. What I suggested as characteristic of the mind is the reflective intelligence of the human animal which can be distinguished from the intelligence of lower forms. If we should try to

[i]In considering the rôle or function of the central nervous system—important though it is—in intelligent human behavior, we must nevertheless keep in mind the fact that such behavior is essentially and fundamentally social; that it involves and presupposes an ever ongoing social life-process; and that the unity of this ongoing social process—or of any one of its component acts—is irreducible, and in particular cannot be adequately analyzed simply into a number of discrete nerve elements. This fact must be recognized by the social psychologist. These discrete nerve elements lie within the unity of this ongoing social process, or within the unity of any one of the social acts in which this process is expressed or embodied; and the analysis which isolates them—the analysis of which they are the results or end-products—does not and cannot destroy that unity.

regard reason as a specific faculty which deals with that which is universal we should find responses in lower forms which are universal. We can also point out that their conduct is purposive, and that types of conduct which do not lead up to certain ends are eliminated. This would seem to answer to what we term "mind" when we talk about the animal mind, but what we refer to as reflective intelligence we generally recognize as belonging only to the human organism. The non-human animal acts with reference to a future in the sense that it has impulses which are seeking expression that can only be satisfied in later experience, and however this is to be explained, this later experience does determine what the present experience shall be. If one accepts a Darwinian explanation he says that only those forms survive whose conduct has a certain relationship to a specific future, such as belongs to the environment of the specific form. The forms whose conduct does insure the future will naturally survive. In such a statement, indirectly at least, one is making the future determine the conduct of the form through the structure of things as they now exist as a result of past happenings.

When, on the other hand, we speak of reflective conduct we very definitely refer to the presence of the future in terms of ideas. The intelligent man as distinguished from the intelligent animal presents to himself what is going to happen. The animal may act in such a way as to insure its food tomorrow. A squirrel hides nuts, but we do not hold that the squirrel has a picture of what is going to happen. The young squirrel is born in the summer time, and has no directions from other forms, but it will start off hiding nuts as well as the older ones. Such action shows that experience could not direct the activity of the specific form. The provident man, however, does definitely pursue a certain course, pictures a certain situation, and directs his own conduct with reference to it. The squirrel follows certain blind impulses, and the carrying-out of its impulses leads to the same result that the storing of grain does for the provident man. It is this picture, however, of what the future is to be as determining our present conduct that is the characteristic of human intelligence—the future as present in terms of ideas.

When we present such a picture it is in terms of our reactions, in terms of what we are going to do. There is some sort of a problem before us, and our statement of the problem is in terms of a future situation which will enable us to meet it by our present reactions. That sort of thinking characterizes the human form and we have endeavored to isolate its mechanism. What is essential

to this mechanism is a way of indicating characters of things which control responses, and which have various values to the form itself, so that such characters will engage the attention of the organism and bring about a desired result. The odor of the victim engages the attention of the beast of prey, and by attention to that odor he does satisfy his hunger and insure his future. What is the difference between such a situation and the conduct of the man who acts, as we say, rationally? The fundamental difference is that the latter individual in some way indicates this character, whatever it may be, to another person and to himself; and the symbolization of it by means of this indicative gesture is what constitutes the mechanism that gives the implements, at least, for intelligent conduct. Thus, one points to a certain footprint, and says that it means bear. Now to identify that sort of a trace by means of some symbol so that it can be utilized by the different members of the group, but particularly by the individual himself later, is the characteristic thing about human intelligence. To be able to identify "this as leading to that," and to get some sort of a gesture, vocal or otherwise, which can be used to indicate the implication to others and to himself so as to make possible the control of conduct with reference to it, is the distinctive thing in human intelligence which is not found in animal intelligence.

What such symbols do is to pick out particular characteristics of the situation so that the response to them can be present in the experience of the individual. We may say they are present in ideal form, as in a tendency to run away, in a sinking of the stomach when we come on the fresh footprints of a bear. The indication that this is a bear calls out the response of avoiding the bear, or if one is on a bear hunt, it indicates the further progress of the hunt. One gets the response into experience before that response is overtly carried out through indicating and emphasizing the stimulus that instigates it. When this symbol is utilized for the thing itself one is, in Watson's terms, conditioning a reflex. The sight of the bear would lead one to run away, the footprint conditioned that reflex, and the word "bear" spoken by one's self or a friend can also condition the reflex, so that the sign comes to stand for the thing so far as action is concerned.

What I have been trying to bring out is the difference between the foregoing type of conduct and the type which I have illustrated by the experiment on the baby with the white rat and the noise behind its head. In the latter situation there is a conditioning of the reflex in which there is no holding apart of the different elements. But when there is a conditioning of the reflex which involves the word "bear," or the sight of the footprint, there is in the experience of the individual the separation of the stimulus and the response. Here the symbol means bear, and that in turn means getting out of the way, or furthering the hunt. Under those circumstances the person who stumbles on the footprints of the bear is not afraid of the footprints—he is afraid of the bear. The footprint means a bear. The child is afraid of the rat, so that the response of fear is to the sight of the white rat; the man is not afraid of the footprint, but of the bear. The footprint and the symbol which refers to the bear in some sense may be said to condition or set off the response, but the bear and not the sign is the object of the fear. The isolation of the symbol, as such, enables one to hold on to these given characters and to isolate them in their relationship to the object, and consequently in their relation to the response. It is that, I think, which characterizes our human intelligence to a peculiar degree. We have a set of symbols by means of which we indicate certain characters, and in indicating those characters hold them apart from their immediate environment, and keep simply one relationship clear. We isolate the footprint of the bear and keep only that relationship to the animal that made it. We are reacting to that, nothing else. One holds on to it as an indication of the bear and of the value that object has in experience as something to be avoided or to be hunted. The ability to isolate these important characters in their relationship to the object and to the response which belongs to the object is, I think, what we generally mean when we speak of a human being thinking a thing out, or having a mind. Such ability makes the worldwide difference between the conditioning of reflexes in the case of the white rat and the human process of thinking by means of symbols.[ii]

What is there in conduct that makes this level of experience possible, this selection of certain characters with their relationship to other characters and to the responses which these call out? My own answer, it is clear, is in terms of such a set of symbols as arise in our social conduct, in the conversation of gestures—in a word, in terms of language. When we get into conduct these symbols which indicate certain characters and their relationship to things and to responses, they enable us to pick out these characters and hold them in so far as they determine our conduct.

A man walking across country comes upon a chasm which he cannot jump. He wants to go ahead but the chasm prevents this tendency from being carried out. In that kind of a situation there arises a sensitivity to all sorts of characters which he has not noticed before. When he stops, mind, we say, is freed. He does not simply look for the indication of the path going ahead. The dog and the man would both try to find a point where they could cross. But what the man could do that the dog could not would be to note that the sides of the chasm seem to be approaching each other in one direction. He picks out the best places to try, and that approach which he indicates to himself determines the way in which he is going to go. If the dog saw at a distance a narrow place he would run to it, but probably he would not be affected by the gradual approach which the human individual symbolically could indicate to himself.

The human individual would see other objects about him, and have other images appear in his experience. He sees a tree which might serve as a bridge across the space ahead of him. He might try various sorts of possible actions which would be suggested to him in such a situation, and present them to himself by means of the symbols he uses. He has not simply conditioned certain

---

[ii]The meanings of things or objects are actual inherent properties or qualities of them; the locus of any given meaning is in the thing which, as we say, "has it." We refer to the meaning of a thing when we make use of the symbol. Symbols stand for the meanings of those things or objects which have meanings; they are given portions of experience which point to, indicate, or represent other portions of experience not directly present or given at the time when, and in the situation in which, any one of them is thus present (or is immediately experienced). The symbol is thus more than a mere substitute stimulus—more than a mere stimulus for a conditioned response or reflex. For the conditioned reflex—the response to a mere substitute stimulus— does not or need not involve consciousness; whereas the response to a symbol does and must involve consciousness. Conditioned reflexes plus consciousness of the attitudes and meanings they involve are what constitute language, and hence lay the basis, or comprise the mechanism for, thought and intelligent conduct. Language is the means whereby individuals can indicate to one another what their responses to objects will be, and hence what the meanings of objects are; it is not a mere system of conditioned reflexes. Rational conduct always involves a reflexive reference to self, that is, an indication to the individual of the significances which his actions or gestures have for other individuals. And the experiential or behavioristic basis for such conduct—the neuro-physiological mechanism of thinking—is to be found, as we have seen, in the central nervous system.

responses by certain stimuli. If he had, he would be bound to those. What he does do by means of these symbols is to indicate certain characters which are present, so that he can have these responses there all ready to go off. He looks down the chasm and thinks he sees the edges drawing together, and he may run toward that point. Or he may stop and ask if there is not some other way in which he can hasten his crossing. What stops him is a variety of other things he may do. He notes all the possibilities of getting across. He can hold on to them by means of symbols, and relate them to each other so that he can get a final action. The beginning of the act is there in his experience. He already has a tendency to go in a certain direction and what he would do is already there determining him. And not only is that determination there in his attitude but he has that which is picked out by means of the term "that is narrow, I can jump it." He is ready to jump, and that reflex is ready to determine what he is doing. These symbols, instead of being a mere conditioning of reflexes, are ways of picking out the stimuli so that the various responses can organize themselves into a form of action.[iii]

The situation in which one seeks conditioning responses is, I think, as far as effective intelligence is concerned, always present in the form of a problem. When a man is just going ahead he seeks the indications of the path but he does it unconsciously. He just sees the path ahead of him; he is not aware of looking for it under those conditions. But when he reaches the chasm, this onward movement is stopped by the very process of drawing back from the chasm. That conflict, so to speak, sets him free to see a whole set of other things. Now, the sort of things he will see will be the characters which represent various possibilities of action under the circumstances. The man holds on to these different possibilities of response in terms of the different stimuli which present themselves, and it is his ability to hold them there that constitutes his mind.

We have no evidence of such a situation in the case of the lower animals, as is made fairly clear by the fact that we do not find in any animal behavior that we can work out in detail any symbol, any method of communication, anything that will answer to these different responses so that they can all be held there in the experience of the individual. It is that which differentiates the action of the reflectively intelligent being from the conduct of the lower forms; and the mechanism that makes that possible is language. We have to recognize that language is a part of conduct. Mind involves, however, a relationship to the characters of things. Those characters are in the things, and while the stimuli call out the response which is in one sense present in the organism, the responses are to things out there. The whole process is not a mental product and you cannot put it inside of the brain. Mentality is that relationship of the organism to the situation which is mediated by sets of symbols. . . .

Mentality on our approach simply comes in when the organism is able to point out meanings to others and to himself. This is the point at which mind appears, or if you like, emerges. What we need to recognize is that we are dealing with the relationship of the organism to the environment selected by its own sensitivity. The psychologist is interested in the mechanism which the human species has evolved to get control over these relationships. The relationships have been there before the indications are made, but the organism has not in its own conduct controlled that relationship. It originally has no mechanism by means of which it can control it. The human animal, however, has worked out a

---

[iii]The reflective act consists in a reconstruction of the perceptual field so that it becomes possible for impulses which were in conflict to inhibit action no longer. This may take place by such a temporal readjustment that one of the conflicting impulses finds a later expression. In this case there has entered into the perceptual field other impulses which postpone the expression of that which had inhibited action. Thus, the width of the ditch inhibits the impulse to jump. There enters into the perceptual field the image of a narrower stretch and the impulse to go ahead finds its place in a combination of impulses, including that of movement toward the narrower stretch.

The reconstruction may take place through the appearance of other sensory characters in the field ignored before. A board long enough to bridge the ditch is recognized. Because the individual has already the complex of impulses which lead to lifting it and placing it across the ditch it becomes a part of the organized group of impulses that carry the man along toward his destination. In neither case would he be ready to respond to the stimulus (in the one case the image of the narrower stretch of the ditch, in the other the sight of the board) if he had not reactions in his nature answering to these objects, nor would these tendencies to response sensitize him to their stimuli if they were not freed from firmly organized habits. It is this freedom, then, that is the prerequisite of reflection, and it is our social self-reflective conduct that gives this freedom to human individuals in their group life (MS).

mechanism of language communication by means of which it can get this control. Now, it is evident that much of that mechanism does not lie in the central nervous system, but in the relation of things to the organism. The ability to pick these meanings out and to indicate them to others and to the organism is an ability which gives peculiar power to the human individual. The control has been made possible by language. It is that mechanism of control over meaning in this sense which has, I say, constituted what we term "mind." The mental processes do not, however, lie in words any more than the intelligence of the organism lies in the elements of the central nervous system. Both are part of a process that is going on between organism and environment. The symbols serve their part in this process, and it is that which makes communication so important. Out of language emerges the field of mind.

It is absurd to look at the mind simply from the standpoint of the individual human organism; for, although it has its focus there, it is essentially a social phenomenon; even its biological functions are primarily social. The subjective experience of the individual must be brought into relation with the natural, socio-biological activities of the brain in order to render an acceptable account of mind possible at all; and this can be done only if the social nature of mind is recognized. The meagerness of individual experience in isolation from the processes of social experience—in isolation from its social environment—should, moreover, be apparent. We must regard mind, then, as arising and developing within the social process, within the empirical matrix of social interactions. We must, that is, get an inner individual experience from the standpoint of social acts which include the experiences of separate individuals in a social context wherein those individuals interact. The processes of experience which the human brain makes possible are made possible only for a group of interacting individuals: only for individual organisms which are members of a society; not for the individual organism in isolation from other individual organisms.

Mind arises in the social process only when that process as a whole enters into, or is present in, the experience of any one of the given individuals involved in that process. When this occurs the individual becomes self-conscious and has a mind; he becomes aware of his relations to that process as a whole, and to the other individuals participating in it with him; he becomes aware of that process as modified by the reactions and interactions of the individuals—including himself—who are carrying it on. The evolutionary appearance of

mind or intelligence takes place when the whole social process of experience and behavior is brought within the experience of any one of the separate individuals implicated therein, and when the individual's adjustment to the process is modified and refined by the awareness or consciousness which he thus has of it. It is by means of reflexiveness—the turning-back of the experience of the individual upon himself—that the whole social process is thus brought into the experience of the individuals involved in it; it is by such means, which enable the individual to take the attitude of the other toward himself, that the individual is able consciously to adjust himself to that process, and to modify the resultant of that process in any given social act in terms of his adjustment to it. Reflexiveness, then, is the essential condition, within the social process, for the development of mind.

Meaning

We are particularly concerned with intelligence on the human level, that is, with the adjustment to one another of the acts of different human individuals within the human social process; an adjustment which takes place through communication: by gestures on the lower planes of human evolution, and by significant symbols (gestures which posses meanings and are hence more than mere substitute stimuli) on the higher planes of human evolution.

The central factor in such adjustment is "meaning." Meaning arises and lies within the field of the relation between the gesture of a given human organism and the subsequent behavior of this organism as indicated to another human organism by that gesture. If that gesture does so indicate to another organism the subsequent (or resultant) behavior of the given organism, then it has meaning. In other words, the relationship between a given stimulus—as a gesture—and the later phases of the social act of which it is an early (if not the initial) phase constitutes the field within which meaning originates and exists. Meaning is thus a development of something objectively there as a relation between certain phases of the social act; it is not a psychical addition to that act and it is not an "idea" as traditionally conceived. A gesture by one organism, the resultant of the social act in which the gesture is an early phase, and the response of another organism to the gesture, are the relata in a triple or threefold relationship of gesture to first organism, of gesture to second organism, and of gesture to subsequent phases of the given social act; and this threefold relationship constitutes the matrix within which meaning arises, or which develops into the field of meaning. The gesture stands for a certain resultant of the social act, a resultant to which there is a definite response on the part of the individuals involved therein; so that meaning is given or stated in terms of response. Meaning is implicit—if not always explicit—in the relationship among the various phases of the social act to which it refers, and out of which it develops. And its development takes place in terms of symbolization at the human evolutionary level.

We have been concerning ourselves, in general, with the social process of experience and behavior as it appears in the calling out by the act of one organism of an adjustment to that act in the responsive act of another organism. We have seen that the nature of meaning is intimately associated with the social process as it thus appears, that meaning involves this three-fold relation among phases of the social act as the context in which it arises and develops: this relation of the gesture of one organism to the adjustive response of another organism (also implicated in the given act), and to the completion of the given act—a relation such that the second organism responds to the gesture of the first as indicating or referring to the completion of the given act. For example, the chick's response to the cluck of the mother hen is a response to the meaning of the cluck; the cluck refers to danger or to food, as the case may be, and has this meaning or connotation for the chick.

The social process, as involving communication, is in a sense responsible for the appearance of new objects in the field of experience of the individual organisms implicated in that process. Organic processes or responses in a sense constitute the objects to which they are responses; that is to say, any given biological organism is in a way responsible for the existence (in the sense of the meanings they have for it) of the objects to which it physiologically and chemically responds. There would, for example, be no food—no edible objects—if there were no organisms which could digest it. And similarly, the social process in a sense constitutes the objects to which it responds, or to which it is an adjustment. That is to say, objects are constituted in terms of meanings within the social process of experience and behavior through the mutual adjustment to one another of the responses or actions of the various individual organisms involved in that process, an adjustment made possible by means of a communication which takes the form of a conversation of gestures in the earlier evolutionary stages of that process, and of language in its later stages.

Awareness or consciousness is not necessary to the presence of meaning in the process of social experience. A gesture on the part of one organism in any given social act calls out a response on the part of another organism which is directly related to the action of the first organism and its outcome; and a gesture is a symbol of the result of the given social act of one organism (the organism making it) in so far as it is responded to by another organism (thereby also involved in that act) as indicating that result. The mechanism of meaning is thus present in the social act before the emergence of consciousness or awareness of meaning occurs. The act or adjustive response of the second organism gives to the gesture of the first organism the meaning which it has.

Symbolization constitutes objects not constituted before, objects which would not exist except for the context of social relationships wherein symbolization occurs. Language does not simply symbolize a situation or object which is already there in advance; it makes possible the existence or the appearance of that situation or object, for it is a part of the mechanism whereby that situation or object is created. The social process relates the responses of one individual to the gestures of another, as the meanings of the latter, and is thus responsible for the rise and existence of new objects in the social situation, objects dependent upon or constituted by these meanings. Meaning is thus not to be conceived, fundamentally, as a state of consciousness, or as a set of organized relations existing or subsisting mentally outside the field of experience into which they enter; on the contrary, it should be conceived objectively, as having its existence entirely within this field itself.[iv] The response of one organism to the gesture of another in any given social act is the meaning of that gesture, and also is in a sense responsible for the appearance or coming into being of the new object—or new content of an old object—to which that gesture refers through the outcome of the given social act in which it is an early phase. For, to repeat, objects are in a genuine sense constituted within the social process of

---

[iv]Nature has meaning and implication but not indication by symbols. The symbol is distinguishable from the meaning it refers to. Meanings are in nature, but symbols are the heritage of man.

experience, by the communication and mutual adjust-ment of behavior among the individual organisms which are involved in that process and which carry it on. Just as in fencing the parry is an interpretation of the thrust, so, in the social act, the adjustive response of one organism to the gesture of another is the interpretation of that gesture by that organism—it is the meaning of that gesture.

At the level of self-consciousness such a gesture becomes a symbol, a significant symbol. But the inter-pretation of gestures is not, basically, a process going on in a mind as such, or one necessarily involving a mind; it is an external, overt, physical, or physiological process going on in the actual field of social experience. Meaning can be described, accounted for, or stated in terms of symbols or language at its highest and most complex stage of development (the stage it reaches in human experience), but language simply lifts out of the social process a situation which is logically or implic-itly there already. The language symbol is simply a significant or conscious gesture.

Two main points are being made here: (1) that the social process, through the communication which it makes possible among the individuals implicated in it, is responsible for the appearance of a whole set of new objects in nature, which exist in relation to it (objects, namely, of "common sense"); and (2) that the gesture of one organism and the adjustive response of another organism to that gesture within any given social act bring out the relationship that exists between the ges-ture as the beginning of the given act and the comple-tion or resultant of the given act, to which the gesture refers. These are the two basic and complementary logical aspects of the social process.

The result of any given social act is definitely sepa-rated from the gesture indicating it by the response of another organism to that gesture, a response which points to the result of that act as indicated by that ges-ture. This situation is all there—is completely given—on the non-mental, non-conscious level, before the analysis of it on the mental or conscious level. Dewey says that meaning arises through communication.[v] It isto the content to which the social process gives rise that this statement refers; not to bare ideas or printed words as such, but to the social process which has been so largely responsible for the objects constituting the daily environment in which we live: a process in which communication plays the main part. That process can give rise to these new objects in nature only in so far as it makes possible communication among the individual organisms involved in it. And the sense in which it is responsible for their existence—indeed for the existence of the whole world of common-sense objects—is the sense in which it determines, conditions, and makes pos-sible their abstraction from the total structure of events, as identities which are relevant for everyday social behavior; and in that sense, or as having that meaning, they are existent only relative to that behavior. In the same way, at a later, more advanced stage of its develop-ment, communication is responsible for the existence of the whole realm of scientific objects as well as identities abstracted from the total structure of events by virtue of their relevance for scientific purposes.

The logical structure of meaning, we have seen, is to be found in the threefold relationship of gesture to adjus-tive response and to the resultant of the given social act. Response on the part of the second organism to the gesture of the first is the interpretation—and brings out the meaning—of that gesture, as indicating the resultant of the social act which it initiates, and in which both organisms are thus involved. This threefold or triadic relation between gesture, adjustive response, and resul-tant of the social act which the gesture initiates is the basis of meaning; for the existence of meaning depends upon the fact that the adjustive response of the second organism is directed toward the resultant of the given social act as initiated and indicated by the gesture of the first organism. The basis of meaning is thus objectively there in social conduct, or in nature in its relation to such conduct. Meaning is a content of an object which is dependent upon the relation of an organism or group of organisms to it. It is not essentially or primarily a psy-chical content (a content of mind or consciousness), for it need not be conscious at all, and is not in fact until significant symbols are evolved in the process of human social experience. Only when it becomes identified with such symbols does meaning become conscious. The meaning of a gesture on the part of one organism is the adjustive response of another organism to it, as indicat-ing the resultant of the social act it initiates, the adjustive response of the second organism being itself directed toward or related to the completion of that act. In other words, meaning involves a reference of the gesture of one organism to the resultant of the social act it indicates

---

[v]See *Experience and Nature*, Chap. V.

or initiates, as adjustively responded to in this reference by another organism; and the adjustive response of the other organism is the meaning of the gesture.

Gestures may be either conscious (significant) or unconscious (non-significant). The conversation of gestures is not significant below the human level, because it is not conscious, that is, not self-conscious (though it is conscious in the sense of involving feelings or sensations). An animal as opposed to a human form, in indicating something to, or bringing out a meaning for, another form, is not at the same time indicating or bringing out the same thing or meaning to or for himself; for he has no mind, no thought, and hence there is no meaning here in the significant or self-conscious sense. A gesture is not significant when the response of another organism to it does not indicate to the organism making it what the other organism is responding to.[vi]

Much subtlety has been wasted on the problem of the meaning of meaning. It is not necessary, in attempting to solve this problem, to have recourse to psychical states, for the nature of meaning, as we have seen, is found to be implicit in the structure of the social act, implicit in the relations among its three basic individual components: namely, in the triadic relation of a gesture of one individual, a response to that gesture by a second individual, and completion of the given social act initiated by the gesture of the first individual. And the fact that the nature of meaning is thus found to be implicit in the structure of the social act provides additional emphasis upon the necessity, in social psychology, of starting off with the initial assumption of an ongoing social process of experience and behavior in which any given group of human individuals is involved, and upon which the existence and development of their minds, selves, and self-consciousness depend.

░░

Introduction to "Self"

In the selections taken from the chapters on the self, Mead presents a number of concepts central to his theory. Here, we present an overview of Mead's notion of the "I," "me," and "generalized other," and their realization through the "play" and "game" stages of the development of self-consciousness. Moreover, in his discussion of the self, you will again encounter two themes that form the core of Mead's program: (1) the interconnectedness between the self and social experiences and (2) language as the tool that mediates this relationship.

For Mead, the self does not exist as a "personality" or an "identity" that answers the question, "Who am I?" Instead, the self exists as self-consciousness, that is, the capacity to be both subject and object to one's self that is made possible solely through interaction. As we project the possible implications of courses of action and attempt to elicit the desired responses from others, we become an "object" to ourselves. However, an individual is aware of his self as an object, "not directly, but only indirectly, from the particular standpoints of other individual [s] . . . or from the generalized standpoint of the social group as a whole to which he belongs" (Mead [1934] 1962:138). In turn, the experience of our self as an object becomes possible only by taking the attitudes of others toward our self.

[vi]There are two characters which belong to that which we term "meanings," one is participation and the other is communicability. Meaning can arise only in so far as some phase of the act which the individual is arousing in the other can be aroused in himself. There is always to this extent participation. And the result of this participation is communicability, i.e., the individual can indicate to himself what he indicates to others. There is communication without significance where the gesture of the individual calls out the response in the other without calling out or tending to call out the same response in the individual himself. Significance from the standpoint of the observer may be said to be present in the gesture which calls out the appropriate response in the other or others within a cooperative act, but it does not become significant to the individuals who are involved in the act unless the tendency to the act is aroused within the individual who makes it, and unless the individual who is directly affected by the gesture puts himself in the attitude of the individual who makes the gesture (MS).

Photo 8.2   The play stage: A "mommy" feeding her "baby"

Viewing ourselves as an object is the phase of the self Mead termed the "**me**."[1] This phase represents a sense of who we are that is created, sustained, and modified through our interaction with others. (In the opening to this chapter, "me" is who you see in the mirror.) It is the "organized set of attitudes of others which one himself assumes" (ibid.:175); the self one is aware of when thinking, or taking the attitude of the other. Thus, the self is a social product that is rooted in our perceptions of how others interpret our behaviors.

That we see ourselves and respond to our conduct as others would is not the only way in which the self is a reflection of social interaction. According to Mead, "We divide ourselves up in all sorts of different selves with reference to our acquaintances" (ibid.:142). In other words, it is interaction and the context within which it takes place that determines who we are for the moment—the side of our self that we show to others. The self you experience as a sibling is different from the one you experience as a coworker or a student. We all have, in a sense, multiple social selves or personalities, as assuming the attitudes of a particular other or community shapes our behavior.

While the "me" is involved in thinking or reflexive role taking, the self enters into another phase during moments of interaction. Here, Mead argues that the "**I**" is in the foreground. The "**I**" reacts or answers to the "me," the phase of the self that one is conscious of. (The "I" is the phase of the self that *looks* in the mirror in our opening example.) It is "the response of the [individual] to the attitudes of the others" (ibid.:175). The individual is aware of the "I" of the present moment only as it passes into memory, into the "me." It represents the here-and-now, creative aspect of one's self, for no matter how much you rehearse your behaviors or try to anticipate the reactions of others, interaction may take an unplanned course. You never know for certain what you will say or how others will interpret your behaviors until after you have spoken and a response has been made. For Mead, it is this spontaneous, unpredictable aspect of the self and social interaction that sparks personal and cultural innovation. Without the "I," social life would be static and relentlessly conformist.

According to Mead, the self (that is, self-consciousness) develops through two cognitive stages: "play" and the "game."[2] The **play stage** is marked by the ability to assume the attitude of only one particular individual at a time. This is the stage of self-consciousness that we find in children until around the age of eight. Here, children perhaps are pretending to be a parent, a superhero, or a princess, moving from one role to the next in an unconnected fashion. In play, the child is able only to switch successively between discrete roles while taking the attitude of the specific other toward herself.

The next phase of development is the **game stage**. Here, the child is able to move beyond simply taking the role of particular others and assume the roles of multiple others simultaneously. Moreover, the

---

[1]In developing his view of the self, Mead borrowed much from William James, including the latter's idea of the "me" and the notion of possessing multiple social selves. See the Significant Others box (page 296) that briefly summarizes James's work.

[2]Contemporary theorists are not in agreement as to how many stages of self-development Mead proposes. Some identify three distinct stages (play, game, and generalized other), while others read Mead as outlining four stages (preplay, play, game, and generalized other). We present a two-stage model, conceiving of the generalized other not as a distinct phase in the development of the self, but as a "community attitude" that is assumed in the game stage. We believe this is a more accurate reading of Mead's ideas.

child has the ability to control his actions on the basis of abstract "rules of the game." This configuration of "roles organized according to rules" brings the attitudes of all participants together in an abstract unity called the "generalized other." The **generalized other** is the organized community or group to which an individual belongs. When taking the attitude of the generalized other towards one's self, one assumes the attitudes that are common to the whole group. (It is the vantage point from which one sees, and thus gives meaning to, the "who" reflected in the mirror from the opening to this chapter.) Responding to ourselves from the point of view of the whole community is the mechanism through which that community controls the behaviors of its members, making possible the coordination of diverse activities in large groups or institutions. Moreover, by assuming the attitude of the generalized other, we are able to orient our behavior according to abstract ideals and principles such as freedom, individual rights, and fairness.

The notion of the generalized other clearly illustrates Mead's view of the dialectical relationship between self and society (a theme echoed in the work of Georg Simmel). As the generalized other develops from its rudimentary form found in the games children play to ever-widening, increasingly abstract social circles, self-consciousness likewise develops more fully as the individual begins to take the imagined attitudes of others toward himself that he does not—and never will—know. As Mead remarks, "We cannot be ourselves unless we are also members in whom there is a community of attitudes which control the attitudes of all. . . . The individual possesses a self only in relation to the selves of the other members of his social group; and the structure of his self expresses or reflects the general behavior pattern of this social group to which he belongs" ([1934] 1962:164).[3] Yet, organized group processes and activities—that is, society itself—are made possible "only in so far as every individual involved in them or belonging to that society can take the general attitudes of all such other individuals with reference to these processes and activities" (ibid.:155).

Early in the game stage, children are preoccupied with rules, often becoming quite literal and rigid in following them. For instance, a batter might not be willing to "take one" for the team, allowing himself to be hit by a pitch, or a player might not move outside her usual position even if it means enhancing the team's chances for success. At this point, children are still unable to abstract themselves from the game as such; they are unable to understand the constructed nature, and thus flexibility, of the rules that organize interaction. With the development of the ability to take the attitude of the generalized other, however, one is not simply following the rules; instead, the player has become conscious of her part in the realization of the rules. No longer is she merely "in" the game; now the player is "above" it, as the rules become subject to self-conscious interpretation and, possibly, to change.

Similarly, consider how a six-year-old might decide whether or not to steal. The child's internal conversation or thinking would entail taking the attitude of a particular other toward his potential lines of conduct. His decision is then based on the imagined response of that other, a parent perhaps, to the action in question. Internally responding to the contemplated action as his parent might, the child's decision to not steal is based on his reaction to the anticipated disapproval of that other. In contrast, an adult's decision would be based also on the imagined response of the generalized other. Here, the same decision is not derived solely from a reaction to the anticipated disapproval of a specific other, but also from an internal response to the community's shared *principle* that it is wrong to steal.

The difference between the play and the game stages is readily apparent if we look at how a soccer game is played by 6-year-olds as opposed to a game played by 15-year-olds. In the former, children play as if they are all magnetically attracted to the ball. They have no sense of position or organized, structured participation. There is little if any passing or cooperation between the players. The game consists mainly in a flock of children running wildly after the ball (though it is not uncommon for some players to lose interest in the game completely—e.g., to stop to pick dandelions, wave at their parents, gaze into space, or examine their socks).

Compare this scenario to a game of soccer played by tenth graders. Here, there is a sense of position and an understanding of when to pass the ball and who to pass it to. Teenagers do not selfishly follow the ball, because *it* no longer is the object of their attention. Instead, the *game* is now the focus of the players.

---

[3]Again, our self is realized only in relation to a wider community of others. One is a teacher only if she has students, or an artist or actor only if she has an audience.

Photo 8.3a Six-year-old children playing soccer

Photo 8.3b Twelve-year-old children playing soccer. Among other things, notice the difference in spacing of players in the age groups.

The ability to cooperate as a team produces a kind of rudimentary society where each player is able simultaneously to take the attitude of all the other players, anticipate their reactions, and adjust her own line of action accordingly.

In taking the attitude of the other (whether a particular individual or generalized other), we, of course, are not capable of getting inside another person's head, nor do we experience our self as a physical, tangible object. Instead, when determining what path of behaviors we should pursue we think and develop ourselves in terms of language. Language is thus the essential medium for the genesis of the mind and self. Moreover, as the medium through which we control our actions and coordinate them with those of others, language makes society itself possible.

"Self" (1934)

George Herbert Mead

THE SELF AND THE ORGANISM

In our statement of the development of intelligence we have already suggested that the language process is essential for the development of the self. The self has a character which is different from that of the physiological organism proper. The self is something which has a development; it is not initially there, at birth, but arises in the process of social experience and activity, that is, develops in the given individual as a result of his relations to that process as a whole and to other individuals within that process. The intelligence of the lower forms of animal life, like a great deal of human intelligence, does not involve a self. In our habitual actions, for example, in our moving about in a world that is simply there and to which we are so adjusted that no thinking is involved, there is a certain amount of sensuous experience such as persons have when they are just waking up, a bare thereness of the world. Such characters about us may exist in experience without taking their place in relationship to the self. One must, of course, under those conditions, distinguish between the experience that immediately takes place and our own organization of it into the experience of the self. One says upon analysis that a certain item had its place in his experience, in the experience of his self. We do inevitably tend at a certain level of sophistication to organize all experience into that of a self. We do so intimately identify our experiences, especially our affective experiences, with the self that it takes a moment's abstraction to realize that pain and pleasure can be there without being the experience of the self. Similarly, we normally organize our memories upon the string of our self. If we date things we always date them from the point of view of our past experiences. We frequently have memories that we cannot date, that we cannot place. A picture comes before us suddenly and we are at a loss to explain when that experience originally took place. We remember perfectly distinctly the picture, but we do not have it definitely placed, and until we can place it in terms of our past experience we are not satisfied. Nevertheless, I think it is obvious when one comes to consider it that the self is not necessarily involved in the life of the organism, nor involved in what we term our sensuous experience, that is, experience in a world about us for which we have habitual reactions.

We can distinguish very definitely between the self and the body. The body can be there and can operate in a very intelligent fashion without there being a self involved in the experience. The self has the characteristic that it is an object to itself, and that characteristic distinguishes it from other objects and from the body. It is perfectly true that the eye can see the foot, but it does not see the body as a whole. We cannot see our backs; we can feel certain portions of them, if we are agile, but we cannot get an experience of our whole body. There are, of course, experiences which are somewhat vague and difficult of location, but the bodily experiences are for us organized about a self. The foot and hand belong to the self. We can see our feet, especially if we look at them from the wrong end of an opera glass, as strange things which we have difficulty in recognizing as our own. The parts of the body are quite distinguishable from the self. We can lose parts of the body without any serious invasion of the self. The mere ability to experience different parts of the body is not different from the experience of a table. The table presents a different feel from what the hand does when one hand feels another, but it is an experience of something with which we come definitely into contact. The body does not experience itself as a whole, in the sense in which the self in some way enters into the experience of the self.

It is the characteristic of the self as an object to itself that I want to bring out. This characteristic is represented in the word "self," which is a reflexive, and indicates that which can be both subject and object. This type of object is essentially different from other objects, and in the past it has been distinguished as conscious, a term which indicates an experience with, an experience of, one's self. It was assumed that consciousness in some way carried this capacity of being an object to itself. In giving a behavioristic statement of consciousness we have to look for some sort of experience in which the physical organism can become an object to itself.[i]

When one is running to get away from someone who is chasing him, he is entirely occupied in this action, and his experience may be swallowed up in the objects about him, so that he has, at the time being, no consciousness of self at all. We must be, of course, very completely occupied to have that take place, but we can, I think, recognize that sort of a possible experience in which the self does not enter. . . . In such instances there is a contrast between an experience that is absolutely wound up in outside activity in which the self as an object does not enter, and an activity of memory and imagination in which the self is the principal object. The self is then entirely distinguishable from an organism that is surrounded by things and acts with reference to things, including parts of its own body. These latter may be objects like other objects, but they are just objects out there in the field, and they do not involve a self that is an object to the organism. This is, I think, frequently overlooked. It is that fact which makes our anthropomorphic reconstructions of animal life so fallacious. How can an individual get outside himself (experientially) in such a way as to become an object to himself? This is the essential psychological problem of selfhood or of self-consciousness; and its solution is to be found by referring to the process of social conduct or activity in which the given person or individual is implicated. The apparatus of reason would not be complete unless it swept itself into its own analysis of the field of experience; or unless the individual brought himself into the same experiential field as that of the other individual selves in relation to whom he acts in any given social situation. Reason cannot become impersonal unless it takes an objective, non-affective attitude toward itself; otherwise we have just consciousness, not *self*-consciousness. And it is necessary to rational conduct that the individual should thus take an objective, impersonal attitude toward himself, that he should become an object to himself. For the individual organism is obviously an essential and important fact or constituent element of the empirical situation in which it acts; and without taking objective account of itself as such, it cannot act intelligently, or rationally.

The individual experiences himself as such, not directly, but only indirectly, from the particular standpoints of other individual members of the same social group, or from the generalized standpoint of the social group as a whole to which he belongs. For he enters his own experience as a self or individual, not directly or immediately, not by becoming a subject to himself, but only in so far as he first becomes an object to himself just as other individuals are objects to him or in his experience; and he becomes an object to himself only by taking the attitudes of other individuals toward himself within a social environment or context of experience and behavior in which both he and they are involved.

The importance of what we term "communication" lies in the fact that it provides a form of behavior in which the organism or the individual may become an object to himself. It is that sort of communication which we have been discussing—not communication in the sense of the cluck of the hen to the chickens, or the bark of a wolf to the pack, or the lowing of a cow, but communication in the sense of significant symbols, communication which is directed not only to others but also to the individual himself. So far as that type of communication is a part of behavior it at least introduces a self. Of course, one may hear without listening; one may see things that he does not realize; do things that he is not really aware of. But it is where one does respond to that which he addresses to another and where that response of his own becomes a part of his conduct, where he not only hears himself but responds to himself, talks and replies to himself as truly as the other person replies to him, that we have behavior in which the individuals become objects to themselves. . . .

The self, as that which can be an object to itself, is essentially a social structure, and it arises in social experience. After a self has arisen, it in a certain sense provides for itself its social experiences, and so we can conceive of an absolutely solitary self. But it is impossible to conceive of a self arising outside of social experience. When it has arisen we can think of a person in solitary confinement for the rest of his life, but who still has himself as a companion, and is able to think and to converse with himself as he had communicated with

[i] Man's behavior is such in his social group that he is able to become an object to himself, a fact which constitutes him a more advanced product of evolutionary development than are the lower animals. Fundamentally it is this social fact—and not his alleged possession of a soul or mind with which he, as an individual, has been mysteriously and supernaturally endowed, and with which the lower animals have not been endowed—that differentiates him from them.

others. That process to which I have just referred, of responding to one's self as another responds to it, taking part in one's own conversation with others, being aware of what one is saying and using that awareness of what one is saying to determine what one is going to say thereafter—that is a process with which we are all familiar. We are continually following up our own address to other persons by an understanding of what we are saying, and using that understanding in the direction of our continued speech. We are finding out what we are going to say, what we are going to do, by saying and doing, and in the process we are continually controlling the process itself. In the conversation of gestures what we say calls out a certain response in another and that in turn changes our own action, so that we shift from what we started to do because of the reply the other makes. The conversation of gestures is the beginning of communication. The individual comes to carry on a conversation of gestures with himself. He says something, and that calls out a certain reply in himself which makes him change what he was going to say. One starts to say something, we will presume an unpleasant something, but when he starts to say it he realizes it is cruel. The effect on himself of what he is saying checks him; there is here a conversation of gestures between the individual and himself. We mean by significant speech that the action is one that affects the individual himself, and that the effect upon the individual himself is part of the intelligent carrying-out of the conversation with others. Now we, so to speak, amputate that social phase and dispense with it for the time being, so that one is talking to one's self as one would talk to another person.[ii]

This process of abstraction cannot be carried on indefinitely. One inevitably seeks an audience, has to pour himself out to somebody. In reflective intelligence one thinks to act, and to act solely so that this action remains a part of a social process. Thinking becomes preparatory to social action. The very process of thinking is, of course, simply an inner conversation that goes on, but it is a conversation of gestures which in its completion implies the expression of that which one thinks to an audience. One separates the significance of what he is saying to others from the actual speech and gets it ready before saying it. He thinks it out, and perhaps writes it in the form of a book; but it is still a part of social intercourse in which one is addressing other persons and at the same time addressing one's self, and in which one controls the address to other persons by the response made to one's own gesture. That the person should be responding to himself is necessary to the self, and it is this sort of social conduct which provides behavior within which that self appears. I know of no other form of behavior than the linguistic in which the individual is an object to himself, and, so far as I can see, the individual is not a self in the reflexive sense unless he is an object to himself. It is this fact that gives a critical importance to communication, since this is a type of behavior in which the individual does so respond to himself.

We realize in everyday conduct and experience that an individual does not mean a great deal of what he is doing and saying. We frequently say that such an individual is not himself. We come away from an interview with a realization that we have left out important things, that there are parts of the self that did not get into what was said. What determines the amount of the self that gets into communication is the social experience itself. Of course, a good deal of the self does not need to get expression. We carry on a whole series of different relationships to different people. We are one thing to one man and another thing to another. There are parts of the self which exist only for the self in relationship to itself. We divide ourselves up in all sorts of different selves with reference to our acquaintances. We discuss politics with one and religion with another. There are all sorts of different selves answering to all sorts of different social reactions. It is the social process itself that is responsible for the appearance of the self; it is not there as a self apart from this type of experience. . . .

The unity and structure of the complete self reflects the unity and structure of the social process as a whole; and each of the elementary selves of which it is composed reflects the unity and structure of one of

[ii]It is generally recognized that the specifically social expressions of intelligence, or the exercise of what is often called "social intelligence," depend upon the given individual's ability to take the rôles of, or "put himself in the place of," the other individuals implicated with him in given social situations; and upon his consequent sensitivity to their attitudes toward himself and toward one another. These specifically social expressions of intelligence, of course, acquire unique significance in terms of our view that the whole nature of intelligence is social to the very core—that this putting of one's self in the places of others, this taking by one's self of their roles or attitudes, is not merely one of the various aspects or expressions of intelligence or of intelligent behavior, but is the very essence of its character. . . .

the various aspects of that process in which the individual is implicated. In other words, the various elementary selves which constitute, or are organized into, a complete self are the various aspects of the structure of that complete self answering to the various aspects of the structure of the social process as a whole; the structure of the complete self is thus a reflection of the complete social process. The organization and unification of a social group is identical with the organization and unification of any one of the selves arising within the social process in which that group is engaged, or which it is carrying on. . . .[iii]

THE BACKGROUND OF THE GENESIS OF THE SELF

The problem now presents itself as to how, in detail, a self arises. We have to note something of the background of its genesis. First of all there is the conversation of gestures between animals involving some sort of co-operative activity. There the beginning of the act of one is a stimulus to the other to respond in a certain way, while the beginning of this response becomes again a stimulus to the first to adjust his action to the oncoming response. Such is the preparation for the completed act, and ultimately it leads up to the conduct which is the outcome of this preparation. The conversation of gestures, however, does not carry with it the reference of the individual, the animal, the organism, to itself. It is not acting in a fashion which calls for a response from the form itself, although it is conduct with reference to the conduct of others. We have seen, however, that there are certain gestures that do affect the organism as they affect other organisms and may, therefore, arouse in the organism responses of the same character as aroused in the other. Here, then, we have a situation in which the individual may at least arouse responses in himself and reply to these responses, the condition being that the social stimuli have an effect on the individual which is like that which they have on the other. That, for example, is what is implied in language; otherwise language as significant symbol would disappear, since the individual would not get the meaning of that which he says.

The peculiar character possessed by our human social environment belongs to it by virtue of the peculiar character of human social activity; and that character, as we have seen, is to be found in the process of communication, and more particularly in the triadic relation on which the existence of meaning is based: the relation of the gesture of one organism to the adjustive response made to it by another organism, in its indicative capacity as pointing to the completion or resultant of the act it initiates (the meaning of the gesture being thus the response of the second organism to it as such, or as a gesture). What, as it were, takes the gesture out of the social act and isolates it as such—what makes it something more than just an early phase of an individual act—is the response of another organism, or of other organisms, to it. Such a response is its meaning, or gives it its meaning. The social situation and process of behavior are here presupposed by the acts of the individual organisms implicated therein. The gesture arises as a separable element in the social act, by virtue of the fact that it is selected out by the sensitivities of other organisms to it; it does not exist as a gesture merely in the experience of the single individual. The meaning of a gesture by one organism, to repeat, is found in the response of another organism to what would be the completion of the act of the first organism which that gesture initiates and indicates.

We sometimes speak as if a person could build up an entire argument in his mind, and then put it into words to convey it to someone else. Actually, our thinking always takes place by means of some sort of symbols. It is possible that one could have the meaning of "chair" in his experience without there being a symbol, but we would not be thinking about it in that case. We may sit down in a chair without thinking about what we are doing, that is, the approach to the chair is presumably already aroused in our experience, so that the meaning is there. But if one is thinking about the chair he must have some sort of a symbol for it. It may be the form of the chair, it may be the attitude that somebody else takes in sitting down, but it is more apt to be some language symbol that arouses this response. In a thought process there has to be some sort of a symbol that can refer to this meaning, that is, tend to call out this

---

[iii]The unity of the mind is not identical with the unity of the self. The unity of the self is constituted by the unity of the entire relational pattern of social behavior and experience in which the individual is implicated, and which is reflected in the structure of the self; but many of the aspects or features of this entire pattern do not enter into consciousness, so that the unity of the mind is in a sense an abstraction from the more inclusive unity of the self.

response, and also serve this purpose for other persons as well. It would not be a thought process if that were not the case.

Our symbols are all universal.[iv] You cannot say anything that is absolutely particular; anything you say that has any meaning at all is universal. You are saying something that calls out a specific response in anybody else provided that the symbol exists for him in this experience as it does for you. There is the language of speech and the language of hands, and there may be the language of the expression of the countenance. One can register grief or joy and call out certain responses. There are primitive people who can carry on elaborate conversations just by expressions of the countenance. Even in these cases the person who communicates is affected by that expression just as the expects somebody else to be affected. Thinking always implies a symbol which will call out the same response in another that it calls out in the thinker. Such a symbol is a universal of discourse; it is universal in its character. We always assume that the symbol we use is one which will call out in the other person the same response, provided it is a part of his mechanism of conduct. A person who is saying something is saying to himself what he says to others; otherwise he does not know what he is talking about.

There is, of course, a great deal in one's conversation with others that does not arouse in one's self the same response it arouses in others. That is particularly true in the case of emotional attitudes. One tries to bully somebody else; he is not trying to bully himself. There is, further, a whole set of values given in speech which are not of a symbolic character. The actor is conscious of these values; that is, if he assumes a certain attitude he is, as we say, aware that this attitude represents grief. If it does he is able to respond to his own gesture in some sense as his audience does. It is not a natural situation; one is not an actor all of the time. We do at times act and consider just what the effect of our attitude is going to be, and we may deliberately use a certain tone of voice to bring about a certain result. Such a tone arouses the same response in ourselves that we want to arouse in somebody else. But a very large part of what goes on in speech has not this symbolic status. . . .

We do not normally use language stimuli to call out in ourselves the emotional response which we are calling out in others. One does, of course, have sympathy in emotional situations; but what one is seeking for there is something which is, after all, that in the other which supports the individual in his own experience. In the case of the poet and actor, the stimulus calls out in the artist that which it calls out in the other, but this is not the natural function of language; we do not assume that the person who is angry is calling out the fear in himself that he is calling out in someone else. The emotional part of our act does not directly call out in us the response it calls out in the other. If a person is hostile the attitude of the other that he is interested in, an attitude which flows naturally from his angered tones, is not one that he definitely recognizes in himself. We are not frightened by a tone which we may use to frighten somebody else. On the emotional side, which is a very large part of the vocal gesture, we do not call out in ourselves in any such degree the response we call out in others as we do in the case of significant speech. Here we should call out in ourselves the type of response we are calling out in others; we must know what we are saying, and the attitude of the other which we arouse in ourselves should control what we do say. Rationality means that the type of the response which we call out in others should be so called out in ourselves, and that this response should in turn take its place in determining what further thing we are going to say and do.

What is essential to communication is that the symbol should arouse in one's self what it arouses in the other individual. It must have that sort of universality to any person who finds himself in the same situation. There is a possibility of language whenever a stimulus can affect the individual as it affects the other. With a blind person such as Helen Keller, it is a contact experience that could be given to another as it is given to herself. It is out of that sort of language that the mind of

---

[iv]Thinking proceeds in terms of or by means of universals. A universal may be interpreted behavioristically as simply the social act as a whole, involving the organization and interrelation of the attitudes of all the individuals implicated in the act, as controlling their overt responses. This organization of the different individual attitudes and interactions in a given social act, with reference to their interrelations as realized by the individuals themselves, is what we mean by a universal; and it determines what the actual overt responses of the individuals involved in the given social act will be, whether that act be concerned with a concrete project of some sort (such as the relation of physical and social means to ends desired) or with some purely abstract discussion, say the theory of relativity or the Platonic ideas.

Helen Keller was built up. As she has recognized, it was not until she could get into communication with other persons through symbols which could arouse in herself the responses they arouse in other people that she could get what we term a mental content, or a self.

Another set of background factors in the genesis of the self is represented in the activities of play and the game.

Among primitive people, as I have said, the necessity of distinguishing the self and the organism was recognized in what we term the "double": the individual has a thing-like self that is affected by the individual as it affects other people and which is distinguished from the immediate organism in that it can leave the body and come back to it. This is the basis for the concept of the soul as a separate entity.

We find in children something that answers to this double, namely, the invisible, imaginary companions which a good many children produce in their own experience. They organize in this way the responses which they call out in other persons and call out also in themselves. Of course, this playing with an imaginary companion is only a peculiarly interesting phase of ordinary play. Play in this sense, especially the stage which precedes the organized games, is a play at something. A child plays at being a mother, at being a teacher, at being a policeman; that is, it is taking different rôles, as we say. We have something that suggests this in what we call the play of animals: a cat will play with her kittens, and dogs play with each other. Two dogs playing with each other will attack and defend, in a process which if carried through would amount to an actual fight. There is a combination of responses which checks the depth of the bite. But we do not have in such a situation the dogs taking a definite rôle in the sense that a child deliberately takes the rôle of another. This tendency on the part of the children is what we are working with in the kindergarten where the rôles which the children assume are made the basis for training. When a child does assume a rôle he has in himself the stimuli which call out that particular response or group of responses. He may, of course, run away when he is chased, as the dog does, or he may turn around and strike back just as the dog does in his play. But that is not the same as playing at something. Children get together to "play Indian." This means that the child has a certain set of stimuli which call out in itself the responses that they would call out in others, and which answer to an Indian. In the play period the child utilizes his own responses to these stimuli which he makes use of in building a self. The response which he has a tendency to make to these stimuli organizes them. He plays that he is, for instance, offering himself something, and he

buys it; he gives a letter to himself and takes it away; he addresses himself as a parent, as a teacher; he arrests himself as a policeman. He has a set of stimuli which call out in himself the sort of responses they call out in others. He takes this group of responses and organizes them into a certain whole. Such is the simplest form of being another to one's self. It involves a temporal situation. The child says something in one character and responds in another character, and then his responding in another character is a stimulus to himself in the first character, and so the conversation goes on. A certain organized structure arises in him and in his other which replies to it, and these carry on the conversation of gestures between themselves.

If we contrast play with the situation in an organized game, we note the essential difference that the child who plays in a game must be ready to take the attitude of everyone else involved in that game, and that these different rôles must have a definite relationship to each other. Taking a very simple game such as hide-and-seek, everyone with the exception of the one who is hiding is a person who is hunting. A child does not require more than the person who is hunted and the one who is hunting. If a child is playing in the first sense he just goes on playing, but there is no basic organization gained. In that early stage he passes from one rôle to another just as a whim takes him. But in a game where a number of individuals are involved, then the child taking one rôle must be ready to take the rôle of everyone else. If he gets in a ball nine he must have the responses of each position involved in his own position. He must know what everyone else is going to do in order to carry out his own play. He has to take all of these rôles. They do not all have to be present in consciousness at the same time, but at some moments he has to have three or four individuals present in his own attitude, such as the one who is going to throw the ball, the one who is going to catch it, and so on. These responses must be, in some degree, present in his own make-up. In the game, then, there is a set of responses of such others so organized that the attitude of one calls out the appropriate attitudes of the other. . . .

PLAY, THE GAME, AND THE GENERALIZED OTHER

The fundamental difference between the game and play is that in the latter the child must have the attitude of all the others involved in that game. The attitudes of the other players which the participant assumes organize into a sort of unit, and it is that organization which controls the response of the individual. The illustration used was of a person playing baseball. Each one of his own acts is determined by his assumption of the action of the others who are playing the game. What he does is controlled by his being everyone else on that team, at least in so far as those attitudes affect his own particular response. We get then an "other" which is an organization of the attitudes of those involved in the same process.

The organized community or social group which gives to the individual his unity of self may be called "the generalized other." The attitude of the generalized other is the attitude of the whole community.[v] Thus, for example, in the case of such a social group as a ball team, the team is the generalized other in so far as it enters—as an organized process or social activity—into the experience of any one of the individual members of it.

If the given human individual is to develop a self in the fullest sense, it is not sufficient for him merely to take the attitudes of other human individuals toward himself and toward one another within the human social process, and to bring that social process as a whole into his individual experience merely in these terms: he must also, in the same way that he takes the attitudes of other individuals toward himself and toward one another, take their attitudes toward the various phases or aspects of the common social activity or set of social undertakings in which, as members of an organized society or social group, they are all engaged; and he must then, by generalizing these individual attitudes of that organized society or social group itself, as a whole, act toward different social projects which at any given time it is carrying out, or toward the various larger phases of the general social process which constitutes

its life and of which these projects are specific manifestations. This getting of the broad activities of any given social whole or organized society as such within the experiential field of any one of the individuals involved or included in that whole is, in other words, the essential basis and prerequisite of the fullest development of that individual's self: only in so far as he takes the attitudes of the organized social group to which he belongs toward the organized, co-operative social activity or set of such activities in which that group as such is engaged, does he develop a complete self or possess the sort of complete self he has developed. And on the other hand, the complex co-operative processes and activities and institutional functionings of organized human society are also possible only in so far as every individual involved in them or belonging to that society can take the general attitudes of all other such individuals with reference to these processes and activities and institutional functionings, and to the organized social whole of experiential relations and interactions thereby constituted—and can direct his own behavior accordingly.

It is in the form of the generalized other that the social process influences the behavior of the individuals involved in it and carrying it on, i.e., that the community exercises control over the conduct of its individual members; for it is in this form that the social process or community enters as a determining factor into the individual's thinking. In abstract thought the individual takes the attitude of the generalized other[vi] toward himself, without reference to its expression in any particular other individuals; and in concrete thought he takes that attitude in so far as it is expressed in the attitudes

---

[v]It is possible for inanimate objects, no less than for other human organisms, to form parts of the generalized and organized—the completely socialized—other for any given human individual, in so far as he responds to such objects socially or in a social fashion (by means of the mechanism of thought, the internalized conversation of gestures). Any thing—any object or set of objects, whether animate or inanimate, human or animal, or merely physical—toward which he acts, or to which he responds, socially, is an element in what for him is the generalized other; by taking the attitudes of which toward himself he becomes conscious of himself as an object or individual, and thus develops a self or personality. Thus, for example, the cult, in its primitive form, is merely the social embodiment of the relation between the given social group or community and its physical environment—an organized social means, adopted by the individual members of that group or community, of entering into social relations with that environment, or (in a sense) of carrying on conversations with it; and in this way that environment becomes part of the total generalized other for each of the individual members of the given social group or community.

[vi]We have said that the internal conversation of the individual with himself in terms of words or significant gestures—the conversation which constitutes the process or activity of thinking—is carried on by the individual from the standpoint of the "generalized other." And the more abstract that conversation is, the more abstract thinking happens to be, the further removed is the generalized other from any connection with particular individuals. It is especially in abstract thinking, that is to say, that the conversation involved is carried on by the individual with the generalized other, rather than with any particular individuals. Thus it is, for example, that abstract concepts are concepts stated in terms of the attitudes of the entire social group or community; they are stated on the basis of the individual's consciousness of the attitudes of the generalized other toward them, as a result of his taking these attitudes of the generalized other and then responding to them. And thus it is also that abstract propositions are stated in a form which anyone—any other intelligent individual—will accept.

toward his behavior of those other individuals with whom he is involved in the given social situation or act. But only by taking the attitude of the generalized other toward himself, in one or another of these ways, can he think at all; for only thus can thinking—or the internalized conversation of gestures which constitutes thinking—occur. And only through the taking by individuals of the attitude or attitudes of the generalized other toward themselves is the existence of a universe of discourse, as that system of common or social meanings which thinking presupposes at its context, rendered possible.

The self-conscious human individual, then, takes or assumes the organized social attitudes of the given social group or community (or of some one section thereof) to which he belongs, toward the social problems of various kinds which confront that group or community at any given time, and which arise in connection with the correspondingly different social projects or organized co-operative enterprises in which that group or community as such is engaged; and as an individual participant in these social projects or co-operative enterprises, he governs his own conduct accordingly. In politics, for example, the individual identifies himself with an entire political party and takes the organized attitudes of that entire party toward the rest of the given social community and toward the problems which confront the party within the given social situation; and he consequently reacts or responds in terms of the organized attitudes of the party as a whole. He thus enters into a special set of social relations with all the other individuals who belong to that political party; and in the same way he enters into various other special sets of social relations, with various other classes of individuals respectively, the individuals of each of these classes being the other members of some one of the particular organized subgroups (determined in socially functional terms) of which he himself is a member within the entire given society or social community. In the most highly developed, organized, and complicated human social communities—those evolved by civilized man—these various socially functional classes or subgroups of individuals to which any given individual belongs (and with the other individual members of which he thus enters into a special set of social relations) are of two kinds. Some of them are concrete social classes or subgroups, such as political parties, clubs, corporations, which are all actually functional social units, in terms of which their individual members are directly related to one another. The others are abstract social classes or subgroups, such as the class of debtors and the class of creditors, in terms of which their individual members are related to one another only more or less indirectly, and which only more or less indirectly function as social units, but which afford or represent unlimited possibilities for the widening and ramifying and enriching of the social relations among all the individual members of the given society as an organized and unified whole. The given individual's membership in several of these abstract social classes or subgroups makes possible his entrance into definite social relations (however indirect) with an almost infinite number of other individuals who also belong to or are included within one or another of these abstract social classes or subgroups cutting across functional lines of demarcation which divide different human social communities from one another, and including individual members from several (in some cases from all) such communities. Of these abstract social classes or subgroups of human individuals the one which is most inclusive and extensive is, of course, the one defined by the logical universe of discourse (or system of universally significant symbols) determined by the participation and communicative interaction of individuals; for of all such classes or subgroups, it is the one which claims the largest number of individual members, and which enables the largest conceivable number of human individuals to enter into some sort of social relation, however indirect or abstract it may be, with one another—a relation arising from the universal functioning of gestures as significant symbols in the general human social process of communication. . . .

The game has a logic, so that such an organization of the self is rendered possible: there is a definite end to be obtained; the actions of the different individuals are all related to each other with reference to that end so that they do not conflict; one is not in conflict with himself in the attitude of another man on the team. If one has the attitude of the person throwing the ball he can also have the response of catching the ball. The two are related so that they further the purpose of the game itself. They are interrelated in a unitary, organic fashion. There is a definite unity, then, which is introduced into the organization of other selves when we reach such a stage as that of the game, as over against the situation of play where there is a simple succession of one rôle after another, a situation which is, of course, characteristic of the child's own personality. The child is one thing at one time and another at another, and what he is at one moment does not determine what he is at another. That is both the charm of childhood as well as its inadequacy. You cannot count on the child; you cannot assume that all the things he does are going to determine what he will do at any moment. He is not organized into a whole. The child has no definite character, no definite personality.

The game is then an illustration of the situation out of which an organized personality arises. In so far as the child does take the attitude of the other and allows that attitude of the other to determine the thing he is going to do with reference to a common end, he is becoming an organic member of society. He is taking over the morale of that society and is becoming an essential member of it. He belongs to it in so far as he does allow the attitude of the other that he takes to control his own immediate expression. What is involved here is some sort of an organized process. That which is expressed in terms of the game is, of course, being continually expressed in the social life of the child, but this wider process goes beyond the immediate experience of the child himself. The importance of the game is that it lies entirely inside of the child's own experience, and the importance of our modern type of education is that it is brought as far as possible within this realm. The different attitudes that a child assumes are so organized that they exercise a definite control over his response, as the attitudes in a game control his own immediate response. In the game we get an organized other, a generalized other, which is found in the nature of the child itself, and finds its expression in the immediate experience of the child. And it is that organized activity in the child's own nature controlling the particular response which gives unity, and which builds up his own self. . . .

We may illustrate our basic concept by a reference to the notion of property. If we say "This is my property, I shall control it," that affirmation calls out a certain set of responses which must be the same in any community in which property exists. It involves an organized attitude with reference to property which is common to all the members of the community. One must have a definite attitude of control of his own property and respect for the property of others. Those attitudes (as organized sets of responses) must be there on the part of all, so that when one says such a thing he calls out in himself the response of the others. He is calling out the response of what I have called a generalized other. That which makes society possible is such common responses, such organized attitudes, with reference to what we term property, the cults of religion, the process of education, and the relations of the family. Of course, the wider the society the more definitely universal these objects must be. In any case there must be a definite set of responses, which we may speak of as abstract, and which can

belong to a very large group. Property is in itself a very abstract concept. It is that which the individual himself can control and nobody else can control. The attitude is different from that of a dog toward a bone. A dog will fight any other dog trying to take the bone. The dog is not taking the attitude of the other dog. A man who says "This is my property" is taking an attitude of the other person. The man is appealing to his rights because he is able to take the attitude which everybody else in the group has with reference to property, thus arousing in himself the attitude of others.

What goes to make up the organized self is the organization of the attitudes which are common to the group. A person is a personality because he belongs to a community, because he takes over the institutions of that community into his own conduct. He takes its language as a medium by which he gets his personality, and then through a process of taking the different rôles that all the others furnish he comes to get the attitude of the members of the community. Such, in a certain sense, is the structure of a man's personality. There are certain common responses which each individual has toward certain common things, and in so far as those common responses are awakened in the individual when he is affecting other persons he arouses his own self. The structure, then, on which the self is built is this response which is common to all, for one has to be a member of a community to be a self. Such responses are abstract attitudes, but they constitute just what we term a man's character. They give him what we term his principles, the acknowledged attitudes of all members of the community toward what are the values of that community. He is putting himself in the place of the generalized other, which represents the organized responses of all the members of the group. It is that which guides conduct controlled by principles, and a person who has such an organized group of responses is a man whom we say has character, in the moral sense. . . .

I have so far emphasized what I have called the structures upon which the self is constructed, the framework of the self, as it were. Of course we are not only what is common to all: each one of the selves is different from everyone else; but there has to be such a common structure as I have sketched in order that we may be members of a community at all. We cannot be ourselves unless we are also members in whom there is a community of attitudes which control the attitudes of all. We cannot have rights unless we have common attitudes. That which we have acquired as self-conscious persons makes us such members of society and gives us selves. Selves can only exist in definite relationships to other selves. No hard-and-fast line can be drawn between our own selves and the selves of others, since our own selves exist and enter as such into our experience only in so far as the selves of others exist and enter as such into our experience also. The individual possesses a self only in relation to the selves of the other members of his social group; and the structure of his self expresses or reflects the general behavior pattern of this social group to which he belongs, just as does the structure of the self of every other individual belonging to this social group. . . .

The Self and the Subjective

There is one other matter which I wish briefly to refer to now. The only way in which we can react against the disapproval of the entire community is by setting up a higher sort of community which in a certain sense outvotes the one we find. A person may reach a point of going against the whole world about him; he may stand out by himself over against it. But to do that he has to speak with the voice of reason to himself. He has to comprehend the voices of the past and of the future. That is the only way in which the self can get a voice which is more than the voice of the community. As a rule we assume that this general voice of the community is identical with the larger community of the past and the future; we assume that an organized custom represents what we call morality. The things one cannot do are those which everybody would condemn. If we take the attitude of the community over against our own responses, that is a true statement, but we must not forget this other capacity, that of replying to the community and insisting on the gesture of the community changing. We can reform the order of things; we can insist on making the community standards better standards. We are not simply bound by the community. We are engaged in a conversation in which what we say is listened to by the community and its response is one which is affected by what we have to say. This is especially true in critical situations. A man rises up and defends himself for what he does; he has his "day in court"; he can present his views. He can perhaps change the attitude of the community toward himself. The process of conversation is one in which the individual has not only the right but the duty of talking to the community of which he is a part, and bringing about those changes which take place through the interaction of individuals. That is the way, of course, in which society gets ahead, by just such interactions as those in which some person thinks a thing out. We are continually changing our social system in some respects, and we are able to do that intelligently because we can think. . . .

The "I" and the "Me"

The "I" is the response of the organism to the attitudes of the others; the "me" is the organized set of attitudes of others which one himself assumes. The attitudes of the others constitute the organized "me," and then one reacts toward that as an "I." I now wish to examine these concepts in greater detail.

There is neither "I" nor "me" in the conversation of gestures; the whole act is not yet carried out, but the preparation takes place in this field of gesture. Now, in so far as the individual arouses in himself the attitudes of the others, there arises an organized group of responses. And it is due to the individual's ability to take the attitudes of these others in so far as they can be organized that he gets self-consciousness. The taking of all of those organized sets of attitudes gives him his "me"; that is the self he is aware of. He can throw the ball to some other member because of the demand made upon him from other members of the team. That is the self that immediately exists for him in his consciousness. He has their attitudes, knows what they want and what the consequence of any act of his will be, and he has assumed responsibility for the situation. Now, it is the presence of those organized sets of attitudes that constitutes that "me" to which he as an "I" is responding. But what that response will be he does not know and nobody else knows. Perhaps he will make a brilliant play or an error. The response to that situation as it appears in his immediate experience is uncertain, and it is that which constitutes the "I."

The "I" is his action over against that social situation within his own conduct, and it gets into his experience only after he has carried out the act. Then he is aware of it. He had to do such a thing and he did it. He fulfils his duty and he may look with pride at the throw which he made. The "me" arises to do that duty—that is the way in which it arises in his experience. He had in him all the attitudes of others, calling for a certain response; that was the "me" of that situation, and his response is the "I."

I want to call attention particularly to the fact that this response of the "I" is something that is more or less uncertain. The attitudes of others which one assumes as affecting his own conduct constitute the "me," and that is something that is there, but the response to it is as yet not given. When one sits down to think anything out, he has certain data that are there. Suppose that it is a social situation which he has to straighten out. He sees himself from the point of view of one individual or another in the group. These individuals, related all together, give him a certain self. Well, what is he going to do? He does not know and nobody else knows. He can get the situation into his experience because he can assume the attitudes of the various individuals involved in it. He knows how they feel about it by the assumption of their attitudes. He says, in effect, "I have done certain things that seem to commit me to a certain course of conduct." Perhaps if he does so act it will place him in a false position with another group. The "I" as a response to this situation, in contrast to the "me" which is involved in the attitudes which he takes, is uncertain. And when the response takes place, then it appears in the field of experience largely as a memory image. . . .

The "I," then, in this relation of the "I" and the "me," is something that is, so to speak, responding to a social situation which is within the experience of the individual. It is the answer which the individual makes to the attitude which others take toward him when he assumes an attitude toward them. Now, the attitudes he is taking toward them are present in his own experience, but his response to them will contain a novel element. The "I" gives the sense of freedom, of initiative. The situation is there for us to act in a self-conscious fashion. We are aware of ourselves, and of what the situation is, but exactly how we will act never gets into experience until after the action takes place.

Such is the basis for the fact that the "I" does not appear in the same sense in experience as does the "me." The "me" represents a definite organization of the community there in our own attitudes, and calling for a response, but the response that takes place is something that just happens. There is no certainty in regard to it. There is a moral necessity but no mechanical necessity for the act. When it does take place then we find what has been done. The above account gives us, I think, the relative position of the "I" and "me" in the situation, and the grounds for the separation of the two in behavior. The two are separated in the process but they belong together in the sense of being parts of a whole. They are separated and yet they belong together. The separation of the "I" and the "me" is not fictitious. They are not identical, for, as I have said, the "I" is something that is never entirely calculable. The "me" does call for a certain sort of an "I" in so far as we meet the obligations that are given in conduct itself, but the "I" is always something different from what the situation itself calls for. So there is always that distinction, if you like, between the "I" and the "me." The "I" both calls out the "me" and responds to it. Taken together they constitute a personality as it appears in social experience. The self is essentially a social process going on with these two distinguishable phases. If it did not have these two phases there could not be conscious responsibility, and there would be nothing novel in experience.

Discussion Questions

1. How does Mead define self-consciousness? What role does language play in the development of self-consciousness? What role does self-consciousness play in social interaction?

2. Compare and contrast Mead's concept of the "I," the "me," and the "generalized other" with Simmel's views on the relationship between the self and group affiliations. How might Simmel's concept of the "stranger" be used to critique Mead's understanding of the "generalized other"?

3. Can television, and mass media more generally, affect the development of the self? If so, how?

4. Given his emphasis on the self, language, thinking, and meaning, what factors might be missing from Mead's view of social interaction?

5. What are some of the implications of Du Bois's discussion on race and Gilman's discussion on gender for Mead's theory of "taking the attitude of the other"?

Part III

Twentieth-Century Sociological Traditions

9 STRUCTURAL FUNCTIONALISM

Talcott C. Parsons

Key Concepts

- Role/Role-Set
- Unit Act
- Social System
- Cultural System
- Personality System
- Pattern Variables
- Socialization
- Internalization
- Institutionalization

Robert K. Merton

Key Concepts

- Manifest Function
- Latent Function
- Deviance
- Dysfunction

A particularly important feature of all systems is the inherent limitation on the compatibility of certain parts or events within the same system. This is indeed simply another way of saying that the relations within the system are determinate and not just anything can happen.

—Parsons and Shils ([1951] 2001:107)

From the 1930s through the 1970s, structural functionalism was the dominant theoretical approach in American sociology. Functionalists coined pivotal concepts, such as "role," "norm," and "social system," that came to form the basic building blocks of contemporary sociology. Moreover, a few functionalist concepts, such as "role model" and "self-fulfilling prophecy," have entered our colloquial vocabulary as well. Yet, structural functionalism is most well known not for the specific concepts that it introduced but rather for the metatheoretical framework on which it is based. Structural functionalists envision society as a system of interrelated parts, and they emphasize how the different parts work together for the good of the system. The classic structural functionalist image of society is as an organism, such as the body, with different parts (e.g., limbs, brain, liver) working together in an interdependent way.

In addition, structural functionalists emphasize "systems within systems." For instance, while each family can be considered its own self-contained "system" or unit, it is also a *component* of society as a whole. Other major components of society include the economy—the *system* for providing goods and services to members of that society; the government (or political realm)—the *system* for determining the rules for that society and the distribution of power; and the religious system—the *system* that provides individuals with core values and a sense of meaning. In short, for structural functionalists, just as the body is a system with specific parts (e.g., limbs, brain, liver) that ensure its overall functioning, so, too, society is a system with specific parts (family, government, economy, religion, etc.) necessary for its very survival. While each of these components can and must be studied separately in order to thoroughly understand them, structural functionalists typically emphasize how the various systems and subsystems work together.

During the 1970s, the image of society as a system of interrelated parts was harshly rejected by those sociologists who emphasized that society was based on conflict, not consensus, among social groups. These sociologists, called "conflict theorists" (such as C. Wright Mills—see Significant Others box on p. 236), underlined how political and economic elites and the organizations they controlled worked for their *own* benefit, rather than for the benefit of society as a whole. Structural functionalism came to be viewed as, at best, an old-fashioned tradition with a conservative bias and, at worst, a perspective that legitimated the exploiters, augmenting social problems rather than helping to correct them.[1]

---

[1]Conflict theorists criticized structural functionalists for evoking a highly idealized image of society. While much of this criticism is valid, conflict theorists mistakenly interpreted the functionalist analogy of society as an organism to mean a *healthy* organism, but this is not necessarily the case. So, too, Parsons is commonly criticized for conflating his theory of society with value judgments as to what is desirable in societies; however, his intent was not to advocate for a particular system but rather to provide an analytical framework for analyzing it. Nevertheless, as you will see in Parsons's "empirical description" of the American family, and as postmodernism (see Chapter 15) today makes eminently clear, social science is not at all a wholly objective, empirical enterprise, and a myriad of gender, racial, cultural, class, and other unacknowledged biases/lenses are readily apparent in the structural functionalism of the 1940s and 1950s.

C. Wright Mills (1916–1962): An American Critic

Though Parsons's variant of structural functionalism dominated American sociology his views were not embraced by all. One of the more vocal critics of Parsons's work was C. Wright Mills. Born in Texas, Mills's foray into academics was launched when he won a research fellowship to attend the University of Wisconsin. After earning his doctorate, Mills went on to Columbia University, where he remained until his untimely death at the age of 46. Despite his passing at a young age, his penetrating analyses of American political, social, and intellectual life have led at least one reputable observer to consider Mills "the greatest sociologist the United States has ever produced" (Horowitz 1963:20).

To what did Mills owe such a reputation? Committed to a vision of a more just and moral society, he was throughout his career a relentless critic of the self-congratulatory hypocrisy that in his view pervaded American culture. For instance, in *The Power Elite* (1958), he detailed the undemocratic character of America's allegedly democratic governance. Far from being subject to the "will of the people,"American politics is becoming dominated more and more by a small interconnected group of political, economic and military leaders. Power has become increasingly centralized within "the big three" institutions, while the public has been transformed into an impotent "mass society," and elected representatives in Congress have become increasingly servile and ineffectual. In American society, important decisions are made by an elite circle of individuals who move from leadership positions in one institutional domain to another. One need only look at the current roster of cabinet members in the executive branch to see how the revolving door works.

In *White Collar* (1951), Mills turns his attention to the plight of the American middle classes who, owing to their "status panic," are unable to realize a meaningful existence. A sense of powerlessness often characterizes the growing ranks of white-collar professionals as their daily lives become increasingly routinized and regimented under the demands of bureaucratic efficiency. While white-collar professionals may have achieved a semblance of economic security, they have traded it for a sense of purpose and the ability to control their own destiny. Alienated from their work and insecure about their status, white-collar professionals often turn to the world of leisure to provide succor. For Mills, this state of affairs has produced psychologically and politically fragmented individuals who are unable to recognize the true sources of their discontent. Akin to the Frankfurt School theorists to be discussed in the next chapter, Mills feared that the malaise of the middle classes and their embracing of a vacuous mass culture left society vulnerable to the rise of authoritarianism.

Not one to avoid controversy, Mills turned his critical outlook onto his own peers. His classic introduction to sociology, *The Sociological Imagination* (1959), not only offers the definitive statement on the task of the discipline, but also reproaches those who are charged with carrying it out. First, Mills exalts sociology as uniquely able to "grasp history and biography and the relations between the two within society" (1959:6). In doing so, the sociological imagination enables its possessor to make the essential distinction between "the personal troubles of milieu" and "the public issues of social structure" (ibid.). Yet, to Mills's dismay, the promise of sociology was being lost, to the extent that academic research is rooted in an "abstracted empiricism" that concerns itself less with addressing pressing human needs than with uncovering scientific facts. Mills saw in his fellow sociologists' quest for "facts" an abdication of their social responsibility to advance a more humane, democratic society.

In this chapter, we explore the basic premises of structural functionalism, an extremely important type of theory not only because it dominated American sociology throughout much of the twentieth century but also because it still informs sociological thinking today. As you will see later in this book, contemporary theorists (such as Jürgen Habermas and Anthony Giddens—see Chapter 16) draw extensively from and/or explicitly "talk back to" central functionalist figures (most important, Talcott Parsons). Sociologists still seek to explain the relations among various parts of society; the extent to which our social world is a product of preexisting structures and conditions as opposed to individual "free will"; and the relation between subjective emotions, symbols, and values and objective, strategic calculation—all vital issues at the heart of structural functionalism. In addition, understanding the basic premises of structural functionalism is necessary for understanding the approaches that sought to upend it in the 1970s.

In the first part of this chapter, we focus on the work of arguably the single most important theorist in the tradition of structural functionalism: Talcott Parsons. In the second part of this chapter, we turn to one of Parsons's most influential students, Robert Merton, who developed and extended Parsons's work.

TALCOTT C. PARSONS (1902–1979): A BIOGRAPHICAL SKETCH ▪▪

Talcott Parsons was born in 1902 in Colorado Springs, Colorado. He was the fifth and last child of Mary Ingersol Parsons, a suffragist and backer of various progressive causes, and Edward Smith Parsons, an ordained Congregational minister who later became president of Marietta College in Ohio. Talcott Parsons acknowledged that his parents' values influenced him considerably. Theirs was a liberal household, in which morality, the modern industrial system, economic individualism, and exploitation of labor were topics of concern. After graduating from high school in 1920, Parsons attended Amherst College (as had his father and two older brothers). He studied the natural sciences, particularly biology, as well as philosophy and the social sciences, graduating magna cum laude in 1924. Thereafter, he attended the London School of Economics as a graduate student, where he met another American economics student, Helen Walker, whom he married in 1927. It was at the London School of Economics that Parsons was introduced both to sociology and to the great functionalist anthropologist Bronislaw Malinowski, with whom he took two courses.

During 1925 and 1926 Parsons studied at the University of Heidelberg, Germany, where he was exposed to and deeply influenced by the work of Max Weber (who had died in 1920). Following Weber, Parsons sought to explore the "relations between economic and sociological theory" (Camic 1991:xxii). He wrote his doctoral dissertation on the role of capitalism in German literature (highlighting both Weber and Marx), and he translated Weber's now-classic *The Protestant Ethic and the Spirit of Capitalism* ([1904] 1958) into English, thereby introducing Weber to the English-speaking academic world.

In 1927, Parsons began teaching at Harvard University, where he would remain until his retirement in 1973. At first, Parsons taught in the department of economics since a sociology department was not founded until 1930. Parsons became a tenured professor of sociology in 1939 and chair of the Harvard Sociology Department in 1944. In 1949 Parsons was elected president of the American Sociological Association. Throughout the 1950s and 1960s, Parsons was a dominant figure in American sociology. He continued to teach and give lectures right up until his death in 1979, although by then an anti-Parsonian backlash was in full swing. C. Wright Mills and other critics charged Parsons with a conservative bias as well as academic elitism. The viciousness of these attacks make Parsons unique not only in terms of the profundity of his influence but also the extent to which his work has been disparaged. As former student, Robin Williams, Jr. (1980:66), notes, "few sociologists of our time have been more subjected to stereotyping, to careless *ad hoc* readings, and to selectively distorted interpretations." By the late 1980s, however, the fervor of anti-Parsonianism had died down, and contemporary theorists—most important, Jeffrey C. Alexander, Jürgen Habermas (see Chapter 16), Anthony Giddens (see Chapter 16), and Richard Münch—began to field a new appreciation for Parsons's work.

⠿ INTELLECTUAL INFLUENCES AND CORE IDEAS

Between his first sole-authored publication in 1928 and his death in 1979, Talcott Parsons enjoyed a long and productive career during which he wrote 17 books and more than 200 articles. This illustrious career can be divided into three phases. In his "early period" (roughly 1928–1937), Parsons focused on bringing classical European theorists into American sociology in order to upend the individualistic and rationalistic utilitarian models prevalent in American sociology at the time. Parsons sought to replace the pragmatic, "grounded" theory dominant in American sociology and most evident in the detailed microlevel work of the Chicago School (see Chapter 12) with an all-encompassing theory of action based on a synthesis of the works of classical European thinkers, particularly Émile Durkheim and Max Weber (see Chapters 3 and 4). This effort culminated in *The Structure of Social Action* (1937). Although it has since become a classic, initially, *The Structure of Social Action* was roundly criticized for its high level of abstraction. One empirically minded critic disparaged Parsons's theory of action as "about as scientifically useful as a sonnet to a skylark" (Bierstadt 1938:18, as cited by Gerhardt 2002:2). For that matter, Parsons's theory of action continues to be maligned for its high level of abstraction even today.

Parsons's second intellectual phase (roughly 1937–1950) is often termed "structural functionalism." It is during this period that Parsons wrote his most famous works, and it is for the ideas developed during this period that he is most well known. This period culminated in 1951 with the publication of two pivotal works: *Toward a General Theory of Action* (coauthored with Edward A. Shils) and *The Social System.* In these works Parsons attempts to resolve the theoretical problems he himself had posed earlier. Specifically, fascinated by Freud and having read Freud thoroughly, Parsons integrated psychoanalytic theory into his general theory of action. In so doing, Parsons added an individualistic bent to what had been a primarily collectivistic orientation (relying as it did on Durkheim and Weber). More generally, Parsons sought a conceptual convergence and synthesis in social psychology, sociocultural anthropology, and sociology. Indeed, *Toward a General Theory of Action* is a collaborative work that originated in a grant from the Carnegie Corporation (a charitable educational trust) that brought social scientists from other universities, including the sociologist Edward Shils (1910–1995) from Chicago, to participate in a regular series of seminars with Parsons and his colleagues in the department of social relations at Harvard in 1949.

In his third and final intellectual phase (roughly 1952–1979), Parsons developed his "interchange model" (dubbed by his students the "AGIL scheme") and focused on the evolution of societies. In this period, Parsons shifted back to his preoccupation with political and economic systems (recall his early interest in Max Weber), epitomized in his book *Politics and Social Structure* (1969).

Throughout these three periods, Parsons wrote on a wide variety of topics—including education, media, politics, and the family—and introduced into sociology a wide variety of terms, including *norms, values, roles, social systems, social structures,* and *social institutions.* However, Parsons (who called himself an "incurable theorist") never lost sight of his central theoretical concern (Rocher [1972] 1974:1). Parsons sought to devise a conceptual and theoretical framework with which to analyze the basis of all action and social organization. It is for this all-encompassing, overarching, metatheoretical framework that Parsons is most well known.

Action Systems and Social Systems

Perhaps Parsons's single most important idea is that action must not be viewed in isolation. Rather, action must be understood as a "process in time," or as a *system.* As Parsons and Shils ([1951] 2001:54) explicitly state, "actions are not empirically discrete but occur in constellations we call systems." To underscore this point, Parsons used the term **unit act** to refer to a hypothetical actor in a hypothetical situation bounded by an array of parameters and conditions (required effort, ends or goals, situation, norms). In other words, instead of construing action in terms of something concrete (such as a business or an individual), Parsons conceptualized action *systems* as a means for analyzing social phenomena.

More generally, he saw social action as composed of four basic elements that distinguish it from isolated, individual behavior:

1. It is oriented toward attainment of ends or goals.

2. It takes place in *situations,* consisting of the physical and social objects to which the actor relates.

3. It is normatively regulated (i.e., regulated by *norms* that guide the orientation of action).

4. It involves expenditure of *effort* or energy.[2]

This model of social action is illustrated in Figure 9.1.

Figure 9.1   Parsons's Model of Social Action

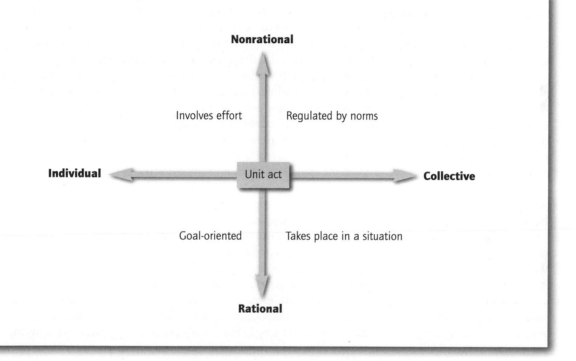

Parsons (1937) and Parsons and Shils ([1951] 2001) further maintain that actions are *organized* into three modes or realms: social systems, personality systems, and cultural systems. These systems are analytically rather than empirically distinct. That is, these systems (and their subsystems) are not physically separate entities but rather a simplified model of society that Parsons and Shils use to explain the organization of action. Put in another way, social systems, personality systems, and cultural systems *undergird* all action and all social life.[3]

---

[2]Parsons and Shils ([1951] 2001:53). Elsewhere, Parsons differentiates *symbols* (by which actors attribute meaning to situations) from *norms* (rules for behavior).

[3]Elsewhere Parsons includes a fourth system, the behavioral organism, which in contemporary terms is equivalent to the "body," or genetic composition. Here, however, he combines the biological (behavioral organism) and the psychological into the personality system, saying that the personality is organized by both "organic" and "emotional" needs.

The **social system** refers to the level of integrated interaction between two or more actors. It also involves cognizance of the other actors' ideas and/or intentions (whether at a relatively conscious or unconscious level), as well as shared norms or expectations and interdependence (Parsons and Shils [1951] 2001:55). In other words, social systems are not material structures or institutions (such as a university) but rather a complex arrangement of interconnected social roles (Alexander 1987:42). Parsons and Shils ([1951] 2001:23) define **roles** as complementary, detailed sets of obligations for interaction. Role-set theory begins with the idea that "each social status involves not a single associated role, but an array of roles." A **role-set** is that complement of interdependent social relationships in which persons are involved simply because they occupy a particular social status (ibid.). For instance, in an everyday situation such as buying groceries, you enter the store, walk around and pick out what you want, and go to the cashier to pay for it. Order and predictability ensue because you, as well as the other actors in the role-set (that is, other shoppers and store employees), know how to play your respective roles (e.g., shopper/cashier).

The **personality system** refers to a system of action organized by *need-dispositions,* both organic (e.g., "drives") and emotional, at the level of the individual. Although the personality system is the source of a distinctive and unique self, Parsons does not construe it as autonomous in the sense that psychologists typically do. Rather, for Parsons, the personality is a distinct level of social life; physical separateness of one's body never entails complete social or cultural differentiation, because personal uniqueness is itself a function of interaction and socialization (Alexander 1987:39). For instance, returning to our example of shopping for groceries, although there is an orderliness and a predictability to the process, not everyone shops for food in exactly the same way. Some people are very organized and follow an exact list (perhaps even outlined according to sections of the grocery store), while others shop more spontaneously, buying on impulse, rather than according to a strict plan. In either case, the particular way that a person shops reflects (to some extent) emotional as well as organic "need-dispositions" at the level of personality.

The **cultural system** is made up of the "values, norms and symbols which guide the choices made by actors and which limit the type of interaction which may occur among actors" (Parsons and Shils [1951] 2001:55). It is composed of intangible ideas and broad symbolic patterns of meaning that establish boundaries of social behavior. In the example of shopping for food, whether you help yourself to fruit or ask a clerk for assistance is not solely a "choice" at the level of personality, nor is it an issue of mere availability at the social system level. Rather, it is set by custom—as anyone who has tried to help themselves in a store that is not "self-help" will know. So, too, even seemingly "personal" likes and dislikes are also a function of one's cultural environment; that is, they are rooted in specific systems of meaning of time and place. Whether your favorite ice cream flavor is *ube* (purple yam), green tea, or peanut butter and chocolate might reflect whether you grew up in Manila (the Philippines), Tokyo (Japan), or Los Angeles (U.S.), respectively. In sum, what we eat is determined, to a large extent, not only by instrumental concerns (e.g., price and convenience) and personal likes and dislikes but also by the taken-for-granted norms and symbols of the particular environment in which we are a part. While this is most evident in the case of sensationalistic regional dietary variations (guinea pig in Ecuador, insects in Africa) and explicitly religious/philosophical variations (e.g., Orthodox Judaism, veganism), it is also true at a more routine, relatively unconscious, habitual level.

As shown in Figure 9.2, the personality, social, and cultural systems interpenetrate each other through socialization, internalization, and institutionalization. **Socialization** refers to the process by which individuals come to regard specific norms as binding. It necessarily involves a community, as it is a process of social learning. **Internalization** refers to the process by which the individual personality system incorporates some specific interpretation of cultural symbols into its need-dispositions. Finally, cultural values and norms are institutionalized at the level of the social system. **Institutionalization** refers to the long-standing processes of communal association that bind actors to particular meanings. Institutionalization privileges particular symbolic constructions and, at the same time, curtails resistance to social norms (Münch 1994:26–28).

Thus, for instance, children in a family of Orthodox Jews are taught, or *socialized* into, the rules for keeping kosher (*kashrut*). Consequently, they acquire, or *internalize,* a specific set of values and identity.

Figure 9.2 Parsons's Model of the Interpenetration of the Cultural, Social, and Personality Systems

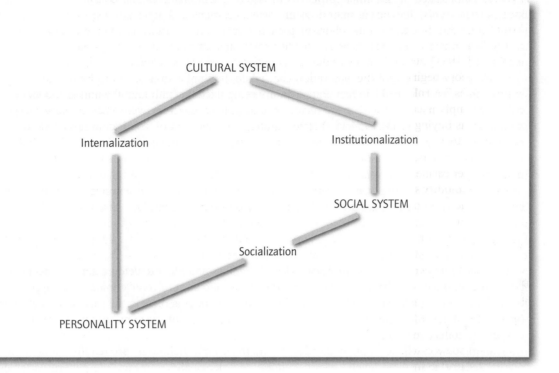

However, what the practice entails is not hashed out anew in each and every Orthodox family. Rather, keeping kosher (which originates in the Torah and is further developed in later rabbinic literature) is *institutionalized* at the level of the social system; it is tradition and social and cultural codes that dictate what is involved in keeping kosher. As this example demonstrates, the processes of socialization, internalization, and institutionalization are intertwined: They involve the personality, cultural, and social realms at once.

One of the most contentious aspects of Parsons's systems theory is the supposition that the personality, social, and cultural systems function together to produce social order and stability. According to Parsons, system equilibrium ensues when the needs of the personality mesh with the resources available in the social system and cultural values and norms (Alexander 1987:47). Thus, for instance, in one essay you will read the Readings at the end of the chapter, Parsons ([1943] 1954) describes the generation gap typical of contemporary American society as a time in which teenagers' need-dispositions are out of sync with the values of their family as well as the wider society. As such, there is a conflict (or gap) between the *personality* needs of the teenager (e.g., to stay out late, wear "weird" clothes) and the prevailing *cultural* values and norms (e.g., to respect authority, aspire toward educational and occupational success) as teenagers adopt identities, systems of meanings, and social roles that contradict the expectations of parents and teachers.

Consequently, many critics contend that Parsons's model of societal equilibrium idealizes social conformity, that it suggests that the goal of society is complete institutionalization of norms and values, and that we should follow the rules for the "good of the system." In fact, however, Parsons argued that equilibrium is only a *theoretical* point of reference. The very notion of "action systems" suggests that systems are never completely stable; action itself is necessarily a disequilibrating factor, since by its very nature it involves change. In other words, although critics have long condemned Parsons for his "static" model of social order that emphasizes consensus, in fact, Parsons's main point is that, like a living organism, social systems are continuously in flux, adjusting, and changing as a result of environmental conditions. Parsons emphasized the continual *process* of managing discrepancies and conflicts between and within systems. He argued that, just as in biological organisms, social and personality systems continuously *seek* equilibrium but that, as the old aphorism goes, the only thing constant is change. Complete and perfect institutionalization is an ideal; it refers to when role demands from the social system complement cultural ideals and when both, in turn, meet the needs of the personality system.

Thus, in the example of the generation gap, while the implication seems to be that this conflict (and conflict in general) is "bad," in fact, Parsons fully recognized that although the norms and values of youth culture may be in conflict with those of wider society, they may very well meet the need-dispositions of youth. As Parsons ([1943] 1954:189) states, among the functions of youth culture is "that of easing the difficult process of adjustment from childhood emotional dependency to full 'maturity.'" Put in another way, because Parsons maintains that each functional subsystem is itself a self-contained system that contains its own subsystems, it is inevitable that changes that are functional for one part of the system will produce changes that are not necessarily functional for other parts of the system. Parsons's student and colleague Robert Merton later deliberately developed this vital point, as is discussed further below.

The Pattern Variables

In *Toward a General Theory of Action* (1951), Parsons and Shils develop a set of concepts called the **pattern variables** (see Table 9.1). The pattern variables are a set of five "choices" that apply not only to the individual level but to the collective level as well. They refer at once to the variant normative priorities of social systems, the dominant modes of orientation in personality systems, and the patterns of values in cultural systems (Coser 1977:569).

Though, as you will see, both ends of each pattern variable dichotomy are readily apparent in contemporary society at all three levels (social, cultural, and personality), in fact, the pattern variables are an extension of a renowned dichotomy first formulated by the German theorist Ferdinand Tönnies in *Gemeinschaft und Gesellschaft* ([1935] 1963). Tönnies's distinction between *Gemeinschaft* (community) and *Gesellschaft* (purposive association) was later reformulated by Émile Durkheim in his conceptualization of "mechanical" versus "organic solidarity." According to these classic dichotomies, modern societies are based on individualistic "purposiveness" and functional interdependence, while traditional societies are rooted in collectivistic "sameness" (or communality) and an intense feeling of community. The relation between Parsons's pattern variables and Tönnies's distinction between Gemeinschaft and Gesellschaft is illustrated in Table 9.2.

As for Parson's pattern variables, as shown in Table 9.1, *affectivity* means that the emotions are considered legitimate in action ("gratification of impulse" in Parsons's terms), while *affective-neutrality* means that emotions are closed out of action ("discipline"; Münch 1994:32–33). For instance, in contemporary societies it is normative to display affectivity in personal relationships but not in bureaucratic relationships: You might kiss or hug a close friend upon greeting him on the street (affectivity), but you would not do the same to the next client in line in your job as a clerk at the department of motor vehicles (affective-neutrality). Indeed, Parsons maintained that affective-neutrality was generally more pervasive in modern Western societies than in traditional societies—which parallels the shift noted by Tönnies as well as Weber from a more religiously oriented (nonrationalistic) society to a more rationalistic, scientifically oriented society.

This shift is particularly apparent when considering the contrast between traditional and modern scientific medicine. Modern medicine typically involves tests and procedures performed by a myriad of specialists who are often strangers to the patient. The relationship between patient and medical specialist (or medical technician) is typically marked by professionalism and affective-neutrality rather than affectivity and emotional involvement (see Photos 9.1a and 9.1b). At the other end of the spectrum, in a traditional society such as the Kamba, together shaman and patient might invoke a drug-induced state in order

Table 9.1   Parsons's Pattern Variables

> **PATTERN VARIABLE: a dichotomy that describes alternatives of action between which each person (or group) has to choose in every situation. The actions are shaped by the three systems: personality, cultural, and social.**

1. **Affectivity/Affective-Neutrality**

   Affectivity: emotional impulses are gratified. For example, a child is allowed to show "love" for his or her parent.

   Affective-neutrality: emotional impulses are inhibited. For example, a bureaucrat in an organization (such as the Department of Motor Vehicles) or a teacher grading papers is expected to be emotionally "neutral."

2. **Self-Orientation/Collectivity-Orientation**

   Self-orientation: action is based on the actor's own interests, needs, and goals. For example, a student decides what to study in college based on his or her own interests.

   Collectivity-orientation: action is based on what is best for the "collectivity." For example, a child quits school to work to support her family.

3. **Universalism/Particularism**

   Universalism: action is based on "general standards" or universal laws and moral rules. For example, the Supreme Court decides cases according to rules that are valid for the whole community.

   Particularism: action is based on the priority and attachment that actors place on relationships and situations. For example, you get a job because of *who* you know rather than *what* you know.

4. **Ascription/Achievement**

   Ascription: action based on given attributes (race, sex, age). For instance, you are eligible for the draft or allowed to buy alcohol or vote because you have reached a specified age.

   Achievement: action based on performance. For example, graduation from college is based on a student's completion of requirements for graduation.

5. **Specificity/Diffuseness**

   Specificity: action based on specific criteria/roles. For example, people who work for bureaucratic organizations, such as the Department of Motor Vehicles, tend to have narrowly and clearly defined roles and criteria for interaction.

   Diffuseness: open guidelines for action. For example, starting a business might entail one person performing a variety of tasks (literally whatever needs doing) from mortgaging ones home for collateral, to writing advertisements, to taking out the trash.

Table 9.2   Parsons's Pattern Variables and Tönnies's Distinction Between Gemeinschaft (Community) and Gesellschaft (Purposive Association)

	TRADITIONAL/Gemeinschaft	MODERN/Gesellschaft
PATTERN VARIABLE	Affectivity	Affective-neutrality
	Collectivity-orientation	Self-orientation
	Particularism	Universalism
	Ascription	Achievement
	Diffuseness	Specificity

Photo 9.1a   Traditional shaman (affectivity): Emotional spontaneity is considered legitimate in shaman/patient interaction..

Affectivity vs. Affective-Neutrality

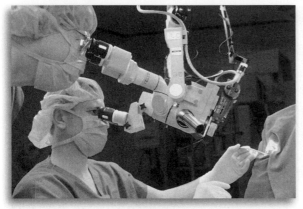

Photo 9.1b   Modern medicine (affective-neutrality): Emotional spontaneity is usually not condoned.

to complete a thorough spiritual and physical "cleansing" of the aggrieved soul. In this case, rather than being excluded from medical treatment, emotions are integral to the healing process.

Self-orientation means that the individual actor prioritizes the needs and goals of the "self," while conversely, the *collectivity-orientation* denotes the prioritization of the needs and goals of the collectivity or group as a whole. Following Tönnies, Parsons suggests that "self-orientation" is considered more legitimate in modern and postmodern than in traditional societies, while in traditional societies, the reverse is true. For instance, as shown in Table 9.1, in contemporary American society, college students typically choose their major based on their own interests and goals ("self-orientation"). So, too, do most Americans choose their own marriage partner, a choice that is based primarily (if not strictly) on personal issues in which the central question is, "Is this the person *I* want to spend the rest of my life with?" Conversely, in many traditional societies, marriages are not so much an individual but, rather, a collective concern. In such societies, marriages are often arranged by elders, who focus on the economic and social benefits of the prospective new kinship formation (a collective consideration) rather than strictly the needs and wants of the bride and groom (who may or may not even know each other).

Universalism means that an action is based on impersonal, universal standards or general rules, such as queuing up, filling out a standardized job application, or having each person's vote count equally in an election. Conversely, *particularism* refers to actions that are guided by the uniqueness of that particular relationship, for instance, inviting only "really good friends" to your party. For example, if you stand by your sister even though she acted badly (or illegally), you are prioritizing the particular relationship you have with her over her action (even though you find it offensive). By contrast, if you turn your sister in to the police because what she did was illegal, you would be applying a universal standard. In this case, your particular relationship is not the central concern. Again, this dichotomy is related to Tönnies's differentiation between traditional and modern society. Action in modern bureaucratic societies typically relies on a myriad of universalistic standards and procedures (student grade point average, or GPA; tax brackets; Supreme Court rulings), while action in traditional societies typically relies on more particularistic criteria (e.g., kinship bonds, rule of elders).

Ascription means that evaluations or interactions are guided by given personal attributes (e.g., race, gender, age), while *achievement* (performance) means that evaluations or interactions are based on an actor's performance with regard to established standards (e.g., college entrance examinations, physical endurance tests for firefighter trainees). Parsons maintained that the achievement orientation was increasingly important in modern Western societies, whereas in traditional societies, ascribed characteristics (e.g., clan, gender, age) were generally given greater priority. Thus, for instance, an individual's tenure and success in her career is determined more by her performance than by her inheritance or privilege.

Of course, this is not to say that ascription is irrelevant in modern societies: In all of the pattern variables, the distinctions between traditional and modern society must not be taken too far. Obviously, upper-class children still inherit significant privileges, making class position one of the most vital attributes in today's world; and race and gender are, to a large extent, ascribed rather than achieved as well. Indeed, although his pattern variables explicitly follow Tönnies's dichotomy between traditional and modern societies, in fact, Parsons fully recognized that each pole of each pattern variable exists in both types of society.

The multivocality of the pattern variables is most readily apparent in Parsons's notion of *subsystems,* or systems within systems, because this means that the pattern-variable continuum is always apparent in full at another level. Thus, for instance, although affective-neutrality is generally considered more legitimate than affectivity in modern societies, as evidenced in economic relationships, affectivity is generally considered a more legitimate basis of action in the context of the family. At the same time, however, even within the economy, there are nevertheless times and places when affectivity is considered a legitimate basis for action; likewise, there are times and places in which affective-neutrality is prescribed within the family as well. So, too, at the level of the actor-ego (personality) as well as social roles, while it is generally considered more legitimate for women to display affectivity than for men, nevertheless, because actions always occur in systems (recall the concept of the unit act), there are situations in which affectivity is normative (or not) for both men and women (e.g., a funeral as opposed to a business conference).

Similarly, although achievement is generally prioritized in the public sector and ascription is prioritized in the family, even within one system (the family, the economy, the legal system), each

Photo 9.2a Small farmer (diffuseness)

Small farmers typically perform a wide range of tasks, from record keeping to operating and servicing heavy machinery, to growing vegetables or raising animals. In Parsons's terms, role-expectations are diffuse. By contrast, duties and role-expectations for fast-food restaurant workers tend to be highly specific; indeed, workers are often mandated to follow finely tuned scripts to maximize efficiency.

Photo 9.2b Fast-food restaurant worker (specificity)

dimension of the pattern variable invariably appears. For instance, parents might base a child's privileges on both achievement (e.g., grades, behavior) and ascription (e.g., age). So, too, ascription and achievement are both readily apparent in the legal system, for instance, in driving laws that say you must both pass a test to get a license (achievement) and be a certain age (ascription). Of course, traditional societies have a parallel continuum, too, in that there are explicit contests or competitions for particular social roles and rewards (achievement), but in order to participate you must possess specific ascribed traits (e.g., be a certain gender or age).

Finally, *diffuseness* means nothing is "closed out" in making a particular choice or undertaking a particular action, while *specificity* means that action is based on a *specific* criterion. For instance, in urban, bureaucratic societies, we are often expected to act within narrowly defined roles in order to maximize efficiency. Impersonal mechanisms, such as queuing up or taking a number, are intended to be based solely on *one* criterion (e.g., who got there first), thereby displaying affective-neutrality and universality (i.e., fairness) rather than emotion or favoritism. So, too, employees at banks, supermarkets, restaurants, insurance companies, and all sorts of other entities are often given a script to follow to ensure that dialogue follows the specific organizational goals for that interaction—and only that interaction (e.g., "Do you want to supersize that?"). By contrast, in smaller, slower-paced, rural, traditional communities in which speed and efficiency are not of such paramount importance, the criteria for action may be more diffuse; less is excluded from the interaction. For instance, a shopper might engage in a significant amount of small-talk before getting to the actual purchase, or a teacher might provide clothes or food for an impoverished student. Here, individuals are acting according to diffuse criteria, rather than according to the specific demands of particular roles.

AGIL

This brings us to another central element in Parsons's theory. Parsons maintains that there are four "functional imperatives" or requirements encountered by all action systems. That is, there are four basic problems that a society, group, or individual must confront in order to survive as a system of action (Parsons 1971:4). Parsons called these four problems or functions adaptation, goal attainment, integration, and latent pattern maintenance (see Table 9.3). Parsons's students used the acronym "AGIL" to refer to this scheme. These four functions or requirements are evident at every level of every system, from entire social systems, to particular subsystems, to the level of the individual actor-ego.

Specifically, as shown in Table 9.3, *adaptation* (A) refers to responses to the physical environment. At the level of the actor-ego, the problem of adaptation is often managed by the behavioral or physiological organism. For example, the body adapts to heat by perspiring, thereby cooling itself down. At the level of the social system, the economy typically fulfills the requirement of adaptation (see Table 9.4). That is, the economy is the subsystem that adapts to the environment for social purposes (providing goods and services). For instance, a coastal island village might revolve around fishing while an inner-island village might revolve around farming, but in either case, it is the *economy* that is most central in adjusting to the environment. So, too, in the economic system of capitalism, the market is continually shifting based on supply and demand for goods and services (the environment).

Table 9.3 Four Functional Requirements of Action Systems: AGIL

Adapation: how well the system adapts to its material environment
Goal attainment: ability of individual or group to identify and pursue goals
Integration: dimensions of cohesion and solidarity
Latent pattern maintenance: sphere of general values

Goal attainment (G) refers to the problem of resolving the discrepancies between "the inertial tendencies of the system and its 'needs' resulting from interchange with the situation" (Parsons 1961:41). At the level of the individual, goal attainment is met primarily by the personality system. As Parsons (1971:6) states, "the personality system is the primary *agency* of action processes," aimed at "the optimization of gratification or satisfaction." At the level of the social system, the requirement of goal attainment is typically met by the *polity,* as it is the realm in which goals and resources are prioritized, and discrepancies are resolved between "the inertial tendencies of the system and its 'needs' resulting from interchange with the situation" (Parsons 1961:41; see Table 9.4). The polity and government establish status and reward systems so that social goals can be attained.

Integration (I) refers to the coordination of a system's or subsystem's constituent parts, since "all social systems are differentiated and segmented into relatively independent units" (Parsons 1961:43). Within the four systems of action (behavioral organism, personality, social system, and cultural system), the function of integration is met primarily by the social system. Integration involves *solidarity,* that is, the feeling of "we-ness" that develops in a social group as distinct roles are carried out; integration depends on interaction and the norms that guide interaction more so than abstract cultural values.

Latent pattern maintenance (L) refers to the "imperative of maintaining the stability of the patterns of institutionalized culture" (Parsons 1961:40). This function is carried out primarily by the cultural system, as it is through culture (made up of shared meanings and values) that specific patterns of behavior are maintained. Within the social system, the function of latent pattern maintenance, that is the maintaining of shared values, is most readily apparent in the realm of religion.

Table 9.4   Parsons's AGIL Scheme at the Level of the Social System

ADAPTATION (A) Economic system (resources)	GOAL ATTAINMENT (G) Polity (goals)
INTEGRATION (I) Social system (norms, interaction)	LATENT PATTERN MAINTENANCE (L) Cultural system (values)

However, as indicated previously, the AGIL scheme refers to the dilemmas or problems faced by *all* systems of action. Thus, for instance, the need for latent pattern maintenance exists within all social units, including the economy, as the Great Depression attests. The great crash of 1929 was rooted in collective panic, a lack of faith in the banking system, or in Parsons's terms, a failure in latent pattern maintenance. Although religion is intimately connected to the problems of latent pattern maintenance (L) and integration (I), nevertheless, religious organizations adapt to the environment (A) and set goals (G). Indeed, as the Protestant Reformation during the 1600s dramatically reflects, it is by adapting to external conditions and setting fresh goals that new religions and religious institutions are born.

PARSONS'S THEORETICAL ORIENTATION ▓

As shown in Figure 9.1 and as previously discussed, Parsons explicitly sought to create a multidimensional metatheoretical model. Indeed, it is Parsons's ideas that gave rise to the "order" and "action" model that we use in this book. Parsons's multidimensionality is readily reflected in his conceptualization of the unit act. As we have seen, according to this model of social action, the hypothetical actor is at once guided by a strategic calculation of ends or goals (which indicates *rational* motivation), as well as *internalized* norms and values (which indicates *nonrational* motivation). At the level of the collective, a *situation* contains both normative (nonrational) and external or environmental (rational) dimensions. That is, it consists of both relatively intangible symbolic phenomena (e.g., societal norms and values) and more tangible constraints (e.g., economic conditions). At the level of the individual, social action involves the *"effort* to conform with *norms"* ([1937] 1967:76–77; emphasis added). "Effort" reflects both organic and nonorganic elements; it is a function of both internal (nonrational) biological constitution and (rational) strategic, conscious choice. Meanwhile, that individual effort is conditioned by norms reflects both the cultural (nonrational) and structural (rational) constraints placed on the individual by society at the collective level.

So, too, Parsons's conceptual trilogy of *institutionalization, internalization,* and *socialization* explicitly integrates individual and collective approaches to order. Norms and values are *institutionalized* in the environment at the collective level in both culture and social structures (e.g., ideologies and organizations) and, at the same time, *internalized* at the level of the individual (nonrational/individual), while *socialization* is the mechanism that bridges these two realms.

Similarly, Parsons's integration of rational and nonrational forms of motivation is also apparent in his AGIL scheme: adaptation and goal attainment speak to the more rationalistic side of action and social systems, while integration and latent pattern maintenance speak to the more nonrational dimensions (see Table 9.4). That is, adaptation and goal attainment reflect how individuals or social groups adapt to, and set goals based on, the environment. At the other end of the action continuum, integration and latent pattern maintenance reflect how action is rooted in norms and ideas (e.g., feelings of belonging and shared values) *internal* to the individual. There is a *commitment* to the norms and ideas of the groups. Thus, from this point of view, of the four functions that every individual or social group must confront, two are largely internal and two are linked to strategic calculations or the environmental conditions to which the individual/social group must adapt.

Finally, Parsons's pattern variables also include both nonrational and rational forms of motivation at the individual and collective levels. Indeed, specific pattern variables speak directly to either action or order. For instance, affectivity versus affective-neutrality is an explicit reference to whether motivation for action is internal/nonrational (based on affectivity) or not; while the self versus collective pattern variable is a specific reference to whether individual decision making or social patterns have priority. Moreover, as discussed previously and shown in Table 9.2, Parsons dichotomized traditional and modern societies along these same lines, construing modern Western societies as geared more toward rational and individualistic motivation and patterns and traditional societies as geared more toward nonrational and collectivistic. Thus, for instance, in traditional societies ascriptive factors, such as kinship ties, may have greater saliency than achievement criteria, such as test scores, in determining vocation and individual opportunity or success. So, too, as indicated previously are institutions such as marriage that tend to be considered intensely personal (individual) in modern societies often rooted more in collective concerns in traditional societies. In this case, marrying someone on the basis of love bespeaks a self-orientation, while marrying for the good of the clan or family is indicative of a collectivity orientation.

That said, as indicated in Chapter 1 and shown in Figures 1.4 and 1.5, overall we situate Parsons and structural functionalism in the collective/nonrational quadrant of our theoretical model. That is because, above all, it is to the *moral* cohesion of the *system* to which structural functionalism and Parsons is most attuned. Perhaps, had Parsons achieved his goal and been *truly* multidimensional (for instance, had he acknowledged the organization and structural barriers to equality that so concerned conflict theorists such as C. Wright Mills), he would not have suffered from the stigma of a conservative purveyor of the "status quo."

Readings

Two very different excerpts from Parsons's extensive oeuvre of works are provided below. The first selection, "Categories of the Orientation and Organization of Action," from *Toward a General Theory of Action* ([1951] 2001), written by Parsons and Edward Shils (1910–1955), a distinguished professor of sociology at the University of Chicago, in the United States, and Cambridge University in the United Kingdom, reflects the quintessence of structural functionalism. Considered "one of the most important theoretical statements in twentieth century sociology" (Smelser 2001:vii), Parsons and Shils set out a conceptual and theoretical framework with which to analyze the basis of all action and social organization. It is for this all-encompassing, overarching, metatheoretical framework that Parsons is most well known. In the second reading, "Sex Roles in the American Kinship System," Parsons applies his theory of action to the empirical topic of the American family.

Introduction to "Categories of the Orientation and Organization of Action"

In this chapter from *Toward a General Theory of Action* ([1951] 2001), Parsons and coauthor Edward Shils delineate their groundbreaking theory of action; outline the basic elements of the social, cultural, and personality systems levels; and discuss the pattern variables. Specifically, in the first section of this reading, "Action and Its Orientation," Parsons and Shils explain their theory of action and the cultural, social, and personality systems. In the second and third sections of this reading, "Dilemmas of Orientation and the Pattern Variables" and "The Definitions of Pattern Variables," Parsons and Shils focus on the basic characteristics of the pattern variables as well as the relation between the pattern variables. In the final sections of this essay, the authors illuminate the interrelations between the pattern variables and the basic structures and interdependencies and interpenetrations of personalities and social systems As noted above, "need-dispositions" refer to the allocative foci for personality systems, while "role-expectations" refer to the allocative foci for social systems. Need-dispositions and role-expectations are the connective tissue between systems of action and the pattern variables. As Parsons and Shils (p. 93) state, "*every* concrete need-disposition of personality, or every role-expectation of social structure, involves a combination of values of the five pattern variables" (emphasis in original).

"Categories of the Orientation and Organization of Action" (1951)

Talcott C. Parsons and Edward A. Shils

ACTION AND ITS ORIENTATION

The theory of action[i] is a conceptual scheme for the analysis of the behavior of living organisms. It conceives of this behavior as oriented to the attainment of ends in situations, by means of the normatively regulated expenditure of energy. There are four points to be noted in this conceptualization of behavior: (1) Behavior is oriented to the attainment of ends or goals or other anticipated states of affairs. (2) It takes place in situations. (3) It is normatively regulated. (4) It involves expenditure of energy or effort or "motivation" (which may be more or less organized independently of its involvement in action). Thus, for example, *a man driving his automobile to a lake to go fishing* might be the behavior to be analyzed. In this case, (1) *to be fishing* is the "end" toward which our man's behavior is oriented; (2) his situation is the road and the car and the place where he is; (3) his energy expenditures are normatively regulated—for example, this driving behavior is an *intelligent*[ii] means of getting to the lake; (4) but he does spend energy to get there; he holds the wheel, presses the accelerator, pays attention, and adapts his action to changing road and traffic conditions. When behavior can be and is so analyzed, it is called "action." This means that any behavior of a living organism might be called action; but to be so called, it must be analyzed in terms of the anticipated states of affairs toward which it is directed, the situation in which it occurs, the normative regulation (e.g., the intelligence)

SOURCE: From T. Parsons and Edward Shils, *Toward a General Theory of Action,* Copyright © 2001. Reprinted with permission of Transaction Publishers.

[i]The present exposition of the theory of action represents in one major respect a revision and extension of the position stated in Parsons, *The Structure of Social Action* (pp. 43–51, 732–733), particularly in the light of psychoanalytic theory, of developments in behavior psychology, and of developments in the anthropological analysis of culture. It has become possible to incorporate these elements effectively, largely because of the conception of a system of action in both the social and psychological spheres and their integration with systems of cultural patterns has been considerably extended and refined in the intervening years.

[ii]Norms of intelligence are one set among several possible sets of norms that function in the regulation of energy expenditure.

of the behavior, and the expenditure of energy or "motivation" involved. Behavior which is reducible to these terms, then, is action.

Each action is the action of an actor, and it takes place in a situation consisting of objects. The objects may be other actors or physical or cultural objects. Each actor has a system of relations-to-objects; this is called his "system of orientations." The objects may be goal objects, resources, means, conditions, obstacles, or symbols. They may become cathected (wanted or not wanted), and they may have different significances attached to them (that is, they may mean different things to different people). Objects, by the significances and cathexes attached to them, become organized into the actor's system of orientations.

The actor's system of orientations is constituted by a great number of specific orientations. Each of these "orientations of action" is a "conception" (explicit or implicit, conscious or unconscious) which the actor has of the situation in terms of what he wants (his ends), what he sees (how the situation looks to him), and how he intends to get from the objects he sees the things he wants (his explicit or implicit, normatively regulated "plan" of action).

Next, let us speak briefly about the sources of energy or motivation. These presumably lie ultimately in the energy potential of the physiological organisms. However, the manner in which the energy is expended is a problem which requires the explicit analysis of the orientation of action, that is, analysis of the normatively regulated relations of the actor to the situation. For, it is the system of orientations which establishes the modes in which this energy becomes attached and distributed among specific goals and objects; it is the system of orientations which regulates its flow and which integrates its many channels of expression into a system.

We have introduced the terms *action* and *actor*. We have said something about the goals of action, the situation of action, the orientation of action, and the motivation of action. Let us now say something about the *organization* of action into systems.

Actions are not empirically discrete but occur in constellations which we call systems. We are concerned with three systems, three modes of organization of the elements of action; these elements are organized as social systems, as personalities, and as cultural systems. Though all three modes are conceptually abstracted from concrete social behavior, the empirical referents of the three abstractions are not on the same plane. Social systems and personalities are conceived as modes of organization of motivated action (social systems are systems of motivated action organized about the relations of actors to each other; personalities are systems of *motivated* action organized about the living organism). Cultural systems, on the other hand, are systems of symbolic patterns (these patterns are created or manifested by individual actors and are transmitted among social systems by diffusion and among personalities by learning).

A social system is a system of action which has the following characteristics: (1) It involves a process of interaction between two or more actors; the interaction process as such is a focus of the observer's attention. (2) The situation toward which the actors are oriented includes other actors. These other actors (alters) are objects of cathexis. Alter's actions are taken cognitively into account as data. Alter's various orientations may be either *goals* to be pursued or *means* for the accomplishment of goals. Alter's orientations may thus be objects for evaluative judgment. (3) There is (in a social system) interdependent and, in part, concerted action in which the concert is a function of collective goal orientation or common values,[iii] and of a consensus of normative and cognitive expectations.

A *personality system* is a system of action which has the following characteristics: (1) It is the system comprising the interconnections of the actions of an individual actor. (2) The actor's actions are organized by a structure of need- dispositions. (3) Just as the actions of a plurality of actors cannot be randomly assorted but must have a determinate organization of compatibility or integration, so the actions of the single actor have a determinate organization of compatibility or integration with one another. Just as the goals or norms which an actor in a social system will pursue or accept will be affected and limited by those pursued or accepted by the other actors, so the goals or norms involved in a single action of one actor will be affected and limited by one another and by other goals and norms of the same actor.

A *cultural system* is a system which has the following characteristics: (1) The system is constituted neither by the organization of interactions nor by the organization

---

[iii]A person is said to have "common values" with another when either (1) he wants the group in which he and the other belong to achieve a certain group goal which the other also wants, or (2) he intrinsically values conformity with the requirements laid down by the other.

of the actions of a single actor (as such), but rather by the organization of the values, norms, and symbols which guide the choices made by actors and which limit the types of interaction which may occur among actors. (2) Thus a cultural system is not an empirical system in the same sense as a personality or social system, because it represents a special kind of abstraction of elements from these systems. These elements, however, may exist separately as physical symbols and be transmitted from one empirical action system to another. (3) In a cultural system the patterns of regulatory norms (and the other cultural elements which guide choices of concrete actors) cannot be made up of random or unrelated elements. If, that is, a system of culture is to be manifest in the organization of an empirical action system it must have a certain degree of consistency. (4) Thus a cultural system is a pattern of culture whose different parts are interrelated to form value systems, belief systems, and systems of expressive symbols.

Social systems, personality systems, and cultural systems are critical subject matter for the theory of action. In the first two cases, the systems themselves are conceived to be actors whose action is conceived as oriented to goals and the gratification of need-dispositions, as occurring in situations, using energy, and as being normatively regulated. Analysis of the third kind of system is essential to the theory of action because systems of value standards (criteria of selection) and other patterns of culture, when *institutionalized* in social systems and *internalized* in personality systems, guide the actor with respect to both the *orientation to ends* and the *normative regulation* of means and of expressive activities, whenever the need-dispositions of the actor allow choices in these matters.

Components of the Frame of Reference of the Theory of Action

1. *The frame of reference of the theory of action* involves actors, a situation of action, and the orientation of the actor to that situation.

   a. One or more *actors* is involved. An actor is an empirical system of action. An actor is an individual or a collectivity which may be taken as a point of reference for the analysis of the modes of its orientation and of its processes of action in relation to objects. Action itself is a process of change of state in such empirical systems of action.

   b. A *situation* of action is involved. It is that part of the external world which means something to the actor whose behavior is being analyzed. It is only part of the whole realm of objects that might be seen. Specifically, it is that part to which the actor is oriented and in which the actor acts. The situation thus consists of objects of orientation.

   c. The *orientation of the actor to the situation* is involved. It is the set of cognitions, cathexes, plans, and relevant standards which relates the actor to the situation.

2. The *actor* is both a system of action and a point of reference. As a system of action the actor may be either an individual or a collectivity. As a point of reference the actor may be either an actor-subject (sometimes called simply *actor*) or a social object.

   a. The *individual-collectivity* distinction is made on the basis of whether the actor in question is a personality system or a social system (a society or subsystem).

   b. The *subject-object distinction* is made on the basis of whether the actor in question occupies a central position (as a point of reference) within a frame of reference or a peripheral position (as an object of orientation for an actor taken as the point of reference. When an actor is taken as the central point of reference, he is an actor-subject. (In an interaction situation, this actor is called *ego*.) When he is taken as an object of orientation for an actor-subject, he is a social object. (In an interaction situation, this actor is called *alter*.) Thus, the actor-subject (*the* actor) is an orienting subject; the social object is the actor who is oriented to. This distinction cross-cuts the individual-collectively distinction. Thus an individual or a collectivity may be either actor-subject or social object in a given analysis.

3. The situation of action may be divided into a class of social objects (individuals and collectivities) and a class of nonsocial (physical and cultural) objects.

   a. *Social objects* include actors as persons and as collectivities (i.e., systems of action composed of a plurality of individual actors in determinate relations to one another). The actor-subject may be oriented to himself as an object as well as to other social objects. A collectivity, when it is considered as a social object, is never constituted by all the action of the participating individual actors; it may, however, be constituted by anything from a specified segment of their actions—for example, their actions in a specific system of roles—to a very inclusive grouping of their actions—for example, all of their many roles in a society. . . .

Neither systems of value-orientation nor systems of culture as a whole are action systems in the same sense as are personalities and social systems. This is because neither motivation nor action is directly attributable to them. They may conjoin with motivation to evoke action in social systems or personalities, but they themselves cannot act, nor are they motivated. It seems desirable to treat them, however, because of the great importance of the particular ways in which they are involved in action systems. . . .

Dilemmas of Orientation and the Pattern Variables

. . . An actor in a situation is confronted by a series of major dilemmas of orientation, a series of choices that the actor must make before the situation has a determinate meaning for him. The objects of the situation do not interact with the cognizing and cathecting organism in such a fashion as to determine automatically the meaning of the situation. Rather, the actor must make a series of choices before the situation will have a determinate meaning. Specifically, we maintain, the actor must make five specific dichotomous choices before any situation will have a determinate meaning. The five dichotomies which formulate these choice alternatives are called the *pattern variables* because any specific orientation (and consequently any action) is characterized by a pattern of the five choices.

Three of the pattern variables derive from the absence of any biologically given hierarchy of primacies among the various modes of orientation. In the first place, the actor must choose whether to accept gratification from the immediately cognized and cathected object or to evaluate such gratification in terms of its consequences for other aspects of the action system. (That is, one must decide whether or not the evaluative mode is to be operative at all in a situation.)[iv] In the second place, if the actor decides to evaluate, he must choose whether or not to give primacy to the moral standards of the social system or subsystem. In the third place, whether or not he decides to grant primacy to such moral standards, he must choose whether cognitive or appreciative standards are to be dominant, the one set with relation to the other. If cognitive standards are dominant over appreciative standards, the actor will tend to locate objects in terms of their relation to some generalized frame of reference; if appreciative standards are dominant over cognitive, the actor will tend to locate objects in terms of their relation to himself, or to his motives.

The other pattern variables emerge from indeterminacies intrinsic to the object situation: social objects as relevant to a given choice situation are either quality complexes or performance complexes, depending on how the actor chooses to see them; social objects are either functionally diffuse (so that the actor grants them every feasible demand) or functionally specific (so that the actor grants them only specifically defined demands), depending on how the actor chooses to see them or how he is culturally expected to see them.

It will be noted now that the three pattern variables which derive from the problems of primacy among the modes of orientation are the first three of the pattern variables as these were listed in our introduction; the two pattern variables which derive from the indeterminacies in the object situation are the last two in that list.

. . . let us restate our definition: a *pattern variable* is a dichotomy, one side of which must be chosen by an actor before the meaning of a situation is determinate for him, and thus before he can act with respect to that situation. We maintain that there are only five *basic* pattern variables (i.e., pattern variables deriving directly from the frame of reference of the theory of action) and that, in the sense that they are *all* of the pattern variables which so derive, they constitute a system. Let us list them and give them names and

---

[iv]In a limited sense the evaluative mode is operative, even when no thought is given to the consequences of immediate gratification; this in the sense that aesthetic (appreciative) standards may be involved to determine the "appropriateness" of the form of gratification chosen. Only in this limited sense, however, does evaluation enter the immediate gratification picture.

numbers so that we can more easily refer to them in the future. They are:

1. Affectivity—Affective neutrality.

2. Self-orientation—Collectivity-orientation.

3. Universalism—Particularism.

4. Ascription—Achievement.

5. Specificity—Diffuseness.

The first concerns the problem of whether or not evaluation is to take place in a given situation. The second concerns the primacy of moral standards in an evaluative procedure. The third concerns the relative primacy of cognitive and cathectic standards. The fourth concerns the seeing of objects as quality or performance complexes. The fifth concerns the scope of significance of the object.

These pattern variables enter the action frame of reference at four different levels. In the first place, they enter at the concrete level as five discrete choices (explicit or implicit) which every actor makes before he can act. In the second place, they enter on the personality level as habits of choice; the person has a set of habits of choosing, ordinarily or relative to certain types of situations, one horn or the other of each of these dilemmas. Since this set of habits is usually a bit of internalized culture, we will list it as a component of the actor's value-orientation standards. In the third place, the pattern variables enter on the collectivity level as aspects of role definition: the definitions of rights and duties of the members of a collectivity which specify the actions of incumbents of roles, and which often specify that the performer shall exhibit a habit of choosing one side or the other of each of these dilemmas. In the fourth place, the variables enter on the cultural level as aspects of value standards; this is because most value standards are rules or recipes for concrete action and thus specify, among other things, that the actor abiding by the standard shall exhibit a habit of choosing one horn or the other of each of the dilemmas.

From the foregoing paragraph, it should be obvious that, except for their integration in concrete acts as discrete choices, the pattern variables are most important as characteristics of value standards (whether these be the value standards of a personality, or the value standards defining the roles of a society, or just value standards in the abstract). In the sense that each concrete act is made up on the basis of a patterning of the choices formulated by the scheme, the pattern variables are not necessarily attributes of value standards, because any specific concrete choice may be a rather discrete and accidental thing. But as soon as a certain consistency of choosing can be inferred from a series of concrete acts, then we can begin to make statements about the value standards involved and the formulation of these standards in terms of the variables of the pattern-variable scheme.

What is the bearing of the pattern variables on our analysis of systems of action and cultural orientation? Basically, the pattern variables are the categories for the description of value-orientations which of course are in various forms integral to all three systems. A given value-orientation or some particular aspect of it may be interpreted as imposing a preference or giving a primacy to one alternative over the other *in a particular type of situation*. The pattern variables therefore delineate the alternative preferences, predispositions, or expectations; in all these forms the common element is the direction of selection in defined situations. In the personality system, the pattern variables describe essentially the predispositions or expectations as evaluatively defined in terms of what will below be called ego-organization[v] and superego-organization. In the case of the social system they are the crucial components in the definition of role-expectations. Culturally, they define patterns of value-orientation.

The pattern variables apply to the *normative* or ideal aspect of the structure of systems of action; they apply to one part of its culture. They are equally useful in the empirical description of the degree of conformity with or divergence of concrete action from the patterns of expectation or aspiration. When they are used to characterize differences of empirical structure of personalities or social systems, they contain an elliptical element. This element appears in such statements as, "The American occupational system is universalistic and achievement- oriented and specific." The more adequate, though still sketchy, statement would be: "Compared to other possible ways of organizing the division of labor, the predominant norms which are institutionalized in the American society and which embody the predominant value-orientation of the culture

---

[v]The term *ego* is here used in the sense current in the theory of personality, not as a point of reference.

give rise to expectations that occupational roles will be treated by their incumbents and those who are associated with them universalistically and specifically and with regard to proficiency of performance."

These categories could equally be employed to describe actual behavior as well as normative expectations and are of sufficient exactitude for first approximations in comparative analysis. For more detailed work, however, much more precise analysis of the degrees and incidence of deviance, with special reference to the magnitude, location, and forms of the tendencies to particularism, to ascriptiveness, and to diffuseness would have to be carried out.

We will now proceed to define the five pattern variables and the problems of alternative selection to which they apply. They are inherently patterns of cultural value-orientation, but they become integrated both in personalities and in social systems. Hence the general definitions will in each case be followed by definitions specific to each of the three types of systems. These definitions will be followed by an analysis of the places of the variables in the frame of reference of the theory of action, the reasons why this list seems to be logically complete on its own level of generality, and certain problems of their systematic interrelations and use in structural analysis.

THE DEFINITIONS OF PATTERN VARIABLES

1. *The dilemma of gratification of impulse versus discipline.* When confronted with situations in which particular impulses press for gratification, an actor faces the problem of whether the impulses should be released or restrained. He can solve the problem by giving primacy, at the relevant selection points, to evaluative considerations, at the cost of interests in the possibility of immediate gratification; or by giving primacy to such interests in immediate gratification, irrespective of evaluative considerations.

   a. Cultural aspect. (1) *Affectivity:* the normative pattern which grants the permission for an actor, in a given type of situation, to take advantage of a given opportunity for immediate gratification without regard to evaluative considerations. (2) *Affective neutrality:* the normative pattern which prescribes for actors in a given type of situation renunciation of certain types of immediate gratification for which opportunity exists, in the interest of evaluative considerations regardless of the content of the latter.

   b. Personality aspect. (1) *Affectivity:* a need-disposition on the part of the actor to permit himself, in a certain situation, to take advantage of an opportunity for a given type of immediate gratification and not to renounce this gratification for evaluative reasons. (2) *Affective neutrality:* a need-disposition on the part of the actor in a certain situation to be guided by evaluative considerations which prohibit his taking advantage of the given opportunity for immediate gratification; in this situation the gratification in question is to be renounced, regardless of the grounds adduced for the renunciation.

   c. Social system aspect. (1) *Affectivity:* the role-expectation[vi] that the incumbent of the role may freely express certain affective reactions to objects in the situation and need not attempt to control them in the interest of discipline. (2) *Affective neutrality:* the role-expectation that the incumbent of the role in question should restrain any impulses to certain affective expressions and subordinate them to considerations of discipline. In both cases the affect may be positive or negative, and the discipline (or permissiveness) may apply only to certain qualitative types of affective expression (e.g., sexual).

2. *The dilemma of private versus collective interests,* or the distribution between private permissiveness and collective obligation. The high frequency of situations in which there is a disharmony of interests creates the problem of choosing between action for private goals or on behalf of collective goals. This dilemma may be resolved by the actor either by giving primacy to interests, goals, and

---

[vi]A role-expectation is, in an institutionally integrated social system (or part of it), an expectation *both* on the part of ego and of the alters with whom he interacts. The same sentiments are shared by both. In a less than perfectly integrated social system, the concept is still useful for describing the expectations of each of the actors, even though they diverge.

values shared with the other members of a given collective unit of which he is a member or by giving primacy to his personal or private interests without considering their bearing on collective interests.

a. Cultural aspect. (1) *Self-orientation:* the normative pattern which prescribes a range of permission for an actor, in a given type of situation, to take advantage of a given opportunity for pursuing a private interest, regardless of the content of the interest or its direct bearing on the interests of other actors. (2) *Collectivity-orientation:* a normative pattern which prescribes the area within which an actor, in a given type of situation, is obliged to take directly into account a given selection of values which he shares with the other members of the collectivity in question. It defines his *responsibility* to this collectivity.

b. Personality aspect. (1) *Self-orientation:* a need-disposition on the part of the actor to permit himself to pursue a given goal or interest of his own—regardless whether from his standpoint it is only cognitive-cathectic or involves evaluative considerations—but without regard to its bearing onc way or another on the interests of a collectivity of which he is a member. (2) *Collectivity-orientation:* a need-disposition on the part of the actor to be guided by the obligation to take directly into account, in the given situation, values which he shares with the other members of the collectivity in question; therefore the actor must accept responsibility for attempting to realize those values in his action. This includes the expectation by ego that in the particular choice in question he will subordinate his private interests, whether cognitive-cathectic or evaluative, and that he will be motivated in superego terms.

c. Social system aspect. (1) *Self-orientation:* the role-expectation by the relevant actors that it is *permissible* for the incumbent of the role in question to give priority in the given situation to his own private interests, whatever their motivational content or quality, independently of their bearing on the interests or values of a given collectivity of which he is a member, or the interests of other actors. (2) *Collectivity-orientation:*

the role-expectation by the relevant actors that the actor is *obliged,* as an incumbent of the role in question, to take directly into account the values and interests of the collectivity of which, in this role, he is a member. When there is a potential conflict with his private interests, he is expected in the particular choice to give priority to the collective interest. This also applies to his action in representative roles on behalf of the collectivity.

3. *The dilemma of transcendence versus immanence.* In confronting any situation, the actor faces the dilemma whether to treat the objects in the situation in accordance with a general norm covering *all* objects in that class or whether to treat them in accordance with their standing in some particular relationship to him or his collectivity, independently of the objects' subsumibility under a general norm. This dilemma can be resolved by giving primacy to norms or value standards which are maximally generalized and which have a basis of validity transcending *any* specific system of relationships in which ego is involved, or by giving primacy to value standards which allot priority to standards *integral* to the *particular* relationship system in which the actor is involved with the object.

a. Cultural aspect. (1) *Universalism:* the normative pattern which obliges an actor in a given situation to be oriented toward objects in the light of general standards rather than in the light of the objects' possession of properties (qualities or performances, classificatory or relational) which have a particular relation to the actor's own properties (traits or statuses). (2) *Particularism:* the normative pattern which obliges an actor in a given type of situation to give priority to criteria of the object's particular relations to the actor's own properties (qualities or performances, classificatory or relational) over generalized attributes, capacities, or performance standards.

b. Personality aspect. (1) *Universalism:* a need-disposition on the part of the actor in a given situation to respond toward objects in conformity with a general standard rather than in the light of their possession of properties (qualities or performances, classificatory or relational) which have a particular relation to the actor's own. (2) *Particularism:* a need-disposition on the part of the actor to be guided by criteria of choice particular to his own and the object's position in an object-relationship system rather than by criteria defined in generalized terms.

c. Social system aspect. (1) *Universalism:* the role-expectation that, in qualifications for memberships and decisions for differential treatment, priority will be given to standards defined in completely generalized terms, independent of the particular relationship of the actor's own statuses (qualities or performances, classificatory or relational) to those of the object. (2) *Particularism:* the role-expectation that, in qualifications for memberships and decisions for differential treatment, priority will be given to standards which assert the primacy of the values attached to objects by their particular relations to the actor's properties (qualities or performances, classificatory or relational) as over against their general universally applicable class properties.

4. *The dilemma of object modalities.* When confronting an object in a situation, the actor faces the dilemma of deciding how to treat it. Is he to treat it in the light of what it is in itself or in the light of what it does or what might flow from its *actions?* This dilemma can be resolved by giving primacy, at the relevant selection points, to the "qualities" aspect of *social objects* as a focus of orientation, or by giving primacy to the objects' performances and their outcomes.

a. Cultural aspect. (1) *Ascription:* the normative pattern which prescribes that an actor in a given type of situation should, in his selections for differential treatment of social objects, give priority to certain attributes that they possess (including collectivity memberships and possessions) over any specific performances (past, present, or prospective) of the objects. (2) *Achievement:* the normative pattern which prescribes that an actor in a given type of situation should, in his selection and differential treatment of social objects, give priority to their specific performances (past, present, or prospective) over their given attributes (including memberships and possessions), insofar as the latter are not significant as direct conditions of the relevant performances.

b. Personality aspect. (1) *Ascription:* the need-disposition on the part of the actor, at a given selection point, to respond to specific given attributes of the social object, rather than to their past, present, or prospective performances. (2) *achievement:* a need-disposition on the part of the actor to respond, at a given selection point, to specific performances (past present, or prospective) of a social object, rather than to its attributes which are not directly involved in the relevant performances as "capacities," "skills," and so forth.

c. Social system aspect. (1) *Ascription:* the role-expectation that the role incumbent, in orienting himself to social objects in the relevant choice situation, will accord priority to the objects' given attributes (whether universalistically or particularistically defined) over their actual or potential performances. (2) *Achievement:* the role-expectation that the role incumbent, in orienting to social objects in the relevant choice situation, will give priority to the objects' actual or expected performances, and to their attributes only as directly relevant to these performances, over attributes which are essentially independent of the specific performances in question.

5. *The dilemma of the scope of significance of the object.* In confronting an object, an actor must choose among the various possible ranges in which he will respond to the object. The dilemma consists in whether he should respond to many aspects of the object or to a restricted range of them—how broadly is he to allow himself to be involved with the object? The dilemma may be resolved by accepting no inherent or prior limitation of the scope of the actor's "concern" with the object, either as an object of interest or of obligations, or by according only a limited and specific type of significance to the object in his system of orientation.

a. Cultural aspect. (1) *Diffuseness:* the normative pattern which prescribes that in a given situation the orientation of an actor to an object should contain no prior specification of the actor's interest in or concern with or for the object, but that the scope should vary with the exigencies of the situation as they arise. (2) *Specificity:* the normative pattern which prescribes that in a given type of situation an actor should confine his concern with a given type of object to a specific sphere and not permit other empirically possible concerns to enter.

b. Personality aspect. (1) *Diffuseness:* the need-disposition to respond to an object in any way which the nature of the actor and the nature of the object and its actual relation to ego require, actual significances varying as occasions arise. (2) *Specificity:* the need-disposition of the actor to respond to a given object in a manner limited to a specific mode or context of significance of a social object, including obligation to it, which is compatible with exclusion of other potential modes of significance of the object.

c. Social system aspect. (1) *Diffuseness:* the role-expectation that the role incumbent, at the relevant choice point, will accept any potential significance of a social object, including obligation to it, which is compatible with his other interests and obligations, and that he will give priority to this expectation over any disposition to confine the role-orientation to a specific range of significance of the object. (2) *Specificity:* the role-expectation that the role incumbent, at the relevant choice point, will be oriented to a social object only within a specific range of its relevance as a cathectic object or as an instrumental means or condition and that he will give priority to this expectation over any disposition to include potential aspects of significance of the object not specifically defined in the expectation pattern.

Of the five pattern variables defined above, the first three are determined by primacies among the interests inherently differentiated within the system of value-orientation itself and in the definition of the limits of its applicability; the other two are determined by the application of value-orientations to the alternatives which are inherent in the structure of the object system, and in the actor's relation to it. . . .

The Interrelations of the Pattern Variables

We hold that the five pattern variables constitute a *system* covering all the fundamental alternatives which can arise directly out of the frame of reference for the theory of action. It should be remembered that the five pattern variables formulate five fundamental choices which must be made by an actor when he is confronted with a situation before that situation can have definitive (unambiguous) meaning for him. We have said that objects do not automatically determine the actors "orientation of action"; rather, a number of choices must be made before the meaning of the objects becomes definite. Now, we maintain that when the situation is social (when one actor is orienting to another), there are but five choices which are completely general (that is, which must always be made) and which derive directly from the action frame of reference; these choices must always be made to give the situation specific defined meaning. Other choices are often necessary to determine the meaning of a situation, but these may be considered accidents of content, rather than genuine alternatives intrinsic to the structure of *all* action. . . .

Classification of Need-Dispositions and Role-Expectations

The pattern variables are tools for the classification of need-dispositions and role-expectations, which, as has been pointed out, represent allocative foci for both personality and social systems. Before we go into the classification of these units, it might be wise to recapitulate briefly the way the allocative and integrative foci fit into the frame of reference of the theory of action. We have said that action systems, as either actors or social objects, may be personalities or collectivities, both of which are abstracted from the same concrete action. The different principles of abstraction used in locating the two systems derive directly from the notion that personalities and collectivities have different kinds of allocative and integrative foci. The integrative foci are, in some sense, the principles of abstraction used in locating or delimiting the system: thus, the individual organism is the integrative focus of a personality system and the interacting social group is the integrative focus of a social system. The integrative foci are therefore used for abstracting social systems themselves from the total realm of possible subject matter.

The allocative foci, on the other hand, are the primary units used for analyzing the action system into elements or parts. The allocative foci of personality systems are need-dispositions. The personality system is in a sense composed of a variety of need-dispositions;

each of these assures that some need of the personality system will be met. The referent of a need-disposition is, in a sense, a set of concrete orientations. That is, a need-disposition is an inferred entity; it is inferred on the basis of a certain consistency of choosing and cathecting in a wide variety of orientations. Thus, when we speak of a need-disposition, we will sometimes seem to be talking about a real entity, causally controlling a wide variety of orientations and rendering them consistent; other times we will seem to be talking about the consistent set of orientations (abstracted on the basis of the postulated entity) themselves. Logicians have shown that it is usually fair to use interchangeably the inferred entity postulated on the basis of a set of data and the whole set of data itself. The postulated entity is, in some sense, a shorthand for the set of data from which it is inferred.

The allocative foci of social systems are roles or role-expectations. The social system is in a sense composed of a variety of roles or role-expectations; each of these assures that some need of the social system will be met. The referent of a role, like that of a need-disposition, is a set of concrete orientations; the role or role-expectation is an inferred entity in exactly the same fashion as is the need-disposition. Each orientation, according to postulate, is a joint function of a role (which partly controls it), a need-disposition (which also partly controls it), and probably of other factors not mentioned here.[vii] When orientations are grouped (or abstracted) according to the need-dispositions that control them, and according to the individual organisms who have these need-dispositions, we are dealing with personality systems. When orientations are grouped (or abstracted) according to the roles or roles-expectations that control them, and according to the interacting groups to which they belong, we are dealing with social systems.

Now, since none of the depth variables (allocative foci, etc.) are effective except as they influence the orientation of action (which is not necessarily either conscious or rational), and since all orientations tend to have not only the allocative foci of both social and personality systems as ingredients but also value standards (which, when internalized, are depth variables similar to need-dispositions and role- expectations), no need-disposition, nor any role-expectation, is effective except in conjunction with certain value-orientations with which it is systematically related (at least in the sense that both control the same orientation for the moment). Hence, in discussing personalities or social systems, using as the primary units of abstraction need-dispositions or role-expectations, we may regard the value-orientation components of the orientations so grouped to be the value- orientation components of the need-dispositions or role-expectations themselves. Thus we can classify the need-dispositions and role-expectations in terms of the value-orientations with which they tend to be linked.

In principle, therefore, *every* concrete need-disposition[viii] of personality, or every role- expectation of social structure, involves a combination of values of the five pattern variables. The cross-classification of each of the five against each of the others, yielding a table of thirty-two cells, will, on the assumption that the list of pattern variables is exhaustive, produce a classification of the basic value patterns. Internalized in the personality system, these value patterns serve as a starting point for a classification of the possible types of need- dispositions; as institutionalized in the system of social action, they are a classification of components of role-expectation definitions.[ix]

[vii]As will be seen in a moment, each orientation is in some sense a function of the value standards which partly control it. Furthermore, each orientation is certainly partly a function of the present object situation.

[viii]A need-disposition as the term is used here always involves a set of dispositions toward objects. In abstraction from objects the concept becomes elliptical. Only for reasons of avoiding even greater terminological cumbersomeness is the more complex term "need-disposition toward objects" usually avoided. However, such a need-disposition and the *particular* objects of its gratification are independently variable. The mechanism of *substitution* links the need-disposition to various objects that are not its "proper" gratifiers.

[ix]The classification of role-expectations and need-dispositions according to value patterns is only a part of the larger problem of classifying concrete need-dispositions and role-expectations. Other components of action must enter the picture before a classification relevant and adequate to the problem of the analysis of systems is attainable. For example, one set of factors entering into need-dispositions, the constitutionally determined components, has been quite explicitly and deliberately excluded from the present analysis. So far as these are essential to an adequate classification of the need-disposition elements of personality, the classification in terms of pattern variables obviously requires adjustment.

It should be clear that the classification of the value components of need-dispositions and of role-expectations in terms of the pattern variables is a *first step* toward the construction of a dynamic theory of systems of action. To advance toward empirical significance, these classifications will have to be related to the functional problems of on-going systems of action.[x]

As a last word before taking up the problem of classification itself, we should mention that of the logically possible combinations of the pattern variables, not all are likely to be of equal empirical significance. Careful analysis of their involvement in a wide variety of phenomena shows that they are all in fact independently variable in some contexts and that there is no tautology in the scheme. Nonetheless there are certainly tendencies for certain combinations to cluster together. The uneven distribution of combinations and the empirical difficulty, or even perhaps impossibility, of the realization of some combinations in systems of action will raise important dynamic problems.

To classify need-dispositions and role-expectations, we must begin by making the cross-classification tables mentioned above. In constructing such tables we find that certain of the pattern-variable dichotomies are of major importance with respect to need-dispositions (and hence personality systems). Similarly, certain pattern-variable dichotomies are of major importance with respect to role-expectations (and hence social systems). Furthermore, the pattern variables of major importance for classification of need-dispositions are not the same as those of major importance for classification of role-expectations. In fact, the two sets are more or less complementary; those of major importance for need-dispositions are the ones of minor importance for role-expectations, and vice versa.

The only one of the pattern variables equally applicable to both need-dispositions and role-expectations is the self-collectivity variable (number two). Of the other four, the first, affectivity-neutrality, and the fifth, specificity-diffuseness, are chiefly important with respect to need-dispositions. The third, universalism-particularism, and the fourth, ascription-achievement, are chiefly important with respect to role-expectations. . . .

THE BASIC STRUCTURE OF THE INTERACTIVE RELATIONSHIP

The interaction of ego and alter is the most elementary form of a social system. The features of this interaction are present in more complex form in all social systems.

In interaction ego and alter are each objects of orientation for the other. The basic differences from orientations to nonsocial objects are two. First, since the outcome of ego's action (e.g., success in the attainment of a goal) is contingent on alter's reaction to what ego does, ego becomes oriented not only to alter's probable *overt* behavior but also to what ego interprets to be alter's expectations relative to ego's behavior, since ego expects that alter's expectations will influence alter's behavior. Second, in an integrated system, this orientation to the expectations of the other is reciprocal or complementary.

Communication through a common system of symbols is the precondition of this reciprocity or complementarity of expectations. The alternatives which are open to alter must have some measure of stability in two respects: first, as realistic possibilities for alter, and second, in their meaning to ego. This stability presupposes generalization from the particularity of the given situations of ego and alter, both of which are continually changing and are never concretely identical over any two moments in time. When such generalization occurs, and actions, gestures, or symbols have more or less the *same* meaning for both ego and alter, we may speak of a common culture existing between them, through which their interaction is mediated.

Furthermore, this common culture, or symbol system, inevitably possesses in certain aspects a normative significance for the actors. Once it is in existence, observance of its conventions is a necessary condition for ego to be "understood" by alter, in the sense of allowing ego to elicit the type of reaction from alter which ego expects. This common set of cultural symbols becomes the medium in which is formed a constellation of the contingent actions of both parties, in such a way that there will simultaneously emerge a definition of a range of *appropriate* reactions on alter's part to each of a range of possible actions ego has taken and

---

[x]This means above all that the motivational *processes* of action must be analyzed as processes in terms of the laws governing them, and as mechanisms in terms of the significance of their outcomes for the functioning of the systems of which they are parts. In due course the attempt to do this will be made. Also, it should be noted that the necessary constitutional factors which are treated as residual in this conceptual scheme will find their place among the functional necessities of systems.

vice versa. It will then be a condition of the stabilization of such a system of complementary expectations, not only that ego and alter should *communicate,* but that they should *react appropriately* to each other's action.

A tendency toward consistent appropriateness of reaction is also a tendency toward conformity with a normative pattern. The culture is not only a set of symbols of communication but a *set of norms* for action.

The motivation of ego and alter become integrated with the normative patterns through interaction. The polarity of gratification and deprivation is crucial here. An appropriate reaction on alter's part is a gratifying one to ego. If ego conforms with the norm, this gratification is in one aspect a reward for his conformity with it; the converse holds for the case of deprivation and deviance. The reactions of alter toward ego's conformity with or deviance from the normative pattern thus become sanctions to ego. Ego's expectations vis-à-vis alter are expectations concerning the roles of ego and of alter; and sanctions reinforce ego's motivation to conform with these role-expectations. Thus the complementarity of expectations brings with it the reciprocal reinforcement of ego's and alter's motivation to conformity with the normative pattern which defines their expectations.

The interactive system also involves the process of generalization, not only in the common culture by which ego and alter communicate but in the interpretation of alter's discrete actions vis-à-vis ego as expressions of alter's *intentions* (that is, as indices of the cathectic-evaluative aspects of alter's motivational orientations toward ego). This "generalization" implies that ego and alter agree that certain actions of alter are indices of the *attitudes* which alter has acquired toward ego (and reciprocally, ego toward alter). Since these attitudes are, in the present paradigm, integrated with the common culture and the latter is internalized in ego's need-dispositions, ego is sensitive not only to alter's overt acts, but to his *attitudes*. He acquires a need not only to obtain specific *rewards* and avoid specific *punishments* but to enjoy the favorable attitudes and avoid the unfavorable ones of alter. Indeed, since he is integrated with the same norms, these are the same as his attitudes toward himself as an object. Thus violation of the norm causes him to feel shame toward alter, guilt toward himself.

It should be clear that as an ideal type this interaction paradigm implies *mutuality* of gratification in a certain sense, though not necessarily equal distribution of gratification. . . . This is also the paradigm of the process of the learning of generalized orientations. Even where special mechanisms of adjustment such as dominance and submission or alienation from normative expectations enter in, the process still must be described and analyzed in relation to the categories of this paradigm. It is thus useful both for the analysis of systems of normative expectations and for that of the actual conformity or deviation regarding these expectations in concrete action.

In summary we may say that this is the basic paradigm for the structure of a solidary interactive relationship. It contains all the fundamental elements of the role structure of the social system and the attachment and security system of the personality. It involves culture in both its communicative and its value-orientation functions. It is the nodal point of the organization of all systems of action.

THE CONCEPT OF SYSTEM AND THE CLASSIFICATION OF TYPES OF SYSTEMS

With our discussion of interaction we have entered upon the analysis of systems. Before we discuss more fully personality and social systems, it is desirable to state explicitly the principal properties of empirical systems which are relevant for the present analysis. The most general and fundamental property of a system is the interdependence of parts or variables. Interdependence consists in the existence of determinate relationships among the parts or variables as contrasted with randomness of variability. In other words, interdependence is *order* in the relationship among the components which enter into a system. This order must have a tendency to self-maintenance, which is very generally expressed in the concept of equilibrium.[xi] It need not, however, be a static self-maintenance or a stable equilibrium. It may be an ordered process of change—a process following a determinate pattern rather than random variability relative to the starting point. This is called a moving equilibrium and is well exemplified by growth. Furthermore, equilibrium, even when stable, by no means implies that process is not going on; process is continual even in stable systems,

---

[xi]That is, if the system is to be permanent enough to be worth study, there must be a tendency to maintenance of order except under exceptional circumstances

the stabilities residing in the interrelations involved in the process.

A particularly important feature of all systems is the inherent limitation on the compatibility of certain parts or events within the same system. This is indeed simply another way of saying that the relations within the system are determinate and that not just anything can happen. Thus, to take an example from the solar system, if the orbit of one of the planets, such as Jupiter, is given, it is no longer possible for the orbits of the other planets to be distributed at random relative to this given orbit. Certain limitations are imposed by the fact that the value of one of the variables is given. This limitation may in turn be looked at from either a negative or a positive point of view. On the one hand, again using the solar system as example, if one of the planets should simply disappear, the fact that no mass was present in that particular orbit would necessitate a change in the equilibrium of the system. It would make necessary a readjustment of the orbits of the other planets in order to bring the system into equilibrium. This may also be expressed in the statement that there is a change in the structure of the system. On the other hand, the same problem may be treated from the standpoint of what would happen in the case of the coexistence of "incompatible" elements or processes within the same system. Incompatibility is always relative to a *given* state of the system. If, for example, the orbits of two of the planets should move closer to each other than is compatible for the maintenance of the current state of the system, one of two things would have to happen. Either processes would be set up which would tend to restore the previous relation by the elimination of the incompatibility; or if the new relation were maintained, there would have to be adjustments in *other* parts of the system, bringing the system into a new state of equilibrium.

These properties are inherent in all systems. A special additional property, however, is of primary significance for the theory of action. This is the tendency to maintain equilibrium, in the most general sense stated above, within certain boundaries relative to an environment—boundaries which are not imposed from outside but which are self-maintained by the properties of the constituent variables as they operate within the system. The most familiar example is the living organism, which is a physicochemical system that is not assimilated to the physicochemical conditions of the environment, but

maintains certain distinct properties in relation to the environment. For example, the maintenance of the constant body temperature of the mammal necessitates processes which mediate the interdependence between the internal and the external systems in respect to temperature; these processes maintain constancy over a wide range of variability in environmental temperatures.

The two fundamental types of processes necessary for the maintenance of a given state of equilibrium of a system we call, in the theory of action, *allocation*[xii] and *integration*. By *allocation* we mean processes which maintain a distribution of the components or parts of the system which is compatible with the maintenance of a given state of equilibrium. By *integration,* we mean the processes by which relations to the environment are mediated in such a way that the distinctive internal properties and boundaries of the system as an entity are maintained in the face of variability in the external situation. It must be realized that self-maintenance of such a system is not only maintenance of boundaries but also maintenance of distinctive relationships of the parts of the system *within* the boundary. The system is in some sense a unity relative to its environment. Also, self-maintenance implies not only control of the environmental variations, but also control of tendencies to change—that is, to alteration of the distinctive state—coming from within the system.

The two types of empirical systems which will be analyzed in the subsequent chapters are personalities and social systems. These systems are, as will be repeatedly pointed out, *different* systems which are not reducible to each other. However, there are certain conceptual continuities or identities between them which derive from two sources. (1) They are both systems built out of the fundamental components of action as these have been discussed in the General Statement and in the present chapter. These components are differently organized to constitute systems in the two cases; nevertheless, they remain the same components. (2) They are both not only systems, but both are systems of the boundary-maintaining, self-maintenance type; therefore, they both have properties which are common to systems in general and the more special properties which are characteristic of this particular type of system. (3) A third basis of their intimate relation to each other is the fact that they *interpenetrate* in the sense that no personality system can exist without *participation* in a social system, by which we mean the integration of *part* of the actor's system of

---

[xii]The term *allocation* is borrowed from the usage of economics, where it has the general meaning here defined. Specifically, economists speak of the allocation of resources in the economy.

action as *part* of the social system. Conversely, there is no social system which is not from one point of view a mode of the integration of parts of the systems of action which constitute the personalities of the members. When we use the term *homology* to refer to certain formal identities between personalities and social systems which are to be understood in terms of the above considerations, it should be clear that we in no way intend to convey the impression that a personality is a microcosm of a social system, or that a social system is a kind of macrocosmic personality.

In spite of the formal similarities and the continuous empirical interdependencies and interpenetrations, both of which are of the greatest importance, personalities and social systems remain two distinct classes of systems.

Introduction to "Sex Roles in the American Kinship System"

In "Sex Roles in the American Kinship System," Parsons applies his theory to the empirical topic of the American family. This descriptive essay did not incite much reaction when it was first published in the 1940s; however, in the 1970s it came to epitomize Parsons's conservatism, interpreted as it was as an explicit endorsement by Parsons of traditional gender roles and the dire consequences that would ensue should they be breeched. Feminists were particularly incensed by Parsons's assertion that "many women succumb to . . . dependency cravings through such channels as neurotic illness or compulsive domesticity" which leads them to "abdicate both their responsibilities and their opportunities for genuine independence" ([1943] 1954:194). In addition, they found Parsons's assumption that "surely the pattern of romantic love which makes his relation to the 'woman he loves' the most important single thing in a man's life, is incompatible with the view that she is an inferior creature, fit only for dependency on him" especially naïve. Yet, interestingly, read in the context of the twenty-first century, one can see that in some respects Parsons got a "bad rap" for this essay. Although there is no question that his description of the ideal, typical, white middle-class family is told from an upper-middle-class white male's point of view, in fact, Parsons did capture important elements of this system. For instance, certainly his assertion that this traditional role structure "serves to concentrate the judgment and valuation of men on their occupational achievements, while the valuation of women is diverted into realms outside the occupationally relevant sphere" rings true (ibid.:122). Indeed, one can even read this statement as an indictment of the traditional kinship system because of its demeaning effect on women. Parsons's most pivotal premise, however—that changes that are functional for one part of the system (e.g., the benefits to women and society as women enter the paid workforce) will produce changes that are not necessarily functional for other parts of the system (e.g., the schools, which relied on the free labor of female "volunteers" for essential tasks—see Dorothy Smith, Chapter 14—and hospitals, who relied heavily on low-paying female positions as nurses)—is not necessarily sexist at all, in the sense that what *is* sexist is the assumption (prevalent in the 1960s and 1970s) that women *could* enter the workforce without significant changes being made to other social structures and systems (e.g., families, schools, the workplace) and, at the least, without a major increase in quality day-care and child-care facilities.

"Sex Roles in the American Kinship System" (1943)

Talcott C. Parsons

Much psychological research has suggested the very great importance to the individual of his affective ties, established in early childhood, to other members of his family of orientation. When strong affective ties have been formed, it seems reasonable to believe that situational pressures which force their drastic modification will impose important strains upon the individual.

Since all known kinship systems impose an incest tabu, the transition from asexual intrafamilial relationships to the sexual relation of marriage—generally to a previously relatively unknown person—is general. But with us this transition is accompanied by a process of "emancipation" from the ties both to parents and to siblings which is considerably more drastic than in most kinship systems, especially in that it applies to both sexes about equally, and includes emancipation from solidarity with *all* members of the family of orientation about equally, so that there is relatively little continuity with *any* kinship ties established by birth for anyone.

The effect of these factors is reinforced by two others. Since the effective kinship unit is normally the small conjugal family, the child's emotional attachments to kin are confined to relatively few persons instead of being distributed more widely. Especially important, perhaps, is the fact that no other adult woman has a role remotely similar to that of the mother. Hence the average intensity of affective involvement in family relations is likely to be high. Secondly, the child's relations outside the family are only to a small extent ascribed. Both in the play group and in the school he must to a large extent "find his own level" in competition with others. Hence the psychological significance of his security within the family is heightened.

We have then a situation where at the same time the inevitable importance of family ties is intensified and a necessity to become emancipated from them is imposed. This situation would seem to have a good deal to do with the fact that with us adolescence—and beyond—is, as has been frequently noted, a "difficult" period in the life cycle. In particular, associated with this situation is the prominence in our society of what has been called a "youth culture," a distinctive pattern of values and attitudes of the age groups between childhood and the assumption of full adult responsibilities. This youth culture, with its irresponsibility, its pleasure-seeking, its "rating and dating," and its intensification of the romantic love pattern, is not a simple matter of "apprenticeship" in adult values and responsibilities. It bears many of the marks of reaction to emotional tension and insecurity, and in all probability has among its functions that of easing the difficult process of adjustment from childhood emotional dependency to full "maturity." In it we find still a third element underlying the prominence of the romantic love complex in American society.

The emphasis which has here been placed on the multilineal symmetry of our kinship structure might be taken to imply that our society was characterized by a correspondingly striking assimilation of the roles of the sexes to each other. It is true that American society manifests a high level of the "emancipation" of women, which in important respects involves relative assimilation to masculine roles, in accessibility to occupational opportunity, in legal rights relative to property holding, and in various other respects. Undoubtedly the kinship system constitutes one of the important sets of factors underlying this emancipation since it does not, as do so many kinship systems, place a structural premium on the role of either sex in the maintenance of the continuity of kinship relations.

But the elements of sex-role assimilation in our society are conspicuously combined with elements of segregation which in many respects are even more striking than in other societies, as for instance in the matter of the much greater attention given by women to style and refinement of taste in dress and personal appearance. This and other aspects of segregation are connected with the structure of kinship, but not so much by itself as in its interrelations with the occupational system.

The members of the conjugal family in our urban society normally share a common basis of economic support in the form of money income, but this income is not derived from the co-operative efforts of the family as a unit—its principal source lies in the remuneration of occupational roles performed by individual members of the family. Status in an occupational role is generally, however, specifically segregated from kinship status—person holds a "job" as an individual, not by virtue of his status in a family.

Among the occupational statuses of members of a family, if there is more than one, much the most important is that of the husband and father, not only because it is usually the primary source of family income, but also because it is the most important single basis of the status of the family in the community at large. To be the main "breadwinner" of his family is a primary role of the normal adult man in our society. The corollary of this role is his far smaller participation than that of his wife in the internal affairs of the household. Consequently, "housekeeping" and the care of children is still the primary functional content of the adult feminine role in the "utilitarian" division of labor. Even if the married woman has a job, it is, at least in the middle classes, in the great majority of cases not one which in status or remuneration competes closely with those held by men of her own class. Hence there is a typically asymmetrical relation of the marriage pair to the occupational structure.

This asymmetrical relation apparently both has exceedingly important positive functional significance and is at the same time an important source of strain in relation to the patterning of sex roles.

On the positive functional side, a high incidence of certain types of patterns is essential to our occupational system and to the institutional complex in such fields as property and exchange which more immediately surround this system. In relatively commonsense terms it requires scope for the valuation of personal achievement, for equality of opportunity, for mobility in response to technical requirements, for devotion to occupational goals and interests relatively unhampered by "personal" consideration. In more technical terms it requires a high incidence of technical competence, of rationality, of universalistic norms, and of functional specificity. All these are drastically different from the patterns which are dominant in the area of kinship relations, where ascription of status by birth play a prominent part, and where roles are defined primarily in particularistic and functionally diffuse terms.

It is quite clear that the type of occupational structure which is so essential to our society requires a far-reaching structural segregation of occupational roles from the kinship roles of the *same* individuals. They must, in the occupational system, be treated primarily as individuals. This is a situation drastically different from that found in practically all non-literate societies and in many that are literate.

At the same time, it cannot be doubted that a solidary kinship unit has functional significance of the highest order, especially in relation to the socialization of individuals and to the deeper aspects of their psychological security. What would appear to have happened is a process of mutual accommodation between these two fundamental aspects of our social structure. On the one hand our kinship system is of a structural type which, broadly speaking, interferes least with the functional needs of the occupational system, above all in that it exerts relatively little pressure for the ascription of an individual's social status—through class affiliation, property, and of course particular "jobs"—by virtue of his kinship status. The conjugal unit can be mobile in status independently of the other kinship ties of its members, that is, those of the spouses to the members of their families of orientation.

But at the same time this small conjugal unit can be a strongly solidary unit. This is facilitated by the prevalence of the pattern that normally only *one* of its members has an occupational role which is of determinate significance for the status of the family as a whole. Minor children, that is, as a rule do not "work," and

when they do, it is already a major step in the process of emancipation from the family of orientation. The wife and mother is either exclusively a "housewife" or at most has a "job" rather than a "career."

There are perhaps two primary functional aspects of this situation. In the first place, by confining the number of status-giving occupational roles of the members of the effective conjugal unit to one, it eliminates any competition for status, especially as between husband and wife, which might be disruptive of the solidarity of marriage. So long as lines of achievement are segregated and not directly comparable, there is less opportunity for jealousy, a sense of inferiority, etc, to develop. Secondly, it aids in clarity of definition of the situation by making the status of the family in the community relatively definite and unequivocal. There is much evidence that this relative definiteness of status is an important fact of in psychological security.

The same structural arrangements which have this positive functional significance also give rise to important strains. What has been said above about the pressure for thoroughgoing emancipation from the family of orientation is a case in point. But in connection with the sex-role problem there is another important source of strain.

Historically, in Western culture, it may perhaps be fairly said that there has been a strong tendency to define the feminine role psychologically as one strongly marked by elements of dependency. One of the best symbols perhaps was the fact that until rather recently the married woman was not *sui juris,* could not hold property, make contracts, or sue in her own right. But in the modern American kinship system, to say nothing of other aspects of the culture and social structure, there are at least two pressures which tend to counteract this dependency and have undoubtedly played a part in the movement for feminine emancipation.

The first, already much discussed, is the multilineal symmetry of the kinship system which gives no basis of sex discrimination, and which in kinship terms favors equal rights and responsibilities for both parties to a marriage. The second is the character of the marriage relationship. Resting as it does primarily on affective attachment for the other person as a concrete human individual, a "personality," rather than on more objective considerations of status, it puts a premium on a certain kind of mutuality and equality. There is no clearly structured superordination-subordination pattern. Each is a fully responsible "partner" with a claim to a voice in decisions, to a certain human dignity, to be "taken seriously." Surely the pattern of romantic love which makes his relation to the "woman he loves" the most important single thing in a man's life, is incompatible with the view that she is an inferior creature, fit only for dependency on him.

In our society, however, occupational status has tremendous weight in the scale of prestige values. The fact that the normal married woman is debarred from testing or demonstrating her fundamental equality with her husband in competitive occupational achievement, creates a demand for a functional equivalent. At least in the middle classes, however, this cannot be found in the utilitarian functions of the role of housewife since these are treated as relatively menial functions. To be, for instance, an excellent cook, does not give a hired maid a moral claim to a higher status than that of domestic servant.

This situation helps perhaps to account for a conspicuous tendency for the feminine role to emphasize broadly humanistic rather than technically specialized achievement values. One of the key patterns is that of "good taste," in personal appearance, house furnishings, cultural things like literature and music. To a large and perhaps increasing extent the more humanistic cultural traditions and amenities of life are carried on by women. Since these things are of high intrinsic importance in the scale of values of our culture, and since by virtue of the system of occupational specialization even many highly superior men are greatly handicapped in respect to them, there is some genuine redressing of the balance between the sexes.

There is also, however, a good deal of direct evidence of tension in the feminine role. In the "glamor girl" pattern, use of specifically feminine devices as an instrument of compulsive search for power and exclusive attention are conspicuous. Many women succumb to their dependency cravings through such channels as neurotic illness or compulsive domesticity and thereby abdicate both their responsibilities and their opportunities for genuine independence. Many of the attempts to excel in approved channels of achievement are marred by garishness of taste, by instability in response to fad and fashion, by a seriousness in community or club activities which is out of proportion to the intrinsic importance of the task. In all of these and other fields there are conspicuous signs of insecurity and ambivalence. Hence it may be concluded that the feminine role is a conspicuous focus of the strains inherent in our social structure, and not the least of the sources of these strains is to be found in the functional difficulties in the integration of our kinship system with the rest of the social structure.

Robert K. Merton (1910–2003): A Biographical Sketch

Robert King Merton was born Meyer R. Schkolnick on July 4, 1910, in South Philadelphia, to working-class, Jewish immigrants from eastern Europe. After a rough-and-tumble early childhood—which allegedly included membership in the local street gang—the teenage Schkolnick began to perform magic tricks around the neighborhood, and at parties and social gatherings, using the stage name "Robert Merlin." After a young friend convinced him that borrowing King Arthur's mentor's name was cliché, he changed it to "Robert King Merton."

From a very early age, Merton showed a passion for learning, and upon graduating from high school he earned a scholarship to Temple University. He graduated from Temple with a degree in sociology in 1931 and went on to earn his doctorate in sociology at Harvard University, where he was one of the first and most important students of Talcott Parsons. Merton first began publishing while still a graduate student, and by the time he was forty he was one of the most influential social scientists in the United States (Calhoun 2003:9). He developed the very idea of studying science sociologically (i.e., the sociology of science), and as early as 1942 his "ethos of science" challenged the common public perception of scientists as eccentric geniuses free of normal social constraints. It was primarily for this work that Merton became the first sociologist to be awarded a National Medal of Science in 1994. Merton continued to study the sociology of science, perhaps the field closest to his heart, publishing his masterpiece on this topic, *On the Shoulders of Giants* (which neatly captures his stance)-in 1965. In 1949 (albeit revised and expanded in 1957 and 1968), Merton published what would become his magnum opus, *Social Theory and Social Structure.*

From 1941 until his death in 2003, Merton was a professor of sociology at Columbia University. He mentored an extraordinary number of students, many of whom would become prominent in their own right, including Robin Williams, Jr.; Jesse Pitts; Peter Blau; James Coleman; Lewis Coser; Rose Coser; Alvin Gouldner; Seymour Martin Lipset; Alice Rossi; and Arthur Stinchcombe. Robert Merton is survived by his wife and collaborator Harriet Zuckerman, three children, nine grandchildren, and nine great-grandchildren and, as Calhoun (2003:1) aptly notes, "by thousands of sociologists whose work is shaped every day by his."

■■ INTELLECTUAL INFLUENCES AND CORE IDEAS

Robert Merton is a major theorist who, as indicated above, coined several pivotal sociological concepts. However, Merton was also a prolific researcher who studied a wide variety of empirical areas and topics, including deviance, drug addiction, friendship formation, medical education, technology, media, and the history of science. In addition, Merton wrote extensively on the discipline of sociology, making vital methodological as well as empirical and theoretical contributions. (For instance, he developed the "focused interview," now called "focus group research.") Indeed, one could argue that Merton's single most important achievement has been to establish connections between theory and research, thereby charting the course of the discipline of sociology.

Merton was influenced by a broad array of theorists, philosophers, and social scientists. He read widely, and one can see traces of Durkheim, Weber, Simmel, and Marx, as well as the structural function-alist anthropologist Bronislaw Malinowski, in his work. Merton also drew inspiration from the work of William I. Thomas and Florian Znaniecki. Above all, however, Merton learned from Pitirim Sorokin, the first chair of the department of sociology at Harvard University, who recruited Merton for graduate school, and Sorokin's young colleague at Harvard, Talcott Parsons.

In contrast to Parsons, who sought to delineate a highly abstract, master conceptual schema, Merton favored what he called middle-range theory: theories that "lie between the minor but necessary working hypotheses that evolve in abundance during day-to-day research and the all-inclusive systematic efforts to develop a unified theory that will explain all the observed uniformities of social behavior, social organization, and social change" (1967:49, as cited in Sztompka 1986:41). Merton's middle-range reformulation of structural functionalism made it eminently more useful. Indeed, Merton has contributed numerous concepts that are now "staples" in sociology, and you will read about these concepts in the following text. Yet, perhaps the single most important contribution that Merton made to functionalism—and sociology—is that he extended Parsons's point that society is a system of interrelated parts and reworked it in order to emphasize that the components of the system may or may not be "in sync" and that the results are not always predictable. This pivotal theoretical contribution is readily apparent not only in Merton's highly influential concepts of manifest and latent function and dysfunction, but also in his oft-cited theory of deviance, discussed below.

Perhaps the most well-known concepts to arise from functionalism are manifest and latent function. For Merton, **manifest function** refers to the overt or intended purpose of action. **Latent function**, on the other hand, refers to implicit or unintended purpose. As Merton readily notes, the terms *manifest* and *latent* come from Freud (1915), who emphasized that almost every behavior, including seemingly mundane ones such as gum chewing, have "latent" functions related to our sexual drive. From a Freudian perspective, gum chewing (and smoking, biting one's nails, etc.) is an "oral fixation" related to the "oral" period of child development (0–15 months), in which gratification revolves around nursing, gumming, sucking, and mouth movement. Oral fixations reflect a failure to complete this stage; the lack of psychological resolution results in a constant "hunger" for activities involving the mouth.

So, too, Merton argues, do social actions sometimes have latent functions far more significant than their more manifest purpose. Consider, for instance, Merton's now classic example of the Hopi rain dance, which you will read about on the following page. While the manifest, or overt, purpose of the dance is to produce rain, the latent function is to reaffirm group identity, solidarity, and/or cohesion. Thus, while from a strictly rationalist point of view, one might consider a rain dance that did not result in rain a failure, in fact, the latent function of the ritual may nevertheless have been fulfilled. Of course, this latent social function—the affirmation of social ties and celebration of shared identity—is among the most important latent functions in all of social life, undergirding a wide variety of social conventions and rituals, such as attending holiday celebrations, conferences, and/or weddings; writing thank-you notes; and joining a gang. (For a more recent discussion of this point, see Randall Collins 2004.) For instance, the "business lunch" often has this same latent function of reinforcing social bonds and asserting or celebrating group identity despite and/or alongside its more obvious (manifest) goals. And just as a rain dance may or may not immediately produce rain, so, too, a business lunch may not achieve immediate (manifest) results (e.g., a deal) but nevertheless fulfill its critical latent function.

In sum, Merton's concept of manifest and latent function greatly enhances the Parsonian notion of society as a system of interrelated parts, not only because it acknowledges that there are multiple functions for any one component but also because it underscores that the various functions within even a single component might not coincide with each other or that they might even conflict. Whereas Parsons's conceptualization of society as a system of interrelated parts seemed to imply that all social institutions were inherently functional—otherwise they would not exist—Merton emphasized that different parts of a system might be at odds with each other and, thus, that even functional or beneficial institutions or subsystems can produce *dysfunctions* or *unintended consequences* as well.

This is precisely the point of Merton's extraordinarily influential theory of **deviance**. In what is "arguably the most cited article" in sociology (Crothers 1987:120), Merton ([1949] 1957) seeks to explain variances in rates of deviance according to social structural location. In sociology, deviance refers to modes of action that do not conform to the dominant norms or values in a social group or society. Following the functionalist notion that society is a system of interrelated parts, Merton hypothesized that deviance results when there is a disconnect between the cultural and social realms. That is, deviance occurs when the *values* of a society are out of sync with the *means* available for achieving them. For Merton, this state of affairs exists when, for instance, an individual internalizes the notion that "success" means having lots of money but is not afforded the opportunity to earn a legitimate, well-paying living. Thus, the individual turns to illegal means (e.g., selling illegal drugs) in order to achieve economic success.

That unintended consequences occur because of a disconnect between the cultural and social realms brings us to the concept of **dysfunction**. It was Émile Durkheim (see Chapter 3) who first emphasized that while positive social changes, such as periods of economic prosperity, might alleviate certain problems, they may also produce significant unanticipated negative consequences (such as an increased likelihood for moral disorder). Merton not only elaborated on Durkheim's point that positive social institutions (or changes) may have unintended negative consequences; he went on to also show that negative (or benign) social institutions (or social changes) might have unanticipated positive consequences as well. But it is for highlighting the "negative" unintended consequences and dysfunctions that Merton is most known.

Significantly, the concept of dysfunction is not incompatible with the functionalist metaphor of the body. The body is not a flawless machine; on the contrary, problems and irregularities often arise, as "subsystems" go awry. For instance, the body typically responds to infection by producing a fever. But while the fever has an important *function* as part of the body's defense, high fevers are dangerous or *dysfunctional* because they

Photo 9.3a   Hopi rain dance

The manifest function of the rain dance is to prove the Hopi worthy enough to receive the much needed thunderstorms, while the latent function is to reinforce group identity and social ties and to celebrate the cosmic order in which rain and person all belong. So, too, is the manifest function of the business lunch to discuss issues or make decisions, while the latent function is to reinforce group identity and social ties (e.g., to "network").

Photo 9.3b   Business lunch

can result in brain damage. Similarly, a drug such as aspirin might be prescribed in order to reduce fever but have the undesirable side effect (or dysfunction) of producing stomach pain.

So, too, Merton points out, laws, can social policies, norms, values, religions, and the like all produce unintended consequences and dysfunctions. For instance, while most people today would not dispute that men and women should be allowed to enter any profession they choose and for which they are qualified, one of the unintended consequences of the rise in gender equality for women in employment has been that it has led to a shortage of nurses and teachers. Now that women have a wide variety of options for employment—many of which are more attractive and lucrative than teaching or nursing— hospitals and schools (especially in poor communities) are hard-pressed to fill these extremely challenging, but still low-status, positions. Indeed, during the past 30 years, the percentage of female physicians rose fivefold, from approximately 8 percent of physicians in 1970 to 34 percent in 2002 (Abdo and Broxterman 2004); however, less than 6 percent of nurses today are male, an increase of only 2 percent in the last 10 years (Burtt 1998).

Similarly, Merton made a significant contribution to role theory by demonstrating that "social status" and "role-sets" are organized in the social structure in a more complex way than Parsons initially supposed. Specifically, Merton (1996:43) defines *social status* as "a position in a social system, with its distinctive array of designated rights and obligations." That is, the status of a role is not fixed, but rather changes in conjunction with the particular role-set involved. Thus, for instance, the status of school teacher might be high in relation to the correlative (low) status of pupil, but low in relation to the school principal and superintendent and the Board of Education (ibid.:44). (See Photos 9.4a and 9.4b.) Yet, the point is that not only role statuses but also role-expectations and role-obligations shift in interaction. What it means to be a "teacher" changes significantly based on the particular role-sets involved. Consequently, not only is role conflict inevitable because any one individual plays multiple roles (e.g., you are a Little League coach on your way to the big game when your elderly mother calls because she needs you to take her to the doctor); even within one role, conflict may occur because of the multiple role-sets involved.

These examples bring us to one of the most important criticisms of functionalism. Critics contend that functionalists rationalize or "explain away" *all* aspects of the social system, including social inequities. In contrast to critical explanations of inequality (which you will read about in the following chapter), functionalism does not expressly indict inequality. Rather, functionalists note that certain policies and institutions can be dysfunctional for entire groups of people but maintained because they are functional for a more powerful social group or the social system itself. For instance, in one of the most famous and controversial functionalist essays, Kingsley Davis and Wilbert E. Moore (1945) argued that a society develops an unequal distribution of rewards, such as money, power, or prestige, because it has to attract the most qualified people to fill its most vital positions. But critics read Davis and Moore's piece as an apology for the "system."

Photo 9.4a   Teacher/student

MERTON'S THEORETICAL ORIENTATION ▪▪

Merton's work is far less abstract than Parsons's, but it is no less theoretically multidimensional. In terms of the question of order, Merton consistently argues that while individuals are constrained by social forces, alternative modes of action are nevertheless possible. This is most readily apparent in Merton's theory of deviance, which is especially attuned to explaining the interchange between the individual and the collective levels. Merton integrates an appreciation for the (collective) structural impediments to acquiring money and status (which reflects a collective approach to order) with an acknowledgment that individuals nevertheless "choose" to follow or abandon specific means and goals (which reflects the level of the individual).

So, too, does Merton's role theory reflect both the individual and collective levels of order in that role-sets are composed of both normative expectations and obligations that preexist the individual actor (thereby reflecting the level of the collective), and also the personality needs of the individual. In other words, that the individual both internalizes role-expectations and, at the same time, takes account of the demands of the situation reflects both the interpenetration of the individual actor-ego and the social systems and of agency at the level of the individual.

In terms of action, on one hand, actors sometimes play their roles quite deliberately and strategically (e.g., in a job interview), and at the level of the *system,* positions and obligations in a highly rationalized society (e.g., in which jobs are eliminated at the drop of a hat with respect to the "bottom line") both reflect the primacy of the rational realm. Above all, however, "role-sets" speak to the nonrational dimension of action in that even the most "strategic" playing of roles inevitably rests on symbols and values (e.g., the "necktie" as a business emblem and sign of professionalism). Moreover, actors often take on roles unreflectively (such as gender roles) because they believe they are

Merton emphasized that the expectations, obligations, and status incumbent in roles is not static. Rather they change systematically depending on the particular *role-set* involved. For instance, the role of teacher may have high status in the teacher-student pair, but low status in relationship with the school board.

Photo 9.4b   Striking teachers

morally "right" or inevitable, which reflects both the nonrational/individual realm as well as that of the collective (since it is the collective level that creates gendered role expectations). This point is also duly reflected in Merton's theory of deviance, which, above all, emphasizes the moral forces and values of the cultural realm.

Significantly, then, as shown in Figure 9.3 and as discussed previously, overall structural functionalism tends to emphasize the collective/nonrational realm. As indicated above, this point was forcefully taken up by the conflict theorists in the 1960s and 1970s who criticized structural functionalists for underemphasizing the influence of preexisting institutions and powerful social groups.

Figure 9.3   Merton's Basic Theoretical Orientation

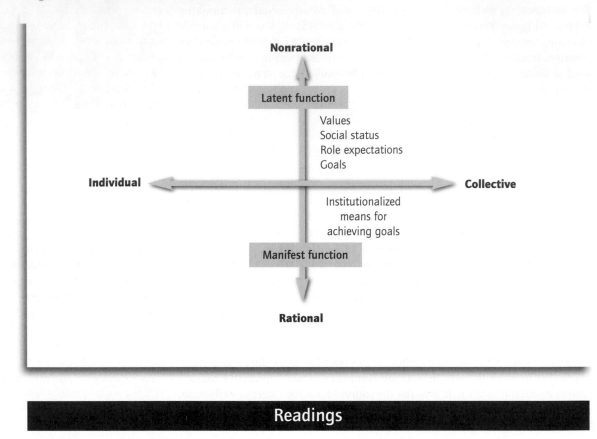

Readings

The two selections that follow reflect the extraordinary breadth and range of Merton's work. The first excerpt demonstrates Merton's pivotal contributions to structural functionalist theory; the second excerpt is one of the most famous and influential essays in the field of deviance.

Introduction to "Manifest and Latent Functions"

"Manifest and Latent Functions" (1949) is a classic functionalist piece that introduces two of Merton's most famous and important sociological concepts: manifest and latent functions, discussed previously. While manifest and latent functions can be found in a wide variety of social situations, interestingly, this excerpt concludes with a pivotal example at the heart of American life: consumer activity. Following Thorsten Veblen (1857–1929), who coined the term *conspicuous consumption,* Merton emphasizes that behind the obvious (manifest) purpose of buying goods in order to satisfy specific needs is the more significant latent function of "heightening or reaffirming social status." The result then is that "people buy expensive goods not so much because they are superior but because they are expensive" (p. 69).

"Manifest and Latent Functions" (1949)

Robert K. Merton

The distinction between manifest and latent functions was devised to preclude the inadvertent confusion, often found in the sociological literature, between conscious *motivations* for social behavior and its *objective consequences.* Our scrutiny of current vocabularies of functional analysis has shown how easily, and how unfortunately, the sociologist may identify *motives* with *functions*. It was further indicated that the motive and the function vary independently and that the failure to register this fact in an established terminology has contributed to the unwitting tendency among sociologists to confuse the subjective categories of motivation with the objective categories of function. This, then, is the central purpose of our succumbing to the not-always-commendable practice of introducing new terms into the rapidly growing technical vocabulary of sociology, a practice regarded by many laymen as an affront to their intelligence and an offense against common intelligibility.

As will be readily recognized, I have adapted the terms "manifest" and "latent" from their use in another context by Freud (although Francis Bacon had long ago spoken of "latent process" and "latent configuration" in connection with processes which are below the threshold of superficial observation).

The distinction itself has been repeatedly drawn by observers of human behavior at irregular intervals over a span of many centuries. Indeed, it would be disconcerting to find that a distinction which we have come to regard as central to functional analysis had not been made by any of that numerous company who have in effect adopted a functional orientation. We need mention only a few of those who have, in recent decades, found it necessary to distinguish in their specific interpretations of behavior between the end-in-view and the functional consequences of action.

George H. Mead[i]: ". . . that attitude of hostility toward the law-breaker has the unique advantage [read: latent function] of uniting all members of the community in the emotional solidarity of aggression. While the most admirable of humanitarian efforts are sure to run counter to the individual interests of very many in the community, or fail to touch the interest and imagination of the multitude and to leave the community divided or indifferent, the cry of thief or murderer is attuned to profound complexes, lying below the surface of competing individual efforts, and citizens who have [been] separated by divergent interests stand together against the common enemy."

Émile Durkheim's[ii] similar analysis of the social functions of punishment is also focused on its latent functions (consequences for the community) rather than confined to manifest functions (consequences for the criminal).

W. G. Sumner[iii]: ". . . from the first acts by which men try to satisfy needs, each act stands by itself, and looks no further than the immediate satisfaction. From recurrent needs arise habits for the individual and customs for the group, but these results are consequences which were never conscious, and never foreseen or intended. They are not noticed until they have long existed, and it is still longer before they are appreciated." Although this fails to locate the latent functions of standardized social actions for a designated social structure, it plainly makes the basic distinction between ends-in-view and objective consequences.

R. M. MacIver[iv]: In addition to the direct effects of institutions, "there are further effects by way of control which lie outside the direct purposes of men . . . this type of reactive form of control . . . may, though unintended, be of profound service to society."

---

[i]George H. Mead, "The psychology of punitive justice," *American Journal of Sociology,* 1918, 23, 577–602, esp. 591.

[ii]As suggested earlier in this chapter, Durkheim adopted a functional orientation throughout his work, and he operates, albeit often without explicit notice, with concepts equivalent to that of latent function in all of his researches. The reference in the text at this point is to his "Deux lois de l'évolution penale," *L'année sociologique,* 1899–1900, 4, 55–95, as well as to his *Division of Labor in Society* (Glencoe, Illinois: The Free Press, 1947).

[iii]This one of his many such observations is of course from W. G. Sumner's *Folkways* (Boston: Ginn & Co., 1906), 3. His collaborator, Albert G. Keller, retained the distinction in his own writings; see, for example, his *Social Evolution* (New York: Macmillan, 1927), at 93–95.

[iv]This is advisedly drawn from one of MacIver's earlier works, *Community* (London: MacMillan, 1915). The distinction takes on greater importance in his later writings, becoming a major element in his *Social Causation* (Boston: Ginn & Co., 1942), esp. at 314–321, and informs the greater part of his *The More Perfect Union* (New York: Macmillan, 1948).

W. I. Thomas and F. Znaniecki[v]: "Although all the new [Polish peasant cooperative] institutions are thus formed with the definite purpose of satisfying certain specific needs, their social function is by no means limited to their explicit and conscious purpose . . . every one of these institutions—commune or agricultural circle, loan and savings bank, or theater—is not merely a mechanism for the management of certain values but also an association of people, each member of which is supposed to participate in the common activities as a living, concrete individual. Whatever is the predominant, official common interest upon which the institution is founded, the association as a concrete group of human personalities unofficially involves many other interests; the social contacts between its members are not limited to their common pursuit, though the latter, of course, constitutes both the main reason for which the association is formed and the most permanent bond which holds it together. Owing to this combination of an abstract political, economic, or rather rational mechanism for the satisfaction of specific needs with the concrete unity of a social group, the new institution is also the best intermediary link between the peasant primary-group and the secondary national system."

These and numerous other sociological observers have, then, from time to time distinguished between categories of subjective disposition ("needs, interest, purposes") and categories of generally unrecognized but objective functional consequences ("unique advantages," "never conscious" consequences, "unintended . . . service to society," "function not limited to conscious and explicit purpose").

Since the occasion for making the distinction arises with great frequency, and since the purpose of a conceptual scheme is to direct observations toward salient elements of a situation and to prevent the inadvertent oversight of these elements, it would seem justifiable to designate this distinction by an appropriate set of terms. This is the rationale for the distinction between manifest functions and latent functions; the first referring to those objective consequences for a specified unit (person, subgroup, social or cultural system) which contribute to its adjustment or adaptation and were so intended; the second referring to unintended and unrecognized consequences of the same order.

There are some indications that the christening of this distinction may serve a heuristic purpose by becoming incorporated into an explicit conceptual apparatus, thus aiding both systematic observation and later analysis. In recent years, for example, the distinction between manifest and latent functions has been utilized in analyses of racial intermarriage,[vi] social stratification,[vii] affective frustration,[viii] Veblen's sociological theories,[ix] prevailing American orientations toward Russia,[x] propaganda as a means of social control,[xi] Malinowski's anthropological theory,[xii] Navajo witchcraft,[xiii] problems in the sociology of knowledge,[xiv] fashion,[xv] the dynamics of personality,[xvi] national security measures,[xvii] the internal social

---

[v]The single excerpt quoted in the text is one of scores which have led to *The Polish Peasant in Europe and America* being deservedly described as a "sociological classic." See pages 1426–7 and 1523 ff. As will be noted later in this chapter, the insights and conceptual distinctions contained in this one passage, and there are many others like it in point of richness of content, were forgotten or never noticed by those industrial sociologists who recently came to develop the notion of "informal organization" in industry.

[vi]Merton, "Intermarriage and the social structure," *op. cit.*

[vii]Kingsley Davis, "A conceptual analysis of stratification," *American Sociological Review,* 1942, 7, 309–321.

[viii]Thorner, *op. cit.,* esp. at 165.

[ix]A. K. Davis, *Thorstein Veblen's Social Theory,* Harvard Ph.D. dissertation, 1941, and "Veblen on the decline of the Protestant Ethic," *Social Forces,* 1944, 22, 282–86; Louis Schneider, *The Freudian Psychology and Veblen's Social Theory* (New York: King's Crown Press, 1948), esp. Chapter 2.

[x]A. K. Davis, "Some sources of American hostility to Russia," *American Journal of Sociology,* 1947, 53, 174–183.

[xi]Talcott Parsons, "Propaganda and social control," in his *Essays in Sociological Theory.*

[xii]Clyde Kluckhohn, "Bronislaw Malinowski, 1884–1942," *Journal of American Folklore,* 1943, 56, 208–219.

dynamics of bureaucracy,[xviii] and a great variety of other sociological problems.

The very diversity of these subject-matters suggests that the theoretic distinction between manifest and latent functions is bound up with a limited and particular range of human behavior. But there still remains the large task of ferreting out the specific uses to which this distinction can be put, and it is to this large task that we devote the remaining pages of this chapter.

Heuristic Purposes of the Distinction

Clarifies the analysis of seemingly irrational social patterns. In the first place, the distinction aids the sociological interpretation of many social practices which persist even though their manifest purpose is clearly not achieved. The time-worn procedure in such instances has been for diverse, particularly lay, observers to refer to these practices as "superstitions," irrationalities," "mere inertia of tradition," *etc.* In other words, when group behavior does not—and, indeed, often cannot—attain its ostensible purpose there is an inclination to attribute its occurrence to lack of intelligence, sheer ignorance, survivals, or so-called inertia. Thus, the Hopi ceremonials designed to produce abundant rainfall may be labelled a superstitious practice of primitive folk and that is assumed to conclude the matter. It should be noted that this in no sense accounts for the group behavior. It is simply a case of name-calling; it substitutes the epithet "superstition" for an analysis of the actual role of this behavior in the life of the group. Given the concept of latent function, however, we are reminded that this behavior *may* perform a function for the group, although this function

may be quite remote from the avowed purpose of the behavior.

The concept of latent function extends the observer's attention beyond the question of whether or not the behavior attains its avowed purpose. Temporarily ignoring these explicit purposes, it directs attention toward another range of consequences: those bearing, for example, upon the individual personalities of Hopi involved in the ceremony and upon the persistence and continuity of the larger group. Were one to confine himself to the problem of whether a manifest (purposed) function occurs, it becomes a problem, not for the sociologist, but for the meteorologist. And to be sure, our meteorologists agree that the rain ceremonial does not produce rain; but this is hardly to the point. It is merely to say that the ceremony does not have this technological use; that this purpose of the ceremony and its actual consequences do not coincide. But with the concept of latent function, we continue our inquiry, examining the consequences of the ceremony not for the rain gods or for meteorological phenomena, but for the groups which conduct the ceremony. And here it may be found, as many observers indicate, that the ceremonial does indeed have functions—but functions which are non-purposed or latent.

Ceremonials may fulfill the latent function of reinforcing the group identity by providing a periodic occasion on which the scattered members of a group assemble to engage in a common activity. As Durkheim among others long since indicated, such ceremonials are a means by which collective expression is afforded the sentiments which, in a further analysis, are found to be a basic source of group unity. Through the systematic application of the concept of latent function, therefore, *apparently* irrational behavior may *at times* be

[xiii]Clyde Kluckhohn, *Navaho Witchcraft, op. cit.,* esp. at 46–47 and ff.

[xiv]Merton, Chapter XII of this volume [*Social Theory and Social Structure,* 1957].

[xv]Bernard Barber and L. S. Lobel, "'Fashion' in women's clothes and the American social system," *Social Forces,* 1952, 31, 124–131.

[xvi]O. H. Mowrer and C. Kluckhohn, "Dynamic theory of personality," in J. M. Hunt, ed., *Personality and the Behavior Disorders* (New York: Ronald Press, 1944), 1, 69–135, esp. at 72.

[xvii]Marie Jahoda and S. W. Cook, "Security measures and freedom of thought: an exploratory study of the impact of loyalty and security programs," *Yale Law Journal,* 1952, 61, 296–333.

[xviii]Philip Selznick, *TVA and the Grass Roots* (University of California Press, 1949); A. W. Gouldner, *Patterns of Industrial Bureaucracy* (Glencoe, Illinois: The Free Press, 1954); P. M. Blau, *The Dynamics of Bureaucracy* (University of Chicago Press, 1955); A. K. Davis, "Bureaucratic patterns in Navy officer corps." *Social Forces* 1948, 27, 142–153.

found to be positively functional for the group. Operating with the concept of latent function, we are not too quick to conclude that if an activity of a group does not achieve its nominal purpose, then its persistence can be described only as an instance of "inertia," "survival," or "manipulation by powerful subgroups in the society."

In point of fact, some conception like that of latent function has very often, almost invariably, been employed by social scientists observing *a standardized practice designed to achieve an objective which one knows from accredited physical science cannot be thus achieved.* This would plainly be the case, for example, with Pueblo rituals dealing with rain or fertility. *But with behavior which is not directed toward a clearly unattainable objective, sociological observers are less likely to examine the collateral or latent functions of the behavior.*

Directs attention to theoretically fruitful fields of inquiry. The distinction between manifest and latent functions serves further to direct the attention of the sociologist to precisely those realms of behavior, attitude and belief where he can most fruitfully apply his special skills. For what is his task if he confines himself to the study of manifest functions? He is then concerned very largely with determining whether a practice instituted for a particular purpose does, in fact, achieve this purpose. He will then inquire, for example, whether a new system of wage-payment achieves its avowed purpose of reducing labor turnover or of increasing output. He will ask whether a propaganda campaign has indeed gained its objective of increasing "willingness to fight" or "willingness to buy war bonds," or "tolerance toward other ethnic groups." Now, these are important, and complex, types of inquiry. But, so long as sociologists *confine* themselves to the study of manifest functions, their inquiry is set for them by practical men of affairs (whether a captain of industry, a trade union leader, or, conceivably, a Navaho chieftain, is for the moment immaterial), rather than by the theoretic problems which are at the core of the discipline. By dealing primarily with the realm of manifest functions, with the key problem of whether deliberately instituted practices or organizations

succeed in achieving their objectives, the sociologist becomes converted into an industrious and skilled recorder of the altogether familiar pattern of behavior. *The terms of appraisal are fixed and limited by the question put to him by the non-theoretic men of affairs, e.g.,* has the new wage-payment program achieved such-and-such purposes?

But armed with the concept of latent function, the sociologist extends his inquiry in those very directions which promise most for the theoretic development of the discipline. He examines the familiar (or planned) social practice to ascertain the latent, and hence generally unrecognized, functions (as well, of course, as the manifest functions). He considers, for example, the consequences of the new wage plan for, say, the trade union in which the workers are organized or the consequences of a propaganda program, not only for increasing its avowed purpose of stirring up patriotic fervor, but also for making large numbers of people reluctant to speak their minds when they differ with official policies, *etc.* In short, it is suggested that the *distinctive* intellectual contributions of the sociologist are found primarily in the study of unintended consequences (among which are latent functions) of social practices, as well as in the study of anticipated consequences (among which are manifest functions).[xix]

There is some evidence that it is precisely at the point where the research attention of sociologists has shifted from the plane of manifest to the plane of latent functions that they have made their *distinctive* and major contributions. . . .

The discovery of latent functions represents significant increments in sociological knowledge. There is another respect in which inquiry into latent functions represents a distinctive contribution of the social scientist. It is precisely the latent functions of a practice or belief which are not common knowledge, for these are unintended and generally unrecognized social and psychological consequences. As a result, findings concerning latent functions represent a greater increment in knowledge than findings concerning manifest functions. They represent, also, greater departures from "common-sense" knowledge about social life. Inasmuch as the latent functions depart, more or less, from the

---

[xix]For a brief illustration of this general proposition, see Robert K. Merton, Marjorie Fiske, and Alberta Curtis, *Mass Persuasion* (New York: Harper, 1946), 185–189; Jahoda and Cook, *op. cit.*

avowed manifest functions, the research which uncovers latent functions very often produces "paradoxical" results. The seeming paradox arises from the sharp modification of a familiar popular preconception which regards a standardized practice or belief *only* in terms of its manifest functions by indicating some of its subsidiary or collateral latent functions. The introduction of the concept of latent function in social research leads to conclusions which show that "social life is not as simple as it first seems." For as long as people confine themselves to *certain* consequences (*e.g.* manifest consequences), it is comparatively simple for them to pass moral judgments upon the practice or belief in question. Moral evaluations, generally based on these manifest consequences, tend to be polarized in terms of black or white. But the perception of further (latent) consequences often complicates the picture. Problems of moral evaluation (which are not our immediate concern) and problems of social engineering (which are our concern[xx]) both take on the additional complexities usually involved in responsible social decisions.

An example of inquiry which implicitly uses the notion of latent function will illustrate the sense in which "paradox"—discrepancy between the apparent, merely manifest, function and the actual, which also includes latent functions—tends to occur as a result of including this concept. Thus, to revert to Veblen's well-known analysis of conspicuous consumption, it is no accident that he has been recognized as a social analyst gifted with an eye for the paradoxical, the ironic, the satiric. For these are frequent, if not inevitable, outcomes of applying the concept of latent function (or its equivalent).

The pattern of conspicuous consumption. The manifest purpose of buying consumption goods is, of course, the satisfaction of the needs for which these goods are explicitly designed. Thus, automobiles are obviously intended to provide a certain kind of transportation; candles to provide light; choice articles of food to provide sustenance; rare art products to provide aesthetic pleasure. Since these products *do* have these uses, it was largely assumed that these encompass the range of socially significant functions. Veblen indeed suggests that this was ordinarily the prevailing view (in the pre-Veblenian era, of course). "The end of acquisition and accumulation is conventionally held to be the consumption of the goods accumulated. . . . This is at least felt to be the economically legitimate end of acquisition, *which alone it is incumbent on the theory to take account of.*"[xxi]

However, says Veblen in effect, as sociologists we must go on to consider the latent functions of acquisition, accumulation and consumption, and these latent functions are remote indeed from the manifest functions. "But, it is only when taken in a sense far removed from its naive meaning [*i.e.* manifest function] that the consumption of goods can be said to afford the incentive from which accumulation invariably proceeds." And among these latent functions, which help explain the persistence and the social location of the pattern of conspicuous consumption, is its symbolization of "pecuniary strength and so of gaining or retaining a good name." The exercise of "punctilious discrimination" in the excellence of "food, drink, shelter, service, ornaments, apparel, amusements" results not merely in direct gratifications derived from the consumption of "superior" to "inferior" articles, but also, and Veblen argues, more importantly, it results in a *heightening or reaffirmation of social status.*

The Veblenian paradox is that people buy expensive goods not so much because they are superior but because they are expensive. For it is the latent equation ("costliness = mark of higher social status") which he singles out in his functional analysis, rather than the manifest equation ("costliness = excellence of the goods"). Not that he denies manifest functions *any* place in buttressing the pattern of conspicuous consumption. These, too, are operative. "What has just been said must not be taken to mean that there are no other incentives to acquisition and accumulation than this desire to excel in pecuniary standing and so gain the esteem and envy of one's fellowmen. The desire for added comfort and security from want is present as a motive at every stage. . . ." Or again: "It would be hazardous to assert that a useful purpose is ever absent from the utility of any article or of any service, however

---

[xx]This is not to deny that social engineering has direct moral implications or that technique and morality are inescapably intertwined, but I do not intend to deal with this range of problems in the present chapter. For some discussion of these problems see Merton, Fiske, and Curtis, *Mass Persuasion, op. cit.*

[xxi]Veblen, *Theory of the Leisure Class, op. cit.,* p. 25.

obviously its prime purpose and chief element is conspicuous waste" and derived social esteem.[xxii] It is only that *these direct, manifest functions do not fully account for the prevailing patterns of consumption. Otherwise put, if the latent functions of status-enhancement or status- reaffirmation were removed from the patterns of conspicuous consumption, these patterns would undergo severe changes of a sort which the "conventional" economist could not foresee.*

In these respects, Veblen's analysis of latent functions departs from the common-sense notion that the end-product of consumption is "of course, the direct satisfaction which it provides": "People eat caviar because they're hungry; buy Cadillacs because they want the best car they can get; have dinner by candlelight because they like the peaceful atmosphere." The common-sense interpretation in terms of selected manifest motives gives way, in Veblen's analysis, to the collateral latent functions which are also, and perhaps more significantly, fulfilled by these practices. To be sure, the Veblenian analysis has, in the last decades, entered so fully into popular thought, that these latent functions are now widely recognized.

▪▪

Introduction to "Social Structure and Anomie"

In this well-known and widely cited excerpt from "Social Structure and Anomie," Merton distinguishes between (1) the cultural norms and goals dominant in a society and (2) the socially structured capacities of members of the group to act in accord with them, and he maintains that the *dysjunction* between cultural goals and the acceptable social or institutional norms to attain them can unintentionally result in pressure to engage in nonconforming rather than conforming behavior. The beauty of Merton's famous typology, which has had a huge impact on the field of deviance, lies in its theoretical generality; it is a simple analytical construct that can be applied to an array of deviant behavior. The actual utility or adequacy of Merton's model is a separate question, and one that has spawned considerable empirical research.

"Social Structure and Anomie"

Robert K. Merton

The framework set out in this essay is designed to provide one systematic approach to the analysis of social and cultural sources of deviant behavior. Our primary aim is to discover how some social *structures exert a definite pressure upon certain persons in the society to engage in non-conforming rather than conforming conduct.* If we can locate groups peculiarly subject to such pressures, we should expect to find fairly high rates of

[xxii]*Ibid.,* 32, 101. It will be noted throughout that Veblen is given to loose terminology. In the marked passages (and repeatedly elsewhere) he uses "incentive," "desire," "purpose," and "function" interchangeably. Since the context usually makes clear the denotation of these terms, no great harm is done. But it is clear that the expressed purposes of conformity to a culture pattern are by no means identical with the latent functions of the conformity. Veblen occasionally recognizes this. For example, "In strict accuracy nothing should be included under the head of conspicuous waste but such expenditure as is incurred on the ground of an invidious pecuniary comparison. But in order to bring any given item or element in under this head *it is not necessary that it should be recognized as waste in this sense by the person incurring the expenditure*" *(Ibid.* 99; italics supplied). Cf. A. K. Davis

deviant behavior in these groups, not because the human beings comprising them are compounded of distinctive biological tendencies hut because they are responding normally to the social situation in which they find themselves. Our perspective is sociological. We look at variations in the *rates* of deviant behavior, not at its incidence.[i] Should our quest be at all successful, some forms of deviant behavior will be found to be as psychologically normal as conformist behavior, and the equation of deviation and psychological abnormality will be put in question.

PATTERNS OF CULTURAL GOALS AND INSTITUTIONAL NORMS

Among the several elements of social and cultural structures, two are of immediate importance. These are analytically separable although they merge in concrete situations. The first consists of culturally defined goals, purposes and interests, held out as legitimate objectives for all or for diversely located members of the society. The goals are more or less integrated—the degree is a question of empirical fact—and roughly ordered in some hierarchy of value. Involving various degrees of sentiment and significance, the prevailing goals comprise a frame of aspirational reference. They are the things "worth striving for." "They are a basic, though not the exclusive, component of what Linton has called "designs for group living." And though some, not all, of these cultural goals are directly related to the biological drives of man, they are not determined by them.

A second element of the cultural structure defines, regulates and controls the acceptable modes of reaching out for these goals. Every social group invariably couples its cultural objectives with regulations, rooted in the mores or institutions, of allowable procedures for moving toward these objectives. These regulatory norms are not necessarily identical with technical or efficiency norms. Many procedures which from the standpoint of particular individuals would be most efficient in securing desired values—the exercise of force, fraud, power—are ruled out of the institutional area of permitted conduct. At times, the disallowed procedures include some which would be efficient for the group itself—e.g., historic taboos on vivisection, on medical experimentation, on the sociological analysis of "sacred" norms—since the criterion of acceptability is not technical efficiency but value-laden sentiments (supported by most members of the group or by those able to promote these sentiments through the composite use of power and propaganda). In all instances, the choice of expedients for striving toward cultural goals is limited by institutionalized norms.

Sociologists often speak of these controls as being "in the mores" or as operating through social institutions. Such elliptical statements are true enough, but they obscure the fact that culturally standardized practices are not all of a piece. They are subject to a wide gamut of control. They may represent definitely prescribed or preferential or permissive or proscribed patterns of behavior. In assessing the operation of social controls, these variations—roughly indicated by the terms *prescription, preference, permission* and proscription—must of course be taken into account.

To say, moreover, that cultural goals and institutionalized norms operate jointly to shape prevailing practices is not to say that they bear a constant relation to one another. The cultural emphasis placed upon certain goals varies independently of the degree of emphasis upon institutionalized means. There may develop a

---

[i]The position, taken here has been perceptively described by Edward Sapir, "" . . . problems of social science differ from problems of individual behavior in degree of specificity, not in kind Every statement about behavior which throws the emphasis, explicitly or implicitly, on the actual, integral experiences of defined personalities or types of personalities is a datum of psychology or psychiatry rather than of social science. Every statement about behavior which aims, not to be accurate about the behavior of an actual individual or individuals or about the expected behavior of a physically and psychologically defined type of individual, but which abstracts from such behavior in order to bring out in clear relief certain expectancies with regard to those aspects of individual behavior which various people share, as an interpersonal or 'social' pattern, is a datum, however crudely expressed, of social science." I have here chosen the second perspective; although I shall have occasion to speak of attitudes, values and function, it will be from the standpoint of how the social structure promotes or inhibits their appearance in specified types of situations. See Sapir, "Why cultural anthropology needs the psychiatrist," *Psychiatry,* 1938, 1, 7–12.

very heavy, at times a virtually exclusive, stress upon the value of particular goals, involving comparatively little concern with the institutionally prescribed means of striving toward these goals. The limiting case of this type is reached when the range of alternative procedures is governed only by technical rather than by institutional norms. Any and all procedures which promise attainment of the all-important goal would be permitted in this hypothetical polar case. This constitutes one type of malintegrated culture. A second polar type is found in groups where activities originally conceived as instrumental are transmuted into self-contained practices, lacking further objectives. The original purposes are forgotten and close adherence to institutionally prescribed conduct becomes a matter of ritual.[ii] Sheer conformity becomes a central value. For a time, social stability is ensured—at the expense of flexibility. Since the range of alternative behaviors permitted by the culture is severely limited, there is little basis for adapting to new conditions. There develops a tradition-bound, 'sacred' society marked by neophobia. Between these extreme types are societies which maintain a rough balance between emphases upon cultural goals and institutionalized practices, and these constitute the integrated and relatively stable, though changing, societies. . . .

TYPES OF INDIVIDUAL ADAPTATION

Turning from these culture patterns, we now examine types of adaptation by individuals within the culture-bearing society. Though our focus is still the cultural and social genesis of varying rates and types of deviant behavior, our perspective shifts from the plane of patterns of cultural values to the plane of types of adaptation to these values among those occupying different positions in the social structure.

We here consider five types of adaptation, as these are schematically set out in the following table, where (+) signifies "acceptance," (−) signifies "rejection," and (±) signifies "rejection of prevailing values and substitution of new values,"

A TYPOLOGY OF MODES OF INDIVIDUAL ADAPTATION[iii]

Modes of Adaptation	Culture Goals	Institutionalized Means
I. Conformity	+	+
II. Innovation	+	−
III. Ritualism	−	+
IV. Retreatism	−	−
V. Rebellion*	±	±

*This fifth alternative is on a plane clearly different from that of the others. It represents a transitional response seeking to *institutionalize* new goals and new procedures to be shared by other members of the society. It thus refers to efforts to *change* the existing cultural and social structure rather than to accommodate efforts *within* this structure.

[ii]This ritualism may be associated with a mythology which rationalizes these practices so that they appear to retain their status as means, but the dominant pressure is toward strict ritualistic conformity, irrespective of the mythology. Ritualism is thus most complete when such rationalizations are not even called forth.

[iii]There is no lack of typologies of alternative modes of response to frustrating conditions. Freud, in his Civilization *and Its Discontents* (p. 30 ff.) supplies one; derivative typologies, often differing in basic details, will be found in Karen Horney, *Neurotic Personality of Our Time* (New York, 1937); S. Rosenzweig, "The experimental measurement of types of reaction, to frustration," in H. A, Murray et al., *Explorations in Personality* (New York, 1938), 585–99; and in the work of John Dollard, Harold Lasswell, Abram Kardiner, Erich Fromm. But particularly in the strictly Freudian typology, the perspective is that of types of individual responses, quite apart from the place of the individual within the social structure. Despite her consistent concern with "culture," for example, Horney does not explore differences in the impact of this culture upon farmer, worker and businessman, upon lower-, middle-, and upper-class individuals, upon members of various ethnic and racial groups, *etc.* As a result, the role of "inconsistencies in culture" is *not* located in its differential impact upon diversely situated groups. Culture becomes a kind of blanket covering all members of the society equally, apart from their idiosyncratic differences of life-history. It is a primary assumption of our typology that these responses occur with different frequency within various sub-groups in our society precisely because members of these groups or strata are differentially subject to cultural stimulation and social restraints. This sociological orientation will be found in the writings of Dollard and, less systematically, in the work of Fromm, Kardiner and Lasswell. On the general point, see note 3 of this chapter.

Examination of how the social structure operates to exert pressure upon individuals for one or another of these alternative modes of behavior must be prefaced by the observation that people may shift from one alternative to another as they engage in different spheres of social activities. These categories refer to role behavior in specific types of situations, not to personality. They are types of more or less enduring response, not types of personality organization, To consider these types of adaptation in several spheres of conduct would introduce a complexity unmanageable within the confines of this chapter. For this reason, we shall be primarily concerned with economic activity in the broad sense of' "the production, exchange, distribution and consumption of goods and services" in our competitive society, where wealth has taken on a highly symbolic cast.

I. Conformity

To the extent that a society is stable, adaptation type I—conformity to both cultural goals and institutionalized means—is the most common and widely diffused. Were this not so, the stability and continuity of the society could not be maintained. The mesh of expectancies constituting every social order is sustained by the modal behavior of its members representing conformity to the established, though perhaps secularly changing, culture patterns. It is, in fact, only because behavior is typically oriented toward the basic values of the society that we may speak of a human aggregate as comprising a society. Unless there is a deposit of values shared by interacting individuals, there exist social relations, if the disorderly interactions may be so called, but no society. It is thus that, at mid-century, one may refer to a Society of Nations primarily as a figure of speech or as an imagined objective, but not as a sociological reality.

Since our primary interest centers on the sources of *deviant* behavior, and since we have briefly examined the mechanisms making for conformity as the modal response in American society, little more need be said regarding this type of adaptation, at this point.

II. Innovation

Great cultural emphasis upon the success-goal invites this mode of adaptation through the use of institutionally proscribed but often effective means of attaining at least the simulacrum of success—wealth and power. This response occurs when the individual has assimilated the cultural emphasis upon the goal without equally internalizing the institutional norms governing ways and means for its attainment.

From the standpoint of psychology, great emotional investment in an objective may be expected to produce a readiness to take risks, and this attitude may be adopted by people in all social strata. From the standpoint of sociology, the question arises, which features of our social structure predispose toward this type of adaptation, thus producing greater frequencies of deviant behavior in one social stratum than in another?

On the top economic levels, the pressure toward innovation not infrequently erases the distinction between business-like strivings this side of the mores and sharp practices beyond the mores. As Veblen observed, "It is not easy in any given case—indeed it is at times impossible until the courts have spoken—to say whether it is an instance of praiseworthy salesmanship or a penitentiary offense." The history of the great American fortunes is threaded with strains toward institutionally dubious innovation as is attested by many tributes to the Robber Barons. The reluctant admiration often expressed privately, and not seldom publicly, of these "shrewd, smart and successful" men Is a product of a cultural structure in which the sacrosanct goal virtually consecrates the means. . . .

III. Ritualism

The ritualistic type of adaptation can be readily identified. It involves the abandoning or scaling down of the lofty cultural goals of great pecuniary success and rapid social mobility to the point where one's aspirations can be satisfied. But though one rejects the cultural obligation to attempt "to get ahead, in the world," though one draws in one's horizons, one continues to abide almost compulsively by institutional norms.

It is something of a terminological quibble to ask whether this represents genuinely deviant behavior. Since the adaptation is, in effect, an internal decision and since the overt behavior is institutionally permitted, though not culturally preferred, it is not generally considered to represent a social problem. Intimates of individuals making this adaptation may pass judgment in terms of prevailing cultural emphases and may "feel sorry for them," they may, in the individual case, feel that "old Jonesy is certainly in a rut" Whether this is described as deviant behavior or no, it clearly represents a departure from the cultural model in which men are obliged to strive actively, preferably through, institutionalized procedures, to move onward and upward in the social hierarchy.

We should expect this type of adaptation to be fairly frequent in a society which makes one's social status largely dependent upon one's achievements. For, as has so often been observed,[iv] this ceaseless competitive struggle produces acute status anxiety. One device for allaying these "anxieties is to lower one's level of aspiration—permanently. Fear produces inaction, or more accurately, routinized action.[v]

The syndrome of the social ritualist is both familiar and instructive. His implicit life-philosophy finds expression in a series of cultural clichés: 'Tm not sticking *my* neck out," "I'm playing safe," "I'm satisfied with what I've got," "Don't aim high and you won't be disappointed." The theme threaded through these attitudes is that high ambitions invite frustration and danger whereas lower aspirations produce satisfaction and security. It is a response to a situation which appears threatening and excites distrust. It is the attitude implicit among workers who carefully regulate their output to a constant quota in an industrial organization where they have occasion to fear that they will "be noticed" by managerial personnel and "something will happen" if their output rises and falls.[vi] It is the perspective of the frightened employee, the zealously conformist bureaucrat in the teller's cage of the private banking enterprise or in the front office of the public works enterprise.[vii] It is, in short, the mode of adaptation of individually seeking a *private* escape from the dangers and frustrations which seem to them inherent in the competition for major cultural goals by abandoning .these goals and clinging all the more closely to the safe routines and the institutional norms.

If we should expect *lower-class* Americans to exhibit Adaptation II—"innovation"—to the frustrations enjoined by the prevailing emphasis on large cultural goals and the fact of small social opportunities, we should expect *lower-middle class* Americans to be heavily represented among those making Adaptation III, "ritualism." For it is in the lower middle class that parents typically exert continuous pressure upon children to abide by the moral mandates of the society, and where the social climb upward is less likely to meet with success than among the upper middle class. The strong disciplining for conformity with mores reduces the likelihood of Adaptation II and promotes the likelihood of Adaptation III. The severe training leads many to carry a heavy burden of anxiety. The socialization patterns of the lower middle class thus promote the very character structure most predisposed toward ritualism,[viii] and it is in this

---

[iv]See, for example, H. S. Sullivan, "Modern conceptions of psychiatry," *Psychiatry,* 1940, 3, 111–12; Margaret Mead, *And Keep Your Powder Dry* (New York, 1942), Chapter VII; Merton, Fiske and Curtis, *Moss Persuasion,* 59–60.

[v]P. Janet, "The fear of action," *Journal of Abnormal Psychology,* 1921, 16, 150–60, and the extraordinary discussion by F. L. Wells, "Social maladjustments: adaptive regression," *op. cit.,* which bears closely on the type of adaptation examined here.

[vi]F. J. Roethlisberger and W. J. Dickson, *Management and the Worker,* Chapter 18 and 531 ff.; and on the more general theme, the typically perspicacious remarks of Gilbert Murray, *op. cit.,* 138–39.

[vii]See the three following chapters.

[viii]See, for example, Allison, Davis and John Dollard, *Children of Bondage* (Washington, 1940), Chapter 12 ("Child Training and Class"), which, though it deals with the lower- and lower-middle class patterns of socialization among Negroes in the Far South, appears applicable, with, slight modification, to file white population as well. On this, see further M. C. Erickson, "Child-rearing and social status," *American Journal of Sociology,* 1946, 53, 190–92; Allison Davis and R. J. Havighurst, "Social class and color differences in child-rearing," *American Sociological Review,* 1946, 11, 698–710: " . . . *the pivotal meaning of social class* to students of human development is that it defines and systematizes different learning environments for children of different classes." "Generalizing from the evidence presented in the tables, we would say that middle-class children [the authors do not distinguish between lower-middle and upper-middle strata] are subjected earlier and more consistently to the influences which make a child an orderly, conscientious, responsible, and tame person. In the course of this training middle-class children probably suffer more frustration of their impulses."

stratum, accordingly, that the adaptive pattern III should most often occur.[ix]

But we should note again, as at the outset of this chapter, that we are here examining *modes of adaptation* to contradictions in the cultural and social structure: we are not focusing on character or personality types. Individuals caught up in these contradictions can and do move from one type of adaptation to another. Thus it may be conjectured that some ritualists, conforming meticulously to the institutional rules, are so steeped in the regulations that they become bureaucratic virtuosos, that they over-conform precisely because they are subject to guilt engendered by previous nonconformity with the rules (*i.e.,* Adaptation II). And the occasional passage from ritualistic adaptation to dramatic kinds of illicit adaptation is well-documented in clinical case-histories and often set forth in insightful

fiction. Defiant outbreaks not infrequently follow upon prolonged periods of over-compliance.[x] But though the psychodynamic mechanisms of this type of adaptation have been fairly well identified and linked with patterns of discipline and socialization in the family, much sociological research is still required to explain why these patterns are presumably more frequent in certain social strata and groups than in others. Our own discussion has merely set out one analytical framework for sociological research focused on this problem.

IV. Retreatism

Just as Adaptation I (conformity) remains the most frequent, Adaptation IV (the rejection of cultural goals and institutional means) is probably the least

---

[ix]This hypothesis still awaits empirical test. Beginnings in this direction have been made with the 'level of aspiration" experiments which explore the determinants of goal-formation and modification in specific, experimentally devised activities. There is, however, a major obstacle, not yet surmounted, in drawing inferences from the laboratory situation, with its relatively slight ego-involvement with the casual task—pencil-and-paper mazes, ring-throwing, arithmetical problems, etc.—which will be applicable to the strong emotional investment with success-goals in the routines of everyday life. Nor have these experiments, with their *ad hoc* group formations, been able to reproduce the acute social pressures obtaining in daily life. (What laboratory experiment reproduces, for example, the querulous nagging of a modern. Xantippe: "The trouble with you is, you've got no ambition; a real man would go out and do things"?) Among studies with a definite though limited relevance, see especially R. Gould, "Some sociological determinants of goal strivings," *Journal of Social Psychology,* 1941, 13, 481–73; L. Festinger, "Wish, expectation and group standards as factors influencing level of aspiration," *Journal of Abnormal and* Social *Psychology,* 1942, 37, 184–200. For a resume of researches, see Kurt Lewin *et al.,* "Level of Aspiration," in J. McV. Hunt, ed., *Personality and the Behavior* Disorders (New York, 1944), I, Chap. 10.

The conception of "success" as a ratio between aspiration and achievement pursued systematically in the level-of-aspiration experiments has, of course, a long history. Gilbert Murray *(op. cit.,* 138–9) notes the prevalence of this conception among the thinkers of fourth century Greece. And in Sartor *Resartus,* Carlyle observes that "happiness" (gratification) can be represented by a fraction in which the numerator represents achievement and the denominator, aspiration. Much the same notion is examined by William James *(The Principles of Psychology* [New York, 1902], I, 310). See also F. L. Wells, *op. cit,* 879, and P. A. Sorokin, Social *and Cultural Dynamics* (New York, 1937), HI, 161–164. The critical question is whether this familiar insight can be subjected to rigorous experimentation in which the contrived laboratory situation adequately reproduces the salient aspects of the real-life situation or whether disciplined observation of routines of behavior in everyday life will prove the more productive method of inquiry.

[x]In her novel, *The Bitter Box* (New York, 1946), Eleanor Clark has portrayed this process with great sensitivity. The discussion by Erich Fromm, *Escape* from *Freedom* (New York, 1941), 185–206, may be cited, without implying acceptance of his concept of "spontaneity" and "man's inherent tendency toward self-development." For an example of a sound sociological formulation: "As long as we assume . . . that the anal character, as it is typical of the European lower middle class, is caused by certain early experiences—in connection with defecation, we have hardly any data that lead us to understand why a specific class should have an anal social character. However, if we understand it as one form of relatedness to others, rooted in the character structure and resulting from the experiences with the outside world, we have a key for understanding why the whole mode of life of the lower middle class, its narrowness, isolation, and hostility, made for the development of this kind of character structure." (293–4) For an example of a formulation stemming from a kind of latter-day benevolent anarchism here judged as dubious: "there are also certain psychological qualities inherent in man that need to be satisfied. . . . The most important seems to be the tendency to grow, to develop and realize potentialities which man has developed in the course of history—as, for instance, the faculty of creative and critical thinking. . . . It also seems that this general tendency to grow—which is the psychological equivalent of the identical biological tendency—results in such specific tendencies as the desire for freedom and the hatred against oppression, since freedom is the fundamental condition for any growth." (287–88)

common. People who adapt (or maladapt) in this fashion are, strictly speaking, *in* the society but not *of* it. Sociologically; these constitute the true aliens. Not sharing the common frame of values, they can be included as members of the *society* (in distinction from the *population)]* only in a fictional sense.

In this category fall some of the adaptive activities of psychotics, autists, pariahs, outcasts, vagrants, vagabonds, tramps, chronic drunkards and drug addicts.[xi] They have relinquished culturally prescribed goals and their behavior does not accord with institutional norms. This is not to say that in some cases the source of their mode of adaptation is not the very social structure which they have in effect repudiated nor that their very existence within an area does not constitute a problem for members of the society.

From the standpoint of its sources in the social structure, this mode of adaptation is most likely to occur when *both* the culture goals and the institutional practices have been thoroughly assimilated by the individual and imbued with affect and high value, but accessible institutional avenues are not productive of success. There results a twofold conflict: the interiorized moral obligation for adopting institutional means conflicts with pressures to resort to illicit means (which may attain the goal) and the individual is shut off from means which are both legitimate and effective. The competitive order is maintained but the frustrated and handicapped individual who cannot cope with this order drops out. Defeatism, quietism and resignation are manifested in escape mechanisms which ultimately lead him to "escape" from the requirements of the society. It is thus an expedient which arises from continued failure to near the goal by legitimate measures and from an inability to use the illegitimate route because of internalized prohibitions, *this process occurring while the supreme value of the success-goal has not yet been renounced.* The conflict is resolved by abandoning *both* precipitating elements, the goals and the means. The escape is complete, the conflict is eliminated and the individual is asocialized.

In public and ceremonial life, this type of deviant behavior is most heartily condemned by conventional representatives of the society. In contrast to the conformist, who keeps the wheels of society running, his deviant is a non-productive liability; in contrast to the innovator who is at least "smart" and actively striving, he sees no value in the success-goal which the culture prizes so highly; in contrast to the ritualist who conforms at least to the mores, he pays scant attention to the institutional practices. . . .

This fourth mode of adaptation, then, is that of the socially disinherited who if they have none of the rewards held out by society also have few of the frustrations attendant upon continuing to seek these rewards. It is, moreover, a privatized rather than a collective mode of adaptation. Although people exhibiting this deviant behavior may gravitate toward centers where they come into contact with other deviants and although they may come to share in the subculture of these deviant groups, their adaptations are largely private and isolated rather than unified under the aegis of a new cultural code. The type of collective adaptation remains to be considered.

V. Rebellion

This adaptation leads men outside the environing social structure to envisage and seek to bring into being a new, that is to say, a greatly modified social structure. It presupposes alienation from reigning goals

---

[xi]Obviously, this is an elliptical statement. These individuals may retain some orientation to the values of their own groupings within the larger society or, occasionally, to the values of the conventional society itself. They may, in other words, shift to other modes of adaptation. But Adaptation IV can be easily detected. Nels Anderson's account of the behavior and attitudes of the bum, for example, can readily be recast in terms of our analytical scheme. See *The Hobo* (Chicago, 1923), 93–98, *et passim.*

[xii]Max Scheler, *L'homme du ressentiment* (Paris, n. d.). This essay first appeared in 1912; revised and completed, it was included in Scheler's *Abhandlungen und Aufsätze,* appearing thereafter in his *Vom Umsturz der Werte* (1919). The last text was used for the French translation. It has had considerable influence in varied intellectual circles. For an excellent and well-balanced discussion of Scheler's essay, indicating some of its limitations and biasses, the respects in which, it prefigured Nazi conceptions, its anti-democratic orientation and, withal, its occasionally brilliant insights, see V. J. McGill, "Sender's theory o£ sympathy and love," *Philosophy and Phenomenological Research,* 1942, 2, 273–9 L For another critical account which properly criticizes Scheler's view that social structure plays only a secondary role in *ressentiment,* see Svend Ranulf, *Moral Indignation and Middle-Class Psychology: A Sociological Study* (Copenhagen, 1938), 199–204.

and standards. These come to be regarded as purely arbitrary. And the arbitrary is precisely that which can neither exact allegiance nor possess legitimacy, for it might as well be otherwise. In our society, organized movements for rebellion apparently aim to introduce a social structure in which the cultural standards of success would be sharply modified and provision would be made for a closer correspondence between merit, effort and reward.

But before examining "rebellion" as a mode of adaptation, we must distinguish it from a superficially similar but essentially different type, *ressentiment*. Introduced in a special technical sense, by Nietzsche, the "concept of *ressentiment* was taken up and developed sociologically by Max Scheler.[xii] This complex sentiment has three interlocking elements. First, diffuse feelings of hate, envy and hostility; second, a sense of being powerless to express these feelings actively against the person or social stratum evoking them; and third, a continual re-experiencing of this impotent hostility.[xiii] The essential point distinguishing *ressentiment* from rebellion is that the former does not involve a genuine change in values. *Ressentiment* involves a sour-grapes pattern which asserts merely that desired but unattainable objectives do not actually embody the prized values—after all, the fox in the fable does not say that he abandons all taste for sweet grapes; he says only that these particular grapes are not sweet. Rebellion, on the other hand, involves a genuine transvaluation, where the direct or vicarious experience of frustration leads to full denunciation of previously prized values—the rebellious fox simply renounces the prevailing taste for sweet grapes. In *ressentiment,* one condemns what one secretly craves; in rebellion, one condemns the craving itself. But though the two are distinct, organized rebellion may draw upon a vast reservoir of the resentful and discontented as institutional dislocations become acute.

When the institutional system is regarded as the barrier to the satisfaction of legitimized goals, the stage is set for rebellion as an adaptive response. To pass into organized political action, allegiance must not only be withdrawn from the prevailing social structure but must be transferred to new groups possessed of a new myth.[xiv] The dual function of the myth is to locate the source of large-scale frustrations in the social structure and to portray an alternative structure which would not, presumably, give rise .to frustration of the deserving. It is a charter for action. In this context, the functions of the counter-myth of the conservatives—briefly sketched in an earlier section of this chapter—become further clarified: whatever the source of mass frustration, it is not to be found in the basic structure of the society. The conservative myth may thus assert that these frustrations are in the nature of things and would occur in *any* social system: "Periodic mass unemployment and business depressions can't be legislated out of existence; it's just like a person who feels good one day and bad the next."[xv] Or, if not the doctrine of inevitability, then the doctrine of gradual and slight adjustment: "A few changes here and there, and we'll have things running as ship-shape as they can possibly be." Or, the doctrine which deflects hostility from the social structure onto the individual who is a "failure" since "every man really gets what's coming to him in this country."

The myths of rebellion and of conservatism both work toward a "monopoly of the imagination" seeking to define the situation in such terms as to move the frustrate toward or away from Adaptation V. It is above all the renegade who, though himself successful, renounces the prevailing values that becomes the target of greatest hostility among those in rebellion. For he not only puts the values in question, as does the out-group, but he signifies that the unity of the group is broken.[xvi] Yet, as has so often been noted, it is typically members of a rising class rather than the most depressed strata who organize the resentful and the rebellious into a revolutionary group.

[xiii]Scheler, op. *cit.,* 55–56. No English word fully reproduces the complex of elements implied by the word ressentiment; its nearest approximation in German would appear to be *Groll*.

[xiv]George S. Pettee, *The Process of Revolution* (New York, 1938), 8–24; see particularly his account of "monopoly of the imagination."

[xv]R. S. and H. M. Lynd, *Middletown in Transition* (New York, 1937), 408, for a series of cultural clichés exemplifying the conservative myth.

[xvi]See the acute observations by Georg Simmel, Soziologie (Leipzig, 1908), 276–77.

Discussion Questions

1. Provide specific examples of each pattern variable in a particular social context, such as the military, church, or school. Then suggest which system—personality, social, and/or cultural—your example primarily reflects. Consider demonstrating the interchange between systems. For instance,

"Affectivity": a soldier is taught that he should be willing to risk his life to save that of his fellow soldier. (CULTURAL SYSTEM—values)

"Affective-neutrality": a soldier is taught to act according to military rules and codes rather than personal emotions. She is taught not to smile when she is saluting an officer. She rides through the streets of occupied territory exhibiting little or no sympathy for the enemy. (SOCIAL SYSTEM—institutionalized in norms)

2. Choose one role that you consistently play (e.g., mother, daughter, teacher, student) and discuss its relation to each of the five pattern variables. For instance,

MOTHER/FATHER ROLE:

Affectivity/affective-neutrality:

It used to be that mothers were expected to display more affectivity and fathers were expected to be more affectively neutral (relegated to the breadwinner role), but now emotional bonding/nurturing is expected/considered legitimate for both parents.

Ascription/achievement:

The mother/father role is mostly ascribed, although interestingly, the saying "It is easy to be a father, hard to be a dad" reflects the notion of performance. Also, because of complicated family situations, one's biological parent might not "perform" the role of parent.

Individual/collective:

Mostly collective, a "good" mother/father acts according to the best interests of the family, not him/herself. Gender is highly relevant, though, as expectation for collective orientation might be even higher for mothers today.

3. Discuss the manifest and latent functions behind everyday activities, using examples from three different aspects of your life: for example, work, school, family, and/or social world. For example, discuss the manifest and latent functions of (1) doing your homework (school life), (2) visiting your parents on the weekend (family life), (3) going to clubs (social life), and/or (4) working.

4. Think about your own patterns of consumption and the extent that they reflect (latent) "status" as opposed to (manifest) "utility" functions. How does this reflect and interrelate with the issue of "rational" as opposed to "nonrational" motivation as well? Using concrete examples of popular advertising campaigns or prominent commercials, discuss the extent to which advertisers speak to rational and nonrational motivation, as well as the "manifest" and "latent" functions of advertising in general.

5. Discuss the extent to which you think that goals are determined by the overarching culture of a society as opposed to smaller social groups or subcultures. What are specific examples of different types of cultural and subcultural means and goals? Do you think that the means and goals of society are as easily identifiable or clear-cut as Merton seems to assume? How so or why not? If not, are there any cultural common denominators that you think transcend various subcultural means and goals? How would you go about researching the means and goals of particular social groups and/or subcultures?

10 Critical Theory

Max Horkheimer

Key Concepts

▓▓ Negate

▓▓ Subjective Reason

▓▓ Objective Reason

Theodor Adorno

Key Concepts

▓▓ Negate

▓▓ Culture Industry

▓▓ Pseudo-individualization

Herbert Marcuse

Key Concepts

▓▓ Negate

▓▓ Individualistic Rationality

▓▓ Technological Rationality

▓▓ Surplus Repression

▓▓ Repressive Desublimation

Culture is only true when implicitly critical, and the mind which forgets this revenges itself in the crit-ics it breeds.

—Theodor Adorno ([1967] 1981:23)

Many of us find living in a modern, industrial society to be pleasant. While such societies certainly are not without their problems, countries like the United States, France, England, and Japan offer their citizens many avenues for educational and occupational success as well as a seemingly unending supply of technological advances and modern conveniences. If ever there was a time and a place to enjoy the "good life," it is now and in democratic nations such as those noted above. For theorists in the critical tradition, however, life in the "land of opportunity" is not all it is cracked up to be. On the contrary, critical theorists find in so-called progress a source of domination and dehumanization.[1] For them, culture, science, and technology are ideological forces that distort consciousness and, thus, prevent individuals from recognizing and satisfying their true human interests. So the next time you tune in to your favorite show on your wide-screen TV, download tunes to your MP3 player, or marvel over the easy access to your cash at the ATM, keep in mind that the pleasures such conveniences bring are not without potential pitfalls. In fact, these technological wonders are in part responsible for perpetuating the very dehumanizing conditions that they are advertised to alleviate.

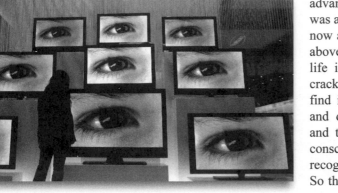

Photo 10.1 The great escape that imprisons

In this chapter, we outline the main arguments of critical theory by emphasizing the work of three of its leading exponents. Before turning to a discussion of the central theoretical ideas, however, we first provide a look into the personal lives of those who were most responsible for shaping this branch of social theory and into the historical conditions that profoundly shaped their outlook.

▪▪ MAX HORKHEIMER, THEODOR ADORNO, AND HERBERT MARCUSE: BIOGRAPHICAL SKETCHES

Max Horkheimer (1895–1973)

Max Horkheimer was born in Stuttgart, Germany, on February 14, 1895. His father was a successful businessman and owner of several textile factories. While his father intended for him to take over the family business, Max's radical political leanings and sympathies with the working class left him ill equipped for embracing the role of a capitalist. Instead, the younger Horkheimer pursued an academic career, taking up graduate studies in philosophy and psychology while attending the University of Frankfurt. During the course of his studies, Horkheimer also spent a year at the University of Freiburg, where he worked with the preeminent philosopher Edmund Husserl and his assistant, Martin Heidegger, who would go on to become a leading intellectual in his own right.

After completing his doctorate in 1925, Horkheimer began working at the University of Frankfurt as an nonsalaried lecturer; in 1928, he was hired as a salaried member of the faculty. The start of his

[1]It is important to note that "critical" is not used here as a generic, derogative term. Critical theory, instead, refers to a specific theoretical tradition that first took shape in Frankfurt, Germany, during the early 1920s.

academic career also saw his marriage to Rose Riekher, with whom he had had a ten-year relationship. Eight years his senior, Rose was the elder Horkheimer's private secretary. Father and son had clashed over the relationship from its beginning, not least because of Rose's humble family background. Yet, Max's attraction to Rose only confirmed his feelings of compassion for the plight of the working class and his distaste for "domineering businessmen like his father" (Wiggershaus [1986] 1994:44). His personal identification with the injustices suffered by those "without money" and his sense of culpability on the part of the capitalist class is revealed in passages such as the following, taken from his personal notes:

> Without money, without any economic security, we are at their mercy. It is certainly a dreadful punishment: having the daily grind wearing you down, being shackled to trivial business, having petty worries day and night, being dependent on the most despicable people. Not just we ourselves, but all of those we love and for whom we are responsible fall with us into this daily treadmill. We become victims of stupidity and sadism. (Quoted in ibid.:48)

Theodor Adorno (1903–1969)

Sharing Horkheimer's deeply felt concerns over social and economic inequalities was Theodor Adorno, whom Horkheimer had met in the early 1920s. Adorno also shared his friend's comfortable family origins. He was born on September 11, 1903, in Frankfurt. Adorno's mother, who was descended from a member of the Corsican nobility, had been a successful singer before her marriage, and his father was the owner of a wholesale wine business. Following in his mother's footsteps as well as those of his aunt, a well-known pianist who lived with his family, the precocious Adorno began his formal studies of music composition at the age of sixteen. The following year he entered the University of Frankfurt, where he studied philosophy, psychology, sociology, and musicology. Although he earned his doctorate in philosophy in 1924, it was in the fields of music criticism and aesthetics that Adorno made his first mark. Indeed, he published nearly a hundred articles on those topics between 1921 and 1932, whereas his first publication in philosophy did not appear until 1933. Perhaps it should come as no surprise, then, that Adorno's original professional ambitions lay with becoming a composer and concert pianist (Wiggershaus:70–72).

Adorno's interests in composing and performing music, however, were tied to his broader philosophy of aesthetics. For Adorno, music—or at least "real" music—offered an expression of truth itself, a truth defined in dialectical terms such that the value of music is measured by its freedom from conformity to existing forms. The rarified place Adorno held for music, and art more generally, stemmed from his view that it alone was capable of transcending the alienated, "soulless" world ushered in by the advance of capitalism and totalitarian political systems. In classic Marxist fashion, Adorno saw in dominant musical forms a reflection of existing material conditions or property relations that fueled the exploitation of one class by another. Thus, in order to perform its socially progressive function, music must escape the restraints imposed by previous forms. It is in this vein that Adorno lavished praise on the work of the pianist Arnold Schoenberg, in whose development of the twelve-tone scale and modernist compositions he saw a critique of corrupted bourgeois society. Meanwhile, he relentlessly attacked popular, "standardized" music for allegedly dumbing down the masses and perpetuating oppressive social conditions.

Herbert Marcuse (1898–1979)

The third figure to be considered in this section is Herbert Marcuse. Born in Berlin on July 19, 1898, Marcuse, like his counterparts, was raised in a well-to-do German family. His mother was the daughter of a factory owner, while his father had worked his way up to become part owner of a textile factory and a real estate entrepreneur. Marcuse served in the military during World War I, although because of vision problems he was spared from combat. As he witnessed the outbreak of strikes, riots, and general social unrest in Berlin during the war years, his views became increasingly politicized and he joined the Social Democratic Party (*Sozialdemokratische Partei Deutschlands,* SPD), a socialist political party that represented working-class interests. Discontented with the actions and policies of the SPD, Marcuse dedicated himself to his studies at Humboldt University in Berlin, where he took up modern Germany history. Soon thereafter, he left for Freiburg, where he studied philosophy and economics, completing his doctorate in 1922. It was here that Marcuse met Horkheimer and, with him, attended Husserl's lectures.

After earning his degree, Marcuse returned to Berlin, where for the next six years he worked in the book business, his father having provided him with a share in a publishing company. Active in literary and artistic circles, Marcuse took a fateful turn upon reading Martin Heidegger's newly published *Being and Time* (1927). He saw in Heidegger's work a crucial addition to Marxist theory, which had come to be the focal point of his intellectual outlook. Determined to pursue an academic career in philosophy, Marcuse abandoned the book business and returned to Freiburg with his wife and young son to study again with Husserl as well as with Heidegger, for whom he served as an assistant.

Finding A Home For the Institute

Having met one another during the course of their respective studies, these three figures' relationships with each other were solidified through their association with the Institute for Social Research at the University of Frankfurt. The Institute was established on February 3, 1923, through the patronage of Felix Weil, the son of a wealthy grain merchant who, to his father's displeasure, sought to advance the ideals of socialism. It is here that the roots of critical theory, a Marxist-inspired social philosophy, were planted. While early on the Institute was home to a number of leading intellectuals, including the psychologist Erich Fromm (1900–1980), it would come to be dominated by Horkheimer, Adorno, and Marcuse. Together, they reshaped the direction of the Institute, changing what was initially a diverse, empirically oriented research program dedicated to reappraising Marxist theory in light of the defeat of the communist revolution throughout central Europe. With Horkheimer's appointment as director in 1930, a new course was established for the Institute. The study of history and economics was replaced with a critical social philosophy that attacked the scientific enterprise as a form of bourgeois ideology. Thus, the empirical sciences were regarded as incapable not only of revealing a true understanding of social life but also of sanctioning the status quo and ushering in new forms of "technocratic" domination (Bottomore 1984:29). Far from leading to progress and the emancipation of humanity, science was leading civilization "into a new kind of barbarism" (Horkheimer and Adorno [1944] 2002:xiv). Thus, the "critical" in critical theory ultimately refers to a critique of empirical sciences and the philosophy of positivism on which they are based.

To account for the critical theorists' rejection of science, and positivism more generally, we need to recall the broader social milieu in which the Institute was formed. Not only had the communist revolution modeled on Marx's scientific prophecy failed to turn Europe into a utopia, but also the Bolsheviks' victory in Russia and Germany's defeat in World War I brought in their wake ruthless, totalitarian regimes. While the communist dream was turning into a nightmare under the dictatorship of Josef Stalin, the crippling of the German economy and sense of national humiliation left that nation vulnerable to the rise of fascism and, with it, the rise of Hitler to political power. Moreover, the rampant anti-Semitism that spread through Germany and elsewhere in Europe had a profound impact on the Frankfurt School, for all of its leading members were Jewish.

Well aware of the impending danger the Nazis posed (his house had been occupied by Nazi storm troops), Horkheimer left for Geneva, Switzerland, where he conducted the affairs of the Institute in exile. In the spring of 1933, the Gestapo (the secret police) closed the Institute and seized its possessions under the charge that it had "encouraged activities hostile to the state" (Wiggershaus [1986] 1994:128). Horkheimer was summarily expelled from his position in Frankfurt. Adorno was fired from his teaching position because he was "half Jewish" (his father was Jewish), at which time he left for London, while Marcuse joined Horkheimer and the Institute in Geneva in 1933. The Institute's relocation to Geneva would be temporary, however. Concerned about the spread of fascism and anti-Semitism throughout Europe, Horkheimer emigrated to New York in May 1934, where he established a new home for the Institute at Columbia University. Marcuse arrived two months later, and Adorno in 1938.

During the next seven years, Horkheimer and his associates worked on a limited number of studies while publishing few essays and articles. Maintaining the academic independence of the Institute while meeting the expectations of the officials of Columbia University was proving to be difficult. For many of the Institute's members, affiliation with Columbia was a mixed blessing. Although it provided needed resources, the type of applied social research expected by the university was not fitting to the temperaments of the ideologically committed critical theorists. Moreover, increasing financial pressures as well as mounting personal and intellectual differences further jeopardized developing productive, collaborative research projects. Sensing the widening gap between his vision for the Institute and the realities that accompanied ties to external institutions, Horkheimer relocated to Pacific Palisades, in Los Angeles, in April 1941. Marcuse arrived the following month, and Adorno in November. Marcuse's stay would be short-lived, however. Receiving only a minimal stipend from the Institute's endowment, he returned to the East Coast, where he began work for the government in the Office of Strategic Services (the forerunner to the Central Intelligence Agency). Here he was assigned the task of determining ways of depicting foreign enemies in the press and in films.

Marcuse's move to Washington, D.C., marked the beginning of his permanent separation from the Institute. Although he desperately wanted to return to Los Angeles and continue his association with Horkheimer, his theoretical views were moving away from Horkheimer's and, even more so, Adorno's. Moreover, the need for steady income kept him working for the State Department until 1954, at which time, at the age of 56, he was offered a full professorship from Brandeis University. In 1965, he took a position at the University of California, San Diego, where he remained until his retirement in 1976. His early years in San Diego sparked much controversy. Not surprisingly, his Marxist views were met with much resistance from the politically conservative community surrounding the university. With the Vietnam War intensifying fears over the spread of communism, some local organizations sought to have Marcuse, nearly 70 years old, expelled from the city, while other residents were content with sending him death threats. Meanwhile, then-Governor Ronald Reagan, who regarded Marcuse as an enemy of the country, and the Board of Regents pressured the university to fire Marcuse, but to no avail. In an important sense, however, Marcuse was a real threat to conservative defenders of the status quo. By the late 1960s, he had become the most widely read social theorist among those fighting for progressive change. (His books *Eros and Civilization* [1955] and *One-Dimensional Man* [1964], excerpted in the Readings for this chapter, garnered the most attention.) With his fame spreading across Europe and America, anti-war activists, feminists, and disciples of the New Left and Black Power movements all saw in his writings an explanation of their discontent and a justification for their political action. And it was in the protests of "the outcasts and outsiders, the exploited and persecuted of other races and other colors, the unemployed and the unemployable" that Marcuse placed a guarded hope for the establishment of a truly democratic society (Marcuse 1964:256). He died of a stroke in 1979, with his hopes unfulfilled.

For their part, while in Los Angeles, Horkheimer and Adorno produced several important works including *The Authoritarian Personality* (Adorno 1950) and *Dialectic of Enlightenment* (Horkheimer and Adorno 1944), the latter a masterpiece of critical theory. The former was the product of a large-scale research project and appeared as part of the Studies in Prejudice series that was edited by Horkheimer and Samuel Flowerman. The series examined anti-Semitism from a range of perspectives, including psychology, sociology, history, and psychoanalysis. In *The Authoritarian Personality*, Adorno and his coauthors administered extensive questionnaires and interviews in an effort to determine the personality traits of persons prone to harboring anti-Semitic views in particular, and fascist attitudes more generally. In developing the "F-scale" (fascism scale), the authors concluded that socialization within the family as well as broader social conditions shape an individual's potential for developing a "fascist" personality—one that is characterized, among other things, by submissiveness, a rigid conformity to dominant values, and contempt for "outsiders" and "deviants."

In *Dialectic of Enlightenment* (1944), Horkheimer and Adorno developed a principal theme of critical theory—namely, that the Enlightenment had produced contradictory developments. Not only did it break the shackles of traditions and religious superstitions, ushering in an epoch of progress through the rise of science, technological advances, and the cultivation of individual freedom, but the Age of Reason also created its own Frankenstein's monster: the "irrationality of rationality." In other words, reason, far from offering a path to human liberation, had been transformed into an irresistible force for new forms of domination. "Enlightenment, understood in the widest sense as the advance of thought, has always aimed at liberating human beings from fear and installing them as masters. Yet the wholly enlightened earth is radiant with triumphant calamity" (ibid.:1). Driven by reason and rationality, social life was becoming increasingly bureaucratized and dehumanized, while ever-more efficient means of death and destruction were being deployed, as demonstrated by the horrors of World War II and the Nazi campaign of genocide.

With the book's translation into English in the 1970s, the authors of *Dialectic of Enlightenment* became iconic figures among the New Left activists. Their claim that the Enlightenment, instead of fostering individual autonomy, is "totalitarian" struck a chord with those challenging the legitimacy of the existing social order. Yet, Horkheimer and Adorno's response to their newfound acclaim contrasted with that of Marcuse. As head of the Institute, Horkheimer had early on established a policy of political abstinence. Not only were associates to refrain from engaging in political activity or supporting the political efforts of others, but even their writings were to contain no direct references to ongoing political events or explicit condemnations of governmental or economic systems, whether fascist or capitalist (Wiggershaus [1986] 1994:133). No doubt Horkheimer's avoidance of activism was shaped, at least in part, by his "outsider" status, first as a Jew in Germany and later as an émigré living in the United States. Nonetheless, Horkheimer became increasingly pessimistic over the prospects for achieving truly progressive change and saw in the social movements of the 1960s no hope for realizing an end to repressive and dehumanizing social conditions. As for Adorno, his views on political engagement paralleled those of his friend and collaborator. Perhaps Adorno's opinion regarding those who invoked his ideas could be considered even bleaker as he saw in their actions and ambitions the very authoritarian tendencies that he himself had sought to expose.

Such were the views of two men who had returned to Frankfurt and reestablished the Institute for Social Research. Horkheimer resumed his duties at the university in 1950 and remained on the faculty until his retirement in 1959. He died in 1973. Adorno returned to Germany in 1953. Although the following decade saw his attitude toward the growing number of student radicals become increasingly hostile, he remained active at the university until his untimely death in 1969 at the age of sixty-six.

▓▓ HORKHEIMER'S, ADORNO'S, AND MARCUSE'S INTELLECTUAL INFLUENCES AND CORE IDEAS

In developing their theoretical perspective, the critical theorists drew from a number of scholars. In addition to those already mentioned (Edmund Husserl and Martin Heidegger), their works owe a debt to the idealist philosophies of Immanuel Kant and Georg Hegel, as well as to their contemporary Georg Lukács (1885–1971). Lukács's groundbreaking book *History and Class Consciousness* (1923) played a pivotal role in the critical theorists' reevaluation of Karl Marx's concept of class consciousness and ideology. In this chapter, however, we confine ourselves to a discussion of the three scholars who arguably made the most decisive impact on the central ideas of the Frankfurt School thinkers: Karl Marx (1818–1893), Max Weber (1864–1920), and Sigmund Freud (1856–1939). To be sure, the critical theorists' mining of the ideas of Marx, Freud, and Weber was by no means identical. Nevertheless, that their perspectives profoundly influenced the work of the critical theorists is undeniable.

Marx, Weber, and the Revolution That Wasn't

The critical theorists developed a framework that at once extends and departs from central Marxist ideas. Like Marx, the critical theorists saw in modern industrial societies an oppressive, dehumanizing social order. And, like Marx, their aim was not simply to construct a theory capable of explaining the real workings of society but also, in Horkheimer's words, to foster "man's emancipation from slavery"

(Horkheimer 1972b:246). Yet, the critical theorists offered a picture of the sources of domination and the struggle to overcome them that was very different from that posited by Marx.

According to Marx's theoretical system, "historical materialism," "the history of all hitherto existing society is the history of class struggles" (Marx and Engels [1848] 1978:473). In other words, the dynamics of historical changes are rooted in opposing class interests that are themselves a product of the distribution of private property. With private property comes economic classes—those who own the means of production and those who work for them—that necessarily are pitted against each other. While it is in the interest of the former to maintain the status quo, it is in the interest of the latter to radically transform the existing distribution of resources and thereby bring an end to their exploitation and subordination. The evolution of history has been shaped by a dialectical process in which each stage of development contains within it the seeds of its own destruction—its own "gravediggers"—as private ownership of the means of production ignites recurring economic conflicts that inevitably sweep away existing social arrangements, giving rise to new classes of oppressors and the oppressed.

In the capitalist stage of development, Marx predicted that society would be divided into two warring classes: the bourgeoisie and the proletariat. However, the class conflict between these two factions was to be like no other previous struggle, for the proletariat heralded the redemption of humankind from the wretchedness, exploitation, and alienation that, for Marx, characterized capitalism. Since the proletariat made up a vast majority of the world's population, their victory over the bourgeoisie was inevitable. In the wake of their victory, the proletariat would abolish private property—the catalyst of class conflict—and thus usher in "the end of prehistory."

As we noted in Chapter 2, Marx argued that the mode of economic production forms the "base" of society which, in turn, determines a society's "social, political, and intellectual life processes in general" (Marx [1859] 1978:4). These noneconomic processes, from a society's legal and educational systems, to its stock of taken-for-granted truths, compose the "superstructure." Thus, an individual's very consciousness, how one makes sense of the world and defines one's interests, is determined by his class position within the mode of production. Moreover, the structural relationship between the superstructure and the base ensures that the dominant economic class will control not only a society's means of material production, but also the production of ideas.

If, as Marx maintained, "the ideas of the ruling class are in every epoch the ruling ideas" ([1846] 1978:172), then a vexing obstacle is raised for prophesizing the downfall of capitalism. For how can the working class develop a revolutionary consciousness when a society's prevailing or dominant ideas (regarding private property rights or market competition, for example) serve to legitimate the very system that they are allegedly destined to overthrow? While Marx portended that historically necessary economic crises would instill the necessary class consciousness, later Marxists noted that changing economic circumstances alone are incapable of mechanically transforming the consciousness of the working class. Thus, Horkheimer concluded, following Lukács that,

> the situation of the proletariat is . . . no guarantee of correct knowledge. The proletariat may indeed have experience of meaninglessness in the form of continuing and increasing wretchedness and injustice in its own life. Yet this awareness is prevented from becoming a social force by the differentiation of social structure which is still imposed from above and by the opposition between personal class interests. . . . Even to the proletariat the world superficially seems quite different than it really is. (1972a:213, 214)

Given this state of affairs, Horkheimer and his colleagues looked elsewhere for the sources of emancipation; not surprisingly, they found them in critical theory. This signaled a major shift from orthodox Marxism, as the critical theorists abandoned analyses of economic conditions and no longer cast the working class in the role of savior of humanity. Moreover, in placing in the hands of critical theoreticians, whatever limited hope they had for the establishment of a just society, they drew inspiration from the work of Georg Hegel and his followers—the "Young Hegelians"—whose philosophy Marx subjected to a scathing critique. In essence, the critical theorists turned away from Marx's doctrine of historical materialism to Hegelian idealism to better focus their analysis of culture and ideology, for it was not economic arrangements that were primarily responsible for the barbarism of humanity but ideas and the irrationality of reason. But if the critical theorists were skeptical of the revolutionary potential of the working class itself, or of the ability of a revolutionary communist party vanguard to advance "correct knowledge" of the proletariat's condition as Lukács theorized, then it would be left to them to realize a free and peaceful society. For critical theory "was not just an extension of proletarian thought, but a means of thinking about the social totality that would aid in the movement from the empirical proletariat's necessarily still partial view of society from its own class position to the achievement of a classless society, one not structured on injustice" (Calhoun 1995:21).

Nevertheless, aside from asserting that the task of the critical theorist is "to reduce the tension between his own insight and oppressed humanity in whose service he thinks" (Horkheimer 1972a:221), or claiming that "truth becomes clearly evident in the person of the theoretician" (ibid.:216), Horkheimer and his associates would fail to offer a specific account of the mechanisms for social change and of who, precisely, the agents of such change would be.[2] Exactly how the critical theorist was to enlighten the class "in whose service he thinks" (ibid.:221) was a question that remained unanswered. This neglect would in turn heighten the sense of pessimism that pervaded the critical theorists' political outlook.

The bleak (if not realistic) picture of the future painted by the critical theorists was not simply a product of the shortcomings of their theoretical model. Instead, it stemmed more directly from what they saw as the changing nature of domination. It was not the exploitation inherent in capitalism that was responsible for the oppression of humanity, but rather forms of thought, and, in particular, the totalitarianism of reason and rationality. Here again the critical theorists draw from the work of Hegel as well as Max Weber. For his part, Hegel argued that the essence of reality lies in thought or ideas because it is only in and through the concepts that order our experiences that experiences, as such, are known. In other words, reality is a product of the ideas we use to make sense of and categorize the otherwise random flow of sensual stimuli we experience. Thus, as our knowledge or concepts change, so does reality itself. Moreover, Hegel combined his idealist philosophy with aspects of Christian theology and Enlightenment philosophy to argue that, as history evolves through a dialectical progression of ideas, humankind's knowledge comes ever closer to the perfected realization of "Spirit" or "Absolute Idea" as revealed by God. The perfectibility of humankind is reached once the utopia of "Truth" and "Reason" resolves the contradictions between our ideas about reality and reality as it is decreed by God. At this point in the evolution of consciousness or ideas, humankind will become truly free as we become fully self-conscious and able to recognize that the objective world is a product of human creation.

In suggesting that history is marked by a separation of the True from the real, it follows that we are alienated from Absolute Idea or Spirit. The alienation of humanity and the obstacles to realizing a perfected social order lie in distorted consciousness. As we noted earlier, the critical theorists located the source of distorted consciousness in the "irrationality of rationality." Reason itself had become corrupted, leaving individuals unable to **negate** or develop a critique of "objective truths," which would alone enable us to resist the domination of the status quo. In short, the power of negative thinking to subvert the established social order and the oppressive conditions it fosters has been lost.

As for Weber, he offered a vision of the future in which "not summer's bloom lies ahead of us, but rather a polar night of icy darkness and hardness" (Weber [1919] 1958:128). His pessimistic outlook stems from his understanding of the development of capitalism in the West. Capitalism was predicated on an instrumental worldview that had ushered in an increasingly impersonal and bureaucratic society. This worldview was born out of the Protestant ethic that originally had promised individuals a path to religious salvation in the form of economic success. Eventually, the religion-based injunction to carry out one's worldly affairs on the basis of methodical and rational planning was shed. The ethical imperative to avoid idleness and enjoyment of luxuries while saving and investing one's wealth in service to God no longer stood as divinely granted proof of one's state of grace. Profit and efficiency were now twin demands that were pursued for their own sakes. The rational and bureaucratic structures necessary to ignite the growth

---

[2]Jürgen Habermas (see Chapter 16), a sometime student of Adorno and leading contemporary theorist, addressed this very issue in his reworking of critical theory.

of modern capitalism rendered obsolete the religious spirit that first had imbued it with meaning. Unleashed from its religious moorings, the process of rationalization transformed the West into a disenchanted "iron cage" from which the modern individual is left with little power to escape. Since it is one of the most compelling passages ever written in sociology, we quote Weber at length:

> The Puritan wanted to work in a calling; we are forced to do so. For when asceticism was carried out of monastic cells into everyday life, and began to dominate worldly morality, it did its part in building the tremendous cosmos of the modern economic order. This order is now bound to the technical and economic conditions of machine production which to-day [sic] determine the lives of all the individuals who are born into this mechanism, not only those directly concerned with economic acquisition, with irresistible force. . . .
>
> Since asceticism undertook to remodel the world and to work out its ideals in the world, material goods have gained an increasing and finally an inexorable power over the lives of men as at no previous period in history. To-day [sic] the spirit of religious asceticism—whether finally, who knows?—has escaped from the cage. But victorious capitalism, since it rests on mechanical foundations, needs its support no longer. . . . For the last stage of this cultural development, it might well be truly said: "Specialists without spirit, sensualists without heart; this nullity imagines that it has attained a level of civilization never before achieved." ([1904] 1958:180, 181)

Along with this transition, "formal rationality" supplanted "substantive rationality" as the motivating force in human action. While substantive rationality provides for an ethic or value principle according to which actions are guided, formal rationality is grounded in rule-bound, matter-of-fact calculations. Thus, substantive rationality establishes ultimate ends that give meaning to our actions, while formal rationality is based on establishing impersonal, calculable procedures. Formal rationality is the lifeblood of the bureaucratic administration of human affairs. And for Weber, it is precisely the bureaucratic form of rationality that is most responsible for creating an oppressive, overly routinized, and depersonalized society. Bureaucracy is the "iron cage" that has stifled individual freedom.

This theme has formed the core of critical theory. For instance, Horkheimer drew a distinction between "subjective reason" and "objective reason."[3] Parallel to Weber's notion of formal or instrumental rationality, **subjective reason** is "essentially concerned with means and ends, with the adequacy of procedures for purposes more or less taken for granted and supposedly self-explanatory. It attaches little importance to the question whether the purposes as such are reasonable" (Horkheimer 1947:3). While subjective reason may allow us to determine the most efficient way of achieving our goals, it cannot in itself offer a guide for determining what is a "reasonable" goal. Subjective reason is the guiding mentality of the technician, the bureaucrat, who, while adept at carrying out functional, procedural rules, are blind to the ethical basis of those rules. It is the form of reason that at its most heinous allows for the callous torturing of others, demonstrated by the Holocaust during Horkheimer's time and the ongoing wars with Iraq and Afghanistan today.

Conversely, **objective reason** speaks to the relative value of the ends of action and thus provides a basis for determining what is ethical, right, and just. It is premised on the notion that

> the existence of reason [is] a force not only in the individual mind but also in the objective world—in relations among human beings, between social classes, in social institutions, and in nature and its manifestations. . . . It aim[s] at evolving a comprehensive system, or hierarchy, of all beings, including man and his aims. The degree of reasonableness of a man's life could be determined according to its harmony with this totality. Its objective structure, and not just man and his purposes, was to be the measuring rod for individual thoughts and actions. (Horkheimer 1947:4)

---

[3]Horkheimer's distinction between subjective and objective reason parallels not only Weber's discussion of rationality, but also the work of his contemporary Karl Mannheim (1893–1947), who drew a contrast between "functional rationality" and "substantial rationality." (See Mannheim 1936.)

For his part, Marcuse (1941) emphasized the distinction between "individualistic rationality" and "technological rationality." (The latter term was adopted by most of the Frankfurt theorists.) Analogous to Horkheimer's notion of objective reason, he defined **individualistic rationality** as "a critical and oppositional attitude that derived freedom of action from the unrestricted liberty of thought and conscience and measured all social standards and relations by the individual's rational self-interest" (ibid.:433). Individualistic rationality allows for negating all that is established in order to critically understand one's world, develop personal objectives, and achieve them through rational methods. The individual's ability to transcend the status quo was "liquidated," however, as the organization of society moved from the era of liberal, competitive capitalism to industrial, corporate capitalism. In the earlier stage, an individual's self-directed efforts enabled him to develop his unique identity and potential while contributing to the needs of society. In the modern era of mechanized, rationalized production, it is society that creates and "administers" all the individual's needs. And "with the disappearance of independent economic subjects, the subject as such disappears" (ibid.:377).

This shift in the form of production altered individual consciousness, as technological rationality replaced individualistic forms of thought.[4] **Technological rationality** is marked by the scientific approach to all human affairs. Social relations as well as humanity's relationship to nature are now understood as "problems" to be efficiently solved—and with solutions comes control. While this form of reason has led to unprecedented material gains, under its sway "individuals are stripped of their individuality, not by external compulsion, but by the very rationality under which they live" (1941:421). As Marcuse notes,

> Autonomy of reason loses its meaning in the same measure as the thoughts, feelings and actions of men are shaped by the technical requirements of the apparatus which they have themselves created. Reason has found its resting place in the system of standardized control, production, and consumption. . . . Rationality here calls for unconditional compliance and coordination, and consequently, the truth values related to this rationality imply the subordination of thought to pregiven external standards. . . .
>
> The point is that today the apparatus to which the individual is to adjust and adapt himself is so rational that individual protest and liberation appear not only as hopeless but as utterly irrational. The system of life created by modern industry is one of the highest expediency, convenience, and efficiency. Reason, once defined in these terms, becomes equivalent to an activity which perpetuates this world. Rational behavior becomes identical with a matter-of-factness which teaches reasonable submissiveness and thus guarantees getting along in the prevailing order. (ibid.:421, 423)

Technological rationality was no longer confined to the productive apparatus—that is, to the system of economic production in advanced industrial society. It represented a worldview that had come to dominate all spheres of life. Unquestioned conformity to the dictates of efficiency, convenience, and profit required by the apparatus now "govern[s] performance not only in the factories and shops, but also in the offices, schools, assemblies, and finally, in the realm of relaxation and entertainment" (Marcuse 1941:421). As a result, it was impossible to escape the attitudes and ways of thinking prescribed by the technical organization of the apparatus, making production not only an economic act, but also an exercise of political power and control. For the critical theorists, this meant "reason [had] liquidated itself as an agency of ethical, moral, and religious insight" (Horkheimer 1947:18). Scientific-technological progress had become the god of modern society, and only a "crank" would refuse to worship before its idols. Paradoxically, then, it was science itself—the bastion of reason—that had promoted the destruction of humanity. Indeed, the critical theorists maintained that science, in claiming that the empirical world can be objectively known, could at best produce a superficial understanding of the natural and social worlds. Based in "traditional theory" and "positivism," science mistook the world of appearances or "facts" for the world of essences.

---

[4]Marcuse's argument here recalls Marx's theory of the relationship between forms of economic production and consciousness. However, unlike Marx, Marcuse and the critical theorists did not contend that consciousness was tied to particular class positions, because no one can escape the numbing effects of technological rationality.

However, "the so-called facts ascertained by quantitative methods, which the positivists are inclined to regard as the only scientific ones, are often surface phenomena that obscure rather than disclose the underlying reality" (Horkheimer 1947:82). Moreover, in treating "surface phenomena" as necessary, inescapable givens, science is unable to transcend the established order and thus reproduces an oppressive social system.

In contrast, the critical theorists claimed that all knowledge is finite; there are no timeless, empirical truths subject to scientific discovery and proof. Because objective knowledge is nothing but a fiction (albeit a comforting one), it is imperative for theorists to recognize that fact and value cannot be separated. Unlike the aloof intellectual who falsely proclaims the truth on the basis of his supposedly neutral and detached reflections on society, the critical theorist found truth "in personal thought and action, in concrete historical activity" (Horkheimer 1972a:222). Departing from Weber's claim (and those of many others, including Karl Mannheim), the critical theorists argued that the notion of a free-floating intellectual who somehow stands above or apart from the object of his investigation is a self-aggrandizing myth. It serves to mask the real, although unintended, social consequences of the scientist's findings—perpetuation of a system of technological domination. For their part, the critical theorists recognized that their ideas were influenced equally by the conditions in which they lived and the concepts that guided their thinking. Thus, they abandoned all pretenses to objectivity and instead sought to develop a theoretical system morally committed to the emancipation of humanity. Their sword was Reason, for it alone cleared the path to the ultimate value: individual freedom.

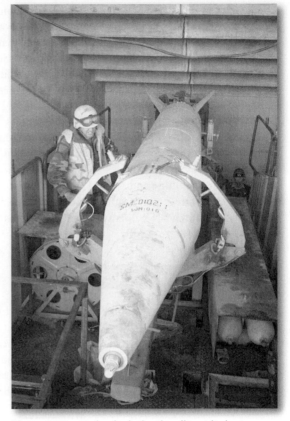

Photo 10.2 Technological rationality at its best . . . or worst. Securing international peace through the threat of mutual annihilation.

Technology and science had reneged on their promise to usher in a just and reasonable world. While industrial and administrative advances had made possible the domination of nature and production of wealth on a scale that would have been unimaginable to the early prophets of the Enlightenment, modern society continued to be plagued by inequality, oppression, destruction, and poverty. While science has made it possible to feed, educate, and care for all the inhabitants of the world, many are left starving, illiterate, poor, and ravaged by preventable illness. And for those whose "needs" are satisfied, "in the unjust state of society the powerlessness and pliability of the masses increase with the quantity of goods allocated to them" (Horkheimer and Adorno [1944] 2002:xvii). It is no longer possible to justify—to find reasonable—the sacrifices of the many for the privileges of the few in the name of progress. Meanwhile, technological advances have produced ever-more lethal and destructive weapons that make war appear to the public as a depersonalized, computer-administered, precision adventure normalized and sanitized through the video games that propagandize mass violence as an "acceptable," everyday occurrence. The root of the inhumanity of the modern condition was not a result of a lack of economic or scientific development. Rather, it was the result of technological or instrumental rationality coming to form the basis of the dominant ideology. In other words, the logic of technological rationality that emphasizes means and aims at answering the question, "How?" was now the authority for determining legitimate courses of action. But this form of rationality is divorced from ethics and value judgments; it is not oriented to answering the question, "Why?" As the dominant ideology, however, the beliefs, concepts, and worldview that inform technological rationality serve as the taken-for-granted reality that shapes social life more generally. It provides the commonsense knowledge by which all groups understand the nature of society and the relationship between individuals. And it is this pervasive ideology, and not class-based exploitation, as Marx contended, that is primarily responsible for sustaining oppression in modern society. Thus, it was to ideology that the critical theorists turned when attempting to articulate a theory of social change that envisioned a free and just society.

This ideology is disseminated in large measure through the **culture industry.** This "industry" encompasses all those sectors involved in the creation and distribution of mass-culture products: television, film, radio, music, magazines, newspapers, books, and the advertisements that sell them. Geared toward entertaining and pacifying the masses, the culture industry "aborts and silences criticism" (Bottomore 1984:19) by peddling "mass deception" through its never-ending supply of mass-produced, standardized commodities. Manufactured movie and television stars act as its leading spokespersons, promoting its superficial, conformist vision of the happy life both in their performances and in their revolving appearances on the cycle of vacuous, ever-the-same talk shows. Suggestive of the effectiveness of the culture industry is the fact that most people seem to prefer its familiar, predictable offerings to alternatives that require active contemplation. For instance, how many theme songs can you hum from TV shows? How many Mozart concertos?

Significant Others

Walter Benjamin (1892–1940): Art in the Age of Mechanical Reproduction

Walter Benjamin was an associate of the Institute for Social Research, his main connection to the group coming in the form of freelance contributions to the Institute's Journal of Social Research. While his philosophical treatises addressed a range of topics, it is his studies on the nature of art in modern societies that have made the most enduring mark in the social sciences. Benjamin met Adorno in 1923 while they were attending the University of Frankfurt, and the two established a close correspondence. However, despite the mutual respect each had for the other's ideas, Benjamin's profound influence on Adorno would be tempered as the two clashed over the revolutionary role assigned to works of art.

While Adorno saw in popular forms of musical and artistic expression a fetishized "standardization" that bred social conformity and the abandonment of reason, Benjamin perceived a potentially revolutionary form of politics. In his classic essay "The Work of Art in the Age of Mechanical Reproduction" (1936), Benjamin contended that technological advances promised to destroy the elitism of art (an elitism that Adorno was himself guilty of advocating). Specifically, lithography, photography, and film, in producing mechanical reproductions of works of art, had removed the distance or separateness that up to that point characterized the relationship between art and those who viewed it. In doing so, it undermined the source of art's aura or sacredness and, in turn, challenged conservative, elitist understandings and uses of art. (The work of Andy Warhol and the debates over its artistic merit illustrate well Benjamin's argument.) Authenticity, considered a hallmark of "true art," is rendered obsolete because images can be endlessly reproduced—popularized—and thus appropriated by "the masses." In Benjamin's own words,

> the technique of reproduction detaches the reproduced object from the domain of tradition. By making many reproductions it substitutes a plurality of copies for a unique existence. And in permitting the reproduction to meet the beholder or listener in his own particular situation, it reactivates the object reproduced. These two processes lead to a tremendous shattering of tradition which is the obverse of the contemporary crisis and renewal of mankind. (1936/1968:223)

The intellectual dispute between Adorno and Benjamin would not be resolved. Tragically, Benjamin, a German of Jewish descent, took his own life when attempting to escape the Nazis through the Pyrenees.

Tied directly to the standardization of products is the **pseudo-individualization** that endows "cultural mass production with the halo of free choice or open market on the basis of standardization itself" (Adorno 1941:25). Here is the "parade of progress," the world of the "new and improved," that masks an eternal sameness. Deodorants and shampoos, hit songs and movies, cars and soft drinks—each product is made to closely resemble its competitors in order to conform to the consumer's pregiven expectations, but offers just the slightest difference in order to capture his attention:

> Although the consumer is, so to speak, given his choice, he does not get a penny's worth too much for his money, whatever the trademark he prefers to possess. The difference in quality between two equally priced popular articles is usually as infinitesimal as the difference in the nicotine content of two brands of cigarettes. Nevertheless, this difference, corroborated by "scientific tests," is dinned into the consumer's mind through posters illuminated by a thousand electric light bulbs, over the radio, and by use of entire pages of newspapers and magazines, as if it represented a revelation altering the entire course of the world rather than an illusory fraction that makes no real difference, even for a chain smoker. (Horkheimer 1947:99)

Despite, or perhaps because of, the superficial differences that distinguish one commodity from the next, the culture industry advertises its products with the promise of an "escape from reality but it really offers an escape from the last thought of resisting that reality" (Horkheimer and Adorno [1944] 2002:116). Such are the movies where the audience can momentarily forget the drudgery and defeat of daily life through the ever-predictable triumph of the good guys; the commercials in which weight-loss pills, whitening toothpaste, and the latest clothing fashions will land you the object of your desires and everlasting happiness; car advertisements that pitch their mass-produced product to a mass audience with the slogan "Engineered for those who never applaud conformity";[5] television shows where viewers can participate in democratic elections by voting for their corporate manufactured, cross-leveraged "American

Photo 10.3 The culture industry fabricates a world where, in Marx's famous aphorism, "all that is solid melts into air." Nowhere is this condition more conspicuously promoted than in advertising. Above, Nissan, a Japanese automaker, "sells" the values of freedom and hope to potential consumers in Harlem, New York, a predominately African American neighborhood that for decades has been denied precisely what is being sold.

---

[5]This was the advertising slogan for Saab's 9-7X SUV, introduced in 2005.

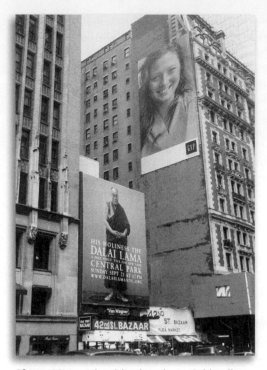

Photo 10.4 Advertising happiness: Spirituality and fashion—it's all the same.

Idol"; and the music industry in which your favorite band's anti-establishment message can be bought for a mere $17. Yet, none of the offerings of the culture industry actually fulfills its promise. In fact, they are not designed to, and although we know this, we are unable to envision an alternative. And this, above all else, marks the power that the culture industry possesses. The essential lesson is that individuals are not to be treated as autonomous, freethinking persons. Rather, in the pursuit of efficient profit making, individuals themselves are to be "created" in order to fit into a standardized model that mirrors the standardized products being sold to them.

> The culture industry endlessly cheats its consumers out of what it endlessly promises. . . . This principle requires that while all needs should be presented to individuals as capable of fulfillment by the culture industry, they should be so set up in advance that individuals experience themselves through their needs only as eternal consumers, as the culture industry's object. Not only does it persuade them that its fraud is satisfaction, it also gives them to understand that they must make do with what is offered, whatever it may be. . . . That is the triumph of advertising in the culture industry: the compulsive imitation by consumers of cultural commodities which, at the same time, they recognize as false. (ibid.:111, 113, 136)

In the end, the culture industry combines with technological rationality to produce a totalitarian social order that transcends any particular economic or political arrangement. Whether a society is organized according to capitalist or socialist, fascist or democratic principles makes little difference, because all advanced, industrial societies are administered alike. Each is rooted to an all-encompassing culture industry that trumpets its conformist products as the avenue for individual success and happiness. No one, regardless of class position, is able to escape its ever-more effective and "pleasant" method of control.

Freud and the "Unhappy Consciousness"

The final influence on the critical theorists considered here is the work of Sigmund Freud, whose ideas had a particularly profound impact on Herbert Marcuse.[6] The critical theorists' turn to psychoanalysis is in large measure a reflection of their emphasis on consciousness, reason, and individual freedom and how ideology distorts human potential. Moreover, Marcuse saw in Freudian theory the basis of an explanation for the continuing repression and unhappiness of individuals and a pathway toward the end of human toil and suffering.

Central to the critical theorists' perspective is Freud's notion of the "pleasure principle" and the "reality principle." The pleasure principle refers to the individual's instinctual drive for the immediate and painless gratification of desires. According to Freud, this unconscious impulse is inimical to the development of civilization, because civilization demands cooperation between individuals in order to achieve social (as opposed to personal) aims, and cooperation entails the delay, if not denial, of self-gratification. Restraining the free play of the pleasure principle is the "reality principle." The reality principle serves as a precondition for entering into associations with others that alone are able to secure basic needs. Essentially, an unconscious trade-off is made as powerful—and at times destructive—instinctual pleasures are exchanged for less-satisfying, sublimated pleasures that, nevertheless, make

---

[6]See especially Marcuse's *Eros and Civilization* (1955) and *Five Lectures: Psychoanalysis, Politics, and Utopia* (1970).

social life possible. As a result, the apparent freedom to do and think and say what one wants actually is based on an essential unfreedom. The necessary repression of the instincts that accompanies the transfer to the reality principle is a form of psychological domination as the socially imposed restraint of instinctual pleasures is internalized within one's own psyche. "The repressive transformation of the instincts becomes the biological constitution of the organism: history rules even in the instinctual structure; culture becomes nature as soon as the individual learns to affirm and to reproduce the reality principle from within himself" (Marcuse 1970:11).

For Freud, the domination of instincts is bound to the development of civilization. Humanity's struggle for existence amid scarcity compels the repression of selfish instincts if civilization is to progress beyond barbarism. Because the progress of civilization requires engaging in unpleasurable tasks and the denial of gratifications, the influence of the pleasure principle must be tamed. Herein lies the "discontent of civilization": the evolution of society is based on the redirecting of instinctual energies from unrestrained, immediate pleasure to the burden of socially useful labor that refuses instinctual desires. The "result is not only the conversion of the organism into an instrument of unpleasurable labor but also and above all the devaluation of happiness and pleasure as ends in themselves, the subordination of happiness and gratification to social productivity without which there is no progress in civilization" (ibid.:35). As Freud himself stated on the matter, "What we call our civilization is largely responsible for our misery" (Freud [1930] 1961:38).

This dynamic results in a "vicious circle of progress" (Marcuse 1970:36) in which the civilizing advances produced through the denial of instinctual satisfaction do not lead to individual happiness. For the very reason that such advances are achieved at the expense of psychological inhibitions, they cannot be fully enjoyed. Yet, frustrated, unfulfilling satisfactions lead to a rise in repressed instinctual energies that find release in ever-more socially approved, although individually malign forms of "progress." (Think, for example, of technologically advanced, violent or sexualized video games that offer an "approved" outlet for our aggressive and sexual instincts.) And so the vicious circle continues.

Over the course of evolution, however, civilization has advanced to such a degree that the repression of instinctual energies in the name of progress is no longer necessary. Humanity's technical control over nature has increased the capacity to fulfill needs to the point where scarcity, not only in regard to life's essentials but also in terms of liberating "free play," is now a choice and not the result of technological obstacles. Enough wealth is produced to ensure the security and comfort of most, if not all, of humankind. Nevertheless, the world's population continues to toil unnecessarily in repressive, alienated labor that fails to provide satisfaction of individual needs and true happiness. With technological advances leading to increased productivity, unfreedom in the form of instinctual denial is increasingly irrational, as scarcity and the struggle for existence can no longer justify the continued repression of individual happiness. To the extent that technology reduces the amount of labor time required to produce life's necessities, it creates the possibility for eliminating repression and satisfying needs beyond those of basic necessity.

> But the closer the real possibility of liberating the individual from the constraints once justified by scarcity and immaturity, the greater the need for maintaining and streamlining these constraints lest the established order of domination dissolve. . . . If society cannot use its growing productivity for reducing repression (because such usage would upset the hierarchy of the *status quo*), productivity must be turned *against* the individuals; it becomes itself an instrument of universal control. Totalitarianism spreads over late industrial civilization wherever the interests of domination prevail upon productivity, arresting and diverting its potentialities (Marcuse 1955:93; emphasis in the original).

Such is a world dominated by the production of waste, where auto lots are filled with new cars that will never be driven and grocery stores are stocked with food that will rot before it is purchased. Meanwhile, workers are subjugated to the technical requirements of the productive apparatus and are left to live paycheck to paycheck.

A question arising from Freud's analysis is whether the conflict between the pleasure principle and the reality principle must inevitably lead to the repression of instinctual pleasure and thus the "misery" of humankind. While Freud himself ambivalently suggested that repression is an inescapable condition of civilization, Marcuse asserted the possibility of a world where "the expanding realm of freedom becomes truly a realm of play—of the free play of individual faculties" (1955:223). Indeed, the real mark of progress is measured by the extent to which the historically conditioned capacity to satisfy human needs is used to advance individual happiness or to meet technological demands for productivity. The degree of freedom and domination is thus not fixed or immutable, nor is it somehow limited by the psychological or instinctual makeup of individuals. In a sense, then, the existing level of freedom can be measured "objectively" by comparing the needs of individuals to the productive capacity for satisfying such needs during a given period of development. The more the available means are organized to minimize instinctual repression in meeting the needs of the individual and of society, the freer a society is.

While the reality principle and the repressive control of instinctual energies that it demands are inherent in the continuing development of civilization, "the specific historical institutions of the reality principle and the specific interests of domination introduce *additional* controls over and above those indispensable for civilized human association" (Marcuse 1955:37; emphasis in original). These additional controls represent **surplus repression:** the portion of repression "which is the result of specific societal conditions sustained in the specific interest of domination" and which unnecessarily impedes the gratification of instinctual desires (ibid.:88). In other words, although the reality principle is intrinsically opposed to the pleasure principle, the specific form that it takes is determined by the prevailing method of social domination—the existing system of social institutions, norms, and values that guides the necessary control of the instincts. Thus, whether a society's mode of production is based on private or collective property, a market or planned economy, or whether all of its members work to secure their survival or only particular groups do so, the content of the reality principle will be affected and, thus, the scope and degree of instinctual repression. The extent of surplus repression, then, provides a standard of measurement according to which the repressiveness of a society can be gauged.

Given that repression is not "natural" but rather is socially conditioned, "psychology in its inner structure must reveal itself to be political" (Marcuse 1970:1). The politics of psychology are, for Marcuse, most evident in the spheres of free time and sexuality. As we noted previously, technological progress has all but eliminated, at least in potential, scarcity and want as a necessary fate of the human condition. Consequently, advanced societies are objectively able to reduce the amount of time individuals spend in burdensome, alienated labor, without compromising the ability to provide the population with a "rational" level of comfort. Yet, advanced societies remain competitive, antagonistic, and enslaved to continued material expansion. As a result, individuals needlessly spend the majority of their time engaged in labor "for an apparatus which they do not control, which operates as an independent power to which individuals must submit if they want to live. . . . Men do not live their own lives but perform pre-established functions. . . . [L]abor time, which is the largest part of the individual's time, is painful time, for alienated labor is absence of gratification, negation of the pleasure principle" (Marcuse 1955:45).

Even when the individual is not laboring for the "apparatus," the limited free time available to him remains tightly controlled. As we discussed previously, the lifeblood of the culture industry is filling individuals' leisure with standardized, conformist commodities, including information. But more than this, leisure is controlled by the very length of the working day and the nature of the dulled, mechanized work experience. Too tired from the day's labor, the individual has only enough energy to passively consume mindless "entertainment" in preparation for the next day's work. If the status quo, the prevailing system of domination, is to be maintained, free time can serve no other purpose because "the individual is not to be left alone. For if left to itself, and supported by a free intelligence aware of the potentialities of liberation from the reality of repression" (ibid.:48), the individual may begin to challenge the legitimacy of the established social order. Herein lies the political dimension to free time: In the name of the sacred values of progress and productivity, humanity has become increasingly repressed and, thus, incapable of realizing true progress in the form of freedom. While earlier stages of civilization required for their development repressive labor, its persistence and the "need" for increasing material wealth lead not to human happiness but to technological unfreedom.

Like his perspective on repression and the politics of leisure, Marcuse's view of sexuality is deeply indebted to Freud. Individuals in their early developmental stages are dominated by the pleasure principle. Thus, before the reality principle is internalized, individuals are controlled by "sexual" instincts that seek polymorphous bodily pleasures. As part of their indoctrination into the repressive social order, individuals must be "desexualized"; that is, sexual energies must be desensualized in "love" in order to sustain the monogamous, patriarchal family structure. For this family structure is itself an essential accomplice to the administered, repressive society. Sex is now tolerated only so long as it furthers the propagation of the species and, thus, existing relations of domination. Meanwhile, sexuality is transformed from the experience of the entire body as a locus of pleasure to a constraining fixation on genitalia. Once tamed by repressive moral codes, the individual's sexuality is no longer a source of uninhibited, total pleasure as it is put into the service of society's needs and not one's own (Marcuse 1970:9).

Marcuse does recognize, however, that, relative to the nineteenth century, sexual mores and behavior have been "desublimated" or liberalized. Yet, the modern liberation of sexuality provides only a false freedom. The conflict between sexuality (a central source of the pleasure principle) and society (the source of the reality principle) has produced a state of **repressive desublimation** or "institutionalized desublimation" "managed by a controlled liberalization which increases satisfaction with the offerings of society" (ibid.:57). As Marcuse noted,

[T]o the degree to which sexuality is sanctioned and even encouraged by society (not "officially," of course, but by the mores and behavior considered as "regular"), it loses the quality which, according to Freud, is its essentially erotic quality, that of freedom from social control. In this sphere was the surreptitious freedom, the dangerous autonomy of the individual under the pleasure principle. . . . Now, with the integration of this sphere into the realm of business and entertainment, the repression itself is repressed: society has enlarged not individual freedom, but its control over the individual. (ibid.:57)

Photo 10.5   Sex, steaks, and "everything" for under a buck. A truly liberated society or one that offers "convenient" repression?

Marcuse saw in the supposed loosening of sexual mores neither a threat to existing civilization, as conservatives may have feared, nor a mark of a freer society. Instead, sexual liberation was but another sign of "business as usual," a sign vividly illustrated today by the growing pornography industry in which "uninhibited" sex is bought and sold. Nevertheless, as an essential force of the pleasure principle, embracing nonrepressive sexuality is necessary to the creation of an authentically liberated society.

▪▪ HORKHEIMER'S, ADORNO'S, AND MARCUSE'S THEORETICAL ORIENTATIONS

From our discussion to this point, it should be clear that critical theory possesses the elements of a multidimensional approach to the central issues of action and order. Drawing principally from Marx, Weber, and Freud, critical theory incorporates concepts that range from an emphasis on institutionalized class relations to the instinctual disposition of individuals. This range is suggested in Figure 10.1, which depicts a number of the concepts outlined in the previous section.

Figure 10.1   Critical Theorists' Basic Concepts and Theoretical Orientation

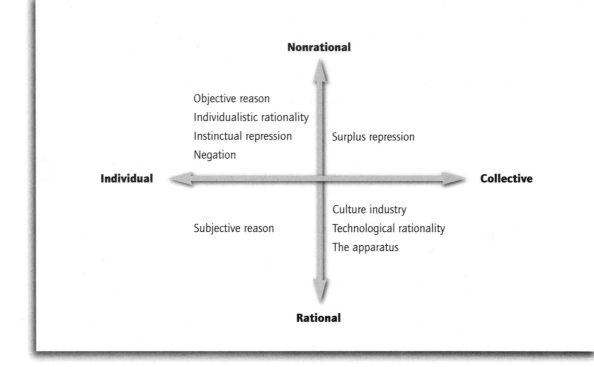

Despite the multidimensionality of their conceptual toolkit, we consider the critical theorists primarily collectivist and rationalist in their theoretical orientation. This view is based on a distinction between the critical theorists' view of the world as it is and as it should be. Undoubtedly, the critical theorists yearn for the creation of a society in which social order is maintained according to dictates of individuals' true needs and desires, where society allows for the realization of goals and aspirations that are determined by the individual's own consciousness and autonomous capacity for reason. But the progress of civilization has not unfolded under such an individualistic dynamic. Instead, repressive, dehumanizing collectivist forces (the culture industry, technology, modes of economic production) have undermined the ability for individuals to develop a society free from domination. Although such forces are the products of human creativity, they have become reified or seemingly immutable. As such, they now confront their creators as alien, abstract powers that operate according to their own logic: The apparatus has a life of its own beyond the control of humans. The "goods and services" spilling off the conveyor belts and assembly lines of the productive apparatus "'sell' or impose the social system as a whole" (Marcuse 1964:12). Horkheimer points to this dynamic when he notes, "the substance of individuality itself, to which the idea of autonomy was bound, did not survive the process of industrialization. . . . With the decline of the ego and its reflective reason, human relationships tend to a point wherein the rule of economy over all personal relationships, the universal control of commodities over the totality of life, turns into a new and naked form of command and obedience" (Horkheimer 1941:376, 379).

Ultimately, then, it is large-scale, institutional forces that account for the creation and re-creation of the social order, albeit one that is founded on oppression. While the critical theorists focus on individual consciousness as the source of human liberation, their account of the social order rests on the broader cultural transformations that have distorted the individual's capacity to reason.

Concerning the issue of action, we again need to distinguish between the critical theorists' understanding of the world as it is and as it should be. In their view, modern society is fueled by a positivist, technological rationality that is incapable of providing an objectively meaningful source of human purpose. Driven by strategic calculations, individuals are motivated to pursue efficiency—a style—above all else in their everyday lives without exercising their own consciousness to determine what the purpose—a substance—of their life should be. While such motivations may prove "rational" insofar as they effectively coordinate means and ends, they fall short of being "reasonable." It is precisely this form of rationality that the critical theorists condemn for supplanting the primacy of morals and values in human affairs. For it is nonrational motivating factors that must guide consciousness and the act of reasoning if modern civilization is to stem the tide of "progress" and create a truly humane society as "the collapse of reason and the collapse of the individual are one and the same thing" (Horkheimer 1941:376).

Readings

The readings that follow explore the central theme of critical theory, namely, how progress has rendered the essence of individuality—negating—obsolete. This argument is based on the premise that individuals in advanced industrial societies live under a radically different set of conditions from those that prevailed during the previous period of competitive capitalism. During the era of "free" capitalism, the family was the center of economic and moral life. The survival of individuals was won and lost through the private family enterprise that simultaneously provided an avenue for individuals to express their own interests and abilities. Moreover, it was the family that trained youth and instilled in them the dominant social values. Thus, the process of socialization and the instinctual conflicts it created were tied to personal relationships between children and their parents. And it is the personal nature of this struggle that allows for the development of individuality.

Under the rule of economic, political, and cultural monopolies, however, the nature of the modern family has changed and with it the nature of the individual. The conflict between the child and the parents as authority figures and administrators of punishment and justice has been replaced. The child learns that it is not his parents that provide for his well-being, but rather the "system" for which they labor. With the parents' loss of power and authority, it is no longer the family but external, standardizing forces to which the individual must adjust and submit. Yet, the reach of such forces is pervasive. The culture industry combines with the economic/political apparatus to erode the distinction between public and private domains. With no autonomous space from which to freely develop their own interests, individuals are unable to confront or oppose the established order. Individuals now identify themselves with and through the interests of the apparatus. Their very consciousness invaded by the apparatus, individuals become objects of technical administration as the whole of their existence has been absorbed by "pleasant" means of domination.

Introduction to Max Horkheimer's *Eclipse of Reason*

In this reading, Horkheimer gives expression to the theme just outlined. In doing so, he argues that modern society is a "totality" that subjects all to its dehumanizing effects; worker and businessperson alike are unable to escape the decline of individuality. This decline is most apparent in the modern form of consciousness that is unwilling and unable to imagine alternative truths. Indeed, "the very idea of truth has been reduced to the purpose of a useful tool in the control of nature, and the realization of the infinite potentialities inherent in man has been relegated to the status of a luxury. Thought that does not serve the interests of any established group or is not pertinent to the business of any industry has no place, is considered vain or superfluous" (Horkheimer 1947:142–43). This is a world in which "usefulness" is defined on technological grounds. Thinking in itself is useless or superfluous. Truth is found in "productivity," and not in critique. While "technocrats maintain that superabundance of goods produced on super-assembly lines will automatically eliminate all economic misery" (ibid.:151), human suffering continues. Yet, such misery and suffering are largely unnecessary, for the technological means to eliminate them are available. Without the ability to reason or find purpose in other than technological terms, humankind is destined to a future in which the "idolization of progress leads to the opposite of progress" (ibid.:153). For it is not technology or productivity in itself that makes for the decline of the individual, but loss of the ability to critically define humane principles according to which real progress must unfold. Perhaps, then, not all hope should be lost; the tide of "progress" may yet still be reversed. As Horkheimer desperately portends,

> Industrial discipline, technological progress, and scientific enlightenment, the very economic and cultural processes that are bringing about the obliteration of individuality, promise—though the augury is faint enough at present—to usher in a new era in which individuality may reemerge as an element in a less ideological and more humane form of existence. (ibid.:160, 161)

Eclipse of Reason (1947)

Max Horkheimer

RISE AND DECLINE OF THE INDIVIDUAL

The crisis of reason is manifested in the crisis of the individual, as whose agency it has developed. The illusion that traditional philosophy has cherished about the individual and about reason—the illusion of their eternity—is being dispelled. The individual once conceived of reason exclusively as an instrument of the self. Now he experiences the reverse of this self-deification. The machine has dropped the driver; it is racing blindly into space. At the moment of consummation, reason has become irrational and stultified. The theme of this time is self-preservation, while there is no self to preserve. In view of this situation, it behooves us to reflect upon the concept of the individual . . .

Individuality presupposes the voluntary sacrifice of immediate satisfaction for the sake of security, material and spiritual maintenance of one's own existence.

When the roads to such a life are blocked, one has little incentive to deny oneself momentary pleasures. Hence, individuality among the masses is far less integrated and enduring than among the so-called elite. On the other hand, the elite have always been more preoccupied with the strategies of gaining and holding power. Social power is today more than ever mediated by power over things. The more intense an individual's concern with power over things, the more will things dominate him, the more will he lack any genuine individual traits, and the more will his mind be transformed into an automaton of formalized reason. . . .

In the era of free enterprise, the so-called era of individualism, individuality was most completely subordinated to self-preserving reason. In that era, the idea of individuality seemed to shake itself loose from metaphysical trappings and to become merely a synthesis of the individual's material interests. That it was not thereby saved from being used as a pawn by ideologists needs no proof. Individualism is the very heart of the theory and practice of bourgeois liberalism, which sees society as progressing through the automatic interaction of divergent interests in a free market. The individual could maintain himself as a social being only by pursuing his own long-term interests at the expense of ephemeral immediate gratifications. The qualities of individuality forged by the ascetic discipline of Christianity were thereby reinforced. The bourgeois individual did not necessarily see himself as opposed to the collectivity, but believed or was prevailed upon to believe himself to be a member of a society that could achieve the highest degree of harmony only through the unrestricted competition of individual interests.

Liberalism may be said to have considered itself the sponsor of a utopia that had come true, needing little more than the smoothing out of a few troublesome wrinkles. These wrinkles were not to be blamed on the liberalistic principle, but on the regrettable nonliberalistic obstacles that impeded its complete fruition. The principle of liberalism has led to conformity through the leveling principle of commerce and exchange which held liberalistic society together. The monad, a seventeenth-century symbol of the atomistic economic individual of bourgeois society, became a social type. All the monads, isolated though they were by moats of self-interest, nevertheless tended to become more and more alike through the pursuit of this very self-interest. In our era of large economic combines and mass culture, the principle of conformity emancipates itself from its individualistic veil, is openly proclaimed, and raised to the rank of an ideal *per se*.

Liberalism at its dawn was characterized by the existence of a multitude of independent entrepreneurs, who took care of their own property and defended it against antagonistic social forces. The movements of the market and the general trend of production were rooted in the economic requirements of their enterprises. Merchant and manufacturer alike had to be prepared for all economic and political eventualities. This need stimulated them to learn what they could from the past and to formulate plans for the future. They had to think for themselves, and although the much-vaunted independence of their thinking was to a certain extent nothing more than an illusion, it had enough objectivity to serve the interests of society in a given form and at a given period. The society of middle-class proprietors, particularly those who acted as middlemen in trade and certain types of manufacturers, had to encourage independent thinking, even though it might be at variance with their particular interests. The enterprise itself, which, it was assumed, would be handed down in the family, gave a businessman's deliberations a horizon that extended far beyond his own life span. His individuality was that of a provider, proud of himself and his kind, convinced that community and state rested upon himself and others like him, all professedly animated by the incentive of material gain. His sense of adequacy to the challenges of an acquisitive world expressed itself in his strong yet sober ego, maintaining interests that transcended his immediate needs.

In this age of big business, the independent entrepreneur is no longer typical. The ordinary man finds it harder and harder to plan for his heirs or even for his own remote future. The contemporary individual may have more opportunities than his ancestors had, but his concrete prospects have an increasingly shorter term. The future does not enter as precisely into his transactions. He simply feels that he will not be entirely lost if he preserves his skill and clings to his corporation, association, or union. Thus the individual subject of reason tends to become a shrunken ego, captive of an evanescent present, forgetting the use of the intellectual functions by which he was once able to transcend his actual position in reality. These functions are now taken over by the great economic and social forces of the era. The future of the individual depends less and less upon his own prudence and more and more upon the national and international struggles among the colossi of power. Individuality loses its economic basis.

There are still some forces of resistance left within man. It is evidence against social pessimism that despite the continuous assault of collective patterns, the spirit of humanity is still alive, if not in the individual as a member of social groups, at least in the individual as far as he is let alone. But the impact of the existing conditions upon the average man's life is such that the submissive type mentioned earlier has become overwhelmingly predominant. From the day of his birth, the individual is made to feel that there is only one way of getting along in this world—that of giving up his hope of ultimate self-realization. This he can achieve solely by imitation. He continuously responds to what he perceives about him, not only consciously but with his whole being, emulating the traits and attitudes represented by all the collectivities that enmesh him—his play group, his classmates, his athletic team, and all the other groups that, as has been pointed out, enforce a more strict conformity, a more radical surrender through complete assimilation, than any father or teacher in the nineteenth century could impose. By echoing, repeating, imitating his surroundings, by adapting himself to all the powerful groups to which he eventually belongs, by transforming himself from a human being into a member of organizations, by sacrificing his potentialities for the sake of readiness and ability to conform to and gain influence in such organizations, he manages to survive. It is survival achieved by the oldest biological means of survival, namely, mimicry.

Just as a child repeats the words of his mother, and the youngster the brutal manners of the elders at whose hands he suffers, so the giant loud-speaker of industrial culture, blaring through commercialized recreation and popular advertising—which become more and more indistinguishable from each other—endlessly reduplicates the surface of reality. All the ingenious devices of the amusement industry reproduce over and over again banal life scenes that are deceptive nevertheless, because the technical exactness of the reproduction veils the falsification of the ideological content or the arbitrariness of the introduction of such content. This reproduction has nothing in common with great realistic art, which portrays reality in order to judge it. Modern mass culture, although drawing freely upon stale cultural values, glorifies the world as it is. Motion pictures, the radio, popular biographies and novels have the same refrain: This is our grove, this is the rut of the great and the would-be great—this is reality as it is and should be and will be.

Even the words that could voice a hope for something besides the fruits of success have been pressed into this service. The idea of eternal bliss and everything relating to the absolute have been reduced to the function of religious edification, conceived as a leisure-time activity; they have been made part of the Sunday-school vernacular. The idea of happiness has similarly been reduced to a banality to coincide with leading the kind of normal life that serious religious thought has often criticized. The very idea of truth has been reduced to the purpose of a useful tool in the control of nature, and the realization of the infinite potentialities inherent in man has been relegated to the status of a luxury. Thought that does not serve the interests of any established group or is not pertinent to the business of any industry has no place, is considered vain or superfluous. Paradoxically, a society that, in the face of starvation in great areas of the world, allows a large part of its machinery to stand idle, that shelves many important inventions, and that devotes innumerable working hours to moronic advertising and to the production of instruments of destruction—a society in which these luxuries are inherent has made usefulness its gospel.

Because modern society is a totality, the decline of individuality affects the lower as well as the higher social groups, the worker no less than the businessman. One of the most important attributes of individuality, that of spontaneous action, which began to decline in capitalism as a result of the partial elimination of competition, played an integral part in socialist theory. But today the spontaneity of the working class has been impaired by the general dissolution of individuality. Labor is increasingly divorced from critical theories as they were formulated by the great political and social thinkers of the nineteenth century. . . . As a matter of fact not theory but its decline furthers surrender to the powers that be, whether they are represented by the controlling agencies of capital or those of labor. However, the masses, despite their pliability, have not capitulated completely to collectivization. Although, under the pressure of the pragmatic reality of today, man's self-expression has become identical with his function in the prevailing system, although he desperately represses any other impulse within himself as well as in others, the rage that seizes him whenever he becomes aware of an unintegrated longing that does not fit into the existing pattern is a sign of his smoldering resentment. This resentment, if repression were abolished, would be turned against the whole social order, which has an intrinsic tendency to prevent its members from gaining insight into the mechanisms of their own repression. Throughout history, physical, organizational, and cultural pressures have always had their role in the integration of the individual into a just or unjust order; today, the labor organizations, in their very effort to improve the status of labor, are inevitably led to contribute to that pressure. . . .

Social theory—reactionary, democratic, or revolutionary—was the heir to the older systems of thought that were supposed to have set the patterns for past totalities. These older systems had vanished because the forms of solidarity postulated by them proved to be deceptive, and the ideologies related to them became hollow and apologetic. The latter-day critique of society for its part refrained from apologetics, and did not glorify its subject—not even Marx exalted the proletariat. He looked upon capitalism as the last form of social injustice; he did not condone the established ideas and superstitions of the dominated class whom his doctrine was supposed to guide. In contrast to the tendencies of mass culture, none of those doctrines undertook to "sell" the people the way of life in which they are fixed and which they unconsciously abhor but overtly acclaim. Social theory offered a critical analysis of reality, including the workers' own warped thoughts. Under the conditions of modern industrialism, however, even political theory is infected with the apologetic trend of the total culture. . . .

As religious and moral ideologies fade, and political theory is abolished by the march of economic and political events, the ideas of the workers tend to be molded by the business ideology of their leaders. The idea of an intrinsic conflict between the laboring masses of the world and the existence of social injustice is superseded by the concepts relating to the strategy of conflicts between the several power groups. It is true that workers of earlier days did not have any conceptual knowledge of the mechanisms unveiled by social theory, and their minds and bodies bore the marks of oppression; yet their misery was still the misery of individual human beings, and therefore linked them with any miserable people in any country and in any sector of society. Their undeveloped minds were not continually being prodded by the techniques of mass culture that hammer the industrialistic behavior patterns into their eyes and ears and muscles during their leisure time as well as during working hours. Workers today, no less than the rest of the population, are intellectually better trained, better informed, and much less naive. They know the details of national affairs and the chicanery of political movements, particularly of those that live by propaganda against corruption. The workers, at least those who have not gone through the hell of fascism, will join in any persecution of a capitalist or politician who has been singled out because he has violated the rules of the game; but they do not question the rules in themselves. They have learned to take social injustice—even inequity within their own group—as a powerful fact, and to take powerful facts as the only things

to be respected. Their minds are closed to dreams of a basically different world and to concepts that, instead of being mere classification of facts, are oriented toward real fulfillment of those dreams. Modern economic conditions make for a positivistic attitude in members as well as in leaders of labor unions, so that they resemble one another more and more. Such a trend, although constantly challenged by contrary tendencies, strengthens labor as a new force in social life.

It is not that inequality has decreased. To the old discrepancies between the social power of single members of different social groups, further differences have been added. While unions dealing in certain categories of labor have been able to raise their prices, the whole weight of oppressive social power is felt by other categories, organized or unorganized. There is, furthermore, the cleavage between members of unions and those who for any one of various reasons are excluded from unions, between the people of privileged nations and those who, in this contracting world, are dominated not only by their own traditional elite, but also by the ruling groups of the industrially more developed countries. The principle has not changed.

At the present time, labor and capital are equally concerned with holding and extending their control. The leaders in both groups contend to an increasing extent that theoretical critique of society has become superfluous as a result of the tremendous technological progress that promises to revolutionize the conditions of human existence. The technocrats maintain that superabundance of goods produced on super-assembly lines will automatically eliminate all economic misery. Efficiency, productivity, and intelligent planning are proclaimed the gods of modern man; so-called "unproductive" groups and "predatory" capital are branded as the enemies of society.

It is true that the engineer, perhaps the symbol of this age, is not so exclusively bent on profitmaking as the industrialist or the merchant. Because his function is more directly connected with the requirements of the production job itself, his commands bear the mark of greater objectivity. His subordinates recognize that at least some of his orders are in the nature of things and therefore rational in a universal sense. But at bottom this rationality, too, pertains to domination, not reason. The engineer is not interested in understanding things for their own sake or for the sake of insight, but in accordance with their being fitted into a scheme, no matter how alien to their own inner structure; this holds for living beings as well as for inanimate things. The engineer's mind is that of industrialism in its streamlined form. His purposeful rule would make men an agglomeration of instruments without a purpose of their own. . . .

It is not technology or the motive of self-preservation that in itself accounts for the decline of the individual; it is not production *per se,* but the forms in which it takes place—the interrelationships of human beings within the specific framework of industrialism. Human toil and research and invention is a response to the challenge of necessity. The pattern becomes absurd only when people make toil, research, and invention into idols. Such an ideology tends to supplant the humanistic foundation of the very civilization it seeks to glorify. While the concepts of complete fulfilment and unrestrained enjoyment fostered a hope that unshackled the forces of progress, the idolization of progress leads to the opposite of progress. Arduous labor for a meaningful end may be enjoyed and even loved. A philosophy that makes labor an end in itself leads eventually to resentment of all labor. The decline of the individual must be charged not to the technical achievements of man or even to man himself—people are usually much better than what they think or say or do—but rather to the present structure and content of the "objective mind," the spirit that pervades social life in all its branches. The patterns of thought and action that people accept ready-made from the agencies of mass culture act in their turn to influence mass culture as though they were the ideas of the people themselves. The objective mind in our era worships industry, technology, and nationality without a principle that could give sense to these categories; it mirrors the pressure of an economic system that admits of no reprieve or escape.

As for the ideal of productivity, it must be observed that economic significance today is measured in terms of usefulness with respect to the structure of power, not with respect to the needs of all. The individual must prove his value to one or other of the groups engaged in the struggle for a greater share of control over the national and the international economy. Moreover, the quantity and quality of the goods or services he contributes to society is merely one of the factors determining his success.

Nor is efficiency, the modern criterion and sole justification for the very existence of any individual, to be confused with real technical or managerial skill. It inheres in the ability to be "one of the boys," to hold one's own, to impress others, to "sell" oneself, to cultivate the right connections—talents that seem to be transmitted through the germ cells of so many persons today. The fallacy of technocratic thinking from St. Simon to Veblen and his followers has lain in underestimating the similarity of the traits that make for success in the various branches of production and business, and in confusing rational use of the means of production with the rational proclivities of certain of its agents.

If modern society tends to negate all the attributes of individuality, are its members not compensated, it may be asked, by the rationality of its organization? The technocrats often maintain that when their theories are put into practice, depressions will become a thing of the past and basic economic disproportions will disappear; the whole productive mechanism will work smoothly according to blueprints. Actually, modern society is not so far from having realized the technocratic dream. The needs of the consumers as well as of the producers, which under the liberal market system made themselves felt in distorted and irrational forms, in a process culminating in depressions, can now to a great extent be forecast and satisfied or negated in accordance with the policies of economic and political leaders. The expression of human needs is no longer distorted by the dubious economic indicators of the market; instead, these needs are determined by statistics, and all kinds of engineers—industrial, technical, political— struggle to keep them under control. But if this new rationality is in one way closer to the idea of reason than the market system, it is in another way farther from it.

Dealings between the members of different social groups under the older system were really determined not by the market but by the unequal distribution of economic power; yet the transformation of human relations into objective economic mechanisms gave the individual, at least in principle, a certain independence. When unsuccessful competitors went to the wall or backward groups were reduced to misery under the liberalistic economy, they could preserve a sense of human dignity even though they were economically cast down, because responsibility for their plight could be thrown upon anonymous economic processes. Today individuals or entire groups may still suffer ruin through blind economic forces; but these are represented by better organized, more powerful elites. Although the interrelations of these dominant groups are subject to vicissitudes, they understand each other well in many respects. When concentration and centralization of industrial forces extinguish political liberalism in its turn, the victims are doomed in their entirety. Under totalitarianism, when an individual or group is singled out by the elite for discrimination, it is not only deprived of the means of livelihood, but its very human essence is attacked. American society may take a different course. However, the dwindling away of individual thinking and resistance, as it is brought about by the economic and cultural mechanisms of modern industrialism, will render evolution toward the humane increasingly difficult.

By making the watchword of production a kind of religious creed, by professing technocratic ideas and branding as "unproductive" such groups as do not have access to the big industrial bastions, industry causes itself and society to forget that production has become to an ever greater extent a means in the struggle for power. The policies of economic leaders, on which society in its present stage more and more directly depends, are dogged and particularistic, and therefore perhaps even blinder with respect to the real needs of society than were the automatic trends that once determined the market. Irrationality still molds the fate of men. . . .

Every instrumentality of mass culture serves to reinforce the social pressures upon individuality, precluding all possibility that the individual will somehow preserve himself in the face of all the atomizing machinery of modern society. The accent on individual heroism and on the self-made man in popular biographics and pseudo-romantic novels and films does not invalidate this observation. These machine-made incentives to self-preservation actually accelerate the dissolution of individuality. Just as the slogans of rugged individualism are politically useful to large trusts in seeking exemption from social control, so in mass culture the rhetoric of individualism, by imposing patterns for collective imitation, disavows the very principle to which it gives lip service. . . .

The objection that the individual, despite everything, does not entirely disappear in the new impersonal institutions, that individualism is as rugged and rampant in modern society as ever before, seems to miss the point. The objection contains a grain of truth, namely, the consideration that man is still better than the world he lives in. Yet his life seems to follow a sequence that will fit any questionnaire he is asked to fill out. His intellectual existence is exhausted in the public opinion polls. Especially the so-called great individuals of today, the idols of the masses, are not genuine individuals, they are simply creatures of their own publicity, enlargements of their own photographs, functions of social processes. The consummate superman, against whom no one has warned more anxiously than Nietzsche himself, is a projection of the oppressed masses, King Kong rather than Caesar Borgia. The hypnotic spell that

such counterfeit supermen as Hitler have exercised derives not so much from what they think or say or do as from their antics, which set a style of behavior for men who, stripped of their spontaneity by the industrial processing, need to be told how to make friends and influence people.

The tendencies described have already led to the greatest catastrophe in European history. Some of the causes were specifically European. Others are traceable to profound changes in man's character under the influence of international trends. Nobody can predict with certainty that these destructive tendencies will be checked in the near future. However, there is increasing awareness that the unbearable pressure upon the individual is not inevitable. It is to be hoped that men will come to see that it springs not directly from the purely technical requirements of the production, but from the social structure. Indeed, the intensification of repression in many parts of the world in itself testifies to fear in face of the imminent possibility of change on the basis of the present development of productive forces. Industrial discipline, technological progress, and scientific enlightenment, the very economic and cultural processes that are bringing about the obliteration of individuality, promise—though the augury is faint enough at present—to usher in a new era in which individuality may reemerge as an element in a less ideological and more humane form of existence.

Fascism used terroristic methods in the effort to reduce conscious human beings to social atoms, because it feared that ever-increasing disillusionment as regards all ideologies might pave the way for men to realize their own and society's deepest potentialities; and indeed, in some cases, social pressure and political terror have tempered the profoundly human resistance to irrationality— a resistance that is always the core of true individuality.

The real individuals of our time are the martyrs who have gone through infernos of suffering and degradation in their resistance to conquest and oppression, not the inflated personalities of popular culture, the conventional dignitaries. These unsung heroes consciously exposed their existence as individuals to the terroristic annihilation that others undergo unconsciously through the social process. The anonymous martyrs of the concentration camps are the symbols of the humanity that is striving to be born. The task of philosophy is to translate what they have done into language that will be heard, even though their finite voices have been silenced by tyranny.

Introduction to Theodor Adorno's "The Culture Industry Reconsidered"

In this selection, Adorno discusses how the culture industry has furthered the collapse of reason. Equating culture with the "high arts," Adorno describes culture as a form of protest "against the petrified relations" (Adorno 1991:100) under which individuals live. The purpose of culture is to render the impossible possible, to offer alternatives to existing social conditions. To the extent that culture (art) is free from the profit motive, it is able to develop according to its own internal logic and thus voice essential social critiques.

In advanced societies, however, culture has become synonymous with industry and hence subject to the rule of efficient production and standardization that is its hallmark. The relationship between mass culture and the individual is one akin to that of seller and buyer. However, "the customer is not king, as the culture industry would like to have us believe, not its subject but its object" (ibid.:99). Individuals, themselves objects of production, are left to consume mass-produced, prepackaged ideas that instill an uncritical consensus that strengthens established authority. Hit songs and movies are not the making of popular tastes but of marketing campaigns that predetermine what will be heard and seen while excluding potentially "disruptive" alternatives. Because culture is now a product of the machine and not the imagination, it is incapable of negating the oppressive conformity churned out by the culture industry. Nor can mass culture critique prevailing patterns of social relations that, likewise, are a reflection of machine production. Culture no longer prods—it pacifies:

> The categorical imperative of the culture industry no longer has anything in common with freedom. It proclaims: You shall conform, without instruction as to what; conform to that which exists anyway, and to that which everyone thinks anyway as a reflex of its power and omnipresence. The power of the culture industry is that conformity has replaced consciousness. The order that springs from it is never confronted with what it claims to be or with the real interests of human beings. (ibid.:104)

While the culture industry claims to be a producer of choice, freedom, and individual identity, it instead provides its customers with a totalitarian, conformist social landscape. It thus "cheats its consumers out of the same happiness which it deceitfully projects" (ibid.:106). So while we are repeatedly instructed to "Just Do It," it is never truthfully revealed what "it" is: BUY.

—————— "The Culture Industry Reconsidered" (1975) ——————

Theodor Adorno

The term culture industry was perhaps used for the first time in the book *Dialectic of Enlightenment*, which Horkheimer and I published in Amsterdam in 1947. In our drafts we spoke of "mass culture." We replaced that expression with "culture industry" in order to exclude from the outset the interpretation agreeable to its advocates: that it is a matter of something like a culture that arises spontaneously from the masses themselves, the contemporary form of popular art. From the latter the culture industry must be distinguished in the extreme. The culture industry fuses the old and familiar into a new quality. In all its branches, products which are tailored for consumption by masses, and which to a great extent determine the nature of that consumption, are manufactured more or less according to plan. The individual branches are similar in structure or at least fit

SOURCE: "The Culture Industry Reconsidered" by Theodor Adorno in *New German Critique*, 6, Fall 1975, pp. 12–19. Copyright © 1975. Reprinted with permission of Taylor & Francis Books, UK.

into each other, ordering themselves into a system almost without a gap. This is made possible by contemporary technical capabilities as well as by economic and administrative concentration. The culture industry intentionally integrates its consumers from above. To the detriment of both it forces together the spheres of high and low art, separated for thousands of years. The seriousness of high art is destroyed in speculation about its efficacy; the seriousness of the lower perishes with the civilizational constraints imposed on the rebellious resistance inherent within it as long as social control was not yet total. Thus, although the culture industry undeniably speculates on the conscious and unconscious state of the millions towards which it is directed, the masses are not primary, but secondary, they are an object of calculation; an appendage of the machinery. The customer is not king, as the culture industry would have us believe, not its subject but its object. The very word mass-media, specially honed for the culture industry, already shifts the accent onto harmless terrain. Neither is it a question of primary concern for the masses, nor of the techniques of communication as such, but of the spirit which sufflates them, their master's voice. The culture industry misuses its concern for the masses in order to duplicate, reinforce and strengthen their mentality, which it presumes is given and unchangeable. How this mentality might be changed is excluded throughout. The masses are not the measure but the ideology of the culture industry, even though the culture industry itself could scarcely exist without adapting to the masses.

The cultural commodities of the industry are governed, as Brecht and Suhrkamp expressed it thirty years ago, by the principle of their realization as value, and not by their own specific content and harmonious formation. The entire practice of the culture industry transfers the profit motive naked onto cultural forms. Ever since these cultural forms first began to earn a living for their creators as commodities in the market-place they had already possessed something of this quality. But then they sought after profit only indirectly, over and above their autonomous essence. New on the part of the culture industry is the direct and undisguised primacy of a precisely and thoroughly calculated efficacy in its most typical products. The autonomy of works of art, which of course rarely ever predominated in an entirely pure form, and was always permeated by a constellation of effects, is tendentially eliminated by the culture industry, with or without the conscious will of those in control. The latter include both those who carry out

directives as well as those who hold the power. In economic terms they are or were in search of new opportunities for the realization of capital in the most economically developed countries. The old opportunities became increasingly more precarious as a result of the same concentration process which alone makes the culture industry possible as an omni-present phenomenon. Culture, in the true sense, did not simply accommodate itself to human beings; but it always simultaneously raised a protest against the petrified relations under which they lived, thereby honouring them. In so far as culture becomes wholly assimilated to and integrated in those petrified relations, human beings are once more debased. Cultural entities typical of the culture industry are no longer *also* commodities, they are commodities through and through. This quantitative shift is so great that it calls forth entirely new phenomena. Ultimately, the culture industry no longer even needs to directly pursue everywhere the profit interests from which it originated. These interests have become objectified in its ideology and have even made themselves independent of the compulsion to sell the cultural commodities which must be swallowed anyway. The culture industry turns into public relations, the manufacturing of "goodwill" per se, without regard for particular firms or saleable objects. Brought to bear is a general uncritical consensus, advertisements produced for the world, so that each product of the culture industry becomes its own advertisement.

Nevertheless, those characteristics which originally stamped the transformation of literature into a commodity are maintained in this process. More than anything in the world, the culture industry has its ontology, a scaffolding of rigidly conservative basic categories which can be gleaned, for example, from the commercial English novels of the late seventeenth and early eighteenth centuries. What parades as progress in the culture industry, as the incessantly new which it offers up, remains the disguise for an eternal sameness; everywhere the changes mask a skeleton which has changed just as little as the profit motive itself since the time it first gained its predominance over culture.

Thus, the expression "industry" is not to be taken too literally. It refers to the standardization of the thing itself—such as that of the Western, familiar to every movie-goer—and to the rationalization of distribution techniques, but not strictly to the production process. Although in film, the central sector of the culture industry, the production process resembles technical modes of operation in the extensive division of labour, the employment of machines and the separation of the labourers from the means of production—expressed in the perennial conflict between artists active in the culture industry and those who control it—individual forms of production are nevertheless maintained. Each product affects an individual air; individuality itself serves to reinforce ideology, in so far as the illusion is conjured up that the completely reified and mediated is a sanctuary from immediacy and life. Now, as ever, the culture industry exists in the "service" of third persons, maintaining its affinity to the declining circulation process of capital, to the commerce from which it came into being. Its ideology above all makes use of the star system, borrowed from individualistic art and its commercial exploitation. The more dehumanized its methods of operation and content, the more diligently and successfully the culture industry propagates supposedly great personalities and operates with heart-throbs. It is industrial more in a sociological sense, in the incorporation of industrial forms of organization even when nothing is manufactured—as in the rationalization of office work—rather than in the sense of anything really and actually produced by technological rationality. Accordingly, the misinvestments of the culture industry are considerable, throwing those branches rendered obsolete by new techniques into crises, which seldom lead to changes for the better . . .

It has recently become customary among cultural officials as well as sociologists to warn against underestimating the culture industry while pointing to its great importance for the development of the consciousness of its consumers. It is to be taken seriously, without cultured snobbism. In actuality the culture industry is important as a moment of the spirit which dominates today. Whoever ignores its influence out of scepticism for what it stuffs into people would be naive. Yet there is a deceptive glitter about the admonition to take it seriously. Because of its social role, disturbing questions about its quality, about truth or untruth, and about the aesthetic niveau of the culture industry's emissions are repressed, or at least excluded from the so-called sociology of communications. The critic is accused of taking refuge in arrogant esoterica. It would be advisable first to indicate the double meaning of importance that slowly worms its way in unnoticed. Even if it touches the lives of innumerable people, the function of something is no guarantee of its particular quality. The blending of aesthetics with its residual communicative aspects leads art, as a social phenomenon, not to its rightful position in opposition to alleged artistic snobbism, but

rather in a variety of ways to the defence of its baneful social consequences. The importance of the culture industry in the spiritual constitution of the masses is no dispensation for reflection on its objective legitimation, its essential being, least of all by a science which thinks itself pragmatic. On the contrary: such reflection becomes necessary precisely for this reason. To take the culture industry as seriously as its unquestioned role demands, means to take it seriously critically, and not to cower in the face of its monopolistic character.

Among those intellectuals anxious to reconcile themselves with the phenomenon and eager to find a common formula to express both their reservations against it and their respect for its power, a tone of ironic toleration prevails unless they have already created a new mythos of the twentieth century from the imposed regression. After all, those intellectuals maintain, everyone knows what pocket novels, films off the rack, family television shows rolled out into serials and hit parades, advice to the lovelorn and horoscope columns are all about. All of this, however, is harmless and, according to them, even democratic since it responds to a demand, albeit a stimulated one. It also bestows all kinds of blessings, they point out, for example, through the dissemination of information, advice and stress reducing patterns of behaviour. Of course, as every sociological study measuring something as elementary as how politically informed the public is has proven, the information is meagre or indifferent. Moreover, the advice to be gained from manifestations of the culture industry is vacuous, banal or worse, and the behaviour patterns are shamelessly conformist.

The two-faced irony in the relationship of servile intellectuals to the culture industry is not restricted to them alone. It may also be supposed that the consciousness of the consumers themselves is split between the prescribed fun which is supplied to them by the culture industry and a not particularly well-hidden doubt about its blessings. The phrase, the world wants to be deceived, has become truer than had ever been intended. People are not only, as the saying goes, falling for the swindle; if it guarantees them even the most fleeting gratification they desire a deception which is nonetheless transparent to them. They force their eyes shut and voice approval, in a kind of self-loathing, for what is meted out to them, knowing fully the purpose for which it is manufactured. Without admitting it they sense that their lives would be completely intolerable as soon as they no longer clung to satisfactions which are none at all.

The most ambitious defence of the culture industry today celebrates its spirit, which might be safely called ideology, as an ordering factor. In a supposedly chaotic world it provides human beings with something like standards for orientation, and that alone seems worthy of approval. However, what its defenders imagine is preserved by the culture industry is in fact all the more thoroughly destroyed by it. The colour film demolishes the genial old tavern to a greater extent than bombs ever could: the film exterminates its *imago*. No homeland can survive being processed by the films which celebrate it, and which thereby turn the unique character on which it thrives into an interchangeable sameness.

That which legitimately could be called culture attempted, as an expression of suffering and contradiction, to maintain a grasp on the idea of the good life. Culture cannot represent either that which merely exists or the conventional and no longer binding categories of order which the culture industry drapes over the idea of the good life as if existing reality were the good life, and as if those categories were its true measure. If the response of the culture industry's representatives is that it docs not dcliver art at all, this is itself the ideology with which they evade responsibility for that from which the business lives. No misdeed is ever righted by explaining it as such.

The appeal to order alone, without concrete specificity, is futile; the appeal to the dissemination of norms, without these ever proving themselves in reality or before consciousness, is equally futile. The idea of an objectively binding order, huckstered to people because it is so lacking for them, has no claims if it does not prove itself internally and in confrontation with human beings. But this is precisely what no product of the culture industry would engage in. The concepts of order which it hammers into human beings are always those of the status quo. They remain unquestioned, unanalysed and undialectically presupposed, even if they no longer have any substance for those who accept them. In contrast to the Kantian, the categorical imperative of the culture industry no longer has anything in common with freedom. It proclaims: you shall conform, without instruction as to what; conform to that which exists anyway, and to that which everyone thinks anyway as a reflex of its power and omnipresence. The power of the culture industry's ideology is such that conformity has replaced consciousness. The order that springs from it is never confronted with what it claims to be or with the real interests of human beings. Order, however, is not good in itself. It would be so only as a good order. The

fact that the culture industry is oblivious to this and extols order *in abstracto,* bears witness to the impotence and untruth of the messages it conveys. While it claims to lead the perplexed, it deludes them with false conflicts which they are to exchange for their own. It solves conflicts for them only in appearance, in a way that they can hardly be solved in their real lives. In the products of the culture industry human beings get into trouble only so that they can be rescued unharmed, usually by representatives of a benevolent collective; and then in empty harmony, they are reconciled with the general, whose demands they had experienced at the outset as irreconcilable with their interests. For this purpose the culture industry has developed formulas which even reach into such non-conceptual areas as light musical entertainment. Here too one gets into a "jam," into rhythmic problems, which can be instantly disentangled by the triumph of the basic beat.

Even its defenders, however, would hardly contradict Plato openly who maintained that what is objectively and intrinsically untrue cannot also be subjectively good and true for human beings. The concoctions of the culture industry are neither guides for a blissful life, nor a new art of moral responsibility, but rather exhortations to toe the line, behind which stand the most powerful interests. The consensus which it propagates strengthens blind, opaque authority. If the culture industry is measured not by its own substance and logic, but by its efficacy, by its position in reality and its explicit pretensions; if the focus of serious concern is with the efficacy to which it always appeals, the potential of its effect becomes twice as weighty. This potential, however, lies in the promotion and exploitation of the ego-weakness to which the powerless members of contemporary society, with its concentration of power, are condemned. Their consciousness is further developed retrogressively. It is no coincidence that cynical American film producers are heard to say that their pictures must take into consideration the level of eleven-year-olds. In doing so they would very much like to make adults into eleven-year-olds.

It is true that thorough research has not, for the time being, produced an airtight case proving the regressive effects of particular products of the culture industry. No doubt an imaginatively designed experiment could achieve this more successfully than the powerful financial interests concerned would find comfortable. In any case, it can be assumed without hesitation that steady drops hollow the stone, especially since the system of the culture industry that surrounds the masses tolerates hardly any deviation and incessantly drills the same formulas on behaviour. Only their deep unconscious mistrust, the last residue of the difference between art and empirical reality in the spiritual make-up of the masses explains why they have not, to a person, long since perceived and accepted the world as it is constructed for them by the culture industry. Even if its messages were as harmless as they are made out to be—on countless occasions they are obviously not harmless, like the movies which chime in with currently popular hate campaigns against intellectuals by portraying them with the usual stereotypes—the attitudes which the culture industry calls forth are anything but harmless. If an astrologer urges his readers to drive carefully on a particular day, that certainly hurts no one; they will, however, be harmed indeed by the stupefication which lies in the claim that advice which is valid every day and which is therefore idiotic, needs the approval of the stars.

Human dependence and servitude, the vanishing point of the culture industry, could scarcely be more faithfully described than by the American interviewee who was of the opinion that the dilemmas of the contemporary epoch would end if people would simply follow the lead of prominent personalities. In so far as the culture industry arouses a feeling of well-being that the world is precisely in that order suggested by the culture industry, the substitute gratification which it prepares for human beings cheats them out of the same happiness which it deceitfully projects. The total effect of the culture industry is one of anti-enlightenment, in which, as Horkheimer and I have noted, enlightenment, that is the progressive technical domination of nature, becomes mass deception and is turned into a means for fettering consciousness. It impedes the development of autonomous, independent individuals who judge and decide consciously for themselves. These, however, would be the precondition for a democratic society which needs adults who have come of age in order to sustain itself and develop. If the masses have been unjustly reviled from above as masses, the culture industry is not among the least responsible for making them into masses and then despising them, while obstructing the emancipation for which human beings are as ripe as the productive forces of the epoch permit.

▪▪
▪▪

Introduction to Herbert Marcuse's *One-Dimensional Man*

Marcuse's *One-Dimensional Man* was one of the most widely read and influential books among advocates for social change during the 1960s. Although his message to the New Left was written more than forty years ago, Marcuse's insights ring just as powerfully today.

As we outlined previously, Marcuse describes contemporary, advanced societies—whether capitalist, communist, or socialist—as totalitarian social orders, for "totalitarian" does not refer only to a particular type of government. It "is not only a terroristic political coordination of society, but also a nonterroristic economic-political coordination which operates through the manipulation of needs by vested interests. It thus precludes the emergence of an effective opposition against the whole" (Marcuse 1964:3). Totalitarian societies are thus characterized by coordinated systems of domination that render all protest obsolete. Instead of being based on fear of external coercion or force, however, the methods of domination in advanced societies are based more on the manipulation of consciousness.

With the development of the industrial capacity to free individuals from want, the working class, the source of revolutionary change under Marxist theory, has been assimilated into the prevailing social order. Under the dominion of technological rationality and the benefits it offers, "the intellectual and emotional refusal 'to go along' appears neurotic and impotent" (ibid.:9). For why would anyone contest the satisfactions that the apparatus delivers through the progress of science and technology? The claim that the working class is alienated now becomes questionable as its members identify with and literally buy into the very system that is the source of their oppression. Once adversaries who harbored conflicting interests, capitalists and workers are united in their unquestioned, welcomed perpetuation of the status quo.

At the root of this social union lies the fabrication of new "needs" that maintain the existing way of life. Such needs do not spring from the consciousness of the individual; instead, they are a product of technological advances. (Do you really "need" a BlackBerry or five-disc CD changer?) While the satisfaction of these false needs is advertised to be a path for happiness, they further the repression of true needs. Technology, then, is not neutral; rather, it is a means for preserving domination. Indeed, its effectiveness as a dominating force resides in the fact that it appears to be neutral while it actually enslaves individuality. The creation of needs and the products dispensed to meet them serve to

indoctrinate and manipulate; they promote a false consciousness which is immune against its falsehood. And as these beneficial products become available to more individuals in more social classes, the indoctrination they carry ceases to be publicity; it becomes a way of life. It is a good way of life—much better than before— and as a good way of life it militates against qualitative change. Thus emerges a pattern of *one-dimensional thought and behavior* in which ideas, aspirations, and objectives that, by their content, transcend the established universe of discourse and action are either repelled or reduced to terms of this universe. (Marcuse 1964:12; emphasis in original)

This one-dimensional society intensifies repressive or institutionalized desublimation in which all opposition, whether political, cultural, or instinctual, is absorbed and thus defused by the very apparatus that it intended to oppose. With the range of alternative ideas and actions reduced to one, the indoctrination of the "Happy Consciousness" leaves individuals unable to grasp the essential unfreedom that characterizes advanced industrial society. Satisfied with the offerings and "liberties" of the established order, their loss of conscience leads to acceptance of the status quo and rampant conformity. As Marcuse argues,

Under the rule of a repressive whole, liberty can be made into a powerful instrument of domination. The range of choice open to the individual is not the decisive factor in determining the degree of human freedom, but *what* can be chosen and what *is* chosen by the individual. . . . Free choice among a wide variety of goods and services does not signify freedom if these goods and services sustain social controls over a life of toil and fear. (ibid.:7, 8; emphasis in original)

--------------------- *One-Dimensional Man* (1964) ---------------------

Herbert Marcuse

THE NEW FORMS OF CONTROL

A comfortable, smooth, reasonable, democratic unfreedom prevails in advanced industrial civilization, a token of technical progress. Indeed, what could be more rational than the suppression of individuality in the mechanization of socially necessary but painful performances; the concentration of individual enterprises in more effective, more productive corporations; the regulation of free competition among unequally equipped economic subjects; the curtailment of prerogatives and national sovereignties which impede the international organization of resources. That this technological order also involves a political and intellectual coordination may be a regrettable and yet promising development.

The rights and liberties which were such vital factors in the origins and earlier stages of industrial society yield to a higher stage of this society: they are losing their traditional rationale and content. Freedom of thought, speech, and conscience were—just as free enterprise, which they served to promote and protect—essentially *critical* ideas, designed to replace an obsolescent material and intellectual culture by a more productive and rational one. Once institutionalized, these rights and liberties shared the fate of the society of which they had become an integral part. The achievement cancels the premises.

To the degree to which freedom from want, the concrete substance of all freedom, is becoming a real possibility, the liberties which pertain to a state of lower productivity are losing their former content. Independence of thought, autonomy, and the right to political opposition are being deprived of their basic critical function in a society which seems increasingly capable of satisfying the needs of the individuals through the way in which it is organized. Such a society may justly demand acceptance of its principles and institutions, and reduce the opposition to the discussion and promotion of alternative policies *within* the status quo. In this respect, it seems to make little difference whether the increasing satisfaction of needs is accomplished by an authoritarian or a non-authoritarian system. Under the conditions of a rising standard of living, non-conformity with the system itself appears to be socially useless, and the more so when it entails tangible economic and political disadvantages and threatens the smooth operation of the whole. Indeed, at least in so far as the necessities of life are involved, there seems to be no reason why the production and distribution of goods and services should proceed through the competitive concurrence of individual liberties.

Freedom of enterprise was from the beginning not altogether a blessing. As the liberty to work or to starve, it spelled toil, insecurity, and fear for the vast majority of the population. If the individual were no longer compelled to prove himself on the market, as a free economic subject, the disappearance of this kind of freedom would be one of the greatest achievements of civilization. The technological processes of mechanization and standardization might release individual energy into a yet uncharted realm of freedom beyond necessity. The very structure of human existence would be altered; the individual would be liberated from the work world's imposing upon him alien needs and alien possibilities. The individual would be free to exert autonomy over a life that would be his own. If the productive apparatus could be organized and directed toward the satisfaction of the vital needs, its control might well be centralized; such control would not prevent individual autonomy, but render it possible.

This is a goal within the capabilities of advanced industrial civilization, the "end" of technological rationality. In actual fact, however, the contrary trend operates: the apparatus imposes its economic and political requirements for defense and expansion on labor time and free time, on the material and intellectual culture. By virtue of the way it has organized its technological base, contemporary industrial society tends to be totalitarian. For "totalitarian" is not only a terroristic political coordination of society, but also a non-terroristic economic-technical coordination which operates through the manipulation of needs by vested interests. It thus precludes the emergence of an effective opposition against the whole. Not only a specific form of government or party rule makes for totalitarianism, but also a specific system of production and distribution which may well be compatible with a "pluralism" of parties, newspapers, "countervailing powers," etc.

Today political power asserts itself through its power over the machine process and over the technical organization of the apparatus. The government of advanced and advancing industrial societies can maintain and secure itself only when it succeeds in mobilizing, organizing, and exploiting the technical, scientific, and mechanical productivity available to industrial civilization. And this productivity mobilizes society as a whole, above and beyond any particular individual or group interests. The brute fact that the machine's physical (only physical?) power surpasses that of the individual, and of any particular group of individuals, makes the machine the most effective political instrument in any society whose basic organization is that of the machine process. But the political trend may be reversed; essentially the power of the machine is only the stored-up and projected power of man. To the extent to which the work world is conceived of as a machine and mechanized accordingly, it becomes the *potential* basis of a new freedom for man.

Contemporary industrial civilization demonstrates that it has reached the stage at which "the free society" can no longer be adequately defined in the traditional terms of economic, political, and intellectual liberties, not because these liberties have become insignificant, but because they are too significant to be confined within the traditional forms. New modes of realization are needed, corresponding to the new capabilities of society.

Such new modes can be indicated only in negative terms because they would amount to the negation of the prevailing modes. Thus economic freedom would mean freedom *from* the economy—from being controlled by economic forces and relationships; freedom from the daily struggle for existence, from earning a living.

Political freedom would mean liberation of the individuals *from* politics over which they have no effective control. Similarly, intellectual freedom would mean the restoration of individual thought now absorbed by mass communication and indoctrination, abolition of "public opinion" together with its makers. The unrealistic sound of these propositions is indicative, not of their utopian character, but of the strength of the forces which prevent their realization. The most effective and enduring form of warfare against liberation is the implanting of material and intellectual needs that perpetuate obsolete forms of the struggle for existence.

The intensity, the satisfaction and even the character of human needs, beyond the biological level, have always been preconditioned. Whether or not the possibility of doing or leaving, enjoying or destroying, possessing or rejecting something is seized as a *need* depends on whether or not it can be seen as desirable and necessary for the prevailing societal institutions and interests. In this sense, human needs are historical needs and, to the extent to which the society demands the repressive development of the individual, his needs themselves and their claim for satisfaction are subject to overriding critical standards.

We may distinguish both true and false needs. "False" are those which are superimposed upon the individual by particular social interests in his repression: the needs which perpetuate toil, aggressiveness, misery, and injustice. Their satisfaction might be most gratifying to the individual, but this happiness is not a condition which has to be maintained and protected if it serves to arrest the development of the ability (his own and others) to recognize the disease of the whole and grasp the chances of curing the disease. The result then is euphoria in unhappiness. Most of the prevailing needs to relax, to have fun, to behave and consume in accordance with the advertisements, to love and hate what others love and hate, belong to this category of false needs.

Such needs have a societal content and function which are determined by external powers over which the individual has no control; the development and satisfaction of these needs is heteronomous. No matter how much such needs may have become the individual's own, reproduced and fortified by the conditions of his existence; no matter how much he identifies himself with them and finds himself in their satisfaction, they continue to be what they were from the beginning—products of a society whose dominant interest demands repression.

The prevalence of repressive needs is an accomplished fact, accepted in ignorance and defeat, but a fact that must be undone in the interest of the happy individual as well as all those whose misery is the price of his satisfaction. The only needs that have an unqualified claim for satisfaction are the vital ones—nourishment, clothing, lodging at the attainable level of culture. The satisfaction of these needs is the prerequisite for the realization of *all* needs, of the unsublimated as well as the sublimated ones.

For any consciousness and conscience, for any experience which does not accept the prevailing societal interest as the supreme law of thought and behavior, the established universe of needs and satisfactions is a fact to be questioned—questioned in terms of truth and falsehood. These terms are historical throughout, and their objectivity is historical. The judgment of needs and their satisfaction, under the given conditions, involves standards of *priority*—standards which refer to the optimal development of the individual, of all individuals, under the optimal utilization of the material and intellectual resources available to man. The resources are calculable. "Truth" and "falsehood" of needs designate objective conditions to the extent to which the universal satisfaction of vital needs and, beyond it, the progressive alleviation of toil and poverty, are universally valid standards. But as historical standards, they do not only vary according to area and stage of development, they also can be defined only in (greater or lesser) *contradiction* to the prevailing ones. What tribunal can possibly claim the authority of decision?

In the last analysis, the question of what are true and false needs must be answered by the individuals themselves, but only in the last analysis; that is, if and when they are free to give their own answer. As long as they are kept incapable of being autonomous, as long as they are indoctrinated and manipulated (down to their very instincts), their answer to this question cannot be taken as their own. By the same token, however, no tribunal can justly arrogate to itself the right to decide which needs should be developed and satisfied. Any such tribunal is reprehensible, although our revulsion does not do away with the question: how can the people who have been the object of effective and productive domination by themselves create the conditions of freedom?

The more rational, productive, technical, and total the repressive administration of society becomes, the more un-imaginable the means and ways by which the administered individuals might break their servitude and seize their own liberation. To be sure, to impose Reason upon an entire society is a paradoxical and scandalous idea—although one might dispute the righteousness of a society which ridicules this idea while making its own population into objects of total administration. All liberation depends on the consciousness of servitude, and the emergence of this consciousness is always hampered by the predominance of needs and satisfactions which, to a great extent, have become the individual's own. The process always replaces one system of preconditioning by another; the optimal goal is the replacement of false needs by true ones, the abandonment of repressive satisfaction.

The distinguishing feature of advanced industrial society is its effective suffocation of those needs which demand liberation—liberation also from that which is tolerable and rewarding and comfortable—while it sustains and absolves the destructive power and repressive function of the affluent society. Here, the social controls exact the overwhelming need for the production and consumption of waste; the need for stupefying work where it is no longer a real necessity; the need for modes of relaxation which soothe and prolong this stupefication; the need for maintaining such deceptive liberties as free competition at administered prices, a free press which censors itself, free choice between brands and gadgets.

Under the rule of a repressive whole, liberty can be made into a powerful instrument of domination. The range of choice open to the individual is not the decisive factor in determining the degree of human freedom, but *what* can be chosen and what *is* chosen by the individual. The criterion for free choice can never be an absolute one, but neither is it entirely relative. Free election of masters does not abolish the masters or the slaves. Free choice among a wide variety of goods and services does not signify freedom if these goods and services sustain social controls over a life of toil and fear—that is, if they sustain alienation. And the spontaneous reproduction of superimposed needs by the individual does not establish autonomy; it only testifies to the efficacy of the controls.

Our insistence on the depth and efficacy of these controls is open to the objection that we overrate greatly the indoctrinating power of the "media," and that by themselves the people would feel and satisfy the needs which are now imposed upon them. The objection misses the point. The preconditioning does not start with the mass production of radio and television and with the centralization of their control. The people enter this stage as preconditioned receptacles of long standing; the decisive difference is in the flattening out of the contrast (or conflict) between the given and the possible, between the satisfied and the unsatisfied needs. Here, the so-called equalization of class distinctions reveals its ideological function. If the worker and his boss enjoy the same television program and visit the same resort places, if the typist is as attractively made up as the daughter of her employer, if the Negro owns a Cadillac, if they all read the same newspaper, then this assimilation indicates not the disappearance of classes, but the extent to which the needs and satisfactions that serve the preservation of the Establishment are shared by the underlying population.

Indeed, in the most highly developed areas of contemporary society, the transplantation of social into individual needs is so effective that the difference between them seems to be purely theoretical. Can one really distinguish between the mass media as instruments of information and entertainment, and as agents of manipulation and indoctrination? Between the automobile as nuisance and as convenience? Between the horrors and the comforts of functional architecture? Between the work for national defense and the work for corporate gain? Between the private pleasure and the commercial and political utility involved in increasing the birth rate?

We are again confronted with one of the most vexing aspects of advanced industrial civilization: the rational character of its irrationality. Its productivity and efficiency, its capacity to increase and spread comforts, to turn waste into need, and destruction into construction, the extent to which this civilization transforms the object world into an extension of man's mind and body makes the very notion of alienation questionable. The people recognize themselves in their commodities; they find their soul in their automobile, hi-fi set, split-level home, kitchen equipment. The very mechanism which ties the individual to his society has changed, and social control is anchored in the new needs which it has produced.

The prevailing forms of social control are technological in a new sense. To be sure, the technical structure and efficacy of the productive and destructive apparatus has been a major instrumentality for subjecting the population to the established social division of labor throughout the modern period. Moreover, such integration has always been accompanied by more obvious forms of compulsion: loss of livelihood, the administration of justice, the police, the armed forces. It still is. But in the contemporary period, the technological controls appear to be the very embodiment of Reason for the benefit of all social groups and interests—to such an extent that all contradiction seems irrational and all counteraction impossible.

No wonder then that, in the most advanced areas of this civilization, the social controls have been introjected to the point where even individual protest is affected at its roots. The intellectual and emotional refusal "to go along" appears neurotic and impotent. This is the socio-psychological aspect of the political event that marks the contemporary period: the passing of the historical forces which, at the preceding stage of industrial society, seemed to represent the possibility of new forms of existence.

But the term "introjection" perhaps no longer describes the way in which the individual by himself reproduces and perpetuates the external controls exercised by this society. Introjection suggests a variety of relatively spontaneous processes by which a Self (Ego) transposes the "outer" into the "inner." Thus introjection implies the existence of an inner dimension distinguished from and even antagonistic to the external exigencies—an individual consciousness and an individual unconscious *apart from* public opinion and behavior.[i] The idea of "inner freedom" here has its reality: it designates the private space in which man may become and remain "himself."

Today this private space has been invaded and whittled down by technological reality. Mass production and mass distribution claim the *entire* individual, and industrial psychology has long since ceased to be confined to the factory. The manifold processes of introjection seem to be ossified in almost mechanical reactions. The result is, not adjustment but *mimesis:* an immediate identification of the individual with *his* society and, through it, with the society as a whole.

This immediate, automatic identification (which may have been characteristic of primitive forms of association) reappears in high industrial civilization; its new "immediacy," however, is the product of a sophisticated, scientific management and organization. In this process, the "inner" dimension of the mind in which opposition to the status quo can take root is whittled down. The loss of this dimension, in which the power of negative thinking—the critical power of Reason—is

at home, is the ideological counterpart to the very material process in which advanced industrial society silences and reconciles the opposition. The impact of progress turns Reason into submission to the facts of life, and to the dynamic capability of producing more and bigger facts of the same sort of life. The efficiency of the system blunts the individuals' recognition that it contains no facts which do not communicate the repressive power of the whole. If the individuals find themselves in the things which shape their life, they do so, not by giving, but by accepting the law of things—not the law of physics but the law of their society.

I have just suggested that the concept of alienation seems to become questionable when the individuals identify themselves with the existence which is imposed upon them and have in it their own development and satisfaction. This identification is not illusion but reality. However, the reality constitutes a more progressive stage of alienation. The latter has become entirely objective; the subject which is alienated is swallowed up by its alienated existence. There is only one dimension, and it is everywhere and in all forms. The achievements of progress defy ideological indictment as well as justification; before their tribunal, the "false consciousness" of their rationality becomes the true consciousness.

This absorption of ideology into reality does not, however, signify the "end of ideology." On the contrary, in a specific sense advanced industrial culture is *more* ideological than its predecessor, inasmuch as today the ideology is in the process of production itself. In a provocative form, this proposition reveals the political aspects of the prevailing technological rationality. The productive apparatus and the goods and services which it produces "sell" or impose the social system as a whole. The means of mass transportation and communication, the commodities of lodging, food, and clothing, the irresistible output of the entertainment and information industry carry with them prescribed attitudes and habits, certain intellectual and emotional reactions which bind the consumers more or less pleasantly to the producers and, through the latter, to the whole. The products indoctrinate and manipulate; they promote a false consciousness which is immune against its falsehood. And as these beneficial products become available to more individuals in more social classes, the

---

[i]The change in the function of the family here plays a decisive role: its "socializing" functions are increasingly taken over by outside groups and media.

indoctrination they carry ceases to be publicity; it becomes a way of life. It is a good way of life—much better than before—and as a good way of life, it militates against qualitative change. Thus emerges a pattern of *one-dimensional thought and behavior* in which ideas, aspirations, and objectives that, by their content, transcend the established universe of discourse and action are either repelled or reduced to terms of this universe. They are redefined by the rationality of the given system and of its quantitative extension. . . .

One-dimensional thought is systematically promoted by the makers of politics and their purveyors of mass information. Their universe of discourse is populated by self-validating hypotheses which, incessantly and monopolistically repeated, become hypnotic definitions or dictations. For example, "free" are the institutions which operate (and are operated on) in the countries of the Free World; other transcending modes of freedom are by definition either anarchism, communism, or propaganda. "Socialistic" are all encroachments on private enterprises not undertaken by private enterprise itself (or by government contracts), such as universal and comprehensive health insurance, or the protection of nature from all too sweeping commercialization, or the establishment of public services which may hurt private profit. This totalitarian logic of accomplished facts has its Eastern counterpart. There, freedom is the way of life instituted by a communist regime, and all other transcending modes of freedom are either capitalistic, or revisionist, or leftist sectarianism. In both camps, non-operational ideas are non-behavioral and subversive. The movement of thought is stopped at barriers which appear as the limits of Reason itself. . . .

However, [earlier] accommodating concepts of Reason were always contradicted by the evident misery and injustice of the "great public bodies" and the effective, more or less conscious rebellion against them. Societal conditions existed which provoked and permitted real dissociation from the established state of affairs; a private as well as political dimension was present in which dissociation could develop into effective opposition, testing its strength and the validity of its objectives.

With the gradual closing of this dimension by the society, the self-limitation of thought assumes a larger significance. The interrelation between scientific-philosophical and societal processes, between theoretical and practical Reason, asserts itself "behind the back" of the scientists and philosophers. The society bars a whole type of oppositional operations and behavior; consequently, the concepts pertaining to them are rendered illusory or meaningless. Historical transcendence appears as metaphysical transcendence, not acceptable to science and scientific thought. The operational and behavioral point of view, practiced as a "habit of thought" at large, becomes the view of the established universe of discourse and action, needs and aspirations. The "cunning of Reason" works, as it so often did, in the interest of the powers that be. The insistence on operational and behavioral concepts turns against the efforts to free thought and behavior *from* the given reality and *for* the suppressed alternatives. Theoretical and practical Reason, academic and social behaviorism meet on common ground: that of an advanced society which makes scientific and technical progress into an instrument of domination.

"Progress" is not a neutral term; it moves toward specific ends, and these ends are defined by the possibilities of ameliorating the human condition. Advanced industrial society is approaching the stage where continued progress would demand the radical subversion of the prevailing direction and organization of progress. This stage would be reached when material production (including the necessary services) becomes automated to the extent that all vital needs can be satisfied while necessary labor time is reduced to marginal time. From this point on, technical progress would transcend the realm of necessity, where it served as the instrument of domination and exploitation which thereby limited its rationality; technology would become subject to the free play of faculties in the struggle for the pacification of nature and of society.

Such a state is envisioned in Marx's notion of the "abolition of labor." The term "pacification of existence" seems better suited to designate the historical alternative of a world which—through an international conflict which transforms and suspends the contradictions within the established societies—advances on the brink of a global war. "Pacification of existence" means the development of man's struggle with man and with nature, under conditions where the competing needs, desires, and aspirations are no longer organized by vested interests in domination and scarcity—an organization which perpetuates the destructive forms of this struggle.

Today's fight against this historical alternative finds a firm mass basis in the underlying population, and finds its ideology in the rigid orientation of thought and behavior to the given universe of facts. Validated by the accomplishments of science and technology, justified by its growing productivity, the status quo defies all transcendence. Faced with the possibility of pacification on the grounds of its technical and intellectual achievements, the mature industrial society closes itself against this alternative. Operationalism, in theory and practice, becomes the theory and practice of *containment*. Underneath its obvious dynamics, this society is a thoroughly static system of life: self-propelling in its oppressive productivity and in its beneficial coordination. Containment of technical progress goes hand in hand with its growth in the established direction. In spite of the political fetters imposed by the status quo, the more technology appears capable of creating the conditions for pacification, the more are the minds and bodies of man organized against this alternative.

The most advanced areas of industrial society exhibit throughout these two features: a trend toward consummation of technological rationality, and intensive efforts to contain this trend within the established institutions. Here is the internal contradiction of this civilization: the irrational element in its rationality. It is the token of its achievements. The industrial society which makes technology and science its own is organized for the ever-more-effective domination of man and nature, for the ever-more-effective utilization of its resources. It becomes irrational when the success of these efforts opens new dimensions of human realization. Organization for peace is different from organization for war; the institutions which served the struggle for existence cannot serve the pacification of existence. Life as an end is qualitatively different from life as a means.

Such a qualitatively new mode of existence can never be envisaged as the mere by-product of economic and political changes, as the more or less spontaneous effect of the new institutions which constitute the necessary prerequisite. Qualitative change also involves a change in the *technical* basis on which this society rests—one which sustains the economic and political institutions through which the "second nature" of man as an aggressive object of administration is stabilized. The techniques of industrialization are political techniques; as such, they prejudge the possibilities of Reason and Freedom.

To be sure, labor must precede the reduction of labor, and industrialization must precede the development of human needs and satisfactions. But as all freedom depends on the conquest of alien necessity, the realization of freedom depends on the *techniques* of this conquest. The highest productivity of labor can be used for the perpetuation of labor, and the most efficient industrialization can serve the restriction and manipulation of needs.

When this point is reached, domination—in the guise of affluence and liberty—extends to all spheres of private and public existence, integrates all authentic opposition, absorbs all alternatives. Technological rationality reveals its political character as it becomes the great vehicle of better domination, creating a truly totalitarian universe in which society and nature, mind and body are kept in a state of permanent mobilization for the defense of this universe. . . .

THE CLOSING OF THE POLITICAL UNIVERSE

The new technological work-world thus enforces a weakening of the negative position of the working class: the latter no longer appears to be the living contradiction to the established society. This trend is strengthened by the effect of the technological organization of production on the other side of the fence: on management and direction. Domination is transfigured into administration. [ii] The capitalist bosses and owners are losing their identity as responsible agents; they are assuming the function of bureaucrats in a corporate machine. Within the vast hierarchy of executive and managerial boards extending far beyond the individual establishment into the scientific laboratory and research institute, the national government and national purpose, the tangible source of exploitation disappears behind the façade of objective rationality. Hatred and frustration are deprived of their specific target, and the technological veil conceals the reproduction of inequality and enslavement. With technical progress as its instrument, unfreedom—in the sense of man's subjection to his productive apparatus—is perpetuated and intensified in the form of many liberties and comforts. The novel feature is the overwhelming rationality in this irrational enterprise, and the depth of the preconditioning which shapes the instinctual drives and aspirations of the individuals and obscures the difference between false and true consciousness. For in reality, neither the utilization of administrative rather than physical controls (hunger, personal dependence, force), nor the change in the character of heavy work, nor the assimilation of occupational classes, nor the equalization in the sphere of consumption compensate

---

[ii] Is it still necessary to denounce the ideology of the "managerial revolution"? Capitalist production proceeds through the investment of private capital for the private extraction and appropriation of surplus value, and capital is a social instrument for the domination of man by man. The essential features of this process are in no way altered by the spread of stock-holdings, the separation of ownership from management, etc.

for the fact that the decisions over life and death, over personal and national security are made at places over which the individuals have no control. The slaves of developed industrial civilization are sublimated slaves, but they are slaves, for slavery is determined "neither by obedience nor by hardness of labor but by the status of being a mere instrument, and the reduction of man to the state of a thing."[iii] This is the pure form of servitude: to exist as an instrument, as a thing. And this mode of existence is not abrogated if the thing is animated and chooses its material and intellectual food, if it does not feel its being-a-thing, if it is a pretty, clean, mobile thing. Conversely, as reification tends to become totalitarian by virtue of its technological form, the organizers and administrators themselves become increasingly dependent on the machinery which they organize and administer. And this mutual dependence is no longer the dialectical relationship between Master and Servant, which has been broken in the struggle for mutual recognition, but rather a vicious circle which encloses both the Master and the Servant. Do the technicians rule, or is their rule that of the others, who rely on the technicians as their planners and executors?

. . . the pressures of today's highly technological arms race have taken the initiative and the power to make the crucial decisions out of the hands of responsible government officials and placed it in the hands of technicians, planners and scientists employed by vast industrial empires and charged with responsibility for their employers' interests. It is their job to dream up new weapons systems and persuade the military that the future of their military profession, as well as the country, depends upon buying what they have dreamed up.[iv]

As the productive establishments rely on the military for self-preservation and growth, so the military relies on the corporations "not only for their weapons, but also for knowledge of what kind of weapons they need, how much they will cost, and how long it will take to get them."[v] A vicious circle seems indeed the proper image of a society which is self-expanding and self- perpetuating in its own preestablished direction—driven by the growing needs which it generates and, at the same time, contains.

[iii]François Perroux, *La Coexistence pacifique* (Paris: Presses Universitaires, 1958), Vol. III, p. 600.

[iv]Stewart Meacham, *Labor and the Cold War* (American Friends Service Committee, Philadelphia 1959), p. 9.

[v]Ibid.

Discussion Questions

1. Given the critical theorists' reservations regarding the scientific pursuit of knowledge, how might a sociologist conduct "valid" research? If the scientific method is rejected, what criteria can be used to assess a researcher's findings? More generally, if not science, what might serve as a basis for accepting anyone's claim to speak the "truth"?

2. Do you think that individuals living in advanced, capitalist societies are as "pacified" as the critical theorists argue? Has individuality been "liquidated" as the critical theorists suggest? What evidence can you point to in support of your view?

3. While the critical theorists presented here deride mass culture and technology for corrupting our ability to reason, what role, if any, might education play in promoting a "totally administered society"? Second, as a form of technology, do you consider the Internet a potential source of liberation or of domination? Why?

4. According to the critical theorists, how, in modern industrial societies, does rationality lead to the oppression or alienation of the individual? Do you agree that the United States is a "totalitarian" society? Why or why not?

5. Many musical groups express in their songs a discontent with existing social conditions and a mistrust of those in positions of authority. What effect, if any, does such music have on the broader society? How might such "protest" songs paradoxically reinforce the very social order they aim to criticize?

11 EXCHANGE AND RATIONAL CHOICE THEORIES

George C. Homans

Key Concepts

- ▪▪ Elementary Social Behavior
- ▪▪ Costs
- ▪▪ Distributive Justice

Peter M. Blau

Key Concepts

- ▪▪ Intrinsic/Extrinsic Rewards
- ▪▪ Power
- ▪▪ Imbalanced Exchange

James S. Coleman

Key Concepts

- ▪▪ Trust
- ▪▪ Norms
  - ❑ Conjoint/Disjoint
- ▪▪ Free Rider
- ▪▪ Social Capital

> *A stable social relationship requires that individuals make some investments to bring it into being and maintain it in existence, and it is to the advantage of each party to have the other or others assume a disproportionate share of the commitments that secure their continuing association. Hence the common interest of individuals in sustaining a relation between them tends to be accompanied by conflicting interests as to whose investment should contribute most to its sustenance.*
>
> —Peter M. Blau (1964:113)

What are the factors that motivate your choice of friends? What is it about you that induces others to strike up an acquaintanceship? Why do relationships—personal or professional—begin and end when they do? To these questions and countless others like them, exchange theorists offer one fundamental answer: social behavior is guided by the *rational* calculation of an *exchange* of rewards and costs. Much like our economic calculations through which we determine whether the value (reward) of a good is worth its price (cost), so in our social interactions we decide to enter into or terminate a relationship (exchange) with another.

For example, why would a person join and remain in a peer group when he is the constant butt of jokes or pressured into conducting embarrassing or perhaps even dangerous acts? Certainly, any familiarity with college fraternities and sororities or even gangs would introduce us to such a person. The explanation is straightforward: The rewards reaped for group membership (perhaps status or companionship) outweigh the costs associated with public humiliation or personal risk. When the value or importance of the rewards diminishes below the costs incurred, the person will exit the group.

In what follows, we outline the work of two sociologists, George C. Homans and Peter M. Blau, who arguably have made the most significant contributions to exchange theory. In doing so, we highlight their points of theoretical convergence and departure—points that are only hinted at in the simplified example drawn above. Equally important, outlining the central elements of their respective contributions also will provide a canvas for sketching the underlying theoretical and methodological assumptions that make exchange theory a distinct approach to understanding social life. For instance, both Homans and Blau (and exchange theorists more generally) portray individuals as strategic actors who use the resources they have at their disposal in an effort to optimize their rewards. Thus, they contend that individuals are motivated to act not on the basis of tradition, unconscious drives, or some type of structural imperative, but rather on the basis of rational considerations: namely, weighing the consequences of alternative lines of conduct in terms of the "profit" they will likely generate.

Despite such fundamental similarities, however, there are key differences in these theorists' work. Homans draws principally from behavioral psychology and neoclassical economics in developing his version of exchange theory. Blau's exchange theory, while sympathetic with economics, evinces a greater indebtedness to the German sociologist Georg Simmel. Consequently, each theorist provides a different pathway—and a different set of answers—to a related set of concerns.

Following our discussion of exchange theory we turn to the work of James S. Coleman, the leading figure in the development of rational choice theory. As you will read, rational choice theory shares with exchange theory many of the same assumptions regarding individual action. Yet, while exchange theory is focused on the individualistic dynamics of decision making and group behavior, rational choice theory introduces collectivist features into the analysis of social life. For rational choice theorists such as Coleman, it is essential to examine not only how strategic actors produce their social environment through interaction, but also how features of that created environment—such as social networks, norms, trust, and access to resources—in turn shape individuals' behavior.

▪▪ GEORGE C. HOMANS (1910–1989): A BIOGRAPHICAL SKETCH

George C. Homans was born in Boston, Massachusetts, in 1910, to one of the city's elite families. He attended Harvard College where, in 1932, he earned a degree in English literature. He was prepared to join the staff of a newspaper after graduating, but his career in journalism ended before it started. Like so many other jobs at the time, Homans's position with the newspaper vanished during the Great Depression. It was through this accident of circumstances that Homans would find his introduction to sociology. Remarking that he had nothing better to do, Homans attended a Harvard seminar on the work of the Italian sociologist Vilfredo Pareto (1848–1923). With another attendee at the seminar, Charles Curtis, he wrote *An Introduction to Pareto,* which in large measure helped earn him admission into the Harvard Society of Fellows in 1934. From that point forward, Homans's academic career would be spent entirely at Harvard, his alma mater. Yet, Homans did not commit himself full-time to the sociology department until 1946 and would wait until 1953 to be appointed full professor—nearly twenty years after being elected to the Society of Fellows. Despite the accidental nature of his introduction into the field and the fact that he never earned a degree in sociology, Homans's work would make an important mark in the discipline. Certainly, his colleagues recognized the significance of his contributions when they elected him president of the American Sociological Association in 1964. During the following twenty-five years, Homans would publish a number of articles and books, including *The Nature of Social Sciences* (1967). The importance of his impact on the discipline was again affirmed when he was awarded the Association's Distinguished Scholarship Award in 1988. Homans died the following year.

HOMANS'S INTELLECTUAL INFLUENCES AND CORE IDEAS ▪▪

Homans developed his exchange theory with the ambition of creating a general framework for understanding individuals' behaviors in face-to-face dealings. As we noted at the beginning of this chapter, his approach is based on a fusion of the principles of behavioral psychology with those of neoclassical economics. Here, we briefly outline the main elements of these two schools and their relevance to Homans's approach.

B. F. Skinner and Behavioral Psychology

In incorporating the insights of behavioral psychology, Homans turned to the work of his friend and colleague at Harvard, B. F. Skinner (1904– 1990). Skinner conducted experiments with pigeons in order to explore the effects of operant conditioning. This conditioning involves a process in which a particular behavior (operant) is reinforced through the use of rewards or punishments that, in turn, condition the likelihood that the behavior will be repeated or avoided. It is in this way, through the provision and removal of rewards and punishments, that all noninstinctual behaviors are learned. In one of the more remarkable examples of the techniques of operant conditioning and their effectiveness, Skinner trained his laboratory pigeons to play table tennis by reinforcing (rewarding) specific behaviors with food.

Skinner, however, was less interested in understanding how pigeons learn reinforced behaviors than he was in discovering the factors that account for changes in the frequency with which learned behaviors are exhibited. This led him to consider two conditions: the state of the pigeon (i.e., whether or not it is hungry) and the rate at which the behavior in question is reinforced. With regard to the former, Skinner noted that the longer a pigeon has gone without food and, thus, the hungrier it is, the more often it will perform a behavior that has been previously reinforced with food. Conversely, the more the pigeon's hunger has been satiated, the less often it will perform the behavior. As for the latter, the basic proposition suggests that the less frequently a behavior is reinforced, the less frequently it will be exhibited. Conversely, the more frequent (and thus reliable) a reinforcement, the more often the conditioned behavior will be exhibited (Homans 1961:19–21).

A final consideration involves the role of punishments in determining the probability that a particular action will be performed. Punishments also take on two forms. First, in laboratory settings, pigeons can be subjected to "aversive conditions" such as an electric shock. In such a setting, a given behavior will be reinforced if it enables the bird to escape from or avoid the punishment. Second, punishments also can come in the form of a withdrawal of positive reinforcements, such as the removal of food if a specified behavior is performed. Parents often resort to both forms of punishment when trying to correct (that is, eliminate) their children's disobedient behavior. Who has not heard, "You're grounded!" or "No dessert for you!"?

Photo 11.1   "Time Out": A favored form of punishment by today's parents.

Skinner's work in behavioral psychology also highlighted a special class of punishments that would become central in Homans's theory on social exchange: **costs**. Costs are unavoidable punishments that are experienced when a behavior also elicits positive reinforcement. Costs have the effect of reducing the frequency of an otherwise rewarding behavior. Costs accompanying a given behavior also have the potential effect of increasing the frequency of an alternative behavior. Similarly, as the costs associated with a given behavior increase, the rewards reaped from pursuing an alternative behavior rise in kind. As a result, the costs entailed in performing a specific act involve the unavoidable punishment of losing or forgoing the rewards that would have been reaped had an alternative behavior been performed (Homans 1961:24–26).

As we noted previously, the preceding discussion of operant conditioning and its attendant concepts such as rewards, punishments, and costs is derived from Skinner's experimentation with pigeons. Thus, Homans had to first acknowledge some basic assumptions when transposing Skinner's work into his own theory, namely, that propositions regarding individual behavior could be adapted to social situations and, perhaps more controversial, that propositions on the behavior of pigeons could be generalized to the behavior of humans (ibid.:31). Taking these two assumptions into account, Homans sought to explain **elementary social behavior**, which he defined as "face-to-face contact between individuals, in which the reward [or punishment] each gets from the behavior of the others is relatively direct and immediate"

(ibid.:7). While individual behavior (of pigeons or humans) involves an "exchange" between an animal and its physical environment, social behavior is based on the type of exchange "where the activity of each of at least two animals reinforces (or punishes) the activity of the other, and where accordingly each influences the other" (ibid.:30). However, it is noteworthy that Homans maintained that the laws of individual behavior developed by Skinner are identical to the laws of social behavior once the complications of mutual reinforcement are considered. Thus, there is nothing unique to social behavior that warrants the introduction of new propositions different from those that explain individual or psychological behavior. Not surprisingly, Homans's position on this issue was (and continues to be) met with much resistance on the part of many sociologists, for, in effect, he was giving over the discipline to psychology.

So what, specifically, are the psychological propositions that explain and predict social behavior? Homans offered five that, when taken together, are said to demonstrate how individuals profit from their actions. In other words, individuals design their conduct such that the value of the rewards gained in an exchange are greater than the costs incurred in forgoing the rewards associated with an alternative line of behavior. We summarize the propositions in Table 11.1.

To illustrate the operation of these propositions, consider the following example adapted from Homans's (1961) work: Andrew is new to his job and needs help in order to complete his assigned tasks. Thus, he values help and finds it rewarding. Andrew turns to Emma, who is more experienced at the job and is able to take some amount of time from her own work in order to assist Andrew. Andrew rewards Emma for her efforts with thanks and expressions of approval, which Emma, we assume, finds worth receiving. In this scenario, help and approval are exchanged.

Considering proposition 1 ("Stimulus"), we can first predict that Emma is more likely to provide Andrew with help if the present situation (i.e., stimulus) resembles previous situations in which her willingness to provide assistance has been rewarded. If in similar past experiences Emma has been "burned"—incurred the costs of taking time from her own work without receiving a sufficient reward for her time and expertise—then she would be less likely to risk her services.

Table 11.1   Homans's Behaviorist Propositions

The Stimulus Proposition	If the previous occurrence of a particular stimulus has been the occasion on which an individual's action has been rewarded, then the more similar the current stimulus is to the past one, the more likely the person is to repeat the action.
The Success Proposition	The more often an action is followed by a reward, the more likely a person will repeat the behavior.
The Value Proposition	The more valuable a particular reward is to a person, the more often he will perform a behavior so rewarded.
The Deprivation-Satiation Proposition	The more often in the recent past an individual has received a particular reward, the less valuable any further unit of that reward becomes (and following the value proposition, the less likely the person is to perform the behavior for which he was so rewarded).
The Frustration-Aggression Proposition	If a person's action receives a punishment he did not expect, or if the person does not receive the reward he did expect, he will become angry and more likely to exhibit aggressive behavior, the results of which will become more valuable to him.

SOURCE: Adapted from Homans (1961), *Social Behavior: Its Elementary Forms,* Harcourt, Brace, and World, Inc., pp. 53–55, 75, and Homans (1987), "Behaviorism and After," in *Social Theory Today,* edited by Anthony Giddens and Jonathan Turner, Stanford University Press, 1987, pp. 58–81.

Taking together propositions 2 and 3 ("Success" and "Value"), we can predict that the more Andrew needs help and finds Emma's assistance valuable, the more he will ask for it, and the more approval he will offer in return. Likewise, the more Emma needs or values approval, the more often she will provide assistance to Andrew. As a result, the frequency of interaction or exchange between the two depends on the frequency with which each rewards the actions of the other and on the value of the reward each receives (1961:55). Conversely, should either of the two begin to offer less-valuable rewards (less-than-adequate help or a lackluster show of approval and thanks), then the frequency of their exchange will decrease and eventually end.

However, proposition 4 ("Deprivation–Satiation") argues that even if Andrew and Emma continue to exchange acceptable rewards, the value of such rewards will begin to decline. For just like the pigeon that stops pecking a lever once its hunger is satiated, at some point Andrew will find that he is no longer in need of help, while Emma will have had her fill of "thank-yous." As a result, following proposition 3 ("Value"), the frequency of their exchange of approval and help will decrease. In short, as the profit of their activities begins to fall, each is more likely to pursue an alternative course of action, even if this simply means not asking for or supplying help.

While the Value Proposition tells us that both Andrew and Emma must have found their exchange more profitable than alternative courses of action (or else they would not have interacted), it does not provide the whole story. Asserting that individuals will enter into and maintain exchange relations as long as they are profitable says little about how the price of rewards and costs is established. In rounding out the account, proposition 5 ("Frustration–Aggression") yields some interesting conclusions regarding evaluations of costs and rewards and, thus, the amount of profit an individual can justifiably expect from an exchange.

To return to our example, when Andrew asks Emma for help, he pays a cost; for instance, he risks being judged incompetent or inferior in status. Accordingly, he forgoes the reward of appearing skilled had he pursued an alternative course and not asked for help. For her part, Emma incurs a cost in the exchange in the form of forgoing whatever rewards she may have reaped had she not spent her time helping Andrew. Perhaps she would have spent more time on her own tasks or used the time to schmooze her boss. The question each is faced with is, "What amount of cost should I expect to incur given the amount of reward I expect to receive?" Striking a fair balance between the rewards and costs each experiences is a problem of **distributive justice**: Each party to the exchange must perceive that he or she is not paying too high a cost relative to the rewards gained. In other words, both Andrew and Emma must be convinced that the profit each derives from the exchange is just.

From where does one's sense of distributive justice arise? According to Homans, it is the history of an individual's past experiences that determines his or her present expectations regarding rewards and costs. Thus, "the more often in the past an activity emitted under particular stimulus-conditions has been rewarded, the more anger [an individual] will display at present when the same activity, emitted under similar conditions, goes without its reward: precedents are always turning into rights" (1961:73). If Emma is interrupted every twenty minutes with a request for help, she is likely to feel that she is being taken advantage of—that with merely a "thank you" in return, the costs are too high to continue the exchange, particularly if similar circumstances in her past required only the occasional interruption. In such a scenario, Emma is likely to feel angry that Andrew is seeking more of a reward than he has the right to expect.

For his part, Andrew is apt to consider whether the help he receives as reward is proportional to the costs that his submission entails. For instance, he may ask himself, "Is Emma expecting me to beg or grovel in order to get her assistance?" If so, then Andrew will gauge Emma's demands (i.e., her price) as unjust: The costs he is being expected to pay far exceed the value of the rewards Emma is able to supply. With exploitation comes the feeling of anger.

In all exchanges, each party evaluates whether he himself profited fairly as well as whether the other's profit was just. In determining if the rewards gained were proportional (i.e., expected) to the costs incurred, a general, although often unstated, rule is applied: "The more valuable to the other (and costly to himself) an activity he gives the other, the more valuable to him (and costly to the other) an activity the other gives him" (ibid.:75). When a balance of expected profits is achieved, distributive justice exists. When either party judges profits to be unjust, then anger (or feelings of guilt if a person thinks he unfairly benefited from the exchange) will follow.

Classical Economics

While Homans draws explicitly on the fundamental principles of Skinner's behavioral psychology, his reliance on classical economics is more implicit. Homans does not turn to the propositions developed by any particular economist. Instead, he asserts that, just as individuals exchange material goods for money in a market, they exchange, for instance, services or activities for social approval in a social market composed of "sellers" and "buyers." Like behavioral psychology, classical economics depicts individual action as being motivated by the pursuit of profit. In attempting to realize our goals, we enter a marketplace of others seeking to maximize our gains and minimize our costs, although in this case our profits are not measured by money.

Take, for instance, the economic Law of Supply states, in part, that the higher the price of a good, the more of it a supplier will seek to sell. This is akin to Homans's value proposition that states that the more valuable a particular reward is to a person, the more often he will perform a behavior so rewarded, and the less often he will perform an alternative activity. Similarly, consider the Law of Demand, which in turn states that the higher the price of a good, the less of it a consumer will purchase. This law is equivalent to the behaviorist principle that the higher the cost incurred by an activity, the less often an individual will perform it, and the more often he will engage in an alternative activity (1961:69).

Although Homans did not incorporate economic analyses into his theory with the same pronouncedness as the principles of behavioral psychology, he was convinced, nevertheless, that economics could shed important light on social behavior. Indeed, Homans felt that his brand of exchange theory had the distinct advantage of "bring[ing] sociology closer to economics—that science of man most advanced, most capable of application, and, intellectually, most isolated" (Homans 1958:598).

Homans's approach to social life is avowedly individualistic and rationalist (see Figure 11.1). Holding himself to be "an ultimate psychological reductionist" (Homans 1958:597), he contends that all social behavior, including that which occurs in large-scale organizations and collectivities, is best explained on the basis of psychological principles. Such an argument is not without controversy, for it runs counter to the very raison d'être of sociology. Homans was quite aware of the critical nature of his theoretical position. Commenting on the state of theory, he remarked, "Much modern sociological theory seems to me to possess every virtue except that of explaining anything" (Homans 1961:10). This salvo was directed in particular toward the dominant sociological theorist of the period, his Harvard colleague Talcott Parsons (see Chapter 9). While Parsons was building abstract conceptual schemes of social systems, Homans was denying that social systems exist, at least not as sui generis social facts independent of the individuals that they comprise. Thus, Homans was not only dismissing structural functionalist theory at the height of its influence; he also was contesting the central ideas of one of the founding figures of the discipline, Émile Durkheim (Homans 1984:297–98, 323–25). On the relationship between individual action and social structures, Homans summed up his position as follows:

> Once [social] structures have been created, they have further effects on the behavior of persons who take part in them or come into contact with them. But these further effects are explained by the same [psychological] propositions as those used to explain the creation and maintenance of the structure in the first place. The structures only provide new given conditions to which the propositions are to be applied. My sociology remains fundamentally individualistic and not collectivistic. (Quoted in Ritzer and Goodman 2004:403)

Figure 11.1   Homans's Basic Concepts and Theoretical Orientation

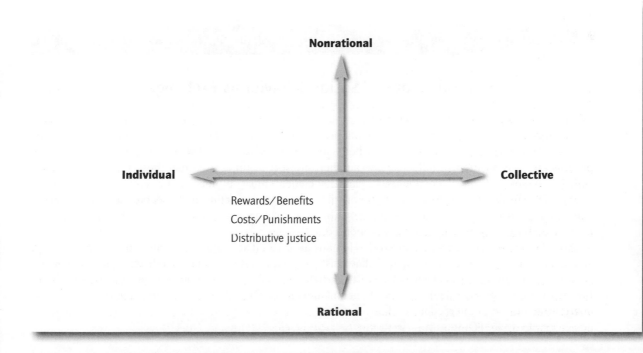

For Homans, then, what accounts for the enduring patterns of behavior and social structures that make up society is basic behavioral psychology, "for structures do not act on individuals automatically. They do so because they establish some of the contingencies under which persons act: their stimuli, rewards and punishments" (Homans 1984:342).

This leads us to Homans's approach to action. If social order is the product of individuals interacting with one another, on what basis is interaction itself motivated? As we noted above, individuals act with the same fundamental goal in mind: maximizing rewards and minimizing costs. In stressing the calculating, strategic aspect of interaction, Homans clearly aligns himself with a rationalist orientation to the problem of action. Yet, Homans acknowledges two limitations on his rationalist position. The first involves the role of values in the decision-making process. Homans assumes as a given the particular values that any individual may seek to maximize. His interest lies in explaining why people *behave* as they do in realizing their values, and not why they value the things that they do. As he puts it, "Our problem is not why they hold [particular] values, but given that they do hold them, what they do about them" (Homans 1961:48). However, this seeming simplification of the problem poses its own puzzles. While it may appear obvious to assume some values—a person new to a job will value the help coworkers can provide—others can confound an observer's expectations. It may well be the case that an individual values his sense of pride more than the help others can provide. Thus, while he is new to a job and in need of help, the cost in pride for asking outweighs the rewards for receiving it. Indeed, values that are their own reward, such as pride and altruism, greatly complicate the ability to predict and explain an individual's behavior (ibid.:45).

The second limitation involves the "bounded" nature of rationality attributed to actors (March and Simon 1958). The rational calculations that guide an individual's behaviors are by no means flawless. Thus, the rewards obtained are, more often than not, less than optimal. The ability to fully maximize the profit on one's values is limited by a number of conditions, including the uniqueness of one's past experiences, the dependence on others for rewards, and the particularities of a given exchange situation. In short, there are boundaries to individuals' decision making such that "All we impute to them in the way of rationality is that they know enough to come in out of the rain unless they enjoy getting wet" (Homans 1961:82).

Reading

Introduction to "Social Behavior as Exchange"

In this article, Homans introduces many of the central ideas discussed above. His individualistic orientation to understanding the basis of social order is made clear in the article's first paragraph. In it, he asserts that in studying elementary social behavior we are studying "what happens when two or three persons are in a position to influence one another, the sort of thing of which those massive structures called 'classes,' 'firms,' 'communities,' and 'societies' must ultimately be composed" (Homans 1958:597). Similarly, he expresses his rationalist orientation to action and indebtedness to Skinner's behavioral psychology when describing interaction as an exchange of goods reinforced by the values that actors attach to rewards sought and costs endured.

After laying the theoretical groundwork, Homans turns to a discussion of a number of experiments and studies that shed empirical light on his behaviorist propositions. Given our own discussion of Homans's work, we will not review here the details of the article. Instead, we leave you to determine the validity of his argument and of the role he assigns to such matters as profit seeking, behavioral reinforcements, and distributive justice in interaction.

"Social Behavior as Exchange" (1958)

George C. Homans

THE PROBLEMS OF SMALL-GROUP RESEARCH

What we are really studying in small groups is elementary social behavior: what happens when two or three persons are in a position to influence one another, the sort of thing of which those massive structures called "classes," "firms," "communities," and "societies" must ultimately be composed. . . .

[Analysis of elementary social behavior] would be furthered by our adopting the view that interaction between persons is an exchange of goods, material and non-material. This is one of the oldest theories of social behavior, and one that we still use every day to interpret our own behavior, as when we say, "I found so-and-so rewarding"; or "I got a great deal out of him"; or, even, "Talking with him took a great deal out of me." But, perhaps just because it is so obvious, this view has been much neglected by social scientists. . . .

An incidental advantage of an exchange theory is that it might bring sociology closer to economics—that science of man most advanced, most capable of application, and, intellectually, most isolated. Economics studies exchange carried out under special circumstances and with a most useful built-in numerical measure of value. What are the laws of the general phenomenon of which economic behavior is one class?

In what follows I shall suggest some reasons for the usefulness of a theory of social behavior as exchange and suggest the nature of the propositions such a theory might contain.

AN EXCHANGE PARADIGM

I start with the link to behavioral psychology and the kind of statement it makes about the behavior of an experimental animal such as the pigeon.[i] As a pigeon explores its cage in the laboratory, it happens to peck a target, whereupon the psychologist feeds it corn. The evidence is that it will peck the target again; it has learned the behavior, or, as my friend Skinner says, the behavior has been reinforced, and the pigeon has undergone *operant conditioning*. This kind of psychologist is not interested in how the behavior was learned: "learning theory" is a poor name for his field. Instead, he is interested in what determines changes in the rate of emission of learned behavior, whether pecks at a target or something else.

The more hungry the pigeon, the less corn or other food it has gotten in the recent past, the more often it will peck. By the same token, if the behavior is often reinforced, if the pigeon is given much corn every time it pecks, the rate of emission will fall off as the pigeon gets *satiated*. If, on the other hand, the behavior is not reinforced at all, then, too, its rate of emission will tend to fall off, though a long time may pass before it stops altogether, before it is *extinguished*. In the emission of many kinds of behavior the pigeon incurs *aversive stimulation,* or what I shall call "cost" for short, and this, too, will lead in time to a decrease in the emission rate. Fatigue is an example of a "cost." Extinction, satiation, and cost, by decreasing the rate of emission of a particular kind of behavior, render more probable the emission of some other kind of behavior, including doing nothing. I shall only add that even a hard-boiled psychologist puts "emotional" behavior, as well as such things as pecking, among the unconditioned responses that may be reinforced in operant conditioning. As a statement of the propositions of behavioral psychology, the foregoing is, of course, inadequate for any purpose except my present one.

We may look on the pigeon as engaged in an exchange—pecks for corn—with the psychologist, but let us not dwell upon that, for the behavior of the pigeon

---

SOURCE: "Social Behavior as Exchange" by George C. Homans from *The American Journal of Sociology, 63*:6, pp. 597–606.

[i]B. F. Skinner, *Science and Human Behavior* (New York: Macmillan Co., 1953).

hardly determines the behavior of the psychologist at all. Let us turn to a situation where the exchange is real, that is, where the determination is mutual. Suppose we are dealing with two men. Each is emitting behavior reinforced to some degree by the behavior of the other. How it was in the past that each learned the behavior he emits and how he learned to find the other's behavior reinforcing we are not concerned with. It is enough that each does find the other's behavior reinforcing, and I shall call the reinforcers—the equivalent of the pigeon's corn—*values,* for this, I think, is what we mean by this term. As he emits behavior, each man may incur costs, and each man has more than one course of behavior open to him.

This seems to me the paradigm of elementary social behavior, and the problem of the elementary sociologist is to state propositions relating the variations in the values and costs of each man to his frequency distribution of behavior among alternatives, where the values (in the mathematical sense) taken by these variable for one man determine in part their values for the other.

I see no reason to believe that the propositions of behavioral psychology do not apply to this situation, though the complexity of their implications in the concrete case may be great indeed. In particular, we must suppose that, with men as with pigeons, an increase in extinction, satiation, or aversive stimulation of any one kind of behavior will increase the probability of emission of some other kind. The problem is not, as it is often stated, merely, what a man's values are, what he has learned in the past to find reinforcing, but how much of any one value his behavior is getting him now. The more he gets, the less valuable any further unit of that value is to him, and the less often he will emit behavior reinforced by it.

THE INFLUENCE PROCESS

We do not, I think, possess the kind of studies of two-person interaction that would either bear out these propositions or fail to do so. But we do have studies of larger numbers of persons that suggest that they may apply, notably the studies by Festinger, Schachter, Back, and their associates on the dynamics of influence. One of the variables they work with they call *cohesiveness,* defined as anything that attracts people to take part in a group. Cohesiveness is a value variable; it refers to the degree of reinforcement people find in the activities of the group. Festinger and his colleagues consider two kinds of reinforcing activity: the symbolic behavior we call "social approval" (sentiment) and activity valuable in other ways, such as doing something interesting.

The other variable they work with they call *communication* and others call *interaction.* This is a frequency variable; it is a measure of the frequency of emission of valuable and costly verbal behavior. We must bear in mind that, in general, the one kind of variable is a function of the other.

Festinger and his co-workers show that the more cohesive a group is, that is, the more valuable the sentiment or activity the members exchange with one another, the greater the average frequency of interaction of the members.[ii] With men, as with pigeons, the greater the reinforcement, the more often is the reinforced behavior emitted. The more cohesive a group, too, the greater the change that members can produce in the behavior of other members in the direction of rendering these activities more valuable.[iii] That is, the more valuable the activities that members get, the more valuable those that they must give. For if a person is emitting behavior of a certain kind, and other people do not find it particularly rewarding, these others will suffer their own production of sentiment and activity, in time, to fall off. But perhaps the first person has found their sentiment and activity rewarding, and, if he is to keep on getting them, he must make his own behavior more valuable to the others. In short, the propositions of behavioral psychology imply a tendency toward a certain proportionality between the value to others of the behavior a man gives them and the value to him of the behavior they give him.

Schachter also studied the behavior of members of a group toward two kinds of other members, "conformers"

[ii]K. W. Back, "The Exertion of Influence through Social Communication," in L. Festinger, K. Back, S. Schachter, H. H. Kelley, and J. Thibaut (eds.), *Theory and Experiment in Social Communication* (Ann Arbor: Research Center for Dynamics, University of Michigan, 1950), pp. 21–36.

[iii]S. Schachter, N. Ellertson, D. McBride, and D. Gregory, "An Experimental Study of Cohesiveness and Productivity," *Human Relations,* IV (1951), 229–38.

and "deviates."[iv] I assume that conformers are people whose activity the other members find valuable. For conformity is behavior that coincides to a degree with some group standard or norm, and the only meaning I can assign to *norm* is "a verbal description of behavior that many members find it valuable for the actual behavior of themselves and others to conform to." By the same token, a deviate is a member whose behavior is not particularly valuable. Now Schachter shows that, as the members of a group come to see another member as a deviate, their interaction with him—communication addressed to getting him to change his behavior—goes up, the faster the more cohesive the group. The members need not talk to the other conformers so much; they are relatively satiated by the conformers' behavior: they have gotten what they want out of them. But if the deviate, by failing to change his behavior, fails to reinforce the members, they start to withhold social approval from him: the deviate gets low sociometric choice at the end of the experiment. And in the most cohesive groups—those Schachter calls "high cohesive-relevant"—interaction with the deviate also falls off in the end and is lowest among those members that rejected him most strongly, as if they had given him up as a bad job. But how plonking can we get? These findings are utterly in line with everyday experience.

PRACTICAL EQUILIBRIUM

. . . [Real-life small groups] often appear to be in practical equilibrium, and by this I mean nothing fancy. I do not mean that all real-life groups are in equilibrium. I certainly do not mean that all groups must tend to equilibrium. I do not mean that groups have built-in antidotes to change: there is no homeostasis here. I do not mean that we assume equilibrium. I mean only that we sometimes *observe* it, that for the time we are with a group—and it is often short—there is no great change in the values of the variables we choose to measure. If, for instance, person A is interacting with B more than with C both at the beginning and at the end of the study, then at least by this crude measure the group is in equilibrium.

Many of the Festinger-Schachter studies are experimental, and their propositions about the process of influence seem to me to imply the kind of proposition that empirically holds good of real-life groups in practical equilibrium. For instance, Festinger *et al.* find that, the more cohesive a group is, the greater the change that members can produce in the behavior of other members. If the influence is exerted in the direction of conformity to group norms, then, when the process of influence has accomplished all the change of which it is capable, the proposition should hold good that, the more cohesive a group is, the larger the number of members that conform to its norms. And it does hold good.[v]

Again, Schachter found, in the experiment I summarized above, that in the most cohesive groups and at the end, when the effort to influence the deviate had failed, members interacted little with the deviate and gave him little in the way of sociometric choice. Now two of the propositions that hold good most often of real-life groups in practical equilibrium are precisely that the more closely a member's activity conforms to the norms the more interaction he receives from other members and the more liking choices he gets from them too. From these main propositions a number of others may be derived that also hold good.

Yet we must ever remember that the truth of the proposition linking conformity to liking may on occasion be masked by the truth of other propositions. If, for instance, the man that conforms to the norms most closely also exerts some authority over the group, this may render liking for him somewhat less than it might otherwise have been. . . .

PROFIT AND SOCIAL CONTROL

Though I have treated equilibrium as an observed fact, it is a fact that cries for explanation. I shall not, as structural-functional sociologists do, use an assumed equilibrium as a means of explaining, or trying to explain, why the other features of a social system should be what they are. Rather, I shall take practical equilibrium as something that is itself to be explained by the other features of the system.

If every member of a group emits at the end of, and during, a period of time much the same kinds of

[iv]S. Schachter, "Deviation, Rejection, and Communication," *Journal of Abnormal and Social Psychology,* XLVI (1951), 190–207.

[v]L. Festinger, S. Schachter, and K. Back, *Social Pressures in Informal Groups* (New York: Harper & Bros., 1950), pp. 72–100.

behavior and in much the same frequencies as he did at the beginning, the group is for that period in equilibrium. Let us then ask why any one member's behavior should persist. Suppose he is emitting behavior of value A_1. Why does he not let his behavior get worse (less valuable or reinforcing to the others) until it stands at $A1\_\Delta A$? True, the sentiments expressed by others toward him are apt to decline in value (become less reinforcing to him), so that what he gets from them may be $S1—\Delta S$. But it is conceivable that, since most activity carries cost, a decline in the value of what he emits will mean a reduction in cost to him that more than offsets his losses in sentiment. Where, then, does he stabilize his behavior? This is the problem of social control.

Mankind has always assumed that a person stabilizes his behavior, at least in the short run, at the point where he is doing the best he can for himself under the circumstances, though his best may not be a "rational" best, and what he can do may not be at all easy to specify, except that he is not apt to think like one of the theoretical antagonists in the *Theory of Games.* Before a sociologist rejects this answer out of hand for its horrid profit-seeking implications, he will do well to ask himself if he can offer any other answer to the question posed. I think he will find that he cannot. Yet experiments designed to test the truth of the answer are extraordinarily rare.

I shall review one that seems to me to provide a little support for the theory, though it was not meant to do so. The experiment is reported by H. B. Gerard, a member of the Festinger-Schachter team, under the title "The Anchorage of Opinions in Face-to-Face Groups."[vi] The experimenter formed artificial groups whose members met to discuss a case in industrial relations and to express their opinions about its probable outcome. The groups were of two kinds: high-attraction groups, whose members were told that they would like one another very much, and low-attraction groups, whose members were told that they would not find one another particularly likable.

At a later time the experimenter called the members in separately, asked them again to express their opinions on the outcome of the case, and counted the number that had changed their opinions to bring them into accord with those of other members of their groups. At the same time, a paid participant entered into a further discussion of the case with each member, always taking, on the probable outcome of the case, a position opposed to that taken by the bulk of the other members of the group to which the person belonged. The experimenter counted the number of persons shifting toward the opinion of the paid participant.

The experiment had many interesting results, from which I choose only those summed up in Tables 1 and 2. The three different agreement classes are made up of people who, at the original sessions, expressed different degrees of agreement with the opinions of other members of their groups. And the figure 44, for instance, means that, of all members of high-attraction groups whose initial opinions were strongly in disagreement with those of other members, 44 per cent shifted their opinion later toward that of others.

Table 1 Percentage of Subjects Changing Toward Someone in the Group

	Agreement	Mild Disagreement	Strong Disagreement
High attraction . . .	0	12	44
Low attraction . . .	0	15	9

Table 2 Percentage of Subjects Changing Toward the Paid Participant

	Agreement	Mild Disagreement	Strong Disagreement
High attraction . . .	7	13	25
Low attraction . . .	20	38	8

[vi]*Human Relations,* VII (1954), 313–25.

In these results the experimenter seems to have been interested only in the differences in the sums of the rows, which show that there is more shifting toward the group, and less shifting toward the paid participant, in the high-attraction than in the low-attraction condition. This is in line with a proposition suggested earlier. If you think that the members of a group can give you much—in this case, liking—you are apt to give them much—in this case, a change to an opinion in accordance with their views—or you will not get the liking. And, by the same token, if the group can give you little of value, you will not be ready to give it much of value. Indeed, you may change your opinion so as to depart from agreement even further, to move, that is, toward the view held by the paid participant.

So far so good, but, when I first scanned these tables, I was less struck by the difference between them than by their similarity. The same classes of people in both tables showed much the same relative propensities to change their opinions, no matter whether the change was toward the group or toward the paid participant. We see, for instance, that those who change least are the high-attraction, agreement people and the low-attraction, strong-disagreement ones. And those who change most are the high-attraction, strong-disagreement people and the low-attraction, mild-disagreement ones.

How am I to interpret these particular results? Since the experimenter did not discuss them, I am free to offer my own explanation. The behavior emitted by the subjects is opinion and changes in opinion. For this behavior they have learned to expect two possible kinds of reinforcement. Agreement with the group gets the subject favorable sentiment (acceptance) from it, and the experiment was designed to give this reinforcement a higher value in the high-attraction condition than in the low-attraction one. The second kind of possible reinforcement is what I shall call the "maintenance of one's personal integrity," which a subject gets by sticking to his own opinion in the face of disagreement with the group. The experimenter does not mention this reward, but I cannot make sense of the results without something much like it. In different degrees for different subjects, depending on their initial positions, these rewards are in competition with one another: they are alternatives. They are not absolutely scarce goods, but some persons cannot get both at once.

Since the rewards are alternatives, let me introduce a familiar assumption from economics—that the cost of a particular course of action is the equivalent of the

foregone value of an alternative—and then add the definition: Profit = Reward – Cost.

Now consider the persons in the corresponding cells of the two tables. The behavior of the high-attraction, agreement people gets them much in the way of acceptance by the group, and for it they must give up little in the way of personal integrity, for their views are from the start in accord with those of the group. Their profit is high, and they are not prone to change their behavior. The low-attraction, strong-disagreement people are getting much in integrity, and they are not giving up for it much in valuable acceptance, for they are members of low-attraction groups. Reward less cost is high for them, too, and they change little. The high-attraction, strong-disagreement people are getting much in the way of integrity, but their costs in doing so are high, too, for they are in high-attraction groups and thus foregoing much valuable acceptance by the group. Their profit is low, and they are very apt to change, either toward the group or toward the paid participant, from whom they think, perhaps, they will get some acceptance while maintaining some integrity. The low-attraction, mild-disagreement people do not get much in the way of integrity, for they are only in mild disagreement with the group, but neither are they giving up much in acceptance, for they are members of low-attraction groups. Their rewards are low; their costs are low too, and their profit—the difference between the two—is also low. In their low profit they resemble the high-attraction, strong-disagreement people, and, like them, they are prone to change their opinions, in this case, more toward the paid participant. The subjects in the other two cells, who have medium profits, display medium propensities to change.

If we define profit as reward less cost, and if cost is value foregone, I suggest that we have here some evidence for the proposition that change in behavior is greatest when perceived profit is least. This constitutes no direct demonstration that change in behavior is least when profit is greatest, but if, whenever a man's behavior brought him a balance of reward and cost, he changed his behavior away from what got him, under the circumstances, the less profit, there might well come a time when his behavior would not change further. That is, his behavior would be stabilized, at least for the time being. And, so far as this were true for every member of a group, the group would have a social organization in equilibrium.

I do not say that a member would stabilize his behavior at the point of greatest conceivable profit to himself, because his profit is partly at the mercy of the behavior of others. It is a commonplace that the short-run pursuit of profit by several persons often lands them in positions where all are worse off than they might conceivably be. I do not say that the paths of behavioral change in which a member pursues his profit under the condition that others are pursuing theirs too are easy to describe or predict; and we can readily conceive that in jockeying for position they might never arrive at any equilibrium at all.

DISTRIBUTIVE JUSTICE

Yet practical equilibrium is often observed, and thus some further condition may make its attainment, under some circumstance, more probable than would the individual pursuit of profit left to itself. I can offer evidence for this further condition only in the behavior of subgroups and not in that of individuals. Suppose that there are two subgroups, working close together in a factory, the job of one being somewhat different from that of the other. And suppose that the members of the first complain and say: "We are getting the same pay as they are. We ought to get just a couple of dollars a week more to show that our work is more responsible." When you ask them what they mean by "more responsible," they say that, if they do their work wrong, more damage can result, and so they are under more pressure to take care.[vii] Something like this is a common feature of industrial behavior. It is at the heart of disputes not over absolute wages but over wage differentials—indeed, at the heart of disputes over rewards other than wages.

In what kind of proposition may we express observations like these? We may say that wages and responsibility give status in the group, in the sense that a man who takes high responsibility and gets high wages is admired, other things equal. Then, if the members of one group score higher on responsibility than do the members of another, there is a felt need on the part of the first to score higher on pay too. There is a pressure, which shows itself in complaints, to bring the *status factors,* as I have called them, into line with one another. If they are in line, a condition of *status congruence* is said to exist.

In this condition the workers may find their jobs dull or irksome, but they will not complain about the relative position of groups.

But there may be a more illuminating way of looking at the matter. In my example I have considered only responsibility and pay, but these may be enough, for they represent the two kinds of thing that come into the problem. Pay is clearly a reward; responsibility may be looked on, less clearly, as a cost. It means constraint and worry—or peace of mind foregone. Then the proposition about status congruence becomes this: If the costs of the members of one group are higher than those of another, distributive justice requires that their rewards should be higher too. But the thing works both ways: If the rewards are higher, the costs should be higher too. This last is the theory of *noblesse oblige,* which we all subscribe to, though we all laugh at it, perhaps because the *noblesse* often fails to *oblige.* To put the matter in terms of profit: though the rewards and costs of two persons or the members of two groups may be different, yet the profits of the two—the excess of reward over cost—should tend to equality. And more than "should." The less-advantaged group will at least try to attain greater equality, as, in the example I have used, the first group tried to increase its profit by increasing its pay.

I have talked of distributive justice. Clearly, this is not the only condition determining the actual distribution of rewards and costs. At the same time, never tell me that notions of justice are not a strong influence on behavior, though we sociologists often neglect them. Distributive justice may be one of the conditions of group equilibrium.

EXCHANGE AND SOCIAL STRUCTURE

I shall end by reviewing almost the only study I am aware of that begins to show in detail how a stable and differentiated social structure in a real-life group might arise out of a process of exchange between members. This is Peter Blau's description of the behavior of sixteen agents in a federal law-enforcement agency.[viii]

The agents had the duty of investigating firms and preparing reports on the firms' compliance with the law. Since the reports might lead to legal action against the

---

[vii]G. C. Homans, "Status among Clerical Workers," *Human Organization,* XII (1953), 5–10.

[viii]Peter M. Blau, *The Dynamics of Bureaucracy* (Chicago: University of Chicago Press, 1955), 99–116.

firms, the agents had to prepare them carefully, in the proper form, and take strict account of the many regulations that might apply. The agents were often in doubt what they should do, and then they were supposed to take the question to their supervisor. This they were reluctant to do, for they naturally believed that thus confessing to him their inability to solve a problem would reflect on their competence, affect the official ratings he made of their work, and so hurt their chances for promotion. So agents often asked other agents for help and advice, and, though this was nominally forbidden, the supervisor usually let it pass.

Blau ascertained the ratings the supervisor made of the agents, and he also asked the agents to rate one another. The two opinions agreed closely. Fewer agents were regarded as highly competent than were regarded as of middle or low competence; competence, or the ability to solve technical problems, was a fairly scarce good. One or two of the more competent agents would not give help and advice when asked, and so received few interactions and little liking. A man that will not exchange, that will not give you what he has when you need it, will not get from you the only thing you are, in this case, able to give him in return, your regard.

But most of the more competent agents were willing to give help, and of them Blau says:

> A consultation can be considered an exchange of values: both participants gain something, and both have to pay a price. The questioning agent is enabled to perform better than he could otherwise have done, without exposing his difficulties to his supervisor. By asking for advice, he implicitly pays his respect to the superior proficiency of his colleague. This acknowledgment of inferiority is the cost of receiving assistance. The consultant gains prestige, in return for which he is willing to devote some time to the consultation and permit it to disrupt his own work. The following remark of an agent illustrates this: "I like giving advice. It's flattering, I suppose, if you feel that others come to you for advice."[ix]

Blau goes on to say: "All agents liked being consulted, but the value of any one of very many consultations became deflated for experts, and the price they paid in frequent interruptions became inflated."[x] This implies that, the more prestige an agent received, the less was the increment of value of that prestige; the more advice an agent gave, the greater was the increment of cost of that advice, the cost lying precisely in the foregone value of time to do his own work. Blau suggests that something of the same sort was true of an agent who went to a more competent colleague for advice: the more often he went, the more costly to him, in feelings of inferiority, became any further request. "The repeated admission of his inability to solve his own problems . . . undermined the self-confidence of the worker and his standing in the group."[xi]

The result was that the less competent agents went to the more competent ones for help less often than they might have done if the costs of repeated admissions of inferiority had been less high and that, while many agents sought out the few highly competent ones, no single agent sought out the latter much. Had they done so (to look at the exchange from the other side), the costs to the highly competent in interruptions to their own work would have become exorbitant. Yet the need of the less competent for help was still not fully satisfied. Under these circumstances they tended to turn for help to agents more nearly like themselves in competence. Though the help they got was not the most valuable, it was of a kind they could themselves return on occasion. With such agents they could exchange help and liking, without the exchange becoming on either side too great a confession of inferiority.

The highly competent agents tended to enter into exchanges, that is, to interact with many others. But, in the more equal exchanges I have just spoken of, less competent agents tended to pair off as partners. That is, they interacted with a smaller number of people, but interacted often with these few. I think I could show why pair relations in these more equal exchanges would be more economical for an agent than a wider distribution of favors. But perhaps I have gone far enough. The final pattern of this social structure was one in which a small number of highly competent agents exchanged advice for prestige with a large

---

[ix]Ibid., p. 108.

[x]Ibid., p. 108.

[xi]Ibid., p. 109.

number of others less competent and in which the less competent agents exchanged, in pairs and in trios, both help and liking on more nearly equal terms.

Blau shows, then, that a social structure in equilibrium might be the result of a process of exchanging behavior rewarding and costly in different degrees, in which the increment of reward and cost varied with the frequency of the behavior, that is, with the frequency of interaction. Note that the behavior of the agents seems also to have satisfied my second condition of equilibrium: the more competent agents took more responsibility for the work, either their own or others,' than did the less competent ones, but they also got more for it in the way of prestige. I suspect that the same kind of explanation could be given for the structure of many "informal" groups.

Summary

The current job of theory in small-group research is to make the connection between experimental and real-life studies, to consolidate the propositions that empirically hold good in the two fields, and to show how these propositions might be derived from a still more general set. One way of doing this job would be to revive and make more rigorous the oldest of theories of social behavior—social behavior as exchange.

Some of the statements of such a theory might be the following. Social behavior is an exchange of goods, material goods but also non-material ones, such as the symbols of approval or prestige. Persons that give much to others try to get much from them, and persons that get much from others are under pressure to give much to them. This process of influence tends to work out at equilibrium to a balance in the exchanges. For a person engaged in exchange, what he gives may be a cost to him, just as what he gets may be a reward, and his behavior changes less as profit, that is, reward less cost, tends to a maximum. Not only does he seek a maximum for himself, but he tries to see to it that no one in his group makes more profit than he does. The cost and the value of what he gives and of what he gets vary with the quantity of what he gives and gets. It is surprising how familiar these propositions are; it is surprising, too, how propositions about the dynamics of exchange can begin to generate the static thing we call "group structure" and, in so doing, generate also some of the propositions about group structure that students of real-life groups have stated.

In our unguarded moments we sociologists find words like "reward" and "cost" slipping into what we say. Human nature will break in upon even our most elaborate theories. But we seldom let it have its way with us and follow up systematically what these words imply. Of all our many "approaches" to social behavior, the one that sees it as an economy is the most neglected, and yet it is the one we use every moment of our lives—except when we write sociology.

▪▪ Peter M. Blau (1918–2002): A Biographical Sketch

Peter M. Blau was born in Vienna, Austria, in 1918. As a teenager, he spoke out against the political repression exacted by Austria's fascist government, then controlled by the National Fascist Party. Writing for an underground journal associated with the Socialist Worker's Party, Blau was imprisoned when the Austrian authorities discovered the journal's circulation. He was convicted of high treason and faced a ten-year sentence until an agreement between Hitler and the Austrian government led to the release of political prisoners.

When Hitler seized control of Austria in 1938, Blau, of Jewish descent, attempted to secure a visa allowing him to emigrate to another country. Unsuccessful, he tried to escape to Czechoslovakia but was captured by Nazi border guards. For the next two months, he faced torture and starvation. When he was released, he made his way to Prague. He was forced to leave only a year later when Hitler invaded Czechoslovakia. Fortunately, arrangements had been made for Blau to emigrate to the United States via France. As fate would have it, he caught what would be the last train out of the country. (His parents remained in Vienna and four years later died in Auschwitz, a Nazi concentration camp.)

Once in France, Blau, who held a German passport, "surrendered" to the Allied forces. The French Army responded by deporting him to a labor camp in Bordeaux. While he was imprisoned, his visa was granted, so he left for the port city of Le Havre and a boat bound for the United States. As fate again would have it, he boarded what turned out to be the last civilian boat departing France. This trip proved doubly miraculous, for also waiting to board were representatives from Elmhurst College who were in Europe offering scholarships for Jewish refugees. Blau was given the address of the son of the college's president, and a remarkable academic career was born.

After receiving his bachelor's degree, Blau served four years in the Army as an interrogation officer. With the end of the war, he began his graduate studies in sociology at Columbia University and, under the tutelage of Robert K. Merton (see Chapter 9), received his PhD in 1952. During the next 50 years, Blau would produce eleven books and more than one hundred articles, in the process becoming one of the leading scholars of his day. From 1953 to 1970, he was a professor at the University of Chicago and then, from 1970 to 1988, a professor at Columbia University. In addition to serving as president of the American Sociological Association in 1973, he was selected to a number of prominent positions. He was a Senior Fellow at King's College as well as a Fellow of the National Academy of Sciences, the American Philosophical Society, and the American Academy of Arts and Sciences. He also was Pitt Professor at Cambridge University and a Distinguished Honorary Professor at the Tianjin Academy of Social Sciences in China. Peter Blau died in March 2002 of adult respiratory distress syndrome.[1]

BLAU'S INTELLECTUAL INFLUENCES AND CORE IDEAS ■■

Like Homans, Blau was interested in examining the processes that guide face-to-face interaction. And like Homans, Blau argued that such interaction is shaped by a reciprocal exchange of rewards, both tangible and intangible. On these points, Homans was an important influence on Blau's work. However, the differences between the two exchange theorists outnumber the similarities. While Homans was interested in studying exchange relations in order to uncover the behaviorist principles that underlie interaction, Blau sought to derive from his analysis of social interaction a better understanding of the complex institutions and organizations that develop out of simpler exchange relations between individuals. Moreover, Blau not only abandoned Homans's brand of behavioral psychology, but—in recognizing that imbalances of rewards and costs often pervade exchange relations—he also emphasized the roles that power, inequality, and norms of legitimation play in interaction.

In extending the work of Homans and fashioning his own brand of exchange theory, Blau drew from a number of scholars. Perhaps most influential in shaping his views on interaction was the German sociologist and philosopher Georg Simmel (1858–1918). In charting a course for the newly created discipline, Simmel argued that sociology should concern itself above all with analyzing the *forms* in which interaction occurs, for it is only through interacting with others that we are able to satisfy our ambitions. Understanding the specific *content* of interactions—what people talk about and why—is of secondary importance. What is of sociological significance, then, is determining the uniformities or shared properties that exist between seemingly diverse social associations. For instance, while a husband and wife may share experiences and discuss issues that are quite different from those expressed among coworkers, both social relations may take on a common form of cooperation or conflict.

Simmel also noted that, while the most dissimilar of contents (individual motivations or interests) may be expressed through an identical form of interaction, the same contents can be expressed in a range of forms of association. For instance, interactions within and between families, gangs, corporations, political organizations, and governments all may take the form of conflict. Thus, despite the varied interests or

---

[1]This biographical account is adapted from the April 2002 edition of *Footnotes,* printed by the American Sociological Association.

purposes that led to interaction in each of these cases, the individuals involved all may find themselves facing an opposing party that hinders the realization of their impulses or desires. Similarly, while giving a wedding present, paying for music lessons, and volunteering at the food co-op are actions motivated by different intentions or "contents," the sociologist sees them all as taking the form of an exchange relation and thus as sharing interaction properties. Conversely, the same drive or interest can be expressed through a number of forms. Attempts to gain economic advantage, for instance, can be realized through cooperative agreements among parties as well as through relations of domination and subordination.

Like Simmel, Blau maintained that the central task of sociology is to uncover the basic forms of interaction through which individuals pursue their interests or satisfy their desires. Blau, moreover, would endorse Simmel's assertion that in exchange relations lie "the purest and most concentrated form of all human interactions" (Simmel [1900b] 1971:43). Both maintained that every interaction (a performance, a conversation, or even a romantic affair) can be understood as a form of exchange in which the participant gives the other "more than he had himself possessed" (ibid.:44). Indeed, at the heart of Blau's theoretical perspective is an attempt to analyze the dynamics of exchange—the interplay of rewards and sacrifice—that are the building blocks of all social relations.

As we noted above, Blau was interested in building a theoretical bridge that would link sociological studies of everyday interactions between individuals and studies that examined the collectivist or structural dimensions of society, such as economic systems, political institutions, or belief systems. While the work of Homans and Simmel (and, to a lesser extent, Erving Goffman) informed the "interactionist" elements of his approach, his analysis of society's structural properties was most influenced by Max Weber (1864–1920) and Talcott Parsons (see Chapter 9). From them he developed his analysis of the role of power and norms of legitimation in shaping group processes.

While beginning with Weber's definition of power as "the probability that one actor within a social relationship will be in a position to carry out his own will despite resistance," Blau stressed the significance of rewards in inducing others to accede to one's wishes.[2] For Blau, then, an individual is able to exercise **power** over others when he alone is able to supply needed rewards to them. If the others are unable to receive the benefits from another source, *and* if they are unable to offer rewards to the individual, they become dependent on the individual. Their only option is to submit to his demands lest he withdraw the needed benefits. In short, power results from an unequal exchange stemming from an individual's or group's monopoly over a desired resource (Blau 1964:115, 118).[3]

In defining power in terms of an inequality of resources and the submission that an **imbalanced exchange** imposes, Blau is led to consider the processes that shape the exercise of power and the rise of opposition to it. These processes, in turn, account for both stability and change in interpersonal and group relations, as well as in more complex social institutions (see Figure 11.2). Of central importance is the role of social norms of fairness and the legitimacy they either confer on or deny those in dominant positions. Following the work of both Weber and Talcott Parsons, Blau argues that legitimate authority—a superior's right to demand compliance from subordinates and their willing obedience—is based on shared norms that constrain an individual's response to issued directives. Thus, imbalanced exchange relations are governed less by individual, rational calculations than they are by shared expectations and the cultural values that legitimate them. As long as the superior meets or exceeds the expectations for rewards deemed acceptable by the group, then the ensuing legitimacy conferred on the superior will foster the stability of the group. That is, the costs incurred by subordinates, both in the services they perform and in the very act of submission, must be judged fair relative to the benefits derived for obedience. Otherwise, opposition to

---

[2]Weber (1947:152), quoted in Blau (1964:115).

[3]In arguing that the dynamics of power and submission are tied to the availability of alternative exchange relations, Blau again draws on Simmel's insights. Of particular importance here is the latter's analysis of the patterns of interaction that distinguish dyads from triads and from other larger groups. See Simmel (1950).

the superior's exercise of power may arise, and with it the potential for change in the structure of existing interpersonal or institutional relations (see Figure 10.2). Yet, this judgment rests, ultimately, on consensual, normative standards of fairness. Interestingly, Blau's emphasis on the normative dimension of group life and the exercise of power in it also represents a metatheoretical shift, a point that we discuss in the following section.

Figure 11.2 Blau's Model of Exchange and the Structure of Social Relations

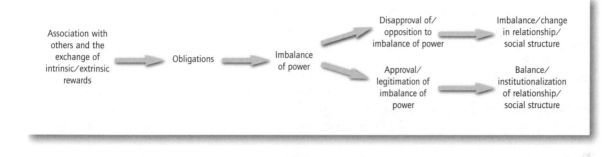

BLAU'S THEORETICAL ORIENTATION ⠿

Like Homans's, Blau's perspective is primarily individualistic and rationalist (see Figure 11.3). In his major work on exchange theory, *Exchange and Power in Social Life* (1964:2), from which the following excerpt is taken, he states his position thusly:

> The [theoretical] problem is to derive the social processes that govern the complex structures of communities and societies from the simpler processes that pervade the daily intercourse among individuals and their interpersonal relations.

In other words, while acknowledging that complex social structures (a family, a university, manufacturing industries, the federal government, etc.) possess properties that are distinct from the individuals that live and work in and through them, such structures, nevertheless, emerge only through individual and group interactions. As such, any attempt to understand complex structures must begin with an analysis of the patterns of daily interactions that guide individual conduct.

Although Blau's emphasis rested on examining the nature of interpersonal interaction, his interests, unlike Homans's, did not lie in searching for the psychological roots of exchange behavior or the patterns of reinforcement that determine the future of such behavior. Instead, reminiscent of Simmel, his objective was to explain the dynamics of exchange *relations* that emerge *between* individuals and groups as they jointly pursue their interests. A simplified model of this dynamic is depicted in Table 10.2. At their root, exchange relations present individuals with four basic interaction options. These options provide the context for an individual's calculation of the possible rewards and costs for initiating an exchange relation. Once an option is realized, it establishes the relational structure for subsequent exchange calculations that may, in turn, alter the very structure that shaped the individual's or group's decision. (You can see the logic of this model in the strategies devised by the participants in currently popular "reality" shows such as *Survivor.* The players' efforts to prove their worth in the challenges and their use of deception and secret alliances are calculated attempts to create a favorable imbalance of power.)

Figure 11.3 Blau's Basic Concepts and Theoretical Orientation

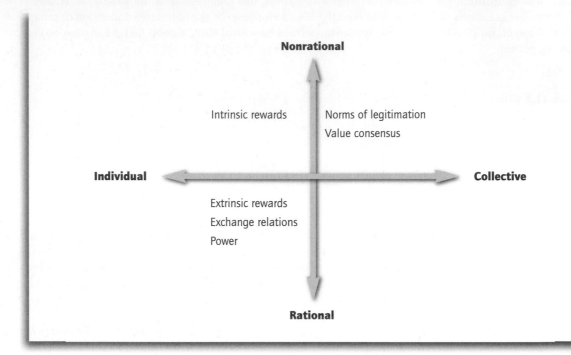

As we alluded earlier, however, Blau would make a theoretical shift that incorporated collectivist (and nonrational) aspects of social life. In attempting to build a theory of social structure on the basis of those "simpler processes" that shape face-to-face interaction, Blau saw in shared norms a generalized mechanism that defines the expectations and governs the reactions of those subjected to power imbalances. More broadly, Blau viewed "value consensus" as a crucial mediating link allowing for "integrative bonds and social solidarity among millions of people in society, most of whom have never met . . . [and] which alone make it possible to transcend personal transactions and develop complex networks of independent exchange" (1964:24). While originally an emergent product of simple exchange relations, norms and values become institutionalized in the broader social fabric as they are transmitted from one generation to the next. Despite this theoretical shift, however, Blau's argument in the following selection underscores the individualistic dimension of social order.[4]

Regarding his general position on the problem of action, Blau adopted a rational orientation. Although the incorporation of norms and values into his model suggests a nonrational perspective, Blau emphasized the goal-directed aspects of human behavior:

> Within the rather broad limits . . . norms impose on social relations, however, human beings tend to be governed in their associations with one another by the desire to obtain social rewards of various sorts, and the resulting exchanges of benefits shape the structure of social relations. (ibid.:18)

---

[4]Blau's theoretical shift also left him open to the charge of ad hoc theorizing. In other words, he resorted to logically inconsistent concepts to plug the holes in his essentially individualistic argument. Interestingly, Blau would later abandon his attempt to develop a theory of complex social structures on the basis of "simple" exchange relations.

Table 11.2 Exchange Relations and Interaction Options

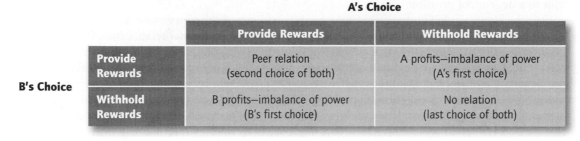

B's Choice		A's Choice	
		Provide Rewards	**Withhold Rewards**
Provide Rewards		Peer relation (second choice of both)	A profits—imbalance of power (A's first choice)
Withhold Rewards		B profits—imbalance of power (B's first choice)	No relation (last choice of both)

SOURCE: Adapted from Peter Blau, *Exchange and Power in Social Life* (1964:45).

Yet, like Homans, he, too, assumed a restricted view on the nature of rational action, contending only that individuals "choose between alternative potential associates or courses of action by evaluating the experiences or expected experiences with each in terms of a preference ranking and then selecting the best alternative" (ibid.). Which objectives an individual pursues and how successful he or she will be in achieving them remains an open question, one that can be answered only inductively after the fact.[5]

As for the "rewards of various sorts," Blau distinguished between two: extrinsic and intrinsic. **Extrinsic rewards** are those that are "detachable" from the association in which they are acquired. In other words, extrinsic benefits are derived not from another person's company itself, but from the external rewards his company will provide. Here, associating with others serves as a means to a further end.[6] Thus, a salesperson is considerate because she wants to make a commission, not because she values the relationship she initiates with any particular customer.

Intrinsic rewards are those things we find pleasurable in and of themselves, not because they provide the means for obtaining other benefits. Examples of intrinsic rewards are celebrating a holiday with one's family, going on a walk with a friend, or love—the purest type of intrinsic reward. In cases such as these, rewards express one's commitment to the relationship and are exchanged in the interest of maintaining it.

Because linking intimate relations to exchange processes runs counter to our conventional understanding of such relations, some additional remarks on the topic perhaps are in order. As Blau notes, "Love appears to make human beings unselfish, since they themselves enjoy giving pleasure to those they love, but this selfless devotion generally rests on an interest in maintaining the other's love" (ibid.:76). Yet, it is often the case that in intimate relations one individual is more in love than the is other. As a result, the "interests" in maintaining the relationship are not equal. This, in turn, creates an imbalance that advantages one partner while disadvantaging the other, as the costs for ending the relationship, as well as the willingness to endure costs to maintain it, are not proportionate. As is the

---

[5]It is important to point out that our theoretical positioning of Blau is based primarily on the selection included in this volume. In fact, the totality of Blau's works spans into each quadrant of our map. For instance, in addition to his contributions to exchange theory, Blau also made a significant impact in the sociology of organizations and social stratification. See, for example, *Structural Contexts of Opportunities* (1994), *Inequality and Heterogeneity* (1977), and *The Dynamics of Bureaucracy* (1955).

[6]Blau saw his distinction between intrinsic and extrinsic rewards as a special case of Parsons's distinction between the pattern variables, particularism and universalism, respectively (see Chapter 9).

case in other types of relationships, the individual who is less committed in an intimate relationship is able to exercise power over the other, whose greater interest in maintaining the relationship resigns him to a dependent position.

It is this dynamic that underlies "playing hard to get," where "the lover who does not express unconditional affection early gains advantages in the established interpersonal relationship. Indeed, the more restrained lover also seems to have a better chance of inspiring another's love for himself or herself" (1964:78). Like other benefits offered, affections that are given too freely decrease in value. Moreover, the more freely one gives his affection, the more he signals that he has few options, thus reducing his value on the "market." From this perspective, dating is a "challenge of conquest" in which those "who are successful in making many conquests . . . validate their attractiveness in their own eyes as well as those of others" (ibid.:81). Conversely, resisting conquests implies that one has many alternatives to choose from, which then enhances a person's desirability in the eyes of others. Just as relationships based on an exchange of extrinsic rewards involve a cost/benefit dilemma, so too do intimate relations. In this instance, though, the dilemma is one in which lovers must express affection for one another in order to maintain the relationship, while at the same time they experience pressure to withhold such expressions in order to make themselves more attractive to each other.

Reading

Introduction to *Exchange and Power in Social Life*

In the following selection, you will encounter the themes discussed to this point: interaction as a reciprocal exchange of rewards, how an imbalance in exchanges can lead to an imbalance of power, and how this inequality can lead to a change in the structure of social relations depending on social norms of fairness that shape the exercise of legitimate authority. It is through such norms that subordinate individuals define the imbalances as just and beneficial or as oppressive and exploitative.

In addition to these themes, Blau also emphasizes the role of "social attraction" in exchange relations. For Blau, social attraction represents the fundamental "force that induces human beings to establish social associations" with others (1964:20). Such attraction rests on whether or not an individual expects to receive rewards for forming an association with another. Thus, it is what the other can offer us that determines whether we will find him "attractive." Our attraction to others often is motivated and sustained by a mix of extrinsic and intrinsic rewards. Indeed, associations we find intrinsically gratifying often were initiated for the extrinsic rewards they provided: the coworker whose advice we sought last year is now a valued friend, the comparative (i.e., extrinsic) good looks and wit of another sparks a lasting courtship. Similarly, in addition to loving one another, spouses also may be attracted to the economic rewards that their marriage brings.

While the forms of social attraction may differ, the same basic exchange processes guide the relations they initiate. Whether the rewards exchanged are intrinsic or extrinsic in nature, the exchange itself produces a "strain toward reciprocity," since individuals are interested in discharging their obligations to provide benefits to those from whom rewards were received. Failure to do so can jeopardize not only the continued supply of benefits, but also the conferring of social approval, a basic and important reward. Yet, the obligation to reciprocate, felt as a blessing or a burden, also produces a strain toward imbalance in the relationship, for individuals seek not merely to stay out of "debt" but also to "accumulate credit that makes their status superior to that of others" (ibid.:26). As you will read in the following excerpt, even love relationships are not free from these twin strains and the power dynamics to which they give rise.

Exchange and Power in Social Life (1964)

Peter M. Blau

THE STRUCTURE OF SOCIAL ASSOCIATIONS

To speak of social life is to speak of the associations between people—their associating together in work and in play, in love and in war, to trade or to worship, to help or to hinder. It is in the social relations men establish that their interests find expression and their desires become realized. As Simmel put it: "Social association refers to the widely varying forms that are generated as the diverse interests of individuals prompt them to develop social units in which they realize these—sensual or ideal, lasting or fleeting, conscious or unconscious, casually impelling or teleologically inducing—interests."[i] Simmel's fundamental postulate, and also that of this book, is that the analysis of social associations, of the processes governing them, and of the forms they assume is the central task of sociology. . . .

The objectives of our investigation are to analyze social associations, the processes that sustain them and the forms they attain, and to proceed to inquire into the complex social forces and structures to which they give rise. Broad as this topic is, it is intended to provide a specific focus that explicitly excludes many sociological problems from consideration. Sociology is defined by Weber as "a science which attempts the interpretative understanding of social action in order thereby to arrive at a causal explanation of its course and effects. . . . Action is social insofar as, by virtue of the subjective meaning attached to it by the acting individual (or individuals), it takes account of the behavior of others and is thereby oriented in its course."[ii] A concern with social action, broadly conceived as any conduct that derives its impetus and meaning from social values, has characterized contemporary theory in sociology for some years. The resulting preoccupation with value orientations has diverted theoretical attention from the study of the actual associations between people and the structures of their associations. While structures of social relations are, of course, profoundly influenced by common values, these structures have a significance of their own, which is ignored if concern is exclusively with the underlying values and norms. Exchange transactions and power relations, in particular, constitute social forces that must be investigated in their own right, not merely in terms of the norms that limit and the values that reinforce them, to arrive at an understanding of the dynamics of social structures. If one purpose of the title of this chapter is to indicate a link with the theoretical tradition of Simmel, another purpose is to distinguish the theoretical orientation in this monograph from that of Weber and Parsons; not "the structure of social action" but the structure of social associations is the focal point of the present inquiry.

After illustrating the concept of social exchange and its manifestations in various social relations, this chapter presents the main theme of how more complex processes of social association evolve out of simpler ones. Forces of social attraction stimulate exchange transactions. Social exchange, in turn, tends to give rise to differentiation of status and power. Further processes emerge in a differentiated status structure that lead to legitimation and organization, on the one hand, and to opposition and change, on the other. Whereas the conception of reciprocity in exchange implies the existence of balancing forces that create a strain toward equilibrium, the simultaneous operations of diverse balancing forces recurrently produce imbalances in social life, and the resulting dialectic between reciprocity and imbalance gives social structures their distinctive nature and dynamics.

The Exchange of Social Rewards

Most human pleasures have their roots in social life. Whether we think of love or power, professional

[i]Georg Simmel, *Soziologie,* Leipzig: Duncker und Humblot, 1908, p. 6 (my translation).

[ii]Max Weber, *The Theory of Social and Economic Organization,* New York: Oxford University Press, 1947, p. 88.

recognition or sociable companionship, the comforts of family life or the challenge of competitive sports, the gratifications experienced by individuals are contingent on actions of others. The same is true for the most selfless and spiritual satisfactions. To work effectively for a good cause requires making converts to it. Even the religious experience is much enriched by communal worship. Physical pleasures that can be experienced in solitude pale in significance by comparison. Enjoyable as a good dinner is, it is the social occasion that gives it its luster. Indeed, there is something pathetic about the person who derives his major gratification from food or drink as such, since it reveals either excessive need or excessive greed; the pauper illustrates the former, the glutton, the latter. To be sure, there are profound solitary enjoyments—reading a good book, creating a piece of art, producing a scholarly work. Yet these, too, derive much of their significance from being later communicated to and shared with others. The lack of such anticipation makes the solitary activity again somewhat pathetic: the recluse who has nobody to talk to about what he reads; the artist or scholar whose works are completely ignored, not only by his contemporaries but also by posterity.

Much of human suffering as well as much of human happiness has its source in the actions of other human beings. One follows from the other, given the facts of group life, where pairs do not exist in complete isolation from other social relations. The same human acts that cause pleasure to some typically cause displeasure to others. For one boy to enjoy the love of a girl who has committed herself to be his steady date, other boys who had gone out with her must suffer the pain of having been rejected. The satisfaction a man derives from exercising power over others requires that they endure the deprivation of being subject to his power. For a professional to command an outstanding reputation in his field, most of his colleagues must get along without such pleasant recognition, since it is the lesser professional esteem of the majority that defines his as outstanding. The joy the victorious team members experience has its counterpart in the disappointment of the losers. In short, the rewards individuals obtain in social associations tend to entail a cost to other individuals. This does not mean that most social associations involve zero-sum games in which the gains of some rest on the losses of others. Quite the contrary, individuals associate with one another because they all profit from their association. But they do not necessarily all profit equally, nor do they share the cost of providing the benefits equally, and even if there are no direct costs to participants, there are often indirect costs born [sic] by those excluded from the association, as the case of the rejected suitors illustrates.

Some social associations are intrinsically rewarding. Friends find pleasure in associating with one another, and the enjoyment of whatever they do together—climbing a mountain, watching a football game—is enhanced by the gratification that inheres in the association itself. The mutual affection between lovers or family members has the same result. It is not what lovers do together but their doing it *together* that is the distinctive source of their special satisfaction—not seeing a play but sharing the experience of seeing it. Social interaction in less intimate relations than those of lovers, family members, or friends, however, may also be inherently rewarding. The sociability at a party or among neighbors or in a work group involves experiences that are not especially profound but are intrinsically gratifying. In these cases, all associates benefit simultaneously from their social interaction, and the only cost they incur is the indirect one of giving up alternative opportunities by devoting time to the association.

Social associations may also be rewarding for a different reason. Individuals often derive specific benefits from social relations because their associates deliberately go to some trouble to provide these benefits for them. Most people like helping others and doing favors for them—to assist not only their friends but also their acquaintances and occasionally even strangers, as the motorist who stops to aid another with his stalled car illustrates. Favors make us grateful, and our expressions of gratitude are social rewards that tend to make doing favors enjoyable, particularly if we express our appreciation and indebtedness publicly and thereby help establish a person's reputation as a generous and competent helper. Besides, one good deed deserves another. If we feel grateful and obligated to an associate for favors received, we shall seek to reciprocate his kindness by doing things for him. He in turn is likely to reciprocate, and the resulting mutual exchange of favors strengthens, often without explicit intent, the social bond between us.

A person who fails to reciprocate favors is accused of ingratitude. This very accusation indicates that reciprocation is expected, and it serves as a social sanction that discourages individuals from forgetting their obligations to associates. Generally, people are grateful for favors and repay their social debts, and both their gratitude and their repayment are social rewards for the associate who has done them favors. The fact that furnishing benefits to others tends to produce these social rewards is, of course, a major reason why people often go to great trouble to help their associates and enjoy doing so. We would not be human if these advantageous consequences of our good deeds were not important inducements for our doing them.[iii] There are, to be sure, some individuals who selflessly work for others without any thought of reward and even without expecting gratitude, but these are virtually saints, and saints are rare. The rest of us also act unselfishly sometimes, but we require some incentive for doing so, if it is only the social acknowledgment that we are unselfish.

An apparent "altruism" pervades social life; people are anxious to benefit one another and to reciprocate for the benefits they receive. But beneath this seeming selflessness an underlying "egoism" can be discovered; the tendency to help others is frequently motivated by the expectation that doing so will bring social rewards. Beyond this self-interested concern with profiting from social associations, however, there is again an "altruistic" element or, at least, one that removes social transactions from simple egoism or psychological hedonism. A basic reward people seek in their associations is social approval, and selfish disregard for others makes it impossible to obtain this important reward.

The social approval of those whose opinions we value is of great significance to us, but its significance depends on its being genuine. We cannot force others to give us their approval, regardless of how much power we have over them, because coercing them to express their admiration or praise would make these expressions worthless. "Action can be coerced, but a coerced show of feeling is only a show."[iv] Simulation robs approval of its significance, but its very importance makes associates reluctant to withhold approval from one another and, in particular, to express disapproval, thus introducing an element of simulation and dissimulation into their communications. As a matter of fact, etiquette prescribes that approval be simulated in disregard of actual opinions under certain circumstances. One does not generally tell a hostess, "Your party was boring," or a neighbor, "What you say is stupid." Since social conventions require complimentary remarks on many occasions, these are habitually discounted as not reflecting genuine approbation, and other evidence that does reflect it is looked for, such as whether guests accept future invitations or whether neighbors draw one into further conversations.

In matters of morality, however, individuals have strong convictions that constrain them to voice their actual judgments more freely. They usually do not hesitate to express disapproval of or, at least, withhold approval from associates who have violated socially accepted standards of conduct. Antisocial disregard for the welfare of the ingroup meets universally with disapprobation regardless of how immoral, in terms of the mores of the wider community, the norms of a particular group may be. The significance of social approval, therefore, discourages conduct that is utterly and crudely selfish. A more profound morality must rest not merely on group pressure and long-run advantage but primarily on internalized normative standards. In the ideal case, an individual unerringly follows the moral commands of his conscience whatever the consequences. While such complete morality is attained only by the saint and the fool, and most men make some compromises, moral standards clearly do guide and restrain human conduct. Within the rather broad limits these norms impose on social relations, however, human beings tend to be governed in their associations with one another by the desire to obtain social rewards of various sorts, and the resulting exchanges of benefits shape the structure of social relations.

The question that arises is whether a rationalistic conception of human behavior underlies this principle that individuals pursue social rewards in their social associations. The only assumption made is that human beings choose between alternative potential associates or courses of action by evaluating the experiences or expected experiences with each in terms of a preference ranking and then selecting the best alternative. Irrational

---

[iii]Once a person has become emotionally committed to a relationship, his identification with the other and his interest in continuing the association provide new independent incentives for supplying benefits to the other. Similarly, firm commitments to an organization lead members to make recurrent contributions to it without expecting reciprocal benefits in every instance.

[iv]Erving Goffman, *Asylums,* Chicago: Aldine, 1962, p. 115.

as well as rational behavior is governed by these considerations, as Boulding has pointed out:

> All behavior, in so far as the very concept of behavior implies doing one thing rather than another, falls into the above pattern, even the behavior of the lunatic and the irrational or irresponsible or erratic person. The distinction between rational and irrational behavior lies in the degree of self-consciousness and the stability of the images involved rather than in any distinction of the principle of optimum.[v]

What is explicitly *not* assumed here is that men have complete information, that they have no social commitments restricting their alternatives, that their preferences are entirely consistent or remain constant, or that they pursue one specific ultimate goal to the exclusion of all others. These more restrictive assumptions, which are not made in the present analysis, characterize rationalistic models of human conduct, such as that of game theory. Of particular importance is the fact that men strive to achieve diverse objectives. The statement that men select the most preferred among available alternatives does not imply that they always choose the one that yields them the greatest material profit. They may, and often do, choose the alternative that requires them to make material sacrifices but contributes the most to the attainment of some lofty ideal, for *this* may be their objective. Even in this choice they may err and select an alternative that actually is not the best means to realize their goal. Indeed, the need to anticipate in advance the social rewards with which others will reciprocate for favors in exchange relations inevitably introduces uncertainty and recurrent errors of judgment that make perfectly rational calculations impossible. Granted these qualifications, the assumption that men seek to adjust social conditions to achieve their ends seems to be quite realistic, indeed inescapable.

BASIC PROCESSES

The basic social processes that govern associations among men have their roots in primitive psychological processes, such as those underlying the feelings of attraction between individuals and their desires for various kinds of rewards. These psychological tendencies are primitive only in respect to our subject matter, that is, they are taken as given without further inquiry into the motivating forces that produce them, for our concern is with the social forces that emanate from them.

The simpler social processes that can be observed in interpersonal associations and that rest directly on psychological dispositions give rise to the more complex social processes that govern structures of interconnected social associations, such as the social organization of a factory or the political relations in a community. New social forces emerge in the increasingly complex social structures that develop in societies, and these dynamic forces are quite removed from the ultimate psychological base of all social life. Although complex social systems have their foundation in simpler ones, they have their own dynamics with emergent properties. In this section, the basic processes of social associations will be presented in broad strokes, to be analyzed subsequently in greater detail, with special attention to their wider implications.

Social attraction is the force that induces human beings to establish social associations on their own initiative and to expand the scope of their associations once they have been formed. Reference here is to social relations into which men enter of their own free will rather than to either those into which they are born (such as kinship groups) or those imposed on them by forces beyond their control (such as the combat teams to which soldiers are assigned), although even in these involuntary relations the extent and intensity of the association depend on the degree of mutual attraction. An individual is attracted to another if he expects associating with him to be in some way rewarding for himself, and his interest in the expected social rewards draws him to the other. The psychological needs and dispositions of individuals determine which rewards are particularly salient for them and thus to whom they will be attracted. Whatever the specific motives, there is an important difference between the expectation that the association will be an intrinsically rewarding experience and the expectation that it will furnish extrinsic benefits, for example, advice. This difference calls attention to two distinct meanings of the term "attraction" and its derivatives. In its narrower sense,

---

[v]Kenneth Boulding, *Conflict and Defense,* New York: Harper, 1962, p. 151.

social attraction refers to liking another person *intrinsically* and having positive feelings toward him; in the broader sense, in which the term is now used, social attraction refers to being drawn to another person for any reason whatsoever. The customer is attracted in this broader sense to the merchant who sells goods of a given quality at the lowest price, but he has no intrinsic feelings of attraction for him, unless they happen to be friends.

A person who is attracted to others is interested in proving himself attractive to them, for his ability to associate with them and reap the benefits expected from the association is contingent on their finding him an attractive associate and thus wanting to interact with him. Their attraction to him, just as his to them, depends on the anticipation that the association will be rewarding. To arouse this anticipation, a person tries to impress others. Attempts to appear impressive are pervasive in the early stages of acquaintance and group formation. Impressive qualities make a person attractive and promise that associating with him will be rewarding. Mutual attraction prompts people to establish an association, and the rewards they provide each other in the course of their social interaction, unless their expectations are disappointed, maintain their mutual attraction and the continuing association.

Processes of social attraction, therefore, lead to processes of social exchange. The nature of the exchange in an association experienced as intrinsically rewarding, such as a love relationship, differs from that between associates primarily concerned with extrinsic benefits, such as neighbors who help one another with various chores, but exchanges do occur in either case. A person who furnishes needed assistance to associates, often at some cost to himself, obligates them to reciprocate his kindness. Whether reference is to instrumental services or to such intangibles as social approval, the benefits each supplies to the others are rewards that serve as inducements to continue to supply benefits, and the integrative bonds created in the process fortify the social relationship.

A situation frequently arises, however, in which one person needs something another has to offer, for example, help from the other in his work, but has nothing the other needs to reciprocate for the help. While the other may be sufficiently rewarded by expressions of gratitude to help him a few times, he can hardly be expected regularly to devote time and effort to providing help without receiving any return to compensate him for his troubles. (In the case of intrinsic

attraction, the only return expected is the willingness to continue the association.) The person in need of recurrent services from an associate to whom he has nothing to offer has several alternatives. First, he may force the other to give him help. Second, he may obtain the help he needs from another source. Third, he may find ways to get along without such help. If he is unable or unwilling to choose any of these alternatives, however, there is only one other course of action left for him; he must subordinate himself to the other and comply with his wishes, thereby rewarding the other with power over himself as an inducement for furnishing the needed help. Willingness to comply with another's demands is a generic social reward, since the power it gives him is a generalized means, parallel to money, which can be used to attain a variety of ends. The power to command compliance is equivalent to credit, which a man can draw on in the future to obtain various benefits at the disposal of those obligated to him. The unilateral supply of important services establishes this kind of credit and thus is a source of power.

Exchange processes, then, give rise to differentiation of power. A person who commands services others need, and who is independent of any at their command, attains power over others by making the satisfaction of their need contingent on their compliance. This principle is held to apply to the most intimate as well as the most distant social relations. The girl with whom a boy is in love has power over him, since his eagerness to spend much time with her prompts him to make their time together especially pleasant for her by acceding to her wishes. The employer can make workers comply with his directives because they are dependent on his wages. To be sure, the superior's power wanes if subordinates can resort to coercion, have equally good alternatives, or are able to do without the benefits at his disposal. But given these limiting conditions, unilateral services that meet basic needs are the penultimate source of power. Its ultimate source, of course, is physical coercion. While the power that rests on coercion is more absolute, however, it is also more limited in scope than the power that derives from met needs.

A person on whom others are dependent for vital benefits has the power to enforce his demands. He may make demands on them that they consider fair and just in relation to the benefits they receive for submitting to his power. On the other hand, he may lack such restraint and make demands that appear excessive to them, arousing feelings of exploitation for having to render more compliance than the rewards received justify.

Social norms define the expectations of subordinates and their evaluations of the superior's demands. The fair exercise of power gives rise to approval of the superior, whereas unfair exploitation promotes disapproval. The greater the resources of a person on which his power rests, the easier it is for him to refrain from exploiting subordinates by making excessive demands, and consequently the better are the chances that subordinates will approve of the fairness of his rule rather than disapprove of its unfairness.

There are fundamental differences between the dynamics of power in a collective situation and the power of one individual over another. The weakness of the isolated subordinate limits the significance of his approval or disapproval of the superior. The agreement that emerges in a collectivity of subordinates concerning their judgment of the superior, on the other hand, has far-reaching implications for developments in the social structure.

Collective approval of power legitimates that power. People who consider that the advantages they gain from a superior's exercise of power outweigh the hardships that compliance with his demands imposes on them tend to communicate to each other their approval of the ruler and their feelings of obligation to him. The consensus that develops as the result of these communications finds expression in group pressures that promote compliance with the ruler's directives, thereby strengthening his power of control and legitimating his authority. "A feeling of obligation to obey the commands of the established public authority is found, varying in liveliness and effectiveness from one individual to another, among the members of any political society."[vi] Legitimate authority is the basis of organization. It makes it possible to organize collective effort to further the achievement of various objectives, some of which could not be attained by individuals separately at all and others that can be attained more effectively by coordinating efforts. Although power that is not legitimated by the approval of subordinates can also be used to organize them, the stability of such an organization is highly precarious.

Collective disapproval of power engenders opposition. People who share the experience of being exploited by the unfair demands of those in positions of power, and by the insufficient rewards they receive for their contributions, are likely to communicate their feelings of anger, frustration, and aggression to each other. There tends to arise a wish to retaliate by striking down the existing powers. "As every man doth, so shall it be done to him, and retaliation seems to be the great law that is dictated to us by nature."[vii] The social support the oppressed give each other in the course of discussing their common grievances and feelings of hostility justifies and reinforces their aggressive opposition against those in power. It is out of such shared discontent that opposition ideologies and movements develop—that men organize a union against their employer or a revolutionary party against their government.

In brief, differentiation of power in a collective situation evokes contrasting dynamic forces: legitimating processes that foster the organization of individuals and groups in common endeavors; and countervailing forces that deny legitimacy to existing powers and promote opposition and cleavage. Under the influence of these forces, the scope of legitimate organization expands to include ever larger collectivities, but opposition and conflict recurrently redivide these collectivities and stimulate reorganization along different lines.

The distinctive characteristic of complex social structures is that their constituent elements are also social structures. We may call these structures of interrelated groups "macrostructures" and those composed of interacting individuals "microstructures." There are some parallels between the social processes in microstructures and macrostructures. Processes of social attraction create integrative bonds between associates, and integrative processes also unite various groups in a community. Exchange processes between individuals give rise to differentiation among them, and intergroup exchanges further differentiation among groups. Individuals become incorporated in legitimate organizations, and these in turn become part of broader bodies of legitimate authority. Opposition and conflict occur not only within collectivities but also between them. These parallels, however, must not conceal the fundamental differences between the processes that govern the interpersonal associations in microstructures and the forces characteristic of the wider and more complex social relations macrostructures.

First, *value consensus* is of crucial significance for social processes that pervade complex social structures,

---

[vi]Bertrand de Jouvenel, *Sovereignty,* University of Chicago Press, 1957, p. 87.

[vii]Adam Smith, *The Theory of Moral Sentiments* (2d ed.), London: A. Millar, 1761, p. 139.

because standards commonly agreed upon serve as mediating links of social transactions between individuals and groups without any direct contact. Sharing basic values creates integrative bonds and solidarity among millions of people in a society, most of whom have never met, and serves as functional equivalent for the feelings of personal attraction that unite pairs of associates and small groups. Common standards of valuation produce media of exchange—money being the prototype but not the only one—which alone make it possible to transcend personal transactions and develop complex networks of indirect exchange. Legitimating values expand the scope of centralized control far beyond the reach of personal influence, as exemplified by the authority of a legitimate government. Opposition ideals serve as rallying points to draw together strangers from widely dispersed places and unite them in a common cause. The study of these problems requires an analysis of the significance of social values and norms that must complement the analysis of exchange transactions and power relations but must not become a substitute for it.

A second emergent property of macrostructures is the complex interplay between the internal forces within substructures and the forces that connect the diverse substructures, some of which may be microstructures composed of individuals while others may themselves be macrostructures composed of subgroups. The processes of integration, differentiation, organization, and opposition formation in the various substructures, which often vary greatly among the substructures, and the corresponding processes in the macrostructure all have repercussions for each other. A systematic analysis of these intricate patterns . . . would have to constitute the core of a general theory of social structures.

Finally, enduring institutions typically develop in macrostructures. Established systems of legitimation raise the question of their perpetuation through time. The strong identification of men with the highest ideals and most sacred beliefs they share makes them desirous to preserve these basic values for succeeding generations. The investments made in establishing and expanding a legitimate organization create an interest in stabilizing it and assuring its survival in the face of opposition attacks. For this purpose, formalized procedures are instituted that make the organization independent of any individual member and permit it to persist beyond the life span or period of tenure of its members. Institutionalization refers to the emergence of social mechanisms through which social values and norms, organizing principles, and knowledge and skills are transmitted from generation to generation. A society's institutions constitute the social matrix in which individuals grow up and are socialized, with the result that some aspects of institutions are reflected in their own personalities, and others appear to them as the inevitable external conditions of human existence. Traditional institutions stabilize social life but also introduce rigidities that make adjustment to changing conditions difficult. Opposition movements may arise to promote such adjustment, yet these movements themselves tend to become institutionalized and rigid in the course of time, creating needs for fresh oppositions.

Reciprocity and Imbalance

There is a strain toward imbalance as well as toward reciprocity in social associations. The term "balance" itself is ambiguous inasmuch as we speak not only of balancing our books but also of a balance in our favor, which refers, of course, to a lack of equality between inputs and outputs. As a matter of fact, the balance of the accounting sheet merely rests, in the typical case, on an underlying imbalance between income and outlays, and so do apparent balances in social life. Individuals and groups are interested in, at least, maintaining a balance between inputs and outputs and staying out of debt in their social transactions; hence the strain toward reciprocity. Their aspirations, however, are to achieve a balance in their favor and accumulate credit that makes their status superior to that of others; hence the strain toward imbalance.

Arguments about equilibrium—that all scientific theories must be conceived in terms of equilibrium models or that any equilibrium model neglects the dynamics of real life—ignore the important point that the forces sustaining equilibrium on one level of social life constitute disequilibrating forces on other levels. For supply and demand to remain in equilibrium in a market, for example, forces must exist that continually disturb the established patterns of exchange. Similarly, the circulation of the elite, an equilibrium model, rests on the operation of forces that create imbalances and disturbances in the various segments of society. The principle suggested is that balanced social states depend on imbalances in other social states; forces that restore equilibrium in one respect do so by creating disequilibrium in others. The processes of association described illustrate this principle.

A person who is attracted to another will seek to prove himself attractive to the other. Thus a boy who is very much attracted to a girl, more so than she is to him, is anxious to make himself more attractive to her. To do so, he will try to impress her and, particularly, go out of his way to make associating with him an especially rewarding experience for her. He may devote a lot of thought to finding ways to please her, spend much money on her, and do the things she likes on their dates rather than those he would prefer. Let us assume that he is successful and she becomes as attracted to him as he is to her, that is, she finds associating with him as rewarding as he finds associating with her, as indicated by the fact that both are equally eager to spend time together.

Attraction is now reciprocal, but the reciprocity has been established by an imbalance in the exchange. To be sure, both obtain satisfactory rewards from the association at this stage, the boy as the result of her willingness to spend as much time with him as he wants, and the girl as the result of his readiness to make their dates enjoyable for her. These reciprocal rewards are the sources of their mutual attraction. The contributions made, however, are in imbalance. Both devote time to the association, which involves giving up alternative opportunities, but the boy contributes in addition special efforts to please her. Her company is sufficient reward by itself, while his is not, which makes her "the more useful or otherwise superior" in terms of their own evaluations, and he must furnish supplementary rewards to produce "equality in a sense between the parties." Although two lovers may, of course, be equally anxious to spend time together and to please one another, it is rare for a perfect balance of mutual affection to develop spontaneously. The reciprocal attraction in most intimate relations—marriages and lasting friendships as well as more temporary attachments—is the result of some imbalance of contributions that compensates for inequalities in spontaneous affection, notably in the form of one partner's greater willingness to defer to the other's wishes. . . .

The theoretical principle that has been advanced is that a given balance in social associations is produced by imbalances in the same associations in other respects. This principle, which has been illustrated with the imbalances that underlie reciprocal attraction, also applies to the process of social differentiation. A person who supplies services in demand to others obligates them to reciprocate. If some fail to reciprocate, he has strong inducements to withhold the needed assistance from them in order to supply it to others who do repay him for his troubles in some form. Those who have nothing else to offer him that would be a satisfactory return for his services, therefore, are under pressure to defer to his wishes and comply with his requests in repayment for his assistance. Their compliance with his demands gives him the power to utilize their resources at his discretion to further his own ends. By providing unilateral benefits to others, a person accumulates a capital of willing compliance on which he can draw whenever it is to his interest to impose his will upon others, within the limits of the significance the continuing supply of his benefits has for them. The general advantages of power enable men who cannot otherwise repay for services they need to obtain them in return for their compliance; although in the extreme case of the person who has much power and whose benefits are in great demand, even an offer of compliance may not suffice to obtain them.

Here, an imbalance of power establishes reciprocity in the exchange. Unilateral services give rise to a differentiation of power that equilibrates the exchange. The exchange balance, in fact, rests on two imbalances: unilateral services and unilateral power. Although these two imbalances make up a balance or equilibrium in terms of one perspective, in terms of another, which is equally valid, the exchange equilibrium reinforces and perpetuates the imbalances of dependence and power that sustain it. Power differences not only are an imbalance by definition but also are actually experienced as such, as indicated by the tendency of men to escape from domination if they can. Indeed, a major impetus for the eagerness of individuals to discharge their obligations and reciprocate for services they receive, by providing services in return, is the threat of becoming otherwise subject to the power of the supplier of the services. While reciprocal services create an interdependence that balances power, unilateral dependence on services maintains an imbalance of power.

Differentiation of power evidently constitutes an imbalance in the sense of an inequality of power; but the question must be raised whether differentiation of power also necessarily constitutes an imbalance in the sense of a strain toward change in the structure of social relations. Power differences as such, analytically conceived and abstracted from other considerations, create such a pressure toward change, because it can be assumed that men experience having to submit to power as a hardship from which they would prefer to escape. The advantages men derive from their ruler or government, however, may outweigh the hardships entailed in submitting to his or its power, with the result that the analytical imbalance or disturbance introduced by power differences is neutralized. The significance of power imbalances for social change depends, therefore, on the reactions of the governed to the exercise of power.

Social reactions to the exercise of power reflect once more the principle of reciprocity and imbalance, although in a new form. Power over others makes it possible to direct and organize their activities. Sufficient resources to command power over large numbers enable a person or group to establish a large organization. The members recruited to the organization receive benefits, such as financial remuneration, in exchange for complying with the directives of superiors and making various contributions to the organization. The leadership exercises power within the organization, and it derives power from the organization for use in relation with other organizations or groups. The clearest illustration of this double power of organizational leadership is the army commander's power over his own soldiers and, through the force of their arms, over the enemy. Another example is the power business management exercises over its own employees and, through the strength of the concern, in the market. The greater the external power of an organization, the greater are its chances of accumulating resources that put rewards at the disposal of the leadership for possible distribution among the members.

The normative expectations of those subject to the exercise of power, which are rooted in their social experience, govern their reactions to it. In terms of these standards, the benefits derived from being part of an organization or political society may outweigh the investments required to obtain them, or the demands made on members may exceed the returns they receive for fulfilling these demands. The exercise of power, therefore, may produce two different kinds of imbalance, a positive imbalance of benefits for subordinates or a negative imbalance of exploitation and oppression.

If the members of an organization, or generally those subject to a governing leadership, commonly agree that the demands made on them are only fair and just in view of the ample rewards the leadership delivers, joint feelings of obligation and loyalty to superiors will arise

and bestow legitimating approval on their authority. A positive imbalance of benefits generates legitimate authority for the leadership and thereby strengthens and extends its controlling influence. By expressing legitimating approval of, and loyalty to, those who govern them subordinates reciprocate for the benefits their leadership provides, but they simultaneously fortify the imbalance of power in the social structure.

If the demands of the men who exercise power are experienced by those subject to it as exploitative and oppressive, and particularly if these subordinates have been unsuccessful in obtaining redress for their grievances, their frustrations tend to promote disapproval of existing powers and antagonism toward them. As the oppressed communicate their anger and aggression to each other, provided there are opportunities for doing so, their mutual support and approval socially justify and reinforce the negative orientation toward the oppressors, and their collective hostility may inspire them to organize an opposition. The exploitative use of coercive power that arouses active opposition is more prevalent in the relations between organizations and groups than within organizations. Two reasons for this are that the advantages of legitimating approval restrain organizational superiors and that the effectiveness of legitimate authority, once established, obviates the need for coercive measures. But the exploitative use of power also occurs within organizations, as unions organized in opposition to exploitative employers show. A negative imbalance for the subjects of power stimulates opposition. The opposition negatively reciprocates, or retaliates, for excessive demands in an attempt to even the score, but it simultaneously creates conflict, disequilibrium, and imbalance in the social structure.[viii]

Even in the relatively simple structures of social association considered here, balances in one respect entail imbalances in others. The interplay between equilibrating and disequilibrating forces is still more evident, if less easy to unravel, in complex macrostructures with their cross-cutting substructures, where forces that sustain reciprocity and balance have disequilibrating and imbalancing repercussions not only on other levels of the same substructure but also on other substructures.

∷∷

[viii]Organized opposition gives expression to latent conflicts and makes them manifest.

▪▪ JAMES S. COLEMAN (1926–1995): A BIOGRAPHICAL SKETCH

James S. Coleman was born in Bedford, Indiana, in 1926. After graduating from high school, he served in the U.S. Navy during World War II. Once discharged from the Navy he attended Purdue University, where he graduated with a bachelor's degree in chemical engineering in 1949. Upon graduating, he was hired by the photography giant Eastman-Kodak, based in Rochester, New York. He soon left his job, however, to purse a PhD in sociology from Columbia University, which he received in 1955. While there, he studied under the tutelage of some of the discipline's most influential figures of the day, most notably Robert Merton, Paul Lazarsfeld, and Seymour Martin Lipset. As a professor at the University of Chicago (1956–1959, 1973–1995) and Johns Hopkins University (1959–1973), Coleman would go on to become one of the most distinguished sociologists in his own right, publishing more than 300 articles and 25 books, including his thousand-page opus, *The Foundations of Social Theory* (1990), in which he laid the intellectual groundwork for rational-choice theory. For Coleman, however, sociology was not to be an academic pursuit carried out by experts in a cloistered ivory tower; instead, it was a science with a unique capacity for informing crucial public policy issues. To that end, much of his work examined the implications of public policy for America's educational system. In 1966, he published the *Equality of Educational Opportunity,* a landmark study conducted for the federal government that examined 600,000 students and 60,000 teachers in 4,000 public schools in order to explore "the lack of availability of equal education opportunities for individuals by reason of race, color, religion, or national origin" (*1964 Civil Rights Act,* Title IV Sec. 402). Based on his findings, he concluded— controversially—that school funding, instructional materials, and teachers' pay did not account for the educational achievement gap between white and black students. In fact, majority black schools were funded comparably to majority white schools by the beginning of the 1960s. (Southern states in particular increased their funding for black schools in the wake of the 1954 *Brown v. Board of Education* Supreme Court decision in an effort to avoid having to comply with its mandates.) Moreover, Coleman found that the academic performance of lower-income black children improved when they attended middle-class, integrated schools, leading him to conclude that the effects of school funding, instructional materials, and the school environment more generally, were mediated through the effects of students' socioeconomic status, race, and family environment. In response, this study, commonly known as the "Coleman Report," helped set the stage for implementing a system of busing to racially integrate schools and create equal educational opportunities for black children. In a subsequent report published in 1975, he revised his policy recommendations as the "white flight" response to busing exacerbated the decline of urban public schools. While a strident opponent of racial segregation, his reversal on the use of busing to achieve racial parity sparked a controversy that saw some civil rights leaders, policy makers, and experts in education vociferously challenging his findings. Indeed, some sociologists called for his expulsion from the discipline's national organization, the American Sociological Association.

Coleman would pursue his interest in the educational system through a number of other works including his coauthored studies, *High School Achievement: Public, Catholic and Private Schools Compared* (1982) and *Public and Private High Schools: The Impact of Communities* (1987). These works were based on a multiwave research program he headed, "High School and Beyond." Based on an initial sample of more than 58,000 sophomore and senior high school students drawn from 1,015 public and private schools, the study is the country's largest longitudinal survey ever conducted on the effects of schooling on individuals' careers and lives.

During his distinguished career, he was a member of the National Academy of Sciences, the American Academy of Arts and Sciences, the Royal Swedish Academy of Sciences, the American Philosophical Society, and the National Academy of Education. His expertise in the American educational system led to his appointment to the President's Science Advisory Committee under President Richard Nixon. In 1989, he was the recipient of the Educational Freedom Award. Coleman also served as president of the American Sociological Association in 1992 and received the Association's

Distinguished Publication Award that year for his *Foundations of Social Theory.* Indeed, it is Coleman's leading role in the development of rational choice theory, a perspective aligned with exchange theory, that established him as a key contemporary theorist. His contribution to the discipline lives on in *Rationality and Society,* a journal he founded in 1989 dedicated to advancing theory and research grounded in a rational choice framework.

COLEMAN'S INTELLECTUAL INFLUENCES AND CORE IDEAS ▓

Rational choice theories share with exchange theories a view of the actor as a rational, purposive agent motivated by maximizing rewards or realizing his interests. Beyond this similarity, however, a subtle yet important shift in emphasis distinguishes the two perspectives. Exchange theorists emphasize the strategic decision making of individuals and how such individual decisions can produce, sustain, modify, or terminate social relationships within small groups, often dyads. Here, the more collectivist aspects of social order—for instance, norms, social networks, institutionalized access to resources—are either left unexamined or incorporated in a superficial manner as social "givens." (Blau's focus on norms was an attempt to rectify such criticism.) Within sociological theory, rational choice theorists, while emphasizing the strategic decisions of individuals, are more inclined to situate such decisions within the context of group dynamics. Thus, rational choice theorists explore how interaction between rationally motivated individuals can produce norms, networks, control of resources, and group solidarity, as well how these factors, once created, orient or constrain individuals' decisions and behavior. In other words, more attention is granted to understanding the "feedback loop" that exists between the purposive action of individuals and the broader social conditions in which they are embedded.

Trust and Norms

Coleman set out to elaborate the premises and promises of rational choice theory in *The Foundations of Social Theory* (1990). To this end, he introduced a number of concepts to explain and predict not only individual decision making and behavior, but also the formation of groups, or, in Coleman's terms, "corporate actors," and the conditions that enable and affect individual and group or corporate decision making. **Trust** is one such concept; as Coleman notes, "actors place trust, in both the judgement and performance of others, based on rational considerations of what is best, given the alternatives they confront" (ibid.:192). Importantly, the alternatives for possible rewards that an individual has at his disposal are determined by the people with whom he associates. As a result, possible rewards are not equally available to all, but instead are distributed along broader social dimensions such as age, gender, class, and race, because each of these influences the amount and type of resources a person has as well as who one associates with (ibid.). Regardless of the relative advantages and disadvantages that stem from one's network of associates, when a person (trustor) places trust in another (trustee), the latter accrues additional resources paid by the trustor. Thus, it is to the advantage of the trustee to be trustworthy in order to reap possible future benefits (whether they be favors from friends or a merchant's continued business with a supplier) that come with establishing a reputation as a trustworthy person. Likewise, it is advantageous to the trustor to place trust in a person when the possible gain for doing so outweighs the possible loss that comes with the risk that the person might prove to be untrustworthy, a risk that increases or decreases depending on the information one has about the trustee and the amount of possible gain and loss. For instance, say Julia (the trustee) asks to borrow $200 from Malika (the trustor) to buy some new clothes, and in return, Julia offers to repay Malika by watching her pets when Malika goes on vacation. Malika now has to decide whether or not to trust Julia to follow through on her promises. Her decision will be based on how valuable the $200 is to her, how important it is to her to avoid boarding her pets when she goes away on vacation, and how well she knows Julia. If

Malika places her trust in Julia, then Julia benefits by getting money that she can use to buy the clothes. If Julia proves to be trustworthy, then Malika benefits by getting her money back and being able to vacation without having to pay the expense of boarding her pets. Julia's trustworthiness allows her to ask for future favors and increases the likelihood of having her requests granted. However, if Julia proves to be untrustworthy, then Malika is out $200 while her pets starve to death while she is away on vacation. On the bright side, Julia has a new wardrobe.

In essence, the decision to place trust in another involves the same rational calculations that are involved in deciding whether or not to place a bet: both entail knowing how much will be lost (the size of the bet), how much can be gained or won, and the chances of winning (the information one has at her disposal). "If the *chance of winning,* relative to the *chance of losing,* is greater than the *amount that would be lost* (if he loses), relative to the *amount that would be won* (if he wins), then by placing the bet he has an expected gain; and if he is rational, he should place it" (Coleman 1990:99; emphasis in the original).

Like trust, Coleman sees norms emerging on the basis of actors' rational considerations. More specifically, they develop out of repeated exchanges between a limited number of individuals who have an interest in maintaining a relationship with one another. The norms that emerge are not assumed to be universally shared across a society, but instead apply only to those individuals involved in the relationship. Decisions to adhere to and enforce norms, like those that led to their initial emergence, are conditioned by the relative costs and benefits derived from doing so. This view fundamentally contrasts with the one taken by collectivist theories such as structural functionalism, where norms are taken as preexisting "givens" more or less internalized universally by individuals within a social system. Here, norms become part of individuals' cognitive makeup as processes of socialization lead to the acceptance and reproduction of particular normative codes of behavior, even those codes that reinforce relations of domination. From such a perspective, the questions of how and why norms initially emerge and are adhered to are left largely unanswered. The same can be said for Homans's incorporation of norms such as distributive justice into his version of exchange theory. He, too, assumes the presence of norms and thus either explains their emergence in an ad hoc fashion or attributes it to psychological principles innate to the individual.

But what, specifically, are norms from Coleman's perspective? **Norms** are socially defined informal rights to control the actions of others. "No norm exists as long as the individual actor holds the right to his own action, and no norm exists if no right [to control others' actions] has come into existence. A norm exists only when others assume the right to affect the direction an actor's action will take" (ibid.:243). Thus, in establishing a norm, individuals are required to forfeit some of their rights to control their own actions even if it prevents them from realizing one or another of their interests. Individuals who hold the right to control the actions of others are the "beneficiaries" of the particular norm as they realize benefits from encouraging or inducing others to perform or refrain from particular actions. The "targets" of the norm are those who give up their rights to control their actions. In some instances, the same individuals are both the beneficiaries and the targets of the norm (a "conjoint" norm). In giving up the right to control some of their own actions—and subordinating their own interests to those of another individual or group—the individuals who are the targets of the norm also receive the benefits that come from having the authority to claim the right to control the same actions of others. Here it is rational for the individual to accept the legitimacy of others' claims to control his action because doing so is a necessary condition for giving him the right to control their actions. In other instances, there is a conflict of interest because some individuals are beneficiaries while others are targets (a "disjoint" norm). In either case, the transferring of rights that is the basis of norms is born out of those circumstances in which individuals see themselves either as benefitting if particular actions are followed or harmed if they are not. To encourage proper behavior and the positive consequences they produce, "prescriptive norms" are established; "proscriptive norms" are established to discourage improper behavior and the negative consequences that follow it. Table 11.5 offers an illustration of the various norms.

Table 11.3 Types of Norms

	Disjoint	Conjoint
Prescriptive	men (beneficiaries) expecting women (targets) to do most of the household chores	an office "policy" requiring that the person who finishes the pot of coffee makes a new pot
Proscriptive	a nonsmoker (beneficiary) inducing a smoker (target) to extinguish his cigarette in a restaurant	members of a group preventing each other from marrying anyone from outside the group

Up to this point, we have discussed what norms are and why some individuals have an interest in controlling the actions of others, namely to encourage behavior that produces positive consequences and to discourage behavior that produces negative consequences. Yet, in order for a demand for a norm to be realized—that is, in order for the norm to be effective—it is not enough that beneficiaries claim the right to control the actions of targets. The beneficiaries also must be able to apply sanctions—rewards and punishments—to the targets to ensure compliance with their demands, particularly in those instances in which the targets have not internalized the desire to conform. Conformity with normative demands, then, can only be sustained when those individuals subjected to them (1) are rewarded by the beneficiaries with positive sanctions for compliance, (2) are punished by the beneficiaries with negative sanctions for engaging in improper behavior, (3) have internalized the demand and thus self-sanction their conduct by experiencing internally generated rewards for performing proper behavior, or (4) have internalized the demand and thus self-sanction their conduct by experiencing internally generated punishments for performing improper actions. In the end, it is the sanctions attached to effective norms that provide the incentive for individuals to cede their rights and coordinate their activities with others in the pursuit of collective goals. With the emergence and realization of norms, group relations and the patterning of behavior that reinforces norms can be sustained over time while limiting wholly self-interested (and potentially socially destructive) conduct.

However, because effective norms require the implementation of sanctions, any number of behaviors are unregulated by others. Those who would benefit from encouraging or preventing particular behaviors may find that the rewards gained for doing so are outweighed by the costs entailed in applying sanctions. An example from Coleman's (Coleman 1990:281) work illustrates this principle:

> [S]uppose all members of a club are expected to clean up after meetings, but one member consistently fails to help. If one person expresses disapproval, this might induce a small effort on the offender's part, but would also have a negative effect on the relationship between the two, an effect that might be more important to the potential sanctioner than the benefit from the offender's effort.

In this case, the rewards gained by the individual for applying "heroic" sanctions are less than the potential costs entailed. As a result, the other member's improper behavior will continue. However, if "incremental" sanctions in the form of collective expressions of disapproval were employed, the offender may be induced to make his full contribution and "the benefits to each would outweigh the costs of each one's worsened relation with the offender" (ibid.). (This same dynamic is likely to arise in our "coffee-pot" example in Table 11.3.)

This example leads to another related issue: how the distribution of power and resources determines who can or cannot impose sanctions and who is or is not subjected to them. Individuals with greater power and resources are less likely to abide by norms and less likely to be sanctioned due to the fact that sanctioning such individuals typically incurs higher costs for those who have less power. Certainly we have all heard of the wealthy businessperson, the politician, or the group leader whose behavior suggests that the rules do not apply to him or her. Conversely, it is less costly to sanction those of lower status, a calculation that perhaps accounts for the fact that often the same violation will exact harsher punishments (fines, prison terms) for lower-status persons than for higher-status persons. Or, to take an example from Table 11.3, a smoker is inclined to give up his right to smoke in particular settings when the beneficiaries of the norm—the nonsmokers—have more power, given their recourse to legally backed negative sanctions. In cases where an unequal distribution of power and resources exists, those subjected to norms, and perhaps even exploited by them (for example, by norms that legitimate racism or sexism), are likely to conform because the costs for compliance are lower than the costs for deviance. In many instances, it is thus in one's self-interest to reproduce his or her own submission.

The Free Rider

As a final consideration, in examining the relationship between individual decisions and group action, Coleman (and many other rational choice theorists) turned his attention to the **free-rider** problem.[7] The term *free rider* refers to the individual's rational decision not to participate in group activity if the goals of the group or the "joint goods" it produces cannot be denied to the individual and if their supply is not reduced by others consuming them. For instance, given that any one individual's efforts in the fight to protect the natural environment border on being inconsequential, why should an individual spend his time and energy fighting for the cause? Surely, one person's taking the time to distribute literature in his community or to write letters to his or her political representatives, or even to donate $50 to an environmental group, will not tip the balance in the fight one way or another. Similarly, common interests derived from a shared structural position (e.g., economic interests of the working class to raise taxes on the wealthy) are not enough to produce group solidarity. That most individuals do not engage in collective action to further their interests is evidenced, for instance, by poor and working-class individuals in the United States who have failed to galvanize as a group to effectively demand a more equitable distribution of the nation's wealth.

But this is not a matter of defeatism; rather, it makes sense (i.e., it is rational) to let others bear the costs of campaigning for cleaner air and water or tax reform, because if the group is successful in achieving its goals, the individual will receive the benefits for free. After all, you cannot stop someone from breathing air, drinking water, or getting a tax rebate, and their supply, while not necessarily unlimited, is not diminished such that noncontributors to the group's cause are unable to consume them. The issue for rational choice theorists, then, is how to explain the formation and success of *groups* when it is rational for *individuals* not to pay the price of participation.

So, why do people participate in the production of joint goods when they cannot be excluded from consuming the benefits the group provides? Rational choice theorists typically offer three solutions to this seeming paradox. First, groups can reward individuals for their participation through "selective incentives"—benefits that are distributed exclusively to those who bear the costs for providing a good. Selective incentives can take on any number of forms, from winning a leadership position in a revolutionary movement to receiving a coffee mug or bumper sticker for making a contribution to a public television station. Second, individuals may work toward achieving a group goal for the intangible benefits that participation itself can provide. For instance, individuals may receive social approval from others or experience a sense of personal satisfaction for having altruistically participated in realizing a collective goal. Thus, a person might even risk imprisonment in order to protest a chemical plant's environmental practices or endure the hardships of a hunger strike to protest political conditions if positive sanctions or the encouragement of others is sufficient to motivate him to bear such extreme costs, even though the possible personal benefits that might come with a successful outcome would pale in comparison. Such a transformation from the rationality of free riding to the rationality of "zeal" is more likely to occur when the social structure in which actors are embedded is "closed." In this case, the actors are connected to

---

[7]See Mancur Olson, *The Logic of Collective Action* (1965); Russell Hardin, *Collective Action* (1982); and Michael Hechter (1987), *Principles of Group Solidarity* for foundational statements on the free-rider problem and its effects on goal-oriented group action.

or know one another such that each is able to influence the behavior of the others by directly offering "low-cost" encouragement or gratitude for their efforts (Coleman 1990:272–78). Third, free riding can be minimized by enforcing participation or negatively sanctioning individuals for not contributing to the public good. This is the rationale of government taxes and the penalties incurred for not paying them.

Michael Hechter (1943–): Rational Choice and Group Solidarity

Michael Hechter graduated with a PhD from Columbia University in 1972. Over his 40-year career, he has taught at the University of Washington, the University of Oxford, the University of Arizona, and Arizona State University, where he currently serves as the director of the School of Politics and Global Studies. A prolific writer, his publications include three single-authored books—*Principles of Group Solidarity* (1987; selected as a *Choice* Outstanding Academic Book in 1988), *Containing Nationalism* (2000), and *Internal Colonialism: The Celtic Fringe in British National Development* ([1975] 1999; selected in 2000 by the International Sociological Association as one of the *Best Books of the Century*)—more than eighty articles and book chapters, and five edited or coedited books. He has received numerous honors and awards, including his election as a Fellow to the American Academy of Arts & Sciences. Hechter has presented his work at colleges and universities across the United States and in a host of other countries (Arizona State University n.d.).

Much of Hechter's research revolves around explaining the development of group solidarity, particularly as it relates to the emergence of nationalist movements. He grounds his examination of the institutional arrangements and processes that shape how, when, and where groups form within the framework of rational choice theory. In doing so, he explicitly seeks to extend the individualist orientation of the framework to an analysis of issues that have traditionally been explored through collectivist approaches. For instance, he argues in *Internal Colonialism* (1975) that the rise of nationalism is tied to the modern development of centralized, direct rule that stratifies the life chances of individuals based on cultural distinctions, typically race or ethnicity. Politicizing such distinctions engenders the rational motivation for culturally subordinated individuals (e.g., the Celts under British rule or the Chechnyans under Russian rule) to engage in nationalist, often violent, political activity.

Hechter's theoretical treatment of the issue of social cohesion is offered in his book *Principles of Group Solidarity* (1987). He begins with the premise that rational actors form or join a group solely to consume the benefits the group jointly (i.e., cooperatively) produces. As a result, groups must ensure the continued production of their benefits and meet the needs of their individual members—otherwise, they will not survive. But how can an individual be compelled to participate in the production of joint goods if it is in her self-interest to free ride? In other words, how can group solidarity be successfully cultivated such that individuals forgo the pursuit of their own selfish goals and interests in order to produce group benefits?

For Hechter, the provision of selective incentives or coercive measures—the conventional rational choice explanations—does not adequately prevent free riding, even in the case of groups such as political parties, professional associations, or food co-ops that produce excludable collective goods (goods produced solely for group members to which they can be denied access). Instead, group solidarity is dependent on guaranteeing that members comply with group rules. The obligation to conform to the will of the group and to forgo one's own interests serves "as a tax that is imposed

(Continued)

(Continued)

upon each member as a condition of access to the joint good" (1987:10). The intensity of obligation, and thus the extensiveness of group solidarity is itself conditioned by two factors: dependency and control mechanisms. The more a member is dependent on the group to provide the goods in question, the greater the tax—in the form of compliance with group obligations—she will bear in order to produce the joint good. Similarly, the fewer alternative sources there are for providing the good, the more costly it will be for her to refuse compliance and do without the group's good. (The range of available alternative sources is typically due to "market" conditions that are not subject to the group's control.) However, dependency alone does not guarantee compliance with group obligations and a solution to the free-rider problem. In addition, compliance—and thus the maintenance of group solidarity—requires the development of control mechanisms in the form of monitoring and sanctioning. Monitoring entails the capacity to collect information about individuals' compliance with group obligations—for instance, by providing opportunities for individuals to discuss the behavior of other group members (gossiping), or increasing the visibility of individuals' performance of their group obligations. Sanctioning refers to a group's capacity to produce and mete out measures that discourage noncompliance. They take the form of rewards such as increased prestige or monetary compensation to members who fulfill their obligations, and punishments such as expulsion, loss of privileges, or public criticism to those who fail to do so. Compliance is increased and group solidarity is enhanced when detection of both conforming and nonconforming behaviors is more likely and when the magnitude of both positive and negative sanctions is increased. Greater dependency on the group for the good it produces, in turn, lowers the group's control costs because individuals are less likely to risk noncompliance for fear of expulsion (ibid.:126).

▓▓ COLEMAN'S THEORETICAL ORIENTATION

On the question of action, Coleman takes a decidedly rationalist view. This is clearly demonstrated when he notes that "[a]ctors have a single principle of action, that of acting so as to maximize their realization of interests" (Coleman 1990:37) as they weigh the potential rewards and punishments accompanying their behavior. In adopting this perspective, however, Coleman does not suggest that nonrational factors such as norms play an inconsequential role in shaping individual behavior. He acknowledges that conduct is subject to normative regulation and that individuals do take norms into account when developing their lines of action: "persons do obey them (though not uniformly), and persons do often act in the interests of others or of a collectivity, 'unselfishly' as we would say" (ibid.: 31). Nevertheless, norms are not "absolute determinants of action," but, rather, "elements which affect [individuals'] decisions about what actions it will be in their interests to carry out" (ibid.: 243). Similarly, trust, while playing a vital role in making cooperative action possible, is not rooted in some type of emotional bond or moral obligation to another. It arises, or not, between individuals based on their rational assessment of the potential costs and gains that come with relying on another. Thus, when someone says, "I feel like he is someone I can trust," it is a calculation of risk, and not the "feeling," that leads to this conclusion.

Coleman's orientation to the question of order, while predominately individualist, recognizes the collectivist aspects to the patterning of social life. This is captured in his view that to best understand social life, theory must be attuned to the feedback process through which individual purposive action is translated into collectivist or "system-level" properties that then work back down to influence the action of individuals. Norms (and rights), for instance, possess this dual feature that is represented by their straddling the "order" line in Figure 11.4. Norms can only originate through the action of individuals, "yet a norm itself is a system-level property which affects the further action of individuals, both the sanctions applied by individuals who hold the norm and the action in conformity with the norm" (Coleman 1990:244). On one hand, then, norms are collectivist constructs that can lead individuals to pursue particular courses of action because the sanctions that are attached to them condition the weighing of benefits and costs and thus the likelihood of undertaking a given path of behavior. Yet, on the other hand, norms exist only as an expression of individual action. To be sure, Coleman's individualist orientation is more a matter of emphasis than of a disavowal of the collectivist or system-level effects on individual action. (Our discussion of social capital below illustrates a collectivist dimension to Coleman's theory.) However, this emphasis is critical to Coleman's theoretical model and understanding of human nature and not a mere academic splitting of hairs. For Coleman, theories that emphasize the collectivist determinants of social life present "a fatalistic view of the future, in which humans are the pawns" of forces outside their control (ibid.:17). Within a collectivist orientation, "the proposed causes of action are not persons' goals or purposes or intents, but some forces outside them or unconscious impulses within them. As a consequence, these theories can do nothing other than describe an inexorable fate; they are useful only to describe the waves of change that wash over us. At the mercy of these uncontrolled external or internal forces, persons are unable to purposefully shape their destiny" (ibid.). Coleman's intent is to counter such a view by asserting that all that is social must be created and maintained—accomplishments that can be effected only through the purposive action of individuals. Normative regulations and moral codes do not originate on their own, and adherence to them is not the product of some sort of "automated conformity." They result from the voluntary action of individuals whose rational decisions determine their birth and development. (See Figure 11.4.)

Figure 11.4 Coleman's Basic Concepts and Theoretical Orientation

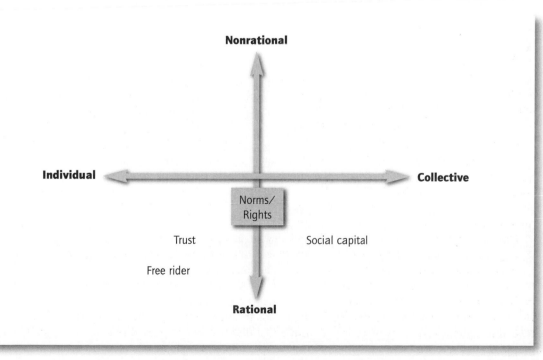

Introduction to "Social Capital in the Creation of Human Capital"

In the reading below, Coleman addresses the role of social capital as an important resource that enables individuals to realize their interests. While it is in a sense something an individual or an organization possesses, **social capital** "inheres in the structure of relations between actors and among actors" (Coleman 1988:S98), particularly those relations that are part of a "closed" social network in which most individuals either directly or indirectly (as in a friend of a friend) know one another. It is thus a product of social relations and not something that an individual or organization alone can develop or use at its own discretion. While social capital comes in many forms, Coleman here emphasizes how authority, trustworthiness (and the obligations and expectations that are tied to trust), information channels, and social norms are types of resources or capital that allow individuals to achieve goals that are otherwise unattainable. For instance, when you do a favor for a friend or colleague, it establishes an obligation on his or her part to repay the favor in order to uphold norms of trustworthiness and maintain the relationship. Much like a "credit slip," you can redeem the obligation owed in order to accomplish a task that might otherwise be impossible or too costly to undertake alone. Similarly, individuals often use their relationships with others in order to gain access to information such as the inside story on a job opening or the latest trends in music or fashion (Coleman 1990:300–21). Here, Coleman explores how the presence of social capital in the family and in the broader community fosters a child's success in school, while its absence increases the likelihood of dropping out. Thus, social capital plays a central role in the creation of human capital.

—— "Social Capital in the Creation of Human Capital" (1988) ——

James S. Coleman

There are two broad intellectual streams in the description and explanation of social action. One, characteristic of the work of most sociologists, sees the actor as socialized and action as governed by social norms, rules, and obligations. The principal virtues of this intellectual stream lie in its ability to describe action in social context and to explain the way action is shaped, constrained, and redirected by the social context.

The other intellectual stream, characteristic of the work of most economists, sees the actor as having goals independently arrived at, as acting independently, and as wholly self-interested. Its principal virtue lies in having a principle of action, that of maximizing utility. This principle of action, together with a single empirical generalization (declining marginal utility) has generated the extensive growth of neoclassical economic theory, as well as the growth of political philosophy of several varieties: utilitarianism, contractarianism, and natural rights.

. . . I have argued for and engaged in the development of a theoretical orientation in sociology that includes components from both these intellectual streams. It accepts the principle of rational or purposive action and attempts to show how that principle, in conjunction with particular social contexts, can account not only for the actions of individuals in particular contexts but also for the development of social organization. In the present paper, I introduce a conceptual tool for use in this theoretical enterprise: social capital. . . .

SOCIAL CAPITAL

. . . If we begin with a theory of rational action, in which each actor has control over certain resources and interests in certain resources and events, then social capital constitutes a particular kind of resource available to an actor.

Social capital is defined by its function. It is not a single entity but a variety of different entities, with two elements in common: they all consist of some aspect of social structures, and they facilitate certain actions of actors—whether persons or corporate actors—within the structure. Like other forms of capital, social capital is productive, making possible the achievement of certain ends that in its absence would not be possible. Like physical capital and human capital, social capital is not completely fungible but may be specific to certain activities. A given form of social capital that is valuable in facilitating certain actions may be useless or even harmful for others.

Unlike other forms of capital, social capital inheres in the structure of relations between actors and among actors. It is not lodged either in the actors themselves or in physical implements of production. Because purposive organizations can be actors ("corporate actors") just as persons can, relations among corporate actors can constitute social capital for them as well (with perhaps the best-known example being the sharing of information that allows price-fixing in an industry). However, in the present paper, the examples and area of application to which I will direct attention concern social capital as a resource for persons. . . .

HUMAN CAPITAL AND SOCIAL CAPITAL

Probably the most important and most original development in the economics of education in the past 30 years has been the idea that the concept of physical capital as embodied in tools, machines, and other productive equipment can be extended to include human capital as well (see Schultz 1961; Becker 1964). Just as physical capital is created by changes in materials to form tools that facilitate production, human capital is created by changes in persons that bring about skills and capabilities that make them able to act in new ways.

Social capital, however, comes about through changes in the relations among persons that facilitate action. If physical capital is wholly tangible, being embodied in observable material form, and human capital is less tangible, being embodied in the skills and knowledge acquired by an individual, social capital is less tangible yet, for it exists in the *relations* among persons. Just as physical capital and human capital facilitate productive activity, social capital does as well. For example, a group within which there is extensive trustworthiness and extensive trust is able to accomplish much more than a comparable group without that trustworthiness and trust.

FORMS OF SOCIAL CAPITAL

The value of the concept of social capital lies first in the fact that it identifies certain aspects of social structure by their functions, just as the concept "chair" identifies certain physical objects by their function, despite differences in form, appearance, and construction. The function identified by the concept of "social capital" is the value of these aspects of social structure to actors as resources that they can use to achieve their interests.

By identifying this function of certain aspects of social structure, the concept of social capital constitutes both an aid in accounting for different outcomes at the level of individual actors and an aid toward making the micro-to-macro transitions without elaborating the social structural details through which this occurs. . . .

Before examining empirically the value of social capital in the creation of human capital, I will go more deeply into an examination of just what it is about social relations that can constitute useful capital resources for individuals.

Obligations, Expectations, and Trustworthiness of Structures

If *A* does something for *B* and trusts *B* to reciprocate in the future, this establishes an expectation in *A* and an obligation on the part of *B*. This obligation can be conceived as a credit slip held by *A* for performance by *B*. If *A* holds a large number of these credit slips, for a number of persons with whom *A* has relations, then the analogy to financial capital is direct. These credit slips constitute a large body of credit that *A* can call in if necessary—unless, of course, the placement of trust has been unwise, and these are bad debts that will not be repaid.

In some social structures, it is said that "people are always doing things for each other." There are a large number of these credit slips outstanding, often on both sides of a relation (for these credit slips appear often not to be completely fungible across areas of activity, so that credit slips of *B* held by *A* and those of *A* held by *B* are not fully used to cancel each other out). . . . In other social structures where individuals are more self-sufficient and depend on each other less, there are fewer of these credit slips outstanding at any time.

This form of social capital depends on two elements: trustworthiness of the social environment, which means that obligations will be repaid, and the actual extent of obligations held. Social structures differ in both these dimensions, and actors within the same structure differ in the second. A case that illustrates the value of the trustworthiness of the environment is that of the rotating-credit associations of Southeast Asia and elsewhere. These associations are groups of friends and neighbors who typically meet monthly, each person contributing to a central fund that is then given to one of the members (through bidding or by lot), until, after a number of months, each of the *n* persons has made *n* contributions and received one payout. As Geertz (1962) points out, these associations serve as efficient institutions for amassing savings for small capital expenditures, an important aid to economic development.

But without a high degree of trustworthiness among the members of the group, the institution could not exist—for a person who receives a payout early in the sequence of meetings could abscond and leave the others with a loss. For example, one could not imagine a rotating-credit association operating successfully in urban areas marked by a high degree of social disorganization—or, in other words, by a lack of social capital.

Differences in social structures in both dimensions may arise for a variety of reasons. There are differences in the actual needs that persons have for help, in the existence of other sources of aid (such as government welfare services), in the degree of affluence (which reduces aid needed from others), in cultural differences in the tendency to lend aid and ask for aid (see Banfield 1967) in the closure of social networks, in the logistics of social contacts (see Festinger, Schachter, and Back 1963), and other factors. Whatever the source, however, individuals in social structures with high levels of obligations outstanding at any time have more social capital on which they can draw. The density of outstanding obligations means, in effect, that the overall usefulness of the tangible resources of that social structure is amplified by their availability to others when needed.

Individual actors in a social system also differ in the number of credit slips outstanding on which they can draw at any time. The most extreme examples are in hierarchically structured extended family settings, in which a patriarch (or "godfather") holds an extraordinarily large set of obligations that he can call in at any time to get what he wants done. Near this extreme are villages in traditional settings that are highly stratified, with certain wealthy families who, because of their wealth, have built up extensive credits that they can call in at any time.

Similarly, in political settings such as a legislature, a legislator in a position with extra resources (such as the Speaker of the House of Representatives or the Majority Leader of the Senate in the U.S. Congress) can, by effective use of resources, build up a set of obligations from other legislators that makes it possible to get legislation passed that would otherwise be stymied. This concentration of obligations constitutes social capital that is useful not only for this powerful legislator but useful also in getting an increased level of action on the part of a legislature. Thus, those members of legislatures among whom such credits are extensive should be more powerful than those without extensive credits and debits because they can use the credits to produce bloc voting on many issues. It is well recognized, for example, that in the U.S. Senate, some senators are members of what is called "the Senate Club," while others are not. This in effect means that some senators are embedded in the system of credits and debits, while others, outside the "Club," are not. It is also well recognized that those in the Club are more powerful than those outside it.

Information Channels

An important form of social capital is the potential for information that inheres in social relations. Information is important in providing a basis for action. But acquisition of information is costly. At a minimum, it requires attention, which is always in scarce supply. One means by which information can be acquired is by use of social relations that are maintained for other purposes. Katz and Lazarsfeld (1955) showed how this operated for women in several areas of life in a midwestern city around 1950. They showed that a woman with an interest in being in fashion, but no interest in being on the leading edge of fashion, used friends who she knew kept up with fashion as sources of information. Similarly, a person who is not greatly interested in current events but who is interested in being informed about important developments can save the time of reading a newspaper by depending on spouse or friends who pay attention to such matters. A social scientist who is interested in being up-to-date on research in related fields can make use of everyday interactions with colleagues to do so, but only in a university in which most colleagues keep up-to-date.

All these are examples of social relations that constitute a form of social capital that provides information that facilitates action. The relations in this case are not valuable for the "credit slips" they provide in the form of obligations that one holds for others' performances or for the trustworthiness of the other party but merely for the information they provide.

Norms and Effective Sanctions

When a norm exists and is effective, it constitutes a powerful, though sometimes fragile, form of social capital. Effective norms that inhibit crime make it possible to walk freely outside at night in a city and enable old persons to leave their houses without fear for their safety. Norms in a community that support and provide effective rewards for high achievement in school greatly facilitate the school's task.

A prescriptive norm within a collectivity that constitutes an especially important form of social capital is the norm that one should forgo self-interest and act in the interests of the collectivity. A norm of this sort, reinforced by social support, status, honor, and other rewards, is the social capital that builds young nations (and then dissipates as they grow older), strengthens families by leading family members to act selflessly in "the family's" interest, facilitates the development of nascent social movements through a small group of dedicated, inward-looking, and mutually rewarding members, and in general leads persons to work for the public good. In some of these cases, the norms are internalized; in others, they are largely supported through external rewards for selfless actions and disapproval for selfish actions. But, whether supported by internal or external sanctions, norms of this sort are important in overcoming the public goods problem that exists in collectivities.

As all these examples suggest, effective norms can constitute a powerful form of social capital. This social capital, however, like the forms described earlier, not only facilitates certain actions; it constrains others. A community with strong and effective norms about young persons' behavior can keep them from "having a

good time." Norms that make it possible to walk alone at night also constrain the activities of criminals (and in some cases of noncriminals as well). Even prescriptive norms that reward certain actions, like the norm in a community that says that a boy who is a good athlete should go out for football, are in effect directing energy away from other activities. Effective norms in an area can reduce innovativeness in an area, not only deviant actions that harm others but also deviant actions that can benefit everyone. (See Merton [1968, pp. 195–203] for a discussion of how this can come about.)

SOCIAL STRUCTURE THAT FACILITATES SOCIAL CAPITAL

All social relations and social structures facilitate some forms of social capital; actors establish relations purposefully and continue them when they continue to provide benefits. Certain kinds of social structure, however, are especially important in facilitating some forms of social capital.

Closure of Social Networks

One property of social relations on which effective norms depend is what I will call closure. In general, one can say that a necessary but not sufficient condition for the emergence of effective norms is action that imposes external effects on others. . . . Norms arise as attempts to limit negative external effects or encourage positive ones. But, in many social structures where these conditions exist, norms do not come into existence. The reason is what can be described as lack of closure of the social structure. Figure 1 illustrates why. In an open structure like that of figure 1*a*, actor *A*, having relations with actors *B* and *C*, can carry out actions that impose negative externalities on *B* or *C* or both. Since they have no relations with one another, but with others instead (*D* and *E*), then they cannot combine forces to sanction *A* in order to constrain the actions. Unless either *B* or *C* alone is sufficiently harmed and sufficiently powerful vis-à-vis *A* to sanction alone, *A*'s actions can continue unabated. In a structure with closure, like that of figure 1*b*, *B* and *C* can combine to provide a collective sanction, or either can reward the other for sanctioning *A*. . . .

In the case of norms imposed by parents on children, closure of the structure requires a slightly more complex structure, which I will call intergenerational closure. Intergenerational closure may be described by a simple diagram that represents relations between parent and child and relations outside the family. Consider the structure of two communities, represented by figure 2. The vertical

Figure 1   Network without (*a*) and with (*b*) closure

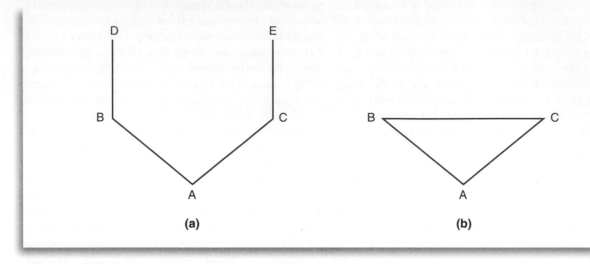

Figure 2   Network involving parents (*A*, *D*) and children (*B*, C) without (*a*) and with (*b*) intergenerational closure

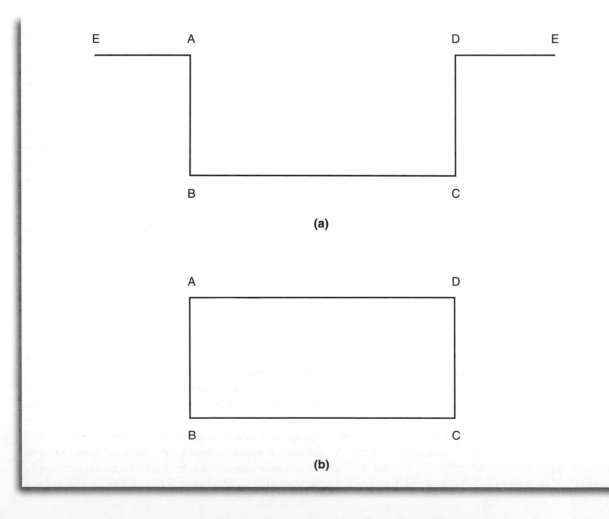

lines represent relations across generations, between parent and child, while the horizontal lines represent relations within a generation. The point labeled *A* in both figure 2*a* and figure 2*b* represents the parent of child *B,* and the point labeled *D* represents the parent of child *C.* The lines between *B* and *C* represent the relations among children that exist within any school. Although the other relations among children within the school are not shown here, there exists a high degree of closure among peers, who see each other daily, have expectations toward each other, and develop norms about each other's behavior.

The two communities differ, however, in the presence or absence of links among the parents of children in the school. For the school represented by figure 2*b,* there is intergenerational closure; for that represented by figure 2*a,* there is not. To put it colloquially, in the lower community represented by 2*b,* the parents' friends are the parents of their children's friends. In the other, they are not.

The consequence of this closure is . . . a set of effective sanctions that can monitor and guide behavior. In the community in figure 2*b,* parents *A* and *D* can discuss their children's activities and come to some consensus about standards and about sanctions, parent *A* is reinforced by parent *D* in sanctioning his child's actions; beyond that, parent *D* constitutes a monitor not only for his own child, *C,* but also for the other child, *B.* Thus, the existence of intergenerational closure provides a quantity of social capital available to each parent in raising his children—not only in matters related to school but in other matters as well.

Closure of the social structure is important not only for the existence of effective norms but also for another form of social capital: the trustworthiness of social structures that allows the proliferation of obligations and expectations. Defection from an obligation is a form of imposing a negative externality on another. Yet, in a structure without closure, it can be effectively sanctioned, if at all, only by the person to whom the obligation *is* owed. Reputation cannot arise in an open structure, and collective sanctions that would ensure trustworthiness cannot be applied. Thus, we may say that closure creates trustworthiness in a social structure. . . .

SOCIAL CAPITAL IN THE CREATION OF HUMAN CAPITAL

The preceding pages have been directed toward defining and illustrating social capital in general. But there is one effect of social capital that is especially important: its effect on the creation of human capital in the next generation. Both social capital in the family and social capital in the community play roles in the creation of human capital in the rising generation. I will examine each of these in turn.

Social Capital in the Family

Ordinarily, in the examination of the effects of various factors on achievement in school, "family background" is considered a single entity, distinguished from schooling in its effects. But there is not merely a single "family background"; family background is analytically separable into at least three different components: financial capital, human capital, and social capital. Financial capital is approximately measured by the family's wealth or income. It provides the physical resources that can aid achievement: a fixed place in the home for studying, materials to aid learning, the financial resources that smooth family problems. Human capital is approximately measured by parents' education and provides the potential for a cognitive environment for the child that aids learning. Social capital within the family is different from either of these. . . . The social capital of the family is the relations between children and parents (and, when families include other members, relationships with them as well). That is, if the human capital possessed by parents is not complemented by social capital embodied in family relations, it is irrelevant to the child's educational growth that the parent has a great deal, or a small amount, of human capital. . . .

Social capital within the family that gives the child access to the adult's human capital depends both on the physical presence of adults in the family and on the attention given by the adults to the child. The physical absence of adults may be described as a structural deficiency in family social capital. The most prominent element of structural deficiency in modern families is the single-parent family. However, the nuclear family itself, in which one or both parents work outside the home, can be seen as structurally deficient, lacking the social capital that comes with the presence of parents during the day, or with grandparents or aunts and uncles in or near the household.

Even if adults are physically present, there is a lack of social capital in the family if there are not strong relations between children and parents. The lack of strong relations can result from the child's embeddedness in a youth community, from the parents' embeddedness in relationships with other adults that do not cross generations, or from other sources. Whatever the source, it means that whatever *human* capital exists in the parents, the child does not profit from it because the *social* capital is missing.

The effects of a lack of social capital within the family differ for different educational outcomes. One for which it appears to be especially important is dropping out of school. With the *High School and Beyond* sample of students in high schools, table 1 shows the expected dropout rates for students in different types of families when various measures of social and human capital in the family and a measure of social capital in the community are controlled statistically. An explanation is necessary for the use of number of siblings as a measure of lack of social capital. The number of siblings represents, in this interpretation, a dilution of adult attention to the child. . . .

Item 1 of table 1 shows that, when other family resources are controlled, the percentage of students who drop out between spring of the sophomore year and spring of the senior year is 6 percentage points higher for children from single-parent families. Item 2 of table 1 shows that the rate is 6.4 percentage points higher for sophomores with four siblings than for those with otherwise equivalent family resources but only one sibling. Or, taking these two together, we can think of the ratio of adults to children as a measure of the social capital in the family available for the education of any one of them. Item 3 of table 1 shows that for a sophomore with four siblings and one parent, and an otherwise average background, the rate is 22.6%; with one sibling and two parents,

Table 1 Dropout Rates between Spring, Grade 10, and Spring, Grade 12, for Students Whose Families Differ in Social Capital, Controlling for Human Capital and Financial Capital in the Family

	Percentage Dropping Out	Difference in Percentage Points
1. Parents' presence:		
Two parents	13.1	
Single parent	19.1	6.0
2. Additional children:		
One sibling	10.8	
Four siblings	17.2	6.4
3. Parents and children:		
Two parents, one sibling	10.1	
One parent, four siblings	22.6	12.5
4. Mother's expectation for child's education:		
Expectation of college	11.6	
No expectation of college	20.2	8.6
5. Three factors together:		
Two parents, one sibling, mother expects college	8.1	
One parent, four siblings, no college expectation	30.6	22.5

the rate is 10.1%—a difference of 12.5 percentage points.

Another indicator of adult attention in the family, although not a pure measure of social capital, is the mother's expectation of the child's going to college. Item 4 of the table shows that, for sophomores without this parental expectation, the rate is 8.6 percentage points higher than for those with it. With the three sources of family social capital taken together, item 5 of the table shows that sophomores with one sibling, two parents, and a mother's expectation for college (still controlling on other resources of family) have an 8.1% dropout rate; with four siblings, one parent, and no expectation of the mother for college, the rate is 30.6%. . . .

Social Capital Outside the Family

The social capital that has value for a young person's development does not reside solely within the family. It can be found outside as well in the community consisting of the social relationships that exist among parents, in the closure exhibited by this structure of relations, and in the parents' relations with the institutions of the community. . . .

In the *High School and Beyond* data set, [a] variation among the schools constitutes a useful indicator of social capital. This is the distinctions among public high schools, religiously based private high schools, and nonreligiously based private high schools. It is the religiously based high schools that are surrounded by a community based on the religious organization. These families have intergenerational closure that is based on a multiplex relation: whatever other relations they have, the adults are members of the same religious body and parents of children in the same school. In contrast, it is the independent private schools that are typically least surrounded by a community, for their student bodies are collections of students, most of whose families have no contact. The choice of private school for most of these parents is an individualistic one, and, although they back their children with extensive human capital, they send their children to these schools denuded of social capital.

In the *High School and Beyond* data set, there are 893 public schools, 84 Catholic schools, and 27 other private schools. Most of the other private schools are independent schools, though a minority have religious

foundations. In this analysis, I will at the outset regard the other private schools as independent private schools to examine the effects of social capital outside the family.

The results of these comparisons are shown in table 2. Item 1 of the table shows that the dropout rates between sophomore and senior years are 14.4% in public schools, 3.4% in Catholic schools, and 11.9% in other private schools. What is most striking is the low dropout rate in Catholic schools. The rate is a fourth of that in the public schools and a third of that in the other private schools.

Adjusting the dropout rates for differences in student-body financial, human, and social capital among the three sets of schools by standardizing the population of the Catholic schools and other private schools to the student-body backgrounds of the public schools shows that the differences are affected only slightly. Furthermore, the differences are not due to the religion of the students or to the degree of religious observance. Catholic students in public school are only slightly less likely to drop out than non-Catholics. Frequency of attendance at religious services, which is itself a measure of social capital through intergenerational closure, is strongly related to dropout rate, with 19.5% of public school students who rarely or never attend dropping out compared with 9.1% of those who attend often. But this effect exists apart from, and in addition to, the effect of the school's religious affiliation. Comparable figures for Catholic school students are 5.9% and 2.6%, respectively (Coleman and Hoffer 1987, p. 138).

The low dropout rates of the Catholic schools, the absence of low dropout rates in the other private schools, and the independent effect of frequency of religious attendance all provide evidence of the importance of social capital outside the school, in the adult community surrounding it, for this outcome of education.

A further test is possible, for there were eight schools in the sample of non-Catholic private schools ("other private" in the analysis above) that have religious foundations and over 50% of the student body of that religion. Three were Baptist schools, two were Jewish, and three from three other denominations. If the inference is correct about the religious community's providing intergenerational closure and thus social capital and about the importance of social

Table 2 Dropout Rates between Spring, Grade 10, and Spring, Grade 12, for Students from Schools with Differing Amounts of Social Capital in the Surrounding Community

	Public	Catholic	Other Private Schools
1. Raw dropout rates	14.4	3.4	11.9
2. Dropout rates standardized to average public school sophomore	14.4	5.2	11.6
	Non-Catholic Religious		Independent
3. Raw dropout rates for students[a] from independent and non-Catholic religious private schools	3.7		10.0

[a] This tabulation is based on unweighted data, which is responsible for the fact that both rates are lower than the rate for other private schools in item 1 of the table, which is based on weighted data.

capital in depressing the chance of dropping out of high school, these schools also should show a lower dropout rate than the independent private schools. Item 3 of table 2 shows that their dropout rate is lower, 3.7%, essentially the same as that of the Catholic schools.[i]

The data presented above indicate the importance of social capital for the education of youth, or, as it might be put, the importance of social capital in the creation of human capital. Yet there is a fundamental difference between social capital and most other forms of capital that has strong implications for the development of youth. It is this difference to which I will turn in the next section.

PUBLIC GOODS ASPECTS OF SOCIAL CAPITAL

Physical capital is ordinarily a private good, and property rights make it possible for the person who invests in physical capital to capture the benefits it produces. Thus, the incentive to invest in physical capital is not depressed; there is not a suboptimal investment in physical capital because those who invest in it are able to capture the benefits of their investments. For human capital also—at least human capital of the sort that is produced in schools—the person who invests the time and resources in building up this capital reaps its benefits in the form of a higher-paying job, more satisfying or higher-status work, or even the pleasure of greater understanding of the surrounding world—in short, all the benefits that schooling brings to a person.

But most forms of social capital are not like this. For example, the kinds of social structures that make possible social norms and the sanctions that enforce them do not benefit primarily the person or persons whose efforts would be necessary to bring them about, but benefit all those who are part of such a structure. For example, in some schools where there exists a dense set of associations among some parents, these are the result of a small number of persons, ordinarily mothers who do not hold full-time jobs outside the home. Yet these mothers themselves experience only a subset of the

[i] It is also true, though not presented here, that the lack of social capital in the family makes little difference in dropout rates in Catholic schools—or, in the terms I have used, social capital in the community compensates in part for its absence in the family. See Coleman and Hoffer (1987, chap. 5).

benefits of this social capital surrounding the school. If one of them decides to abandon these activities—for example, to take a full-time job—this may be an entirely reasonable action from a personal point of view and even from the point of view of that household with its children. The benefits of the new activity may far outweigh the losses that arise from the decline in associations with other parents whose children are in the school. But the withdrawal of these activities constitutes a loss to all those other parents whose associations and contacts were dependent on them. . . .

It is not merely voluntary associations, such as a PTA, in which underinvestment of this sort occurs. When an individual asks a favor from another, thus incurring an obligation, he does so because it brings him a needed benefit; he does not consider that it does the other a benefit as well by adding to a drawing fund of social capital available in a time of need. If the first individual can satisfy his need through self-sufficiency, or through aid from some official source without incurring an obligation, he will do so—and thus fail to add to the social capital outstanding in the community.

Similar statements can be made with respect to trustworthiness as social capital. An actor choosing to keep trust or not (or choosing whether to devote resources to an attempt to keep trust) is doing so on the basis of costs and benefits he himself will experience. That his trustworthiness will facilitate others' actions or that his lack of trustworthiness will inhibit others' actions does not enter into his decision. A similar but more qualified statement can be made for information as a form of social capital. An individual who serves as a source of information for another because he is well informed ordinarily acquires that information for his own benefit, not for the others who make use of him. . . .

For norms also, the statement must be qualified. Norms are intentionally established, indeed as means of reducing externalities, and their benefits are ordinarily captured by those who are responsible for establishing them. But the capability of establishing and maintaining effective norms depends on properties of the social structure (such as closure) over which one actor does not have control yet are affected by one actor's action. These are properties that affect the structure's capacity to sustain effective norms, yet properties that ordinarily do not enter into an individual's decision that affects them.

Some forms of social capital have the property that their benefits can be captured by those who invest in them; consequently, rational actors will not underinvest in this type of social capital. Organizations that produce a private good constitute the outstanding example. The result is that there will be in society an imbalance in the relative investment in organizations that produce private goods for a market and those associations and relationships in which the benefits are not captured—an imbalance in the sense that, if the positive externalities created by the latter form of social capital could be internalized, it would come to exist in greater quantity.

The public goods quality of most social capital means that it is in a fundamentally different position with respect to purposive action than are most other forms of capital. It is an important resource for individuals and may affect greatly their ability to act and their perceived quality of life. They have the capability of bringing it into being. Yet, because the benefits of actions that bring social capital into being are largely experienced by persons other than the actor, it is often not in his interest to bring it into being. The result is that most forms of social capital are created or destroyed as by-products of other activities. This social capital arises or disappears without anyone's willing it into or out of being and is thus even less recognized and taken account of in social action than its already intangible character would warrant.

There are important implications of this public goods aspect of social capital that play a part in the development of children and youth. Because the social structural conditions that overcome the problems of supplying these public goods—that is, strong families and strong communities—are much less often present now than in the past, and promise to be even less present in the future, we can expect that, ceteris paribus, we confront a declining quantity of human capital embodied in each successive generation. The obvious solution appears to be to attempt to find ways of overcoming the problem of supply of these public goods, that is, social capital employed for the benefit of children and youth. This very likely means the substitution of some kind of formal organization for the voluntary and spontaneous social organization that has in the past been the major source of social capital available to the young. . . .

Discussion Questions

1. Homans claimed that the behaviorist propositions used to explain individual conduct are capable of explaining social or group behavior. Do you think that individual and group behaviors are rooted in the same set of processes? If not, what processes or mechanisms differentiate the two forms of conduct? Do you think that social systems or structures can produce behavioral effects independent of the individuals that compose them? If so, what might be some of these effects?

2. Compare Parsons's notion of system equilibrium to Blau's view of social interaction producing a strain toward reciprocity.

3. Following Blau's discussion, what is the link between social attraction and power? What role does social approval play in mediating this link?

4. Homans and Blau suggest that, in calculating the potential profits derived from interacting with others, individuals attempt to establish a position of dominance. As such, maximizing our own benefits is contingent on having others dependent on us for the rewards they seek. Do you think this motive characterizes all relationships? If not, what other considerations might guide our interactions with others? What types of relationships are more likely to generate such considerations?

5. Based on Coleman's approach to norms, why is there a demand on the part of some individuals and groups for a prescriptive norm that assigns women the responsibility for taking care of the home? What dynamics lead to the persistence of and challenges to this norm?

6. How does the structure of social networks contribute to the production of social capital? How does social capital affect individuals' and groups' actions and the likelihood that their goals will be realized? How might the concept of social capital help to explain the existence and persistence of social inequality?

12 SYMBOLIC INTERACTIONISM AND DRAMATURGY

Erving Goffman

Key Concepts

- :: Impression Management
- :: Definition of the Situation
- :: Backstage
- :: Front
- :: Self
  - ❏ Performer/Character

- :: Merchants of Morality
- :: Total Institutions
- :: Secondary Adjustments
- :: Deference/Demeanor

Arlie Russell Hochschild

Key Concepts

- :: Feeling Rules
- :: Emotion Work
- :: Emotional Labor
- :: Commodification of Feeling

Universal human nature is not a very human thing. By acquiring it, the person becomes a kind of construct, built up not from inner psychic propensities but from moral rules impressed upon him from without.

—Goffman (1967:45)

Consider the following commonplace scenario: You hop on a bus only to find that most of the seats are already taken. However, you notice an empty seat on which a passenger has placed her bags. As you approach the seat, you try to make eye contact with the passenger seated next to the bags, only to find that she pretends not to notice you. Pausing for moment, you spot another empty seat at the rear of the bus. What do you do?

a. Attempt to make eye contact, say, "Excuse me; may I sit here?" and wait for the passenger to move her bags.

b. Pick up the passenger's bags and politely hand them to her as you proceed to sit down.

c. Stomp your foot and angrily blurt out, "Hey lady, I paid for this ride. Move your bags or I'll do it for you!" and then sit down with a sigh of disdain next to the passenger.

d. Mutter some less-than-kind words as you pass the rider on your way to the back of the bus.

The primary purpose of this quiz lies not in determining your answer, but rather in calling your attention to the richness and complexity of social life that can be found under the surface of your response.

Photo 12.1 May I have this dance? Jockeying for a bus seat

Nothing short of a well-choreographed ballet is performed during this encounter. From determining the "definition of situation" as you read the meaning of the passenger's bags placed on the empty seat and her attempt to avoid eye contact, to controlling (or not) your emotional reaction, the scene is a give-and-take dance of gestures. While it may seem to unfold spontaneously, such an encounter illustrates the constructed and ritualized nature of everyday interaction. No one answer above is "right," but each sheds light on how the meaning of gestures arises out of interaction and on the expectation that one's self is to be treated with ceremonial care.

In this chapter we take up such issues through the works of two sociologists who have made significant contributions to interactionist approaches to social life. Such approaches highlight how the self and society are created and re-created during the course of interaction. We begin with a brief overview of symbolic interactionism. We then turn to a discussion of Erving Goffman and his development of "dramaturgy," a perspective that is informed not only by symbolic interactionism but also by Émile Durkheim's insights into the ritual and moral realms of society. We conclude the chapter with an examination of the work of Arlie Russell Hochschild, whose analyses of emotions expanded the theoretical and empirical terrain of symbolic interactionism. Moreover, Hochschild's study of gender and family dynamics in contemporary American society has provided important insights into the relationship between self-identity and broader social conditions.

▪▪ SYMBOLIC INTERACTIONISM: AN OVERVIEW

The birth of symbolic interactionism as a distinct theoretical perspective dates to 1937, when Herbert Blumer (1900–1987) first coined the term and outlined the central concepts that would form the foundation of the perspective. The inspiration for Blumer's work lay principally with the ideas developed by George Herbert Mead. Blumer was a student of Mead's at the University of Chicago during the period in which the sociology department was solidifying its reputation as a leading voice in the discipline. Known as the "Chicago School," the department was at the forefront in developing a sociology that, through detailed, empirical studies, explores how individuals understand and negotiate their everyday life.[1] For his part, Blumer advanced the intellectual agenda of the Chicago School through his codification of Mead's analyses of the self and its relationship to language.

Because we examined Mead's social psychology in Chapter 8, we confine ourselves here to a brief discussion of those aspects of Blumer's thought that, based on Mead's insights, proved pivotal to the development of symbolic interactionism. We begin with the three premises that form the basis of Blumer's understanding of a symbolic interactionist approach to social life:

> The first premise is that human beings act toward things on the basis of the meanings that the things have for them. Such things include everything that the human being may note in his world—physical objects, such as trees or chairs; other human beings, such as a mother or a store clerk; categories of human beings, such as friends or enemies; institutions, such as a school or a government; guiding ideals, such as individual independence or honesty; activities of others, such as their commands or requests; and such situations as an individual encounters in his daily life. The second premise is that the meaning of such things is derived from, or arises out of, the social interaction that one has with one's fellows. The third premise is that these meanings are handled in, and modified through, an interpretative process used by the person in dealing with the things he encounters. (Blumer 1969:2)

Like Mead, then, Blumer highlighted the significance of meaning, interaction, and interpretation to the fitting together of individuals' lines of action. Interpretation entails *constructing* the meaning of objects or another's actions, for meaning is not "released" by, or inherent in, things or the actions themselves. It is on the basis of one's interpretation or definition that one then responds to her physical and social surroundings. Thus, your response to another's gesture—to copy your class notes, for instance—is dependent on how you define or give meaning to the request. But the meaning is not a product solely of your mind or inherent in the specific words used by the other person; instead, it is developed out of the interaction itself. While you may let some classmates copy your notes, you may refuse others depending on your interpretation of the motives (i.e., meaning) underlying their request. Was the person absent from class because of an illness, or is he trying to avoid doing the work? Perhaps he's asking for the notes as a way to strike up a friendship. Knowing the answer to such questions, however, does not determine the decision for each and every individual. For some, being sick is not a valid reason for missing class, while others are more than happy to help out a slacker.

Following Mead, Blumer notes that the process of interpretation is carried out through a conversation of gestures involving significant symbols (words and gestures whose meaning is shared by those engaged in interaction). During this internal dialogue the individual takes the attitude of the other, experiences herself as an object, and calls out in herself the same response to her actions that is called out in others. Self-consciousness, the ability to see oneself as others do and thus anticipate responses to one's behavior has a most profound impact on social life, for it makes possible "joint action." Blumer defines joint action as "the larger collective form of action that is constituted by the fitting together of the lines of behavior of the separate participants. . . . Joint actions range from a simple collaboration of two individuals to a complex alignment of the acts of huge organizations or institutions" (1969:70). Whether it is a family gathering, a corporate merger, or a nation conducting foreign policy, each is the outcome of a reflexive, interpretative process in which participants assign meaning to the separate acts that together constitute the joint action. In other words, in order to align their behaviors with one another, individuals must construct a shared interpretation—by becoming an object to one's self—of each other's gestures. This alone allows for the formation of joint action, the building block of society.

---

[1]There are a number of works that examine the Chicago School and its impact on sociology. Three such books that may be of interest to readers are Martin Blumer's (1984) *The Chicago School of Sociology,* Robert Faris's (1967) *Chicago Sociology, 1920–1932,* and Lewis and Smith's (1980) *American Sociology and Pragmatism.*

One of Blumer's central theoretical objectives was to underscore the differences between symbolic interactionism and the more collectivist approaches such as functionalism that, to his mind, presented social structure as a "straitjacket" that determines the behavior of individuals and groups. While remarking that societal factors such as norms, values, culture, roles, and status positions play an important part in organizing social life, he nevertheless argued that they are significant "only as they enter into the process of interpretation and definition out of which joint actions are formed" (1969:75). For Blumer, actors are not vessels for the mechanical expression of conduct prescribed by roles and norms. Instead, social life is seen as a dynamic process in which people, through interpreting the gestures of others as well as their own, are at every moment creating and recreating the patterns of behavior that form the basis for the social order. As Blumer noted, "It is the social process in group life that creates and upholds the rules, not the rules that create and uphold group life" (ibid:19). That joint actions, and the institutions that they sustain, take on repetitive and stable forms is a function not of an organization's "inner dynamics" or "system requirements," but of the recurring use of schemes of interpretation and definition. Thus, even established patterns of group life are constantly "formed anew" as they are "just as much a result of an interpretive process as is a new form of joint action that is being developed for the first time" (ibid.:18).

Significant Others

Sheldon Stryker and Identity Theory

Sheldon Stryker served on the faculty at Indiana University from 1951 until his retirement in 2002. Much of his work is dedicated to developing and testing identity theory, a perspective that draws from and extends the ideas set out by George Herbert Mead. In fact, Stryker's theory offers a counter to the dominant strand of symbolic interactionism that was formulated by Blumer. As a critic of Blumer's views of interactionist theory and methods, Stryker's own perspective was ill received by many and created an antagonistic schism within the field (Stryker 2003).

Stryker's identity theory offers a social structural version of symbolic interactionism centered on specifying the reciprocal relationship between self and society—how each is a product and producer of the other—which was a central theme in Mead's social psychology. As he remarked, "A satisfactory theoretical framework must bridge social structure and person, must be able to move from the level of the person to that of large-scale social structure and back again" (Stryker 1980:53). The "bridge" between these two levels is found in the concept of roles, or the behavioral expectations and meanings that are attached to positions located in the social structure. Roles allow us to predict the behavior of others and to orient our own conduct, thus making possible coordinated, organized interaction. Moreover, the concept of roles sheds light on the reciprocal relationship between self and society. "Asking how . . . society is patterned has led sociology to an image of society in terms of positions and roles. Given that the person is the other side of the society coin, then this view of society leads to an image of the person as a structure of positions and roles which, internalized, is the self" (ibid.:79).

This view leads to an understanding of the self that Mead (as well as Blumer) recognized but did not fully explore: A person has as many selves as he or she has patterned relations with others. To capture the complexity of the self and its connection to role making, Stryker introduced three concepts: identity, identity salience, and commitment. An identity is a "part" of one's self that is "called up" in the course of interacting with others. The number of identities a person possesses corresponds to the number of structured role relationships in which he participates. Thus, a person's self can be composed of multiple identities including child, parent, worker, friend, spouse, and so on (ibid.:60).

Identity salience refers to how the self is organized according to a hierarchy of identities. Not every identity has the same salience or importance to the individual, "such that the higher the identity in [the salience] hierarchy, the more likely that the identity will be invoked in a given situation or in many situations" (ibid.:61). When a situation structurally overlaps with other situations (for instance, your boss asks you to work a shift that conflicts with your class schedule), different identities are invoked. Which identity most influences an individual's subsequent behavior, and how it does so, depends on its location in the salience hierarchy. As a result, social structural characteristics—the presence or absence of situational overlaps—are directly tied to the self that is called up during interaction.

Rounding out Stryker's identity theory is the concept of commitment. Here he notes that, to the extent that a person's relationship to others is dependent on being a specific kind of person, "one is committed to being that kind of person" (ibid.:61). Similarly, to the extent that maintaining one's ties to a particular group of others is important to an individual, he or she is committed to being a member of that group. In sum, because the forging of relationships with others is determined, in part, on the basis of identities and their salience for an individual, commitment is itself a reflection of identity salience (ibid.:61, 62). Yet, identities are themselves tied to socially structured role relationships that have been internalized as parts of one's self. Thus, social structure (society) creates identities (self) that (re-)create social structure that creates identities that (re-)create. . . .

ERVING GOFFMAN (1922–1982): A BIOGRAPHICAL SKETCH ▪▪

Erving Goffman was born in Mannville, Alberta, Canada, in 1922. He graduated with a bachelor's degree from the University of Toronto in 1945. He then attended the University of Chicago, where he received his master's degree in 1949 and his doctorate in 1953. After completing his doctorate, Goffman spent the next three years as a visiting scientist at the National Institute of Mental Health (NIMH) in Bethesda, Maryland. During that period, he conducted fieldwork at St. Elizabeth's Hospital in Washington, D.C., a federal mental institution with more than seven thousand patients in its charge at the time. His research there led to one of the landmark books in sociology, *Asylums* (1961). Following his stint at the NIMH, Goffman took a position in the sociology department at the University of California, Berkeley, where he remained until 1968. He then left for the University of Pennsylvania, where he assumed an endowed chair, the Benjamin Franklin Professor of Anthropology and Psychology. He served on the University of Pennsylvania faculty until his untimely death in 1982 at the age of sixty.

Goffman produced a number of works that have made their way into the canon of contemporary sociology. In addition to *Asylums,* his more notable books include *The Presentation of Self in Everyday Life* (1959), *Encounters* (1961), *Behavior in Public Places* (1963), *Stigma* (1963), *Interaction Ritual* (1967), *Strategic Interaction* (1969), *Frame Analysis* (1974), and *Gender Advertisements* (1979). His contributions to the discipline notwithstanding, Goffman maintained a reputation as an iconoclast: never content to be conventional, he charted a rebellious intellectual course (R. Collins 1986; Lemert 1997). His unique, if not visionary, position within the profession was developed in several ways. For starters, Goffman refused to situate his perspective within a specific theoretical tradition, although that did not stop others from trying to do so. Indeed, while he is often considered a symbolic interactionist (and for good reason), Goffman himself found the label wanting. Denying an allegiance to that tradition or even to the more general label of "theorist," he was more prone to refer to himself as simply an "empiricist" or a "social psychologist" (Lemert 1997:xxi). In some respects, Goffman's self-description may be the more accurate since his work drew from a number of distinct approaches that he fashioned together in forming his own novel account of everyday life.

Goffman's writing style also fueled his iconoclasm. Writing with the flair of a literary stylist, his was not the dry prose all too common among scientists. Instead of adopting the standard practice of situating one's analyses within a particular intellectual lineage or reigning contemporary debates, Goffman was busy inventing his own terminology as he set out to "raise questions that no one else had ever asked and to look at data that no one had ever examined before" (R. Collins 1986:110). As a result, Goffman was at the forefront of important movements within sociology—for instance, doing ethnomethodology before the ethnomethodologists laid claim to it, and exploring the central role of language in social life (the "linguistic turn") well ahead of most of his brethren in the field.

That Goffman stood with one foot in sociology and the other out can be read in his posthumously published article "The Interaction Order" (1983). Commenting on the state of the discipline, he remarked,

> We have not been given the credence and weight that economists lately have acquired, but we can almost match them when it comes to the failure of rigorously calculated predictions. Certainly our systematic theories are every bit as vacuous as theirs: we manage to ignore almost as many critical variables as they do. We do not have the esprit that anthropologists have, but our subject matter at least has not been obliterated by the spread of the world economy. So we have an undiminished opportunity to overlook the relevant facts with our very own eyes. We can't get graduate students who score as high as those who go into Psychology, and at its best the training the latter gets seems more professional and more thorough than what we provide. So we haven't managed to produce in our students the high level of trained incompetence that psychologists have achieved in theirs, although, God knows, we're working on it. . . .
>
> From the perspective of the physical and natural sciences, human life is only a small irregular scab on the face of nature, not particularly amenable to deep systematic analysis. And so it is. . . . I'm not one to think that so far our claims can be based on magnificent accomplishment. Indeed I've heard it said that we should be glad to trade what we've so far produced for a few really good conceptual distinctions and a cold beer. But there's nothing in the world we should trade for what we do have: the bent to sustain in regard to all elements of social life a spirit of unfettered, unsponsored inquiry, and the wisdom not to look elsewhere but ourselves and our discipline for this mandate. That is our inheritance and that so far is what we have to bequeath. (ibid.:2, 17)

Such were the words Goffman had written in 1982 for his address to the American Sociological Association. The occasion was his election to the office of president. However, he was unable to deliver his speech, too ill to attend the conference. He died from cancer shortly thereafter. But perhaps that is how it should have been, for the maverick had entered the mainstream.

▪▪ GOFFMAN'S INTELLECTUAL INFLUENCES AND CORE IDEAS

In this section and those that follow, we largely confine our discussion to Goffman's earlier works, for arguably they have had the most significant impact on the discipline. As we noted previously, Goffman resisted the typical tendency to position his work in one or another theoretical camp. For that matter, his citations and footnotes seemingly were designed to obscure the identities of those whose ideas he drew from, whether it was to champion or to challenge their approaches (R. Collins 1986:108). Not surprisingly, this complicates efforts to trace connections to those to whom Goffman is most indebted. Despite these obstacles, we nevertheless can identify a number of theorists whose work played a central role in the development of his dramaturgical perspective.

Symbolic Interactionism: George Herbert Mead and William I. Thomas

Notwithstanding his avoidance of the interactionist label, one figure who had an important impact on Goffman's work was George Herbert Mead. As discussed in Chapter 8, Mead envisioned the self as essentially a social construct rooted in our ability to see ourselves as an object. We experience, or are aware of, ourselves not directly, but, rather, indirectly through taking the attitude of others toward ourselves. In this way the self is a product of social interaction during which individuals engage in an internal conversation and interpret others' responses to their conduct. Thus, who we *think* we are is a reflection of the organized set of attitudes of others that the individual assumes. As the others that inform our reflexive role-taking change, so does our sense of who we are. In turn, we do not have *a* self, but, instead, multiple selves. As Mead asserted, "[w]e divide ourselves up in all sorts of different selves with reference to our acquaintances. . . . There are all sorts of different selves answering to all sorts of different social creations" ([1934] 1962:142). It is interaction that determines the "sort" of self that appears as we imaginatively rehearse the potential effects of our behavior in order to elicit the desired responses from others.

The notion that we see ourselves as objects, as others see us, forms the basis for one of Goffman's central concepts: **impression management**. Impression management refers to the verbal and nonverbal practices we employ in an attempt to present an acceptable image of our self to others. Much of Goffman's work is dedicated to describing the subtle ways in which we carry out such performances, for instance, concealing information about ourselves that is incompatible with the image we are trying to project, or ensuring "audience segregation" so that those for whom we play one of our parts will not be present for our performance of a different, potentially incompatible, role. It is this controlling of what we say and do and don't say and do, and how we do and don't say and do it, that speaks to taking the attitude of the other. For if our attempts to project an image of our self were not in some measure guided by the imagined responses of others to our actions, then we would be unable to coordinate our behavior with others and engage in relatively predictable and smoothly functioning interaction.

Yet, it is in the "some measure" clause that Goffman makes one of his breaks with the traditional symbolic interaction of Mead and Blumer. While he by no means denies that individuals see themselves as others see them, he is far less interested in exploring the internal conversations that individuals engage in as they map out their conduct. Instead, Goffman's genius lies in exploring how social arrangements themselves and the actual, physical copresence of individuals—the "interaction order"—shapes the organization of the self. Commenting on Mead's approach, Goffman remarks,

> The Meadian notion that the individual takes toward himself the attitude others take toward him seems very much an oversimplification. Rather the individual must rely on others to complete the picture of him of which he himself is allowed to paint only certain parts. Each individual is responsible for the demeanor image of himself and the deference image of others, so that for a complete man to be expressed, individuals must hold hands in a chain of ceremony, each giving deferentially with proper demeanor to the one on the right what will be received deferentially from the one on the left. While it may be true that the individual has a unique self all his own, evidence of this possession is thoroughly a product of joint ceremonial labor. (1967:84, 85)

While Mead contends that the central elements of social interaction are rooted in one's imagination, Goffman looks to the "scene" within which individuals orient their actions to one another. For Goffman, then, the essence of the self is found not in the interior, cognitive deliberations of the individual, but rather in interaction itself. In the end, both the dynamics of social encounters and the image of one's self are dependent on the willingness of others to "go along" with the particular impression that an individual is seeking to present.

The implications of these remarks extend beyond capturing the underpinnings of social interaction to expose the very nature of the self. But before addressing this latter issue, let us first turn to another concept central to Goffman's work. Closely connected to Goffman's notion of impression management is the **definition of the situation**, a concept derived from another early contributor to the Chicago School of

Sociology, William I. Thomas (1863–1947). Like many of the early interactionists, Thomas developed his perspective in part out of a critique of psychological behaviorism. His notion of the definition of the situation countered the view that individuals respond to stimuli on the basis of conditioned or instinctual reflexes. Instead, Thomas argued, in one of the most oft-cited concepts in sociology, that "preliminary to any self-determined act of behavior there is always a stage of examination and deliberation which we may call the definition of the situation" (1923:41). Thus, an individual's behavior or reaction to a situation is not automatic, but rather is constructed on the basis of the meanings that are attributed to the situation. In turn, those interested in understanding individuals and their behaviors must attend to the subjective meanings that they attach to their actions.

There are two additional aspects to Thomas's notion that are noteworthy. First, Thomas points out that a person "is always born into a group of people among whom all the general types of situation which may arise have already been defined and corresponding rules of conduct developed, and where he has not the slightest chance of making his definitions and following his wishes without interference" (ibid.:42). More often than not, then, individuals do not create definitions of the situations so much as they select among preexisting definitions when determining the meaning of an event or encounter. Second, Thomas, in a phrase that would become one of the most famous in sociology, noted the link between meaning and action: "If men define situations as real, they are real in their consequences" (Thomas 1928:571–72). In other words, behavior is fundamentally shaped by, or a consequence of, the definition of the situation assigned by an individual. Much like Mead, then, Thomas contended that we act on the basis of the meanings we ascribe to the situations or stimuli that we confront. Moreover, Thomas's dictum suggests that reality itself is created through the definition of the situation, because it lays the foundation on which individuals will interpret others' and their own actions. In short, individuals create reality as they define it. Thus, if you define a class as boring, then it will be, and your efforts will be different from what they would be if you had defined it as interesting.

So, how does Goffman's work connect with Thomas's insights? First, Goffman concurs that the act of defining the situation establishes the basis on which an encounter unfolds. He notes,

> When an individual enters the presence of others, they commonly seek to acquire information about him or to bring into play information about him already possessed. . . . Information about the individual helps to define the situation, enabling others to know in advance what he will expect of them and what they may expect of him. Informed in this way, the others will know how best to act in order to call forth a desired response from him. (1959:1)

In defining the situation, actors are then able to practice the arts of impression management more effectively, which, as we discussed previously, is a central ingredient of social interaction. For knowing which "self" a person is obliged to present and how to present it—knowing what to say and how to say it, what to do and how to do it—is largely determined by defining what is going on in a situation. Knowing that you are, for the purposes of a given encounter, a student, a worker, or a friend, and whether you should present yourself as studious, compliant, or caring, requires knowing first what the situation itself demands of the actors. Goffman points to this relationship between the definition of the situation and impression management when he remarks,

> Regardless of the particular objective which the individual has in mind and of his motive for having this objective, it will be in his interests to control the conduct of others, especially their responsive treatment of him. This control is achieved largely by influencing the definition of the situation which the others come to formulate, and he can influence this definition by expressing himself in such a way as to give them the kind of impression that will lead them to act voluntarily in accordance with his own plans. (ibid.:3, 4)

While arguing that the definition of the situation plays a central role in interaction, particularly as it affects our attempts to manage impressions, Goffman nevertheless parts with Thomas's original understanding of the concept in an important way. For Goffman, central to the definition of the situation is its infusion with moral obligations that structure the interaction between the participants in an encounter. As a structural (i.e., collectivist) component of encounters, definitions and the moral obligations attached to them both constrain and enable actors' self-presentations in particular ways. It is in his study of the structural and, more specifically, moral and ritual dimensions of interaction where Goffman most clearly parts with the symbolic interactionists and takes up the charge of social anthropologists.

Social Anthropology: Émile Durkheim, A. R. Radcliffe-Brown, and W. Lloyd Warner

While he was earning his master's and doctorate degrees at the University of Chicago, one of Goffman's major influences was the anthropologist W. Lloyd Warner (1898–1970) (R. Collins 1986:109). For his part, Warner studied with the noted British social anthropologist A. R. Radcliffe-Brown (1881–1955), who, in turn, was largely responsible for introducing the work of Émile Durkheim (1858–1917) into British anthropology during the early decades of the twentieth century. While an extensive review of the works and legacy of these three figures is beyond the scope of this volume, we outline here the commonalties that span their approaches and their implications for Goffman's own approach to the study of social life.

It is Durkheim's classic study of tribal religions, *The Elementary Forms of Religious Life* ([1912] 1995), that arguably had the most impact on Goffman's unique view of social interaction. In this work, Durkheim sought to uncover the origins of religion and, in doing so, to demonstrate the inevitable presence of religion in all societies, including allegedly secular, modern ones. According to Durkheim, the worshipping of gods is, in fact, a worshipping of society itself. For while religious individuals experience and commune with a power greater than themselves, an external force that inspires awe and demands respect, this greater power is in fact society. The sacrifices we make in the name of gods, the feelings of dependence on their wisdom and mercy, and the willingness to submit to their commands are nothing other than our offerings of allegiance to the society of which we are a part and through which our individual nature and fate are determined. It is not a supernatural deity that we depend on for our well-being and from which we draw our strength, but rather society, and the benefits that civilization provides.

If "the idea of society is the soul of religion" (ibid.:466), then religious rituals do more than pay tribute to or placate the gods. Indeed, the importance of such rituals for Durkheim lies in their capacity to bind participants to a common experience. By unifying participants' focus through a shared practice, rituals reaffirm a society's collective conscience (the totality of beliefs and sentiments commonly held in a society) and thus play a central role in preserving social solidarity. Thus, in a very real sense the continued existence of any society requires that its members periodically reunite in ritual ceremonies in order to rekindle their adherence to the collective conscience and their commitment to upholding the moral order. In this way, not only are communions, baptisms, and bar mitzvahs religious rituals, but so too are college graduations, family reunions, annual professional meetings, July Fourth celebrations, and recitations of the Pledge of Allegiance at the start of the school day. Each marks an occasion during which participants "reaffirm in common their common sentiments" and strengthen their ties to the community (ibid.:475).

That such secular ceremonies serve "religious" purposes raises interesting questions: What is sacred in a society, and how is sacredness recognized? Neither objects nor animals are intrinsically sacred, yet every culture assigns to one or the other extraordinary qualities that set them apart from the everyday world of profane things. Above all else, such extraordinary qualities grant a moral authority to that which is deemed sacred. To those things that are sacred, that possess a moral authority or power, we yield unquestioningly our own interests and desires. And because, as Durkheim maintains, our interests are unlimited and we alone are incapable of restraining our passions and desires, we must obey something "bigger" than ourselves, lest life be as Thomas Hobbes warned, "solitary, poor, nasty, brutish, and short" ([1651] 1904:185). Given that sacred objects and beings allegedly possess the power to make our lives possible and contented, it is no wonder that we celebrate them through ritual practices and ceremonies. For their part, rituals are designed to protect what is sacred and, in doing so, protect us.

Both Durkheim and Radcliffe-Brown confined their studies of rituals and their role in sustaining the stability of group life to "simple," traditional societies, primarily Australian aboriginal communities. However, Durkheim noted that as societies became larger and increasingly complex, what was imbued with sacred qualities changed as well. Inhabitants of small, traditional societies are tied together by "mechanical solidarity" or feelings of likeness or "oneness" that come with each member participating in the same round of activities and following the same system of beliefs. Indeed, the survival of such societies is dependent on each of its members engaging in a variety of tasks that contribute to the maintenance of the group. As a result, the collective conscience is marked by shared beliefs and sentiments that encompass the individual's entire existence, and what is sacred above all else is the group.

Modern, complex societies, by contrast, are bonded together by "organic solidarity" or interdependence that comes from their members performing specialized functions that, taken together, form a stable, cohesive whole. What is shared in societies such as our own are not similarities, but differences. Modern societies are populated by individuals who hold different occupations, who maintain different religious and political beliefs, and who affirm different ethnic and racial identities. And if specialization and cultivating individual differences marks modern societies, then that which is sacred must take on a different character, for likeness and "oneness" can no longer ensure the stability and survival of the society. The one thing we do share in modern societies, paradoxically, is our individuality. Thus, it is the individual who is deemed sacred, for he alone provides the common basis on which social solidarity can be maintained. Worshipping the group is replaced by the "cult of personality."

That the individual is sacred in modern societies can be seen in the plethora of laws that have been established to protect and to deny the "inalienable rights" of the individual. From the right to vote, the right to hold property, and the right to privacy, to incarceration and the denial of freedoms as a favored mode of punishment, the legal system is, in a sense, the church of the individual, while judges are its priests. For that matter, setting aside the veiled racism and sexism that underlie much of the backlash against affirmative action, those who are opposed to this policy could be seen as defending the sacredness of the individual who in their eyes has been profaned at the expense of group-based claims. Yet, it is not to the formal rule of law that Goffman sets his sights. Instead, it is to the interaction rituals that pervade everyday encounters, the ceremonial chain of "hellos" and "excuse mes," that he turns his perceptive eye. For it is through such perfunctory, conventionalized acts that the individual affirms the sacred status of others while receiving his due in kind.

Goffman's interest in exploring the rituals that structure social life in modern societies stems in part from the influence of W. Lloyd Warner, with whom he studied while earning his graduate degree at the University of Chicago. In his "Yankee City" series (1941–1959), Warner creatively adapted the anthropological study of tribal societies to the study of class and ethnic stratification in urban society. Conducted in a Massachusetts town, Warner's community studies examined the ritual practices that shape modern status systems and the inequalities they sustain. For instance, he argued that secular, patriotic ceremonies served not only to revivify group solidarity, as Durkheim maintained, but also to reinforce class domination. "[T]hey suppress feelings of class conflict and dissension by emphasizing group unity, while implicitly conferring legitimacy on the class [WASPs in 'Yankee City'] that leads the rituals and exemplifies the culture expressed in them" (R. Collins 1994:217).

Goffman followed Warner's lead in applying the insights and methods of anthropology to urban society; however, he largely moved away from his mentor's emphasis on stratification and turned to a Durkheimian-inspired, functionalist interpretation of the ritual practices that guide interaction (R. Collins 1986:109). Defining interaction rituals as "acts through whose symbolic component the actor shows how worthy he is of respect or how worthy he feels others are of it" (Goffman 1967:19), Goffman illuminated the significance of seemingly insignificant acts. Of particular import are a person's **demeanor** (conduct, dress) and the **deference** (honor, dignity, respect) that his demeanor symbolically accords to others. By expressing oneself as a well-demeaned or poorly demeaned person, an individual simultaneously bestows or withholds deference to others. The reciprocal nature of defer-ence and demeanor is such that maintaining a well-demeaned image allows those present to do likewise as the deference they receive obligates them to confer proper deference in kind. Each is rewarded for his or her good behavior by the deference that person reaps in turn. Yet, whether or not an individual is judged to be well demeaned is determined not by the individual himself but, rather, by the interpre-tations others make of his behavior during interaction. Indeed, claiming oneself to be well demeaned is a sign of poor demeanor.

The ceremonial shows of deference and demeanor of which Goffman speaks are more commonly described as manners or rules of etiquette. In his hands, however, the perfunctory acts through which politeness is expressed are seen in an entirely new light. Through the tactful observance of "avoidance rituals" and "presentational rituals," the individual is able to proffer the deference owed to the recipient. Avoidance rituals take the "form of proscriptions, interdictions, and taboos, which imply acts the actor must refrain from doing lest he violate the right of the recipient to keep him at a distance" (Goffman 1967:73). They entail, for instance, avoidance of conversational topics that may be embarrassing or pain-ful to one or more of those present, or refraining from trespassing in another's "personal" space. Additionally, they grant the individual the privacy necessary to conceal himself from others should he momentarily be unable to present himself in a dignified state (ibid.:83). Presentational rituals, conversely, take the form of prescriptions, not proscriptions. In specifying what is to be done, presentational rituals encompass "acts through which the individual makes specific attestations to recipients concerning how he regards them and how he will treat them in the on-coming interaction" (ibid.: 71). Thus, we knock on a person's door before invading his private sphere, congratulate a colleague on her promotion, and rave about a friend's new hairdo. In short, the greetings and salutations we offer others, the disclosure or con-cealment of personal information, the tactful "nonobservance" we display in response to another's gaffe, the closing or granting of physical space we afford others, and countless other acts, if carried out properly, all serve as ceremonial indulgences that mark an individual as a well-demeaned person and thus deserving of the deference only others can provide him. We rely on such avoidance and presentational rituals to steer our way through the potentially hazardous conditions that always lurk beneath the surface of an encoun-ter, for much of interaction is guided by an attempt to avoid embarrassment and the discrediting of one another's self as an object worthy of receiving deference. In a way, interaction is like driving a car: the object is to get where you want to go without having an accident.

Consider the ritual act of wearing a suit for a job interview. Even if the employees do not wear suits at work, if you show up in more casual attire for your interview you may ruin your chances of getting the job. Your failure to be hired in this case may have little to do with your ability to fulfill the requirements of the position. More critical is the fact that your demeanor paid insufficient deference to the interviewer and the company. A strip of cloth tied in a knot under one's throat speaks volumes about one's character! By suggesting that the interviewer is unworthy of your donning "sacred" clothes, you likewise demon-strate a lack of commitment to or knowledge of basic rules of interaction, the most important of which is to treat others, and to allow them to tactfully present themselves, as "gods." Reminiscent of Durkheim's "cult of personality," Goffman remarked on the connections between demeanor, displays of deference, and the self,

> [T]he self is in part a ceremonial thing, a sacred object which must be treated with proper ritual care and in turn must be presented in a proper light to others. As a means through which this self is established, the individual acts with proper demeanor while in contact with others and is treated by others with deference. . . . The implication is that in one sense this secular world is not so irreligious as we may think. Many gods have been done away with, but the individual himself stubbornly remains a deity of considerable importance. . . . Perhaps the individual is so viable a god because he can actually understand the ceremonial significance of the way he is treated, and quite on his own can respond dramatically to what is proffered him. In contacts between such deities there is no need for middlemen; each of these gods is able to serve as his own priest. (1967:91, 95)

Before, During, and After: The Dinner Party

Photo 12.2a   Preparing for guests
In anticipation of their arrival, we prepare the ceremonial "red carpet" for our sacred guests— emptying trashcans, vacuuming the carpet, mopping the floors.

Photo 12.2b   The dinner party
During the festive ritual, however, all the stress and glamourless labor that went into its preparation is momentarily forgotten as hosts and guests alike revel in the celebration of their solidarity.

Photo 12.2c   The cleanup

Yet not all gods are created equal. While some rules of interaction lead all involved to share "symmetrical" obligations and expectations, the proper expression of deference and demeanor may at times entail "asymmetrical" treatment. Symmetrical rules make up the collection of common courtesies that any well-demeaned person is obligated to perform and can expect to receive. Asymmetrical rules require an individual to treat others differently from the way he is treated by them. Such rules symbolically convey the status differences that obtain between certain groupings within society (Goffman 1967:53). Thus, the CEO of a company treats and is treated by the mailroom clerk differently from how the mailroom clerk treats and is treated by the CEO. While the CEO is free to give orders to or interrupt the conversations of the clerk, this same license is not reciprocated. Similarly, the use of first names or the telling of jokes is a freedom granted to those whose positions within a given social hierarchy accord greater deference. For a subordinate to do the same would be a sign of bad demeanor because it violates the deferential preserve that guards the superior against contamination by an inferior. The higher one goes up the social ladder, the greater the distance—both physical and verbal— and the more extensive and elaborate the taboos that protect the individual from intrusions of his sacred self (ibid.:63).

Dramaturgy: A Synthesis

With his Durkheimian focus on rituals leaving him in, but not of, the symbolic interactionist tradition, Goffman developed his dramaturgical approach to the study of social life. Inspired in part by the literary critic and theorist Kenneth Burke (1897–1993), who viewed language as a symbolically enacted drama, Goffman analyzed interaction through analogy with the theater. For his part, Burke constructed a philosophy of rhetoric—"dramatism"—that sought to uncover the motives for action symbolically revealed in the formal structure underlying all instances of speech and writing. Goffman, however, claiming that "life itself is a dramatically enacted thing" (Goffman 1959:72), turned his attention to the symbolic dimensions of social encounters in his effort to explore the nature of the self and its relation to the broader moral code that shapes interaction performances.

To this end, Goffman introduced a vocabulary normally associated with the world of the theater: front, backstage, setting, audience, performance, and, perhaps most provocatively, performer and character are all part of his repertoire of terms used to examine the often unspoken and taken-for-granted subtleties that structure the interaction order. Consider, for instance, Goffman's notion of the **front** that he describes as

> that part of the individual's performance which regularly functions in a general and fixed fashion to define the situation for those who observe the performance. Front, then, is the expressive equipment of a standard kind intentionally or unwittingly employed by the individual during his performance. (ibid.:22)

Moreover, fronts tend to become "institutionalized" as performances conducted in similar settings and by similar actors give rise to "stereotyped expectations" that transcend and shape any particular presentation. Thus, "when an actor takes on an established social role, usually he finds that a particular front has already been established for it" (ibid.:27). As "facts in their own right," fronts, then, are typically selected, not created, by performers.

Goffman divides the front into two parts: the setting and the personal front. The setting consists of the scenery and props that make up the physical space where a performance is conducted. A professor needs a classroom and a lifeguard needs a body of water if they are to perform their roles. Presenting one's self as a high-powered executive, for instance, requires a spacious office, not a cubicle, adorned with expensive furniture, works of art, and a majestic view. It is much harder to convince an audience that you are an important "mover and shaker" if your office overlooks the Dumpster in the back alley and the cushions on your furniture are stained or patched with duct tape. The personal front refers to those items of "expressive equipment" that the audience identifies with the performer himself. These items consist of "insignia of office or rank; clothing; sex, age, and racial characteristics; size and looks; posture; speech patterns; facial expressions; bodily gestures; and the like" (Goffman 1959:24). Such things provide the resources that make it possible for a person to carry himself and appear before others in a particular light. Certainly, audiences attribute traits, fairly or unfairly, to performers on the basis of their age, sex, race, and looks, and performers are often aware of this fact. For instance, all too often audiences equate intelligence with looks and personality traits with sex. Thus, males are seen as "naturally" aggressive and in turn have license to present themselves as such, while women are considered "naturally" passive and often face negative repercussions if they do not conform to this gendered expectation.

The front is contrasted with the **backstage**, the region of the performance normally unobserved by, and restricted from, members of the audience. Backstage is

> where the impression fostered by a performance is knowingly contradicted as a matter of course . . . [where] illusions and impressions are openly constructed. . . . Here costumes and other parts of the personal front may be adjusted and scrutinized for flaws. . . . Here the performer can relax; he can drop his front, forgo speaking his lines, and step out of character. (ibid.:112)

Restaurants illustrate well the distinction between the front and backstage and their facilitation of performances. While managing his impression in the front region as a courteous, deft, and hygienic server, a waiter often can be found in the backstage of the kitchen cursing a customer, sneezing atop someone's meal, or assembling an assortment of previously tabled bread into a basket for the next diners.[2] Of course, the front and the backstage are not solely the province of business establishments. Our homes are likewise divided up into distinct regions, the separation of which takes on particular import with the arrival of guests.

In addition to shedding light on how space functions in managing performances and impressions, the front and backstage also call attention to an issue that we alluded to earlier: the nature of the self. Implied in the passage above, Goffman draws a distinction between the self as **character** and the self as **performer**, and in doing so comes to a radical, antipsychological conclusion. As a character, the self is not an organic thing that has a specific location, whose fundamental fate is to be born, to mature and to die; it is a dramatic effect arising diffusely from a scene that is presented, and the characteristic issue, the crucial concern, is whether it will be credited or discredited. (ibid.:253)

In other words, the self is in reality an image, a managed impression, that is fabricated in concert with others during an encounter. While we typically see one's performed self as "something housed within the body of possessor . . . in the psychobiology of the personality" (ibid.:252), in actuality the self is imputed

---

[2]In one episode of *Seinfeld,* Jerry finds himself standing at the urinal next to the chef of a restaurant. After relieving himself, the chef heads back to the kitchen without first washing his hands. Upon returning to his table, Jerry watches with dismay as the chef energetically kneads the dough for a "special" pizza he is preparing for Jerry and his guest: the chef's daughter.

by others such that it "does not derive from its possessor but from the whole scene of his action. . . . [This] imputation—this self—is a *product* of a scene that comes off, and is not a *cause* of it" (ibid.). Goffman sums up his notion of the self as character with the following remarks:

> In analyzing the self . . . we are drawn from its possessor, from the person who will profit or lose most by it, for he and his body merely provide the peg on which something of collaborative manufacture will be hung for a time. And the means for producing and maintaining selves do not reside inside the peg; in fact these means are often bolted down in social establishments. (ibid.:253)

However, when we turn to the self as performer, Goffman offers a different view, one that suggests that the individual does indeed possess a self that is uniquely his own. For while we are presenting a contrived image to an audience in the front, in the backstage we can relax, forgo speaking our lines, and step out of character. But if we step out of character, to what do we step in? Here the self is not a fabrication, but rather a performer, "a fabricator of impressions . . . [who] has a capacity to learn, this being exercised in the task of training for a part" (Goffman 1959:252, 253). The self as performer is more in keeping with our conventional understanding of selfhood, which maintains that behind whatever part may be played or impression cast, there lies a thinking, feeling "person," a core being that is who we *really* are. Yet, in the end, perhaps this self is really nothing more than an idea, a comforting myth. For if the front you present to audiences is what they know of you, and if you develop your sense of self through interaction with others, then is your self not an image realized in performances? Truth, in reality, is a fiction.

■■ GOFFMAN'S THEORETICAL ORIENTATION

Goffman's theoretical orientation is particularly difficult to decipher. Not only is his overall approach multidimensional in the sense that it speaks to the different quadrants of our action/order framework, but in many instances the substance of specific concepts also ranges across the presuppositions that inform them (see Figure 12.1). With that said, we begin our outline of Goffman's orientation by taking up first the issue of action, in which, far from positing a single motivating force, he asserts that when in the presence of others,

> Sometimes the individual will act in a thoroughly calculating manner, expressing himself in a given way solely in order to give the kind of impression to others that is likely to evoke from them a specific response he is concerned to obtain. Sometimes the individual will be calculating in his activity but be relatively unaware that this is the case. Sometimes he will intentionally and consciously express himself in a particular way but chiefly because the tradition of his group or social status requires this kind of expression and not because of any particular response . . . that is likely to be evoked from those impressed by his expressions. Sometimes the traditions of an individual's role will lead him to give a well-designed impression of a particular kind and yet he may be neither consciously nor unconsciously disposed to create such an impression. (ibid.:6)

Here we clearly see that impression management, the fundamental dynamic underlying interaction, is carried out according to any number of motives. Thus, we can be motivated by rational, calculating self-interest, the nonrational commitment to tradition or status requirements, or even by purposes of which we are not consciously aware. (The multidimensionality of impression management is reflected in Figure 12.1 through its placement in two quadrants of our action/order framework.) This "having it both ways" approach can be unsatisfying if not frustrating, at least to the extent that the conditions are not specified under which we should expect rational or nonrational factors to be most salient.[3] Conversely, Goffman's explicit inclusion of both forms of motivating factors offers a more realistic portrayal of action. After all, behavior is rarely, if

---

[3]Like his equivocal view on the nature of the self, his position on the issue of action says everything and thus arguably says nothing.

Figure 12.1 Goffman's Basic Concepts and Theoretical Orientation

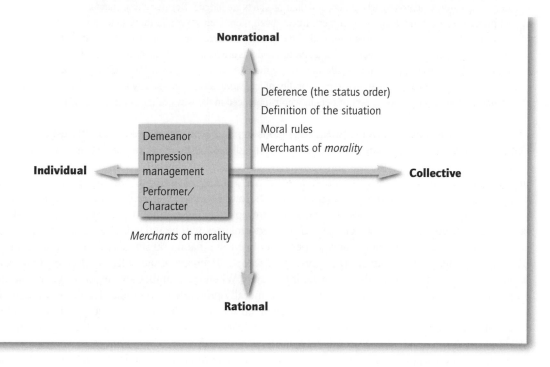

ever, motivated by a single consideration. This claim is all the more credible if we are interested in describing behavior in general and not as a solitary slice of conduct.

To further illustrate Goffman's theoretical orientation, consider the following question: What is to prevent an individual from deceiving his audience into thinking that he is someone he is in fact not? Given that as sacred gods we have the moral "right" to present an image of our self and to expect its acceptance without intrusive or violating questions from the audience, we have the freedom to be anything we want. Or do we? While it is true that we have license to portray who we "are" to our own advantage, this holds only so long as we abide by the moral rules of interaction. Foremost among these rules is that we are who we claim to be. Because audiences are left to infer traits to the individual on the basis of impressions, they have little choice but to accept a performance on faith. Thus, impressions are "treated as claims and promises that have implicitly been made, and claims and promises tend to have a moral character" (1959:249). This, too, accounts in part for the destructive nature of lies, for they breach the fundamental trust on which relationships are based.[4] This may also account for the legal challenges that often arise when law-enforcement agents pose as someone they are not to (en)trap a suspected criminal. And finally, notwithstanding the heinousness of the crime, perhaps this also accounts for the shock and feelings of betrayal that come with learning that your pillar-of-the-community, apple-pie-eating neighbor is actually a serial murderer.

This connection between impression management and the moral standards to which it is subject is captured in Goffman's notion of individuals as **merchants of morality**. Here we see his dialectical understanding not only of rational and nonrational motivating forces, but also of the individualist and collectivist dimensions of social order.

---

[4]An individual who knowingly presents a false impression runs the risk not only of losing face during the interaction, but also of having his face "destroyed." Once he is found guilty of being an imposter, mindful audiences may never again be willing to submit on his behalf to the moral vulnerability that is required of all interactions, leaving the individual in need of finding new congregations willing to worship at his altar (1959:62).

In their capacity as performers, individuals will be concerned with maintaining the impression that they are living up to the many standards by which they and their products are judged. Because these standards are so numerous and so pervasive, the individuals who are performers dwell more than we might think in a moral world. But, qua performers, individuals are concerned not with the moral issue of realizing these standards, but with the amoral issue of engineering a convincing impression that these standards are realized. Our activity, then, is largely concerned with moral matters, but as performers we do not have a moral concern with them. As performers we are merchants of morality . . . [T]he very obligation and profitability of appearing always in a steady moral light, of being a socialized character, forces one to be the sort of person who is practiced in the ways of the stage. (ibid.:251)

We present ourselves as well-demeaned persons in part because it is in our best interest to do so (a rationalist motivation), but in doing so, we announce our adherence to the moral standards that ritually organize social encounters. The notion that individuals are "merchants of morality" suggests that we are calculating in our attempts to "engineer" a properly enacted performance in order to reap the profits (*merchant*) that may be earned from delivering a convincing impression. However, we are in an important sense compelled by the force of moral obligations (*morality*) to live up to the standards of propriety—or at least offer the appearance of doing so—by which we and our performances will be judged. Our actions are not guided by a conscious effort to *be* a morally upright person or to ensure that moral standards are actually being upheld. Rather, they are guided by an effort to present an acceptable image of our self, and it is *this* that requires us to abide by the moral standards that shape interaction. When the plumber arrives to fix your sink, he is not concerned with convincing you that he is truly a morally principled individual: He does not recite the Bill of Rights or religious scripture, nor does he recount episodes from his past as evidence of the morally sound decisions he has made. Indeed, doing so would violate the rules of interaction and thus call his moral character into question. Instead, he conducts himself in a well-demeaned way—that is, as someone "who is practiced in the ways of stage," in order to engineer the impression that he is competent in his work and is not at your house to steal your television and silverware so that he can earn his living. His adherence to moral standards is implied by his performance, but this adherence, whether sincere or fabricated, must be expressed if he is to craft a successful performance. As a result, the interaction order is produced and reproduced not by individuals coordinating their activities through the mutual process of interpretation as, for instance, Blumer's view of the social order suggests, but rather through the moral rules around which interaction is itself organized (a collectivist approach to order). Like impression management, the multidimensionality of this concept is reflected in Figure 12.1 through its placement in two quadrants of our theoretical framework.

Figure 12.1 depicts a number of concepts central to Goffman's perspective in addition to those just discussed. (Again, several appear in more than one quadrant to reflect their multidimensional character.) Because we addressed several concepts in the previous section, we direct your attention to those earlier comments as you develop your understanding of Goffman's theoretical orientation. Yet from the figure it is clear that, despite his multidimensional approach, Goffman does not consider concepts that would be positioned in the rational/collective quadrant of our model. Here we might find an emphasis on such matters as political institutions, legal institutions, or economic systems. Instead, most of Goffman's central concepts are used to examine how *individuals* navigate their everyday life. Thus, we end this section with a question: What impact, if any, does the omission of these aspects of society have on Goffman's ability to construct a theory of interaction that accurately portrays social life?

Readings

In the following selections, Goffman pursues three central themes. Taken from *The Presentation of Self in Everyday Life* (1959), the first excerpts represent his most sustained use of theatrical metaphors as he advances his dramaturgical approach to impression management and its centrality to self and society. The second selection is from *Asylums* (1961), Goffman's study of total institutions, wherein he examines the ways in which social establishments produce the self. Both works evince Goffman's indebtedness to Durkheimian social anthropology as he sheds light on the ceremonial dimensions of interaction and the sacredness that must envelop the individual if he is to possess an acceptable self.

Introduction to *The Presentation of Self in Everyday Life*

In this modern classic, Goffman develops his dramaturgical approach to social life and the self. It is in this work that Goffman introduces many of the central concepts discussed in the preceding sections: impression management, definition of the situation, front and backstage, and performer and character. In a sense, his book is an analysis of the familiar saying, "actions speak louder than words," but after reading Goffman you will never be able to adjust the volume again.

He begins by distinguishing between two types of "sign-vehicles"—expressions one *gives* and expressions one *gives off*—that are used by actors and audiences as means for both managing impressions and defining the situation. The former refers to "verbal symbols or their substitutes which [the individual] uses admittedly and solely to convey the information that he and the others are known to attach to these symbols." The latter "involves a wide range of action that others can treat as symptomatic of the actor, the expectation being that the action was performed for reasons other than the information conveyed in this way" (Goffman 1959:2). With this distinction, Goffman sets the backdrop for his analysis of the subtle techniques that actors and audiences employ, sometimes with consciously strategic intent and sometimes without, as they go about the business of dramatically staging acceptable images of their self. Yet, whether or not one's staged performance is intentionally manipulative, and regardless of whether the performer is sincere or cynical in playing his role, all participants in the encounter are obligated to abide by the moral rules that establish just what "acceptable" means. Here, participants in an interaction will establish a "working consensus" or surface agreement regarding the definition of the situation in which, in an effort to avoid conflict, embarrassment, or the discrediting of self-performances, each "is expected to suppress his immediate heartfelt feelings, conveying a view of the situation which he feels the others will be able to find at least temporarily acceptable" (ibid.:9). Under the unwritten rule to uphold a working consensus, we often find that in order to present a viable image of one's self, the definition of the situation compels us to tell "white lies." This explains our tactful refusals of invitations for interaction from people that we do not like: "Sorry, I'd love to come to your party but my cousin is visiting from out of town." You can only hope that on the night of the party no one sees you without your "cousin." Or, better yet, perhaps you can convince a friend to join your "team" for the evening and play the part of your favorite relative as you show her a night on the town.

The Presentation of Self in Everyday Life (1959)

Erving Goffman

INTRODUCTION

When an individual enters the presence of others, they commonly seek to acquire information about him or to bring into play information about him already possessed.

They will be interested in his general socio-economic status, his conception of self, his attitude toward them, his competence, his trustworthiness, etc. Although some of this information seems to be sought almost as an end in itself, there are usually quite practical reasons for acquiring it. Information about the individual helps to define the situation, enabling others to know in advance what he will expect of them and what they may expect of him. Informed in these ways, the others will know how best to act in order to call forth a desired response from him.

For those present, many sources of information become accessible and many carriers (or "sign-vehicles") become available for conveying this information. If unacquainted with the individual, observers can glean clues from his conduct and appearance which allow them to apply their previous experience with individuals roughly similar to the one before them or, more important, to apply untested stereotypes to him. They can also assume from past experience that only individuals of a particular kind are likely to be found in a given social setting. They can rely on what the individual says about himself or on documentary evidence he provides as to who and what he is. If they know, or know of, the individual by virtue of experience prior to the interaction, they can rely on assumptions as to the persistence and generality of psychological traits as a means of predicting his present and future behavior.

However, during the period in which the individual is in the immediate presence of the others, few events may occur which directly provide the others with the conclusive information they will need if they are to direct wisely their own activity. Many crucial facts lie beyond the time and place of interaction or lie concealed within it. For example, the "true" or "real" attitudes, beliefs, and emotions of the individual can be ascertained only indirectly, through his avowals or through what appears to be involuntary expressive behavior. Similarly, if the individual offers the others a product or service, they will often find that during the interaction there will be no time and place immediately available for eating the pudding that the proof can be found in. They will be forced to accept some events as conventional or natural signs of something not directly available to the senses. In Ichheiser's terms,[i] the individual will have to act so that he intentionally or unintentionally *expresses* himself, and the others will in turn have to be *impressed* in some way by him.

The expressiveness of the individual (and therefore his capacity to give impressions) appears to involve two radically different kinds of sign activity: the expression that he *gives,* and the expression that he *gives off.* The first involves verbal symbols or their substitutes which he uses admittedly and solely to convey the information that he and the others are known to attach to these symbols. This is communication in the traditional and narrow sense. The second involves a wide range of action that others can treat as symptomatic of the actor, the expectation being that the action was performed for reasons other than the information conveyed in this way. As we shall have to see, this distinction has an only initial validity. The individual does of course intentionally convey misinformation by means of both of these types of communication, the first involving deceit, the second feigning.

Taking communication in both its narrow and broad sense, one finds that when the individual is in the immediate presence of others, his activity will have a promissory character. The others are likely to find that they must accept the individual on faith, offering him a just return while he is present before them in exchange for something whose true value will not be established until after he has left their presence. (Of course, the others also live by inference in their dealings with the physical world, but it is only in the world of social interaction that the objects about which they make inferences will purposely facilitate and hinder this inferential process.) The security that they justifiably feel in making inferences about the individual will vary, of course, depending on such factors as the amount of information they already possess about him, but no amount of such past evidence can entirely obviate the necessity of acting on the basis of inferences. As William I. Thomas suggested:

> It is also highly important for us to realize that we do not as a matter of fact lead our lives, make our decisions, and reach our goals in everyday life either statistically or scientifically. We live by inference. I am, let us say, your guest. You do not know, you cannot determine scientifically, that I will not steal your money or your spoons. But inferentially I will not, and inferentially you have me as a guest.[ii]

---

[i]Gustav Ichheiser, "Misunderstandings in Human Relations," Supplement to *The American Journal of Sociology,* LV (September, 1949), pp. 6–7.

[ii]Quoted in E. H. Volkart, editor, *Social Behavior and Personality,* Contributions of W. I. Thomas to Theory and Social Research (New York: Social Science Research Council, 1951), p. 5.

Let us now turn from the others to the point of view of the individual who presents himself before them. He may wish them to think highly of him, or to think that he thinks highly of them, or to perceive how in fact he feels toward them, or to obtain no clear-cut impression; he may wish to ensure sufficient harmony so that the interaction can be sustained, or to defraud, get rid of, confuse, mislead, antagonize, or insult them. Regardless of the particular objective which the individual has in mind and of his motive for having this objective, it will be in his interests to control the conduct of the others, especially their responsive treatment of him. This control is achieved largely by influencing the definition of the situation which the others come to formulate, and he can influence this definition by expressing himself in such a way as to give them the kind of impression that will lead them to act voluntarily in accordance with his own plan. Thus, when an individual appears in the presence of others, there will usually be some reason for him to mobilize his activity so that it will convey an impression to others which it is in his interests to convey. Since a girl's dormitory mates will glean evidence of her popularity from the calls she receives on the phone, we can suspect that some girls will arrange for calls to be made, and Willard Waller's finding can be anticipated:

> It has been reported by many observers that a girl who is called to the telephone in the dormitories will often allow herself to be called several times, in order to give all the other girls ample opportunity to hear her paged.[iii]

Of the two kinds of communication—expressions given and expressions given off—this report will be primarily concerned with the latter, with the more theatrical and contextual kind, the non-verbal, presumably unintentional kind, whether this communication be purposely engineered or not. As an example of what we must try to examine, I would like to cite at length a novelistic incident in which Preedy, a vacationing Englishman, makes his first appearance on the beach of his summer hotel in Spain:

> But in any case he took care to avoid catching anyone's eye. First of all, he had to make it clear to those potential companions of his holiday that they were of no concern to him whatsoever. He stared through them, round them, over them—eyes lost in space. The beach might have been empty. If by chance a ball was thrown his way, he looked surprised; then let a smile of amusement lighten his face (Kindly Preedy), looked round dazed to see that there *were* people on the beach, tossed it back with a smile to himself and not a smile *at* the people, and then resumed carelessly his nonchalant survey of space.
>
> But it was time to institute a little parade, the parade of the Ideal Preedy. By devious handlings he gave any who wanted to look a chance to see the title of his book—a Spanish translation of Homer, classic thus, but not daring, cosmopolitan too—and then gathered together his beach-wrap and bag into a neat sand-resistant pile (Methodical and Sensible Preedy), rose slowly to stretch at ease his huge frame (Big-Cat Preedy), and tossed aside his sandals (Carefree Preedy, after all).
>
> The marriage of Preedy and the sea! There were alternative rituals. The first involved the stroll that turns into a run and a dive straight into the water, thereafter smoothing into a strong splashless crawl towards the horizon. But of course not really to the horizon. Quite suddenly he would turn on to his back and thrash great white splashes with his legs, somehow thus showing that he could have swum further had he wanted to, and then would stand up a quarter out of water for all to see who it was.
>
> The alternative course was simpler, it avoided the cold-water shock and it avoided the risk of appearing too high-spirited. The point was to appear to be so used to the sea, the Mediterranean, and this particular beach, that one might as well be in the sea as out of it. It involved a slow stroll down and into the edge of the water—not even noticing his toes were wet, land and water all the same to *him!*—with his eyes up at the sky gravely surveying portents, invisible to others, of the weather (Local Fisherman Preedy).[iv]

The novelist means us to see that Preedy is improperly concerned with the extensive impressions he feels his sheer bodily action is giving off to those around him. We can malign Preedy further by assuming that he has acted merely in order to give a particular impression, that this is a false impression, and that the others present

---

[iii]Willard Waller, "The Rating and Dating Complex," *American Sociological Review,* II, p. 730.

[iv]William Sansom, *A Contest of Ladies* (London: Hogarth, 1956), pp. 230–32.

receive either no impression at all, or, worse still, the impression that Preedy is affectedly trying to cause them to receive this particular impression. But the important point for us here is that the kind of impression Preedy thinks he is making is in fact the kind of impression that others correctly and incorrectly glean from someone in their midst.

I have said that when an individual appears before others his actions will influence the definition of the situation which they come to have. Sometimes the individual will act in a thoroughly calculating manner, expressing himself in a given way solely in order to give the kind of impression to others that is likely to evoke from them a specific response he is concerned to obtain. Sometimes the individual will be calculating in his activity but be relatively unaware that this is the case. Sometimes he will intentionally and consciously express himself in a particular way, but chiefly because the traditions of his group or social status require this kind of expression and not because of any particular response (other than vague acceptance or approval) that is likely to be evoked from those impressed by the expression. Sometimes the traditions of an individual's role will lead him to give a well-designed impression of a particular kind and yet he may be neither consciously nor unconsciously disposed to create such an impression. The others, in their turn, may be suitably impressed by the individual's efforts to convey something, or may misunderstand the situation and come to conclusions that are warranted neither by the individual's intent nor by the facts. In any case, in so far as the others act *as if* the individual had conveyed a particular impression, we may take a functional or pragmatic view and say that the individual has "effectively" projected a given definition of the situation and "effectively" fostered the understanding that a given state of affairs obtains.

There is one aspect of the others' response that bears special comment here. Knowing that the individual is likely to present himself in a light that is favorable to him, the others may divide what they witness into two parts; a part that is relatively easy for the individual to manipulate at will, being chiefly his verbal assertions, and a part in regard to which he seems to have little concern or control, being chiefly derived from the expressions he gives off. The others may then use what are considered to be the ungovernable aspects of his expressive behavior as a check upon the validity of what is conveyed by the governable aspects. In this a fundamental asymmetry is demonstrated in the communication process, the individual presumably being aware of only one stream of his communication, the witnesses of this stream and one other. For example, in Shetland Isle one crofter's wife, in serving native dishes to a visitor from the mainland of Britain, would listen with a polite smile to his polite claims of liking what he was eating; at the same time she would take note of the rapidity with which the visitor lifted his fork or spoon to his mouth, the eagerness with which he passed food into his mouth, and the gusto expressed in chewing the food, using these signs as a check on the stated feelings of the eater. The same woman, in order to discover what one acquaintance (A) "actually" thought of another acquaintance (B), would wait until B was in the presence of A but engaged in conversation with still another person (C). She would then covertly examine the facial expressions of A as he regarded B in conversation with C. Not being in conversation with B, and not being directly observed by him, A would sometimes relax usual constraints and tactful deceptions, and freely express what he was "actually" feeling about B. This Shetlander, in short, would observe the unobserved observer.

Now given the fact that others are likely to check up on the more controllable aspects of behavior by means of the less controllable, one can expect that sometimes the individual will try to exploit this very possibility, guiding the impression he makes through behavior felt to be reliably informing. For example, in gaining admission to a tight social circle, the participant observer may not only wear an accepting look while listening to an informant, but may also be careful to wear the same look when observing the informant talking to others; observers of the observer will then not as easily discover where he actually stands. A specific illustration may be cited from Shetland Isle. When a neighbor dropped in to have a cup of tea, he would ordinarily wear at least a hint of an expectant warm smile as he passed through the door into the cottage. Since lack of physical obstructions outside the cottage and lack of light within it usually made it possible to observe the visitor unobserved as he approached the house, islanders sometimes took pleasure in watching the visitor drop whatever expression he was manifesting and replace it with a sociable one just before reaching the door. However, some visitors, in appreciating that this examination was occurring, would blindly adopt a social face a long distance from the house, thus ensuring the projection of a constant image.

This kind of control upon the part of the individual reinstates the symmetry of the communication process, and sets the stage for a kind of information game—a potentially infinite cycle of concealment, discovery, false revelation, and rediscovery. It should be added that since the others are likely to be relatively unsuspicious of the presumably unguided aspect of the individual's conduct, he can gain much by controlling it. The others of course may sense that the individual is manipulating the presumably spontaneous aspects of his behavior, and seek in this very act of manipulation some shading of conduct that the individual has not managed to control. This again provides a check upon the individual's behavior, this time his presumably uncalculated behavior, thus re-establishing the asymmetry of the communication process. Here I would like only to add the suggestion that the arts of piercing an individual's effort at calculated unintentionality seem better developed than our capacity to manipulate our own behavior, so that regardless of how many steps have occurred in the information game, the witness is likely to have the advantage over the actor, and the initial asymmetry of the communication process is likely to be retained.

When we allow that the individual projects a definition of the situation when he appears before others, we must also see that the others, however passive their role may seem to be, will themselves effectively project a definition of the situation by virtue of their response to the individual and by virtue of any lines of action they initiate to him. Ordinarily the definitions of the situation projected by the several different participants are sufficiently attuned to one another so that open contradiction will not occur. I do not mean that there will be the kind of consensus that arises when each individual present candidly expresses what he really feels and honestly agrees with the expressed feelings of the others present. This kind of harmony is an optimistic ideal and in any case not necessary for the smooth working of society. Rather, each participant is expected to suppress his immediate heartfelt feelings, conveying a view of the situation which he feels the others will be able to find at least temporarily acceptable. The maintenance of this surface of agreement, this veneer of consensus, is facilitated by each participant concealing his own wants behind statements which assert values to which everyone present feels obliged to give lip service. Further, there is usually a kind of division of definitional labor. Each participant is allowed to establish the tentative official ruling regarding matters which are vital to him but not immediately important to others, e.g., the rationalizations and justifications by which he

accounts for his past activity. In exchange for this courtesy he remains silent or non-committal on matters important to others but not immediately important to him. We have then a kind of interactional *modus vivendi*. Together the participants contribute to a single over-all definition of the situation which involves not so much a real agreement as to what exists but rather a real agreement as to whose claims concerning what issues will be temporarily honored. Real agreement will also exist concerning the desirability of avoiding an open conflict of definitions of the situation. I will refer to this level of agreement as a "working consensus." It is to be understood that the working consensus established in one interaction setting will be quite different in content from the working consensus established in a different type of setting. Thus, between two friends at lunch, a reciprocal show of affection, respect, and concern for the other is maintained. In service occupations, on the other hand, the specialist often maintains an image of disinterested involvement in the problem of the client, while the client responds with a show of respect for the competence and integrity of the specialist. Regardless of such differences in content, however, the general form of these working arrangements is the same.

In noting the tendency for a participant to accept the definitional claims made by the others present, we can appreciate the crucial importance of the information that the individual *initially* possesses or acquires concerning his fellow participants, for it is on the basis of this initial information that the individual starts to define the situation and starts to build up lines of responsive action. The individual's initial projection commits him to what he is proposing to be and requires him to drop all pretenses of being other things. As the interaction among the participants progresses, additions and modifications in this initial informational state will of course occur, but it is essential that these later developments be related without contradiction to, and even built up from, the initial positions taken by the several participants. It would seem that an individual can more easily make a choice as to what line of treatment to demand from and extend to the others present at the beginning of an encounter than he can alter the line of treatment that is being pursued once the interaction is underway.

In everyday life, of course, there is a clear understanding that first impressions are important. Thus, the work adjustment of those in service occupations will often hinge upon a capacity to seize and hold the initiative in the service relation, a capacity that will require subtle aggressiveness on the part of the server when he is of lower socio-economic status than his client. W. F. Whyte suggests the waitress as an example:

> The first point that stands out is that the waitress who bears up under pressure does not simply respond to her customers. She acts with some skill to control their behavior. The first question to ask when we look at the customer relationship is, "Does the waitress get the jump on the customer, or does the customer get the jump on the waitress?" The skilled waitress realizes the crucial nature of this question....
>
> The skilled waitress tackles the customer with confidence and without hesitation. For example, she may find that a new customer has seated himself before she could clear off the dirty dishes and change the cloth. He is now leaning on the table studying the menu. She greets him, says, "May I change the cover, please?" and, without waiting for an answer, takes his menu away from him so that he moves back from the table, and she goes about her work. The relationship is handled politely but firmly, and there is never any question as to who is in charge.[v]

When the interaction that is initiated by "first impressions" is itself merely the initial interaction in an extended series of interactions involving the same participants, we speak of "getting off on the right foot" and feel that it is crucial that we do so. Thus, one learns that some teachers take the following view:

> You can't ever let them get the upper hand on you or you're through. So I start out tough. The first day I get a new class in, I let them know who's boss.... You've got to start off tough, then you can ease up as you go along. If you start out easy-going, when you try to get tough, they'll just look at you and laugh.[vi]

Similarly, attendants in mental institutions may feel that if the new patient is sharply put in his place the first day on the ward and made to see who is boss, much future difficulty will be prevented.

Given the fact that the individual effectively projects a definition of the situation when he enters the presence of others, we can assume that events may occur within the interaction which contradict, discredit, or otherwise throw doubt upon this projection. When these disruptive events occur, the interaction itself may come to a confused and embarrassed halt. Some of the assumptions upon which the responses of the participants had been predicated become untenable, and the participants find themselves lodged in an interaction for which the situation has been wrongly defined and is now no longer defined. At such moments the individual whose presentation has been discredited may feel ashamed while the others present may feel hostile, and all the participants may come to feel ill at ease, nonplussed, out of countenance, embarrassed, experiencing the kind of anomy that is generated when the minute social system of face-to-face interaction breaks down.

In stressing the fact that the initial definition of the situation projected by an individual tends to provide a plan for the co-operative activity that follows—in stressing this action point of view—we must not overlook the crucial fact that any projected definition of the situation also has a distinctive moral character. It is this moral character of projections that will chiefly concern us in this report. Society is organized on the principle that any individual who possesses certain social characteristics has a moral right to expect that others will value and treat him in an appropriate way. Connected with this principle is a second, namely that an individual who implicitly or explicitly signifies that he has certain social characteristics ought in fact to be what he claims he is. In consequence, when an individual projects a definition of the situation and thereby makes an implicit or explicit claim to be a person of a particular kind, he automatically exerts a moral demand upon the others, obliging them to value and treat him in the manner that persons of his kind have a right to expect. He also implicitly forgoes all claims to be things he does not appear to be and hence forgoes the treatment that would be appropriate for such individuals. The others find, then, that the individual has informed them as to what is and as to what they *ought* to see as the "is."

One cannot judge the importance of definitional disruptions by the frequency with which they occur, for

---

[v]W. F. Whyte, "When Workers and Customers Meet," Chap. VII, *Industry and Society,* ed. W. F. Whyte (New York: McGraw-Hill, 1946), pp. 132–33.

[vi]Teacher interview quoted by Howard S. Becker, "Social Class Variations in the Teacher-Pupil Relationship," *Journal of Educational Sociology,* XXV, p. 459.

apparently they would occur more frequently were not constant precautions taken. We find that preventive practices are constantly employed to avoid these embarrassments and that corrective practices are constantly employed to compensate for discrediting occurrences that have not been successfully avoided. When the individual employs these strategies and tactics to protect his own projections, we may refer to them as "defensive practices"; when a participant employs them to save the definition of the situation projected by another, we speak of "protective practices" or "tact." Together, defensive and protective practices comprise the techniques employed to safeguard the impression fostered by an individual during his presence before others. It should be added that while we may be ready to see that no fostered impression would survive if defensive practices were not employed, we are less ready perhaps to see that few impressions could survive if those who received the impression did not exert tact in their reception of it.

In addition to the fact that precautions are taken to prevent disruption of projected definitions, we may also note that an intense interest in these disruptions comes to play a significant role in the social life of the group. Practical jokes and social games are played in which embarrassments which are to be taken unseriously are purposely engineered. Fantasies are created in which devastating exposures occur. Anecdotes from the past—real, embroidered, or fictitious—are told and retold, detailing disruptions which occurred, almost occurred, or occurred and were admirably resolved. There seems to be no grouping which does not have a ready supply of these games, reveries, and cautionary tales, to be used as a source of humor, a catharsis for anxieties, and a sanction for inducing individuals to be modest in their claims and reasonable in their projected expectations. The individual may tell himself through dreams of getting into impossible positions. Families tell of the time a guest got his dates mixed and arrived when neither the house nor anyone in it was ready for him. Journalists tell of times when an all-too-meaningful misprint occurred, and the paper's assumption of objectivity or decorum was humorously discredited. Public servants tell of times a client ridiculously misunderstood form instructions, giving answers which implied an unanticipated and bizarre definition of the situation. Seamen, whose home away from home is rigorously he-man, tell stories of coming back home and inadvertently asking mother to "pass the fucking butter." Diplomats tell of the time a near-sighted queen asked a republican ambassador about the health of his king.

To summarize, then, I assume that when an individual appears before others he will have many motives for trying to control the impression they receive of the situation. This report is concerned with some of the common techniques that persons employ to sustain such impressions and with some of the common contingencies associated with the employment of these techniques. The specific content of any activity presented by the individual participant, or the role it plays in the interdependent activities of an on-going social system, will not be at issue; I shall be concerned only with the participant's dramaturgical problems of presenting the activity before others. The issues dealt with by stagecraft and stage management are sometimes trivial but they are quite general; they seem to occur everywhere in social life, providing a clear-cut dimension for formal sociological analysis.

It will be convenient to end this introduction with some definitions that are implied in what has gone before and required for what is to follow. For the purpose of this report, interaction (that is, face-to-face interaction) may be roughly defined as the reciprocal influence of individuals upon one another's actions when in one another's immediate physical presence. *An* interaction may be defined as all the interaction which occurs throughout any one occasion when a given set of individuals are in one another's continuous presence; the term "an encounter" would do as well. A "performance" may be defined as all the activity of a given participant on a given occasion which serves to influence in any way any of the other participants. Taking a particular participant and his performance as a basic point of reference, we may refer to those who contribute the other performances as the audience, observers, or co-participants. The pre-established pattern of action which is unfolded during a performance and which may be presented or played through on other occasions may be called a "part" or "routine." These situational terms can easily be related to conventional structural ones. When an individual or performer plays the same part to the same audience on different occasions, a social relationship is likely to arise. Defining social role as the enactment of rights and duties attached to a given status, we can say that a social role will involve one or more parts and that each of these different parts may be presented by the performer on a series of occasions to the same kinds of audience or to an audience of the same persons.

PERFORMANCES

Belief in the Part One Is Playing

When an individual plays a part he implicitly requests his observers to take seriously the impression that is fostered before them. They are asked to believe that the character they see actually possesses the attributes he appears to possess, that the task he performs will have the consequences that are implicitly claimed for it, and that, in general, matters are what they appear to be. In line with this, there is the popular view that the individual offers his performance and puts on his show "for the benefit of other people." It will be convenient to begin a consideration of performances by turning the question around and looking at the individual's own belief in the impression of reality that he attempts to engender in those among whom he finds himself.

At one extreme, one finds that the performer can be fully taken in by his own act; he can be sincerely convinced that the impression of reality which he stages is the real reality. When his audience is also convinced in this way about the show he puts on—and this seems to be the typical case—then for the moment at least, only the sociologist or the socially disgruntled will have any doubts about the "realness" of what is presented.

At the other extreme, we find that the performer may not be taken in at all by his own routine. This possibility is understandable, since no one is in quite as good an observational position to see through the act as the person who puts it on. Coupled with this, the performer may be moved to guide the conviction of his audience only as a means to other ends, having no ultimate concern in the conception that they have of him or of the situation. When the individual has no belief in his own act and no ultimate concern with the beliefs of his audience, we may call him cynical, reserving the term "sincere" for individuals who believe in the impression fostered by their own performance. It should be understood that the cynic, with all his professional disinvolvement, may obtain unprofessional pleasures from his masquerade, experiencing a kind of gleeful spiritual aggression from the fact that he can toy at will with something his audience must take seriously.[vii]

It is not assumed, of course, that all cynical performers are interested in deluding their audiences for purposes of what is called "self-interest" or private gain. A cynical individual may delude his audience for what he considers to be their own good, or for the good of the community, etc. . . . We know that in service occupations practitioners who may otherwise be sincere are sometimes forced to delude their customers because their customers show such a heartfelt demand for it. Doctors who are led into giving placebos, filling station attendants who resignedly check and recheck tire pressures for anxious women motorists, shoe clerks who sell a shoe that fits but tell the customer it is the size she wants to hear—these are cynical performers whose audiences will not allow them to be sincere. Similarly, it seems that sympathetic patients in mental wards will sometimes feign bizarre symptoms so that student nurses will not be subjected to a disappointingly sane performance. So also, when inferiors extend their most lavish reception for visiting superiors, the selfish desire to win favor may not be the chief motive; the inferior may be tactfully attempting to put the superior at ease by simulating the kind of world the superior is thought to take for granted.

I have suggested two extremes: an individual may be taken in by his own act or be cynical about it. These extremes are something a little more than just the ends of a continuum. Each provides the individual with a position which has its own particular securities and defenses, so there will be a tendency for those who have traveled close to one of these poles to complete the voyage. Starting with lack of inward belief in one's role, the individual may follow the natural movement described by Park:

It is probably no mere historical accident that the word person, in its first meaning, is a mask. It is rather a recognition of the fact that everyone is always and everywhere, more or less consciously, playing a role. . . . It is in these roles that we know each other; it is in these roles that we know ourselves.[viii]

In a sense, and in so far as this mask represents the conception we have formed of ourselves—the role we are striving to live up to—this mask is our truer self, the self we would like to be. In the end, our conception

---

[vii]Perhaps the real crime of the confidence man is not that he takes money from his victims but that he robs all of us of the belief that middle-class manners and appearance can be sustained only by middle-class people. A disabused professional can be cynically hostile to the service relation his clients expect him to extend to them, the confidence man is in a position to hold the whole "legit" world in this contempt.

[viii]Robert Ezra Park, *Race and Culture* (Glencoe, Ill.: The Free Press, 1950), p. 249.

of our role becomes second nature and an integral part of our personality. We come into the world as individuals, achieve character, and become persons.[ix]

. . . While we can expect to find natural movement back and forth between cynicism and sincerity, still we must not rule out the kind of transitional point that can be sustained on the strength of a little self-illusion. We find that the individual may attempt to induce the audience to judge him and the situation in a particular way, and he may seek this judgment as an ultimate end in itself, and yet he may not completely believe that he deserves the valuation of self which he asks for or that the impression of reality which he fosters is valid. . . .

Front

I have been using the term "performance" to refer to all the activity of an individual which occurs during a period marked by his continuous presence before a particular set of observers and which has some influence on the observers. It will be convenient to label as "front" that part of the individual's performance which regularly functions in a general and fixed fashion to define the situation for those who observe the performance. Front, then, is the expressive equipment of a standard kind intentionally or unwittingly employed by the individual during his performance. For preliminary purposes, it will be convenient to distinguish and label what seem to be the standard parts of front.

First, there is the "setting," involving furniture, décor, physical layout, and other background items which supply the scenery and stage props for the spate of human action played out before, within, or upon it. A setting tends to stay put, geographically speaking, so that those who would use a particular setting as part of their performance cannot begin their act until they have brought themselves to the appropriate place and must terminate their performance when they leave it. It is only in exceptional circumstances that the setting follows along with the performers; we see this in the funeral cortège, the civic parade, and the dreamlike processions that kings and queens are made of. In the main, these exceptions seem to offer some kind of extra protection for performers who are, or who have momentarily become, highly sacred. These worthies are to be distinguished, of course, from quite profane performers

of the peddler class who move their place of work between performances, often being forced to do so. In the matter of having one fixed place for one's setting, a ruler may be too sacred, a peddler too profane. . . .

If we take the term "setting" to refer to the scenic parts of expressive equipment, one may take the term "personal front" to refer to the other items of expressive equipment, the items that we most intimately identify with the performer himself and that we naturally expect will follow the performer wherever he goes. As part of personal front we may include: insignia of office or rank; clothing; sex, age, and racial characteristics; size and looks; posture; speech patterns; facial expressions; bodily gestures; and the like. Some of these vehicles for conveying signs, such as racial characteristics, are relatively fixed and over a span of time do not vary for the individual from one situation to another. On the other hand, some of these sign vehicles are relatively mobile or transitory, such as facial expression, and can vary during a performance from one moment to the next.

It is sometimes convenient to divide the stimuli which make up personal front into "appearance" and "manner," according to the function performed by the information that these stimuli convey. "Appearance" may be taken to refer to those stimuli which function at the time to tell us of the performer's social statuses. These stimuli also tell us of the individual's temporary ritual state, that is, whether he is engaging in formal social activity, work, or informal recreation, whether or not he is celebrating a new phase in the season cycle or in his life-cycle. "Manner" may be taken to refer to those stimuli which function at the time to warn us of the interaction role the performer will expect to play in the oncoming situation. Thus a haughty, aggressive manner may give the impression that the performer expects to be the one who will initiate the verbal interaction and direct its course. A meek, apologetic manner may give the impression that the performer expects to follow the lead of others, or at least that he can be led to do so. . . .

In order to explore more fully the relations among the several parts of social front, it will be convenient to consider here a significant characteristic of the information conveyed by front, namely, its abstractness and generality.

However specialized and unique a routine is, its social front, with certain exceptions will tend to claim facts that can be equally claimed and asserted of other,

---

[ix]Ibid., p. 250.

somewhat different routines. For example many service occupations offer their clients a performance that is illuminated with dramatic expressions of cleanliness, modernity, competence, and integrity. While in fact these abstract standards have a different significance in different occupational performances the observer is encouraged to stress the abstract similarities. For the observer this is a wonderful, though sometime disastrous, convenience. Instead of having to maintain a different pattern of expectation and responsive treatment for each slightly different performer and performance he can place the situation in a broad category around which it is easy for him to mobilize his past experience and stereotypical thinking. Observers then need only be familiar with a small and hence manageable vocabulary of fronts, and know how to respond to them, in order to orient themselves in a wide variety of situations. . . .

In addition to the fact that different routines may employ the same front, it is to be noted that a given social front tends to become institutionalized in terms of the abstract stereotyped expectations to which it gives rise, and tends to take on a meaning and stability apart from the specific tasks which happen at the time to be performed in its name. The front becomes a "collective representation" and a fact in its own right.

When an actor takes on an established social role, usually he finds that a particular front has already been established for it. Whether his acquisition of the role was primarily motivated by a desire to perform the given task or by a desire to maintain the corresponding front, the actor will find that he must do both.

Further, if the individual takes on a task that is not only new to him but also unestablished in the society, or if he attempts to change the light in which his task is viewed, he is likely to find that there are already several well-established fronts among which he must choose. Thus, when a task is given a new front we seldom find that the front it is given is itself new. . . .

Reality and Contrivance

In our own Anglo-American culture there seems to be two common-sense models according to which we formulate our conceptions of behavior: the real, sincere, or honest performance; and the false one that thorough fabricators assemble for us, whether meant to be taken unseriously, as in the work of stage actors, or seriously, as in the work of confidence men. We tend to see real performances as something not purposely put together

at all, being an unintentional product of the individual's unselfconscious response to the facts in his situation. And contrived performances we tend to see as something painstakingly pasted together, one false item on another, since there is no reality to which the items of behavior could be a direct response. It will be necessary to see now that these dichotomous conceptions are by way of being the ideology of honest performers, providing strength to the show they put on, but a poor analysis of it.

First, let it be said that there are many individuals who sincerely believe that the definition of the situation they habitually project is the real reality. In this report I do not mean to question their proportion in the population but rather the structural relation of their sincerity to the performances they offer. If a performance is to come off, the witnesses by and large must be able to believe that the performers are sincere. This is the structural place of sincerity in the drama of events. Performers may be sincere—or be insincere but sincerely convinced of their own sincerity—but this kind of affection for one's part is not necessary for its convincing performance. There are not many French cooks who are really Russian spies, and perhaps there are not many women who play the part of wife to one man and mistress to another but these duplicities do occur, often being sustained successfully for long periods of time. This suggests that while persons usually are what they appear to be, such appearances could still have been managed. There is, then, a statistical relation between appearances and reality, not an intrinsic or necessary one. In fact, given the unanticipated threats that play upon a performance, and given the need . . . to maintain solidarity with one's fellow performers and some distance from the witnesses, we find that a rigid incapacity to depart from one's inward view of reality may at times endanger one's performance. Some performances are carried off successfully with complete dishonesty, others with complete honesty; but for performances in general neither of these extremes is essential and neither, perhaps, is dramaturgically advisable.

The implication here is that an honest, sincere, serious performance is less firmly connected with the solid world than one might first assume. And this implication will be strengthened if we look again at the distance usually placed between quite honest performances and quite contrived ones. In this connection take, for example, the remarkable phenomenon of stage acting. It does take deep skill, long training, and psychological capacity to become a good stage actor. But this fact should not blind us to another one: that almost anyone can quickly learn a script well enough to give a charitable audience some sense of realness in what is being contrived before them. And it seems this is so because ordinary social intercourse is itself put together as a scene is put together, by the exchange of dramatically inflated actions, counteractions, and terminating replies. Scripts even in the hands of unpracticed players can come to life because life itself is a dramatically enacted thing. All the world is not, of course, a stage, but the crucial ways in which it isn't are not easy to specify. . . .

When the individual does move into a new position in society and obtains a new part to perform, he is not likely to be told in full detail how to conduct himself, nor will the facts of his new situation press sufficiently on him from the start to determine his conduct without his further giving thought to it. Ordinarily he will be given only a few cues, hints, and stage directions, and it will be assumed that he already has in his repertoire a large number of bits and pieces of performances that will be required in the new setting. The individual will already have a fair idea of what modesty, deference, or righteous indignation looks like, and can make a pass at playing these bits when necessary. He may even be able to play out the part of a hypnotic subject or commit a "compulsive" crime on the basis of models for these activities that he is already familiar with.

A theatrical performance or a staged confidence game requires a thorough scripting of the spoken content of the routine; but the vast part involving "expression given off" is often determined by meager stage directions. It is expected that the performer of illusions will already know a good deal about how to manage his voice, his face, and his body, although he—as well as any person who directs him—may find it difficult indeed to provide a detailed verbal statement of this kind of knowledge. And in this, of course, we approach the situation of the straightforward man in the street. Socialization may not so much involve a learning of the many specific details of a single concrete part—often there could not be enough time or energy for this. What does seem to be required of the individual is that he learn enough pieces of expression to be able to "fill in" and manage, more or less, any part that he is likely to be given. The legitimate performances of everyday life are not "acted" or "put on" in the sense that the performer knows in advance just what he is going to do, and does this solely because of the effect it is likely to have. The expressions it is felt he is giving off will be especially "inaccessible" to him. But as in the case of

less legitimate performers, the incapacity of the ordinary individual to formulate in advance the movements of his eyes and body does not mean that he will not express himself through these devices in a way that is dramatized and pre-formed in his repertoire of actions. In short, we all act better than we know how. . . .

[W]hen we observe a young American middle-class girl playing dumb for the benefit of her boy friend, we are ready to point to items of guile and contrivance in her behavior. But like herself and her boy friend, we accept as an unperformed fact that this performer *is* a young American middle-class girl. But surely here we neglect the greater part of the performance. It is commonplace to say that different social groupings express in different ways such attributes as age, sex, territory, and class status, and that in each case these bare attributes are elaborated by means of a distinctive complex cultural configuration of proper ways of conducting oneself. To *be* a given kind of person, then, is not merely to possess the required attributes, but also to sustain the standards of conduct and appearance that one's social grouping attaches thereto. The unthinking ease with which performers consistently carry off such standard-maintaining routines does not deny that a performance has occurred, merely that the participants have been aware of it.

A status, a position, a social place is not a material thing, to be possessed and then displayed; it is a pattern of appropriate conduct, coherent, embellished, and well articulated. Performed with ease or clumsiness, awareness or not, guile or good faith, it is none the less something that must be enacted and portrayed, something that must be realized. . . .

CONCLUSION

The Role of Expression Is Conveying Impressions of Self

Underlying all social interaction there seems to be a fundamental dialectic. When one individual enters the presence of others, he will want to discover the facts of the situation. Were he to possess this information, he could know, and make allowances for, what will come to happen and he could give the others present as much of their due as is consistent with his enlightened self-interest. To uncover fully the factual nature of the situation, it would be necessary for the individual to know all the relevant social data about the others. It would also be necessary for the individual to know the actual outcome or end product of the activity of the others during the interaction, as well as their innermost feelings concerning him. Full information of this order is rarely available; in its absence, the individual tends to employ substitutes—cues, tests, hints, expressive gestures, status symbols, etc.—as predictive devices. In short, since the reality that the individual is concerned with is unperceivable at the moment, appearances must be relied upon in its stead. And, paradoxically, the more the individual is concerned with the reality that is not available to perception, the more must he concentrate his attention on appearances.

The individual tends to treat the others present on the basis of the impression they give now about the past and the future. It is here that communicative acts are translated into moral ones. The impressions that the others give tend to be treated as claims and promises they have implicitly made, and claims and promises tend to have a moral character. In his mind the individual says: "I am using these impressions of you as a way of checking up on you and your activity, and you ought not to lead me astray." The peculiar thing about this is that the individual tends to take this stand even though he expects the others to be unconscious of many of their expressive behaviors and even though he may expect to exploit the others on the basis of the information he gleans about them. Since the sources of impression used by the observing individual involve a multitude of standards pertaining to politeness and decorum, pertaining both to social intercourse and task-performance, we can appreciate afresh how daily life is enmeshed in moral lines of discrimination.

Let us shift now to the point of view of the others. If they are to be gentlemanly, and play the individual's game, they will give little conscious heed to the fact that impressions are being formed about them but rather act without guile or contrivance, enabling the individual to receive valid impressions about them and their efforts. And if they happen to give thought to the fact that they are being observed, they will not allow this to influence them unduly, content in the belief that the individual will obtain a correct impression and give them their due because of it. Should they be concerned with influencing the treatment that the individual gives them, and this is properly to be expected, then a gentlemanly means will be available to them. They need only guide their action in the present so that its future consequences will be the kind that would lead a just individual to treat them now in a way they want to be treated; once this is done, they have only to rely on the perceptiveness and justness of the individual who observes them.

Sometimes those who are observed do, of course, employ these proper means of influencing the way in which the observer treats them. But there is another way, a shorter and more efficient way, in which the observed can influence the observer. Instead of allowing an impression of their activity to arise as an incidental by-product of their activity, they can reorient their frame of reference and devote their efforts to the creation of desired impressions. Instead of attempting to achieve certain ends by acceptable means, they can attempt to achieve the impression that they are achieving certain ends by acceptable means. It is always possible to manipulate the impression the observer uses as a substitute for reality because a sign for the presence of a thing, not being that thing, can be employed in the absence of it. The observer's need to rely on representations of things itself creates the possibility of misrepresentation.

There are many sets of persons who feel they could not stay in business, whatever their business, if they limited themselves to the gentlemanly means of influencing the individual who observes them. At some point or other in the round of their activity they feel it is necessary to band together and directly manipulate the impression that they give. The observed become a performing team and the observers become the audience. Actions which appear to be done on objects become gestures addressed to the audience. The round of activity becomes dramatized.

We come now to the basic dialectic. In their capacity as performers, individuals will be concerned with maintaining the impression that they are living up to the many standards by which they and their products are judged. Because these standards are so numerous and so pervasive, the individuals who are performers dwell more than we might think in a moral world. But, *qua* performers, individuals are concerned not with the moral issue of realizing these standards, but with the amoral issue of engineering a convincing impression that these standards are being realized. Our activity, then, is largely concerned with moral matters, but as performers we do not have a moral concern with them. As performers we are merchants of morality. Our day is given over to intimate contact with the goods we display and our minds are filled with intimate understandings of them; but it may well be that the more attention we give to these goods, then the more distant we feel from them and from those who are believing enough to buy them. To use a different imagery, the very obligation and profitability of appearing always in a steady moral light, of being a socialized character, forces one to be the sort of person who is practiced in the ways of the stage.

Staging and the Self

The general notion that we make a presentation of ourselves to others is hardly novel; what ought to be stressed in conclusion is that the very structure of the self can be seen in terms of how we arrange for such performances in our Anglo-American society.

In this report, the individual was divided by implication into two basic parts: he was viewed as a *performer,* a harried fabricator of impressions involved in the all-too-human task of staging a performance; he was viewed as a *character,* a figure, typically a fine one, whose spirit, strength, and other sterling qualities the performance was designed to evoke. The attributes of a performer and the attributes of a character are of a different order, quite basically so, yet both sets have their meaning in terms of the show that must go on.

First, character. In our society the character one performs and one's self are somewhat equated, and this self-as-character is usually seen as something housed within the body of its possessor, especially the upper parts thereof, being a nodule, somehow, in the psychobiology of personality. I suggest that this view is an implied part of what we are all trying to present, but provides, just because of this, a bad analysis of the presentation. In this report the performed self was seen as some kind of image, usually creditable, which the individual on stage and in character effectively attempts to induce others to hold in regard to him. While this image is entertained *concerning* the individual, so that a self is imputed to him, this self itself does not derive from its possessor, but from the whole scene of his action, being generated by that attribute of local events which renders them interpretable by witnesses. A correctly staged and performed scene leads the audience to impute a self to a performed character, but this imputation—this self—is a *product* of a scene that comes off, and is not a *cause* of it. The self, then, as a performed character, is not an organic thing that has a specific location, whose fundamental fate is to be born, to mature, and to die; it is a dramatic effect arising diffusely from a scene that is presented, and the characteristic issue, the crucial concern, is whether it will be credited or discredited.

In analyzing the self then we are drawn from its possessor, from the person who will profit or lose most by it, for he and his body merely provide the peg on which something of collaborative manufacture will be hung for a time. And the means for producing and maintaining selves do not reside inside the peg; in fact these means are often bolted down in social establishments. There will be a back region with its tools for shaping the body, and a front region with its fixed props. There will be a team of persons whose activity on stage in conjunction with available props will constitute the scene from which the performed character's self will emerge, and another team, the audience, whose interpretive activity will be necessary for this emergence. The self is a product of all of these arrangements, and in all of its parts bears the marks of this genesis.

The whole machinery of self-production is cumbersome, of course, and sometimes breaks down, exposing its separate components: back region control; team collusion; audience tact; and so forth. But, well oiled, impressions will flow from it fast enough to put us in the grips of one of our types of reality—the performance will come off and the firm self accorded each performed character will appear to emanate intrinsically from its performer.

Let us turn now from the individual as character performed to the individual as performer. He has a capacity to learn, this being exercised in the task of training for a part. He is given to having fantasies and dreams, some that pleasurably unfold a triumphant performance, others full of anxiety and dread that nervously deal with vital discreditings in a public front region. He often manifests a gregarious desire for teammates and audiences, a tactful considerateness for their concerns; and he has a capacity for deeply felt shame, leading him to minimize the chances he takes of exposure.

These attributes of the individual *qua* performer are not merely a depicted effect of particular performances; they are psychobiological in nature, and yet they seem to arise out of intimate interaction with the contingencies of staging performances.

And now a final comment. In developing the conceptual framework employed in this report, some language of the stage was used. I spoke of performers and audiences; of routines and parts; of performances coming off or falling flat; of cues, stage settings and backstage; of dramaturgical needs, dramaturgical skills, and dramaturgical strategies. Now it should be admitted that this attempt to press a mere analogy so far was in part a rhetoric and a maneuver.

The claim that all the world's a stage is sufficiently commonplace for readers to be familiar with its limitations and tolerant of its presentation, knowing that at any time they will easily be able to demonstrate to themselves that it is not to be taken too seriously. An action staged in a theater is a relatively contrived illusion and an admitted one; unlike ordinary life, nothing real or actual can happen to the performed characters—although at another level of course something real and actual can happen to the reputation of performers *qua* professionals whose everyday job is to put on theatrical performances.

And so here the language and mask of the stage will be dropped. Scaffolds, after all, are to build other things with, and should be erected with an eye to taking them down. This report is not concerned with aspects of theater that creep into everyday life. It is concerned with the structure of social encounters—the structure of those entities in social life that come into being whenever persons enter one another's immediate physical presence. The key factor in this structure is the maintenance of a single definition of the situation, this definition having to be expressed, and this expression sustained in the face of a multitude of potential disruptions.

A character staged in a theater is not in some ways real, nor does it have the same kind of real consequences as does the thoroughly contrived character performed by a confidence man; but the *successful* staging of either of these types of false figures involves use of *real* techniques—the same techniques by which everyday persons sustain their real social situations. Those who conduct face to face interaction on a theater's stage must meet the key requirement of real situations; they must expressively sustain a definition of the situation: but this they do in circumstances that have facilitated their developing an apt terminology for the interactional tasks that all of us share.

⊞

Introduction to *Asylums*

In *Asylums* (1961), Goffman again takes up the issue of the social sources of this self. However, in this study he reports on daily life as it transpires in mental hospitals and prisons. Yet his aim is not solely to understand how these specific establishments shape the self, but also to explore how such places shed light on the nature of the self as it is experienced in more ordinary, civilian settings.

Mental hospitals, prisons, monasteries, convents, boarding schools, and the military all have one thing in common: They are all total institutions. **Total institutions** are places of "residence and work where a large number of like-situated individuals, cut off from the wider society for an appreciable period of time, together lead an enclosed, formally administered round of life" (Goffman 1961:xiii). It is here where "under one roof and according to one rational plan, all spheres of individuals' lives—sleeping, eating, playing, and working—are regulated" (Branaman 1997:lv). To one degree or another, inhabitants of such facilities are stripped of the freedoms and resources to manage their self-presentation that are normally provided by social arrangements. As a result, they are subjected to the **mortification of self**: the process of "killing off" the multiple selves possessed prior to one's entrance into the total institution and replacing them with one totalizing identity over which the person exercises little, if any, control. Here is the life of the prison inmate or military recruit: shaven head, dressed in institutional clothing, a number or insult substituted for one's name, dispossessed of personal property, endless degradation, and complete loss of privacy over intimate information and matters of personal hygiene. All work together to construct a self radically different from the one that entered the establishment.

To counteract this profanation of self, inhabitants of total institutions engage in the types of practices we all do when we sense that our sacredness is threatened, only here they appear more extreme. Thus, when patients in a mental institution repeatedly bang a chair against the floor or smear their feces on the wall of the ward, their "recalcitrance" is less a symptom of their psychosis and more a sign of the conditions under which they are forced to live. Such actions serve the function of preserving one's self-autonomy, as do our calling in sick to work or devising (unconvincing) excuses for missing an exam in order to recuperate from a "festive" weekend, or the sarcastic enthusiasm we display in response to our boss's directives. The failure to cooperate with institutional demands that is expressed in each case is a "normal" attempt to preserve control over one's self. Each represents a **secondary adjustment** or a way "of taking leave of a place without moving from it" (Goffman 1961:308). Secondary adjustments, then, are "ways in which the individual stands apart from the role and the self that were taken for granted for him by the institution" (ibid.:189). They are oppositional practices through which we refuse the "official" view of what we should be and thus distance ourselves from an organization.

Goffman's notion of secondary adjustments recalls the distinction he made in *The Presentation of Self in Everyday Life* between the "performer" and the "character." While he arguably emphasized the "fictional" nature of the self in that work, here he contends that behind the performer's mask lies a solid, "stance-taking entity." Nevertheless, the self, contrary to the claims of psychologists and psychiatrists, is realized only in and through the social arrangements that alone create the conditions for its expression. And contrary to the claims of sociologists, the self is much more than a simple reflection of the groups to which it belongs:

> Without something to belong to, we have no stable self, and yet total commitment and attachment to any social unit implies a kind of selflessness. Our sense of being a person can come from being drawn into a wider social unit; our sense of selfhood can rise through the little ways we resist the pull. Our status is backed by the solid building of the world, while our sense of personal identity often resides in the cracks. (ibid.:320)[1]

---

[1]Georg Simmel (1858–1918), a German sociologist and philosopher whose ideas influenced the development of the Chicago School of symbolic interactionism, as well as Goffman's own perspective on the social order, made a similar observation half a century before. Commenting on the relationship between self and society, Simmel noted that while who you are as an individual is in an important sense defined and made possible by the groups to which you belong, preserving your individuality demands that your identity not be completely submerged into or engulfed by group membership. Otherwise, you have no self that you can call your own. Society, in his words, is

> a structure which consists of beings who stand inside and outside of it at the same time. This fact forms the basis for one of the most important sociological phenomena, namely, that between a society and its component individuals a relation may exist as if between two parties. . . . [T]he individual can never stay within a unit which he does not at the same time stay outside of, that he is not incorporated into any order without also confronting it. (Simmel [1908] 1971:14, 15)

Asylums (1961)

Erving Goffman

CHARACTERISTICS OF TOTAL INSTITUTIONS

I

A basic social arrangement in modern society is that the individual tends to sleep, play, and work in different places, with different co-participants, under different authorities, and without an over-all rational plan. The central feature of total institutions can be described as a breakdown of the barriers ordinarily separating these three spheres of life. First, all aspects of life are conducted in the same place and under the same single authority. Second, each phase of the member's daily activity is carried on in the immediate company of a large batch of others, all of whom are treated alike and required to do the same thing together. Third, all phases

---

of the day's activities are tightly scheduled, with one activity leading at a prearranged time into the next, the whole sequence of activities being imposed from above by a system of explicit formal rulings and a body of officials. Finally, the various enforced activities are brought together into a single rational plan purportedly designed to fulfill the official aims of the institution.

Individually, these features are found in places other than total institutions. For example, our large commercial, industrial, and educational establishments are increasingly providing cafeterias and free-time recreation for their members; use of these extended facilities remains voluntary in many particulars, however, and special care is taken to see that the ordinary line of authority does not extend to them. Similarly, housewives or farm families may have all their major spheres of life within the same fenced-in area, but these persons are not collectively regimented and do not march through the day's activities in the immediate company of a batch of similar others.

The handling of many human needs by the bureaucratic organization of whole blocks of people—whether or not this is a necessary or effective means of social organization in the circumstances—is the key fact of total institutions. From this follow certain important implications.

When persons are moved in blocks, they can be supervised by personnel whose chief activity is not guidance or periodic inspection (as in many employer-employee relations) but rather surveillance—a seeing to it that everyone does what he has been clearly told is required of him, under conditions where one person's infraction is likely to stand out in relief against the visible, constantly examined compliance of the others. Which comes first, the large blocks of managed people, or the small supervisory staff, is not here at issue; the point is that each is made for the other.

In total institutions there is a basic split between a large managed group, conveniently called inmates, and a small supervisory staff. Inmates typically live in the institution and have restricted contact with the world outside the walls; staff often operate on an eight-hour day and are socially integrated into the outside world. Each grouping tends to conceive of the other in terms of narrow hostile stereotypes, staff often seeing inmates as bitter, secretive, and untrustworthy, while inmates often see staff as condescending, highhanded, and mean.

Staff tends to feel superior and righteous; inmates tend, in some ways at least, to feel inferior, weak, blameworthy, and guilty.

Social mobility between the two strata is grossly restricted; social distance is typically great and often formally prescribed. Even talk across the boundaries may be conducted in a special tone of voice. . . . Just as talk across the boundary is restricted, so, too, is the passage of information, especially information about the staff's plans for inmates. Characteristically, the inmate is excluded from knowledge of the decisions taken regarding his fate. Whether the official grounds are military, as in concealing travel destination from enlisted men, or medical, as in concealing diagnosis, plan of treatment, and approximate length of stay from tuberculosis patients, such exclusion gives staff a special basis of distance from and control over inmates. . . .

The total institution is a social hybrid, part residential community, part formal organization; therein lies its special sociological interest. There are other reasons for being interested in these establishments, too. In our society, they are the forcing houses for changing persons; each is a natural experiment on what can be done to the self.

II

The recruit comes into the establishment with a conception of himself made possible by certain stable social arrangements in his home world. Upon entrance, he is immediately stripped of the support provided by these arrangements. In the accurate language of some of our oldest total institutions, he begins a series of abasements, degradations, humiliations, and profanations of self. His self is systematically, if often unintentionally, mortified. He begins some radical shifts in his *moral career,* a career composed of the progressive changes that occur in the beliefs that he has concerning himself and significant others.

The processes by which a person's self is mortified are fairly standard in total institutions;[i] analysis of these processes can help us to see the arrangements that ordinary establishments must guarantee if members are to preserve their civilian selves.

The barrier that total institutions place between the inmate and the wider world marks the first curtailment of self. In civil life, the sequential scheduling of the

---

[i]An example of the description of these processes may be found in Gresham M. Sykes, *The Society of Captives* (Princeton: Princeton University Press, 1958), ch. iv, "The Pains of Imprisonment," pp. 63–83.

individual's roles, both in the life cycle and in the repeated daily round, ensures that no one role he plays will block his performance and ties in another. In total institutions, in contrast, membership automatically disrupts role scheduling, since the inmate's separation from the wider world lasts around the clock and may continue for years. Role dispossession therefore occurs. In many total institutions the privilege of having visitors or of visiting away from the establishment is completely withheld at first, ensuring a deep initial break with past roles and an appreciation of role dispossession. A report on cadet life in a military academy provides an illustration:

> This clean break with the past must be achieved in a relatively short period. For two months, therefore, the swab is not allowed to leave the base or to engage in social intercourse with non-cadets. This complete isolation helps to produce a unified group of swabs, rather than a heterogeneous collection of persons of high and low status. Uniforms are issued on the first day, and discussions of wealth and family background are taboo. Although the pay of the cadet is very low, he is not permitted to receive money from home. The role of the cadet must supersede other roles the individual has been accustomed to play. There are few clues left which will reveal social status in the outside world.[ii]

I might add that when entrance is voluntary, the recruit has already partially withdrawn from his home world; what is cleanly severed by the institution is something that had already started to decay....

The inmate, then, finds certain roles are lost to him by virtue of the barrier that separates him from the outside world. The process of entrance typically brings other kinds of loss and mortification as well. We very generally find staff employing what are called admission procedures, such as taking a life history, photographing, weighing, fingerprinting, assigning numbers, searching, listing personal possessions for storage, undressing, bathing, disinfecting, haircutting, issuing institutional clothing, instructing as to rules, and assigning to quarters. Admission procedures might better be called "trimming" or "programming" because in thus being squared away the new arrival allows himself to be shaped and coded into an object that can be fed into the administrative machinery of the establishment, to be worked on smoothly by routine operations. Many of these procedures depend upon attributes such as weight or fingerprints that the individual possesses merely because he is a member of the largest and most abstract of social categories, that of human being. Action taken on the basis of such attributes necessarily ignores most of his previous bases of self-identification. ...

The admission procedure can be characterized as a leaving off and a taking on, with the midpoint marked by physical nakedness. Leaving off of course entails a dispossession of property, important because persons invest self feelings in their possessions. Perhaps the most significant of these possessions is not physical at all, one's full name; whatever one is thereafter called, loss of one's name can be a great curtailment of the self.

Once the inmate is stripped of his possessions, at least some replacements must be made by the establishment, but these take the form of standard issue, uniform in character and uniformly distributed. These substitute possessions are clearly marked as really belonging to the institution and in some cases are recalled at regular intervals to be, as it were, disinfected of identifications. With objects that can be used up—for example, pencils—the inmate may be required to return the remnants before obtaining a reissue. Failure to provide inmates with individual lockers and periodic searches and confiscations of accumulated personal property reinforce property dispossession. Religious orders have appreciated the implications for self of such separation from belongings. Inmates may be required to change their cells once a year so as not to become attached to them. ...

One set of the individual's possessions has a special relation to self. The individual ordinarily expects to exert some control over the guise in which he appears before others. For this he needs cosmetic and clothing supplies, tools for applying, arranging, and repairing them, and an accessible, secure place to store these supplies and tools—in short, the individual will need an "identity kit" for the management of his personal front. He will also need access to decoration specialists such as barbers and clothiers.

On admission to a total institution, however, the individual is likely to be stripped of his usual appearance and of the equipment and services by which he maintains it, thus suffering a personal defacement. Clothing, combs, needle and thread, cosmetics, towels, soap, shaving sets, bathing facilities—all these may be taken away or denied him, although some may be kept

---

[ii]Sanford M. Dornbusch, "The Military Academy as an Assimilating Institution," *Social Forces,* XXXIII (1955), p. 317.

in inaccessible storage, to be returned if and when he leaves. In the words of St. Benedict's Holy Rule:

> Then forthwith he shall, there in the oratory, be divested of his own garments with which he is clothed and be clad in those of the monastery. Those garments of which he is divested shall be placed in the wardrobe, there to be kept, so that if, perchance, he should ever be persuaded by the devil to leave the monastery (which God forbid), he may be stripped of the monastic habit and cast forth.[iii]

. . . At admission, loss of identity equipment can prevent the individual from presenting his usual image of himself to others. After admission, the image of himself he presents is attacked in another way. Given the expressive idiom of a particular civil society, certain movements, postures, and stances will convey lowly images of the individual and be avoided as demeaning. Any regulation, command, or task that forces the individual to adopt these movements or postures may mortify his self. In total institutions, such physical indignities abound. In mental hospitals, for example, patients may be forced to eat all food with a spoon. In military prisons, inmates may be required to stand at attention whenever an officer enters the compound. In religious institutions, there are such classic gestures of penance as the kissing of feet, and the posture recommended to an erring monk that he

> . . . lie prostrate at the door of the oratory in silence; and thus, with his face to the ground and his body prone, let him cast himself at the feet of all as they go forth from the oratory.[iv]

In some penal institutions we find the humiliation of bending over to receive a birching.

Just as the individual can be required to hold his body in a humiliating pose, so he may have to provide humiliating verbal responses. An important instance of this is the forced deference pattern of total institutions; inmates are often required to punctuate their social interaction with staff by verbal acts of deference, such as saying "sir." Another instance is the necessity to beg, importune, or humbly ask for little things such as a light for a cigarette, a drink of water, or permission to use the telephone.

Corresponding to the indignities of speech and action required of the inmate are the indignities of treatment others accord him. The standard examples here are verbal or gestural profanations: staff or fellow inmates call the individual obscene names, curse him, point out his negative attributes, tease him, or talk about him or his fellow inmates as if he were not present.

Whatever the form or the source of these various indignities, the individual has to engage in activity whose symbolic implications are incompatible with his conception of self. A more diffuse example of this kind of mortification occurs when the individual is required to undertake a daily round of life that he considers alien to him—to take on a disidentifying role. In prisons, denial of heterosexual opportunities can induce fear of losing one's masculinity. In military establishments, the patently useless make-work forced on fatigue details can make men feel their time and effort are worthless. In religious institutions there are special arrangements to ensure that all inmates take a turn performing the more menial aspects of the servant role. An extreme is the concentration-camp practice requiring prisoners to administer whippings to other prisoners.

There is another form of mortification in total institutions; beginning with admission a kind of contaminative exposure occurs. On the outside, the individual can hold objects of self-feeling—such as his body, his immediate actions, his thoughts, and some of his possessions—clear of contact with alien and contaminating things. But in total institutions these territories of the self are violated; the boundary that the individual places between his being and the environment is invaded and the embodiments of self profaned.

There is, first, a violation of one's informational preserve regarding self. During admission, facts about the inmate's social statuses and past behavior—especially discreditable facts—are collected and recorded in a dossier available to staff. Later, in so far as the establishment officially expects to alter the self-regulating inner tendencies of the inmate, there may be group or individual confession—psychiatric, political, military, or religious, according to the type of institution. On

---

[iii]*The Holy Rule of Saint Benedict,* Ch. 58.

[iv]*The Holy Rule of Saint Benedict,* Ch. 44.

these occasions the inmate has to expose facts and feelings about self to new kinds of audiences. . . .

New audiences not only learn discreditable facts about oneself that are ordinarily concealed but are also in a position to perceive some of these facts directly. Prisoners and mental patients cannot prevent their visitors from seeing them in humiliating circumstances. Another example is the shoulder patch of ethnic identification worn by concentration-camp inmates. Medical and security examinations often expose the inmate physically, sometimes to persons of both sexes; a similar exposure follows from collective sleeping arrangements and door-less toilets. An extreme here, perhaps, is the situation of a self-destructive mental patient who is stripped naked for what is felt to be his own protection and placed in a constantly lit seclusion room, into whose Judas window any person passing on the ward can peer. In general, of course, the inmate is never fully alone; he is always within sight and often earshot of someone, if only his fellow inmates. Prison cages with bars for walls fully realize such exposure.

Perhaps the most obvious type of contaminative exposure is the directly physical kind—the besmearing and defiling of the body or of other objects closely identified with the self. Sometimes this involves a breakdown of the usual environmental arrangements for insulating oneself from one's own source of contamination, as in having to empty one's own slops or having to subject one's evacuation to regimentation. . . . A very common form of physical contamination is reflected in complaints about unclean food, messy quarters, soiled towels, shoes and clothing impregnated with previous users' sweat, toilets without seats, and dirty bath facilities. . . . Finally, in some total institutions the inmate is obliged to take oral or intravenous medications, whether desired or not, and to eat his food, however unpalatable. When an inmate refuses to eat, there may be forcible contamination of his innards by "forced feeding."

I have suggested that the inmate undergoes mortification of the self by contaminative exposure of a physical kind, but this must be amplified: when the agency of contamination is another human being, the inmate is in addition contaminated by forced interpersonal contact and, in consequence, a forced social relationship. (Similarly, when the inmate loses control over who observes him in his predicament or knows about his past, he is being contaminated by a forced relationship to these people—for it is through such perception and knowledge that relations are expressed.) . . .

THE MORAL CAREER OF THE MENTAL PATIENT

I

In general . . . mental hospitals systematically provide for circulation about each patient the kind of information that the patient is likely to try to hide. And in various degrees of detail this information is used daily to puncture his claims. At the admission and diagnostic conferences, he will be asked questions to which he must give wrong answers in order to maintain his self-respect, and then the true answer may be shot back at him. An attendant whom he tells a version of his past and his reason for being in the hospital may smile disbelievingly, or say, "That's not the way I heard it," in line with the practical psychiatry of bringing the patient down to reality. When he accosts a physician or nurse on the ward and presents his claims for more privileges or for discharge, this may be countered by a question which he cannot answer truthfully without calling up a time in his past when he acted disgracefully. When he gives his view of his situation during group psychotherapy, the therapist, taking the role of interrogator, may attempt to disabuse him of his face-saving interpretations and encourage an interpretation suggesting that it is he himself who is to blame and who must change. When he claims to staff or fellow patients that he is well and has never been really sick, someone may give him graphic details of how, only one month ago, he was prancing around like a girl, or claiming that he was God, or declining to talk or eat, or putting gum in his hair.

Each time the staff deflates the patient's claims, his sense of what a person ought to be and the rules of peer-group social intercourse press him to reconstruct his stories; and each time he does this, the custodial and psychiatric interests of the staff may lead them to discredit these tales again. . . .

Learning to live under conditions of imminent exposure and wide fluctuation in regard, with little control over the granting or withholding of this regard, is an important step in the socialization of the patient, a step that tells something important about what it is like to be an inmate in a mental hospital. Having one's past mistakes and present progress under constant moral review seems to make for a special adaptation consisting of a less than moral attitude to ego ideals. One's shortcomings and successes become too central and fluctuating an issue in life to allow the usual commitment of concern for other persons' views of them. It is not very practicable to try to sustain solid claims about oneself. The inmate tends to learn that degradations and reconstructions of the self need not be given too much weight, at the same time learning that staff and inmates are ready to view an inflation or deflation of a self with some indifference. He learns that a defensible picture of self can be seen as something outside oneself that can be constructed, lost, and rebuilt, all with great speed and some equanimity. He learns about the viability of taking up a standpoint—and hence a self—that is outside the one which the hospital can give and take away from him.

The setting, then, seems to engender a kind of cosmopolitan sophistication, a kind of civic apathy. In this unserious yet oddly exaggerated moral context, building up a self or having it destroyed becomes something of a shameless game, and learning to view this process as a game seems to make for some demoralization, the game being such a fundamental one. In the hospital, then, the inmate can learn that the self is not a fortress, but rather a small open city; he can become weary of having to show pleasure when held by troops of his own, and weary of having to show displeasure when held by the enemy. Once he learns what it is like to be defined by society as not having a viable self, this threatening definition—the threat that helps attach people to the self society accords them—is weakened. The patient seems to gain a new plateau when he learns that he can survive while acting in a way that society sees as destructive of him.

[An] . . . illustration of this moral loosening and moral fatigue might be given. . . . On the worst ward level, discreditings seem to occur the most frequently, in part because of lack of facilities, in part through the mockery and sarcasm that seem to be the occupational norm of social control for the attendants and nurses who administer these places. At the same time, the paucity of equipment and rights means that not much self can be built up. The patient finds himself constantly toppled, therefore, but with very little distance to fall. A kind of jaunty gallows humor seems to develop in some of these wards, with considerable freedom to stand up to the staff and return insult for insult. While these patients can be punished, they cannot, for example, be easily slighted, for they are accorded as a matter of course few of the niceties that people must enjoy before they can suffer subtle abuse. Like prostitutes in connection with sex, inmates on these wards have very little reputation or rights to lose and can therefore take certain liberties. As the person moves up the ward system, he can

manage more and more to avoid incidents which discredit his claim to be a human being, and acquire more and more of the varied ingredients of self-respect; yet when eventually he does get toppled—and he does—there is a much farther distance to fall. For instance, the privileged patient lives in a world wider than the ward, containing recreation workers who, on request, can dole out cake, cards, table-tennis balls, tickets to the movies, and writing materials. But in the absence of the social control of payment which is typically exerted by a recipient on the outside, the patient runs the risk that even a warmhearted functionary may, on occasion, tell him to wait until she has finished an informal chat, or teasingly ask why he wants what he has asked for, or respond with a dead pause and a cold look of appraisal.

Moving up and down the ward system means, then, not only a shift in self-constructive equipment, a shift in reflected status, but also a change in the calculus of risks. Appreciation of risks to his self-conception is part of everyone's moral experience, but an appreciation that a given risk level is itself merely a social arrangement is a rarer kind of experience, and one that seems to help to disenchant the person who undergoes it. . . .

Each moral career, and behind this, each self, occurs within the confines of an institutional system, whether a social establishment such as a mental hospital or a complex of personal and professional relationships. The self, then, can be seen as something that resides in the arrangements prevailing in a social system for its members. The self in this sense is not a property of the person to whom it is attributed, but dwells rather in the pattern of social control that is exerted in connection with the person by himself and those around him. This special kind of institutional arrangement does not so much support the self as constitute it. . . .

In the usual cycle of adult socialization one expects to find alienation and mortification followed by a new set of beliefs about the world and a new way of conceiving of selves. In the case of the mental-hospital patient, this rebirth does sometimes occur, taking the form of a strong belief in the psychiatric perspective, or, briefly at least, a devotion to the social cause of better treatment for mental patients. The moral career of the mental patient has unique interest, however; it can illustrate the possibility that in casting off the raiments of the old self—or in having this cover torn away—the person need not seek a new robe and a new audience before which to cower. Instead he can learn, at least for a time, to practise before all groups the amoral arts of shamelessness.

THE UNDERLIFE OF A PUBLIC INSTITUTION

I

In every social establishment, there are official expectations as to what the participant owes the establishment. Even in cases where there is no specific task, as in some night-watchman jobs, the organization will require some presence of mind, some awareness of the current situation, and some readiness for unanticipated events; as long as an establishment demands that its participants not sleep on the job, it asks them to be awake to certain matters. And where sleeping is part of the expectation, as in a home or a hotel, then there will be limits on where and when the sleeping is to occur, with whom, and with what bed manners. And behind these claims on the individual, be they great or small, the managers of every establishment will have a widely embracing implicit conception of what the individual's character must be for these claims on him to be appropriate.

Whenever we look at a social establishment, we find a counter to this first theme: we find that participants decline in some way to accept the official view of what they should be putting into and getting out of the organization and, behind this, of what sort of self and world they are to accept for themselves. Where enthusiasm is expected, there will be apathy; where loyalty, there will be disaffection; where attendance, absenteeism; where robustness, some kind of illness; where deeds are to be done, varieties of inactivity. We find a multitude of homely little histories, each in its way a movement of liberty. Whenever worlds are laid on, underlives develop.

II

The study of underlife in restrictive total institutions has some special interest. When existence is cut to the bone, we can learn what people do to flesh out their lives. Stashes, means of transportation, free places, territories, supplies for economic and social exchange—these apparently are some of the minimal requirements for building up a life. Ordinarily these arrangements are taken for granted as part of one's primary adjustment; seeing them twisted out of official existence through bargains, wit, force, and cunning, we can see their significance anew. The study of total institutions also suggests that formal organizations have standard places of vulnerability, such as supply rooms, sick bays, kitchens, or scenes of highly technical labor. These are the damp corners where secondary adjustments breed and start to infest the establishment.

The mental hospital represents a peculiar instance of those establishments in which underlife is likely to proliferate. Mental patients are persons who caused the kind of trouble on the outside that led someone physically, if not socially, close to them to take psychiatric action against them. Often this trouble was associated with the "prepatient" having indulged in situational improprieties of some kind, conduct out of place in the setting. It is just such misconduct that conveys a moral rejection of the communities, establishments, and relationships that have a claim to one's attachment.

Stigmatization as mentally ill and involuntary hospitalization are the means by which we answer these offenses against propriety. The individual's persistence in manifesting symptoms after entering the hospital, and his tendency to develop additional symptoms during his initial response to the hospital, can now no longer serve him well as expressions of disaffection. From the patient's point of view, to decline to exchange a word with the staff or with his fellow patients may be ample evidence of rejecting the institution's view of what and who he is; yet higher management may construe this alienative expression as just the sort of symptomatology the institution was established to deal with and as the best kind of evidence that the patient properly belongs where he now finds himself. In short, mental hospitalization outmaneuvers the patient, tending to rob him of the common expressions through which people hold off the embrace of organizations—insolence, silence, *sotto voce* remarks, unco-operativeness, malicious destruction of interior decorations, and so forth; these signs of disaffiliation are now read as signs of their maker's proper affiliation. Under these conditions all adjustments are primary.

Furthermore, there is a vicious-circle process at work. Persons who are lodged on "bad" wards find that very little equipment of any kind is given them—clothes may be taken from them each night, recreational materials may be withheld, and only heavy wooden chairs and benches provided for furniture. Acts of hostility against the institution have to rely on limited, ill-designed devices, such as banging a chair against the floor or striking a sheet of newspaper sharply so as to make an annoying explosive sound. And the more inadequate this equipment is to convey rejection of the hospital, the more the act appears as a psychotic symptom, and the more likely it is that management feels justified in assigning the patient to a bad ward. When a patient finds himself in seclusion, naked and without visible means of expression, he may have to rely on tearing up his mattress, if he can, or writing with feces on the wall—actions management takes to be in keeping with the kind of person who warrants seclusion.

We can also see this circular process at work in the small, illicit, talisman-like possessions that inmates use as symbolic devices for separating themselves from the position they are supposed to be in. What I think is a typical example may be cited from prison literature:

> Prison clothing is anonymous. One's possessions are limited to toothbrush, comb, upper or lower cot, half the space upon a narrow table, a razor. As in jail, the urge to collect possessions is carried to preposterous extents. Rocks, string, knives—anything made by man and forbidden in man's institution—anything,—a red comb, a different kind of toothbrush, a belt—these things are assiduously gathered, jealously hidden or triumphantly displayed.[v]

But when a patient, whose clothes are taken from him each night, fills his pockets with bits of string and rolled up paper, and when he fights to keep these possessions in spite of the consequent inconvenience to those who must regularly go through his pockets, he is usually seen as engaging in symptomatic behavior befitting a very sick patient, not as someone who is attempting to stand apart from the place accorded him.

Official psychiatric doctrine tends to define alienative acts as psychotic ones—this view being reinforced by the circular processes that lead the patient to exhibit alienation in a more and more bizarre form—but the hospital cannot be run according to this doctrine. The hospital cannot decline to demand from its members exactly what other organizations must insist on; psychiatric doctrine is supple enough to do this, but institutions are not. Given the standards of the institution's environing society, there have to be at least the minimum routines connected with feeding, washing, dressing, bedding the patients, and protecting them from physical harm. Given these routines, there have to be inducements and exhortations to get patients to follow them. Demands must be made, and disappointment is shown when a patient does not live up to what is expected of him. Interest in seeing

psychiatric "movement" or "improvement" after an initial stay on the wards leads the staff to encourage "proper" conduct and to express disappointment when a patient backslides into "psychosis." The patient is thus re-established as someone whom others are depending on, someone who ought to know enough to act correctly. Some improprieties, especially ones like muteness and apathy that do not obstruct and even ease ward routines, may continue to be perceived naturalistically as symptoms, but on the whole the hospital operates semi-officially on the assumption that the patient ought to act in a manageable way and be respectful of psychiatry, and that he who does will be rewarded by improvement in life conditions and he who doesn't will be punished by a reduction of amenities. Within this semi-official reinstatement of ordinary organizational practices, the patient finds that many of the traditional ways of taking leave of a place without moving from it have retained their validity; secondary adjustments are therefore possible.

III

Of the many different kinds of secondary adjustment, some are of particular interest because they bring into the clear the general theme of involvement and disaffection, characteristic of all these practices.

One of these special types of secondary adjustment is "removal activities" (or "kicks"), namely, undertakings that provide something for the individual to lose himself in, temporarily blotting out all sense of the environment which, and in which, he must abide. In total institutions a useful exemplary case is provided by Robert Stroud, the "Birdman," who, from watching birds out his cell window, through a spectacular career of finagling and make-do, fabricated a laboratory and became a leading ornithological contributor to medical literature, all from within prison. Language courses in prisoner-of-war camps and art courses in prisons can provide the same release.

Central Hospital provided several of these escape worlds for inmates.[vi] One, for example, was sports. Some of the baseball players and a few tennis players seemed

---

[v]Cantine and Rainer, *op. cit.,* p. 78. Compare the things that small boys stash in their pockets; some of these items also seem to provide a wedge between the boy and the domestic establishment.

[vi]Behind informal social typing and informal group formation in prisons there is often to be seen a removal activity. Caldwell, *op. cit.,* pp. 651–53, provides some interesting examples of prisoners on such kicks: those involved in securing and using drugs; those focused on leatherwork for sale; and "Spartans," those involved in the glorification of their bodies, the prison locker room apparently serving as a muscle beach; the homosexuals; the gamblers, etc. The point about these activities is that each is world-building for the person caught up in it, thereby displacing the prison.

to become so caught up in their sport, and in the daily record of their efforts in competition, that at least for the summer months this became their overriding interest. In the case of baseball this was further strengthened by the fact that, within the hospital, parole patients could follow national baseball as readily as could many persons on the outside. For some young patients, who never failed to go, when allowed, to a dance held in their service or in the recreation building, it was possible to live for the chance of meeting someone "interesting" or remeeting someone interesting who had already been met—in much the same way that college students are able to survive their studies by looking forward to the new "dates" that may be found in extracurricular activities. The "marriage moratorium" in Central Hospital, effectively freeing a patient from his marital obligations to a non-patient, enhanced this removal activity. For a handful of patients, the semi-annual theatrical production was an extremely effective removal activity: tryouts, rehearsals, costuming, scenery-making, staging, writing and rewriting, performing—all these seemed as successful as on the outside in building a world apart for the participants. Another kick, important to some patients—and a worrisome concern for the hospital chaplains—was the enthusiastic espousal of religion. Still another, for a few patients, was gambling.

Portable ways of getting away were much favored in Central Hospital, paper-back murder mysteries, cards, and even jigsaw puzzles being carried around on one's person. Not only could leave be taken of the ward and grounds be taken leave of through these means, but if one had to wait for an hour or so upon an official, or the serving of a meal, or the opening of the recreation building, the self-implication of this subordination could be dealt with by immediately bringing forth one's own world-making equipment. . . .

IV

If a function of secondary adjustments is to place a barrier between the individual and the social unit in which he is supposed to be participating, we should expect some secondary adjustments to be empty of intrinsic gain and to function solely to express unauthorized distance—a self-preserving "rejection of one's

rejectors."[vii] This seems to happen with the very common forms of ritual insubordination, for example, griping or bitching, where this behavior is not realistically expected to bring about change. Through direct insolence that does not meet with immediate correction, or remarks passed half out of hearing of authority, or gestures performed behind the back of authority, subordinates express some detachment from the place officially accorded them. . . . Some of these ways of openly but safely taking a stand outside the authorized one are beautiful, especially when carried out collectively. Again, prisons provide ready examples:

> . . . When the sky pilot got up in the pulpit to give us our weekly pep talk each Sunday he would always make some feeble joke which we always laughed at as loud and as long as possible, although he must have known that we were sending him up. He still used to make some mildly funny remark and every time he did the whole church would be filled with rawcous [sic] laughter, even though only half the audience had heard what had been said.[viii]

. . . Beyond irony, however, there is an even more subtle and telling kind of ritual insubordination. There is a special stance that can be taken to alien authority; it combines stiffness, dignity, and coolness in a particular mixture that conveys insufficient insolence to call forth immediate punishment and yet expresses that one is entirely one's own man. Since this communication is made through the way in which the body and face are held, it can be constantly conveyed wherever the inmate finds himself. Illustrations can be found in prison society:

> "Rightness" implies bravery, fearlessness, loyalty to peers, avoidance of exploitation, adamant refusal to concede the superiority of the official value system, and repudiation of the notion that the inmate is of a lower order. It consists principally in the reassertion of one's basic integrity, dignity, and worth in an essentially degrading situation, and the exhibition of these personal qualities regardless of any show of force by the official system.[ix]

[vii]Lloyd W. McCorkle and Richard Korn, "Resocialization Within Walls," *The Annals,* CCXCIII (1954), p. 88.

[viii]J. F. N., *op. cit.,* pp. 15–16. See also Goffman, *Presentation of Self,* "derisive collusion," pp. 186–88.

[ix]Richard Cloward, "Social Control in the Prison," S.S.R.C. Pamphlet No. 15, *op. cit.,* p. 40. See also Sykes and Messinger, *op. cit.,* pp. 10–11.

Similarly, in Central Hospital, in the "tough" punishment wards of maximum security, where inmates had very little more to lose, fine examples could be found of patients not going out of their way to make trouble but by their very posture conveying unconcern and mild contempt for all levels of the staff, combined with utter self-possession.

V

It would be easy to account for the development of secondary adjustments by assuming that the individual possessed an array of needs, native or cultivated, and that when lodged in a milieu that denied these needs the individual simply responded by developing makeshift means of satisfaction. I think this explanation fails to do justice to the importance of these undercover adaptations for the structure of the self.

The practice of reserving something of oneself from the clutch of an institution is very visible in mental hospitals and prisons but can be found in more benign and less totalistic institutions, too. I want to argue that this recalcitrance is not an incidental mechanism of defense but rather an essential constituent of the self.

Sociologists have always had a vested interest in pointing to the ways in which the individual is formed by groups, identifies with groups, and wilts away unless he obtains emotional support from groups. But when we closely observe what goes on in a social role, a spate of sociable interaction, a social establishment—or in any other unit of social organization—embracement of the unit is not all that we see. We always find the individual employing methods to keep some distance, some elbow room, between himself and that with which others assume he should be identified. No doubt a state-type mental hospital provides an overly lush soil for the growth of these secondary adjustments, but in fact, like weeds, they spring up in any kind of social organization. If we find, then, that in all situations actually studied the participant has erected defenses against his social bondedness, why should we base our conception of the self upon how the individual would act were conditions "just right"?

The simplest sociological view of the individual and his self is that he is to himself what his place in an organization defines him to be. When pressed, a sociologist modifies this model by granting certain complications: the self may be not yet formed or may exhibit conflicting dedications. Perhaps we should further complicate the construct by elevating these qualifications to a central place, initially defining the individual, for sociological purposes, as a stance-taking entity, a something that takes up a position somewhere between identification with an organization and opposition to it, and is ready at the slightest pressure to regain its balance by shifting its involvement in either direction. It is thus *against something* that the self can emerge. . . .

Without something to belong to, we have no stable self, and yet total commitment and attachment to any social unit implies a kind of selflessness. Our sense of being a person can come from being drawn into a wider social unit; our sense of selfhood can arise through the little ways in which we resist the pull. Our status is backed by the solid buildings of the world, while our sense of personal identity often resides in the cracks.

▪▪ ARLIE RUSSELL HOCHSCHILD (1940–): A BIOGRAPHICAL SKETCH

Born in 1940 and raised in a Maryland suburb, Arlie Russell's interest in "emotion management" was kindled at the age of twelve when her parents joined the U.S. Foreign Service. As her family hosted parties for foreign diplomats, she became precociously attuned to the subtleties of emotional displays and the importance of controlling, if not manipulating, them. This led her to question early on whether the "self" we present to others is a reality or fiction, as she found herself wondering, "Had I passed the peanuts to a person . . . or to an actor?" (Hochschild 1983:ix). The insightfulness of her innocent childhood musings has been confirmed: Today Hochschild has earned an international audience as a leading social psychologist.

Hochschild completed her master's and doctorate degrees at the University of California, Berkeley. While a student there, she became increasingly attuned to the struggles women faced as they attempted to straddle two worlds: the "male" world of professional life and the "female" world of caregiver. Moreover, her studies led her to the conclusion that sociology was a patently male discipline whose points of reference, theories, and methods were derived from men's experiences. This realization would, along with her fleeting childhood encounters with her parents' colleagues, inspire her to develop a branch of sociology that addressed not only the experiences of men but also the experiences of women; that informed the public not through "objective," statistical facts but through stories of everyday life told by those living it; and that sought to explore not only what people *think* but also how people *feel:* the sociology of emotions.

During the past thirty years, Hochschild has written five books, coedited another, and published more than fifty scholarly articles, book chapters, and reviews. Her academic works have been translated into ten languages. The American Sociological Association has twice recognized her contributions to the discipline by presenting her with both a Lifetime Achievement Award and the Award for Public Understanding of Sociology. Hochschild has also contributed significantly as a public intellectual. She established the Center for Working Families at the University of California, Berkeley, for which she currently serves as the director, and her expertise on the impact of government policy and business/employee relations on the family has led her into the political arena as well. She has presented her work to California's Child Development Policy Advisory Committee and, on the national level, to the Democratic Leadership Council, the National Council on Family Relations, and the White House's Domestic Policy Council. She also advised former vice president Al Gore on the development of work–family policies. Hochschild currently teaches at the University of California, Berkeley.

HOCHSCHILD'S INTELLECTUAL INFLUENCES AND CORE IDEAS

In developing her theory of emotions, Hochschild draws from two distinct approaches. The first, which she labels the "organismic model," focuses on how emotions are rooted in an individual's biological or psychological makeup. The second, the "interactional model," stresses the role of social processes in shaping emotive experiences. As you will shortly see, Hochschild fashions a third approach, the "emotion-management perspective," by weaving together the insights of these two approaches; in doing so, she significantly expands the terrain of symbolic interactionism (Hochschild 2003).

The Organismic Model

The organismic model is derived primarily from the writings of Charles Darwin (1809–1882), Sigmund Freud (1856–1939), and William James (1842–1910). Darwin argued that emotive "gestures" are vestiges of humankind's earlier, direct experiences. Thus, baring one's teeth in an expression of rage is a vestige of the instinctual act of biting used by primitive humans to attack or defend themselves against another. Similarly, our expression of love is a holdover from what was once a direct act of sexual intercourse. For Darwin, then, emotive gestures—that is, emotional expressions such as anger or fear—are largely universal, since all humans can be traced to the same ancestral gene pool. Biology, not social factors, is said to shape emotions; that, in turn, accounts for why there allegedly is little cross-cultural variation in emotive experience. From this perspective, emotions are viewed as instinctual or automatic responses to a stimulus that produces physiological changes in our bodies. Our conscious awareness of such changes and the meanings we attribute to them are considered extrinsic to the emotions themselves. For example, the emotion we call fear is fundamentally a biologically driven response marked by an increased heart rate, heightened adrenaline levels, and a constriction of muscles. Fear is simply the label we use to describe the combination of these biological changes.

For his part, Freud connects emotions to instinctual drives, namely, Eros (love or creation) and Thanatos (death or destruction). When our ability to directly satisfy the demands of these two psychobiological instincts is hindered—when we cannot get what we want nor do what we want—we experience anxiety, the single most powerful emotion. Yet, civilization itself is predicated on the individual repressing or subli-mating these basic drives because cooperative activity—the lifeblood of all civilizations—requires us to forgo or at least compromise our own needs and demands. Anxiety is thus inescapably part of the human condition, tied as it is to the necessary denial of instinctual gratification.[5] In our attempt to avoid or minimize unpleasant or painful psychological distress we develop defense mechanisms through which we uncon-sciously repress and redefine the real source of our anxiety, the blocking of our instinctual drives. Displacing thoughts and behaviors that are socially defined as inappropriate or impermissible shields our ego from what would otherwise be endless psychological turmoil caused by the conflict between the demands of our instincts and the demands of society. Such mechanisms, however, can lead to neurotic behavior or misfitted emotional expressions (for instance, laughing at a funeral or feeling depressed at one's own wedding) that to the psychoanalyst are a reflection of unconscious psychic processes.

Freud also assigns an important "signal function" to anxiety, a function that would become central to Hochschild's theory of emotions. Here, Freud asserts that the experience of anxiety serves to alert the individual to impending dangers emanating either from the immediate environment or from inside the individual's own psychic state. For instance, anxiety is likely to increase when we find ourselves in a potentially embarrassing situation or when we feel as though we are emotionally "flooded out" from a particularly intense situation and unable to control our behaviors. However, what Freud reserves primar-ily for anxiety, Hochschild extends to all emotions, arguing that the entire range of emotive experiences—from joy to sadness, pride to shame—serve as signals through which we fit our prior expectations to our present situation. Envy, for example, is the name for the "signal" we receive when our expectations regarding what we think we deserve are not met, while another person has what it is we desire.

The Interactional Model

The "interactional model" is expressed in the work of a number of figures including John Dewey (1859–1952), George Herbert Mead, Hans Gerth (1908–1978) and C. Wright Mills (1916–1962), Herbert Blumer, and Erving Goffman. These theorists all share a conception of the individual as an active, con-scious participant in the production and reproduction of social life—a view that informs Hochschild's own emotion-management perspective. Although they pursued different lines of inquiry, each called attention to the relationship between thinking (a social act rooted in taking the attitude of others toward our own conduct) and feeling. Here, emotive experiences thus are not tied to biologically driven responses or to the unconscious workings of the psyche. Emotions are viewed, instead, as intimately connected to our conscious perceptions and interpretations of the situations in which we are involved. In other words, the act of assigning meaning to an emotion is intrinsic to—not an aftereffect of—the emotive experience itself. Importantly, the active production of emotions is itself a social process since the interpretation and labeling of feeling are based both on our reaction to the imagined responses of others as well as on our awareness of the normative expectations that infuse a given situation. Thus, we are sad at funerals not only because of the loss of a loved one, but also because we are *supposed* to be sad. Similarly, we feel a sense of pride in our accomplishments in part because we reflect back on the obstacles confronted along the path to our successes and realize that we deserve to feel proud.

The model of the actor and his or her relation to the broader society posed by the interactional perspec-tive is instructive on several fronts. Emphasizing the social dimensions of self-consciousness provides a vantage point for linking emotive experiences to cognitive processes and for recognizing the connections between self- and social-control mechanisms. From this perspective, emotions can be understood as something other than irrational, displaced, or instinctual physiological responses to events. Moreover, recognizing that an individual's "private" emotions are produced and shaped within a public setting opens the door for examining how broader social structures can impinge on emotive experiences, for instance, how the institutionalization of class and gender relations shapes both the inner experience and the outer expression of emotions, an issue that Hochschild addresses in the readings that follow.

---

[5]For additional comments on Freud's theory, see our discussion of his influence on critical theory in Chapter 10.

Goffman and Impression Management

Above all, it is Erving Goffman who had the greatest impact on Hochschild's conceptualization of emotions. Although Goffman did not explore explicitly the full range of emotions, his notion of impression management is central to Hochschild's own emotion-management model. As we discussed earlier in connection to Goffman's dramaturgical perspective, impression management refers to the technique of controlling one's behavior in order to present an acceptable image of one's self to others. But, according to Hochschild, Goffman's singular focus on outward expressions, or "surface acting," leads to a blind spot in his analysis. The notion that individuals have inner feelings that, like outward appearances, must be actively managed is scarcely entertained by Goffman. This assumption, however, is fundamental to Hochschild's own model. Indeed, Goffman's own ambivalent position on the nature of self (whether it is primarily an imputed "dramatic effect" arising from a scene or a "psychobiological" organism that stands behind the performance) makes it difficult to determine how *inner* feelings might be managed or shaped.

While Goffman did not make the study of emotions a cornerstone to his approach, he did devote attention to embarrassment and its role in sustaining both the informal rules that govern social encounters and the self-identity of the actors involved (Goffman 1967). In his analysis, we can glimpse the blind spot that Hochschild sought to rectify. Goffman defines embarrassment as a feeling that arises when one or more actors in an encounter are unable to fulfill the social expectations required to project an acceptable self-image to others. Certainly, no one has fully escaped the feeling of embarrassment (or "flustering") that signals an interaction gone awry. Goffman's description will no doubt ring familiar:

> [A] completely flustered individual is one who cannot for the time being mobilize his muscular and intellectual resources for the task at hand, although he would like to; he cannot volunteer a response to those around him that will allow them to sustain the conversation smoothly. He and his flustered actions block the line of activity the others have been pursuing. He is present with them, but he is not "in play." The others may be forced to stop and turn their attention to the impediment; the topic of conversation is neglected, and energies are directed to the task of re-establishing the flustered individual, of studiously ignoring him, or of withdrawing from his presence . . . [F]lustering threatens the encounter itself by disrupting the smooth transmission and reception by which encounters are sustained. (ibid.:100 102)

Here, embarrassment is studied less as an inner emotional state and more for what it contributes to the flow of interaction. The point is not that Goffman ignores the play and power of emotions, but rather that his situationalism leads him to view emotions as something "done" to the actor: Embarrassment happens *to* him. Thus, essential features of emotions are not to be found in the mind or body of the individual as the organismic and emotion-management models contend, but rather in their ramifications for the encounter and for the actor's attempt to maintain a viable presentation of self. As Goffman puts it,

> By showing embarrassment . . . [the individual's] role in the current interaction may be sacrificed, and even the encounter itself, but he demonstrates that, while he cannot present a sustainable and coherent self on this occasion, he is at least disturbed by the fact and may prove worthy another time. To this extent, embarrassment is not an irrational impulse breaking through socially prescribed behavior but part of the orderly behavior itself. (ibid.:111)

Hochschild's Emotion-Management Model

Having presented in broad strokes the models from which Hochschild draws, we turn now to a brief overview of her own emotion-management perspective. To begin, Hochschild (1983), borrowing from Darwin, defines emotion as a "biologically-given sense" (ibid.:219), much like hearing, sight, and smell, that communicates information about the world around us. Thus, emotions are fundamentally connected to biological processes. Unlike our other senses, however, emotions are directly tied to behavior; they are experienced as the body physiologically readies itself to engage in action. Emotions also possess a cognitive component in the form of a "signal function." In this way, emotions are tied to our attempts to reconcile our expectations with the actuality of the situations in which we find ourselves.

Yet, unlike the picture that is painted by the organismic model, emotions are not viewed as biological or psychological givens that are immune to social and cultural influences. Instead, Hochschild maintains that emotions are actively produced and managed by individuals in the context of interaction: they are not simply experienced, they are created. In an important sense, we "do" emotions in the form of **emotion work**. This refers to efforts to alter (i.e., manage) the intensity or type of feelings one is, or should be, experiencing. Efforts to evoke or suppress emotions involve three types of work: cognitive, bodily, and expressive. The cognitive dimension refers to "the attempt to change images, ideas, or thoughts in the service of changing the feelings associated with them." Bodily work entails efforts to alter the physical effects of an emotion; for instance, trying to control one's shaking when angry or afraid. The expressive aspect of emotion work involves attempts to alter the public display of an emotional state in order to realize a specific feeling (2003:96). For instance, we may cry in an effort to feel sad. Here we see both Hochschild's indebtedness to and extension of Goffman's dramaturgical perspective. For while Goffman's notion of impression management emphasizes surface acting—the *outward* behavioral and verbal expressions we "put on" in order to convincingly play a role—Hochschild's emotion management focuses on "deep acting"—our *inner* efforts to produce not the appearance of feeling, but rather a "real feeling that has been self-induced" (ibid.:35). Thus, after a stressful day, we *try* to be happy at our friend's birthday party. Similarly, we often must work to suppress the nervousness or anxiety we experience during a job interview or first date. It is important to point out that, in arguing that individuals actively attempt to evoke or suppress their emotions, Hochschild is not claiming that emotion work necessarily is engaged in for purposes of deceit or manipulation. Instead, the fact that we try to shape our emotional experiences speaks to the power that the definition of the situation and the feeling rules that help frame situations have over our expressive life. This is a theme we will return to in the following section.

Directly related to the notion of emotion work is **emotional labor** in which one's deep acting is sold for a wage. Here, inner feelings are managed in order to produce an outward display as part of one's job. For example, in her research on flight attendants, Hochschild found that manufacturing feelings of caring and cheerfulness while suppressing anger or boredom was a common feature of the job. Of course, emotional labor does not pertain only to flight attendants. Rather, Hochschild points out that it is required by an array of jobs that share three characteristics: (1) face-to-face or voice-to-voice interaction with the public, (2) a directive to produce specified emotional states in clients, and (3) through training and supervision, employer control over workers' emotional activities (ibid.:147). From this description, it is easy to see that emotional labor is a pervasive feature in a service economy such as that in the United States. Waiters and waitresses, front-desk attendants in hotels, and just about anyone who works in a retail store are all subject to the supervised control of their own emotions because a major component of their job is to produce a sense of satisfaction or self-importance in their customers. Likewise, police officers engage in emotional labor in their dealings with the public. Trained to keep their own emotions in check under dangerous and threatening conditions—and

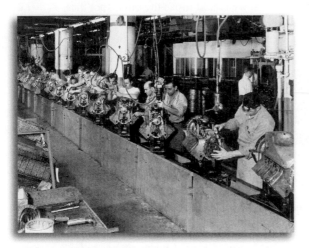

Photo 12.3a   Factory Workers

Equally alienated? The factory worker (Photo 12.3a) sells his physical labor, finding its embodiment in a tangible commodity. The flight attendant (Photo 12.3b) sells her emotions and smile to produce desired feeling states (contentment, safety) in others.

Photo 12.3b   Flight Attendant

subject to a range of penalties if they fail to do so—police officers also are instructed in techniques designed to produce specified emotional states in suspects and witnesses. Indeed, emotional labor is essential to performing the often-used "good cop/bad cop" routine, and to eliciting the required emotions, such as fear or trust, from those they are questioning.

Emotional labor, however, often exacts a high cost from the worker. Following the work of Gerth and Mills (and through them, the work of Karl Marx and Max Weber), Hochschild draws a parallel between the alienation experienced by the factory worker and that experienced by those employed in service industries. If the factory worker's body is bought and controlled by his employer, it is the service worker's feelings that are subject to the dictates of another. Though manifested in different ways, both are detached or alienated from something that is vital to their self. As Hochschild describes it,

> Those who perform emotional labor in the course of giving service are like those who perform physical labor in the course of making things: both are subject to the demands of mass production. But when the product—the thing to be engineered, mass-produced, and subjected to speed-up and slow-down—is a smile, a mood, a feeling, or a relationship, it comes to belong more to the organization and less to the self. (2003:198)

When our natural capacity to engage in emotion work is sold for a wage and bought to serve the profit motive, our feelings become a commodity engineered to further corporate and organizational interests. "Transmuted" from a private act controlled by the individual herself to one that is publicly administered by a supervisor and codified in training manuals and company policies, emotion work becomes rationalized to better serve instrumental purposes. However, the commercial reshaping of emotions, or **commodification of feelings**, is not experienced alike by all. The demands for emotion management are shaped by both class and gender relations such that middle-class workers and women are more susceptible to the commodification of their emotive experiences. Success within the "people professions" requires diligently controlling one's emotions in order to express the appropriate feeling state (trustworthy, dedicated, ambitious, caring, etc.) associated with the occupation. Conversely, performance in blue-collar jobs is graded more in terms of the quality of one's manual labor. An irritable or abrasive carpenter or assembly-line worker would likely keep his job as long as his finished product met with approval. In a corporate office, however, the same display of feelings would probably earn a pink slip.

If members of the lower and working classes tend more to things than to people, and thus are less practiced in the skills of emotional labor, the hierarchical patterning of managing emotions is reversed when it comes to gender. In other words, while the occupations associated with the more advantaged classes (corporate executives, sales workers, teachers, health care providers, etc.) are more likely to require the control of personal feelings, it is women, the less advantaged gender, who more often find it necessary to be skilled emotion managers and are thus more susceptible to the commodification of their feelings.

HOCHSCHILD'S THEORETICAL ORIENTATION ◼◼

In the preceding discussion, we introduced several concepts that suggest the contours of Hochschild's theoretical orientation. Given the theoretical wells from which she draws and the influence of Goffman's work in particular, it is not surprising that Hochschild has developed a multidimensional approach to the study of emotion that spans the four quadrants of our framework (see Figure 12.2). Consider first the twin concepts of emotion work and deep acting that, in speaking to the actor's inner experiences, are rooted in individualist and nonrational presuppositions. Here, Hochschild focuses on the efforts of individuals to actively shape their *real* feelings in the course of interacting with others. Thus, we may find ourselves trying to get excited about a concert we've been invited to attend or trying to suppress our feelings of disappointment over not getting a pay raise. Together, these two concepts emphasize the role that individuals play in creating their emotive experiences.

In contrast, surface acting refers to the individual's managing of her outward appearance in order to present a convincing image of her self. The emphasis is less on emotions per se, and more on the actor's

Figure 12.2   Hochschild's Basic Concepts and Theoretical Orientation

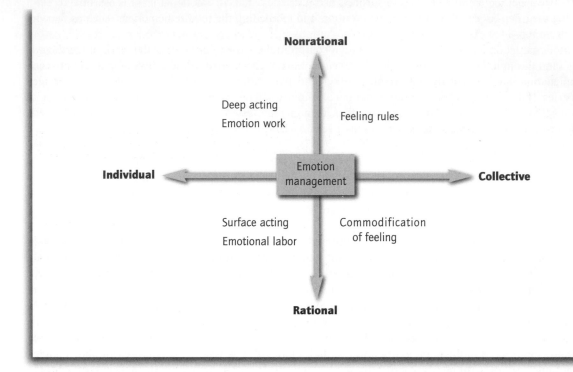

interest in maintaining an acceptable impression, which may involve faking feelings. For instance, we may pretend to thoroughly enjoy another's company when in actuality we are more interested in having him help us with our work. Thus, a more strategic or calculating motive often forms the backdrop of surface acting. This rationalist dimension of action is carried one step farther with the notion of emotional labor where an employee tries to evoke or suppress real emotions not because she wants to, but because she is paid to do so. While the feelings experienced may be real, they are nevertheless manufactured for "artificial" rational purposes—namely, to keep one's job.

Closely related to emotional labor is the concept of the commodification of feelings, a concept that points to the collectivist and rational dimension of emotive experience. In this case, the instrumental, strategic stance toward emotions is adopted and executed through corporate design. It is not the interests of the individual that are being served here, but rather those of the company. Like a car or a pair of shoes, emotions are bought and sold under the logic of the profit motive and the impersonal market forces it infuses. The notion of the commodification of feelings thus turns our attention to the institutional or structural pressures that shape the individual's experience of emotions.

Hochschild's multidimensional approach is rounded out by her notion of feeling rules, which captures the collectivist, nonrational dimension of emotions. **Feeling rules** are the shared, social (collectivist) conventions that determine what we should properly feel in a given situation (the "direction" of emotions), how intensely we should feel it (the "extent"), and how long we should feel it (the "duration") (Hochschild 2003:97). They form the taken-for-granted (nonrational) backdrop according to which we manage our emotions and assess the emotive expressions of others. Feeling rules account for the social patterning of emotive experiences, and in doing so establish the "standards used in emotional conversation to determine what is rightly owed and owing in the currency of feeling" (Hochschild 1983:18). It is on the basis of feeling rules, and the sense of emotional entitlement or "rights" and obligations or duties they establish, that we guide our private emotion work (ibid.:56). For just as there are social expectations that set the boundaries of acceptable behavior, so, too, there are rules that set limits to our feelings. In extreme cases, feeling rules may even have the power of life and death over the individual. Consider a court trial where the defendant is found guilty of murder in part because he expressed the "wrong amount" of grief over the victim's death, thus suggesting his guilt.

Moreover, feeling rules are themselves embedded in a broader structural context. For instance, as we noted previously, existing class relations subject middle-class workers more than others to the demands of emotional labor and the commodification of feelings. For it is the middle-class job that is more likely to entail personal interaction with the public, the requirement to produce specific feeling-states in others, and the surveillance of emotional labor by superiors.

Reading

Introduction to *The Managed Heart*

In this chapter from *The Managed Heart,* Hochschild turns her attention to the effects of gender relations on emotion management. If members of the lower and working classes tend more to things than to people, and thus are less practiced in the skills of emotional labor, the hierarchical patterning of managing emotions is reversed when it comes to gender. In other words, while the occupations associated with the more advantaged classes (flight attendants, sales workers, teachers, lawyers, health-care providers, etc.) are more likely to require the manipulation of personal feelings, it is women, the less-advantaged gender, who more often find it necessary to be skilled emotion managers and thus who are more susceptible to the commodification of their feelings. As you will read, Hochschild attributes these differences in the emotional lives of men and women to the unequal distribution of money, power, authority, and status. As a result, in their private lives, "women make a resource out of feeling and offer it to men as a gift in return for the more material resources they lack" (ibid.:163). Emotion work is central to how to be, and what it means to be, a wife, a mother, and a woman. Meanwhile, men and women are typically called on to perform different types of emotional labor because of the gendered nature of occupations. This, too, carries with it a number of consequences that make the managing of feelings a different business for women and for men.

The Managed Heart (1983)

Arlie Russell Hochschild

GENDER, STATUS, AND FEELING

More emotion management goes on in the families and jobs of the upper classes than in those of the lower classes. That is, in the class system, social conditions conspire to make it more prevalent at the top. In the gender system, on the other hand, the reverse is true: social conditions make it more prevalent, and prevalent in different ways, for those at the bottom—women. In what sense is this so? And why?

Both men and women do emotion work, in private life and at work. In all kinds of ways, men as well as women get into the spirit of the party, try to escape the grip of hopeless love, try to pull themselves out of depression, try to allow grief. But in the whole realm of emotional experience, is emotion work as important for men as it is for women? And is it important in the same ways? I believe that the answer to both questions is No. The reason, at bottom, is the fact that women in general have far less independent access to money, power, authority, or status in society. They are a subordinate social stratum, and this has four consequences.

First, lacking other resources, women make a resource out of feeling and offer it to men as a gift in return for the more material resources they lack. . . . Thus their capacity to manage feeling and to do "relational" work is for them a more important resource. Second, emotion work is important in different ways for men and for women. This is because each gender tends to be called on to do different kinds of this work. . . . This specialization of emotional labor in the marketplace rests on the different childhood training of the heart that is given to girls and to boys. ("What are little girls made of? Sugar and spice and everything nice. What are little boys made of? Snips and snails and puppy dog tails.") Moreover, each specialization presents men and women with different emotional tasks. Women are more likely to be presented with the task of mastering anger and aggression in the service of "being nice." To men, the socially assigned task of aggressing against those that break rules of various sorts creates the private task of mastering fear and vulnerability.

Third, and less noticed, the general subordination of women leaves every individual woman with a weaker "status shield" against the displaced feelings of others. . . . The fourth consequence of the power difference between the sexes is that for each gender a different portion of the managed heart is enlisted for commercial use. Women more often react to subordination by making defensive use of sexual beauty, charm, and relational skills. For them, it is these capacities that become most vulnerable to commercial exploitation, and so it is these capacities that they are most likely to become estranged from. For male workers in "male" jobs, it is more often the capacity to wield anger and make threats that is delivered over to the company, and so it is this sort of capacity that they are more likely to feel estranged from. After the great transmutation, then, men and women come to experience emotion work in different ways. . . .

Women as Emotion Managers

Middle-class American women, tradition suggests, feel emotion more than men do. The definitions of "emotional" and "cogitation" in the *Random House Dictionary of the English Language* reflect a deeply rooted cultural idea. Yet women are also thought to command "feminine wiles," to have the capacity to premeditate a sigh, an outburst of tears, or a flight of joy. In general, they are thought to *manage* expression and feeling not only better but more often than men do.

How much the conscious feelings of women and men may differ is an issue I leave aside here. However, the evidence seems clear that women do *more* emotion managing than men. And because the well-managed feeling has an outside resemblance to spontaneous feeling, it is possible to confuse the condition of being more "easily affected by emotion" with the action of willfully managing emotion when the occasion calls for it.

Especially in the American middle class, women tend to manage feeling more because in general they depend on men for money, and one of the various ways of repaying their debt is to do extra emotion work—*especially emotion work that affirms, enhances, and celebrates the well-being and status of others.* When the emotional skills that children learn and practice at home move into the marketplace, the emotional labor of women becomes more prominent because men in general have not been trained to make their emotions a resource and are therefore less likely to develop their capacity for managing feeling.

There is also a difference in the kind of emotion work that men and women tend to do. Many studies have told us that women adapt more to the needs of others and cooperate more than men do. These studies often imply the existence of gender-specific characteristics that are inevitable if not innate. But do these characteristics simply exist passively in women? Or are they signs of a social work that women *do*—the work of affirming, enhancing, and celebrating the well-being and status of others? I believe that much of the time, the adaptive, cooperative woman is actively working at showing deference. This deference requires her to make an outward display of what Leslie Fiedler has called the "seriously" good girl in her and to support this effort by evoking feelings that make the "nice" display seem natural.[i] Women who want to put their own feelings less at the service of others must still confront the idea that if they do so, they will be considered less "feminine."

. . . The emotional arts that women have cultivated are analogous to the art of feigning that Lionel Trilling has noted among those whose wishes outdistance their opportunities for class advancement. As for many others of lower status, it has been in the woman's interest to be the better actor.[ii] As the psychologists would say,

---

[i]Fiedler (1960) suggests that girls are trained to be "seriously" good and to be ashamed of being bad whereas boys are asked to be good in formalistic ways but covertly invited to be ashamed of being "too" good. Oversocialization into "sugar-and-spice" demeanor produces feminine skills in delivering deference.

[ii]Other researchers have found men to have a more "romantic" orientation to love, women a more "realistic" orientation. That is, males may find cultural support for a passive construction of love, for seeing themselves as "falling head over heels," or "walking on air." According to Kephart, "the female is not pushed hither and yon by her romantic compulsions. On the contrary, she seems to have a greater measure of rational control over her romantic inclinations than the male" (1967, p. 473).

the techniques of deep acting have unusually high "secondary gains." Yet these skills have long been mislabeled "natural," a part of woman's "being" rather than something of her own making.[iii]

Sensitivity to nonverbal communication and to the micropolitical significance of feeling gives women something like an ethnic language, which men can speak too, but on the whole less well. It is a language women share offstage in their talk "about feelings." This talk is not, as it is for men offstage, the score-keeping of conquistadors. It is the talk of the artful prey, the language of tips on how to make him want her, how to psyche him out, how to put him on or turn him off. Within the traditional female subculture, subordination at close quarters is understood, especially in adolescence, as a "fact of life." Women accommodate, then, but not passively. They actively adapt feeling to a need or a purpose at hand, and they do it so that it *seems* to express a passive state of agreement, the chance occurrence of coinciding needs. Being becomes a way of doing. Acting is the needed art, and emotion work is the tool. . . .

Almost everyone does the emotion work that produces what we might, broadly speaking, call deference. But women are expected to do more of it. A study by Wikler (1976) comparing male with female university professors found that students expected women professors to be warmer and more supportive than male professors; given these expectations, proportionally more women professors were perceived as cold. In another study, Broverman, Broverman, and Clarkson (1970) asked clinically trained psychologists, psychiatrists, and social workers to match various characteristics with "normal adult men" and "normal adult women"; they more often associated "very tactful, very gentle, and very aware of feelings of others" with their ideas of the normal adult woman. In being adaptive, cooperative, and helpful, the woman is on a private stage behind the public stage, and as a consequence she is often seen as less good at arguing, telling jokes, and teaching than she is at expressing appreciation of these activities. She is the conversational cheerleader. She actively enhances other people—usually men, but also other women to whom she plays woman. The more she seems natural at it, the more her labor does not show as labor, the more successfully it is disguised as the *absence* of other, more prized qualities. As a *woman* she may be praised for out-enhancing the best enhancer, but as a *person* in comparison with comics, teachers, and argument-builders, she usually lives outside the climate of enhancement that men tend to inhabit. Men, of course, pay court to certain other men and women and thus also do the emotion work that keeps deference sincere. The difference between men and women is a difference in the psychological effects of having or not having power.

Racism and sexism share this general pattern, but the two systems differ in the avenues available for the translation of economic inequality into private terms. The white manager and the black factory worker leave work and go home, one to a generally white neighborhood and family and the other to a generally black neighborhood and family. But in the case of women and men, the larger economic inequality is filtered into the intimate daily exchanges between wife and husband. Unlike other subordinates, women seek *primary* ties with a supplier. In marriage, the principle of reciprocity applies to wider arenas of each self: there is more to choose from in how we pay and are paid, and the paying between economically unequal parties goes on morning, noon, and night. The larger inequities find intimate expression.

Wherever it goes, the bargain of wages-for-other-things travels in disguise. Marriage both bridges and obscures the gap between the resources available to men and those available to women.[iv] Because men and women do try to love one another—to cooperate in making love, making babies, and making a life together—the very closeness of the bond they accept calls for some disguise of subordination. There will be talk in the "we" mode, joint bank accounts and joint decisions, and the idea among women that they are equal in the ways that "really count." But underlying this pattern will be *different potential futures outside the marriage* and the effect of that on the patterning of life.[v] The woman may thus become especially assertive about certain secondary decisions, or especially active in certain limited domains,

[iii]The use of feminine wiles (including flattery) is felt to be a psychopolitical style of the subordinate; it is therefore disapproved of by women who have gained a foothold in the man's world and can afford to disparage what they do not need to use.

[iv]Because women have less access to money and status than their male class peers do, they are more motivated than men to marry in order to win access to a much higher "male wage."

[v]Zick Rubin's study of young men and women in love relationships (generally middle-class persons of about the same age) found that the women tended to admire their male loved ones more than they were, in turn, admired by them. The women also felt "more like" their loved ones than the men did. (See Rubin 1970; Reiss 1960.)

in order to experience a sense of equality that is missing from the overall relationship.

Women who understand their ultimate disadvantage and feel that their position cannot change may jealously guard the covertness of their traditional emotional resources, in the understandable fear that if the secret were told, their immediate situation would get worse. For to confess that their social charms are the product of secret work might make them less valuable, just as the sexual revolution has made sexual contact less "valuable" by lowering its bargaining power without promoting the advance of women into better-paying jobs. In fact, of course, when we redefine "adaptability" and "cooperativeness" as a form of shadow labor, we are pointing to a hidden cost for which some recompense is due and suggesting that a general reordering of female-male relationships is desirable.

There is one further reason why women may offer more emotion work of this sort than men: more women at all class levels do unpaid labor of a highly interpersonal sort. They nurture, manage, and befriend children. More "adaptive" and "cooperative," they address themselves better to the needs of those who are not yet able to adapt and cooperate much themselves. Then, according to Jourard (1968), because they are seen as members of the category from which mothers come, women in general are asked to look out for psychological needs more than men are. The world turns to women for mothering, and this fact silently attaches itself to many a job description.

Women at Work

With the growth of large organizations calling for skills in personal relations, the womanly art of status enhancement and the emotion work that it requires has been made more public, more systematized, and more standardized. It is performed by largely middle-class women in largely public-contact jobs. . . . Jobs involving emotional labor comprise over a third of all jobs. But they form only a *quarter* of all jobs that men do, and over *half* of all jobs that women do.

Many of the jobs that call for public contact also call for giving service to the public. Richard Sennett and Jonathan Cobb, in *The Hidden Injuries of Class,* comment on how people tend to rank service jobs in relation to other kinds of jobs: "At the bottom end of the scale are found not factory jobs but service jobs where the individual has to perform personally for someone else. A bartender is listed below a coal miner, a taxi driver below a truck driver; we believe this occurs because their functions *are felt to be more dependent on and more at the mercy of others*" [my emphasis].[vi] Because there are more women than men in service jobs (21 percent compared with 9 percent), there are "hidden injuries" of gender attached to those of class.

Once women are at work in public-contact jobs, a new pattern unfolds: they receive less basic deference. That is, although some women are still elbow-guided through doors, chauffeured in cars, and protected from rain puddles, they are not shielded from one fundamental consequence of their lower status: their feelings are accorded less weight than the feelings of men. . . .

How, then, does a woman's lower status influence how she is treated by others? More basically, what is the prior link between status and the treatment of feeling? High-status people tend to enjoy the privilege of having their feelings noticed and considered important. The lower one's status, the more one's feelings are not noticed or treated as inconsequential. H. E. Dale, in *The Higher Civil Service of Great Britain,* reports the existence of a "doctrine of feelings":

> The doctrine of feelings was expounded to me many years ago by a very eminent civil servant. . . . He explained that the importance of feelings varies in close correspondence with the importance of the person who feels. If the public interest requires that a junior clerk should be removed from his post, no regard need be paid to his feelings; if it is the case of an assistant secretary, they must be carefully considered, within reason; if it is a permanent secretary, feelings are a principal element in the situation, and only imperative public interest can override their requirements.[vii]

Working women are to working men as junior clerks are to permanent secretaries. Between executive and secretary, doctor and nurse, psychiatrist and social worker, dentist and dental assistant, a power difference is reflected as a gender difference. The "doctrine of feelings" is another double standard between the two sexes.[viii]

[vi]Sennett and Cobb (1973), p. 236.

[vii]Quoted in Goffman (1967), p. 10.

[viii]The code of chivalry is said to require protection of the weaker *by* the stronger. Yet a boss may bring flowers to his secretary or open the door for her only to make up for the fact that he gets openly angry at her more often than he does at a male equal or superior, and more often than she does at him. The flowers symbolize redress, even as they obscure the basic maldistribution of respect and its psychic cost.

The feelings of the lower-status party may be discounted in two ways: by considering them rational but unimportant or by considering them irrational and hence dismissable. An article entitled "On Aggression in Politics: Are Women Judged by a Double Standard?" presented the results of a survey of female politicians. All those surveyed said they believed there was an affective double standard. As Frances Farenthold, the president of Wells College in Aurora, New York, put it: "You certainly see to it that you don't throw any tantrums. Henry Kissinger can have his scenes—remember the way he acted in Salzburg? But for women, we're still in the stage that if you don't hold in your emotions, you're pegged as emotional, unstable, and all those terms that have always been used to describe women."[ix] These women in public life were agreed on the following points. When a man expresses anger, it is deemed "rational" or understandable anger, anger that indicates not weakness of character but deeply held conviction. When women express an equivalent degree of anger, it is more likely to be interpreted as a sign of personal instability. It is believed that women are more emotional, and this very belief is used to invalidate their feelings. That is, the women's feelings are seen not as a response to real events but as reflections of themselves as "emotional" women.

Here we discover a corollary of the "doctrine of feelings": the lower our status, the more our manner of seeing and feeling is subject to being discredited, and the less believable it becomes. An "irrational" feeling is the twin of an invalidated perception. A person of lower status has a weaker claim to the right to define what is going on; less trust is placed in her judgments; and less respect is accorded to what she feels. Relatively speaking, it more often becomes the burden of women, as with other lower-status persons, to uphold a minority viewpoint, a discredited opinion.

Medical responses to male and female illness provide a case in point. One study of how doctors respond to the physical complaints of back pain, headache, dizziness, chest pain, and fatigue—symptoms for which a doctor must take the patient's word—showed that among fifty-two married couples, the complaints of the husbands elicited more medical response than those of the wives. The authors conclude: "The data may bear out . . . that the physicians . . . tend to take illness more seriously in men than in women."[x] Another study of physician interactions with 184 male and 130 female patients concluded that "doctors were more likely to consider the psychological component of the patient's illness important when the patient was a woman."[xi] The female's assertion that she was physically sick was more likely to be invalidated as something "she just imagined," something "subjective," not a response to anything real.

To make up for either way of weighing the feelings of the two sexes unequally, many women urge their feelings forward, trying to express them with more force, so as to get them treated with seriousness. But from there the spiral moves down. For the harder women try to oppose the "doctrine of feeling" by expressing their feelings more, the more they come to fit the image awaiting them as "emotional." Their efforts are discounted as one more example of emotionalism. The only way to counter the doctrine of feelings is to eliminate the more fundamental tie between gender and status.

The Status Shield at Work

Given this relation between status and the treatment of feeling, it follows that persons in low-status categories—women, people of color, children—lack a status shield against poorer treatment of their feelings. This simple fact has the power to utterly transform the content of a job. The job of flight attendant, for example, is not the *same job* for a woman as it is for a man. A day's accumulation of passenger abuse for a woman differs from a day's accumulation of it for a man. Women tend to be more exposed than men to rude or surly speech, to tirades against the service, the airline, and airplanes in general. As the company's main shock absorbers against "mishandled" passengers, their own feelings are more frequently subjected to rough treatment. In addition, a day's exposure to people who resist authority in women is a different experience for a woman than it is for a man. Because her gender is accorded lower status, a woman's shield against abuse is weaker, and the importance of what she herself might be feeling—when faced with

---

[ix]*New York Times,* February 12, 1979.

[x]More women than men go to doctors, and this might seem to explain why doctors take them less seriously. But here it is hard to tell cause from effect, for if a woman's complaints are not taken seriously, she may have to make several visits to doctors before a remedy is found (Armitage et al. 1979).

[xi]Wallens et al. (1979), p. 143.

blame for an airline delay, for example—is correspondingly reduced. Thus the job for a man differs in essential ways from the same job for a woman.

In this respect, it is a disadvantage to be a woman—as 85 percent of all flight attendants are. And in this case, they are not simply women in the biological sense. They are also a highly visible distillation of middle-class American notions of femininity. They symbolize Woman. Insofar as the category "female" is mentally associated with having less status and authority, female flight attendants are more readily classified as "really" female than other females are. And as a result their emotional lives are even less protected by the status shield.

More than female accountants, bus drivers, or gardeners, female flight attendants mingle with people who expect them to *enact* two leading roles of Womanhood: the loving wife and mother (serving food, tending the needs of others) and the glamorous "career woman" (dressed to be seen, in contact with strange men, professional and controlled in manner, and literally very far from home). They do the job of symbolizing the transfer of homespun femininity into the impersonal marketplace, announcing, in effect, "I work in the public eye, but I'm still a woman at heart."

Passengers borrow their expectations about gender biographies from home and from the wider culture and then base their demands on this borrowing. The different fictive biographies they attribute to male and female workers make sense out of what they expect to receive in the currency of caretaking and authority. One male flight attendant noted:

> They always ask about my work plans. "Why are you doing this?" That's one question we get all the time from passengers. "Are you planning to go into management?" Most guys come in expecting to do it for a year or so and see how they like it, but we keep getting asked about the management training program. I don't know any guy that's gone into management from here.

In contrast, a female flight attendant said:

> Men ask me why I'm not married. They don't ask the guys that. Or else passengers will say, "Oh, when you have kids, you'll quit this job. I know you will." And I say, "Well, no, I'm not going to have kids." "Oh yes you will," they say. "No I'm not," I say, and I don't want to get more personal than that. They may expect me to have kids because of my gender, but I'm not, no matter what they say.

If a female flight attendant is seen as a protomother, then it is natural that the work of nurturing should fall to her. As one female attendant said: "The guys bow out of it more and we pick up the slack. I mean the handling of babies, the handling of children, the coddling of the old folks. The guys don't get involved in that quite as much." Confirming this, one male flight attendant noted casually, "Nine times out of ten, when I go out of my way to talk, it will be to attractive gal passengers." In this regard, females generally appreciated gay male flight attendants who, while trying deftly to sidestep the biography test, still gravitate more toward nurturing work than straight males are reputed to do.

Gender makes two jobs out of one in yet another sense. Females are asked more often than males to appreciate jokes, listen to stories, and give psychological advice. Female specialization in these offerings takes on meaning only in light of the fact that flight attendants of both sexes are required to be both deferential and authoritative; they have to be able to appreciate a joke nicely, but they must also be firm in enforcing the rules about oversized luggage. But because more deference is generally expected from a woman, she has a weaker grasp on passenger respect for her authority and a harder time enforcing rules.

In fact, passengers generally assume that men have *more* authority than women and that men exercise authority *over* women. For males in the corporate world to whom air travel is a way of life, this assumption has more than a distant relation to fact. As one flight attendant put it: "Say you've got a businessman sitting over there in aisle five. He's got a wife who takes his suit to the cleaners and makes the hors d'oeuvres for his business guests. He's got an executive secretary with horn-rimmed glasses who types 140 million words a minute and knows more about his airline ticket than he does. There's no woman in his life over him." This assumption of male authority allows ordinary twenty-year-old male flight attendants to be mistaken for the "managers" or "superintendents" of older female flight attendants. A uniformed male among women, passengers assume, must have authority over women. In fact, because males were excluded from this job until after a long "discrimination" suit in the mid-1960s and few were hired until the early 1970s, most male flight attendants are younger and have less seniority than most female attendants.

The assumption of male authority has two results. First, authority, like status, acts as a shield against scapegoating. Since the women workers on the plane were thought to have less authority and therefore less status, they were more susceptible to scapegoating. When the plane was late, the steaks gone, or the ice out, frustrations were vented more openly toward female workers. Females were expected to "take it" better, it being more their role to absorb an expression of displeasure and less their role to put a stop to it.

In addition, both male and female workers adapted to this fictional redistribution of authority. Both, in different ways, made it more real. Male flight attendants tended to react to passengers *as if they had more authority* than they really did. This made them less tolerant of abuse and firmer in handling it. They conveyed the message that *as authorities* they expected compliance without loud complaint. Passengers sensing this message were discouraged from pursuing complaints and stopped sooner. Female flight attendants, on the other hand, assuming that passengers would honor their authority less, used more tactful and deferential means of handling abuse. They were more deferential toward male passengers (from whom they expected less respect) than toward female passengers (whose own fund of respect was expected to be lower). And they were less successful in preventing the escalation of abuse. As one male flight attendant observed: "I think the gals tend to get more intimidated if a man is crabby at them than if a woman is."

Some workers understood this as merely a difference of style. As one woman reflected:

The guys have a low level of tolerance and their own male way of asserting themselves with the passenger that I'm not able to use. I told a guy who had a piece of luggage in front of him that wouldn't fit under the seat, I told him, "It won't fit, we'll have to do something with it." He came back with, "Oh, but it's been here the whole trip, I've had it with me all the time, blah, blah, blah." He gave me some guff. I thought to myself, I'll finish this later, I'll walk away right now. I intended to come back to him. A flying partner of mine, a young man, came by this passenger, without knowing about our conversation, and said to him, "Sir, that bag is too big for your seat. We're going to have to take it away." "Oh, here you are," the guy says, and he hands it over to him. . . . You don't see the male flight attendants being physically abused or verbally abused nearly as much as we are.

The females' supposed "higher tolerance for abuse" amounted to a combination of higher exposure to it and less ammunition—in the currency of respect—to use against it.

This pattern set in motion another one: female workers often went to their male co-workers to get them to "cast a heavier glance." As one woman who had resigned herself to this explained wearily: "I used to fight it and assert myself. Now I'm just too overworked. It's simpler to just go get the male purser. One look at him and the troublemaker shuts up. Ultimately it comes down to the fact that I don't have time for a big confrontation. The job is so stressful these days, you don't go out of your way to make it more stressful. A look from a male carries more weight." Thus the greater the respect males could command, the more they were called on to claim it.

This only increased the amount of deference that male workers felt their female co-workers owed them, and women found it harder to supervise junior males than females. One young male attendant said that certain conditions had to be met—and deference offered—before he would obey a woman's orders: "If it's an order without a human element to it, then I'll balk. I think sometimes it's a little easier for a man to be an authority figure and command respect and cooperation. I think it depends on how the gal handles herself. If she doesn't have much confidence or if she goes the other way and gets puffed out of shape, then in that case I think she could have more trouble with the stewards *than with the gals*" [my emphasis]. Workers tended to agree that females took orders better than males, no matter how "puffed out of shape" the attendant in charge might be, and that women in charge had to be nicer in exercising their authority than men did.

This attitude toward status and authority inspired compensatory reactions among some female workers. One response was to adopt the crisply cheerful but no-nonsense style of a Cub Scout den mother—a model of female authority borrowed from domestic life and used here to make it acceptable for women to tell adult men what to do. In this way a woman might avoid being criticized as "bossy" or "puffed out of shape" by placing her behavior within the boundaries of the gender expectations of passengers and co-workers.

Another response to displaced anger and challenged authority was to make small tokens of respect a matter of great concern. Terms of address, for example, were seen as an indicator of status, a promise of the right to politeness which those deprived of status unfortunately lack. The term "girl," for example, was recognized by female workers as the moral equivalent of calling black men "boys." Although in private and among themselves, the women flight attendants I knew usually called themselves "girls," many were opposed to the use of the term in principle. They saw it not only as a question of social or moral importance but as a *practical matter.* To be addressed as a "girl" was to be subjected to more on-the-job stress. The order, "Girl, get me some cream" has a different effect than the request "Oh miss, could I please have some cream?" And if the cream has run out because the commissary didn't provide enough, it will be the "girls" who get the direct expressions of disappointment, exasperation, and blame. Tokens of respect can be exchanged to make a bargain: "I'll manage my unpleasant feelings for you if you'll manage yours for me." When outrageously rude people occasionally enter a plane, it reminds all concerned why the flimsy status shield against abuse is worth struggling over.

Schooled in emotion management at home, women have entered in disproportionate numbers those jobs that call for emotional labor outside the home. Once they enter the marketplace, a certain social logic unfolds. Because of the division of labor in the society at large, women *in any particular job* are assigned lower status and less authority than men. As a result, they lack a shield against the "doctrine of feelings." Much more often than men, they become the complaint department, the ones to whom dissatisfaction is fearlessly expressed. Their own feelings tend to be treated as less important. In ways that the advertising smiles obscure, the job has different contents for women and men.

Estrangement From Sexual Identity

Regardless of gender, the job poses problems of identity. What is my work role and what is "me"? How can I do deep acting without "feeling phony" and losing self-esteem? How can I redefine the job as "illusion making" without becoming cynical?

But there are other psychological issues a flight attendant faces if she is a woman. In response to her relative lack of power and her exposure to the "doctrine of feelings," she may seek to improve her position by making use of two traditionally "feminine" qualities—those of the supportive mother and those of the sexually desirable mate. Thus, some women *are* motherly; they support and enhance the well-being and status of others. But in *being* motherly, they may also *act* motherly and may sometimes experience themselves using the motherly act to win regard from others. In the same way, some women are sexually attractive and may act in ways that are sexually alluring. For example, one flight attendant who played the sexual queen—swaying slowly down the aisle with exquisitely understated suggestiveness—described herself as using her sexual attractiveness to secure interest and favors from male passengers. In each case, the woman is using a feminine quality for private purposes. But it is also true, for the flight attendant, that both "motherly" behavior and a "sexy" look and manner are partly an achievement of corporate engineering—a result of the company's emphasis on the weight and (former) age requirements, grooming classes, and letters from passengers regarding the looks and demeanor of flight attendants. In its training and supervisory roles, the company may play the part of the protective duenna. But in its commercial role as an advertiser of sexy and glamorous service, it acts more like a backstage matchmaker. Some early United Airlines ads said, "And she might even make a good wife." The company, of course, has always maintained that it does not meddle in personal affairs.

Thus the two ways in which women traditionally try to improve their lot—by using their motherly capacity to enhance the status and well-being of others, and by using their sexual attractiveness—have come under company management. Most flight attendants I spoke with agreed that companies used and attached profit to these qualities. . . .

Estrangement from aspects of oneself are, in one light, a means of defense. On the job, the acceptance of a division between the "real" self and the self in a company uniform is often a way to avoid stress, a wise realization, a saving grace. But this solution also poses serious problems. For in dividing up our sense of self, in order to save the "real" self from unwelcome intrusions, we necessarily relinquish a healthy sense of wholeness. We come to accept as normal the tension we feel between our "real" and our "on-stage" selves.

More women than men go into public-contact work and especially into work in which status enhancement is the essential social-psychological task. In some jobs, such as that of the flight attendant, women may perform this task by playing the Woman. Such women are more vulnerable, on this account, to feeling estranged from their capacity to perform and enjoy two traditional feminine roles—offering status enhancement and sexual attractiveness to others. These capacities are now under corporate as well as personal management.

Perhaps this realization accounts for the laughter at a joke I heard surreptitiously passed around the Delta Training Office, as if for an audience of insiders. It went like this: A male passenger came across a woman flight attendant seated in the galley, legs apart, elbows on knees, her chin resting in one hand and a lighted cigarette in the other—held between thumb and forefinger. "Why are you holding your cigarette like that?" the man asked. Without looking up or smiling, the woman took another puff and said, "If I had balls, I'd be driving this plane." Inside the feminine uniform and feminine "act" was a would-be man. It was an estrangement joke, a poignant behind-the-scenes protest at a commercial logic that standardizes and trivializes the dignity of women.

Discussion Questions

1. Symbolic interactionism emphasizes the constructed and negotiated aspects of the self and social life more generally. As a result, it shifts attention from analyses of the structural and institutional features of society. How might some of the key concepts in this framework (for instance, meaning, impression management, definition of the situation, interaction rituals, front and backstage, secondary adjustments, feeling rules, emotion work) be used to explore stratification systems and the relations of domination and subordination that they sustain?

2. Symbolic interactionism is an inductive theoretical perspective that examines the interpretive nature of social life. For their part, symbolic interactionists are themselves inescapably engaged in an interpretation of their observations (as all observers are). As a result, their findings are, in a sense, interpretations of interpretations. What implications might this have on producing generalizable conclusions and, with them, theory itself? For that matter, how does one empirically study the self, which, after all, is the central concept within this theoretical tradition? Similarly, how might you observe and scientifically analyze the operation of feeling rules?

3. A common theme expressed throughout Goffman's work, and particularly in *The Presentation of Self in Everyday Life,* is the notion that impression management is shaped by both strategic and moral concerns. How is this related to Goffman's distinction between the self as a "performer" and a "character"? Are there some contexts in which behavior might be guided more by strategic than by moral concerns, or vice versa? If so, what commonalities might such contexts share?

4. Frederick Wiseman's *Titicut Follies* (1967), Ken Kesey's *One Flew Over the Cuckoo's Nest* (1962), and Philip Zimbardo's *Stanford Prison Experiment* (1991) are classic works that explore the psychological effects of being confined within total institutions. View one of these films and discuss its implications for Goffman's analysis of the relationship between the self and social arrangements. In what way does the film confirm or challenge Goffman's perspective?

5. In general, how important do you think emotions are in shaping an individual's (or your own) experiences and decisions? To what extent are emotions seen as a legitimate basis for action? In which domains of life (work, school, family, religion, politics, etc.) are emotive experiences encouraged or discouraged? Following Hochschild's insights, what factors might account for the differing types and intensity of emotions that are experienced within particular domains?

13 PHENOMENOLOGY

Alfred Schutz

Key Concepts

- Bracketing
- Intersubjectivity
- Lifeworld
- Stocks of Knowledge
- Typifications
- Recipes

Peter Berger

Key Concepts

- Habitualizaton
- Institutionalization
- Externalization/Objectivation/Reification
- Internalization
- Socialization
  - Primary
  - Secondary
- Social Construction of Reality

Thomas Luckmann

It is the meaning of our experiences, and not the ontological structure of the objects, which constitutes reality.

—Alfred Schutz (1970:252)

H ave you ever thought about just how many intricate habits, rules, and principles there are in every-day life? For instance, have you ever been talking to an acquaintance on the phone, trying to end the conversation by saying things like, "So, well, thanks for calling . . ." but he or she just prattles on as if you hadn't said anything (or, even worse, starts a whole new topic of conversation)? Or have you ever said your good-byes at a party and left, only to get to your car and realize you forgot your keys? Why does it feel so extremely awkward to go back through the door? These small, unexpected disruptions to routine social interactions exemplify just how nuanced social life is—and how many aspects of it are taken for granted. One classic example of this is elevator-riding: Everyone everywhere rides an elevator in pretty much the same way (by turning around and facing the door through which you just entered). How would people react if you were to get on an elevator and then face someone already riding in the elevator, looking straight at him or her, rather than at the door? Ethnomethodologists actually do experiments like this and see the results.

In this chapter, we focus on the phenomena of everyday life by examining phenomenology and, to a lesser extent, ethnomethodology. These two interrelated traditions analyze the taken-for-granted everyday world that is the basis for all human conduct. Phenomenology is a philosophy, a method, and an approach that has deeply influenced psychology and psychiatry as well as sociology. Focusing particularly on the work of Alfred Schutz and his students Peter Berger and Thomas Luckmann, we will see that phenomenologists seek to explain how people actively produce and sustain meaning. We conclude this chapter with a brief look at ethnomethodology, which is a more recent tradition developed principally by Harold Garfinkel. Ethnomethodologists focus less on meaning and subjectivity and more on the actual *methods* people use to accomplish their everyday lives. In contrast to phenomenology, which as indicated above has close ties to psychology and philosophy, ethnomethodology has close ties to linguistics and mainstream sociology. Ethnomethodologists are more interested in how actors assure each other that meaning is shared than the actual meaning structures themselves, although obviously the subject matter of phenomenology and ethnomethodology is very much intertwined.

ALFRED SCHUTZ (1899–1959): A BIOGRAPHICAL SKETCH ▓▓

Alfred Schutz was born in Vienna, Austria, in 1899 to an upper-middle-class Jewish family. He studied law and the social sciences at the University of Vienna and after graduating began a career in banking. Although he was not an academic by profession, Schutz was intrigued by phenomenological sociology and participated in informal lecture and discussion circles. He was particularly interested in the work of the German sociologist Max Weber (1864–1920; see Chapter 4 of this book)) and philosopher Edmund Husserl (1859–1938), who Schutz often visited in Freiburg and who even invited Schutz to be his assistant. Although he was only informally connected to academic life, in 1932 Schutz published what would come to be an essential work in the field: *Phenomenology of the Social World*. This groundbreaking work was not widely appreciated at first, however, both because Schutz was not an established academic with a university position and because it would not be translated into English until more than 35 years later. With the outbreak of World War II, Schutz fled to Paris and then to the United States shortly thereafter. Although he continued to work in banking, he also began teaching at the New School for Social Research in New York City in 1943, an institution that became home to a number of exiled German intellectuals. Among those he met in the United States and whose work significantly influenced his was the eminent pragmatist philosopher George Herbert Mead (see Chapter 8 of this book). Schutz finally gave up his banking career in 1956 and devoted himself full time to teaching and research. At the time of his death in 1959, Schutz was working on a comprehensive statement as to the development of his ideas since publication of *Phenomenology of the Social World*. This work (discussed later in the chapter) was finished by Schutz's student Thomas Luckmann and published as *The Structures of the Life World* (*Strukturen der Lebenswelt*) by Schutz and Luckmann (1973).

▋▋ SCHUTZ'S INTELLECTUAL INFLUENCES AND CORE IDEAS

Above all, two key thinkers were pivotal in shaping the work of Alfred Schutz: Edmund Husserl and Max Weber.[1] As indicated previously, Edmund Husserl is commonly considered the founder of phenomenology. A student of mathematics, physics, and philosophy and influenced himself by Rene Descartes, David Hume, and Immanuel Kant, Husserl developed what he called "transcendental phenomenology," which holds that there is no pure subjective subject or pure objective object. Rather, all consciousness is consciousness of something, and objects do not have appearances independent of the beings that perceive them. Thus, the human experience in the world is as "a world of *meaningful* objectives and relations" (Wagner 1973:14, emphasis in original). Husserl used the term **lifeworld** (*Lebenswelt*) to refer to the world of existing assumptions as they are experienced and made meaningful in consciousness (Wagner 1973:63). Husserl ([1913] 1982) explains how intentional consciousness, that is, directing our attention in one way or another, enables the phenomenologist to reconstruct or bracket his basic views on the world and himself and explore their interconnections. In doing so, Husserl made the lifeworld, or "thinking as usual" in everyday life situations, a legitimate object of investigation. Phenomenology investigates the systematic **bracketing** of all existing assumptions regarding the external world.

Following Husserl, Schutz emphasized that humans do not experience the world as an "objective" reality; rather, they experience the world as made of meaningful objects and relations. Understanding and explaining "reality," then, requires paying attention to the meaning structures by and through which individuals perceive the world. Extending Husserl's idea that the human actor, as a socialized member of society, operates within a lifeworld that is pregiven and already organized, Schutz emphasized that this taken-for-granted world includes whatever prejudices and typical interpretations may derive from it. Within the standpoint of this "natural attitude," the individual does not question the structures of meaning that comprise her lifeworld. Instead, the individual is guided by practical concerns, and her task is to live in rather than to reflect on the lifeworld. As Schutz (1970:136) states, the lifeworld is "the unproblematic ground for the emergence of all possible problems, the prerequisite for transforming any unclarified situation into warranted ascertainability." The lifeworld is the taken-for-granted backdrop within which all situations are measured and given meaning.

Most important, the lifeworld is an intersubjective world, known and experienced by others. **Intersubjectivity** refers not so much to the fact that we share the same empirical or material world as others, but that we share the same consciousness. It is intersubjectivity—that you know that I know that you know—that allows human beings from a wide variety of personal and social backgrounds to function and interact.

On one hand, this emphasis on shared consciousness and meaning recalls Émile Durkheim's conceptualization of "collective conscience." Durkheim used this term to refer to the "totality of beliefs and sentiments common to average citizens of the same society" that "forms a determinate system which has its own life" ([1893] 1984:38–39 see Chapter 3 of this book).[2] However, in contrast to Durkheim, Schutz does not conceptualize the preorganized and pregiven elements of the lifeworld as acting on the individual with the external power of constraint. Rather, in accordance with the basic premises of symbolic interactionism (see Chapter 12), Schutz views one's "natural attitude" as based on the acceptance, interpretation, redefinition,

---

[1]Schutz was also particularly influenced by the sociologist and philosopher Max Scheler (1874–1928), who wrote on intersubjectivity, the intentionality of the Other, and sympathy (see *The Nature of Sympathy*, [1922] 1945), as well as the French philosopher Henri Bergson, whose conception of *durée* provided the recognition of subjective experience of time in distinction from the mechanical notion of clock time (Wagner 1973:61).

[2]As discussed in Chapter 3, for Durkheim, the task of sociology is to analyze "social facts," that is, the conditions and circumstances external to the individual that, nevertheless, determine one's course of action.

and modification of cultural elements by the individual (Wagner 1973:64). That is, the individual gives the elements of the lifeworld a personal note and subtly changes, enlarges, or reduces their meaning in distinctive ways. A unique accumulation of lived experiences results, as the individual goes from situation to situation that makes up her everyday world.

This brings us to another important influence on Schutz: Max Weber, one of the leading figures in the development of modern sociology. As discussed in Chapter 4, Weber was one of the first to differentiate sociology from philosophy by proclaiming that the social sciences must abstain from value judgments. He envisioned sociology as the science of human behavior and its consequences, and he sought a method that would prevent the intrusion of political and moral ideologies that all too easily influence the judgment of the social scientist (whether this influence is conscious or not). However, Weber was not a crass positivist who simply ignored subjectivity in the quest for objectivity and timeless truths. His position on the proper method and subject of sociology is laid out in his definition of the discipline: sociology is "a science which attempts the *interpretive understanding* of social action in order thereby to arrive at a causal explanation of its course and effects" (Weber 1947:88, emphasis added). In casting interpretive understanding, or *Verstehen,* as the principal objective of sociology, Weber offered a distinctive counter to those who sought to base sociology on the effort to uncover universal laws applicable to all societies. In short, Weber sought to explain both the objective dimensions of social life (e.g., historical conditions, politico-economic structures, etc.) and the subjective dimension of social life, particularly the states of mind or motivations that guide individuals' behavior.

Schutz sought to expand on Weber's conceptualization of *Verstehen* and interpretive sociology (*verstedhende Soziologie*) by formulating his own concept of meaning. Starting with Weber's conceptualization of action as behavior to which a subjective meaning is attached, and drawing heavily on Husserl (as well as Bergson), Schutz envisions social action as an action oriented toward the past, present, or future behavior of another person or persons. In other words, while agreeing with Weber that social science must be interpretive, Schutz finds that Weber had failed to state clearly the essential characteristics of understanding (*Verstehen*), of subjective meaning (*gemeinter Sinn*), or of action (*Handeln*) (Walsh:xxi). As Schutz (1967:8) maintains,

> Weber makes no distinction between the *action,* considered as something in progress, and the completed *act,* between the meaning of the producer of a cultural object and the meaning of the object produced, between the meaning of my own action and the meaning of another's action, between my own experience and that of someone else, between my self-understanding and my understanding of another person. He does not ask how an actor's meaning is constituted or what modifications this meaning undergoes for his partners in the social world or for a nonparticipating observer. He does not try to identify the unique and fundamental relation existing between the self and the other self, that relation whose clarification is essential to a precise understanding of what it is to know another person. (emphasis in original)

Schutz sets out several interrelated concepts that help clarify the Weberian notion of social action and interpretive understanding. These concepts include lifeworld and intersubjectivity, discussed previously, and **stocks of knowledge**, **recipes**, and **typifications**.

Stocks of knowledge (*Erfahrung*) provide actors with rules for interpreting interactions, social relationships, organizations, institutions, and the physical world. This is the "lower strata" of consciousness that does not receive a "reflective glance"; it consists of what has already been experienced and is thus taken for granted (Schutz 1967:80). For instance, the actor has a stock of knowledge about physical things and fellow creatures and artifacts, including cultural objects. She likewise has a stock of knowledge made up of "syntheses of inner experience." As Schutz states,

> The ordinary man in every moment of his lived experience lights upon past experiences in the storehouse of his consciousness. He knows about the world and he knows what to expect. With every moment of conscious life a new item is filed away in this vast storehouse. At a minimum, this is due to the fact that, with the arrival of a new moment, things are seen in a slightly different light. All of this is involved in the conception of a duration that is manifold, continuous, and irreversible in direction. (1973:81)

Moreover, it is only with a stock of knowledge that one is able to imaginatively explore courses of action other than those he already knows (Schutz 1964–97: 229); although, to be sure, Schutz insists that it is only when the unexpected happens or new situations occur and the taken-for-granted is thrown into question that one is forced to consider alternative schemes of interpretation.

Schutz (1970:98) also refers to stocks of knowledge as "cookery-book knowledge." Just as a cookbook has recipes and lists of ingredients and formulas for making something to eat, so, too, do we all have a "cookbook" of recipes, or implicit instructions, for accomplishing everyday life. Indeed, according to Schutz (1970:99), most of our daily activities, from rising to going to bed, "are performed by following recipes reduced to automatic habits or unquestioned platitudes."

Although Schutz sometimes uses the terms "recipe" and "typification" interchangeably, typification is the process of constructing personal "ideal-types" based on the typical function of people or things rather than their unique features. On one hand, typifications are akin to stereotypes (or stereotypes can be considered a form of typification). In both cases, perceptions are formed based on preconceived categories rather than on all possible information. However, Schutz's conceptualization of typification is more individualistic and interactive than the collectivistic sociological notion of "stereotype." While stereotypes are, by definition, pregiven and somewhat stagnant or fixed, building on Weber's notion of "ideal-type," Schutz emphasizes that typification is a process through which actors isolate the generic characteristics that are relevant for their particular interactive goal. As Schutz (1970:107) states,

> We typify in daily life, human activities which interest us only as appropriate means of bringing about intended effects, but not as emanations of the personality of our fellow-men. . . . In short, the ideal type is but a model of a conscious mind without the faculty of spontaneity and without a will of its own. In typical situations of our daily life, we too, assume certain typical roles. By isolating one of our activities from its interrelations with all the other manifestations of our personality, we disguise ourselves as consumers or taxpayers, citizens, members of a church or of a club, clients, smokers, bystanders, etc. The traveler, for instance, has to behave in the specific way he believes the type "railway agent" to expect from a typical passenger.

Thus, for instance, we use typifications and recipes to know who and how to ask for help in the grocery store. We look around for a "kind-and-knowledgeable-looking shopper" or a "sales clerk," and then we initiate an interaction by saying, "Excuse me, do you know where I can find . . ." (as opposed to just blurting out the item that we need). Most important, the point is that the successful accomplishment of this everyday interaction relies not only on an abstract perception and knowledge of who and how to ask for help, but on our own actualization of recipes and typification as well. That is, we "bracket" or set aside other aspects of our personality, life, and experience in order to carry out the typified role of the "generic" shopper looking for an item in the store. For instance, if I can't find something in the store, I don't begin an interaction by saying, "Hi, I'm Laura, and I teach sociology"—even though that is an important biographical element of my life. I "bracket" that information and dimension of myself in order to carry out the shopper role.

Schutz's point is not only that typifications and recipes are economical and efficient, but also that typification is vital to the very accomplishment of social life. Conscious life (as well as social life) itself relies on us having relatively unreflexive or routine typifications and recipes at our disposal. We cannot "carry on" without bracketing certain features of the social world because otherwise our consciousness would be overloaded. Indeed, Asperger syndrome (a type of autism) can be viewed as an example of the painful consequences that ensue for people unable to typify and/or bracket irrelevant information and stimuli. Asperger syndrome is a neurological disorder marked by deficiencies in social skills. Individuals with Asperger syndrome are often sensitive to sounds, tastes, smells, and sights that seem not to affect others. They often have a great deal of difficulty reading nonverbal cues and determining proper body space. In social interaction, people with Asperger syndrome may demonstrate gaze avoidance or turn away at the same moment as greeting another. They often want to interact with others, but they have trouble knowing how to make it work. From the perspective of a person with

Asperger syndrome, the world is an inconstant, chaotic place, a deafening cacophony of overwhelming sound and an unintelligible jumble of symbols, as Alison Hale, in her autobiography, *My World Is Not Your World* (1998), makes clear:

> My first vivid memories of life are the senseless confusion at play school and the terror of being unable to make sense of my surroundings. I was bewildered by "the vast place of deafening, confusing mush of sound" where "the multi-coloured blurs [children] rushed past sometimes knocking into me" . . .
>
> I spent my primary school life tormented by confusion, bemused by the shapes on a page of printed text "should I read the black bits or the white bits?," puzzled by the riddle of "Why high notes on a piano are not further from the ground than low notes" and baffled by the way people treated me and why I was considered unintelligent.

Hale eventually learned basic social skills in the same way that other people learn to play the piano: with effort and practice. She consciously learned to "make sense" of the world, that is, to decipher and respond to social cues that people without Asperger syndrome pick up intuitively (reflexively) and take for granted simply by being a member of the social group.[3]

In sum, the language we learn and the social structures within which we live provide us with a stockpile of typifications and recipes that make the world both intelligible and manageable. This does not mean, however, that specific elements of the cultural realm are the same for every person. Rather, stocks of knowledge are "biographically articulated"; every individual has a unique stock of knowledge because no two individuals have the same biographical or subjective experience. Thus, for instance, even identical twins, whom obviously have significant genetic and environmental commonalities, have distinct stocks of knowledge and subjectivity.

It is important to note, too, that not only does each of us have our own biographically articulated stock of knowledge but also the elements in our stock of knowledge do not contain the same weight or value in every situation. Schutz uses the terms *umwelt* and *mitwelt* to differentiate various realms of social experience, based on the level of intimacy/immediacy. Specifically, the umwelt is the realm of directly experienced social reality. Umwelt experiences (we relations) are a product of face-to-face relationships and are defined by a high degree of intimacy, as actors are in one another's immediate copresence. Because the pure umwelt (or we relation) is rooted in intimate interaction, it includes indicators of the subjectivity of the other. That is, each person must be aware of the other's body as a field of expression of inner processes of consciousness in a face-to-face situation (Psathas 1989:55). This, in turn, fosters the development of intersubjectivity.

By contrast, the mitwelt (world of contemporaries) is the realm of indirectly experienced social reality. In mitwelt relations, people are experienced only as "types," or within larger social structures, rather than individual actors (e.g., the postal worker who we've never met who delivers our mail). As people grow together (from mitwelt to umwelt relations), typifications and recipes are modified as they are infused with intimate bits of knowledge and actual experience. Conversely, as people grow apart (from mitwelt to umwelt relations), they move from less anonymous to more anonymous interactions, and as a result they become less able to know what goes on in the other person's mind.[4]

Another way to discuss the relationship between our biographically articulated stock of knowledge, typifications, and recipes is to say that individuals do not actually have identical experiences, but rather that we act as if our experiences are, for all practical purposes, equivalent. For instance, even though you and I may go to the same football game, our actual experience of the game (i.e., what we actually see and experience) is different. First, since we inhabit different bodies, it is impossible for us to look at exactly

---

[3]See Barbara L. Kirby, *What Is Asperger Syndrome?* Founder of the OASIS Web site (www.asperger syndrome.org), coauthor of *The Oasis Guide to Asperger Syndrome* (Crown, 2001, revised 2005).

[4]As you saw in Chapter 12, Erving Goffman (1961, 1963, 1969) brilliantly describes and analyzes the characteristics of "face-to-face" versus less personal interaction.

the same things at each and every moment. Moreover, even if we both happen to look at exactly the same thing at exactly the same moment, we are not in the same seat or physical location, thus we are actually seeing the "same" thing but from slightly different perspectives. Most important, however, we do not have the same experience of the game because each of us brings our own biographically articulated stock of knowledge with us to the game. Thus, even if we were to look at exactly the same thing at the same moment from exactly the same vantage point, we would not in fact "see" exactly the same thing. Rather, we would be interpreting and bracketing the action we are witnessing in different ways, based on our unique biographically articulated lifeworld. This is most readily apparent in the difference between how experts and novices view the "same" phenomenon. The individual who really knows the game of football (and who perhaps has actually played football himself) will "see" entirely different categories of phenomena than the uninitiated spectator. While the experienced fan sees attempted versus actualized plays, standard versus unexpected formations, and the like the unversed spectator sees only who's on what side (based on the color of the uniforms) or who has the ball. So, too, the professional musician hears individual instruments, errors, interpretive styles, and patterns of syncopation that the inexperienced listener does not, even though they might be sitting side by side at the "same" recital.

Despite the fact that individuals never experience exactly the same thing, it is still possible to develop intersubjectivity, or shared meaning. Indeed, the very existence of society is dependent on individuals developing a shared meaning of their environment, for this alone allows for the fitting together of actions. In short, intersubjectivity is itself made possible through the process of typification, intercommunication, and language. As Schutz states,

Photo 13.1 Individuals do not have identical experiences of even the "same" phenomena because stocks of knowledge are biographically articulated. Thus, for instance, a dedicated connoisseur will often see (or hear) entirely different categories of phenomena than the uninitiated viewer. Here, an opera connoisseur literally sees more with the help of opera glasses.

> From the outset, we, the actors on the social scene, experience the world we live in as a world both of nature and of culture, not as private but as an intersubjective one, that is, as a world common to all of us, either actually given or potentially accessible to everyone; and this involves intercommunication and language. (1973:53)

In sum, this is the importance of language and shared interpretive schemes: they unite individuals who inevitably have their own experiences into a cohesive whole. In other words, because we cannot access other people's experiences or get inside each other's heads, social life is possible only via shared interpretive schemes and language.

▪▪ SCHUTZ'S THEORETICAL ORIENTATION

As shown in Figure 13.1, Schutz pointedly seeks to illuminate the relationship between the individual and collective levels. On one hand, in the spirit of Durkheim, Schutz emphasizes the pregiven nature of the lifeworld, which reflects a collective approach to order. Schutz maintains that the elements of the lifeworld exist long before our birth, that we are born into particular cultural systems, such that, for instance, our "native language" and "natural attitude" are somewhat predetermined. As Schutz states, "I find myself in my everyday life within a world not of my own making . . . I was born into a pre-organized social world which will survive me, a world shared from the outset with fellow-men who are organized in groups" (1973:329).

On the other hand, however, Schutz emphasizes that we do not merely "internalize" elements of the lifeworld, rather, we experience them; we interpret them, thereby reflecting an individual approach regarding order. We are not simply "vessels" for pregiven cultural forms. In other words, while the inter-subjective dimension of the lifeworld is essential for social relationships and interaction, it nevertheless possesses a "private component." This dual collectivistic/individualistic bent regarding the question of order is also readily apparent in Schutz's notion that stocks of knowledge are "biographically articulated"; that is, they vary from individual to individual because personal and subjective experiences are never exactly the same.

As shown in Figure 13.1, in emphasizing that social life is accomplished only via the intersubjective nature of stocks of knowledge, Schutz takes a primarily nonrational approach regarding action. Schutz stresses how each biographically articulated experience is part of a relatively unconscious, taken-for-granted (and thereby nonrational) lifeworld. We experience the world as one of common and shared objects, events, values, goals, and recipes for acting, and it is this taken-for-granted intersubjectivity that sustains social life, as well as each of our biographically articulated lifeworlds.

Figure 13.1 Schutz's Basic Concepts and Theoretical Orientation

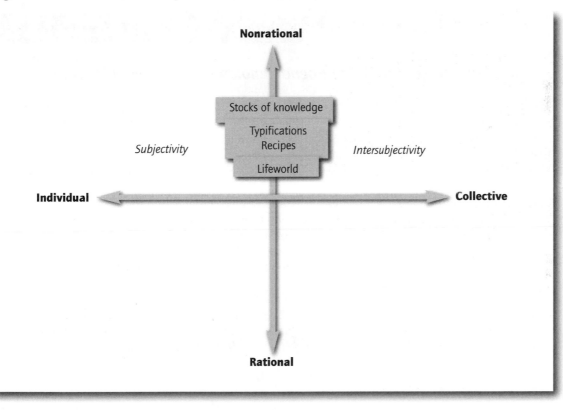

To be sure, in addition to the unreflective, habitual nature of everyday action is the problem-solving aspect of stocks of knowledge. There is a strategic (and thereby rational) dimension to typifications and recipes in that they enable us to "save face" and effectively manage new developments and situations. In short, we use our existing stocks of knowledge to muddle through life. Indeed, as you will shortly see, understanding the methods at our disposal for muddling through everyday situations is a central point taken up by ethnomethodologists such as Harold Garfinkel. For instance, even if we make mistakes in using recipes (e.g., we embrace someone as a greeting when they are not expecting or prepared for that because we are in, for instance, Japan), typifications and recipes enable us to do *something* in new situations.

Nevertheless, overall Schutz clearly emphasizes the taken-for-granted character of the lifeworld much more than the ability of actors to use the elements of the lifeworld to maximize rewards and minimize costs. Indeed, as indicated above, Schutz suggests that it is only when taken-for-granted recipes and typifications don't work that new and creative ideas are then applied. Moreover, Schutz's understanding of rational motivation remains at the level of the individual and not the collective. In fact, Schutz does not have much to say about the collective/rational dimensions of social life at all. As you will see in the following section, this theoretical lacuna (the collective/rational quadrant) is to some extent taken up by Peter Berger and Thomas Luckmann, who add the concept of institutionalization to the phenomenological focus on habitualization and consciousness in order to address social structural concerns. Nevertheless, one of the most important criticisms of both ethnomethodology and phenomenology is that they ignore the social structural constraints that impact both collective cultural life and biographically articulated lifeworlds.

Reading

Introduction to *The Phenomenology of the Social World*

The following selection from *The Phenomenology of the Social World* begins with a discussion of intersubjectivity. Here Schutz explains how it is that we can come to "know" another person even though their "stream of consciousness" is not open to us. How is it that we can come to say that we "understand" him or her? Schutz's answer is twofold. First, he points out how we "put ourselves in the place of the actor and identify our lived experiences with his." Second, he notes that we interpret our own and others' experience using shared signs and interpretive schemes. (As you read in the previous chapter, these two themes are central to symbolic interactionism as well.) In the second part of this excerpt, Schutz focuses on the distinctive characteristics of face-to-face (we) relations versus indirect (they) relations. He argues that the pure "they-orientation" is based on the supposition of characteristics "in the form of a type," and hence actions between contemporaries are only mutually related. By contrast, in the pure "we" relation, actions are mutually interlocked, rooted as they are not in "types" but in immediate experience.

The Phenomenology of the Social World (1967)

Alfred Schutz

INTERSUBJECTIVE UNDERSTANDING

Your whole stream of lived experience is *not* open to me. To be sure, your stream of lived experience is also a continuum, but I can catch sight of only disconnected segments of it. We have already made this point. If I

could be aware of your whole experience, you and I would be the same person. But we must go beyond this. You and I differ from each other not merely with respect to how much of each other's lived experiences we can observe. We also differ in this: When I become aware of a segment of your lived experience, I arrange what I see within my own meaning-context. But meanwhile you have arranged it in yours. Thus I am always interpreting your lived experiences from my own standpoint. Even if I had ideal knowledge of all your meaning-contexts at a given moment and so were able to arrange your whole supply of experience, I should still not be able to determine whether the particular meaning-contexts of yours in which *I* arranged your lived experiences were the same as those which *you* were using. This is because your manner of attending to your experiences would be different from my manner of attending to them. However, if I look at my whole stock of knowledge of your lived experiences and ask about the structure of this knowledge, one thing becomes clear: *This is that everything I know about your conscious life is really based on my knowledge of my own lived experiences. My* lived experiences of you are constituted in simultaneity or quasisimultaneity with *your* lived experiences, to which they are intentionally related. It is only because of this that, when I look backward, I am able to synchronize *my* past experiences of you with *your* past experiences. . . .

BEFORE WE PROCEED FURTHER, it would be well to note that there are ambiguities in the ordinary notion of understanding another person. Sometimes what is meant is intentional Acts directed toward the other self; in other words, my lived experiences of you. At other times what is in question is *your* subjective experiences. Then, the arrangements of all such experiences into meaning-contexts (Weber's comprehension of intended meaning) is sometimes called "understanding of the other self," as is the classification of others' behavior into motivation contexts. The number of ambiguities associated with the notion of "understanding another person" becomes even greater when we bring in the question of understanding the signs he is using. On the one hand, what is understood is the sign itself, then again *what* the other person means by using this sign, and finally the significance of the fact *that* he is using the sign, here, now, and in this particular context. . . .

The man in the natural attitude, then, understands the world by interpreting his own lived experiences of it, whether these experiences be of inanimate things, of animals, or of his fellow human beings. And so our initial concept of the understanding of the other self is simply the concept "our explication of our lived experiences *of* our fellow human beings as such." The fact that the Thou who confronts me is a fellow man and not a shadow on a movie screen—in other words, that he has duration and consciousness—is something I

discover by explicating my own lived experiences of him.

Furthermore, the man in the natural attitude perceives changes in that external object which is known to him as the other's body. He interprets these changes just as he interprets changes in inanimate objects, namely, by interpretation of his own lived experiences of the events and processes in question. Even this second phase does not go beyond the bestowing of meaning within the sphere of the solitary consciousness.

The transcending of this sphere becomes possible only when the perceived processes come to be regarded as lived experiences belonging to another consciousness, which, in accordance with the general thesis of the other self, exhibits the same structure as my own. The perceived bodily movements of the other will then be grasped not merely as *my* lived experience of these movements within *my* stream of consciousness. Rather it will be understood that, simultaneous with *my* lived experience of you, there is *your* lived experience which belongs to you and is part of your stream of consciousness. Meanwhile, the specific nature of your experience is quite unknown to me, that is, I do not know the meaning-contexts you are using to classify those lived experiences of yours, provided, indeed, you are even aware of the movements of your body.

However, I can know the meaning-context into which I classify my own lived experiences of you. We have already seen that this is not your intended meaning in the true sense of the term. What can be comprehended is always only an "approximate value" of the limiting concept "the other's intended meaning." . . .

But, of course, by "understanding the other person" much more is meant, as a rule. This additional something, which is really the only strict meaning of the term, involves grasping what is really going on in the other person's mind, grasping those things of which the external manifestations are mere indications. To be sure, interpretation of such external indications and signs in terms of interpretation of one's own experiences must come first. But the interpreter will not be satisfied with this. He knows perfectly well from the total context of his own experience that, corresponding to the outer objective and public meaning which he has just deciphered, there is this other, inner, subjective meaning. He asks, then, . . ." What does this person mean by speaking to me in this manner, at this particular moment? For the sake of what does he do this (what is his in-order-to motive)? What circumstance does he give as the reason for it (that is, what is his genuine because-motive)? What does the choice of these words indicate?" Questions like these point to the other person's *own* meaning-contexts, to the complex ways in which his own lived experiences have been constituted polythetically and also to the monothetic glance with which he attends to them. . . .

Having established that all genuine understanding of the other person must start out from acts of explication performed by the observer on his own lived experience, we must now proceed to a precise analysis of this genuine understanding itself. From the examples we have already given, it is clear that our inquiry must take two different directions. First we must study the genuine understanding of actions which are performed *without any communicative intent*. . . . Second we would examine cases where such communicative intent was present. The latter type of action involves a whole new dimension, the using[i] and interpreting of signs.

Let us first take actions performed without any communicative intent. We are watching a man in the act of cutting wood and wondering what is going on in his mind. Questioning him is ruled out, because that would require entering into a social relationship[ii] with him, which in turn would involve the use of signs.

Let us further suppose that we know nothing about our woodcutter except what we see before our eyes. By subjecting our own perceptions to interpretation, we know that we are in the presence of a fellow human being and that his bodily movements indicate he is engaged in an action which we recognize as that of cutting wood.

Now how do we know what is going on in the woodcutter's mind? Taking this interpretation of our own perceptual data as a starting point, we can plot out in our mind's eye exactly how *we* would carry out the action in question. Then we can actually imagine ourselves doing so. In cases like this, then, we project the other person's goal as if it were our own and fancy ourselves carrying it out. Observe also that we here project the action in the future perfect tense as completed and that our imagined execution of the action is accompanied by the usual retentions and reproductions of the project, although, of course, only in fancy. Further, let us note that the imagined execution may fulfill or fail to fulfill the imagined project.

Or, instead of imagining for ourselves an action wherein we carry out the other person's goal, we may recall in concrete detail how we once carried out a similar action ourselves. Such a procedure would be merely a variation on the same principle.

In both these cases, we put ourselves in the place of the actor and identify our lived experiences with his. It might seem that we are here repeating the error of the well-known "projective" theory of empathy. For here we are reading our own lived experiences into the other person's mind and are therefore only discovering our own experiences. But, if we look more closely, we will see that our theory has nothing in common with the empathy theory except for one point. This is the general thesis of the Thou as the "other I," the one whose experiences are constituted in the same fashion as mine. But even this similarity is only apparent, for we start out from the general thesis of the other person's flow of duration, while the projective theory of empathy jumps from the mere fact of empathy to the belief in other minds by an act of blind faith. Our theory only brings out the implications of what is already present in the self-explicative judgment "I am experiencing a fellow human being." We know with certainty that the other person's subjective experience of his own action is in principle different from our own imagined picture of what we would do in the same situation. The reason, as we have already pointed out, is that the intended meaning of an action is always in principle subjective and accessible only to the actor. The error in the empathy theory is two-fold. First, it naïvely tries to trace back the constitution of the other self within the ego's consciousness to empathy, so that the latter becomes the direct source of knowledge of the other.[iii] Actually, such a task of discovering the constitution of the other self can only be carried out in a transcendentally phenomenological manner. Second, it pretends to a knowledge of the other person's mind that goes far beyond the establishment of a structural parallelism between that mind and my own. In fact, however, when we are dealing with actions having no communicative intent, all that we can assert about their meaning is already contained in the general thesis of the alter ego.

---

[i][*Setzung;* literally, "positing" or "establishing."]

[ii]The term "social relationship" is here being used in Weber's vague colloquial sense.

[iii]For a critique of the empathy theory see Scheler, *Wesen und Formen der Sympathie,* pp. 277 ff. [E.T., Heath, p. 241].

It is clear, then, that we imaginatively project the in-order-to motive of the other person as if it were our own and then use the fancied carrying-out of such an action as a scheme in which to interpret his lived experiences. However, to prevent misunderstanding, it should be added that what is involved here is only a reflective analysis of another person's completed act. It is an interpretation carried out after the fact. When an observer is directly watching someone else to whom he is attuned in simultaneity, the situation is different. Then the observer's living intentionality carries him along without having to make constant playbacks of his own past or imaginary experiences. The other person's action unfolds step by step before his eyes. In such a situation, the identification of the observer with the observed person is not carried out by starting with the goal of the act as already given and then proceeding to reconstruct the lived experiences which must have accompanied it. Instead, the observer keeps pace, as it were, with each step of the observed person's action, identifying himself with the latter's experiences within a common "we-relationship." We shall have much more to say about this later.

So far we have assumed the other person's bodily movement as the only datum given to the observer. It must be emphasized that, if the bodily movement is taken by itself in this way, it is necessarily isolated from its place within the stream of the observed person's living experience. And this context is important not only to the observed person but to the observer as well. He can, of course, if he lacks other data, take a mental snapshot of the observed bodily movement and then try to fit it into a phantasied filmstrip in accordance with the way he thinks he would act and feel in a similar situation. However, the observer can draw much more reliable conclusions about his subject if he knows something about his past and something about the over-all plan into which this action fits. To come back to Max Weber's example, it would be important for the observer to know whether the woodcutter was at his regular job or just chopping wood for physical exercise. An adequate model of the observed person's subjective experiences calls for just this wider context. We have already seen, indeed, that the unity of the action is a function of the project's span. From the observed bodily movement, all the observer can infer is the single course of action which has directly led to it. If, however, I as the observer wish to avoid an inadequate interpretation of what I see another person doing, I must "make my own" all those meaning-contexts which make sense of this action on the basis of my past knowledge of this particular person. We shall come back later on to this concept of "inadequacy" and show its significance for the theory of the understanding of the other person. . . .

At this point we shall . . . try to recapitulate the complex structures involved in understanding another person insofar as these bear on communication and the use of signs. For to say, as we do, that for the user of the sign the sign stands in a meaning-context involves a number of separate facts which must be disentangled.

First of all, whenever I make use of a sign, those lived experiences signified by that sign stand for me in a meaning-context. For they have already been constituted into a synthesis, and I look upon them as a unit.

In the second place, for me the sign must already be part of a sign system. Otherwise I would not be able to use it. A sign must already have been interpreted before it can be used. But the understanding of a sign is a complicated synthesis of lived experiences resulting in a special kind of meaning-context. This meaning-context is a configuration involving two elements: the sign as object in itself and the *signatum,* each of which, of course, involves separate meaning-contexts in its own right. The total new meaning-context embracing them both we have called the "coordinating scheme" of the sign.

Third, the act of selecting and using the sign is a special meaning-context for the sign-user to the extent that each use of a sign is an expressive action. Since every action comprises a meaning-context by virtue of the fact that the actor visualizes all the successive lived experiences of that action as one unified act, it follows that every expressive action is therefore a meaning-context. This does not mean that every case of sign-using is *ipso facto* a case of communication. A person may, talking to himself for instance, use a sign purely as an act of self-expression without any intention of communication.

Fourth, the meaning-context "sign-using as act" can serve as the basis for a superimposed meaning-context "sign-using as communicative act" without in any way taking into account the particular person addressed.

Fifth, however, this superimposed meaning-context can enter into a still higher and wider meaning-context in which the addressee *is* taken into account. In this case the communicating act has as its goal not merely that someone take cognizance of it but that its message should motivate the person cognizing to a particular attitude or piece of behavior.

Sixth, the fact that this particular addressee is communicated with *here, now,* and *in this way* can be placed within a still broader context of meaning by finding the in-order-to motive of that communicative act.

All these meaning-contexts are in principle open to the interpreter and can be uncovered systematically by him. Just which ones he does seek to inquire into will depend upon the kind of interest he has in the sign.

However, the statement that all these meaning-contexts in principle lie open to interpretation requires some modification. As we have said repeatedly, the structure of the social world is by no means homogeneous. Our fellow men and the signs they use can be given to us in different ways. There are different approaches to the sign and to the subjective experience it expresses. Indeed, we do not even need a sign in order to gain access to another person's mind; a mere indication can offer us the opening. This is what happens, for instance, when we draw inferences from artifacts concerning the experiences of people who lived in the past. . . .

The Structure of the Social World

In the face-to-face situation, directness of experience is essential, regardless of whether our apprehension of the Other is central or peripheral and regardless of how adequate our grasp of him is. I am still "Thou-oriented" even to the man standing next to me in the subway. When we speak of "pure" Thou-orientation or "pure" We-relationship, we are ordinarily using these as limiting concepts referring to the simple givenness of the Other in abstraction from any specification of the degree of concreteness involved. But we can also use these terms for the lower limits of experience obtainable in the face-to-face relationship, in other words, for the most peripheral and fleeting kind of awareness of the other person.

We make the transition from direct to indirect social experience simply by following this spectrum of decreasing vividness. The first steps beyond the realm of immediacy are marked by a decrease in the number of perceptions I have of the other person and a narrowing of the perspectives within which I view him. At one moment I am exchanging smiles with my friend, shaking hands with him, and bidding him farewell. At the

next moment he is walking away. Then from the far distance I hear a faint good-by, a moment later I see a vanishing figure give a last wave, and then he is gone. It is quite impossible to fix the exact instant at which my friend left the world of my direct experience and entered the shadowy realm of those who are merely my contemporaries. As another example, imagine a face-to-face conversation, followed by a telephone call, followed by an exchange of letters, and finally messages exchanged through a third party. Here too we have a gradual progression from the world of immediately experienced social reality to the world of contemporaries. In both examples the total number of the other person's reactions open to my observation is progressively diminished until it reaches a minimum point. It is clear, then, that the world of contemporaries is itself a variant function of the face-to-face situation. They may even be spoken of as two poles between which stretches a continuous series of experiences. . . .

We tend to picture marriage or friendship as primarily face-to-face relationships, especially intimate ones at that. We do this because of a tendency we have to conceive the actions of the partners as integrated into the larger unity of the relationship and goal-directed toward that unity.

In actual life, however, a marriage or a friendship is made up of many separate events occurring over a long period of time. Some of these events involve face-to-face situations, in others the partners simply exist side by side as contemporaries. To call such social relationships as these "continuous" is erroneous in the extreme,[iv] since discontinuity and repeatability are included in their very definition. What, then, do friends mean when they speak of their "friendship"? We can distinguish three different meanings they may have in mind.

1. When *A* speaks of his friendship with *B,* he may be thinking of a series of past face-to-face relationships which he shared with *B.* We say "series," because *A* does remember that during the course of his friendship with *B* he did spend some time alone or with other people.

2. When *A* speaks of his friendship with *B,* he may mean that, over and above such face-to-face situations, his behavior is oriented to *B*'s expected

---

[iv][There is an unfortunate linguistic ambiguity here. A friendship, it is true, is (happily) not a continu*ous* series of contacts in the Cantorian sense that between any two contacts there is another. It is a series of continu*al* or recurring contacts. But, although it is not a continuous series, it can be spoken of as a continu*ous relationship* unless every *au revoir* is a temporary "breaking-off" of the friendship.]

behavior or to the fact that B exists—that he is the kind of man he is. In this case, A is oriented toward B as a contemporary, and their relationship is the kind that exists between contemporaries. This relationship can be either one of orientation or of social interaction.[v] For instance, A may perform a certain action because he thinks it will please B as soon as the latter finds out about it. Whereas in the face-to-face situation he would literally see B's reaction, here he is confined to merely imagining it. Within the "friendship" such contemporary- oriented acts are inserted between consociate-oriented acts. Face-to-face interaction involves mutual engagement in which the partners can witness the literal coming-to-birth of each other's experiences. Interaction between contemporaries, however, merely involves the expectation on the part of each partner that the other will respond in a relevant way. But this expectation is always a shot in the dark compared to the knowledge one has of one's consociate in the face-to-face situation. Actions between contemporaries are only mutually *related,* whereas actions between consociates are mutually *interlocked.*[vi] The being related to each other of contemporaries occurs in imagination, whereas the interlocking mutual engagement of the We-relationship is a matter of immediate experience. Between these two situations we find many intermediate degrees. For instance, think of the gradually decreasing immediacy of the following: (a) carrying on an imagined conversation with a friend, (b) wondering what my friend would say if I were to do such and such, (c) doing something "for him."

3. When A speaks of his friendship with $B,$ he may be referring to the fact that, external obstacles aside, they can always get together again and begin where they have left off. This is parallel to what happens in the sphere of judgment. We showed in our analysis of the concept "knowledge" that the latter refers to a sum of already constituted objectified judgments [or judgment-objectivities—*Urteilsgegen ständlichkeiten*]. Knowledge, then, is a storehouse which can be drawn on at any time by the reactivation of the judgments in question. In the same way, when A speaks of his friendship with $B,$ he is referring to a storehouse of past experiences of $B.$ But he is assuming at the same time that these experiences can be reactivated in a revived We-relationship and that, on that basis, both parties can proceed as before. What is here revived, of course, is not so much the specific lived experiences that previously occurred within the We-relationship but the lived experience of the We-relationship itself.

In the last few pages we have been describing the intermediate zone between the face-to-face situation and the situation involving mere contemporaries. Let us continue our journey. As we approach the outlying world of contemporaries, our experience of others becomes more and more remote and anonymous. Entering the world of contemporaries itself, we pass through one region after another: (1) the region of those whom I once encountered face to face and could encounter again (for instance, my absent friend); then (2) comes the region of those once encountered by the person I am now talking to (for instance, your friend, whom you are promising to introduce to me); next (3) the region of those who are as yet *pure* contemporaries but whom I will soon meet (such as the colleague whose books I have read and whom I am now on my way to visit); then (4) those contemporaries of whose existence I know, not as concrete individuals, but as points in social space as defined by a certain function (for instance, the postal employee who will process my letter); then (5) those collective entities whose function and organization I know while not being able to name any of their members, such as the Canadian Parliament; then (6) collective entities which are by their very nature anonymous and of which I could never in principle have direct experience, such as "state" and "nation"; then (7) objective configurations of meaning which have been instituted in the world of my contemporaries and which live a kind of anonymous life of their own, such as the interstate commerce clause and the rules of French grammar; and finally (8) artifacts of any kind which bear witness to the subjective meaning-context of some unknown person. The farther out we get into the world of contemporaries, the more anonymous its inhabitants become, starting with the innermost region, where

---

[v] The different forms of orientation relationships and social interaction in the world of contemporaries remain to be described exactly.

[vi] ["Aufeinanderbezogen . . . aufeinander eingestellt."]

they can almost be seen, and ending with the region where they are by definition forever inaccessible to experience. . . .

MY MERE CONTEMPORARY (or "contemporary"), then, is one whom I know coexists with me in time but whom I do not experience immediately. This kind of knowledge is, accordingly, always indirect and impersonal. I cannot call my contemporary "Thou" in the rich sense that this term has within the We-relationship. Of course, my contemporary may once have been my consociate or may yet become one, but this in no way alters his present status.

Let us now examine the ways in which the world of contemporaries is constituted and the modifications which the concepts "Other- orientation" and "social relationship" undergo in that world. These modifications are necessitated by the fact that the contemporary is only indirectly accessible and that his subjective experiences can only be known in the form of *general types* of subjective experience.

That this should be the case is easy to understand if we consider the difference between the two modes of social experience. When I encounter you face to face I know you as a person in one unique moment of experience. While this We-relationship remains unbroken, we are open and accessible to each other's intentional Acts. For a little while we grow older together, experiencing each other's flow of consciousness in a kind of intimate mutual possession.

It is quite otherwise when I experience you as my contemporary. Here you are not prepredicatively given to me at all. I do not even directly apprehend your existence (*Dasein*). My whole knowledge of you is mediate and descriptive. In this kind of knowledge your "characteristics" are established for me by inference. From such knowledge results the indirect We-relationship. . . .

THE THEY-ORIENTATION is the pure form[vii] of understanding the contemporary in a predicative fashion, that is, in terms of his typical characteristics. Acts of They-orientation are, therefore, intentionally directed toward another person imagined as existing at the same time as oneself but conceived in terms of an ideal type. And just as in the cases of the Thou-orientation and the We-relationship, so also with the They- orientation can

we speak of different *stages of concretization* and *actualization.*

In order to distinguish from one another the various stages of concretization of the We-relationship, we established as our criterion the degree of closeness to direct experience. We cannot use this criterion within the They- orientation. The reason is that the latter possesses by definition a high degree of remoteness from direct experience, and the other self which is its object possesses a correspondingly higher degree of anonymity.

It is precisely this degree of anonymity which we now offer as the criterion for distinguishing between the different levels of concretization and actualization that occur in the They-orientation. The more anonymous the personal ideal type applied in the They-orientation, the greater is the use made of objective meaning-contexts instead of subjective ones, and likewise, we shall find, the more are lower- level personal ideal types and objective meaning-contexts pregiven. (The latter have in turn been derived from other stages of concretization of the They-orientation.)

Let us get clear as to just what we mean by the anonymity of the ideal type in the world of contemporaries. The pure Thou-orientation consists of mere awareness of the existence of the other person, leaving aside all questions concerning the characteristics of that person. On the other hand, the pure They-orientation is based on the presupposition of such characteristics in the form of a type. Since these characteristics are genuinely typical, they can in principle be presupposed again and again. Of course, whenever I posit such typical characteristics, I assume that they now exist or did once exist. However, this does not mean that I am thinking of them as existing in a particular person in a particular time and place. The contemporary alter ego is therefore anonymous in the sense that its existence is only the individuation of a type, an individuation which is merely supposable or possible. Now since the very existence of my contemporary is always less than certain, any attempt on my part to reach out to him or influence him may fall short of its mark, and, of course, I am aware of this fact.

The concept which we have been analyzing is the concept of the anonymity of the partner in the world of contemporaries. It is crucial to the understanding of the nature of the indirect social relationship.

[vii][*Die Leerform,* literally, "the empty form."]

PETER BERGER (1929–) AND THOMAS LUCKMANN (1927–): BIOGRAPHICAL SKETCHES

Peter Berger

Peter Berger was born in Vienna in 1929. He studied for a short time in London and then emigrated to the United States shortly after World War II. He received his bachelor's degree in philosophy from Wagner College on Staten Island, New York, in 1949, and both his master's and doctorate degrees in sociology from the New School for Social Research. The New School faculty included at the time a number of illustrious German expatriates including Alfred Schutz. These men were not only well versed in the works of Karl Marx and Max Weber but also fluent in a number of languages and an array of fields that stretched from classical to modern thought (Fernandez 2003:237). In addition to his exposure to a broad liberal arts education, Berger was a fervent student of religion. He continuously contemplated his own Christian beliefs and spent a "very happy" year at the Lutheran Theological Seminary in Philadelphia, where he studied to be a minister (ibid.). Indeed, Berger is just as well known for his work in the sociology of religion as in phenomenology and the sociology of knowledge.[5] His now-classic *Invitation to Sociology* (1963) continues to be one of the most acclaimed and inspiring introductions to the discipline of sociology today. Professor Berger has taught at a number of universities including The New School; Evangelische Akademie in Bad Boll, Germany; the University of North Carolina; the Hartford Theological Seminary; Rutgers University; and Boston College. Since 1981, Professor Berger has been professor of sociology and theology at Boston University. He is also currently director of the Institute on Culture, Religion and World Affairs at Boston University.

Thomas Luckmann

Thomas Luckmann was born in 1927 in Slovenia. He was educated both in Europe and in the United States, studying at the Universities of Vienna and Innsbruck and the New School for Social Research. Luckmann's first major sole-authored publication, *The Invisible Religion* (original title *Das problem der Religion,* 1963), did not appear in English until the year after the publication of Berger and Luckmann's groundbreaking *The Social Construction of Reality* (1966), and he never became quite as well known in the United States as either his teacher, Alfred Schutz, or his collaborator, Peter Berger. Luckmann's "unequal" relationship with Schutz is duly noted by Luckmann himself in the preface to *The Structures of the Life World* (Schutz and Luckmann 1973), which he finished editing after Schutz's death:

> The completion of the *Strukturen der Lebenswelt* combined the difficulties of the posthumous editing of the manuscript of a great teacher by his student with the problems of collaboration between two unequal authors (one dead, the other living), one looking back at the results of many years of singularly concentrated efforts devoted to the resolution of the problems that were to be dealt with in the book, the other the beneficiary of these efforts; one a master, always ready to revise his analyses but now incapable of doing so, the other a pupil, hesitant to revise what the master had written but forced by the exigencies of the analyses that he continued in the direction indicated by the master to go back, occasionally, to the beginning. (1973:xvii ff.)

Like Berger, Luckmann has taught at a variety of universities throughout the world, including in New York, Massachusetts, California, Frankfurt, and Constance, Germany. Almost 30 years after Peter Berger and Thomas Luckmann wrote *The Social Construction of Reality,* they coauthored a second book, *Modernity, Pluralism and the Crisis of Meaning* (1995).

---

[5] Among Berger's prodigious work in the sociology of religion are *The Precarious Vision* (1961), *The Sacred Canopy* (1967), *A Far Glory* (1992), *The Desecularization of the World* (1999), and *Questions of Faith* (2003).

Berger and Luckmann extend Schutz's phenomenology by melding it with the pivotal ideas of Émile Durkheim, Max Weber, Karl Marx, and George Herbert Mead. Berger and Luckmann also draw extensively from philosophical anthropology, especially the work of Arnold Gehlen and Helmuth Plessner. The result is a theoretically multidimensional and provocative phenomenological sociology, captured, above all, in Berger and Luckmann's famous assertion that "*society is a human product. Society is an objective reality. Man is a social product*" (1966:61, emphasis in original).

From Gehlen and Plessner, Berger and Luckmann borrow the idea that humans, unlike other animals, are "instinctually deprived" or biologically underdeveloped (Berger and Luckmann 1966:48). Important organismic developments that take place in the womb in other animals take place in humans' first year of life. This means not only that the survival of the human infant is dependent on certain social arrangements, but that lacking an instinctual basis for action, human beings have to create a world that ensures social stability. Commonsense knowledge and social institutions compensate for biological underdevelopment. They provide a "base" that operates "automatically" (analogous to the instincts that guide other animals' behavior). "Commonsense knowledge is the knowledge that I share with others in the normal, self-evident routines of everyday life" (Berger and Luckmann 1966:23). It is what allows us to perceive the reality of everyday life as "reality," to suspend our doubts so that we can act in the world. Social institutions are the bridges between humans and their physical environments (B. Turner 2001:109). Following Schutz, Berger and Luckmann emphasize that it is the intersubjective character of commonsense knowledge that enables human institutions and culture to produce stability. It is because "most of the time, my encounters with others in everyday life are typical in a double sense—I apprehend the other *as* a type and I interact with him in a situation that is itself typical" that social interaction is successful (Berger and Luckmann 1966:31, emphasis in original). Without intersubjectivity—that you know that I know that we both know—social order and interaction would break down, as we would be left to doubt the most fundamental aspects of communication.

This brings us to the issue of **habitualization**, that is, the process by which the flexibility of human actions is limited. All activity is subject to habitualization, as repeated actions inevitably become routinized. Habitualization carries with it the psychological advantage that choices are narrowed. That an action may be "performed again in the future in the same manner and with the same economical effort" provides a stable background from which human activity can proceed (Berger and Luckmann 1966:53–54). In other words, from the time we wake up in the morning until we go to bed at night, we can direct our minds and bodies to constructive action only because we take most actions for granted.

Moreover, habitualized actions set the stage for **institutionalization**, for "institutionalization occurs whenever there is a reciprocal typification of habitualized action by types of actors" (Berger and Luckmann 1966:54). That is, it is when habitualized actions are shared and/or "available to all members of the particular social group" (ibid.) that institutions are born. Akin to habits that function at the level of the individual, then, institutions control human conduct by setting up predefined patterns of conduct that channel in one direction as opposed to another theoretically possible direction. Of course, institutions are not created instantaneously, but rather are "built up in the course of a shared history" (ibid.). In other words, over time, shared habitualized actions become institutions that are taken for granted and therefore limiting for the individuals who are subject to them. Thus, it is through institutions that human life becomes coherent, meaningful, and continuous (ibid.).

Thus, it is through the dual processes of habitualization and institutionalization that the systems of meaning and social institutions that humans create act back on them as a fixed "objective reality." Interestingly, however, we seldom ponder the fact that our institutions are arbitrarily rooted in shared typifications; it is only when the taken-for-granted elements of our lifeworld are disrupted (e.g., when we travel abroad) that we think about them at all. Most often, the system appears to us (and we act) as if it is a fixed, "objective" reality.

Consider, for instance, driving a car. At the level of the individual, learning to drive entails learning to habitualize specific actions, one of the most important of which is knowing which pedal is the gas and which is the brake and where the brake is in the car. You must automatically know this, for if you had to stop and think, "Which pedal is the brake?" every time you needed to brake, or if you had to stop and think "Where is the brake?" (and consider all theoretically possible locations for the brake in the vehicle), you would most certainly crash, for you don't have time for hemming and hawing and contemplation while driving a car. Of course, in terms of institutionalization, if we each had our own solitary driving habits that shared no common features, the result would be chaos. It is only to the extent that our typifications and recipes for driving are shared (or "reciprocal") that the traffic system works and that order is maintained at all.

Berger and Luckmann use the terms **externalization**, **objectivation**, and **reification** to refer to the process by which human activity and society attain the character of objectivity. Externalization and objectivation enable the actor to confront the social world as something outside of herself. For instance, institutions appear external to the individual, as historical and objective facticities. They confront the individual as undeniable facts (Berger and Luckmann 1966:60). Reification is "an extreme step" in the process of objectivation. In reification, "the real relationship between man and his world is reversed in consciousness. Man, the producer of a world, is apprehended as its product, and human activity as an epiphenomenon of non-human processes." That is, reification is the apprehension of human phenomena as if they were "non-human or possibly suprahuman" things (Berger and Luckmann 1966:89). For instance, we reify our social roles in such a way that we say, "I have no choice in the matter. I have to act this way" (ibid.:91). This is what Berger refers to as "bad faith." Of course, history is full of examples of the horrendous consequences that ensue from such reification. The Nazi concentration camps relied on guards who are said to have merely "taken orders." A parallel also can be drawn with the recent example of torture in the Abu Ghraib prison.

Objectivation and reification are related to the Marxist concept of alienation (Berger and Luckmann 1966:197, 200). Here Berger and Luckmann are consciously drawing on Marx in order to underscore the dialectical relationship between substructure and superstructure, that is, the intimate relationship between the material and ideal realms. For instance, in a famous passage in *The German Ideology* (1845), Marx maintains that the material or economic basis of religious systems is obscured for believers much like "the inversion of objects on the retina."[6] For Marx, religion is rooted in material processes (namely, the ruling class's interest in maintaining its dominance) that are experienced as reified "Facts" and fixed "Truths" by the proletariat. Similarly, Berger and Luckmann emphasize that human beings create systems of meaning and ideologies that they then experience as "objective" and "suprahuman." And just as Marx perceived religious beliefs as having the same effect as an opiate—providing the masses with "comfort" and a cushion to shield them from the harsh realities of everyday life—so, too, Berger and Luckmann argue that "human actors prefer reification to anomie, because the former offers comfort through amnesia"; the "burden of the homeless mind cannot be easily endured" (B. Turner 2001:111).

However, in stark contrast to Marx, Berger and Luckmann perceive the process of reification as not necessarily alienating. While Marx likened reification to a delusion that must be dispelled in order to create a fully human, liberated society, Berger and Luckmann view reification as inherent to the human condition. Institutionalization, objectivation, and reification are simply the processes by which man becomes "capable of producing a world that he then experiences as something other than a human product" (B. Turner 2001:61). They are what enable social life to be both meaningful and continuous.

The notion that objectivation is tied to shared cultural systems vital to social life reflects Berger and Luckmann's indebtedness to Durkheim (rather than to Marx). Durkheim ([1895] 1966:13) used the term "social fact" to refer to any manner of action capable of exercising "exterior constraint" over the individual, that is, "*every way of acting which is general throughout a given society, while at the same time existing in its own right independent of its individual manifestations*" (ibid., emphasis in original), and he defined sociology as the study of social facts. In other words, just as a game has rules that are "coercive"

---

[6]Marx states, "If in all ideology men and their circumstances appear upside-down as in a camera obscura, this phenomenon arises just as much from historical life-process as the inversion of objects on the retina does from their physical life-process" (in Robert C. Tucker, ed., *The Marx-Engels Reader,* 2nd ed. New York: Norton, 1978:154).

for its players, so, too, are societies made up of "social facts" that are "coercive." In the process of participating in society (or playing a game), the rules/social structures are iterated and learned and unwittingly upheld. Thus, for instance, in the case of learning to drive a car discussed above, once we learn to drive we take the rules of the road (e.g., which side of the road to drive on) for granted as we go about our business (e.g., where it is we're going). However, if we ever do "forget" and get a ticket for not following the law, this rule/law for us is again reaffirmed.

Indeed, in his later work, Durkheim ([1912] 1965) maintains that the primary function of religion is to encode the system of relations of the group, and that through the dual processes of ritualization and symbolization collective sentiments are reaffirmed. As Durkheim ([1912] 1965:474–75) states,

> There can be no society which does not feel the need of upholding and reaffirming at regular intervals the collective sentiments and the collective ideas which make its unity and personality. Now this moral remaking cannot be achieved except by the means of reunions, assemblies, and meetings where the individuals, being closely united to one another, reaffirm in common their common sentiments.

Thus, highly routinized acts, such as taking communion or saluting the flag, are rituals through which participants are bound together and their collective beliefs and sentiments are reaffirmed. In addition, collective sentiments and beliefs are reaffirmed through symbolization. Symbols (defined as something that stands for something else), such as the Christian cross, are markers that are capable of calling up and reaffirming shared meaning and the feeling of community in between periodic ritual acts (such as religious celebrations and church services). As Durkheim ([1912] 1965:232) states, "without symbols, social sentiments could have only a precarious existence."

In the spirit of Durkheim, Berger and Luckmann underscore that shared meanings and categories are a "sacred canopy" vital to social existence (Berger 1967). And just as Durkheim emphasized how systems of meaning tie individuals together in a system that they experience as "objective" reality (i.e., social facts), so, too, do Berger and Luckmann (1966:58) emphasize that institutions are "experienced as possessing a reality of their own, a reality that confronts the individual as an external and coercive fact." Moreover, akin to Durkheim's emphasis on symbolization, so, too, Berger and Luckmann (1966:35) stress that signification (the human production of signs) is simply one of the most important examples of objectivation. As Berger (1967:18–19) states,

> Language confronts the individual as an objective facticity. He subjectively appropriates it by engaging in linguistic interaction with others. In the course of this interaction, however, he inevitably modifies the language. . . . Furthermore, his continuing participation in the language is part of the human activity that is the only ontological base for the language in question. The language exists because he, along with others, continues to employ it. In other words, both with regard to language and to the socially objectivated world as a whole, it may be said that the individual keeps "talking back" to the world that formed him and thereby continues to maintain the latter as reality.

Nevertheless, as this discussion of language reflects, in contrast to both Marx and Durkheim, Berger and Luckmann stress that "the individual is not molded as an inert and passive thing" (Berger 1967:18). They emphasize the active role that individuals take in maintaining the world that they experience as an objective reality; that is, they stress the experience of institutions over their external, empirical "reality." This focus on everyday life reflects Berger and Luckmann's indebtedness not only to Schutz but also to symbolic interactionism, particularly to the work of George Herbert Mead (see the previous chapter). In conjunction with the microlevel emphasis of Mead, Berger and Luckmann focus on the dynamics of face-to-face interaction. They explain, for instance, how during interaction actors "A and B . . . play roles vis-à-vis each other. . . . That is, A will inwardly appropriate B's reiterated roles and make them the models for his own role-playing." As Berger and Luckmann (1966:22) state, "The reality of everyday life is organized around the 'here' of my body and the 'now' of my present. This 'here' and 'now' is the focus of my attention to the reality of everyday life."

This brings us to the final step in the dynamic process of the social construction of reality: **internalization**. Internalization is "the immediate apprehension or interpretation of an objective event as expressing meaning" (1966:129), that is, the process through which individual subjectivity is attained. Internalization means that "the objectivated social world is retrojected into consciousness in the course of socialization" (ibid.:61). As such, internalization is the "beginning point" in the process of becoming a member of society (ibid.:129), as well as the "end point" in institutionalization (see Figure 13.2).

These moments of externalization, objectivation, and internalization are not to be understood "as occurring in a temporal sequence," but rather as a simultaneous, dialectical process. Nevertheless, it is in intergenerational transmission that the process of internalization is complete. As Berger and Luckmann (1966:61) maintain,

> only with the transmission of the social world to a new generation (that is, internalization as effectuated in socialization) does the fundamental social dialectic appear in its totality. To repeat, only with the appearance of a new generation can one properly speak of a social world.

In other words, every individual is born into an environment within which she encounters the significant others who are in charge of her socialization. One does not choose one's own significant others; rather, they are imposed on her. In the process of socialization, the stocks of knowledge that the individual experiences as a preexisting objective reality are imposed on her. The individual is thereby "born into not only an objective social structure but also an objective social world" (1966:131).

Berger and Luckmann differentiate two types of socialization based on the extent to which individuals are active and conscious of the process of internalization. **Primary socialization** refers to "the first socialization an individual undergoes in childhood, through which he becomes a member of society" (1966:130–31). On the other hand, **secondary socialization** refers to subsequent processes of socialization that induct "an already socialized individual into new sectors of the objective world of his society"

Figure 13.2 Berger and Luckmann's Basic Concepts and Theoretical Orientation

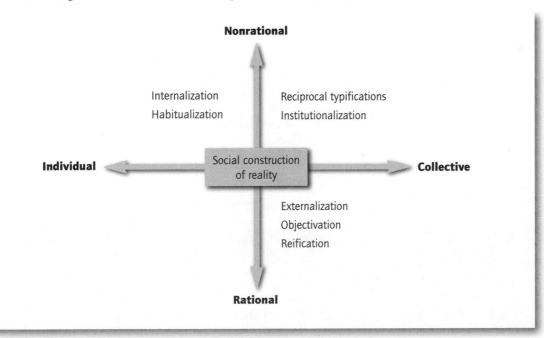

(ibid.). Whereas primary socialization is predefined and taken for granted, secondary socialization is acquired in a more conscious way (e.g., training for a new job). It is for this reason that primary socialization has so much more of an impact on the individual than secondary socialization. As Berger and Luckmann (1966:134, emphasis in original) state,

> The child does not internalize the world of his significant others as one of many possible worlds. He internalizes it as *the* world, the only existent and only conceivable world, the world *tout court*. It is for this reason that the world internalized in primary socialization is so much more firmly entrenched in consciousness than worlds internalized in secondary socialization.

Furthermore, primary socialization is distinguished by the fact that it cannot take place without an emotionally shared identification of the child with his significant others: you have to love your mother, but not your teacher (1966:141). This distinction between the more intimate (primary) and less intimate (secondary) types of socialization recalls Schutz's more abstract discussion of umwelt versus mitwelt relations. Each type of relationship is distinguished by a different level of intersubjectivity and typification. Primary socialization and significant others (essential to "we relations") are far more central to the maintenance of "identity" than are secondary relationships/socialization (Berger and Luckmann 1966:151).

▪▪ BERGER AND LUCKMANN'S THEORETICAL ORIENTATION

Berger and Luckmann (1966:18) explicitly seek to illuminate "the dual character of society in terms of objective facticity *and* subjective meaning" (emphasis in original). In so doing, they move back and forth between individual and collective approaches to order. Specifically, as shown in Figure 13.2, Berger and Luckmann emphasize both how individual actors shape social patterns and how social patterns structure individual behavior and consciousness. They argue that institutions "control human conduct by setting up predefined patterns of conduct" (ibid.:55), but at the same time, they assert that it is only to the extent that such patterns of conduct are internalized and taken for granted by the individual that they shape consciousness and behavior.

Of course, this dual approach to order is a function of Berger and Luckmann's reliance both on Schutz's phenomenology and on the classical sociologists Marx, Weber, Durkheim, and Mead. While as we have seen, Schutz and Mead are most attuned to the level of the individual, Durkheim, Marx, and Weber are particularly attuned to the collective realm. In addition, from Mead and symbolic interactionism, Berger and Luckmann borrow an appreciation for the pragmatic dimensions of everyday consciousness and action, which also reflects a more individualistic approach, as it emphasizes individuals' attempts to make sense of or give meaning to the situations that comprise their everyday life.

In terms of action, above all Berger and Luckmann emphasize the *taken-for-granted* workings of society, which reflects a primarily nonrational theoretical approach. As discussed previously, at the level of the individual, this emphasis on nonrational motivation reflects Berger and Luckmann's indebtedness to Schutz and symbolic interactionism. Rather than stress the strategic, calculative aspects of human behavior, they assert the extent to which we act according to internalized typifications, recipes, and lifeworlds (see Figure 13.2). At the level of the collective, Berger and Luckmann emphasize the shared, intersubjective dimensions of symbolic life (à la Durkheim). Indeed, Berger and Luckmann insist that it is only by using the preexisting, intersubjective elements of language and culture that individuals are able to act at all. This emphasis on preexisting, taken-for-granted systems of meaning can be said to reflect a primary interest in the collective/nonrational realm.

This is not to say that Berger and Luckmann ignore the rational dimension of action entirely, however. As shown in Figure 13.2, objectivation and reification are concepts rooted in Marx that exude a rationalistic bent at the collective level. That is, Berger and Luckmann's whole point is that the power and force of social as well as culture structures is that they are perceived as forces outside and external to the individual.

Reading

Introduction to *The Social Construction of Reality*

Berger and Luckmann's *The Social Construction of Reality* (1966) is one of the most widely read and influential books in contemporary sociology. In the following excerpts, you will see many of the themes discussed above, including the Schutzian concepts of intersubjectivity, typification, and language, as well as Berger and Luckmann's own analyses of habitualization, institutionalization, externalization, objectivation, reification, internalization, and socialization. The selection begins with Berger and Luckmann's fundamental point that "the reality of everyday life appears already objectified" (ibid.:21) and that this objectivation is *intersubjective,* as is most readily apparent in language and sign systems. In the next section, Berger and Luckmann focus on the dual processes of habitualization and institutionalization, emphasizing that it is the lack of human instinctual parameters that necessitates the development of sociocultural limits on social behavior. Berger and Luckmann then move on to viewing society as "subjective reality," emphasizing the processes of internalization and socialization at the level of the individual. Here Berger and Luckmann stress that "no individual internalizes the totality of what is objectivated as reality in his society . . . [and that] there are always elements of subjective reality that have not originated in socialization, such as the awareness of one's own body prior to and apart from any socially learned apprehension of it" (ibid.:134). Here Berger and Luckmann also insist that "the symmetry between objective and subjective reality is never a static, one-for-all state of affairs. It must always be produced and reproduce *in actu*" (ibid.), which reflects their dual individualistic and collectivistic approach.

The Social Construction of Reality (1966)

Peter L. Berger and Thomas Luckmann

FOUNDATIONS OF KNOWLEDGE IN EVERYDAY LIFE

I apprehend the reality of everyday life as an ordered reality. Its phenomena are prearranged in patterns that seem to be independent of my apprehension of them and that impose themselves upon the latter. The reality of everyday life appears already objectified, that is, constituted by an order of objects that have been designated *as* objects before my appearance on the scene. The language used in everyday life continuously provides me with the necessary objectifications and posits the order within which these make sense and within which everyday life has meaning for me. I live in a place that is geographically designated; I employ tools, from can openers to sports cars, which are designated in the technical vocabulary of my society; I live within a web of human relationships, from my chess club to the United States of America, which are also ordered by means of vocabulary. In this manner language marks the co-ordinates of my life in society and fills that life with meaningful objects.

The reality of everyday life is organized around the "here" of my body and the "now" of my present. This "here and now" is the focus of my attention to the reality of everyday life. What is "here and now" presented to me in everyday life is the *realissimum* of my consciousness. The reality of everyday life is not, however, exhausted by these immediate presences, but embraces phenomena that are not present "here and now." This means that I experience everyday life in terms of differing degrees of closeness and remoteness, both spatially and temporally. Closest to me is the zone of everyday life that is directly accessible to my bodily manipulation. This zone contains the world within my reach, the world in which I act so as to modify its reality, or the world in which I work. In this world of working my consciousness is dominated by the pragmatic motive, that is, my attention to this world is mainly determined by what I am doing, have done or plan to do in it. In this way it is *my* world par excellence. I know, of course, that the reality of everyday life contains zones that are not accessible to me in this manner. But either I have no pragmatic interest in these zones or my interest in them is indirect insofar as they may be, potentially, manipulative zones for me. Typically, my interest in the far zones is less intense and certainly less urgent. I am intensely interested in the cluster of objects involved in my daily occupation—say, the world of the garage, if I am a mechanic. I am interested, though less directly, in what goes on in the testing laboratories of the automobile industry in Detroit—I am unlikely ever to be in one of these laboratories, but the work done there will eventually affect my everyday life. I may also be interested in what goes on at Cape Kennedy or in outer space, but this interest is a matter of private, "leisure-time" choice rather than an urgent necessity of my everyday life.

The reality of everyday life further presents itself to me as an intersubjective world, a world that I share with others. This intersubjectivity sharply differentiates everyday life from other realities of which I am conscious. I am alone in the world of my dreams, but I know that the world of everyday life is as real to others as it is to myself. Indeed, I cannot exist in everyday life without continually interacting and communicating with others. I know that my natural attitude to this world corresponds to the natural attitude of others, that they also comprehend the objectifications by which this world is ordered, that they also organize this world around the "here and now" of *their* being in it and have projects for working in it. I also know, of course, that the others have a perspective on this common world that is not identical with mine. My "here" is their "there." My "now" does not fully overlap with theirs. My projects differ from and may even conflict with theirs. All the same, I know that I live with them in a common world. Most importantly, I know that there is an ongoing correspondence between *my* meanings and *their* meanings in this world, that we share a common

sense about its reality. The natural attitude is the attitude of commonsense consciousness precisely because it refers to a world that is common to many men. Commonsense knowledge is the knowledge I share with others in the normal, self-evident routines of everyday life.

The reality of everyday life is taken for granted *as* reality. It does not require additional verification over and beyond its simple presence. It is simply *there,* as self-evident and compelling facticity. I *know* that it is real. While I am capable of engaging in doubt about its reality, I am obliged to suspend such doubt as I routinely exist in everyday life. This suspension of doubt is so firm that to abandon it, as I might want to do, say, in theoretical or religious contemplation, I have to make an extreme transition. The world of everyday life proclaims itself and, when I want to challenge the proclamation, I must engage in a deliberate, by no means easy effort. The transition from the natural attitude to the theoretical attitude of the philosopher or scientist illustrates this point. But not all aspects of this reality are equally unproblematic. Everyday life is divided into sectors that are apprehended routinely, and others that present me with problems of one kind or another. Suppose that I am an automobile mechanic who is highly knowledgeable about all American-made cars. Everything that pertains to the latter is a routine, unproblematic facet of my everyday life. But one day someone appears in the garage and asks me to repair his Volkswagen. I am now compelled to enter the problematic world of foreign-made cars. I may do so reluctantly or with professional curiosity, but in either case I am now faced with problems that I have not yet routinized. At the same time, of course, I do not leave the reality of everyday life. Indeed, the latter becomes enriched as I begin to incorporate into it the knowledge and skills required for the repair of foreign-made cars. The reality of everyday life encompasses both kinds of sectors, as long as what appears as a problem does not pertain to a different reality altogether (say, the reality of theoretical physics, or of nightmares). As long as the routines of everyday life continue without interruption they are apprehended as unproblematic.

But even the unproblematic sector of everyday reality is so only until further notice, that is, until its continuity is interrupted by the appearance of a problem. When this happens, the reality of everyday life seeks to integrate the problematic sector into what is already unproblematic. Commonsense knowledge contains a variety of instructions as to how this is to be done. . . .

SOCIAL INTERACTION IN EVERYDAY LIFE

The reality of everyday life is shared with others. But how are these others themselves experienced in everyday life? Again, it is possible to differentiate between several modes of such experience.

The most important experience of others takes place in the face-to-face situation, which is the prototypical case of social interaction. All other cases are derivatives of it.

In the face-to-face situation the other is appresented to me in a vivid present shared by both of us. I know that in the same vivid present I am appresented to him. My and his "here and now" continuously impinge on each other as long as the face-to-face situation continues. As a result, there is a continuous interchange of my expressivity and his. I see him smile, then react to my frown by stopping the smile, then smiling again as I smile, and so on. Every expression of mine is oriented toward him, and vice versa, and this continuous reciprocity of expressive acts is simultaneously available to both of us. This means that, in the face-to-face situation, the other's subjectivity is available to me through a maximum of symptoms. To be sure, I may misinterpret some of these symptoms. I may think that the other is smiling while in fact he is smirking. Nevertheless, no other form of social relating can reproduce the plenitude of symptoms of subjectivity present in the face-to-face situation. Only here is the other's subjectivity emphatically "close." All other forms of relating to the other are, in varying degrees, "remote."

In the face-to-face situation the other is fully real. This reality is part of the overall reality of everyday life, and as such massive and compelling. To be sure, another may be real to me without my having encountered him face to face—by reputation, say, or by having corresponded with him. Nevertheless, he becomes real to me in the fullest sense of the word only when I meet him face to face. Indeed, it may be argued that the other in the face-to-face situation is more real to me than I myself. Of course I "know myself better" than I can ever know him. My subjectivity is accessible to me in a way his can never be, no matter how "close" our relationship. My past is available to me in memory in a fullness with which I can never reconstruct his, however much he may tell me about it. But this "better knowledge" of myself requires reflection. It is not immediately appresented to me. The other, however, *is* so appresented to me in the face-to-face situation. "What he is," therefore, is ongoingly available to me. This availability is continuous and prereflective. On the other hand, "What I am" is *not* so available. To make it available requires that I stop, arrest the continuous spontaneity of my experience, and deliberately turn my attention back upon myself. What is more, such reflection about myself is typically occasioned by the attitude toward me that *the other* exhibits. It is typically a "mirror" response to attitudes of the other.

It follows that relations with others in the face-to-face situation are highly flexible. Put negatively, it is comparatively difficult to impose rigid patterns upon face-to-face interaction. Whatever patterns are introduced will be continuously modified through the exceedingly variegated and subtle interchange of subjective meanings that goes on. For instance, I may view the other as someone inherently unfriendly to me and act toward him within a pattern of "unfriendly relations" as understood by me. In the face-to-face situation, however, the other may confront me with attitudes and acts that contradict this pattern, perhaps up to a point where I am led to abandon the pattern as inapplicable and to view him as friendly. In other words, the pattern cannot sustain the massive evidence of the other's subjectivity that is available to me in the face-to-face situation. By contrast, it is much easier for me to ignore such evidence as long as I do not encounter the other face to face. Even in such a relatively "close" relation as may be maintained by correspondence I can more successfully dismiss the other's protestations of friendship as not actually representing his subjective attitude to me, simply because in correspondence I lack the immediate, continuous and massively real presence of his expressivity. It is, to be sure, possible for me to misinterpret the other's meanings even in the face-to-face situation, as it is possible for him "hypocritically" to hide his meanings. All the same, both misinterpretation and "hypocrisy" are more difficult to sustain in face-to-face interaction than in less "close" forms of social relations.

On the other hand, I apprehend the other by means of typificatory schemes even in the face-to-face situation, although these schemes are more "vulnerable" to his interference than in "remoter" forms of interaction. Put differently, while it is comparatively difficult to impose rigid patterns on face-to-face interaction, even it is patterned from the beginning if it takes place within the routines of everyday life. (We can leave aside for later consideration cases of interaction between complete strangers who have no common background of everyday life.) The reality of everyday life contains typificatory schemes in terms of which others are apprehended and "dealt with" in face-to-face encounters. Thus I apprehend the other as "a man," "a European," "a buyer," "a jovial type," and so on. All these typifications ongoingly affect my interaction with him as, say, I decide to show him a good time on the town before trying to sell him my product. Our face-to-face interaction will be patterned by these typifications as long as they do not become problematic through interference on his part. Thus he may come up with evidence that, although "a man," "a European" and "a buyer," he is also a self- righteous moralist, and that what appeared first as joviality is actually an expression of contempt for Americans in general and American salesmen in particular. At this point, of course, my typificatory scheme will have to be modified, and the evening planned differently in accordance with this modification. Unless thus challenged, though, the typifications will hold until further notice and will determine my actions in the situation.

The typificatory schemes entering into face-to-face situations are, of course, reciprocal. The other also apprehends me in a typified way—as "a man," "an American," "a salesman," "an ingratiating fellow," and so on. The other's typifications are as susceptible to my interference as mine are to his. In other words, the two typificatory schemes enter into an ongoing "negotiation" in the face-to-face situation. In everyday life such "negotiation" is itself likely to be prearranged in a typical manner—as in the typical bargaining process between buyers and salesmen. Thus, most of the time, my encounters with others in everyday life are typical in a double sense—I apprehend the other *as* a type and I interact with him in a situation that is itself typical.

The typifications of social interaction become progressively anonymous the farther away they are from the face-to-face situation. Every typification, of course, entails incipient anonymity. If I typify my friend Henry as a member of category X (say, as an Englishman), I *ipso facto* interpret at least certain aspects of his conduct as resulting from this typification—for instance, his tastes in food are typical of Englishmen, as are his manners, certain of his emotional reactions, and so on. This implies, though, that these characteristics and actions of my friend Henry appertain to *anyone* in the category of Englishman, that is, I apprehend these aspects of his being in anonymous terms. Nevertheless, as long as my friend Henry is available in the plenitude of expressivity of the face-to-face situation, he will constantly break through my type of anonymous

Englishman and manifest himself as a unique and therefore atypical individual—to wit, as my friend Henry. The anonymity of the type is obviously less susceptible to this kind of individualization when face-to-face interaction is a matter of the past (my friend Henry, *the Englishman,* whom I knew when I was a college student), or is of a superficial and transient kind (the Englishman with whom I have a brief conversation on a train), or has never taken place (my business competitors in England).

An important aspect of the experience of others in everyday life is thus the directness or indirectness of such experience. At any given time it is possible to distinguish between consociates with whom I interact in face-to-face situations and others who are mere contemporaries, of whom I have only more or less detailed recollections, or of whom I know merely by hearsay. In face-to-face situations I have direct evidence of my fellowman, of his actions, his attributes, and so on. Not so in the case of contemporaries—of them I have more or less reliable knowledge. Furthermore, I must take account of my fellowmen in face-to-face situations, while I may, but need not, turn my thoughts to mere contemporaries. Anonymity increases as I go from the former to the latter, because the anonymity of the typifications by means of which I apprehend fellowmen in face-to-face situations is constantly "filled in" by the multiplicity of vivid symptoms referring to a concrete human being.

This, of course, is not the whole story. There are obvious differences in my experiences of mere contemporaries. Some I have experienced again and again in face-to-face situations and expect to meet again regularly (my friend Henry); others I *recollect* as concrete human beings from a past meeting (the blonde I passed on the street), but the meeting was brief and, most likely, will not be repeated. Still others I *know* of as concrete human beings, but I can apprehend them only by means of more or less anonymous intersecting typifications (my British business competitors, the Queen of England). Among the latter one could again distinguish between likely partners in face-to-face situations (my British business competitors), and potential but unlikely partners (the Queen of England).

The degree of anonymity characterizing the experience of others in everyday life depends, however, upon another factor too. I see the newspaper vendor on the street corner as regularly as I see my wife. But he is less important to me and I am not on intimate terms with him. He may remain relatively anonymous to me. The degree of interest and the degree of intimacy may combine to increase or decrease anonymity of experience. They may also influence it independently. I can be on fairly intimate terms with a number of the fellow-members of a tennis club and on very formal terms with my boss. Yet the former, while by no means completely anonymous, may merge into "that bunch at the courts" while the latter stands out as a unique individual. And finally, anonymity may become near-total with certain typifications that are not intended ever to become individualized—such as the "typical reader of the London *Times*." Finally, the "scope" of the typification—and thereby its anonymity—can be further increased by speaking of "British public opinion."

The social reality of everyday life is thus apprehended in a continuum of typifications, which are progressively anonymous as they are removed from the "here and now" of the face-to-face situation. At one pole of the continuum are those others with whom I frequently and intensively interact in face-to-face situations—my "inner circle," as it were. At the other pole are highly anonymous abstractions, which by their very nature can never be available in face-to-face interaction. Social structure is the sum total of these typifications and of the recurrent patterns of interaction established by means of them. As such, social structure is an essential element of the reality of everyday life.

One further point ought to be made here, though we cannot elaborate it. My relations with others are not limited to consociates and contemporaries. I also relate to predecessors and successors, to those others who have preceded and will follow me in the encompassing history of my society. . . .

LANGUAGE AND KNOWLEDGE IN EVERYDAY LIFE

The reality of everyday life is not only filled with objectivations; it is only possible because of them. I am constantly surrounded by objects that "proclaim" the subjective intentions of my fellowmen, although I may sometimes have difficulty being quite sure just what it is that a particular object is "proclaiming," especially if it was produced by men whom I have not known well or at all in face-to-face situations. Every ethnologist or archaeologist will readily testify to such difficulties, but the very fact that he *can* overcome them and reconstruct from an artifact the subjective intentions of men whose society may have been extinct for millennia is eloquent proof of the enduring power of human objectivations.

A special but crucially important case of objectivation is significant, that is, the human production of signs. A sign may be distinguished from other objectivations by its explicit intention to serve as an index of subjective meanings. . . .

Language, which may be defined here as a system of vocal signs, is the most important sign system of human society. Its foundation is, of course, in the intrinsic capacity of the human organism for vocal expressivity, but we can begin to speak of language only when vocal expressions have become capable of detachment from the immediate "here and now" of subjective states. It is not yet language if I snarl, grunt, howl, or hiss, although these vocal expressions are capable of becoming linguistic insofar as they are integrated into an objectively available sign system. The common objectivations of everyday life are maintained primarily by linguistic signification. Everyday life is, above all, life with and by means of the language I share with my fellowmen. An understanding of language is thus essential for any understanding of the reality of everyday life. . . .

As a sign system, language has the quality of objectivity. I encounter language as a facticity external to myself and it is coercive in its effect on me. Language forces me into its patterns. I cannot use the rules of German syntax when I speak English; I cannot use words invented by my three-year-old son if I want to communicate outside the family; I must take into account prevailing standards of proper speech for various occasions, even if I would prefer my private "improper" ones. Language provides me with a ready-made possibility for the ongoing objectification of my unfolding experience. Put differently, language is pliantly expansive so as to allow me to objectify a great variety of experiences coming my way in the course of my life. Language also typifies experiences, allowing me to subsume them under broad categories in terms of which they have meaning not only to myself but also to my fellowmen. As it typifies, it also anonymizes experiences, for the typified experience can, in principle, be duplicated by anyone falling into the category in question. For instance, I have a quarrel with my mother-in-law. This concrete and subjectively unique experience is typified linguistically under the category of "mother-in-law trouble." In this typification it makes sense to myself, to others, and, presumably, to my mother-in-law. The same typification, however, entails anonymity. Not only I but *anyone* (more accurately, anyone in the category of son-in-law) can have "mother-in-law trouble." In this way, my biographical experiences are ongoingly subsumed under general orders of meaning that are both objectively and subjectively real.

Because of its capacity to transcend the "here and now," language bridges different zones within the reality of everyday life and integrates them into a meaningful whole. The transcendences have spatial, temporal and social dimensions. Through language I can transcend the gap between my manipulatory zone and that of the other; I can synchronize my biographical time sequence with his; and I can converse with him about individuals and collectivities with whom we are not at present in face-to-face interaction. As a result of these transcendences language is capable of "making present" a variety of objects that are spatially, temporally and socially absent from the "here and now." *Ipso facto* a vast accumulation of experiences and meanings can become objectified in the "here and now." Put simply, through language an entire world can be actualized at any moment. This transcending and integrating power of language is retained when I am not actually conversing with another. Through linguistic objectification, even when "talking to myself" in solitary thought, an entire world can be appresented to me at any moment. As far as social relations are concerned, language "makes present" for me not only fellowmen who are physically absent at the moment, but fellowmen in the remembered or reconstructed past, as well as fellowmen projected as imaginary figures into the future. All these "presences" can be highly meaningful, of course, in the ongoing reality of everyday life.

Moreover, language is capable of transcending the reality of everyday life altogether. It can refer to experiences pertaining to finite provinces of meaning, and it can span discrete spheres of reality. For instance, I can interpret "the meaning" of a dream by integrating it linguistically within the order of everyday life. Such integration transposes the discrete reality of the dream into the reality of everyday life by making it an enclave within the latter. The dream is now meaningful in terms of the reality of everyday life rather than of its own discrete reality. Enclaves produced by such transposition belong, in a sense, to both spheres of reality. They are "located" in one reality, but "refer" to another. . . .

Society as Objective Reality

Man's instinctual organization may be described as underdeveloped, compared with that of the other higher mammals. Man does have drives, of course. But these drives are highly unspecialized and undirected. This means that the human organism is capable of applying its constitutionally given equipment to a very wide and, in addition, constantly variable and varying range of activities. This peculiarity of the human organism is grounded in its ontogenetic development. Indeed, if one looks at the matter in terms of organismic development, it is possible to say that the fetal period in the human being extends through about the first year after birth. Important organismic developments, which in the animal are completed in the mother's body, take place in the human infant after its separation from the womb. At this time, however, the human infant is not only *in* the outside world, but interrelating with it in a number of complex ways.

The human organism is thus still developing biologically while already standing in a relationship to its environment. In other words, the process of becoming man takes place in an interrelationship with an environment. This statement gains significance if one reflects that this environment is both a natural and a human one. That is, the developing human being not only interrelates with a particular natural environment, but with a specific cultural and social order, which is mediated to him by the significant others who have charge of him. Not only is the survival of the human infant dependent upon certain social arrangements, the direction of his organismic development is socially determined. From the moment of birth, man's organismic development, and indeed a large part of his biological being as such, are subjected to continuing socially determined interference.

Despite the obvious physiological limits to the range of possible and different ways of becoming man in this double environmental interrelationship the human organism manifests an immense plasticity in its response to the environmental forces at work on it. This is particularly clear when one observes the flexibility of man's biological constitution as it is subjected to a variety of socio-cultural determinations. It is an ethnological commonplace that the ways of becoming and being human are as numerous as man's cultures. Humanness is socio-culturally variable. In other words, there is no human nature in the sense of a biologically fixed substratum determining the variability of socio-cultural formations. There is only human nature in the sense of anthropological constants (for example, world-openness and plasticity of instinctual structure) that delimit and permit man's socio-cultural formations. But the specific shape into which this humanness is molded is determined by those socio-cultural formations and is relative to their numerous variations. While it is possible to say that man has a nature, it is more significant to say that man constructs his own nature, or more simply, that man produces himself.

The plasticity of the human organism and its susceptibility to socially determined interference is best illustrated by the ethnological evidence concerning sexuality. While man possesses sexual drives that are comparable to those of the other higher mammals, human sexuality is characterized by a very high degree of pliability. It is not only relatively independent of temporal rhythms, it is pliable both in the objects toward which it may be directed and in its modalities of expression. Ethnological evidence shows that, in sexual matters, man is capable of almost anything. . . .

Social order is not part of the "nature of things," and it cannot be derived from the "laws of nature." Social order exists *only* as a product of human activity. No other ontological status may be ascribed to it without hopelessly obfuscating its empirical manifestations. Both in its genesis (social order is the result of past human activity) and its existence in any instant of time (social order exists only and insofar as human activity continues to produce it) it is a human product. . . .

All human activity is subject to habitualization. Any action that is repeated frequently becomes cast into a pattern, which can then be reproduced with an economy of effort and which, *ipso facto,* is apprehended by its performer *as* that pattern. Habitualization further implies that the action in question may be performed again in the future in the same manner and with the same economical effort. This is true of non-social as well as of social activity. Even the solitary individual on the proverbial desert island habitualizes his activity. When he wakes up in the morning and resumes his attempts to construct a canoe out of matchsticks, he may mumble to himself. "There I go again," as he starts on step one of an operating procedure consisting of, say, ten steps. In other words, even solitary man has at least the company of his operating procedures.

Habitualized actions, of course, retain their meaningful character for the individual although the meanings involved become embedded as routines in his general stock of knowledge, taken for granted by him and at hand for his projects into the future. Habitualization carries with it the important psychological gain that choices are narrowed. While in theory there may be a hundred ways to go about the project of building a canoe out of matchsticks, habitualization narrows these down to one. This frees the individual from the burden of "all those decisions," providing a psychological relief that has its basis in man's undirected

instinctual structure. Habitualization provides the direction and the specialization of activity that is lacking in man's biological equipment, thus relieving the accumulation of tensions that result from undirected drives. And by providing a stable background in which human activity may proceed with a minimum of decision-making most of the time, it frees energy for such decisions as may be necessary on certain occasions. In other words, the background of habitualized activity opens up a foreground for deliberation and innovation.

In terms of the meanings bestowed by man upon his activity, habitualization makes it unnecessary for each situation to be defined anew, step by step. A large variety of situations may be subsumed under its predefinitions. The activity to be undertaken in these situations can then be anticipated. Even alternatives of conduct can be assigned standard weights.

These processes of habitualization precede any institutionalization, indeed can be made to apply to a hypothetical solitary individual detached from any social interaction. The fact that even such a solitary individual, assuming that he has been formed as a self (as we would have to assume in the case of our matchstick-canoe builder), will habitualize his activity in accordance with biographical experience of a world of social institutions preceding his solitude need not concern us at the moment. Empirically, the more important part of the habitualization of human activity is coextensive with the latter's institutionalization. The question then becomes how do institutions arise.

Institutionalization occurs whenever there is a reciprocal typification of habitualized actions by types of actors. Put differently, any such typification is an institution. What must be stressed is the reciprocity of institutional typifications and the typicality of not only the actions but also the actors in institutions. The typifications of habitualized actions that constitute institutions are always shared ones. They are *available* to all the members of the particular social group in question, and the institution itself typifies individual actors as well as individual actions. The institution posits that actions of type X will be performed by actors of type X. For example, the institution of the law posits that heads shall be chopped off in specific ways under specific circumstances, and that specific types of individuals shall do the chopping (executioners, say, or members of an impure caste, or virgins under a certain age, or those who have been designated by an oracle).

Institutions further imply historicity and control. Reciprocal typifications of actions are built up in the course of a shared history. They cannot be created instantaneously. Institutions always have a history, of which they are the products. It is impossible to understand an institution adequately without an understanding of the historical process in which it was produced. Institutions also, by the very fact of their existence, control human conduct by setting up predefined patterns of conduct, which channel it in one direction as against the many other directions that would theoretically be possible. It is important to stress that this controlling character is inherent in institutionalization as such, prior to or apart from any mechanisms of sanctions specifically set up to support an institution. These mechanisms (the sum of which constitute what is generally called a system of social control) do, of course, exist in many institutions and in all the agglomerations of institutions that we call societies. Their controlling efficacy, however, is of a secondary or supplementary kind. As we shall see again later, the primary social control is given in the existence of an institution as such. To say that a segment of human activity has been institutionalized is already to say that this segment of human activity has been subsumed under social control. Additional control mechanisms are required only insofar as the processes of institutionalization are less than completely successful. Thus, for instance, the law may provide that anyone who breaks the incest taboo will have his head chopped off. This provision may be necessary because there have been cases when individuals offended against the taboo. It is unlikely that this sanction will have to be invoked continuously (unless the institution delineated by the incest taboo is itself in the course of disintegration, a special case that we need not elaborate here). It makes little sense, therefore, to say that human sexuality is socially controlled by beheading certain individuals. Rather, human sexuality is socially controlled by its institutionalization in the course of the particular history in question. One may add, of course, that the incest taboo itself is nothing but the negative side of an assemblage of typifications, which define in the first place which sexual conduct is incestuous and which is not. . . .

An institutional world, then, is experienced as an objective reality. It has a history that antedates the individual's birth and is not accessible to his biographical recollection. It was there before he was born, and it will be there after his death. This history itself, as the tradition of the existing institutions, has the character of objectivity. The individual's biography is apprehended as an episode located within the objective history of the society. The institutions, as historical and objective facticities, confront the individual as undeniable facts.

The institutions are *there,* external to him, persistent in their reality, whether he likes it or not. He cannot wish them away. They resist his attempts to change or evade them. They have coercive power over him, both in themselves, by the sheer force of their facticity, and through the control mechanisms that are usually attached to the most important of them. The objective reality of institutions is not diminished if the individual does not understand their purpose or their mode of operation. He may experience large sectors of the social world as incomprehensible, perhaps oppressive in their opaqueness, but real nonetheless. Since institutions exist as external reality, the individual cannot understand them by introspection. He must "go out" and learn about them, just as he must to learn about nature. This remains true even though the social world, as a humanly produced reality, is potentially understandable in a way not possible in the case of the natural world.

It is important to keep in mind that the objectivity of the institutional world, however massive it may appear to the individual, is a humanly produced, constructed objectivity. The process by which the externalized products of human activity attain the character of objectivity is objectivation. The institutional world is objectivated human activity, and so is every single institution. In other words, despite the objectivity that marks the social world in human experience, it does not thereby acquire an ontological status apart from the human activity that produced it. The paradox that man is capable of producing a world that he then experiences as something other than a human product will concern us later on. At the moment, it is important to emphasize that the relationship between man, the producer, and the social world, his product, is and remains a dialectical one. That is, man (not, of course, in isolation but in his collectivities) and his social world interact with each other. The product acts back upon the producer. Externalization and objectivation are moments in a continuing dialectical process. The third moment in this process, which is internalization constitution of subject (by which the objectivated social world is retrojected into consciousness in the course of socialization), will occupy us in considerable detail later on. It is already possible, however, to see the fundamental relationship of these three dialectical moments in social reality. Each of them corresponds to an essential characterization of the social world. *Society is a human product. Society is an objective reality. Man is a social product.* It may also already be evident than an analysis of the social world that leaves out any one of these three moments will be distortive. One may further add that only with the transmission of the social world to a new generation (that is, internalization as effectuated in socialization) does the fundamental social dialectic appear in its totality. To repeat, only with the appearance of a new generation can one properly speak of a social world.

At the same point, the institutional world requires legitimation, that is, ways by which it can be "explained" and justified. This is not because it appears less real. As we have seen, the reality of the social world gains in massivity in the course of its transmission. This reality, however, is a historical one, which comes to the new generation as a tradition rather than as a biographical memory. . . .

A final question of great theoretical interest arising from the historical variability of institutionalization has to do with the manner in which the institutional order is objectified: To what extent is an institutional order, or any part of it, apprehended as a non-human facticity? This is the question of the reification of social reality.

Reification is the apprehension of human phenomena as if they were things, that is, in non-human or possibly supra-human terms. Another way of saying this is that reification is the apprehension of the products of human activity *as if* they were something else than human products—such as facts of nature, results of cosmic laws, or manifestations of divine will. Reification implies that man is capable of forgetting his own authorship of the human world, and further, that the dialectic between man, the producer, and his products is lost to consciousness. The reified world is, by definition, a dehumanized world. It is experienced by man as a strange facticity, an *opus alienum* over which he has no control rather than as the *opus proprium* of his own productive activity.

It will be clear from our previous discussion of objectivation that, as soon as an objective social world is established, the possibility of reification is never far away. The objectivity of the social world means that it confronts man as something outside of himself. The decisive question is whether he still retains the awareness that, however objectivated, the social world was made by men—and, therefore, can be remade by them. In other words, reification can be described as an extreme step in the process of objectivation, whereby the objectivated world loses its comprehensibility as a human enterprise and becomes fixated as a non-human, non-humanizable, inert facticity. Typically, the real relationship between man and his world is reversed in consciousness. Man, the producer of a world, is apprehended as its product, and human activity as an epiphenomenon of non-human processes. Human meanings are no longer understood as world-producing but as being, in their turn, products of the "nature of things." It must be emphasized that reification is a modality of consciousness, more precisely, a modality of man's objectification of the human world. Even while apprehending the world in reified terms, man continues to produce it. That is, man is capable paradoxically of producing a reality that denies him.

Reification is possible on both the pretheoretical and theoretical levels of consciousness. Complex theoretical systems can be described as reifications, though presumably they have their roots in pretheoretical reifications established in this or that social situation. Thus it would be an error to limit the concept of reification to the mental constructions of intellectuals. Reification exists in the consciousness of the man in the street and, indeed, the latter presence is more practically significant. It would also be a mistake to look at reification as a perversion of an originally non-reified apprehension of the social world, a sort of cognitive fall from grace, On the contrary, the available ethnological and psychological evidence seems to indicate the opposite, namely, that the original apprehension of the social world is highly reified both phylogenetically and ontogenetically. This implies that an apprehension of reification *as* a modality of consciousness is dependent upon an at least relative dereification of consciousness, which is a comparatively late development in history and in any individual biography.

Both the institutional order as a whole and segments of it may be apprehended in reified terms. For example, the entire order of society may be conceived of as a microcosm reflecting the macrocosm of the total universe as made by the gods. Whatever happens "here below" is but a pale reflection of what takes place "up above." Particular institutions may be apprehended in similar ways. The basic "recipe" for the reification of institutions is to bestow on them an ontological status independent of human activity and signification. Specific reifications are variations on this general theme. Marriage, for instance, may be reified as an imitation of divine acts of creativity, as a universal mandate of natural law, as the necessary consequence of biological or psychological forces, or, for that matter, as a functional imperative of the social system. What all these reifications have in common is their obfuscation of marriage as an ongoing human production. As can be readily seen in this example, the reification may occur both theoretically and pretheoretically. Thus the mystagogue can concoct a highly sophisticated theory reaching out from the concrete human event to the farthest corners of the divine cosmos, but an illiterate peasant couple being married may apprehend the event with a similarly reifying shudder of metaphysical dread. Through reification, the world of institutions appears to merge with the world of nature. It becomes necessity and fate, and is lived through as such, happily *or* unhappily as the case may be.

Roles may be reified in the same manner as institutions. The sector of self-consciousness that has been objectified in the role is then also apprehended as an inevitable fate, for which the individual may disclaim responsibility. The paradigmatic formula for this kind of reification is the statement "I have no choice in the matter, I have to act this way because of my position"—as husband, father, general, archbishop, chairman of the board, gangster, or hangman, as the case may be. This means that the reification of roles narrows the subjective distance that the individual may establish between himself and his role-playing. The distance implied in all objectification remains, of course, but the distance brought about by disidentification shrinks to the vanishing point. Finally, identity itself (the total self, if one prefers) may be reified, both one's own and that of others. There is then a total identification of the individual with his socially assigned typifications. He is apprehended as *nothing but* that type. This apprehension may be positively or negatively accented in terms of values or emotions. The identification of "Jew" may be equally reifying for the anti-Semite and the Jew himself, except that the latter will accent the identification positively and the former negatively. Both reifications bestow an ontological and total status on a typification that is humanly produced and that, even as it is internalized, objectifies but a segment of the self. Once more, such reifications may range from the pretheoretical level of "what everybody knows about Jews" to the most complex theories of Jewishness as a manifestation of biology ("Jewish blood"), psychology ("the Jewish soul") or metaphysics ("the mystery of Israel"). . . .

SOCIETY AS SUBJECTIVE REALITY

Since society exists as both objective and subjective reality, any adequate theoretical understanding of it must comprehend both these aspects. As we have already argued, these aspects receive their proper recognition if

society is understood in terms of an ongoing dialectical process composed of the three moments of externalization, objectivation, and internalization. As far as the societal phenomenon is concerned, these moments are *not* to be thought of as occurring in a temporal sequence. Rather society and each part of it are simultaneously characterized by these three moments, so that any analysis in terms of only one or two of them falls short. The same is true of the individual member of society, who simultaneously externalizes his own being into the social world and internalizes it as an objective reality. In other words, to be in society is to participate in its dialectic.

The individual, however, is not born a member of society. He is born with a predisposition toward sociality, and he becomes a member of society. In the life of every individual, therefore, there *is* a temporal sequence, in the course of which he is inducted into participation in the societal dialectic. The beginning point of this process is internalization: the immediate apprehension or interpretation of an objective event as expressing meaning, that is, as a manifestation of another's subjective processes which thereby becomes subjectively meaningful to myself. . . .

Only when he has achieved this degree of internalization is an individual a member of society. The ontogenetic process by which this is brought about is socialization, which may thus be defined as the comprehensive and consistent induction of an individual into the objective world of a society or a sector of it. Primary socialization is the first socialization an individual undergoes in childhood, through which he becomes a member of society. Secondary socialization is any subsequent process that inducts an already socialized individual into new sectors of the objective world of his society. We may leave aside here the special question of the acquisition of knowledge about the objective world of societies other than the one of which we first became a member, and the process of internalizing such a world as reality—a process that exhibits, at least superficially, certain similarities with both primary and secondary socialization, yet is structurally identical with neither.

It is at once evident that primary socialization is usually the most important one for an individual, and that the basic structure of all secondary socialization has to resemble that of primary socialization. Every individual is born into an objective social structure within which he encounters the significant others who are in charge of his socialization. These significant others are imposed upon him. Their definitions of his situation are posited for him as objective reality. He is thus born into not only an objective social structure but also an objective social world. The significant others who mediate this world to him modify it in the course of mediating it. They select aspects of it in accordance with their own location in the social structure, and also by virtue of their individual, biographically rooted idiosyncrasies. The social world is "filtered" to the individual through this double selectivity. Thus the lower-class child not only absorbs a lower-class perspective on the social world, he absorbs it in the idiosyncratic coloration given it by his parents (or whatever other individuals are in charge of his primary socialization). The same lower-class perspective may induce a mood of contentment, resignation, bitter resentment, or seething rebelliousness. Consequently, the lower-class child will not only come to inhabit a world greatly different from that of an upper-class child, but may do so in a manner quite different from the lower-class child next door. . . .

The formation within consciousness of the generalized other marks a decisive phase in socialization. It implies the internalization of society as such and of the objective reality established therein, and, at the same time, the subjective establishment of a coherent and continuous identity. Society, identity *and* reality are subjectively crystallized in the same process of internalization. This crystallization is concurrent with the internalization of language. Indeed, for reasons evident from the foregoing observations on language, language constitutes both the most important content and the most important instrument of socialization.

When the generalized other has been crystallized in consciousness, a symmetrical relationship is established between objective and subjective reality. What is real "outside" corresponds to what is real "within." Objective reality can readily be "translated" into subjective reality, and vice versa. Language, of course, is the principal vehicle of this ongoing translating process in both directions. It should, however, be stressed that the symmetry between objective and subjective reality cannot be complete. The two realities correspond to each other, but they are not coextensive. There is always more objective reality "available" than is actually internalized in any individual consciousness, simply because the contents of socialization are determined by the social distribution of knowledge. No individual internalizes the totality of what is objectivated as reality in his society, not even if the society and its world are relatively simple ones. On the other hand, there are always elements of subjective reality that have not originated in socialization, such as the awareness of one's own body prior to and apart from any socially learned apprehension of it. Subjective biography is not fully social. The individual apprehends himself as being both inside *and* outside society. This implies that the symmetry between objective and subjective reality is never a static, once-for-all state of affairs. It must always be produced and reproduced *in actu*. In other words, the relationship between the individual and the objective social world is like an ongoing balancing act. The anthropological roots of this are, of course, the same as those we discussed in connection with the peculiar position of man in the animal kingdom.

In primary socialization there is no *problem* of identification. There is no choice of significant others. Society presents the candidate for socialization with a predefined set of significant others, whom he must accept as such with no possibility of opting for another arrangement. *Hic Rhodus, hic salta.* One must make do with the parents that fate has regaled one with. This unfair disadvantage inherent in the situation of being a child has the obvious consequence that, although the child is not simply passive in the process of his socialization, it is the adults who set the rules of the game. The child can play the game with enthusiasm or with sullen resistance. But, alas, there is no other game around. This has an important corollary. Since the child has no choice in the selection of his significant others, his identification with them is quasi-automatic. For the same reason, his internalization of their particular reality is quasi-inevitable. The child does not internalize the world of his significant others as one of many possible worlds. He internalizes it as *the* world, the only existent and only conceivable world, the world *tout court*. It is for this reason that the world internalized in primary socialization is so much more firmly entrenched in consciousness than worlds internalized in secondary socializations. However much the original sense of inevitability may be weakened in subsequent disenchantments, the recollection of a never-to-be- repeated certainty— the certainty of the first dawn of reality—still adheres to the first world of childhood. Primary socialization thus accomplishes what (in hindsight, of course) may be seen as the most important confidence trick that society plays on the individual—to make appear as necessity what is in fact a bundle of contingencies, and thus to make meaningful the accident of his birth.

▪▪ ETHNOMETHODOLOGY: AN OVERVIEW

Ethnomethodology literally means the study of the methods people use to accomplish their everyday lives. The basic premise of ethnomethodology is that "people do what they do, right there and then, to be reasonable and effective and they do so for pervasively practical reasons and under unavoidably local conditions of knowledge, action and material resources" (Boden 1990:189).

Like phenomenology, ethnomethodology is concerned with how individuals make sense of their everyday circumstances. Both emphasize that "we accept as *unquestionable* the world of facts which surrounds us" (Schutz 1970:58, emphasis added), and both pay attention to the methods or procedures that individuals use to interpretively produce the recognizable, intelligible forms of action that, paradoxically, they treat as "facts." In short, both are intrigued by the "suspension of doubt" that sustains our everyday world.

The central difference between phenomenology and ethnomethodology is that while phenomenology has deeply influenced and been influenced by psychology, ethnomethodology is resolutely sociological. Indeed, as the subtitle of one of Garfinkel's books—*Ethnomethodology's Program: Working Out Durkheim's Aphorism* (2002)—makes clear, this perspective is based on a direct response to Durkheim's famous dictum discussed earlier, that the concreteness of social facts is sociology's most fundamental phenomenon (Rawls 2002:2). Ethnomethodologists seek to explain how "people make joint sense of their social world *together* and [how] they do so *methodically using social procedures or methods* that they share" (Heritage 1998:176, emphasis in original). In other words, instead of focusing on the cognitive or psychological aspects of sense-making and perceptions, ethnomethodologists attribute the very existence of an apparently real world to the interactional and interpretive skills of individual actors. In so doing, they shift the focus from "the real world" to "reality-constituting procedures" that are carried out during social interaction (Gubrium and Holstein 1997:39–41). Ethnomethodologists are particularly intrigued by accounting practices, that is, by the way in which actors "offer accounts" in order to make sense out of events.

Consider, for instance, a visit to the doctor. When the doctor comes in, she may begin with a friendly hello and chat about the upcoming holiday. Then she asks, "So, what brings you here today?" and you begin to describe your stomach pains. This completely routine, but nevertheless complex, interaction involves much more than a simple internalization and acting out of preprogrammed norms or an automatic exchange of gestures. Rather, the success of this interaction depends on both actors' ability to read and respond to verbal and nonverbal cues on the spot. In this way everyday interaction is more like a dance using shared methods (or steps) than an instance of predetermined socialization. For instance, you (and the doctor) must gain a sense of exactly how much small talk in which to engage before getting to the point. (You won't spend 20 minutes answering the doctor's question about the holidays.) So, too, is your response based on a quick implicit assessment as to whether the doctor seems friendly and sincere or rushed and mechanical.

While ethnomethodologists focus on social interaction and behavior rather than abstract mental phenomenon, and thus share much with other sociologists, they are highly critical of work associated with "conventional" forms of sociological theory. First, ethnomethodologists reject the fundamental sociological conceit (especially apparent in structural functionalism and the Frankfurt School—see Chapters 9 and 10) that actors' views of their social worlds are somehow flawed or marginal to a full understanding of social phenomena (Cicourel 1964). In other words, ethnomethodologists condemn those perspectives that construe actors as "judgmental dopes" (Garfinkel 1967). Second, ethnomethodologists criticize conventional approaches for naïvely taking for granted the same skills, practices, and suppositions as the "unenlightened" members of the society they are studying (Pollner 1987:xi). Ethnomethodologists strive for "ethnomethodological indifference," an attitude of detachment that is rooted in neither intellectual naïveté nor condescension (Garfinkel and Sacks 1970:346). They seek to suspend belief in a rule-governed order in order to observe how the regular, coherent, connected patterns of social life are described and explained in ways that create that order itself (Zimmerman and Wieder 1970:289). That is, they seek to understand how people see, describe, and jointly develop a definition of the situation (ibid.).

University of California Professor Harold Garfinkel (see Significant Others box below) developed a method for identifying the building blocks of everyday interaction, which he called breaching experiments. The point of breaching experiments is to "modify the *objective* structure of the familiar, known-in-common environment by rendering the background expectancies inoperative" (Garfinkel 1967:54, emphasis in original); in other words, to disrupt normal procedures in order to expose them. According to Garfinkel, an effective breaching experiment (1) makes it impossible for the subject to interpret the situation as a game or experiment or deception, (2) makes it necessary for the actor to reconstruct the "natural facts" of the encounter, and (3) deprives the actor of consensual support for an alternative definition of the situation (ibid.:58). In short, breaching experiments work much in the same way that "culture shock" does: by thrusting into consciousness the basic elements of a previously taken-for-granted world.

Significant Others

Harold Garfinkel (1917– 2011): The Father of Ethnomethodology

Harold Garfinkel was born on October 29, 1917, in Newark, New Jersey. His father, Abraham Garfinkel, owned a small furniture business, and after Harold graduated from high school the family decided that he would take business courses at the University of Newark and go into the furniture business as well. According to Garfinkel, the business courses he took in college—particularly one called "theory of accounts"—influenced his later work as much as or more than did the work of C. Wright Mills and Kenneth Burke, whose social theories he would later study (Rawls 2002:10). Garfinkel was most interested in how formal analytic theorizing creates an orderly social world out of "indicators," that is, how actors use signs and categories methodically to make sense of the world (Rawls 2002:10).

After he graduated from the University of Newark in 1939, Garfinkel decided to not join the family's furniture business. Instead, he enrolled in graduate school at the University of North Carolina at Chapel Hill. In 1942, Garfinkel completed a master's thesis on intraracial and interracial homicide. Based on observations in court cases and court records on homicides, Garfinkel concluded that the formulation and processing of cases within races was different from that between races (Rawls 2002:12). After serving in the Air Force in World War II, Garfinkel decided to attend Harvard to study with Talcott Parsons (see Chapter 9) rather than go back to Chapel Hill. In 1952, Garfinkel completed his doctoral thesis, which was essentially an extended debate with Parsons as to the relative importance of empirical detail versus conceptual categories and generalizations. Most renowned for setting out the fundamentals of ethnomethodology in his book of essays, *Studies in Ethnomethodology* (1967), Garfinkel was on the faculty at the University of California, Los Angeles from 1954 to 1987. He remained active in the discipline until his death on April 21, 2011.

For instance, in one well-known experiment, Garfinkel asked his students to spend 15 minutes to an hour interacting with their family members as if they were boarders and not intimates. "They were instructed to conduct themselves in a circumspect and polite fashion. They were to avoid getting personal, to use formal address, to speak only when spoken to" (1967:47). Garfinkel reports that in four-fifths of the cases, "family members were stupefied. They vigorously sought to make the strange actions intelligible and to restore the situation to normal appearances" (ibid.). The point is that the families were stupefied because their taken-for-granted world was challenged, and as such, their normal procedures for making sense of things were rendered useless or at least questionable, which produced more than a little anxiety. Indeed, while always controversial, today such stress-inducing experiments are no longer considered ethical.

Photo 13.2   Comedians such as Stephen Colbert earn their living uncovering taken-for-granted stocks of knowledge and recipes.

Comedy "reality" shows such as *Punk'd* and *Boiling Point* and, long before that, *Candid Camera* have no compunction carrying out their own versions of Garfinkel's stress-inducing breaching experiments, however. *Candid Camera,* which first premiered in 1948, used hidden cameras to catch unsuspecting guests trying to figure out and deal with hoaxes such as mailboxes that talked or moved.[7] Similarly, the short-lived cable show *Boiling Point* placed unsuspecting victims in situations that tried their patience (such as a florist who kept on cutting and cutting flowers out of an already-paid-for bouquet). The television audience watched as the clock ticked down, and if the victim was still involved in the interaction (without using profanity) after a given time limit, their patience was rewarded with $100. In the spirit of Garfinkel's breaching experiments, these shows are based on observing how unsuspecting victims respond to slight (or not-so-slight) disruptions in their taken-for-granted world.

Yet, it is not only reality-based comedy that is related to ethnomethodology. Identifying the intricate, taken-for-granted elements of everyday life—and then commenting on, satirizing, or subverting them—is central to stand-up and slapstick comedy as well. For example, comedian George Carlin's famous routine on the "seven words you can't say on television" epitomizes how comedians expose and then violate core but unspoken recipes. As Carlin quips, "I think it's the duty of the comedian to find out where the line is drawn and cross it deliberately" (http://www.quotationspage.com/quotes/George_Carlin).[8] So, too, is

---

[7]Interestingly, *Candid Camera* (which is considered to be the first and longest-running reality-based comedy program) was inspired in part by host Allen Funt's background as a research assistant at Cornell University. Funt aided psychologist Kurt Lewin in experiments on the behaviors of mothers and children (http://www.museum.tv/archives/etv/C/htmlC/candidcamera/candidcamera.htm, maintained by Amy Loomis).

[8]Akin to many comedians, Carlin is a student of language and words. Here is the beginning of the original "Seven Dirty Words" routine that caused such a fracas: "I love words. I thank you for hearing my words. I want to tell you something about words that I uh, I think is important. I love . . . as I say, they're my work, they're my play, they're my passion. Words are all we have, really. We have thoughts, but thoughts are fluid. You know, [humming]. And, then we assign a word to a thought, [clicks tongue]. And we're stuck with that word for that thought. So be careful with words. I like to think, yeah, the same words that hurt can heal. It's a matter of how you pick them . . ." (http://www.erenkrantz.com/Humor/SevenDirtyWords.shtml).

Saturday Night Live's "Weekend Update" based on an exact imitation of newscasters' actual comments, mannerisms, demeanor, and so on coupled with an unexpected jolt from not following the "script," as in Chevy Chase's famous recurring line, "Good evening, I'm Chevy Chase, and you're not." Comedy Central's *The Daily Show with Jon Stewart* and *The Colbert Report* hosted by Stephen Colbert (Photo 13.2), are rooted in a similar tack, parodying both newscasters and cocky "correspondents in the field," as well as political pundit programs.

One of the central methods that actors use to order and make sense of their everyday world (and that breaching experiments were explicitly designed to expose) is accounting practices. For instance, in Garfinkel's experiment discussed above, parents typically responded to the unexpected behavior of their child by constructing and attributing motives that looked to the past to explain the present: "the student was 'working too hard' in school; the student was 'ill'; there had been 'another fight' with a fiancée" (Garfinkel 1967:48). Here, parents are trying to make "accountable" the behavior observed. Garfinkel emphasizes that the process of accounting is vital to the sustenance of reality because it is what enables us to believe that social life is coherent and that meaning is shared.

One of the most important features of accounting practices is indexicality. Taken from linguistics, the term "indexical" refers to the fact that just as words have different meanings in different contexts, so, too, all expressions and practical actions are interpreted in a particular context.

Consider, for instance, the following conversation:

A: I have a fourteen-year-old son.

B: Well, that's all right.

A: I also have a dog.

B: Oh, I'm sorry. (Sacks 1984, as cited in Heritage 1984:237)

At first glance this conversation appears incoherent or nonsensical. However, once we are told that "A" is a would-be tenant of an apartment and "B" is the landlord, the conversation makes perfect sense. Then we recognize that the conversation is an exchange about issues that might disqualify the potential tenant from the rental. This is indexicality.

The centrality of both accounting practices and indexicality to ethnomethodology is evident in Garfinkel's very definition of ethnomethodology as "the investigation of the rational properties of index-ical expressions and other practical actions as contingent on ongoing accomplishments of organized artful practices of everyday life" (1967:11). Everyday life is made up of not only determining the meaning of objects or practices within a given context, but also inferring meaning by creating or attributing a context or "index" when one is not readily apparent (as the experiment above makes clear). Thus, for example, staring out at pouring rain, a speaker observes: "What a wonderful summer we're having!" and the hearer concludes that the remark was intended to be ironical (Heritage 1984:152).

In sum, the everyday world is created and sustained by our ability to make "definite sense with indefinite resources" (Heritage 1984:144). Human descriptive resources, our abilities to understand the situations that we face, are only "approximate"; they do not reflect or correspond directly to "real" states of affairs, but rather provide a field of possibilities. Individuals must actively perform contextualizing work in order to see what the descriptions used in encounters mean (ibid.:148).

This brings us back to the concept of typification. Recall that for Schutz (1967:14), language is the principal tool for typifying the elements in all encounters; that is, it is a kind of "treasure house of ready-made pre-constituted types and characteristics all socially derived and carrying along an open horizon of unexplored content." As John Heritage (1984:145) states,

> Language is the medium through which . . . common-sense equivalence classes are constituted and communicated. It embodies a continual compromise between generality and specificity. There is thus an inherently approximate relationship between a descriptor and the range of states of affairs it may be used to describe.

In the 1960s, ethnomethodologists' emphasis on language as a constituting practice combined with new technology to produce a new direction in the field, called conversation analysis. Conversation analysis infuses the ethnomethodological interest in the details of mundane everyday action and the production of order with a rigorous methodology and focus on the fundamental, taken-for-granted structures of conversational interaction. Using audio- and videotape for recording conversations as they naturally occur and linguistic conventions for meticulously coding and analyzing them, conversation analysts investigate not only the use of words but also "the hesitations, cut-offs, restarts, silences, breathing noises, throat clearings, sniffles, laughter and laughterlike noises, prosody, and the like, not to mention the 'nonverbal' behaviors available on video records that are usually closely integrated with the stream of activity captured on the audiotape" (Zimmerman 1988:413). Conversation analysts have been quite prolific in the last 30 years, successfully uncovering the characteristic features of such phenomena as conversational openings and closings, turn-taking, interruptions, as well as applause and booing (Clayman 1993; Heritage and Greatbatch 1986; Jefferson 1978; Sacks, Schegloff, and Jefferson 1974).

Yet, ironically, by taking a more positivistic approach to conversational structures, conversation analysts have veered from ethnomethodology's phenomenological roots. In contrast to Berger and Luckmann and Schutz, conversation analysts are not interested in studying the actor's consciousness or a community of "shared understandings," but rather only studying observable, measurable behaviors. Put in another way, conversation analysts are interested in the "architecture of intersubjectivity" (Heritage 1984:254)—rather than intersubjectivity itself.

Discussion Questions

1. What would happen if you got on an elevator and then faced the person already riding in the elevator (looking straight at him or her) rather than the door? Try it a few times. Report your findings to the class. Most important, you should not only explain exactly what you did and how the other person responded, but also analyze and explain how you felt during this experiment.

2. Discuss three specific recipes that you use in a typical classroom situation. Be as precise as possible about the exact conditions under which you use these recipes (indexicality) and how you acquired them.

3. According to Berger and Luckmann, "Habitualization carries with it the important psychological gain that choices are narrowed." Discuss this point using a concrete example from your everyday life. How does this process result in objectivation?

4. What impact might the insights of phenomenology and ethnomethodology have on doing cross-cultural research? More generally, what types of research questions do you think these two perspectives can address? To what types are they not well suited?

5. Ethnomethodologists explore the methods people use to construct their everyday world. What methods do we use to construct such things as race or gender? In other words, how do we "do" race or gender? For instance, how does an individual act like a "white" person or a woman? How, or on what basis, do you "read" such behaviors?

Part IV

BROADENING SOCIOLOGICAL THEORY

14 Feminist and Gender Theories

Dorothy E. Smith

Key Concepts

- Relations of Ruling
- Bifurcation of Consciousness
- Institutional Ethnography
- Standpoint Theory

Patricia Hill Collins

Key Concepts

- Standpoint Epistemology
- Black Feminist Thought
- Matrix of Domination

Raewyn Connell

Key Concepts

- Hegemonic Masculinity
- Patriarchal Dividend

Judith Butler

Key Concepts

▪▪ Queer Theory

▪▪ Heterosexual Matrix

▪▪ Performativity

There is no original or primary gender a drag imitates, but gender is a kind of imitation for which there is no original.

—Judith Butler

A Brief History of Women's Rights in the United States

1700s

American colonial law held that "by marriage, the husband and wife are one person in the law. The very being and legal existence of the woman is suspended during the marriage, or at least is incorporated into that of her husband under whose wing and protection she performs everything."

By 1777, women are denied the right to vote in all states in the United States.

1800s

In *Missouri v. Celia* (1855), a slave, a black woman, is declared to be property without the right to defend herself against a master's act of rape.

In 1866, the Fourteenth Amendment is passed by Congress (ratified by the states in 1868). It is the first time "citizens" and "voters" are defined as male in the U.S. Constitution.

1900s

In 1920, the Nineteenth Amendment to the U.S. Constitution is ratified. It declares, "The right of citizens of the United States to vote shall not be denied or abridged by the United States or by any State on account of sex."

In 1923, the Equal Rights Amendment is introduced in Congress in the United States.

In 1963, the Equal Pay Act is passed by the U.S. Congress, promising equitable wages for the same work, regardless of the race, color, religion, national origin, or sex of the worker. In 1982, the Equal Rights Amendment (which had languished in Congress for fifty years) is defeated, falling three states short of the thirty-eight needed for ratification.

SOURCE: http://www.legacy98.org/timeline.html; Jo Freeman, *American Journal of Sociology,* in Goodwin and Jasper (2004).

The brief timeline above underscores an obvious but all too often overlooked point: the experience of women in society is not the same as that of men. In the United States, women's rights have expanded considerably since the nineteenth century, when women were denied access to higher education and the right to own property and vote. Despite major advances, there are still some troubling gender gaps in the United States, however. Women still suffer disproportionately, leading to what sociologists refer to as the "feminization of poverty," where two out of every three poor adults are women. In addition, in contrast to countries such as Sweden where more than 45 percent of elected officials in the national parliament are women, in the United States only about 15 percent of the politicians in the House or Senate are women, placing the United States a lowly sixtieth worldwide in the global ranking of women in politics (www.iwdc.org/resources/factsheet.htm; www.ipu.org/wmn-e/classif.htm).

Yet, it was not until 2005 that women in Kuwait were granted the right to vote and stand for election (see Table 14.1), and sadly, as of this writing, women in Saudi Arabia do not yet have those political freedoms. Indeed, in a recent study by Freedom House, Saudi Arabia ranked dead last in all five categories analyzed in terms of women's equality, although in none of the 17 societies of the Arab Middle East and Northern Africa (MEDA) studied do women enjoy the same citizenship and nationality rights as men.[1] In Saudi Arabia, women are segregated in public places, cannot drive cars, and must be covered from head to toe when in public. Men are entitled to divorce without explanation simply by registering a statement to the court and repeating it three times. By contrast, most women not only lack the right of divorce, but because their children legally belong to the father, to leave their husband means giving up their children.

What these latter cases also demonstrate is that the expansion of women's rights does not proceed automatically and must not be taken for granted. Laws that discriminate against women were instituted in the United States in the nineteenth century that had not existed in the previous decades. On a global scale, nowhere was the precariousness of women's rights more evident than when the Taliban radically rescinded them in Afghanistan (1996–2002). Under the rule of the Taliban, women who had previously enjoyed many rights were banished from the workforce, forbidden an education, and prohibited from leaving their homes unless accompanied by a close male relative (www.pbs.org/wgbh/globalconnections/mideast/timeline/text/qwomen.html).

In this chapter, we explore the works of four different analysts who take seriously the distinct social situation of women and examine it from various theoretical viewpoints. We begin with the Canadian sociologist Dorothy Smith, who provocatively blends neo-Marxist, phenomenological, and ethnomethodological concepts and ideas. We then turn to the work of African American sociologist Patricia Hill Collins, who extends the work of Smith by formally situating the variable of race into the critical/phenomenological exploration of class and gender, while also borrowing significantly from postmodernism and recent work on the body and sexuality. The final two theorists featured in this chapter challenge the prevailing sex/gender dichotomy, that is, the notion that "sex" is the *biological* difference between male and female human animals, while "gender" is the *social* difference between male and female roles or men's and women's personalities (Connell 2002:33). Australian sociologist Raewyn (R. W.) Connell explains how in many ways men and boys are

Photo 14.1 Kuwaiti women press for their full political rights amid crucial parliamentary meeting in March 2005.

---

[1]For instance, in no country in the region is domestic violence outlawed, and some laws, such as those that encourage men who rape women to marry their victims, even condone violence against women.

gatekeepers of gender equality. While in accordance with postmodern lines of thought, the American philosopher Judith Butler challenges the very binary categories that we use to think about both gender and sexual orientation.[2]

Table 14.1 International Women's Suffrage Timeline

• 1893 New Zealand	• 1950 India
• 1902 Australia[1]	• 1954 Colombia
• 1906 Finland	• 1957 Malaysia, Zimbabwe
• 1913 Norway	• 1962 Algeria
• 1915 Denmark	• 1963 Iran, Morocco
• 1917 Canada[2]	• 1964 Libya
• 1918 Austria, Germany, Poland, Russia	• 1967 Ecuador
• 1919 Netherlands	• 1971 Switzerland
• 1920 United States	• 1972 Bangladesh
• 1921 Sweden	• 1974 Jordan
• 1928 Britain, Ireland	• 1976 Portugal
• 1931 Spain	• 1989 Namibia
• 1944 France	• 1990 Western Samoa
• 1945 Italy	• 1993 Kazakhstan, Moldova
• 1947 Argentina, Japan, Mexico, Pakistan	• 1994 South Africa
• 1949 China	• 2005 Kuwait

SOURCE: Information Please® Database, © 2005 Pearson Education, Inc. All rights reserved. (http://www.infoplease.com/ipa/A0931343.html)

Note: Two countries do not allow their people, male or female, to vote: Brunei and the United Arab Emirates. Saudi Arabia is the only country with suffrage that does not allow its women to vote.

[1]Australian women, with the exception of aboriginal women, won the vote in 1902. Aboriginals, male and female, did not have the right to vote until 1962.

[2]Canadian women, with the exception of Canadian Indian women, won the vote in 1917. Canadian Indians, male and female, did not win the vote until 1960. *Source: The New York Times,* May 22, 2005.

---

[2]To be sure, feminism has never been a unified body of thought, and there are various ways that feminisms and feminist theorists can be contemplated. One of the most common is according to political/ideological orientation. According to this approach (which typically equates "feminism" with "feminist theory"), "liberal feminists," such as Betty Friedan, focus on how political, economic, and social rights can be fully extended to women within contemporary society; while "radical feminists," such as Andrea Dworkin (1946–2005) and Catharine MacKinnon (1946–), most famous for their proposal for a law that defined pornography as a violation of women's civil rights (thereby allowing women to sue the producers and distributors of pornography in a civil court for damages), view women as an oppressed group, who, like other oppressed peoples, must struggle for their liberation against their oppressors—in this case, men. However, here we consider feminists largely in terms of their theoretical orientation rather than political/ideological commitment, because we view the former as *prior to* the latter (Alexander 1987:7). As discussed in Chapter 1, theoretical presuppositions are, by definition, simply the most basic assumptions that theorists make as they go about thinking and writing about the world (ibid.:12).

▓▓ DOROTHY E. SMITH (1926–): A BIOGRAPHICAL SKETCH

Dorothy E. Smith was born in Great Britain in 1926. She worked at a variety of jobs and was a secretary at a publishing company before she decided to enhance her employment prospects by attaining a college degree. She began college at the London School of Economics in 1951, and she received her bachelor's degree in sociology from the London School of Economics in 1955. She and her husband then decided to both go on to graduate school at the University of California, Berkeley. Smith maintains that although her years at Berkeley were in many ways the unhappiest of her life, she learned a lot, both inside and outside the classroom (http://sociology.berkeley.edu/alumni/). Through "the experience of marriage . . . and the arrival of children, and the departure of her husband rather early one morning," she learned about the discrepancy between social scientific description and lived experience (1979:151). Through courses in survey methods and mathematical sociology, she learned a type of sociological methodology that she would come to reject, but with which would come to formulate her own opposing methodology. Through a wonderful course taught by Tamotsu Shibutani, she gained a deep appreciation for George Herbert Mead, which "laid the groundwork for a later deep involvement with the phenomenology of Maurice Merleau-Ponty" (http://faculty.maxwell.syr.edu/mdevault/dorothy_smith.htm).

After completing her doctorate in sociology in 1963, Smith worked as a research sociologist and lecturer at the University of California, Berkeley. At times, she was the only woman in the UC Berkeley Department of Sociology. Deeply moved by the newly emerging women's movement, Smith organized a session for graduate students to "tell their stories" about gender inequities in academia (of which "there were many") (http://faculty .maxwell.syr.edu/mdevault/dorothy_smith.htm).

By the late 1960s, Smith's marriage had fallen apart, and lacking day care and family support, she returned home to England to raise her children and teach. She became a lecturer in sociology at the University of Essex, Colchester. Several years later, Smith accepted a full-time position at the University of British Columbia, and it was here that Smith's feminist transformation (which had begun in Berkeley) deepened. Smith taught one of the first women's studies courses, and the lack of existing materials gave her impetus to "go from the kind of deep changes in my psyche that accompanied the women's movement to writing those changes into the social" (ibid.). Smith also helped create a women's action group that worked to improve the status of women "at all levels of the university," and she was involved in establishing a women's research center in Vancouver outside the university that would provide action-relevant research to women's organizations (ibid.). Smith also edited a volume providing a feminist critique of psychiatry (*Women Look at Psychiatry: I'm Not Mad, I'm Angry,* 1975), and she began to reread Marx and integrate Marxist ideas into her work, as is reflected in her pamphlet *Feminism and Marxism, a Place to Begin, a Way to Go* (1977).[3]

In 1977, Smith became a professor in the Department of Sociology and Equity Studies in Education at the Ontario Institute for Studies in Education at the University of Toronto. Here Smith published the works for which she is most well known, including *A Sociology for Women* (1979), *The Everyday World as Problematic* (1987), *The Conceptual Practices of Power* (1990a), *Texts, Facts, and Femininity* (1990b), *Writing the Social* (1999), and most recently, *Institutional Ethnography: A Sociology for People* (2005). In these works, Smith exhorts a powerful feminist theory of what she calls relations of ruling, and she sets out her own approach, which she calls institutional ethnography, as a means for building knowledges as to how the relations of ruling operate from the standpoints of the people participating in them. These pivotal ideas are discussed further later in the chapter.

Smith continues to be an active teacher and scholar. As professor emerita in the Department of Sociology and Equity Studies in Education at the Ontario Institute for Studies in Education at the University of Toronto and an adjunct professor in the Department of Sociology, University of Victoria, British Columbia, she continues to educate and inspire a new generation of scholars dedicated to

---

[3]Interestingly, Smith (1977:9) maintains that although she worked as a socialist when she was a young woman in England, it was not until she reread Marx in the 1970s that she came to really understand what Marx meant.

institutional ethnography (see, for instance, Marie Campbell and Ann Manicom, eds., *Experience, Knowledge and Ruling Relations: Explorations in the Social Organization of Knowledge,* University of Toronto Press, 1995).

SMITH'S INTELLECTUAL INFLUENCES AND CORE IDEAS

Although Dorothy Smith has written on a wide variety of topics, including education, Marxism, the family, mental illness, and textual analysis, she is most well known as one of the originators of **standpoint theory**.[4] Smith uses the notion of standpoint to emphasize that what one knows is affected by where one stands (one's subject position) in society. We begin from the world as we actually experience it, and what we know of the world and of the "other" is conditional upon that location (Smith 1987). Yet, Smith's argument is not that we cannot look at the world in any way other than our given standpoint. Rather, her point is that (1) no one can have complete, objective knowledge; (2) no two people have exactly the same standpoint; and (3) we must not take the standpoint from which we speak for granted. Instead, we must recognize it, be reflexive about it, and problematize it. Our situated, everyday experience should serve as a "point of entry" of investigation (Smith 2005:10).

Put in another way, the goal of Smith's feminist sociology is to explicitly reformulate sociological theory by fully accounting for the standpoint of gender and its effects on our experience of reality. Interestingly, it was Smith's particular standpoint as a female in a male-dominated world, and specifically as simultaneously a wife and mother and a sociology graduate student in the 1960s, that led her to the formulation of her notion of standpoint. By overtly recognizing the particular standpoint from which she spoke, Smith was bringing to the fore the extent to which the issue of standpoint had been unacknowledged in sociology. This point is quite ironic, really. Sociology was explicitly set out as the "scientific" and "objective" study of society when it first emerged as a discipline in the nineteenth century; but because its first practitioners were almost exclusively men, it implicitly assumed and reflected the relevancies, interests, and perspectives of (white, middle-class) European males.[5] "Its method, conceptual schemes and theories had been based on and built up within the male social universe" (Smith 1990a:23).

The failure to recognize the particular standpoints from which they spoke not only left sociologists unaware of the biases inherent to their position. In addition, by implicitly making the discipline of sociology a masculine sociology—that is, by focusing on the world of paid labor, politics, and formal organizations (spheres of influence from which women have historically been excluded) and erasing or ignoring women's world of sexual reproduction, children, household labor, and affective ties—sociology unwittingly served as a vehicle for alienating women from their own lives (Seidman 1994:212–13). This is the irony mentioned previously: at the same time that sociology emerged as provocative new discipline dedicated to explaining the inequalities and systems of stratification at the heart of various societies (especially apparent, for instance, in Marx and Weber), it created its own version of domination by shifting attention almost exclusively to one particular dimension of human social life—the masculine-dominated macrolevel public sphere—at the expense of another (the world of women).

In short, Smith underscores not only that the standpoint of men is consistently privileged and that of women devalued, but that the standpoint of the (white) male upper-class pervades and *dominates* other

---

[4]The term "feminist standpoint theory" was actually not coined by Smith. Rather, feminist standpoint theory (and hence "standpoint theory") is traced to Sandra Harding (1986), who, based on her reading of the work of feminist theorists—most important, Dorothy Smith, Nancy Hartsock, and Hilary Rose—used the term to describe a feminist critique beyond the strictly empirical one of claiming a special privilege for women's knowledge, and emphasizing that knowledge is always rooted in a particular position and that women are privileged epistemologically by being members of any oppressed group (Smith 2005:8). See also Harding (2004).

[5]Although Smith did not focus on race, as you will shortly see, Patricia Hill Collins built on Smith's work by illuminating how race is intertwined with gender and class standpoints.

worldviews. This idea, that not all standpoints are equally valued and accessed in society, clearly reflects Smith's critical/Marxist roots. As discussed previously, beginning with her pamphlet *Feminism and Marxism* (1977), Smith explicitly links her feminism with Marxism. She explains how "objective social, economic and political relations . . . shape and determine women's oppression" (ibid.:12). She focuses on "the relations between patriarchy and class in the context of the capitalist mode of production" (1985:1) and emphasizes how "the inner experiences which also involved our exercise of oppression against ourselves were ones that had their location in the society outside and originated there" (1977:10).

Yet, Smith's feminist theory is not just derived from an application of Marx to the issue of gender; rather, it reflects Smith's phenomenological roots (see Chapter 13) as well. Specifically, Smith links a neo-Marxist concern about structures of domination with a phenomenological emphasis on consciousness and the active construction of the taken-for-granted world. She explicitly demonstrates the extent to which men and women bracket and view the world in distinctive ways, in conjunction with their distinct, biographically articulated lifeworlds. In her own case, for instance, Smith recognizes that she experienced "two subjectivities, home and university" that could not be blended, for "they ran on separate tracks with distinct phenomenal organization" (Smith 2005:11). "Home was organized around the particularities of my children's bodies, faces, movements, the sound of their voices, the smell of their hair . . . and the multitudes of the everyday that cannot be enumerated," while the "practice of subjectivity in the university excluded the local and bodily from its field" (ibid.:12). In this way, Smith (1987:83–84) notes that female-dominated work in the concrete world of the everyday demands one to be attuned to the sensory experiences of the body. "Here there are textures and smells. . . . It has to happen here somehow if she is to experience it at all" (1987:82). The abstract world of the professions, on the other hand, requires an individual to take this level of experience for granted.

Smith is particularly indebted to the phenomenologist Alfred Schutz (see Chapter 13). Recall that it was Schutz (1970:11, as cited in Smith 1987:83) who argued that we put various levels of our personality "in play" in various provinces of reality. Schutz used the term *mitwelt relations* to refer to relations in which individuals are experienced as "types" (e.g., the relationship between you and the person who delivers your mail), and he used the term *umwelt relations* to refer to more intimate face-to-face relations. According to Schutz, in contrast to mitwelt relations, in umwelt relations each person must be aware of the other's body as a field of expression that fosters the development of intersubjectivity.

Smith (1987:83) extends Schutz's distinction between umwelt and mitwelt relations by asserting that "if men are to participate fully in the abstract mode of action, they must be liberated from having to attend to their needs in the concrete and particular." That is, traditionally not only are umwelt relations more central in women's lives, but men *relegate* their umwelt relations to women (for instance, a boss has his secretary shop for an anniversary present for his wife and make his personal calls). Thus, Smith argues that "women's work conceals from men the actual concrete forms on which their work depends" (ibid.:83–84).

This brings us to Smith's concept of **bifurcation of consciousness**. Smith uses this term to refer to a separation or split between the world as you actually experience it and the dominant view to which you must adapt (e.g., a masculine point of view). The notion of bifurcation of consciousness underscores that subordinate groups are conditioned to view the world from the perspective of the dominant group, since the perspective of the latter is *embedded* in the institutions and practices of that world, while the dominant group, on the other hand, enjoys the privilege of remaining oblivious to the worldview of the Other, or subordinate group, since the Other is fully expected to accommodate to them. The "governing mode" of the professions, then, creates a bifurcation of consciousness in the actor: "It establishes two modes of knowing, experiencing, and acting—one located in the body and in the space that it occupies and moves into, the other passing beyond it" (1987:82).

Of course, bifurcation of consciousness reflects Smith's own experience of living in "two worlds": the dominant, masculine-oriented, "abstract" world of the sociologist and the "concrete" world of wife and mother. The key point, as Smith (2005:11) notes, is that "the two subjectivities, home and university, could not be blended." In this way, Smith's concept of bifurcation of consciousness recalls W. E. B. Du Bois's

Thus far we have discussed Smith's dual neo-Marxist and phenomenological roots. There is also an

concept of "double consciousness," which he used to describe the experiential condition of black Americans.[6] In both cases, it is the oppressed person who must adapt to the "rules of the game" that do not reflect her interests or desires, even though, in both cases, the dual subjectivities provide a uniquely "clairvoyant" vantage point (in Du Bois's terms). Thus, for instance, women in male-dominated professions (e.g., law enforcement, construction) acclimate themselves to sexist and even misogynistic talk about the female body that is a normal part of their everyday work environment. Not only do they learn to ignore the banter; indeed, they might even chime in. However, because they must continually accommodate themselves to the dominant group in order to gain acceptance in a world that is not theirs, members of oppressed or minority groups become alienated from their "true" selves.

Thus far we have discussed Smith's dual neo-Marxist and phenomenological roots. There is also an important discursive bent in Smith's work that has become especially apparent in the last decade, however. In conjunction with the poststructuralist turn (see Chapter 15), Smith emphasizes that in modern, Western societies social domination operates through *texts* (such as medical records, census reports, psychiatric evaluations, employment files) that facilitate social control. Thus, Smith (1990b:6) describes **relations of ruling** as including not only forms such as "bureaucracy, administration, management, professional organization and media" but also "the complex of discourses, scientific, technical, and cultural, that intersect, interpenetrate, and coordinate" them. Smith (1987:4) maintains that behind and within the "apparently neutral and impersonal rationality of the ruling apparatus" is concealed a "male subtext." Women are "excluded from the practices of power within textually mediated relations of ruling" (ibid.). Thus, for instance, official psychiatric evaluations replace the individual's actual lived experience with a means for interpreting it; the individual becomes a case history, a type, a disease, a syndrome, and a treatment possibility (Seidman 1994:216).

Smith goes on to suggest that because sociology too relies on these same kinds of texts, it too is part and parcel of the relations of ruling. The subject matter and topics of sociology are those of the ruling powers. Sociological knowledge receives its shape less from actualities and the lived experiences of real individuals than from the interests in control and regulation, by the state, professional associations, and bureaucratization (Seidman 1994:216).

Most important, Smith does not just criticize modern, "masculinist" sociology; she provides an alternative to it. Inspired by Marx's historical realism but also drawing on ethnomethodology—which, as discussed in Chapter 13, considers that practical activities, practical circumstances, and practical sociological reasoning must not be taken for granted but rather be topics of empirical study (Garfinkel 1967:1)—Smith advocates a "sociology for women" that begins "where women are situated": in the "particularities of an actual, everyday world" (Smith 1987:109). Smith's sociology for women aims not to "transform people into objects" but to "preserve their presence as subjects" (1987:151). Smith (1987:143) argues that the "only route to a faithful telling that does not privilege the perspectives arising in the sites of her sociological project and her participation in a sociological discourse is to commit herself to an inquiry that is ontologically faithful, faithful to the presence and activity of her subjects and faithful to the actualities of the world that arises for her, for them, for all of us, in the ongoing co-ordering of our actual practices."[7]

Smith calls her particular approach **institutional ethnography**. Institutional ethnography is a method of elucidating and examining the relationship between everyday activities and experiences and larger institutional imperatives. Interestingly, the very term "institutional ethnography" explicitly couples an

---

[6]See Edles and Appelrouth (2005:323–25).

[7]In her most recent book, Smith (2005) updates her terminology by replacing the notion of "a sociology for women" with that of "a sociology for people." In other words, the notion of "a sociology for women" can be understood as reflecting a particular historical era in which feminists called attention to the fact that the standpoint of women was absent in the academy; however, today the more pertinent (and more postmodern) point is that we must begin wherever we are, that is, not only in terms of "gender" but also class, race, sexual orientation, ablebodiedness, and so on. This is institutional ethnography.

emphasis on structures of power ("institutions") with the microlevel practices that make up everyday life ("ethnography"). Smith's point, of course, is that it is in microlevel, everyday practices at the level of the individual that collective, hierarchical patterns of social structure are experienced, shaped, and reaffirmed. For instance, in one passage you will read, Smith explains how the seemingly benign, everyday act of walking her dog actually reaffirms the class system. As Smith "keeps an eye on her dog" so that it does its business on some lawns as opposed to others, she is, in fact, "observing some of the niceties of different forms of property ownership" (renters versus owners) (1987:155); she is participating in the existing relations of ruling. This point is illustrated in Figure 14.1.

Figure 14.1   Smith's Concept of Institutional Ethnography: Walking the Dog

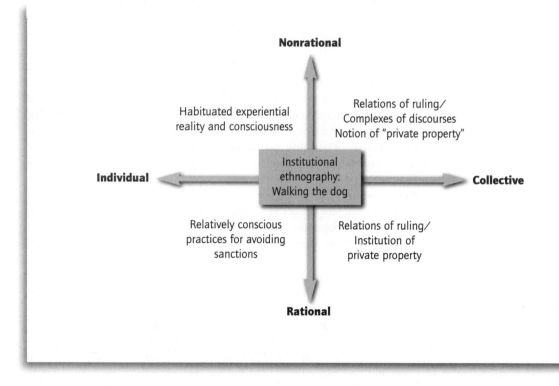

▪▪ SMITH'S THEORETICAL ORIENTATION

Smith's theoretical approach is explicitly multidimensional, as can be readily seen in her central concepts (see Figures 14.1 and 14.2). For instance, as discussed above, the term "institutional ethnography" explicitly reflects Smith's dual emphasis on collective structures of ruling and the institutionalization of power and their actual workings at the level of the individual in everyday life. In terms of action, as shown in Figure 14.1, institutional ethnography can be said to reflect a rationalistic emphasis on practical action both at the level of the individual and at the collective level of the institution; however, clearly Smith's phenomenological roots lead her to appreciate the nonrational motivation for action as well. Above all, Smith emphasizes that taken-for-granted, subjective categories provide the backdrop for the pragmatic performances that constitute the everyday world and, in doing so, reaffirm the existing structural order. For instance, in the example above, it is only because of her internalization of taken-for-granted notions of class and private property that Smith knows how and where to walk her dog. In Schutz's terms, she uses specific "recipes" (see Chapter 13) and taken-for-granted habits, which, by definition, work at the individual/nonrational level.

Figure 14.2 Smith's Concepts of "Standpoint" and "Relations of Ruling"

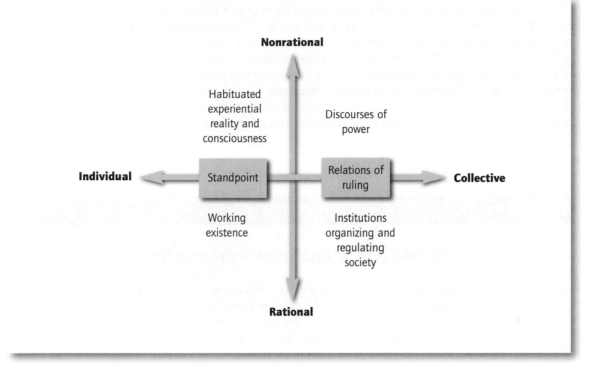

So, too, the term "standpoint" reflects Smith's dual rational and nonrational approach to action and individual and collective approach to order, in that "standpoint" refers both to our objective (rational) position and subjective (nonrational) position in the (collective) social hierarchy, and to our unique biographical (individual) situation. For instance, as shown in Figure 14.2, my "standpoint" as a mother is rooted at once in the *meaning* (including social status or honor) accorded to "mothers" in general in our society, as determined by the complexes of discourses that are part of relations of ruling (collective/ nonrational), and the specific reward structure accrued to that position by the (collective/rational) institutions organizing and regulating society (collective/ rational) as well. That said, above all, "standpoint" reflects the specific attitudes, emotions, and values that I experience and internalize at the level of the individual (individual/nonrational) as well as the habituated day-to-day experience, and the particular strategic advantages and disadvantages I am able to accrue through this position and my mundane working existence (individual/rational).

Put in another way, Smith articulates not only how individuals unthinkingly "do" gender (and class) in daily life at the individual/nonrational level, but also the subjective categories that make this possible, that is, the taken-for-granted subjective understandings of what it means to be a "boy" or a "girl" that reflect the collective, nonrational realm. Akin to Schutz and Berger and Luckmann (see Chapter 13) as well as the poststructuralists who emphasize discourse and are discussed in the next chapter, Smith continually emphasizes that gender cannot be "done" at the individual level in everyday life without taken-for-granted conceptualizations at the collective level.

In a similar vein, that Smith's concept "relations of ruling" encompasses both such forms as "bureaucracy, administration, management, professional organization and media" and scientific, technical, and cultural discourses reflects the collective/rational and collective/nonrational realms, respectively (see Figure 14.2). Specifically, that Smith defines ruling relations as "objectified forms of consciousness and organization, *constituted externally to particular places and people*" (2005:227, emphasis added) clearly reflects her collectivistic orientation to order. And although Smith also underscores that ruling relations refer to "that total complex of activities, differentiated into many spheres . . . through which we are ruled and through which we, and I emphasize this *we,* participate in ruling" (Smith 1990a, as cited in Calhoun 2003:316), which indicates an acknowledgment of individual agency, that "forms of consciousness are created that are properties of organization or discourse *rather than of individual subjects*" (1987:3, emphasis added) clearly reflects a collectivistic approach to order. This dual rational/nonrational approach to action and collectivistic approach to order inherent in Smith's concept of relations of ruling is illustrated in Figure 14.2. Interestingly, then, taken together Figures 14.1 and 14.2 illustrate that the multidimensionality of the concept of institutional ethnography is a function of its incorporation of the more individualistic concept of standpoint and the more collectivistic concept of ruling relations.

Readings

Introduction to *Institutional Ethnography*

In this excerpt from her most recent book, *Institutional Ethnography* (2005), Smith explicitly defines "institutional ethnography" and explains how she came to formulate this unique method of inquiry. In addition, Smith explains the historical trajectory of gender and relations of ruling, that is, how the radical division between spheres of action and of consciousness of middle-class men and women came to emerge. As indicated previously, it is precisely this conceptualization of relations of ruling (or ruling relations) as not simply modes of domination but also forms of *consciousness* that forms the crux of Smith's work.

———— *Institutional Ethnography* (2005) ————

Dorothy Smith

WOMEN'S STANDPOINT: EMBODIED KNOWING VS. THE RULING RELATIONS

It's hard to recall just how radical the experience of the women's movement was at its inception for those of us who had lived and thought within the masculinist regime against which the movement struggled. For us, the struggle was as much within ourselves, with what we knew how to do and think and feel, as with that regime as an enemy outside us. Indeed we ourselves had participated however passively in that regime. There was no developed discourse in which the experiences that were spoken originally as everyday experience could be translated into a public language and become political in the ways distinctive to the women's movement. We learned in talking with other women about experiences that we had and about others that we had not had. We began to name "oppression," "rape," "harassment," "sexism," "violence," and others. These were terms that did more than name. They gave shared experiences a political presence.

Starting with our experiences as we talked and thought about them, we discovered depths of alienation and anger that were astonishing. Where had all these feelings been? How extraordinary were the transformations we experienced as we discovered with other women how to speak with one another about such experiences and then how to bring them forward publicly, which meant exposing them to men. Finally, how extraordinary were the transformations of ourselves in this process. Talking our experience was a means of discovery. What we did not know and did not know how to think about, we could examine as we found what we had in common. The approach that I have taken in developing an alternative sociology takes up women's standpoint in a way that is modeled on these early adventures of the women's movement. It takes up women's standpoint not as a given and finalized form of knowledge but as a ground in experience from which discoveries are to be made.

It is this active and shared process of speaking from our experience, as well as acting and organizing to change how those experiences had been created, that has been translated in feminist thinking into the concept of a feminist standpoint–or, for me, women's standpoint. However the concept originated, Sandra Harding (1988) drew together the social scientific thinking by feminists, particularly Nancy Hartsock, Hilary Rose, and myself, that had as a common project taking up a standpoint in women's experience. Harding argued that feminist empiricists who claimed both a special privilege for women's knowledge and an objectivity were stuck in an irresolvable paradox. Those she described as "feminist standpoint theorists" moved the feminist critique a step beyond feminist empiricism by claiming that knowledge of society must always be from a position in it and that women are privileged epistemologically by being members of an oppressed group. Like the slave in Hegel's parable of the master-slave relationship, they can see more, further, and better than the master precisely because of their marginalized and oppressed condition. She was, however, critical of the way in which experience in the women's movement had come to hold authority as a ground for speaking, and claiming to speak truly, that challenged the rational and objectified forms of knowledge and their secret masculine subject. Furthermore, feminist standpoint theory, according to Harding, implicitly reproduced the universalized subject and claims to objective truth of traditional philosophical discourse, an implicit return to the empiricism we claimed to have gone beyond.

The notion of women's standpoint—or indeed the notion that women's experience has special authority—has also been challenged by feminist theorists. It fails to take into account diversities of class and race as well as the various forms and modulations of gender. White middleclass heterosexual women dominated the early phases of the women's movement in the 1960s and 1970s, but soon our, and I speak as one, assumptions about what would hold for women in general were challenged and undermined, first by working-class women and lesbians, then by African-North American, Hispanic, and Native women. The implicit presence of class, sexuality, and colonialism began to be exposed. Our assumptions were also challenged by women in other societies who experience wasn't North American, by women such as those with disabilities and older women whose experience was not adequately represented and, as the women's movement evolved over time, by younger women who have found the issues of older feminists either alien or irrelevant.

The theoretical challenge to the notion of women's standpoint has been made in terms of its alleged essentialism. It has been seen as essentialist because it excludes other bases of oppression and inequity that intersect with the category "women." The critique of essentialism, however, assumes the use of the category "women" or "woman" to identify shared and defining attributes. While essentialism has been a problem in the theorizing of *woman,* it cannot be extended to all uses of such categories. In practice in the women's movement, the category has worked politically rather than referentially. As a political concept, it coordinates struggle against the masculinist forms of oppressing women that those forms themselves explicitly or implicitly universalize. Perhaps most important, it creates for women what had been missing, a subject position in the public sphere and, more generally, one in the political, intellectual, and cultural life of the society.

Claiming a subject position within the public sphere in the name of women was a central enterprise of the women's movement in its early days in the 1970s and 1980s. A powerful dynamic was created. While those making the claim first were white middle-class women, the new subject position in public discourse opened the way for others who had found themselves excluded by those who'd gone before. Their claims were positioned and centered differently, and their own experience became authoritative. It is indeed one of the extraordinary characteristics of the women's movement that its continual disruption, its internal struggles against racism and white cultural dominance, its internal quarrels and angers, have been far from destructive to the movement. On the contrary, these struggles in North

America and Europe have expanded and diversified the movement as women other than those with whom it originated gave their own experiences voice.

WOMEN'S STANDPOINT AND THE RULING RELATIONS

Standpoint is a term lifted out of the vernacular, largely through Harding's innovative thinking and her critique (1988), and it is used for doing new discursive work. Harding identifies standpoint in terms of the social positioning of the subject of knowledge, the knower and creator of knowledge. Her own subsequent work develops an epistemology that relies on a diversity of subject positions in the sociopolitical-economic regimes of colonialism and imperialism. The version of standpoint that I have worked with, after I had adopted the term from Harding (previously I'd written of "perspective"; D. E. Smith (1974a), is rather different. It differs also from the concept of a feminist standpoint that has been put forward by Nancy Hartsock (1998) in that it does not identify a socially determined position or category of position in society (or political economy).[i] Rather, my notion of women's (rather than feminist) standpoint is integral to the design of what I originally called "a sociology for women," which has necessarily been transformed into "a sociology for people." It does not identify a position or a category of position, gender, class, or race within the society, but it does establish as a subject position for institutional ethnography as a method of inquiry, a site for the knower that is open to anyone.

As a method of inquiry, institutional ethnography is designed to create an alternate to the objectified subject of knowledge of established social scientific discourse. The latter conforms to and is integrated with what I have come to call the "ruling relations"—that extraordinary yet ordinary complex of relations that are textually mediated, that connect us across space and time and organize our everyday lives—the corporations, government bureaucracies, academic and professional discourses, mass media, and the complex of relations that interconnect them. At the inception of this early stage of late- twentieth-century women's movement, women were excluded from appearing as agents or subjects with the ruling relations. However we might have been at work in them, we were subordinates. We were women whose work as mothers reproduced the same gendered organization that subordinated us; we were the support staff, store clerks, nurses, social workers doing casework and not administration, and so on. In the university itself, we were few and mostly marginal (two distinguished women in the department where I first worked in Canada had never had more than annual lectureships).

"Standpoint" as the design of a subject position in institutional ethnography creates a point of entry into discovering the social that does not subordinate the knowing subject to objectified forms of knowledge of society or political economy. It is a method of inquiry that works from the actualities of people's everyday lives and experience to discover the social as it extends beyond experience. A standpoint in people's everyday lives is integral to that method. It is integral to a sociology creating a subject position within its discourse, which anyone can occupy. The institutional ethnographer works from the social in people's experience to discover its presence and organization in their lives and to explicate or map that organization beyond the local of the everyday.

EXAMINING SOCIOLOGY FROM A WOMAN'S STANDPOINT

The project of developing a sociology that does not objectify originated, as did so much in the women's

---

[i]Hartsock's concern is to reframe historical materialism so that women's experience and interests are fully integrated. Of particular importance to her is the adequate recognition of the forms of power that the women's movement has named "patriarchal." Women's marginal position, structured as it is around the work associated with reproduction and the direct production of subsistence, locates women distinctively in the mode of production in general. For her, taking a feminist standpoint introduces a dimension into historical materialism neglected by Marx and his successors. She designs a feminist standpoint that has a specifically political import. It might, I suppose, be criticized as essentialist, but, if we consider not just North America and not just white middle-class professional North America, it's hard to deny that Hartsock is characterizing a reality for women worldwide. In Canada a recent census report shows that while women's participation in the paid labor force has increased substantially over the past thirty years, "women remain more than twice as likely as men to do at least 30 hours a week of cooking and cleaning" (Andersen 2003, A7) and are move involved in child care than men, particularly care of younger children.

movement, in exploring experiences in my life as a woman. That exploration put into question the fundamentals of the sociology I had learned at length and sometimes painfully as an undergraduate and graduate school student. I was, in those early times, a sociologist teaching at the University of British Columbia, on the west coast of Canada, and a single parent with two small boys. My experience was of contradictory modes of working existence: on the one hand was the work of the home and of being a mother; on the other, the work of the academy, preparing for classes, teaching, faculty meetings, writing papers, and so on. I could not see my work at home in relation to the sociology I taught, in part, of course, because that sociology had almost nothing to say about it.

I learned from the women's movement to begin in my own experience and start there in finding the voice that asserted the buried woman. I started to explore what it might mean to think sociologically from the place where I was in-body, living with my children in my home and with those cares and consciousness that are integral to that work. Here were the particularities of my relationships with my children, my neighbors, my friends, their friends, our rabbit (surprisingly fierce and destructive—my copy of George Herbert Mead's *Mind, Self, and Society* bars scores inflicted by our long-eared pet's teeth and claws), our two dogs, and an occasional hamster. In this mode, I was attentive to the varieties of demands that housekeeping, cooking, child care, and the multiple minor tasks of our local settings made on me. When I went to work in the university, I did not, of course, step out of my body, but the focus of my work was not on the local particularities of relationships and setting but on sociological discourse read and taught or on the administrative work of a university department. Body, of course, was there as it had to be to get the work done, but the work was not organized by and in relation to it.

The two subjectivities, home and university, could not be blended. They ran on separate tracks with distinct phenomenal organization. Memory, attention, reasoning, and response were organized quite differently. Remembering a dental appointment for one of the children wasn't part of my academic consciousness, and if I wasn't careful to find some way of reminding myself that didn't depend on memory, I might have well forgot it. My experiences uncovered radical differences between home and academy in how they were situated, and how they situated me, in the society. Home was organized around the particularities of my children's

bodies, faces, movements, the sounds of their voices, the smell of their hair, the arguments, the play, the evening rituals of reading, the stress of getting them off to school in the morning, cooking, and serving meals, and the multitudes of the everyday that cannot be enumerated, an intense, preoccupying world of work that also cannot really be defined. My work at the university was quite differently articulated; the sociology I thought and taught was embedded in the texts that linked me into a discourse extending indefinitely into only very partially known networks of others, some just names of the dead; some the heroes and masters of the contemporary discipline; some just names on books or articles; and others known as teachers, colleagues, and contemporaries in graduate school. The administrative work done by faculty tied into the administration of the university, known at that time only vaguely as powers such as dean or president or as offices such as the registrar, all of whom regulated the work we did with students. My first act on arriving in the department office, after greeting the secretaries, was to open my mail and thus to enter a world of action in texts.

I knew a practice of subjectivity in the university that excluded the local and bodily from its field. Learning from the women's movement to start from where I was as a woman, I began to attend to the university and my work there from the standpoint of "home" subjectivity. I started to notice what I had not seen before. How odd, as I am walking down the central mall of that university that opens up to the dark blue of the humped islands and the further snowy mountains to the north, to see on my left a large hole where before there had been a building! In the mode of the everyday you can find the connections, though you may not always understand them. In a house with children and dogs and rabbits, the connection between the destruction of the spine of my copy of *Mind, Self, and Society* and that rabbit hanging around in my workspace was obvious. But the hole where once there'd been a building couldn't be connected to any obvious agent. The peculiar consciousness I practiced in the university began to emerge for me as a puzzlingly strange form of organization. If I traced the provenance of that hole, I'd be climbing up into an order of relations linking administrative process with whatever construction company was actually responsible for the making of the hole; I'd be climbing into a web of budgets, administrative decisions, provincial and federal government funding, and so on and so on. I'd be climbing into that order of relations that institutional ethnographers call the "ruling relations." These could be seen as relations that divorced the subject from the particularized settings and relationships of her life and work as mother and housewife. They created subject positions that elevated consciousness into a universalized mode, whether of the social relations mediated by money or of those organized as objectivity in academic or professional discourse. Practicing embodiment on the terrain of the disembodied of those relations brought them into view. I became aware of them as I became aware of their presence and power in the everyday, and, going beyond that hole in the ground, I also began to think of the sociology I practiced in the everyday working world of the university as an organization of discursive relations fully integrated with them.

Introduction to *The Everyday World as Problematic*

In this reading taken from *The Everyday World as Problematic* (1987), Smith further elucidates institutional ethnography using concrete examples from her own experience. As you will see, by starting from her own experience, Smith does *not* mean that she engages only in a self-indulgent inner exploration with herself as sole focus and object. Rather, Smith means that she begins from her own original but tacit knowledge as well as from the acts by which she brings this knowledge into her grasp (Calhoun 2003:320). As Smith states, "we can never escape the circles of our own heads if we accept that as our territory . . . We aim not at a reiteration of what we already (tacitly) know, but at an exploration of what passes beyond that knowledge and is deeply implicated in how it is" (ibid).

The Everyday World as Problematic (1987)

Dorothy Smith

INSTITUTIONAL ETHNOGRAPHY: A FEMINIST RESEARCH STRATEGY

Institutional Relations as Generalizers of Actual Local Experience

Let me give an everyday example of what I mean by the "problematic of the everyday world." When I take my dog for a walk in the morning, I observe a number of what we might call "conventions." I myself walk on the sidewalk; I do not walk on the neighbors' lawns. My dog, however, freely runs over the lawns. My dog also, if I am not careful, may shit on a neighbor's lawn, and there are certainly some neighbors who do not like this. I am, of course, aware of this problem, and I try to arrange for my dog to do his business in places that are appropriate. I am particularly careful to see that he avoids the well-kept lawns because those are the ones I know I am most likely to be in trouble over should I/he slip up—which does happen occasionally. The neighborhood I live in is a mixture of single-family residences and rental units, and the differences between the well- and ill-kept lawns are related to this. On the whole, those living in rental units do not care so much about the appearance of their front lawn, whereas those who own their own residences are more likely to give care and attention to the grass and sometimes to the flower beds in front of the house.

So as I walk down the street keeping an eye on my dog I am observing some of the niceties of different forms of property ownership. I try to regulate my dog's behavior with particular scrupulousness in relation to the property rights of the owners of single-family dwellings and am a little more casual where I know the house consists of rented apartments or bachelor units, or, as in one case, a fraternity house.[i]

Customarily in sociology we talk about this behavior in terms of norms. Then we see my selection of a path of behavior for my dog as guided by certain norms held in common by myself and my neighbors. But something important escapes this. The notion of "norm" provides for the surface properties of my behavior, what I can be seen to be doing—in general preventing my dog from shitting on others' lawns and being particularly careful where negative sanctions are more likely to be incurred. A description of the kind I have given is in this way transposed into a normative statement.

As a norm it is represented as governing the observed behavior. What is missing, however, is an account of the constitutive work that is going on. This account arises from a process of practical reasoning. How I walk my dog attends to and constitutes in an active way different forms of property as a locally realized organization. The normative analysis misses how this local course of action is articulated to social relations. Social relations here mean concerted sequences or courses of social action implicating more than one individual whose participants are not necessarily present or known to one another. There are social relations that are not encompassed by the setting in which my dog is walked, but they nonetheless enter in and organize it. The existence of single-family dwellings, of rental units, and the like has reference to and depends upon the organization of the state at various levels, its local by-laws, zoning laws, and so forth determining the "real estate" character of the neighborhood; it has reference to and depends upon the organization of a real estate market in houses and apartments, and the work of the legal profession and others; it has reference to and organizes the ways in which individual ownership is expressed in local practices that maintain the value of the property both in itself and as part of a respectable neighborhood. Thus this ordinary daily scene, doubtless enacted by many in various forms and settings, has an implicit organization tying each particular local setting to a larger generalized complex of social relations. . . .

The language of the everyday world as it is incorporated into the description of that world is rooted in social relations beyond it and expresses relations not peculiar to the particular setting it describes. In my account of walking the dog, there are categories anchored in and depending for their meaning on a larger complex of social relations. The meaning of such terms as "single-family residence" and "rental units," for example, resides in social relations organizing local settings but not fully present in them. The particularizing description gives access to that which is not particular since it is embedded in categories whose meaning reaches into the complex of social relations our inquiry would explicate. Ordinary descriptions, ordinary talk, trail along with them as a property of the meaning of their terms, the extended social relations they name as phenomena.

Thus taking the everyday world as problematic does not confine us to particular descriptions of local settings without possibility of generalization. This has been seen to be the problem with sociological ethnographies, which, however fascinating as accounts of people's lived worlds, cannot stand as general or typical statements about society and social relations. They have been seen in themselves as only a way station to the development of systematic research procedures that would establish the level of generality or typicality of what has been observed of such-and-such categories of persons. Or they may be read as instances of a general sociological principle. This procedure has been turned on its head in an ingenious fashion in "grounded theory," which proposes a method of distilling generalizing concepts from the social organization of the local setting observed whereupon the latter becomes an instance of the general principles distilled from it.[ii] The popularity of this device testifies to the extent to which the problem of generalizability is felt by sociologists. The single case has no significance unless it can in some way or another be extrapolated to some general statement either about society or some subgroup represented methodologically as a population of individuals, or connecting the local and particular with a generalizing concept of sociological discourse.

Beginning with the everyday world as problematic bypasses this issue. The relation of the local and particular to generalized social relations is not a conceptual

---

[i]The more tender and civic-minded of my readers may like to know that two things have changed in my life since I wrote this. One is that I no longer have a dog of my own. I do, however, sometimes dog-sit my two sons' dogs. The second is that we now have "poop 'n' scoop" laws in Toronto, so I have learned to overcome my rural-bred tendencies to let the shit lie where it falls.

[ii]Barney Glaser and Anselm L. Strauss, *The Discovery of Grounded Theory: Strategies for Qualitative Research* (Chicago: Aldine Press, 1967).

or methodological issue, it is a property of social organization. The particular "case" is not particular in the aspects that are of concern to the inquirer. Indeed, it is not a "case" for it presents itself to us rather as a point of entry, the locus of an experiencing subject or subjects, into a larger social and economic process. The problematic of the everyday world arises precisely at the juncture of particular experience, with generalizing and abstracted forms of social relations organizing a division of labor in society at large. . . .

I am using the terms "institutional" and "institution" to identify a complex of relations forming part of the ruling apparatus, organized around a distinctive function—education, health care, law, and the like. In contrast to such concepts as bureaucracy, "institution" does not identify a determinate form of social organization, but rather the intersection and coordination of more than one relational mode of the ruling apparatus. Characteristically, state agencies are tied in with professional forms of organization, and both are interpenetrated by relations of discourse of more than one order. We might imagine institutions as nodes or knots in the relations of the ruling apparatus to class, coordinating multiple strands of action into a functional complex. Integral to the coordinating process are ideologies systematically developed to provide categories and concepts expressing the relation of local courses of action to the institutional function (a point to be elaborated later), providing a currency or currencies enabling interchange between different specialized parts of the complex and a common conceptual organization coordinating its diverse sites. The notion of ethnography is introduced to commit us to an exploration, description, and analysis of such a complex of relations, not conceived in the abstract but from the entry point of some particular person or persons whose everyday world of working is organized thereby. . . .

Institutional ethnography explores the social relations individuals bring into being in and through their actual practices. Its methods, whether of observation, interviewing, recollection of work experience, use of archives, textual analysis, or other, are constrained by the practicalities of investigation of social relations as actual practices. Note however that the institutional

ethnography as a way of investigating the problematic of the everyday world does not involve substituting the analysis, the perspectives and views of subjects, for the investigation by the sociologist. Though women are indeed the expert practitioners of their everyday worlds, the notion of the everyday world as problematic assumes that disclosure of the extralocal determinations of our experience does not lie within the scope of everyday practices. We can see only so much without specialized investigation, and the latter should be the sociologist's special business.

Ideology, Institutions, and the Concept of Work as Ethnographic Ground

The coordination of institutional processes is mediated ideologically. The categories and concepts of ideology express the relation of member's actual practices—their work—to the institutional function. Ethnomethodology has developed the notion of accountability to identify members' methods accomplishing the orderliness and sense of local processes.[iii] Members themselves and for themselves constitute the observability and reportability of what has happened or is going on, in how they take it up as a matter for anyone to find and recognize. Members make use of categories and concepts to analyze settings for features thus made observable. The apparently referential operation of locally applied categories and concepts is constitutive of the reference itself.[iv] When applied to the institutional context, the notion of accountability locates practices tying local settings to the nonlocal organization of the ruling apparatus. Indeed, the institutional process itself can be seen as a dialectic between what members do intending the categories and concepts of institutional ideology and the analytic and descriptive practices of those categories and concepts deployed in accomplishing the observability of what is done, has happened, is going on, and so forth. Thus local practices in their historical particularity and irreversibility are made accountable in terms of categories and concepts expressing the function of the institution. Members' interpretive practices analyzing the work processes that bring the institutional process into

---

[iii]Harold Garfinkel, *Studies in Ethnomethodology* (Englewood Cliffs, NJ: Prentice Hall, 1967).

[iv]D.L. Wieder, *Language and Social Reality: The C ase of Telling the Convict Code* (The Hague: Moulton, 1974).

being in actuality constitute those work processes as institutional courses of action.[v]

Institutional ideologies are acquired by members as methods of analyzing experiences located in the work process of the institution. Professional training in particular teaches people how to recycle the actualities of their experience into the forms in which it is recognizable within institutional discourse. For example, when teachers are in training they learn a vocabulary and analytic procedures that accomplish the classroom in the institutional mode. They learn to analyze and name the behavior of students as "appropriate" or "inappropriate" and to analyze and name their own (and others') responses. In responding to "inappropriate" behavior, they have been taught to avoid "undermining the student's ego" and hence to avoid such practices as "sarcasm." They should, rather, be "supportive." This ideological package provides a procedure for subsuming what goes on in the classroom under professional educational discourse, making classroom processes observable-reportable within an institutional order.[vi] In this way the work and practical reasoning of individuals and the locally accomplished order that is their product become an expression of the non-local relations of the professional and bureaucratic discourse of the ruling apparatus.

The accountability procedures of institutions make some things visible, while others as much a part of the overall work organization that performs the institution do not come into view at all or as other than themselves. Local practices glossed by the categories of the discourse are provided with boundaries of observability beneath which a subterranean life continues. What is observable does not appear as the work of individuals, and not all the work and practices of individuals become observable. When my son was in elementary school, his homework one day was to write up an experiment he had done in science class that day. He asked me how to do it and I replied (not very helpfully), "Well, just write down everything you did." He told me not to be so stupid. "Of course," he said, "they don't mean you write about *everything,* like about filling the jar with water from the tap and taking it to the bench." Clearly there were things done around the doing of an experiment that were essential to, but not entered into or made accountable within, the "experimental procedure." Its boundaries were organized conceptually to select from a locally indivisible work process, some aspects to be taken as part of the experiment and others to be discounted. All were done. All were necessary. But only some were to be made observable-reportable within the textual mode of the teaching of science. In like ways, institutional ideologies analyze local settings, drawing boundaries and the like. They provide analytic procedures for those settings that attend selectively to work processes, thus making only selective aspects of them accountable within the institutional order.

---

[v]Dorothy E. Smith, "No one commits suicide: Textual analyses of ideological practices," *Human Studies* 6 (1983):309–359.

[vi]See Garfinkel, *Studies in Ethnomethodology.*

PATRICIA HILL COLLINS (1948–): A BIOGRAPHICAL SKETCH ■■

Patricia Hill Collins was born in 1948 and grew up in a working-class family in Philadelphia. She earned her BA from Brandeis University in 1969 and her MAT from Harvard University in 1970. Collins worked as a school teacher and curriculum specialist before returning to graduate school and receiving her PhD in sociology from Brandeis University in 1984. It was in teaching a course called "The Black Woman" to middle-school girls in 1970 that Collins realized not only the dearth of teaching materials by and about black women, but also the *significance* of this dearth. The exclusion of black women from intellectual discourses became the subject of her first book, *Black Feminist Thought: Knowledge, Consciousness, and the Politics of Empowerment* ([1990] 2000), which won the Jessie Bernard Award of the American Sociological Association for significant scholarship in gender as well as the C. Wright Mills Award from the Society for the Study of Social Problems. In this highly acclaimed book (excerpts from which you will read), Collins illuminates the rich, self-defined intellectual tradition of black women, which, she argues, has persisted despite formal discursive exclusion. By positioning itself as documenting a tradition or canon, *Black Feminist Thought* legitimates black women's intellectual production as critical social theory (Collins 1998:8).

Collins further explores black feminist thought in *Fighting Words: Black Women and the Search for Justice* (1998) and *Black Sexual Politics* (2004). In *Fighting Words,* Collins shows not only how elite discourses present a view of social reality that elevates the ideas and actions of highly educated white men as normative and superior (ibid.:45), but also how black feminist thought has remained dynamic and oppositional under changing social conditions. In *Black Sexual Politics* (2004), Collins continues to firmly situate black feminist thought in the critical tradition by underscoring that antiracist African American politics in the post–civil rights era must soundly address questions of gender and sexuality.

Collins has taught at a number of universities, including Northern Kentucky University, Tufts University, Boston College, and the University of Cincinnati, where she is Charles Phelps Taft Emeritus Professor of Sociology within the Department of African American Studies. Since 2005, she has also been a professor of sociology at the University of Maryland. Collins's most recent book, *From Black Power to Hip Hop: Essays on Racism, Nationalism, and Feminism* (2006), explores how black nationalism works today in the wake of changing black youth identity.

▪▪ Collins's Intellectual Influences and Core Ideas

Patricia Hill Collins's work integrates elements of feminist theory, standpoint theory, critical theory, Afrocentrism, poststructuralism, and postmodernism. Collins was particularly influenced by Dorothy Smith, as is evident in her concept of **standpoint epistemology**, which she defines as the philosophic viewpoint that what one knows is affected by the standpoint (or position) one has in society ("epistemology" means how we know what we know, how we decide what is valid knowledge). Collins extends the critical/phenomenological/feminist ideas of Dorothy Smith by illuminating the particular epistemological standpoint of black women. Yet, Collins does not merely add the empirical dimension of "race" to Smith's feminist, critical/phenomenological framework. Rather, taking a poststructural/ postmodern turn, Collins emphasizes the "interlocking" nature of the wide variety of statuses—for example, race, class, gender, nationality, sexual orientation—that make up our standpoint, and, in the spirit of Foucault (see Chapter 15), she stresses that where there are sites of domination, there are also potential sites of resistance.

Specifically, Collins (1998, 2004) explicitly situates her work within the critical tradition (indeed, she conceptualizes standpoint theory and postmodernism as "*examples* of critical theory" [1998:254], emphasis added).[8] For Collins (2004:350), what makes critical theory "critical" is its commitment to "justice, for one's own group and/or for that of other groups." Critical social theory illuminates the "bodies of knowledge and sets of institutional practices that actively grapple with the central questions facing groups of people differently placed in specific political, social, and historical contexts characterized by injustice" (ibid.). Yet, Collins rejects "additive" models of oppression that reflect a dichotomous ("top-down") way of thinking about domination rooted specifically in European masculinist thought. Rather than simply elevate one group's suffering over that of another, Collins maps "differences in penalty and privilege that accompany race, class, gender and similar systems of social injustice" (2004:3).

Collins uses the term **matrix of domination** to underscore that one's position in society is made up of multiple contiguous standpoints rather than just one essentialist standpoint. Thus, in contrast to earlier

---

[8]By following Craig Calhoun (1995) in considering postmodernism and standpoint theory as "examples of critical theory," Collins (1998:254n4) is rejecting the narrower (but perhaps more well known) definition of critical theory as simply the Frankfurt School tradition (see Chapter 10) or the style of theorizing of Jürgen Habermas (see Chapter 16).

critical accounts (e.g., the Frankfurt School—see Chapter 10) that assume that power operates from the top down by forcing and controlling unwilling victims to bend to the will of more powerful superiors, Collins ([1990] 2000:226) asserts that "depending on the context, an individual may be an oppressor, a member of an oppressed group, or simultaneously oppressor and oppressed. . . . Each individual derives varying amounts of penalty and privilege from the multiple systems of oppression which frame everyone's lives."[9]

In addition, Collins emphasizes "that people simultaneously experience and resist oppression on three levels: the level of personal biography; the group or community level of the cultural context created by race, class, and gender; and the systemic level of social institutions." (Collins [1990] 2000:227). At the level of the individual, she insists on "the power of the self-definition" (Collins 2004:306) and "self-defined standpoint" (1998:47) and that "each individual has a unique personal biography made up of concrete experiences, values, motivations, and emotions," thereby reasserting both the subjectivity and agency absent in earlier critical models (e.g., the Frankfurt School). For Collins (1998:50), breaking silence represents a moment of insubordination in relations of power—"a direct, blatant insult delivered before an audience."

The group or community level of the cultural context created by race, class, and gender is vital to Collins's conceptualization of black feminist thought, which like all specialized thought reflects the interests and standpoint of its creators. Collins locates black feminist thought in the unique literary traditions forged by black women such as bell hooks, Audre Lorde, and Alice Walker, as well as in the everyday experience of ordinary black women. In addition, black feminist thought is rooted in black women's intellectual tradition nurtured by black women's community. As Collins ([1990] 2000:253) maintains,

> When white men control the knowledge validation process, both political criteria (contextual credibility and evaluation of knowledge claims) can work to suppress Black feminist thought. Therefore, Black women are more likely to choose an alternative epistemology for assessing knowledge claims, one using different standards that are consistent with Black women's criteria for substantiated knowledge and with our criteria for methodology adequacy. . . .

In other words, Collins maintains that the experience of multiple oppressions makes black women particularly skeptical of and vulnerable to dominant paradigms of knowledge and thus more reliant on their own experiential sources of information. Black women "come to voice" and break the silence of oppression by drawing both from their own experience and from the "collective secret knowledge generated by groups on either side of power," that is, the black community and the black female community in particular (Collins 1998:48–49). Black feminist thought offers individual African American women the conceptual tools to resist oppression. Black women have historically resisted, and continue to resist, oppression at individual, community, and institutional levels. A women's blues tradition, the voices of contemporary African American woman writers and thinkers, and women's everyday relationships with each other speak to the outpouring of contemporary black feminist thought in history and literature despite exclusion and/or marginalization in the hegemonic framework.[10]

---

[9]In her recent *Black Sexual Politics* (2004:9–10), Collins takes an even more radical postmodern stance. Here she sees the complexity of "mutually constructing," intertwined dimensions of race, class, gender, and sexuality as so great that she sets her sights not on "untangling the effects" of race, class, gender, sexuality, ethnicity, age, and the like, but rather on simply illuminating them. The point of *Black Sexual Politics*, she says, is not "to tell readers what to think," but rather "it examines what we might think about" (ibid.).

[10]For instance, Alice Walker's *The Color Purple* epitomizes black feminist thought. Told from the perspective of the 14-year-old Celie, a semiliterate black girl brutalized first by her stepfather and then by her husband, *The Color Purple* supplants the typical patriarchal concerns of the historical novel—"the taking of lands, or the birth, battles, and deaths of Great Men"—with the scene of "one woman asking another for her underwear" (Berlant 2000:4).

By articulating the powerful but hidden dynamics of black feminist thought, Collins highlights the underlying assumed whiteness of both feminism and academia and reminds white women in particular that they are not the only feminists. In addition, however, black feminist thought disrupts the masculinist underpinnings of Afrocentrism. Collins maintains that in the same way that European theorists have historically prioritized class over race or gender, and feminists have prioritized gender over either race or class, Afrocentric scholarship, although formally acknowledging the significance of gender, relegates it secondary to the more-pressing fight against racism.

To be sure, Collins (1998:174) readily appreciates the guiding principles at the heart of Afrocentrism— most important, the emphases on reconstructing black culture, reconstituting black identity, using racial solidarity to build black community, and fostering an ethic of service to black community development. Yet, she is highly critical of the "unexamined yet powerful" gender ideology in black nationalist projects, particularly that of Afrocentrists such as Molefi Kete Asante (1942–) who seek to replace Eurocentric systems of knowledge with African-centered ways of knowing.[11]

▪▪ COLLINS'S THEORETICAL ORIENTATION

As indicated earlier, the terms "matrix of domination" and "standpoint epistemology" are explicitly devised so as to reflect a multidimensional approach to order; that is, they pointedly work at the level of the social structure or group and the individual. However, above all, in the spirit of the critical tradition, it is to the collective level that Collins's work is most attuned. For instance, while on one hand Collins's term "self-defined standpoint" seems to readily reflect agency at the level of the individual, interestingly, Collins (1998:47) maintains that she favors this term over bell hooks's term "self-reflexive speech" because self-defined standpoint "ties Black women's speech communities much more closely to institutionalized power relations." Clearly, that "standpoint" refers to "historically shared *group-based* experiences" and that "groups have a degree of permanence over time such that group realities transcend individual experiences" reflects a prioritization of the collective realm ([1990] 2000:247, emphasis in original). As Collins ([1990] 2000:249) states,

> Groups who share common placement in hierarchical power relations also share common experiences in such power relations. Shared angles of vision lead those in similar social locations to be predisposed to interpret in comparable fashion.

To be sure, Collins readily acknowledges that the individual has "unique" experiences that are rooted in her inimitable social location, which reflects her cognizance of the level of the individual (see Figure 14.3). Here we see that the individual is not a proxy for the group, that in contrast to what Marx supposed, oppressed groups do not possess a fixed or stagnant (or "essential") identity. As Collins (1998:249) contends, "using the group as the focal point provides space for individual agency." Nevertheless, Collins never loses sight of the "collective secret knowledge generated by groups on either side of power" from within which individual self-definition ensues (1998:49).

In terms of action, overall Collins's theory reflects a collective/rationalistic view of power characteristic of critical theory, in that relations of power are perceived as a preexisting hierarchical structure external to the individual. However, at the same time, by emphasizing that these are *relations*

---

[11]Asante asserts that Afrocentricity can be done only via complete separation, and that Afrocentrism is vital to combat the Eurocentric arrogance that necessarily obliterates others, for Eurocentrism is nothing less than "symbolic imperialism." However, Collins identifies several specific ways in which gender assumptions undergird black cultural nationalism.

of power and that this involves both collective, discursive codes and their internalization at the nonrational/individual level, Collins presents a nonrational approach to action as well. Of course, her emphasis on "shared angles of vision" as well as "self-defined standpoint" reflects the collective/nonrational realm and the individual/ nonrational realms, respectively. Here we see the significance of "group consciousness, group self-definition and 'voice'" (ibid.:251), that is, the collective/nonrational realm. Explicitly challenging the materialist, structural Marxist point of view, standpoint theorists such as Collins argue that "ideas matter in systems of power" (ibid.:252). This multidimensional approach is illustrated in Figure 14.3.

Figure 14.3   Collins's Black Feminist Thought

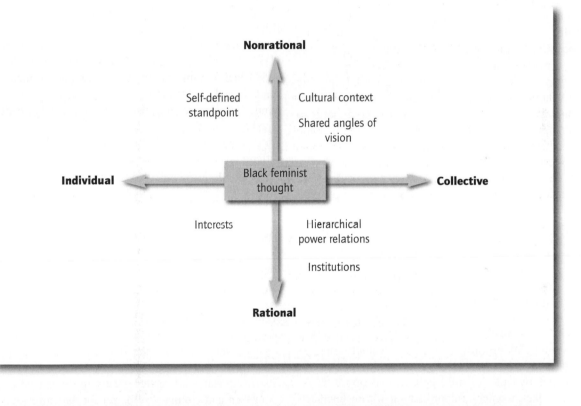

Reading

Introduction to *Black Feminist Thought*

In the following selection from Collins's most highly acclaimed book, *Black Feminist Thought,* Collins exposes and discusses the tension for black women as agents of knowledge, acknowledging that "Black culture and many of its traditions oppress women" (p. 230). However, she also warns against portraying black women either "solely as passive, unfortunate recipients of racial and sexual abuses" or "heroic figures who easily engage in resisting oppression" (p. 238). In sum, Collins continually emphasizes the complexity of both systems of domination and resistance.

Black Feminist Thought ([1990] 2000)

Patricia Hill Collins

DISTINGUISHING FEATURES OF BLACK FEMINIST THOUGHT

Widely used yet increasingly difficult to define, U.S. Black feminist thought encompasses diverse and often contradictory meanings. . . .

Rather than developing definitions and arguing over naming practices—for example, whether this thought should be called Black feminism, womanism, Afrocentric feminism, Africana womanism, and the like—a more useful approach lies in revisiting the reasons why Black feminist thought exists at all. Exploring six distinguishing features that characterize Black feminist thought may provide the common ground that is so sorely needed both among African-American women, and between African-American women and all others whose collective knowledge or thought has a similar purpose. Black feminist thought's distinguishing features need not be unique and may share much with other bodies of knowledge. Rather, it is the *convergence* of these distinguishing features that gives U.S. Black feminist thought its distinctive contours.

Why U.S. Black Feminist Thought?

Black feminism remains important because U.S. Black women constitute an oppressed group. As a collectivity, U.S. Black women participate in a *dialectical* relationship linking African-American women's oppression and activism. Dialectical relationships of this sort mean that two parties are opposed and opposite. As long as Black women's subordination within intersecting oppressions of race, class, gender, sexuality, and nation persists, Black feminism as an activist response to that oppression will remain needed.

In a similar fashion, the overarching purpose of U.S. Black feminist thought is also to resist oppression, both its practices and the ideas that justify it. If intersecting oppressions did not exist, Black feminist thought and similar oppositional knowledges would be unnecessary. As a critical social theory, Black feminist thought aims to empower African-American women within the context of social injustice sustained by intersecting oppressions. Since Black women cannot be fully empowered unless intersecting oppressions themselves are eliminated, Black feminist thought supports broad principles of social justice that transcend U.S. Black women's particular needs.

Because so much of U.S. Black feminism has been filtered through the prism of the U.S. context, its contours have been greatly affected by the specificity of American multiculturalism (Takaki 1993). In particular, U.S. Black feminist thought and practice respond to a fundamental contradiction of U.S. society. On the one hand, democratic promises of individual freedom, equality under the law, and social justice are made to all American citizens. Yet on the other hand, the reality of differential group treatment based on race, class, gender, sexuality, and citizenship status persists. Groups organized around race, class, and gender in and of themselves are not inherently a problem. However, when African-Americans, poor people, women, and other groups discriminated against see little hope for group-based advancement, this situation constitutes social injustice.

Within this overarching contradiction, U.S. Black women encounter a distinctive set of social practices that accompany our particular history within a unique matrix of domination characterized by intersecting oppressions. Race is far from being the only significant marker of group difference—class, gender, sexuality, religion, and citizenship status all matter greatly in the United States (Andersen and Collins 1998). Yet for African-American women, the effects of institutionalized racism remain visible and palpable. Moreover, the institutionalized racism that African-American women encounter relies heavily on racial segregation and accompanying discriminatory practices designed to deny U.S. Blacks equitable treatment. Despite important strides to desegregate U.S. society since 1970, racial segregation remains deeply entrenched in housing, schooling, and employment (Massey and Denton 1993). For many African-American women, racism is not something that exists in the distance. We encounter racism in everyday situations in workplaces, stores, schools, housing, and daily social interaction (St. Jean and Feagin 1998). Most Black women do not have the opportunity to befriend White women and men as neighbors, nor do their children attend school with White children. Racial segregation remains a fundamental feature of the U.S. social landscape, leaving many African-Americans with the belief that "the more things change, the more they stay the same" (Collins 1998a, 11–43). Overlaying these persisting inequalities is a rhetoric of color blindness designed to render these social inequalities invisible. In a context where many believe that to talk of race fosters racism, equality allegedly lies in treating everyone the same. Yet as Kimberle Crenshaw (1997) points out, "it is fairly obvious that treating different things the same can generate as much inequality as treating the same things differently" (p. 285).

Although racial segregation is now organized differently than in prior eras (Collins 1998a, 11–43), being Black and female in the United States continues to expose African-American women to certain common experiences. U.S. Black women's similar work and family experiences as well as our participation in diverse expressions of African-American culture mean that, overall, U.S. Black women as a group live in a different world from that of people who are not Black and female. For individual women, the particular experiences that accrue to living as a Black woman in the United States can stimulate a distinctive consciousness concerning our own experiences and society overall. Many African-American women grasp this connection between what one does and how one thinks. Hannah Nelson, an elderly Black domestic worker, discusses how work shapes the perspectives of African-American and White women: "Since I have to work, I don't really have to worry about most of the things that most of the white women I have worked for are worrying about. And if these women did their own work, they would think just like I do—about this, anyway" (Gwaltney 1980, 4). Ruth Shays, a Black inner-city resident, points out how variations in men's and women's experiences lead to differences in perspective. "The mind of the man and the mind of the woman is the same" she notes, "but this business of living makes women use their minds in ways that men don't even have to think about" (Gwaltney 1980, 33).

A recognition of this connection between experience and consciousness that shapes the everyday lives of individual African-American women often pervades the works of Black women activists and scholars. In her autobiography, Ida B. Wells-Barnett describes how the lynching of her friends had such an impact on her worldview that she subsequently devoted much of her life to the anti-lynching cause (Duster 1970). Sociologist Joyce Ladner's discomfort with the disparity between the teachings of mainstream scholarship and her experiences as a young Black woman in the South led her to write *Tomorrow's Tomorrow* (1972), a groundbreaking study of Black female adolescence. Similarly, the transformed consciousness experienced by Janie, the light-skinned heroine of Zora Neale Hurston's (1937) classic *Their Eyes Were Watching God,* from obedient granddaughter and wife to a self-defined African-American woman, can be directly traced to her experiences with each of her three husbands. In one scene Janie's second husband, angry because she served him a dinner of scorched rice, underdone fish, and soggy bread, hits her. That incident stimulates Janie to stand "where he left her for unmeasured time" and think. And in her thinking "her image of Jody tumbled down and shattered. . . . [S]he had an inside and an outside now and suddenly she knew how not to mix them" (p. 63).

Overall, these ties between what one does and what one thinks illustrated by *individual* Black women can also characterize Black women's experiences and ideas as a *group*. Historically, racial segregation in housing, education, and employment fostered group commonalities that encouraged the formation of a group-based, collective standpoint. For example, the heavy concentration of U.S. Black women in domestic work coupled with racial segregation in housing and schools meant that U.S. Black women had common organizational networks that enabled them to share experiences and construct a collective body of wisdom. This collective wisdom on how to survive as U.S. Black women constituted a distinctive Black women's standpoint on gender-specific patterns of racial segregation and its accompanying economic penalties.

The presence of Black women's collective wisdom challenges two prevailing interpretations of the consciousness of oppressed groups. One approach claims that subordinate groups identify with the powerful and have no valid independent interpretation of their own oppression. The second assumes the oppressed are less human than their rulers, and are therefore less capable of interpreting their own experiences (Rollins 1985; Scott 1985). Both approaches see any independent consciousness expressed by African-American women and other oppressed groups as being either not of our own making or inferior to that of dominant groups. More importantly, both explanations suggest that the alleged lack of political activism on the part of oppressed groups stems from our flawed consciousness of our own subordination.

Historically, Black women's group location in intersecting oppressions produced commonalities among individual African-American women. At the same time, while common experiences may predispose Black women to develop a distinctive group consciousness, they guarantee neither that such a consciousness will develop among all women nor that it will be articulated as such by the group. As historical conditions change, so do the links among the types of experiences Black women will have and any ensuing group consciousness concerning those experiences. Because group standpoints are situated in, reflect, and help shape unjust power relations, standpoints are not static (Collins 1998a, 201–28). Thus, common challenges may foster similar angles of vision leading to a group knowledge or standpoint among African-American women. Or they may not.

Diverse Responses to Common Challenges Within Black Feminism

A second distinguishing feature of U.S. Black feminist thought emerges from a tension linking experiences and ideas. On the one hand, all African-American women face similar challenges that result from living in a society that historically and routinely derogates women of African descent. Despite the fact that U.S. Black women face common challenges, this neither means that individual African-American women have all had the same experiences nor that we agree on the significance of our varying experiences. Thus, on the other hand, despite the common challenges confronting U.S. Black women as a group, diverse responses to these core themes characterize U.S. Black women's group knowledge or standpoint.

Despite differences of age, sexual orientation, social class, region, and religion, U.S. Black women encounter societal practices that restrict us to inferior housing, neighborhoods, schools, jobs, and public treatment and hide this differential consideration behind an array of common beliefs about Black women's intelligence, work habits, and sexuality. These common challenges in turn result in recurring patterns of experiences for individual group members. For example, African-American women

from quite diverse backgrounds report similar treatment in stores. Not every *individual* Black woman consumer need experience being followed in a store as a potential shoplifter, ignored while others are waited on first, or seated near restaurant kitchens and rest rooms, for African-American women as a collectivity to recognize that differential *group* treatment is operating.

Since standpoints refer to group knowledge, recurring patterns of differential treatment such as these suggest that certain themes will characterize U.S. Black women's group knowledge or standpoint. For example, one core theme concerns multifaceted legacies of struggle, especially in response to forms of violence that accompany intersecting oppressions (Collins 1998d). Katie Cannon observes, "[T]hroughout the history of the United States, the interrelationship of white supremacy and male superiority has characterized the Black woman's reality as a situation of struggle—a struggle to survive in two contradictory worlds simultaneously, one white, privileged, and oppressive, the other black, exploited, and oppressed" (1985, 30). Black women's vulnerability to assaults in the workplace, on the street, at home, and in media representations has been one factor fostering this legacy of struggle.

Despite differences created by historical era, age, social class, sexual orientation, skin color, or ethnicity, the legacy of struggle against the violence that permeates U.S. social structures is a common thread binding African-American women. Anna Julia Cooper, an educated, nineteenth-century Black woman intellectual, describes Black women's vulnerability to sexual violence:

> I would beg . . . to add my plea for the *Colored Girls* of the South:—that large, bright, promising fatally beautiful class . . . so full of promise and possibilities, yet so sure of destruction; often without a father to whom they dare apply the loving term, often without a stronger brother to espouse their cause and defend their honor with his life's blood; in the midst of pitfalls and snares, waylaid by the lower classes of white men, with no shelter, no protection. (Cooper 1892, 240)

Yet during this period Cooper and other middle-class U.S. Black women built a powerful club movement and numerous community organizations (Giddings 1984, 1988; Gilkes 1985).

Stating that a legacy of struggle exists does not mean that all U.S. Black women share its benefits or even recognize it. For example, for African-American girls, age often offers little protection from assaults. Far too many young Black girls inhabit hazardous and hostile environments (Carroll 1997). In 1975 I received an essay titled "My World" from Sandra, a sixth-grade student who was a resident of one of the most dangerous public housing projects in Boston. Sandra wrote, "My world is full of people getting rape. People shooting on another. Kids and grownups fighting over girlsfriends. And people without jobs who can't afford to get a education so they can get a job . . . winos on the streets raping and killing little girls." Her words poignantly express a growing Black feminist sensibility that she may be victimized by racism, misogyny, and poverty. They reveal her awareness that she is vulnerable to rape as a form of sexual violence. Despite her feelings about her neighborhood, Sandra not only walked the streets daily but managed safely to deliver three siblings to school. In doing so she participated in a Black women's legacy of struggle. Sandra prevailed, but at a cost. Unlike Sandra, others simply quit.

This legacy of struggle constitutes one of several core themes of a Black women's standpoint. Efforts to reclaim U.S. Black women's intellectual traditions have revealed Black women's long-standing attention to additional core themes first recorded by Maria W. Stewart (Richardson 1987). Stewart's perspective on intersecting oppressions, her call for replacing derogated images of Black womanhood with self-defined images, her belief in Black women's activism as mothers, teachers, and Black community leaders, and her sensitivity to sexual politics are all core themes advanced by a variety of Black feminist intellectuals.

Despite the common challenges confronting African-American women as a group, individual Black women neither have identical experiences nor interpret experiences in a similar fashion. The existence of core themes does not mean that African-American women respond to these themes in the same way. Differences among individual Black women produce different patterns of experiential knowledge that in turn shape individual reactions to the core themes. For example, when faced with controlling images of Black women as being ugly and unfeminine, some women—such as Sojourner Truth—demand, "Ain't I a woman?" By deconstructing the conceptual apparatus of the dominant group, they challenge notions of Barbie-doll femininity premised on middle-class White women's experiences (duCille 1996, 8–59). In contrast, other women internalize the controlling images and come to believe that they are the

stereotypes (Brown-Collins and Sussewell 1986). Still others aim to transgress the boundaries that frame the images themselves. Jaminica, a 14-year-old Black girl, describes her strategies: "Unless you want to get into a big activist battle, you accept the stereotypes given to you and just try and reshape them along the way. So in a way, this gives me a lot of freedom. I can't be looked at any worse in society than I already am—black and female is pretty high on the list of things not to be" (Carroll 1997, 94–95).

Many factors explain these diverse responses. For example, although all African-American women encounter institutionalized racism, social class differences among African American women influence patterns of racism in housing, education, and employment. Middle-class Blacks are more likely to encounter a pernicious form of racism that has left many angry and disappointed (Cose 1993; Feagin and Sikes 1994). A young manager who graduated with honors from the University of Maryland describes the specific form racism can take for middle-class Blacks. Before she flew to Cleveland to explain a marketing plan for her company, her manager made her go over it three or four times in front of him so that she would not forget *her* marketing plan. Then he explained how to check luggage at an airport and how to reclaim it. "I just sat at lunch listening to this man talking to me like I was a monkey who could remember but couldn't think," she recalled. When she had had enough, "I asked him if he wanted to tie my money up in a handkerchief and put a note on me saying that I was an employee of this company. In case I got lost I would be picked up by Traveler's Aid, and Traveler's Aid would send me back" (Davis and Watson 1985, 86). Most middle-class Black women do not encounter such blatant incidents, but many working-class Blacks do. Historically, working-class Blacks have struggled with forms of institutionalized racism directly organized by White institutions and by forms mediated by some segments of the Black middle class. Thus, while it shares much with middle-class Black women, the legacy of struggle by working-class Blacks (Kelley 1994) and by working-class Black women in particular will express a distinctive character (Fordham 1993).

Sexuality signals another important factor that influences African-American women's varying responses to common challenges. Black lesbians have identified heterosexism as a form of oppression and the issues they face living in homophobic communities as shaping their interpretations of everyday events (Shockley

1974; Lorde 1982, 1984; Clarke et al. 1983; Barbara Smith 1983, 1998; Williams 1997). Beverly Smith describes how being a lesbian affected her perceptions of the wedding of one of her closest friends: "God, I wish I had one friend here. Someone who knew me and would understand how I feel. I am masquerading as a nice, straight, middle-class Black 'girl'" (1983, 172). While the majority of those attending the wedding saw only a festive event, Beverly Smith felt that her friend was being sent into a form of bondage. In a similar fashion, varying ethnic and citizenship statuses within the U.S. nation-state as well also shape differences among Black women in the United States. For example, Black Puerto Ricans constitute a group that combines categories of race, nationality, and ethnicity in distinctive ways. Black Puerto Rican women thus must negotiate a distinctive set of experiences that accrue to being racially Black, holding a special form of American citizenship, and being ethnically Latino.

Given how these factors influence diverse response to common challenges, it is important to stress that no homogeneous Black *woman's* standpoint exists. There is no essential or archetypal Black woman whose experiences stand as normal, normative, and thereby authentic. An essentialist understanding of a Black woman's standpoint suppresses differences among Black women in search of an elusive group unity. Instead, it may be more accurate to say that a Black *women's* collective standpoint does exist, one characterized by the tensions that accrue to different responses to common challenges. Because it both recognizes and aims to incorporate heterogeneity in crafting Black women's oppositional knowledge, this Black *women's* standpoint eschews essentialism in favor of democracy. Since Black feminist thought both arises within and aims to articulate a Black *women's* group standpoint regarding experiences associated with intersecting oppressions, stressing this group standpoint's heterogeneous composition is significant.

Moreover in thinking through the contours of a Black women's standpoint it is equally important to recognize that U.S. Black women also encounter the same challenges (and correspondingly different expressions) as women of African descent within a Black diasporic context. This context in turn is situated within a transnational, global context. The term *diaspora* describes the experiences of people who, through slavery, colonialism, imperialism, and migration, have been forced to leave their native lands (Funani 1998, 417). For U.S. Black women and other people of African descent, a diasporic framework suggests a dispersal from Africa to societies in the Caribbean, South America, North America, and Europe.

Understandings of African-American womanhood thus reflect a distinctive pattern of dispersal associated with forced immigration to the United States and subsequent enslavement (Pala 1995). Since a diasporic framework is not normative, it should not be used to assess the authenticity of people of African descent in reference to an assumed African norm. Rather, Black diasporic frameworks center analyses of Black women within the context of common challenges experienced transnationally.

The version of Black feminism that U.S. Black women have developed certainly must be understood in the context of U.S. nation-state politics. At the same time, U.S. Black feminism as a social justice project shares much with comparable social justice projects advanced not only by other U.S. racial/ethnic groups (see, e.g., Takaki 1993), but by women of African descent across quite diverse societies. In the context of an "intercontinental Black women's consciousness movement" (McLaughlin 1995, 73), women of African descent are dispersed globally, yet the issues we face may be similar. Transnationally, women encounter recurring social issues such as poverty, violence, reproductive concerns, lack of education, sex work, and susceptibility to disease (*Rights of Women* 1998). Placing African-American women's experiences, thought, and practice in a transnational, Black diasporic context reveals these and other commonalities of women of African descent while specifying what is particular to African-American women.

Black Feminist Practice and Black Feminist Thought

A third distinguishing feature of Black feminist thought concerns the connections between U.S. Black women's experiences as a heterogeneous collectivity and any ensuing group knowledge or standpoint. . . .

As members of an oppressed group, U.S. Black women have generated alternative practices and knowledges that have been designed to foster U.S. Black women's group empowerment. In contrast to the dialectical relationship linking oppression and activism, a *dialogical* relationship characterizes Black women's collective experiences and group knowledge. On both the individual and the group level, a dialogical relationship suggests that changes in thinking may be accompanied by changed actions and that altered experiences may in turn stimulate a changed consciousness. For U.S. Black women as a collectivity, the struggle for a self-defined Black feminism occurs through an ongoing dialogue whereby action and thought inform one another.

U.S. Black feminism itself illustrates this dialogical relationship. On the one hand, there is U.S. Black feminist practice that emerges in the context of lived experience. When organized and visible, such practice has taken the form of overtly Black feminist social movements dedicated to the empowerment of U.S. Black women. Two especially prominent moments characterize Black feminism's visibility. Providing many of the guiding ideas for today, the first occurred at the turn of the century via the Black women's club movement. The second or modern Black feminist movement was stimulated by the antiracist and women's social justice movements of the 1960s and 1970s and continues to the present. However, these periods of overt political activism where African-American women lobbied in our own behalf remain unusual. They appear to be unusual when juxtaposed to more typical patterns of quiescence regarding Black women's advocacy.

Given the history of U.S. racial segregation, Black feminist activism demonstrates distinctive patterns. Because African-Americans have long been relegated to racially segregated environments, U.S. Black feminist practice has often occurred within a context of Black community development efforts and other Black nationalist-inspired projects. Black nationalism emerges in conjunction with racial segregation—U.S. Blacks living in a racially integrated society would most likely see less need for Black nationalism. As a political philosophy, Black nationalism is based on the belief that Black people constitute a people or "nation" with a common history and destiny. Black solidarity, the belief that Blacks have common interests and should support one another, has long permeated Black women's political philosophy. Thus, Black women's path to a "feminist" consciousness often occurs within the context of antiracist social justice projects, many of them influenced by Black nationalist ideologies. In describing how this phenomenon affects Black women in global context, Andree Nicola McLaughlin contends, "[A]mong activist Black women, it is generally recognized that nationalist struggle provides a rich arena for developing a woman's consciousness" (McLaughlin 1995, 80). To look for Black feminism by searching for U.S. Black women who self-identify as "Black feminists" misses the complexity of how Black feminist practice actually operates (Collins 1993a). . . .

As critical social theory, Black feminist thought encompasses bodies of knowledge and sets of institutional practices that actively grapple with the central questions facing U.S. Black women as a group. Such theory recognizes that U.S. Black women constitute one group among many that are differently placed within situations of injustice. What makes critical social theory "critical" is its commitment to justice, for one's own group and for other groups.

Within these parameters, knowledge for knowledge's sake is not enough—Black feminist thought must both be tied to Black women's lived experiences and aim to better those experiences in some fashion. When such thought is sufficiently grounded in Black feminist practice, it reflects this dialogical relationship. Black feminist thought encompasses general knowledge that helps U.S. Black women survive in, cope with, and resist our differential treatment. It also includes more specialized knowledge that investigates the specific themes and challenges of any given period of time. Conversely, when U.S. Black women cannot see the connections among themes that permeate Black feminist thought and those that influence Black women's everyday lives, it is appropriate to question the strength of this dialogical relationship. Moreover, it is also reasonable to question the validity of that particular expression of Black feminist thought. For example, during slavery, a special theme within Black feminist thought was how the institutionalized rape of enslaved Black women operated as a mechanism of social control. During the period when Black women worked primarily in agriculture and service, countering the sexual harassment of live-in domestic workers gained special importance. Clear connections could be drawn between the content and purpose of Black feminist thought and important issues in Black women's lives.

The potential significance of Black feminist thought goes far beyond demonstrating that African-American women can be theorists. Like Black feminist practice, which it reflects and which it seeks to foster, Black feminist thought can create collective identity among African-American women about the dimensions of a Black women's standpoint. Through the process of *rearticulation,* Black feminist thought can offer African-American women a different view of ourselves and our worlds (Omi and Winant 1994, 99). By taking the core themes of a Black women's standpoint and infusing them with new meaning. Black feminist thought can stimulate a new consciousness that utilizes Black women's everyday, taken-for-granted knowledge. Rather than raising consciousness, Black feminist thought affirms, rearticulates, and provides a vehicle for expressing in public a consciousness that quite often already exists. More important, this rearticulated consciousness aims to empower African-American women and stimulate resistance. . . .

Dialogical Practices and Black Women Intellectuals

A fourth distinguishing feature of Black feminist thought concerns the essential contributions of African-American women intellectuals. The existence of a Black women's standpoint does not mean that African-American women, academic or otherwise, appreciate its content, see its significance, or recognize its potential as a catalyst for social change. One key task for Black women intellectuals of diverse ages, social classes, educational backgrounds, and occupations consists of asking the right questions and investigating all dimensions of a Black women's standpoint with and for African-American women. Historically, Black women intellectuals stood in a special relationship to the larger community of African-American women, a relationship that framed Black feminist thought's contours as critical social theory. . . .

This special relationship of Black women intellectuals to the community of African-American women parallels the existence of two interrelated levels of knowledge (Berger and Luckmann 1966). The commonplace, taken-for-granted knowledge shared by African-American women growing from our everyday thoughts and actions constitutes a first and most fundamental level of knowledge. The ideas that Black women share with one another on an informal, daily basis about topics such as how to style our hair, characteristics of "good" Black men, strategies for dealing with White folks, and skills of how to "get over" provide the foundations for this taken-for-granted knowledge.

Experts or specialists who participate in and emerge from a group produce a second, more specialized type of knowledge. Whether working-class or middle-class, educated or not, famous or everyday, the range of Black women intellectuals discussed in Chapter 1 are examples of these specialists. Their theories that facilitate the expression of a Black women's standpoint form the specialized knowledge of Black feminist thought. The two types of knowledge are interdependent. While Black feminist thought articulates the often taken-for-granted knowledge shared by African-American women as a group, the consciousness of Black women may be transformed by such thought. Many Black women blues singers have long sung about taken-for-granted situations that affect U.S. Black women. Through their music, they not only depict Black women's realities, they aim to shape them.

Because they have had greater opportunities to achieve literacy, middle-class Black women have also had greater access to the resources to engage in Black feminist scholarship. Education need not mean alienation from this dialogical relationship. The actions of educated Black women within the Black women's club movement typify this special relationship between one segment of Black women intellectuals and the wider community of African-American women:

> It is important to recognize that black women like Frances Harper, Anna Julia Cooper, and Ida B. Wells were not isolated figures of intellectual genius; they were shaped by and helped to shape a wider movement of Afro-American women. This is not to claim that they were representative of all black women; they and their counterparts formed an educated, intellectual elite, but an elite that tried to develop a cultural and historical perspective that was organic to the wider condition of black womanhood. (Carby 1987, 115)

The work of these women is important because it illustrates a tradition of joining scholarship and activism. Because they often lived in the same neighborhoods as working-class Blacks, turn-of-the-century club women lived in a Black civil society where this dialogical relationship was easier to establish. They saw the problems. They participated in social institutions that encouraged solutions. They fostered the development of a "cultural and historical perspective that was organic to the wider condition of black womanhood." Contemporary Black women intellectuals face similar challenges of fostering dialogues, but do so under greatly changed social conditions. Whereas racial segregation was designed to keep U.S. Blacks oppressed, it fostered a form of racial solidarity that flourished in all-Black neighborhoods. In contrast, now that Blacks live in economically heterogeneous neighborhoods, achieving the same racial solidarity raises new challenges. . . .

Black Feminism as Dynamic and Changing

A fifth distinguishing feature of U.S. Black feminist thought concerns the significance of change. In order for Black feminist thought to operate effectively within Black feminism as a social justice project, both must remain dynamic. Neither Black feminist thought as a critical social theory nor Black feminist practice can be static; as social conditions change, so must the knowledge and practices designed to resist them. For example, stressing the importance of Black women's centrality to Black feminist thought does not mean that all African-American women desire, are positioned, or are qualified to exert this type of intellectual leadership. Under current conditions, some Black women thinkers have lost contact with Black feminist practice. Conversely, the changed social conditions under which U.S. Black women now come to womanhood—class-segregated neighborhoods, some integrated, far more not—place Black women of different social classes in entirely new relationships with one another. . . .

The changing social conditions that confront African-American women stimulate the need for new Black feminist analyses of the common differences that characterize U.S. Black womanhood. Some Black women thinkers are already engaged in this process. Take, for example, Barbara Omolade's (1994) insightful analysis of Black women's historical and contemporary participation in mammy work. Most can understand mammy work's historical context, one where Black women were confined to domestic service, with Aunt Jemima created as a controlling image designed to hide Black women's exploitation. Understanding the limitations of domestic service, much of Black women's progress in the labor market has been measured by the move out of domestic service. Currently, few U.S. Black women work in domestic service in private homes. Instead, a good deal of this work in private homes is now done by undocumented immigrant women of color who lack U.S. citizenship; their exploitation resembles that long visited upon African-American women (Chang 1994). But, as Omolade points out, these changes do not mean that U.S. Black women have escaped mammy work. Even though few Aunt Jemimas exist today, and those that do have been cosmetically altered, leading to the impression that mammy work has disappeared, Omolade reminds us that mammy work has assumed new forms. Within each segment of the labor market—the low-paid jobs at fast-food establishments, nursing homes, day-care centers, and dry cleaners that characterize the secondary sector, the secretaries and clerical workers of the primary lower tier sector, or the teachers, social workers, nurses, and administrators of the primary upper tier sector—U.S. Black women still do a remarkable share of the emotional nurturing and cleaning up after other people, often for lower pay. In this context the task for contemporary Black feminist thought lies in explicating these changing relationships and developing analyses of how these commonalities are experienced differently.

The changing conditions of Black women's work overall has important implications for Black women's intellectual work. Historically, the suppression of Black feminist thought has meant that Black women intellectuals have traditionally relied on alternative institutional locations to produce specialized knowledge about a Black women's standpoint. Many Black women scholars, writers, and artists have worked either alone, as was the case with Maria W. Stewart, or within African-American community organizations, the case for Black women in the club movement and in Black churches. The grudging incorporation of work on Black women into curricular offerings of historically White colleges and universities, coupled with the creation of a critical mass of African-American women writers such as Toni Morrison, Alice Walker, and Gloria Naylor within these institutional locations, means that Black women intellectuals can now find employment within academia. Black women's history and Black feminist literary criticism constitute two focal points of this renaissance in Black women's intelectual work (Carby 1987). Moreover, U.S. Black women's access to the media remains unprecedented, as talk show hostess Oprah Winfrey's long- running television show and forays into film production suggest.

The visibility provided U.S. Black women and our ideas via these new institutional locations has been immense. However, one danger facing African-American women intellectuals working in these new locations concerns the potential isolation of individual thinkers from Black women's collective experiences—lack of access to other U.S. Black women and to Black women's communities. Another is the pressure to separate thought from action—particularly political activism—that typically accompanies training in standard academic disciplines or participating in allegedly neutral spheres like the "free" press. Yet another involves the inability of some Black women "superstars" to critique the terms of their own participation in these new relations. Blinded by their self-proclaimed Black feminist diva aspirations, they feel that they owe no one, especially other Black women. Instead, they become trapped within their own impoverished Black feminist universes. Despite these dangers, these new institutional locations provide a multitude of opportunities for enhancing Black feminist thought's visibility. In this new context, the challenge lies in remaining dynamic, all the while keeping in mind that a moving target is more difficult to hit.

U.S. Black Feminism and Other Social Justice Projects

A final distinguishing feature of Black feminist thought concerns its relationship to other projects for social justice. A broad range of African-American women intellectuals have advanced the view that Black women's struggles are part of a wider struggle for human dignity, empowerment, and social justice. In an 1893 speech to women, Anna Julia Cooper cogently expressed this worldview:

> We take our stand on the solidarity of humanity, the
> oneness of life, and the unnaturalness and injustice of
> all special favoritisms, whether of sex, race, country, or

condition. . . . The colored woman feels that woman's cause is one and universal; and that . . . not till race, color, sex, and condition are seen as accidents, and not the substance of life; not till the universal title of humanity to life, liberty, and the pursuit of happiness is conceded to be inalienable to all; not till then is woman's lesson taught and woman's cause won—not the white woman's nor the black woman's, not the red woman's but the cause of every man and of every woman who has writhed silently under a mighty wrong. (Loewenberg and Bogin 1976, 330–31)

Like Cooper, many African-American women intellectuals embrace this perspective regardless of particular political solutions we propose, our educational backgrounds, our fields of study, or our historical periods. Whether we advocate working through autonomous Black women's organizations, becoming part of women's organizations, running for political office, or supporting Black community institutions, African-American women intellectuals repeatedly identify political actions such as these as a *means* for human empowerment rather than ends in and of themselves. Thus one important guiding principle of Black feminism is a recurring humanist vision (Steady 1981, 1987). . . .

Perhaps the most succinct version of the humanist vision in U.S. Black feminist thought is offered by Fannie Lou Hamer, the daughter of sharecroppers and a Mississippi civil rights activist. While sitting on her porch, Ms. Hamer observed, "Ain' no such thing as I can hate anybody and hope to see God's face" (Jordan 1981, xi).

Significant Others

Nancy Chodorow (1944–): Turning Freud on His Head

Arguably the most important psychoanalytic feminist and reinterpreter of Freud, Nancy Chodorow (1944–) is a practicing clinical psychoanalyst and psychotherapist as well as a sociologist. Although she accepts the basic Freudian idea that unconscious and innate erotic and aggressive drives exist, Chodorow is intensely critical of Freud's phallocentric approach. In contrast to Freud, Chodorow situates innate erotic drives in the context of interpersonal relations, focusing not so much on sexuality per se as on intimacy and separation, primarily in the family and especially between mother and child.

Specifically, in her award-winning book, *The Reproduction of Mothering* (1978), Chodorow turns Freud on his head. Chodorow maintains that because of the allocation of work roles, infants usually originally identify with the female parent. That is, the infant first develops a sense of his or her own selfhood in a close, one-on-one relationship with the mother, and qualities possessed by the mother are internalized by the infant to form the beginnings of the child's personality. The particularly strong bonds formed with the mother and the relative absence of the father has important implications for the development of "normal" adult heterosexual identity. Boys achieve their adult sexual identity (i.e., become "men") only by separating themselves from their mothers. This separation entails denying the world of emotional intimacy that she represents. That is, boys become men by defining themselves in opposition to the femininity of their mother. According to Chodorow, the consequence of this is that men have difficulty in dealing with emotional matters: They see acknowledgment of emotions as a sign of vulnerability and weakness. Moreover, social contempt for women (and, in its extreme form, misogyny) arises as boys deny their earliest emotional experiences with their mother and particularly the sexually charged nature of their oedipal love for their mothers. The acknowledgment of emotions, in particular feelings of vulnerability, is considered "femininizing" and is threatening to their status as "real" men (Alsop, Fitzsimons, and Lennon 2002:59).

Girls, on the other hand, are never required to make a complete break with their mothers in order to achieve their adult sexual identity (become "women"). Rather, society fosters the continuation of intense mother-daughter bonds into adulthood. However, not having been forced to emotionally separate from their mothers, women continue to long for the emotional intimacy provided by close relationships. This unconscious desire to form attachments to others leads women to suffer greater dependency needs, as their self-identity is tied to their relationships with others. According to Chodorow, this lack of differentiation explains why women become preoccupied with the very relational issues at the heart of motherhood: intimacy and a lack of ego separation. Women find their self-in-relation (in intimate relations with others), but because of their socialization into adult heterosexuality men lack the emotional capabilities that women need in order to be fulfilled in relationships. Because masculinity is defined by separation and distance, women turn not to men but to motherhood to fulfill their unconscious desire for intimacy; they re-create the early infant–mother relationship by becoming mothers themselves. Of course, as women again mother (and fathers continue to eschew intimacy), the cycle continues on in another generation: a female self that is fundamentally a self-in-relation and a male self that is fundamentally a self in denial of relations (Gerhenson and Williams 2001:282).

Since 1986, Chodorow has been a professor of sociology at the University of California, Berkeley. Her more recent books include *Feminism and Psychoanalytic Theory* (1989), *Femininities Masculinities, Sexualities: Freud and Beyond* (1994), and *The Power of Feelings: Personal Meaning in Psychoanalysis, Gender and Culture* (1999).

RAEWYN CONNELL (1944–): A BIOGRAPHICAL SKETCH ▪▪

Raewyn Connell (formerly R.W. or Bob Connell) is one of Australia's most highly acclaimed sociologists. She has authored or coauthored a number of books, including *Ruling Class Ruling Culture* (1977), *Class Structure in Australian History* (1980), *Making the Difference* (1982), *Gender and Power* (1987), and *Masculinities* (1995), which has been translated into 13 languages and is among the most-cited research publication in the field. Raewyn's most recent book, *Southern Theory* (2007) discusses theorists unfamiliar in the European canon of social science, and explores the possibility of a genuinely global social science. Her ongoing work explores the relation between masculinities and neoliberal globalization, combining, in characteristic form, her concern for large-scale social structures with recognition of personal experience and collective agency.

Connell received her doctorate in sociology from the University of Sydney, where she currently holds a University Chair. She has also taught at the University of California at Santa Cruz, Macquarie University in Sydney, Flinders University in Adelaide, and held visiting posts at the University of Toronto, Harvard University, and Ruhr-Universität Bochum. Connell's work is widely cited in social science and humanities publications internationally. Four of her books were listed among the 10 most influential books in Australian sociology. She is frequently invited to give keynote addresses at conferences and seminars, including events in Canada, Switzerland, Germany, Senegal, and Britain. Connell has received the American Sociological Association's award for distinguished contribution to the study of sex and gender, as well as the Australian Sociological Association's award for distinguished service to sociology.

CONNELL'S INTELLECTUAL INFLUENCES AND CORE IDEAS ▪▪

Akin to Chodorow (see Significant Others Box), Connell is concerned about the resiliency of gender roles and the pattern of practices that allows men's dominance over women. However, rather than use object relations theory to explain these practices, Connell expands on the work of the Italian journalist, communist and political activist, Antonio Gramsci (1891–1937), who coined the concept of cultural hegemony (See Significant Others Box, Chapter 2, p. 30). Building on Marx's notion that "the ruling ideas are the ideas of the ruling class," and fascinated by the extraordinary ideological power of the Catholic Church in Italy, Gramsci used the term "cultural hegemony" to refer to how the ruling class maintains its dominance not primarily through force or coercion, but through the willing, "spontaneous" consent of the ruled. In a similar vein, Connell uses the term **hegemonic masculinity** to refer to the pattern of practices that allows men's dominance over women to continue (Connell and Messerschmidt 2005:832). Connell maintains that there are many kinds of masculinities but always there is one which is hegemonic to the rest and marginalize others in a gender system. This does not mean that hegemonic masculinity is either monolithic or static, but rather, that it is the kind of masculinity which is in a superior level. No matter what, each culture will prefer one kind of masculinity over others. Significantly, however, Connell maintains that most men do *not* live in the model of hegemonic masculinity and that masculinity (as femininity) has internal contradictions and historical ruptures, because what is hegemonic is determined in a mobile relation. Above all, Connell is concerned with the changing patterns of "hegemony," that is to say that the dominance of particular patterns of masculinity over others.

Connell's conceptualization of **hegemonic masculinity** has the central advantage of locating male dominance not solely to the microlevel and the inter-personal dynamics of the family, but to the macro-level and the public sphere. **Hegemonic masculinity** recognizes not only the gendered character or bureaucracies and workplaces as well as educational institutions, including classroom dynamics and patterns of bullying, but also the media, for instance, the interplay of sports and war imagery, as well as the virtual monopoly of men in certain forms of crime, including syndicated and white collar crime. In theoretical terms, Connell explicitly accounts for both the more "rational" dimensions of dominance (institutionalized bureaucracies) as well as the "nonrational" dimensions (e.g. sports and war imagery), as shown in Figure 14.4. As Connell and Messerschmidt (ibid, p. 846) state, "Cultural consent, discursive centrality, institutionalization and the marginalization or delegitimation of alternatives are widely documented features of socially dominant masculinities . . . Hegemony works in part through the production of exemplars of masculinity (e.g. professional sports stars), symbols that have authority despite the fact that most men and boys do not fully live up to them."

Figure 14.4   Connell's Basic Theoretical Orientation

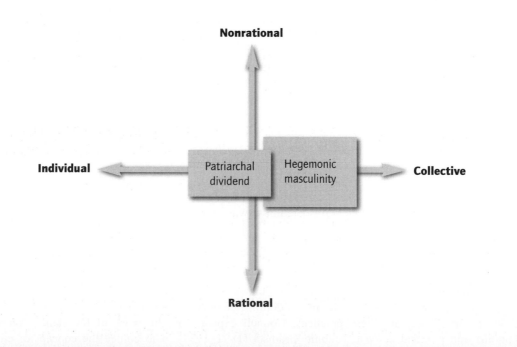

This brings us to a second vital concept in Connell's work: **patriarchal dividend.** Connell uses this term to refer both to the honor and prestige and the more material dividends men accrue under patriarchy, the point being that this dividend is not *uniformly* distributed among men, but it is, nevertheless, universally distributed among them. In other words, though men as a whole may gain from living in a patriarchal gender order, not all gain in the same way or to the same degree. Patriarchal systems are intertwined with a wide variety of other hierarchical relations (e.g. class, race, nation, region, generation, sexual orientation); consequently, not all men receive the same share of the patriarchal dividend.

CONNELL'S THEORETICAL ORIENTATION

As illustrated in Figure 14.4, in terms of the theoretical model used in this book, "patriarchal dividend" and "hegemonic masculinity" might be conceptualized as twin terms, the former highlighting the costs and benefits of the gender order as played out at the level of the individual; the latter highlighting, as indicated previously, dominant *patterns* of masculinity, at both the cultural and social structural levels. As Connell (2000, p. 11) states, "Masculinities are defined collectively in culture, and are sustained in institutions." In other words, in terms of the question of order, Connell's work is thoroughly multidimensional. As a sociologist and historian, Connell is most interested in "collective masculinities," which she defines as "the patterns of conduct our society defines as masculine" (ibid). She emphasizes not only that there are different types of masculinities in different cultures and periods of history, but also "multiple masculinities" in any particular place and time (ibid). At the same time, however, Connell goes to great lengths to explain "the active construction" of masculinity at the level of the individual (ibid). Connell asserts that "the hegemonic form need not be the most common form of masculinity," that masculinities are not fixed, and that significant contradictions exist not only at the level of the collective, but at the level of the individual—for instance, in contradictory desires.

In terms of the question of action, as indicated previously, Connell explicitly accounts for both the more "rational" and "nonrational" dimensions of dominance at the level of the collective (for instance, institutionalized bureaucracies, and sports and war imagery), as well as both the conscious and relatively unconscious costs and benefits that accrue from the patriarchal dividend at the level of the individual (for instance, intricate maneuvering in peer groups and competitive sports). Most importantly, Connell's theoretical multidimensionality is rooted not only in her comprehensive analysis of distinct sorts of variables (e.g. the economy, the body, media), but also in her comprehensive analysis of a single variable across space and time. Thus, in the essay you will read below, she maintains that the disadvantages to men that accrue in the current gender order are "the conditions of the advantages. For instance, men cannot be the beneficiaries of women's domestic labor and 'emotion work' without many of them losing intimate connections, for instance, with young children" (Connell 2005: 2809).

Reading

Introduction to "Change Among the Gatekeepers"

In this essay, Connell makes three pivotal points regarding gender equality in the global arena. First, Connell argues that men are the "gatekeepers" to equality between men and women in many ways; that is, they have access to resources, authority, and skills that may all be important in social change. The point is that men who believe in gender equality can do a great deal. Second, Connell illuminates the *diversity* of masculinities and men's movements worldwide. For instance, on one hand, homosexual men are mobilizing in antidiscrimination campaigns, the gay liberation movement, and community responses to the HIV/AIDS pandemic. At the same time, however, there are very large numbers of men engaged in preserving gender inequality. For instance, conservative religious organizations (Christian, Islamic, Buddhist) controlled by men sometimes completely exclude women, and these organizations have often been used to oppose the emancipation of women; while "transnational media organizations, such as Rupert Murdoch's conglomerate are equally active in promoting conservative gender ideology," and "neoliberalism can function as a form of masculinity politics largely because of the powerful role of the state in the gender order" as well (Connell, p. 1816). Finally, Connell (p. 1803) points out that "we now have a far more sophisticated and detailed scientific understanding of issues about men, masculinities, and gender than ever before." Such that, though clearly given the diversity of masculinity politics it is unrealistic to expect worldwide consensus for gender equality, it is possible that gender equality might someday become *hegemonic* among men.

"Change Among the Gatekeepers: Men, Masculinities, and Gender Equality in the Global Arena" (2005)

R. W. Connell

Equality between women and men has been a doctrine well recognized in international law since the adoption of the 1948 *Universal Declaration of Human Rights* (United Nations 1958), and as a principle it enjoys popular support in many countries. The idea of gender equal rights has provided the formal basis for the international discussion of the position of women since the 1975–85 UN Decade for Women, which has been a key element in the story of global feminism (Bulbeck 1988). The idea that men might have a specific role in relation to this principle has emerged only recently.

The issue of gender equality was placed on the policy agenda by women. The reason is obvious: it is women who are disadvantaged by the main patterns of gender inequality and who therefore have the claim for redress. Men are, however, necessarily involved in gender-equality reform. Gender inequalities are embedded in a multidimensional structure of relationships between women and men, which, as the modern sociology of gender shows, operates at every level of human experience, from economic arrangements, culture, and the state to interpersonal relationships and individual emotions (Holter 1997; Walby 1997; Connell 2002). Moving toward a gender-equal society involves profound institutional change as well as change in everyday life and personal conduct. To move far in this direction requires

widespread social support, including significant support from men and boys.

Further, the very gender inequalities in economic assets, political power, and cultural authority, as well as the means of coercion, that gender reforms intend to change, currently mean that men (often specific groups of men) control most of the resources required to implement women's claims for justice. Men and boys are thus in significant ways gatekeepers for gender equality. Whether they are willing to open the gates for major reforms is an important strategic question.

In this article, I will trace the emergence of a worldwide discussion of men and gender-equality reform and will try to assess the prospects of reform strategies involving men. To make such an assessment, it is necessary to set recent policy discussions in the wider context of the cultural problematization of men and boys, the politics of "men's movements," the divided interests of men and boys in gender relations, and the growing research evidence about the changing and conflict-ridden social construction of masculinities.

In an article of this scope, it is not possible to address particular national agendas in detail. I will refer to a number of texts where these stories can be found. Because my primary concern is with the global character of the debate, I will give particular attention to

SOURCE: From *Signs: Journal of Women in Culture and Society,* Vol. 30, no. 3, pp. 1801–1826. Copyright © 2005 The University of Chicago. Reprinted with permission of The University of Chicago Press.

policy discussions in UN forums. These discussions culminated in the 2004 meeting of the UN Commission on the Status of Women, which produced the first world-level policy document on the role of men and boys in relation to gender equality (UN Commission on the Status of Women 2004).

MEN AND MASCULINITIES IN THE WORLD GENDER ORDER

In the last fifteen years, in the "developed" countries of the global metropole, there has been a great deal of popular concern with issues about men and boys. Readers in the United States may recall a volume by the poet Robert Bly, *Iron John: A Book about Men* (1990), which became a huge best seller in the early 1990s, setting off a wave of imitations. This book became popular because it offered, in prophetic language, simple solutions to problems that were increasingly troubling the culture. A therapeutic movement was then developing in the United States, mainly though not exclusively among middle-class men, addressing problems in relationships, sexuality, and identity (Kupers 1993; Schwalbe 1996).

More specific issues about men and boys have also attracted public attention in the developed countries. Men's responses to feminism, and to gender-equality measures taken by government, have long been the subject of debate in Germany and Scandinavia (Metz-Göckel and Müller 1985; Holter 2003). In anglophone countries there has been much discussion of "the new fatherhood" and of supposed changes in men's involvement in families (McMahon 1999). There has been public agonizing about boys' "failure" in school, and in Australia there are many proposals for special programs for boys (Kenway 1997; Lingard 2003). Men's violence toward women has been the subject of practical interventions and extensive debate (Hearn 1998). There has also been increasing debate about men's health and illness from a gender perspective (Hurrelmann and Kolip 2002).

Accompanying these debates has been a remarkable growth of research about men's gender identities and practices, masculinities and the social processes by which they are constructed, cultural and media images of men, and related matters. Academic journals have been founded for specialized research on men and masculinities, there have been many research conferences, and there is a rapidly growing international literature.

We now have a far more sophisticated and detailed scientific understanding of issues about men, masculinities, and gender than ever before (Connell 2003a).

This set of concerns, though first articulated in the developed countries, can now be found worldwide (Connell 2000; Pease and Pringle 2001). Debates on violence, patriarchy, and ways of changing men's conduct have occurred in countries as diverse as Germany, Canada, and South Africa (Hagemann-White 1992; Kaufman 1993; Morrell 2001a). Issues about masculine sexuality and fatherhood have been debated and researched in Brazil, Mexico, and many other countries (Arilha, Unbehaum Ridenti, and Medrado 1998; Lerner 1998). A men's center with a reform agenda has been established in Japan, where conferences have been held and media debates about traditional patterns of masculinity and family life continue (Menzu Senta 1997; Roberson and Suzuki 2003). A "traveling seminar" discussing issues about men, masculinities, and gender equality has recently been touring in India (Roy 2003). Debates about boys' education, men's identities, and gender change are active from New Zealand to Denmark (Law, Campbell, and Dolan 1999; Reinicke 2002). Debates about men's sexuality, and changing sexual identities, are also international (Altman 2001).

The research effort is also worldwide. Documentation of the diverse social constructions of masculinity has been undertaken in countries as far apart as Peru (Fuller 2001), Japan (Taga 2001), and Turkey (Sinclair-Webb 2000). The first large-scale comparative study of men and gender relations has recently been completed in ten European countries (Hearn et al. 2002). The first global synthesis, in the form of a world handbook of research on men and masculinities, has now appeared (Kimmel, Hearn, and Connell 2005).

The rapid internationalization of these debates reflects the fact—increasingly recognized in feminist thought (Bulbeck 1998; Marchand and Runyan 2000)—that gender relations themselves have an international dimension. Each of the substructures of gender relations can be shown to have a global dimension, growing out of the history of imperialism and seen in the contemporary process of globalization (Connell 2002). Change in gender relations occurs on a world scale, though not always in the same direction or at the same pace.

The complexity of the patterns follows from the fact that gender change occurs in several different modes. Most dramatic is the direct colonization of the gender order of regions beyond the metropole. There has also been a more gradual recomposition of gender orders, both those of the colonizing society and the colonized, in the process of colonial interaction. The hybrid gender identities and sexualities now much discussed in the context of postcolonial societies are neither unusual nor new. They are a feature of the whole history of imperialism and are visible in many contemporary studies (e.g., Valdes and Olavarria 1998).

Imperialism and globalization change the conditions of existence for gender orders. For instance, the linking of previously separate production systems changes the flow of goods and services in the gendered division of labor, as seen in the impact of industrially produced foods and textiles on household economies. Colonialism itself often confronted local patriarchies with colonizing patriarchies, producing a turbulent and sometimes very violent aftermath, as in southern Africa (Morrell 1998). Pressure from contemporary Western commercial culture has destabilized gender arrangements, and models of masculinity, in Japan (Ito 1992), the Arab world (Ghoussoub 2000), and elsewhere.

Finally, the emergence of new arenas of social relationship on a world scale creates new patterns of gender relations. Transnational corporations, international communications systems, global mass media, and international state structures (from the United Nations to the European Union) are such arenas. These institutions have their own gender regimes and may form the basis for new configurations of masculinity, as has recently been argued for transnational business (Connell 2000) and the international relations system (Hooper 2001). Local gender orders now interact not only with the gender orders of other local societies but also with the gender order of the global arena.

The dynamics of the world gender order affect men as profoundly as they do women, though this fact has been less discussed. The best contemporary research on men and masculinity, such as Matthew C. Gutmann's (2002) ethnographic work in Mexico, shows in fine detail how the lives of particular groups of men are shaped by globally acting economic and political dynamics.

Different groups of men are positioned very differently in such processes. There is no single formula that accounts for men and globalization. There is, indeed, a growing polarization among men on a world scale. Studies of the "super-rich" (Haseler 2000) show a privileged minority reaching astonishing heights of wealth and power while much larger numbers face poverty, cultural dislocation, disruption of family relationships, and forced renegotiation of the meanings of masculinity.

Masculinities, as socially constructed configurations of gender practice, are also created through a historical process with a global dimension. The old-style ethnographic research that located gender patterns purely in a local context is inadequate to the reality. Historical research, such as Robert Morrell's (2001b) study of the masculinities of the colonizers in South Africa and T. Dunbar Moodie's (1994) study of the colonized, shows how a gendered culture is created and transformed in relation to the international economy and the political system of empire. There is every reason to think this principle holds for contemporary masculinities.

SHIFTING GROUND: MEN AND BOYS IN GENDER-EQUALITY DEBATES

Because of the way they came onto the agenda of public debate, gender issues have been widely regarded as women's business and of little concern to men and boys. In almost all policy discussions, to adopt a gender perspective substantially means to address women's concerns.

In both national and international policy documents concerned with gender equality, women are the subjects of the policy discourse. The agencies or meetings that formulate, implement, or monitor gender policies usually have names referring to women, such as Department for Women, Women's Equity Bureau, Prefectural Women's Centre, or Commission on the Status of Women. Such bodies have a clear mandate to act for women. They do not have an equally clear mandate to act with respect to men. The major policy documents concerned with gender equality, such as the UN *Convention on the Elimination of all Forms of Discrimination against Women* (United Nations [1979] 1989), often do not name men as a group and rarely discuss men in concrete terms.

However, men are present as background throughout these documents. In every statement about women's disadvantages, there is an implied comparison with men as the advantaged group. In the discussions of violence against women, men are implied, and sometimes named, as the perpetrators. In discussions of gender and HIV/AIDS, men are commonly construed as being "the problem," the agents of infection. In discussions of women's exclusion from power and decision making, men are implicitly present as the power holders.

When men are present only as a background category in a policy discourse about women, it is difficult to raise issues about men's and boys' interests, problems, or differences. This could be done only by falling into a backlash posture and affirming "men's rights" or by moving outside a gender framework altogether.

The structure of gender-equality policy, therefore, created an opportunity for antifeminist politics. Opponents of feminism have now found issues about boys and men to be fertile ground. This is most clearly seen in the United States, where authors such as Warren Farreh (1993) and Christina Hoff Sommers (2000), purporting to speak on behalf of men and boys, bitterly accuse feminism of injustice. Men and boys, they argue, are the truly disadvantaged group and need supportive programs in education and health, in situations of family breakup, and so forth. These ideas have not stimulated a social movement, with the exception of a small-scale (though active and sometimes violent) "father's rights" movement in relation to divorce. The arguments have, however, strongly appealed to the neoconservative mass media, which have given them international circulation. They now form part of the broad neoconservative repertoire of opposition to "political correctness" and to social justice measures.

Some policy makers have attempted to straddle this divide by restructuring gender-equality policy in the form of parallel policies for women and men. For instance, some recent health policy initiatives in Australia have added a "men's health" document to a "women's health" document (Schofield 2004). Similarly, in some school systems a "boys' education" strategy has been added to a "girls' education" strategy (Lingard 2003).

This approach acknowledges the wider scope of gender issues. But it also risks weakening the equality rationale of the original policy. It forgets the relational character of gender and therefore tends to redefine women and men, or girls and boys, simply as different market segments for some service. Ironically, the result may be to promote more gender segregation, not less. This has certainly happened in education, where some privileged boys' schools have jumped on the "gender equality" bandwagon and now market themselves as experts in catering to the special needs of boys.

On the other hand, bringing men's problems into an existing framework of policies for women may weaken the authority that women have so far gathered in that policy area. In the field of gender and development, for instance, some specialists argue that "bringing men in"—given the larger context in which men still control most of the wealth and institutional authority—may undermine, not help, the drive for gender equality (White 2000). . . .

DIVIDED INTERESTS: SUPPORT AND RESISTANCE

There is something surprising about the worldwide problematizing of men and masculinities, because in many ways the position of men has not greatly changed. For instance, men remain a very large majority of corporate executives, top professionals, and holders of public office. Worldwide, men hold nine out of ten cabinet-level posts in national governments, nearly as many of the parliamentary seats, and most top positions in international agencies. Men, collectively, receive approximately twice the income that women receive and also receive the benefits of a great deal of unpaid household labor, not to mention emotional support, from women (Gierycz 1999; Godenzi 2000; Inter-Parliamentary Union 2003).

The UN Development Program (2003) now regularly incorporates a selection of such statistics into its annual report on world human development, combining them into a "gender-related development index" and a "gender empowerment measure." This produces a dramatic outcome, a league table of countries ranked in terms of gender equality, which shows most countries in the world to be far from gender-equal. It is clear that, globally, men have a lot to lose from pursuing gender equality because men, collectively, continue to receive a patriarchal dividend.

But this way of picturing inequality may conceal as much as it reveals. There are multiple dimensions in gender relations, and the patterns of inequality in these dimensions may be qualitatively different. If we look separately at each of the substructures of gender, we find a pattern of advantages for men but also a linked pattern of disadvantages or toxicity (Connell 2003c).

For instance, in relation to the gender division of labor, men collectively receive the bulk of income in the money economy and occupy most of the managerial positions. But men also provide the workforce for the most dangerous occupations, suffer most industrial injuries, pay most of the taxation, and are under heavier social pressure to remain employed. In the domain of power men collectively control the institutions of coercion and the means of violence (e.g., weapons). But men are also the main targets of military violence and criminal assault, and many more men than women are imprisoned or executed. Men's authority receives more social recognition (e.g., in religion), but men and boys are underrepresented in important learning experiences (e.g., in humanistic studies) and important dimensions of human relations (e.g., with young children).

One could draw up a balance sheet of the costs and benefits to men from the current gender order. But this balance sheet would not be like a corporate accounting exercise where there is a bottom line, subtracting costs from income. The disadvantages listed above are, broadly speaking, the conditions of the advantages. For instance, men cannot hold state power without some men becoming the agents of violence. Men cannot be the beneficiaries of women's domestic labor and "emotion work" without many of them losing intimate connections, for instance, with young children.

Equally important, the men who receive most of the benefits and the men who pay most of the costs are not the same individuals. As the old saying puts it, generals die in bed. On a global scale, the men who benefit from corporate wealth, physical security, and expensive health care are a very different group from the men who provide the workforce of developing countries. Class, race, national, regional, and generational differences cross-cut the category "men," spreading the gains and costs of gender relations very unevenly among men. There are many situations where groups of men may see their interest as more closely aligned with the women in their communities than with other men. It is not surprising that men respond very diversely to gender-equality politics.

There is, in fact, a considerable history of support for gender equality among men. There is certainly a tradition of advocacy by male intellectuals. In Europe, well before modern gender-equality documents were written, the British philosopher John Stuart Mill published "The Subjection of Women" (1912), which established the presumption of equal rights; and the Norwegian dramatist Henrik Ibsen, in plays like *A Doll's House* ([1923] 1995), made gender oppression an important cultural theme. In the following generation, the pioneering Austrian psychoanalyst Alfred Adler established a powerful psychological argument for gender equality (Connell 1995). A similar tradition of men's advocacy exists in the United States (Kimmel and Mosmiller 1992). . . .

There is, however, also significant evidence of men's and boys' resistance to change in gender relations. The survey research reveals substantial levels of doubt and

opposition, especially among older men. Research on workplaces and on corporate management has documented many cases where men maintain an organizational culture that is heavily masculinized and unwelcoming to women. In some cases there is active opposition to gender-equality measures or quiet undermining of them (Cockburn 1991; Collinson and Hearn 1996). Research on schools has also found cases where boys assert control of informal social life and direct hostility against girls and against boys perceived as being different. The status quo can be defended even in the details of classroom life, for instance, when a particular group of boys used misogynist language to resist study of a poem that questioned Australian gender stereotypes (Kenworthy 1994; Holland et al. 1998).

Some men accept change in principle but in practice still act in ways that sustain men's dominance of the public sphere and assign domestic labor and child care to women. In strongly gender segregated societies, it may be difficult for men to recognize alternatives or to understand women's experiences (Kandiyoti 1994; Fuller 2001; Meuser 2003). Another type of opposition to reform, more common among men in business and government, rejects gender-equality measures because it rejects all government action in support of equality, in favor of the unfettered action of the market.

The reasons for men's resistance include the patriarchal dividend discussed above and threats to identity that occur with change. If social definitions of masculinity include being the breadwinner and being "strong," then men may be offended by women's professional progress because it makes men seem less worthy of respect. Resistance may also reflect ideological defense of male supremacy. Research on domestic violence suggests that male batterers often hold very conservative views of women's role in the family (Ptacek 1988). In many parts of the world, there exist ideologies that justify men's supremacy on grounds of religion, biology, cultural tradition, or organizational mission (e.g., in the military). It is a mistake to regard these ideas as simply outmoded. They may be actively modernized and renewed.

GROUNDS FOR OPTIMISM: CAPACITIES FOR EQUALITY AND REASONS FOR CHANGE

The public debates about men and boys have often been inconclusive. But they have gone a long way, together with the research, to shatter one widespread belief that has hindered gender reform. This obstacle is the belief that men *cannot* change their ways, that "boys will be boys," that rape, war, sexism, domestic violence, aggression, and self-centeredness are natural to men.

We now have many documented examples of the diversity of masculinities and of men's and boys' capacity for equality. For instance, life-history research in Chile has shown that there is no unitary Chilean masculinity, despite the cultural homogeneity of the country. While a hegemonic model is widely diffused across social strata, there are many men who depart from it, and there is significant discontent with traditional roles (Valdes and Olavarria 1998). Though groups of boys in schools often have a dominant or hegemonic pattern of masculinity, there are usually also other patterns present, some of which involve more equal and respectful relations with girls.

Research in Britain, for instance, shows how boys encounter and explore alternative models of masculinity as they grow up (Mac an Ghaill 1994; O'Donnell and Sharpe 2000).

Psychological and educational research shows personal flexibility in the face of gender stereotypes. Men and boys can vary, or strategically use, conventional definitions of masculinity. It is even possible to teach boys (and girls) how to do this in school, as experiments in Australian classrooms have shown (Davies 1993; Wetherell and Edley 1999).

Changes have occurred in men's practices within certain families, where there has been a conscious shift toward more equal sharing of housework and child care. The sociologist Barbara J. Risman (1998), who has documented such cases in one region of the United States, calls them "fair families." It is clear from her research that the change has required a challenge to traditional models of masculinity. In the Shanghai region of China, there is an established local tradition of relative gender equality, and men are demonstrably willing to be involved in domestic work. Research by Da Wei Wei (Da 2004) shows this tradition persisting among Shanghai men even after migration to another country.

Perhaps the most extensive social action involving men in gender change has occurred in Scandinavia. This includes provisions for paternity leave that have had high rates of take-up, among the most dramatic of all demonstrations of men's willingness to change gender practices. Øystein Holter sums up the research and practical experience: "The Nordic 'experiment' has shown that a *majority* of men can change their practice when circumstances are favorable.... When reforms

or support policies are well-designed and targeted towards an on-going cultural process of change, men's active support for gender-equal status increases" (1997, 126). Many groups of men, it is clear, have a capacity for equality and for gender change. But what reasons for change are men likely to see?

Early statements often assumed that men had the same interest as women in escaping from restrictive sex roles (e.g., Palme 1972). Later experience has not confirmed this view. Yet men and boys often do have substantial reasons to support change, which can readily be listed.

First, men are not isolated individuals. Men and boys live in social relationships, many with women and girls: wives, partners, mothers, aunts, daughters, nieces, friends, classmates, workmates, professional colleagues, neighbors, and so on. The quality of every man's life depends to a large extent on the quality of those relationships. We may therefore speak of men's relational interests in gender equality.

For instance, very large numbers of men are fathers, and about half of their children are girls. Some men are sole parents and are then deeply involved in caregiving—an important demonstration of men's capacity for care (Risman 1986). Even in intact partnerships with women, many men have close relationships with their children, and psychological research shows the importance of these relationships (Kindler 2002). In several parts of the world, young men are exploring more engaged patterns of fatherhood (Olavarria 2001). To make sure that daughters grow up in a world that offers young women security, freedom, and opportunities to fulfil their talents is a powerful reason for many men to support gender equality.

Second, men may wish to avoid the toxic effects that the gender order has for them. James Harrison long ago issued a "Warning: The Male Sex Role May Be Dangerous to Your Health" (1978). Since then health research has documented specific problems for men and boys. Among them are premature death from accident, homicide, and suicide; occupational injury; higher levels of drug abuse, especially of alcohol and tobacco; and in some countries at least, a relative unwillingness by men to seek medical help when it is needed. Attempts to assert a tough and dominant masculinity sustain some of these patterns (Sabo and Gordon 1995; Hurrelmann and Kolip 2002).

Social and economic pressures on men to compete in the workplace, to increase their hours of paid work, and sometimes to take second jobs are among the most powerful constraints on gender reform. Desire for a better balance between work and life is widespread among employed men. On the other hand, where unemployment is high the lack of a paid job can be a damaging pressure on men who have grown up with the expectation of being breadwinners. This is, for instance, an important gender issue in postapartheid South Africa. Opening alternative economic paths and moving toward what German discussions have called "multioptional masculinities" may do much to improve men's well-being (*Widersfruche* 1998; Morrell 2001a).

Third, men may support gender change because they see its relevance to the well-being of the community they live in. In situations of mass poverty and underemployment, for instance in cities in developing countries, flexibility in the gender division of labor may be crucial to a household that requires women's earnings as well as men's. Reducing the rigidity of masculinities may also yield benefits in security. Civil and international violence is strongly associated with dominating patterns of masculinity and with marked gender inequality in the state. Movement away from these patterns makes it easier for men to adopt historically "feminine" styles of nonviolent negotiation and conflict resolution (Zalewski and Parpart 1998; Breines, Connell, and Eide 2000; Gockburn 2003). This may also reduce the toxic effects of policing and incarceration (Sabo, Rupees, and London 2001).

Finally, men may support gender reform because gender equality follows from their political or ethical principles. These may be religious, socialist, or broad democratic beliefs. Mill argued a case based on classical liberal principles a century and a half ago, and the idea of equal human rights still has purchase among large groups of men.

GROUNDS FOR PESSIMISM: THE SHAPE OF MASCULINITY POLITICS

The diversity among men and masculinities is reflected in a diversity of men's movements in the developed countries. A study of the United States found multiple movements, with different agendas for the remaking of masculinity. They operated on the varying terrains of gender equality, men's rights, and ethnic or religious identities (Messner 1997). There is no unified political position for men and no authoritative representative of men's interests.

Men's movements specifically concerned with gender equality exist in a number of countries. A well-known example is the White Ribbon Campaign, dedicated to mobilizing public opinion and educating men and boys for the prevention of men's violence against women. Originating in Canada, in response to the massacre of women in Montreal in 1989, the White Ribbon Campaign achieved very high visibility in that country, with support from political and community leaders and considerable outreach in schools and mass media. More recently, it has spread to other countries. Groups concerned with violence prevention have appeared in other countries, such as Men against Sexual Assault in Australia and Men Overcoming Violence (MOVE) in the United States. These have not achieved the visibility of the White Ribbon Campaign but have built up a valuable body of knowledge about the successes and difficulties of organizing among men (Lichterman 1989; Pease 1997; Kaufman 1999).

The most extensive experience of any group of men organizing around issues of gender and sexual politics is that of homosexual men, in antidiscrimination campaigns, the gay liberation movement, and community responses to the HIV/AIDS pandemic. Gay men have pioneered in areas such as community care for the sick, community education for responsible sexual practices, representation in the public sector, and overcoming social exclusion, which are important for all groups of men concerned with gender equality (Kippax et al. 1993; Altman 1994).

Explicit backlash movements also exist but have not generally had a great deal of influence. Men mobilizing as men to oppose women tend to be seen as cranks or fanatics. They constantly exaggerate women's power. And by defining men's interests in opposition to women's, they get into cultural difficulties, since they have to violate a main tenet of modern patriarchal ideology— the idea that "opposites attract" and that men's and women's needs, interests, and choices are complementary.

Much more important for the defense of gender inequality are movements in which men's interests are a side effect—nationalist, ethnic, religious, and economic movements. Of these, the most influential on a world scale is contemporary neoliberalism—the political and cultural promotion of free-market principles and individualism and the rejection of state control.

Neoliberalism is in principle gender neutral. The "individual" has no gender, and the market delivers advantage to the smartest entrepreneur, not to men or women as such. But neoliberalism does not pursue social justice in relation to gender. In Eastern Europe, the restoration of capitalism and the arrival of neoliberal politics have been followed by a sharp deterioration in the position of women. In rich Western countries, neoliberalism from the 1980s on has attacked the welfare state, on which far more women than men depend; supported deregulation of labor markets, resulting in increased casualization of women workers; shrunk public sector employment, the sector of the economy where women predominate; lowered rates of personal taxation, the main basis of tax transfers to women; and squeezed public education, the key pathway to labor market advancement for women. However, the same period saw an expansion of the human-rights agenda, which is, on the whole, an asset for gender equality.

The contemporary version of neoliberalism, known as neoconservatism in the United States, also has some gender complexities. George W. Bush was the first U.S. president to place a woman in the very heart of the state security apparatus, as national security adviser to the president. And some of the regime's actions, such as the attack on the Taliban regime in Afghanistan, were defended as a means of emancipating women.

Yet neoconservatism and state power in the United States and its satellites such as Australia remain overwhelmingly the province of men—indeed, men of a particular character: power oriented and ruthless, restrained by little more than calculations of likely opposition. There has been a sharp remasculinization of political rhetoric and a turn to the use of force as a primary instrument in policy. The human rights discourse is muted and sometimes completely abandoned (as in the U.S. prison camp for Muslim captives at Guantanamo Bay and the Australian prison camps for refugees in the central desert and Pacific islands).

Neoliberalism can function as a form of masculinity politics largely because of the powerful role of the state in the gender order. The state constitutes gender relations in multiple ways, and all of its gender policies affect men. Many mainstream policies (e.g., in economic and security affairs) are substantially about men without acknowledging this fact (Nagel 1998; O'Connor, Orloff, and Shaver 1999; Connell 2003b).

This points to a realm of institutional politics where men's and women's interests are very much at stake, without the publicity created by social movements. Public-sector agencies (Jensen 1998; Mackay and Bilton 2000; Schofield, forthcoming), private-sector corporations (Marcband and Runyan 2000; Hearn and Parkin 2001), and unions (Gorman et al. 1993; Pranzway 2001) are all sites of masculinized power and struggles

for gender equality. In each of these sites, some men can be found with a commitment to gender equality, but in each case that is an embattled position. For gender-equality outcomes, it is important to have support from men in the top organizational levels, but this is not often reliably forthcoming.

One reason for the difficulty in expanding men's opposition to sexism is the role of highly conservative men as cultural authorities and managers. Major religious organizations, in Christianity, Islam, and Buddhism, are controlled by men who sometimes completely exclude women, and these organizations have often been used to oppose the emancipation of women. Transnational media organizations such as Rupert Murdoch's conglomerate are equally active in promoting conservative gender ideology.

A specific address to men is found in the growing institutional, media, and business complex of commercial sports. With its overwhelming focus on male athletes; its celebration of force, domination, and competitive success; its valorization of male commentators and executives; and its marginalization and frequent ridicule of women, the sports/business complex has become an increasingly important site for representing and defining gender. This is not traditional patriarchy. It is something new, welding exemplary bodies to entrepreneurial culture. Michael Messner (2002), one of the leading analysts of contemporary sports, formulates the effect well by saying that commercial sports define the renewed centrality of men and of a particular version of masculinity.

On a world scale, explicit backlash movements are of limited importance, but very large numbers of men are nevertheless engaged in preserving gender inequality. Patriarchy is defended diffusely. There is support for change from equally large numbers of men, but it is an uphill battle to articulate that support. That is the political context with which new gender-equality initiatives have to deal.

WAYS FORWARD: TOWARD A GLOBAL FRAMEWORK

Inviting men to end men's privileges, and to remake masculinities to sustain gender equality, strikes many people as a strange or Utopian project. Yet this project is already under way. Many men around the world are engaged in gender reforms, for the good reasons discussed above.

The diversity of masculinities complicates the process but is also an important asset. As this diversity becomes better known, men and boys can more easily see a range of possibilities for their own lives, and both men and women are less likely to think of gender inequality as unchangeable. It also becomes possible to identify specific groups of men who might engage in alliances for change.

The international policy documents discussed above rely on the concept of an alliance between men and women for achieving equality. Since the growth of an autonomous women's movement, the main impetus for reform has been located in women's groups. Some groups within the women's movement, especially those concerned with men's violence, are reluctant to work with men or are deeply skeptical of men's willingness to change. Other feminists argue that alliances between women and men are possible, even crucial. In some social movements, for instance, environmentalism, there is a strong ideology of gender equality and a favorable environment for men to support gender change (Connell 1995; Segal 1997).

In local and central government, practical alliances between women and men have been important in achieving equal-opportunity measures and other gender-equality reforms. Even in the field of men's violence against women, there has been cooperation between women's groups and men's groups, for instance, in prevention work. This cooperation can be an inspiration to grassroots workers and a powerful demonstration of women and men's common interest in a peaceful and equal society (Pease 1997; Schofield, forthcoming). The concept of alliance is itself important, in preserving autonomy for women's groups, in preempting a tendency for any one group to speak for others, and in defining a political role for men that has some dignity and might attract widespread support.

Given the spectrum of masculinity politics, we cannot expect worldwide consensus for gender equality. What is possible is that support for gender equality might become hegemonic among men. In that case it would be groups supporting equality that provide the agenda for public discussion about men's lives and patterns of masculinity. . . .

▪▪ Judith Butler (1957–): A Biographical Sketch

Judith Butler was born in 1956. She received her BA in philosophy from Bennington College in 1978 and her PhD in philosophy from Yale University in 1984. Butler has taught at Wesleyan and Johns Hopkins Universities and is currently professor of rhetoric and comparative literature at the University of California at Berkeley. Butler's books include *Subjects of Desire* (1999), *Gender Trouble: Feminism and the Subversion of Identity* (1989), *Bodies That Matter: On the Discursive Limits of "Sex"* (1993), and *Excitable Speech: Politics of the Performance* (1997), which analyzes name-calling as both a social injury and the way in which individuals are called into action for political purposes.

▪▪ Butler's Intellectual Influences and Core Ideas

Whereas feminists committed to modern ideas about gender ask the question "And what about women?" postmodern feminists, such as Judith Butler, ask "And what do you *mean* by 'women'?" Butler ([1990] 1999:145–47) rejects the very idea that "women" can be understood as a concrete category at all, construing gender identity instead as an unstable "fiction." She criticizes modern feminists for remaining within the confines of traditional binary categories that in her view necessarily perpetuate sexism. In keeping with Foucault (see Chapter 15), Butler provides a "critical genealogy of gender categories in . . . different discursive domains" ([1990] 1999:xxx). In short, while modern feminists had in separating (biologically determined) "sex" from (socially constructed) "gender" helped rupture the idea of a stable or essential self, Butler takes this rupture to an extreme by upending the alleged "biological" dimensions of sexuality. Far from seeing "desire" as a biological given, Butler ([1990] 1999:70) maintains that "which pleasures shall live and which shall die is often a matter of which serve the legitimating practices of identity formation that take place within the matrix of gender norms."

Specifically, Butler conceptualizes gendered subjectivity as a fluid identity and contends that the individual subject is never exclusively "male" or "female" but rather, is always in a state of contextually dependent flux. That is, gendered subjectivity is not something "fixed" or "essential," but a continual performance. Gender is not "a singular act, but a representation and a ritual" (1999:xv). Consequently, Butler (1993) seeks to explain the conditions of performances; that is, she seeks to explain "the practice by which gendering occurs," for the embodying of norms is "a compulsory practice, a forcible production," repeated, although "not fully determining."

Indeed, it is the sustained, continual nature of gender performance that compels Butler to use the term **performativity** rather than "performance." Performativity contests the very notion of a subject. Whereas the noun "performance" implies distinct, concrete, finished events, the term "performativity" reflects "culturally sustained temporal duration." As Butler (1999:xv) states,

> The view that gender is performative show[s] that what we take to be an internal essence of gender is manufactured through a sustained set of acts, posited through the gendered stylization of the body. In this way, it show[s] that what we take to be an "internal" feature of ourselves is one that we anticipate and produce through certain bodily acts, at an extreme, an hallucinatory effect of naturalized gestures.

So, too, is the "culturally sustained" (rather than essentialist) nature of gender performances evident in Butler's discussion of performative *acts,* which she conceptualizes as "forms of authoritative speech . . . [or] statements that, "*in uttering* . . . exercise a binding power" (1993:224, emphasis added); as Butler maintains,

> Implicated in a network of authorization and punishment, performatives tend to include legal sentences, baptisms, inaugurations, declarations of ownership, statements which not only perform an action, but confer a binding power on the action performed. If the power of discourse to produce that which it names is linked with the question of performativity, then the performative is one domain in which power acts *as* discourse. (1993:224, emphasis added)

In other words, for Butler, "what we take to be an internal essence of gender is manufactured through a sustained set of acts, posited through the gendered stylization of the body" (ibid.:xv). "Gender is a kind of persistent impersonation that passes as the real" (ibid.:xxviii). Just as in Kafka's "Before the Law," where one sits before the door of the law awaiting that authority to be distributed, so, too, gender is "an expectation that ends up producing the very phenomenon that it anticipates" (1999:xiv).

This brings us to the issue of **queer theory.** In addition to being a leading feminist theorist, Butler is one of the most important figures in queer theory. Queer theory emerged from gay/lesbian studies (which emerged from gender studies) in the 1980s. Up until the 1980s, the term "queer" had a derogatory connotation, meaning "odd" or "peculiar" or "out of the ordinary." However, queer theorists, including Butler, appropriated this term, insisting that all sexual behaviors, all concepts linking sexual behaviors to sexual identities, and all categories of normative and deviant sexualities are social constructs, which create certain types of social meaning. In short, "sex is a norm" (Osborne and Segel, 1993).

Thus, the undergirding emphasis in all these projects (gay/lesbian, queer, feminist) is that the categories of normative and deviant sexual behavior are not biologically but rather socially constructed. In contrast to those who see sexuality as biological and gender as a social construction, Butler sees sex as no more a natural category than gender. She conceptualizes gender norms as structuring biology and not the reverse, which informs the more conventional view.

Butler does not deny certain kinds of biological differences, but she seeks to explain under what discursive and institutional conditions do certain arbitrary biological differences become salient characteristics of sex (ibid.). She emphasizes that sexuality is a complex array of individual activity and institutional power, of social codes and forces, which interact to shape the ideas of what is normative and what is deviant at any particular moment, and which then result in categories as to "natural," "essential," "biological," or "god-given." She seeks to show how a norm can actually materialize a body, that is, how the body is not only invested with a norm but in some sense is also animated by a norm or contoured by a norm (ibid.).

Photo 14.2 Harris Glenn Milstead (1945–1988), better known by his drag persona, Divine, who starred in several of John Waters's films, including *Hairspray*, exemplifies performativity.

Specifically, Butler describes a **heterosexual matrix** in which "proper men" and "proper women" are identified as heterosexual. She shows that the essential unity between biological sex, gender identification, and heterosexuality is not dictated by nature; indeed, this unity is an illusion mediated through cultural systems of meaning that underlie our understanding of material, anatomical differences. According to Butler, heterosexual normativity "ought *not* to order gender" ([1990] 1999:xiv, emphasis in original). The subversion of gender performances (e.g., drag performances) indicates nothing about sexuality or sexual practice. "Gender can be rendered ambiguous without disturbing or reorientating normative sexuality at all" (ibid.).

Thus, for instance, Butler points out that discrimination against gays is a function not of their sexuality but of their failure to perform heterosexual gender norms. Heterosexuality is a highly unstable system that relies on carefully constructed individual performances of identity, and the exclusion of homosexuality. Because heterosexuality is based on a binary difference between male and female (a person is either one or the other), there is a socially constructed gender in which heterosexuality is central, which informs our understanding of biology.

Interestingly, then, akin to Harold Garfinkel's "breaching" experiments, which exposed taken-for-granted normative expectations (see Chapter 13), cross-dressing, "kiss-ins," gender parodies, and so on can be used to transgress and rebel against existing sexual categories. In short, queer politics seeks to explicitly challenge gender norms to show their lack of naturalness and inevitability and to celebrate transgressions from them (Alsop, et al. 2002:96), while postmodern queer theorists seek to upend and "resignify" our gender expectations.

╟ BUTLER'S THEORETICAL ORIENTATION

As will be discussed further in the next chapter, postmodernists tend to eschew metatheoretical frameworks as "essentializing." However, it is difficult not to see postmodernists, including Butler, as nonrationalistic in their approach to action. That "there is no reality" anymore (only "hyperreality"—Baudrillard—see Chapter 15); that sex is not a "natural" category but constituted through social discourse; that performances create subjectivities (see Butler, this chapter) seems a profoundly nonrationalistic orientation to action. In contrast to Goffman, who, as we have seen (see Chapter 12), also at times used the term "performance" in a more rationalistic way (wittingly constructed, via calculation and even rehearsal), Butler argues that we become subjects *from* our performances. Subjectivity is a process of submitting ourselves to socially constituted norms and practices (Alsop, et al. 2002:98). This speaks to the nonrational realm.

In terms of order, on one hand postmodernists such as Butler emphasize the role of structured "scripts," discourses, and preexisting symbolic patterns that reflect a collective orientation. In addition, Butler exudes a neo-Marxist emphasis on hierarchical (class, gender, racial) structures, oppression, and corporate control, which also speaks to the collective realm (see Chapters 3 and 10). She shows how gender

Figure 14.5 Butler's Basic Theoretical Orientation

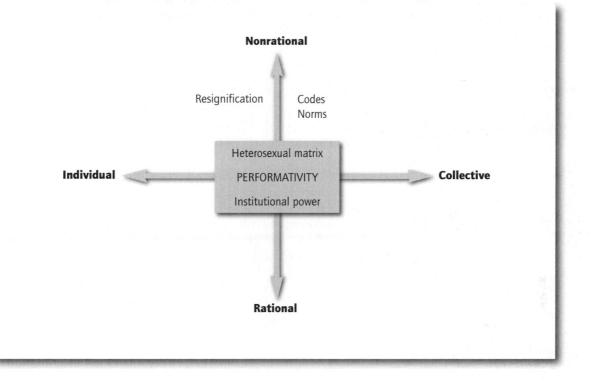

performances are tied to relations of ruling, in Smith's terms. On the other hand, however, like Foucault (see Chapter 15), Butler insists that regulatory norms and discourses are never wholly determining. One could argue that in the end, Butler's work seems individualistic because she emphasizes that it is in interaction that subjectivities are formed. Moreover, in contrast to cultural Marxists (e.g., the Frankfurt School), Butler optimistically asserts that because of the multiplicity of symbols that cannot all be obeyed coherently, we can "reconfigure" and "redeploy" symbols. However, again it must be emphasized that Butler would undoubtedly chafe at this label. First, because Butler goes to great lengths to show that performances are never isolated acts, but occur only *within* specific discursive contexts, and second, because she would chafe at any sort of metatheoretical label at all. The whole point of postmodernism is to do away with this kind of academic theoretical scaffolding. In keeping with the spirit of Butler, then, we place the notion of "performativity" at the center of our theoretical map, thereby indicating its fluid, multidimensional nature, while nevertheless acknowledging its nonrational bent (see Figure 14.5).

Reading

Introduction to "Subjects of Sex/Gender/Desire"

The following excerpt is from Butler's most widely read and celebrated book, *Gender Trouble* (1990), which has sold more than 100,000 copies. Here you will see Butler challenge the presumed universality and unity of the concept of "woman" in feminist theory and, drawing on Foucault, dispute the predominant binary opposition of sex as a "biological" and gender as a "cultural" category.

"Subjects of Sex/Gender/Desire" (1990)

Judith Butler

One is not born a woman, but rather becomes one.
—Simone de Beauvoir

Strictly speaking, "women" cannot be said to exist.
—Julia Kristeva

Woman does not have a sex.
—Luce Irigaray

The deployment of sexuality . . . established this notion of sex.
—Michel Foucault

The category of sex is the political category that founds society as heterosexual.
—Monique Wittig

I. "WOMEN" AS THE SUBJECT OF FEMINISM

For the most part, feminist theory has assumed that there is some existing identity, understood through the category of women, who not only initiates feminist interests and goals within discourse, but constitutes the subject for whom political representation is pursued. But *politics* and *representation* are controversial terms. On the one hand, *representation* serves as the operative term within a political process that seeks to extend visibility and legitimacy to women as political subjects; on the other hand, representation is the normative function of a language which is said either to reveal or to distort what is assumed to be true about the category of women. For feminist theory, the development of a language that fully or adequately represents women has seemed necessary to foster the political visibility of women. This has seemed obviously important considering the pervasive cultural condition in which women's lives were either misrepresented or not represented at all.

Recently, this prevailing conception of the relation between feminist theory and politics has come under challenge from within feminist discourse. The very subject of women is no longer understood in stable or abiding terms. There is a great deal of material that not only questions the viability of "the subject" as the ultimate candidate for representation or, indeed, liberation, but there is very little agreement after all on what it is that constitutes, or ought to constitute, the category of women. The domains of political and linguistic "representation" set out in advance the criterion by which subjects themselves are formed, with the result that representation is extended only to what can be acknowledged as a subject. In other words, the qualifications for being a subject must first be met before representation can be extended.

Foucault points out that juridical systems of power *produce* the subjects they subsequently come to represent. Juridical notions of power appear to regulate political life in purely negative terms—that is, through the limitation, prohibition, regulation, control, and even "protection" of individuals related to that political structure through the contingent and retractable operation of choice. But the subjects regulated by such structures are, by virtue of being subjected to them, formed, defined, and reproduced in accordance with the requirements of those structures. If this analysis is right, then the juridical formation of language and politics that represents women as "the subject" of feminism is itself a discursive formation and effect of a given version of representational politics. And the feminist subject turns out to be discursively constituted by the very political system that is supposed to facilitate its emancipation. This becomes politically problematic if that system can be shown to produce gendered subjects along a differential axis of domination or to produce subjects who are presumed to be masculine. In such cases, an uncritical appeal to such a system for the emancipation of "women" will be clearly self-defeating.

The question of "the subject" is crucial for politics, and for feminist politics in particular, because juridical subjects are invariably produced through certain exclusionary practices that do not "show" once the juridical structure of politics has been established. In other words, the political construction of the subject proceeds with certain legitimating and exclusionary aims, and these political operations are effectively concealed and naturalized by a political analysis that takes juridical structures as their foundation. Juridical power inevitably "produces" what it claims merely to represent; hence, politics must be concerned with this dual function of power: the juridical and the productive. In effect, the law produces and then conceals the notion of "a subject before the law" in order to invoke that discursive formation as a naturalized foundational premise that subsequently legitimates that law's own regulatory hegemony. It is not enough to inquire into how women might become more fully represented in language and politics. Feminist critique ought also to understand how the category of "women," the subject of feminism, is produced and restrained by the very structures of power through which emancipation is sought.

Indeed, the question of women as the subject of feminism raises the possibility that there may not be a subject who stands "before" the law, awaiting representation in or by the law. Perhaps the subject, as well as the invocation of a temporal "before," is constituted by the law as the fictive foundation of its own claim to legitimacy. The prevailing assumption of the ontological integrity of the subject before the law might be understood as the contemporary trace of the state of nature hypothesis, that foundationalist fable constitutive of the juridical structures of classical liberalism. The performative invocation of a nonhistorical "before" becomes the foundational premise that guarantees a presocial ontology of persons who freely consent to be governed and, thereby, constitute the legitimacy of the social contract.

Apart from the foundationalist fictions that support the notion of the subject, however, there is the political problem that feminism encounters in the assumption that the term *women* denotes a common identity. Rather than a stable signifier that commands the assent of those whom it purports to describe and represent, *women,* even in the plural, has become a troublesome term, a site of contest, a cause for anxiety. As Denise Riley's title suggests, *Am I That Name?* is a question produced by the very possibility of the name's multiple significations. If one "is" a woman, that is surely not all one is; the term fails to be exhaustive, not because a pregendered "person" transcends the specific paraphernalia of its gender, but because gender is not always constituted coherently or consistently in different historical

contexts, and because gender intersects with racial, class, ethnic, sexual, and regional modalities of discursively constituted identities. As a result, it becomes impossible to separate out "gender" from the political and cultural intersections in which it is invariably produced and maintained.

The political assumption that there must be a universal basis for feminism, one which must be found in an identity assumed to exist cross- culturally, often accompanies the notion that the oppression of women has some singular form discernible in the universal or hegemonic structure of patriarchy or masculine domination. The notion of a universal patriarchy has been widely criticized in recent years for its failure to account for the workings of gender oppression in the concrete cultural contexts in which it exists. Where those various contexts have been consulted within such theories, it has been to find "examples" or "illustrations" of a universal principle that is assumed from the start. That form of feminist theorizing has come under criticism for its efforts to colonize and appropriate non-Western cultures to support highly Western notions of oppression, but because they tend as well to construct a "Third World" or even an "Orient" in which gender oppression is subtly explained as symptomatic of an essential, non-Western barbarism. The urgency of feminism to establish a universal status for patriarchy in order to strengthen the appearance of feminism's own claims to be representative has occasionally motivated the shortcut to a categorial or fictive universality of the structure of domination, held to produce women's common subjugated experience.

Although the claim of universal patriarchy no longer enjoys the kind of credibility it once did, the notion of a generally shared conception of "women," the corollary to that framework, has been much more difficult to displace. Certainly, there have been plenty of debates: Is there some commonality among "women" that preexists their oppression, or do "women" have a bond by virtue of their oppression alone? Is there a specificity to women's cultures that is independent of their subordination by hegemonic, masculinist cultures? Are the specificity and integrity of women's cultural or linguistic practices always specified against and, hence, within the terms of some more dominant cultural formation? Is there is a region of the "specifically feminine," one that is both differentiated from the masculine as such and recognizable in its difference by an unmarked and, hence, presumed universality of "women"? The masculine/feminine binary constitutes not only the exclusive framework in which that specificity can be recognized, but in every other way the "specificity" of the feminine is once again fully decontextualized and separated off analytically and politically from the constitution of class, race, ethnicity, and other axes of power relations that both constitute "identity" and make the singular notion of identity a misnomer.

My suggestion is that the presumed universality and unity of the subject of feminism is effectively undermined by the constraints of the representational discourse in which it functions. Indeed, the premature insistence on a stable subject of feminism, understood as a seamless category of women, inevitably generates multiple refusals to accept the category. These domains of exclusion reveal the coercive and regulatory consequences of that construction, even when the construction has been elaborated for emancipatory purposes. Indeed, the fragmentation within feminism and the paradoxical opposition to feminism from "women" whom feminism claims to represent suggest the necessary limits of identity politics. The suggestion that feminism can seek wider representation for a subject that it itself constructs has the ironic consequence that feminist goals risk failure by refusing to take account of the constitutive powers of their own representational claims. This problem is not ameliorated through an appeal to the category of women for merely "strategic" purposes, for strategies always have meanings that exceed the purposes for which they are intended. In this case, exclusion itself might qualify as such an unintended yet consequential meaning. By conforming to a requirement of representational politics that feminism articulate a stable subject, feminism thus opens itself to charges of gross misrepresentation.

Obviously, the political task is not to refuse representational politics—as if we could. The juridical structures of language and politics constitute the contemporary field of power; hence, there is no position outside this field, but only a critical genealogy of its own legitimating practices. As such, the critical point of departure is *the historical present,* as Marx put it. And the task is to formulate within this constituted frame a critique of the categories of identity that contemporary juridical structures engender, naturalize, and immobilize.

Perhaps there is an opportunity at this juncture of cultural politics, a period that some would call "postfeminist," to reflect from within a feminist perspective on the injunction to construct a subject to feminism. Within feminist political practice, a radical rethinking of the ontological constructions of identity appears to be necessary in order to formulate a representational politics that might revive feminism on other grounds. On the other hand, it may be time to entertain a radical critique that seeks to free feminist theory from the necessity of having to construct a single or abiding ground which is invariably contested by those identity positions or anti-identity positions that it invariably excludes. Do the exclusionary practices that ground feminist theory in a notion of "women" as subject paradoxically undercut feminist goals to extend its claims to "representation"?

Perhaps the problem is even more serious. Is the construction of the category of women as a coherent and stable subject an unwitting regulation and reification of gender relations? And is not such a reification precisely contrary to feminist aims? To what extent does the category of women achieve stability and coherence only in the context of the heterosexual matrix? If a stable notion of gender no longer proves to be the foundational premise of feminist politics, perhaps a new sort of feminist politics is now desirable to contest the very reifications of gender and identity, one that will take the variable construction of identity as both a methodological and normative prerequisite, if not a political goal.

To trace the political operations that produce and conceal what qualifies as the juridical subject of feminism is precisely the task of *a feminist genealogy* of the category of women. In the course of this effort to question "women" as the subject of feminism, the unproblematic invocation of that category may prove to *preclude* the possibility of feminism as a representational politics. What sense does it make to extend representation to subjects who are constructed through the exclusion of those who fail to conform to unspoken normative requirements of the subject? What relations of domination and exclusion are inadvertently sustained when representation becomes the sole focus of politics? The identity of the feminist subject ought not to be the foundation of feminist politics, if the formation of the subject takes place within a field of power regularly buried through the assertion of that foundation. Perhaps, paradoxically, "representation" will be shown to make sense for feminism only when the subject of "women" is nowhere presumed.

II. THE COMPULSORY ORDER OF SEX/GENDER/DESIRE

Although the unproblematic unity of "women" is often invoked to construct a solidarity of identity, a split is introduced in the feminist subject by the distinction between sex and gender. Originally intended to dispute the biology-is-destiny formulation, the distinction between sex and gender serves the argument that whatever biological intractability sex appears to have, gender is culturally constructed: hence, gender is neither the causal result of sex nor as seemingly fixed as sex. The unity of the subject is thus already potentially contested by the distinction the permits of gender as a multiple interpretation of sex.

If gender is the cultural meanings that the sexed body assumes, then a gender cannot be said to follow from a sex in any one way. Taken to its logical limit, the sex/gender distinction suggests a radical discontinuity between sexed bodies and culturally constructed genders. Assuming for the moment the stability of binary sex, it does not follow that the construction of "men" will accrue exclusively to the bodies of males or that "women" will interpret only female bodies. Further, even if the sexes appear to be unproblematically binary in their morphology and constitution (which will become a question), there is no reason to assume that genders ought also to remain as two. The presumption of a binary gender system implicitly retains the belief in a mimetic relation of gender to sex whereby gender mirrors sex or is otherwise restricted by it. When the constructed status of gender is theorized as radically independent of sex, gender itself becomes a free-floating artifice, with the consequence that *man* and *masculine* might just as easily signify a female body as a male one, and *woman* and *feminine* a male body as easily as a female one.

This radical splitting of the gendered subject poses yet another set of problems. Can we refer to a "given" sex or a "given" gender without first inquiring into how sex and/or gender is given, through what means? And what is "sex" anyway? Is it natural, anatomical, chromosomal, or hormonal, and how is a feminist critic to assess the scientific discourses which purport to establish such "facts" for us? Does sex have a history? Does each sex have a different history, or histories? Is there a history of how the duality of sex was established, a genealogy that might expose the binary options as a variable construction? Are the ostensibly natural facts of sex discursively produced by various scientific discourses in the service of other political and social interests? If the immutable character of sex is contested, perhaps this construct called "sex" is as culturally constructed as gender; indeed, perhaps it was always already gender, with the consequence that the distinction between sex and gender turns out to be no distinction at all.

It would make no sense, then, to define gender as the cultural interpretation of sex, if sex itself is a gendered category. Gender ought not to be conceived merely as the cultural inscription of meaning on a pregiven sex (a juridical conception); gender must also designate the very apparatus of production whereby the sexes themselves are established. As a result, gender is not to culture as sex is to nature; gender is also the discursive/cultural means by which "sexed nature" or "a natural sex" is produced and established as "prediscursive," prior to culture, a politically neutral surface *on which* culture acts. This construction of "sex" as the radically unconstructed will concern us again in the discussion of Lévi-Strauss and structuralism, [but] at this juncture it is already clear that one way the internal stability and binary frame for sex is effectively secured is by casting the duality of sex in a prediscursive domain. This production of sex as the prediscursive ought to be understood as the effect of the apparatus of cultural construction designated by *gender.* How, then, does gender need to be reformulated to encompass the power relations that produce the effect of a prediscursive sex and so conceal that very operation of discursive production?

Discussion Questions

1. Dorothy Smith (1987:64) states, "So as I walk down the street keeping an eye on my dog I am observing some of the niceties of different forms of property ownership." In what ways do you "do class" in your everyday life? In what ways do you "do gender"? How do your class and gender performances reaffirm forms of domination? To what extent do your gender performances reflect, reaffirm, and/or challenge normative heterosexuality or what Butler calls the heterosexual matrix?

(Continued)

(Continued)

2. According to Smith (1987:68), "The role of women is central both in the work that is done and in the management of its routine daily order . . . whatever the relations between school achievement, career success, and the 'intricate psychosocial processes' of the family, the conscious, planned thoughtful work of women as mothers has been part of its actuality." Discuss the extent to which "behind-the-scenes" women's work is still taken for granted in both schools and the workplace today. Do you think that this aspect of gender roles has changed in the past 20 years? How so or why not? Do you think full gender equality can be achieved? Why or why not?

3. According to Collins (1990:228), "A matrix of domination contains few pure victims or oppressors. Each individual derives varying amounts of penalty and privilege from the multiple systems of oppression which frame everyone's lives." Give concrete examples of moments or situations in which you have found yourself a "victim" and concrete examples of moments or situations in which you found yourself an "oppressor." Explain how your examples reflect the matrix of domination at the level of personal biography, the community, and the systemic level of social institutions.

4. Discuss the neo-Marxist or critical dimensions of Smith, Collins, Connell and Butler. In addition to critical theory, what other traditions and concepts does each draw from to produce their own distinct perspective?

5. Compare and contrast Butler's conceptualization of "performativity" with Goffman's dramaturgical theory (Chapter 12).

15 POSTSTRUCTURAL AND POSTMODERN THEORIES

With David Boyns

Michel Foucault

Key Concepts

- Archaeology
- Discipline/Disciplinary Society
- Episteme
- Genealogy

- Panopticon
- Power/Knowledge
- Sign
- Surveillance
- Metanarratives

Jean Baudrillard

Key Concepts

- Hyperreality
- Simulation/Simulacrum

The judges of normality are present everywhere. We are in the society of the teacher-judge, the doctor-judge, the educator-judge, the "social worker" judge.

—Michel Foucault

Television knows no night. It is perpetual day. TV embodies our fear of the dark, of night, of the other side of things.

—Jean Baudrillard

Have you ever felt that you were being watched—imagined that unbeknownst to you, someone, somewhere was inspecting you, maybe through the lens of a camera or through some information tracking system, monitoring your purchases, collecting information on your "identity," measuring your rate of speed, observing your every move? Such is the alarming experience of computer programmer Thomas Anderson ("Neo"), the central character in the popular film *The Matrix*. Through a series of cryptic messages from a computer hacker named Trinity, Neo discovers that someone is monitoring his computer activities. As he learns more about this surveillance from Trinity, Neo's ordinary life begins to unravel. He finds that a group of mysterious "agents" have him under observation and that technologies of surveillance are being used to record and scrutinize his daily activities. Neo soon recognizes that these are just the beginning of his problems. Rescued from the agents by Trinity and a faction of rogue revolutionaries from the "real world," Neo discovers that the entirety of his life experiences has been false, that he has been "living" inside a simulated world generated by an elaborate computer program. All of his memories and relationships, loves and fears, have been delusions designed to keep him unaware of the reality of his virtual deception. Bewildered by these discoveries, Neo is forced to confront two overwhelming contemporary truths. First, he lives in a social world that is saturated with technologies of **surveillance** that monitor his daily activities. Second, the authenticity of his experiences is distorted by the existence of **simulations** of reality. "Simulation" means a model or reproduction, and in postmodernism simulation refers particularly to when an image or model becomes more "real" than "reality" itself.

While Neo's story is certainly the invention of science fiction, the film *The Matrix* and other films like it[1] bring into awareness some important changes in the contemporary social world. We are increasingly both the objects of surveillance and the inhabitants of a world permeated by simulations. Like Neo, our daily lives are increasingly monitored. We are watched by cameras in our workplaces and as we patronize shopping centers; our consumer activities are "recorded" to document our spending habits; we are monitored by police radar guns and photo-radar cameras as we drive our cars; our financial history is recorded and scrutinized. Such processes allow a great deal of information about us to be gathered; they allow our lives to be documented and recorded in very personal ways. Strangely, and perhaps frighteningly, we have no real idea who has access to this information, what they are doing with it, and exactly what is known about us. Also like Neo, we increasingly find that our daily experiences are fused with simulations. Think for a moment about what you know about the world in which you live, and more important, *how* you have come to know what you know. If we take an honest inventory of our knowledge about the world, many of us would discover that much of what we "know" we haven't directly experienced for ourselves. Instead, we "know" much of our world secondhand, through conversation, books, television, radio, newspaper, and the Internet. Reflect on all the people that you know of but have never actually met—celebrities, newscasters, political figures, athletes, cartoon characters, and pop stars; or consider to how many places—even fictional places—you have traveled only through photographs, film, and television. Increasingly, as the French postmodern theorist Jean Baudrillard argues, the social world we inhabit has become **hyperreal**, filled with simulations of reality that replace "reality" itself.

Neo's world is one of ubiquitous surveillance and simulation. The key poststructural and postmodern theorists profiled in this chapter—Michel Foucault and Jean Baudrillard—both develop these two central themes, exploring them as a dominant condition of the contemporary world. The ideas of these three writers have become greatly influential in the development of poststructural and postmodern theories. As mentioned previously, Jean Baudrillard, who is one of the first and most important sociological theorists to be associated with postmodern theory, is particularly concerned about the issue of simulation. Michel Foucault, who is commonly identified as a chief progenitor of poststructuralist theory as well as one of the most influential intellectuals of the twentieth century, has written extensively on the issue of surveillance. Although Foucault died in 1984, his work still reverberates throughout many scholarly fields (e.g., history,

---

[1]For instance, *Matrix 3, Existenz, The Thirteenth Floor, Blade Runner, The Truman Show, Edtv,* and *Donnie Darko.*

philosophy, literary criticism, feminist studies, psychology, gay and lesbian studies, and sociology) and has become an important reference point for activists around the world.

Before we turn to the specific ideas of Foucault and Baudrillard, however, it is essential to define the two terms that frame this chapter: poststructuralism and postmodernism. Poststructuralism and postmodernism are among the most controversial and puzzling theories within sociology. These "post" theories radically oppose many of the established understandings of sociology and call for social thinkers to interpret the world in new, and often startling, ways. Indeed, they attack the ability of sociology to develop claims to the truth of social reality, as well as the very idea of the "social" upon which sociology is based. It should come as no surprise, then, that not all sociologists accept the legitimacy of poststructural and postmodern analysis. Many see these "post" theories as academic fads created by social theorists who have sought to carve new intellectual territory by making wild and exaggerated claims. Moreover, even among those who do recognize their validity, clear, consistent, and succinct definitions of poststructuralism and postmodernism are rarely offered. In fact, many of the theorists primarily identified with poststructuralism and postmodernism—including the theorists profiled in this chapter—never, or at least rarely, use the terms in their own work. Thus, before we go any further, it is essential to clarify these ambiguous and controversial terms.

Photo 15.1 *The Matrix* exemplifies two critical features of the postmodern world: technological surveillance and simulation.

Defining Poststructuralism ▪▪

Poststructuralism is not so much a coherent theory as an assemblage of converging themes developed by theorists in the 1960s. These common themes (to be discussed shortly) include the fragmentation of meaning, the localization of politics, the decline of the idea of Truth, and the decentering of the subject. Put in another way, poststructuralism is better understood as a theoretical trend, as a loosely articulated set of ideas that find common expression in the works of influential French writers of the second half of the twentieth century.[2]

Poststructuralism is so named because it represents a challenge to the central European and American tradition of theory dominant in the mid-twentieth century—structuralism. Structuralist thought emphasizes that there are forces in social life that emerge out of human activity but stand outside of human agency or intervention. Often called "emergent properties," these social forces are the external and constraining "social facts" that the classical sociological theorist Émile Durkheim described in his work, and a similar structuralist emphasis is echoed in many of the works of other classical theorists in sociology, particularly Karl Marx and Max Weber. Indeed, it was by drawing from these classical sources that Talcott Parsons would crystallize his structural functionalist theory that dominated American sociology during the middle years of the twentieth century (see Chapter 9).

---

[2]In addition to Foucault, whose work is profiled in this chapter, these writers include Roland Barthes, Jacques Derrida, Julia Kristeva, Jacques Lacan, Gilles Deleuze, and Felix Guattari. Interestingly, although most of the writers identified with poststructuralism are French, poststructuralism has its general origins in American traditions of literary criticism (Poster 1989:6). It is the way in which American literary scholars adopted and unified the ideas of French thinkers that created what is known today as poststructural theory. Ironically but not surprisingly, then, most of those French theorists who were canonized and identified as "poststructuralists" do not write under the label of "poststructuralism," and they do not identify themselves with poststructuralism as a theoretical movement.

A different version of structuralism emerged during the mid-twentieth century in France.[3] French structuralism is most distinct from American structural functionalism in its particular emphasis on language. In contrast to Parsons's structural functionalism which emphasizes the patterns of human social organization, French structuralism focuses on meaning and the role of language in the organization of systems of ideas. French structuralists seek to demonstrate how language has formal properties (such as grammar) that provide the structure not only for communication but also for broader aspects of human existence. For French structuralists, myths, kinship systems, religious rituals, advertising messages, fashion, class relationships, and the human psyche all have formal elements that can be understood through an analysis of the structures of meaning, especially those organized by the properties of language. That is, just as there are grammatical rules and linguistic structures in our native language that speakers follow without being cognizant of them, there is a rigid formula for virtually every aspect of social life that structures understanding, although social actors are not necessarily cognizant of its impact.

In most accounts, the story of French structuralism begins with Swiss linguist Ferdinand de Saussure's (1857–1913) development of what is today known as *semiotics*.[4] Semiotics is the study of signs, those aspects of human communication that are used to indicate and convey meaning. In *A Course in General Linguistics* (1966), Saussure develops his theory by making a key distinction between the formal structure of language (*langue*) and the everyday use of speech (*parole*). For Saussure, *langue* is the structural dimension of language that is nstitutionalized and patterned through the establishment of social conventions. According to Saussure, language has several characteristics: it is collective, well defined, concrete, composed of systems of shared meanings, and subject to objective study. Speech (*parole*), on the other hand, is the active use of language in everyday interaction and is more transitory and individualized.

From this distinction, Saussure developed a description of the formal structure of the **sign**. For Saussure, a sign is that which is socially designated to represent the meaning of objects and experiences. Anything that carries meaning—language, nonverbal gestures, street signs, clothing, pictures—can function as a sign. Each sign in turn is constructed out of two interrelated elements, the *signified* and the *signifier*. The signified is that idea, object, experience, belief, concept, or feeling that one individual wishes to express to another (e.g., greeting a person upon meeting them). The signifier is that deeply inculcated representation designated to stand for the signified (e.g., shaking hands, kissing on both cheeks, or saying hello). The *signifie–signifier* pair combines to form two inseparable dimensions of a sign that has two fundamental characteristics. First, the relationship that connects a signifier to a signified is essentially arbitrary, produced only through social conventions established by a community of speakers. Second, the signifier–signified pair has a relatively stable meaning because signs exist only in the context of a stable, institutionalized structure of language. Thus, for Saussure, language has both static and dynamic properties, as it embodies the structured nature of a social institution but is also subject to shifts and changes as a consequence of its arbitrary nature.

The semiotic model developed by Saussure is the foundation for the development of French structuralism and for the poststructural critique of the sign. While structuralist theorists are inclined to develop theories based on the assumption of formal, patterned, and commonly shared meaning, the poststructural position expresses extreme doubt about the existence of universal patterns of meaning and culture. Thus, one of the guiding themes that unify the various poststructural thinkers is their general skepticism toward the universality of shared meaning as conveyed by signs. Instead, in one of their most influential arguments, poststructuralists argue that the meaning of signs has fragmented, resulting in "floating signifiers." They contend that the links between signifieds and their signifiers have become destabilized. Signifiers

---

[3]French structuralism emerged throughout a variety of disciplines and theoretical traditions, including the anthropology of Claude Lévi-Strauss, the Marxist sociology of Louis Althusser, the psychoanalytic psychology of Jacques Lacan, the literary criticism of Roland Barthes, and the linguistics of Ferdinand de Saussure.

[4]A parallel set of ideas is found in the works of American philosopher Charles Sanders Peirce, who is credited with coining the term "semiotics." Saussure preferred the term "semiology."

are no longer connected to only one signified, nor are signifieds represented by only one signifier. The internal structure of the sign has collapsed, with signifiers disconnected from any stable signified, making meaning multiplicative, open-ended, and fragmented. The cultural world described by the poststructuralists is one of inherent fragmentation, instability and confusion. But language, meaning, and signs are not the only objects of the poststructuralist critique. The notions of Truth, knowledge, power, and identity also are challenged. From the perspective of poststructuralism, social life is chaotic and radically relative without any potential for unity, consensus, or coherent analysis. This extreme relativism is one of the primary reasons that poststructural theories are thus named: The analysis of the social world is based on the assumption that the patterns, routines, and conventions of social life are inherently unstable and thus only temporarily structured. As a result, existing notions of truth and knowledge are not seen as universal claims to a total understanding of reality, but instead are seen as derived from the perspective of individuals who inhabit positions of privilege (such as high levels of status and power). Truth, with a capital *T*, is no longer seen as "the Truth" but is instead "a truth" resulting from a particular, privileged point of view. The poststructuralists contend that the importance of social structures as constraining forces upon individual action has been reduced and replaced by a more fluid exercise of power that manifests itself in multiple forms within local contexts, and without the coercive force of overarching social institutions. In short, from the perspective of poststructuralism, a sociological understanding of contemporary societies requires moving beyond the analysis of structures and unified perspectives and toward an investigation of local, everyday practices.

Significant Others

Jean-François Lyotard (1924–1998): The Postmodern Condition

Jean-François Lyotard is a French philosopher and self-conscious proponent of the postmodern position, whose writings span from philosophical and political treatises to critical discussions of art, literature, and aesthetics. Although his work covers a broad range of topics and disciplines, Lyotard is most famous for his commissioned report *The Postmodern Condition: A Report on Knowledge* (1979), which Lyotard developed on behalf of the *Conseil des Universities* of the government of Quebec, Canada. The publication of *The Postmodern Condition* earned Lyotard worldwide acclaim and laid the groundwork for key developments in postmodern theory.

As its subtitle suggests, *The Postmodern Condition: A Report on Knowledge* is primarily concerned with knowledge and its changing organization in contemporary societies. Lyotard argues that the structure of knowledge has changed dramatically since the Second World War. Prior to World War II, knowledge was legitimized by reference to what Lyotard calls *metanarratives*. Metanarratives, or grand narratives, are paradigmatic systems of knowledge that contain established and credible worldviews and describe a total picture of society; they provide the basis on which truth claims are made and through which the validity of knowledge is judged. Examples of modern metanarratives are Marxism and its socialist variants, the democratic and progressive ideologies of the European Enlightenment, science and the quest for empirical truth, and systems of religious thought like Christianity. Lyotard suggests that changes in the structure of contemporary societies have initiated a profound skepticism toward the legitimacy of metanarratives—not only the dominant metanarratives of the Western world (Marxism, science, Christianity, Enlightenment progress, etc.) but *all* metanarratives. Lyotard ([1979] 1984:xxiv) argues that this suspicious attitude toward metanarratives is the hallmark of the postmodern condition, defining postmodernism simply and famously as "incredulity toward metanarratives."

(Continued)

(Continued)

Specifically, Lyotard contends that with the *computerization of society;* that is, as computers have come to play a major role in the development, manipulation, storage, and rationalization of knowledge, knowledge has become commercialized and subjected to the whims of the capitalist marketplace. With the advent of computer-based information technology, knowledge is easily fragmented, and often sold piecemeal. Circumscribed by economic forces, it is no longer important *what* one knows but whether one can *afford* to buy what one needs to know. This *mercantilization of knowledge* should be of great concern, suggests Lyotard, even though by his own analysis not much could be done about it.

According to Lyotard, gone are the days when knowledge was cumulative and consisted of the simple "truths" validated by scientific discovery. In the postmodern condition, "scientific knowledge" (and all metanarratives, for that matter) has been replaced by a new structure of knowing in which the truth claims are deeply intertwined with power. The legitimacy of science does not derive from its truth, but derives from the support it obtains from the state or the corporation, which saturate the public with narratives hailing them as the chief bastions of knowledge. Thus, in Lyotard's analysis, scientific knowledge has become a form of "narrative knowledge," a story that is told to affirm a local and particularistic point of view.

For Lyotard, this inability to form universal or "totalizing" truth is the *sine qua non* of the postmodern condition. Knowledge is forced to take refuge in localized forms with only limited possibilities for consensual understanding and collective politics. The Western metanarratives that supported the modern world have collapsed and been replaced by a postmodern world which pivots on pluralism and multivocality. Lyotard applies his argument directly to the revisionist theories of Jürgen Habermas (see Chapter 16). Lyotard criticizes Habermas's emphasis on the role of language and communication in the development of a universal discourse. Lyotard contends that such a project necessarily entails the development of shared rules for a singular "language game" and, in effect, the reinscription of a new metanarrative. In sum, Lyotard finds a return to meta-narratives not only largely unfeasible but also undesirable in the postmodern condition, attainable as it is only through force and the suppression of alternative voices.

In 1987, Lyotard became a professor emeritus at the University of Paris VIII at Vincennes, and throughout the late 1980s and the 1990s he accepted professorships at universities in the United States, Canada, Brazil, Germany, and France. Suffering from leukemia, Jean-François Lyotard died in Paris on April 21, 1998.

▪▪ Defining Postmodernism

Like poststructuralism, there are many definitions of postmodernism and many approaches to understanding what is described as "postmodern." Unlike poststructuralism, however, the term "postmodern" has achieved wide public use, especially in popular forms of culture. Newscasters, talk-show hosts, pop-song lyricists, magazine and newspaper columnists, and television program script-writers have all been conspicuous in their use of the term. MTV even ran a "Postmodern Hour" in the early 1990s. Despite its pervasive use, however, there is considerable ambiguity and range in the meaning of the term "postmodern."[5]

---

[5]While there is some debate over the true origins of the term "postmodern," Best and Kellner (1991) trace initial uses to the late nineteenth and early twentieth centuries. They find that the first applications of the term "postmodern" were used by the English painter John Watkins Chapman in the 1870s to describe a burgeoning artistic movement in painting and by Arnold Toynbee (1957) in his study of history to describe a new stage of social development beginning with the rise of an urbanized industrial working class in the late nineteenth century. However, as Bertens (1995:20) appropriately points out, there is little in the way of continuity between these early uses of the term and its appearance in social and cultural theory today.

As the name suggests, postmodernism questions the adequacy of the designation "modern" with reference to contemporary societies, engaging in debates on the importance and sense of the changes in the surrounding world. In very general terms, "postmodernism" points to a basic skepticism about the methods, goals, and ideals of modern society. The advent of modern society, typically considered to be an outgrowth of the Industrial Revolution and the eighteenth-century Enlightenment period in Europe, was premised upon the potential for scientific knowledge, the universal emancipation and institution of dignity of the individual, democratic equality, the economic effectiveness of a capitalist division of labor, and the security of the rational organization of society (see Chapter 1). Looking back over the last three centuries of human history and especially over the twentieth century, postmodern theorists have arrived at a profoundly skeptical attitude toward the promises of modern society. The emergence of international conflict, weapons of mass destruction, environmental threats, fascism and totalitarianism, hyperrationalization, global inequality, and rampant consumerism all incline to pessimism about the prospects of the modern world.

However, while some postmodernists see the postmodern shift in contemporary societies as harkening a new type of sociocultural environment that represents a firm rupture with, and implosion of, the modern, others argue that the postmodern is a natural extension and development of the modern world, an intensive expansion of advanced (or "high") modern tendencies. Yet, both positions in this debate are premised upon the idea that the contemporary world is significantly different from what preceded it and thus requires novel theoretical approaches to capture the dynamics of the new social order.

One of the most important changes giving rise to postmodernism as a historical moment is the rise of new media and interactive technology.[6] While media in the modern period was understood as having clear boundaries between "fact" and "fiction," postmodernism is characterized by a complex *fusion* of "reality" and "unreality." Instead of television programs being labeled *either* "drama" or "documentary" we now have a preponderance of "docudramas" (based on or inspired by a true story), "reality" shows with all their staged enactments and simulations, and "live" crimes in action such as car chases and shootouts. In the postmodern era, *virtual reality* and *simulation* are the order of the day.[7]

Postmodernism, as a more or less coherent perspective, became most visible in the artistic and cultural criticism of the 1960s. Its early uses were in the field of American literary criticism. In these works, the possibility for arriving at transcendental truths and unbiased readings, indicative of a "modernist" approach to art, were denied and replaced by an emphasis on antirationality, antirepresentation, and the visceral experience of art. In general, localized and idiosyncratic meaning and interpretation were celebrated over transcendent and objective truth. The cut-up literature of William S. Burroughs, the jump-cut fragmentation of Jean Luc Goddard's "French New Wave" films, and the "found object" collages of Robert Rauschenberg were seen to be exemplars of these antimodern aesthetics. These artistic forms emphasized discontinuity, multivocality, indeterminacy, and the immanence of consciousness as its own reality. Ihab Hassan's (1987) famous definition of postmodernism is a useful tool for clarifying this distinction between modernism and postmodernism. An abbreviated and modified version of Hassan's table can be found in Table 15.1.

---

[6]Traditionally, media analysts have used the term "mass media" to describe a one-way dissemination of information to a mass, anonymous audience (e.g., newspapers) and "interpersonal media" to describe a two-way process between specific individuals (e.g., the telephone). However, today "mass" and "interpersonal" media are merged: One can respond to mass-generated messages, set up one's own Web site for a mass, anonymous audience, or engage in virtual relationships and even virtual sex, which clearly epitomizes the melding of intimate, personal worlds and mass communications (see Edles 2002:57).

[7]For instance, on the Internet site voyeurdorm.com you can watch live pictures of "college women" as they go about their daily lives, 24 hours a day. For merely $34 a month, subscribers can see women take showers, put on their makeup, study, sleep, and (so the advertising promises) sunbathe nude and have lingerie parties (see Edles 2002:80–88). Interestingly, however, although the site describes the women as "sexy, young college girls," none of those interviewed for a story in the *Los Angeles Times* about the site was in fact a college student. See *Los Angeles Times* (Sept. 16, 1999:E2).

Table 15.1   Distinctions Between Modernism and Postmodernism

Modernism	Postmodernism
romanticism/symbolism	paraphysics/dadaism
form (conjunctive, closed)	antiform (disjunctive, open)
purpose	play
design	chance
hierarchy	anarchy
mastery/logos	exhaustion/silence
art object/finished work	process/performance/happening
distance	participation
creation/totalization/synthesis	decreation/deconstruction/antithesis
presence	absence
centering	dispersal
genre/boundary	text/intertext
semantics	rhetoric
paradigm	syntagm
hypotaxis	parataxis
metaphor	metonymy
selection	combination
root/depth	rhizome/surface
interpretation/reading	against interpretation/misreading
signified	signifier
lisible (readerly)	scriptable (writerly)
narrative/grand histoire	anti-narrative/petite histoire
master code	idiolect
symptom	desire
type	mutant
genital/phallic	polymorphous/androgynous
paranoia	schizophrenia
god the father	the holy ghost
metaphysics	irony
determinacy	indeterminacy
transcendence	immanence

SOURCE: Adapted from Hassan (1987); used with permission.

Whereas criticism within the artistic and literary worlds would announce the emergence of postmodernism, it would really be the field of architecture that ensconced postmodernism as a field of cultural inquiry. Beginning in the 1970s, a substantial body of work emerged that introduced the term "postmodernism" as an analytical construct used to describe a new, iconoclastic architectural movement characterized not only by bric-a-brac forms but also by its playful appropriation of existing stylistic conventions. Postmodern architecture threw out the design principles of modernism altogether in favor of a collagelike eclecticism. Perhaps the best large-scale example of postmodern architecture is Las Vegas (Venturi, Scott Brown, and Izenour 1977), where a patchwork of architectural styles—ranging from ancient Egypt, to classical Rome, to the early American Southwest, to the streets of New York and Monte Carlo—are combined to create a discontinuous architectural landscape. Many other architectural examples have been recognized as postmodern: for example, Philip Johnson's AT&T (Sony) building in New York City, Charles Moore's Piazza d'Italia in New Orleans, shopping malls like Horton Plaza in San Diego, Marriott's Bonaventure Hotel in Los Angeles, and the Mall of America in Minneapolis.

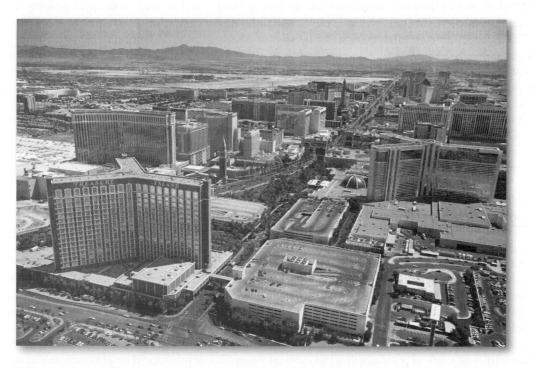

Photo 15.2   Las Vegas architecture epitomizes postmodernism, a patchwork of simulated styles ranging from Ancient Egypt to classical Rome, to the early American West, to the streets of New York and Monte Carlo.

With the development of postmodernist analyses of literature, art, and architecture, it was not long before social scientists began developing an interest in the theme of postmodernism. In fact, early sociological uses of the term "postmodern" can be found in the works of several prominent sociological theorists: C. Wright Mills (1959), Talcott Parsons (1971), and Ralf Dahrendorf (1979). In his famous work *The Sociological Imagination,* C. Wright Mills (1959:165–67) argues that the "Modern Age is being succeeded by a post-modern period" in which the liberating notion of progress celebrated by the Enlightenment is challenged and that "the ideas of freedom and reason have become moot." This point would be later contested by Talcott Parsons (1971:143) in his claim that "anything like a 'culminating' phase of modern development is a good way off—very likely a century or more. Talk of 'postmodern' society is thus decidedly premature . . . the main trend of the next century or more will be toward completion of the type of society we have called 'modern.'" Debates of this nature would come to the surface during the 1950s and 1960s but with little impact on sociology. It would not be until the 1980s that the idea of postmodernism would make important inroads into the discipline with the incorporation of elements of French poststructuralism.

■■ MICHEL FOUCAULT (1926–1984): A BIOGRAPHICAL SKETCH

Paul-Michel Foucault was born on October 15, 1926, in the town of Poitiers in central France. The son of a prominent surgeon, it was hoped that the younger Foucault would follow his father in his occupational aspirations. Michel, however, found himself drawn toward philosophy, history, and the social sciences. After graduating from the Jesuit College Saint-Stanislaus, Foucault enrolled in the prestigious *lycée* Henri-IV in Paris, where he studied under the well-known French philosopher Jean Hyppolite (1907–1968), a noted Hegelian scholar. In 1946, Foucault passed the entrance exams for the École Normale Supérieure, one of the most prestigious universities in France, and entered as the fourth-highest ranked student in his class. His time at the École was difficult as he suffered from acute depression and even attempted suicide. However, Foucault was able to recover, and studying under the famous French phenomenologist Maurice Merleau-Ponty (1908–1961) he completed his studies in 1950, earning diplomas in both philosophy and psychology. He would receive an additional academic degree in psychopathology in 1952 from the Institute de Psychologie in Paris.

Strongly influenced by the structuralist Marxism of Louis Althusser (1918–1990), Foucault became politically radicalized during the 1950s and joined the Parti Communiste Français (the French Communist Party). (He would later leave the party in 1953.) At the invitation of Althusser, Foucault became a member of the faculty at the École Normale Supérieure, where he taught psychology. Between 1952 and 1969, Foucault would be invited to teach at a number of universities throughout Europe (in France, Sweden, and Germany and in Tunisia in Northern Africa. It was also during this time, in 1960, that Foucault met Daniel Defert, the man who would become his lover and companion throughout the rest of his life. In 1969, at the age of 43, Foucault was elected to the Collège de France, France's most prestigious research university, and accepted the chair of the newly developed "History of Systems of Thought."

As Foucault established himself as an eminent lecturer and scholar during the 1960s, the publication of a number of works secured him considerable intellectual prominence. The first of these, *Madness and Civilization: A History of Insanity in the Age of Reason,* published in 1960, was a historical account of the rise of Western conceptualizations of mental illness and the means by which systems of knowledge increasingly served to socially construct "madness." This work was followed by *The Birth of the Clinic: An Archaeology of Medical Perception* (1963), an analysis of the institutionalization of clinical practices and the medical "gaze." Foucault's most celebrated book came with the publication of *The Order of Things: An Archaeology of the Human Sciences* (1966), a historical treatment of the socially constituted basis of knowledge, a work that surprisingly became a national bestseller in France. Foucault closed the 1960s with the publication of the *Archaeology of Knowledge and the Discourse on Language* (1969), investigating the links between systems of discourse and the production of knowledge.

The 1970s marked a notable shift in Foucault's lifestyle and work. After his election to the Collège de France, Foucault became a sought-after lecturer and traveled widely throughout Europe, North and South America, and the Far East. His political activities increased, especially in the area of prison reform. Throughout his travels, Foucault had visited many prison facilities, including Attica State Prison in New York State, which was the site of a massive prison riot in 1971. In an effort to initiate prison reforms, Foucault helped establish the Groupe d'Information sur les Prisons (the Prison Information Group) and called for a radical reorganization of the care and treatment of prisoners. Also during this period, Foucault began to incorporate personal experiences into his philosophical investigations. Foucault experimented with LSD on a trip to Death Valley, California, in 1975 and described the experience as one of the most profound of his life.[8] In addition, he immersed himself in anonymous sex in San Francisco's S-and-M and gay subcultures, creating experiential ground for the exploration of their theoretical possibilities.

In the mid-1970s, Foucault published two of his most influential works, *Discipline and Punish: The Birth of the Prison* (1975) and the first volume of his trilogy on sexuality, *The History of Sexuality* (1976). These two works incorporated a new dimension to Foucault's analysis: the interpenetration of knowledge, discourse, and, most important, power. Foucault published the second and third volumes of his study of sexuality, *The Use of Pleasure* and *The Care of the Self,* shortly before his death in 1984. At the time of his death, Foucault was at work on a fourth volume in his history of sexuality.

In his life and in his work, Foucault pushed the boundaries of conventional worldviews. He succumbed to AIDS-related illness in a Paris hospital on June 24, 1984, at the age of 57.

FOUCAULT'S INTELLECTUAL INFLUENCES AND CORE IDEAS ▓▓

Michel Foucault emerged on the French intellectual scene amid an intense debate between two mid-twentieth-century philosophical perspectives: existentialism and French structuralism. Rather than taking a side in this debate, however, Foucault sought to simultaneously challenge them both. On the one hand, existentialism—spearheaded by one of France's most famous intellectuals, Jean-Paul Sartre (1905–1980)—emphasized the importance of individual autonomy and personal responsibility. Premised upon the idea that individual human activity is a more potent and meaningful force than any underlying and determining structure, existentialism argues that human beings are charged with discovering the significance of their lives while confronted with the finality of their own death. In short, as Sartre puts it, human beings are "condemned to be free," responsible for their every action and left alone with the daunting task of providing meaning to their own lives. In contrast, French structuralism emphasized the latent social, psychological, and linguistic structures that shape and mold human existence and activity.

In challenging these reigning perspectives, Foucault adopted two distinct methodologies: **archaeology** and **genealogy**. As Foucault uses the term, "archaeology" is a historical method whereby discursive practices are "unearthed" much like the artifacts of past civilizations. This makes it possible to expose the evolution or history of human understanding. By excavating forms of discourse, the knowledge that is embedded in them can be revealed along with the means by which humans have come to construct particular meanings about reality and themselves. Central here are Foucault's works of the 1960s in which he systematically investigated the origins of madness (1960), medical practices (1963), and the human sciences (1966). Foucault argues that the pattern of knowledge changes over time via discursive shifts in what he calls *episteme*. For Foucault, **episteme** is a framework of knowledge (such as religion or science) that shapes discourse, that collection of linguistic tools, rules, descriptions, and habits of logic that make possible specific understandings of the world. Guided by his archaeological investigations, Foucault seeks to uncover the development of episteme and the social practices that they set in motion. Thus, for example, in *Madness and Civilization* Foucault explores how the emergence of psychiatric labels and treatments served not to cure or increase the knowledge of mental illness, but to define and refine definitions of insanity and the identities of those who allegedly suffered from the affliction.

The mid-1970s reflects a notable shift in Foucault's methodology. Now, his investigations borrow heavily from the nineteenth-century German philosopher Friedrich Nietzsche (1844–1900) and his method of "genealogy." The move from archaeology to genealogy does not simply involve a terminological change but primarily reflects an important shift in Foucault's focus of analysis. For Foucault, "genealogy" is a method of sociohistorical analysis of the impact of power on discourse. Unlike archaeology, which seeks to examine

---

[8]This account is taken from one of Foucault's biographers, James Miller (1993). Although it is clear that Foucault did experiment with psychedelic drugs and heightened sexual experiences, Miller claims that these events fundamentally shifted the course of Foucault's philosophical investigations. The importance of Foucault's experimentation with drugs and sexual experience on his philosophical work has been met with some controversy. See Halperin (1995) for a critical engagement of these ideas.

the role of discourse in the production of knowledge, genealogy articulates the dependence of the production of knowledge on relationships of power. Nietzsche had used the term "genealogy" to explore the changing nature of moral norms and value standards, revealing the intersection of ethics and power in the formation of morals and law.[9] Foucault explores similar correlations focusing on the topics of punishment and sexuality, continuing and expanding Nietzschean motifs with sociological analyses.

Popular understanding of the interrelationship between knowledge and power is frequently expressed through the phrase "knowledge is power." Foucault, in his genealogical studies, reverses the logic of this expression. He contends that it is not the acquisition of knowledge that gives one power. Instead, knowledge is already always deeply invested with power in such a way that it must be said that "power is knowledge." Thus, in Foucault's analysis, knowledge is never separate from power but is instead a specific means for exercising power. In this way, power is not simply something embodied within an individual or a social structure and expressed by brute coercion or punishment. Power appears in its most potent form when successfully translated into systems of "knowledge" and thus removed from reflection under the veil of obvious truths. The inseparability of power and knowledge is so thoroughgoing, according to Foucault, that he often conjoins the two into the phrase **power/knowledge**.

Discipline and Punish (1975)—from which reading selections in this chapter are drawn—is Foucault's first, and perhaps best, genealogical investigation. In what is among the most memorable openings of any academic work, Foucault begins *Discipline and Punish* with a description of the gruesome 1757 execution of Frenchman Robert-François Damiens (1715–1757). Damiens, who was commonly known during his day as "Robert the Devil," was condemned to death for the attempted assassination of the French monarch Louis XV. Although he failed in his attempt, inflicting only a single stab wound, Damiens was mercilessly and publicly tortured, dismembered, and finally burned—all to the delight and appreciation of the crowd that had gathered to watch.

Foucault contends that public executions like that of Damiens, although condoned in eighteenth-century Europe as an established form of punishment, are today considered abhorrent not because they violate the boundaries of moral acceptability. Rather, it is because punishment and the power that guides it have taken new, more acceptable forms. Specifically, in his analysis of penal practices, Foucault argues that punishment has moved through three stages since the eighteenth century. The first stage of punishment, according to Foucault, is characterized by corporal practices such as torture, bodily mutilation, and disfigurement, such as those exemplified in the grisly execution of Damiens. Here, punishment is orchestrated and meted out by a centralized authority, like a king, and is clearly visible among, and given legitimacy by, the general population. In addition, there are explicit social structures and hierarchies through which the exercise of power is channeled. In short, the first phase of punishment is characterized by the existence of clear and unambiguous sources of power that adopt severe physical punitive practices typically performed for public demonstration and repressive in intent.

The second phase of punishment moves Foucault's investigation into the nineteenth century and toward the contemporary era. Foucault presents this phase of punishment as marked by a noted decline in the ferocity and publicity of penal force accompanied by the rise of humanistic concerns surrounding the implementation of punishment. However, Foucault suggests that the contemporary decline of torture and corporal punishment is much less an indication of a growing humaneness concerning penal practices than a transformation in the technologies of power. This new stage is marked by the introduction of penal practices based on **surveillance** and **discipline**. The physical harm as the central focus of punishment is replaced by technologies of observation and discipline aimed at the mind rather than the body. While surveillance-based systems of punishment are certainly less physically severe, Foucault argues that they actually may be more insidious and sinister. In an era where penal practices are organized by surveillance and discipline, there is no longer a need for a sole, centralized authority around which punishment is legitimized and carried out. Instead, surveillance and discipline become penal techniques integral to à wide range of social institutions— schools, hospitals, military camps, asylums, workplaces, churches, and, most notably, prisons. (For instance,

---

[9]Nietzsche's genealogical method is especially illustrated in his works *Beyond Good and Evil* (1866) and *On the Genealogy of Morals* (1887).

a dossier or record follows a student throughout his or her scholastic career.) Foucault contends that such social institutions do not need to rely on external forms of authority, like the state, in order to surveil and discipline their members and thus produce punishment. On the contrary, institutions have come to increasingly develop and incorporate technologies of surveillance and discipline into their own operations as normalized and taken-for-granted operating procedures. For instance, students and workers are routinely kept in line by the possibility of notes or warnings being placed in their files/records regarding violations of school or workplace policy or unsatisfactory progress in their work.

Pivotal in Foucault's investigation of surveillance and discipline is his analysis of the architectural form of an ideal prison called the **Panopticon**. The blueprint for the Panopticon was designed by the British philosopher Jeremy Bentham (1748–1832) in the 1790s and was intended to be the model for a perfectly rational and efficient prison. The architecture of the Panopticon was based on two unique elements: a central guard tower surrounded by a circular gallery of cells. The inward faces of each cell within the Panopticon were barred but left unobstructed, such that their contents could be completely visible from a single vantage point in the central tower. Thus, while hundreds of inmates might be confined within the circular architecture of the Panopticon, only a single, watchful guard would be required to surveil the inmates at any given time. However, with a bit of architectural ingenuity, Bentham added an important element to the design: He made it impossible for inmates to see the guard in the central tower, making it difficult to discern at any given moment whether or not they were being observed or whether or not a guard was even present in the tower. Thus, in the Panopticon it was simply the threat of surveillance that provided the basis for the self-sanctioning of behavior, where inmates would develop a psychology of self-discipline motivated by the fear of being under constant observation.

Foucault argues that the architecture of the Panopticon is a metaphor for the general emergence of a new type of penal system that he calls the **disciplinary society**. Because the anxiety produced by being continually watched tends to put one on his "best behavior"— as Foucault suggests, to be "visible" is to be "trapped"—the intimidation created by possible surveillance tends to normalize human activity and create a self-induced complicity with the rules. Foucault suggests, however, that surveillance need not take the form of the continual observation exemplified by the Panopticon. Instead, he argues that there are many methods by which the unrelenting "gaze" of surveillance has been operationalized in the contemporary world: routine inspections in the workplace, rigorous training in procedures and protocol, registration and credentialing, self-reports of daily activities, standardized work quotas, measurement of effort, grades, performance evaluations, and so on. Such technologies of surveillance have become part and parcel of normal and routine daily activity. In fact, Panopticonesque surveillance has become so effective that individuals now sanction and normalize their own behavior without any prompting, surveilling and disciplining themselves as if they were simultaneously the inmate and guard of their own self-produced Panopticon.

Photo 15.3a Foucault's first phase of punishment: Public torture

Photo 15.3b The second phase: The Panopticon

Photo 15.3c The third phase: Disciplinary society—A virtual fortress

Table 15.2 Foucault's Three Phases of Punishment

Phase	Period	Basis of Authority/Power	Methods
1st	18th C	Orchestrated by a central authority, e.g., king	Public corporal punishment, e.g., torture
2nd	19th–20th C	Orchestrated by decentralized institutions	Surveillance, discipline, e.g., Panopticon
3rd	21st C	Diffuse, multiple self-regulation	Surveillance, disciplinary individuals

The systematic internalization of the Panopticon, where individuals continually surveil and discipline themselves, has resulted in the third and most recent phase of punishment. Foucault characterizes this phase by its increasingly diffuse and polymorphic locations of power. While the first phase of punishment was based on the centralization of power within a specific social institution, the second phase witnessed the expansion of power into surveillance and disciplinary practices of manifold institutions. This trend is intensified in the third phase of punishment, which is exemplified by a generalized and multiple physics of "micro-power" creating what Foucault describes as a "disciplinary individual." No longer are social structures or specific institutions necessary for the exercise of power and the meting out of punishment. Power has become destructured and individualized, free-floating within society in multiple manifestations. These three distinct phases of power are summarized in Table 15.2.

It is with this last phase of punishment that Foucault makes an intriguing connection between knowledge and power. He argues that while the development of new forms of knowledge (through scientific investigations, educational practices, psychotherapeutic techniques, pharmacological and medical discoveries, etc.) creates broader and deeper understandings of the social and physical world, it simultaneously generates new locations for the application of power. As that knowledge expands, so does the number of sites in which power is exercised. Thus, as knowledge grows, the techniques of discipline and surveillance multiply such that power takes on an ever-increasing number of forms and circulates throughout society everywhere without originating in any single location or source.

■■ FOUCAULT'S THEORETICAL ORIENTATION

As shown in Figure 15.1, Foucault's provocative poststructural approach can be understood as explicitly confronting and seeking to correct the theoretical weaknesses of structuralism. That is, whereas structuralism is predominantly collectivistic and nonrationalistic, Foucault's theory adds both individualistic and rationalistic dimensions to this theoretical orientation without undermining it. The emphasis on how preexisting social patterns shape individual behavior reflects the collective approach to order in structuralism. This is most evident in French structuralists' analysis of language and sign systems as pregiven structures that shape and mold (or work *down* on) human activity and behavior; in contrast, the ethnomethodologists and phenomenologists, highlighted in Chapter 13, focus on individual agency and the creativity entailed in interpretation. In terms of action, French structuralists emphasize that language works in a relatively unconscious and habitual, rather than a strategically calculative, way. For the most part, we do not manipulate linguistic codes to maximize our self-interest; rather, we simply use them in a taken-for-granted, unreflective way.[10]

---

[10]As foreign-language students well know, trying to construe your thoughts in a language that you have to consciously think about and manipulate in order to use is exhausting. One of the most important cognitive leaps in language acquisition occurs when one stops translating from one's first language into one's second language and can automatically use the new one. Many people recognize having made this transition by noting that they have begun to dream in their new tongue.

Figure 15.1 Foucault's Basic Concepts, Intellectual Influences, and Theoretical Orientation

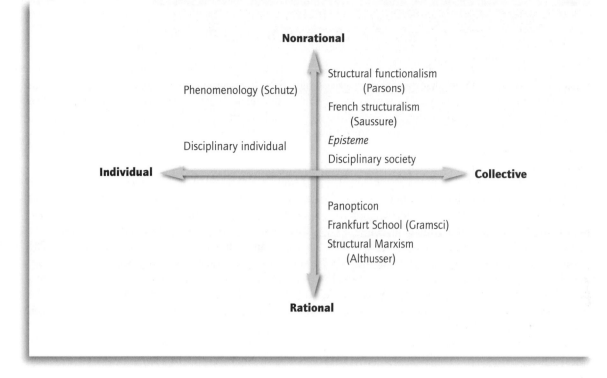

In sum, Foucault's theoretical approach is best understood not as subverting the structuralist approach but as making it more multidimensional. In terms of order, Foucault does not at all abandon structuralists' emphasis on the determining force of linguistic, social, and psychological codes; instead, he emphasizes that, in addition, in the process of *using* these codes, the codes themselves are reshaped and changed. Thus, rather than adopting a unidimensional conceptualization of preexisting codes working down on individuals as French structuralists do, like the phenomenologists outlined in Chapter 13, Foucault emphasizes that it is *only* in their use at the level of the individual that their existence takes form. In this sense, codes do not really "preexist" the subject at all.

In terms of action, as his work on punishment makes clear, Foucault emphasizes not only the relatively unconscious, and thereby nonrational, workings of semiotic, social, and psychological codes, but their strategic and thereby rational dimensions as well. Thus, Foucault identifies a historical shift in the methods of punishment from external (e.g., physical torture) to unconscious and internal (e.g., self-regulation of the disciplinary society), which reflects a shift in emphasis from the rational to the nonrational realm. Yet, Foucault never loses sight of the strategic, manipulative premises of both discipline and punishment. Discipline and punishment suit the interests of power and governability.

So, too, does Foucault's power/knowledge problematic reflect his stimulating multidimensional approach. Power is not something embodied within either an individual or a social structure and executed in a singularly coercive manner; rather, power is most potent when appearing in the guise of knowledge at both the individual and the collective levels. Power is at once profoundly strategic and manipulative (i.e., rational) and relatively unconscious and unreflective (i.e., nonrational). Power is not so much structured (collectivistic) or unstructured (individualistic) as *de*-structured.

Reading

Introduction to *Discipline and Punish*

The following excerpts from *Discipline and Punish* (1975) highlight the distinction between the first and third phases of punishment discussed earlier (see Table 15.2). In the first excerpt taken from the "body of the condemned," we see Foucault dramatically enticing the reader with voyeuristic descriptions of the torture spectacle. Here scintillating images of the tearing and burning of flesh, the pulling away of limbs, and so on, are contrasted with placid descriptions from a mere 80 years later describing a prisoner's typical day. Foucault's point, of course, is that this shift in the methods of punishment reflects a profound cultural as well as social change. Rather than being based on physical, brute force, discipline and punishment are managed in a discursive, but from Foucault's perspective, no less insidious, way.

In the second excerpt, titled "Panopticism," Foucault develops his ideas regarding surveillance. Here Foucault draws a parallel between the aggressive mechanisms used by plague-stricken cities in the late seventeenth century and Bentham's Panopticon, which was intended to be the model for the perfectly rational and efficient prison, and the unrelenting "gaze" of surveillance in the contemporary world. The point of this genealogical investigation is that although surveillance-based regulatory systems are certainly physically less severe than earlier methods of punishment, they actually may be more insidious and effective. Foucault's theme, so popular in science fiction, is the internalization of the Panopticon. The fear of being "watched" in contemporary disciplinary society normalizes and "traps" complicity within the individual's own psyche.

Discipline and Punish (1975)

Michel Foucault

TORTURE

The Body of the Condemned

On 2 March 1757 Damiens the regicide was condemned "to make the *amende honorable* before the main door of the Church of Paris," where he was to be "taken and conveyed in a cart, wearing nothing but a shirt, holding a torch of burning wax weighing two pounds"; then, "in the said cart, to the Place de Grève, where, on a scaffold that will be erected there, the flesh will be torn from his breasts, arms, thighs and calves with red-hot pincers, his right hand, holding the knife with which he committed the said parricide, burnt with sulphur, and, on those places where the flesh will be torn away, poured molten lead, boiling oil, burning resin, wax and sulphur melted together and then his body drawn and quartered by four horses and his limbs and body consumed by fire, reduced to ashes and his ashes thrown to the winds" (*Pièces originales . . .*, 372–4).

"Finally, he was quartered," recounts *the Gazette d'Amsterdam* of 1 April 1757. "This last operation was very long, because the horses used were not accustomed to drawing; consequently, instead of four, six were needed; and when that did not suffice, they were forced, in order to cut off the wretch's thighs, to sever the sinews and hack at the joints. . . .

SOURCE: From *Discipline and Punish* by Michel Foucault. English translation copyright © 1977 by Alan Sheridan (New York: Pantheon). Originally published in French as *Surveiller et Punir: Naissance de la prison* by Editions Gallimard, Allen Lane 1975. Copyright © 1975 by Editions Gallimard. Reprinted by permission of George Borchardt, Inc. for Editions Gallimard and Penguin Books, Ltd.

"It is said that, though he was always a great swearer, no blasphemy escaped his lips; but the excessive pain made him utter horrible cries, and he often repeated: 'My God, have pity on me! Jesus, help me!' The spectators were all edified by the solicitude of the parish priest of St Paul's who despite his great age did not spare himself in offering consolation to the patient."

Bouton, an officer of the watch, left us his account: "The sulphur was lit, but the flame was so poor that only the top skin of the hand was burnt, and that only slightly. Then the executioner, his sleeves rolled up, took the steel pincers, which had been especially made for the occasion, and which were about a foot and a half long, and pulled first at the calf of the right leg, then at the thigh, and from there at the two fleshy parts of the right arm; then at the breasts. Though a strong, sturdy fellow, this executioner found it so difficult to tear away the pieces of flesh that he set about the same spot two or three times, twisting the pincers as he did so, and what he took away formed at each part a wound about the size of a six-pound crown piece.

"After these tearings with the pincers, Damiens, who cried out profusely, though without swearing, raised his head and looked at himself; the same executioner dipped an iron spoon in the pot containing the boiling potion, which he poured liberally over each wound. Then the ropes that were to be harnessed to the horses were attached with cords to the patient's body; the horses were then harnessed and placed alongside the arms and legs, one at each limb.

"Monsieur Le Breton, the clerk of the court, went up to the patient several times and asked him if he had anything to say. He said he had not; at each torment, he cried out, as the damned in hell are supposed to cry out, 'Pardon, my God! Pardon, my Lord.' Despite all this pain, he raised his head from time to time and looked at himself boldly. The cords had been tied so tightly by the men who pulled the ends that they caused him indescribable pain. Monsieur le [sic] Breton went up to him again and asked him if he had anything to say; he said no. Several confessors went up to him and spoke to him at length; he willingly kissed the crucifix that was held out to him; he opened his lips and repeated: 'Pardon, Lord.'

"The horses tugged hard, each pulling straight on a limb, each horse held by an executioner. After a quarter of an hour, the same ceremony was repeated and finally, after several attempts, the direction of the horses had to changed, thus: those at the arms were made to pull towards the head, those at the thighs towards the arms, which broke the arms at the joints. This was repeated several times without success. He raised his head and looked at himself. Two more horses had to be added to those harnessed to the thighs, which made six horses in all. Without success.

"Finally, the executioner, Samson, said to Monsieur Le Breton that there was no way or hope of succeeding, and told him to ask their Lordships if they wished him to have the prisoner cut into pieces. Monsieur Le Breton, who had come down from the town, ordered that renewed efforts be made, and this was done; but the horses gave up and one of those harnessed to the thighs fell to the ground. The confessors returned and spoke to him again. He said to them (I heard him): 'Kiss me, gentlemen.' The parish priest of St Paul's did not dare to, so Monsieur de Marsilly slipped under the rope holding the left arm and kissed him on the forehead. The executioners gathered round and Damiens told them not to swear, to carry out their task and that he did not think ill of them; he begged them to pray to God for him, and asked the parish priest of St Paul's to pray for him at the first mass.

"After two or three attempts, the executioner Samson and he who had used the pincers each drew out a knife from his pocket and cut the body at the thighs instead of severing the legs at the joints; the four horses gave a tug and carried off the two thighs after them, namely, that of the right side first, the other following; then the same was done to the arms, the shoulders, the arm-pits and the four limbs; the flesh had to be cut almost to the bone, the horses pulling hard carried off the right arm first and the other afterwards.

"When the four limbs had been pulled away, the confessors came to speak to him; but his executioner told them that he was dead, though the truth was that I saw the man move, his lower jaw moving from side to side as if he were talking. One of the executioners even said shortly afterwards that when they had lifted the trunk to throw it on the stake, he was still alive. The four limbs were untied from the ropes and thrown on the stake set up in the enclosure in line with the scaffold, then the trunk and the rest were covered with logs and faggots, and fire was put to the straw mixed with this wood.

". . . In accordance with the decree, the whole was reduced to ashes. The last piece to be found in the embers was still burning at half-past ten in the evening. The pieces of flesh and the trunk had taken about four hours to burn. The officers of whom I was one, as also was my son, and a detachment of archers remained in the square until nearly eleven o'clock.

"There were those who made something of the fact that a dog had lain the day before on the grass where the fire had been, chased away several times, and had always returned. But it is not difficult to understand that an animal found this place warmer than elsewhere" (quoted in Zevaes, 201–14).

Eighty years later, Léon Faucher drew up his rules "for the House of young prisoners in Paris":

"Art. 17. The prisoners' day will begin at six in the morning in winter and at five in summer. They will work for nine hours a day throughout the year. Two hours a day will be devoted to instruction. Work and the day will end at nine o'clock in winter and at eight in summer.

Art. 18. *Rising.* At the first drum-roll, the prisoners must rise and dress in silence, as the supervisor opens the cell doors. At the second drum-roll, they must be dressed and make their beds. At the third, they must line up and proceed to the chapel for morning prayer. There is a five-minute interval between each drum-roll.

Art. 19. The prayers are conducted by the chaplain and followed by a moral or religious reading. This exercise must not last more than half an hour.

Art. 20. *Work.* At a quarter to six in the summer, a quarter to seven in winter, the prisoners go down into the courtyard where they must wash their hands and faces, and receive their first ration of bread. Immediately afterwards, they form into work-teams and go off to work, which must begin at six in summer and seven in winter.

Art. 21. *Meal.* At ten o'clock the prisoners leave their work and go to the refectory; they wash their hands in their courtyards and assemble in divisions. After the dinner, there is recreation until twenty minutes to eleven.

Art. 22. *School.* At twenty minutes to eleven, at the drum-roll, the prisoners form into ranks, and proceed in divisions to the school. The class lasts two hours and consists alternately of reading, writing, drawing and arithmetic.

Art. 23. At twenty minutes to one, the prisoners leave the school, in divisions, and return to their courtyards for recreation. At five minutes to one, at the drum-roll, they form into workteams.

Art. 24. At one o'clock they must be back in the workshops: they work until four o'clock.

Art. 25. At four o'clock the prisoners leave their workshops and go into the courtyards where they wash their hands and form into divisions for the refectory.

Art. 26. Supper and the recreation that follows it last until five o'clock: the prisoners then return to the workshops.

Art. 27. At seven o'clock in the summer, at eight in winter, work stops; bread is distributed for the last time in the workshops. For a quarter of an hour one of the prisoners or supervisors reads a passage from some instructive or uplifting work. This is followed by evening prayer.

Art. 28. At half-past seven in summer, half-past eight in winter, the prisoners must be back in their cells after the washing of hands and the inspection of clothes in the courtyard; at the first drum-roll, they must undress, and at the second get into bed. The cell doors are closed and the supervisors go the rounds in the corridors, to ensure order and silence" (Faucher, 274, 82).

We have, then, a public execution and a time-table. They do not punish the same crimes or the same type of delinquent. But they each define a certain penal style. Less than a century separates them. It was a time when, in Europe and in the United States, the entire economy of punishment was redistributed. It was a time of great "scandals" for traditional justice, a time of innumerable projects for reform. It saw a new theory of law and crime, a new moral or political justification of the right to punish; old laws were abolished, old customs died out. "Modern" codes were planned or drawn up: Russia, 1769; Prussia, 1780; Pennsylvania and Tuscany, 1786; Austria, 1788; France, 1791, Year IV, 1808 and 1810. It was a new age for penal justice.

Among so many changes, I shall consider one: the disappearance of torture as a public spectacle. Today we are rather inclined to ignore it; perhaps, in its time, it gave rise to too much inflated rhetoric; perhaps it has been attributed too readily and too emphatically to a process of "humanization," thus dispensing with the need for further analysis. And, in any case, how important is such a change, when compared with the great institutional transformations, the formulation of explicit, general codes and unified rules of procedure; with the almost universal adoption of the jury system, the definition of the essentially corrective character of the penalty and the tendency, which has become increasingly marked since the nineteenth century, to adapt punishment to the individual offender? Punishment of a less immediately physical kind, a certain discretion in the art of inflicting pain, a combination of more subtle, more subdued sufferings, deprived of their visible display, should not all this be treated as a special case, an incidental effect of deeper changes? And yet the fact remains that a few decades saw the disappearance of the tortured, dismembered, amputated body, symbolically branded on face or shoulder, exposed alive or dead to public view. The body as the major target of penal repression disappeared.

DISCIPLINE

Panopticism

The following, according to an order published at the end of the seventeenth century, were the measures to be taken when the plague appeared in a town.

First, a strict spatial partitioning: the closing of the town and its outlying districts, a prohibition to leave the town on pain of death, the killing of all stray animals; the division of the town into distinct quarters, each governed by an intendant. Each street is placed under the authority of a syndic, who keeps it under surveillance; if he leaves the street, he will be condemned to death. On the appointed day, everyone is ordered to stay indoors: it is forbidden to leave on pain of death. The syndic himself comes to lock the door of each house from the outside; he takes the key with him and hands it over to the intendant of the quarter; the intendant keeps it until the end of the quarantine. Each family will have made its own provisions; but, for bread and wine, small wooden canals are set up between the street and the interior of the houses, thus allowing each person to receive his ration without communicating with the suppliers and other residents; meat, fish and herbs will be hoisted up into the houses with pulleys and baskets. If it is absolutely necessary to leave the house, it will be done in turn, avoiding any meeting. Only the intendants, syndics and guards will move about the streets and also, between the infected houses, from one corpse to another, the 'crows,' who can be left to die: these are 'people of little substance who carry the sick, bury the dead, clean and do many vile and abject offices.' It is a segmented, immobile, frozen space. Each individual is fixed in his place.

And, if he moves, he does so at the risk of his life, contagion or punishment.

Inspection functions ceaselessly. The gaze is alert everywhere: 'A considerable body of militia, commanded by good officers and men of substance,' guards at the gates, at the town hall and in every quarter to ensure the prompt obedience of the people and the most absolute authority of the magistrates, 'as also to observe all disorder, theft and extortion.' At each of the town gates there will be an observation post; at the end of each street sentinels. Every day, the intendant visits the quarter in his charge, inquires whether the syndics have carried out their tasks, whether the inhabitants have anything to complain of; they 'observe their actions.' Every day, too, the syndic goes into the street for which he is responsible; stops before each house: gets all the inhabitants to appear at the windows (those who live overlooking the courtyard will be allocated a window looking onto the street at which no one but they may show themselves); he calls each of them by name; informs himself as to the state of each and every one of them—'in which respect the inhabitants will be compelled to speak the truth under pain of death'; if someone does not appear at the window, the syndic must ask why: 'In this way he will find out easily enough whether dead or sick are being concealed.' Everyone locked up in his cage, everyone at his window, answering to his name and showing himself when asked—it is the great review of the living and the dead.

This surveillance is based on a system of permanent registration: reports from the syndics to the intendants, from the intendants to the magistrates or mayor. At the beginning of the 'lock up,' the role of each of the inhabitants present in the town is laid down, one by one; this document bears 'the name, age, sex of everyone, notwithstanding his condition': a copy is sent to the intendant of the quarter, another to the office of the town hall, another to enable the syndic to make his daily roll call. Everything that may be observed during the course of the visits—deaths, illnesses, complaints, irregularities—is noted down and transmitted to the intendants and magistrates. The magistrates have complete control over medical treatment; they have appointed a physician in charge; no other practitioner may treat, no apothecary prepare medicine, no confessor visit a sick person without having received from him a written note 'to prevent anyone from concealing and dealing with those sick of the contagion, unknown to the magistrates.' The registration of the pathological must be constantly centralized. The relation of each individual to his disease and to his death passes through the representatives of power, the registration they make of it, the decisions they take on it.

Five or six days after the beginning of the quarantine, the process of purifying the houses one by one is begun. All the inhabitants are made to leave; in each room 'the furniture and goods' are raised from the ground or suspended from the air; perfume is poured around the room; after carefully sealing the windows, doors and even the keyholes with wax, the perfume is set alight. Finally, the entire house is closed while the perfume is consumed; those who have carried out the work are searched, as they were on entry, 'in the presence of the residents of the house, to see that they did not have something on their persons as they left that they did not have on entering.' For hours later, the residents are allowed to re-enter their homes.

This enclosed, segmented space, observed at every point, in which the individuals are inserted in a fixed place, in which the slightest movements are supervised, in which all events are recorded, in which an uninterrupted work of writing links the centre and periphery, in which power is exercised without division, according to a continuous hierarchical figure, in which each individual is constantly located, examined and distributed among the living beings, the sick and the dead—all this constitutes a compact model of the disciplinary mechanism. The plague is met by order; its function is to sort out every possible confusion: that of the disease, which is transmitted when bodies are mixed together; that of the evil, which is increased when fear and death overcome prohibitions. It lays down for each individual his place, his body, his disease and his death, his well-being, by means of an omnipresent and omniscient power that subdivides itself in a regular, uninterrupted way even to the ultimate determination of the individual, of what characterizes him, of what belongs to him, of what happens to him. Against the plague, which is a mixture, discipline brings into play its power, which is one of analysis. A whole literary fiction of the festival grew up around the plaque: suspended laws, lifted prohibitions, the frenzy of passing time, bodies mingling together without respect, individuals unmasked, abandoning their statutory identity and the figure under which they had been recognized, allowing a quite different truth to appear. But there was also a political dream of the plague, which was exactly its reverse: not the collective festival, but strict divisions; not laws transgressed, but the penetration of regulation into even the smallest details of everyday life through the mediation of the complete hierarchy that assured the capillary functioning of power; not masks that were put on and taken off, but the assignment to each individual of his 'true' name, his 'true' place, his 'true' body, his 'true' disease. The plague as a form, at once real and imaginary, of disorder had as its medical and political correlative discipline. Behind the disciplinary mechanisms can be read the haunting memory of 'contagions,' of the plague, of rebellions, crimes, vagabondage, desertions, people who appear and disappear, live and die in disorder.

If it is true that the leper gave rise to rituals of exclusion, which to a certain extent provided the model for and general form of the great Confinement, then the plague gave rise to disciplinary projects. Rather than the massive, binary division between one set of people and another, it called for multiple separations, individualizing distributions, an organization in depth of surveillance and control, an intensification and a ramification of power. The leper was caught up in a practice of rejection, of exile-enclosure; he was left to his doom in a mass among which it was useless to differentiate; those sick of the plague were caught up in a meticulous tactical partitioning in which individual differentiations were the constricting effects of a power that multiplied, articulated and subdivided itself; the great confinement on the one hand; the correct training on the other. The leper and his separation; the plague and its segmentations. The first is marked; the second analysed and distributed. The exile of the leper and the arrest of the plague do not bring with them the same political dream. The first is that of a pure community, the second that of a disciplined society. Two ways of exercising power over men, of controlling their relations, of separating out their dangerous mixtures. The plague-stricken town, traversed throughout with hierarchy, surveillance, observation, writing; the town immobilized by the functioning of an extensive power that bears in a distinct way over all individual bodies—this is the utopia of the perfectly governed city. The plague (envisaged as a possibility at least) is the trial in the course of which one may define ideally the exercise of disciplinary power. In order to make rights and laws function according to pure theory, the jurists place themselves in imagination in the state of nature; in order to see perfect disciplines functioning, rulers dreamt of the state of plague. Underlying disciplinary projects the image of the plague stands for all forms of confusion and disorder; just as the image of the leper, cut off from all human contact, underlies projects of exclusion.

They are different projects, then, but not incompatible ones. We see them coming slowly together, and it is the peculiarity of the nineteenth century that it applied to the space of exclusion of which the leper was the symbolic inhabitant (beggars, vagabonds, madmen and the disorderly formed the real population) the technique of power proper to disciplinary partitioning. Treat 'lepers' as 'plague victims,' project the subtle segmentations of discipline onto the confused space of internment, combine it with the methods of analytical distribution proper to power, individualize the excluded, but use procedures of individualization to mark exclusion—this is what was operated regularly by disciplinary power from the beginning of the nineteenth century in the psychiatric asylum, the penitentiary, the reformatory, the approved school and, to some extent, the hospital. Generally speaking, all the authorities exercising individual control function according to a double mode; that of binary division and branding (mad/sane; dangerous/harmless; normal/abnormal); and that of coercive assignment of differential distribution who he is; where he must be; how he is to be characterized; how he is to be recognized; how a constant surveillance is to be exercised over him in an individual way, etc.). On the one hand, the lepers are treated as plague victims; the tactics of individualizing disciplines are imposed on the excluded; and, on the other hand, the universality of disciplinary controls makes it possible to brand the 'leper' and to bring into play against him the dualistic mechanisms of exclusion. The constant division between the normal and the abnormal, to which every individual is subjected, brings us back to our own time, by applying the binary branding and exile of the leper to quite different objects; the existence of a whole set of techniques and institutions for measuring, supervising and correcting the abnormal brings into play the disciplinary mechanisms to which the fear of the plague gave rise. All the mechanisms of power which, even today, are disposed around the abnormal individual, to brand him and to alter him, are composed of those two forms from which they distantly derive.

Bentham's Panopticon is the architectural figure of this composition. We know the principle on which it was based: at the periphery, an annular building; at the centre, a tower; this tower is pierced with wide windows that open onto the inner side of the ring; the peripheric building is divided into cells, each of which extends the whole width of the building; they have two windows, one on the inside, corresponding to the windows of the tower; the other, on the outside, allows the light to cross the cell from one end to the other. All that is needed, then, is to place a supervisor in a central tower and to shut up in each cell a madman, a patient, a condemned man, a worker or a schoolboy. By the effect of backlighting, one can observe from the tower, standing out precisely against the light, the small captive shadows in the cells of the periphery. They are like so many cages, so many small theatres, in which each actor is alone, perfectly individualized and constantly visible. The panoptic mechanism arranges spatial unities that make it possible to see constantly and to recognize immediately. In short, it reverses the principle of the dungeon; or rather of its three functions—to enclose, to deprive of light and to hide—it preserves only the first and eliminates the other two. Full lighting and the eye of a supervisor capture better than darkness, which ultimately protected. Visibility is a trap.

To begin with, this made it possible—as a negative effect—to avoid those compact, swarming, howling masses that were to be found in places of confinement, those painted by Goya or described by Howard. Each individual, in his place, is securely confined to a cell from which he is seen from the front by the supervisor; but the side walls prevent him from coming into contact with his companions. He is seen, but he does not see; he is the object of information, never a subject in communication. The arrangement of his room, opposite the central tower, imposes on him an axial visibility; but the divisions of the ring, those separated cells, imply a lateral invisibility. And this invisibility is a guarantee of order. If the inmates are convicts, there is no danger of a plot, an attempt at collective escape, the planning of new crimes for the future, bad reciprocal influences; if they are patients, there is no danger of contagion; if they are madmen there is no risk of their committing violence upon one another; if they are schoolchildren, there is no copying, no noise, no chatter, no waste of time; if they are workers, there are no disorders, no theft, no coalitions, none of those distractions that slow down the rate of work, make it less perfect or cause accidents. The crowd, a compact mass, a locus of multiple exchanges, individualities merging together, a collective effect, is abolished and replaced by a collection of separated individualities. From the point of view of the guardian, it is replaced by a multiplicity that can be numbered and supervised; from the point of view of the inmates, by a sequestered and observed solitude (Bentham, 60–64).

Hence the major effect of the Panopticon: to induce in the inmate a state of conscious and permanent visibility that assures the automatic functioning of power. So to arrange things that the surveillance is permanent in its effects, even if it is discontinuous in its action; that the perfection of power should tend to render its actual exercise unnecessary; that this architectural apparatus should be a machine for creating and sustaining a power relation independent of the person who exercises it; in short, that the inmates should be caught up in a power situation of which they are themselves the bearers. To achieve this, it is at once too much and too little that the prisoner should be constantly observed/by an inspector: too little, for what matters is that he knows himself to be observed; too much, because he has no need in fact of being so. In view of this, Bentham laid down the principle that power should be visible and unverifiable. Visible: the inmate will constantly have before his eyes the tall outline of the central tower from which he is spied upon. Unverifiable: the inmate must never know whether he is being looked at any one moment; but he must be sure that he may always be so. In order to make the presence or absence of the inspector unverifiable, so that the prisoners, in their cells, cannot even see a shadow, Bentham envisaged not only venetian blinds on the windows of the central observation hall, but, on the inside, partitions that intersected the hall at right angles and, in order to pass from one quarter to the other, not doors but zig-zag openings; for the slightest noise, a gleam of light, a brightness in a half-opened door would betray the presence of the guardian. The Panopticon is a machine for dissociating the see/being seen dyad: in the peripheric ring, one is totally seen, without ever seeing; in the central tower, one sees everything without ever being seen.

It is an important mechanism, for it automatizes and disindividualizes power. Power has its principle not so much in a person as in a certain concerted distribution of bodies, surfaces, lights, gazes; in an arrangement whose internal mechanisms produce the relation in which individuals are caught up. The ceremonies, the rituals, the marks by which the sovereign's surplus power was manifested are useless. There is a machinery that assures dissymmetry, disequilibrium, difference. Consequently, it does not matter who exercises power. Any individual, taken almost at random, can operate the machine: in the absence of the director, his family, his friends, his visitors, even his servants (Bentham, 45). Similarly, it does not matter what motive animates him: the curiosity of the indiscreet, the malice of a child, the thirst for knowledge of a philosopher who wishes to visit this museum of human nature, or the perversity of those who take pleasure in spying and punishing. The more numerous those anonymous and temporary observers are, the greater the risk for the inmate of being surprised and the greater his anxious awareness of being observed. The Panopticon is a marvellous machine which, whatever use one may wish to put it to, produces homogeneous effects of power.

A real subjection is born mechanically from a fictitious relation. So it is not necessary to use force to constrain the convict to good behaviour, the madman to calm, the worker to work, the schoolboy to application, the patient to the observation of the regulations. Bentham was surprised that panoptic institutions could be so light: there were no more bars, no more chains, no more heavy locks; all that was needed was that the separations should be clear and the openings well arranged. The heaviness of the old 'houses of security,' with their fortress-like architecture, could be replaced by the simple, economic geometry of a 'house of certainty.' The efficiency of power, its constraining force have, in a sense, passed over to the other side—to the side of its surface of application. He who is subjected to a field of visibility, and who knows it, assumes responsibility for the constraints of power; he makes them play spontaneously upon himself; he inscribes in himself the power relation which he simultaneously plays both roles; he becomes the principle of his own subjection. By this very fact, the external power may throw off its physical weight; it tends to the non-corporal; and, the more it approaches this limit, the more constant, profound and permanent are its effects: it is a perpetual victory that avoids any physical confrontation and which is always decided in advance.

Bentham does not say whether he was inspired, in his project, by Le Vaux's menagerie at Versailles: the first menagerie in which the different elements are not, as they traditionally were, distributed in a park (Loisel, 104–7). At the centre was an octagonal pavilion which, on the first floor, consisted of only a single room, the king's salon; on every side large windows looked out onto seven cages (the eighth side was reserved for the entrance), containing different species of animals. By Bentham's time, this menagerie had disappeared. But one finds in the programme of the Panopticon a similar concern with individualizing observation, with characterization and classification, with the analytical arrangement of space. The Panopticon is a royal menagerie; the animal is replaced by man, individual distribution by specific grouping and the king by the machinery of a furtive power. With this exception, the Panopticon also does the work of a naturalist. It makes it possible to draw up differences: among patients, to observe the symptoms of each individual, without the proximity of beds, the circulation of miasmas, the effects of contagion confusing the clinical tables; among schoolchildren, it makes it possible to observe performances (without there being any imitation or copying), to map aptitudes, to assess characters, to draw up rigorous classifications and, in relation to normal development, to distinguish 'laziness and stubbornness' from 'incurable imbecility'; among workers, it makes it possible to note theaptitudes of each worker, compare the time he takes to perform a task, and if they are paid by the day, to calculate their wages (Bentham, 60–64).

So much for the question of observation. But the Panopticon was also a laboratory; it could be used as a machine to carry out experiments, to alter behaviour, to train or correct individuals. To experiment with medicines and monitor their effects. To try out different punishments on prisoners, according to their crimes and character, and to seek the most effective ones. To teach different techniques simultaneously to the workers, to decide which is the best. To try out pedagogical experiments—and in particular to take up once again the well-debated problem of secluded education, by using orphans. One would see what would happen when, in the sixteenth or eighteenth year, they were presented with other boys or girls; one could verify whether, as Helvetius thought, anyone could learn anything; one would follow 'the genealogy of every observable idea'; one could bring up different children according to different systems of thought, making certain children believe that two and two do not make four or that the moon is a cheese, then put them together when they are twenty or twenty-five years old; one would then have discussions that would be worth a great deal more than the sermons or lectures on which so much money is spent; one would have at least an opportunity of making discoveries in the domain of metaphysics. The Panopticon is a privileged place for experiments on men, and for analysing with complete certainty the transformations that may be obtained from them. The Panopticon may even provide an apparatus for supervising its own mechanisms. In this central tower, the director may spy on all the employees that he has under his orders: nurses, doctors, foremen, teachers, warders; he will be able to judge them continuously, alter their behaviour, impose upon them the methods he thinks best; and it will even be possible to observe the director himself. An inspector arriving unexpectedly at the centre of the Panopticon will be able to judge at a glance, without anything being concealed from him, how the entire establishment is functioning. And, in any case, enclosed as he is in the middle of this architectural mechanism, is not the director's own fate entirely bound up with it? The incompetent physician who has allowed contagion to spread, the incompetent prison governor or workshop manager will be the first victims of an epidemic or a revolt.' 'By every tie I could devise,' said the master of the Panopticon, 'my own fate had been bound up by me with theirs' (Bentham, 177). The Panopticon functions as a kind of laboratory of power. Thanks to its mechanisms of observation, it gains in efficiency and in the ability to penetrate into men's behaviour; knowledge follows the advances of power, discovering new objects of knowledge over all the surfaces on which power is exercised.

The plague-stricken town, the panoptic establishment—the differences are important. They mark, at a distance of a century and a half, the transformations of the disciplinary programme. In the first case, there is an exceptional situation: against an extraordinary evil, power is mobilized; it makes itself everywhere present and visible; it invents new mechanisms; it separates, it immobilizes, it partitions constructs for a time what is both a counter-city and the perfect society; it imposes an ideal functioning, but one that is reduced, in the final analysis, like the evil that it combats, to a simple dualism of life and death: that which moves brings death, and one kills that which moves. The Panopticon, on the other hand, must be understood as a generalizable model of functioning; a way of defining power relations in terms of the everyday life of men. No doubt Bentham presents it as a particular institution, closed in upon itself. Utopias, perfectly closed in upon themselves, are common enough. As opposed to the ruined prisons, littered with mechanisms of torture, to be seen in Piranese's engravings, the Panopticon presents a cruel, ingenious cage. The fact that it should have given rise, even in our own time, to so many variations, projected or realized, is evidence of the imaginary intensity that it has possessed for almost two hundred years. But the Panopticon must not be understood as a dream building: it is the diagram of a mechanism of power reduced to its ideal form; its functioning, abstracted from any obstacle, resistance or friction, must be represented as a pure architectural and optical system: it is in fact a figure of political technology that may and must be detached from any specific use.

It is polyvalent in its applications; it serves to reform prisoners, but also to treat patients, to instruct schoolchildren, to confine the insane, to supervise workers, to put beggars and idlers to work. It is a type of location of bodies in space, of distribution of individuals in relation to one another, of hierarchical organization, of disposition of centres and channels of power, of definition of the instruments and modes of intervention of power, which can be implemented in hospitals, workshops, schools, prisons. Whenever one is dealing with a multiplicity of individuals on whom a task or a particular form of behaviour must be imposed, the panoptic schema may be used. It is—necessary modifications apart—applicable 'to all establishments whatsoever, in which, within a space not too large to be covered or commanded by buildings, a number of persons are meant to be kept under inspection' (Bentham, 40; although Bentham takes the penitentiary house as his prime example, it is because it has many different functions to fulfill—safe custody, confinement, solitude, forced labour and instruction).

In each of its applications, it makes it possible to perfect the exercise of power. It does this in several ways: because it can reduce the number of those who exercise it, while increasing the number of those on whom it is exercised. Because it is possible to intervene at any moment and because the constant pressure acts even before the offences, mistakes or crimes have been committed. Because, in these conditions, its strength is that it never intervenes, it is exercised spontaneously and without noise, it constitutes a mechanism whose effects follow from one another. Because, without any physical instrument other than architecture and geometry, it acts directly on individuals; it gives 'power of mind over mind.' The panoptic schema makes any apparatus of power more intense: it assures its economy (in material, in personnel, in time); it assures its efficacity by its preventative character, its continuous functioning and its automatic mechanisms. It is a way of obtaining from power 'in hitherto unexampled quantity,' 'a great and new instrument of government . . . ; its great excellence consists in the great strength it is capable of giving to any institution it may be thought proper to apply it to' (Bentham, 66).

It's a case of 'it's easy once you've thought of it' in the political sphere. It can in fact be integrated into any function (education, medical treatment, production, punishment); it can increase the effect of this function, by being linked closely with it; it can constitute a mixed mechanism in which relations of power (and of knowledge) may be precisely adjusted, in the smallest detail, to the processes that are to be supervised; it can establish

a direct proportion between 'surplus power' and 'surplus production.' In short, it arranges things in such a way that the exercise of power is not added on from the outside, like a rigid, heavy constraint to the functions it invests, but is so subtly present in them as to increase their efficiency by itself increasing its own points of contact. The panoptic mechanism is not simply a hinge, a point of exchange between a mechanism of power and a function; it is a way of making power relations function in a function, and of making a function through these power relations. Bentham's Preface to Panopticon opens with a list of the benefits to be obtained from his 'inspection-house': 'Morals reformed—health preserved—industry invigorated—instruction diffused—public burthens lightened—Economy seated, as it were, upon a rock—the gordian knot of the Poor-Laws not cut, but untied—all by a simple idea in architecture!' (Bentham, 39)

Furthermore, the arrangement of this machine is such that its enclosed nature does not preclude a permanent presence from the outside: we have seen that anyone may come and exercise in the central tower the functions of surveillance, and that, this being the case, he can gain a clear idea of the way in which the surveillance is practised. In fact, any panoptic institution, even if it is as rigorously closed as a penitentiary, may without difficulty be subjected to such irregular and constant inspections: and not only by the appointed inspectors, but also by the public; any member of society will have the right to come and see with his own eyes how the schools, hospitals, factories, prisons function. There is no risk, therefore, that the increase of power created by the panoptic machine may degenerate into tyranny; the disciplinary mechanism will be democratically controlled, since it will be constantly accessible 'to the great tribunal committee of the world.' This Panopticon, subtly arranged so that an observer may observe, at a glance, so many different individuals, also enables everyone to come and observe any of the observers. The seeing machine was once a sort of dark room into which individuals spied; it has become a transparent building in which the exercise of power may be supervised by society as a whole.

The panoptic schema, without disappearing as such or losing any of its properties, was destined to spread throughout the social body; its vocation was to become a generalized function. The plague-stricken town provided an exceptional disciplinary model: perfect, but absolutely violent; to the disease that brought death, power opposed its perpetual threat of death; life inside it was reduced to its simplest expression; it was, against the power of death, the meticulous exercise of the right of the sword. The Panopticon, on the other hand, has a role of amplification; although it arranges power, although it is intended to make it more economic and more effective, it does so not for power itself, nor for the immediate salvation of a threatened society: its aim is to strengthen the social forces—to increase production, to develop the economy, spread education, raise the level of public morality; to increase and multiply. . . .

There are two images, then, or discipline. At one extreme, the discipline-blockade, the enclosed institution, established on the edges of society, turned inwards towards negative functions: arresting evil, breaking communications, suspending time. At the other extreme, with panopticism, is the discipline-mechanism: a functional mechanism that must improve the exercise of power by making it lighter, more rapid, more effective, a design of subtle coercion for a society to come. The movement from one project to the other, from a schema of exceptional discipline to one of a generalized surveillance, rests on a historical transformation: the gradual extension of the mechanisms of discipline throughout the seventeenth and eighteenth centuries, their spread throughout the whole social body, the formation of what might be called in general the disciplinary society.

A whole disciplinary generalization—the Benthamite physics of power represents an acknowledgement of this—had operated throughout the classical age. The spread of disciplinary institutions, whose network was beginning to cover an ever larger surface and occupying above all a less and less marginal position, testifies to this: what was an islet, a privileged place, a circumstantial measure, or a singular model, became a general formula; the regulations characteristic of the Protestant and pious armies of William of Orange or of Gustavus Adolphus were transformed into regulations for all the armies of Europe; the model colleges of the Jesuits, or the schools of Batencour or Demia, following the example set by Sturm, provided the outlines for the general forms of educational discipline; the ordering of the naval and military hospitals provided the model for the entire reorganization of hospitals in the eighteenth century.

But this extension of the disciplinary institutions was no doubt only the most visible aspect of various, more profound processes.

1. The functional inversion of the disciplines. At first, they were expected to neutralize dangers, to fix

useless or disturbed populations, to avoid the inconveniences of over-large assemblies; now they were being asked to play a positive role, for they were becoming able to do so, to increase the possible utility of individuals. Military discipline is no longer a mere means of preventing looting, desertion or failure to obey orders among the troops; it has become a basic technique to enable the army to exist, not as an assembled crowd, but as a unity that derives from this very unity an increase in its forces; discipline increases the skill of each individual, coordinates these skills, accelerates movements, increases fire power, broadens the fronts of attack without reducing their vigour, increases the capacity for resistance, etc. The discipline of the workshop, while remaining a way of enforcing respect for the regulations and authorities, of preventing thefts or losses, tends to increase aptitudes, speeds, output and therefore profits: it still exerts a moral influence over behaviour, but more and more it treats actions in terms of their results, introduces bodies into a machinery, forces into an economy. When, in the seventeenth century, the provincial schools or the Christian elementary schools were founded, the justifications given for them were above all negative: those poor who were unable to bring up their children left them 'in ignorance of their obligations: given the difficulties they have in earning a living, and themselves having been badly brought up, they are unable to communicate a sound upbringing that they themselves never had'; this involves three major inconveniences: ignorance of God, idleness (with its consequent drunkenness, impurity, larceny, brigandage); and the formation of those gangs of beggars, always ready to stir up public disorder and 'virtually to exhaust the funds of the Hotel-Dieu' (Demia, 60–61). Now, at the beginning of the Revolution, the end laid down for primary education was to be, among other things, to 'fortify,' to 'develop the body,' to prepare the child 'for a future in some mechanical work,' to give him 'an observant eye, a sure hand and prompt habits' (Talleyrand's Report to the Constituent Assembly, 10 September 1791, quoted by Leon, 106). The disciplines function increasingly as techniques for making useful individuals. Hence their emergence from a marginal position on the confines of society, and detachment from the forms of exclusion or expiation, confinement or retreat. Hence the slow loosening of their kinship with religious regularities and enclosures. Hence also their rooting in the most important, most central and most productive sectors of society. They become attached to some of the great essential functions: factory production, the transmission of knowledge, the diffusion of aptitudes and skills, the war-machine. Hence, too, the double tendency one sees developing throughout the eighteenth century to increase the number of disciplinary institutions and to discipline the existing apparatuses.

2. The swarming of disciplinary mechanisms. While, on the one hand, the disciplinary establishments increase, their mechanisms have a certain tendency to become 'de-institutionalized,' to emerge from the closed fortresses in which they once functioned and to circulate in a 'free' state; the massive, compact disciplines are broken down into flexible methods of control, which may be transferred and adapted. Sometimes the closed apparatuses add to their internal and specific function a role of external surveillance, developing around themselves a whole margin of lateral controls. Thus the Christian School must not simply train docile children; it must also make it possible to supervise the parents, to gain information as to their way of life, their resources, their piety, their morals. The school tends to constitute minute social observatories that penetrate even to the adults and exercise regular supervision over them: the bad behaviour of the child, or his absence, is a legitimate pretext, according to Demia, for one to go and question the neighbours, especially if there is any reason to believe that the family will not tell the truth; one can then go and question the parents themselves, to find out whether they know their catechism and the prayers, whether they are determined to root out the vices of their children, how many beds there are in the house and what the sleeping arrangements are; the visit may end with the giving of alms, the present of a religious picture, or the provision of additional beds (Demia, 39–40). . . .

3. The state-control of the mechanisms of discipline. In England, it was private religious groups that carried out, for a long time, the functions of social discipline (cf. Radzinovitz, 203–14); in France, although a part of this role remained in the hands of parish guilds or charity associations, another—and no doubt the most important part—was very soon taken over by the police apparatus.

The organization of a centralized police had long been regarded, even by contemporaries, as the most direct expression of absolutism; the sovereign had wished to have 'his own magistrate to whom he might directly entrust his orders, his commissions, intentions, and who was entrusted with the execution of orders and orders under the King's private seal' (a note by Duval, first secretary at the police magistrature, quoted in Funck-Brentano, 1). In effect, in taking over a number of pre-existing functions—the search for criminals, urban surveillance, economic and political supervision—the police magistratures and the magistrature-general that presided over them in Paris transposed them into a single, strict, administrative machine: 'All the radiations of force and information that spread from the circumference culminate in the magistrate-general. . . . It is he who operates all the wheels that together produce order and harmony. The effects of his administration cannot be better compared than to the movement of the celestial bodies' (Des Essarts, 344 and 528).

But, although the police as an institution were certainly organized in the form of a state apparatus, and although this was certainly lined directly to the centre of political sovereignty, the type of power that it exercises, the mechanisms it operates and the elements to which it applies them are specific. It is an apparatus that must be coextensive with the entire social body and not only by the extreme limits that it embraces, but by the minuteness of the details it is concerned with. Police power must bear 'over everything': it is not however the totality of the state nor of the kingdom as visible and invisible body of the monarch; it is the dust of events, actions, behaviour, opinions—'everything that happens'; the police are concerned with 'those things of every moment,' those 'unimportant things,' of which Catherine II spoke in her Great Instruction (Supplement to the Instruction for the drawing up of a new code, 1769, article 535). With the police, one is in the indefinite world of a supervision that seeks ideally to reach the most elementary particle, the most passing phenomenon of the social body: 'The ministry of the magistrates and police officers is of the greatest importance; the objects that it embraces are in a sense definite, one may perceive them only by a sufficiently detailed examination' (Delamare, unnumbered Preface): the infinitely small of political power.

And, in order to be exercised, this power had to be given the instrument of permanent, exhaustive, omnipresent surveillance, capable of making all visible, as long as it could itself remain invisible. It had to be like a faceless gaze that transformed the whole social body into a field of perception: thousands of eyes posted everywhere, mobile attentions ever on the alert, a long, hierarchized network which, according to Le Maire, comprised for Paris the forty-eight commissaires, the twenty inspecteurs, then the 'observers,' who were paid regularly, the 'basses mouches,' or secret agents, who were paid by the day, then the informers, paid according to the job done, and finally the prostitutes. And this unceasing observation had to be accumulated in a series of reports and registers; throughout the eighteenth century, an immense police text increasingly covered society by means of a complex documentary organization (on the police registers in the eighteenth century, cf. Chassaigne). And, unlike the methods of judicial or administrative writing, what was registered in this way were forms of behaviour, attitudes, possibilities, suspicions—a permanent account of individuals' behaviour.

Now, it should be noted that, although this police supervision was entirely 'in the hands of the king,' it did not function in a single direction. It was in fact a double-entry system: it had to correspond, by manipulating the machinery of justice, to the immediate wishes of the king, but it was also capable of responding to solicitations from below; the celebrated lettres de cachet, or orders under the king's private seal, which were long the symbol of arbitrary royal rule and which brought detention into disrepute on political grounds, were in fact demanded by families, masters, local notables, neighbours, parish priests; and their function was to punish by confinement a whole infra-penality, that of disorder, agitation, disobedience, bad conduct; those things that Ledoux wanted to exclude from his architecturally perfect city and which he called 'offences of non-surveillance.' In short, the eighteenth-century police added a disciplinary function to its role as the auxiliary of justice in the pursuit of criminals and as an instrument for the political supervision of plots, opposition movements or revolts. It was a complex function since it linked the absolute power of the monarch to the lowest levels of power disseminated in society; since, between these different, enclosed institutions of discipline (workshops, armies, schools), it extended an intermediary network, acting where they could not intervene, disciplining the non-disciplinary spaces; but it filled in the gaps, linked them together, guaranteed with its armed forces an interstitial discipline and a meta-discipline. 'By means of a wise police, the sovereign accustoms the people to order and obedience' (Vattel, 162).

The organization of the police apparatus in the eighteenth century sanctioned a generalization of the disciplines that became co-extensive with the state itself. Although it was linked in the most explicit way with everything in the royal power that exceeded the exercise of regular justice, it is understandable why the police offered such slight resistance to the rearrangement of the judicial power; and why it has not ceased to impose its prerogatives upon it, with everincreasing weight, right up to the present day; this is no doubt because it is the secular arm of the judiciary; but it is also because to a far greater degree than the judicial institution, it is identified, by reason of its extent and mechanisms, with a society of the disciplinary type. Yet it would be wrong to believe that the disciplinary functions were confiscated and absorbed once and for all by a state apparatus.

'Discipline' may be identified neither with an institution nor with an apparatus; it is a type of power, a modality for its exercise, comprising a whole set of instruments, techniques, procedures, levels of application, targets; it is a 'physics' or an 'anatomy' of power, a technology. And it may be taken over either by 'specialized' institutions (the penitentiaries or 'houses of correction' of the nineteenth century), or by institutions that use it as an essential instrument for a particular end (schools, hospitals), or by pre-existing authorities that find in it a means of reinforcing or reorganizing their internal mechanisms of power (one day we should show how intra-familial relations, essentially in the parents-children cell, have become 'disciplined,' absorbing since the classical age external schemata, first educational and military, then medical, psychiatric, psychological, which have made the family the privileged locus of emergence for the disciplinary question of the normal and the abnormal); or by apparatuses that have made discipline their principle of internal functioning (the disciplinarization of the administrative apparatus from the Napoleonic period), or finally by state apparatuses whose major, if not exclusive, function is to assure that discipline reigns over society as a whole (the police).

On the whole, therefore, one can speak of the formation of a disciplinary society in this movement that stretches from the enclosed disciplines, a sort of social 'quarantine,' to an indefinitely generalizable mechanism of 'panopticism.' Not because the disciplinary modality of power has replaced all the others; but because it has infiltrated the others, sometimes undermining them, but serving as an intermediary between them, linking them together, extending them and above all making it possible to bring the effects of power to the most minute and distant elements. It assures an infinitesimal distribution of the power relations. . . .

The formation of the disciplinary society is connected with a number of broad historical processes—economic, juridico-political and, lastly, scientific—of which it forms part.

1. Generally speaking, it might be said that the disciplines are techniques for assuring the ordering of human multiplicities. It is true that there is nothing exceptional or even characteristic in this; every system of power is presented with the same problem. But the peculiarity of the disciplines is that they try to define in relation to the multiplicities a tactics of power that fulfils three criteria: firstly, to obtain the exercise of power at the lowest possible cost (economically, by the low expenditure it involves; politically, by its discretion, its low exteriorization, its relative invisibility, the little resistance it arouses); secondly, to bring the effects of this social power to their maximum intensity and to extend them as far as possible, without either failure or interval; thirdly, to link this 'economic' growth of power with the output of the apparatuses (educational, military, industrial or medical) within which it is exercised; in short, to increase both the docility and the utility of all the elements of the system. This triple objective of the disciplines corresponds to a well-known historical conjuncture. One aspect of this conjuncture was the large demographic thrust of the eighteenth century; an increase in the floating population (one of the primary objects of discipline is to fix; it is an anti-nomadic technique); a change of quantitative scale in the groups to be supervised or manipulated (from the beginning of the seventeenth century to the eve of the French Revolution, the school population had been increasing rapidly, as had no doubt the hospital population; by the end of the eighteenth century, the peace-time army exceeded 200,000 men). The other aspect of the conjuncture was the growth in the apparatus of production, which was becoming more and more extended and complex, it was also becoming more costly and its profitability had to be increased. The development of the disciplinary methods corresponded to these two processes, or rather, no doubt, to the new need to adjust their correlation. . . .

In short, to substitute for a power that is manifested through the brilliance of those who exercise it, a power that insidiously objectifies those on whom it is applied; to form a body of knowledge about these individuals, rather than to deploy the ostentatious signs of sovereignty. In a word, the disciplines are the ensemble of minute technical inventions that made it possible to increase the useful size of multiplicities by decreasing the inconveniences of the power which, in order to make them useful, must control them. A multiplicity, whether in a workshop or a nation, an army or a school, reaches the threshold of a discipline when the relation of the one or the other becomes favourable. . . .

2. The panoptic modality of power—at the elementary, technical, merely physical level at which it is situated—is not under the immediate dependence or a direct extension of the great juridico-political structures of a society; it is nonetheless not absolutely independent. Historically, the process by which the bourgeoisie became in the course of the eighteenth century the politically dominant class was masked by the establishment of an explicit, coded and formally egalitarian juridical framework, made possible by the organization of a parliamentary, representative regime. But the development and generalization of disciplinary mechanisms constituted the other, dark side of these processes. The general juridical form that guaranteed a system of rights that were egalitarian in principle was supported by these tiny, everyday, physical mechanisms, by all those systems of micro-power that are essentially non-egalitarian and asymmetrical that we call the disciplines. And although, in a formal way, the representative regime makes it possible, directly or indirectly, with or without relays, for the will of all to form the fundamental authority of sovereignty, the disciplines provide, at the base, a guarantee of the submission of forces and bodies. The real, corporal disciplines constituted the foundation of the formal, juridical liberties. The contract may have been regarded as the ideal foundation of law and political power; panopticism constituted the technique, universally widespread, of coercion. It continued to work in depth on the juridical structures of society, in order to make the effective mechanisms of power function in opposition to the formal framework that it had acquired. The 'Enlightenment,' which discovered the liberties, also invented the disciplines.

In appearance, the disciplines constitute nothing more than an infra-law. They seem to extend the general forms defined by law to the infinitesimal level of individual lives; or they appear as methods of training that enable individuals to become integrated into these general demands. They seem to constitute the same type of law on a different scale, thereby making it more meticulous and more indulgent. The disciplines should be regarded as a sort of counter-law. They have the precise role of introducing insuperable asymmetries and excluding reciprocities. First, because discipline creates between individuals a 'private' link, which is a relation of constraints entirely different from contractual obligation; the acceptance of a discipline may be underwritten by contract; the way in which it is imposed, the mechanisms it brings into play, the non-reversible subordination of one group of people by another, the 'surplus' power that is always fixed on the same side, the inequality of position of the different 'partners' in relation to the common regulation, all these distinguish the disciplinary link from the contractual link, and make it possible to distor the contractual link systematically from the moment it has as its content a mechanism of discipline. We know, for example, how many real procedures undermine the legal fiction of the work contract: workshop discipline is not the least important. Moreover, whereas the juridical systems define juridical subjects according to universal norms, the disciplines characterize, classify, specialize; they distribute along a scale, around a norm, hierarchize individuals in relation to one another and, if necessary, disqualify and invalidate. In any case, in the space and during the time in which they exercise their control and bring into play the asymmetries of their power, they effect a suspension of the law that is never total, but is never annulled either. Regular and institutional as it may be, the discipline, in its mechanism, is a 'counter-law.' And, although the universal juridicism of modern society seems to fix limits on the exercise of power, its universally widespread panopticism enables it to operate, on the underside of the law, a machinery that is both immense and minute, which supports, reinforces, multiplies the asymmetry of power and undermines the limits that are traced around the law. The minute disciplines, the panopticisms of every day may well be below the level of emergence of the great apparatuses and the great political struggles. But, in the genealogy of modern society, they have been, with the class domination that traverses it, the political counterpart of the juridical norms according to which power was redistributed. Hence, no doubt, the importance that has been given for so long to the small techniques of discipline, to those apparently insignificant tricks that it has invented, and even to those 'sciences' that give it a respectable face; hence the fear of abandoning them if one cannot find any substitute; hence the affirmation that they are at the very foundation of society, and an element in its equilibrium, whereas they are a series of mechanisms for unbalancing power relations definitively and everywhere; hence the persistence in regarding them as the humble, but concrete form of every morality, whereas they are a set of physico-political techniques.

To return to the problem of legal punishments, the prison with all the corrective technology at its disposal is to be resituated at the point where the codified power to punish turns into a disciplinary power to observe; at the point where the universal punishments of the law are applied selectively to certain individuals and always the same ones; at the point where the redefinition of the juridical subject by the penalty becomes a useful training of the criminal; at the point where the law is inverted and passes outside itself, and where the counter-law becomes the effective and institutionalized content of the juridical forms. What generalizes the power to punish, then, is not the universal consciousness of the law in each juridical subject; it is the regular extension, the infinitely minute web of panoptic techniques.

3. Taken one by one, most of these techniques have a long history behind them. But what was new, in the eighteenth century, was that, by being combined and generalized, they attained a level at which the formation of knowledge and the increase of power regularly reinforce one another in a circular proccss. At this point, the disciplines crossed the 'technological' threshold. First the hospital, then the school, then, later, the workshop were not simply 'reordered' by the disciplines; they became, thanks to them, apparatuses such that any mechanism of objectification could be used in them as an instrument of subjection, and any growth of power could give rise in them to possible branches of knowledge; it was this link, proper to the technological systems, that made possible within the disciplinary element the formation of clinical medicine, psychiatry, child psychology, educational psychology, the rationalization of labour. It is a double process, then: an epistemological 'thaw' through a refinement of power relations; a multiplication of the effects of power through the formation and accumulation of new forms of knowledge. . . .

In the Middle Ages, the procedure of investigation gradually superseded the old accusatory justice, by a process initiated from above; the disciplinary technique, on the other hand, insidiously and as if from below, has invaded a penal justice that is still, in principle, inquisitorial. All the great movements of extension that characterize modern penality—the problematization of the criminal behind his crime, the concern with a punishment that is a correction, a therapy, a normalization, the division of the act of judgment between various authorities that are supposed to measure, assess, diagnose, cure, transform individuals—all this betrays the penetration of the disciplinary examination into the judicial inquisition.

What is now imposed on penal justice as its point of application, its 'useful' object, will no longer be the body of the guilty man set up against the body of the king; nor will it be the juridical subject of an ideal contract; it will be the disciplinary individual. The extreme point of penal justice under the Ancien Regime was the infinite segmentation of the body of the regicide: a manifestation of the strongest power over the body of the greatest criminal, whose total destruction made the crime explode into its truth. The ideal point of penalty today would be an indefinite discipline: an interrogation without end, an investigation that would be extended without limit to a meticulous and ever more analytical observation, a judgment that would at the same time be the constitution of a file that was never closed, the calculated leniency of a penalty that would be interlaced with the ruthless curiosity of an examination, a procedure that would be at the same time the permanent measure of a gap in relation to an inaccessible norm and the asymptotic movement that strives to meet in infinity. The public execution was the logical culmination of a procedure governed by the Inquisition. The practice of placing individuals under 'observation' is a natural extension of a justice imbued with disciplinary methods and examination procedures. Is it surprising that the cellular prison, with its regular chronologies, forced labour, its authorities of surveillance and registration, its experts in normality, who continue and multiply the functions of the judge, should have become the modern instrument of penalty? Is it surprising that prisons resemble factories, schools, barracks, hospitals, which all resemble prisons?

▪▪ JEAN BAUDRILLARD (1929–2007): A BIOGRAPHICAL SKETCH

Jean Baudrillard was born in 1929, in Reims, France, the capital of the Champagne region. The son of civil servants, Baudrillard was the first in his family to attend the university, where he studied German. After graduation, he taught German in a high school from 1958 to 1966 before completing his doctoral work in sociology under the advisement of Henri Lefebvre. In 1966, Baudrillard completed his doctoral thesis "Thèse de Troisième Cycle: Le Système des Objets" ("Third Cycle Thesis: The System of Objects"), which would later be published as his first major book, *The System of Objects*. He began his university career as an assistant professor in 1966 at University of Paris X, Nanterre. He would stay at Nanterre until 1987.

The turmoil of the 1950s and 1960s politicized Baudrillard's outlook. Of particular import was the rising opposition to the Algerian War voiced by Jean-Paul Sartre and other French intellectuals. At Nanterre, Baudrillard became embroiled in the student revolts in 1968 and began contributing essays to the Marxist-inspired journal *Utopie*. While many of his early writings were works of literary criticism and translations of German playwrights into French, the Marxist influence on Baudrillard's early work is nevertheless clear. But with the rise of the "poststructural" movement in France, he would become critical of the economic focus of Marxist theory and later abandoned Marxism altogether. In the mid-1970s and continuing into the late 1980s, Baudrillard's work would move away from economic analyses almost entirely and toward the study of culture, particularly the examination of semiotics and the changing nature of signs in late-twentieth-century consumer capitalism. It is during this period that Baudrillard developed his central ideas of *hyperreality* and *simulations* that have become widely influential in postmodern theory.

In 1987, Baudrillard left Nanterre University to become the scientific director of the Institut de Recherche et d'Information Socio-Économique at the University of Paris IX in Dauphine. During the

1980s and 1990s, Baudrillard increasingly gained both academic and public celebrity. In *America* (1988) and the *Cool Memories* series (1990), Baudrillard chronicled his travels throughout Europe, Australia, and the United States, expressing a cynical attitude toward consumer culture, especially as he witnessed it in America. In the early 1990s, during the first U.S.–Iraq conflict, Baudrillard published a series of widely contested essays about why the Gulf War would, could, and did not take place (Baudrillard 1995). Baudrillard proclaimed that the Gulf War "did not take place"; rather, what took place was a carefully scripted media event, the outcome of which was predetermined. In this media spectacle, Saddam Hussein emerged as a villain of extravagant, exaggerated proportions; equally inflated were American media constructs of an effortless victory (see Dixon 1997:54; Edles 2002:81). These essays brought him much criticism but also served to increase his stature as a controversial intellectual. His November 1996 appearance at Whiskey Pete's Hotel and Casino in Nevada at the event "Chance: 3 Days in the Desert," an eclectic mixture of an academic conference and celebratory rave, increased his currency as an eccentric scholar and public icon. However, perhaps the pinnacle of Baudrillard's claim to fame is the use of several of his key ideas in the storyline for the film *The Matrix* (1999). Not only are passages of his text *Simulacra and Simulation* (1981) referenced verbatim in the film, but a copy of the book is shown in a key early sequence.

Baudrillard continued to produce numerous books and essays on a wide variety of subjects, including terrorism, politics, television, symbolic exchange, and virtual reality, right up until his death in 2007.

BAUDRILLARD'S INTELLECTUAL INFLUENCES AND CORE IDEAS

Jean Baudrillard is arguably the most prominent postmodern theorist. He assumes a unique position in the debates over postmodernism because his ideas provide much of the groundwork for the discussion. His conceptions of **simulacra**—or copies of objects for which there is no true original—and **hyperreality**—"reality" that has always already been reproduced—are among the most important contributions to postmodern theory. While Baudrillard did not begin to write explicitly under the label "postmodern" until the late 1980s, it is the works of the late 1970s and early 1980s that have become central to the theory of postmodernism.

Baudrillard began his career influenced by Marxist theory, but his early writing investigated the capitalist mode of consumption rather than the mode of production emphasized by Marx. His first book, *The System of Objects* (1968), initiated a series of studies in which Baudrillard developed a critique of consumerism and commodity fetishism. In this first book, Baudrillard explores everyday home furnishings and their role in the creation of domestic environments. He examines how mundane elements of home decor create environments of "modern" functionality (through furnishings like the "kitchen set" or the "living room set"), signify "keeping up" with technological progress (through the increasing accumulation of "gadgets"), and contribute to the overall process of conspicuous consumption (especially through the accumulation of "collections"). Baudrillard's indebtedness to Saussure and semiotics is clear as he explores the systems of signs through which everyday objects are attributed cultural meaning and symbolic value.

Baudrillard expanded this line of analysis in his next series of books, *Consumer Society* (1970), *For a Critique of the Political Economy of the Sign* (1972), and *The Mirror of Production* (1973), in which he developed a critique of the Marxist analytical framework, moving away from an emphasis on capitalist economic production toward the increasing relevance of consumption practices. These works are representative of an important shift in his approach to cultural analysis. Baudrillard no longer investigates the obvious, overt meaning of household objects but instead analyzes the multiple meanings attached to objects of consumption. For example, Baudrillard explores department stores, shopping malls, television programs, and leisure activities and the role they play in creating a culture of consumption. With a theoretical perspective that parallels that developed by the Frankfurt School (see Chapter 10), Baudrillard is critical of mass consumption and its ability to create false needs and to tie individual identities to commodities.

In these works Baudrillard also adumbrated the ideas underlying the theory of simulations. What marks this development is Baudrillard's shift in the way he explores the role of the sign in conveying meaning. Gone, he argues, are the days when signs conveyed simple and clear meaning. Today, consumer goods, and in particular mass media, have created a uniquely complex cultural world where signs routinely take on multiple meanings or have no stable meaning at all.

However, in postmodern society meaning is obscured not only because of the multiplicity of connotations but also because symbols no longer refer directly to a concrete reality. Nowhere is this trend more readily apparent than in the rise of "imagistic" advertising. Today, statements of explicit value (buy this car because it is well made) have given way to statements regarding implicit values and lifestyle images (buy this car because it will make you "cool"). Rather than focus on clear-cut evaluations as to the price and quality of a particular product, imagistic advertising encourages emotional engagement as well as unconscious desires and compels consumers to organize and find identity in cultural products. We come to understand who we (and you) are via symbols in the same way we come to understand the commercial products at hand. Thus, we docilely observe and order our lives using symbols (do they sport baggy pants or Armani, drive a Honda CRX or a Hummer). The result is not only that the world of true needs has been subordinated to the world of false needs but that "the realm of needing has become a function of the field of communication" (Kline and Leiss 1978:18).

Consider, for instance, a favorite object of postmodern analysis: the Nike "swoosh." While you are probably familiar with this symbol, you might have a difficult time identifying its specific meaning. What makes the semiotics of the "swoosh" so interesting is that it doesn't have any particular denotative meaning. In fact, the "swoosh" means multiple things: a powerful company, athleticism, status, sweatshop labor, and connections to celebrities like Tiger Woods, Mia Hamm, and Michael Jordan. It is difficult to pin down the clear, denotative meaning of the Nike "swoosh" because it does not have *one;* instead, it has multiple connotative associations and in some ways no real meaning at all. In Baudrillard's terms, the "swoosh" has meaning not through its specific connections to any particular signified; instead, its meaning stems from its multiplicity as a signifier. The Nike "swoosh" is a "floating signifier" that is able to move from one signified to another.

Photo 15.4 Nike "swoosh": The epitome of a floating signifier

As we have seen, Baudrillard uses the term "hyperreality" to refer to this state when the distinction between "reality" and the model or simulation is completely dissolved. In the condition of hyperreality, simulations stand in for—they are more "real" than—reality; the map of the territory is taken for the territory itself. Baudrillard argues that the development of hyperreality can be seen through history. He suggests that the last six centuries can be described by changes in the strategies with which human beings have come to simulate reality. He argues that prior to the fifteenth-century Renaissance in Europe, the symbolic order of society was built upon the obligation of signs to represent an authentic reality. This was particularly true for feudal societies in which the nature and number of signs were extremely limited. The meaning of such signs was hierarchically bound by the obligations of a rigid social structure, an effect of religious and politic hegemony. In other words, it was not only perfectly clear who was a lord and who was a serf, but which markers (clothing, style, language, etc.) connoted each position. The meaning of signs was "sure," nonarbitrary, and denotative, as it was based on what was considered to be an irrefutable social order.

The first true order of simulacra, Baudrillard suggests, began with the Renaissance and culminated with the birth of the Industrial

Revolution. Here, the order of obligated signs is replaced by the first "modern signs" in an era of the "counterfeit," where the meaning of signs is no longer bound to the obligations of a feudal order. In this phase, signs begin to multiply with the expansion of strategies by which human beings come to represent the world (primarily painting and the printing press). Nevertheless, although signs remain largely unrestricted during this phase, they are still tied to representations of the natural world. Thus, this order of signs is based on what Baudrillard calls a "natural law of value," as signs acquire their value based on their ability, or inability, to represent the natural world. In this order, signs do not have to represent directly the material world; they may be flagrantly counterfeit, but their meaning is judged with respect to an empirical reality. An example of this order of simulacra is an artistic reproduction of a famous painting that confirms rather than calls into question the value of the "original" piece.

Baudrillard's second order of simulacra begins with the Industrial Revolution and ends in the early part of the twentieth century. A newfound logic of technological reproduction creates a mass proliferation of signs serially produced on a massive scale (through photography, film, print, etc.). A novel order of signs emerges not only because of increases in quantitative scale, but also as a result of the production of signs as reproductions of other signs. Mass advertising not only represents products and their uses but also begins to refer to lifestyles that are found only in other ads. The value of objects is now determined by the logic of the marketplace of signs, where they take on a second-order of signification and their meaning is severed (!) from their commodity-like "exchange value."

The third order of simulacra, beginning in the late twentieth century, is the order of simulation proper and describes the contemporary period. The dominant scheme is one in which signs no longer have any contact with material reality but, instead, are completely liberated from objective signifieds. Pure simulacra emerge as copies of things that have always only been copies. For Baudrillard, this order is organized around what he calls a "structural law of value," where the meaning of a particular sign is not derived from a connection to a real signified but instead to its structural relationship to a semiotic code that is divorced from reality. For instance, the popularity of Nike shoes is based on an emulation of a "street style," an attitude of self-assertion and even rebellion associated with African American inner-city culture, a "prestige from below." However, not only is this nothing more than *hyperreality* since Nike is a multinational corporation *not* rooted in inner-city street culture at all, paradoxically, inner-city youth who buy Nike themselves are victims of this imaging. In a sense, they "buy themselves back" by sporting the expensive, popular Nike brand (see Edles 2002:88; Lipsitz 1994:123).

Baudrillard's description of the historical progression of simulacra has become widely influential and is frequently recounted as a seminal statement on postmodern culture. A summary of Baudrillard's "order of simulacra" is presented in Table 15.3.

Throughout his works, Baudrillard offers many examples of the simulacra he describes. Below we summarize a few of these examples:

- Theme parks like Disneyland now simulate the experiences of space travel, submarine voyages, driving on the Autobahn, strolling through the French Quarter in New Orleans, expeditions through the "Western Frontier," nostalgic walks down "Main Street" America, and journeys into fairy tales and children's stories. Baudrillard argues that Disneyland is a perfect model of a simulation. Disneyland is not just simply a simulated fantasyland, but instead it is also presented as fantasy, which, ironically, leads us to believe that the rest of America is real. Thus for Baudrillard, in contrast to the simulated malls and theme restaurants of everyday "*real* life," Disneyland is more authentic than "reality" because it does not purport to be "real."

- Television, which invites viewers into a counterfeit world rich with the intimate and personal details of the lives of "fictional" and "nonfictional" characters, delivers individuals into a fantasy world of consumerism and consumer goods, allows masses of people to simultaneously attend public events without leaving the couch, and ultimately simulates travel through both time and space. There is perhaps no better example of the role of television as a simulation than the way that television actors and their characters

Table 15.3   Baudrillard's Theory of the "Orders of Simulacra"

Dominant Scheme	Law of Value	Nature of Semiotic Value	Semiotic Focus	Historical Era
Obligation	Prestige	Non-arbitrary	Sure-signs	Pre-Renaissance
Counterfeit	Natural	Use-value	Natural object	Renaissance to Industrial
Production	Market	Exchange-value	Logic of commodity	Industrial to early 20th century
Simulation	Structural	Sign-value	Codes and models	Late 20th century to contemporary period

are confused by the viewing public. It is often reported that television characters (not the actors) will receive "get well" cards from viewers when they (the *characters*) become sick and that television actors will be chastised in public for something their *characters have* done on the show. The "Trekkies" phenomenon, those devoted fans of *Star Trek* who are passionately committed to the show, is another characteristic example of the role of television in producing simulations.

• The first Gulf War, which Baudrillard claims did not really occur, was more of a simulation of the much-feared nuclear war between the Soviet Union and the United States. In addition, the war was entirely simulated for viewers through highly regulated news reports and tightly controlled military briefings. For Americans, Baudrillard claims, much of the actual experience of the war resembled a flight simulation video game in which planes dropped bombs on targets marked by crosshairs and casualties were largely removed from the equation.

• Watergate, Baudrillard argues, was a simulated political scandal in which reporters and government officials revived a dying political process by creating an indignity as a reminder of the existence of American politics. Baudrillard argues that the Watergate scandal was a hyperreal event created in order to simulate political credibility in an era of political cynicism.

Demographic polls, statistical research, and voting have become simulated attempts to represent the idea of "the social" in the era of the "mass" society. Borrowing a phrase used by Richard Nixon to manufacture consensus as a political tool, Baudrillard argues that the social is now "in the shadow of the silent majorities." He suggests that contemporary mass society is "frozen," as individuals demonstrate their social connections only "silently" in simulated forms, through surveys, demographic counts, and political polls. This has a dual effect: On one hand, it continually reproduces a simulation of the social, and on the other hand, it neutralizes the need for real social relationships (i.e., "we have been counted; therefore we exist").

Photo 15.5   Picture perfect: Downtown celebration

Additional examples of simulations abound in our contemporary world—flight simulators, interactive video games, virtual-reality technology, IMAX films, robotic pets, virtual celebrities, computer-generated animation, Barbie dolls, museums, zoos, aquariums, and so on. There are even copies of Disneyland in France and Japan. For Baudrillard, it is the mass media with its endless ability to create and reproduce images and signs that drives this process, becoming the great simulation-producing machine of the twentieth and twenty-first centuries. The result is that ours is a thoroughly hyperreal world, as images and signs become our primary references. For instance, Disneyland's Main Street has become the benchmark for what America *is*, such that new towns are created emulating *it*. Indeed, Disney "opened" its own neotraditional town of "Celebration" in the swamplands of Florida in 1996. In addition to an instantly functioning downtown modeled after Disneyland's Main Street, the town sports a man-made lake and cutting-edge technology. Disney expects Celebration to grow to a population of 20,000 (Edles 2002:87; Ehrenhalt 1999:8).

BAUDRILLARD'S THEORETICAL ORIENTATION ▪▪

As discussed, Baudrillard seeks to complement Marxist theories of economic production with parallel studies of consumerism. Theoretically, Baudrillard's orientation can best be understood as preserving the Frankfurt School's collectivistic approach to order while moving it in a nonrationalistic direction (see Figure 15.2). Theoretically, this means that Baudrillard's collectivist and nonrationalist approach parallels both American structural functionalism (see Chapter 9) and French structuralism—which is no surprise. As discussed, both these versions of structuralism emphasize the salience of culture, as does Baudrillard; and of course, Baudrillard's work is explicitly based on the semiotic model. In other words, by definition, theoretical approaches that emphasize how *systems of meaning* shape action necessarily prioritize the nonrational realm.

In terms of order, Baudrillard's emphasis is almost entirely collectivist. He has little to say about individuals and their social interaction, except the degree to which they are ensconced in a system of hyperreality saturated by simulations. Like the theories associated with functionalism, the Frankfurt School,

Figure 15.2 Baudrillard's Basic Concepts and Theoretical Orientation

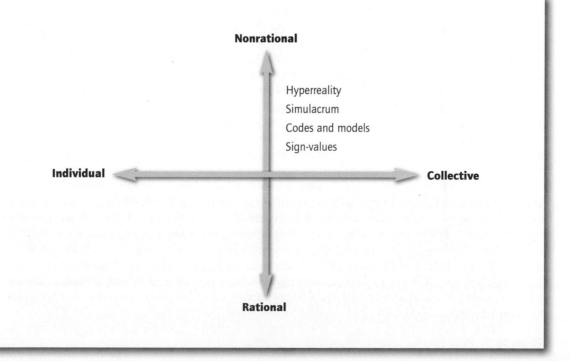

Nonrational

Hyperreality
Simulacrum
Codes and models
Sign-values

Individual ←——————————→ **Collective**

Rational

and French structuralism, his analytical focus is at the level of the collective, underscoring the role of simulations and their transformative effects on culture. Individuals are entirely absent as producers of simulations; instead they are simply subject to the ability of simulations to distort reality. In Baudrillard's analysis, it is culture that reigns supreme, filled with simulacra that construct an artificial world and minimize the efficacy of individual agency.

In terms of action, although Baudrillard shares with structural functionalists an emphasis on the nonrationalist workings of the cultural realm, his emphasis on the changing nature of culture in the contemporary era of mass media–generated simulations, and the blurring of the boundaries of reality and unreality, make his conceptualization of culture far less coherent and integrative than that of Parsons (see Chapter 9). Instead, culture is depicted as a cacophony of signs and symbols that circulate without any necessarily rational logic. Despite its transcendence over individuals, culture is not seen as a standardized system of domination created by a culture industry, as described by the theorists of the Frankfurt School (see Chapter 10). For Baudrillard, the signs that circulate through culture are ever fluctuating, unstable, and potentially incoherent. Without a concrete, empirical reality from which to draw, even an allegedly "strategic" calculation of "costs" and "benefits" is itself an illusion.

Reading

Introduction to *Simulacra and Simulations*

In the following essay from *Simulacra and Simulations* (1981), Baudrillard sets out his key terms "simulacra" and "hyperreality" and explains how today the circulation of signs within the mass media creates a semiotic "code" that is based not on "reality" but, rather, on a symbolic logic of its own. To help us understand this idea, Baudrillard begins by borrowing an allegory from the Argentinean writer Jorge Luis Borges. Borges describes an imaginary empire where a group of cartographers set out to construct a map with so much detail that it becomes a perfect replication of the original territory it charts. This map is such a seamless simulation of the territory that it is mistaken for the real territory itself. However, as fate would have it, the map begins to disintegrate and the citizens of the empire who inhabit the map, mistaking it for the territory itself, find themselves disoriented and uncertain of the authenticity of their reality.

Baudrillard suggests that a similar condition of confusion is present in contemporary societies today. We have increasingly filled our world with simulations of reality and, mistaking them for reality itself, cannot distinguish between the simulation and the real. Such is the condition of hyperreality. In semiotic terms, Baudrillard argues that hyperreality is a product of the separation of signifiers from signifieds. If signifieds are some object, thought, or feeling that one wishes to represent, and a signifier is that symbol chosen to stand for the signified, then, drawing a parallel to Borges's allegory, signifieds are like the territory and signifiers are like the map. In Baudrillard's theory of the simulation, we have increasingly come to create and use signifiers (i.e., maps) that have lost their connection to a stable signified (i.e., the territory). In short, simulations now stand in for reality, where the map of the territory is taken for the territory itself, the signifier is mistaken for the signified, and reproductions are experienced as real—all creating a condition of hyperreality . . . a "reality" that is rooted in reproduction rather than in "reality" itself. In this essay, Baudrillard gives several other examples of this "divine irreference of images" characteristic of contemporary society, including Disneyland and Watergate discussed previously. Most provocatively, in the final section of this essay Baudrillard reiterates that since "reality" does not exist anymore (only hyperreality), neither does "illusion" and, in accordance with Foucault's conceptualization of disciplinary society, he emphasizes that power has been replaced by *signs* of power.

Simulacra and Simulations (1981)

Jean Baudrillard

The simulacrum is never that which conceals the truth—it is the truth which conceals that there is none.

The simulacrum is true.

— *Ecclesiastes*

If we were able to take as the finest allegory of simulation the Borges tale where the cartographers of the Empire draw up a map so detailed that it ends up exactly covering the territory (but where, with the decline of the Empire this map becomes frayed and finally ruined, a few shreds still discernible in the deserts—the metaphysical beauty of this ruined abstraction, bearing witness to an imperial pride and rotting like a carcass, returning to the substance of the soil, rather as an aging double ends up being confused with the real thing), this fable would then have come full circle for us, and now has nothing but the discrete charm of second-order simulacra.[i]

Abstraction today is no longer that of the map, the double, the mirror or the concept. Simulation is no longer that of a territory, a referential being or a substance. It is the generation by models of a real without origin or reality: a hyperreal. The territory no longer precedes the map, nor survives it. Henceforth, it is the map that precedes the territory—*precession of simulacra*—it is the map that engenders the territory and if we were to revive the fable today, it would be the territory whose shreds are slowly rotting across the map. It is

the real, and not the map, whose vestiges subsist here and there, in the deserts which are no longer those of the Empire, but our own. *The desert of the real itself.*

In fact, even inverted, the fable is useless. Perhaps only the allegory of the Empire remains. For it is with the same imperialism that present-day simulators try to make the real, all the real, coincide with their simulation models. But it is no longer a question of either maps or territory. Something has disappeared: the sovereign difference between them that was the abstraction's charm. For it is the difference which forms the poetry of the map and the charm of the territory, the magic of the concept and the charm of the real. This representational imaginary, which both culminates in and is engulfed by the cartographer's mad project of an ideal coextensivity between the map and the territory, disappears with simulation, whose operation is nuclear and genetic, and no longer specular and discursive. With it goes all of metaphysics. No more mirror of being and appearances, of the real and its concept; no more imaginary coextensivity: rather, genetic miniaturization is the dimension of simulation. The real is produced from miniaturized units, from matrices, memory banks and command models—and with these it can be reproduced an indefinite number of times. It no longer has to be rational, since it is no longer measured against some ideal or negative instance. It is nothing more than operational. In fact, since it is no longer enveloped by an imaginary, it is no longer real at all. It is a hyperreal: the

---

[i]Counterfeit and reproduction imply always an anguish, a disquieting foreignness: the uneasiness before the photograph, considered like a witch's trick—and more generally before any technical apparatus, which is always an apparatus of reproduction, is related by Benjamin to the uneasiness before the mirror-image. There is already sorcery at work in the mirror. But how much more so when this image can be detached from the mirror and be transported, stocked, reproduced at will (cf. The Student of Prague, where the devil detaches the image of the student from the mirror and harrasses him to death by the intermediary of this image). All reproduction implies therefore a kind of black magic, from the fact of being seduced by one's own image in the water, like Narcissus, to being haunted by the double and, who knows, to the mortal turning back of this vast technical apparatus secreted today by man as his own image (the narcissistic mirage of technique, McLuhan) and that returns to him, cancelled and distorted-endless reproduction of himself and his power to the limits of the world. Reproduction is diabolical in its very essence; it makes something fundamental vacillate. This has hardly changed for us: simulation (that we describe here as the operation of the code) is still and always the place of a gigantic enterprise of manipulation, of control and of death, just like the imitative object (primitive statuette, image of photo) always had as objective an operation of black image.

product of an irradiating synthesis of combinatory models in a hyperspace without atmosphere.

In this passage to a space whose curvature is no longer that of the real, nor of truth, the age of simulation thus begins with a liquidation of all referentials—worse: by their artificial resurrection in systems of signs, which are a more ductile material than meaning, in that they lend themselves to all systems of equivalence, all binary oppositions and all combinatory algebra. It is no longer a question of imitation, nor of reduplication, nor even of parody. It is rather a question of substituting signs of the real for the real itself; that is, an operation to deter every real process by its operational double, a metastable, programmatic, perfect descriptive machine which provides all the signs of the real and short-circuits all its vicissitudes. Never again will the real have to be produced: this is the vital function of the model in a system of death, or rather of anticipated resurrection which no longer leaves any chance even in the event of death. A hyperreal henceforth sheltered from the imaginary, and from any distinction between the real and the imaginary, leaving room only for the orbital recurrence of models and the simulated generation of difference.

THE DIVINE IRREFERENCE OF IMAGES

To dissimulate is to feign not to have what one has. To simulate is to feign to have what one hasn't. One implies a presence, the other an absence. But the matter is more complicated, since to simulate is not simply to feign: "Someone who feigns an illness can simply go to bed and pretend he is ill. Someone who simulates an illness produces in himself some of the symptoms" (Littre). Thus, feigning or dissimulating leaves the reality principle intact: the difference is always clear, it is only masked; whereas simulation threatens the difference between "true" and "false," between "real" and "imaginary." Since the simulator produces "true" symptoms, is he or she ill or not? The simulator cannot be treated objectively either as ill, or as not ill. Psychology and medicine stop at this point, before a thereafter undiscoverable truth of the illness. For if any symptom can be "produced," and can no longer be accepted as a fact of nature, then every illness may be considered as simulatable and simulated, and medicine loses its meaning since it only knows how to treat "true" illnesses by their objective causes. Psychosomatics evolves in a dubious way on the edge of the illness principle. As for psychoanalysis, it transfers the symptom from the organic to the unconscious order: once again, the latter is held to be real, more real than the former; but why should simulation stop at the portals of the unconscious? Why couldn't the "work" of the unconscious be "produced" in the same way as any other symptom in classical medicine? Dreams already are.

The alienist, of course, claims that "for each form of the mental alienation there is a particular order in the succession of symptoms, of which the stimulator is unaware and in the absence of which the alienist is unlikely to be deceived." This (which dates from 1865) in order to save at all cost the truth principle, and to escape the specter raised by simulation: namely that truth, reference and objective causes have ceased to exist. What can medicine do with something which floats on either side of illness, on either side of health, or with the reduplication of illness in a discourse that is no longer true or false? What can psychoanalysis do with the reduplication of the discourse of the unconscious in a discourse of simulation that can never be unmasked, since it isn't false either?[ii]

What can the army do with simulators? Traditionally, following a direct principle of identification, it unmasks and punishes them. Today, it can reform an excellent simulator as though he were equivalent to a "real" homosexual, heart-case or lunatic. Even military psychology retreats from the Cartesian clarifies and hesitates to draw the distinction between true and false, between the "produced" symptom and the authentic symptom. "If he acts crazy so well, then he must be mad." Nor is it mistaken: in the sense that all lunatics are simulators, and this lack of distinction is the worst form of subversion. Against it, classical reason armed itself with all its categories. But it is this today which again outflanks them, submerging the truth principle.

Outside of medicine and the army, favored terrains of simulation, the affair goes back to religion and the simulacrum of divinity: "I forbade any simulacrum in the temples because the divinity that breathes life into nature cannot be represented." Indeed it can. But what becomes of the divinity when it reveals itself in icons,

---

[ii]There is furthermore in Monod's book a flagrant contradiction, which reflects the ambiguity of all current science. His discourse concerns the code, that is the third-order simulacra, but it does so still according to "scientific" schemes of the second-order—objectiveness, "scientific" ethic of knowledge, science's principle of truth and transcendence. All things incompatible with the indeterminable models of the third-order.

when it is multiplied in simulacra? Does it remain the supreme authority, simply incarnated in images as a visible theology? Or is it volatilized into simulacra which alone deploy their pomp and power of fascination—the visible machinery of icons being substituted for the pure and intelligible Idea of God? This is precisely what was feared by the Iconoclasts, whose millennial quarrel is still with us today.[iii] Their rage to destroy images rose precisely because they sensed this omnipotence of simulacra, this facility they have of erasing God from the consciousnesses of people, and the overwhelming, destructive truth which they suggest: that ultimately there has never been any God; that only simulacra exist; indeed that God himself has only ever been his own simulacrum. Had they been able to believe that images only occulted or masked the Platonic idea of God, there would have been no reason to destroy them. One can live with the idea of a distorted truth. But their metaphysical despair came from the idea that the images concealed nothing at all, and that in fact they were not images, such as the original model would have made them, but actually perfect simulacra forever radiant with their own fascination. But this death of the divine referential has to be exorcised at all cost.

It can be seen that the iconoclasts, who are often accused of despising and denying images, were in fact the ones who accorded them their actual worth, unlike the iconolaters, who saw in them only reflections and were content to venerate God at one remove. But the converse can also be said, namely that the iconolaters possessed the most modern and adventurous minds, since, underneath the idea of the apparition of God in the mirror of images, they already enacted his death and his disappearance in the epiphany of his representations (which they perhaps knew no longer represented anything, and that they were purely a game, but that this was precisely the greatest game—knowing also that it is dangerous to unmask images, since they dissimulate the fact that there is nothing behind them).

This was the approach of the Jesuits, who based their politics on the virtual disappearance of God and on the worldly and spectacular manipulation of consciences—the evanescence of God in the epiphany of power—the end of transcendence, which no longer serves as alibi for a strategy completely free of influences and signs. Behind the baroque of images hides the grey eminence of politics.

Thus perhaps at stake has always been the murderous capacity of images: murderers of the real; murderers of their own model as the Byzantine icons could murder the devine identity. To this murderous capacity is opposed the dialectical capacity of representations as a visible and intelligible mediation of the real. All of Western faith and good faith was engaged in this wager on representation: that a sign could refer to the depth of meaning, that a sign could exchange for meaning and that something could guarantee this exchange God, of course. But what if God himself can be simulated, that is to say, reduced to the signs which attest his existence? Then the whole system becomes weightless; it is no longer anything but a gigantic simulacrum: not unreal, but a simulacrum, never again exchanging for what is real, but exchanging in itself, in an uninterrupted circuit without reference or circumference.

So it is with simulation, insofar as it is opposed to representation. Representation starts from the principle that the sign and the real are equivalent (even if this equivalence is Utopian, it is a fundamental axiom). Conversely, simulation starts from the Utopia of this principle of equivalence, *from the radical negation of the sign as value,* from the sign as reversion and death sentence of every reference. Whereas representation tries to absorb simulation by interpreting it as false representation, simulation envelops the whole edifice of representation as itself a simulacrum.

These would be the successive phases of the image:

1. It is the reflection of a basic reality.

2. It masks and perverts a basic reality.

3. It masks the *absence* of a basic reality.

4. It bears no relation to any reality whatever: it is its own pure simulacrum.

In the first case, the image is a *good* appearance: the representation is of the order of sacrament. In the second, it is an evil appearance: of the order of malefice.

[iii]"It's the feeble 'definition' of TV which condemns its spectator to rearranging the few points retained into a kind of abstract work. He participates suddenly in the creation of a reality that was only just presented to him in dots: the television watcher is in the position of an individual who is asked to project his own fantasies on inkblots that are not supposed to represent anything." TV as perpetual Rorshach test. And furthermore: "The TV image requires each instant that we 'close' the spaces in the mesh by a convulsive sensuous participation that is profoundly kinetic and tactile."

In the third, it *plays at being* an appearance: it is of the order of sorcery. In the fourth, it is no longer in the order of appearance at all, but of simulation.

The transition from signs which dissimulate something to signs which dissimulate that there is nothing, marks the decisive turning point. The first implies a theology of truth and secrecy (to which the notmn of ideology still belongs). The second inaugurates an age of simulacra and simulation, in which there is no longer any God to recognize his own, nor any last judgment to separate truth from false, the real from its artificial resurrection, since everything is already dead and risen in advance.

When the real is no longer what it used to be, nostalgia assumes its full meaning. There is a proliferation of myths of origin and signs of reality; of second-hand truth, objectivity and authenticity. There is an escalation of the true, of the lived experience; a resurrection of the figurative where the object and substance have disappeared. And there is a panic-stricken production of the real and the referential, above and parallel to the panic of material production. This is how simulation appears in the phase that concerns us: a strategy of the real, neo-real and hyperreal, whose universal double is a strategy of deterrence.

HYPERREAL AND IMAGINARY

Disneyland is a perfect model of all the entangled orders of simulation. To begin with it is a play of illusions and phantasms: pirates, the frontier, future world, etc. This imaginary world is supposed to be what makes the operation successful. But, what draws the crowds is undoubtedly much more the social microcosm, the miniaturized and *religious* revelling in real America, in its delights and drawbacks. You park outside, queue up inside, and are totally abandoned at the exit. In this imaginary world the only phantasmagoria is in the inherent warmth and affection of the crowd, and in that aufficiently excessive number of gadgets used there to specifically maintain the multitudinous affect. The contrast with the absolute solitude of the parking lot—a veritable concentration camp—is total. Or rather: inside, a whole range of gadgets magnetize the crowd into direct flows; outside, solitude is directed onto a single gadget: the automobile. By an extraordinary coincidence (one that undoubtedly belongs to the peculiar enchantment of this universe), this deep-frozen infantile world happens to have been conceived and realized by a man who is himself now cryogenized; Walt Disney, who awaits his resurrection at minus 180 degrees centigrade.

The objective profile of the United States, then, may be traced throughout Disneyland, even down to the morphology of individuals and the crowd. All its values are exalted here, in miniature and comic-strip form. Embalmed and pacified. Whence the possibility of an ideological analysis of Disneyland (L. Marin does it well in Utopies, jeux d'espaces): digest of the American way of life, panegyric to American values, idealized transposition of a contradictory reality. To be sure. But this conceals something else, and that "ideological" blanket exactly serves to cover over a third-order simulation: Disneyland is there to conceal the fact that it is the "real" country, all of "real" America, which is Disneyland (just as prisons are there to conceal the fact that it is the social in its entirety, in its banal omnipresence, which is carceral). Disneyland is presented as imaginary in order to make us believe that the rest is real, when in fact all Los Angeles and the America surrounding it are no longer real, but of the order of the hyperreal and of simulation. It is no longer a question of a false representation of reality (ideology), but of concealing the fact that the real is no longer real, and thus of saving the reality principle.

The Disneyland imaginary is neither true nor false: it is a deterrence machine set up in order to rejuvenate in reverse the fiction of the real. Whence the debility, the infantile degeneration of this imaginary. It's meant to be an infantile world, in order to make us believe that the adults are elsewhere, in the "real" world, and to conceal the fact that real childishness is everywhere, particularly among those adults who go there to act the child in order to foster illusions of their real childishness.

Moreover, Disneyland is not the only one. Enchanted Village, Magic Mountain, Marine World: Los Angeles is encircled by these "imaginary stations" which feed reality, reality-energy, to a town whose mystery is precisely that it is nothing more than a network of endless, unreal circulation: a town of fabulous proportions, but without space or dimensions. As much as electrical and nuclear power stations, as much as film studios, this town, which is nothing more than an immense script and a perpetual motion picture, needs this old imaginary made up of childhood signals and faked phantasms for its sympathetic nervous system. . . .

Strategy of the Real

Of the same order as the impossibility of rediscovering an absolute level of the real, is the impossibility of staging an illusion. Illusion is no longer possible, because the real is no longer possible. It is the whole *political* problem of the parody, of hypersimulation or offensive simulation, which is posed here.

For example: it would be interesting to see whether the repressive apparatus would not react more violently to a simulated hold up than to a real one? For a real hold up only upsets the order of things, the right of property, whereas a simulated hold up interferes with the very principle of reality. Transgression and violence are less serious, for they only contest the *distribution* of the real. Simulation is infinitely more dangerous since it always suggests, over and above its object, that *law and order themselves might really be nothing more than a simulation.*

But the difficulty is in proportion to the peril. How to feign a violation and put it to the test? Go and simulate a theft in a large department store: how do you convince the security guards that it is a simulated theft? There is no "objective" difference: the same gestures and the same signs exist as for a real theft; in fact the signs incline neither to one side nor the other. As far as the established order is concerned, they are always of the order of the real.

Go and organize a fake hold up. Be sure to check that your weapons are harmless, and take the most trustworthy hostage, so that no life is in danger (otherwise you risk committing an offence). Demand ransom, and arrange it so that the operation creates the greatest commotion possible. In brief, stay close to the "truth," so as to test the reaction of the apparatus to a perfect simulation. But you won't succeed: the web of artificial signs will be inextricably mixed up with real elements (a police officer will really shoot on sight; a bank customer will faint and die of a heart attack; they will really turn the phoney ransom over to you). In brief, you will unwittingly find yourself immediately in the real, one of whose functions is precisely to devour every attempt at simulation, to reduce everything to some reality: that's exactly how the established order is, well before institutions and justice come into play.

In this impossibility of isolating the process of simulation must be seen the whole thrust of an order that can only see and understand in terms of some reality, because it can function nowhere else. The simulation of an offence, if it is patent, will either be punished more lightly (because it has no "consequences") or be punished as an offence to public office (for example, if one triggered off a police operation "for nothing")—but *never as simulation,* since it is precisely as such that no equivalence with the real is possible, and hence no repression either. The challenge of simulation is irreceivable by power. How can you punish the simulation of virtue? Yet as such it is as serious as the simulation of crime. Parody makes obedience and transgression equivalent, and that is the most serious crime, since it *cancels out the difference upon which the law is based.* The established order can do nothing against it, for the law is a second-order simulacrum whereas simulation is a third-order simulacrum, beyond true and false, beyond equivalences, beyond the rational distinctions upon which function all power and the entire social stratum. Hence, *failing the real,* it is here that we must aim at order.

This is why order always opts for the real. In a state of uncertainty, it always prefers this assumption (thus in the army they would rather take the simulator as a true madman). But this becomes more and more difficult, for it is practically impossible to isolate the process of simulation; through the force of inertia of the real which surrounds us, the inverse is also true (and this very reversibility forms part of the apparatus of simulation and of power's impotency): namely, *it is now impossible to isolate the process of the real,* or to prove the real.

Thus all hold ups, hijacks and the like are now as it were simulation hold ups, in the sense that they are inscribed in advance in the decoding and orchestration rituals of the media, anticipated in their mode of presentation and possible consequences. In brief, where they function as a set of signs dedicated exclusively to their recurrence as signs, and no longer to their "real" goal at all. But this does not make them inoffensive. On the contrary, it is as hyperreal events, no longer having any particular contents or aims, but indefinitely refracted by each other (for that matter like so-called historical events: strikes, demonstrations, crises, etc.[iv]), that they are precisely unverifiable by an order which can only exert itself on the real and the rational, on ends and means: a referential order which can

---

[iv]The entire current "psychological" situation is characterized by this shortcircuit.

Doesn't emancipation of children and teenagers, once the initial phase of revolt is passed and once there has been established the principle of the right to emancipation, seem like the real emancipation of parents? And the young (students, high-schoolers,

(Continued)

only dominate referentials, a determinate power which can only dominate a determined world, but which can do nothing about that indefinite recurrence of simulation, about that weightless nebula no longer obeying the law of gravitation of the real—power itself eventually breaking apart in this space and becoming a simulation of power (disconnected from its aims and objectives, and dedicated to *power effects* and mass simulation).

The only weapon of power, its only strategy against this defection, is to reinject realness and referentiality everywhere, in order to convince us of the reality of the social, of the gravity of the economy and the finalities of production. For that purpose it prefers the discourse of crisis, but also—why not?—the discourse of desire. "Take your desires for reality!" can be understood as the ultimate slogan of power, for in a nonreferential world even the confusian of the reality principle with the desire principle is less dangerous than contagious hyperreality. One remains among principles, and there power is always right.

Hyperreality and simulation are deterrents of every principle and of every objective; they turn against power this deterrence which is so well utilized for a long time itself. For, finally, it was capital which was the first to feed throughout its history on the destruction of every referential, of every human goal, which shattered

---

(Continued)

adolescents) seem to sense it in their always more insistent demand (though still as paradoxical) for the presence and advice of parents or of teachers. Alone at least, free and responsible, it seemed to them suddenly that other people possibly have absconded with their true liberty. Therefore, there is no question of "leaving them be." They're going to hassle them, not with any emotional or material spontaneous demand, but with an exigency that has been premediated and corrected by an implicit oedipal knowledge. Hyperdependence (much greater than before) distorted by irony and refusal, parody of libidinous original mechanisms. Demand without content, without referent, unjustified, but for all that all the more severe—naked demand with no possible answer. The contents of knowledge (teaching) or of affective relations, the pedagogical or familial referent having been eliminated in the act of emancipation, there remains only a demand linked to the empty form of the institution—perverse demand, and for that reason all the more obstinate. "Transferable" desire (that is to say non-referential, un-referential), desire that has been fed by lack, by the place left vacant, "liberated," desire captured in its own vertiginous image, desire of desire, as pure form, hyperreal. Deprived of symbolic substance, it doubles back upon itself, draws its energy from its own reflection and its disappointment with itself. This is literally today the "demand," and it is obvious that unlike the "classical" objective or transferable relations this one here is insoluble and interminable.

Simulated Oedipus.

Francois Richard: "Students asked to be seduced either bodily or verbally. But also they are aware of this and they play the game, ironically. 'Give us your knowledge, your presence, you have the word, speak, you are there for that.' Contestation certainly, but not only: the more authority is contested, vilified, the greater the need for authority as such. They play at Oedipus also, to deny it all the more vehemently. The 'teach,' 'he's Daddy, they say' it's fun, you play at incest, malaise, the untouchable, at being a tease—in order to de-sexualize finally." Like one under analysis who asks for Oedipus back again, who tells the "oedipal" stories, who has the "analytical" dreams to satisfy the supposed request of the analyst, or to resist him? In the same way the student goes through his oedipal number, his seduction number, gets chummy, close, approaches, dominates—but this isn't desire, it's simulation. Oedipal psychodrama of simulation (neither less real nor less dramatic for all that). Very difference from the real libidinal stakes of knowledge and power or even of a real mourning for the absence of same (as could have happened after 1968 in the universities). Now we've reached the phase of desperate reproduction, and where the stakes are nil, the simulacrum is maximal—exacerbated and parodied simulation at one and the same time—as interminable as psychoanalysis and for the same reasons.

The interminable psychoanalysis.

There is a whole chapter to add to the history of transference and countertransference: that of their liquidation by simulation, of the impossible psychoanalysis because it is itself, from now on, that produces and reproduces the unconscious as its institutional substance. Psychoanalysis dies also of the exchange of the signs of the unconscious. Just as revolution dies of the exchange of the critical signs of political economy. This short-circuit was well known to Freud in the form of the gift of the analytic dream, or with the "uninformed" patients, in the form of the gift of their analytic knowledge. But this was still interpreted as resistance, as detour, and did not put fundamentally into question either the process of analysis or the principle of transference. It is another thing entirely when the unconscious itself, the discourse of the unconscious becomes unfindable—according to the same scenario of simulative anticipation that we have seen at work on all levels with the machines of the third order. The analysis then can no longer end, it becomes logically and historically interminable, since it stabilizes on a puppetsubstance of reproduction, an unconscious programmed on demand—an impossible-to-breakthrough point around which the whole analysis is rearranged. The messages of the unconscious have been shortcircuited by the psychoanalysis "medium." This is libidinal hyperrealism. To the famous categories of the real, the symbolic and the imaginary, it is going to be necessary to add the hyperreal, which captures and obstructs the functioning of the three orders.

every ideal distinction between true and false, good and evil, in order to establish a radical law of equivalence and exchange, the iron law of its power. It was the first to practice deterrence, abstraction, disconnection, deterritorialization, etc.; and if it was capital which fostered reality, the reality principle, it was also the first to liquidate it in the extermination of every use value, of every real equivalence, of production and wealth, in the very sensation we have of the unreality of the stakes and the omnipotence of manipulation. Now, it is this very logic which is today hardened even more *against* it. And when it wants to fight this catastrophic spiral by secreting one last glimmer of reality, on which to found one last glimmer of power, it only multiplies the *signs* and accelerates the play of simulation.

As long as it was historically threatened by the real, power risked deterrence and simulation, disintegrating every contradiction by means of the production of equivalent signs. When it is threatened today by simulation (the threat of vanishing in the play of signs), power risks the real, risks crisis, it gambles on remanufacturing artificial, social, economic, political stakes. This is a question of life or death for it. But is too late.

Whence the characteristic hysteria of our time: the hysteria of production and reproduction of the real. The other production, that of goods and commodities, that of *la belle epoque* of political economy, no longer makes any sense of its own, and has not for some time. What society seeks through production, and overproduction, is the restoration of the real which escapes it. That is why *contemporary "material" production is itself hyperreal.* It retains all the features, the whole discourse of traditional production, but it is nothing more than its scaled-down refraction (thus the hyperrealists fasten in a striking resemblance a real from which has fled all meaning and charm, all the profundity and energy of representation). Thus the hyperrealism of simulation is expressed everywhere by the real's striking resemblance to itself.

Power, too, for some time now produces nothing but signs of its resemblance. And at the same time, another figure of power comes into play: that of a collective demand for *signs* of power—a holy union which forms around the disappearance of power. Everybody belongs to it more or less in fear of the collapse of the political. And in the end the game of power comes down to nothing more than the *critical* obsession with power: an obsession with its death; an obsession with its survival which becomes greater the more it disappears. When it

has totally disappeared, logically we will be under the total spell of power—a haunting memory already foreshadowed everywhere, manifesting at one and the same time the satisfaction of having got rid of it (nobody wants it any more, everybody unloads it on others) and grieving its loss. Melancholy for societies without power: this has already given rise to fascism, that overdose of a powerful referential in a society which cannot terminate its mourning.

But we are still in the same boat: none of our societies know how to manage their mourning for the real, for power, for the *social itself,* which is implicated in this same breakdown. And it is by an artificial revitalization of all this that we try to escape it. *Undoubtedly this will even end up in socialism.* By an unforeseen twist of events and an irony which no longer belongs to history, it is through the death of the social that socialism will emerge—as it is through the death of God that religions emerge. A twisted coming, a perverse event, an unintelligible reversion to the logic of reason. As is the fact that power is no longer present except to conceal that there is none. A simulation which can go on indefinitely, since—unlike "true" power which is, or was, a structure, a strategy, a relation of force, a stake—this is nothing but the object of a social *demand,* and hence subject to the law of supply and demand, rather than to violence and death. Completely expunged from the *political* dimension, it is dependent, like any other commodity, on production and mass consumption. Its spark has disappeared; only the fiction of a political universe is saved.

Likewise with work. The spark of production, the violence of its stake no longer exists. Everybody still produces, and more and more, but work has subtly become something else: a need (as Marx ideally envisaged it, but not at all in the same sense), the object of a social "demand," like leisure, to which it is equivalent in the general run of life's options. A demand exactly proportional to the loss of stake in the work process.[v] The same change in fortune as for power: the *scenario* of work is there to conceal the fact that the work-real, the production-real, has disappeared. And for that matter so has the strike-real too, which is no longer a stoppage of work, but its alternative pole in the ritual scansion of the social calendar. It is as if everyone has "occupied" their work place or work post, after declaring the strike, and resumed production, as is the custom in a "self-managed" job, in exactly the same terms as before, by declaring themselves (and virtually being) in a state or permanent strike.

---

[v]Athenian democracy, much more advanced than our own, had reached the point where the vote was considered as payment for a service, after all other repressive solutions had been tried and found wanting in order to insure a quorum.

This isn't a science-fiction dream: everywhere it is a question of a doubling of the work process. And of a double or locum for the strike process—strikes which are incorporated like obsolescence in objects, like crises in production. Then there are no longer any strikes or work, but both simultaneously, that is to say something else entirely: a wizardry of work, a trompe l'oeil, a scenodrama (not to say melodrama) of production, collective dramaturgy upon the empty stage of the social.

It is no longer a question of the *ideology* of work—of the traditional ethic that obscures the "real" labour process and the "objective" process of exploitation—but of the scenario of work. Likewise,

it is no longer a question of the ideology of power, but of the *scenario* of power. Ideology only corresponds to a betrayal of reality by signs; simulation corresponds to a short-circuit of reality and to its reduplication by signs. It is always the aim of ideological analysis to restore the objective process; it is always a false problem to want to restore the truth beneath the simulacrum.

This is ultimately why power is so in accord with ideological discourses and discourses on ideology, for these are all discourses of *truth*—always good, even and especially if they are revolutionary, to counter the mortal blows of simulation.

Discussion Questions

1. Compare and contrast critical theories/the Frankfurt School (see Chapter 10) and poststructural and postmodern theories, paying particular attention to the themes of standardization, rationality, power, and the impact of media in contemporary social life. How are the respective theoretical orientations reflected in their similarities and differences?

2. In her famous essay, "Foucault on Power: A Theory for Women?" (1990:163–64), feminist Nancy Hartsock criticizes both Foucault and postmodernism in the following way:

Why is it that just at the moment when so many of us who have been silenced begin to demand the right to name ourselves, to act as subjects rather than objects of history, that just then the concept of subjecthood becomes problematic? Just when we are forming our own theories about the world, uncertainty emerges about whether the world can be theorized. Just when we are talking about the changes we want, ideas of progress and the possibility of systematically and rationally organizing human society become dubious and suspect. Why is it only now that critiques are made of the will to power inherent in the effort to create theory? I contend that these intellectual moves are no accident (but no conspiracy either). They represent the transcendental voice of the Enlightenment attempting to come to grips with the social and historical changes of the middle-to-late twentieth century.

What are the essential issues Hartsock raises in this eloquent critique? To what extent do you agree and/or disagree with her position?

3. Discuss Foucault's genealogy of punishment with an emphasis on contemporary society. Do you think that the continued existence of torture in contemporary societies today (e.g., the Abu Ghraib scandal) undermines Foucault's historical trajectory of discipline and punishment? Or do you think that the contemporary prohibitions against torture and the moral attitude toward it confirm his position? To what extent do you think that Foucault's notion of a disciplinary society rings true? Using concrete examples, discuss the effect of surveillance in your own life and in contemporary society as a whole.

4. Keep a 24-hour log documenting the extent to which your life is mediated. That is, write down *every* instance of your use of media, including radio, newspapers, television, Internet, telephone, and so on, over a 24-hour period. Then analyze your log and discuss the extent to which your life reflects postmodern "hyperreality" as opposed to modern "reality." To what extent is your knowledge of the world a function of representations and the media rather than personal experience? Do you think the fact that contemporary life is experienced *via* representations (going to the movies, watching TV, etc.) negates Baudrillard's point that representations *replace* experience? Why not or how so?

5. Discuss the methodological implications of postmodernism. If there are no "objective" truths to be uncovered, if sociology is always told from a particular perspective, can sociologists "document" particular social phenomena or uncover "new" information? What kinds of claims *can* sociologists make in the postmodern condition?

16 CONTEMPORARY THEORETICAL SYNTHESES

Pierre Bourdieu

Jürgen Habermas

Anthony Giddens

Key Concepts

- Habitus
- Fields
- Cultural Capital
- Social Capital
- Economic Capital
- Symbolic Capital
- Symbolic Violence

Key Concepts

- System
- Lifeworld
- Colonization of the Lifeworld
- Communicative Action
- Public Sphere
- Steering Media
  - Money/Power

Key Concepts

- Agency
- Structure
  - Rules/Resources
- System
- Duality of Structure
- Discursive Consciousness
- Practical Consciousness
- Trust
- Time–Space Distanciation
- Ontological Security

651

Of all the oppositions that artificially divide social science, the most fundamental, and the most ruinous, is the one that is set up between subjectivism and objectivism.

—Pierre Bourdieu (1990b:25)

In this chapter, we consider the work of the three leading theorists in contemporary sociology: Pierre Bourdieu was engaged, and Jürgen Habermas and Anthony Giddens are engaged in a similar project: developing a multidimensional theoretical approach to understanding social life. While most of the theorists whose work we have discussed thus far emphasize one or another approach to the questions of order and action, the theorists examined here all seek to bridge the divide in social theory between individualist (subjectivist) and collectivist (objectivist) orientations, on one hand, and rational and nonrational orientations on the other. This long-standing theoretical divide was tempered for some thirty years when Parsonian structural-functionalism (see Chapter 9) dominated sociological theorizing. The period of the "orthodox consensus," however, was fractured during the 1960s as theorists began to incorporate ideas from an array of disciplines and perspectives—psychoanalysis, economics, comparative literature, linguistics, Marxism, phenomenology, and anthropology, to name but a few. While the injection of various approaches ushered in a period of astounding creativity, it also left many in the field to talk past one another. The proliferation of competing approaches made for little common ground between those who emphasized one aspect of social order and action and those who emphasized another dimension of these fundamental theoretical questions. Equally important, this division necessarily produced incomplete portraits of social life, since the unidimensional theoretical models were ill equipped to capture the complexities of the relationship between the individual and society.

Addressing this state of the field, each of the three theorists discussed in this chapter has developed a perspective that explicitly incorporates elements from all sides of the theoretical divide. Because each uniquely melds the work of a host of scholars, their various approaches cast different shades of light on their common effort to bridge the oppositions separating the different theoretical dimensions. Along the way, they also introduce a unique set of concerns. While Bourdieu develops his project through an emphasis on the reproduction of class relations, Habermas examines the prospects for democracy in the modern world. For his part, Giddens explores the effects of modern society on the nature of trust, risk, and the self.

■■ PIERRE BOURDIEU (1930–2002): A BIOGRAPHICAL SKETCH

Pierre Bourdieu was born in 1930 in Denguin, a small village in southwestern France. Bourdieu was raised in a family of modest means; his father was a farmer turned town postman. Bourdieu proved to be a talented and hardworking student, enabling him to gain entry into the elite lycées (public secondary schools) in Paris. In 1951, Bourdieu entered the prestigious École Normale Supériéure, long a training ground for French intellectuals, including Maurice Merleau-Ponty, Michel Foucault, Jean-Paul Sartre, and Louis Althusser, all of whom developed a decidedly Marxist orientation to their studies, as would Bourdieu. However, as a provincial outsider to the Parisian intellectual and cultural elite, Bourdieu became acutely aware of the advantages that his upper-class schoolmates possessed. This fostered within Bourdieu an antiestablishment sentiment that some have construed as an "extraordinary desire for revenge" against the Parisian intellectual world (Dufay and Dufort 1993:196, cited in Swartz 1997:18). Even though he himself was a successful product of it, Bourdieu would become a relentless critic of the French educational establishment and the aura of meritocracy with which it legitimated itself. He wrote extensively about the hostility harbored by the upper classes toward the middle and lower classes, a hostility that was embedded not only in the educational system, but also in matters such as the arts, sports, and food. In a number of his works, including *The Inheritors* (1979), *Distinction* (1979), and *The State Nobility* (1989), Bourdieu makes clear that education and cultural tastes are central to creating differences between social classes and to the reproduction of those class differences.

Bourdieu completed his *agrégation* (a highly competitive exam for teachers) in philosophy in 1955 and began teaching philosophy in the French provinces at the secondary level. Three years later, however, he was called into military service and sent to Algeria to combat the Algerians' armed struggle to end French rule. Like many French intellectuals, Bourdieu opposed the French colonial war effort, and his military experience had a profound effect on him, leading to his lifelong commitment to producing socially relevant empirical research. After completing his military duties, Bourdieu remained in Algeria to lecture at the University of Algiers and to carry out ethnographic studies of tribal life, farming communities, and industrial laborers. Bourdieu left for Algeria as a philosopher, but after conducting extensive interviews with local inhabitants and migrant workers and taking hundreds of photographs, he returned to France in 1960 as a self-taught anthropologist.

Upon his return to France, Bourdieu attended the Sorbonne (University of Paris, Sorbonne), where he studied anthropology and sociology under the direction of Raymond Aron, one of France's leading sociologists. In 1962, he took a position at the University of Lille; two years later, he joined the faculty of the Practical School of Advanced Studies in Paris (École pratique des hautes études). Ever the agent provocateur, Bourdieu refused to complete the state doctorate degree, which is the standard requirement for those seeking chairs in French universities (Swartz and Zolberg 2004:19).[1]

In 1968, in the midst of the student political protests that swept through much of western Europe, Bourdieu split with Aron and set up his own research center, the Center of Sociology for Education and Culture (Centre de Sociologie de l'éducation et de la culture). Thereafter followed a spate of groundbreaking publications. In 1970, Bourdieu coauthored with Jean-Claude Passeron *Reproduction: In Education, Society and Culture,* now considered a classic in the sociology of education. This was followed by *Distinction* (1979), which was named the sixth most important social scientific work of the twentieth century in a survey by the International Sociological Association (Swartz and Zolberg 2004:17), while his *The Logic of Practice* (1980) placed fourth in the same survey. In addition, he published *Outline of a Theory of Practice* (1972), *Sociology in Question* (1984), *Language and Symbolic Power* (1982), *Homo Academicus* (1984), and dozens of articles addressing a wide variety of methodological, philosophical, and other issues. Meanwhile, in 1975, Bourdieu founded the review journal *Proceedings in Social Science Research;* in 1981 he was elected chair in sociology at the prestigious College of France, a position formerly held by Aron.

Bourdieu dominated French public intellectual life during the 1980s and 1990s. He continued to produce work at a prolific pace, publishing *The Rules of Art* (1992) and the widely read *An Invitation to Reflexive Sociology* (coauthored with Loïc Wacquant) in 1992. Two more books were published in 1993: *The Field of Cultural Production* and *The Weight of the World.* In 1993, he was awarded the Gold Medal from the National Center for Scientific Research in France—the highest accolade to be awarded to an intellectual (Grenfell 2004:8). His last publications dealt with such topics as masculine domination, neoliberal newspeak, globalization, and television. Apt to be recognized in the streets or cafés throughout France, particularly after he was featured in the documentary film *Sociology Is a Martial Art* (2000), Bourdieu's fame was unusual for an intellectual (Calhoun and Wacquant 2002). Bourdieu died from cancer in January 2002, with a lifetime achievement of more than forty books and hundreds of articles published in a variety of languages; his influence on sociology will undoubtedly be felt for years to come.

BOURDIEU'S INTELLECTUAL INFLUENCES AND CORE IDEAS ▪▪▪

Bourdieu's work largely can be understood as an attempt to overcome the "dualism" that plagues much of social theory. Indeed, he notes, "I can say that all of my thinking started from this point: how can

---

[1]Grenfell (2004:3) subtitles his intellectual biography of Bourdieu "agent provocateur" because this "term seems best to sum up Bourdieu as iconoclast, as someone who was ready to challenge established orthodoxies and incite action against the violence (both symbolic and real) of the world."

behaviour be regulated without being the product of obedience to rules?" (quoted in Swartz 1997:95). In seeking to answer this question, he developed an approach that takes into consideration the objective or external social forces that shape attitudes and behaviors as well as an individual's subjectivity or perception of and action in the world. Along the way, he has introduced a number of concepts into the lexicon of sociology: cultural and symbolic capital, fields, habitus, and symbolic violence, to name but a few. Moreover, his conceptual vocabulary evinces his cross-disciplinary influences: He draws not only from sociology but also from anthropology and philosophy. In addition to Marx, Durkheim, and Weber, his work bears the imprint of Durkheim's nephew, the anthropologist Marcel Mauss (1872–1950), the structural anthropologist Claude Lévi-Strauss (1908–2009), and a range of philosophers, including the "father" of phenomenology Edmund Husserl (1859–1938; see Chapter 13) and many of his intellectual descendents, most notably Martin Heidegger (1889–1976), Alfred Schutz (1899–1959; see Chapter 13), Maurice Merleau-Ponty (1908–1961), and the existentialist Jean-Paul Sartre (1905–1980). Finally, important aspects of Bourdieu's work were developed in reaction to the structural Marxist philosophy of Louis Althusser (1918–1990), his teacher at the École Normale Superiéure.[2]

Bourdieu draws from these scholars but avoids adopting their ideas wholesale, opting instead to extend the themes they addressed by way of his own set of conceptual tools. In this way, the concepts he devises often fuse together those aspects of others' ideas that he finds most fruitful for his own empirical analyses, while casting off those ideas deemed least instructive or useful. The result is a highly creative and complex approach to the study of social life that defies easy summary. Nevertheless, in the remainder of this section we attempt the impossible, discussing several of the concepts that are central to Bourdieu's unique perspective while pointing to the intellectual sources from which he drew inspiration.

Habitus

The cornerstone of Bourdieu's efforts to link objectivist and subjectivist approaches is his notion of **habitus**, a concept he began to develop during his anthropological studies of Kabyle society in Algeria. The habitus is a mental filter that structures an individual's perceptions, experiences, and practices such that the world takes on a taken-for-granted, commonsense appearance. It refers to an individual's "dispositions" or "mental structures" through which the social world is apprehended and expressed through both verbal and bodily language (Bourdieu 1990a:131). "As an acquired system of generative schemes, the habitus makes possible the free production of all the thoughts, perceptions and actions inherent in the condition of its production" (Bourdieu 1990b:55). In short, it is through the habitus that one acquires a "sense of one's place" in the world or a "point of view" from which one is able to interpret one's own actions as well as the actions of others. As a "way of being," however, the habitus shapes not only interpretive schemes and thoughts—the mind—but also the body, by molding one's "natural propensity" for a wide range of movements including posture, gait, and agility.[3]

Bourdieu's notion of habitus draws from Husserl's phenomenological philosophy and thus speaks to an understanding of social life that emphasizes its active construction as part and parcel of individuals' attempts to navigate their everyday world. Yet, Bourdieu goes to great lengths to demonstrate that individuals do not create their dispositions—rather, they *acquire* them. In this way, his notion of habitus offers a critique of not only Husserl's work but also of that of Jean-Paul Sartre, a towering figure in French

---

[2]Because Bourdieu, Habermas, and Giddens draw from numerous perspectives, discussing their intellectual influences at any length would exceed the space limitations of this chapter. However, we have outlined the work of many of their influences in previous chapters. Thus, to help readers better situate the approaches of the three theorists examined here, we reference the relevant chapters whenever possible.

[3]Bourdieu's emphasis on the corporeal dimensions of the habitus stems from the work of Maurice Merleau-Ponty, whose phenomenology emphasized the centrality of embodied practice to everyday experience. See Jeremy Lane (2000).

intellectual and political life. A Marxist philosopher, novelist, and playwright, Sartre's work and political activism made him the model "total intellectual," for which he served as a standard for some three decades. Most noted for his existentialist philosophy, Sartre maintained that "man is condemned to be free" (Sartre 1947:27) as a result of living in a world without meaning or design. However, this inherent meaninglessness provided the opportunity for the individual to develop an "authentic" self, a potential that he notes as the "first principle of existentialism": "Man is nothing else but what he makes of himself" (ibid.:18). For Sartre, consciousness is not determined by one's social environment; rather, individuals possess the capacity to will themselves an existence that is freely chosen. Ultimately, individuals are solely responsible for being what they have willed themselves to be.

While drawing on phenomenology and existentialism, Bourdieu maintains that the habitus is not simply a mental or internal compass that shapes one's attitudes, perceptions, tastes, and "inclinations," nor does it refer to one's will or undetermined consciousness; it is instead an "internalization of externality" (Bourdieu 1980/1990:55). Bourdieu introduces this objectivist or structural element into the workings of the habitus by critically incorporating the insights of Marx, Weber, Durkheim, and Claude Lévi-Strauss, whose structural anthropology was a landmark in the social sciences. Lévi-Strauss sought, in part through a critique of Sartre's subjectivist orientation and emphasis on free will, to uncover the "fundamental structures of the human mind" (Lévi-Strauss 1969:75) in the binary codes that structure a society's cultural system. In essence, Lévi-Strauss was expanding Ferdinand de Saussure's structural linguistics (see Chapter 15) to explain what he took as the universal foundation of human society—namely, the binary opposites (good/evil, weak/strong, masculine/feminine, etc.) that structure all human ways of thinking, kinship systems, and myths. On the basis of his richly detailed anthropological studies, Lévi-Strauss created quasi-mathematical formulas to map out the structural relations according to which all cultures are said to be organized like so many variations on a common theme.

Lévi-Strauss's commitment to a scientific, as opposed to a philosophical, examination of social life had a major impact on Bourdieu's empirical research. It was in the "confrontation" between the works of Sartre and Lévi-Strauss that he saw the possibility "of reconciling theoretical and practical intentions, bringing together the scientific and the ethical or political vocation . . . in a humbler and more responsible way of performing [one's] task as researcher" (ibid.:2). Nevertheless, Bourdieu did not adopt Lévi-Strauss's structuralism wholesale. Most troubling for Bourdieu is structuralism's "mechanical" view of action whereby individuals' consciousness or reasons for acting are of secondary interest, since they are held to be reflections of underlying cultural codes of which they are unaware. Through his conception of the habitus, Bourdieu sought "to reintroduce agents that Lévi-Strauss and the structuralists . . . tended to abolish, making them into simple epiphenomena of structure" (1990a:9). Contrary to the structuralists' portrayal of conduct, Bourdieu contended, "action is not the mere carrying out of a rule, or obedience to a rule. Social agents . . . are not automata regulated like clocks, in accordance with laws they do not understand" (ibid.).

Despite Bourdieu's criticism of the mechanistic nature of Lévi-Strauss's structural anthropology, there is a decidedly structuralist element in Bourdieu's notion of the habitus, since an individual's dispositions are a product of the "internalization of externality." The externality that shapes the habitus is readily apparent if we compare it to a "point of view." As the term itself suggests, points of view "are views taken from a certain point, that is, from a given position within social space. And we know too that there will be different or even antagonistic points of view, since points of view depend on the point from which they are taken, since the vision that every agent has of space depends on his or her position in that space" (ibid.:131). For example, your vision of geographic space is dependent on your position or the point you occupy within that space. If you were to stand facing north, your surroundings would appear to you in a particular way that is quite different from how they would appear to someone facing south. You each will see things that the other is unaware of, because it is outside of your (his) field of vision. Moreover, those things that you both see will not look the same, because your distinctive points of view will determine their appearance. Nevertheless, both of you will perceive the world as natural or self-evident despite the fact that your perceptions are dependent not on the world, but rather on the point of view from which you apprehend the world.

Now extend the notion of geographic space to social space. In this case, your point of view or disposition is determined by your position within a space that is structured by two "principles of differentation": economic capital and cultural capital (Bourdieu 1994/1998:6). **Economic capital** refers to the material resources—wealth, land, money—that one controls or possesses. **Cultural capital** refers to nonmaterial goods such as educational credentials, types of knowledge and expertise, verbal skills, and aesthetic preferences that can be converted into economic capital. It is these two forms of capital that constitute the "externality" that is internalized via the habitus by forming the social space within which points of view are taken. Within this social space, individuals are positioned relative to one another, first according to the overall volume of the capital they possess and, second, according to the relative amount of economic and cultural capital they possess. Moreover, the closer individuals are to one another in terms of the amount and types of capital they possess, the more they have in common (the more their lifestyles, tastes, and aspirations coincide), while the farther apart they are in social space (that is, the less similar the composition of their capital), the less they have in common (ibid.:6, 7). Thus, you are more apt to see "eye to eye" or to "hit it off" with someone the more your positions in social space overlaps. But this is not a direct result of possessing similar amounts of money and types of educational credentials, per se, but rather a result of the similarity of the habitus each has acquired by virtue of being similarly positioned in social space.

The habitus, then, is a structured structure that structures how one views and acts in the world. First, as a scheme or structure of perceptions, dispositions, and actions, the habitus "generates and organizes practices and representations"—it *structures* an individual's experience of and orientation to the social world. An individual's early childhood socialization has a particularly strong effect in this regard because it provides the basis for apprehending and structuring all future experiences. As Bourdieu notes, the habitus is a system of durable dispositions that "tends to ensure its own constancy and defence against change through the selection it makes within new information by rejecting information capable of calling into question its accumulated information. . . . Through the systematic 'choices' it makes among the places, events and people that might be frequented, the habitus tends to protect itself from crises and critical challenges by providing itself with a milieu to which it is as pre-adapted as possible, that is, a relatively constant universe of situations tending to reinforce its dispositions" (Bourdieu 1980/1990:60, 61).

However, as a structure, the habitus is itself *structured* by one's position in social space, which is determined by the volume and types of capital possessed. As a product of an objective position within social space, the habitus encompasses a system of objectively determined practices that reflect the possibilities or "life chances" that are tied to a given social position. In this way, the habitus is an "embodiment of history" or a "present past" born out of the long accumulation of life experiences distinctive to a given social position.

Bourdieu sees in the habitus the union of structures and practices: "objective structures tend to produce structured subjective dispositions that produce structured actions which, in turn, tend to reproduce objective structure" (Bourdieu and Passeron [1970] 1977:203, quoted in Swartz 1997:103). For Bourdieu, however, a central unintended consequence of this circular process is the legitimation and reproduction of a stratified social order that advantages some groups while disadvantaging others. Insofar as an individual's habitus is structured by the existing social distribution of economic and cultural capital, the aspirations and expectations it engenders will conform to what is objectively accessible, given one's class position. Objective probabilities for success or failure are translated into a subjective appraisal of a "sense of one's place." On one hand, a person whose parents are successful, college-educated professionals or corporate executives will grow up expecting to be a successful, college-educated professional or corporate executive; more than likely, he or she will not aspire to be a worker in a furniture assembly plant. On the other hand, a person whose family has worked as manual laborers for generations will be more likely to expect to do the same while assuming that a career as a corporate executive or physician is outside of his reach. As a result, an individual's internalized estimation of what is objectively possible or impossible, reasonable or unreasonable to accomplish, fosters aspirations and practices that reproduce the objective structures that generate the world of unequal possibilities. Like a self-fulfilling prophecy, the habitus perpetuates structural inequality across generations by adapting individuals' expectations and behaviors to a social space that is constructed on an unequal distribution of resources (Swartz 1997:103–107).

The discriminatory effects of the habitus apply not only to life chances and career aspirations, but also to a wide range of lifestyle "choices." For example, persons who possess relatively little economic capital but possess a relatively high volume of cultural capital (the "dominated fraction of the dominant class," in Bourdieu's terms) are likely to enjoy the same types of food, read the same types of books, listen to the same types of music, speak with a similar vocabulary and accent, participate in the same types of sports, and share similar political views. These affinities are the product of a shared point of view that enables those who occupy this space to acquire the code necessary for understanding, for exam-

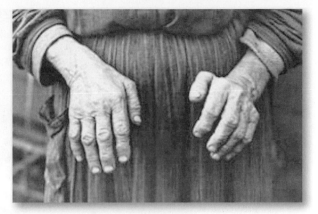

Photo 16.1 An old woman's hands

ple, the distinctions that make Beethoven different from Mozart, James Joyce different from Joseph Conrad, a Monet different from a Manet, or a bottle of Bordeaux different from a bottle of Gallo. Those situated differently in social space—for instance, those who possess little in the way of economic and cultural capital—are unlikely to be exposed to the necessary socialization that would endow them with the categorical schemes required to appreciate these distinctions. Being unable to "understand" such types of music, novels, paintings, or wines, those so situated would be little interested in partaking of them.

In his groundbreaking book, *Distinction,* Bourdieu (1979) analyzed the schemes of perception individuals bring to bear on the social world, including works of art. Based on interviews and questionnaires, he documented distinctive modes of apprehending art that corresponded to the class position of respondents. For instance, when asked to comment on the photo of an old woman's hands (see Photo 16.1), working-class respondents—those who possessed little economic and cultural capital—replied in every day terms that lacked an explicitly aesthetic judgment: "Oh, she's got terribly deformed hands! . . . Looks like she's got arthritis. . . . I really feel sorry seeing that poor old woman's hands, they're all knotted." Respondents from the more privileged classes, however, described the photograph in abstract terms that reflect their own distance from the necessities of everyday life and their possession of the aesthetic code required for properly appreciating art: "I find this a very beautiful photograph. It's the very symbol of toil. It puts me in mind of Flaubert's old servant-woman. . . . That woman's gesture, [is] at once very humble. . . . It's terrible that work and poverty are so deforming" (Bourdieu 1984/1990:44, 45).

Like their comments on the photo of the woman's hands, when responding to a "modernist" photo similar to Photo 16.2, those possessing little cultural capital were unable to apply an aesthetic code to the photograph that would allow them to perceive it as a form of artistic expression. Neither, however, could the image be decoded through common, "realist" perceptual schemes. Perplexed and defeated, manual workers offered the following descriptions: "At first sight it's a construction in metal but I can't make head or tail of it. It might be something used in an electric power station. . . . I can't make out what it really is, it's a mystery to me." "Now, that one really bothers me, I haven't got anything to say about it. . . . I can't see what it could be, apart from lighting. It isn't car headlights, it wouldn't be all straight lines like that." "That's something to do with electronics, I don't know anything about that." Similarly situated in social space, small shopkeepers and craftsmen offered similar verdicts: "That is of no interest, it may be all very fine, but not for me. . . . Personally that stuff leaves me cold." " I wouldn't know what to do with a photo like that. Perhaps it suits modern tastes." By contrast, members of the dominant class were able to "appreciate" the image by applying an abstract, aesthetic code in which judgments of form are freed from judgments of contents: "It's inhuman but aesthetically beautiful because of the contrasts" (ibid.:46, 47).

Bourdieu's analysis reveals how possessing the code—that is, the cultural capital—to properly understand works of art, haute cuisine, or the finer points of golf is not simply a matter of an individual's

Photo 16.2   Nighttime industrial scene

aesthetic preferences and tastes. Because differences in lifestyles and knowledge are the product of hier-archically ordered positions within social space, they, too, are ordered hierarchically. In class-based societies, it matters whether or not an individual prefers to visit museums or watch wrestling on TV, read *The New Yorker* or *The National Enquirer,* attend a tennis match or a stock car race. While the former are generally considered "highbrow" and refined, the latter are considered "lowbrow" and vulgar. As a result, tastes symbolically mark the social positions they are connected to—and the individuals who inhabit them—as either dignified or inferior, and serve to legitimate existing social inequalities. Even how a person handles his silverware at mealtime can signify the cultural capital he possesses; "cultured" parents who are attuned to the symbolic significance of such behavior often stress to their young children the need to refrain from holding their forks and spoons like a "caveman."

Social Reproduction

While Bourdieu's notion of the habitus speaks to a central debate that has long framed the social sciences—namely, the relationship between agency and structure—he combines the concept of the habitus with his discussion of economic and cultural capital to shed light on another longstanding problematic: how are societies reproduced from one generation to the next such that social stability is preserved? In answering this question, Bourdieu draws heavily from Marx, Durkheim, and Weber to fashion a novel account of the process of social reproduction. The resulting analyses of cultural capital and its accumulation through social-ization and education represent one of his most important contributions to social theory.

Like Marx, Bourdieu sees modern societies as based fundamentally on relations of power. Arguing that "economic capital is at the root of all the other types of capital" (Bourdieu 1990a:252), Bourdieu shares with Marx's work, and Marxism more generally, an emphasis on class-based forms of dominance. Economic capital (money and property) provides the means for acquiring other forms of capital, including cultural capital, by providing an escape from "necessity." The more wealth a person possesses, the more he is able to remove himself from the necessary daily concerns over physical survival. By freeing one from pressing concerns over one's material existence, economic capital grants opportunities for traveling to foreign lands, frequenting the bastions of high culture (museums, operas, etc.), and otherwise indulging one's tastes. In short, economic capital affords an individual the possibility for becoming "worldly"—that is, "knowing" the world—and thus deserving of the privileged status that is bestowed on him or her by others. Most important for the purposes of social reproduction, economic capital can be readily transferred to succeeding generations through inheritance. Certainly it is easier to give your money to your family members than it is to give them your cultural capital in the form of knowledge and aesthetic preferences, no matter how "superior" they may be. Moreover, the value of specific tastes can fluctuate within a relatively short time, while the value of money remains an enduring feature of advanced societies.

Although Bourdieu contends that class relations form the basis of modern, hierarchical societies and that economic capital is the most valuable form of capital, his analysis of social reproduction parts in significant ways from traditional Marxist interpretations. Marxist social theory typically assigns a derivative role to noneconomic domains. Thus, political, legal, and cultural systems compose a "superstructure" that reflects the organization of production within the more primary economic "base." However, as our discussion of cultural capital suggests, Bourdieu contends that economic resources alone do not form the social space of positions. Nor are money and property the only avenues for expressing and sustaining relations of domination. Moreover, in addition to cultural capital, social positions also are endowed with varying degrees of **social capital** or networks of contacts and acquaintances that can be used to secure or advance one's position. For instance, a friend's father may write a letter of recommendation on your behalf to attend a prestigious university of which he is an alumnus, or perhaps someone you know knows someone who can arrange an interview for you at a well-respected software firm. Of course, irrespective of the amount of economic and cultural capital we may possess, we all have friends and acquaintances that can help us in one way or another. Bourdieu's point, however, is that social capital circulates within defined boundaries of social space and thus serves to reproduce existing relations of domination. If one is raised in a poor family living in a rural community, it is unlikely that one will have connections to corporate executives working in a major metropolitan center. Thus, social capital promotes the perpetuation of class position across generations by providing access to opportunities denied to those who do not possess such resources.

Social reproduction, then, essentially consists of the reproduction of stratified, hierarchical relations that deflect or resist calls for radical change by those positioned in dominated positions in social space. Typically, this requires that dominated social groups sanction the legitimacy of the existing system of relations, thus perpetuating their own domination. Like Marx, who argued that "the ruling ideas are the ideas of the ruling class," Bourdieu sees relations of domination disguised through a false consciousness that renders the social system immune from challenges. This is made possible once the real sources of individuals' domination are "misrecognized" as stemming from personal failings or from causes beyond the control of their society and its leaders.

According to Bourdieu, no institution does more to ensure the reproduction of class relations than education. Misrecognized as a meritocratic institution that rewards individual aptitudes over hereditary privileges, the educational system "maintains the preexisting social order, that is, the gap between pupils endowed with unequal amounts of cultural capital. More precisely, by a series of selection operations, the system separates holders of inherited cultural capital from those who lack it. Differences of aptitude being inseparable from social differences according to inherited capital, the system thus tends to maintain preexisting social differences" (Bourdieu 1994/1998:20). As guardians of the dominant culture—what is worth knowing and what the worthy should know—universities are charged with separating the cultured from the uncultured. Their mission is effected through an application process and competitive entrance exams (e.g., SATs) that establish a "true magical threshold separating the last candidate to have passed from the first to have failed" and thus a boundary separating the sacred from the profane (ibid.:21). Applicants are judged, and, once admitted, grades are assigned, as much, if not more, on the basis of what students already know than on what they will have learned in class. And what successful students already know is the "feel for the game" that comes with possessing the same types of cultural capital that are enshrined in institutions of higher education. Success in school is not simply tied to writing and speaking well; rather, it is a matter of writing and speaking in a particular way. Similarly, knowing how to install home plumbing, repair a television, or provide emotional support to those who are distressed is of little value in university classrooms unless it is coupled with an understanding of, and ability to articulate, the *science* that underlies such skills. What the best schools do best—the Grandes Écoles in France or the Ivy

League in the United States—is conceal the links between scholastic aptitude and inherited cultural capital in a diploma that consecrates a social difference in the guise of a technical competence (ibid.:22). Ostensibly, meritocratic qualifications serve to perpetuate social inequalities by transforming their effects into the fate of personal traits.

Symbolic Struggles

While economic, cultural, and social capital are crucial resources that shape an individual's position in social space and thus his life chances, Bourdieu argues that these forms of capital are not directly responsible for charting one's destiny or for the reproduction of social relations more generally. Instead, these forms of capital are realized symbolically through a "war of words," the stakes of which are establishing a monopoly over the "legitimate principle of vision and division" on which social reality is constructed (Bourdieu 1990a:134). In this "war," opposing sides struggle to define what is just and unjust, good and bad, right and wrong, pure and corrupt. Both the stability of and challenges to the prevailing social order are the consequence of individuals and groups strategically marshalling the capital at their disposal in an effort to advance their particular interests.

However, the struggle to impose a particular vision of the world in the name of a universal "truth" cannot be waged through naked coercion or under the banner that "might makes right" if that vision is to retain its legitimacy. To be effective, capital must be deployed in a "disinterested" fashion such that all strategies, interests, and calculations appear to be untainted by self-serving ambition. This requires that individuals and groups possess **symbolic capital**. Commonly labeled prestige, honor, reputation, or charisma, symbolic capital is converted economic or cultural capital denied as capital, "recognized as legitimate, that is, misrecognized as capital" (Bourdieu 1980/1990:118). Here, Bourdieu is extending Weber's insights into charisma as a source of legitimacy and authority (Swartz 1997:43). In his studies of world religions, Weber examined how the charisma of prophets served as a basis for exercising authority over their followers. Charisma—a state of "grace"—endows religious prophets with a virtual magical power that instills in believers an unquestioned obedience to their commands. While Weber confined his discussion of charisma to the religious field, Bourdieu argues that charisma operates within the secular world as well. Artists, scientists, politicians, university professors, journalists, and others can use their charisma or reputation, accumulated within their respective areas of expertise, as a source of power to legitimately demand obedience from others. Those who possess symbolic capital possess the authority to make the world through a "magical power" that transforms and disguises what is the result of the self-interested exercise of economic and political domination into the naturally inevitable.

Bourdieu uses the term **symbolic violence** to refer to acts leading to the misrecognition of reality or distortion of underlying power relations. "Symbolic violence rests on the adjustment between the structures constitutive of the habitus of the dominated and the structure of the relation of domination to which they apply: the dominated perceive the dominant through the categories that the relation of domination has produced and which are thus identical to the interests of the dominant" (Bourdieu 1994/1998:121). In this way, relations of domination take on a "naturalness" that is inscribed in the habitus—the schemes of perception and apprehension—of both the dominant and the dominated. Through committing acts of symbolic violence, a misrecognized vision of the social world is legitimated—a vision that reproduces, with the complicity of the dominated, a stratified social order. Thus, for instance, through acts of symbolic violence the beneficiaries of the educational system appear intrinsically worthy of their success, while the less successful appear intrinsically unworthy (Bourdieu 1982/1991:24, 25).

The struggle to impose the legitimate categories according to which social life is understood is, for Bourdieu, at the root of all action. This struggle takes place within "fields," a concept central in Bourdieu's theoretical scheme. **Fields** are relatively autonomous arenas within which actors and institutions mobilize their capital in an effort to capture the stakes—the distribution of capital—that are specific to it. Examples of fields include art, literature, science, religion, the family, and education. The evolving history of each of these arenas is determined by the struggles that take place between its dominant and subordinate factions as each attempts to either defend or subvert the legitimacy of existing practices—the status quo—and the meanings assigned to them. For instance, painters and critics within the field of art are engaged in a continual struggle over what qualifies as "art" and whose works deserve to be so consecrated. As is the case in all such struggles, participants here adopt strategies that correspond to the amount and types of capital they possess. Thus, artists, critics, dealers, and gallery owners whose interests and positions are aligned with established, orthodox styles will put into play their symbolic capital to denounce avant-garde or heretical works as "pretentious," "undisciplined," or, the worst of all charges in the artistic field, "commercial." For their part, the "newcomers" will challenge the legitimacy of the establishment by classifying works associated with it as "stale," "dated," or "hackneyed." While the former seek to fix the current boundaries of the field, the latter, who can make a name for themselves only by being different or distinctive from consecrated producers, seek to rupture those boundaries (Bourdieu 1984/1993).

No matter the field, those who have an interest or stake in it take a position—that is, they pursue a strategy that corresponds to the position they occupy in that field. In this light, all action is self-interested, although in order to be perceived as legitimate it must be misrecognized as disinterested. Because strategies of misrecognition are carried out symbolically, the words used to classify or assign meaning to acts and others are instruments of power. Symbolic power is a power of "worldmaking," "a power of creating things with words" exercised in the struggle to impose the legitimate vision of the social world (Bourdieu 1980/1990:137, 138). Of course, not every individual or group is able to exercise symbolic power and define the "truth." The power to name the world—to "make people see and believe"—is dependent on the amount of symbolic capital (charisma, authority, recognition) one possesses, which, in turn, is dependent on the outcome of previous struggles as well as one's position within the relevant field.

Consider the controversies surrounding the definition of obscenity, attempts to form an employee union in a retail store chain, or efforts to redraw electoral districts. In each instance, the interested parties (artists and religious leaders, workers and business owners, politicians and community activists) seek to impose their particular vision on the issue as the universal, and thus legitimate, truth. When is a painting or song lyric a form of artistic expression and when is it an obscenity? What is the best way to strike a balance between workers' rights and a corporation's ability to turn a profit? When does redistricting serve to enhance democratic representation or unfairly distort the voting process? Seldom do such questions yield a simple, obvious answer. Nevertheless, they are answered, when one faction in the controversy is able to mobilize belief in the legitimacy of its words through "an almost magical power which enables [it] to obtain the equivalent of what is obtained through force (whether physical or economic)"—namely, the ability to establish the social order (Bourdieu 1982/1991:170).

We conclude this section by briefly noting several of the influences Bourdieu draws on in developing his analyses of symbolic struggles and their relevance to the reproduction of social relations. Like Durkheim and Lévi-Strauss, Bourdieu views all symbolic systems as grounded in fundamental binary categories.[4] Evincing the imprint of both scholars, he notes,

> All the agents in a given social formation share a set of basic perceptual schemes, which receive the beginnings of objectification in the pairs of antagonistic adjectives commonly used to classify and qualify persons or objects in the most varied areas of practice. The network of oppositions between high (sublime, elevated, pure) and low (vulgar, low, modest), spiritual and material, fine (refined, elegant) and coarse (heavy, fat, crude, brutal), light (subtle, lively, sharp, adroit) and heavy (slow, thick, blunt, laborious, clumsy), free and forced, broad and narrow, or, in another dimension, between unique (rare, different, distinguished, exclusive,

---

[4]Durkheim [1912] 1995 saw in the early religious distinction drawn between the sacred and the profane the basis for all other systems of thought. This distinction was born out of focused group activity in tribal societies that produced in participants an altered state of consciousness. Durkheim referred to this group energy as "mana" or "collective effervescence" and argued that tribal members attributed their newly found feelings of efficacy to the influence of supernatural spirits and gods. These feelings, in turn, provided the foundation for group solidarity and a sense of community. Lévi-Strauss would later develop Durkheim's ideas regarding binary classifications into the basis for his structural anthropology.

exceptional, singular, novel) and common (ordinary, banal, commonplace, trivial, routine), brilliant (intelligent) and dull (obscure, grey, mediocre), is the matrix of all the commonplaces which find such ready acceptance because behind them lies the whole social order. (Bourdieu 1984/1990:468)

Bourdieu parts company with Durkheim and Lévi-Strauss, however, when he argues that binary classifications, while serving as the foundation for appending the everyday world and creating shared meaning and social solidarity, also provide the conceptual basis for social domination (Swartz 1997). The world of common sense produced by the logic of binary opposites is a world in which only some individuals and groups are entitled to the privileged axis of opposing terms. The binary categories we use to distinguish practices and things classify at the same time those individuals and groups associated with them. As an outcome of symbolic struggles, what is consecrated as sacred, pure, strong, or brilliant reflects and legitimates the underlying hierarchical social order that pits the dominant against the dominated (ibid.:85, 86).

This leads us to consider the influence of Weber's work on Bourdieu. Like Weber, Bourdieu sees interests and power as deriving from a number of sources; his analyses of multiple forms of capital represent an important extension of Weber's earlier work in this regard.[5] Bourdieu's analyses of lifestyle patterns as signs of distinction that disguise class-based domination are particularly indebted to Weber's conceptualization of class and status groups (ibid.:45). Bourdieu also adopts a view similar to Weber's regarding the existence of social classes. For Weber, classes, contrary to Marxian definitions of the concept, are not real groups whose composition reflects existing property relations. Instead, classes refer to individuals who share life chances that are determined by "economic interests in the possession of goods and opportunity for income" (Weber [1925b] 1978:927). As you will read in the first selection excerpted, Bourdieu argues that classes exist only "on paper," as individuals who are "related" to one another in social space, and not as real groups. This view has important ramifications for understanding how the social order and relations of domination are maintained and challenged, particularly by calling attention to the shortcomings of Marxist perspectives.

Significant Others

Randall Collins (1941–): Bridging the Micro and Macro

No theorist has been more engaged in bridging the divide between micro and macro accounts of social life than Randall Collins. Collins's distinctive multidimensional approach can be traced to his educational training. As an undergraduate at Harvard University, Collins was significantly influenced by the work of Talcott Parsons (see Chapter 9). Although there are many points on which Parsons and Collins sharply disagree, they share in interest in building a comprehensive sociological theory that works on both the micro and macro levels, as well as a particular appreciation for the works of Weber and Durkheim. Collins refined his interests in microsociology as a graduate student at the University of California, Berkeley, in the 1960s, where he worked with Erving Goffman and Herbert Blumer (see Chapter 12). The turbulent political climate of that time and place also shaped his views by making clear the continuing real-world relevance of the works of Marx and Weber.

In his most recent book, *Interaction Ritual Chains* (2004), Collins draws on Durkheim, Mead, and Goffman to further develop his multidimensional perspective. Here, Collins investigates variations in the emotional intensity of social rituals in order to explain patterns of group membership. He argues that successful rituals create feelings of group membership and "pump up"

---

[5] In his essay "Class, Status, Party," Weber ([1925b] 1978) explored how interests and the power to realize them flow not only from one's class position, but from one's position in the status hierarchy as well. Moreover, Weber notes that "parties" can be formed in an effort to strategically pursue group goals or otherwise influence communal action that is not directly tied to specific class or status group interests.

individuals with emotional energy, while failed rituals drain emotional energy, making continued interaction between the individuals involved less likely. Moreover, indicative of his conflict approach to understanding social life, Collins emphasizes that privilege and power are not only tied to an unequal distribution of material and cultural resources, but are also derived from a "flow of emotional energy across situations that makes some individuals more impressive, more attractive or dominant" while narrowing the sources of emotional energy for others (ibid.:xiii). Describing the emotional and symbolic exchanges at the heart of such activities as sex and smoking, Collins argues that individuals are drawn to interactions in which their economic and cultural capital promises them the best emotional payoff.

Collins's earlier work includes *Conflict Sociology* (1975), a Durkheimian reworking of the conflict tradition, and *Credential Society* (1979). In the latter he argues that educational credentials (university degrees) not only serve as barriers to the labor market that lead to further inflationary effects (the need for more and more degrees), but also are mired in inequality. In order for individuals to receive fair treatment in the labor market, then, this system of barriers would have to be removed. Collins argues that "credential abolitionism" is necessary because "the prospects of continuing to expand the credential system indefinitely, to let job requirements inflate to the point where 4 years of college is needed for a manual laborer or 20 years of postdoctoral study is required for a technical profession, would be exceedingly alienating to all concerned. Moreover, it would not affect the rate of mobility, nor change the order of stratification among ethnic groups; it would simply reproduce their order at higher and higher levels of education" (ibid.:197). Evidencing his disdain for the credential society, at various points in his life Collins has quit the academic world because of his disapproval of the credential-seeking process.

BOURDIEU'S THEORETICAL ORIENTATION ▓▓

As noted at the outset, Bourdieu developed his theoretical model with the intention of overcoming the theoretical dualism that, in his estimation, is responsible for producing only partial understandings of social life. Bourdieu's concerns in this matter speak directly to the questions of both action and order. Specifically, with regard to the question of order, Bourdieu's concept of the habitus is a theoretical device that makes possible an encompassing answer to the question, "What accounts for the patterns of behavior that lead us to experience social life as routine and predictable?" Bourdieu's response is centered on the notion that the habitus, as the "internalization of externality," structures an individual's perception of the social world at the same time that it is itself a product of social and historical conditions.

Understood as "schemes of perception, thought and action," the habitus incorporates both individualist and collectivist dimensions of the nature of social order (see Figure 16.1). The habitus both enables and constrains our actions that then reproduce the very social conditions that structure the habitus. Bourdieu goes to great lengths to argue that the system of dispositions produced by the habitus does not lead to "mechanical" conduct that simply reflects the external conditions that structure the habitus. In this light, individuals actively make and remake the social order as they "take positions" or advance particular points of view, constructing their vision of the world in the process. However, position-taking is not unconstrained: Individuals are not free to adopt any stance they choose. Points of view and the "strategies" individuals construct as they go about fitting their conduct to others' are structured by the point they occupy in the space of social positions. The distribution of capital, in particular economic and cultural capital, structures the relation of positions within social space as well as within the specific fields in which individuals seek to impose the legitimate categories of perception. This is crucial, because the distribution of capital, in defining the space of positions, defines the space of possible position-takings—that is, the dispositions or habitus internalized by those who are located in one point or another.

Figure 16.1   Bourdieu's Basic Concepts and Theoretical Orientation

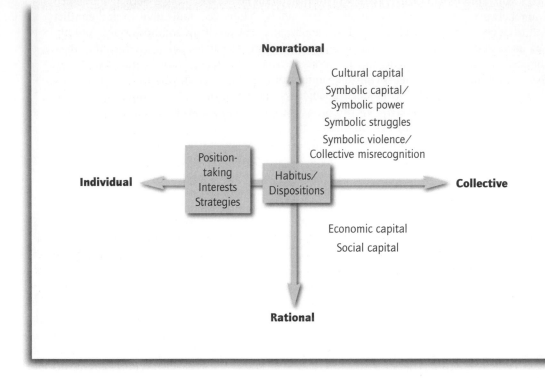

Concerning the question of action, Bourdieu has developed a framework that likewise seeks to incorporate both rational and nonrational dimensions. (This is indicated by the boxed terms that span the rational/nonrational axis in Figure 16.1.) His efforts are based on adopting a conceptual language linked to economic and utilitarian perspectives ("interests," "strategies," "calculation," "profit") that underscores the rational aspects of motivation while arguing that action is typically carried out in a prereflective, nonrational manner. Bourdieu argues that all action is "interested" and thus oriented to the maximization of profit. Indeed, Bourdieu extends the notion of interests and profits to incorporate economic or material gains and advantages as well as the pursuit of symbolic "goods" (Swartz 1997:42). Thus, Bourdieu contends that actors adopt strategies that maximize their economic, cultural, social, and symbolic capital. In this light, we might take Bourdieu to mean that action is motivated by a conscious, rational calculation of costs and benefits. Yet, this is the very interpretation of action he means to dispel. Action is not guided by the logic of profit: It is driven by a *practical* logic that Bourdieu likens to a prereflective "feel for the game." Similar to how a good tennis player will position himself "not where the ball is but where it will be, one invests oneself and one invests not where the profit is but where it will be" (Bourdieu 1994/1998:79). Bourdieu argues that the habitus embodies actors with a "know-how" that enables them to strategically orient their conduct without necessarily having a strategic intention or goal that can be put into words. Actions are more adaptive than purposive in nature (Robbins 2000:29).

Put in another way, in taking a position, actors deploy their capital so as to realize their interests. However, conduct is not consciously chosen as if alternative means were weighed in terms of their potential for achieving specific ends. Rather, conduct is inscribed in the very nature of the "game" in which actors are involved and in the habitus through which a taken-for-granted world is constructed. Although the strategies that actors pursue are rational—that is, fitting to the field in which they are "absorbed"—they are not pursued rationally. While a politician may find it "natural" to publicly express his religious commitments on the campaign trail, a scientist would not be predisposed to such a strategy when trying to convince the public of the significance of her findings.

In the first selection, "Social Space and the Genesis of Groups," Bourdieu outlines his relational model for the study of social life. He conceptualizes social space as an arena in which individuals and groups are positioned in relation to one another based on the amount and types of capital they have at their disposal. It is on the basis of such relations that struggles are waged to classify the world. In the second selection, "Outline of a Sociological Theory of Art Perception," Bourdieu argues that artistic competence, far from being the product of an individual's natural aesthetic preferences, is dependent on the possession of cultural capital. Unrecognized as such, artistic tastes function as a sign of distinction that legitimates the dominance of those who possess the code for deciphering works of art.

Introduction to "Social Space and the Genesis of Groups"

In this reading, Bourdieu addresses many of the themes discussed above. After describing social space as structured according to the distribution of capital, he embarks on a critique of "substantialist" approaches to social classes by emphasizing the space of relations within which ongoing struggles are waged to name the social world. In particular, Bourdieu takes aim at Marxist theory for treating classes as real groups mobilized solely on the basis of the conditions of economic production. He counters by presenting the view that classes and groups do not "exist"; rather, they are an outcome of symbolic struggles that seek to pronounce a reality that exists only because it is pronounced by those in an authorized position to declare its existence—that is, by those who possess the necessary symbolic capital. It is above all the spokesperson "who, speaking about a group, speaking on behalf of a group, surreptitiously posits the existence of the group in question, institutes the group, through that magical operation which is inherent in any act of naming" (Bourdieu 1982/1991:250).

Yet, the struggle for the "monopoly of legitimate naming" is not an entirely symbolic affair. Imposing the legitimate vision of the social world reproduces the relations of power that structure the social world. Thus, to fully understand the classifications, the principles of vision and division, according to which the world is named, one must account for the positions occupied by those engaged in the struggle to establish meaning, because the categories through which the world is perceived "are the product of the incorporation of the very structures to which they are applied" (ibid.:238). Different points of view, and the categories of perception that are used to express them, are but expressions, mediated through the habitus, of different points or positions within the space of objective relations. Any "science of classifications" must acknowledge this fundamental reality if it is to offer sound analyses of social life.

————— "Social Space and the Genesis of Groups" (1982) —————

Pierre Bourdieu

THE SOCIAL SPACE

Initially, sociology presents itself as a *social topology.* Thus, the social world can be represented as a space (with several dimensions) constructed on the basis of principles of differentiation or distribution constituted by the set of properties active within the social universe in question, i.e. capable of conferring strength, power within that universe, on their holder. Agents and groups of agents are thus defined by their relative positions

SOURCE: "Social Space and the Genesis of Groups" is reprinted by permission of the publisher from *Language and Symbolic Power* by Pierre Bourdieu, pp. 229–250, edited and introduced by John B. Thompson, trans. by Gino Raymond and Matthew Adamson, Cambridge, MA: Harvard University Press. Copyright © Pierre Bourdieu 1983, 1977, 1981, 1984. This English translation copyright © 1991 by Polity Press.

within that space. Each of them is assigned to a position or a precise class of neighbouring positions (i.e. a particular region in this space) and one cannot really—even if one can in thought—occupy two opposite regions of the space. Inasmuch as the properties selected to construct this space are active properties, one can also describe it as a field of forces, i.e. as a set of objective power relations which impose themselves on all who enter the field and which are irreducible to the intentions of the individual agents or even to the direct *interactions* among the agents.

The active properties which are selected as principles of construction of the social space are the different kinds of power or capital which are current in the different fields. Capital, which may exist in objectified form—in the form of material properties—or, in the case of cultural capital, in the embodied state, and which may be legally guaranteed, represents a power over the field (at a given moment) and, more precisely, over the accumulated products of past labour (in particular over the set of instruments of production) and thereby over the mechanisms tending to ensure the production of a particular category of goods and so over a set of incomes and profits. The kinds of capital, like the aces in a game of cards, are powers which define the chances of profit in a given field (in fact, to each field or sub-field there corresponds a particular kind of capital, which is current, as a power or stake, in that game). For example, the volume of cultural capital (the same thing would be true, *mutatis mutandis,* of the economic game) determines the aggregate chances of profit in all the games in which cultural capital is effective, thereby helping to determine position in social space (to the extent that this is determined by success in the cultural field).

The position of a given agent within the social space can thus be defined by the positions he occupies in the different fields, that is, in the distribution of the powers which are active within each of them. These are, principally, economic capital (in its different kinds), cultural capital and social capital, as well as symbolic capital, commonly called prestige, reputation, renown, etc., which is the form in which the different forms of capital are perceived and recognized as legitimate. One can thus construct a simplified model of the social field as a whole which makes it possible to conceptualize, for each agent, his position in all possible spaces of competition (it being understood that, while each field has its own logic and its own hierarchy, the hierarchy which prevails among the different kinds of capital and the statistical link between the different types of assets tends to impose its own logic on the other fields).

The social field can be described as a multi-dimensional space of positions such that every actual position can be defined in terms of a multi-dimensional system of co-ordinates whose values correspond to the values of the different pertinent variables. Thus, agents are distributed within it, in the first dimension, according to the overall volume of the capital they possess and, in the second dimension, according to the composition of their capital—i.e. according to the relative weight of the different kinds of assets within their total assets. . . .

Thus, in the first dimension, which is undoubtedly the most important, the holders of a great volume of overall capital, such as industrial employers, members of liberal professions, and university professors are opposed, in the mass, to those who are most deprived of economic and cultural capital, such as unskilled workers. But from another point of view, that is, from the point of view of the relative weight of economic capital and cultural capital in their patrimony, professors (relatively wealthier in cultural capital than in economic capital) are strongly opposed to industrial employers (relatively wealthier in economic capital than in cultural capital). . . . The second opposition, like the first, is the source of differences in dispositions and, therefore, in position-takings. This is the case of the opposition between intellectuals and industrial employers or, on a lower level of the social hierarchy, between, [for instance,] primary school teachers and small merchants. . . .

In a more general sense, the space of social positions is retranslated into a space of position-takings through the mediation of the space of dispositions (or habitus). In other words, the system of differential deviations which defines the different positions in the two major dimensions of social space corresponds to the system of differential deviations in agents' properties (or in the properties of constructed classes of agents), that is, in their practices and in the goods they possess. To each class of positions there corresponds a class of habitus (or *tastes*) produced by the social conditioning associated with the corresponding condition and, through the mediation of the habitus and its generative capability, a systematic set of goods and properties, which are united by an affinity of style.

One of the functions of the notion of habitus is to account for the unity of style, which unites the practices and goods of a single agent or a class of agents. . . . The habitus is this generative and unifying principle which retranslates the intrinsic and relational characteristics of a position into a unitary lifestyle, that is, a unitary set of choices of persons, goods, practices.

Like the positions of which they are the product, habitus are differentiated, but they are also differentiating. Being distinct and distinguished, they are also distinction operators, implementing different principles of differentiation or using differently the common principles of differentiation.

Habitus are generative principles of distinct and distinctive practices—what the worker eats, and especially the way he eats it, the sport he practices and the way he practices it, his political opinions and the way he expresses them are systematically different from the industrial owner's corresponding activities. But habitus are also classificatory schemes, principles of classification, principles of vision and division, different tastes. They make distinctions between what is good and what is bad, between what is right and what is wrong, between what is distinguished and what is vulgar, and so forth, but the distinctions are not identical. Thus, for instance, the same behavior or even the same good can appear distinguished to one person, pretentious to someone else, and cheap or showy to yet another.

But the essential point is that, when perceived through these social categories of perception, these principles of vision and division, the differences in practices, in the goods possessed, or in the opinions expressed become symbolic differences and constitute a veritable *language*. Differences associated with different positions, that is, goods, practices, and especially *manners,* function, in each society, in the same way as differences which constitute symbolic systems, such as the set of phonemes of a language or the set of distinctive features and of differential "*écarts*" that constitute a mythical system, that is, as *distinctive signs. . . .*

The form that is taken, at every moment, in each social field, by the set of distributions of the different kinds of capital (embodied or materialized), as instruments for the appropriation of the objectified product of accumulated social labour, defines the state of the power relations, institutionalized in long-lasting social statuses, socially recognized or legally guaranteed, between social agents objectively defined by their position in these relations; it determines the actual or potential powers within the different fields and the chances of access to the specific profits that they offer.[i]

Knowledge of the position occupied in this space contains information as to the agents' intrinsic properties (their condition) and their relational properties (their position). This is seen particularly clearly in the case of the occupants of the intermediate or middle positions, who, in addition to the average or median values of their properties, owe a number of their most typical properties to the fact that they are situated *between* the two poles of the field, in the *neutral* point of the space, and that they are balanced between the two extreme positions.

Classes on Paper

On the basis of knowledge of the space of positions, one can separate out *classes,* in the logical sense of the word, i.e. sets of agents who occupy similar positions and who, being placed in similar conditions and subjected to similar conditionings, have every likelihood of having similar dispositions and interests and therefore of producing similar practices and adopting similar stances. This "class on paper" has the *theoretical* existence which is that of theories: insofar as it is the product of an explanatory classification, entirely similar to those of zoologists or botanists, it makes it possible to *explain* and predict the practices and properties of the things classified—including their group-forming practices. It is not really a class, an *actual* class, in the sense of a group, a group mobilized for struggle; at most, it might be called a *probable class,* inasmuch as it is a set of agents which will present fewer hindrances to efforts at mobilization than any other set of agents.

Thus, contrary to the *nominalist relativism* which cancels out social differences by reducing them to pure theoretical artefacts, one must therefore assert the existence of an objective space determining compatibilities and incompatibilities, proximities and distances.

---

[i]In some social universes, the principles of division which, like volume and structure of capital, determine the structure of the social space, are reinforced by principles of division relatively independent of economic or cultural properties, such as ethnic or religious affiliation. In such cases, the distribution of the agents appears as the product of the intersection of two spaces which are partially independent: an ethnic group situated in a lower position in the space of the ethnic groups may occupy positions in all the fields, including the highest, but with rates of representation inferior to those of an ethnic group situated in a higher position. Each ethnic group may thus be characterized by the social positions of its members, by the rate of dispersion of these positions, and by its degree of social integration despite this dispersion. (Ethnic solidarity may have the effect of ensuring a form of collective mobility.)

Contrary to the *realism of the intelligible* (or the reification of concepts), one must assert that the classes which can be separated out in social space (for example, for the purposes of the statistical analysis which is the only means of manifesting the structure of the social space) do not exist as real groups although they explain the probability of individuals constituting themselves as practical groups, in families (homogamy), clubs, associations and even trade-union or political "movement." What does exist is a *space of relationships* which is as real as a geographical space, in which movements are paid for in work, in efforts and above all in time (moving up means raising oneself, climbing, and acquiring the marks, the stigmata, of this effort). Distances within it are also measured in time (time taken to rise or to convert capital, for example). And the probability of mobilization into organized movements, equipped with an apparatus and spokesmen, etc. (precisely that which leads one to talk of a "class") will be inverse ratio to distance in this space. While the probability of assembling a set of agents, really or nominally—through the power of the delegate—rises when they are closer in social space and belong to a more restricted and therefore more homogeneous constructed class, alliance between those who are closest is never *necessary,* inevitable (because the effects of immediate competition may act as a screen), and alliance between those most distant from each other is never *impossible.* Though there is more chance of mobilizing the set of workers than the set composed of workers and bosses, it is possible, in an international crisis, for example, to provoke a grouping on the basis of links of national identity (partly because, by virtue of its specific history, each national social space has its specific structure—e.g. as regards hierarchical distances within the economic field).

Like "being," according to Aristotle, the social world can be uttered and constructed in different ways. It may be practically perceived, uttered, constructed, according to different principles of vision and division—for example, ethnic division. But groupings grounded in the structure of the space constructed in terms of capital distribution are more likely to be stable and durable, while other forms of grouping are always threatened by the splits and oppositions linked to distances in social space. To speak of a social space means that one cannot group just *anyone* with *anyone* while ignoring the fundamental differences, particularly economic and cultural ones. But this never entirely excludes the possibility of organizing agents in accordance with other principles of division—ethnic or national ones, for example,—though it has to be remembered that these are generally linked to the fundamental principles, with ethnic groups themselves being at least roughly hierarchized in the social space, in the USA for example (through seniority in immigration).[ii]

This marks a first break with the Marxist tradition. More often than not, Marxism either summarily identifies constructed class with real class (in other words, as Marx complained about Hegel, it confuses the things of logic with the logic of things); or, when it does make the distinction, with the opposition between "class-in-itself," defined in terms of a set of objective conditions, and "class-for-itself," based on subjective factors, it describes the movement from one to the other (which is always celebrated as nothing less than an ontological promotion) in terms of a logic which is either totally determinist or totally voluntarist. In the former case, the transition is seen as a logical, mechanical or organic necessity (the transformation of the proletariat from class-in-itself to class-for-itself is presented as an inevitable effect of time, of the "maturing of the objective conditions"); in the latter case, it is seen as the effect of an "awakening of consciousness" (*prise de conscience*) conceived as a "taking cognizance" (*prise de connaissance*) of theory, performed under the enlightened guidance of the Party. In all cases, there is no mention of the mysterious alchemy whereby a "group in struggle," a personalized collective, a historical agent assigning itself its own ends, arises from the objective economic conditions. . . .

PERCEPTION OF THE SOCIAL WORLD AND POLITICAL STRUGGLE

The most resolutely objectivist theory has to integrate the agents' representation of the social world; more precisely, it must take account of the contribution that

---

[ii]The same thing would be true of the relationship between geographical space and social space. These two spaces never coincide completely, but a number of differences that are generally attributed to the effect of geographical space, e.g. the opposition between centre and periphery, are the effect of distance in social space, i.e. the unequal distribution of the different kinds of capital in geographical space.

agents make towards constructing the view of the social world, and through this, towards constructing this world, by means of the *work of representation* (in all senses of the word) that they constantly perform in order to impose their view of the world or the view of their own position in this world—their social identity. Perception of the social world is the product of a double social structuration: on the "objective" side, it is socially structured because the properties attached to agents or institutions do not offer themselves independently to perception, but in combinations that are very unequally probable (and, just as animals with feathers are more likely to have wings than are animals with fur, so the possessors of a substantial cultural capital are more likely to be museum-goers than those who lack such capital); on the "subjective" side, it is structured because the schemes of perception and appreciation available for use at the moment in question, especially those that are deposited in language, are the product of previous symbolic struggles and express the state of the symbolic power relations, in a more or less transformed form. The objects of the social world can be perceived and uttered in different ways because, like objects in the natural world, they always include a degree of indeterminacy and fuzziness—owing to the fact, for example, that even the most constant combinations of properties are only founded on statistical connections between interchangeable features; and also because, as historical objects, they are subject to variations in time so that their meaning, insofar as it depends on the future, is itself in suspense, in waiting, dangling, and therefore relatively indeterminate. This element of play, or uncertainty, is what provides a basis for the plurality of world views, itself linked to the plurality of points of view, and to all the symbolic struggles for the power to produce and impose the legitimate world view and, more precisely, to all the cognitive "filling-in" strategies which produced the meaning of the objects of the social world by going beyond the directly visible attributes by reference to the future or the past. This reference may be implicit and tacit, through what Husserl calls protention and retention, practical forms of prospection or retrospection without a positing of the future and the past as such; or it may be explicit, as in political struggles, in which the past—with retrospective reconstruction of a past tailored to the needs of the present—and especially the future, with creative forecasting, are endlessly invoked, to determine, delimit and define the always open meaning of the present.

To point out that perception of the social world implies an act of construction in no way entails acceptance of an intellectualist theory of knowledge: the essential part of the experience of the social world and of the act of construction that it implies takes place in practice, below the level of explicit representation and verbal expression. More like a class unconscious than a "class consciousness" in the Marxist sense, the sense of the position occupied in social space (what Erving Goffman calls the "sense of one's place") is the practical mastery of the social structure as a whole that reveals itself through the sense of the position occupied within that structure. The categories of perception of the social world are, as regards their most essential features, the product of the internalization, the incorporation, of the objective structures of social space. Consequently, they incline agents to accept the social world as it is, to take it for granted, rather than to rebel against it, to counterpose to it different, even antagonistic, possibles. The sense of one's place as a sense of what one can or cannot "permit oneself," implies a tacit acceptance of one's place, a sense of limits ("that's not for the likes of us," etc.), or, which amounts to the same thing, a sense of distances, to be marked and kept, respected or expected. It does so all the more strongly where the conditions of existence are most rigorous and where the reality principle most rigorously asserts itself. . . .

If objective power relations tend to reproduce themselves in views of the social world which contribute to the permanence of these relations, this is therefore because the structuring principles of a world view are rooted in the objective structures of the social world: power relations are also present in people's minds, in the form of the categories of perception of these relations. However, the degree of indeterminacy and fuzziness in the objects of the social world, together with the practical, pre-reflexive and implicit nature of the schemes of perception and appreciation that are applied to them, is the Archimedean leverage point that is objectively offered for political action proper. Knowledge of the social world and, more precisely, the categories which make it possible, are the stakes, par excellence, of political struggle, the inextricably theoretical and practical struggle for the power to conserve or transform the social world by conserving or transforming the categories through which it is perceived.

The capacity to make entities exist in the explicit state, to publish, make public (i.e. render objectified, visible, and even official) that which had not previously attained objective and collective existence and had therefore remained in the state of individual or serial existence—people's malaise, anxiety, disquiet, expectations—represents a formidable social power, the power to make groups by making the *common sense,* the explicit consensus, of the whole group. In fact, this work of categorization, i.e. of making-explicit and of classification, is performed incessantly, at every moment of ordinary existence, in the struggles in which agents clash over the meaning of the social world and of their position within it, the meaning of their social identity, through all the forms of benediction or malediction, eulogy, praise, congratulations, compliments, or insults, reproaches, criticisms, accusations, slanders, etc. . . .

It becomes clear why one of the elementary forms of political power, in many archaic societies, consisted in the quasi-magical power to *name* and to make-exist by virtue of naming. Thus in traditional Kabylia, the function of making-explicit and the work of symbolic production that the poets performed, particularly in crisis situations, when the meaning of the world slips away, conferred on them major political functions, those of the warlord or ambassador. But with the growing differentiation of the social world and the constitution of relatively autonomous fields, the work of producing and imposing meaning is carried on in and through the struggles within the field of cultural production (particularly the political sub-field); it becomes the particular concern, the specific interest, of the professional producers of objectified representations of the social world or, more precisely, of methods of objectification.

If the legitimate mode of perception is such an important prize at stake in social struggles, this is partly because the shift from the implicit to the explicit is in no way automatic: the same experience of the social may be uttered in very different expressions. And partly it is because the most marked objective differences may be masked by more immediately visible differences (e.g. those between ethnic groups). It is true that perceptual configurations, social *Gestalten,* exist objectively, and that the proximity of conditions, and therefore of dispositions, tends to be translated into durable linkages and groupings, immediately perceptible social units, such as socially distinct regions or neighbourhoods (with spatial segregation), or sets of agents endowed with entirely similar visible properties, such as Weber's *Stände.* But the fact remains that socially known and recognized differences only exist for a subject capable not only of perceiving differences but of recognizing them as significant, interesting, i.e. only for a subject endowed with the capacity and inclination to *make* the distinctions that are regarded as significant in the social universe in question.

Thus, particularly through properties and their distributions, the social world achieves, objectively, the status of a *symbolic system,* which, like the system of phonemes, is organized according to the logic of difference, differential deviation, thereby constituted as significant *distinction.* The social space, and the differences that "spontaneously" emerge within it, tends to function symbolically as a *space of life-styles* or as a set of *Stände,* of groups characterized by different life-styles.

Distinction does not necessarily imply the pursuit of distinction, as is often supposed, following Veblen and his theory of conspicuous consumption. All consumption and, more generally, all practice, is "conspicuous," visible, whether or not it is performed *in order to be seen;* it is distinctive, whether or not it springs from the intention of being "conspicuous," standing out, or distinguishing oneself or behaving with distinction. As such, it inevitably functions as a *distinctive sign* and, when the difference is recognized, legitimate and approved, as a *sign of distinction* (in all senses of the phrase). However, because social agents are capable of perceiving as significant distinctions the "spontaneous" distinctions that their categories of perception lead them to regard as pertinent, it follows that they are also capable of intentionally underscoring these spontaneous differences in life-style by what Weber calls "the stylization of life" (*die Stilisierung des Lebens*). The pursuit of distinction—which may be expressed in ways of speaking or the refusal of misalliances— produces separations intended to be perceived or, more precisely, known and recognized, as legitimate differences, which most often means differences in nature ("natural distinction").

Distinction—in the ordinary sense of the word—is the difference inscribed in the very structure of the social space when perceived through categories adapted to that structure; and the Weberian *Stände,* which is often contrasted with the Marxist class, is the class constructed by an adequate division of social space, when perceived through categories derived from the structure of that space. Symbolic capital—another name for distinction—is nothing other than capital, in whatever form, when perceived by an agent endowed with categories of perception arising from the internalization (embodiment) of the structure of its distribution, i.e., when it is known and recognized as self-evident. Distinctions, as symbolic transfigurations of de facto differences and, more generally, ranks, orders, grades, and all other symbolic hierarchies, are the product of the application of schemes of construction which, like (for example) the pairs of adjectives used to utter most social judgements, are the product of the internalization of the structures to which they are applied; and the most absolute recognition of legitimacy is nothing other than the apprehension of the everyday world as self-evident which results from the quasi-perfect coincidence of objective structures and embodied structures.

It follows, among other things, that symbolic capital goes to symbolic capital, and that the—real—autonomy of the field of symbolic production does not prevent it being dominated, in its functioning, by the constraints which dominate the social field, so that objective power relations tend to reproduce themselves in symbolic power relations, in views of the social world which help to ensure the permanence of these power relations. In the struggle to impose the legitimate view of the social world, in which science itself is inevitably involved, agents wield a power proportionate to their symbolic capital, i.e. to the recognition they receive from a group. The authority which underlies the performative efficacy of discourse about the social world, the symbolic strength of the views and forecasts aimed at imposing principles of vision and division of the social world, is a *percipi,* a being-known and being-recognized (this is the etymology of *nobilis*), which makes it possible to impose a *percipere.* Those most *visible* in terms of the prevailing categories of perception are those best placed to change the vision by changing the categories of perception. But also, on the whole, those least inclined to do so.

The Symbolic Order and the Power to Name

In the symbolic struggle over the production of common sense or, more precisely, for the monopoly of legitimate *naming,* that is to say, official—i.e. explicit and public—imposition of the legitimate vision of the social world, agents engage the symbolic capital they have acquired in previous struggles, in particular, all the power they possess over the instituted taxonomies, inscribed in minds or in objectivity, such as qualifications. Thus, all the symbolic strategies through which agents seek to impose their vision of the divisions of the social world and their position within it, can be located between two extremes: the insult, and *idios logos* with which an individual tries to impose his point of view while taking the risk of reciprocity, and *official nomination,* an act of symbolic imposition which has behind it all the strength of the collective, the consensus, the common sense, because it is performed by a delegated agent of the state, the holder of the *monopoly of legitimate symbolic violence.* On the one hand, there is the

world of particular perspectives, singular agents who, from their individual viewpoint, their personal position, produce particular, self-interested namings of themselves and others (nicknames, by-names, insults, even accusations, slanders) that lack the capacity to force recognition, and therefore to exert a symbolic effect, to the extent that their authors are less authorized and have a more direct interest in forcing recognition of the viewpoint they seek to impose. On the other hand, there is the authorized viewpoint of an agent authorized, in his personal capacity, such as a "major critic," a prestigious prefacer or a consecrated author (cf. Zola's *J'accuse*); and, above all, the legitimate viewpoint of the authorized spokesman of the mandated representative of the state, the "plane of all perspectives," in Leibniz's phrase—official nomination, the "entitlement" (*titre*) which, like the academic qualification (*titre scolaire*) is valid on all markets and which, as an official definition of official identity, rescues its holders from the symbolic struggle of all against all, by uttering the authorized, universally recognized perspective on all social agents. The state, which produces the official classifications, is in a sense the supreme tribunal to which Kafka was referring in *The Trial* when he had Block say of the advocate and his claim to be one of the "great advocates": "Naturally anyone can call himself 'great' if he wants to, but in such matters it is the practices of the court that decide." The fact is that scientific analysis does not have to choose between perspectivism and what has to be called absolutism; the truth of the social world is the stake in a struggle between agents very unequally equipped to achieve absolute, i.e. self-fulfilling, vision and pre-vision.

One could analyse in this light the functioning of an institution like the French national statistics office, INSEE, a state institute which produces official taxonomies, invested with quasi-legal authority, particularly, in relations between employers and employees, that of the *title,* capable of conferring rights independent of actual productive activity. In so doing, it tends to fix the hierarchies and thus to sanction and consecrate a power relationship between the agents with respect to the names of trades and occupations, an essential component of social identity. The management of names is one of the ways of managing material scarcity, and the names of groups, especially occupational groups, record a state of the struggles and bargaining over official designations and the material and symbolic advantages associated with them. The occupational name which is conferred on agents, the title they are given, is one of the positive or negative retributions (on the same footing as their salary), inasmuch as it is a *distinctive mark* (an emblem or stigma) which receives its *value* from its position in a hierarchically organized system of titles and which thereby helps to determine the relative positions of agents and groups. Consequently, agents have recourse to practical or symbolic strategies aimed at maximizing the symbolic profit of naming: for example, they may decline the economic gratifications provided by one job in order to occupy a less well-paid but more prestigiously named position; or they may try to move towards positions whose designation is less precise and so escape the effects of symbolic devaluation. Similarly, in stating their personal identity, they may give themselves a name which includes them in a class sufficiently broad to include agents occupying positions superior to their own: for example, in France, a primary school teacher, an *instituteur,* may refer to himself as an *enseignant,* thereby implying that he might be a *lycée* teacher or a university teacher. More generally, they always have a choice between several names and they can play on the uncertainties and the effects of vagueness linked to the plurality of perspectives so as to try to escape the verdict of the official taxonomy.

But the logic of official naming is most clearly seen in the case of all the symbolic property rights which in French are called *titres*—titles of nobility, educational qualifications, professional titles. Titles are symbolic capital, socially and even legally recognized. The noble is not just someone who is known (*nobilis*), noteworthy, well-regarded, recognized, he is someone recognized by an *official,* "universal" tribunal, in other words known and recognized by all. The professional or academic title is a kind of legal rule of social perception, a "being- perceived" guaranteed as a right. It is symbolic capital in an institutionalized, legal (and no longer merely legitimate) form. Increasingly inseparable from the academic qualification, since the educational system increasingly tends to represent the ultimate and only guarantee of professional titles, it has a value in itself and, although it is a "common noun," it functions like a "great name" (the name of a great family or a proper name), securing all sorts of symbolic profits (and assets that cannot be obtained directly with money).[iii] It is the symbolic scarcity of the title in the space of the names of professions that tends to govern the rewards of the occupation (and not the relationship between the supply of and demand for a particular form of labour). It follows from this that the rewards of the title tend to acquire autonomy with respect to the rewards of labour. Thus the same work may receive different remuneration depending on the titles of the person who does it (e.g. tenured, official post-holder (*titulaire*) as opposed to a part-timer (*intérimaire*) or someone "acting" (*faisant fonction*) in that capacity, etc.). Since the title is in itself an institution (like language) that is more durable than the intrinsic characteristics of the work, the rewards of the title may be maintained despite changes in the work and its relative value. It is not relative value of the work that determines the value of the name, but the institutionalized value of the title that can be used as a means of defending or maintaining the value of the work.[iv]

This means that one cannot conduct a science of classifications without conducting a science of the struggle over classifications and without taking account of the position occupied, in this struggle over the power of knowledge, for power through knowledge, for the monopoly of legitimate symbolic violence, by each of the agents or groups of agents who are involved in it, whether they be ordinary individuals, exposed to the vicissitudes of the everyday symbolic struggle, or authorized (and full-time) professionals, which includes all those who speak or write about the social classes, and who are distinguished according to the greater or lesser extent to which their classifications commit the authority of the state, the holder of the monopoly of *official naming,* correct classification, the correct order.

While the structure of the social world is defined at every moment by the structure of the distribution of the capital and profits characteristic of the different particular fields, the fact remains that in each of these arenas, the very definition of the stakes and of the "trump

---

[iii]Entry into an occupation endowed with a title is increasingly subordinated to possession of an educational qualification (*titre scolaire*), and there is a close relationship between educational qualifications and remuneration, in contrast to untitled occupations in which agents doing the same work may have very different qualifications.

[iv]The possessors of the same title tend to constitute themselves into a group and to equip themselves with permanent organizations—medical associations, alumni associations, etc. intended to ensure the group's cohesion—periodic meetings, etc. and to promote its material and symbolic interests.

cards" can be called into question. Every field is the site of more or less overt struggle over the definition of the legitimate principles of division of the field. The question of legitimacy arises from the very possibility of this questioning, of a break with the *doxa* which takes the ordinary order for granted. Having said this, the symbolic strength of the participants in this struggle is never completely independent of their position in the game, even if the specifically symbolic power of nomination constitutes a strength relatively independent of the other forms of social power. The constraints of the necessity inscribed in the very structure of the different fields continue to bear on the symbolic struggles aimed at conserving or transforming that structure. The social world is, to a large extent, what the agents make of it, at each moment; but they have no chance of un-making and re-making it except on the basis of realistic knowledge of what it is and of what they can do with it from the position they occupy within it.

In short, scientific work aims to establish adequate knowledge both of the space of objective relations between the different positions constituting the field and of the necessary relations set up, through the mediation of the *habitus* of their occupants, between these positions and their corresponding stances (*prises de position*), that is to say, between the points occupied within that space and the points of view on that very space, which play a part in the reality and the evolution of that space. In other words, the objective delimitation of constructed classes, i.e. of *regions* of the constructed space of positions, makes it possible to understand the principle and the efficacy of the classificatory strategies by means of which agents seek to conserve or modify this space, in the forefront of which is the constituting of groups organized with a view to defending their members' interests.

Analysis of the struggle over classifications brings to light the political ambition which pervades the epistemic ambition of producing the correct classification—the ambition which defines the rex, to whom it falls, according to Emile Benveniste, to *regere fines* and *regere sacra,* to set forth the frontier between the sacred and the profane, good and evil, the vulgar and the distinguished. If he is not to make social science merely a way of pursuing politics by other means, the sociologist must take as his object the intention of assigning others to classes and of telling them thereby what they are and what they have to be (this is the whole ambiguity of forecasting); he must analyse, in order to repudiate, the ambition of the creative world view, a kind of *intuitus*

originarius that would make things exist in accordance with its vision (this is the whole ambiguity of the Marxist conception of class, which is inextricably an "is" and an "ought"). He must objectify the ambition of objectifying, of classifying from outside, objectively, agents who struggle to classify others and to classify themselves. If he does classify—by making divisions, for the purposes of statistical analysis, in the continuous space of social positions—he does so precisely so as to be able to objectify *all* forms of objectification, from the particular insult to the official nomination, not forgetting the claim, characteristic of science in its positivist, bureaucratic, definition, to arbitrate in these struggles in the name of "axiological neutrality." The symbolic power of agents, understood as the power to make things seen—*theorein*—and to make things believed, to produce and impose the legitimate or legal classification, in fact depends, as the case of the *rex* reminds us, on the position occupied in the space (and in the classifications potentially inscribed in it). But objectifying objectification means, above all, objectifying the field of production of objectified representations of the social world, in particular, of the law-making taxonomies, in a word, the field of cultural or ideological production, a space and a game in which the social scientist himself is caught, like all those who argue about the social classes (and who else talks about them?).

THE POLITICAL FIELD AND THE EFFECT OF THE HOMOLOGIES

It is this field of political struggles, in which the professional practitioners of representation, in all senses of the word, clash with one another over another field of struggles, that has to be analysed if one wants to understand (without subscribing to the mythology of the "awakening of consciousness") the shift from the practical sense of the position occupied, itself amenable to being made explicit in different ways, to specifically political manifestations. Those who occupy the dominated positions within the social space are also located in dominated positions in the field of symbolic production, and it is not clear where they could obtain the instruments of symbolic production that are needed in order to express their specific viewpoint on the social space, were it not that the specific logic of the field of cultural production, and the particular interests that are generated within it, have the effect of inclining a fraction of the professionals involved in this field to supply the dominated, on the basis of homology of position, with the means of challenging the representations that arise from the immediate complicity between social structures and mental structures, and which tend to ensure the continuous reproduction of the distribution of symbolic capital. . . .

The inadequacies of the Marxist theory of classes, in particular its inability to explain the set of objectively observed differences, stems from the fact that, in reducing the social world solely to the economic field, it is forced to define social position solely in terms of position in the relations of economic production and consequently ignores positions in the different fields and sub-fields, particularly in the relations of cultural production, as well as all the oppositions that structure the social field, which are irreducible to the opposition between owners and non-owners of the means of economic production. It thereby secures a one-dimensional social world, simply organized around the opposition between two blocs (and one of the major questions is then that of the *boundary* between these two blocs, with all the associated, endlessly debated, questions of the "labour aristocracy," the "embourgeoisement" of the working class, etc.). In reality, the social space is a multi-dimensional space, an open set of fields that are relatively autonomous, i.e. more or less strongly and directly subordinated, in their functioning and their transformations, to the field of the economic production. Within each of these sub-spaces, the occupants of the dominated positions are constantly engaged in struggles of different forms (without necessarily constituting themselves into antagonistic groups).

But the most important thing, from the standpoint of the problem of breaking the circle of symbolic production, is the fact that, on the basis of the homologies between positions within different fields (and the invariant, or indeed universal, content of the relationship between the dominant and the dominated), *alliances* can be set up which are more or less lasting and always based on a more or less conscious misunderstanding. Homology of position between intellectuals and industrial workers—with the former occupying within the field of power, i.e. vis-à-vis industrial and commercial employers, positions which are homologous to those which industrial workers occupy within the social space as a whole—is the basis of an ambiguous alliance, in which the cultural producers, dominated agents among the dominant, divert their accumulated cultural capital so as to offer to the dominated the means of objectively constituting their view of the world and the representation of their interest in an explicit theory and in institutionalized instruments of representation—trade-union organizations, parties, social technologies for mobilization and demonstration, etc. . . .

But above all it would become clear that the effect of the specific interests associated with the position they occupy in the field and in the competition to impose views of the social world, inclines professional theoreticians and spokesmen, i.e. all those who are called in everyday language "full-time" or "permanent" officials, to produce differentiated, distinctive products which, because of the homology between the field of professional producers and the field of the consumers of opinions, are quasi-automatically adjusted to the different forms of demand—this demand being defined, especially in this case, as a demand for difference, for opposition, which they actually help to produce by helping it to find expression. It is the structure of the political field, in other words the objective relationship to the occupants of the other positions, and the relationship to the competing stances which they offer, which, as much as the direct relationship to their mandators, determines the stances they take, i.e. the supply of political products. Because the interests directly involved in the struggle for the monopoly of the legitimate expression of the truth of the social world tend to be the specific equivalent of the interests of the occupants of homologous positions in the social field, political discourses have a sort of structural duplicity. They seem to be directly addressed to the mandators, but in reality they are aimed at competitors within the field. . . .

CLASS AS REPRESENTATION AND AS WILL

But to establish how the power to constitute and institute that is held by the authorized spokesman—a party leader or trade-union leader, for example—is itself constituted and instituted, it is not sufficient to give an account of the specific interests of the theoreticians or spokesmen and of the structural affinities that link them to their mandators. One must also analyze the logic of the process of institution, which is ordinarily perceived and described as a process of delegation, in which the mandated representative receives from the group the power to make the group. Here, making the necessary transpositions, we may follow the historians of law (Kantorowicz, Post and others), when they describe the mystery of "ministry"—the *mysterium* of *ministerium,* a play on words much favoured by the canonists. The mystery of the process of transubstantiation whereby the spokesman becomes the group that he expresses can only be understood through a historical analysis of the genesis and functioning of *representation,* through which the representative makes the group he represents. The spokesman endowed with full power to speak and act in the name of the group, and first of all to act on the group through the magic of the slogan, the password (*mot d'ordre*), is the substitute of the group which exists only through this surrogacy. Personifying a fictitious person, a social fiction, he raises those whom he represents from the state of separate individuals, enabling them to act and speak, through him, as one man. In exchange, he receives the right to take himself for the group, to speak and act as if he were the group made man: "*status est magistratus,*" "*l'etat c'est moi,*" "the union thinks that . . . ," etc.

The mystery of ministry is one of those cases of social magic in which a thing or a person becomes something other than what it or he is, so that a man (a government minister, a bishop, a delegate, a member of parliament, a general secretary, etc.) can identify himself, and be identified, with a set of men, the People, the Workers, etc. or a social entity, the Nation, the State, the Church, the Party. The mystery of ministry culminates when the group can only exist through delegation to a spokesman who will make it exist by speaking for it, i.e. on its behalf and in its place. The circle is then complete: the group is made by the man who speaks in its name, who thus appears as the source of the power which he exerts on those who are its real source. This circular relationship is the root of the charismatic illusion in which, in extreme cases, the spokesman can appear to himself or herself and others as *causa sui.* Political alienation arises from the fact that isolated agents—the more so the less strong they are symbolically—cannot constitute themselves as a group, i.e. as a force capable of making itself heard in the political field, except by dispossessing themselves in favour of an apparatus; in other words from the fact that one always has to risk political dispossession in order to escape political dispossession. Fetishism, according to Marx, is what happens when "the products of the human brain appear as autonomous figures endowed with a life of their own"; political fetishism lies precisely in the fact that the value of the hypostatized individual, a product of the human brain, appears as charisma, a mysterious objective property of the person, an impalpable charm, an unnameable mystery. The minister—a minister of religion or a minister of

state—is related metonymically to the group; a part of the group, he functions as a sign in place of the whole of the group. It is he who, as an entirely real for an entirely symbolic being, induces a "category mistake," as Ryle would have said, rather like that of the child who, after seeing the soldiers composing a regiment march past, asks where the regiment is. By his mere visible existence, he constitutes the pure serial diversity of the separate individuals (*collectio personarum plurium*) into an "artificial person" (*une personne morale*), a *corporatio,* a constituted body, and, through the effect of mobilization and demonstration, he may even make it appear as a social agent.

Politics is the site par excellence of symbolic efficacy, the action that is performed through signs capable of producing social things and, in particular, groups. Through the potency of the oldest of the metaphysical effects linked to the existence of a symbolism, the one that enables one to regard as really existing everything that can be *symbolized* (God or non-being), political representation produces and reproduces at every moment a derived form of the case of the bald king of France, so dear to the logicians: any predicative proposition having "the working class" as its subject disguises an existential proposition (*there* is a working class). More generally, all utterances which have as their subject a collective noun—people, class, university, school, state, etc.—presuppose the existence of the group in question and conceal the same sort of metaphysical boot-strapping that was denounced in the ontological argument. The spokesman is he who, in speaking of a group, on behalf of a group, surreptitiously posits the existence of the group in question, institutes the group, through the magical operation that is inherent in any act of naming. That is why one must perform a critique of political reason, which is intrinsically inclined to abuses of language that are also abuses of power, if one wants to pose the question with which all sociology ought to begin, that of the existence and the mode of existence of collectives.

A class exists insofar—and only insofar—as mandated representatives endowed with *plena potestas agendi* can be and feel authorized to speak in its *name*—in accordance with the question "the Party is the working class," or "the working class is the Party," a formula which reproduces the canonists' equation: "The Church is the Pope (or the Bishops), the Pope is (or the Bishops are) the Church"—and so to make it exist as a real force within the political field. The mode of existence of what is nowadays called, in many societies (with variations, of course), "the working class," is entirely paradoxical: it is a sort of *existence in thought,* an existence in the thinking of a large proportion of those whom the taxonomies designate as workers, but also in the thinking of the occupants of the positions remotest from the workers in the social space. This almost universally recognized existence is itself based on the existence of a *working class in representation,* i.e. of political and trade-union apparatuses and professional spokesmen, vitally interested in believing that it exists and in having this believed both by those who identify with it and those who exclude themselves from it, and capable of making the "working class" *speak,* and with one voice, of invoking it, as one invokes gods or patron saints, even of symbolically manifesting it through *demonstration,* a sort of theatrical deployment of the class-in-representation, with on the one hand the corps of professional representatives and all the symbolism constitutive of its existence, and on the other the most convinced fraction of the believers who, through their presence, enable the representatives to manifest their representativeness. This working class "as will and representation" (in the words of Schopenhauer's famous title) is not the self-enacting class, a real group really mobilized, which is evoked in the Marxist tradition. But it is no less real, with the magical reality which (as Durkheim and Mauss maintained) defines institutions as social fictions. It is a "mystical body," created through an immense historical labour of theoretical and practical invention, starting with that of Marx himself, and endlessly re-created through the countless, constantly renewed, efforts and energies that are needed to produce and reproduce belief and the institution designed to ensure the reproduction of belief. It exists in and through the corps of mandated representatives who give it material speech and visible presence, and in the belief in its existence which this corps of plenipotentiaries manages to enforce, by its sheer existence and by its representations, on the basis of the affinities objectively uniting the members of the same "class on paper" as a probable group.

Introduction to "Outline of a Sociological Theory of Art Perception"

In this essay, Bourdieu examines artistic competence as a symbolic asset that legitimates relations of social domination. As an asset, the ability to appropriately decipher and appreciate works of art functions as a sign of distinction that separates those who possess the ability from those who do not. Individuals are not born with this competence; rather, they acquire it through exposure to and internalization of uniquely artistic interpretive schemes. These schemes allow the viewer to understand works of art in an aesthetic manner that is freed from the "necessities" of everyday reality. However, the connection between "being" cultured and "acquiring" culture is seldom recognized. Instead, the link between artistic competence and education is denied by a "charismatic ideology" that misrecognizes social privilege as a gift of nature. Thus, while art is made accessible to all, not everyone is equally comfortable in its presence. To the extent that we understand the "unfortunate" as simply being born without the blessings of grace that allow an individual's spirit to be touched by works of art, we fail to uncover the ideological functions of culture. Bourdieu speaks directly to the ideological or political dimensions of art when he notes,

> By symbolically shifting the essential of what sets them apart from other classes from the economic field to that of culture, or rather, by adding to strictly economic differences, namely those created by the simple possession of material wealth, differences created by the possession of symbolic wealth such as works of art, or by the pursuit of symbolic distinctions in the manner of using such wealth (economic or symbolic) . . . the privileged members of bourgeois society replace the difference between two cultures, historic products of social conditions, by the essential difference between two natures, a naturally cultivated nature and a naturally

natural nature. Thus, the sacralization of culture and art fulfills a vital function of contributing to the consecration of the social order: to enable educated people to believe in barbarism and persuade the barbarians within the gates of their own barbarity, all they must and need do is manage to conceal themselves and to conceal the social conditions which render possible not only culture as a second nature in which society recognizes human excellence or "good form" as the "realization" in a habitus of the aesthetics of the ruling classes, but also the legitimized dominance . . . of a particular definition of culture. (Bourdieu 1968/1993:236)

To reproduce their privileged position, the "civilized" need only convince the "barbarians" that the conditions that produce the culturally gifted and ungifted are but the expression of a state of nature that condemns both to their destinies.

"Outline of a Sociological Theory of Art Perception" (1968)

Pierre Bourdieu

I

Any art perception involves a conscious or unconscious deciphering operation.

An act of deciphering *unrecognized as such,* immediate and adequate "comprehension," is possible and effective only in the special case in which the cultural code which makes the act of deciphering possible is immediately and completely mastered by the observer (in the form of cultivated ability or inclination) and merges with the cultural code which has rendered the work perceived possible. . . .

The question of the conditions that make it possible to experience the work of art (and, in a more general way, all cultural objects) as at once endowed with meaning is totally excluded from the experience itself, because the recapturing of the work's objective meaning (which may have nothing to do with the author's intention) is completely adequate and immediately effected in the case— and only in the case—where the culture that the originator puts into the work is identical with the culture or, more accurately, the *artistic competence* which the beholder brings to the deciphering of the work. In this case, everything is a matter of course and the question of the meaning, of the deciphering of the meaning and of the conditions of this deciphering does not arise.

Whenever these specific conditions are not fulfilled, misunderstanding is inevitable: the illusion of immediate comprehension leads to an illusory comprehension based on a mistaken code. In the absence of the perception that the works are coded, and coded in another code, one unconsciously applies the code which is good for everyday perception, for the deciphering of familiar objects, to works in a foreign tradition. There is no perception which does not involve an unconscious code and it is essential to dismiss the myth of the "fresh eye," considered a virtue attributed to naïveté and innocence. One of the reasons why the less educated beholders in our societies are so strongly inclined to demand a realistic representation is that, being devoid of specific categories of perception, they cannot apply any other code to works of scholarly culture than that which enables them to apprehend as meaningful objects of their everyday environment. Minimum, and apparently immediate, comprehension, accessible to the simplest observers and enabling them to recognize a house or a tree, still presupposes partial (unconscious) agreement between artist and beholder concerning categories that define the representation of the real that a historic society holds to be "realistic." . . .

The spontaneous theory of art perception is founded on the experience of familiarity and immediate comprehension—an unrecognized special case.

Educated people are at home with scholarly culture. They are consequently carried towards that kind of ethnocentrism which may be called class-centrism and which consists in considering as natural (in other words, both as a matter of course and based on nature) a way of perceiving which is but one among other possible ways and which is acquired through education that may be diffuse or specific, conscious or unconscious, institutionalized or non-institutionalized. "When, for instance, a man wears a pair of spectacles which are so close to him physically that they are 'sitting on his nose,' they are environmentally more remote from him than the picture on the opposite wall. Their proximity is normally so weakly perceived as to go unnoticed." Taking Heidegger's analysis metaphorically, it can be said that the illusion of the "fresh eye" as a "naked eye" is an attribute of those who wear the spectacles of culture and who do not see that which enables them to see, any more than they see what they would not see if they were deprived of what enables them to see.

Conversely, faced with scholarly culture, the least sophisticated are in a position identical with that of ethnologists who find themselves in a foreign society and present, for instance, at a ritual to which they do not hold the key. The disorientation and cultural blindness of the less-educated beholders are an objective reminder of the objective truth that art perception is a mediate deciphering operation. Since the information presented by the works exhibited exceeds the deciphering capabilities of the beholder, he perceives them as devoid of signification—or, to be more precise, of structuration and organization—because he cannot "decode" them, i.e. reduce them to an intelligible form.

Scientific knowledge is distinguished from naïve experience (whether this is shown by disconcertment or by immediate comprehension) in that it involves an awareness of the conditions permitting adequate perception. The object of the science of the work of art is that which renders possible both this science and the immediate comprehension of the work of art, that is, culture. It therefore includes, implicitly at least, the science of the difference between scientific knowledge and naïve perception. . . .

II

The work of art considered as a symbolic good (and not as an economic asset, which it may also be) only exists as such for a person who has the means to appropriate it, or in other words, to decipher it.[i]

The degree of an agent's art competence is measured by the degree to which he or she masters the set of instruments for the appropriation of the work of art, available at a given time, that is to say, the interpretation schemes which are the prerequisite for the appropriation of art capital or, in other words, the prerequisite for the deciphering of works of art offered to a given society at a given moment.

Art competence can be provisionally defined as the preliminary knowledge of the possible divisions into complementary classes of a universe of representations. A mastery of this kind of system is necessarily determined in relation to another class, itself constituted by all the art representations consciously or unconsciously taken into consideration which do not belong to the class in question. The *style* proper to a period and to a social group is none other than such a class defined in relation to all the works of the same universe which it excludes and which are complementary to it. The *recognition* (or, as the art historians say when using the vocabulary of logic, the *attribution*) proceeds by *successive elimination* of the possibilities to which the class is—negatively—related and to which the possibility which has become a reality in the work concerned belongs. It is immediately evident that the uncertainty concerning the different characteristics likely to be attributed to the work under consideration (authors, schools, periods, styles, subjects, etc.) can be removed by employing different codes, functioning as classification systems; it may be a case of a properly artistic code which, by permitting the deciphering of specifically stylistic characteristics, enables the work concerned to be assigned to the class formed by the whole of the works of a period, a society, a school or an author ("that's a Cézanne"), or a code from everyday life which, in the form of previous knowledge of the possible divisions into complementary classes of the universe of signifiers and of the universe of signifieds, and of the correlations between the divisions of the one and the divisions of the other, enables the particular representation, treated as a sign, to be assigned to a class of

---

[i]The laws governing the reception of works of art are a special case of the laws of cultural diffusion: whatever may be the nature of the message—religious prophecy, political speech, publicity image, technical object—reception depends on the categories of perception, thought and action of those who receive it. In a differentiated society, a close relationship is therefore established between the nature and quality of the information transmitted and the structure of the public, its "readability" and its effectiveness being all the greater when it meets as directly as possible the expectations, implicit or explicit, which the receivers owe chiefly to their family upbringing and social circumstances (and also, in the matter of scholarly culture at least, to their school education) and which the diffuse pressure of the reference group maintains, sustains and reinforces by constant recourse to the norm.

signifiers and consequently makes it possible to know, by means of the correlations with the universe of signifieds, that the corresponding signified belongs to a certain class of signifieds ("that's a forest"). In the first case the beholder is paying attention to the manner of treating the leaves or the clouds, that is to say to the stylistic indications, locating the possibility realized, characteristic of one class of works, by reference to the universe of stylistic possibilities; in the other case, she is treating the leaves or the clouds as indications or signals associated, according to the logic set forth above, with significations transcendent to the representation itself ("that's a poplar," "that's a storm").

Artistic competence is therefore defined as the previous knowledge of the strictly artistic principles of division which enable a representation to be located, through the classification of the stylistic indications which it contains, among the possibilities of representation constituting the universe of art and not among the possibilities of representation constituting the universe of everyday objects or the universe of signs, which would amount to treating it as a mere monument, i.e. as a mere means of communication used to transmit a transcendent signification. The perception of the work of art in a truly aesthetic manner, that is, as a signifier which signifies nothing other than itself, does not consist of considering it "without connecting it with anything other than itself, either emotionally or intellectually," in short of giving oneself up to the work apprehended in its irreducible singularity, but rather of noting its *distinctive stylistic features* by relating it to the ensemble of the works forming the class to which it belongs, and to these works only. On the contrary, the taste of the working classes is determined, after the manner of what Kant describes in his *Critique of Judgement* as "barbarous taste," by the refusal or the impossibility (one should say the impossibility-refusal) of operating the distinction between "what is liked" and "what pleases" and, more generally, between "disinterestedness," the only guarantee of the aesthetic quality of contemplation, and "the interest of the senses" which defines "the agreeable" or "the interest of reason": it requires that every image shall fulfil a function, if only that of a sign. This "functionalist" representation of the work of art is based on the refusal of gratuitousness, the idolatry of work or the placing of value on what is "instructive" (as opposed to what is "interesting") and also on the impossibility of placing each individual work in the universe of representations, in the absence of strictly stylistic principles of classification. . . .

Being an historically constituted system, founded on social reality, this set of instruments of perception whereby a particular society, at a given time, appropriates artistic goods (and, more generally, cultural goods) does not depend on individual wills and consciousnesses and forces itself upon individuals, often without their knowledge, defining the distinctions they can make and those which escape them. Every period arranges artistic representations as a whole according to an institutional system of classification of its own, bringing together works which other periods separated, or distinguishing between works which other periods placed together, and individuals have difficulty in imagining differences other than those which the available system of classification allows them to imagine. . . .

III

Since the work of art only exists as such to the extent that it is perceived, or, in other words, deciphered, it goes without saying that the satisfactions attached to this perception—whether it be a matter of purely aesthetic enjoyment or of more indirect gratification, such as the *effect of distinction*—are only accessible to those who are disposed to appropriate them because they *attribute a value to them,* it being understood that they can do this only if they have the means to appropriate them. Consequently, the need to appropriate goods which, like cultural goods, only exist as such for those who have received the means to appropriate them from their family environment and school, can appear only in those who can satisfy it, and it can be satisfied as soon as it appears. . . .

The disposition to appropriate cultural goods is the product of general or specific education, institutionalized or not, which creates (or cultivates) art competence as a mastery of the instruments for appropriation of these goods, and which creates the "cultural need" by giving the means to satisfy it.

The repeated perception of works of a certain style encourages the unconscious internalization of the rules that govern the production of these works. Like rules of grammar, these rules are not apprehended as such, and are still less explicitly formulated and capable of being formulated; for instance, lovers of classical music may have neither awareness nor knowledge of the laws obeyed by the sound-making art to which they are accustomed, but their auditive education is such that, having heard a dominant chord, they are induced urgently to await the tonic which seems to them the "natural" resolution of this chord, and they have difficulty in apprehending the internal coherence of music founded on other principles. The unconscious mastery of the instruments of appropriation which are the basis of familiarity with cultural works is acquired by slow familiarization, a long succession of "little perceptions," in the sense in which Leibniz uses the expression. Connoisseurship is an "art" which, like the art of thinking or the art of living, cannot be imparted entirely in the form of precepts or instruction, and apprenticeship to it presupposes the equivalent of prolonged contact between disciple and initiate in traditional education, i.e. repeated contact with the work (or with works of the same class). And, just as students or disciples can *unconsciously* absorb the rules of the art—including those which are not explicitly known to the initiates themselves—by giving themselves up to it, excluding analysis and the selection of elements of exemplary conduct, so art-lovers can, by abandoning themselves in some way to the work, internalize the principles and rules of its construction without there ever being brought to their consciousness and formulated as such. This constitutes the difference between the art theorist and the connoisseur, who is usually incapable of explicating the principles on which his judgements are based. In this field as in others (learning the grammar of one's native tongue, for instance), school education tends to encourage the conscious reflection of patterns of thought, perception or expression which have already been mastered unconsciously by formulating explicitly the principles of the creative grammar, for example, the laws of harmony and counterpoint or the rules of pictorial composition, and by providing the verbal and conceptual material essential for naming differences previously experienced in a purely intuitive way. . . .

Even when the educational institution makes little provision for art training proper (as is the case in France and many other countries), even when, therefore, it gives neither specific encouragement to cultural activities nor a body of concepts specifically adapted to the plastic arts, it tends on the one hand to inspire a certain *familiarity*—conferring a feeling of belonging to the cultivated class—with the world of art, in which people feel at home and among themselves as the appointed addressees of works which do not deliver their message to the first-comer; and on the other to inculcate (at least in France and in the majority of European countries, at the level of secondary education) a *cultivated disposition* as a durable and generalized attitude which implies recognition of the value of works of art and the ability to appropriate them by means of generic categories.[ii]

Although it deals almost exclusively with literary works, in-school learning tends to create on the one hand a transposable inclination to admire works approved by the school and a duty to admire and to love certain works or, rather, certain classes of works which gradually seem to become linked to a certain educational and social status; and, on the other hand, an equally generalized and transposable aptitude for categorizing by authors, by genres, by schools and by periods, for the handling of educational categories of literary analysis and for the mastery of the code which governs the use of the different codes, giving at least a tendency to acquire equivalent categories in other fields and to store away the typical knowledge which, even though extrinsic and anecdotal, makes possible at least an elementary form of apprehension, however inadequate it may be. Thus, the first degree of strictly pictorial competence shows itself in the mastery of an arsenal of words making it possible to name differences and to apprehend them while naming them: these are the proper names of famous painters—da Vinci, Picasso, Van Gogh—which function as generic categories, because one can say about any painting or non-figurative object "that suggests Picasso," or, about any work recalling nearly or distantly the manner of the Florentine painter, "that looks like a da Vinci"; there are also broad categories, like "the Impressionists" (a school commonly considered to include Gaugin, Cézanne and Degas), "the Dutch School," "the Renaissance." It is particularly significant that the proportion of subjects who think in terms of schools very clearly grows as the level of education rises and that, more generally, generic knowledge which is required for the perception of differences and consequently for memorizing—proper names and historical, technical or aesthetic concepts—becomes increasingly specific as we go towards the more educated beholders, so that the most adequate perception differs only from the least adequate in so far as the specificity, richness and subtlety of the categories employed are concerned. By no means contradicting these arguments is the fact that the less educated visitors to museums—who tend to prefer the most famous paintings and those sanctioned by school teaching, whereas modern painters who have the least chance of being mentioned in schools are quoted only by those with the highest educational qualifications—live in large cities. To be able to form discerning or so-called "personal" opinions is again a result of the education received: the ability to go beyond school constraints is the

---

[ii]School instruction always fulfils a function of legitimation, if only by giving its blessing to works which it sets up as worthy of being admired, and thus helps to define the hierarchy of cultural goods valid in a particular society at a given time.

privilege of those who have sufficiently assimilated school education to make their own the free attitude towards scholastic culture taught by a school so deeply impregnated with the values of the ruling classes that it accepts the fashionable depreciation of school instruction. The contrast between accepted, stereotyped and, as Max Weber would say, "routinized" culture, and genuine culture, freed from school discourse, has meaning only for an infinitely small minority of educated people for whom culture is second nature, endowed with all the appearances of talent, and the full assimilation of school culture is a prerequisite for going beyond it towards this "free culture"—free, that is to say, from its school origins—which the bourgeois class and its school regard as the value of values.

But the best proof that the general principles for the transfer of training also hold for school training lies in the fact that the practices of one single individual and, *a fortiori,* of individuals belonging to one social category or having a specific level of education, tend to constitute a system, so that a certain type of practice in any field of culture very probably implies a corresponding type of practice in all the other fields; thus, frequent visits to museums are almost necessarily associated with an equal amount of theatre-going and, to a lesser degree, attendance at concerts. Similarly, everything seems to indicate that knowledge and preferences tend to form into constellations that are strictly linked to the level of education, so that a typical structure of preferences in painting is most likely to be linked to a structure of preferences of the same type in music or literature.

Owing to the particular status of the work of art and the specific logic of the training which it implies, art education which is reduced to a discourse (historical, aesthetic or other) on the works is necessarily at a secondary level; like the teaching of the native tongue, literary or art education (that is to say "the humanities" of traditional education) necessarily presupposes, without ever, or hardly ever, being organized in the light of this principle, that individuals are endowed with a previously acquired competence and with a whole capital of experience unequally distributed among the various social classes (visits to museums or monuments, attending concerts, lectures, etc.).

In the absence of a methodical and systematic effort, involving the mobilization of all available means from the earliest years of school onwards, to procure for all those attending school a direct contact with the works or, at least, an approximate substitute for that experience (by showing reproductions or reading texts, organizing visits to museums or playing records, etc.), art education can be of full benefit only to those who owe the competence acquired by slow and imperceptible familiarization to their family milieu, because it does not explicitly give to all what it implicitly demands from all. While it is true that only the school can give the continuous and prolonged, methodical and uniform training capable of *mass production,* if I may use that expression, of competent individuals, provided with schemes of perception, thought and expression which are prerequisites for the appropriation of cultural goods, and endowed with that generalized and permanent inclination to appropriate them which is the mark of devotion to culture, the fact remains that the effectiveness of this formative action is directly dependent upon the degree to which those undergoing it fulfil the preliminary conditions for adequate reception: the influence of school activity is all the stronger and more lasting when it is carried on for a longer time (as is shown by the fact that the decrease of cultural activity with age is less marked when the duration of schooling was longer), when those upon whom it is exercised have greater previous competence, acquired through early and direct contact with works (which is well known to be more frequent always as one goes higher up the social scale) and finally when a propitious cultural atmosphere sustains and relays its effectiveness. Thus, humanities students who have received a homogeneous and homogenizing training for a number of years, and who have been constantly selected according to the degree to which they conform to school requirements, remain separated by systematic differences, both in their pursuit of cultural activities and in their cultural preferences, depending upon whether they come from a more or less cultivated milieu and for how long this has been so; their knowledge of the theatre (measured according to the average number of plays that they have seen on the stage) or of painting is greater if their father or grandfather (or, *a fortiori,* both of them) belongs to a higher occupational category; and, furthermore, if one of these variables (the category of the father or of the grandfather) has a fixed value, the other tends, by itself, to hierarchize the scores. Because of the slowness of the acculturation process, subtle differences linked with the length of time that they have been in contact with culture thus continue to separate individuals who are apparently equal with regard to social success and even educational success. Cultural nobility also has its quarterings.

Only an institution like the school, the specific function of which is methodically to develop or create the dispositions which produce an educated person and which lay the foundations, quantitatively and consequently qualitatively, of a constant and intense pursuit of culture, could offset (at least partially) the initial disadvantage of those who do not receive from their family circle the encouragement to undertake cultural activities and the competence presupposed in any discourse on works, on the condition—and only on the condition—that it employs every available means to break down the endless series of cumulative processes to which any cultural education is condemned. For if the apprehension of a work of art depends, in its intensity, its modality and in its very existence, on the beholders' mastery of the generic and specific code of the work, i.e. on their competence, which they owe partly to school training, the same thing applies to the pedagogic communication which is responsible, among its other functions, for transmitting the code of works of scholarly culture (and also the code according to which it effects this transmission). Thus the intensity and modality of the communication are here again a function of culture (as a system of schemes of perception, expression and historically constituted and socially conditioned thinking) which the receiver owes to his or her family milieu and which is more or less close to scholarly culture and the linguistic and cultural models according to which the school effects the transmission of this culture. Considering that the direct experience of works of scholarly culture and the institutionally organized acquisition of culture which is a prerequisite for adequate experience of such works are subject to the same laws, it is obvious how difficult it is to break the sequence of the cumulative effects which cause cultural capital to attract cultural capital. In fact, the school has only to give free play to the objective machinery of cultural diffusion without working systematically to give to all, in and through the pedagogical message itself, what is given to some through family inheritance—that is, the instruments which condition the adequate reception of the school message—for it to redouble and consecrate by its approval the socially conditioned inequalities of cultural competence, by treating them as natural inequalities or, in other words, as inequalities of gifts or natural talents.

Charismatic ideology is based on parenthesizing the relationship, evident as soon as it is revealed, between art competence and education, which alone is capable of creating both the disposition to recognize a value in cultural goods and the competence which gives a meaning to this disposition by making it possible to appropriate such goods. Since their art competence is the product of an imperceptible familiarization and an automatic transferring of aptitudes, members of the privileged classes are naturally inclined to regard as a gift of nature a cultural heritage

which is transmitted by a process of unconscious training. But, in addition, the contradictions and ambiguities of the relationship which the most cultured among them maintain with their culture are both encouraged and permitted by the paradox which defines the "realization" of culture as *becoming natural*. Culture is thus achieved only by negating itself as such, that is, as artificial and artificially acquired, so as to become second nature, a habitus, a possession turned into being; the virtuosi of the judgement of taste seem to reach an experience of aesthetic grace so completely freed from the constraints of culture and so little marked by the long, patient training of which it is the product that any reminder of the conditions and the social conditioning which have rendered it possible seems to be at once obvious and scandalous. It follows that the most experienced connoisseurs are the natural champions of charismatic ideology, which attributes to the work of art a magical power of conversion capable of awakening the potentialities latent in a few of the elect, and which contrasts authentic experience of a work of art as an "affection" of the heart or immediate enlightenment of the intuition with the laborious proceedings and cold comments of the intelligence, ignoring the social and cultural conditions underlying such an experience, and at the same time treating as a birthright the virtuosity acquired through long familiarization or through the exercises of a methodical training; silence concerning the social prerequisites for the appropriation of culture or, to be more exact, for the acquisition of art competence in the sense of mastery of all the means for the specific appropriation of works of art is a self-seeking silence because it is what makes it possible to legitimatize a social privilege by pretending that it is a gift of nature.

To remember that culture is not what one is but what one has, or rather, what one has become; to remember the social conditions which render possible aesthetic experience and the existence of those beings—art lovers or "people of taste"—for whom it is possible; to remember that the work of art is given only to those who have received the means to acquire the means to appropriate it and who could not seek to possess it if they did not already possess it, in and through the possession of means of possession as an actual possibility of effecting the taking of possession; to remember, finally, that only a few have the real possibility of benefitting from the theoretical possibility, generously offered to all, of taking advantage of the works exhibited in museums—all this is to bring to light the hidden force of the effects of the majority of culture's social uses.

The parenthesizing of the social conditions which render possible culture and culture become nature, cultivated nature, having all the appearances of grace or a gift and yet acquired, so therefore "deserved," is the precedent condition of charismatic ideology which makes it possible to confer on culture and in particular on "love of art" the all-important place which they occupy in bourgeois "sociodicy." The bourgeoisie find naturally in culture as cultivated nature and culture that has become nature the only possible principle for the legitimation of their privilege. Being unable to invoke the right of birth (which their class, through the ages, has refused to the aristocracy) or nature which, according to "democratic" ideology, represents universality, i.e. the ground on which all distinctions are abolished, or the aesthetic virtues which enabled the first generation of bourgeois to invoke their merit, they can resort to cultivated nature and culture become nature, to what is sometimes called "class," through a kind of tell-tale slip, to "education," in the sense of a product of education which seems to owe nothing to education,[iii] to distinction, grace which is merit and merit which is grace, an unacquired merit which justifies unmerited acquisitions, that is to say, inheritance. To enable culture to fulfil its primary ideological function of class co-optation and legitimation of this mode of selection, it is necessary and sufficient that the link between culture and education, which is simultaneously obvious and hidden, be forgotten, disguised and denied. The unnatural idea of inborn culture, of a gift of culture, bestowed on certain people by nature, is inseparable from blindness to the functions of the institution which ensures the profitability of the cultural heritage and legitimizes its transmission while concealing that it fulfils this function. The school in fact is the institution which, through its outwardly irreproachable verdicts, transforms socially conditioned inequalities in regard to culture into inequalities of success, interpreted as inequalities of gifts which are also inequalities of merit. Plato records, towards the end of *The Republic,* that the souls who are to begin another life must themselves choose their lot among "patterns of life" of all kinds and that, when the choice has been made, they must drink of the water of the river Lethe before returning to earth. The function which Plato attributes to the water of forgetfulness falls, in our societies, on the university which, in its impartiality, through pretending to recognize students as equal in rights and duties, divided only by inequalities

of gifts and of merit, in fact confers on individuals degrees judged according to their cultural heritage, and therefore according to their social status.

By symbolically shifting the essence of what sets them apart from other classes from the economic field to that of culture, or rather, by adding to strictly economic differences, namely those created by the simple possession of material goods, differences created by the possession of symbolic goods such as works of art, or by the pursuit of symbolic distinctions in the manner of using such goods (economic or symbolic), in short, by turning into a fact of nature everything which determines their "value," or to take the word in the linguistic sense, their *distinction*—a mark of difference which, according to the Littré, sets people apart from the common herd "by the characteristics of elegance, nobility and good form"—the privileged members of bourgeois society replace the difference between two cultures, historic products of social conditions, by the essential difference between two natures, a naturally cultivated nature and a naturally natural nature. Thus, the sacralization of culture and art fulfils a vital function by contributing to the consecration of the social order: to enable educated people to believe in barbarism and persuade the barbarians within the gates of their own barbarity, all they must and need do is to manage to conceal themselves and to conceal the social conditions which render possible not only culture as a second nature in which society recognizes human excellence or "good form" as the "realization" in a habitus of the aesthetics of the ruling classes, but also the legitimized dominance (or, if you like, the legitimacy) of a particular definition of culture. And in order that the ideological circle may be completely closed, all they have to do is to find in an essentialist representation of the bipartition of society into barbarians and civilized people the justification of their right to conditions which produce the possession of culture and the dispossession of culture, a state of "nature" destined to appear based on the nature of the men who are condemned to it.

If such is the function of culture and if it is love of art which really determines the choice that separates, as by an invisible and insuperable barrier, those who have from those who have not received this grace, it can be seen that museums betray, in the smallest details of their morphology and their organization, their true

---

[iii]It was understood thus by a very cultivated old man who declared during a conversation: "Education, Sir, is inborn."

function, which is to strengthen the feeling of belonging in some and the feeling of exclusion in others.[iv] Everything, in these civic temples in which bourgeois society deposits its most sacred possessions, that is, the relics inherited from a past which is not its own, in these holy places of art, in which the chosen few come to nurture a faith of virtuosi while conformists and bogus devotees come and perform a class ritual, old palaces or great historic homes to which the nineteenth century added imposing edifices, built often in the Greco-Roman style of civic sanctuaries, everything combines to indicate that the world of art is as contrary to the world of everyday life as the sacred is to the profane. The prohibition against touching the objects, the religious silence which is forced upon visitors, the puritan asceticism of the facilities, always scarce and uncomfortable, the almost systematic refusal of any instruction, the grandiose solemnity of the decoration and the decorum, colonnades, vast galleries, decorated ceilings, monumental staircases both outside and inside, everything seems done to remind people that the transition from the profane world to the sacred world presupposes, as Durkheim says, "a genuine metamorphosis," a radical spiritual change, that the bringing together of the worlds "is always, in itself, a delicate operation which calls for precaution and a more or less complicated initiation," that "it is not even possible unless the profane lose their specific characteristics, unless they themselves become sacred to some extent and to some degree."[v] Although the work of art, owing to its sacred character, calls for particular dispositions or predispositions, it brings in return its consecration to those who satisfy its demands, to the small elite who are self-chosen by their aptitude to respond to its appeal.

The museum gives to all, as a public legacy, the monuments of a splendid past, instruments of the sumptuous glorification of the great figures of bygone ages, but this is false generosity, because free entrance is also optional entrance, reserved for those who, endowed with the ability to appropriate the works, have the privilege of using this freedom and who find themselves consequently legitimized in their privilege, that is, in the possession of the means of appropriating cultural goods or, to borrow an expression of Max Weber, in the *monopoly* of the handling of cultural goods and of the institutional signs of cultural salvation (awarded by the school). Being the keystone of a system which can function only by concealing its true function, the charismatic representation of art experience never fulfills its function of mystifying so well as when it resorts to a "democratic" language: to claim that works of art have power to awaken the grace of aesthetic enlightenment in anyone, however culturally uninitiated he or she may be, to presume in all cases to ascribe to the unfathomable accidents of grace or to the arbitrary bestowal of "gifts" aptitudes which are always the product of unevenly distributed education, and therefore to treat inherited aptitudes as personal virtues which are both natural and meritorious. Charismatic ideology would not be so strong if it were not the only outwardly irreproachable means of justifying the right of the heirs to the inheritance without being inconsistent with the ideal of formal democracy, and if, in this particular case, it did not aim at establishing in nature the sole right of the bourgeoisie to appropriate art treasures to itself, to appropriate them to itself *symbolically,* that is to say, in the only legitimate manner, in a society which pretends to yield to all, "democratically," the relics of an aristocratic past.

[iv]It is not infrequent that working-class visitors explicitly express the feeling of exclusion which, in any case, is evident in their whole behaviour. Thus, they sometimes see in the absence of any indication which might facilitate the visit—arrows showing the direction to follow, explanatory panels, etc.—the signs of a deliberate intention to exclude the uninitiated. The provision of teaching and didactic aids would not, in fact, really make up for the lack of schooling, but it would at least proclaim the right not to know, the right to be there in ignorance, the right of the ignorant to be there, a right which everything in the presentation of works and in the organization of the museum combines to challenge, as this remark overheard in the Chateau of Versailles testifies: "This chateau was not made for the people, and it has not changed."

[v]E. Durkheim, *Les formes élémentaires de la vie religieuse,* 6th edn (Paris: Presses Universitaires de France, 1960), 55–6. The holding of a Danish exhibition showing modern furniture and utensils in the old ceramic rooms of the Lille museum brought about such a "conversion" in the visitors as can be summarized in the following contrasts, the very ones which exist between a department store and a museum: noise/silence; touch/see; quick, haphazard exploration, in no particular order/leisurely, methodical inspection, according to a fixed arrangement; freedom/constraint; economic assessment of works which may be purchased/aesthetic appreciation of "priceless" works. However, despite these differences, bound up with the things exhibited, the solemnizing (and distancing) effect of the museum no less continued to be felt, contrary to expectations, for the structure of the public at the Danish exhibition was more "aristocratic" (in respect of level of education) than the ordinary public of the museum. The mere fact that works are consecrated by being exhibited in a consecrated place is sufficient, in itself, profoundly to change their signification and, more precisely, to raise the level of their emission; were they presented in a more familiar place, a large emporium for instance, they would be more accessible.

JÜRGEN HABERMAS (1929–): A BIOGRAPHICAL SKETCH ▦

Jürgen Habermas was born in Dusseldorf, Germany, in 1929. He grew up in the small town of Gummersbach, where his father, a successful businessman, was the head of the local chamber of industry and commerce. Ten years old at the outbreak of World War II, Habermas served in the Hitler Youth, a compulsory organization that trained young boys for military service and young girls for motherhood. With the defeat of Germany and fall of Nazism, however, Habermas was forced to confront the atrocities committed under the Nazis' reign of terror. Horrified by the genocide carried out under Hitler's rule, Habermas viewed the end of the war as "a liberation, both historical and personal," as he came to realize that "it was a politically criminal system in which we lived" (interview, cited in Horster 1992:78–79).

Between 1949 and 1954, Habermas studied philosophy, history, and psychology at the universities of Göttingen, Zurich, and Bonn. In 1953, he published a provocative critique of the influential philosopher Martin Heidegger (1889–1976) that questioned how Heidegger's allegiance to the Nazi effort shaped his development of existential philosophy (Lalonde 1999:37, 38). The following year, he completed his dissertation, "The Absolute in History," on the work of Friedrich Wilhelm Schelling (1775–1854), a renowned German idealist philosopher. Toward the end of his studies, Habermas turned his attention to the writings of Marx and developed a fervent interest in the work associated with the Frankfurt School of critical theory. Greatly influenced by Georg Lukács' *History and Class Consciousness* and Horkheimer and Adorno's *Dialectic of Enlightenment* (see Chapter 10), Habermas served as Adorno's assistant from 1956 to 1959 at the Institute for Social Research, which by then had relocated to Frankfurt, Germany.

Despite his interests in critical theory, Habermas's intellectual position began to shift away from key tenets espoused by his mentors. He was particularly disillusioned by the earlier critical theorists' rejection of the public sphere of debate as a possible arena for democratic, progressive change and their despairing view of modern society more generally. In 1959, Habermas left the Institute and began a second doctorate ("Habilitation") at the University of Marburg. The following year, he joined the faculty at the University of Heidelberg, where he served as a professor of philosophy until 1964. He then returned to Frankfurt as professor of philosophy and sociology. He would eventually leave Frankfurt after disagreements with the student protest movement over what he believed to be its politically and intellectually extremist positions. Between 1971 and 1981, he lived in Starnberg, Bavaria, where he served as director of the Max Planck Institute for the Study of the Conditions of Life in the Scientific-Technical World. During this time, he wrote several works that were highly acclaimed, including *Knowledge and Human Interests* (1968) and *Legitimation Crisis* (1973). In 1980, Habermas won the Adorno Prize from the city of Frankfurt, and in 1982 he was named Extraordinary Professor of Philosophy in Heidelberg. That same year, he accepted a professorship at the University of Frankfurt, where he remained until his retirement in 1994.

Considered one of Europe's most important public intellectuals and described by *Der Spiegel* magazine as "the most intellectually powerful" philosopher in Germany (Horster 1992:3), Habermas is read not only by sociologists, philosophers, and political scientists, but also by politicians throughout the world. His most important works, *The Theory of Communicative Action, v. 1: Reason and the Rationalization of Society* (1984) and *The Theory of Communicative Action, v. 2: Lifeworld and System* (1987), continue to inspire democratic activists who seek to increase the scope and power of the public sphere.

HABERMAS'S INTELLECTUAL INFLUENCES AND CORE IDEAS ▦

Jürgen Habermas is commonly regarded as one of sociology's greatest synthetic thinkers (McCarthy 1978). He is also the most celebrated inheritor of the tradition of thought developed by the Frankfurt School (Kellner 2000; McCarthy 1978; Seidman 1989). As Habermas was a student of two of the School's founders—Max Horkheimer and Theodor Adorno (see Chapter 10)—his work is centrally informed by critical theory and the issues with which its leading figures grappled, particularly their inquiries into the role of reason in promoting a free and just society. However, Habermas's influences run much deeper into German intellectual history, for, like the critical theorists, he draws on the legacy of German Enlightenment thinkers whose roots date back to the eighteenth century. Incorporating the insights of Kant, Hegel, and Marx, he fashions a contemporary approach to the emancipatory project of social philosophy. This project envisions the creation of a social order in which reason and rationality provide the pathway for establishing economic equality, political democracy, and the liberation of the individual from systems of domination. These themes form the core of Habermas's own theoretical perspective, a perspective that, in addition to Marxist-inspired writers, incorporates the insights of a range of additional scholars including Weber, Durkheim, Mead, Husserl, Schutz, and Parsons, as well as linguistic philosophers and developmental psychologists.

Like Weber and the Frankfurt School theorists, Habermas (1981) contends that the modern world is characterized by increasing instrumental rationalization—that is, by the spread of methodical procedures and calculable rules into more and more domains of social and personal life. However, whereas the earlier critical theorists saw rationalization leading only to the corruption of the human spirit and the decay of civilization, Habermas finds modernity yielding mixed results. For while it is true that the possibilities for democracy and human emancipation have been distorted in modern societies, modernity has also brought the rule of law, the expansion of political and civil rights, and, if not the full realization of democracy, at least the hallowing of democratic principles of government. Moreover, rationalizing processes have led to advances in a variety of fields, from medical sciences and food production to architecture and the arts. Nevertheless, the growth of bureaucratic structures of social organization, complex political processes, sophisticated economic systems, and the rationalized means of mass-mediated communication has not been without its drawbacks. It is precisely this duality embodied in the rationalization of society—both facilitating and restricting the development of human knowledge, democratic organization, and ultimately liberation—that Habermas seeks to address.

In stark contrast to the Frankfurt School theorists, whose antimodernist stance and loss of faith in reason led them to abandon Marx's utopian commitment to emancipation, Habermas holds out great hope for the power of reason to combat the dehumanizing consequences stemming from the rationalization of society. His rekindled utopian vision, however, does not mark a return to "orthodox" Marxism's emphasis on capitalism as the primary cause for humanity's unfreedom. Instead, drawing from phenomenology, Habermas reconstructs Marxism to examine the *intersubjective* and *normative* dimensions of social life that Marx had largely ignored. For according to Habermas, it is not the form of *economic* or material reproduction of society (the relationship between individuals and their physical environment) but, rather, the form of its *symbolic* reproduction (processes of socialization, identity formation, and social integration) that poses the greatest threat to freedom and progress in modern society.

Lifeworld and System

Habermas's efforts to chart the historical development of societies and the course for emancipation begins with his discussion of the lifeworld and system. Drawing from and extending the work of Husserl and Schutz (see Chapter 13), Habermas conceives of the **lifeworld** as a prereflexive framework of background assumptions, a "network of shared meanings that individuals draw from to construct identities, to negotiate situational definitions, or to create social solidarity" (Seidman 1989:18). It consists of taken-for-granted cultural know-how, customs, and norms through which we are able to construct common understandings of our social world. In addition, the lifeworld provides for the socialization of society's members and the internalization of the norms and values essential to the stability of the social order.

The lifeworld serves as the backdrop for social integration, identity formation, and the construction of meaning in two domains: the private sphere of the family and the public sphere of open civic debate (see Figure 16.2). (We address the issue of the public sphere more fully later in this chapter.) Interaction within these domains is mediated through language and the development of shared meanings that in turn serves to reinforce or challenge the legitimacy of existing social norms and values.

A key component of modern societies, however, is the ongoing *rationalization* of the lifeworld in which the discourses regarding "truth, goodness, and beauty" are differentiated into separate spheres of

Figure 16.2   The Domains of Lifeworld and System

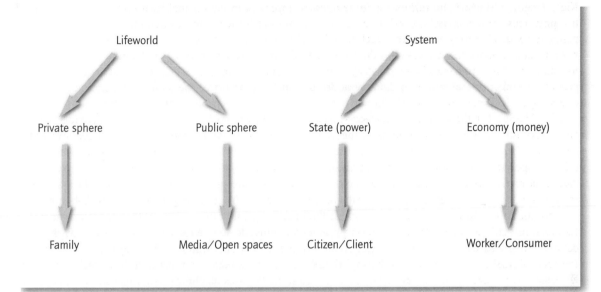

knowledge (Seidman 1989:24). Rationalization produces a greater potential to question our actions and the actions of others and to question more generally the conditions of the world around us. In the process, highly abstract concepts like democracy, equality, freedom, and universal rights circulate in debates that, in turn, create the conditions for a consensual understanding of the forces that shape social life.

Habermas maintains that his notion of the lifeworld helps correct Marx's reductive, one-sided theory of society. According to Habermas, Marx and his successors failed to recognize the significance of the symbolic and communicative domains of society, opting instead to emphasize the role of economic production and property relations both in generating conditions of exploitation and in sparking the eventual communist revolution. Nevertheless, to avoid developing his own one-sided, and thus incomplete, theory, Habermas (1981/1984, 1981/1987) introduces the notion of the "system" to address some of the concerns that Marx had earlier studied. The **system** comprises a society's political and economic structures that are responsible for the organization of power relations and the production and distribution of material resources. As societies evolve, both the state and the economy develop their own formal structure and mechanisms for self-organization. Habermas calls these organizational mechanisms **steering media** and argues that two primary forms emerge: **power** and **money**. As shown in Figure 16.2, while the state is organized around the distribution of power, the economy is organized around the circulation of money. What makes both power and money unique is that they are abstract, "delinguistified" media; they operate according to a goal-oriented, instrumental logic that lies outside of language and norm-confirming, consensus-producing debate.

In modern societies, individuals are related to the system in their roles as formal citizens and clients of the state and as economic workers and consumers. To the extent that the state (power) and the economy (money) shape social relations, they do so without the benefit of negotiated, consensual understandings of their impact on individuals and the broader society. This is because such understandings are made possible only through interactions that are grounded in the norm- and value-rich environment of the lifeworld. That power and money are impersonal, "meaningless" steering media without normative value is of crucial significance both for the maintenance of social stability and for the emancipation of individuals from domination. In advanced capitalist societies, families are linked to the economy through the medium of money: Monetary wages are paid for work, which are then spent to consume goods and services. At no point during this process do families exercise significant control over the terms of these exchanges. Meanwhile, individuals are related to the state less as informed citizens capable of effectively influencing political decisions than as "clients" who use the services (education, health care, social security, etc.) it provides. This client relation is reinforced as the state enters into more domains that formerly were free from political interference. Not only does the state intervene in the economy to minimize destabilizing fluctuations in the market (e.g., by creating a public sector or raising interest rates), but it also increasingly penetrates into private matters ordinarily addressed in the lifeworld. The state now regulates virtually every aspect of personal "choice," from marriage and biological reproduction to euthanasia.

In exchange for the goods and services purchased in the economy and the benefits provided by the state, individuals offer their obedience to the system. Mass loyalty and compliance are thus secured at the cost of forestalling genuine democratic participation in decisions that profoundly affect the most fundamental aspects of one's life. However, because loyalty and compliance can never be completely ensured, "legitimation crises" are endemic to the system. These crises are prone to erupt when the class biases of the political system are revealed. While the state is charged with promoting the general well-being of all its citizens, it often effects policies that further the interests of only the most privileged (for example, the financial reform bill recently signed into law); indeed, it must do so to ensure the continued growth of the economy. Nevertheless, such policies run counter to the democratic principles of egalitarianism underlying the political system, principles that politicians and state bureaucrats must uphold in order to confer legitimacy on state actions. To avoid such crises, the state provides its clients with material rewards in the form of various social welfare programs such as minimum wage increases, unemployment benefits, and Social Security. And by restricting the political participation of citizens to periodic voting, state officials are afforded "substantial freedom from public accountability" while securing the legitimacy of the state (Seidman 1989:21, 22).

This dynamic is vividly expressed in two current areas of government policy. First, the ongoing economic crisis, the worst in the country since the Great Depression, has left a record high of more than 14 million workers unemployed as of mid-2010. Moreover, many of those still employed are left with no choice but to accept steep cuts in pay and benefits, while federal, state, and local budgets for public programs such as education and health care are being dramatically reduced. The principal causes of the economic crisis lie in the unscrupulous investment and accounting practices of the nation's major banks and investment firms. To stem the tide of the fallout and prevent what some observers allege would be a complete collapse of the nation's (and the world's) economy, the government has intervened in the form of loaning billions of taxpayer dollars to the very financial institutions that are largely responsible for creating the crisis, with little oversight for how the funds will be used. The actions of government officials, which they claimed to have been "necessary," have nevertheless sparked public uproar because the bailout of Wall Street has come at the expense of Main Street. The bailout has thus unambiguously exposed the class bias of government policy, since the typical taxpaying citizen has seen little relief from the fragility of her own economic situation. That these actions have ignited a legitimation crisis is made visible by the numerous public protests that have occurred in the wake of the government's actions. Potentially intensifying this crisis is the "return to normalcy" within the financial industry aided by elected federal officials who refused to pass stringent regulatory policies that would prevent the use of banking and investment practices that could precipitate another collapse. Meanwhile, executive bonuses at firms such as Goldman Sachs have quickly returned to their precrisis, multimillion-dollar level, despite the fact that the firms' solvency is due solely to the massive infusion of taxpayers' funds—taxpayers who continue to face economic hardships. To the degree that democratic governments are premised on fostering the well-being of all their citizens, such actions certainly jeopardize their legitimacy.

As Habermas notes, states seek to avoid or mitigate legitimation crises by providing their "clients" with material rewards in the form of social programs. The recent health care reform debate can be viewed as a stage on which an attempt to "buy" the public's loyalty was played out. While it is in the interest of all government officials to secure the obedience of the citizenry, efforts to expand health care coverage, nevertheless, were opposed by many members of Congress as well as by a health insurance industry that would lose billions of dollars in profits if a wholesale overhaul of the system (in the form of establishing single-payer universal care or a public health insurance option) was enacted. With the passage of the health care reform bill it is clear that a dramatic shift away from a for-profit healthcare model will not occur, thus blunting the reform's potential to buttress the state's legitimacy. Moreover, during the debate, press reports disclosed the efforts of lobbyists for the insurance industry to prevent significant reform, efforts that included making campaign contributions to sympathetic lawmakers, publishing industry reports that deceptively inflated the costs of healthcare reform, and taking an active role in writing proposed health care legislation. Their success in thwarting reform not only compromises the state's legitimacy but also demonstrates the encroachment of the steering media of money and power into the essential lifeworld requirement to advance democratic, normatively sanctioned debate.

The relationship between lifeworld and system represents for Habermas the crucial dynamic shaping the development of human society. Although the system has its origins in the lifeworld, highly differentiated and complex societies have developed a set of emergent structures that become increasingly segmented from the communicative and integrative processes of the lifeworld. In smaller and less-differentiated societies, the system and the lifeworld are closely coupled in a tight, mutual interrelationship. In such societies, the weight of tradition presents its members with a world that is unquestioned. Marriage arrangements, productive tasks, and the distribution of wealth and power confront members as natural and unchanging conditions of social life. However, as societies become larger and more complex, system and lifeworld begin to grow increasingly detached. Economic and political institutions become functionally independent of kinship structures that previously provided the basis for societal organization (Seidman 1989:19). On one hand, this "uncoupling" is an essential condition for the development of modern societies and the greater potential for individual freedom they offer. It enables societies to more efficiently organize and control their productive and administrative requirements. Moreover, with the growing rationalization of the lifeworld, individuals come to possess the symbolic resources enabling them to question the legitimacy of the existing social conditions and to demand that they be explicitly justified according to shared principles and standards for action. Indeed, for Habermas, the rationalization of the lifeworld is the "pacesetter" of social evolution.

On the other hand, as the complexity, power, and differentiation of the system grows, it eventually becomes scaled off from the lifeworld and ultimately comes to engulf it. In one of his more famous expressions, Habermas (1981/1987) describes this process as the **colonization of the lifeworld**. In this process, system steering media (money and power) and technical/instrumental logic come to replace the consensual negotiation of shared meanings as the foundation for social integration and the reproduction of the lifeworld. The result is a "totally administered" society in which social relationships are increasingly mediated by power and money, and the interpersonal debates and discussions within the lifeworld come to have less and less impact on the constitution of the system. (The examples above concerning the economic crisis and healthcare reform speak to this development.) Thus, while modern societies have witnessed a phenomenal expansion of productive capacity and material wealth, they have yet to fulfill the promise of the Enlightenment. To more fully understand Habermas's ambivalent view of modernity, we must explore his conception of rationality. It is to this task that we now turn.

Habermas and Rational Action

Like that of the critical theorists, Habermas's work is grounded in an investigation of the increasing rationalization of society, a subject first explored by Max Weber. Habermas's discussion of modern forms of rationality parallels the "moments of reason" outlined by Weber. To Weber's three moments of reason—science, law, and art—Habermas posits three complementary modes of rationality—instrumental, moral-practical, and aesthetic-expressive—that individuals use to give meaning to their everyday lives. As illustrated in Table 16.1, each form of rationality corresponds to a specific mode of reasoning; addresses specific philosophical, ethical, or interpersonal questions; speaks to a unique dimension of the world of experience; and has distinct consequences for sociocultural reproduction.

Following Weber, Habermas conceptualizes instrumental rationality as a purposive and goal-oriented attitude underlying the pursuit of objective truth and technical, empirical knowledge. Instrumental rationality reflects the moment of reason captured by science as it serves to quantify, organize, and systematically control the *objective* or *external* world of physical objects. Moral-practical rationality is a form of reasoning that shapes our understanding of cultural norms and values, ethics and law, justice, and the exercise of authority—in short, the "rightness" and legitimacy of action. Tied to law, it serves as a normative and moral guide within the *social* world of interpersonal relations. Finally, aesthetic-expressive rationality informs our evaluations of taste and the sincerity or deceitfulness of managed self-presentations and expressions (Goffman 1959). It is oriented to the *subjective* or *internal* world of personal experience and artistic judgment (see Table 16.1).

Table 16.1 Summary of Habermas's Theory of Action

Rationality	Moment of Reason	Questions Addressed	World	Action
Instrumental	Science	Truth/Knowledge	Objective	Teleological
Moral-Practical	Law	Justice/Morality	Social	Normatively regulated
Aesthetic-Expressive	Art	Aesthetics/Sincerity	Subjective	Dramaturgical
Communicative	Critical	Mutual understanding Coordination of action Socialization	Lifeworld	Transmission of cultural knowledge Social integration Formation of identity

While his analysis of rationality shares much with Weber's moments of reason, Habermas nevertheless contends that Weber (as well as Marx, the earlier critical theorists, and Parsons, all of whose ideas he incorporates in some fashion) failed to recognize a vital dimension of human rationality, one that is embodied in intersubjective communication. To fill this gap, Habermas draws on the work of symbolic interactionists and phenomenologists (most notably George Herbert Mead and Alfred Schutz) and linguistic philosophers to develop his notion of "communicative rationality" or "communicative action." Habermas defines **communicative action** as the process in which individuals come to mutual understanding and consensus through open, noncoercive debate and discussion freed from the corrosive effects of money, power, and manipulation. Encompassing the other forms of rationality and addressing simultaneously the worlds of objective, social, and subjective experience, communicative action embodies a critical stance that allows for the negotiation of shared meanings, the coordination of action, and the socialization of individuals. In the process, communicative action—itself an outgrowth of the evolutionary rationalization of the lifeworld—reproduces the lifeworld by transmitting the cultural stock of knowledge, integrating individuals into the community, and securing the formation of personal and social identities. For Habermas, then, the rationality embedded within communicative action provides the foundation for renewing the social order according to the simultaneous demands for proof, justice, and authenticity.

Significantly, communicative rationality is achieved only when individuals are confident that the claims being made by others meet the criteria for proof specific to the other three worlds. For instance, while a person may convince us of the truth of her statements (objective world), we may still doubt the sincerity of her intentions (subjective world) or the normative appropriateness, the "rightness," of what she says (social world). Indeed, determining whether someone is attempting to reach a shared understanding or trying to manipulate us is a crucial aspect of communication. Communicative action is the crucial means through which technical, moral, and aesthetic critiques can be advanced, providing in turn the basis for mutual agreement and a thriving democracy.

To shed light on how these forms of rationality and the worlds they connect to shape interaction, consider the following set of statements that could come up in a discussion. "Did you know that in America, women are paid less than men for performing the same work? I think America is a sexist country; otherwise, we wouldn't accept this. We must do something to change gender relations and stop the discriminatory treatment of women." To these statements we would apply the various forms of rationality in order to determine whether or not an agreement can be reached on the issues at hand. The first statement primarily addresses the objective world of facts. Here we might seek relevant scientific statistics in order to be convinced of the truth of the claim. The second statement does not address the world of facts; rather, it speaks to the social world of normative expectations that regulate behavior and interpersonal relations in American society. This statement implies that the truth is not "right," that in a society that promises equality for all, such treatment is not morally legitimate and "ought" not exist. In response, we would consider whether or not the person is entitled or justified to assert this claim, given our understanding of society's normative code. The third statement is directed toward the subjective world and leads us to consider whether or not the speaker is being sincere or deceitful in making her claim. In determining if she really means what she says, we would gauge her self-presentation and management of expressions to assess whether ulterior motives might be at play. Communicative rationality is obtained when uncoerced, mutual agreement is reached in each "world" of interaction, and the truth, rightness, and sincerity of all claims are established.

Faith in Reason: The Public Sphere and "New" Social Movements

One of the more controversial elements of Habermas's theory is his enduring faith in the power of reason and critical reflection to rescue humanity from systems of domination and oppression. His conviction not only places him at odds with the founders of critical theory, but also runs against the most basic claims of postmodern theorists. While postmodern theory (see Chapter 15) challenges the very existence of truth and contends that all knowledge is inherently relative, Habermas asserts that universal standards for establishing both facts and normative judgments not only are possible but also mark the road toward advancing individual freedom and creating a progressive society. In order to win the mantle of "truth," universal standards must be born out of the democratic expression of communicative rationality. In modern society, communicative action that is directed toward broad social issues is most visible in the public sphere and in the protests mounted by new social movements.

Habermas (1962) argues that the world of public discussion and debate that first emerged within European bourgeois social circles in the eighteenth and nineteenth centuries held great promise for the rise of democratic government. This world of public discussion and debate, or **public sphere**, is composed of an array of social spaces where, ideally, private individuals can publicly congregate and freely debate political, ethical, and social issues in a noncoercive and "undistorted" manner. The public sphere is not an institution, an organization, or a system; rather,

> it is a network for communicating information and points of view (i.e., opinions expressing affirmative or negative attitudes); the streams of communication are, in the process, filtered and synthesized in such a way that they coalesce into bundles of topically specified *public* opinions. Like the lifeworld as a whole, so, too, is the public sphere reproduced through communicative action. . . .
>
> The public sphere is differentiated into levels according to the density of communication, organizational complexity, and range—from the *episodic* publics found in taverns, coffee houses, or on the streets; through the *occasional* or "arranged" publics of particular presentations and events, such as theater performances, rock concerts, party assemblies, or church congresses; up to the *abstract* public sphere of isolated readers, listeners, and viewers scattered across large geographic areas, or even around the globe, and brought together only through the mass media. (Habermas 1992/1996:360, 374)

If left free from distorting influences and guided by its own logic of communicative rationality, the public sphere makes possible the vigorous, compelling debate necessary for the unrestricted growth of democratic processes and the legitimation of the political system. In the public sphere, debate ideally is won on the merits of the best argument, and opinions are formed without consideration of the status or power of speakers or the intrusions of manipulation or coercion.

While Habermas's notion of the public sphere has been roundly criticized for being overly idyllic (e.g., see Calhoun 1992; Kellner 2000; McCarthy 1978), he nevertheless maintains that the potential for creating a truly democratic society based on universal reason lies within this domain. He is, however, not entirely blind to his critics. He notes that the public sphere of open debate was deformed by the scale of its growth over the past century and by the rationalization of economic and political institutions whose technical requirements and forms of rationality have seeped into the public sphere. With the extension of democratic rights, more citizens demanded inclusion in the public sphere, and the resulting "structural transformation" impeded its own rational development. "Social decisions were increasingly removed from the rational-critical discourse of citizens in the political public sphere and made the province of negotiation (rather than discourse proper) among bureaucrats, credentialed experts, and interest group elites" (Calhoun 1995:31). This is particularly evident in commercial mass media, whose dissemination of *public* news and information has been altered by "an influx of *private* interests that achieved privileged representation within it" (Habermas 1989:235, emphasis added). It is likewise evident in the manufacturing of "publicity" on the part of organized interest groups that manipulate the public sphere to carry out their "secret policies" (ibid.:236). The public sphere is now less a space for free debate than it is a market for the sale of prepackaged opinions.

Habermas's argument reflects real concerns that we encounter every day. Think for a moment about the last time that you had a rigorous political or philosophical debate. This debate may have occurred within any number of arenas: your family; a friendship network; your school, workplace, or neighborhood; a community association devoted to sociopolitical concerns or leisure activities; or perhaps the Internet. Yet, consider how much impact this debate may have had on the political processes that shape your social world. For many of us, such encounters happen infrequently, and when they do, it is rarer that they have anything but a trivial impact on the political and economic life of our society. Habermas's theory helps us understand why public debate seldom occurs and why it has little influence on the institutions that most dramatically affect our lives. He contends that the space for open, democratic, public discussion is limited to the extent that resources like money and power interfere in the process of opinion formation. Public debates have become all too infrequent because, in the face of political and economic processes that seem to be beyond our control, we recognize the limitations of their efficacy. Thus, we experience a type of communicative alienation within a public sphere that has become increasingly dominated by powerful and rationalized social forces (like political and economic interest groups) that are largely divorced from the influence of public deliberation. The result is that the public sphere idealized by Habermas has been deformed, making it all the more essential to reconstruct the vision of the emancipatory project that began with the Enlightenment.

Nevertheless, Habermas refuses to give up on the promise of communicative rationality and its expression in the public sphere. He places his hopes for freedom and progress in "new" social movements whose efforts signal a resistance to the colonization of the lifeworld and the deformation of the public sphere. New social movements are composed of an array of groups whose disparate aims are united by the "critique of growth." They emerge from the margins of established economic and political structures to give voice to a "crisis consciousness" that calls attention to the social and environmental pathologies created in the wake of advanced industrial capitalism. Theirs are not the struggles of class-based labor movements seeking to effect a more egalitarian distribution

Photo 16.3 New social movements and the struggle over the "grammar of forms of life": Protesting for civil rights

of wealth or to secure more humane working conditions: The movements in which Marxist theorists placed their hopes and disappointments are no longer the vanguard for progressive change. It is not economic interests or issues concerning material reproduction that form the nucleus of their causes; rather, it is "quality of life" issues that are embedded in the symbolic reproduction of the lifeworld. These concerns are expressed in the form of "green" groups fighting to stop the destruction of timberlands or reduce the reliance on polluting energy sources, antinuclear movements, squatters protesting the commercial buyout of affordable residential housing, efforts to end discrimination against homosexuals (such as recent efforts to legalize gay marriage), protests against the consolidation of the media, and demands for greater personal control over decisions regarding one's health and life (such as efforts to legalize assisted suicide or the use of medical marijuana).

These movements are potentially effective avenues for invigorating the public sphere because they inject claims into the political process that are largely peripheral to the interests of bureaucratic political organizations and their leaders. New social movements and their counterinstitutional political tactics, like civil disobedience, are indicators that the emancipatory project of the Enlightenment is still "unfinished." By nurturing the mechanisms for communicative action, new social movements hold out the promise for democracy and social equality that can be fulfilled only through struggles to defend the lifeworld against the intrusion of money and power and the corrupting influences of instrumental rationality.

HABERMAS'S THEORETICAL ORIENTATION ▪▪

Like Bourdieu, Habermas draws from a diverse array of influences to create a system of thought that is broad, complex, and theoretically multidimensional. His work bears the imprint not only of classical sociological theorists (particularly Marx, Weber, Durkheim, and Mead) but also of more contemporary thinkers in sociology, psychology, political science, phenomenology, and philosophy. The result of this synthesis is a theoretical system that blends individualist and collectivist levels of analysis, addresses concerns of individual agency and social structures, and explores issues of rational and nonrational dimensions of social life (see Figure 16.3).

Figure 16.3   Habermas's Basic Concepts and Theoretical Orientation

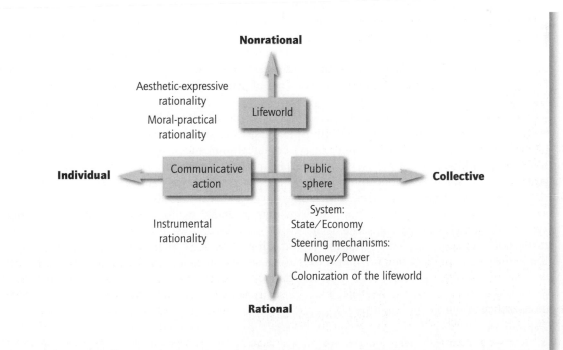

With regard to the issue of action, Habermas's multidimensional approach is most apparent in his discussion of rationality (see Table 16.1). This model clearly reflects that action can be motivated by a number of factors. First, instrumental rationality refers to the means/ends calculations in which individuals seek to optimize the benefits that are likely to be reaped from pursuing a particular course of action. This entails a rational or technical approach to the social and physical world that asks *how* a given goal can be achieved most efficiently. Conversely, Habermas's discussion of moral-practical rationality and aesthetic-expressive rationality speaks to nonrational motivating forces. A different set of motivating questions is raised to the extent that these forms of reasoning guide our actions. Instead of asking "how" or "if" a line of action can be profitably pursued, our behaviors are governed by normative considerations that lead us to ask *why* we should carry out a course of action. Our determinations of whether or not we *ought* to pursue a given line of conduct are grounded not in an evaluation of facts but, rather, in normative prescriptions for behavior that are not subject to empirical proof. Such prescriptions are instead embedded in the traditions and customs of a given culture.

Habermas's notion of communicative action is premised on the conviction that human emancipation is not possible as long as one form of reasoning dominates another. In an ideal, democratic society, both rational and nonrational forms of reasoning would motivate an individual's actions. That is, in order to be truly progressive, our actions must satisfy technical requirements for efficiency as well as normative requirements for "rightness" and authenticity. Yet, modern societies have fallen short of the democratic ideal of mutual understanding based on the full range of motivating forces. Individuals remain dominated by seemingly autonomous forces that have corrupted their capacity to reason.

In terms of the question of order (that is, "What accounts for the routines and patterns of behavior through which social life appears, and is experienced as, orderly and predictable?"), Habermas again devises a multidimensional approach. While he holds on to the promise of a future society ordered on the basis of individually negotiated communication action, this condition, as well as all other forms of interaction, is dependent on the evolutionary stage of the relationship between the system and the lifeworld. Thus, Habermas contends that the distortion of consciousness—the unfreedom—that plagues modern individuals is a result of the steering media of economic and political systems colonizing the lifeworld and its framework of shared meanings and norms. Economic and political systems operate according to their own impersonal logic that confronts individuals as an abstract, amorphous force that largely defies their ability to control their own destinies. In short, modern society is ordered less by the ongoing negotiation of meanings through which the world is made and remade than by the imperatives of collectivist structures that shape the nature of social interaction. Indeed, it is this very condition that Habermas seeks to dismantle through his reviving of critical theory and reinvigorating of the public sphere.

It is important to note, however, that Habermas's perspective by no means overlooks the individualist dimension. Incorporating both collectivist and individualist dimensions is an essential element in his theoretical system. His twin notions of system integration and social integration and his emphasis on the lifeworld are aimed precisely at understanding the (distorted) ways in which individuals construct their personal and collective identities, negotiate meanings, and, in the process, create and re-create the social order. The social order is produced and reproduced as individuals coordinate their actions through processes of reaching understanding mediated within the lifeworld. In providing the symbolic resources for making sense of the world, the lifeworld establishes a horizon of possibilities that places "internal limitations" on the reproduction of society. Such interpretive possibilities and limitations are formed within the minds of individuals and thus are not determined by the functional "needs" of the system. At the same time, however, actions are interconnected by system mechanisms that "are not intended by [individuals] and are usually not even perceived within the horizon of everyday practice" (ibid.).

In the selections that follow, you are presented with two key works from Jürgen Habermas. The first is an essay titled "Civil Society, Public Opinion, and Communicative Power," in which Habermas outlines the key characteristics of the public sphere, its role in promoting discussion and debate, and the forces that impair its efficacy and corrupt the communicative processes ushering from the lifeworld. The second selection, "The Tasks of a Critical Theory of Society," finds Habermas extending these ideas, exploring the growing tensions that emerge between the system and the lifeworld and the colonization of the latter by delinguistified steering media rooted in the system. Together, these selections offer a set of ideas that examine the obstacles that jeopardize the realization of the emancipatory project embraced by Enlightenment thinkers, and a vision of hope for democracy in contemporary societies.

Introduction to "Civil Society,
Public Opinion, and Communicative Power"

In the essay "Civil Society, Public Opinion, and Communicative Power," Habermas investigates the connections between the public sphere, the formation of public opinion, and the effectiveness of democratic politics. The public sphere serves as a "sounding board" or "warning system" that alerts the political system to pressing social problems. It is not an organization or institution, but rather a "social space" or "network" that is created and reproduced through communicative action. These spaces can take the form of "episodic" publics found, for instance, in cafés and on the street; "occasional" publics that meet for specific events, such as church group meetings, annual conventions, or political protests and rallies; or "abstract" publics consisting of isolated individuals who are "brought together" through their shared consumption of mass media.

As a space for opinion formation, the public sphere is not itself invested with political power, yet within its borders actors engage in a struggle over the ability to shape public opinion in order to catapult an issue into official, institutionalized political bodies. For public opinion to be transformed into effective political power it must first pass "through the filters of institutionalized *procedures* of democratic opinion- and will-formation and [enter] through parliamentary debates into legitimate lawmaking" (Habermas 1992/1996:371). These procedures are, in effect, "rules of a *shared* practice of communication" that determine the "quality" or "success" of public opinion. They require that consensus on an issue "*develops* only as a result of more or less exhaustive controversy in which proposals, information, and reasons can be more or less rationally dealt with" (ibid.:362, emphasis in original). As a result, not all opinions are equally qualified to *convince* the public of their relevance. The public sphere is thus a crucial space for fostering democracy, since its egalitarian public alone possesses the final authority to certify the salience and comprehensibility of publicly voiced opinions.

However, the democratic potential inherent in the public sphere is subject to distortion from at least two sources. First, money and organizational power are capable of manipulating the process of public opinion formation so long as their use remains hidden from public view. Once the general public becomes aware of a previously undeclared infusion of money and power, public opinions developed and made visible by such resources lose their credibility. Certainly, the public uproar sparked by news stories detailing the illicit actions of interest groups and lobbyists and the behind-the-scene efforts of corporations and government officials attempting to shape public opinion reminds us that "public opinion can be manipulated, but neither publicly bought nor publicly blackmailed" (ibid.: 364).

A second threat to the democratically formed public sphere stems from journalism and the mass media. In a mass media–dominated public sphere such as our own, journalists and media executives act as gatekeepers for the flow of information. Through their decisions about what is newsworthy, they determine the topics and the viewpoints that are disseminated to the mass audience. As a result, they have become increasingly powerful in the world of public opinion formation. Moreover, as the mass media become increasingly complex and expensive, avenues of communication become more centralized and consolidated within a few media conglomerates. Groups that are peripheral to the organized political system or that do not possess significant resources have become less able to influence media programming. As Habermas notes, this problem is compounded for those groups whose messages "do not fall inside the 'balanced,' that is, the centrist and rather narrowly defined, spectrum of 'established opinions' dominating the programs of the electronic media" (ibid.:377). In addition, the profit-driven market strategies that steer media programming lead to an oversimplification or watering down of information—infotainment—that "works to depoliticize public communication" (ibid.). Yet, Habermas claims that despite the public sphere's apparent undermining, as soon as a "crisis consciousness" develops within the public sphere, the balance of power shifts as actors press for solutions to relevant social problems.

The crucial question now is whether or not a crisis consciousness is capable of developing in light of recent political and legal events that have amplified the dangers posed by these two threats, further eroding what remains of the public sphere and our democracy. Perhaps most significant in this regard are the Federal Communications Commission's December 2007 decision that overturned a thirty-year restriction preventing media corporations from owning both a major TV or radio station and a major daily newspaper in one city, and the Supreme Court's January 2010 ruling that struck down certain limits on corporate spending in political elections as violations of corporations' First Amendment right to free speech. With the Court's new ruling, corporations now have the right as "individuals" to advocate directly for or against specific candidates through the unrestricted purchasing of advertising time and space. Such decisions suggest that the distortion of the democratic potential inherent in the public sphere is being championed by our own "democratic" government through its expansion of the role that corporate money and power can play in shaping the "will of the people."

"Civil Society, Public Opinion, and Communicative Power" (1996)

Jürgen Habermas

I have described the political public sphere as a sounding board for problems that must be processed by the political system because they cannot be solved elsewhere. To this extent, the public sphere is a warning system with sensors that, though unspecialized, are sensitive throughout society. From the perspective of democratic theory, the public sphere must, in addition, amplify the pressure of problems, that is not only detect and identify problems but also convincingly and *influentially* thematize them, furnish them with possible solutions, and dramatize them in such a way that they are taken up and dealt with by parliamentary complexes. Besides the "signal" function, there must be an effective problematization. The capacity of the public sphere to solve problems *on its own* is limited. But this capacity must be utilized to oversee the further treatment of problems that takes place inside the political system. . . .

SOURCE: From *Between Facts and Norms: Contributions to a Discourse Theory of Law and Democracy* by Jurgen Harbermas, translated by William Rehg. Copyright © 1996 Massachusetts Institute of Technology. Reprinted with permission of The MIT Press and Polity Press Ltd.

I

The public sphere is a social phenomenon just as elementary as action, actor, association, or collectivity, but it eludes the conventional sociological concepts of "social order." The public sphere cannot be conceived as an institution and certainly not as an organization. It is not even a framework of norms with differentiated competences and roles, membership regulations, and so on. Just as little does it represent a system; although it permits one to draw international boundaries, outwardly it is characterized by open, permeable, and shifting horizons. The public sphere can best be described as a network for communicating information and points of view (i.e., opinions expressing affirmative or negative attitudes); the streams of communication are, in the process, filtered and synthesized in such a way that they coalesce into bundles of topically specified *public* opinions. Like the lifeworld as a whole, so, too, the public sphere is reproduced through communicative action, for which mastery of a natural language suffices; it is tailored to the *general comprehensibility* of everyday communicative practice. We have become acquainted with the "lifeworld" as a reservoir for simple interactions; specialized systems of action and knowledge that are differentiated within the lifeworld remain tied to these interactions. These systems fall into one of two categories. Systems like religion, education, and the family become associated with general reproductive functions of the lifeworld (that is, with cultural reproduction, social integration, or socialization). Systems like science, morality, and art take up different validity aspects of everyday communicative action (truth, rightness, or veracity). The public sphere, however, is specialized in neither of these two ways; to the extent that it extends to politically relevant questions, it leaves their specialized treatment to the political system. Rather, the public sphere distinguishes itself through a *communication structure* that is related to a third feature of communicative action: it refers neither to the *functions* nor to the *contents* of everyday communication but to the *social space* generated in communicative action.

Unlike success-oriented actors who mutually observe each other as one observes something in the objective world, persons acting communicatively encounter each other in a *situation* they at the same time constitute with their cooperatively negotiated interpretations. The intersubjectively shared space of a speech situation is disclosed when the participants enter into interpersonal relationships by taking positions on mutual speech-act offers and assuming illocutionary obligations. Every encounter in which actors do not just observe each other but take a second-person attitude, reciprocally attributing communicative freedom to each other, unfolds in a linguistically constituted public space. This space stands open, in principle, for potential dialogue partners who are present as bystanders or could come on the scene and join those present. That is, special measures would be required to prevent a third party from entering such a linguistically constituted space. Founded in communicative action, this spatial structure of simple and episodic encounters can be expanded and rendered more permanent in an abstract form for a larger public of present persons. For the public infrastructure of such *assemblies,* performances, presentations, and so on, architectural metaphors of structured spaces recommend themselves: we speak of forums, stages, arenas, and the like. These public spheres still cling to the concrete locales where an audience is physically gathered. The more they detach themselves from the public's physical presence and extend to the virtual presence of scattered readers, listeners, or viewers linked by public media, the clearer becomes the abstraction that enters when the spatial structure of simple interactions is expanded into a public sphere.

When generalized in this way, communication structures contract to informational content and points of view that are uncoupled from the thick context of simple interactions, from specific persons, and from practical obligations. At the same time, context generalization, inclusion, and growing anonymity demand a higher degree of explication that must dispense with technical vocabularies and special codes. Whereas the *orientation to laypersons* implies a certain loss in differentiation, uncoupling communicated opinions from concrete practical obligations tends to have an *intellectualizing* effect. Processes of opinion-formation, especially when they have to do with political questions, certainly cannot be separated from the transformation of the participants' preferences and attitudes, but they can be separated from putting these dispositions into action. To this extent, the communication structures of the public sphere *relieve* the public *of the burden of decision making;* the postponed decisions are reserved for the institutionalized political process. In the public sphere, utterances are sorted according to issue and contribution, whereas the contributions are weighted by the affirmative versus negative responses they receive. Information and arguments are thus worked into focused opinions. What makes such "bundled" opinions into *public opinion* is both the controversial way it comes about and the amount of approval that "carries" it. Public opinion is not representative in the statistical sense. It is not an aggregate of individually gathered, privately expressed opinions held by isolated persons. Hence it must not be confused with survey results. Political opinion polls provide a certain reflection of "public opinion" only if they have been preceded by a focused public debate and a corresponding opinion-formation in a mobilized public sphere.

The diffusion of information and points of view via effective broadcasting media is not the only thing that matters in public processes of communication, nor is it the most important. True, only the broad circulation of comprehensible, attention-grabbing messages arouses a sufficiently inclusive participation. But the rules of a *shared* practice of communication are of greater significance for structuring public opinion. Agreement on issues and contributions develops only as the result of more or less exhaustive controversy in which proposals, information, and reasons can be more or less rationally dealt with. In general terms, the *discursive level* of opinion-formation and the "quality" of the outcome vary with this "more or less" in the "rational" processing of "exhaustive" proposals, information, and reasons. Thus the success of public communication is not intrinsically measured by the requirement of inclusion either but by the formal criteria governing how a qualified public opinion comes about. The structures of a power-ridden, oppressed public sphere exclude fruitful and clarifying discussions. The "quality" of public opinion, insofar as it is measured by the procedural properties of its process of generation, is an empirical variable. From a normative perspective, this provides a basis for measuring the legitimacy of the influence that public opinion has on the political system. Of course, actual influence coincides with legitimate influence just as little as the belief in legitimacy coincides with legitimacy. But conceiving things this way at least opens a perspective from which the relation between actual influence and the procedurally grounded quality of public opinion can be empirically investigated.

Parsons introduced "influence" as a symbolically generalized form of communication that facilitates interactions in virtue of conviction or persuasion. For example, persons or institutions can enjoy a reputation that allows their utterances to have an influence on others' beliefs without having to demonstrate authority or to give explanations in the situation. "Influence" feeds on the resource of mutual understanding, but it is based on advancing trust in beliefs that are not currently tested. In this sense, public opinion represents political potentials that can be used for influencing the voting behavior of citizens or the will-formation in parliamentary bodies, administrative agencies, and courts. Naturally, political *influence* supported by public opinion is converted into political *power*—into a potential for rendering binding decisions—only when it affects the beliefs and decisions of *authorized* members of the political system and determines the behavior of voters, legislators, officials, and so forth. Just like social power, political influence based on public opinion can be transformed into political power only through institutionalized procedures.

Influence develops in the public sphere and becomes the object of struggle there. This struggle involves not only the political influence that has already been acquired (such as that enjoyed by experienced political leaders and officeholders, established parties, and well-known groups like Greenpeace and Amnesty International). The reputation of groups of persons and experts who have acquired their influence in special public spheres also comes into play (for example, the authority of religious leaders, the public visibility of literary figures and artists, the reputation of scientists, and the popularity of sports figures and movie stars). For as soon as the public space has expanded beyond the context of simple interactions, a differentiation sets in among organizers, speakers, and hearers; arenas and galleries; stage and viewing space. The *actors' roles* that increasingly professionalize and multiply with organizational complexity and range of media are, of course, furnished with unequal opportunities for exerting influence. But the political influence that the actors gain through public communication must *ultimately* rest on the resonance and indeed the approval of a lay public whose composition is egalitarian. The public of citizens must be *convinced* by comprehensible and broadly interesting contributions to issues it finds relevant. The public audience possesses final authority, because it is *constitutive* for the internal structure and reproduction of the public sphere, the *only* place where actors can appear. There can be no public sphere without a public.

To be sure, we must distinguish the actors who, so to speak, emerge from the public and take part in the reproduction of the public sphere itself from actors who occupy an already constituted public domain in order to use it. This is true, for example, of the large and well-organized interest groups that are anchored in various social subsystems and affect the political system *through* the public sphere. They cannot make any manifest use in the public sphere of the sanctions and rewards they rely on in bargaining or in nonpublic attempts at pressure. They can capitalize on their social power and convert it into political power only insofar as they can advertise their interests in a language that can mobilize convincing reasons and shared value orientations—as, for example, when parties to wage negotiations inform the public about demands, strategies, or outcomes. The contributions of interest groups are, in any case, vulnerable to a kind of criticism to which contributions from other sources are not exposed. Public opinions that can acquire visibility only because of an undeclared infusion of money or organizational power lose their credibility as soon as these sources of social power are made public. Public opinion can be manipulated but neither publicly bought nor publicly blackmailed. This is due to the fact that a public sphere cannot be "manufactured" as one pleases. Before it can be captured by actors with strategic intent, the public sphere together with its public must have developed as a structure that stands on its own and reproduces itself *out of itself.* This lawlike regularity governing the formation of a public sphere remains latent in the constituted public sphere—and takes effect again only in moments when the public sphere is mobilized.

The political public sphere can fulfill its function of perceiving and thematizing encompassing social problems only insofar as it develops out of the communication taking place among *those who are potentially affected.* It is carried by a public recruited from the entire citizenry. But in the diverse voices of this public, one hears the echo of private experiences that are caused throughout society by the externalities (and internal disturbances) of various functional systems—and even by the very state apparatus on whose regulatory activities the complex and poorly coordinated sub-systems depend. Systemic deficiencies are experienced in the context of individual life histories; such burdens accumulate in the lifeworld. The latter has the appropriate antennae, for in its horizon are intermeshed the private life histories of the "clients" of functional systems that might be failing in their delivery of services. It is only for those who are immediately affected that such services are paid in the currency of "use values." Besides religion, art, and literature, only the spheres of "private" life have an existential language at their disposal, in which such socially generated problems can be *assessed in terms of one's own life history.* Problems voiced in the public sphere first become visible when they are mirrored in personal life experiences. To the extent that these experiences find their concise expression in the languages of religion, art, and literature, the "literary" public sphere in the broader sense, which is specialized for the articulation of values and world disclosure, is intertwined with the political public sphere.

As both bearers of the political public sphere and as *members of society,* citizens occupy two positions at once. As members of society, they occupy the roles of employees and consumers, insured persons and patients, taxpayers and clients of bureaucracies, as well as the roles of students, tourists, commuters, and the like; in such complementary roles, they are especially exposed to the specific requirements and failures of the corresponding service systems. Such experiences are first assimilated "privately," that is are interpreted within the horizon of a life history intermeshed with other life histories in the contexts of shared lifeworlds. The communication channels of the public sphere are linked to private spheres—to the thick networks of interaction found in families and circles of friends as well as to the looser contacts with neighbors, work colleagues, acquaintances, and so on—and indeed they are linked in such a way that the spatial structures of simple interactions are expanded and abstracted but not destroyed. Thus the orientation to reaching understanding that is predominant in everyday practice is also preserved for a *communication among strangers* that is conducted over great distances in public spheres whose branches are quite complex. The threshold separating the private sphere from the public is not marked by a fixed set of issues or relationships but by *different conditions of communication.* Certainly these conditions lead to differences in the accessibility of the two spheres, safeguarding the intimacy of the one sphere and publicity of the other. However, they do not seal off the private from the public but only channel the flow of topics from the one sphere into the other. For the public sphere draws its impulses from the private handling of social problems that resonate in life histories. It is symptomatic of this close connection, incidentally, that a modern bourgeois public sphere developed in the European societies of the seventeenth and eighteenth centuries as the "sphere of private persons come together as a public." Viewed historically, the connection between the public and the private spheres is manifested in the clubs and organizational forms of a reading public composed of bourgeois private persons and crystallizing around newspapers and journals.

II

This sphere of civil society has been rediscovered today in wholly new historical constellations. The expression "civil society" has in the meantime taken on a meaning different from that of the "bourgeois society" of the liberal tradition, which Hegel conceptualized as a "system of needs," that is, as a market system involving social labor and commodity exchange. What is meant by "civil society" today, in contrast to its usage in the Marxist tradition, no longer includes the economy as constituted by private law and steered through markets in labor, capital, and commodities. Rather, its institutional core comprises those nongovernmental and noneconomic connections and voluntary associations that anchor the communication structures of the public sphere in the society component of the lifeworld. Civil society is composed of those more or less spontaneously emergent associations, organizations, and movements that, attuned to how societal problems resonate in the private life spheres, distill and transmit such reactions in amplified form to the public sphere. The core of civil society comprises a network of associations that institutionalizes problem-solving discourses on questions of general interest inside the framework of organized public spheres. These "discursive designs" have an egalitarian, open form of organization that mirrors essential features of the kind of communication around which they crystallize and to which they lend continuity and permanence.

Such associations certainly do not represent the most conspicuous element of a public sphere dominated by mass media and large agencies, observed by market and opinion research, and inundated by the public relations work, propaganda, and advertising of political parties and groups. All the same, they do form the organizational substratum of the general public of citizens. More or less emerging from the private sphere, this public is made of citizens who seek acceptable interpretations for their social interests and experiences and who want to have an influence on institutionalized opinion- and will-formation.

One searches the literature in vain for clear definitions of civil society that would go beyond such descriptive characterizations. . . . Jean Cohen and Andrew Arato, who have presented the most comprehensive study on this topic, provide a catalog of features characterizing the civil society that is demarcated from the state, the economy, and other functional systems but coupled with the core private spheres of the lifeworld:

(1) *Plurality:* families, informal groups, and voluntary associations whose plurality and autonomy allow for a variety of forms of life; (2) *Publicity:* institutions of culture and communication; (3) *Privacy:* a domain of individual self-development and moral choice; (4) *Legality:* structures of publicity from at least the state and, tendentially, the economy. Together, these structures secure the institutional existence of a modern differentiated civil society[i]. . . .

The *constitution of this sphere through basic rights* provides some indicators for its social structure. Freedom of assembly and freedom of association, when linked with freedom of speech, define the scope for various types of associations and societies: for voluntary associations that intervene in the formation of public opinion, push topics of general interest, and act as advocates for neglected issues and underrepresented groups; for groups that are difficult to organize or that pursue cultural, religious, or humanitarian aims; and for ethnical communities, religious denominations, and so on. Freedom of the press, radio, and television, as well as the right to engage in these areas, safeguards the media infrastructure of public communication; such liberties are thereby supposed to preserve an openness for competing opinions and a representative diversity of voices. The political system, which must remain sensitive to the influence of public opinion, is intertwined with the public sphere and civil society through the activity of political parties and general elections. This intermeshing is guaranteed by the right of parties to "collaborate" in the political will-formation of the people, as well as by the citizens' and passive voting rights and other participatory rights. Finally, the network of associations can assert its autonomy and preserve its spontaneity only insofar as it can draw support from a mature pluralism of forms of life, subcultures, and worldviews. The constitutional protection of "privacy" promotes the integrity of private life spheres: rights of personality, freedom of belief and of conscience, freedom of movement, the privacy of letters, mail, and telecommunications, the inviolability

---

[i]J. L. Cohen and A. Arato, *Civil Society and Political Theory* (Cambridge, Mass., 1992), p. 346.

of one's residence, and the protection of families circumscribe an untouchable zone of personal integrity and independent judgment.

The tight connection between an autonomous civil society and an integral private sphere stands out even more clearly when contrasted with totalitarian societies of bureaucratic socialism. Here a panoptic state not only directly controls the bureaucratically desiccated public sphere, it also undermines the private basis of this public sphere. Administrative intrusions and constant supervision corrode the communicative structure of everyday contacts in families and schools, neighborhoods and local municipalities. The destruction of solidary living conditions and the paralysis of initiative and independent engagement in overregulated yet legally uncertain sectors go hand in hand with the crushing of social groups, associations, and networks: with indoctrination and the dissolution of cultural identities; with the suffocation of spontaneous public communication. Communicative rationality is thus destroyed *simultaneously* in both public and private contexts of communication. The more the bonding force of communicative action wanes in private life spheres and the embers of communicative freedom die out, the easier it is for someone who monopolizes the public sphere to align the mutually estranged and isolated actors into a mass that can be directed and mobilized in a plebiscitarian manner.

Basic constitutional guarantees alone, of course, cannot preserve the public sphere and civil society from deformations. The communication structures of the public sphere must rather be kept intact by an energetic civil society. That the political public sphere must in a certain sense reproduce and stabilize itself from its own resources is shown by the odd *self-referential character of the practice of communication in civil society*. Those actors who are the carriers of the public sphere put forward "texts" that always reveal the same subtext, which refers to the critical function of the public sphere in general. Whatever the manifest content of their public utterances, the performative meaning of such public discourse at the same time actualizes the function of an undistorted political public sphere as such. Thus, the institutions and legal guarantees of free and open opinion-formation rest on the unsteady ground of the political communication of actors who, in making use of them, at the same time interpret, defend, and radicalize their normative content. Actors

who know they are involved in the *common* enterprise of reconstituting and maintaining structures of the public sphere as they contest opinions and strive for influence differ from actors who merely use forums that already exist. More specifically, actors who support the public sphere are distinguished by the *dual orientation* of their political engagement: with their programs, they directly influence the political system, but at the same time they are also reflexively concerned with revitalizing and enlarging civil society and the public sphere as well as with confirming their own identities and capacities to act.

Cohen and Arato see this kind of "dual politics" especially in the "new" social movements that simultaneously pursue offensive and defensive goals. "Offensively," these movements attempt to bring up issues relevant to the entire society, to define ways of approaching problems, to propose possible solutions, to supply new information, to interpret values differently, to mobilize good reasons and criticize bad ones. Such initiatives are intended to produce a broad shift in public opinion, to alter the parameters of organized political will-formation, and to exert pressure on parliaments, courts, and administrations in favor of specific policies. "Defensively," they attempt to maintain existing structures of association and public influence, to generate subcultural counterpublics and counterinstitutions, to consolidate new collective identities, and to win new terrain in the form of expanded rights and reformed institutions:

> On this account, the "defensive" aspect of the movements involves preserving *and developing* the communicative infrastructure of the lifeworld. . . . This is the sine qua non for successful efforts to redefine identities, to reinterpret norms, and to develop egalitarian, democratic associational forms. The expressive, normative and communicative modes of collective action . . . [also involve] efforts to secure *institutional* changes within civil society that correspond to the new meanings, identities, and norms that are created.[ii]

In the self-referential mode of reproducing the public sphere, as well as in the Janus-faced politics aimed at the political system and the self-stabilization of public sphere and civil society, the space is provided for the extension and radicalization of existing rights: "The combination of associations, publics, and rights, when

---

[ii]Cohen and Arato, *Civil Society,* p. 531.

supported by a political culture in which independent initiatives and movements represent an ever-renewable, legitimate, political option, represents, in our opinion, an effective set of bulwarks around civil society within whose limits much of the program of radical democracy can be reformulated.[iii]

In fact, the *interplay* of a public sphere based in civil society with the opinion- and will-formation institutionalized in parliamentary bodies and courts offers a good starting point for translating the concept of deliberative politics into sociological terms. However, we must not look on civil society as a focal point where the lines of societal self-organization as a whole would converge. Cohen and Arato rightly emphasize the *limited scope for action* that civil society and the public sphere afford to noninstitutionalized political movements and forms of political expression. They speak of a structurally necessary "self-limitation" of radical-democratic practice:

First, a robust civil society can develop only in the context of a liberal political culture and the corresponding patterns of socialization, and on the basis of an integral private sphere; it can blossom only in an already rationalized lifeworld. Otherwise, populist movements arise that blindly defend the frozen traditions of a lifeworld endangered by capitalist modernization. In their forms of mobilization, these fundamentalist movements are as modern as they are antidemocratic.

Second, within the boundaries of the public sphere, or at least of a liberal public sphere, actors can acquire only influence, not political power. The influence of a public opinion generated more or less discursively in open controversies is certainly an empirical variable that can make a difference. But public influence is transformed into communicative power only after it passes through the filters of the institutionalized *procedures* of democratic opinion- and will-formation and enters through parliamentary debates into legitimate lawmaking. The informal flow of public opinion issues in beliefs that have been *tested* from the standpoint of the generalizability of interests. Not influence per se, but influence transformed into communicative power legitimates political decisions. The popular sovereignty set communicatively aflow cannot make itself felt *solely* in the influence of informal public discourses— not even when these discourses arise from autonomous public spheres. To generate political power, their influence must have an effect on the democratically regulated deliberations of democratically elected assemblies and assume an authorized form in formal decisions.

This also holds, mutatis mutandis, for courts that decide politically relevant cases.

Third, and finally, the instruments that politics have available in law and administrative power have limited effectiveness in functionally differentiated societies. Politics indeed continues to be the addressee for all unmanaged integration problems. But political steering can often take only an indirect approach and must, as we have seen, leave intact the modes of operation internal to functional systems and other highly organized spheres of action. As a result, democratic movements emerging from civil society must give up holistic aspirations to a self-organizing society, aspirations that also undergirded Marxist ideas of social revolution. Civil society can directly transform only itself, and it can have at most an indirect effect on the self-transformation of the political system; generally, it has an influence only on the personnel and programming of this system. But in no way does it occupy *the position* of a macrosubject supposed to bring society as a whole under control and simultaneously act for it. Besides these limitations, one must bear in mind that the administrative power deployed for purposes of social planning and supervision is not a suitable medium for fostering emancipated forms of life. These can *develop* in the wake of democratization processes but they cannot be *brought about* through intervention. . . .

III

In complex societies, the public sphere consists of an intermediary structure between the political system, on the one hand, and the private sectors of the lifeworld and functional systems, on the other. It represents a highly complex network that branches out into a multitude of overlapping international, national, regional, local, and subcultural arenas. Functional specifications, thematic foci, policy fields, and so forth, provide the points of reference for a substantive differentiation of public spheres that are, however, still accessible to laypersons (for example, popular science and literary publics, religious and artistic publics, feminist and "alternative" publics, publics concerned with health-care issues, social welfare, or environmental policy). Moreover, the public sphere is differentiated into levels according to the density of communication, organizational complexity, and range—from the *episodic* publics found in taverns, coffee houses, or on the streets; through the *occasional* or

---

[iii]Cohen and Arato, *Civil Society,* p. 474.

"arranged" publics of particular presentations and events, such as theater performances, rock concerts, party assemblies, or church congresses; up to the *abstract* public sphere of isolated readers, listeners, and viewers scattered across large geographic areas, or even around the globe, and brought together only through the mass media. Despite these manifold differentiations, however, all the partial publics constituted by ordinary language remain porous to one another. The one text of "the" public sphere, a text continually extrapolated and extending radially in all directions, is divided by internal boundaries into arbitrarily small texts for which everything else is context; yet one can always build hermeneutical bridges from one text to the next. Segmented public spheres are constituted with the help of exclusion mechanisms; however, because publics cannot harden into organizations or systems, there is no exclusion rule without a proviso for its abolishment. . . .

The more the audience is widened through mass communications, the more inclusive and the more abstract in form it becomes. Correspondingly, the *roles of the actors* appearing in the arenas are, to an increasing degree, sharply separated from the roles of the spectators in the galleries. Although the "success of the actors in the arena is ultimately decided in the galleries, the question arises of how autonomous the public is when it takes a position on an issue, whether its affirmative or negative stand reflects a process of becoming informed or in fact only a more or less concealed game of power. Despite the wealth of empirical investigations, we still do not have a well-established answer to this cardinal question. But one can at least pose the question more precisely by assuming that public processes of communication can take place with less distortion the more they are left to the internal dynamic of a civil society that emerges from the lifeworld.

One can distinguish, at least tentatively, the more loosely organized actors who "emerge from" the public, as it were, from other actors merely "appearing before" the public. The latter have organizational power, resources, and sanctions available *from the start*. Naturally, the actors who are more firmly anchored in civil society and participate in the reproduction of the public sphere also depend on the support of "sponsors" who supply the necessary resources of money, organization, knowledge, and social capital. But patrons or "like-minded" sponsors do not necessarily reduce the authenticity of the public actors they support. By contrast, the collective actors who merely enter the public sphere from, and utilize it for, a specific organization or functional system have *their own* basis of support.

Among these political and social actors who do not have to obtain their resources from other spheres, I primarily include the large interest groups that enjoy social power, as well as the established parties that have largely become arms of the political system. They draw on market studies and opinion surveys and conduct their own professional public-relations campaigns.

In and of themselves, organizational complexity, resources, professionalization, and so on, are admittedly insufficient indicators for the difference between "indigenous" actors and mere users. Nor can an actor's pedigree be read directly from the interests actually represented. Other indicators are more reliable. Thus actors differ in how they can be identified. Some actors one can easily identify from their functional background; that is, they represent political parties or pressure groups; unions or professional associations; consumer-protection groups or rent-control organizations, and so on. Other actors, by contrast, must first *produce* identifying features. This is especially evident with social movements that initially go through a phase of self-identification and self-legitimation; even after that, they still pursue a self-referential "identity politics" parallel to their goal-directed politics—they must continually reassure themselves of their identity. Whether actors merely use an already constituted public sphere or whether they are involved in reproducing its structures is, moreover, evident in the abovementioned sensitivity to threats to communication rights. It is also shown in the actors' willingness to go beyond an interest in self-defense and take a universalist stand against the open or concealed exclusion of minorities or marginal groups. The very existence of social movements, one might add, depends on whether they find organizational forms that produce solidarities and publics, forms that allow them to fully utilize and radicalize existing communication rights and structures as they pursue special goals.

A third group of actors are the journalists, publicity agents, and members of the press (i.e., in the broad sense of *Publizisten*) who collect information, make decisions about the selection and presentation of "programs," and to a certain extent control the entry of topics, contributions, and authors into the mass-media-dominated public sphere. As the mass media become more complex and more expensive, the effective channels of communication become more centralized. To the degree this occurs, the mass media face an increasing pressure of selection, on both the supply side and the demand side. These selection processes become the source of a new sort of power. This *power of the media* is not sufficiently reined in by professional standards, but today, by fits and starts, the "fourth branch of government" is being subjected to constitutional regulation. In the Federal Republic, for example, it is both the legal form and the institutional structure of television networks that determine whether they depend more on the influence of political parties and public interest groups or more on private firms with large advertising outlays. In general, one can say that the image of politics presented on television is predominantly made up of issues and contributions that are professionally produced as media input and then fed in via press conferences, news agencies, public-relations campaigns, and the like. These official producers of information are all the more successful the more they can rely on trained personnel, on financial and technical resources, and in general on a professional infrastructure. Collective actors operating outside the political system or outside large organizations normally have fewer opportunities to influence the content and views presented by the media. This is especially true for messages that do not fall inside the "balanced," that is, the centrist and rather narrowly defined, spectrum of "established options" dominating the programs of the electronic media.

Moreover, before messages selected in this way are broadcast, they are subject to *information-processing strategies* within the media. These are oriented by reception conditions as perceived by media experts, program directors, and the press. Because the public's receptiveness, cognitive capacity, and attention represent unusually scarce resources for which the programs of numerous "stations" compete, the presentation of news and commentaries for the most part follows market strategies. Reporting facts as human-interest stores, mixing information with entertainment, arranging material episodically, and breaking down complex relationships into smaller fragments—all of this comes together to form a syndrome that works to depoliticize public communication. This is the kernel of truth in the theory of the culture industry. The research literature provides fairly reliable information on the institutional framework and structure of the media, as well as on the way they work, organize programs, and are utilized. But, even a generation after Paul Lazarsfeld, propositions concerning the *effects of the media* remain controversial. The research on effect and reception has at least done away with the image of passive consumers as "cultural dopes" who are manipulated by the programs offered to them. It directs our attention to the *strategies of interpretation* employed by viewers, who communicate with one another, and who in fact can be provoked to criticize or reject what programs offer or to synthesize it with judgments of their own.

Even if we know something about the internal operation and impact of the mass media, as well as about the distribution of roles among the public and various actors, and even if we can make some reasonable conjectures about who has privileged access to the media and who has a share in media power, it is by no means clear how the mass media intervene in the diffuse circuits of communication in the political public sphere. The *normative reactions* to the relatively new phenomenon of the mass media's powerful position in the competition for public influence are clearer. Michael Gurevitch and Jay G. Blumler have summarized the tasks that the media *ought* to fulfill in democratic political systems:

1. surveillance of the sociopolitical environment, reporting developments likely to impinge, positively or negatively, on the welfare of citizens;

2. meaningful agenda-setting, identifying the key issues of the day, including the forces that have formed and may resolve them;

3. platforms for an intelligible and illuminating advocacy by politicians and spokespersons of other causes and interest groups;

4. dialogue across a diverse range of views, as well as between power-holders (actual and prospective) and mass publics;

5. mechanisms for holding officials to account for how they have exercised power;

6. incentives for citizens to learn, choose, and become involved, rather than merely to follow and kibitz over the political process;

7. a principled resistance to the efforts of forces outside the media to subvert their independence, integrity and ability to serve the audience;

8. a sense of respect for the audience member, as potentially concerned and able to make sense of his or her political environment.[iv]

[iv]M. Gurevitch and J. G. Blumler, "Political Communication Systems and Democratic Values," in J. Lichtenberg, ed., *Democracy and the Mass Media* (Cambridge, Mass., 1990), p. 270.

Such principles orient the professional code of journalism and the profession's ethical self-understanding, on the one hand, and the formal organization of a free press by laws governing mass communication, on the other. In agreement with the concept of deliberative politics, these principles express a simple idea: the mass media ought to understand themselves as the mandatary of an enlightened public whose willingness to learn and capacity for criticism they at once presuppose, demand, and reinforce; like the judiciary, they ought to preserve their independence from political and social pressure; they ought to be receptive to the public's concerns and proposals, take up these issues and contributions impartially, augment criticisms, and confront the political process with articulate demands for legitimation. The power of the media should thus be neutralized and the tacit conversion of administrative or social power into political influence blocked. According to this idea, political and social actors would be allowed to "use" the public sphere only insofar as they make convincing contributions to the solution of problems that have been perceived by the public or have been put on the public agenda with the public's consent. In a similar vein, political parties would have to participate in the opinion- and will-formation from the public's own perspective, rather than patronizing the public and extracting mass loyalty from the public sphere for the purposes of maintaining their own power. . . .

IV

With this I return to the central question of who can place issues on the agenda and determine what direction the lines of communication take. Roger Cobb, Jennie-Keith Ross, and Marc Howard Ross have constructed models that depict how new and compelling issues develop, from the first initiative up to formal proceedings in bodies that have the power to decide.[v] If one suitably modifies the proposed models—inside access model, mobilization model, outside initiative model—from the viewpoint of democratic theory, they present basic alternatives in how the public sphere and the political system influence each other. In the first case, the initiative comes from office

holders or political leaders, and the issue continues to circulate inside the political system all the way to its formal treatment, while the broader public is either excluded from the process or does not have any influence on it. In the second case, the initiative again starts inside the political system, but the proponents of the issue must mobilize the public sphere, because they need the support of certain groups, either to obtain formal consideration or to implement an adopted program successfully. Only in the third case does the initiative lie with forces at the periphery, outside the purview of the political system. With the help of the mobilized public sphere, that is, the pressure of public opinion, such forces compel formal consideration of the issue:

> The outside initiative model applies to the situation in which a group outside the government structure 1) articulates a grievance, 2) tries to expand interest in the issue enough to other groups in the population to gain a place on the public agenda, in order to 3) create sufficient pressure on decision makers to force the issue onto the formal agenda for their serious consideration. This model of agenda building is likely to predominate in more egalitarian societies. Formal agenda status, . . . however, does not necessarily mean that the final decisions of the authorities or the actual policy implementation will be what the grievance group originally sought.[vi]

In the normal case, issues and proposals have a history whose course corresponds more to the first or second model than to the third. As long as the informal circulation of power dominates the political system, the initiative and power to put problems on the agenda and bring them to a decision lies more with the Government leaders and administration than with the parliamentary complex. As long as in the public sphere the mass media prefer, contrary to their normative self-understanding, to draw their material from powerful, well-organized information producers and as long as they prefer media strategies that lower rather than raise the discursive level of public communication, issues will tend to start in, and be managed from, the center, rather than follow a spontaneous course originating in the periphery. At least, the skeptical findings on problem articulation in public arenas accord with this view. In the present

---

[v]R. Cobb, J. K. Ross, and M. H. Ross, "Agenda Building as a Comparative Political Process," *American Political Science Review* 70 (1976): 126–38; R. Cobb and C. Elder, "The Politics of Agenda-Building," *Journal of Politics* (1971): 892–915.

[vi]Cobb, Ross, and Ross, "Agenda Building as a Comparative Political Process," p. 132.

context, of course, there can be no question of a conclusive empirical evaluation of the mutual influence that politics and public have on each other. For our purposes, it suffices to make it plausible that in a perceived crisis situation, the *actors in civil society* thus far neglected in our scenario *can* assume a surprisingly active and momentous role. In spite of a lesser organizational complexity and a weaker capacity for action, and despite the structural disadvantages mentioned earlier, at the critical moments of an accelerated history, these actors get the chance to *reverse* the normal circuits of communication in the political system and the public sphere. In this way they can shift the entire system's mode of problem solving.

The communication structures of the public sphere are linked with the private life spheres in a way that gives the civil-social periphery, in contrast to the political center, the advantage of greater sensitivity in detecting and identifying new problem situations. The great issues of the last decades give evidence for this. Consider, for example, the spiraling nuclear-arms race; consider the risks involved in the peaceful use of atomic energy or in other large-scale technological projects and scientific experimentation, such as genetic engineering; consider the ecological threats involved in an overstrained natural environment (acid rain, water pollution, species extinction, etc.); consider the dramatically progressing impoverishment of the Third World and problems of the world economic order; or consider such issues as feminism, increasing immigration, and the associated problems of multiculturalism. Hardly any of these topics were *initially* brought up by exponents of the state apparatus, large organizations, or functional systems. Instead, they were broached by intellectuals, concerned citizens, radical professionals, self-proclaimed "advocates," and the like. Moving in from this outermost periphery, such issues force their way into newspapers and interested associations, clubs, professional organizations, academies, and universities. They find forums, citizen initiatives, and other platforms before they catalyze the growth of social movements and new subcultures. The latter can in turn dramatize contributions, presenting them so effectively that the mass media take up the matter. Only through their controversial presentation in the media do such topics reach the larger public and subsequently gain a place on the "public agenda." Sometimes the support of sensational actions, mass protests, and incessant campaigning is required before an issue can make its way via the surprising election of marginal candidates or radical parties, expanded platforms of "established" parties, important court decisions, and so on, into the core of the political system and there receive formal consideration.

Naturally, there are other ways in which issues develop, other paths from the periphery to the center, and other patterns involving complex branchings and feedback loops. But, in general, one can say that even in more or less power-ridden public spheres, the power relations shift as soon as the perception of relevant social problems evokes a *crisis consciousness* at the periphery. If actors from civil society then join together, formulate the relevant issue, and promote it in the public sphere, their efforts can be successful, because the endogenous mobilization of the public sphere activates an otherwise latent dependency built into the internal structure of every public sphere, a dependency also present in the normative self-understanding of the mass media: the players in the arena owe their influence to the approval of those in the gallery. At the very least, one can say that insofar as a rationalized lifeworld supports the development of a liberal public sphere by furnishing it with a solid foundation in civil society, the authority of a position-taking public is strengthened in the course of escalating public controversies. Under the conditions of a *liberal* public sphere, informal public communication accomplishes two things in cases in which mobilization depends on crisis. On the one hand, it prevents the accumulation of indoctrinated masses that are seduced by populist leaders. On the other hand, it pulls together the scattered critical potentials of a public that was only abstractly held together through the public media, and it helps this public have a political influence on institutionalized opinion- and will-formation. Only in *liberal* public spheres, of course, do subinstitutional political movements—which abandon the conventional paths of interest politics in order to boost the constitutionally regulated circulation of power in the political system—take this direction. By contrast, an authoritarian, distorted public sphere that is brought into alignment merely provides a forum for plebiscitary legitimation.

This sense of reinforced demand for legitimation becomes especially clear when subinstitutional protest movements reach a high point by escalating their protests. The last means for obtaining more of a hearing and greater media influence for oppositional arguments are acts of civil disobedience. These acts of nonviolent, symbolic rule violation are meant as expressions of protest against binding decisions that, their legality notwithstanding, the actors consider illegitimate in the light of valid constitutional principles. Acts of civil disobedience are directed simultaneously to two addressees. On the one hand, they appeal to officeholders and parliamentary representatives to reopen formally concluded political deliberations so that their decisions may possibly be revised in view of the continuing public criticism. On the other hand, they appeal "to the sense of justice of the majority of the community," as Rawls puts it,[vii] and thus to the critical judgment of a public of citizens that is to be mobilized with exceptional means. Independently of the current object of controversy, civil disobedience is also always an implicit appeal to connect organized political will-formation with the communicative processes of the public sphere. The message of this subtext is aimed at a political system that, as constitutionally organized, may not detach itself from civil society and make itself independent vis-à-vis the periphery. Civil disobedience thereby refers to its own origins in a civil society that in crisis situations actualizes the normative contents of constitutional democracy in the medium of public opinion and summons it against the systemic inertia of institutional politics. . . .

This interpretation of civil disobedience manifests the self-consciousness of a civil society confident that at least in a crisis it can increase the pressure of a mobilized public on the political system to the point where the latter switches into the conflict mode and neutralizes the unofficial countercirculation of power.

Beyond this, the justification of civil disobedience relies on a *dynamic understanding* of the constitution as an unfinished project. From this long-term perspective, the constitutional state does not represent a finished structure but a delicate and sensitive—above all fallible and revisable—enterprise, whose purpose is to realize the system of rights *anew* in changing circumstances, that is, to interpret the system of rights better, to institutionalize it more appropriately, and to draw out its contents more radically. This is the perspective of citizens who are actively engaged in realizing the system of rights. Aware of, and referring to, changed contexts, such citizens want to overcome in practice the tension between social facticity and validity. Although legal theory cannot adopt this participant perspective as its own, it can reconstruct the paradigmatic *understanding* of law and democracy that guides citizens whenever they form an idea of the structural constraints on the self-organization of the legal community in their society.

■■

Introduction to "The Tasks of a Critical Theory of Society"

In this selection, Habermas extends his analysis of the public sphere into a more general theory of communication in modern societies. Central to the essay are several key concepts: system, lifeworld, uncoupling, the colonization of the lifeworld, and communicative action. In Habermas's account, all societies can be understood theoretically as composed of two primary components—the system and the lifeworld. The system is that set of interrelated social structures, like the state and economy, that form the primary basis for the organization of social life. The lifeworld is the arena of everyday, human interaction carved out of ongoing and negotiated interpersonal communication.

Habermas suggests that an optimally functioning society is one in which systemic processes are neatly coupled with the lifeworld dynamics of everyday communication and consensus formation. Such interrelations allow for rationalized and differentiated system-level structures to emerge that are grounded in and guided by the critical and communicative demands of the lifeworld. Habermas argues, however, that in the modern world, system and lifeworld have become "uncoupled," resulting in the emergence of "steering crises" and "pathologies in the lifeworld." The most striking disturbance has been within the lifeworld, which has been "colonized" by reified and increasingly self-insulated systemic processes. The

---

[vii]J. Rawls, *A Theory of Justice* (Cambridge, Mass., 1971), p. 364.

result of this "colonization of the lifeworld" is that system-level institutions, most notably economic institutions and governments, operate with virtual autonomy from the normative checks and balances that can only emanate from the lifeworld, and that the lifeworld itself has become radically deformed. Here, the "steering media" that guide systemic institutional processes (money and power) penetrate the lifeworld and displace meaningful social interaction and symbolic communication as a primary basis for social relationships. It is the "task of a critical theory of society" to critique existing social structures in order to rebalance the relationship between system and lifeworld and thus create the conditions necessary for producing a free and democratic social order.

Habermas places his hopes for the emancipation of humanity in the counterweight of social movements and, in particular, the power of communicative action. While the system is steered by money and power and a form of instrumental rationality that emphasizes efficiency and means over ends, communicative action emanates from within the lifeworld. As such, communicative action is grounded in a distinct form of rationality (i.e., "communicative rationality") that makes possible a discourse capable of generating social solidarity, mutual understandings, and personal and collective identities. For their part, new social movements mount protests and give air to public opinion in an effort to defend and restore "endangered ways of life" that cannot be secured through money and power. Rather, the conflicts they give voice to "arise in domains of cultural reproduction, social integration, and socialization." Their aims speak not to economic or political transformations but, rather, to reestablishing control over "the grammar of forms of life."

"The Tasks of a Critical Theory of Society" (1987)

Jürgen Habermas

In respects a theory of capitalist modernization developed by some means of a theory of communicative action does follow the Marxian model. It is *critical* both of contemporary social sciences and of the social reality they are supposed to grasp. It is critical of the reality of developed societies inasmuch as they do not make full use of the learning potential culturally available to them, but deliver themselves over to an uncontrolled growth of complexity. . . . This increasing system complexity encroaches upon nonrenewable supplies like a quasi-natural force; not only does it out-flank traditional forms of life, it attacks the communicative infrastructure of largely rationalized lifeworlds. But the theory is also critical of social-scientific approaches that are incapable of deciphering the paradoxes of societal rationalization because they make complex social systems their object only from one or another abstract point of view, without accounting for the historical constitution of their object domain (in the sense of a reflexive sociology). Critical social theory does not relate to established lines of research as a competitor; starting from its concept of the rise of modern societies, it attempts to explain the specific limitations and the relative rights of those approaches.

If we leave to one side the insufficiently complex approach of behaviorism, there are today three main lines of inquiry occupied with the phenomenon of modern societies. We cannot even say that they are in competition, for they scarcely have anything to say to one another. Efforts at theory comparison do not issue in reciprocal critique; fruitful critique that might foster a common undertaking can hardly be developed across these distances, but at most within one or another camp. There is a good reason for this mutual incomprehension: the object domains of the competing approaches do not come into contact, for they are the result of one-sided abstractions that unconsciously cut the ties between system and lifeworld constitutive for modern societies. . . .

Whereas the theory of structural differentiation does not sufficiently separate systemic and lifeworld aspects, systems theory and action theory, each isolates and overgeneralizes one of the two aspects. The methodological abstractions have the same result in all three cases. The theories of modernity made possible by these approaches remain insensitive to what Marx called "real abstractions"; the latter can be gotten at through an analysis that at once traces the rationalization of lifeworlds *and* the growth in complexity of media-steered subsystems, and that keeps the paradoxical nature of their interference in sight. As we have seen, it is possible to speak in a nonmetaphorical sense of paradoxical conditions of life if the structural differentiation of lifeworlds is described as rationalization. Social pathologies are not to be measured against "biological" goal states but in relation to the contradictions in which communicatively intermeshed interaction can get caught because deception and self-deception can gain objective power in an everyday practice reliant on the facticity of validity claims.

By "real abstractions" Marx was referring not only to paradoxes experienced by those involved as deformations of their lifeworld, but above all to paradoxes that could be gotten at only through an analysis of reification (or of rationalization). It is in this latter sense that we call "paradoxical" those situations in which systemic relief mechanisms made possible by the rationalization of the lifeworld turn around and overburden the communicative infrastructure of the lifeworld. After attempting to render a fourth approach to inquiry—the *genetic structuralism* of developmental psychology—fruitful for appropriating Weber's sociology of religion, Mead's theory of communication, and Durkheim's theory of social integration, I proposed that we read the Weberian rationalization thesis in that way. The basic conceptual framework I developed by these means was, naturally, not meant to be an end in itself; rather, it has to prove itself against the task of explaining those pathologies of modernity that other approaches pass right by for methodological reasons. . . .

With these illustrative remarks I also intend to emphasize the fully open character and the flexibility of an approach to social theory whose fruitfulness can be confirmed only in the ramifications of social and philosophical research. As to what social theory can accomplish in and of itself—it resembles the focusing power of a magnifying glass. Only when the social sciences no longer sparked a single thought would the time for social theory be past.

(a) On the forms of integration in postliberal societies. Occidental rationalism arose within the framework of bourgeois capitalist societies. For this reason, following Marx and Weber I have examined the initial conditions of modernization in connection with societies of this type and have traced the capitalist path of development. In postliberal societies there is a fork in this path: modernization pushes forward in one direction through endogenously produced problems of economic accumulation, in the other through problems arising from the state's efforts at rationalization. Along the developmental path of organized capitalism, a political order of welfare-state mass democracy took shape. In some places, however, under the pressure of economic crises, the mode of production, threatened by social disintegration, could be maintained for a time only in the political form of authoritarian or fascist orders. Along the developmental path of bureaucratic socialism a political order of dictatorship by state parties took shape. . . . If we permit ourselves to simplify in an ideal-typical manner and limit ourselves to the two dominant variants of postliberal societies, and if we start from the assumption that alienation phenomena arise as systemically induced deformations of the lifeworld, then we can take a few steps toward a comparative analysis of principles of societal organizations, kinds of crisis tendencies, and forms of social pathology.

On our assumption, a considerably rationalized lifeworld is one of the initial conditions for modernization processes. It must be possible to anchor money and power in the lifeworld as media, that is, to institutionalize them by means of positive law. If these conditions are met, economic and administrative systems can be differentiated out, systems that have a complementary relation to one another and enter into interchanges with their environments via steering media. At this level of system differentiation modern societies arise, first capitalist societies, and later—setting themselves off from those—bureaucratic-socialist societies. A capitalist path of modernization opens up as soon as the economic system develops its own intrinsic dynamic of growth and, with its endogenously produced problems, takes the lead, that is, the evolutionary primacy, for society as a whole. The path of modernization runs in another direction when, on the basis of state ownership of most of the means of production and an institutionalized one-party rule, the administrative action system gains a like autonomy in relation to the economic system.

To the extent that these organizational principles are established, there arise interchange relations between the two functionally interlocked subsystems and the societal components of the lifeworld in locked subsystems and the societal components of the lifeworld in which the media are anchored. The lifeworld, more or less relieved of tasks of material reproduction, can in turn become more differentiated in its symbolic structures and can set free the inner logic of development of cultural modernity. At the same time, the private and public spheres are now set off as the environments of the system. According to whether the economic system or the state apparatus attains evolutionary primacy, either private households or politically relevant memberships are the points of entry for crises that are shifted from the subsystems to the lifeworld. In modernized societies disturbances in the material reproduction of the lifeworld take the form of stubborn systemic disequilibria; the latter either take effect directly as *crises* or they call forth *pathologies* in the lifeworld.

Steering crises were first studied in connection with the business cycle of market economies. In bureaucratic socialism, crisis tendencies spring from self-blocking mechanisms in planning administrations, as they do on the other side from endogenous interruptions of accumulation processes. Like the paradoxes of exchange rationality, the paradoxes of planning rationality can be explained by the fact that rational action orientations come into contradiction with themselves through unintended systemic effects. These crisis tendencies are worked through not only in the subsystem in which they arise, but also in the complementary action system into which they can be shifted. Just as the capitalist economy relies on organizational performances of the state, the socialist planning bureaucracy has to rely on self-steering performances of the economy. Developed capitalism swings between the contrary policies of "the market's self-healing powers" and state interventionism. The structural dilemma is even clearer on the other side, where policy oscillates hopelessly between increased central planning and decentralization, between orienting economic programs toward investment and toward consumption.

These *systemic disequilibria* become *crises* only when the performances of economy and state remain manifestly below an established level of aspiration and harm the symbolic reproduction of the lifeworld by calling forth conflicts and reactions of resistance there. It is the societal components of the lifeworld that are directly affected by this. Before such conflicts threaten core domains of social integration, they are pushed to the periphery—before anomic conditions arise there are appearances of withdrawal of legitimation or motivation. But when steering crises—that is, perceived disturbances of material reproduction—are successfully intercepted by having recourse to lifeworld resources, pathologies arise in the lifeworld. These resources appear as contributions to cultural reproduction, social integration, and socialization. For the continued existence of the economy and the state, it is the resources that contribute to the maintenance of society that are relevant, for it is here, in the institutional orders of the lifeworld, that subsystems are anchored.

We can represent the replacement of steering crises with lifeworld pathologies as follows: anomic conditions are avoided, and legitimations and motivations important for maintaining institutional orders are secured, at the expense of, and through the ruthless exploitation of, other resources. Culture and personality come under attack for the sake of warding off crises and stabilizing society. Instead of manifestations of anomie (and instead of the withdrawal of legitimation and motivation in place of anomie), phenomena of alienation and the unsettling of collective identity emerge. I have traced such phenomena back to a colonization of the lifeworld and characterized them as a reification of the communicative practice of everyday life.

However, deformations of the lifeworld take the form of a reification of communicative relations only in capitalist societies, that is, only where the private household is the point of incursion for the displacement of crises into the lifeworld. This is not a question of the overextension of a single medium but of the monetarization and bureaucratization of the sphere of action of employees and of consumers, of citizens and of clients of state bureaucracies. Deformations of the lifeworld take a different form in societies in which the points of incursion for the penetration of crises into the lifeworld are politically relevant memberships. There too, in bureaucratic-socialist societies, domains of action that are dependent on social integration are switched over to mechanisms of system integration. But instead of the reification of communicative relations we find the shamming of communicative relations in bureaucratically desiccated, forcibly "humanized" domains of pseudopolitical intercourse in an overextended and administered public sphere. This pseudopoliticization is symmetrical to reifying privatization in certain respects. The lifeworld is not directly assimilated to the system, that is, to legally regulated, formally organized domains of action; rather, systemically self- sufficient organizations are fictively put back into a simulated horizon of the lifeworld. While the system is draped out as the lifeworld, the lifeworld is absorbed by the system.

(b) Family socialization and ego development. The diagnosis of an uncoupling of system and lifeworld also offers a different perspective for judging the structural change in family, education, and personality development. For a psychoanalysis viewed from a Marxist standpoint, the theory of the Oedipus complex, interpreted sociologically, was pivotal for explaining how the functional imperatives of the economic system could establish themselves in the superego structures of the dominant social character. Thus, for example, Löwenthal's studies of drama and fiction in the nineteenth century served to show in detail that the constraints of the economic system—concentrated in status hierarchies, occupational roles, and gender stereotypes—penetrated into the innermost aspects of life history via intra-familial dependencies and patterns of socialization. The intimacy of highly personalized relations merely concealed the blind force of economic inter-dependencies that had become autonomous in relation to the private sphere—a force that was experienced as "fate."

Thus the family was viewed as the agency through which systemic imperatives influenced our instinctual vicissitudes; its communicative internal structure was not taken seriously. Because the family was always viewed only from functionalist standpoints and was never given its own weight from structuralist points of view, the epochal changes in the bourgeois family could be misunderstood; in particular, the results of the leveling out of paternal authority could be interpreted wrongly. It seemed as if systemic imperatives now had the chance—by way of a mediatized family—to take hold directly of intrapsychic events, a process that the soft medium of mass culture could at most slow down. If, by contrast, we *also* recognize in the structural transformation of the bourgeois family the inherent rationalization of the lifeworld; if we see that, in egalitarian patterns of relationship, in individual forms of intercourse, and in liberalized child-rearing practices, some of the potential for rationality ingrained in communicative action is *also* released; then the changed conditions of socialization in the middle-class nuclear family appear in a different light.

Empirical indicators suggest the growing autonomy of a nuclear family in which socialization processes take place through the medium of largely deinstitutionalized communicative action. Communicative infrastructures are developing that have freed themselves from latent entanglements in systemic dependencies. The contrast between the *homme* who is educated to freedom and humanity in the intimate sphere and the *citoyen* who obeys functional necessities in the sphere of social labor was always an ideology. But it has now taken on a different meaning. Familial lifeworlds see the imperatives of the economic and administrative systems coming at them from outside, instead of being mediatized by them from behind. In the families and their environments we can observe a polarization between communicatively structured and formally organized domains of action; this places socialization processes under different conditions and exposes them to a different type of danger. This view is supported by two rough sociopsychological clues: the diminishing significance of the Oedipal problematic and the growing significance of adolescent crises.

For some time now, psychoanalytically trained physicians have observed a symptomatic change in the typical manifestations of illness. Classical hysterias have almost died out; the number of compulsion neuroses is drastically reduced; on the other hand, narcissistic disturbances are on the increase. Christopher Lasch has taken this symptomatic change as the occasion for a diagnosis of the times that goes beyond the clinical domain.[i] It confirms the fact that the significant changes in the present escape sociopsychological explanations that start from the Oedipal problematic, from an internalization of societal repression which is simply masked by parental authority. The better explanations start from the premise that the communications structures that have been set free in the family provide conditions for socialization that are as demanding as they are vulnerable. The potential for irritability grows, and with it the probability that instabilities in parental behavior will have a comparatively strong effect—a subtle neglect.

The other phenomenon, a sharpening of the adolescence problematic, also speaks for the socializatory significance of the uncoupling of system and lifeworld. Systemic imperatives do not so much insinuate themselves into the family, establish themselves in systematically distorted communication, and inconspicuously intervene in the formation of the self as, rather, openly

---

[i]Christopher Lasch, *The Culture of Narcissism* (New York, 1978).

come at the family from outside. As a result, there is a tendency toward disparities between competences, attitudes, and motives, on the one hand, and the functional requirements of adult roles on the other. The problem of detaching oneself from the family and forming one's own identity has in any case turned adolescent development (which is scarcely safeguarded by institutions anymore) into a critical test for the ability of the coming generation to connect up with the preceding one. When the conditions of socialization in the family are no longer functionally in tune with the organizational membership conditions that the growing child will one day have to meet, the problems that young people have to solve in their adolescence become insoluble for more and more of them. One indication of this is the social and even political significance that youth protest and withdrawal cultures have gained since the end of the 1960s.

This new problem situation cannot be handled with the old theoretical means. If we connect the epochal changes in family socialization with the rationalization of the lifeworld, socializatory interaction becomes the point of reference for the analysis of ego development, and systematically distorted communication—the reification of interpersonal relations—the point of reference for investigating pathogenesis. The theory of communicative action provides a framework within which the structural model of ego, id, and superego can be recast. Instead of an instinct theory that represents the relation of ego to inner nature in terms of a philosophy of consciousness— on the model of relations between subject and object— we have a theory of socialization that connects Freud with Mead, gives structures of intersubjectivity their due, and replaces hypotheses about instinctual vicissitudes with assumptions about identity formation. This approach can (i) appropriate more recent developments in psychoanalytic research, particularly the theory of object relations and ego psychology, (ii) take up the theory of defense mechanisms in such a way that the interconnections between intrapsychic communication barriers and communication disturbances at the interpersonal level become comprehensible, and (iii) use the assumptions about mechanisms of conscious and unconscious mastery to establish a connection between orthogenesis and pathogenesis. The cognitive and sociomoral development studied in the Piagetian tradition takes place in accord with structural patterns that provide a reliable foil for intuitively recorded clinical deviations.

(c) Mass media and mass culture. With its distinction between system and lifeworld, the theory of communicative action brings out the independent logic of socializatory interaction; the corresponding distinction between two contrary types of communication media makes us sensitive to the ambivalent potential of mass communications. The theory makes us skeptical of the thesis that the essence of the public sphere has been liquidated in postliberal societies. According to Horkheimer and Adorno, the communication flows steered via mass media *take the place of* those communication structures that had once made possible public discussion and self-understanding by citizens and private individuals. With the shift from writing to images and sounds, the electronic media—first film and radio, later television—present themselves as an apparatus that completely permeates and dominates the language of everyday communication. On the one hand, it transforms the authentic content of modern culture into the sterilized and ideologically effective stereotypes of a mass culture that merely replicates what exists; on the other hand, it uses up a culture cleansed of all subversive and transcending elements for an encompassing system of social controls, which is spread over individuals, in part reinforcing their weakened internal behavioral controls, in part replacing them. The mode of functioning of the culture industry is said to be a mirror image of the psychic apparatus, which, as long as the internalization of paternal authority was still functioning, had subjected instinctual nature to the control of the superego in the way that technology had subjected outer nature to its domination. . . .

I distinguished two sorts of media that can ease the burden of the (risky and demanding) coordinating mechanism of reaching understanding: on the one hand, steering media, via which subsystems are differentiated out of the lifeworld; on the other hand, generalized forms of communication, which do not replace reaching agreement in language but merely condense it, and thus remain tied to lifeworld contexts. Steering media uncouple the coordination of action from building consensus in language altogether and neutralize it in regard to the alternative of coming to an agreement or failing to do so. In the other case we are dealing with a specialization of linguistic processes of consensus formation that remains dependent on recourse to the resources of the lifeworld background. The mass media belong to these generalized forms of communication. They free communication processes from the provinciality of spatiotemporally restricted contexts and permit public spheres to emerge, through establishing the abstract simultaneity of a virtually present network of communication contents far removed in space and time and through keeping messages available for manifold contexts.

These media publics hierarchize and at the same time remove restrictions on the horizon of possible communication. The one aspect cannot be separated from the other—and therein lies their ambivalent potential. Insofar as mass media one-sidedly channel communication flows in a centralized network—from the center to the periphery or from above to below—they considerably strengthen the efficiency of social controls. But tapping this authoritarian potential is always precarious because there is a counterweight of emancipatory potential built into communication structures themselves. Mass media can simultaneously contextualize and concentrate processes of reaching understanding, but it is only in the first instance that they relieve interaction from yes/no responses to criticizable validity claims. Abstracted and clustered though they are, these communications cannot be reliably shielded from the possibility of opposition by responsible actors.

When communications research is not abridged in an empiricist manner and allows for dimensions of reification in communicative everyday practice, it confirms this ambivalence. Again and again reception research and program analysis have provided illustrations of the theses in culture criticism that Adorno, above all, developed with a certain overstatement. In the meantime, the same energy has been put into working out the contradictions resulting from the facts that

- the broadcasting networks are exposed to competing interests; they are not able to smoothly integrate economic, political and ideological, professional and aesthetic viewpoints;
- normally the mass media cannot, without generating conflict, avoid the obligations that accrue to them from their journalistic mission and the professional code of journalism;
- the programs do not only, or even for the most part, reflect the standards of mass culture; even when they take the trivial forms of popular entertainment, they may contain critical messages—"popular culture as popular revenge";[ii]
- ideological messages miss their audience because the intended meaning is turned into its opposite

under conditions of being received against a certain subcultural background;
- the inner logic of everyday communicative practice sets up defenses against the direct manipulative intervention of the mass media; and
- the technical development of electronic media does not necessarily move in the direction of centralizing networks, even though "video pluralism" and "television democracy" are at the moment not much more than anarchist visions.

(d) Potentials for protest. My thesis concerning the colonization of the lifeworld, for which Weber's theory of societal rationalization served as a point of departure, is based on a critique of functionalist reason, which agrees with the critique of instrumental reason only in its intention and in its ironic use of the word reason. One major difference is that the theory of communicative action conceives of the lifeworld as a sphere in which processes of reification do not appear as mere reflexes—as manifestations of a repressive integration emanating from an oligopolistic economy and an authoritarian state. In this respect, the earlier critical theory merely repeated the errors of Marxist functionalism. My references to the socializatory relevance of the uncoupling of system and lifeworld and my remarks on the ambivalent potentials of mass media and mass culture show the private and public spheres in the light of a rationalized lifeworld in which system imperatives *clash with* independent communication structures. The transposition of communicative action to media-steered interactions and the deformation of the structures of a damaged intersubjectivity are by no means predecided processes that might be distilled from a few global concepts. The analysis of lifeworld pathologies calls for an (unbiased) investigation of tendencies *and* contradictions. The fact that in welfare-state mass democracies class conflict has been institutionalized and thereby pacified does not mean that protest potential has been altogether laid to rest. But the potentials for protest emerge now along different lines of conflict—just where we would expect them to emerge if the thesis of the colonization of the lifeworld were correct.

In the past decade or two, conflicts have developed in advanced Western societies that deviate in various ways from the welfare-state pattern of institutionalized conflict over distribution. They no longer flare up in

---

[ii]D. Kellner, "TV, Ideology and Emancipatory Popular Culture," *Socialist Review* 45 (1979): 13ff.

domains of material reproduction; they are no longer channeled through parties and associations; and they can no longer be allayed by compensations. Rather, these new conflicts arise in domains of cultural reproduction, social integration, and socialization; they are carried out in subinstitutional—or at least extraparliamentary—forms of protest; and the underlying deficits reflect a reification of communicatively structured domains of action that will not respond to the media of money and power. The issue is not primarily one of compensations that the welfare state can provide, but of defending and restoring endangered ways of life. In short, the new conflicts are not ignited by distribution problems but by questions having to do with the grammar of forms of life.

This new type of conflict is an expression of the "silent revolution" in values and attitudes that R. Inglehart has observed in entire populations. Studies by Hildebrandt and Dalton, and by Barnes Kaase, confirm the change in themes from the "old politics" (which turns on questions of economic and social security, internal and military security) to a "new politics." The new problems have to do with quality of life, equal rights, individual self-realization, participation, and human rights. In terms of social statistics, the "old politics" is more strongly supported by employers, workers, and middle-class tradesmen, whereas the new politics finds stronger support in the new middle classes, among the younger generation, and in groups with more formal education. These phenomena tally with my thesis regarding internal colonization.

If we take the view that the growth of the economic-administrative complex sets off processes of erosion in the lifeworld, then we would expect old conflicts to be overlaid with new ones. A line of conflict forms between, on the one hand, a center composed of strata *directly* involved in the production process and interested in maintaining capitalist growth as the basis of the welfare-state compromise, and, on the other hand, a periphery composed of a variegated array of groups that are lumped together. Among the latter are those groups that are further removed from the "productivist core of performance" in late capitalist societies, that have been more strongly sensitized to the self-destructive consequences of the growth in complexity or have been more strongly affected by them. The bond that unites these heterogeneous groups is the critique of growth. Neither the bourgeois emancipation movements nor the struggles of the organized labor movement can serve as a model for this protest. Historical parallels are more likely to be found in the social-romantic movements of the early industrial period, which were supported by craftsmen, plebians, and workers, in the defensive movements of the populist middle class, in the escapist movements (nourished by bourgeois critiques of civilization) undertaken by reformers, the *Wandervögel,* and the like.

The current potentials for protest are very difficult to classify, because scenes, grouping, and topics change very rapidly. To the extent that organizational nuclei are formed at the level of parties or associations, members are recruited from the same diffuse reservoir. The following catchphrases serve at the moment to identify the various currents in the Federal Republic of Germany: the antinuclear and environmental movements; the peace movement (including the theme of north-south conflict); single-issue and local movements; the alternative movement (which encompasses the urban "scene," with its squatters and alternative projects, as well as the rural communes); the minorities (the elderly, gays, handicapped, and so forth); the psychoscene, with support groups and youth sects; religious fundamentalism; the tax-protest movement, school protest by parents' associations, resistance to "modernist" reforms; and, finally, the women's movement. Of international significance are the autonomy movements struggling for regional, linguistic, cultural, and also religious independence.

In this spectrum I will differentiate emancipatory potentials from potentials for resistance and withdrawal. After the American civil rights movement—which has since issued in a particularistic self-affirmation of black subcultures—only the feminist movement stands in the tradition of bourgeois-socialist liberation movements. The struggle against patriarchal oppression and for the redemption of a promise that has long been anchored in the acknowledged universalistic foundations of morality and law gives feminism the impetus of an offensive movement, whereas the other movements have a more defensive character. The resistance and withdrawal movements aim at stemming formally organized domains of action for the sake of communicatively structured domains, and not at conquering new territory. There is an element of particularism that connects feminism with these movements; the emancipation of women means not only establishing formal equality and eliminating male privilege, but overturning concrete forms of life marked by male monopolies. Furthermore, the historical legacy of the sexual division of labor to which women were subjected in the bourgeois nuclear family has given them access to contrasting virtues, to a register of values complementary to those of the male world and opposed to a one-sidedly rationalized everyday practice.

Within resistance movements we can distinguish further between the defense of traditional and social rank (based on property) and a defense that already operates on the basis of a rationalized lifeworld and tries out new ways of cooperating and living together. This criterion makes it possible to demarcate the protest of the traditional middle classes against threats to neighborhoods by large technical projects, the protest of parents against comprehensive schools, the protest against taxes (patterned after the movement in support of Proposition 13 in California), and most of the movements for autonomy, on the one side, from the core of a new conflict potential, on the other: youth and alternative movements for which a critique of growth sparked by themes of ecology and peace is the common focus. It is possible to conceive of these conflicts in terms of resistance to tendencies toward a colonization of the lifeworld, as I hope now to indicate, at least in a cursory way. The objectives, attitudes, and ways of acting prevalent in youth protest groups can be understood, to begin with, as reactions to certain problem situations that are perceived with great sensitivity.

"Green" problems. The intervention of large-scale industry into ecological balances, the growing scarcity of nonrenewable natural resources, as well as demographic developments present industrially developed societies with major problems; but these challenges are abstract at first and call for technical and economic solutions, which must in turn be globally planned and implemented by administrative means. What sets off the protest is rather the tangible destruction of the urban environment; the despoliation of the countryside through housing developments, industrialization, and pollution; the impairment of health through the ravages of civilization, pharmaceutical side-effects, and the like—that is, developments that noticeably affect the organic foundations of the lifeworld and make us drastically aware of standards of livability, of inflexible limits to the deprivation of sensual-aesthetic background needs.

Problems of excessive complexity. There are certainly good reasons to fear military potentials for destruction, nuclear power plants, atomic waste, genetic engineering, the storage and central utilization of private data, and the like. These real anxieties are combined, however, with the terror of a new category of risks that are literally invisible and are comprehensible only from the perspective of the system. These risks invade the lifeworld and at the same time burst its dimensions. The anxieties function as catalysts for a feeling of being overwhelmed in view of the possible consequences of processes for which we are morally accountable—since we do set them in motion technically and politically—and yet for which we can no longer take moral responsibility—since their scale has put them beyond our control. Here resistance is directed against abstractions that are forced upon the lifeworld, although they go beyond the spatial, temporal, and social limits of complexity of even highly differentiated lifeworlds, centered as these are around the senses.

Overburdening the communicative infrastructure. Something that is expressed rather blatantly in the manifestations of the psychomovement and renewed religious fundamentalism is also a motivating force behind most alternative projects and many citizens' action groups—the painful manifestations of deprivation in a culturally impoverished and one-sidedly rationalized practice of everyday life. For this reason, ascriptive characteristics such as gender, age, skin color, neighborhood or locality, and religious affiliation serve to build up and separate off communities, to establish subculturally protected communities supportive of the search for personal and collective identity. The revaluation of the particular, the natural, the provincial, of social spaces that are small enough to be familiar, of decentralized forms of commerce and despecialized activities, of segmented pubs, simple interactions and dedifferentiated public spheres—all this is meant to foster the revitalization of possibilities for expression and communication that have been buried alive. Resistance to reformist interventions that turn into their opposite, because the means by which they are implemented run counter to the declared aims of social integration, also belongs in this context.

The new conflicts arise along the seams between system and lifeworld. Earlier I described how the interchange between the private and public spheres, on the one hand, and the economic and administrative action systems, on the other, takes place via the media of money and power, and how it is institutionalized in the roles of employees and consumers, citizens and clients of the state. It is just these roles that are the targets of protest. Alternative practice is directed against the profit-dependent instrumentalization of work in one's vocation, the market-dependent mobilization of labor power, against the extension of pressures of competition and performance all the way down into elementary school. It also takes aim at the monetarization of services, relationships, and time, at the consumerist redefinition of private spheres of life and personal lifestyles. Furthermore, the relation of clients to public service agencies is to be opened up and reorganized in a participatory mode, along the lines of self-help organizations. It is above all in the domains of social policy and health policy (e.g., in connection with psychiatric care) that models of reform point in this direction. Finally, certain forms of protest negate the definitions of the role of citizen and the routines for pursuing interests in a purposive-rational manner—forms ranging from the undirected explosion of disturbances by youth ("Zurich is burning!"), through calculated or surrealistic violations of rules (after the pattern of the American civil rights movement and student protests), to violent provocation and intimidation.

According to the programmatic conceptions of some theoreticians, a partial disintegration of the social roles of employees and consumers, of clients and citizens of the state, is supposed to clear the way for counterinstitutions that develop from within the lifeworld in order to set limits to the inner dynamics of the economic and political-administrative action systems. These institutions are supposed, on the one hand, to divert out of the economic system a second, informal sector that is no longer oriented to profit and, on the other hand, to oppose to the party system new forms of a "politics in the first person," a politics that is expressive and at the same time has a democratic base. Such institutions would reverse just those abstractions and neutralizations by which in modern societies labor and political will-formation have been tied to media-steered interaction. The capitalist enterprise and the mass party (as an "ideology-neutral organization for acquiring power") generalize their points of social entry via labor markets and manufactured public spheres; they treat their employees and voters as abstract labor power and voting subjects; and they keep at a distance— as environments of the system—those spheres in which personal and collective identities can alone take shape. By contrast, the counterinstitutions are intended to dedifferentiate some parts of the formally organized domains

of action, remove them from the clutches of the steering media, and return these "liberated areas" to the action-coordinating mechanisms of reaching understanding.

However unrealistic these ideas may be, they are important for the polemical significance of the new resistance and withdrawal movements reacting to the colonization of the lifeworld. This significance is obscured, both in the self-understanding of those involved and in the ideological imputations of their opponents, if the communicative rationality of cultural modernity is rashly equated with the functionalist rationality of self-maintaining economic and administrative action systems—that is, whenever the rationalization of the lifeworld is not carefully distinguished from the increasing complexity of the social system. This confusion explains the fronts—which are out of place and obscure the real political oppositions—between the antimodernism of the Young Conservatives and the neoconservative defense of postmodernity that robs a modernity at variance itself of its rational content and its perspectives on the future. . . .

The theory of modernity that I have here sketched in broad strokes permits us to recognize the following: In modern societies there is such an expansion of the scope of contingency for interaction loosed from normative contexts that the inner logic of communicative action "becomes practically true" in the deinstitutionalized forms of intercourse of the familial private sphere as well as in a public sphere stamped by the mass media. At the same time, the systemic imperatives of autonomous subsystems penetrate into the lifeworld and, through monetarization and bureaucratization, force an assimilation of communicative action to formally organized domains of action—even in areas where the action-coordinating mechanism of reaching understanding is functionally necessary. It may be that this provocative threat, this challenge that places the symbolic structures of the lifeworld as a whole in question, can account for why they have become accessible to us.

▩

▩ ANTHONY GIDDENS (1938–): A BIOGRAPHICAL SKETCH

Anthony Giddens was born in 1938 in Edmonton, a suburb of London England. His father was a clerk who worked for London Transport, and his mother was a housewife. Theirs was a typical middle-class family, and Giddens graduated from grammar school with below-average marks. He later attended the University of Hull, where he intended to study philosophy. However, Giddens found that he excelled in psychology and sociology. Inspired by a lively sociology department led by Peter Worsley, Giddens graduated with a first-class honors degree in sociology in 1959 (Kaspersen 2000:1). He then went on to study at the prestigious London School of Economics and Political Science, where he wrote a Weber-inspired master's thesis on the development of sport in England in the nineteenth century. Unfortunately for Giddens, sport was a marginal topic in sociology at the time, leaving him to feel that his supervisors did not take his work all that seriously (Elliot 2001:293).

From 1961 to 1969, Giddens worked as a lecturer at the University of Leicester alongside several of Europe's top sociologists, including Norbert Elias and Ilya Neustadt, and held guest lectureships at Simon Fraser University near Vancouver, Canada, and the University of California, Los Angeles. During this time, Giddens began to focus on the connections and divergences between the core classical figures in sociology (Marx, Weber, and Durkheim) and drew up plans for his first book, *Capitalism and Modern Social Theory* (1971). This book was followed by a spate of others, including *New Rules of Sociological Method* (1976), *Studies in Social and Political Theory* (1977), *Central Problems in Social Theory* (1979), and *A Contemporary Critique of Historical Materialism* (1981a).[6] His most ambitious work, *The Constitution of Society,* was published in 1984. In this book, Giddens outlined his theory of "structuration," which has made a lasting impact on the field.

In 1970, Giddens became a lecturer at King's College, Cambridge, and attained a professorship there in 1985. During this time, he continued to focus on structuration theory and historical sociology in such works as *The Nation-State and Violence* (1985) and *Social Theory and Modern Sociology* (1987). In *The Consequences of Modernity* (1990), he turned his attention to the debates over postmodern theory while examining the effects of contemporary society on social relations and the role of trust in everyday life. In 1992, Giddens was named winner of the prestigious Spanish prize the Prince of Asturias Award for Social Sciences, which recognizes the "scientific, technical, cultural, social and humanistic work performed by individuals, groups or institutions worldwide." In 1996, Giddens left Cambridge to become director of the London School of Economics and Political Science, a position that he still holds.

In the mid-1990s, Giddens's career took a striking new turn as he became an advisor to British Prime Minister Tony Blair, helping popularize the ideas of left-of-center politics known as the "Third Way." His recent books, *The Third Way* (1998) and *Beyond Left and Right* (1994), have influenced debates about the future of democracy in many countries across the world. In the 1990s, his work also took a decidedly social psychological turn, as Giddens addressed such issues as the self, identity, intimacy, and sexuality in *Modernity and Self-Identity* (1991) and *The Transformation of Intimacy* (1992). Although Giddens's ongoing political and publishing activities (he is cofounder of Polity Press and consulting editor for two publishing companies) reflect a departure from more formal theoretical concerns, his query into the sociology of the self is not necessarily inconsistent with his earlier work: Giddens has always held that sociology is "not only about the big institutions, such as government organizations, business firms or societies as a whole. It is very much about the individual and our individual experiences. We come to understand ourselves much better through grasping the wider social forces that influence our lives" (an interview with Anthony Giddens, 1999, http://www.polity.co.uk/giddens6/about/interview.asp).

GIDDENS'S INTELLECTUAL INFLUENCES AND CORE IDEAS ▪▪

In this section, we provide an overview of Giddens's general theory of social action: the theory of structuration. In developing his approach, Giddens creatively combines the insights of linguistic philosophy, phenomenology, ethnomethodology, psychoanalysis, hermeneutics, structuralism, poststructuralism, dramaturgy, and existentialism (Craib 1992:chap. 2) to fashion a multidimensional account of social life.

---

[6] The title of Giddens's book connotes the influence of Émile Durkheim, whose book *The Rules of Sociological Method* ([1895] 1966) was one of the first systematic statements outlining sociology as a distinct field of scientific study.

At the heart of his theory is the notion that the individual and society form an interdependent duality that he terms the **duality of structure**. By this, Giddens means that social structures are both the "medium and outcome" of the practices they organize. To understand what Giddens means by the duality of structure, consider the use of language. Every time an individual communicates with someone, she relies on language to convey the information she wishes to impart. In this way, language serves as the medium or structure that allows the individual to communicate. Moreover, in intentionally using language to communicate, the language itself and the rules that govern its use are unintentionally reproduced as an outcome or consequence of the process of communication (Giddens 1981b:171, 172). The significance of Giddens's notion of the duality of structure lies in his attempt to overcome the dualism that plagues social theory—the divide between subjectivist/individualist and objectivist/collectivist approaches that, in fostering either/or views of social life, present only half of the story.

To more fully appreciate Giddens's position, we must define several core ideas that inform his structuration theory. First is **structure**. Giddens defines structure as rules and resources drawn on by individuals "in the production and reproduction of social action" (Giddens 1984:19). Evidencing his indebtedness to the work of Alfred Schutz and Harold Garfinkel (see Chapter 13, "Phenomenology and Ethnomethodology"), **rules** are defined as "techniques or generalizable procedures applied in the enactment/reproduction of social practices" (ibid.:21). Rules are commonsense social recipes that allow us to make sense of the words and conduct of others as well as to form our own words and behavior. **Resources** refer to "'the capabilities of making things happen,' of bringing about particular states of affairs" (Giddens 1981b:170). Often this capability is tied to an individual's command over objects, goods, or other persons that then allows him to affect the world around him. In this light, resources are "bases" or "vehicles" for exercising power (Giddens 1979:69) and are by no means equally shared.

Expanding on the work of Ferdinand de Saussure, the founder of linguistic structuralism (see Chapter 15), Giddens argues that structures are properties that exist only in the moment of their use by actors. They have no existence outside of the time and space in which actors draw on them during social encounters, save as "know-how" or "memory traces." Yet, structures are reproduced at the very moment individuals draw on them to make sense of and act in the social world—thus the duality of structure. Recalling de Saussure's distinction between speech (*parole*) and language (*langue*), the duality of structure illustrates how "just as each speech act implies and draws upon the whole structure of language, so each social action implies and draws upon the structure it instantiates" (Craib 1992:42).

Giddens's understanding of structure is a significant departure from conventional sociological uses of the term, particularly those uses that are aligned with functionalist modes of analysis. Functionalist approaches view social structures as external forces that determine or constrain individual action.[7] Like the scaffolding of a building or the skeleton of an organism that is invisible to the observer, social structures are said to work "behind the backs" of individuals, since such structures exist to fulfill the needs of larger social systems. From this viewpoint, functionalists emphasize how the patterned actions of individuals have the unintended consequence or "latent function" of preserving the stability of the existing social order. What is of importance here is not understanding the motives and reasons individuals themselves may possess for carrying out certain behaviors, but, rather, uncovering "society's reasons" for

---

[7]In *The Rules of Sociological Method* ([1895] 1966:1), Émile Durkheim offered an early functionalist account of the constraints imposed on individual action by society. He noted,

> When I fulfill my obligations as brother, husband, or citizen, when I execute my contracts, I perform duties which are defined, externally to myself and my acts, in law and custom. Even if they conform to my own sentiments and I feel their reality subjectively, such reality is still objective, for I did not create them. . . . The system of signs I use to express my thought, the system of currency I employ to pay my debts, the instruments of credit I utilize in my commercial relations, the practices followed in my profession, etc., function independently of my own use of them. . . . These types of conduct or thought are not only external to the individual, but are, moreover, endowed with coercive power, by virtue of which they impose themselves upon him, independent of his individual will.

requiring such behavior on the part of its members (Giddens 1979:210–15). Ultimately, such reasons entail a society's need to survive and to reproduce the structures—or parts—that compose it.

For Giddens, this account of structure is problematic on a number of fronts. Most important, it ignores the fact that individuals more often than not "know what they are doing and why they are doing it" (Giddens 1981b:163). Individuals are "purposeful" actors who draw on the rules and resources at their disposal in order to make their way through everyday life. Structure, then, is both constraining *and* enabling: while it establishes limitations on action, it is also the medium through which one is able to affect his surroundings. Returning to our example of language, without the words (resources) and the know-how to use them correctly (rules) to describe an event or object, we cannot communicate their meaning to others. In this way, structure is constraining; it prevents us from acting. Conversely, structure is enabling to the extent that possessing both the words and the recipes for their use allows us to communicate the meaning of situations and things to others.

Claiming that individuals are "purposeful" and that they "know what they are doing" regarding their use of rules and resources leads Giddens to consider the issue of **agency**. Agency refers to a person's capability or power to act, to "make a difference," or to intentionally intervene in her world. "Agency concerns events of which an individual is the perpetrator, in the sense that the individual could, at any phase in a given sequence of conduct, have acted differently" (Giddens 1984:9). This notion of agency likewise entails assumptions regarding the "knowledgeability" or consciousness of actors. Here, Giddens draws an important distinction between what we "know" about our actions and our ability to put this knowledge into words—that is, our ability to explain or give reasons for our conduct. This distinction represents different modes of consciousness that Giddens depicts in Figure 16.4.

Figure 16.4   Modes of Consciousness

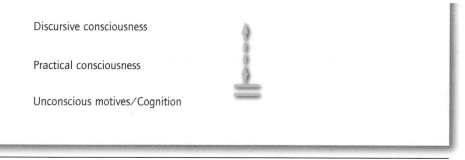

Discursive consciousness

Practical consciousness

Unconscious motives/Cognition

SOURCE: Figure reproduced from Giddens (1984:7).

Drawing again on phenomenology and ethnomethodology, as well as on Freudian psychoanalysis and Goffman's dramaturgy, Giddens notes that, although we often are unable to explicitly state the reasons for our behaviors, it does not necessarily follow that we do not know why we carried them out. Thus, much of our conduct is carried out with only our tacit, unspoken awareness of how and why we do what we do. This tacit awareness is lodged in "practical consciousness" and consists of shared stocks of knowledge, or "mutual knowledge," on the basis of which we are able to coordinate our actions with others. Parallel to Bourdieu's notion of the habitus, **practical consciousness** refers to what individuals know "about social conditions, including especially the conditions of their own action, but cannot express" in words (ibid.:375). For instance, when conversing, we employ any number of rules regarding turn-taking, eye contact, body positioning, choice of topics, and speed and volume of speech. Typically these behaviors are simply done; we are not asked to supply explicit reasons or explanations for doing them, nor would it be easy for us to do so should we be asked. The knowledge that informs these behaviors is not immediately accessible to the individual's consciousness; rather, it is practical in nature, "inherent in the capability to 'go on' within the routines of social life" (ibid.:4). While social interaction is shaped largely by tacit adherence to rules governing behaviors such as these, that does not mean that individuals do not understand why they act as they do, nor does it suggest that such behaviors are unintentional.

Discursive consciousness refers to the capacity to explicitly state reasons or explanations for our conduct. It is "what actors are able to say . . . about social conditions, including especially the conditions of their own action" (ibid.:374). Thus, if you were asked why you gave someone a gift, you might reply, "Because it's his birthday." Or when asked to explain why you refused to lend your notes to another student, you might respond, "Because the last time I did, I never got them back." Yet as Giddens points out, there is no "bar" between discursive and practical consciousness such that one is impermeable to the other. The differences between the two lie only in "what can be said and what is characteristically done," differences that can be modified through ongoing socialization (ibid.:7).

While no rigid boundary exists between discursive and practical consciousness, there are barriers between discursive consciousness and the individual's unconscious motives for action (represented by the double line in Figure 16.4). By definition, unconscious motives are repressed from conscious awareness or seep into consciousness only in distorted form. Actors, then, typically are unable to offer a discursive account of their unconscious motivations. Nevertheless, Giddens maintains that a complete theory of action must incorporate an analysis of the unconscious, because it is the source of our wishes and desires that in turn often moves us to pursue particular lines of action.

Giddens's discussion of the conscious/unconscious is offered as a corrective to structural-functionalist and psychoanalytic accounts of agency. From the perspective of the former, individuals' actions stem from system imperatives, and thus, their significance lies in how they serve societal purposes of which the individuals themselves are unaware. Learning what individuals say or think about their conduct is of little importance, given that the impetus for action operates behind the backs of individuals who are seen as vessels for the play of larger social forces. For their part, psychoanalytic theories, while focusing on the individual and related concepts such as personality and identity, likewise suggest that conduct often is not consciously guided; in this case it is driven by unknown psychic forces. For Giddens, both accounts underestimate the degree to which individuals are aware of and control their conduct.

To counter these approaches, Giddens offers a "stratification model of action" (Giddens 1979:53–59, 1984:5–14), which is graphically represented in Figure 16.5. In this model, Giddens combines his perspective on consciousness with a reworking of key structural-functionalist ideas. During social encounters, individuals routinely and continuously monitor their own activity, the behavior of others with whom they are interacting, and the setting in which the interaction is taking place. Moreover, during social encounters, individuals routinely maintain a taken-for-granted practical (that is, nondiscursive) understanding or rationalization of the

Figure 16.5 Giddens's Stratification Model of Action

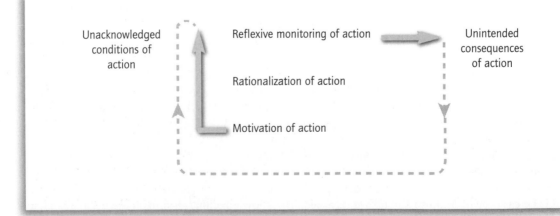

SOURCE: Figure reproduced from Giddens (1984:5).

interaction. The motivation of action refers to the wants and desires—often unconscious—that impel us to act. Yet, the individual's intentional action is bounded by two dimensions of conduct that remain largely outside of his awareness but that nevertheless profoundly shape the ability to act. Giddens notes that actions are situated within unacknowledged conditions and produce unintended consequences that, in turn, reproduce the unacknowledged conditions. In other words, our intentional actions are often motivated by unconscious wants, while these same actions often have the unintended consequence of reproducing the social institutions in which they are embedded. As such, the unintended consequences "loop back" to create the unacknowledged conditions of action. Thus, "every process of action is a production of something new, a fresh act; but at the same time all action exists in continuity with the past, which supplies the means of its initiation" (Giddens 1979:70).

As we noted above, Giddens's view of agency and structure is developed out of a critique of structural-functionalist analyses represented by the work of notable scholars such as Talcott Parsons and Robert Merton, and the structural Marxism of Louis Althusser. In Figure 16.6, Giddens contrasts his version of agency and structure with that put forward by functionalist interpretations.

Figure 16.6   Agency and Structure

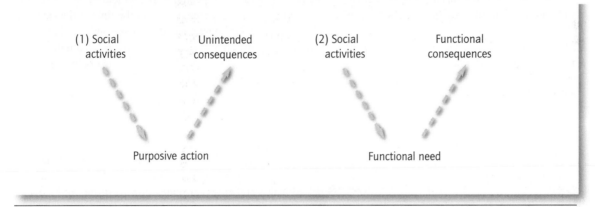

(1) Social activities   Unintended consequences   (2) Social activities   Functional consequences

Purposive action   Functional need

SOURCE: Figure reproduced from Giddens (1984:294).

Aligned with structuration theory, the first version (1) views social activities (for instance, attending religious services) as purposive action intentionally initiated by individuals for reasons of which they are aware (for instance, searching for moral and spiritual guidance). However, while a degree of rationality is imputed to the actors, this version recognizes that actors' knowledgeability of the conditions of their conduct and its consequences is not perfect or total. This presents the sociologist with an opportunity to uncover unintended consequences that stem from the actors' intentional conduct. Thus, attending religious services does more than provide a sense of moral or spiritual direction to worshippers. For example, it may reinforce feelings of community or solidarity among participants while producing tensions or hostilities directed toward those who do not maintain similar beliefs. Yet, whatever the unintended consequences may be, they do not exist as a result of societal "needs" that somehow filter down into the behavior of individuals. Instead, they are a result of intentional activity pursued, albeit for different purposes.

Giddens associates the second version (2) with functionalist analysis, which does posit the existence of system needs that operate independently of the individuals who fulfill them. Giddens illustrates this version through a critique of Robert Merton's analysis of the Hopi rain dance in which Merton asks us to consider why a rain dance ceremony (social activity) would persist when it fails to produce its manifest or intended function—to make rain. Merton's answer is that, far from a mere "superstition" or "irrational" practice, the rain dance serves crucial societal needs, namely, to ensure the survival of the group by providing an occasion for its members to participate in a common activity and thereby rekindle their attachment to the group. From this perspective, to understand the significance of social activities, it is less important to examine the manifest or intended consequences of the actors involved because such activities are driven by functional needs unknown to the individuals but, nevertheless, accomplished through their conduct. While Merton is right to point out that unintended consequences often issue from social activities, Giddens maintains that he is wrong to suggest that they do so because of "society's reasons" or some "invisible" social need. Thus, while participation in a religious ceremony may produce the unintended consequence of fostering group unity, this consequence does not explain why such activity persists, because societies have no needs—only people do. Social activities, "irrational" or otherwise, persist only because they satisfy the desires or interests of the individuals who engage in them.

Giddens's understanding of structure and agency is rounded out with his notion of **system**. Here, again, Giddens is attempting to separate his theory from varieties of structuralism and functionalism that, he argues, in defining a system as a "functioning structure" (Giddens 1981b:169), fail to adequately distinguish between the two concepts. In Giddens's structuration theory, social systems are patterned or "reproduced relations between individuals or collectivities, organized as regular social practices" (Giddens 1984:25). As recurring patterns of interaction, social systems do not consist of structures, but rather have "structural properties" that are reproduced across time and space as an unintended consequence of interaction (ibid.:17). Unlike structures, which exist only in the moment of their instantiation (that is, the moment of their use or appearance in interaction) or as enabling and constraining memory traces possessed by actors, social systems do stretch beyond the time and space of a given interaction. Social systems, then, exist "outside" the particular interactions they shape and are shaped by. Any ongoing, patterned social relation constitutes a social system. Thus, a friendship between two individuals, a family, a hospital, or a university are all forms of systems, as are recurring relations that exist between two or more nations. As Craib notes, "The scope of a system has to do with the amount of time-space that it binds, or rather the extent of its ability to bind time and space" (Craib 1992:60, 61). In other words, in "binding" time and space (what Giddens terms "time-space distanciation"), social systems consist of similar practices that are repeated at different times and in different places. Social systems thus make the experience of social life orderly and predictable. When two friends meet after not having seen each other for a month, they do not have to remake their relationship anew. Likewise, the system of market relations provides for recurring, predictable practices that bind the activities of individuals and groups across large expanses of time and space. From my home, I can invest in the future profitability of a Russian corporation or purchase a car manufactured with parts from four different countries. In both cases, my activities are embedded in a larger system of relations and practices that extends far beyond the time and space of my involvement.

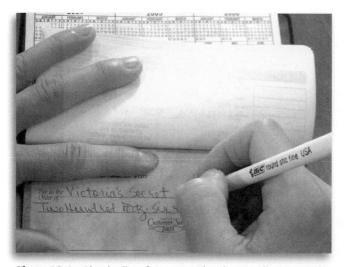

Photo 16.4 The duality of structure: There's more than meets the eye in writing a check

To illustrate the foregoing discussion, consider the example of writing a check. The ability to write a check is dependent on our commonsense understanding or know-how of the rules regarding the uses of them. More often than not, this knowledge is part of our practical consciousness. When asked, few of us can articulate exactly how a slip of paper can be converted into a monetary value, which is then transferred to the business to which we are giving the slip of paper. What we do know is that the process "works." However, if we were to take a course in economics or read a few books on the subject, that know-how would become explicit and "converted" to our discursive consciousness. In addition, the rules that govern check writing structure the activity in ways that make it predictable and orderly. These rules, or structure, are both constraining and enabling. We cannot use just any scrap of paper to write a check, and without sufficient funds in a bank account, checks are of little use (unless one is willing to pay a penalty for overdrawing funds—another rule). Conversely, if we possess the necessary *resources,* the rules governing the use of checks can enable us to purchase goods and services that we would otherwise have to do without. Here we see the duality of structure operating as well, for not only do the rules provide the medium for our action; they are also an unintended outcome, reproduced at the same moment we draw on them. This unintended consequence is not the product of a structural need working behind our backs that must be met in order to ensure the stability of the economy. Rather, it is the unintended consequence of actions purposively pursued for reasons of which we are consciously aware. Finally, each instance of writing a check involves a practice that is reproduced countless times and in countless places. As a "regular social practice," check writing is a structural property of a broader system of economic relations that binds the activities of numerous individuals and groups across time and space. The recurring interactions (either face-to-face or by mail) that involve check writing are patterned by the system of economic relations that is reproduced each time a check is written.

GIDDENS'S THEORETICAL ORIENTATION ▪

As discussed previously and as illustrated in Figure 16.7, Giddens's theory of structuration is an overt attempt to overcome the theoretical dualism that has divided individualist from collectivist perspectives. In this attempt, he has created a synthetic general theory of social action that seeks to incorporate elements from each of the quadrants making up our action/order framework. Taking up the issue of action first, Giddens draws from the work of Erving Goffman, phenomenology, and ethnomethodology (see Chapters 12 and 13) to develop a view of agency that highlights how individuals make sense of their everyday world. While he by no means rules out the rational dimension of action (indicated by the "Agency" box spanning both dimensions in Figure 16.7), his emphasis lies in exploring the practical knowledgeability of actors—that is, how actors put their commonsense know-how to use as they go about constructing their daily lives. Perhaps, to the extent

Figure 16.7   Giddens's Basic Concepts and Theoretical Orientation

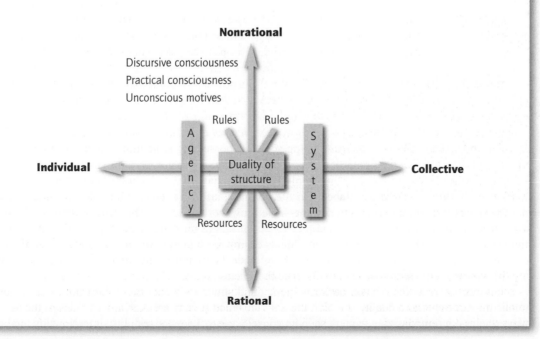

that Giddens defines resources in terms of the power that actors are capable of exercising over people and things, conduct can then be understood as motivated by calculated attempts to secure advantages. However, Giddens's view of rational motivation speaks less to a strategic weighing of costs and benefits and more to an individual's pragmatic and routine interventions in the world. It is not that an individual is seen as acting in an unreflexive or unthinking manner. Rather, the thinking that guides an individual's conduct is prefaced on the ability to "go on" with the routines making up everyday life.

As for the issue of order, Giddens again develops a number of concepts that incorporate both dimensions of the continuum. Consider first his notion of the duality of structure. Giddens defines rules as shared (i.e., collective) recipes of know-how that enable individuals to meaningfully navigate their everyday lives. As generalized procedures for "getting on," rules are not the creation or "property" of isolated individuals. Instead, their existence transcends individuals, as we are born into a culture whose rules and procedures for making sense of the world predate our birth. Yet, these recipes do not exist outside of their use by individuals; they exist only insofar as people apply them. Conversely, because they are not of our own making, rules constrain as well as enable our ability to act. Moreover, an unintended consequence of our individual conduct is the reproduction of the very rules that both enable and constrain our actions. And in reproducing the rules, we are at the same time reproducing the broader social systems in which these rules are embedded and in which our action takes place.

Further insight into Giddens's perspective on the relationship between agency and structure and the theoretical question of order is provided in his discussion of the limits or constraints on freedom of choice and action (Giddens 1984). Giddens identifies three forms of limits: material constraints, sanctions, and structural constraints. *Material constraints* refer to limitations on choice and action imposed by the "physical capacities of the human body, plus relevant features of the physical environment" (ibid.:174). There are some things we cannot do simply because our bodies have finite abilities. For instance, we cannot live 175 years or survive without water for more than a few days.

Sanctions refer to constraints on choice and action that are derived from another's exercise of power. (As noted previously, the exercise of power is itself tied to control over resources.) Sanctions range from "the direct application of force or violence, or the threat of such application, to the mild expression of disapproval" (ibid.:175). However, because Giddens maintains that resistance to power is almost always an option, the effectiveness of sanctions, regardless of how oppressive they may be, rests on the compliance of those who are subjected to them. As Giddens remarks,

> Even the threat of death carries no weight unless it is the case that the individual so threatened in some way values life. To say that an individual "had no choice but to act in such and such a way," in a situation of this sort evidently means "Given his/her desire not to die, the only alternative open was to act in the way he or she did." (ibid.:175)

More often than not, sanctions are "implied" in power relations that are themselves "often most profoundly embedded in modes of conduct which are taken for granted by those who follow them, most especially in routinized behavior, which is only diffusely motivated" (ibid.:175).

Finally, limits on action also take the form of *structural constraints*. Such constraints derive from the fact that individuals are born into preexisting societies possessing structural properties. As we noted previously, although structural properties possess an "objectivity" that constrains (and enables) the conduct of individuals, this objectivity does not imply that individuals "have no choice" but to pursue a particular line of action in a given circumstance. On the contrary, structural constraints "do not operate independently of the motives and reasons that agents have for what they do. . . . The structural properties of social systems do not act, or 'act on,' anyone like forces of nature to 'compel' him or her to behave in a particular way" (ibid.:181). To argue that individuals cannot resist or alter the course of a structural constraint only means "that they are unable to do anything other than conform to whatever the trends in question are, given the motives or goals which underlie their action" (ibid.:178).

Thus, for Giddens the relation between agency and structure is understood best not as an either/or—a dualism—but rather as a duality in which the individual and society are seen as two sides of the same coin. Nevertheless, in emphasizing matters such as consciousness, knowledgeability, intentionality, and choice, his analysis privileges the role of the individual in shaping and reshaping the social world. For Giddens, social life does not move according to some mechanical formula, despite the efforts of those sociologists who seek to locate general laws of social behavior in one or another variant of structural constraint that allegedly governs conduct. For, in the end, "the only moving objects in human social relations are individual agents, who employ resources to make things happen, intentionally or otherwise" (ibid.:181).

Readings

In the first excerpt you will read, taken from *The Constitution of Society,* Giddens lays out his general theory of action: the theory of structuration. In the second selection, taken from *The Consequences of Modernity,* Giddens explores the nature of modern society. Here he outlines a number of concepts including trust, risk, the transformation of intimacy, and ontological security. In doing so, he examines how the development of modern institutions has radically changed our relationship with others and our experience of the world around us.

Introduction to *The Constitution of Society*

In this selection from *The Constitution of Society,* "Elements of the Theory of Structuration," you will encounter the key concepts discussed in previous sections. Giddens begins by outlining the "empire-building endeavors" associated with two opposing theoretical traditions: structuralism and interpretative sociology. He then moves to an explication of the theory of structuration, offering his framework as a corrective to the one-sided portraits presented by approaches that unduly emphasize either collectivist or individualist dimensions of social life. To this end, he provides an account of individual agency, taking into consideration the different layers of consciousness and the role of intention in action. He then introduces the central concepts of structuration theory: structure, system, and the duality of structure—rules and resources that are both the medium and the outcome of the behavior they recursively organize. The reading concludes with Giddens's discussion of "forms of institution." Here he considers how interaction is linked to structures via the paired "modalities" of communication/signification, power/domination, and sanction/legitimation. The first term in each pair refers to the modalities that shape everyday interaction, while the second term refers to structural dimensions of various institutional orders that are produced in the course of interaction.

The Constitution of Society (1984)

Anthony Giddens

ELEMENTS OF THE THEORY OF STRUCTURATION

In offering a preliminary exposition of the main concepts of structuration theory it will be useful to begin from the divisions which have separated functionalism (including systems theory) and structuralism on the one hand from hermeneutics and the various forms of "interpretative sociology" on the other. Functionalism and structuralism have some notable similarities, in spite of the otherwise marked contrasts that exist between them. Both tend to express a naturalistic stand-point, and both are inclined towards objectivism. Functionalist thought, from Comte onwards, has looked particularly towards biology as the science providing the closest and most compatible model for social science. Biology has been taken to provide a guide to conceptualizing the structure and the functioning of social systems and to analysing processes of evolution via mechanisms of adaptation. Structuralist thought, especially in the writings of Lévi-Strauss, has been hostile to evolutionism and free from biological analogies. Here the homology between social and natural science is primarily a cognitive one in so far as each is supposed to express similar features of the overall constitution of mind. Both structuralism and functionalism strongly emphasize the pre-eminence of the social whole over its individual parts (i.e., its constituent actors, human subjects).

In hermeneutic traditions of thought, of course, the social and natural sciences are regarded as radically discrepant. Hermeneutics has been the home of that "humanism" to which structuralists have been so strongly and persistently opposed. In hermeneutic thought, such as presented by Dilthey, the gulf between subject and social object is at its widest. Subjectivity is the preconstituted centre of the experience of culture and history and as such provides the basic foundation of the social or human sciences. Outside the realm of subjective experience, and alien to it, lies the material world governed by impersonal relations of cause and effect. Whereas for those schools of thought which tend towards naturalism subjectivity has been regarded as something of a mystery, or almost a residual phenomenon, for hermeneutics it is the world of nature which is opaque—which, unlike human activity, can be grasped only from the outside. In interpretative sociologies, action and meaning are accorded primacy in the explication of human conduct; structural concepts are not notably prominent, and there is not much talk of constraint. For functionalism and structuralism, however, structure (in the divergent senses attributed to that concept) has primacy over action, and the constraining qualities of structure are strongly accentuated.

The differences between these perspectives on social science have often been taken to be epistemological, whereas they are in fact also ontological. What is at issue is how the concepts of action, meaning and subjectivity should be specified and how they might relate to notions of structure and constraint. If interpretative sociologies are founded, as it were, upon an imperialism of the subject, functionalism and structuralism propose an imperialism of the social object. One of my principal ambitions in the formulation of structuration theory is to put an end to each of these empire-building endeavours. The basic domain of study of the social sciences, according to the theory of structuration, is neither the experience of the individual actor, nor the existence of any form of society totality, but social practices ordered across space and time. Human social activities, like some self-reproducing items in nature, are recursive. That is to say, they are not brought into being by social actors but continually recreated by them via the very means whereby they express themselves *as* actors. In and through their activities agents reproduce the conditions that make these activities possible. However, the sort of "knowledgeability" displayed in nature, in the form of coded programmes, is distant from the cognitive skills displayed by human agents. It is in the conceptualizing of human knowledgeability and its involvement in action that I seek to appropriate some of the major contributions of interpretative sociologies. In structuration theory a hermeneutic starting-point is accepted in so far as it is acknowledged that the description of human activities demands a familiarity with the forms of life expressed in those activities.

It is the specifically reflexive form of the knowledgeability of human agents that is most deeply involved in the recursive ordering of social practices. Continuity of practices presumes reflexivity, but reflexivity in turn is possible only because of the continuity of practices that makes them distinctively "the same" across space and time. "Reflexivity" hence should be understood not merely as "self-consciousness" but as the monitored character of the ongoing flow of social life. To be a human being is to be a purposive agent, who both has reasons for his or her activities and is able, if asked, to elaborate discursively upon those reasons (including lying about them). But terms such as "purpose" or "intention," "reason," "motive" and so on have to be treated with caution, since their usage in the philosophical literature has very often been associated with a hermeneutical voluntarism, and because they extricate human action from the contextuality of time-space. Human action occurs as a durée, a continuous flow of conduct, as does cognition. Purposive action is not composed of an aggregate or series of separate intentions, reasons and motives. Thus it is useful to speak of reflexivity as grounded in the continuous monitoring of action which human beings display and expect others to display. The reflexive monitoring of action depends upon rationalization, understood here as a process rather than a state and as inherently involved in the competence of agents. An ontology of time-space as constitutive of social practices is basic to the conception of structuration, which *begins* from temporality and thus, in one sense, "history." . . .

Structure, Structuration

Let me now move to the core of structuration theory: the concepts of "structure," "system" and "duality of structure." The notion of structure (or "social structure"), of course, is very prominent in the writings of most functionalist authors and has lent its name to the traditions of "structuralism." But in neither instance is this conceptualized in a fashion best suited to the demands of social theory. Functionalist authors and their critics have given much more attention to the idea of "function" than to that of "structure," and consequently the latter has tended to be used as a received notion. But there can be no doubt about how "structure" is usually understood by functionalists and, indeed, by the vast majority of social analysts—as some kind of "patterning" of social relations or social phenomena. This is often naively conceived of in terms of visual imagery, akin to the skeleton or morphology of an organism or to the girders of a building. Such conceptions are closely connected to the dualism of subject and social object: "structure" here appears as "external" to human action, as a source of constraint on the free initiative of the independently constituted subject. As conceptualized in structuralist and post-structuralist thought, on the other hand, the notion of structure is more interesting. Here it is characteristically thought of not as a patterning of presences but as an intersection of presence and absence; underlying codes have to be inferred from surface manifestations.

These two ideas of structure might seem at first sight to have nothing to do with one another, but in fact each relates to important aspects of the structuring of social relations, aspects which, in the theory of structuration, are grasped by recognizing a differentiation between the concepts of "structure" and "system." In analysing social relations we have to acknowledge both a syntagmatic dimension, the patterning of social relations in time-space involving the reproduction of situated practices, and a paradigmatic dimension involving a virtual order of "modes of structuring" recursively implicated in such reproduction. In structuralist traditions there is usually ambiguity over whether structures refer to a matrix of admissible transformations within a set or to rules of transformation governing the matrix. I treat structure, in its most elemental meaning at least, as referring to such rules (and resources). It is misleading, however, to speak of "rules of transformation" because all rules are inherently transformational. Structure thus refers, in social analysis, to the structuring properties allowing the "binding" of time-space in social systems, the properties which make it possible for discernibly similar social practices to exist across varying spans of time and space and which lend them "systemic" form. To say that structure is a "virtual order" of transformative relations means that social systems, as reproduced social practices, do not have "structures" but rather exhibit "structural properties" and that structure exists, as time-space presence, only in its instantiations in such practices and as memory traces orienting the conduct of knowledgeable human agents. This does not prevent us from conceiving of structural properties as hierarchically organized in terms of the time-space extension of the practices they recursively organize. The most deeply embedded structural properties, implicated in the reproduction of societal totalities, I call *structural principles*. Those practices which have the greatest time-space extension within such totalities can be referred to as *institutions*.

To speak of structure as "rules" and resources, and of structures as isolable sets of rules and resources, runs a distinct risk of misinterpretation because of certain dominant uses of "rules" in the philosophical literature.

1. Rules are often thought of in connection with games, as formalized prescriptions. The rules implicated in the reproduction of social systems are not generally like this. Even those which are codified as laws are characteristically subject to a far greater diversity of contestations than the rules of games. Although the use of the rules of games such as chess, etc. as prototypical of the rule-governed properties of social systems is frequently associated with Wittgenstein, more relevant is what Wittgenstein has to say about children's play as exemplifying the routines of social life.

2. Rules are frequently treated in the singular, as if they could be related to specific instances or pieces of conduct. But this is highly misleading if regarded as analogous to the operation of social life, in which practices are sustained in conjunction with more or less loosely organized sets.

728 ■ BROADENING SOCIOLOGICAL THEORY

3. Rules cannot be conceptualized apart from resources, which refer to the modes whereby transformative relations are actually incorporated into the production and reproduction of social practices. Structural properties thus express forms of *domination* and *power.*

4. Rules imply "methodical procedures" of social inter-action, as Garfinkel in particular has made clear. Rules typically intersect with practices in the contex-tuality of situated encounters: the range of "ad hoc" considerations which he identifies are chronically involved with the instantiation of rules and are fun-damental to the form of those rules. Every competent social actor, it should be added, is *ipso facto* a social theorist on the level of discursive consciousness and a "methodological specialist" on the levels of both discursive and practical consciousness.

5. Rules have two aspects to them, and it is essential to distinguish these conceptually, since a number of philosophical writers (such as Winch) have tended to conflate them. Rules relate on the one hand to the constitution of *meaning,* and on the other to the *sanctioning* of modes of social conduct.

I have introduced the above usage of "structure" to help break with the fixed or mechanical character which the term tends to have in orthodox sociological usage. The concepts of system and structuration do much of the work that "structure" is ordinarily called upon to perform. In proposing a usage of "structure" that might appear at first sight to be remote from conventional interpretations of the term, I do not mean to hold that looser versions be abandoned altogether. "Society," "culture" and a range of other forms of sociological terminology can have double usages that are embarrassing only in contexts where a difference is made in the nature of the statements employing them. Similarly, I see no particular objection to speaking of "class structure," "the structure of the industrialized societies" and so on, where these terms are meant to indicate in a general way relevant institutional features of a society or range of societies.

One of the main propositions of structuration theory is that the rules and resources drawn upon in the pro-duction and reproduction of social action are at the same time the means of system reproduction (the dual-ity of structure). But how is one to interpret such a claim? In what sense is it the case that when I go about my daily affairs my activities incorporate and repro-duce, say, the overall institutions of modern capitalism? What rules are being invoked here in any case? . . .

Let us regard the rules of social life . . . as techniques or generalizable procedures applied in the enactment/repro-duction of social practices. Formulated rules—those that are given verbal expression as canons of law, bureaucratic rules, rules of games and so on—are thus codified interpre-tations of rules rather than rules as such. They should be taken not as exemplifying rules in general but as specific types of formulated rule, which, by virtue of their overt formulation, take on various specific qualities.

So far these considerations offer only a preliminary approach to the problem. How do formulae relate to the practices in which social actors engage, and what kinds of formulae are we most interested in for general pur-poses of social analysis? As regards the first part of the question, we can say that awareness of social rules, expressed first and foremost in practical consciousness, is the very core of that "knowledgeability" which spe-cifically characterizes human agents. As social actors, all human beings are highly "learned" in respect of knowledge which they possess, and apply, in the pro-duction and reproduction of day-to-day social encoun-ters; the vast bulk of such knowledge is practical rather than theoretical in character. As Schutz and many others have pointed out, actors employ typified schemes (for-mulae) in the course of their daily activities to negotiate routinely the situations of social life. Knowledge of procedure, or mastery of the techniques of "doing" social activity, is by definition methodological. That is to say, such knowledge does not specify all the situa-tions which an actor might meet with, nor could it do so; rather, it provides for the generalized capacity to respond to and influence an indeterminate range of social circumstances.

Those types of rule which are of most significance for social theory are locked into the reproduction of institutionalized practices, that is, practices most deeply sedimented in time-space. The main characteristics of rules relevant to general questions of social analysis can be described as follows:

intensive	tacit	informal	weakly sanctioned
:	:	:	
shallow	discursive	formalized	strongly sanctioned

By rules that are intensive in nature, I mean formulae that are constantly invoked in the course of day-to-day activities, that enter into the structuring of much of the texture of everyday life. Rules of language are of this char-acter. But so also, for example, are the procedures utilized by actors in organizing turn-taking in conversations or in interaction. They may be contrasted with rules which, although perhaps wide in scope, have only a superficial impact upon much of the texture of social life. The contrast is an important one, if only because it is commonly taken for granted among social analysts that the more abstract rules—e.g., codified law—are the most influential in the structuring of social activity. I would propose, however, that many seemingly trivial procedures followed in daily life have a more profound influence upon the generality of social conduct. The remaining categories should be more or less self-explanatory. Most of the rules implicated in the production and reproduction of social practices are only tacitly grasped by actors: they know how to "go on." *The discursive formulation of a rule is already an interpretation of it,* and, as I have noted, may in and of itself alter the form of its application. Among rules that are not just discursively formulated but are formally codified, the type case is that of laws. Laws, of course, are among the most strongly sanctioned types of social rules and in modern societies have formally prescribed gradations of retribution. However, it would be a serious mistake to underestimate the strength of informally applied sanctions in respect of a variety of mundane daily practices. Whatever else Garfinkel's "experiments with trust" might be thought to demonstrate, they do show the extraordinarily compelling force with which apparently minor features of conversational response are invested.

The structuring qualities of rules can be studied in respect, first of all, of the forming, sustaining, termination and reforming of encounters. Although a dazzling variety of procedures and tactics are used by agents in the constitution and reconstitution of encounters, probably particularly significant are those involved in the sustaining of ontological security. Garfinkel's "experiments" are certainly relevant in this respect. They indicate that the prescriptions involved in the structuring of daily interaction are much more fixed and constraining than might appear from the ease with which they are ordinarily followed. This is surely because the deviant responses or acts that Garfinkel instructed his "experimenters" to perform disturbed the sense of ontological security of the "subjects" by undermining the intelligibility of discourse. Breaking or ignoring rules is not, of course, the only way in which the constitutive and sanctioning properties of intensively invoked rules can be studied. But there is no doubt that Garfinkel has helped to disclose a

remarkably rich field of study—performing the "sociologist's alchemy," the "transmutation of any patch of ordinary social activity into an illuminating publication."

I distinguish "structure" as a generic term from "structures" in the plural and both from the "structural properties of social systems." "Structure" refers not only to rules implicated in the production and reproduction of social systems but also to resources (about which I have so far not said much but will do so shortly). As ordinarily used in the social sciences, "structure" tends to be employed with the more enduring aspects of social systems in mind, and I do not want to lose this connotation. The most important aspects of structure are rules and resources recursively involved in institutions. Institutions by definition are the more enduring features of social life. In speaking of the structural properties of social systems I mean their institutionalized features, giving "solidity" across time and space. I use the concept of "structures" to get at relations of transformation and mediation which are the "circuit switches" underlying observed conditions of system reproduction.

Let me now answer the question I originally posed: in what manner can it be said that the conduct of individual actors reproduces the structural properties of larger collectivities? The question is both easier and more difficult to answer than it appears. On a logical level, the answer to it is nothing more than a truism. That is to say, while the continued existence of large collectivities or societies evidently does not depend upon the activities of any particular individual, such collectivities or societies manifestly would cease to be if all the agents involved disappeared. On a substantive level, the answer to the question depends upon issues yet to be broached—those concerning the mechanisms of integration of different types of societal totality. It is always the case that the day-to-day activity of social actors draws upon and reproduces structural features of wider social systems. But "societies"—as I shall make clear—are not necessarily unified collectivities. "Social reproduction" must not be equated with the consolidation of social cohesion. The location of actors and of collectivities in different sectors or regions of more encompassing social systems strongly influences the impact of even their habitual conduct upon the integration of societal totalities. Here we reach the limits of linguistic examples which might be used to illustrate the concept of the duality of structure. Considerable illumination of problems of social analysis can be derived from studying the recursive qualities of speech and language. When I produce a grammatical utterance, I draw upon the same syntactical rules as those that utterance helps to produce. But I speak the "same" language as the other speakers in my language community; we all share the same rules and linguistic practices, give or take a range of relatively minor variations. Such is not necessarily the case with the structural properties of social systems in general. But this is not a problem to do with the concept of the duality of structure as such. It is to do with how social systems, especially "societies," should be conceptualized.

The Duality of Structure

Let me summarize the argument thus far. Structure, as recursively organized sets of rules and resources, is out of time and space, save in its instantiations and co-ordination as memory traces, and is marked by an "absence of the subject." The social systems in which structure is recursively implicated, on the contrary, comprise the situated activities of human agents, reproduced across time and space. Analysing the structuration of social systems means studying the modes in which such systems, grounded in the knowledgeable activities of situated actors who draw upon rules and resources in the diversity of action contexts, are produced and reproduced in interaction [see figure below]. Crucial to the idea of structuration is the theorem of the duality of structure, which is logically implied in the arguments portrayed above. The constitution of agents and structures are not two independently given sets of phenomena, a dualism, but represent a duality. According to the notion of the duality of structure, the structural properties of social systems are both medium and outcome of the practices they recursively organize. Structure is not "external" to individuals: as memory traces, and as instantiated in social practices, it is in a certain sense more "internal" than exterior to their activities in a Durkheimian sense. Structure is not to be equated with constraint but is always both constraining and enabling. This, of course, does not prevent the structured properties of social systems from stretching away, in time and space, beyond the control of any individual actors. Nor does it compromise the possibility that actors' own theories of the social systems which they help to constitute and reconstitute in their activities may reify those systems. The reification of social relations, or the discursive "naturalization" of the historically contingent circumstances and products of human action, is one of the main dimensions of ideology in social life.

Even the crudest forms of reified thought, however, leave untouched the fundamental significance of the knowledgeability of human actors. For knowledgeability is founded less upon discursive than practical consciousness. The knowledge of social conventions, of oneself and of other human beings, presumed in being able to "go on" in the diversity of contexts of social life is detailed and dazzling. All competent members of society are vastly skilled in the practical accomplishments of social activities and are expert "sociologists." The knowledge they possess is not incidental to the persistent patterning of social life but is integral to it. This stress is absolutely essential if the mistakes of functionalism and structuralism are to be avoided, mistakes which, suppressing or discounting agents' reasons—the rationalization of action as chronically involved in the structuration of social practices—look for the origins of their activities in phenomena of which these agents are ignorant. But it is equally important to avoid tumbling into the opposing error of hermeneutic approaches and of various versions of phenomenology, which tend to regard society as the plastic creation of human subjects. Each of these is an illegitimate form of reduction, deriving from a failure adequately to conceptualize the duality of structure. According to structuration theory, the moment of the production of action is also one of reproduction in the contexts of the day-to-day enactment of social life. This is so even during the most violent upheavals or most radical forms of social change. It is not accurate to see the structural properties of social systems as "social products" because this tends to imply that pre-constituted actors somehow come together to create them. In reproducing structural properties to repeat a phrase used earlier, agents

Structure(s)	System(s)	Structuration
Rules and resources, or sets of transformation relations, organized as properties of social systems	Reproduced relations between actors or collectivities, organized as regular social practices	Conditions governing the continuity of transmutation of structures, and therefore the reproduction of social systems

also reproduce the conditions that make such action possible. Structure has no existence independent of the knowledge that agents have about what they do in their day-to-day activity. Human agents always know what they are doing on the level of discursive consciousness under some description. However, what they do may be quite unfamiliar under other descriptions, and they may know little of the ramified consequences of the activities in which they engage.

The duality of structure is always the main grounding of continuities in social reproduction across time-space. It in turn presupposes the reflexive monitoring of agents in, and as constituting, the durée of daily social activity. But human knowledgeability is always bounded. The flow of action continually produces consequences which are unintended by actors, and these unintended consequences also may form unacknowledged conditions of action in a feedback fashion. Human history is created by intentional activities but is not an intended project; it persistently eludes efforts to bring it under conscious direction. However, such attempts are continually made by human beings, who operate under the threat and the promise of the circumstance that they are the only creatures who make their "history" in cognizance of that fact.

The theorizing of human beings about their action means that just as social theory was not an invention of professional social theorists, so the ideas produced by those theorists inevitably tend to be fed back into social life itself. One aspect of this is the attempt to monitor, and thereby control, highly generalized conditions of system reproduction—a phenomenon of massive importance in the contemporary world. To grasp such monitored processes of reproduction conceptually, we have to make certain distinctions relevant to what social systems "are" as reproduced practices in interaction settings. The relations implied or actualized in social systems are, of course, widely variable in terms of their degree of "looseness" and permeability. But, this being accepted, we can recognize two levels in respect of the means whereby some element of "systemness" is achieved in interaction. One is that generally prominent in functionalism, as referred to earlier, where interdependence is conceived of as a homeostatic process akin to mechanisms of self-regulation operating within an organism. There can be no objection to this as long as it is acknowledged that the "looseness" of most social systems makes the organic parallel a very remote one and that this relatively "mechanized" mode of system reproduction is not the only one found in human societies. Homeostatic system reproduction in human

society can be regarded as involving the operation of causal loops, in which a range of unintended consequences of action feed back to reconstitute the initiating circumstances. But in many contexts of social life there occur processes of selective "information filtering" whereby strategically placed actors seek reflexively to regulate the overall conditions of system reproduction either to keep things as they are or to change them.

The distinction between homeostatic causal loops and reflexive self-regulation in system reproduction must be complemented by one further, and final, one: that between social and system integration. "Integration" may be understood as involving reciprocity of practices (of autonomy and dependence) between actors or collectivities. Social integration then means systemness on the level of face-to-face interaction. System integration refers to connections with those who are physically absent in time or space. The mechanisms of system integration certainly presuppose those of social integration, but such mechanisms are also distinct in some key respects from those involved in relations of co-presence.

Social Integration	System Integration
Reciprocity between actors in contexts of co-presence	Reciprocity between actors or collectivities across extended time-space

Forms of Institution

The division of rules into modes of signifying or meaning constitution and normative sanctions, together with the concept of resources—fundamental to the conceptualization of power—carries various implications which need to be spelled out. What I call the "modalities" of structuration serve to clarify the main dimensions of the duality of structure in interaction, relating the knowledgeable capacities of agents to structural features. Actors draw upon the modalities of structuration in the reproduction of systems of interaction, by the same token reconstituting their structural properties. The communication of meaning in interaction, it should be stressed, is separable only analytically from the operation of normative sanctions. This is obvious, for example, in so far as language use is itself sanctioned by the very nature of its "public" character. The very identification of acts or of aspects of interaction—their accurate description, as grounded hermeneutically in the capability of an observer to "go on" in a form of life—implies the interlacing of meaning, normative elements and power. This is most evident in the not infrequent contexts of social life where what social phenomena "are," how they are aptly described, is contested. Awareness of such contestation, of divergent and overlapping characterizations of activity, is an essential part of "knowing a form of life." . . .

The dimensions of the duality of structure are portrayed in [Figure 1 below]. Human actors are not only able to monitor their activities and those of others in the regularity of day-to-day conduct; they are also able to "monitor that monitoring" in discursive consciousness. "Interpretative schemes" are the modes of typification incorporated within actors' stocks of knowledge, applied reflexively in the sustaining of communication. The stocks of knowledge which actors draw upon in the production and reproduction of interaction are the same as those whereby they are able to make accounts, offer reasons, etc. The communication of meaning, as with all aspects of the contextuality of action, does not have to be seen merely as happening "in" timespace. Agents routinely incorporate temporal and spatial features of encounters in processes of meaning constitution. Communication, as a general element of interaction, is a more inclusive concept than communicative intent (i.e. what an actor "means" to say or do). There are once more

two forms of reductionism to be avoided here. Some philosophers have tried to derive overall theories of meaning or communication from communicative intent; others, by contrast, have supposed that communicative intent is at best marginal to the constitution of the meaningful qualities of interaction, "meaning" being governed by the structural ordering of sign systems. In the theory of structuration, however, these are regarded as of equivalent interest and importance, aspects of a duality rather than a mutually exclusive dualism.

The idea of "accountability" in everyday English gives cogent expression to the intersection of interpretative schemes and norms. To be "accountable" for one's activities is both to explicate the reasons for them and to supply the normative grounds whereby they may be "justified." Normative components of interaction always centre upon relations between the rights and obligations "expected" of those participating in a range of interaction contexts. Formal codes of conduct, as, for example, those enshrined in law (in contemporary societies at least), usually express some sort of claimed symmetry between rights and obligations, the one being the justification of the other. But no such symmetry

Figure 1

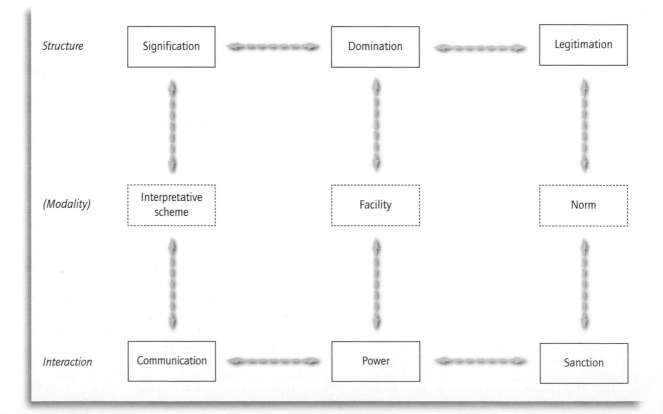

Structure	Signification	←·······→	Domination	←·······→	Legitimation
(Modality)	Interpretative scheme		Facility		Norm
Interaction	Communication	←·······→	Power	←·······→	Sanction

Structure(s)	Theoretical Domain	Institutional Order
Signification	Theory of coding	Symbolic orders/modes of discourse
Domination	Theory of resource authorization Theory of resource allocation	Political institutions Economic institutions
Legitimation	Theory of normative regulation	Legal institutions

necessarily exists in practice, a phenomenon which it is important to emphasize, since both the "normative functionalism" of Parsons and the "structuralist Marxism" of Althusser exaggerates the degree to which normative obligations are "internalized" by the members of societies. Neither standpoint incorporates a theory of action which recognizes human beings as knowledgeable agents, reflexively monitoring the flow of interaction with one another. When social systems are conceived of primarily from the point of view of the "social object," the emphasis comes to be placed upon the pervasive influence of a normatively co-ordinated legitimate order as an overall determinant or "programmer" of social conduct. Such a perspective masks the fact that the normative elements of social systems are contingent claims which have to be sustained and "made to count" through the effective mobilization of sanctions in the contexts of actual encounters. Normative sanctions express structural asymmetries of domination, and the relations of those nominally subject to them may be of various sorts other than expressions of the commitments those norms supposedly engender.

Concentration upon the analysis of the structural properties of social systems, it should be stressed, is a valid procedure only if it is recognized as placing an *epoché* upon—holding in suspension—reflexively monitored social conduct. Under such an *epoché* we may distinguish three structural dimensions of social systems: signification, domination and legitimation. The connotations of the analysis of these structural properties are indicated in the table below. . . ." Domination" and "power" cannot be thought of only in terms of asymmetries of distribution but have to be recognized as inherent in social association (or, I would say, in human action as such). Thus . . . power is not an inherently noxious phenomenon, not just the capacity to "say no"; nor can domination be "transcended" in some kind of putative society of the future, as has been the characteristic aspiration of at least some strands of socialist thought. . . .

In the terminology indicated in the table above the "signs" implied in "signification" should not be equated with "symbols." Many writers treat the two terms as equivalent, but I regard symbols, interpolated within symbolic orders, as one main dimension of the "clustering" of institutions. Symbols coagulate the "surpluses of meaning" implied in the polyvalent character of signs; they conjoin those intersections of codes which are especially rich in diverse forms of meaning association, operating along the axes of metaphor and metonymy. Symbolic orders and associated modes of discourse are a major institutional locus of ideology. However, in the theory of structuration ideology is not a particular "type" of symbolic order or form of discourse. One cannot separate off "ideological discourse" from "science," for example. "Ideology" refers only to those asymmetries of domination which connect signification to the legitimation of sectional interests.

We can see from the case of ideology that structures of signification are separable only analytically either from domination or from legitimation. Domination depends upon the mobilization of two distinguishable types of resource. Allocative resources refer to capabilities—or, more accurately, to forms of transformative capacity—generating command over objects, goods or material phenomena. Authoritative resources refer to types of transformative capacity generating command over persons or actors. Some forms of allocative resources (such as raw materials, land, etc.) might seem to have a "real existence" in a way which I have claimed that structural properties as a whole do not. In the sense of having a time-space "presence," in a certain way such is obviously the case. But their "materiality" does not affect the fact that such phenomena become resources, in the manner in which I apply that term here, only when incorporated within processes of structuration. The transformational character of resources is logically equivalent to, as well as inherently bound up with the instantiation of, that of codes and normative sanctions. . . .

Introduction to *The Consequences of Modernity*

In addition to developing his theory of structuration, Giddens has undertaken a number of investigations into the nature of modern society and the interpersonal relationships that are forged within the context of everyday, modern life. In the excerpt provided below, taken from *The Consequences of Modernity,* Giddens argues that modern societies are marked by two paradoxical developments: *security* versus *danger* and *trust* versus *risk.* Inhabitants of modern societies, by and large, are able to avoid the disease, famine, and other harsh realities of daily life that were all too common in civilizations past. While not everyone is able to enjoy all the conveniences modern society has to offer, what we now consider basic necessities would be unfathomable for the vast majority of the populations in traditional or premodern societies. However, the security that modernity affords us is offset by the unparalleled dangers that we face today. Ongoing destruction of our ecology, the rise of totalitarian regimes, the outbreak of horrific wars, and the proliferation of nuclear and biological weapons have all played their part in creating a world that is more dangerous than ever before.

The interplay between trust and risk is also novel to modern societies. Giddens defines **trust** as "confidence in the reliability of a person or system, regarding a given set of outcomes or events, where that confidence expresses a faith in the probity [honor] or love of another, or in the correctness of abstract principles (technical knowledge)" (Giddens 1990:34). Trust inevitably involves risk: The people and systems in which we place our trust may not perform as we had anticipated; chance, incompetence, or unforeseen events can all lead to undesired outcomes. Nevertheless, trust is an inescapable feature of modern society, in large measure because **time–space distanciation**—system and social integration linking individuals across large spans of time and space—has "disembedded" or "lifted out" social relations from the immediate contexts of face-to-face interaction (ibid.:21). As Giddens notes, "There would be no need to trust anyone whose activities were continually visible and whose thought processes were transparent, or to trust any system whose workings were wholly known or understood . . . [T]he prime condition of requirements for trust is . . . lack of full information" (ibid.:33). Because we do not "know" the pilot of the airplane we are flying in or understand how planes can defy gravity, we have little choice but to trust the expertise of the pilot and all those whose efforts make air travel a manageable risk. The same can be said for those who monitor the systems that purify our drinking water or that deliver highly combustible natural gas to our homes. From the car mechanic, traffic engineer, and architect, to the electrician, doctor, and nuclear power plant technician, modern life is predicated on trust in the expertise of others and the abstract systems whose workings we do not fully understand.

Disembedding mechanisms are of two types: "symbolic tokens" and "expert systems" (sometimes referred to as "abstract systems," as above). Giddens defines the former as "media of interchange which can be 'passed around' without regard to the specific characteristics of individuals or groups that handle them at any particular juncture" (ibid.:22). Drawing on the work of Georg Simmel (1858–1918), Giddens finds in money the paramount symbolic token because it makes possible transactions between individuals who are distanced from one another in terms of both time and space. Expert systems are "systems of technical accomplishment or professional expertise that organize large areas of the material and social environments in which we live" (ibid.:27). Despite the fact that we do not regularly seek expert knowledge, our lives are continuously linked to the expert or abstract systems in which that knowledge is integrated. Given that individuals are linked across increasing spans of time and space, a hallmark of modern society is our routine placing of trust or faith in such abstract, disembedding mechanisms and the expertise that defines them. Yet, expert knowledge, like scientific knowledge more generally, is never certain, but instead always open to refutation and revision.

Giddens next considers how the spreading of abstract systems and our trust in them has led to a "transformation of intimacy." Here he notes the impact of modernity on our personal relationships with others, particularly how in modern society trust is a "project" that requires work in the form of self-disclosure in order to be "won." The need for self-disclosure is itself bound up with a reflexive search for self-identity in which the construction of the self and a concern for self-fulfillment likewise become ongoing projects. However, these projects are carried out in an environment fraught with risks of unprecedented proportions. The disembedding mechanisms associated with modernity have altered both the scope and the types of risks that currently confront individuals. The potential for global catastrophe is now part of everyday life. The threat of nuclear war, the collapse of the global economy, destruction of the planet's resources, and global warming are just a few of the risks that are forever looming on the horizon. No one can completely escape the effects of these modern global dangers, almost all of which are the product of humans' intervention in the world. Living daily with the knowledge that risks of high consequence are always present affects feelings of ontological security. **Ontological security** refers to unconscious feelings regarding the continuity of one's self-identity and of the broader social and physical world. It is a sense of security rooted in a trust that people and the surrounding world are as they appear. Yet, precisely because modern society poses an array of catastrophic risks of which individuals are aware, and because the experts and abstract systems charged with controlling the sources of potential danger are known to be imperfect, the sense of confidence, reliability, and trust that is basic to feelings of ontological security is jeopardized (ibid.:92). Such is the "juggernaut of modernity."

The Consequences of Modernity (1990)

Anthony Giddens

THE DISCONTINUITIES OF MODERNITY

The modes of life brought into being by modernity have swept us away from *all* traditional types of social order, in quite unprecedented fashion. In both their extensionality and their intensionality the transformations involved in modernity are more profound than most sorts of change characteristic of prior periods. On the extensional plane they have served to establish forms of social interconnection which span the globe; in intensional terms they have come to alter some of the most intimate and personal features of our day-to-day existence. Obviously there are continuities between the traditional and the modern, and neither is cut of whole cloth; it is well known how misleading it can be to contrast these two in too gross a fashion. But the changes occurring over the past three or four centuries—a tiny period of historical time—have been so dramatic and so comprehensive in their impact that we get only limited assistance from our knowledge of prior periods of transition in trying to interpret them. . . .

How should we identify the discontinuities which separate modern social institutions from the traditional social orders? Several features are involved. One is the sheer *pace of change* which the era of modernity sets into motion. Traditional civilisations may have been considerably more dynamic than other pre-modern systems, but the rapidity of change in conditions of modernity is extreme. If this is perhaps most obvious in respect of technology, it also pervades all other spheres. A second discontinuity is the *scope of change*. As different areas of the globe are drawn into interconnection with one another, waves of social transformation crash across virtually the whole of the earth's surface. A third feature concerns the intrinsic *nature of modern institutions*. Some modern social forms are simply not found in prior historical periods—such as the political system of the nation-state, the wholesale dependence of production upon inanimate power sources, or the thoroughgoing commodification of products and wage labour. Others only have a specious continuity with pre-existing social orders. An example is the city. Modern urban settlements often incorporate the sites of traditional cities, and it may look as though they have merely spread out from them. In fact, modern urbanism is ordered according to quite different principles from those which set off the pre-modern city from the countryside in prior periods.

SECURITY AND DANGER, TRUST AND RISK

In pursuing my enquiry into the character of modernity, I want to concentrate a substantial portion of the discussion upon the themes of *security versus danger* and *trust versus risk*. Modernity, as everyone living in the closing

---

years of the twentieth century can see, is a double-edge phenomenon. The development of modern social institutions and their worldwide spread have created vastly greater opportunities for human beings to enjoy a secure and rewarding existence than any type of pre-modern system. But modernity also has a sombre side, which has become very apparent in the present century.

On the whole, the "opportunity side" of modernity was stressed most strongly by the classical founders of sociology. Marx and Durkheim both saw the modern era as a troubled one. But each believed that the beneficent possibilities opened up by the modern era outweighed its negative characteristics. Marx saw class struggle as the source of fundamental schisms in the capitalistic order, but at the same time envisaged the emergence of a more humane social system. Durkheim believed the further expansion of industrialism would establish a harmonious and fulfilling social life, integrated through a combination of the division of labour and moral individualism. Max Weber was the most pessimistic among the three founding fathers, seeing the modern world as a paradoxical one in which material progress was obtained only at the cost of an expansion of bureaucracy that crushed individual creativity and autonomy. Yet even he did not fully anticipate how extensive the darker side of modernity would turn out to be.

To take an example, all three authors saw that modern industrial work had degrading consequences, subjecting many human beings to the discipline of dull, repetitive labour. But it was not foreseen that the furthering of the "forces of production" would have large-scale destructive potential in relation to the material environment. Ecological concerns do not brook large in the traditions of thought incorporated into sociology, and it is not surprising that sociologists today find it hard to develop a systematic appraisal of them.

A second example is the consolidated use of political power, particularly as demonstrated in episodes of totalitarianism. The arbitrary use of political power seemed to the sociological founders to belong primarily to the past (although sometimes having echoes in the present, as indicated in Marx's analysis of the rule of Louis Napoleon). "Despotism" appeared to be mainly characteristic of pre-modern states. In the wake of the rise of fascism, the Holocaust, Stalinism, and other episodes of twentieth-century history, we can see that totalitarian possibilities are contained within the institutional parameters of modernity rather than being foreclosed by them. Totalitarianism is distinct from traditional despotism, but is all the more frightening as a result. Totalitarian rule connects political, military, and ideological power in more concentrated form than was ever possible before the emergence of modern nation-states.

The development of military power as a general phenomenon provides a further case in point. . . . None of the classical founders of sociology gave systematic attention to the phenomenon of the "industrialisation of war."

Social thinkers writing in the late nineteenth and early twentieth centuries could not have foreseen the invention of nuclear weaponry.[i] But the connecting of industrial innovation and organisation to military power is a process that dates back to the early origins of modern industrialisation itself. That this went largely unanalysed in sociology is an indication of the strength of the view that the newly emergent order of modernity would be essentially pacific, in contrast to the militarism that had characterised previous ages. Not just the threat of nuclear confrontation, but the actuality of military conflict, form a basic part of the "dark side" of modernity in the current century. The twentieth century is the century of war, with the number of serious military engagements involving substantial loss of life being considerably higher than in either of the two preceding centuries. In the present century thus far, over 100 million people have been killed in wars, a higher proportion of the world's population than in the nineteenth century, even allowing for overall population increase. Should even a limited nuclear engagement be fought, the loss of life would be staggering, and a full superpower conflict might eradicate humanity altogether.

The world in which we live today is a fraught and dangerous one. This has served to do more than simply blunt or force us to qualify the assumption that the emergence of modernity would lead to the formation of a happier and more secure social order. Loss of a belief in "progress," of course, is one of the factors that underlies the dissolution of "narratives" of history. Yet there is much more at stake here than the conclusion that history "goes nowhere." We have to develop an institutional analysis of the double-edged character of modernity. . . .

---

[i]Yet, writing in 1914, just before the outbreak of the Great War, H. G. Wells did make such a prediction, influenced by the physicist Frederick Soddy, a collaborator of Ernest Rutherford. Wells's book, *The World Set Free,* recounts the story of a war which erupts in Europe in 1958, from there spreading throughout the world. In the war, a terrible weapon is used, constructed from a radioactive substance called Carolinum. Hundreds of these bombs, which Wells called "atomic bombs," are dropped on the world's cities, causing immense devastation. A time of mass starvation and political chaos follows, after which a new world republic is set up, in which war is forever prohibited.

ABSTRACT SYSTEMS AND THE TRANSFORMATION OF INTIMACY

Abstract systems have provided a great deal of security in day-to-day life which was absent in pre-modern orders. A person can board a plane in London and reach Los Angeles some ten hours later and be fairly certain that not only will the journey be made safely, but that the plane will arrive quite close to a predetermined time. The passenger may perhaps only have a vague idea of where Los Angeles is, in terms of a global map. Only minimal preparations need to be made for the journey (obtaining passport, visa, air-ticket, and money)—no knowledge of the actual trajectory is necessary. A large amount of "surrounding" knowledge is required to be able to get on the plane, and this is knowledge which has been filtered back from expert systems to lay discourse and action. One has to know what an airport is, what an air-ticket is, and very many other things besides. But security on the journey itself does not depend upon mastery of the technical paraphernalia which make it possible.

Compare this with the task of an adventurer who undertook the same journey no more than three or four centuries ago. Although he would be the "expert," he might have little idea of where he was traveling *to*—and the very notion of "traveling" sounds oddly inapplicable. The journey would be fraught with dangers, and the risk of disaster or death very considerable. No one could participate in such an expedition who was not physically tough, resilient, and possessed of skills relevant to the conduct of the voyage.

Every time someone gets cash out of the bank or makes a deposit, casually turns on a light or a tap, sends a letter or makes a call on the telephone, she or he implicitly recognises the large areas of secure, coordinated actions and events that make modern social life possible. Of course, all sorts of hitches and breakdowns can also happen, and attitudes of scepticism or antagonism develop which produce the disengagement of individuals from one or more of these systems. But most of the time the taken-for-granted way in which everyday actions are geared into abstract systems bears witness to the effectiveness with which they operate (within the contexts of what is expected from them, because they also produce many kinds of unintended consequences).

Trust in abstract systems is the condition of time-space distanciation and of the large areas of security in day-to-day life which modern institutions offer as compared to the traditional world. The routines which are integrated with abstract systems are central to ontological security in conditions of modernity. Yet this situation also creates novel forms of psychological vulnerability, and trust in abstract systems is not psychologically rewarding in the way in which trust in person is. I shall concentrate on the second of these points here, returning to the first later. To begin, I want to advance the following theorems: that there is a direct (although dialectical) connection between the globalising tendencies of modernity and what I shall call the *transformation of intimacy* in contexts of day-to-day life; that the transformation of intimacy can be analysed in terms of the building of trust mechanisms; and that personal trust relations, in such circumstances, are closely bound up with a situation in which the construction of the self becomes a reflexive project. . . .

TRUST AND PERSONAL IDENTITY

With the development of abstract systems, trust in impersonal principles, as well as in anonymous others, becomes indispensable to social existence. Nonpersonalized trust of this sort is discrepant from basic trust. There is a strong psychological need to find others to trust, but institutionally organised personal connections are lacking, relative to pre-modern social situations. The point here is *not* primarily that many social characteristics which were previously part of everyday life or the "life-world" become drawn off and incorporated into abstract systems. Rather, the tissue and form of day-to-day life become reshaped in conjunction with wider social changes. Routines which are structured by abstract systems have an empty, unmoralised character—this much is valid in the idea that the impersonal increasingly swamps the personal. But this is not simply a diminishment of personal life in favour of impersonally organized systems—it is a genuine transformation of the nature of the personal itself. Personal relations whose main objective is sociability, informed by loyalty and authenticity, become as much a part of the social situations of modernity as the encompassing institutions of time-space distanciation.

It is quite wrong, however, to set off the impersonality of abstract systems against the intimacies of personal life as most existing sociological accounts tend to do. Personal life and the social ties it involves are deeply intertwined with the most far-reaching of abstract systems. It has long been the case, for example, that Western diets reflect global economic interchanges: "every cup of coffee contains within it the whole history of Western imperialism." With the accelerating globalisation of the past fifty years or so, the connections between personal life of the most intimate kind and disembedding mechanisms have intensified. As Ulrich Beck has observed, "The most intimate—say, nursing a child—and the most distant, most general—say a reactor accident in the Ukraine, energy politics—are now suddenly *directly* connected."

What does this mean in terms of personal trust? The answer to this question is fundamental to the transformation of intimacy in the twentieth century. Trust in persons is not focused by personalised connections within the local community and kinship networks. Trust on a personal level becomes a project, to be "worked at" by the parties involved, and demands the *opening out of the individual to the other*. Where it cannot be controlled by fixed normative codes, trust has to be *won*, and the means of doing this is demonstrable warmth and openness. Our peculiar concern with "relationships," in the sense which that word has now taken on, is expressive of this phenomenon. Relationships are ties based upon trust, where trust is not pre-given but worked upon, and where the work involved means a *mutual process of self-disclosure*.

Given the strength of the emotions associated with sexuality, it is scarcely surprising that erotic involvements become a focal point for such self-disclosure. The transition to modern forms of erotic relations is generally thought to be associated with the formation of an ethos of romantic love, or with what Lawrence Stone calls "affective individualism." The ideal of romantic love is aptly described by Stone in the following way:

the notion that there is only one person in the world with whom one can unite at all levels; the personality of that person is so idealised that the normal faults and follies of human nature disappear from view; love is like a thunderbolt and strikes at first sight; love is the most important thing in the world, to which all other considerations, particularly material ones, should be sacrificed; and lastly, the giving of full rein to personal emotions is admirable, no matter how exaggerated and absurd the resulting conduct might appear to others.[ii]

Characterised in this way, romantic love incorporates a cluster of values scarcely ever realisable in their totality. Rather than being an ethos associated in a continuous way with the rise of modern instructions, it seems essentially to have been a transitional phenomenon, bound up with a relatively early phase in the dissolution of the older forms of arranged marriage. Aspects of the "romantic love complex" as described by

Stone have proved quite durable, but these have become increasingly meshed with the dynamics of personal trust described above. Erotic relations involve a progressive path of mutual discovery, in which a process of self-realisation on the part of the lover is as much a part of the experience as increasing intimacy with the loved one. Personal trust, therefore, has to be established through the process of self-enquiry; the discovery of oneself becomes a project directly involved with the reflexivity of modernity.

Interpretations of the quest for self-identity tend to divide in much the same way as views of the decline of community, to which they are often linked. Some see a preoccupation with self-development as an offshoot of the fact that the old communal orders have broken down, producing a narcissistic, hedonistic concern with the ego. Others reach much the same conclusion, but trace this end result to forms of social manipulation. Exclusion of the majority from the arenas where the most consequential policies are forged and decisions taken forces a concentration upon the self; this is a result of the powerlessness most people feel. In the words of Christopher Lasch:

As the world takes on a more and more menacing appearance, life becomes a never-ending search for health and well-being through exercise, dieting, drugs, spiritual regimens of various kinds, psychic self-help, and psychiatry. For those who have withdrawn interest from the outside world except in so far as it remains a source of gratification and frustration, the state of their own health becomes an all-absorbing concern.[iii]

Is the search for self-identity a form of somewhat pathetic narcissism, or is it, in some part at least, a subversive force in respect of modern institutions? Most of the debate about the issue has concentrated upon this question, . . . but for the moment we should see that there is something awry in Lasch's statement. A "search for health and well-being" hardly sounds compatible with a "withdrawal of interest in the outside world." The benefits of exercise or dieting are not personal discoveries but came from the lay reception of expert knowledge, as does the appeal of therapy or psychiatry. The spiritual regimens in question may be an eclectic assemblage, but include religions and cults from around the world. The outside world not only enters

---

[ii]Lawrence Stone, *The Family, Sex and Marriage in England 1500–1800,* London: Weidenfeld 1977, p. 282.

[iii]Christopher Lasch, *Haven in a Heartless World,* New York: Basic Books 1977, p. 140.

in here; it is an outside world vastly more extensive in character than anyone would have had contact with in the pre-modern era.

To summarise all this, the transformation of intimacy involves the following:

1. An intrinsic relation between the *globalising tendencies* of modernity and *localised events* in day-to-day life—a complicated, dialectical connection between the "extensional" and the "intensional."

2. The construction of the self as a *reflexive project,* an elemental part of the reflexivity of modernity; an individual must find her or his identity amid the strategies and options provided by abstract systems.

3. A drive towards self-actualisation, founded upon *basic trust,* which in personalised contexts can only be established by an "opening out" of the self to the other.

4. The formation of personal and erotic ties as "relationships," guided by the *mutuality of self-disclosure.*

5. *A concern for self-fulfillment,* which is not just a narcissistic defense against an externally threatening world, over which individuals have little control, but also in part a *positive appropriation* of circumstances in which globalised influences impinge upon everyday life.

Risk and Danger in the Modern World

How should we seek to analyze the "menacing appearance" of the contemporary world of which Lasch speaks? To do so means looking in more detail at the specific risk profile of modernity, which may be outlined in the following way:

1. *Globalisation of risk* in the sense of *intensity:* for example, nuclear war can threaten the survival of humanity.

2. *Globalisation of risk* in the sense of the *expanding number of contingent events* which affect everyone or at least large numbers of people on the planet: for example, changes in the global division of labour.

3. Risk stemming from the *created environment,* or *socialised nature;* the infusion of human knowledge into the material environment.

4. The development of *institutionalised risk environments* affecting the life-chances of millions: for example, investment markets.

5. *Awareness of risk as risk:* the "knowledge gaps" in risks cannot be converted into "certainties" by religious or magical knowledge.

6. The *well-distributed awareness of risk:* many of the dangers we face collectively are known to wide publics.

7. *Awareness of the limitations of expertise:* no expert system can be wholly expert in terms of the consequences of the adoption of expert principles.

If the disembedding mechanisms have provided large areas of security in the present-day world, the new array of risks which have thereby been brought into being are truly formidable. The main forms I have listed can be separated out into those that alter the objective distribution of risks (the first four items listed) and those that alter the experience of risk or the perception of perceived risk (the remaining three items).

What I have termed the intensity of risk is surely the basic element in the "menacing appearance" of the circumstances in which we live today. The possibility of nuclear war, ecological calamity, uncontainable population explosion, the collapse of global economic exchange, and other potential global catastrophes provide an unnerving horizon of dangers for everyone. As Beck has commented, globalised risks of this sort do not respect divisions between rich and poor or between regions of the world. The fact that "Chernobyl is everywhere" spells what he calls "the end of 'others'"— boundaries between those who are privileged and those who are not. The global intensity of certain kinds of risk transcends all social and economic differentials. (Of course, this should not blind us to the fact that, in conditions of modernity, as in the pre-modern world, many risks are differentially distributed between the privileged and the underprivileged. Differential risk—in relation, for example, to levels of nutrition and susceptibility to illness—is a large part of what is actually meant by "privilege" and "underprivilege.")

Nuclear war is plainly the most potentially immediate and catastrophic of all current global dangers. Since the early 1980's it has been recognised that the climatic and environmental effects of a quite limited nuclear engagement could be very far-reaching. The detonation of a small number of warheads might produce irreversible environmental damage which could threaten the life of all complex animal species. The threshold for the occurrence of a "nuclear winter" has been calculated at between 500 and 2,000 warheads—less than 10 percent of the total held by the nuclear nations. It is even below the number possessed during the 1950's. This circumstance wholly justifies the assertion that in such a context, there are no longer "others": the combatants and those uninvolved would all suffer.

The second category of globalised risks concerns the world-wide extension of risk environments, rather than the intensification of risk. All disembedding mechanisms take things out of the hands of any specific individuals or groups; and the more such mechanisms are of global scope, the more this tends to be so. Despite the high levels of security which globalised mechanisms can provide, the other side of the coin is that novel risks come into being: resources or services are no longer under local control and therefore cannot be locally refocused to meet unexpected contingencies, and there is a risk that the mechanism as a whole can falter, thus affecting everyone who characteristically makes use of it. Thus someone who has oil-fired central heating and no fireplaces is particularly vulnerable to changes in the price of oil. In circumstances such as the "oil crisis" of 1973, produced as a result of the actions of the OPEC cartel, all consumers of petroleum products are affected.

The first two categories in the risk profile concern the scope of risk environments; the next two are to do with changes in the type of risk environment. The category of the created environment, or "socialised nature" refers to the altered character of the relation between human beings and the physical environment. The variety of ecological dangers in such a category derive from the transformation of nature by human knowledge systems. The sheer number of serious risks in respect of socialised nature is quite daunting: radiation from major accidents at nuclear power-stations or from nuclear waste; chemical pollution of the seas sufficient to destroy the phytoplankton that renews much of the oxygen in the atmosphere; a "greenhouse effect" deriving from atmospheric pollutants which attack the ozone layer, melting part of the ice caps and flooding vast areas; the destruction of large areas of rain forest which are a basic source of renewable oxygen; and the exhaustion of millions of acres of topsoil as a result of widespread use of artificial fertilizers. . . .

In terms of the experience of risk, far more could be said than I have the opportunity to analyse here. The three aspects of the awareness of risk indicated in the risk profile above, however, are immediately relevant to the arguments developed in this study thus far and to subsequent sections. The fact that risks—including in this regard many different forms of activity—are generally accepted by the lay population to *be* risks is a major aspect of the disjuncture between the pre-modern and the modern worlds. High-risk enterprises undertaken in traditional cultures may sometimes have occurred in a secular domain, but more typically were carried out under the auspices of religion or magic. How far individuals may have been prepared to vest trust in particular religious or magical prescriptions in specific risk domains was no doubt widely variable. But religion and magic very often provided a way of sealing over the uncertainties entailed in risky endeavours, thus translating the experience of risk into feelings of relative security. Where risk is known *as* risk, this mode of generating confidence in hazardous actions is by definition unavailable. In a predominantly secular milieu, there are various ways of trying to transmute risk into providential *fortuna,* but they remain half-hearted superstitions rather than truly effective psychological supports. People in occupations entailing life-threatening risk, such as steeplejacks, or in enterprises where the outcome is structurally indetermined, like sports players, quite often have recourse to charms or superstitious rituals, to "influence" the outcomes of what they do. But they might very well be scorned by others if they make these practices too public.

We can take the final two points in the risk profile together. Widespread lay knowledge of modern risk environments leads to awareness of the limits of expertise and forms one of the "public relations" problems that has to be faced by those who seek to sustain lay trust in expert systems. The faith that supports trust in expert systems involves a blocking off of the ignorance of the lay person when faced with the claims of expertise; but realisation of the areas of ignorance which confront the experts themselves, as individual practitioners and in terms of overall fields of knowledge, may weaken or undermine that faith on the part of lay individuals. Experts often take risks "on behalf" of lay clients while concealing, or fudging over, the true nature of those risks or even the fact that there are risks at all. More damaging than the lay discovery of this kind of concealment is the circumstance where the full extent of a particular set of dangers and the risks associated with them is not realised by the experts. For in this case what is in question is not only the limits of, or the gaps in, expert knowledge, but an inadequacy which compromises the very idea of expertise.

RISK AND ONTOLOGICAL SECURITY

In what ways does this array of risks impinge upon lay trust in expert systems and feelings of ontological security? The baseline for analysis has to be the *inevitability* of living with dangers which are *remote* from the control not only of individuals, but also of large organisations, including states; and which are of *high intensity* and *life-threatening* for millions of human beings and potentially for the whole of humanity. The facts that these are not risks anyone *chooses* to run and that there are, in Beck's terms, no "others" who could be held responsible, attacked, or blamed reinforce the sense of foreboding which so many have noted as a characteristic of the current age. Nor is it surprising that some of those who hold to religious beliefs are inclined to see the potential for global disaster as an expression of the wrath of God. For the high consequence global risks which we all now run are key elements of the runaway, juggernaut character of modernity, and no specific individuals or groups are responsible for them or can be constrained to "set things right."

How can we constantly keep in the forefront of our minds dangers which are enormously threatening, yet so remote from individual control? The answer is that most of us cannot. People who worry all day, every day, about the possibility of nuclear war, as was noted earlier, are liable to be thought disturbed. While it would be difficult to deem irrational someone who was constantly and consciously anxious in this way, this outlook would paralyse ordinary day-to-day life. . . .

The large majority of people do not spend much of their time, on a conscious level at least, worrying about nuclear war or about the other major hazards for which it may or may not be a metaphor. The need to get on with the more local practicalities of day-to-day life is no doubt one reason, but much more is involved psychologically. In a secular environment, low-probability high-consequence risks tend to conjure up anew a sense of *fortuna* closer to the pre-modern outlook than that cultivated by minor superstitions. A sense of "fate," whether positively or negatively tinged—a vague and generalised sense of trust in distant events over which one has no control—relieves the individual of the burden of engagement with an existential situation which might otherwise be chronically disturbing. Fate, a feeling that things will take their own course anyway, thus reappears at the core of a world which is supposedly taking rational control of its own affairs. Moreover, this surely exacts a price on the level of the unconscious, since it essentially presumes the repression of anxiety. The sense of dread which is the antithesis of basis trust is likely to infuse unconscious sentiments about the uncertainties faced by humanity as a whole.

Low-probability high-consequence risks will not disappear in the modern world, although in an optimal scenario they could be minimised. Thus, were it to be the case that all existing nuclear weapons were done away with, no other weapons of comparable destructive force were invented, and no comparably catastrophic disturbances of socialised

nature were to loom, a profile of global danger would still exist. For if it is accepted that the eradication of established technical knowledge could not be achieved, nuclear weaponry could be reconstructed at any point. Moreover, any major technological initiative could thoroughly disturb the overall orientation of global affairs. The juggernaut effect is inherent in modernity, for reasons I shall amplify in the next section of this work.

The heavily counterfactual character of the most consequential risks is closely bound up with the numbness that a listing of them tends to promote. In mediaeval times, the invention of hell and damnation as the fate of the unbeliever in the afterlife was "real." Yet things are different with the most catastrophic dangers which face us today. The greater the danger, measured not in terms of probability of occurrence but in terms of its generalised threat to human life, the more thoroughly counterfactual it is. The risks involved are necessarily "unreal," because we could only have clear demonstration of them if events occurred that are too terrible to contemplate. Relatively small-scale events, such as the dropping of atomic bombs on Hiroshima and Nagasaki or the accidents at Three Mile Island or Chernobyl, give us some sense of what could happen. But these do not in any way bear upon the necessarily counterfactual character of other, more cataclysmic happenings— the main basis of their "unreality" and the narcotising effects produced by the repeated listing of risks. As Susan Sontag remarks, "A permanent modern scenario: apocalypse looms—and it doesn't occur. And still it looms. . . . Apocalypse is now a long-running serial: not 'Apocalypse Now,' but 'Apocalypse from now on.'"[iv] . . .

A PHENOMENOLOGY OF MODERNITY

Two images of what it feels like to live in the world of modernity have dominated the sociological literature, yet both of them seem less than adequate. One is that of Weber, according to which the bonds of rationality are drawn tighter and tighter, imprisoning us in a featureless cage of bureaucratic routine. Among the three major founders of modern sociology, Weber saw most clearly the significance of expertise in modern social development and used it to outline a phenomenology of modernity. Everyday experience, according to Weber, retains its colour and spontaneity, but only on the perimeter of the "steel-hard" cage of bureaucratic rationality. . . . There are many contexts of modern institutions which are marked by bureaucratic fixity. But they are far from all-pervasive, and even in the core settings of its application, namely,

---

[iv]Susan Sontag, *AIDS and Its Metaphors,* Harmondsworth: Penguin, 1989.

large-scale organisations, Weber's characterisation of bureaucracy is inadequate. Rather than tending inevitably towards rigidity, organisations produce areas of autonomy and spontaneity—which are actually often less easy to achieve in smaller groups. . . . The closed climate of opinion within some small groups and the modes of direct sanction available to their members fix the horizons of action much more narrowly and firmly than in larger organisational settings.

The second is the image of Marx—and of many others, whether they regard themselves as Marxist or not. According to this portrayal, modernity is seen as a monster. More limpidly perhaps than any of his contemporaries, Marx perceived how shattering the impact of modernity would be, and how irreversible. At the same time, modernity was for Marx what Habermas has aptly called an "unfinished project." The monster can be tamed, since what human beings have created they can always subject to their own control. Capitalism, simply, is an irrational way to run the modern world, because it substitutes the whims of the market for the controlled fulfillment of human need.

For these images I suggest we should substitute that of the juggernaut[v]—a runaway engine of enormous power which, collectively as human beings, we can drive to some extent but which also threatens to rush out of our control and which could rend itself asunder. The juggernaut crushes those who resist it, and while it sometimes seems to have a steady path, there are times when it veers away erratically in directions we cannot foresee. The ride is by no means wholly unpleasant or unrewarding; it can often be exhilarating and charged with hopeful anticipation. But, so long as the institutions of modernity endure, we shall never be able to control completely either the path or the pace of the journey. In turn, we shall never be able to feel entirely secure, because the terrain across which it runs is fraught with risks of high consequence. Feelings of ontological security and existential anxiety will coexist in ambivalence.

The juggernaut of modernity is not all of one piece, and here the imagery lapses, as does any talk of a single path which it runs. It is not an engine made up of integrated machinery, but one in which there is a tensionful, contradictory, push-and-pull of different influences. Any attempt to capture the experience of modernity must begin from this view, which derives ultimately from the dialectics of space and time, as expressed in the time-space constitution of modern institutions. I shall sketch a phenomenology of modernity in terms of four dialectically related frameworks of experience, each of which connects in an integral way with the preceding discussion in this study:

Displacement and reembedding: the intersection of estrangement and familiarity.

Intimacy and impersonality: the intersection of personal trust and impersonal ties.

Expertise and reappropriation: the intersection of abstract systems and day-to-day knowledgeability.

Privatism and engagement: the intersection of pragmatic acceptance and activism.

Modernity "dis-places" in the sense previously analyzed—place becomes phantasmagoric. Yet this is a double-layered, or ambivalent, experience rather than simply a loss of community. We can see this clearly only if we keep in mind the contrasts between the pre-modern and the modern described earlier. What happens is not simply that localised influences drain away into the more impersonalised relations of abstract systems. Instead, the very tissue of spatial experience alters, conjoining proximity and distance in ways that have few close parallels in prior ages. There is a complex relation here between familiarity and estrangement. Many aspects of life in local contexts continue to have a familiarity and ease to them, grounded in the day-to-day routines individuals follow. But the sense of the familiar is one often mediated by time-space distanciation. It does not derive from the particularities of localised place. And this experience, so far as it seeps into general awareness, is simultaneously disturbing and rewarding. The reassurance of the familiar, so important to a sense of ontological security, is coupled with the realisation that what is comfortable and nearby is actually an expression of distant events and was "placed into" the local environment rather than forming an organic development within it. The local shopping mall is a milieu in which a sense of ease and security is cultivated by the layout of the buildings and the careful planning of public places. Yet everyone who shops there is aware that most of the shops are chain stores, which one might find in any city, and indeed that innumerable shopping malls of similar design exist elsewhere.

A feature of displacement is our insertion into globalised cultural and information settings, which means that familiarity and place are much less consistently connected than hitherto. This is less a phenomenon of estrangement from the local than one of integration within globalised "communities" of shared experience. The boundaries of concealment and disclosure become altered, since many erstwhile quite distinct activities are juxtaposed in unitary public

[v]The term comes from the Hindi *Jaganna-th,* "lord of the world," and is a title of Krishna; an idol of this deity was taken each year through the streets on a huge car, which followers are said to have thrown themselves under, to be crushed beneath the wheels.

domains. The newspaper and the sequence of television programmes over the day are the most obvious concrete examples of this phenomenon, but it is generic to the time-space organisation of modernity. We are all familiar with events, with actions, and with the visible appearance of physical settings thousands of miles away from where we happen to live. The coming of electronic media has undoubtedly accentuated these aspects of displacement, since they override presence so instantaneously and at such distance. As Joshua Meyrowitz points out, a person on the telephone to another, perhaps on the opposite side of the world, is more closely bound to that distant other than to another individual in the same room (who may be asking, "Who is it? What's she saying?" and so forth).

The counterpart of displacement is reembedding. The disembedding mechanisms lift social relations and the exchange of information out of specific time-space contexts, but at the same time provide new opportunities for their reinsertion. This is another reason why it is a mistake to see the modern world as one in which large, impersonal systems increasingly swallow up most of personal life. The self-same processes that lead to the destruction of older city neighbourhoods and their replacement by towering office-blocks and skyscrapers often permit the gentrification of other areas and a recreation of locality. Although the picture of tall, impersonal clusters of city-centre buildings is often presented as the epitome of the landscape of modernity, this is a mistake. Equally characteristic is the recreation of places of relative smallness and informality. The very means of transportation which help to dissolve the connection between locality and kinship provide the possibility for reembedding, by making it easy to visit "close" relatives who are far away.

Parallel comments can be made about the intersection of intimacy and impersonality in modern contexts of action. It is simply not true that in conditions of modernity we live increasingly in a "world of strangers." We are not required more and more to exchange intimacy for impersonality in the contacts with others we routinely make in the course of our day-to-day lives. Something much more complex and subtle is involved. Day-to-day contacts with others in pre-modern settings were normally based upon a familiarity stemming in part from the nature of place. Yet contacts with familiar others probably rarely facilitated the level of intimacy we associate with personal and sexual relations today. The "transformation of intimacy" of which I have spoken is contingent upon the very distancing which the disembedding mechanisms bring about, combined with the altered environments of trust which they presuppose. There are some very obvious ways in which intimacy and abstract systems interact. Money, for example, can be spent to purchase the expert services of a psychologist who guides the individual in an exploration of the inner universe of the intimate and the personal.

A person walks the streets of a city and encounters perhaps thousands of people in the course of a day, people she or he has never met before—"strangers" in the modern sense of that term. Or perhaps that individual strolls along less crowded thoroughfares, idly scrutinising passers by and the diversity of products for sale in the shops—Baudelaire's *flâneur*. Who could deny that these experiences are an integral element of modernity? Yet the world "out there"—the world that shades off into indefinite time-space from the familiarity of the home and the local neighbourhood—is not at all a purely impersonal one. On the contrary, intimate relationships can be sustained at a distance (regular and sustained contact can be made with other individuals at virtually any point on the earth's surface—as well as some below and above), and personal ties are continually forged with others with whom one was previously unacquainted. We live in a *peopled* world, not merely one of anonymous, blank faces, and the interpolation of abstract systems into our activities is intrinsic to bringing this about.

In relations of intimacy of the modern type, trust is always ambivalent, and the possibility of severance is more or less ever present. Personal ties can be ruptured, and ties of intimacy returned to the sphere of impersonal contacts—in the broken love affair, the intimate suddenly becomes again a stranger. The demand of "opening oneself up" to the other which personal trust relations now presume, the injunction to hide nothing from the other, mix reassurance and deep anxiety. Personal trust demands a level of self-understanding and self-expression which must itself be a source of psychological tension. For mutual self-revelation is combined with the need for reciprocity and support; yet the two are frequently incompatible. Torment and frustration interweave themselves with the need for trust in the other as the provider of care and support.

DESKILLING AND RESKILLING IN EVERYDAY LIFE

Expertise is part of intimacy in conditions of modernity, as is shown not just by the huge variety of forms of psychotherapy and counseling available, but by the plurality of books, articles, and television programmes providing technical information about "relationships." Does this mean that, as Habermas puts it, abstract systems "colonise" a pre-existing "life-world," subordinating personal decisions to technical expertise? It does not. The reasons are twofold. One is that modern institutions do not just implant themselves into a "life-world," the residues of which remain much the same as they always were. Changes in the nature of day-to-day life also affect the disembedding mechanisms in a dialectical interplay. The second reason is that technical expertise is continuously reappropriated by lay agents as part of their routine dealings with abstract systems. No one can become an expert, in the sense of the possession either of full expert knowledge or of the appropriate formal credentials, in more than a few small sectors of the immensely complicated knowledge systems which now exist. Yet no one can interact with abstract systems without mastering some of the rudiments of the principles upon which they are based.

Sociologists often suppose that, in contrast to the pre-modern era, where many things were mysteries, today we live in a world from which mystery has retreated and where the way "the world works" can (in principle) be exhaustively known. But this is not true for either the lay person or the expert, if we consider their experience as individuals. To all of us living in the modern world things are specifically *opaque,* in a way that was not the case previously. In pre-modern environments the "local knowledge," to adapt a phrase from Clifford Geertz, which individuals possessed was rich, varied, and adapted to the requirements of living in the local milieu. But how many of us today when we switch on the light know much about where the electricity supply comes from or even, in a technical sense, what electricity actually is?

Yet, although "local knowledge" cannot be of the same order as it once was, the sieving off of knowledge and skill from everyday life is not a one-way process. Nor are individuals in modern contexts less knowledgeable about their local milieux than their counterparts in pre-modern cultures. Modern social life is a complex affair, and there are many "filter-back" processes whereby technical knowledge, in one shape or another, is reappropriated by lay persons and routinely applied in the course of their day-to-day activities. . . . Economic factors may decide whether a person learns to fix her or his car engine, rewire the electrical system of the house, or fix the roof; but so do the levels of trust that an individual vests in the particular expert systems and known experts involved. Processes of reappropriation relate to all aspects of social life—for example, medical treatment, child-rearing, or sexual pleasure.

For the ordinary individual, all this does not add up to feelings of secure control over day-to-day life circumstances. Modernity expands the arenas of personal fulfillment and of security in respect of large swathes of day-to-day life. But the lay person—and *all* of us are lay persons in respect of the vast majority of expert systems—must ride the juggernaut. The lack of control which many of us feel about some of the circumstances of our lives is real. . . .

Balanced against the deep anxieties which such circumstances must produce in virtually everyone is the psychological prop of the feeling that "there's nothing that I as an individual can do," and that at any rate the risk must be very slight. Business-as-usual, as I have pointed out, is a prime element in the establishing of trust and ontological security, and this no doubt applies in respect of high consequence risks just as it does in other areas of trust relations.

Yet obviously even high-consequence risks are not only remote contingencies, which can be ignored in daily life, albeit at some probable psychological cost. Some such risks, and many others which are potentially life-threatening for individuals or otherwise significantly affect them, intrude right into the core of day-to-day activities. This is true, for example, of any pollution damage which affects the health of adults or children, and anything which produces toxic contents in food or affects its nutritional properties. It is also true of a multitude of technological changes that influence life chances, such as reproductive technologies. The mix of risk and opportunity is so complex in many of the circumstances involved that it is extremely difficult for individuals to know how far to vest trust in particular prescriptions or systems, and how far to suspend it. How can one manage to eat "healthily," for example, when all kinds of food are said to have toxic qualities of one sort or another and when what is held to be "good for you" by nutritional experts varies with the sifting state of scientific knowledge?

Trust and risk, opportunity and danger—these polar, paradoxical features of modernity permeate all aspects of day-to-day life, once more reflecting an extraordinary interpolation of the local and the global. Pragmatic acceptance can be sustained towards most of the abstract systems that impinge on individuals' lives, but by its very nature such an attitude cannot be carried on all the while and in respect of all areas of activity. For incoming expert information is often fragmentary or inconsistent,[vi] as is

---

[vi]Consider, as one among an indefinite range of examples, the case of cyclamate, an artificial sweetener, and the U.S. authorities. Cyclamate was widely used in the United States until 1970, and the Food and Drug Administration classified it as "generally recognised as safe." The attitude of the FDA changed when scientific research concluded that rats given large doses of the substance were prone to certain types of cancer. Cyclamate was banned from use in foodstuffs. As more and people began to drink low-calorie beverages in the 1970's and early 1980's, however, manufacturers exerted pressure on the FDA to change its stance. In 1984, a committee of the FDA decided that cyclamate was not after all a cancer-producing agent. A year later, the National Academy of Sciences intervened, reaching yet a different conclusion. In its report on the subject, the Academy declared that cyclamate is unsafe when used with sacharin, although probably harmless when used on its own as a sweetener. See James Bellini, *High Tech Holocaust* (London: Tarrant, 1986).

the recycled knowledge which colleagues, friends, and intimates pass on to one another. On a personal level, decisions must be taken and policies forged. Privatism, the avoidance of contestatory engagement—which can be supported by attitudes of basic optimism, pessimism, or pragmatic acceptance—can serve the purposes of day-to-day "survival" in many respects. But it is likely to be interspersed with phases of active engagement, even on the part of those most prone to attitudes of indifference or cynicism. For, to repeat, in respect of the balance of security and danger which modernity introduces into our lives, there are no longer "others"—no one can be completely outside. Conditions of modernity, in many circumstances, provoke activism rather than privatism, because of modernity's inherent reflexivity

and because there are many opportunities for collective organisation within the polyarchic systems of modern nation-states.

OBJECTIONS TO POST-MODERNITY

. . . I have sought to develop an interpretation of the current era which challenges the usual views of the emergence of post-modernity. As ordinarily understood, conceptions of post-modernity—which mostly have their origin in post-structuralist thought—involve a number of distinct strands. I compare this conception of post-modernity (PM) with my alternative position, which I shall call radicalised modernity (RM), in Table 1.

Table 1 A Comparison of Conceptions of "Post-Modernity" (PM) and "Radicalised Modernity" (RM)

PM	RM
1. Understands current transitions in epistemological terms or as dissolving epistemology altogether.	1. Identifies the institutional developments which create a sense of fragmentation and dispersal.
2. Focuses upon the centrifugal tendencies of current social transformation and their dislocating character.	2. Sees high modernity as a set of circumstances in which dispersal is dialectically connected to profound tendencies towards global integration.
3. Sees the self as dissolved or dismembered by the fragmenting of experience.	3. Sees the self as more than just a site of intersecting forces; active processes of reflexive self-identity are made possible by modernity.
4. Argues for the contextuality of truth claims or sees them as "historical."	4. Argues that the universal features of truth claims force themselves upon us in an irresistible way given the primacy of problems of a global kind. Systematic knowledge about these developments is not precluded by the reflexivity of modernity.
5. Theorizes powerlessness which individuals feel in the face of globalising tendencies.	5. Analyses a dialectic of powerlessness and empowerment, in terms of both experience and action.
6. Sees the "emptying" of day-to-day life as a result of the intrusion of abstract systems.	6. Sees day-to-day life as an active complex of reactions to abstract systems, involving appropriation as well as loss.
7. Regards coordinated political engagement as precluded by the primacy of contextuality and dispersal.	7. Regards coordinated political engagement as both possible and necessary, on a global level as well as locally.
8. Defines post-modernity as the end of epistemology/the individual/ethics.	8. Defines post-modernity as possible transformations moving "beyond" the institutions of modernity.

Discussion Questions

1. Using Giddens's theory of structuration, how might intergenerational poverty be explained?

2. Bourdieu notes how class relations shape the habitus such that "the dominated perceive the dominant through the categories that the relation of domination has produced and which are thus identical to the interests of the dominant" (Bourdieu 1994/1998:121). In addition to economic classes, how might relations of domination based on gender or race shape the categories of perception that induce the dominated to reproduce their own subordinate position?

3. For Habermas, the strength and vibrancy of democracy lie in the public sphere—that is, spaces where ideas and reasoned arguments can be freely debated. What role do you think the mass media play in shaping the public sphere? How might the corporate consolidation of media outlets and growing use of the Internet affect public debate?

4. Each of the theorists discussed in this chapter has sought to develop a multidimensional framework that bridges the divide between individualist and collectivist perspectives. Which framework do you think provides the most accurate and balanced account of social life? What are the strengths of this framework relative to the others?

5. Given that each of the theorists discussed in this chapter explores social life through a unique set of concepts, what types of research strategies might be best suited to each approach? In other words, what types of data would you need to collect in order to empirically examine, for instance, the duality of structure, the reproduction of class relations, or the colonization of the lifeworld?

17 THE GLOBAL SOCIETY

Immanuel Wallerstein

Key Concepts

- World-System
- Core
- Periphery
- Semiperiphery

George Ritzer

Key Concepts

- McDonaldization
- Grobalization
- Americanization
- Nothing
- Something

Edward Said

Key Concepts

- Orientalism
- Imaginative Geography

It was with the emergence of the modern world-economy in sixteenth century Europe that we saw the full development and economic predominance of market trade. This was the system called capitalism. Capitalism and a world- economy . . . are obverse sides of the same coin.

—Immanuel Wallerstein (1979:6)

Several years ago, I spent some time on an ecotour in the Amazon rain forest. While the trip was spectacular in every sense, it also provided a firsthand experience of the shrinking of space that increasingly characterizes our (post)modern world. After a full day of travel from New York, which included four different modes of transportation, I had the sense of being far away but, yet, not removed from the world as I know it. During the four-hour trip down the Napo River in a dugout canoe, we came across an outpost that resembled a small town like those depicted in Western movies. It had a general store, one or two other small, wooden buildings, and what appeared to be a brothel. Just past the outpost was an area of deforestation or clear-cutting where mud roads crisscrossed. The sights and sounds of bulldozers, dump trucks, and the like dominated the landscape. And off the bank of the river a large sign emblazoned with a single word—TEXACO—stood perched some thirty feet off the ground. Here, deep in the Amazonian rain forest, was a transnational corporation, headquartered in a dominant, global superpower, exploiting the natural resources of an "underdeveloped" country in the name of private profit.

The reach of the modern global society is not simply economic, however. Indeed, international commerce preceded both capitalism and colonization and thus by itself represents nothing new. Rather, it is the scope and speed at which economic capital flows across borders that differentiates the contemporary global society from that of earlier eras. Moreover, it is technological innovations in communications and transportation that are directly responsible for shrinking the size of the globe. For while it took me a day to travel some three thousand miles to the rain forest from New York, it took me *only* a day. In addition, I was able to make all of the necessary travel plans and plot out my adventure long before I left the comforts of my home.

Yet, my trip highlighted another dimension of the contemporary global society. One day as I ventured from my hut to the part of the encampment where the local guides were housed, I happened upon a generator that ran to one of their huts. As I approached the hut I saw the guides huddled around a generator-powered television watching a video of Sylvester Stallone's *Rambo*. And here was a glimpse into how not only Western corporations, but also Western culture penetrates even the most remote nooks of our planet. To understand the forces that shape today's global society, then, we must not only be attuned to economic, technological, and political dynamics: we also need to recognize the role played by the spread of Western culture and images. For both individuals and larger collectivities around the globe, the impetus for action is in many ways shaped by encounters with and reactions to the "West," understood not as a place on a map or as a particular form of economy, but as an *idea*. And, of course, those living in the West also are affected by globalization as more and more of the products we consume, whether it be clothes, food, toys, computers, medicine, or cars and the gas to drive them—the list is endless—are produced by people working and living in non-Western countries. Similarly, the everyday lives of Westerners are affected by their governments' policies in matters ranging from tariffs, currency value, access to natural resources, environmental conditions, and military strategy that are in large measure crafted in response to the political realities in non-Western countries.

In this chapter, we explore three perspectives pertaining to the contemporary global society. We begin with a discussion of Immanuel Wallerstein's world-systems analysis, which focuses on the economic processes that have shaped the modern world. This is followed by a discussion of George Ritzer's insights into the types of practices, places, and people spread through globalization. We conclude with a discussion of postcolonial theory and the work of Edward Said, whose analysis of the social and cultural construction of the "Oriental" continues to provoke debate throughout the social sciences and the humanities. These perspectives focus neither on the dynamics of interpersonal interaction nor on the internal forces that give form to a single, self-contained society. Rather, they emphasize how such aspects of social life are themselves embedded in a global context, and how what happens in any given country (or geographical zone) is a function of its interconnections with other geographical regions. Certainly, what happens in Ecuador's rain forest is not solely, or most importantly, a result of internal economic, political, or cultural factors. Rather, Ecuador, like the rest of the world, is but a part of the global society.

DEFINING GLOBALIZATION ▪▪

Before we turn to the work of these three theorists, we need to clarify further a central concept in this chapter: globalization. In this section, we outline some of the major themes that occupy studies of globalization and, in doing so, provide a fuller context for examining the work of the theorists that follows later in the chapter. At the outset, however, it is important to note that despite, or perhaps because of, the vast literature that has been written on the subject, there is no one theory of globalization. Instead, there are "many theoretical discourses . . . that tend to be grounded in broader theoretical traditions and perspectives, such as Marxism, Weberianism, functionalism, post-modernism, critical and feminist theory, and involve a number of distinct approaches to social inquiry, such as cultural studies, international relations, postcolonial studies, literature, and so on" (Robinson 2007).[1]

Globalization can be defined as "the rapidly developing and ever-densening network of interconnections and interdependencies that characterize modern life" (Tomlinson 1999:2). In a very real sense, the modern world is undergoing processes of compression or "shrinking," as geographical distances become less constraining of the flow of information, technology, products, and people across territorial boundaries. Coupled with the increasing interconnectedness is the growing awareness and experience of the world as a "whole" (Robertson 1992:8; Waters 2001:5). While the physical distance separating, say, Japan and England still exists, today the inhabitants of the respective nations experience this distance in ways very different from those living only a century ago. With advances in transportation and communications technology, the six thousand miles between the two nations is routinely traversed through an airplane ride, television show, or click of a mouse (Tomlinson 1999:4). Thus, whether through economic trade, corporate expansion, political alliances, military conquest, the Internet, or TV, the world has become more structurally intertwined, while individuals' everyday lives are increasingly shaped by events taking place outside of their local environment. The world has become, in Marshall McLuhan's words, a "global village."

An endless supply of examples could be used to illustrate this point: through the technology of the Internet, individuals develop virtual relationships with people who live on the other side of the globe; satellites enable us to watch and listen to events—wars, music concerts, the Olympics, the passing of popes—taking place thousands of miles away yet providing us with a sense of connection to distant lands and their people. Or consider the ripple effects of rising gas and oil prices in the United States, triggered in part by China's and India's increasing consumption of petroleum. A rise in gas prices can lead to higher costs for most other goods, since their distribution is dependent on a transportation system fueled by gasoline. Thus, for example, the trucking industry, a key transporter of goods to warehouses and stores across the nation, will have to pass off to consumers the higher fuel costs that otherwise would cut into its profits. If fuel prices remain high, this in turn could produce a significant slowing of the national economy, as many consumers become less willing and less able to spend their wages on anything other than essential items. If prolonged, this slowdown can lead to growing unemployment across the nation as businesses begin to lay off workers in an effort to compensate for declining sales. Moreover, because America is a major trading partner with numerous countries and, for that matter, the leading consumer of countless commodities, a slowdown of its economy could have a devastating impact on virtually every nation. What may result is not a local recession or even a national one but, rather, a global recession that

---

[1]This same point also applies to feminist theories and postmodern and poststructural theories (see Chapters 14 and 15).

affects every region of the world. Certainly, the interlinking of the world's economies can be no more clearly illustrated than through the current global economic crisis sparked in large measure by the collapse of the U.S. housing market and the "toxic" mortgage-backed securities that have infested banks and other financial institutions worldwide. The resulting Great Recession has crippled not only the U.S. economy but also the economies of countries across Europe (most notably Greece) and Asia.

When, What, and Where?

While most globalization theorists would agree that increasing connectivity and interdependency—a growing compression of the world—are essential features of the globalization process, disagreements remain over determining some of its fundamental aspects. In particular, theorists have offered competing views concerning *when* the process of globalization began, *what* factors are primarily responsible for shaping its development, and *where* the process ultimately is leading.

Concerning the issue of when the globalization process began, theorists generally subscribe to one of three competing views (Robinson, 2007; Waters 2001:6, 7). The first view argues that globalizing forces have been evolving since the beginning of human history; thus, globalization has been occurring for well over five thousand years. A second view, whose adherents include Immanuel Wallerstein, contends that globalization began with the rise of the modern era and the development of capitalism. From this perspective, globalization was born some five hundred years ago as European nations raced to establish colonies and an international network of economic trade relations. The third view argues that globalization is in its infancy, having begun in the second half of the twentieth century. Globalization is here associated with the development of postindustrial economies and the contemporary restructuring of capitalism. Despite the different historical dimensions that scholars assign to globalization, most agree that the pace of the process increased markedly in the last decades of the twentieth century.

Perhaps more thorny is sorting out the factors that are most significant in shaping the process of globalization. While there is a general consensus that globalization involves economic, political, and cultural transformations, theorizing how these domains intersect, overlap, and influence one another remains contentious. A glimpse into the complexity of such transformations is offered in Arjun Appadurai's notion of the "global cultural economy." In his widely cited essay, "Disjuncture and Difference in the Global Cultural Economy," Appadurai (1992) notes that the contemporary global order is marked by "disjunctures" between and within five dimensions or "scapes" of global cultural flows. These scapes, in turn, form the basis of the fluidity of today's world and the multiple perspectives, images, and meanings that inform the worldviews of social actors, whether they be individuals, communities, corporations, or nation-states.[2]

Ethnoscapes are created through the flow of individuals and groups "who constitute the shifting world in which we live" (ibid.:33). Ethnoscapes comprise the growing number of exiles, guest workers, tourists, immigrants, and refugees whose movements affect the economic, political, and cultural landscapes of an increasing number of, and increasingly distant, nations. *Technoscapes* refer to the "global configuration . . . of technology and the fact that technology, both high and low, both mechanical and informational, now moves at high speeds across" previously impermeable corporate and national boundaries (ibid.:34). Technologies are now bought and sold and moved across borders at an unprecedented rate. For instance, an automobile plant in the United States may be using parts and technologies imported from Mexico, Japan, and Germany. *Financescapes* represent the rapid flow of massive amounts of currencies and stocks across national boundaries. More than ever, national economies are linked globally as the rise and fall of major financial markets, the patterns of foreign investment and the purchasing of foreign debt, and the value of national currencies impact the economic stability of not just one country, but of many countries simultaneously.

Appadurai also notes the growing importance of *mediascapes,* the ability to electronically produce and disseminate information and images on a global scale through television, newspapers, magazines, and films. According to Appadurai, what is most important about mediascapes is that "they provide . . . large

---

[2]Nation-states are defined as self-governing territories demarcated by recognized spatial boundaries.

and complex repertoires of images, narratives, and ethnoscapes to viewers throughout the world, in which the world of commodities and the world of news and politics are profoundly mixed" (1992:35). In turn, the lines between the real and the fictional are blurred, while the audiences for such images often are farther and farther removed from the direct experience of the mediated narratives that they observe. Thus, from the strips of information presented on the news, viewers form opinions about people and places from around the globe, as they come to "know" about, for instance, civil unrest in France or a tsunami in Indonesia. Meanwhile, the line between news and propaganda, reality and fantasy, is becoming increasingly blurred as television subjects viewers to a steady dose of "infomercials" and "infotainment."

Closely related to mediascapes are *ideoscapes,* which likewise entail the dissemination of images. In the case of ideoscapes, however, the images "are often directly political and frequently have to do with the ideologies of states and the counterideologies of movements explicitly oriented to capturing state power or a piece of it" (ibid.:36). Such politically charged images are tied to an Enlightenment worldview that trumpets notions of rights, freedom, sovereignty, and democracy. It is these terms and the ideas that they encompass that often form the discourse through which political struggles are cast. For instance, the meanings of elections, military interventions, legal rulings, and protest movements are formed out of the attempts by proponents and opponents to affirm or contest their adherence to the principles of freedom, equality, and democracy. Taken together, these five scapes represent the global flows of people, machines, money, and images. While previous periods in history experienced such flows across national boundaries, Appadurai contends that the contemporary era of globalization is witness to an extreme intensification of their speed, scale, and volume such that the disjunctures between and within these scapes are a major component of the global economic, political, and cultural landscapes. No longer are people, machines, money, and images more or less fixed to an identifiable territory; instead, they traverse the planet with unprecedented ease, creating truly global ideas and practices.

As we noted at the outset of our discussion, the world we live in today is shrinking. National economies are becoming increasingly integrated through expanding trade relationships, the spread of transnational corporations, and organizations such as the World Trade Organization (WTO) and the International Monetary Fund (IMF). Transnational political institutions including the International Court of Justice, the European Union, the South American Community of Nations, and the United Nations are furthering international, if not global, forms of governance. A "global culture" consisting of common practices, ideas, and images now flows across national boundaries as more people, products, and information move to more places. That nations have become increasingly subject to the push and pull of foreign economic, political, and cultural interests begs what is perhaps the most salient and controversial question: Where is the world headed?

There are two broad positions into which most scholars fall on this question, depending on whether they view globalization as producing an increasingly homogeneous or heterogeneous world. These two broad positions can be further divided into "weak" and "strong" versions. (See Table 17.1 for a summary of these positions.) For those who view globalization as a homogenizing force, technological advances in communication and transportation are seen as leading to the creation of a unified, interconnected "world society" where the world's population becomes integrated into a common culture. Yet, to the extent that globalization is tied to the expansion of Western capitalism and culture, this raises the issue of specifying what the West is actually exporting around the globe. For some observers, Westernization primarily implies a sense of universalism in which people around the world come to embrace a common cultural ground rooted in the Western ideals of democracy, human rights, and individual freedom. This could be termed a "weak" form of homogenization in which what is shared is a common appreciation and respect for differences (Tomlinson 1999:69). This openness to difference can in turn lead to a fusion of distinct cultures and practices, giving birth to such things as world music, world cuisine, or world cinema. In essence, a universal trend toward global pluralism is created that nurtures an ever-expanding mixture of cultural practices, meanings, tastes, and personal identities. (See Robertson 1992.) The crucial point to this notion of homogenization is the sense that globalization is not based on the systematic exploitation or domination of one region of the world by another, but rather on a developing, more humane, and tolerant world-as-a-whole.

Table 17.1   Where Is Globalization Leading? Four Views on the Consequences of Globalization

	Homogeneity	Heterogeneity
"Weak" Version	World democracy Universalism via fusion Global pluralism	Particular colored with the global Glocalism McArabia
"Strong" Version	Universalism via conquest Western cultural imperialism McDonaldization "Grobalization" Coca-colonization	Ethnic and religious conflicts Rising fundamentalism Polarization of cultural identities "Clash of civilizations"

The "strong" form of the homogenization thesis, however, contends that globalization does indeed involve the destruction of indigenous or national cultures as the West, pursuing "grobalization" practices (see the following discussion), continues its expansion through both economic and cultural imperialism. (Wallerstein and Ritzer adopt this "strong" position, although they emphasize different dynamics.) This is a homogenization or universalism bred from conquest and the perpetuation of global inequalities as corporate capitalism and the ideology of consumerism continue their invasion of national and local cultures and ways of life. Thus, homogenization does not usher in a global democracy, the establishment of universal human rights, or a pastiche of cultural practices but, rather, a "McWorld" or "Coca-colonization," in which Western nations (most notably the United States) remodel the world in their image while siphoning off the profits that come with the increasing flows of technology, goods, money, and people across national borders.

One adherent to this view is the Scottish sociologist Leslie Sklair (1940–). Sklair traces the emergence of globalization to the second half of the twentieth century. It was at this time that "transnational practices" (TNPs) became the dominant mechanism for ordering social life across the boundaries of nation-states, effectively supplanting territorial borders and national governments as the basic elements for structuring contemporary life. TNPs are defined as "practices that cross state borders but do not originate with state actors, agencies, or institutions" (Sklair 2002:10). To understand better the modern world, then, one must move beyond studying the actions of, and relations between, particular states and instead examine the "global system." More specifically, this requires recognizing that the most important global force in the world today is the capitalist global system.

Sklair distinguishes between three types of TNPs: economic, political, and culture-ideology. Forming the institutional foundation for transnational economic practices are major "transnational corporations" (TNCs) whose extensive powers have made capitalism the dominant organizing principle of the global system. (General Motors, Monsanto, and Sony are but three examples.) Those who own and oversee TNCs control the manufacturing of goods and the labor necessary to produce and sell them. The "transnational capitalist class" (TCC) is the central shaper of transnational political practices. This class is made up of those who control the major TNCs and media outlets, as well as an elite segment of politicians and state bureaucrats (think Alan Greenspan and his successor, Ben Bernanke), "who see their own interests and/or interests of their social and/or ethnic group . . . as best served by an identification with the interests of the capitalist global system" (Sklair 2002:9). The "state fraction" of this class is charged with stabilizing the political environment in order to ensure the profitable marketing of goods and services across the globe.

The primary "transnational culture-ideology" practices are found in consumerism and the meanings attached to the globally marketed goods we purchase. The culture-ideology of consumerism is based on values and attitudes aimed at instilling artificially created "needs" for the products produced by the TNCs.

Following the work of Antonio Gramsci (see Chapter 2), Sklair contends that class domination and the stability of the capitalist global society is not predicated on overt or threatened uses of force, nor through economic coercion. Instead, long-term stability requires popular support and a belief in the legitimacy of existing social arrangements, arrangements that are rooted in fundamental inequalities. This allegiance to the status quo among subordinated classes is secured ideologically, through the development of a shared culture that leads subordinate groups to identify their own interests with those that reinforce the power of dominant groups. In the global capitalist society, this is effected through the dominant culture-ideology of consumerism, expressed via the mass media and advertising, that induces individuals to equate personal satisfaction and happiness with the continual purchase of commodities. Identifying themselves in the products they purchase, individuals lend their monetized consensual support to the very corporations that are responsible for perpetuating both global inequalities and the exploitation of human labor and natural resources. Although the rewards experienced by individuals are at best fleeting, consumerism generates lasting profits and power for transnational corporations and the trans-

Photo 17.1 Transnational practices: Capitalism without borders Starbucks, a U.S.-based corporation, has stores in more than 30 countries, including one in Shanghai, China. The coffee giant uses beans imported from more than half a dozen countries across four continents. Of course, the company sells more than a drink. Its Web site speaks to the culture-ideology of consumerism as it touts the creation of "The Starbucks Experience" based on a "passion for a quality product, excellent customer service, and people. With more than 1,500 coffeehouses in 31 markets outside North America, it is clear that our passion transcends language and culture." We should not overlook the fact, however, that underlying the "passion" and commitment to "bringing the Starbucks Experience to customers worldwide" is a highly successful branding campaign.

national capitalist class. Sklair's notion of the culture-ideology of consumerism underscores how the global expansion of capitalism is dependent on creating legions of consumers who find in the goods they purchase much more than a practical satisfaction of needs. More important, consumers must see in commodities an avenue for *becoming* something they otherwise cannot be.

Another pertinent description of modern homogenization is George Ritzer's **McDonaldization** of society. Ritzer defines this as a process of growing standardization "by which the principles of the fast-food restaurant are coming to dominate more and more sectors of American society as well as the rest of the world" (Ritzer 2000b:1). Yet, as Mike Featherstone (1995) points out, the homogenizing effects of exporting McDonald's business model means more than spreading a particular method for producing and consuming goods and services. With restaurants in more than sixty countries, McDonald's serves as an icon of America and an American "way of life." Those who dine at McDonald's in South Africa, Thailand, India, Uruguay, and other countries around the world are eating more than just a burger: they are "consuming" America and images associated with it: power, wealth, beauty, freedom, youth. This is true not only for McDonald's but also for other American products that have penetrated global markets such as Nike, Hollywood movies, Coca-Cola, and pop music. In the process, indigenous types of food, clothing, and music are displaced as a country's population adopts "American style."

We must not be too quick to assume however, that McDonald's, Hollywood, and American CEOs have carte blanche to remake the world as they see fit. The consumption of goods and images is not a simple, unidirectional process, but instead one that often involves a reconfiguring of products on the part of consumers. There are a number of reasons to doubt that the sights, sounds, and tastes of America—and the West more generally—are sweeping the globe in some monolithic fashion. While American television shows and movies are beamed across the world, audiences do not assign identical meanings to them: *CSI* and *Desperate Housewives* can mean one thing (or many things, for that matter) to white, suburban viewers in America, and something entirely different to an audience in Korea. This holds true even more so when considering the meaning of ideals such as democracy, justice, and human rights. There is much variation in how particular countries define these terms and in how they are implemented in political institutions. What a Saudi might see as evidence of democracy, a Swede might view as promoting dictatorship. Or consider how Coke, a "simple" soft drink, is consumed throughout the world. As Howes notes,

> No imported object, Coca-Cola included, is completely immune from creolization. Indeed, one finds that Coke is often attributed with meanings and uses within particular cultures that are different from those imagined by the manufacturer. These include that it can smooth wrinkles (Russia), that it can revive a person from the dead (Haiti), and that it can turn copper into silver (Barbados). . . . Coke is also indigenised through being mixed with other drinks, such as rum in the Caribbean to make Cuba Libre or augadiente in Bolivia to produce Ponche Negro. Finally it seems that Coke is perceived as a "native product" in many places—that is you will often find people who believe the drink originated in their country not in the United States. (Howes 1996, cited in Tomlinson 1999:84)

For some final food for thought, in addition to offering the standard menu items found in their "home" country, fast-food chains often provide options tailored to the local culture. Thus, in Norway, McDonald's features the "McLaks" (a grilled salmon sandwich with dill sauce); in the Philippines, "McSpaghetti"; in Uruguay, the McHuevo (hamburgers with poached egg); in Japan, a "Teriyaki McBurger" (sausage patty with teriyaki sauce); in the Middle East, the McArabia (grilled chicken wrapped in flatbread); and in India, where neither beef nor pork products are eaten, the Chicken Maharaja Mac and the Chicken McCurry Pan. These examples suggest that, at the very least, it is premature to portend a homogeneous global order in which the economies, political systems, and cultures of countries around the world are becoming exact duplicates of one another.

The second broad position adopted by scholars concerning where globalization is leading gives us even more cause to be cautious about such projections. For some, globalization is linked more to an increasing "glocalism" or the hybridization of existing cultural differences. The world society is thus seen as characterized by a growing heterogeneity that highlights the distinctiveness of local cultures and ways of life, albeit a distinctiveness that is transformed by its merger with "outside" influences. For example, proponents of this view would see in McDonald's global expansion not an invasion of the Big Mac, but rather a force that has created novel products that incorporate local ingredients and tastes like the McLak. Thus, while the global mass-marketing of products leads to homogenization in the sense that the same goods can be found in virtually every corner of the globe, this also means that in any one place there exists the possibility for increased choice and, with it, greater heterogeneity at the local level (Waters 2001:196). This view, emphasizing the combining of particular, local practices and ideas with those of foreign origin, represents what we call here a "weak" form of the heterogeneity thesis. In short, the process of globalization is seen as penetrating the regions of the world without, however, producing a uniform global culture. Regional distinctions remain, albeit colored by a constant interweaving with "outside" cultural and economic influences.

Last, we can speak of a "strong" form of heterogeneity. (Edward Said's work is aligned most closely with this perspective.) Those who adopt this position maintain that globalization is leading to an increasingly fragmented world where local communities organize to resist the homogenizing tendencies associated with the spread of global—and more often than not Western—influences. Far from promoting global unity or uniformity, scholars here note that the interconnectedness spawned by globalization sparks religious, ethnic, and cultural conflicts as people fight to preserve their identity and particular way of life. For instance, in response to the increasing encroachment of Western culture and ideals, Asian leaders have sought to articulate a unique model of human rights based on Confucianism to counter Western conceptions of individualism and liberty. Moreover, globalization is seen as providing fertile ground not only for the rise of Islamic fundamentalism in the Middle East but also for an intensifying ethnocentrism witnessed in many countries such as Germany, where Turkish immigrants are subjected to violent attacks; England, where South Asians face similar hostilities; and the United States, where, in order to stem the "tide" of illegal immigrants, armed "Minutemen" have taken to patrolling the border with Mexico and the governor of Arizona has signed a bill making it a crime for immigrants not to carry their immigration papers that gives the police the power to detain anyone suspected of being in the country illegally.[3] Rejecting globalization's trend toward universalism, these reactions are symptomatic of a "clash of civilizations" that pits the West against the Rest (Huntington 1993).

Photo 17.2 Democracy, Iranian style

While they have popular elections for both the office of the president and the national assembly, Iran is a theocracy governed according to Shari'a or Islamic law. The "Supreme Leader" of the country is a cleric who is appointed for life. In this "democracy," there is no separation between church and state, and under Islamic law women can be stoned to death for adultery, have unequal rights to divorce and child custody, can be married off at the age of nine, are able to work or travel only with the permission of their father or husband, and, in a court of law, evidence and testimony offered by a female witness is valued at half that offered by a male. Pictured above is Supreme Leader of the Islamic Revolution Ayatollah Seyed Ali Khamenei, confirming Mahmoud Ahmadinejad as the sixth president of the Islamic Republic of Iran in a special ceremony.

Outside of Europe and the United States, this resistance to global homogeneity is in large part a response to the history of Western colonialism (a theme of Said's addressed later in the chapter). Many nations that were once colonies or "client states" of Western powers have sought to combine the economic modernization that globalization has brought with a rekindling of their traditional cultural identities.[4] This entails embracing the economic advantages and technological advances flowing from the West without embracing Western culture. A similar stance is often adopted by non-Western nations (for instance, China, Japan, and Saudi Arabia) that remained largely free from direct or indirect rule by Western powers. As these nations struggle for and win greater political, economic, and military power, the clash between their religious and cultural orientations and those of the West will likewise increase. No longer mere pawns on the world stage, nations whose religious and political cultures are based on Islamic, Hindu, and Buddhist traditions are gaining greater voice and are using it to nurture a way of life that contrasts in fundamental ways with the "universal" values and ideals of the West. "For the relevant future, there will be no universal civilization, but instead a world of different civilizations," where the West seeks to maintain its dominance as other nations attempt to assert their own cultural identities and priorities (Huntington 1993:49).

---

[3]It is also true that religious fundamentalism has an expansionist, homogenizing dimension to it as adherents seek to instill their own particular set of beliefs and way of life on others. At times, this ambition has been carried out by way of force or conquest.

[4]The list of such nations is long. Some that students may be more familiar with are Algeria, Egypt, Hong Kong, India, Iran, Kuwait, Pakistan, and Vietnam.

Having provided an overview of some of the central themes that inform studies of the global society, we now turn to our discussion of the work of Immanuel Wallerstein, George Ritzer, and Edward Said.

▪▪ IMMANUEL WALLERSTEIN (1930–): A BIOGRAPHICAL SKETCH

Immanuel Wallerstein was born in New York City in 1930. He spent his childhood and early adult years in the city, attending a local high school and going on to Columbia University, where he received his bachelor's, master's, and doctorate degrees. While finishing his doctorate, he took a position as an instructor in Columbia's sociology department. He remained there for the next thirteen years, advancing to the position of associate professor; then, in 1971, he took a position at McGill University in Montreal. In 1976, he returned to New York to become director of the newly established Fernand Braudel Center for the Study of Economics, Historical Systems, and Civilizations at the State University of New York at Binghamton. While he continues to be the director of the Center, Wallerstein now holds a position at Yale University.

Wallerstein grew up in a politically minded family, and early socialization into political discussions played an important role in shaping his intellectual interests. He became involved both politically and as a scholar in the national liberation movements spreading through Africa during the early 1950s. During the next two decades, Wallerstein would become a leading specialist in the study of African political and economic affairs, producing numerous books and articles on the subject. His early experiences in Africa led to his "gut feeling . . . that the most important thing that was happening in the twentieth-century world was the struggle to overcome the control by the Western world of the rest of the world" (Wallerstein, quoted in Yale University, N.d.). His solidarity with those struggling for independence from colonial exploitation also shaped his development of "world-systems analysis" in the 1970s.

One additional aspect of Wallerstein's intellectual position is worth noting here. While objectivity is often proclaimed as a hallmark of scientific inquiry, Wallerstein argues that the intellectual detachment necessary to achieving such a perspective is, in fact, impossible to maintain. Moreover, by attempting to detach one's self from the subject of study, the "disengaged" scholar will produce knowledge that is less capable of fostering positive social change. Wallerstein, however, does not suggest that scholars should produce knowledge to serve the interests of particular political parties or to serve their own political loyalties. Rather, his point is that there is not nor could there "be such a thing as value-free historical social science. Every choice of conceptual framework is a political option. Every assertion of 'truth,' even if one qualifies it as transitory truth or heuristic theory, is an assertion of value" (Wallerstein 1979:x). Scholars who are convinced otherwise are merely attempting to hide their value assertions behind the transparent cloak of scientific neutrality and objectivity. Wallerstein states his position on the relationship between politics and scientific research thusly:

> It seems to me that it is the duty of the scholar to be politically and intellectually subversive of received truths, but that the only way this subversion can be socially useful is if it reflects a serious attempt to engage with and understand the real world as best we can. . . . I have argued that world-systems analysis is not a theory but a protest against neglected issues and deceptive epistemologies. It is a call for intellectual change, indeed for "unthinking" the premises of nineteenth-century social science. It is an intellectual task that is and has to be a political task as well, because—I insist—the search for the true and the search for the good is but a single quest. (Wallerstein, quoted in Yale University, N.d.)

Not surprisingly, Wallerstein's criticisms of mainstream theories and research practices and his own emphasis on qualitative and historical methods were not enthusiastically received by all social scientists. Nevertheless, he has established himself as a leading intellectual with an international reputation. In addition to serving as president of the International Sociological Association from 1994 to 1998, he has over the years produced an amazing corpus of work comprising more than fifty authored and coauthored books and nearly three hundred articles. However, it is his three-volume study, *The Modern World-System* (1974, 1980, 1989), for which he is most noted. Translated into fourteen languages, the first volume attracted worldwide attention and placed Wallerstein squarely in the pantheon of contemporary intellectual giants.

Wallerstein lists a number of thinkers whose ideas have impacted his intellectual development. Among them are Karl Marx, the economists Joseph Schumpeter (1883–1950) and Karl Polanyi (1886–1964), and Fernand Braudel (1902–1985), the French historian for whom the Fernand Braudel Center for the Study of Economics, Historical Systems, and Civilizations was named. In what follows, we briefly introduce the central ideas associated with these thinkers as they pertain to Wallerstein's own perspective.

Fernand Braudel was a leading figure in the Annales School of history, established in France in the late 1920s. Members of the Annales School revolutionized the study of history by moving from events-based research to the study of *la longue durée,* or the long term. Up to this point, historians had focused on detailing single accounts of war and diplomacy and the great deeds of great men, but the Annales historians sought to integrate geography, sociology, economics, and, in later generations, a focus on the "mentalities" or psychology of the period into their analyses. This approach produced more sweeping and in-depth social and economic histories such as Braudel's monumental two-thousand-page work *The Mediterranean and the Mediterranean World in the Age of Philip II* (1949). Written largely during his captivity in a German prisoner-of-war camp, this master-piece of historical reconstruction, produced in the most inhospitable conditions with the barest of resources (including few books), confirmed his reputation as one of the world's foremost historians.

In this work and his subsequent publications, Braudel adopted the *longue durée* to complete enormously detailed studies that incorporated analyses of everything from a region's climate, geography, natural catastrophes, and demographics to its social customs, fashions, food, technology, and economic and political institutions.[5] Examining such an extensive array of factors across continents and over centuries, Braudel concluded that historical development is a "messy" project, incapable of being subsumed under one or another general theory. Nevertheless, Braudel maintained that it was the life of the peasant, the craftsman, and the merchant, not the heroics of the prince, the statesman, and the rich, that largely shaped the march of history.

Three aspects of Wallerstein's research demonstrate a particular debt to Braudel. First, like the French historian, Wallerstein's work covers large expanses of time. His three-volume *The Modern World-System* spans some four hundred years, while offering detailed accounts of the economic development of several regions of the world. Second, akin to Braudel, Wallerstein sets out to explain the origins of modern capitalism and the causes and consequences of its spread throughout the world. The third Braudelian stamp is evidenced by Wallerstein's rejection of traditional methodological and disciplinary boundaries, particularly those separating history from sociology.[6] While history has typically been devoted to analyses of unique or particular events, sociology (or more accurately, some branches of sociology) has been devoted to the development of abstract or general causal statements. For his part, Wallerstein believed this to be a false dichotomy and instead advocated building on both the idiographic accounts associated with the humanities and the nomothetic explanations that characterize the sciences. For Wallerstein, both methods are necessary for any attempt to accurately describe and explain the real world.

Wallerstein's history of economic development is also influenced by the work of two renowned economists: Joseph Schumpeter and Karl Polanyi. Schumpeter was an Austrian economist who rose to prominence at the age of twenty-eight with the publication of his *Theory of Economic Development* (1911).[7] His subsequent book, *Capitalism, Socialism and Democracy* (1942), further solidified his reputation in the field

---

[5]His most notable works include the three-volume *Civilization and Capitalism, 15th–18th Century; A History of Civilizations;* and *On History.*

[6]Max Weber, one of the founding figures in sociology, took a similar position on this debate, developing what is now commonly referred to as historical-comparative methods.

[7]Apart from working in academia, Schumpeter also served briefly as Austria's finance minister in 1919, his tenure short-lived in part because of the hyperinflation that was strangling his country's economy. With Hitler's rise to power, however, Schumpeter left Europe and emigrated to the United States in 1932. That same year he accepted a position at Harvard University, where he remained until his retirement in 1949.

while expanding his audience to include, among other scholars, sociologists. In this work, Schumpeter mounted a masterful defense of the virtues of capitalism—a particularly important accomplishment, given the prevailing geopolitical climate. He championed the individual entrepreneur as the source of great innovations that not only advanced capitalism as a system of production and economic organization but also improved the living standards of the general population. However, like Marx, Schumpeter imagined that the very success of capitalism would ultimately lead to its collapse and to the rise of socialism. Yet, in stark contrast to Marx, who viewed capitalism as inherently exploitative and destined to be overthrown by a proletarian revolution, Schumpeter saw capitalism imploding on itself because of moments of "creative destruction" in which the constant, beneficial innovations sparked by the entrepreneurial spirit would in time lead to the rise of monopolies that are better able to withstand the costs associated with "progress." In short, only monopolies and large corporations that possess the necessary capital and control of the market can bear the rising costs and risks associated with the continual supplanting of "obsolete" technologies, products, and skills by new innovations. With the gradual eclipse of competitive capitalism, the individual entrepreneur is himself rendered obsolete as bureaucratic managers take on the administrative task of steering the economy. However, the slow extinction of the entrepreneur produces a workforce increasingly uncertain about its prospects for employment. Coupled with the intellectual class's growing protests against the social inequities stemming from capitalist production, the stage has been set for the emergence of socialism.

The other economist Wallerstein credits for shaping his approach is Karl Polanyi.[8] Here we note two of Polanyi's most influential works. First is *The Great Transformation* (1944), a now-classic study of the emergence of the contemporary market economy in the nineteenth century and its connection to the development of the modern nation-state. In this book, Polanyi details the role that governments played in supporting the economic interests of the rising bourgeoisie. The second work is his coauthored *Trade and Market in the Early Empires* (1957). In his contribution to this volume, Polanyi explored the development of nonmarket forms of economic systems, which he classified as reciprocal, redistributive, and householding. Polanyi's investigations led him to construct a "substantive" approach to the study of economics. Akin to anthropological studies, this approach was based on the premise that economic systems and modes of exchange are not separate from but, rather, are embedded in broader social and cultural relations.

The imprint of these two economists on Wallerstein's world-systems analysis is apparent on several fronts. Like Schumpeter and Polanyi, Wallerstein provides rich, historical accounts of economic developments and connects them to transformations taking place in other domains of society. Similarly, all three view capitalism as an economic system whose reign will eventually come to a close. In Polanyi's perspective, capitalism will collapse not because of the system's successes, as Schumpeter argued, but because of its failures. More specifically, Polanyi asserted that market-based forms of exchange cannot be sustained ecologically over the long term, because the logic of market relations destroys other forms of social relations, such as kinship ties, that are better suited to creating a sense of solidarity or community within populations.

Making the deepest impact on Wallerstein's work, however, is Karl Marx, whose ideas figure prominently also in the work of each of the scholars discussed previously. At its most fundamental level, Wallerstein's model is based on the assumption that all social systems, whether a particular nation-state or the more encompassing world-system, are formed, maintained, and destroyed through conflict. To understand the history of social development thus requires an analysis of the struggles that are waged between factions competing over resources. Moreover, in the spirit of Marx's historical materialism, Wallerstein maintained that, while the conflicts that form the basis of social systems can be contained through legal or political institutions, they can never be entirely eliminated. For Marx saw all history as

---

[8]Polanyi was born in Vienna and raised in Budapest, Hungary. He emigrated to the United States in 1940 and later took a position at Columbia University, where he served on the faculty from 1947 to 1953. However, although he worked in New York City, Polanyi was unable to live there or anywhere else in the United States. Because his wife was at one time a communist, she was unable to secure a visa to enter the United States, leaving Karl to commute between New York and Canada, where they resided.

the history of class struggle, a struggle against inequality that is inevitable as long as a society's means of production can be privately owned. (See our overview of Marx's work in Chapter 3 as well as additional remarks in Chapter 9.) Wallerstein expresses a similar view when he remarks that "the history of the world is one of a constant series of revolts against inequality—whether that of one people or nation vis-à-vis another or of one class within a geographical area against another" (Wallerstein 1979:49).

To connect the range of intellectual influences outlined to this point, we now turn to an overview of Wallerstein's world-systems analysis. At its root, world-systems analysis is a product of Wallerstein's rejection of the conventional sociological practice of equating states (e.g., France, China) with societies. This intellectual position, according to Wallerstein, leads to erroneous explanations of social history and misguided predictions of future developments. To counter these pitfalls, he insists that the only unit of analysis capable of advancing the knowledge of the social sciences is the "world-system." In essence, social scientists must approach their subject—society—with the understanding that there is only one social system as opposed to viewing the world as made up of a collection of more or less independent societies or nation-states. He defines his central concept, **world-system**, as a

> social system, one that has boundaries, structures, member groups, rules of legitimation, and coherence. Its life is made up of conflicting forces which hold it together by tension and tear it apart as each group seeks eternally to remold it to its advantage. It has the characteristics of an organism, in that it has a life-span over which its characteristics change in some respects and remain stable in others (Wallerstein 1979:229).

Such "total" world-systems contain within their boundaries a diversity of cultures and a single division of labor that links together the smaller social units to form a coherent whole. The constituent units (for instance, nations) depend on economic exchanges with each other in order to meet their own needs. By contrast, minisystems—such as small agricultural or hunting and gathering societies—are culturally homogeneous and economically self-sustaining (Wallerstein 1979:4).

According to Wallerstein, to this point in history there have been only two types of world-systems in existence: world-empires and world-economies. World-empires, such as those that existed in ancient Rome or Egypt between 8000 B.C. and A.D. 1500, are ruled by, and integrated through, political and military domination. Those regions making up the empire pay taxes in the form of "tributes" to those who control the empire. World-economies, conversely, are not structured according to political rule but, instead, are based on economic exchanges that extend beyond the politically defined boundaries and control of any given state. Thus, it is not a political system but economics that is the integrating force within world-economies, an integration that nevertheless remains based on relations of domination. And from roughly the sixteenth century to the present, only one world-system has been in existence: the capitalist world-economy. This economy is the only mode of production in which the never-ending drive for profit is the central objective for participating in economic production. Thus, the capitalist's interest in producing goods or services is determined not by their use, but rather by the potential value of their sale in a market. Moreover, the logic of capitalism demands that those who profit from the market are required to constantly expand production in an effort to further increase their economic reward (Wallerstein 1979:85).

Wallerstein (1974) traces the origins of the modern world-capitalist system during the sixteenth century to the convergence of three factors: (1) an expansion of the world's geography due to exploration and colonization, (2) the development of different methods of labor control in different geographical zones, and (3) the development of strong states able to dictate the terms of international economic trade to their relative advantage. As ruling elites in European nations vied to maintain their power, they sought to exploit the natural and human resources of foreign lands. In need of gold, raw materials for production and trade, and the manpower to extract them, factions within the European nobility looked to overseas expansion as a means for securing their positions of dominance. This created a regional and hierarchical division of labor and system of economic specialization that to this day serves as the foundation for the world capitalist system, whose essential feature "is production for sale in a market in which the object is to realize the maximum profit" (Wallerstein 1979:15).

In the attempt to consolidate their economic power, emerging capitalist classes appealed to political leaders to intervene in the market on their behalf. Those capitalists who were able to turn to a strong state bureaucracy—which is itself largely a product of large tax revenues—capable of dictating international trade policy and possessing the military power to secure its demands, while also curbing the economic demands of their own national labor force, were able to reap a greater share of the world's wealth. Indeed, in the world-economy, strong, relatively autonomous states "serve primarily to distort the 'free' workings of the capitalist market" in order to increase one or another group's potential for profit (ibid.:223).[9] This dynamic led to the creation of three hierarchically structured positions within the capitalist world-economy: core, periphery, and semiperiphery. As a result, the stability of the modern world-system came to be based on a fundamental inequality in which some regions of the globe accumulate wealth at the expense of the continuing impoverishment of other regions.

The **core** region of the modern world-system first emerged in northwestern Europe and now includes the United States, Canada, Japan, and other similarly industrialized nation-states. Together, this "area" controls a vast majority of the world's wealth while producing a highly skilled workforce that is controlled through wage payment. Using its economic and political superiority, the core is able to dominate and control the periphery. The **periphery** is exploited for both its cheap labor and raw materials, such as cotton, sugar, rubber, and gold, which are then exported to the core. Often "possessions" or colonies of core states, the periphery is made up of weak nations lacking the economic, political, and military clout necessary for pressing their own agendas. The workforce in the periphery historically has been controlled through coercion and slavery, and today it still breeds the worst of labor conditions. At the outset, eastern Europe and parts of the Western Hemisphere served as the periphery, while today it comprises Africa, the Caribbean, Central America, and other "third world" regions. The lack of progress that is often said to characterize peripheral areas is in large measure a result of their economic and political dependence on the core. While the core areas have been enriched through their exploitation of the natural and human resources of the periphery, the periphery has been subject to the "development of underdevelopment" (Frank 1967). In other words, the "backwardness" of the periphery is the result of economic policies strategically intended to maintain its status as an exporter of resources while being prevented from appropriating the profits that the exporting of commodities would otherwise generate. An example of this phenomenon was provided at the beginning of the chapter when we described the multinational corporation Texaco exploiting the rain forests of Ecuador. Unfortunately, the list of examples is exceedingly long: diamonds and coal in Africa, rice and rubber in Cambodia, coffee and sugarcane from Haiti. In each case, core areas institute international economic trade policies and subsidize local governments that perpetuate the underdevelopment of the regions making up the periphery.

The **semiperiphery** occupies a position between the core and the periphery—economically and politically weaker than the former, but stronger than the latter. This area, with a labor force historically controlled through sharecropping, first emerged in the Mediterranean region of Europe, a position that today includes Eastern Europe, Mexico, and parts of South America. Moreover, semiperipheral areas play an essential role in maintaining the political stability of the modern world-system by deflecting some of the protests and political pressures that groups located in the periphery would otherwise direct toward the core states and the economic interests that are aligned with them. These areas prevent a completely polarized world-system from developing such that "the upper stratum is not faced with the *unified* opposition of all the others because the *middle* stratum is both exploited and exploiter" (Wallerstein 1979:23).

---

[9]Wallerstein defines a "strong state" as one that possesses "strength vis-à-vis other states within the world economy including other core states, and strong vis-à-vis local political units within the boundaries of the state. . . . A strong state then is a partially autonomous entity in the sense that it has a margin of action available to it wherein it reflects the compromises of multiple interests, even if the bounds of these margins are set by some groups of primordial strength. To be a partially autonomous entity, there must be a group of people whose direct interests are served by such an entity: state managers and a state bureaucracy" (Wallerstein 1974:232).

In one sense, Wallerstein's notion of the core and periphery resembles Marx's discussion of the struggle between the "two great classes" under capitalism: the bourgeoisie—those who own the means of production—and the proletariat—those who own only their labor power and thus must work for the bourgeoisie. However, a critical distinction separates these two views. While Marx emphasized the conflict between the two classes that occurs within a particular nation, Wallerstein contends that "capitalism involves not only appropriation of the surplus value [profit] by an owner from a laborer, but an appropriation of surplus of the whole world-economy by core areas" (ibid.:19). In other words, the reality of the capitalist world-economy is that the surplus value produced by the proletariat is routinely appropriated by a bourgeoisie that is located in a different region of the globe. With the development of the world-economy, no single nation-state or

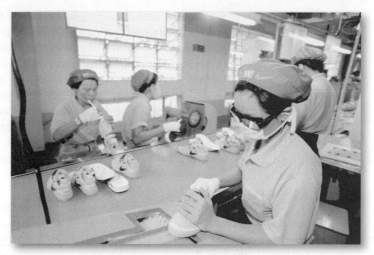

Photo 17.3   Nike in Vietnam

Although a communist country, Vietnam is not isolated from the market forces of the capitalist world-economy. With some fifty thousand employees, Nike is Vietnam's largest private employer, exporting twenty-two million pairs of shoes annually.

central political system exists that is capable of controlling the flow of capital across national boundaries and its unequal accumulation into "private" hands. Capitalism has no determinate "external" boundaries in the modern world-system, as it has on the basis of its own logic come to expand the globe, absorbing all other social systems (Wallerstein 1987:318). This expansion, however, has served to increase the wealth of some regions while perpetuating the poverty of others.

Yet, that the world-capitalist economy has evolved over the course of several centuries demonstrates that nations are subject to both advancement and loss of status within the world-system. One need only think of the United States to recognize that a colony located in the periphery can, over time, become a dominant force located in the core region. Similarly, at the present, other nations are actively vying for improving their status in the world-capitalist economy—for instance, the semiperipheral nations of India, Brazil, and Singapore. And as Wallerstein points out, history is filled with nations whose trajectories during certain historical periods have been less fortunate: Russia, Poland, Turkey, Spain, and even Britain, among others (Wallerstein 1979:23–31). Indeed, while the status of the United States as a dominant core nation is secure, its position as *the* global economic power is in jeopardy as China, despite its impoverished standard of living, may have the world's largest economy as early as 2020 and its currency, the renminbi, may soon rival the dollar as the world's reserve currency. Additionally, with its massive reserves of foreign currency, China is able to flex its political clout by flouting American and European diplomatic efforts regarding countries such as Iran. China (along with India and other developing countries) has also exercised its growing political and economic power by blocking Western, industrialized nations' efforts to reverse global warming that involve imposing taxes on nations that do not lower their carbon emissions. Such proposals, if enacted, would restrict the economic development of industrializing nations, particularly China (Benhold 2010).

This "shuffling of the deck" in no way suggests the impending collapse of the capitalist world-economy. Rather, the essential point here is that the fate of any one nation or region is tied to the economic dynamics of the overall world-system that experiences cycles of geographical expansion and contraction according to the balance between world supply of raw materials and production of goods, and world demand for those goods.

There are two other important differences between Marx's analysis and that offered by Wallerstein. First, while Marx envisioned a future utopia brought on by the inevitable revolution of the proletariat and the establishment of communism, Wallerstein's vision of the future is much more cautious. Unlike Marx, Wallerstein does not ascribe a necessary trajectory or path of progress to history. There is no utopian final stage or "end of prehistory" toward which the capitalist world-economy inexorably marches. At best, an entirely new form of world-system—a socialist world-government—may emerge out of the capitalist world-economy as "antisystemic movements" primarily organized around class and ethno-national identities struggle to resist the inequities and injustices resulting from the unchecked expansion of capitalism (Wallerstein 1982). Such a world-system may not offer humankind full redemption from the exploitation and destruction wrought by capitalism or the end of class conflict, but it would lead to a more equitable distribution of the world's wealth by maintaining a high level of economic productivity managed more through collective planning decisions than through market forces. The freedom currently exercised by capitalists and core states to direct profits to their sole advantage would be curtailed, since a single political system whose boundaries encompass (as opposed to being exceeded by) the whole of economic production would be established. This would in turn stabilize economic development by curbing fluctuations in supply and demand. The tighter control exercised over the market would then reduce the level of exploitation and inequality currently tied to the capitalist world-economy, while meeting the needs of the world's population.

Second, Wallerstein's studies of the political and economic conditions in Africa, and in the modern world-system more generally, led him to identify not only class, but also race as a major locus of conflict within the capitalist world-economy. Indeed, he contends that gender, ethnicity, and nationality are likewise all major sources of divisiveness in the modern world. This emphasis represents a break from more orthodox variants of Marxism that argue that class conflict is the singular site of exploitation and the motor force of all social change. Wallerstein argues that it is not by coincidence that the vast majority of the periphery is peopled by the "darker races," thus directly tying racism to the expansion of the capitalist world-economy. It is not by coincidence that the technologically advanced, wealthy core states compose an area where "whites" constitute a majority of the population. As you will see in the final portion of this chapter when we discuss the work of Edward Said, this point is also taken up in postcolonial theories. Yet, Wallerstein's argument on this point remains ambiguous, since he contends that ethnic and racial struggles and nationalist movements ultimately are generated by the dynamics of the *capitalist* world-economy. (This must be the case given that the modern world-system is fundamentally organized around the logic of capitalism.) As a result, such struggles are at their root expressions of economic, not racial or nationalist, conflict. Thus, Wallerstein notes that despite the success of some ethno-nationalist movements to win political independence for their countries, they remain dependent and subordinate because of the overwhelming economic power possessed by core states. Moreover, such movements serve to mute the underlying reality of capitalist exploitation to the extent that calls for ethnic equality and national liberation are the principal battleground for social change (Wallerstein 1979:200).

Table 17.2   Comparing the Theories of Marx and Wallerstein on Modern Society

Central Themes	Marx	Wallerstein
Primary social entity	Nation-state	World-system
Basis of society	Capitalist economy	Capitalist economy
Primary social relations	Bourgeoisie and Proletariat	Core, Periphery, Semiperiphery
Basis of social relations	Conflict/Exploitation	Conflict/Exploitation
Future of society	Inevitable communist revolution	Possible socialist world-government

Consider, for instance, the case of South Africa. First colonized by the Dutch, South Africa became a British colony in the late eighteenth century. After the Anglo-Boer Wars, the Union of South Africa was established in 1910, and with it a political and social environment that promoted the systematic repression of the indigenous nonwhite population. The repression was intensified under the brutally oppressive system of apartheid formally established in 1948. South Africa declared its independence from British rule in 1961, and the Republic of South Africa was created. It would take several decades of domestic protests from the black population and mounting pressure from the international community to end the country's overtly racist policies. Yet, even with the dismantling of apartheid in the early 1990s and the granting of political equality for the nonwhite population, South Africa's economic picture remains bleak. With estimates of unemployment ranging from 30 to 40 percent of the population, an annual per capita income of $2,600, and the implementation of socially punishing austerity measures to attract foreign (i.e., core) investment, South Africa remains economically weak in the world capitalist-economy. In the end, the morally motivated battle to end racial injustice may have been won, but the vast majority of South Africans are still not free from the ravages of poverty.

WALLERSTEIN'S THEORETICAL ORIENTATION ▪▪

The presuppositions underlying Wallerstein's perspective can be gleaned from our preceding discussion. Employing concepts such as core, periphery, and semiperiphery, Wallerstein adopts a decidedly collectivist approach to the issue of order. Indeed, what could be more encompassing than a world-system! From this vantage point, social life is fundamentally patterned according to the economic relationships existing between regions of the globe. As such, individuals and their interactions do not figure prominently in Wallerstein's theoretical framework.

Figure 17.1 Wallerstein's Basic Concepts and Theoretical Orientation

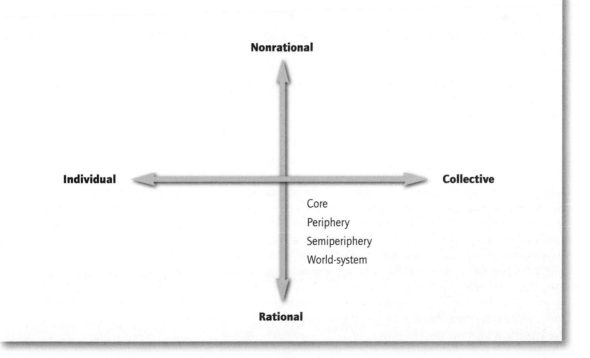

Regarding the issue of action, Wallerstein's emphasis on matters such as capitalist markets, class interests, and the pursuit of profit points to a rationalist orientation. It is important to note, however, that accounting for motivational factors, which is what the question of action addresses, is problematic within world-systems analysis. This is because the central "actors" are not individuals, or even groups of individuals, but rather regions of the globe. At the very least, we must be cautious about projecting motives onto spatial areas. Nevertheless, if we set aside this issue, it is clear that what "motivates" the core is the strategic attempt to maintain its dominant position in the modern world-system. Similarly, the peripheral and semiperipheral regions seek to improve their position through strategic use of the limited resources they possess. Ultimately, the world is shaped by the play of interests, not ideas, as the competing regions struggle to maximize their economic advantage while perpetuating the inequalities on which the capitalist world-economy is based.

The collectivistic and rationalistic dimensions of Wallerstein's theoretical orientation are reflected in the following remarks, in which he describes the dynamics of the capitalist world-economy:

> The capitalist world-economy functions, as do most . . . historical systems by means of a pattern of cyclical rhythms. . . . One part, however, of the process is that, periodically, the capitalist world-economy has seen the need to expand the geographic boundaries of the system as a whole, creating thereby new loci of production to participate in its axial division of labor. Over 400 years, these successive expansions have transformed the capitalist world-economy from a system located primarily in Europe to one that covers the entire globe. (Wallerstein 1990:36)

Here Wallerstein posits the existence of an economic system with its own "needs" and that experiences necessary periods of expansion according to its own internal logic, and in classic functionalist—and Marxist—fashion, Wallerstein offers a picture of society in which the primary mover of history is not people but rather an abstract force that carries them along in its own wake.

Reading

Introduction to "The Modern World-System as a Capitalist World-Economy"

A consistent theme throughout this reading is Wallerstein's analysis of the economic and political dynamics that led to and sustain the division of the modern world-system into core, peripheral, and semiperipheral regions. After outlining the interconnections between the world-economy and capitalism, Wallerstein goes on to describe the institutions that form the basis of the capitalist world-economy: markets, firms, states, households, classes, and status groups. To this end, he discusses the efforts of firms to develop market quasi-monopolies and the ability or inability of states to assist those entities in reaping the profits that such market control makes possible. The global cycles of economic expansion and stagnation are largely a result of the relocation of what were once quasi-monopolies from core region production processes to semiperipheral and peripheral production processes. A firm that has a quasi-monopoly on the market is able to generate extensive capital that in turn can lead to growing employment, higher wages, and a general expansion of the world-economy. Yet, as more firms enter the market and break up the quasi-monopoly, overproduction and increased price competition result. The ensuing buildup of unsold goods leads to the slowdown and relocation of production, lower wages, rising unemployment, and weakening consumer demand; in short, economic stagnation. This fifty- to sixty-year cycle—an A-phase of expansion followed by a B-phase of stagnation—is called a Kondratieff cycle (named after the Russian economist who described the process). Counted among the effects of these recurring cycles is the increasing proletarianization of households and the workforce whose now-higher wages, while stimulating demand for goods, create pressures on firms to accumulate profits. This, too, contributes to the cycles of global economic expansion and stagnation and the shifting of a nation's position within the world-system.

Wallerstein concludes this reading by taking up the normative dimension of the world-system in the form of socialization and norms of universalism and anti-universalism. Here he notes the importance of households as the principal arena for socializing individuals and thus for instilling beliefs and values that legitimate the world-system and the hierarchies of domination on which it is based. In a similar vein, Wallerstein notes the importance of norms of universalism that lead to the equal application of rules to individuals within the range of institutions that make up the modern world-system. Based on meritocratic distinctions, universalism promotes greater efficiency and productivity within the various institutional domains. Yet, norms of anti-universalism (for instance, racism and sexism) also play an important role in justifying and enforcing the reality of the unequal distribution of work and power within the world-system. By offering biological explanations for a given group's inferior status, the active social construction of the polarized institutions making up the world-system remains hidden.

"The Modern World-System as a Capitalist World-Economy: Production, Surplus, Value, and Polarization" (2004)

Immanuel Wallerstein

The world in which we are now living, the modern world-system, had its origins in the sixteenth century. This world-system was then located in only a part of the globe, primarily in parts of Europe and the Americas. It expanded over time to cover the whole globe. It is and has always been a *world-economy*. It is and has always been a *capitalist* world-economy. We should begin by explaining what these two terms, world-economy and capitalism, denote. It will then be easier to appreciate the historical contours of the modern world-system—its origins, its geography, its temporal development, and its contemporary structural crisis.

What we mean by a world-economy (Braudel's *èconomie-monde*) is a large geographic zone within which there is a division of labor and hence significant internal exchange of basic or essential goods as well as flows of capital and labor. A defining feature of a world-economy is that it is *not* bounded by a unitary political structure. Rather, there are many political units inside the world-economy, loosely tied together in our modern world-system in an interstate system. And a world-economy contains many cultures and groups—practicing many religions, speaking many languages,

differing in their everyday patterns. This does not mean that they do not evolve some common cultural patterns, what we shall be calling a geoculture. It does mean that neither political nor cultural homogeneity is to be expected or found in a world-economy. What unifies the structure most is the division of labor which is constituted within it.

Capitalism is not the mere existence of persons or firms producing for sale on the market with the intention of obtaining a profit. Such persons or firms have existed for thousands of years all across the world. Nor is the existence of persons working for wages sufficient as a definition. Wage-labor has also been known for thousands of years. We are in a capitalist system only when the system gives priority to the *endless* accumulating of capital. Using such a definition, only the modem world-system has been a capitalist system. Endless accumulation is a quite simple concept: it means that people and firms are accumulating capital in order to accumulate still more capital, a process that is continual and endless. If we say that a system "gives priority" to such endless accumulation, it means that there exist structural mechanisms by which those who act with other motivations are penalized in some way, and are eventually eliminated from the social scene, whereas those who act with the appropriate motivations are rewarded and, if successful, enriched.

A world-economy and a capitalist system go together. Since world-economies lack the unifying cement of an overall political structure or a homogeneous culture, what holds them together is the efficacy of the division of labor. And this efficacy is a function of the constantly expanding wealth that a capitalist system provides. Until modern times, the world-economies that had been constructed either fell apart or were transformed *manu militari* into world-empires. Historically, the only world-economy to have survived for a long time has been the modern world-system, and that is because the capitalist system took root and became consolidated as its defining feature.

Conversely, a capitalist system cannot exist within any framework except that of a world-economy. We shall see that a capitalist system requires a very special relationship between economic producers and the holders of political power. If the latter are too strong, as in a world-empire, their interests will override those of the economic producers, and the endless accumulation of capital will cease to be a priority. Capitalists need a large market (hence minisystems are too narrow for them) but they also need a multiplicity of states, so that they can gain the advantages of working with states but also can circumvent states hostile to their interests in favor of states friendly to their interests. Only the existence of a multiplicity of states within the overall division of labor assures this possibility.

A capitalist world-economy is a collection of many institutions, the combination of which accounts for its processes, and all of which are intertwined with each other. The basic institutions are the market, or rather the markets; the firms that compete in the markets; the multiple states, within an interstate system; the households; the classes; and the status-groups (to use Weber's term, which some people in recent years have renamed the "identities"). They are all institutions that have been created within the framework of the capitalist world-economy. Of course, such institutions have some similarities to institutions that existed in prior historical systems to which we have given the same or similar names. But using the same name to describe institutions located in different historical systems quite often confuses rather than clarifies analysis. It is better to think of the set of institutions of the modern world-system as contextually specific to it.

Let us start with markets, since these are normally considered the essential feature of a capitalist system. A market is both a concrete local structure in which individuals or firms sell and buy goods, and a virtual institution across space where the same kind of exchange occurs. How large and widespread any virtual market is depends on the realistic alternatives that sellers and buyers have at a given time. In principle, in a capitalist world-economy the virtual market exists in the world-economy as a whole. But as we shall see, there are often interferences with these boundaries, creating narrower and more "protected" markets. There are of course separate virtual markets for all commodities as well as for capital and different kinds of labor. But over time, there can also be said to exist a single virtual world market for all the factors of production combined, despite all the barriers that exist to its free functioning. One can think of this complete virtual market as a magnet for all producers and buyers, whose pull is a constant political factor in the decision-making of everyone—the states, the firms, the households, the classes, and the status-groups (or identities). This complete virtual world market is a reality in that it influences all decision making, but it never functions fully and freely (that is, without interference). The totally free market functions as an ideology, a myth, and a constraining influence, but never as a day-to-day reality.

One of the reasons it is not a day-to-day reality is that a totally free market, were it ever to exist, would make impossible the endless accumulation of capital. This may seem a paradox because it is surely true that capitalism cannot function without markets, and it is also true that capitalists regularly say that they favor free markets. But capitalists in fact need not totally free markets but rather markets that are only partially free. The reason is clear. Suppose there really existed a world market in which all the factors of production were totally free, as our textbooks in economics usually define this—that is, one in which the factors flowed without restriction, in which there were a very large number of buyers and a very large number of sellers, and in which there was perfect information (meaning that all sellers and all buyers knew the exact state of all costs of production). In such a perfect market, it would always be possible for the buyers to bargain down the sellers to an absolutely minuscule level of profit (let us think of it as a penny), and this low level of profit would make the capitalist game entirely uninteresting to producers, removing the basic social underpinnings of such a system.

What sellers always prefer is a monopoly, for then they can create a relatively wide margin between the costs of production and the sales price, and thus realize high rates of profit. Of course, perfect monopolies are extremely difficult to create, and rare, but quasi-monopolies are not. What one needs most of all is the support of the machinery of a relatively strong state, one which can enforce a quasi-monopoly. There are many ways of doing this. One of the most fundamental is the system of patents which reserves rights in an "invention" for a specified number of years. This is what basically makes "new" products the most expensive for consumers and the most profitable for their producers. Of course, patents are often violated and in any case they eventually expire, but by and large they protect a quasi-monopoly for a time. Even so, production protected by patents usually remains only a quasi-monopoly, since there may be other similar products on the market that are not covered by the patent. This is why the normal situation for so-called leading products (that is, products that are both new and have an important share of the overall world market for commodities) is an oligopoly rather than an absolute monopoly. Oligopolies are however good enough to realize the desired high rate of profits, especially since the various firms often collude to minimize price competition.

Patents are not the only way in which states can create quasi-monopolies. State restrictions on imports and exports (so-called protectionist measures) are another. State subsidies and tax benefits are a third. The ability of strong states to use their muscle to prevent weaker states from creating counter-protectionist measures is still another. The role of the states as large-scale buyers of certain products willing to pay excessive prices is still another. Finally, regulations which impose a burden on producers may be relatively easy to absorb by large producers but crippling to smaller producers, an asymmetry which results in the elimination of the smaller producers from the market and thus increases the degree of oligopoly. The modalities by which states interfere with the virtual market are so extensive that they constitute a fundamental factor in determining prices and profits. Without such interferences, the capitalist system could not thrive and therefore could not survive.

Nonetheless, there are two inbuilt anti-monopolistic features in a capitalist world-economy. First of all, one producer's monopolistic advantage is another producer's loss. The losers will of course struggle politically to remove the advantages of the winners. They can do this by political struggle within the states where the monopolistic producers are located, appealing to doctrines of a free market and offering support to political leaders inclined to end a particular monopolistic advantage. Or they do this by persuading other states to defy the world market monopoly by using their state power to sustain competitive producers. Both methods are used. Therefore, over time, every quasi-monopoly is undone by the entry of further producers into the market.

Quasi-monopolies are thus self-liquidating. But they last long enough (say thirty years) to ensure considerable accumulation of capital by those who control the quasi-monopolies. When a quasi-monopoly does cease to exist, the large accumulators of capital simply move their capital to new leading products or whole new leading industries. The result is a cycle of leading products. Leading products have moderately short lives, but they are constantly succeeded by other leading industries. Thus the game continues. As for the once-leading industries past their prime, they become more and more "competitive," that is, less and less profitable. We see this pattern in action all the time.

Firms are the main actors in the market. Firms are normally the competitors of other firms operating in the same virtual market. They are also in conflict with those firms from whom they purchase inputs and those firms to which they sell their products. Fierce intercapitalist rivalry is the name of the game. And only the strongest and the most agile survive. One must remember that bankruptcy, or absorption by a more powerful firm, is the daily bread of capitalist enterprises. Not all capitalist entrepreneurs succeed in accumulating capital. Far from it. If they all succeeded, each would be likely to obtain very little capital. So, the repeated "failures" of firms not only weed out the weak competitors but are a condition sine qua non of the endless accumulation of capital. That is what explains the constant process of the concentration of capital.

To be sure, there is a downside to the growth of firms, either horizontally (in the same product), vertically (in the different steps in the chain of production), or what might be thought of as orthogonally (into other products not closely related). Size brings down costs through so-called economies of scale. But size adds costs of administration and coordination, and multiplies the risks of managerial inefficiencies. As a result of this contradiction, there has been a repeated zigzag process of firms getting larger and then getting smaller. But it has not at all been a simple up-and-down cycle. Rather, worldwide there has been a secular increase in the size of firms, the whole historical process taking the form of a ratchet, two steps up then one step back, continuously. The size of firms also has direct political implications. Large size gives firms more political clout but also makes them more vulnerable to political assault—by their competitors, their employees, and their consumers. But here too the bottom line is an upward ratchet, toward more political influence over time.

The axial division of labor of a capitalist world-economy divides production into core-like products and peripheral products. Core-periphery is a relational concept. What we mean by core-periphery is the degree of profitability of the production processes. Since profitability is directly related to the degree of monopolization, what we essentially mean by core-like production processes is those that are controlled by quasi-monopolies. Peripheral processes are then those that are truly competitive. When exchange occurs, competitive products are in a weak position and quasi-monopolized products are in a strong position. As a result, there is a constant flow of surplus-value from the producers of peripheral products to the producers of core-like products. This has been called unequal exchange.

To be sure, unequal exchange is not the only way of moving accumulated capital from politically weak regions to politically strong regions. There is also plunder, often used extensively during the early days of incorporating new regions into the world-economy (consider, for example, the conquistadores and gold in the Americas). But plunder is self-liquidating. It is a case of killing the goose that lays the golden eggs. Still, since the consequences are middle-term and the advantages short-term, there still exists much plunder in the modern world- system, although we are often "scandalized" when we learn of it. When Enron goes bankrupt, after procedures that have moved enormous sums into the hands of a few managers, that is in fact plunder. When "privatizations" of erstwhile state property lead to its being garnered by mafia-like businessmen who quickly leave the country with destroyed enterprises in their wake, that is plunder. Self-liquidating, yes, but only after much damage has been done to the world's productive system, and indeed to the health of the capitalist world-economy.

Since quasi-monopolies depend on the patronage of strong states, they are largely located—juridically, physically, and in terms of ownership—within such states. There is therefore a geographical consequence of the core-peripheral relationship. Core-like processes tend to group themselves in a few states and to constitute the bulk of the production activity in such states.

Peripheral processes tend to be scattered among a large number of states and to constitute the bulk of the production activity in these states. Thus, for shorthand purposes we can talk of core states and peripheral states, so long as we remember that we are really talking of a relationship between production processes. Some states have a near even mix of core-like and peripheral products. We may call them semiperipheral states. They have, as we shall see, special political properties. It is however not meaningful to speak of semiperipheral production processes.

Since, as we have seen, quasi-monopolies exhaust themselves, what is a core-like process today will become a peripheral process tomorrow. The economic history of the modern world-system is replete with the shift, or downgrading, of products, first to semiperipheral countries, and then to peripheral ones. If circa 1800 the production of textiles was possibly the preeminent core-like production process, by 2000 it was manifestly one of the least profitable peripheral production processes. In 1800 these textiles were produced primarily in a very few countries (notably England and some other countries of northwestern Europe); in 2000 textiles were produced in virtually every part of the world-system, especially cheap textiles. The process has been repeated with many other products. Think of steel, or automobiles, or even computers. This kind of shift has no effect on the structure of the system itself. In 2000 there were other core-like processes (e.g. aircraft production or genetic engineering) which were concentrated in a few countries. There have always been new core-like processes to replace those which become more competitive and then move out of the states in which they were originally located.

The role of each state is very different vis-à-vis productive processes depending on the mix of core-peripheral processes within it. The strong states, which contain a disproportionate share of core-like processes, tend to emphasize their role of protecting the quasi-monopolies of the core-like processes. The very weak states, which contain a disproportionate share of peripheral production processes, are usually unable to do very much to affect the axial division of labor, and in effect are largely forced to accept the lot that has been given them.

The semiperipheral states which have a relatively even mix of production processes find themselves in the most difficult situation. Under pressure from core states and putting pressure on peripheral states, their major concern is to keep themselves from slipping into the periphery and to do what they can to advance themselves toward the core. Neither is easy, and both require considerable state interference with the world market. These semi-peripheral states are the ones that put forward most aggressively and most publicly so-called protectionist policies. They hope thereby to "protect" their production processes from the competition of stronger firms outside, while trying to improve the efficiency of the firms inside so as to compete better in the world market. They are eager recipients of the relocation of erstwhile leading products, which they define these days as achieving "economic development." In this effort, their competition comes not from the core states but from other semiperipheral states, equally eager to be the recipients of relocation which cannot go to all the eager aspirants simultaneously and to the same degree. In the beginning of the twenty-first century, some obvious countries to be labeled semiperipheral are South Korea, Brazil, and India—countries with strong enterprises that export products (for example steel, automobiles, pharmaceuticals) to peripheral zones, but that also regularly relate to core zones as importers of more "advanced" products.

The normal evolution of the leading industries—the slow dissolution of the quasi-monopolies—is what accounts for the cyclical rhythms of the world-economy. A major leading industry will be a major stimulus to the expansion of the world-economy and will result in considerable accumulation of capital. But it also normally leads to more extensive employment in the world-economy, higher wage-levels, and a general sense of relative prosperity. As more and more firms enter the market of the erstwhile quasi-monopoly, there will be "overproduction" (that is, too much production for the real effective demand at a given time) and consequently increased price competition (because of the demand squeeze), thus lowering the rates of profit. At some point, a buildup of unsold products results, and consequently a slowdown in further production.

When this happens, we tend to see a reversal of the cyclical curve of the world-economy. We talk of stagnation or recession in the world-economy. Rates of unemployment rise worldwide. Producers seek to reduce costs in order to maintain their share of the world market. One of the mechanisms is relocation of the production processes to zones that have historically lower wages, that is, to semiperipheral countries. This shift puts pressure on the wage levels in the processes still remaining in core zones, and wages there tend to become lower as well. Effective demand which was at

first lacking because of overproduction now becomes lacking because of a reduction in earnings of the consumers. In such a situation, not all producers necessarily lose out. There is obviously acutely increased competition among the diluted oligopoly that is now engaged in these production processes. They fight each other furiously, usually with the aid of their state machineries. Some states and some producers succeed in "exporting unemployment" from one core state to the others. Systemically, there is contraction, but certain core states and especially certain semiperipheral states may seem to be doing quite well.

The process we have been describing—expansion of the world-economy when there are quasi-monopolistic leading industries and contraction in the world-economy when there is a lowering of the intensity of quasi-monopoly—can be drawn as an up-and-down curve of so-called A- (expansion) and B- (stagnation) phases. A cycle consisting of an A-phase followed by a B-phase is sometimes referred to as a Kondratieff cycle, after the economist who described this phenomenon with clarity in the beginning of the twentieth century. Kondratieff cycles have up to now been more or less fifty to sixty years in length. Their exact length depends on the political measures taken by the states to avert a B-phase, and especially the measures to achieve recuperation from a B-phase on the basis of new leading industries that can stimulate a new A-phase.

A Kondratieff cycle, when it ends, never returns the situation to where it was at the beginning of the cycle. That is because what is done in the B-phase in order to get out of it and return to an A-phase changes in some important way the parameters of the world-system. The changes that solve the immediate (or short-run) problem of inadequate expansion of the world-economy (an essential element in maintaining the possibility of the endless accumulation of capital) restore a middle-run equilibrium but begin to create problems for the structure in the long run. The result is what we may call a secular trend. A secular trend should be thought of as a curve whose abscissa (or x-axis) records time and whose ordinate (or y-axis) measures a phenomenon by recording the proportion of some group that has a certain characteristic. If over time the percentage is moving upward in an overall linear fashion, it means by definition (since the ordinate is in percentages) that at some point it cannot continue to do so. We call this reaching the asymptote, or 100 percent point. No characteristic can be ascribed to more than 100 percent of any group. This means that as we solve the middle-run problems by moving up on the curve, we will eventually run into the long-run problem of approaching the asymptote.

Let us suggest one example of how this works in a capitalist world-economy. One of the problems we noted in the Kondratieff cycles is that at a certain point major production processes become less profitable, and these processes begin to relocate in order to reduce costs. Meanwhile, there is increasing unemployment in core zones, and this affects global effective demand. Individual firms reduce their costs, but the collectivity of firms finds it more difficult to find sufficient customers. One way to restore a sufficient level of world effective demand is to increase the pay levels of ordinary workers in core zones, something which has frequently occurred at the latter end of Kondratieff B- periods. This thereby creates the kind of effective demand that is necessary to provide sufficient customers for new leading products. But of course higher pay levels may mean lesser profits for the entrepreneurs. At a world level this can be compensated for by expanding the pool of wage workers elsewhere in the world, who are willing to work at a lower level of wages. This can be done by drawing new persons into the wage-labor pool, for whom the lower wage represents in fact an increase in real income. But of course every time one draws "new" persons into the wage-labor pool, one reduces the number of persons remaining outside the wage-labor pool. There will come a time when the pool is diminished to the point where it no longer exists effectively. We are reaching the asymptote. . . .

A typical household consists of three to ten persons who, over a long period (say thirty years or so), pool multiple sources of income in order to survive collectively. Households are not usually egalitarian structures internally nor are they unchanging structures (persons are born and die, enter or leave households, and in any case grow older and thus tend to alter their economic role). What distinguishes a household structure is some form of obligation to provide income for the group and to share in the consumption resulting from this income. Households are quite different from clans or tribes or other quite large and extended entities, which often share obligations of mutual security and identity but do not regularly share income. Or if there exist such large entities which are income-pooling, they are dysfunctional for the capitalist system.

We first must look at what the term "income" covers. There are in fact generically five kinds of income in the modern world-system. And almost all households seek and obtain all five kinds, although in different proportions (which turns out to be very important). One obvious form is wage-income, by which is meant payment (usually in money form) by persons outside the household for work of a member of the household that is performed outside the household in some production process. Wage-income may be occasional or regular. It may be payment by time employed or by work accomplished (piecework). Wage-income has the advantage to the employer that it is "flexible" (that is, continued work is a function of the employer's need), although the trade union, other forms of syndical action by workers, and state legislation have often limited employers' flexibility in many ways. Still, employers are almost never obligated to provide lifetime support to particular workers. Conversely, this system has the disadvantage to the employer that when more workers are needed, they may not be readily available for employment, especially if the economy is expanding. That is, in a system of wage-labor, the employer is trading not being required to pay workers in periods when they are not needed for the guarantee that the workers are available when they are needed.

A second obvious source of household income is subsistence activity. We usually define this type of work too narrowly, taking it to mean only the efforts of rural persons to grow food and produce necessities for their own consumption without passing through a market. This is indeed a form of subsistence production, and this kind of work has of course been on a sharp decline in the modern world-system, which is why we often say that subsistence production is disappearing. By using such a narrow definition, we are however neglecting the numerous ways in which subsistence activity is actually increasing in the modern world. When someone cooks a meal or washes dishes at home, this is subsistence production. When a homeowner assembles furniture bought from a store, this is subsistence production. And when a professional uses a computer to send an e-mail which, in an earlier day, a (paid) secretary would have typed, he or she is engaged in subsistence production. Subsistence production is a large part of household income today in the most economically wealthy zones of the capitalist world-economy.

A third kind of household income we might generically call petty commodity production. A petty commodity is defined as a product produced within the confines of the household but sold for cash on a wider market. Obviously, this sort of production continues to be very widespread in the poorer zones of the world-economy but is not totally absent anywhere. In richer zones we often call it free-lancing. This kind of activity involves not only the marketing of produced goods (including of course intellectual goods) but also petty marketing. When a small boy sells on the street cigarettes or matches one by one to consumers who cannot afford to buy them in the normal quantity that is packaged, this boy is engaged in petty-commodity production, the production activity being simply the disassembly of the larger package and its transport to the street market.

A fourth kind of income is what we can generically call rent. Rent can be drawn from some major capital investment (offering urban apartments for rent, or rooms within apartments) or from locational advantage (collecting a toll on a private bridge) or from capital ownership (clipping coupons on bonds, earning interest on a savings account). What makes it rent is that it is ownership and not work of any kind that makes possible the income.

Finally, there is a fifth kind of income, which in the modern world we call transfer payments. These may be defined as income that comes to an individual by virtue of a defined obligation of someone else to provide this income. The transfer payments may come from persons close to the household, as when gifts or loans are given from one generation to the other at the time of birth, marriage, or death. Such transfer payments between households may be made on the basis of reciprocity (which in theory ensures no extra income over a lifetime but tends to smooth out liquidity needs). Or transfer payments may occur through the efforts of the state (in which case one's own money may simply be returning at a different moment in time), or through an insurance scheme (in which one may in the end benefit or lose), or through redistribution from one economic class to another. . . .

As we have already noted, the relative importance of the various forms of income in particular households has varied widely. Let us distinguish two major varieties: the household where wage-income accounts for 50 percent or more of the total lifetime income, and the household where it accounts for less. Let us call the former a "proletarian household" (because it seems to be heavily dependent on wage-income, which is what the term proletarian is supposed to invoke); and let us then call the latter a "semiproletarian household" (because there is doubtless at least some wage-income for most members of it). If we do this, we can see that an employer has an advantage in employing those wage-laborers who are in a semiproletarian household.

Whenever wage-labor constitutes a substantial component of household income, there is necessarily a floor for how much the wage-earner can be paid. It must be an amount that represents at least a proportionate share of the reproduction costs of the household. This is what we can think of as an *absolute* minimum wage. If, however, the wage-earner is ensconced in a household that is only semiproletarian, the wage-earner can be paid a wage *below* the absolute minimum wage, without necessarily endangering the survival of the household. The difference can be made up by additional income provided from other sources and usually by other members of the household. What we see happening in such cases is that the other producers of income in the household are in effect transferring surplus-value to the employer of the wage-earner over and above whatever surplus-value the wage-earner himself is transferring, by permitting the employer to pay less than the absolute minimum wage.

It follows that in a capitalist system employers would in general prefer to employ wage-workers coming from semiproletarian households. There are however two pressures working in the other direction. One is the pressure of the wage-workers themselves who seek to be "proletarianized," because that in effect means being better paid. And one is the contradictory pressure on the employers themselves. Against their individual need to lower wages, there is their collective longer-term need to have a large enough effective demand in the world-economy to sustain the market for their products. So over time, as a result of these two very different pressures, there is a slow increase in the number of households that are proletarianized. Nonetheless, this description of the long-term trend is contrary to the traditional social science picture that capitalism as a system requires primarily proletarians as workers. If this were so, it would be difficult to explain why, after four to five hundred years, the proportion of proletarian workers is not much higher than it is. Rather than think of proletarianization as a capitalist necessity, it would be more useful to think of it as a locus of struggle, whose outcome has been a slow if steady increase, a secular trend moving toward its asymptote.

There are classes in a capitalist system, since there are clearly persons who are differently located in the economic system with different levels of income who have differing interests. For example, it is obviously in the interest of workers to seek an increase in their wages, and it is equally obviously in the interest of employers to resist these increases, at least in general. But, as we have just seen, wage-workers are ensconced in households. It makes no sense to think of the workers belonging to one class and other members of their household to another. It is obviously households, not individuals, that are located within classes. Individuals who wish to be class-mobile often find that they must withdraw from the households in which they are located and locate themselves in other households, in order to achieve such an objective. This is not easy but it is by no means impossible.

Classes however are not the only groups within which households locate themselves. They are also members of status-groups or identities. (If one calls them status-groups, one is emphasizing how they are perceived by others, a sort of objective criterion. If one calls them identities, one is emphasizing how they perceive themselves, a sort of subjective criterion. But under one name or the other, they are an institutional reality of the modern world-system.) Status-groups or identities are ascribed labels, since we are born into them, or at least we usually think we are born into them. It is on the whole rather difficult to join such groups voluntarily, although not impossible. These status-groups or identities are the numerous "peoples" of which all of us are members—nations, races, ethnic groups, religious communities, but also genders and categories of sexual preferences. Most of these categories are often alleged to be anachronistic leftovers of pre-modern times. This is quite wrong as a premise. Membership in status-groups or identities is very much a part of modernity. Far from dying out, they are actually growing in importance as the logic of a capitalist system unfolds further and consumes us more and more intensively. . . .

Why is it so important for households to maintain singular class and status-group identities, or at least pretend to maintain them? Such a homogenization of course aids in maintaining the unity of a household as an income-pooling unit and in overcoming any centrifugal tendencies that might arise because of internal inequalities in the distribution of consumption and decision making. It would however be a mistake to see this tendency as primarily an internal group defensive mechanism. There are important benefits to the overall world-system from the homogenizing trends within household structures.

Households serve as the primary socializing agencies of the world-system. They seek to teach us, and particularly the young, knowledge of and respect for the social rules by which we are supposed to abide. They are of course seconded by state agencies such as schools and armies as well as by religious institutions and the media. But none of these come close to the households in actual impact. What however determines how the households will socialize their members? Largely how the secondary institutions frame the issues for the households, and their ability to do so effectively depends on the relative homogeneity of the households—that is, they have and see themselves as having a defined role in the historical social system. A household that is certain of its status-group identity—its nationality, its race, its religion, its ethnicity, its code of sexuality—knows exactly how to socialize its members. One whose identity is less certain but that tries to create a homogenized, even if novel, identity can do almost as well. A household that would openly avow a permanently split identity would find the socialization function almost impossible to do, and might find it difficult to survive as a group.

Of course, the powers that be in a social system always hope that socialization results in the acceptance of the very real hierarchies that are the product of the system. They also hope that socialization results in the internalization of the myths, the rhetoric, and the theorizing of the system. This does happen in part but never in full. Households also socialize members into rebellion, withdrawal, and deviance. To be sure, up to a point even such antisystemic socialization can be useful to the system by offering an outlet for restless spirits, provided that the overall system is in relative equilibrium. In that case, one can anticipate that the negative socializations may have at most a limited impact on the functioning of the system. But when the historical system comes into structural crisis, suddenly such antisystemic socializations can play a profoundly unsettling role for the system. . . .

We must look at the pressures on households coming from outside. Most of the status-groups have some kind of trans-household institutional expression. And these institutions place direct pressure on the households not merely to conform to their norms and their collective strategies but to give them priority. Of the trans-household institutions, the states are the most successful in influencing the households because they have the most immediate weapons of pressure (the law, substantial benefits to distribute, the capacity to mobilize media). But wherever the state is less strong, the religious structures, the ethnic organizations, and similar groups may become the strongest voices insisting on the priorities of the households. Even when status-groups or identities describe themselves as antisystemic, they may still be in rivalry with other antisystemic status-groups or identities, demanding priority in allegiance. It is this complicated turmoil of household identities that underlies the roller coaster of political struggle within the modern world-system.

The complex relationships of the world-economy, the firms, the states, the households, and the trans-household institutions that link members of classes and status-groups are beset by two opposite—but symbiotic—ideological themes: universalism on the one hand and racism and sexism on the other.

Universalism is a theme prominently associated with the modern world-system. It is in many ways one of its boasts. Universalism means in general the priority to general rules applying equally to all persons, and therefore the rejection of particularistic preferences in most spheres. The only rules that are considered permissible within the framework of universalism are those which can be shown to apply directly to the narrowly defined proper functioning of the world-system.

The expressions of universalism are manifold. If we translate universalism to the level of the firm or the school, it means for example the assigning of persons to positions on the basis of their training and capacities (a practice otherwise known as meritocracy). If we translate it to the level of the household, it implies among other things that marriage should be contracted for reasons of "love" but not those of wealth or ethnicity or any other general particularism. If we translate it to the level of the state, it means such rules as universal suffrage and equality before the law. We are all familiar with the mantras, since they are repeated with some regularity in public discourse. They are supposed to be the central focus of our socialization. Of course, we know that these mantras are unevenly advocated in various locales of the world-system (and we shall want to discuss why this is so), and we know that they are far from fully observed in practice. But they have become the official gospel of modernity.

Universalism is a positive norm, which means that most people assert their belief in it, and almost everyone claims that it is a virtue. Racism and sexism are just the opposite. They too are norms, but they are negative norms, in that most people deny their belief in them. Almost everyone declares that they are vices, yet nonetheless they are norms. What is more, the degree to which the negative norms of racism and sexism are observed is at least as high as, in fact for the most part

much higher than, the virtuous norm of universalism. This may seem to be an anomaly. But it is not.

Let us look at what we mean by racism and sexism. Actually these are terms that came into widespread use only in the second half of the twentieth century. Racism and sexism are instances of a far wider phenomenon that has no convenient name, but that might be thought of as anti-universalism, or the active institutional discrimination against all the persons in a given status-group or identity. For each kind of identity, there is a social ranking. It can be a crude ranking, with two categories, or elaborate, with a whole ladder. But there is always a group on top in the ranking, and one or several groups at the bottom. These rankings are both worldwide and more local, and both kinds of ranking have enormous consequences in the lives of people and in the operation of the capitalist world-economy.

We are all quite familiar with the worldwide rankings within the modern world-system: men over women, Whites over Blacks (or non-Whites), adults over children (or the aged), educated over less educated, heterosexuals over gays and lesbians, the bourgeois and professionals over workers, urbanites over rural dwellers. Ethnic rankings are more local, but in every country, there is a dominant ethnicity and then the others. Religious rankings vary across the world, but in any particular zone everyone is aware of what they are. Nationalism often takes the form of constructing links between one side of each of the antinomies into fused categories, so that, for example, one might create the norm that adult White heterosexual males of particular ethnicities and religions are the only ones who would be considered "true" nationals.

There are several questions which this description brings to our attention. What is the point of professing universalism and practicing anti-universalism simultaneously? Why should there be so many varieties of anti-universalism? Is this contradictory antinomy a necessary part of the modern world-system? Universalism and anti-universalism are in fact both operative day to day, but they operate in different arenas. Universalism tends to be the operative principle most strongly for what we could call the cadres of the world-system—neither those who are at the very top in terms of power and wealth, nor those who provide the large majority of the world's workers and ordinary people in all fields of work and all across the world, but rather an in-between group of people who have leadership or supervisory roles in various institutions. It is a norm that spells out the optimal recruitment mode for such technical, professional, and scientific personnel. This in-between group may be larger or smaller according to a country's location in the world-system and the local political situation. The stronger the country's economic position, the larger the group. Whenever universalism loses its hold even among the cadres in particular parts of the world-system, however, observers tend to see dysfunction, and quite immediately there emerge political pressures (both from within the country and from the rest of the world) to restore some degree of universalistic criteria.

There are two quite different reasons for this. On the one hand, universalism is believed to ensure relatively competent performance and thus make for a more efficient world-economy, which in turn improves the ability to accumulate capital. Hence, normally those who control production processes push for such universalistic criteria. . . . [Moreover,] the norm of universalism is an enormous comfort to those who are benefitting from the system. It makes them feel they deserve what they have.

On the other hand, racism, sexism, and other anti-universalistic norms perform equally important tasks in allocating work, power, and privilege within the modern world-system. They seem to imply exclusions from the social arena. Actually they are really modes of inclusion, but of inclusion at inferior ranks. These norms exist to justify the lower ranking, to enforce the lower ranking, and perversely even to make it somewhat palatable to those who have the lower ranking. Anti-universalistic norms are presented as codifications of natural, eternal verities not subject to social modification. They are presented not merely as cultural verities but, implicitly or even explicitly, as biologically rooted necessities of the functioning of the human animal.

They become norms for the state, the workplace, the social arena. But they also become norms into which households are pushed to socialize their members, an effort that has been quite successful on the whole. They justify the polarization of the world-system. Since polarization has been increasing over time, racism, sexism, and other forms of anti-universalism have become ever more important, even though the political struggle against such forms of anti-universalism has also become more central to the functioning of the world-system.

The bottom line is that the modern world-system has made as a central, basic feature of its structure the simultaneous existence, propagation, and practice of both universalism and anti-universalism. This antinomic duo is as fundamental to the system as is the core-peripheral axial division of labor.

▪▪ GEORGE RITZER (1940–): A BIOGRAPHICAL SKETCH

George Ritzer was born in 1940 to a working-class Jewish family in upper Manhattan. His father worked as a taxicab driver and his mother was employed as a secretary in order to support Ritzer and his younger brother. He attended the prestigious Bronx High School of Science, then went on to earn a B.A. at the City College of New York. He received his PhD from Cornell University in 1968. In 1974, after working briefly at Tulane University and the University of Kansas, he joined the faculty at the University of Maryland, where he is currently a Distinguished University Professor. His primary interests center on social theory and analyses of consumption and globalization. His theoretical works include *Sociology: A Multiple Paradigm Science* (1975), *Toward an Integrated Sociological Paradigm* (1981) and *Metatheorizing in Sociology* (1991). In addition, he has written a number of textbooks on classical and contemporary sociological theory. His books on consumption and globalization include *The McDonaldization of Society* (1993), *Expressing America: A Critique of the Global Credit-Card Society* (1995), *Enchanting a Disenchanted World* (1999), and *The Globalization of Nothing* (2004). His books have been translated into twenty-five languages. He also has published nearly one hundred scholarly articles in his areas of expertise and is a founding editor of the *Journal of Consumer Culture*. Ritzer has presented his work at universities and conferences across the United States, as well as in some twenty foreign nations, including Taiwan, China, Dubai, Australia, New Zealand, England, Italy, Germany, Ireland, and Denmark, to name but a few. In 2000, he received the American Sociological Association's Distinguished Contributions to Teaching Award (http://www.georgeritzer.com/).

RITZER'S INTELLECTUAL INFLUENCES AND CORE IDEAS

Ritzer's work in the areas of consumption and globalization is influenced by Karl Marx, Max Weber, and critical theory. From Marx, he takes the view that Western capitalism is an exploitative economic system driven by the pursuit of profit—a pursuit that inevitably leads to the expansion of capitalism across all corners of the globe. Indeed, this expansion is well under way: The collapse of state socialism under the former Soviet Union and the continuing growth of capitalistic enterprises in China have left few if any economic alternatives to capitalism and the access to cheap resources, cheap labor, and favorable political conditions that are its lifeblood. Yet, the long-term success of capitalist economic expansion depends not only on these three factors, vital as they are to developing more profitable technologies for the production and distribution of goods, but also on establishing cultural hegemony or a system of "universal" beliefs, values, and ideas that serve the "particular" interests of the powerful. Cultural hegemony is created and sustained both domestically and internationally through the ideas, values, and information promoted through schools, television programming, news media, government policy and diplomatic efforts, and, of course, corporate advertising (Ritzer [2004] 2007:16). Under present-day capitalism, "the ruling ideas of the ruling class" take the form of a free market democracy that equates political and social freedom with the freedom to consume.

Thus, to complement Marx's focus on capitalism as a system of commodity production, Ritzer turns his attention to capitalism as a global system of consumption that necessarily requires increasing numbers of individuals to be inculcated with the desire and need to consume. Yet, intensifying desires to consume mass-produced commodities is not enough to sustain the continued expansion of global capitalism and ensure corporate profits. Consumers the world over must also be given the means and places to pursue their desires in the marketplace of commodities. To this end, the global spread of credit cards such as Visa and MasterCard have opened up unparalleled possibilities for consumption by allowing individuals to spend more money than they actually have, in turn making otherwise impossible purchases as easy as swiping a card (Ritzer 1995). Indeed, in America alone credit card spending between 1990 and 2005 rose from $213 billion to $1.3 trillion (1.3 million million). Not surprisingly, during this same period the number of major credit cards in use more than doubled from 213 million to 568 million (the Board of Governors of the Federal Reserve System, cited in Ritzer [2004] 2007:37). Meanwhile the development of "cathedrals of consumption" in the form of chain stores, theme parks, shopping malls, casino-hotels, cruise ships, and a host of other locales provide consumers with ample sites in which they can swipe their cards (Ritzer [2004] 2007).

The global expansion of capitalism as a system of production and consumption is itself intertwined with the spread of Western rationalization. Drawing from Max Weber's notion of the "iron cage" and the bureaucratization of modern societies, and the critical theorists' ideas on the "irrationality of rationality" (see Chapter 4), Ritzer offers a critique of the spread of Western rationalization that distinguishes his analysis from those of the other theorists presented in this chapter. He first captured this trend by exploring the practices of the epitome of the "cathedrals": McDonald's. Ritzer argues that the operating procedures perfected (but not originated) by McDonald's have been duplicated in countless arenas from amusement parks and grocery stores to childcare companies, health care organizations, religious institutions, educational institutions, and entertainment industries; as a result, those procedures have seeped into the very fabric of everyday life. Specifically, Ritzer notes five dimensions to the "McDonaldization of society": (1) Efficiency—optimizing the method for achieving goals, whether it be satisfying one's hunger or mass-producing laptop computers. (2) Calculability—an emphasis on the quantitative aspects of goods and services—for instance, how *much* they cost or how *long* it takes to receive them. As a result, the importance of quality becomes secondary. Big Macs and Domino's Pizza may not taste as good as a homemade burger or pizza, but they are relatively inexpensive and you get them quickly. (3) Predictability—the comforting promise that goods and services will be the same whenever or wherever they are produced or consumed, and the profits gained in the cost controls derived from high-volume, standardized production. (4) Control through nonhuman technology—the use of machines to replace workers or, at the very least, to routinize their tasks, while controlling the behavior of consumers through techniques such as offering drive-through windows and limited menu options, clearly marking queues for ordering, and providing uncomfortable seats and inhospitable lighting. All are designed to encourage individuals to leave as quickly as possible in order to make room for new paying customers (Ritzer 2000:12–15). (5) Irrationality of rationality—the dehumanization of individuals and social relationships that result from anonymous, superficial, scripted interactions between employees and customers (Ritzer [2004] 2007:25). In the end, McDonaldization and the practices that underlie it promote a standardized, homogenized approach to production, consumption, and social relationships. Yet the rationalizing processes perfected by McDonald's offer immense competitive advantages in the efficient pursuit of profits. And to the extent they have been adopted globally and across an array of institutional arenas, everyday life has become the same the world over. Cultural distinctiveness, the particular, and the unique are all lost in the irrational quest for rationalized, profit-driven expediency.

Ritzer connects his interests in the globalizing pressures of capitalism and the rationalization of societies through his concept of grobalization. **Grobalization** calls attention to the "imperialistic ambitions of nations, corporations, organizations and other entities and their desire—indeed, their need—to impose themselves on various geographic areas" (Ritzer 2003:194). Such entities are primarily interested in "seeing their power, influence, and in many cases profits *grow* (hence the term *gro*balization) throughout the world" (Ritzer [2004] 2007:16). While grobalization is tied generally to the practices of Western nations and corporations, it is the United States government and American corporations that play the leading role in advancing the process. Thus, grobalization is tied not only to the spread of capitalism and McDonaldization, but also to **Americanization**—that is, to the "propagation of American ideas, customs, social patterns, industry, and capital around the world" (ibid.:27). Americanization "involves a commitment to the *growth* in American influence in all realms throughout the world [by] overwhelm[ing] competing processes (e.g., Japanization) as well as the strength of local (and glocal) forces that might resist, modify, or transform American models into hybrid forms" (ibid., emphasis in the original). Examples of Americanization are endless: the worldwide exporting of American sports (for instance, football, basketball, and baseball), popular music, television shows, Hollywood films, soft drinks, fast-food chains, computer software programs and platforms (for instance, Microsoft Windows, Internet Explorer, and Google); efforts to spread the American model of democracy to other countries through wide-ranging diplomatic and military efforts; and the extraction of natural resources and use of labor markets around the globe by American corporations (ibid.:28). The issue, however, is not simply the exporting of American commodities, business models, political interests, and culture. Rather, it is the displacement and subsequent loss of local and national industries, brands, business practices, political practices, customs, and cultures that occur in the wake of exporting America.

While grobalization may be the dominant trend behind globalization, Ritzer contrasts it with the parallel process of glocalization that, as we noted previously, underscores the intertwining of global and local practices and cultures. Glocalization, and the heterogeneity it produces, signals that local actors are not powerless in the face of Western—and more specifically, American—economic, political, and cultural imperialism. Instead, globalization offers them the opportunity to creatively adapt to global "intrusions."

This points to an essential difference between grobalization and glocalization, namely, the effects these processes are producing. To address this issue, Ritzer turns to the related concepts of "nothing" and "something." He defines **nothing** as "a social form that is generally centrally conceived, controlled, and comparatively devoid of distinctive content" (Ritzer [2007] 2007:36). Such social forms are the staple of grobalizing tendencies. He contrasts this with **something**, which he defines as those glocalized social forms that are "generally indigenously conceived, controlled, and comparatively rich in distinctive substantive content" (ibid.:38). For Ritzer, modern life, particularly within the realm of consumption, is characterized by a long-term trend in which something is being replaced by nothing—in which the indigenous and unique is being increasingly marginalized by the central and generic.

Grobalization is thus creating a world in which our lives are dominated more and more by "nullities": "*non*places," "*non*things," "*non*people," and "*non*services" (ibid.). For instance, compare the eating experiences that one is likely to find in a chain restaurant, such as Chili's or T.G.I. Friday's, to one that may be found in a privately owned family restaurant. Any given Chili's or Friday's is identical to another. The signage, menus, and décor are the same in every restaurant, and indeed they must be in order to maximize profit through economies of scale. As a result, these establishments have little if any connection to the local environments in which they operate. There will be no artwork created by local artists hanging on the walls, nor will the architecture reflect the heritage indigenous to the area. Eating in a *non*place like Chili's, one would never know if she is in the southwest or the northeast of the United States. Similarly, the food one eats in chain restaurants is not made to order from local ingredients, but rather usually involves simply heating frozen, prepackaged portions that are shipped from a central location. These *non*things are purposely bland and lacking in complexity in order to reach as wide a customer base as possible. Offering distinctive food runs the risk of repelling potential customers who might be put off by unique flavors or ingredients. When the goal is to maximize profits, this is an experience that must be avoided. Compare this to the "things" one might eat in a "place" like a local, family-owned restaurant where people go to eat precisely because the food and décor are unique.

Moreover, in a chain restaurant setting, customers are likely to interact with *non*people, that is, with waiters, staff, and other customers with whom one has no connection. Here, all involved are interchangeable and anonymous, save for the "flair" the servers wear on their wardrobe or the color of their ties.[10] The workers are themselves easily replaced by management and come to see the job as simply a temporary means for making money. These factors culminate in the generic *non*service offered to customers. Waiters in chain restaurants have scripts and corporate protocol to follow when interacting with customers, and likewise, those who are making the food have little if any opportunity to be creative. Instead, they, too, follow standardized recipes that they did not develop, making variation virtually impossible. For both waiters and cooks, little in the way of skill, know-how, or specialized training is required. One does not make a reservation at Chili's in order to taste a chef's (a "person" who offers a "service") inventive culinary creation or a one-of-a-kind family recipe while enjoying the distinctive service provided by the owner or one of his family members (again, "persons" who offer "services"). In fact, both the food and the dining atmosphere are secondary to the experience—and purposively so—to encourage customers to eat as quickly as possible, thus freeing a table for the next patrons and generating greater profits. Local establishments do not survive for long when they treat their clientele as impersonal means to making money. On the contrary, their success depends on developing a reputation for providing distinctive food and personable service that are "something."

RITZER'S THEORETICAL ORIENTATION ▪▪

Ritzer's theoretical orientation is primarily collectivist and rational. Regarding the question of order, grobalization, and glocalization, McDonaldization and Americanization all speak to broad structural processes that are understood to be centrally responsible for shaping the social order. To the degree that individuals figure into Ritzer's analysis, they are conceived as reacting to, and not directly determining, the development of these processes. For instance, Ritzer's definition of grobalization—the dominant trend shaping the global society—identifies the key actors as nations, corporations, and organizations, particularly the United States government and American corporations, and not as individuals per se. It is these collectivist actors and the actions that they take that determine the course of contemporary society and the fate of the individuals within it.

Addressing the question of action, Ritzer notes that these institutions operate according to a rational logic rooted in the principles of efficiency, predictability, calculability, and control, because it is these organizational features that offer the best advantage in the quest for power and profits that is the hallmark of globalization. The imperialistic motivations that drive the actions of major economic, political, and cultural institutions are realized through intensive planning. While their decisions may prove to be wrong,

---

[10]The movie *Office Space* (1999) offers a comical critique of the *non*places in which *non*people work. In one scene Joanna (played by Jennifer Aniston) quits her job at a chain restaurant over her refusal to add more pieces of "flair" to her uniform.

if not disastrous, the consequences of corporate mergers are not left to chance, investments in foreign countries or new markets are not determined by "gut feeling," treaties and free trade agreements are not signed without deliberation, and movies and TV shows are not aired without first screening them before a test audience. Centralized, rationalized decision-making molded by the imperatives of efficiency, predictability, calculability, and control are the lifeblood of grobalizing entities.

Yet, Ritzer does acknowledge that both individualist and nonrational forces play a role in contemporary society, albeit a role that is of secondary importance in his model. Specifically, his discussion of glocalization makes it clear that the juggernaut of grobalization gives rise to counterpressures in the form of local actors (individuals) who are committed to creating or retaining space for the unique, indigenous, and heterogeneous. While these efforts are increasingly marginalized, they nevertheless demonstrate that individuals are not powerless in their struggle against the forces of grobalization: Local farmers' markets still exist and play a vital role in some communities; individuals still practice and consume forms of alternative medicine; some families choose to camp in national parks for their vacation instead of going to Disneyland. For Ritzer ([2004] 2007), such efforts to produce and consume something speak to the desire to cultivate the "magical" and "enchanted"—or, in our terms, nonrational—aspects of social life. However, he insists that they represent reactions to and an escape from the overriding process of grobalization and the spreading production and consumption of nothing that it carries in its wake. To capture this dynamic, the concepts of glocalization and something straddle the action/order axes in Figure 17.2. The play of individual, magical forces in social life, while not extinct, is itself subjected to the homogeneity and "disenchantment" spurred by centrally controlled rationalizing processes that predominantly color our world. Yet glocalization, as "the last outpost of most lingering . . . forms of the local," represents the "most realistic and viable" measure of hope for those opposed to the growing sway of grobalization (Ritzer 2003:207).

Figure 17.2   Ritzer's Basic Concepts and Theoretical Orientation

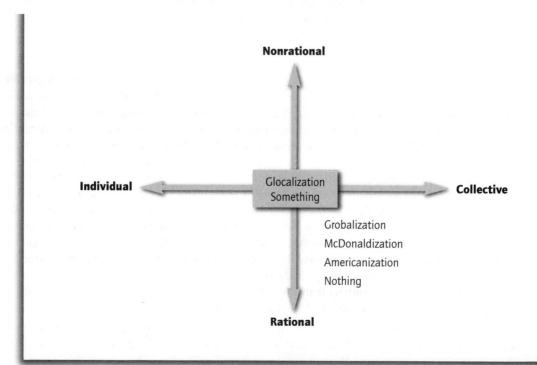

Reading

Introduction to "Rethinking Globalization"

In the following article, Ritzer introduces his notion of grobalization as the central dynamic shaping the world today. To this end, he analyzes the production and consumption of nothing, as well as the production and consumption of something that is associated with the counterprocess of glocalization. However, he repeatedly points out that the elective affinities that exist between grobalization and nothing on the one hand, and glocalization and something on the other, represent developments that lie on a continuum: They do not presuppose a necessary, singular relationship between the various elements. In other words, grobalization can lead to the production and consumption of something and, conversely, glocalization can lead to the production and consumption of nothing. Yet, such possibilities are of comparatively minor significance, the reasons for which Ritzer outlines in the following excerpt.

Finally, the reader should note the parallels between Wallerstein's discussion of core, peripheral, and semiperipheral regions and Ritzer's analysis of the "economics of nothing." As you will read, the production and consumption of nothing and something are not spread equally around the globe. Those who produce many of the *non*things in the world have little access to them (though they are surrounded by them at their workplace) and cannot afford to purchase them. Workers living within the peripheral and semiperipheral regions labor long hours for low wages, often in unsafe conditions, so that those of us living in the core can enjoy nothing.

"Rethinking Globalization: Glocalization/Grobalization and Something/Nothing" (2003)

George Ritzer

This essay seeks to offer a unique theoretical perspective by reflecting on and integrating some well-known ideas in sociology (and the social sciences) on globalization and a body of thinking, virtually unknown in sociology, on the concept of nothing (and, implicitly, something). The substantive focus will be on consumption, and all of the examples will be drawn from it. However, the implications of this analysis extend far beyond that realm, or even the economy more generally.

It is beyond the scope of this discussion to deal fully with globalization, but two centrally important processes—glocalization and grobalization—will be of focal concern. Glocalization (and related ideas such as hybridity and creolization) gets to the heart of what many—perhaps most—contemporary globalization theorists think about the nature of transnational processes (Appadurai 1996; Garcia Canclini 1995; Hannerz 1987; Pieterse 1995; Robertson 1992, 1995, 2001; Tomlinson 1999). *Glocalization* can be defined as the interpenetration of the global and the local, resulting in unique outcomes in different geographic areas. This view emphasizes global heterogeneity and tends to reject the idea that forces emanating from the

SOURCE: "Rethinking Globalization: Glocalization/Grobalization/and Something/Nothing" by George Ritzer in *Sociological Theory,* Vol. 21, No. 3 (September 2003), pp. 193–209. Reprinted with permission of the American Sociological Association and author George Ritzer.

West in general and the United States in particular (Featherstone 1995:8–9) are leading to economic, political, institutional, and—most importantly—cultural homogeneity.

One of the reasons for the popularity of theories of glocalization is that they stand in stark contrast to the much hated and maligned *modernization theory* that had such a wide following in sociology and the social sciences for many years (Rostow 1960). Some of the defining characteristics of this theory were its orientation to issues of central concern in the West, the preeminence it accorded to developments there, and the idea that the rest of the world had little choice but to become increasingly like it (more democratic, more capitalistic, more consumption-oriented, and so on). While there were good reasons to question and to reject modernization theory and to develop the notion of glocalization, there are elements of that theory that remain relevant to thinking about globalization today.

In fact, some of those associated with globalization theory have adhered to and further developed perspectives that, while rejecting most of modernization theory, retain an emphasis on the role of Westernization and Americanization in global processes (Giddens 2000; Kuisel 1993; Ritzer 1995). Such concerns point to the need for a concept—*grobalization*—coined here for the first time as a much-needed companion to the notion of glocalization. While it does *not* deny the importance of glocalization and, in fact, complements it, grobalization focuses on the imperialistic ambitions of nations, corporations, organizations, and other entities and their desire—indeed, their need—to impose themselves on various geographic areas. Their main interest is in seeing their power, influence, and (in some cases) profits *grow* (hence the term "*gro*balization") throughout the world. It will be argued that grobalization tends to be associated with the proliferation of nothing, while glocalization tends to be tied more to something and therefore stands opposed, at least partially (and along with the local itself), to the spread of nothing. Globalization as a whole is not unidirectional, because these two processes coexist under that broad heading and because they are, at least to some degree, in conflict in terms of their implications for the spread of nothingness around the world.

Having already begun to use the concepts of nothing and something, we need to define them as they will be used here. Actually, it is the concept of nothing that is of central interest here (as well as to earlier scholars); the idea of something enters the discussion mainly because nothing is meaningless without a sense of something. However, nothing is a notoriously obscure concept: "Nothing is an awe-inspiring yet essentially undigested concept, highly esteemed by writers of a mystical or existentialist tendency, but by most others regarded with anxiety, nausea, or panic" (Heath 1967:524). . . .

Conceptualizing Nothing (and Something)

Nothing is defined here as *a social form that is generally centrally conceived, controlled, and comparatively devoid of distinctive substantive content*. This leads to a definition of *something as a social form that is generally indigenously conceived, controlled, and comparatively rich in distinctive substantive content*. This definition of nothing's companion term makes it clear that neither nothing nor something exists independently of the other: *each makes sense only when paired with and contrasted to the other*. While presented as a dichotomy, this implies a *continuum* from something to nothing, and that is precisely the way the concepts will be employed here—as the two poles of that continuum.

A major and far more specific source of the interest here in nothing—especially conceptually—is the work in social geography by anthropologist Marc Auge (1995) on the concept of nonplaces (see also Morse [1990] on "nonspaces"; Relph 1976). To Auge, nonplaces are "the real measure of our time" (Auge 1995:79). This can be generalized to say that nothing is, in many ways, the true measure of our time! The present work extends the idea of nonplaces to nonthings, nonpeople, and nonservices and, following the logic used above, none of these make sense without their polar opposites—places, things, people, and services. In addition, they need to be seen as the poles of four subtypes that are subsumed under the broader heading of the something/nothing continuum. Figure 1 offers an overview of the overarching something/nothing continuum and these four subtypes, as well as an example of each.

Figure 1 The Four Major Subtypes of Something/Nothing (with examples) Presented as Subcontinua under the Broad Something/Nothing Continuum

Something	Nothing
Place (community bank)	Nonplace (credit card company)
Thing (personal loan)	Nothing (credit card loan)
Person (personal banker)	Nonperson (telemarketer)
Service (individualized assistance)	Nonservice (automated, dial-up aid)

Following the definition of nothing, it can be argued that a credit card is nothing (or at least lies toward that end of the something/nothing continuum) because it is centrally conceived and controlled by the credit card company and there is little to distinguish one credit card (except a few numbers and a name) from any other (they all do just about the same things). Extending this logic, a contemporary credit card company, especially its telephone center, is a nonplace, the highly programmed and scripted individuals who answer the phones are nonpeople, and the often automated functions can be thought of as nonservices. Those entities that are to be found at the something end of each continuum are locally conceived and controlled forms that are rich in distinctive substance. Thus, a traditional line of credit negotiated by local bankers and personal clients is a thing; a place is the community bank to which people can go and deal with bank employees in person and obtain from them individualized services.

NOTHING/SOMETHING AND GROBALIZATION/GLOCALIZATION

We turn now to a discussion of the relationship between grobalization/glocalization and something/nothing. Figure 2 offers the four basic possibilities that emerge when we cross-cut the grobalization/glocalization and something/nothing continua (along with representative examples of places/nonplaces, things/nonthings, people/nonpeople, and services/nonservices for each of the four possibilities and quadrants). It should be noted that while this yields four "ideal types," there are no hard and fast lines between them. This is reflected in the use of both dotted lines and multidirectional arrows in Figure 2.

Quadrants one and four in Figure 2 are of greatest importance, at least for the purposes of this analysis. They represent a key point of tension and conflict in the world today. Clearly, there is great pressure to grobalize nothing, and often all that stands in its way in terms of achieving global hegemony is the glocalization of something. We will return to this conflict and its implications below.

While the other two quadrants (two and three) are clearly residual in nature and of secondary importance, it is necessary to recognize that there is, at least to some degree, a glocalization of nothing (quadrant two) and a grobalization of something (quadrant three). Whatever tensions may exist between them are of far less significance than that between the grobalization of nothing and the glocalization of something. However, a discussion of the glocalization of nothing and the grobalization of something makes it clear that grobalization is not an unmitigated source of nothing (it can involve something) and that glocalization is not to be seen solely as a source of something (it can involve nothing).

The close and centrally important relationship between (1) grobalization and nothing and (2) glocalization and something leads to the view that there is an *elective affinity* between the two elements of each of these pairs. The idea of elective affinity, derived from the historical comparative sociology of Max Weber, is meant to imply that there is *not* a necessary, law-like causal relationship between these elements. That is, neither in the case of grobalization and nothing nor that of glocalization and something does one of these elements "cause" the other to come into existence. Rather, the development and diffusion of one tends to go hand in hand with the other. Another way of putting this is that grobalization/nothing and glocalization/something tend to mutually favor one another; they are inclined to combine with one another (Howe 1978). Thus, it is far easier to grobalize nothing than something: the development of grobalization creates a favorable ground for the development and spread of nothing (and nothing is easily grobalized). Similarly, it is far easier to glocalize something than nothing: the development of glocalization creates a favorable ground for the development and proliferation of something (and something is easily glocalized).

Figure 2 The Relationship between Glocal-Grobal and Something-Nothing with Exemplary (Non-)Places, (Non-)Things, (Non-)Persons, and (Non-)Services

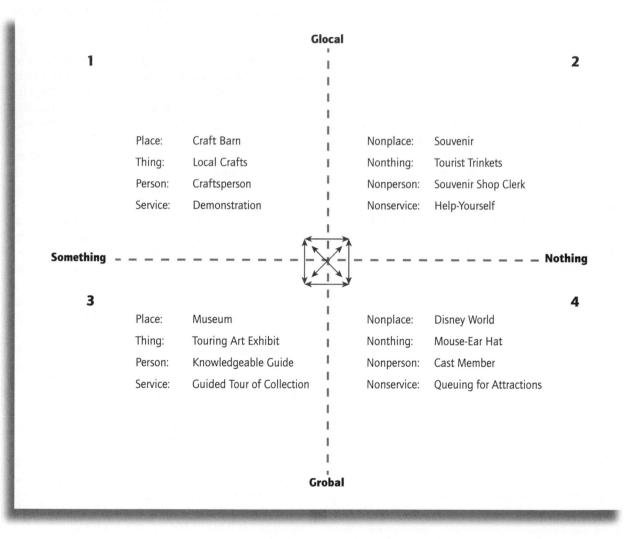

However, the situation is more complex than this, since we can also see support for the argument that grobalization can, at times, involve something (e.g., art exhibits that move among art galleries throughout the world; Italian exports of food such as Parmiagiano Reggiano and Culatella ham; touring symphony orchestras and rock bands that perform in venues throughout the world) and that glocalization can sometimes involve nothing (e.g., the production of local souvenirs and trinkets for tourists from around the world). However, we would *not* argue that there is an elective affinity between grobalization and something and between glocalization and nothing. The existence of examples of the grobalization of something and the glocalization of nothing makes it clear why we need to think in terms of elective affinities and not law-like relationships.

THE GROBALIZATION OF SOMETHING

Some types of something have been grobalized to a considerable degree. For example, gourmet foods, handmade crafts, custom-made clothes, and Rolling Stones concerts are now much more available throughout the world, and more likely to move transnationally, than ever in history. In a very specific example in the arts, a touring series of "Silk Road" concerts recently brought together Persian artists and music, an American symphony orchestra, and Rimsky-Korsakov's (Russian) "Scheherezade" (Delacoma 2002).

Returning to Figure 2, we have used as examples of the grobalization of something touring art exhibitions (thing) of the works of Vincent van Gogh, the museums throughout the world in which such exhibitions occur (place), the knowledgeable guides who show visitors the highlights of the exhibition (person),[i] and the detailed information and insights they are able to impart in response to questions from gallery visitors (service).

In spite of the existence of examples like these, why is there comparatively little affinity between grobalization and something? First, there is simply far less demand throughout the world for most forms of something, at least in comparison to the demand for nothing. One reason for this is that the distinctiveness of something tends to appeal to far more limited tastes than nothing, be it gourmet foods, handmade crafts, or Rolling Stones or Silk Road concerts. Second, the complexity of something, especially the fact that it is likely to have many different elements, means that it is more likely that it will have at least some characteristics that will be off-putting for or will even offend large numbers of people in many different cultures. For example, a Russian audience at a Silk Road concert might be bothered by the juxtaposition of Persian music with that of Rimsky-Korsakov. Third, the various forms of something are usually more expensive—frequently much more expensive—than competing forms of nothing (gourmet food is much more costly than fast food).

Higher cost means, of course, that far fewer people can afford something. As a result, the global demand for expensive forms of something is minuscule in comparison to that for the inexpensive varieties of nothing. Fourth, because the prices are high and the demand is comparatively low, far less can be spent on the advertising and marketing of something, which serves to keep demand low. Fifth, something is far more difficult to mass-manufacture and, in some cases (Silk Road concerts, van Gogh exhibitions), impossible to produce in this way. Sixth, since the demand for something is less price-sensitive than nothing (the relatively small number of people who can afford it are willing, and often able, to pay almost any price), there is less need to mass-manufacture it (assuming it could be produced in this way) in order to lower prices. Seventh, the costs of shipping (insurance, careful packing and packaging, special transports) of something (gourmet foods, the van Gogh paintings) are usually very high, adding to the price and thereby reducing the demand.

It could also be argued that the fact that the grobalization of something (compared to nothing) occurs to a lesser degree helps to distinguish something from nothing. Because it is relatively scarce, something retains its status and its distinction from nothing. If something came to be mass-produced and grobalized, it is likely that it would move toward the nothing end of the continuum. This raises the intriguing question of what comes first—nothing, or grobalization and the associated mass production. That is, does a phenomenon start out as nothing? Or is it transformed into nothing by mass production and grobalization? We will return to this issue below.

THE GROBALIZATION OF NOTHING

The example of the grobalization of nothing in Figure 2 is a trip to one of Disney's worlds. Any of Disney's worlds is a nonplace, awash with a wide range of nonthings (such as mouse-ear hats), staffed largely by nonpeople (the "cast members," in costume or out), who offer nonservices (what is offered is often dictated by rules, regulations, and the scripts followed by employees).

The main reasons for the strong elective affinity between grobalization and nothing are basically the

---

[i] An interesting example of the trend toward nothingness is the increasing use of audio guides and rented tape players at such shows, and at museums more generally.

inverse of the reasons for the lack of such affinity between grobalization and something. Above all, there is a far greater demand throughout the world for nothing than something. This is the case because nothing tends to be less expensive than something (although this is not always true), with the result that more people can afford the former than the latter. Large numbers of people are also far more likely to want the various forms of nothing, because their comparative simplicity and lack of distinctiveness appeals to a wide range of tastes. In addition, as pointed out earlier, that which is nothing—largely devoid of distinctive content—is far less likely to bother or offend those in other cultures. Finally, because of the far greater potential sales, much more money can be—and is—devoted to the advertising and marketing of nothing, thereby creating a still greater demand for it than for something.

Given the great demand, it is far easier to mass-produce and mass-distribute the empty forms of nothing than the substantively rich forms of something. Indeed, many forms of something lend themselves best to limited, if not one-of-a-kind, production. A skilled potter may produce a few dozen pieces of pottery and an artist a painting or two in, perhaps, a week, a month, or even (a) year(s). While these craft and artworks may, over time, move from owner to owner in various parts of the world, this traffic barely registers in the total of global trade and commerce. Of course, there are the rare masterpieces that may bring millions of dollars, but in the main these are small-ticket items. In contrast, thousands, even many millions, and sometimes billions of varieties of nothing are mass-produced and sold throughout the globe. Thus, the global sale of Coca-Cola, Whoppers, Benetton sweaters, Gucci bags, and even Rolex watches is a far greater factor in grobalization than is the international sale of pieces of high art or of tickets to the Rolling Stones' most recent world tour. Furthermore, the various forms of nothing can range in cost from a dollar or two to thousands, even tens of thousands of dollars. The cumulative total is enormous and infinitely greater than the global trade in something.

Furthermore, the economics of the marketplace demands that the massive amount of nothing that is produced be marketed and sold on a grobal basis. For one thing, the economies of scale mean that the more that is produced and sold, the lower the price. This means that, almost inevitably, American producers of nothing (and they are, by far, the world leaders in this) must become dissatisfied with the American market, no matter how vast it is, and aggressively pursue a world market for their consumer products. The greater the grobal market, the lower the price that can be charged. This, in turn, means that even greater numbers of nothing can be sold and farther reaches of the globe in less-developed countries can be reached. Another economic factor stems from the demand of the stock market that corporations that produce and sell nothing (indeed, all corporations) increase sales and profits from one year to the next. Those corporations that simply meet the previous year's profitability or experience a decline are likely to be punished in the stock market and see their stock prices fall, sometimes precipitously. In order to increase profits continually, the corporation is forced, as Marx understood long ago, to continue to search out new markets. One way of doing that is constantly to expand globally. In contrast, since something is less likely to be produced by corporations—certainly by the large corporations listed in the stock market—there is far less pressure to expand the market for it. In any case, as we saw above, given the limited number of these things that can be produced by artisans, skilled chefs, artists, and so on, there are profound limits on such expansion. This, in turn, brings us back to the pricing issue and relates to the price advantage that nothing ordinarily has over something. As a general rule, the various types of nothing cost far less than something. The result, obviously, is that nothing can be marketed globally far more aggressively than something.

Also, nothing has an advantage in terms of transportation around the world. These are things that generally can be easily and efficiently packaged and moved, often over vast areas. Lunchables, for example, are compact, prepackaged lunch foods, largely for schoolchildren, that require no refrigeration and have a long shelf life. Furthermore, because the unit cost of such items is low, it is of no great consequence if they go awry, are lost, or are stolen. In contrast, it is more difficult and expensive to package something—say, a piece of handmade pottery or an antique vase—and losing such things or having them stolen or broken is a disaster. As a result, it is far more expensive to insure something than nothing, and this difference is another reason for the cost advantage that nothing has over something. It is these sorts of things that serve to greatly limit the global trade in items that can be included under the heading of something.

It is important to remember that while most of our examples in this section are nonthings, it is the case that nonplaces (franchises), nonpeople (counterpeople in fast-food chains), and nonservices (automatic teller machines—ATMs) are also being grobalized.

While the grobalization of nothing dominates in the arena of consumption as it is generally defined, we find domains—medicine, science, pharmaceuticals (*Financial Times* 2001), biotechnology (Abate 2002), education, and others—in which the grobalization of something is of far greater importance. While these areas have experienced their share of the grobalization of nothing, they are also characterized by a high degree of the grobalization of something. For example, the worldwide scientific community benefits from the almost instantaneous distribution of important scientific findings, often, these days, via new journals on the Internet. Thus, our focus on the grobalization of nothing should not blind us to the existence and importance—especially in areas such as these—of the grobalization of something.

The Glocalization of Nothing

Just as there has historically been a tendency to romanticize and glorify the local, there has been a similar trend in recent years among globalization theorists to overestimate the glocal (Tam, Dissanayake, and Siuhuan Yip 2002). It is seen by many as not only the alternative to the evils of grobalization, but also a key source of much that is worthwhile in the world today. Theorists often privilege the glocal something over the grobal nothing (as well as over the glocal nothing, which rarely appears in their analyses). For example, Jonathan Friedman (1994) associates cultural pluralism with "a dehegemonizing, dehomogenizing world incapable of a formerly enforced politics of assimilation or cultural hierarchy." Later, he links the "decline of hegemony" to "a liberation of the world arena to the free play of already extant but suppressed projects and potential new projects" (Friedman 1994:252). Then there are the essays in James Watson's (1997) *McDonald's in East Asia,* which, in the main, focus on glocal adaptations (and generally downplay grobal impositions) and tend to describe them in positive terms.

While most globalization theorists are not postmodernists (Featherstone 1995 is one exception), the widescale acceptance of various postmodern ideas (and rejection of many modern positions) has helped lead to positive attitudes toward glocalization among many globalization theorists. Friedman is one who explicitly links "cultural pluralism" and the "postmodernization of the world" (Friedman 1994:100). The postmodern

perspective is linked to glocalization theory in a number of ways. For example, the work of de Certeau and others on the power of the agent in the face of larger powers (such as grobalization) fits with the view that indigenous actors can create unique phenomena out of the interaction of the global and the local. De Certeau talks of actors as "unrecognized producers, poets of their own affairs, trailblazers in the jungles of functionalist rationality" (de Certeau 1984:34). A similar focus on the local community (Seidman 1991) gives it the power to create unique glocal realities. More generally, a postmodern perspective is tied to hybridity, which, in turn, is "subversive" of such modern perspectives as essentialism and homogeneity.

While there are good reasons for the interest in and preference for glocalization among globalization theorists, such interest is clearly overdone. For one thing, grobalization (especially of nothing) is far more prevalent and powerful than glocalization (especially of something). For another, glocalization itself is a significant source of nothing.

One of the best examples of the glocalization of nothing is to be found in the realm of tourism (Wahab and Cooper 2001), especially where the grobal tourist meets the local manufacturer and retailer (where they still exist) in the production and sale of glocal goods and services (this is illustrated in quadrant two of Figure 2). There are certainly instances—perhaps even many of them—in which tourism stimulates the production of something: well-made, high-quality craft products made for discerning tourists; meals lovingly prepared by local chefs using traditional recipes and the best of local ingredients. However, far more often—and increasingly, as time goes by—grobal tourism leads to the glocalization of nothing. Souvenir shops are likely to be bursting at the seams with trinkets reflecting a bit of the local culture. Such souvenirs are increasingly likely to be mass-manufactured—perhaps using components from other parts of the world—in local factories. If demand grows great enough and the possibilities of profitability high enough, low-priced souvenirs may be manufactured by the thousands or millions elsewhere in the world and then shipped back to the local area to be sold to tourists (who may not notice, or care about, the "made in China" label embossed on their souvenir replicas of the Eiffel Tower). The clerks in these souvenir shops are likely to act like nonpeople, and tourists are highly likely to serve themselves. Similarly, large numbers of meals slapped together by semiskilled chefs to suggest vaguely local cooking are far more likely than authentic meals that are true to the region, or that truly integrate local elements. Such meals are likely to be offered in "touristy" restaurants that are close to the nonplace end of the continuum and to be served by nonpeople who offer little in the way of service.

Another major example involves the production of native shows—often involving traditional costumes, dances, and music—for grobal tourists. While these could be something, there is a very strong tendency for them to be transformed into nothing to satisfy grobal tour operators and their clientele. Hence these shows are examples of the glocalization of nothing, because they become centrally conceived and controlled empty forms. They are often watered down, if not eviscerated, with esoteric or possibly offensive elements removed. The performances are designed to please the throngs of tourists and to put off as few of them as possible. They take place with great frequency, and interchangeable performers often seem as if they are going through the motions in a desultory fashion. For their part, this is about all the grobal tourists want in their rush (and that of the tour operator) to see a performance, to eat an ersatz local meal, and then to move on to the next stop on the tour. Thus, in the area of tourism—in souvenirs, performances, and meals—we are far more likely to see the glocalization of nothing than of something.

THE GLOCALIZATION OF SOMETHING

The example of the glocalization of something in Figure 2 (quadrant 1) is in the realm of indigenous crafts such as pottery or weaving. Such craft products are things, and they are likely to be displayed and sold in places such as craft barns. The craftperson who makes and demonstrates his or her wares is a person, and customers are apt to be offered a great deal of service.

Such glocal products are likely to remain something, although there are certainly innumerable examples of glocal forms of something that have been transformed into glocal—and in some cases grobal—forms of nothing (see below for a discussion of Kokopelli figures and matryoshka dolls). In fact, there is often a kind of progression here, from glocal something to glocal nothing

as demand grows, and then to grobal nothing[ii] if some entrepreneur believes that there might be a global market for such products. However, some glocal forms of something are able to resist this process.

Glocal forms of something tend to remain as such for various reasons. For one thing, they tend to be costly, at least in comparison to mass-manufactured competitors. High price tends to keep demand down locally, let alone globally. Second, glocal forms of something are loaded with distinctive content. Among other things, this means that they are harder and more expensive to produce and that consumers, especially in other cultures, find them harder to understand and appreciate. Furthermore, their idiosyncratic and complex character make it more likely that those in other cultures will find something about them they do not like or even find offensive. Third, unlike larger manufacturers of nothing, those who create glocal forms of something are not pushed to expand their business and increase profits to satisfy stockholders and the stock market. While craftspeople are not immune to the desire to earn more money, the pressure to do so is more internal than external, and it is not nearly as great or inexorable. In any case, the desire to earn more money is tempered by the fact that the production of each craft product is time-consuming and only so many of them can be produced in a given time. Further, craft products are even less likely to lend themselves to mass marketing and advertising than they are to mass manufacture.

WHICH COMES FIRST: NOTHING, OR ITS GROBALIZATION?

At this point, we need to deal with a difficult issue: Is it possible to determine which comes first—nothing or its grobalization? The key components of the definition of nothing—central conception and control, lack of distinctive content—tend to lead us to associate nothing with the modern era of mass production. After all, the system of mass production is characterized by centralized conception and control, and it is uniquely able to turn out large numbers of products lacking in distinctive content. While there undoubtedly were isolated examples of nothing prior to the Industrial Revolution, it is hard to find many that fit our basic definition of nothing.

[ii]Grobal forms of nothing (e.g., McDonald's toys) can be transformed into something (either grobal or glocal) when, for example, they become collector's items.

Thus, as a general rule, nothing requires the prior existence of mass production. However, that which emanates from mass-production systems need not necessarily be distributed and sold globally. Nevertheless, as we have discussed, there are great pressures on those who mass-produce nothing to market it globally. Thus, there is now a very close relationship between mass production and grobalization; the view here is that *both* precede nothing and are prerequisites to it.

Take, for example, such historic examples of something in the realm of folk art as Kokopellis from the southwestern United States and matryoshka dolls from Russia. At their points of origin long ago in local cultures, these were clearly hand-made products that one would have had to put close to the something end of the continuum. For example, the Kokopelli, usually depicted as an arch-backed flute player, can be traced back to at least 800 A.D. and to rock art in the mountains and deserts of the southwestern United States (Acacia Artisans 2002; Malotki 2000). Such rock art is clearly something. But in recent years, Kokopellis have become popular among tourists to the area and have come to be produced in huge numbers in innumerable forms (figurines, lamps, keychains, light-switch covers, Christmas ornaments, and so on), with increasingly less attention to the craftsmanship involved in producing them. Indeed, they are increasingly likely to be mass-produced in large factories. Furthermore, offending elements are removed in order not to put off potential consumers anywhere in the world. For example, the exposed genitals that usually accompanied the arched back and the flute have been removed. More recently, Kokopellis have moved out of their locales of origin in the Southwest and come to be sold globally. In order for them to be marketed globally at a low price, much of the distinctive character and craftsmanship involved in producing the Kokopelli is removed. That is, the grobalization of Kokopellis has moved them even closer to the nothing end of the continuum.

A similar scenario has occurred in the case of the matryoshka doll (from five to as many as 30 dolls of increasingly small size nested within one another) (Gift to Give 2002), although its roots in Russian culture are not nearly as deep (little more than a century) as that of the Kokopelli in the culture of the southwestern United States. Originally hand-made and hand-painted by skilled craftspeople and made from seasoned birch (or lime), the traditional matryoshka doll was (and is) rich in detail. With the fall of communism and the Soviet Union, Russia has grown as a tourist destination, and the matryoshka doll has become a popular souvenir. In order to supply the increasing demand of tourists, and even to distribute matryoshka dolls around the world, they are now far more likely to be machine-made: automatically painted; made of poor quality, unseasoned wood; and greatly reduced in detail. In many cases, the matryoshka doll has been reduced to the lowest level of schlock and kitsch in order to enhance sales. For example, the traditional designs depicting precommunist nobles and merchants have been supplemented with caricatures of global celebrities such as Bill Clinton, Mikhail Gorbachev, and—post-September 11—Osama bin Laden (Korchagina 2002). Such mass-produced and mass-distributed matryoshka dolls bear little resemblance to the folk art that is at their root. The mass production and grobalization of these dolls has transformed that which was something into nothing. Many other products have followed that course, and still more will do so in the future.

While we have focused here on nonthings that were things at one time, much the same argument can be made about places, people, and services. That is, they, too, have come to be mass-manufactured and grobalized, especially in the realm of consumption. This is most obvious in virtually all franchises for which settings are much the same throughout the world (using many mass-manufactured components), people are trained and scripted to work in much the same way, and the same "services" are offered in much the same way. They all have been centrally conceived, are centrally controlled, and are lacking in distinctive content.

GROBALIZATION AND LOSS

Grobalization has brought with it a proliferation of nothing around the world. While it carries with it many advantages (as does the grobalization of something), it has also led to a loss, as local (and glocal) forms of something are progressively threatened and replaced by grobalized (and glocalized) forms of nothing.

This reality and sense of loss are far greater in much of the rest of the world than they are in the United States. As the center and source of much nothingness, the United States has also progressed furthest in the direction of nothing and away from something. Thus, Americans are long accustomed to nothing and have fewer and fewer forms of something with which to compare it. Each new form of or advance in nothing barely creates a ripple in American society.

However, the situation is different in much of the rest of the world. Myriad forms of something remain well entrenched and actively supported. The various forms of nothing—often, at least initially, imports from the United States—are quickly and easily perceived as nothing, since alternative forms of something, and the standards they provide, are alive and well. Certainly, large numbers of people in these countries demand and flock to nothing in its various forms, but many others are critical of it and on guard against it. The various forms of something thriving in these countries give supporters places, things, people, and services to rally around in the face of the onslaught of nothing. Thus, it is not surprising that the Slow Food Movement, oriented to the defense of "slow food" against the incursion of fast food, began in Italy (in fact, the origin of this movement was a battle to prevent McDonald's from opening a restaurant at the foot of the Spanish Steps in Rome) and has its greatest support throughout Europe (Kummer 2002).

THE INCREASE IN NOTHING! THE DECLINE IN SOMETHING?

A basic idea—even a grand narrative—in this essay is the idea that there is a long-term trend in the social world in general, and in the realm of consumption in particular, in the direction of nothing. More specifically, there is an historic movement from something to nothing. Recall that this is simply an argument about the increase in forms that are centrally conceived and controlled and are largely devoid of distinctive content. In other words, we have witnessed a long-term trend *from* a world in which indigenously conceived and controlled forms laden with distinctive content predominated *to* one where centrally conceived and controlled forms largely lacking in distinctive content are increasingly predominant.

There is no question that there has been an increase in nothing and a relative decline in something, but many forms of something have not experienced a decline in any absolute sense. In fact, in many cases, forms of something have increased; they have simply not increased at anything like the pace of the increase in nothing. For example, while the number of fast-food restaurants (nonplaces) has increased astronomically since the founding of the McDonald's chain in 1955, the number of independent gourmet and ethnic restaurants (places) has also increased, although at not nearly the pace of fast-food restaurants (Nelson 2001). This helps to account for the fact that a city such as Washington, DC (to take an example I know well) has, over the last half century, witnessed a massive increase in fast-food restaurants *at the same time* that there has been a substantial expansion of gourmet and ethnic restaurants. In fact, it could be argued that there is a dialectic here—that the absolute increase in nothing sometimes serves to spur at least some increase in something. That is, as people are increasingly surrounded by nothing, at least some are driven to search out or create something. However, the grand narrative presented here is more about the relative ascendancy of nothing and the relative decline in something than about absolute change.

Nonetheless, at least some forms of something (e.g., local groceries, cafeterias) have suffered absolute declines and may have disappeared or be on the verge of disappearance. It could be argued that all of these have been victims of what Joseph Schumpeter (1950) called "creative destruction." That is, while they have largely disappeared, in their place have arisen successors such as the fast-food restaurant, the supermarket, and the "dinner-house" (e.g., the Cheesecake Factory) (Jones 2002). While there is no question that extensive destruction of older forms has occurred, and that considerable creativity has gone into the new forms, one must question Schumpeter's one-sidedly positive view of this process. Perhaps some things—even some measure of creativity—have been lost with the passing of these older forms. It may be that the destruction has not always been so creative.

However, no overall value judgment needs to be made here; forms laden with content are not inherently better than those devoid of content, or vice versa. In fact, there were and are many forms rich in content that are among the most heinous of the world's creations. We could think, for example, of the pogroms that were so common in Russia, Poland, and elsewhere (Klier and Lambroza 1992). These were largely locally conceived and controlled and were awash in distinctive content (anti-Semitism, nationalism, and so on). Conversely, forms largely devoid of content are not necessarily harmful. For example, the bureaucracy, as Weber ([1921] 1968) pointed out, is a form (and ideal type) that is largely lacking in content. As such, it is able to operate in a way that other, more content-laden forms of organization—those associated with traditional and charismatic forms of organization—could not. That is, it was set up to be impartial—to *not* (at least theoretically) discriminate against anyone.

There is very strong support for the argument, especially in the realm of consumption, that we are in the midst of a long-term trend away from something and in the direction of nothing. By the way, this implies a forecast for the future: we will see further increases in nothing and further erosions of something in the years to come.

THE ECONOMICS OF NOTHING

Several points can be made about the economics of nothing. First, it is clear that, in general, there is an inverse relationship between income and nothing. That is, those with money can still afford to acquire various forms of something, whereas those with little money are largely restricted to nothing.[iii] Thus, only the affluent can afford expensive bottles of complex wine, or gourmet French meals with truffles. Those with little means are largely restricted to Coca-Cola, Lunchables, microwave meals, and McDonald's fries.

Second, there is an economic floor to this: those below a certain income level cannot even afford much of that which is categorized here as nothing. Thus, there are those near or below the poverty line in America who often cannot afford a meal at McDonald's or a six-pack of Coca Cola. More importantly, there are many more people in the less-developed parts of the world who do not have access to and cannot afford such forms of nothing. Interestingly, extreme poverty relegates people to something—homemade meals and home brews made from whatever is available. However, in this case it is hard to make the argument for something. These forms of something are often meager, and those who are restricted to them would love to have access to that which has been defined here, as well as by many people throughout the world, as nothing.

Third, thinking of society as a whole, some minimum level of affluence and prosperity must be reached before it can afford nothing. That is, there are few ATMs, fast-food restaurants, and Victoria's Secret boutiques in the truly impoverished nations of the world. There simply is not enough income and wealth for people to be able to afford nothing; people in these societies are, ironically, doomed—at least for the time being—to something. Thus, they are more oriented to barter, preparing food at home from scratch, and making their own nightgowns. It is not that they would not readily trade their something for the forms of nothing described above, but that they are unable to do so. It seems clear that as soon as the level of wealth in such a country reaches some minimal level, the various forms of nothing will be welcomed and, for their part, the companies that produce them will enter eagerly.

Fourth, even the wealthiest of people often consume nothing. For one thing, as has been pointed out previously, nothing is not restricted to inexpensive (non) places, (non)things, (non)people, and (non)services. Some forms of nothing—a Four Seasons hotel room, a Dolce and Gabbana frock, the salesperson at Gucci, and the service of a waiter at a Morton's steakhouse—are very costly, but they still qualify as nothing as that term is used here: relatively empty forms that are centrally conceived and controlled. The consumption of these very expensive forms of nothing is obviously restricted to the uppermost reaches of the economic ladder.

Fifth, the wealthy are drawn to many of the same low-priced forms of nothing that cater to the mass of the population, even those who would be considered poor or very close to it. A credit card knows no income barriers—at least at the high end of the spectrum—and the same is true of ATMs. The wealthy, especially wealthy teenagers, are just as likely to be attracted to fast-food restaurants as are those from virtually every other income group.

There is no simple relationship between wealth and nothingness.

GROBALIZATION VERSUS GLOCALIZATION

Returning to the issue with which we began this discussion, one of the key contributions here is the argument that the/a key dynamic under the broad heading of globalization is the conflict between grobalization and glocalization. This is a very different view than *any* of the conventional perspectives on global conflict. For example, I think a large number of observers have tended to see the defining conflict, where one is seen to exist, as that between globalization and the local. However, the perspective offered here differs from that perspective on several crucial points.

[iii]But not exclusively. There are certainly many forms of something—a homemade soup or stew, a hand-knitted ski cap, homemade ice cream—that are inexpensive; indeed, they are often far less costly than comparable store-bought products.

First, globalization does not represent one side in the central conflict. It is far too broad a concept, encompassing as it does all transnational processes. It needs further refinement to be useful in this context, such as the distinction between grobalization and glocalization. When that differentiation is made, it is clear that the broad process of globalization already encompasses important conflicting processes. Since globalization contains the key poles in the conflict, it therefore is not, and cannot be, one position in that conflict.

Second, the other side of the traditional view of that conflict—the local—is relegated to secondary importance in this conceptualization. That is, to the degree that the local continues to exist, it is seen as increasingly insignificant and a marginal player in the dynamics of globalization. Little of the local remains that has been untouched by the global. Thus, much of what we often think of as the local is, in reality, the glocal. As the grobal increasingly penetrates the local, less and less of the latter will remain free of grobal influences. That which does will be relegated to the peripheries and interstices of the local community. The bulk of that which remains is much better described as glocal than local. In community after community, the real struggle is between the more purely grobal versus the glocal.

One absolutely crucial implication of this is that *it is increasingly difficult to find anything in the world untouched by globalization.* Ironically, then, the hope for those opposed to globalization, especially the grobalization of nothing, seems to lie in an alternative form of globalization—glocalization. This is hardly a stirring hope as far as most opponents of grobalization are concerned, but it is the most realistic and viable one available. The implication is that those who wish to oppose globalization, and specifically grobalization, must support and align themselves with the other major form of globalization—glocalization.

Yet glocalization does represent some measure of hope. For one thing, it is the last outpost of most lingering (if already adulterated by grobalization) forms of the local. That is, important vestiges of the local remain in the glocal. For another, the interaction of the grobal and the local produces unique phenomena that are not reducible to either the grobal or the local. If the local alone is no longer the source that it once was of uniqueness, at least some of the slack has been picked up by the glocal. It is even conceivable that the glocal and the interaction among various glocalities are—or at least can be—a significant source of uniqueness and innovation.

■■ EDWARD SAID (1935–2003): A BIOGRAPHICAL SKETCH

Edward Said was born in Jerusalem in 1935 to a Christian Arab family. At that time, Jerusalem was a Palestinian city under British colonial rule. Said spent his formative years living in both Jerusalem and Cairo until 1948, when the state of Israel was established and his family became refugees. In 1951, Said emigrated to the United States to attend an elite boarding school in Massachusetts while his family stayed behind in the Middle East. He proved to be an exceptional student in his new country, graduating at the top of his class. The school, however, withheld his deserved title as valedictorian or salutatorian (Said does not recall graduating first or second) on the dubious moral grounds of being unfit for the honor (Said 2000:559). Nevertheless, he would go on to receive his bachelor's degree from Princeton University and his master's and doctorate from Harvard, where he studied Western literature, music, and philosophy.

In 1963, Said joined the faculty of Columbia University, where he would remain for the next four decades as professor of English and comparative literature. During this time, he also taught courses at Yale, Johns Hopkins, and Harvard universities, and delivered lectures in Canada, Europe, and the Middle East, among other locations. He received more than half a dozen honorary doctorates from universities in some eight countries. He published more than a dozen books (with translations in thirty-six languages) and wrote countless articles and essays appearing in both scholarly journals and the popular media. Said's writings run the gamut from literary criticism and music criticism (he was a Juilliard-trained classical pianist) to the cultural dynamics of colonialism, the Arab-Israeli conflict, media depictions of Arabs and

Islam, and personal memoirs. Said was a member of the Council on Foreign Relations, the American Academy of Arts and Sciences, and the Royal Society of Literature. He served as president of the Modern Language Association in 1999.

Said is arguably best known for his political views on the "Palestinian question." An ardent supporter of the Palestinians' right of return, he remained an outspoken critic of the Israeli occupation of the West Bank and Gaza Strip and the United States' complicity in the continued degradation of the Palestinian people. He was a regular contributor to newspapers and magazines published in England, France, and the United States and across the Arab world. For more than a decade, he served on the Palestinian National Council, until a falling out with Yasser Arafat and his leadership of the Palestinian Liberation Organization (PLO) led him to resign his position. Such was Said's vocal opposition to Arafat's leadership that the Palestinian Authority banned the sale of his books. Once an official spokesperson for the Palestinian struggle for independence, Said was now considered a traitor to the cause. His disagreement with the PLO came largely in response to Arafat's signing of the Oslo Peace Accord in 1993. For Said, the Accord was fundamentally flawed, because it did not establish the right of Palestinian refugees to return to their homes in Israel and left unchecked the expansion of Israeli settlements into the occupied territories. In his view, a lasting Middle East peace could be secured only through the creation of a single, binational state in which Israelis and Palestinians shared political authority. A coexistence between Israeli Jews and Palestinian Arabs based on a mutual equality was for Said the only "road map" capable of ending the continuing oppression of refugees and the violence that occupation breeds. Said summed up his position in this way:

> I've been consistent in my belief that no military option exists for either side, that only a process of peaceful reconciliation, and justice for what the Palestinians have had to endure by way of dispossession and military occupation, would work. (2000:564)

As for his views on the United States and the media's role in the politics of the Middle East, Said noted all too presciently in 1980,

> So far as the United States seems to be concerned, it is only a slight overstatement to say that Moslems and Arabs are essentially seen as either oil suppliers or potential terrorists. Very little of the detail, the human density, the passion of Arab-Moslem life has entered the awareness of even those people whose profession it is to report the Arab world. What we have instead is a series of crude, essentialized caricatures of the Islamic world presented in such a way as to make that world vulnerable to military aggression. ("Islam Through Western Eyes," The Nation, April 26, 1980)

Over the decades, Said's position earned him his share of enemies on both the political right and left. His office at Columbia University was set on fire, his family was subjected to countless death threats, and in 1985 he was labeled a Nazi by the Jewish Defense League. His more recent criticisms of the ongoing Iraq War also made him a popular target for those in the media and government who support the invasion and subsequent occupation. Nevertheless, Said remained steadfastly committed to his principles and vision of a humane world until his death at the age of sixty-seven in September 2003.

SAID'S INTELLECTUAL INFLUENCES AND CORE IDEAS ▪▪

While Wallerstein and Ritzer emphasize broad, structural aspects of globalization, Said, as a postcolonial theorist, underscores the subjective dimensions of Western expansion into non-Western societies. Developing largely out of English and comparative literature departments, postcolonial studies examines the relationship between the colonizer and the colonized as it is inscribed in language. More specifically, postcolonial theorists analyze how scientists, philosophers, novelists, and political officials from imperialist powers (most notably England, France, and the United States) construct, understand, and subjugate the populations of colonized nations through the written word. In addition, postcolonial theory explores how writers living under colonial rule seek to reconstruct their cultural identity and resist domination through their own appropriation or use of the colonial language (i.e., English or French).

While the term *"post"colonial* implies a specific relationship between the colonizer and the colonized—after colonial rule has ended—it should not be taken literally. Postcolonial scholars explore writings produced prior to and during colonial rule, as well as after a nation has gained its political independence. Indeed, it is often the case that political independence does not bring with it complete autonomy or freedom from foreign intervention but leads, instead, to neocolonial relations aimed at "modernizing" underdeveloped nations. Under these conditions, Western powers install "puppet regimes" in their former colonies to ensure their continued economic (if not political or military) exploitation of the region. For their part, the leaders of these regimes all too often resort to ruthless repression of their own native populations in order to maintain the favor of Western investors. Unfortunately, the list of such dictators is quite long: Saddam Hussein of Iraq, King Fahd in Saudi Arabia, Papa Doc and Baby Doc Duvalier of Haiti, Manuel Noriega of Panama, Rafael Trujillo in the Dominican Republic, Ngo Dinh Diem of South Vietnam, and Reza Muhammed Shah Pahlavi, the former Shah of Iran, to name but a few.

At its root, postcolonial studies explore relations of power that have defined geopolitics from the age of European colonial expansion to the present-day era of declining Western dominance. Although spearheaded by scholars associated with the field of comparative literature, this emphasis has led to a host of questions that sociologists have recently taken up, a sign of the increasing cross-disciplinary encounters between the humanities and the sciences. Perhaps the most compelling question involves how colonizing powers use language to fix the meaning of the colonized **"Other."** In other words, how do colonizing powers use texts—the written word—to simplify complex civilizations, to erase the existence of a rich cultural heritage, and to deny the humanity of a population as part of the effort to legitimate the subjugation of those nations they seek to control? Such efforts demand that a population be torn from its history in order to create a new future—a future literally written by the West. For those seeking to fashion an alternative, "postcolonial" means of expression freed from the dehumanizing effects of the colonial language, a central task involves reclaiming the culture, the history, and the language of marginalized populations in order to speak as and for themselves. Examining the promises and pitfalls tied to this struggle for self-representation is a primary concern for Gayatri Chakravorty Spivak, a leading postcolonial theorist whose work we highlight in the Significant Others box (page 793), and who, as an Indian woman living in the United States, knows firsthand what it means to be the "Other."

The notion of the "Other" raises several interesting issues, two of which we address here. First, colonization is premised on the notion that the colonized Other is inherently inferior, weak, and evil. This understanding of the native population allows for the implementation of brutal and repressive tactics as a means for "civilizing the savages." Yet, such an understanding must first be constructed, and it is colonial discourse that accomplishes this very objective: "to construe the colonized as a population of degenerate types on the basis of racial origin, in order to justify conquest and to establish systems of administration and instruction. . . . The colonized population is then deemed to be both the cause and effect of the system, imprisoned in the circle of interpretation" (Bhabha 1986:154, 171). Thus, people throughout the Arab world, the Caribbean islands, Africa, and elsewhere in the developing world have been defined in Western literary texts and in scientific works as simultaneously irrational, bloodthirsty, infantile, and deceitful and in need of "correct training." And who better to provide this correct training than those in the colonizing West, who, by virtue of not being the inferior and degenerate Other, are superior, strong, righteous, and pure? To the colonizers, then, the colonized are an untamed group who, despite their inferiority, are to be feared because of their potential to corrupt civilized Western culture, hence the need to tame those who are less civilized. Such is the "white man's burden."[11]

---

[11]It is important to note that a similar colonial discourse is often used to subjugate segments of a population within a nation, and not only as a means for justifying the oppression of those living in a foreign country. The history of racism within the United States, as well as its contemporary manifestations, provides ample evidence of this. Black Americans in particular have been subjected to a litany of negative depictions, from violent, oversexed, and conniving to childish, inept, and feebleminded.

Gayatri Chakravorty Spivak (1942–): Can the Subaltern Speak?

Alongside Edward Said, Gayatri Chakravorty Spivak is one of the leading postcolonial scholars. Spivak was born in Calcutta, West Bengal, to an urban middle-class family. After graduating from the University of Calcutta with a degree in English, she moved to the United States to pursue a graduate education in comparative literature at Cornell University. After completing her degree, Spivak achieved notoriety with the publication of *Of Grammatology* (1976), an English translation of *De la Grammatologie* (1967) written by the French poststructuralist theorist Jacques Derrida. This work is most notable for Spivak's contribution to "deconstruction," a branch of philosophy that seeks to deconstruct the hierarchical binary oppositions (such as black and white, male and female, good and evil) on which Western thought is based. In doing so, deconstructionists expose how otherwise unquestioned "truths" are produced. However, the practice of deconstruction necessarily involves adopting the existing system of binary codes in order to articulate a critique of that system. Unable to escape from this hierarchical system, the deconstructionist unavoidably reproduces the prevailing patterns of social domination that are understood and legitimated through it.

Spivak's work in large measure addresses this problematic, a problematic that for her takes on a particularly personal dimension. On one hand, as a woman from a "third world" nation, Spivak inhabits a marginalized social position. Yet, on the other hand, she is a member of the educated elite who enjoys a privileged career as a professor at one of the most prestigious institutions, Columbia University. In what is her most oft-cited piece, "Can the Subaltern Speak?" (1988), Spivak borrows Antonio Gramsci's notion of the "subaltern" to argue that those groups who remain shut out from the dominant social institutions and whose voices do not conform to the assumptions of the dominant discourse are effectively mute. While such groups can indeed speak, they will not be heard by those who occupy positions of privilege. Yet intellectuals committed to dismantling Western imperialism and advocating the independence of colonized nations all too often do a disservice to those oppressed groups on whose behalf they speak. Speaking from a vantage point of privilege, intellectuals cannot help but distort the lived truth of those who, because of their marginalized status, are unable to represent themselves.

In the end, Spivak calls into question the very foundation of postcolonial studies by asking whether intellectuals are necessarily complicit in perpetuating the colonizing forces and institutions they are ostensibly seeking to unmask. Aligned with male-dominated educational institutions (many of which have dubious records with regard to foreign investment practices) located in imperialist countries, Spivak asks if it is possible for Western intellectuals to develop a postcolonial discourse that is not, however inadvertently, colonizing. Speaking *for* the subaltern, whether it is Iraqi women, Indian women, or East Timorese peasants, intellectuals simultaneously construct the subaltern through a discourse that essentializes what is in reality always a diverse, heterogeneous group. Spivak warns that such a discursive practice is a form of colonization, a way of erasing and remaking the identity and culture of a group that is analogous to the economic exploitation and political domination exercised by colonial powers. For those engaged with postcolonial studies, their best hope is not to "give" the subaltern a voice in hopes of protecting or fighting for "them," for this only reinforces their subaltern status. Rather, they should work toward institutionalizing avenues that allow subalterns to speak for themselves.

This relationship between the Other and the West leads to a central paradox in the development of cultural identity. In a very real sense, the West, despite its alleged superiority and strength, is dependent on the inferior, colonized Other who alone is capable of producing its sublime identity. In other words, one cannot be "superior" unless there is an "inferior" against which the necessary comparison can be made. As a result, the source of the West's superiority lies not in its own "advanced" civilization but, rather, in constructing non-Western cultures through negative terms that project onto the colonized all those traits that the West cannot possess if it is to legitimate its position as the center of progress and the beacon of humanity. The purity of the West is thus intertwined with the impurity of the non-West. And such oversimplified, homogenized depictions of both Western and non-Western cultures and peoples must be maintained in order to perpetuate existing relations based on domination and subordination.[12] Thus, when a politician remarks that Muslims "hate us for our freedom," he or she is making a claim dependent on a purified notion of what Western freedom has promoted around the world, in turn justifying an ennobled aggression against those who are "against freedom."

The second set of issues raised by the notion of the Other turns our attention from the colonizers to the colonized. For colonized groups, or those who have won political independence from colonial rule, the question remains as to how they can shed their identity as an inferior Other. Educated in schools established by the colonizers and adopting the language of their rulers, the colonized are effectively cut off from their history and cultural traditions. What the colonized come to learn about themselves, their identity, and their heritage is shaped by the images and meanings that the colonizers project back onto their subject populations. In this context, reclaiming their past and their precolonized identity becomes a near-impossible task. Yet, doing so is a necessary and fundamental step in the struggle for the freedom to "be." In this regard, the colonized share a parallel fate with their oppressors, for they, too, are unable to escape the duality of their identity. Every attempt to shed their "otherness" entails incorporating the very terms that define them as an Other. Thus, even after their independence is won, the colonized are condemned to an identity that is defined at least in part through the vision of their colonizers.

The psychological turmoil this situation exacts was described by Frantz Fanon, whose penetrating work represents an important precursor to what would later become postcolonial studies. As a black colonial subject, Fanon, like Spivak, draws from his personal experiences as an Other to describe the psyche of the subjugated. We outline Fanon's contributions in the Significant Others on page 795.

However, we turn here to an example of the structural consequences of this "othering" by considering the 1994 genocide in Rwanda, in which hundreds of thousands of Tutsis were slaughtered by Hutus with the encouragement and assistance of the Rwandan government. The intense hatred for the Tutsis has its roots in Belgium's colonization of the country. The different ethnicities of these two groups were politicized by the Belgians, who established a caste system in which the Hutus were defined as an inferior, lower-class people and the lighter-skinned Tutsis as an upper-class people worthy of colonial privileges such as an education, good jobs, and access to political positions. While the two groups had coexisted relatively peacefully for centuries prior to the colonization of the region, Hutu resentment and hatred toward the Tutsi mounted under Belgium's colonial rule and the system of discrimination it created. When Rwanda was granted its independence from Belgium in the early 1960s, the stage was set for a massive civil war. The Hutus gained control of the government, and decades of ensuing violence culminated in the atrocities committed during the early months of 1994. The ethnic genocide was the tragic consequence of a racially informed caste system installed by a colonial power whose own definitions of the relative worth of the two groups became the reality through which the groups came to define themselves.

---

[12]The use of the term "non-West" itself points to the oversimplified "othering" that we are discussing: The non-West refers to everything that is "not" West. On a parallel note, in discussing the notion of the "other" and the role of identity construction, postcolonial scholars often draw on psychoanalysis, particularly the work of Sigmund Freud and Jacques Lacan.

Frantz Fanon (1925–1961): The Father of Postcolonial Studies

Frantz Fanon was born in Martinique, a French colony in the Caribbean. After serving in the French army during World War II, he completed his degree in psychiatry at the University of Lyon. In 1953, he was appointed to head the psychiatry department in an Algerian hospital. The following year, however, the war for Algeria's independence from French colonial rule broke out. The horrors and torture associated with the struggle served to amplify Fanon's own experiences as a black, colonized subject growing up in a white, colonial world. Fanon formally left his hospital position (a move that signaled a symbolic renunciation of his French upbringing) in 1956 to join the Algerian liberation movement, and in 1960 the Algerian Provisional Government appointed him ambassador to Ghana. In addition to his diplomatic post, Fanon served the revolution through publishing a number of political essays supportive of its cause.

Although he was a trained psychiatrist, Fanon rose to prominence as a result of his political activism. He was a leading intellectual in the struggle to end colonial rule in Africa and throughout the world. In his now-classic works, *Black Skin, White Masks* (1952) and *The Wretched of the Earth* (1961), Fanon emphasized that colonization has both physical and psychological dimensions. That colonial rule is established on repressive legal and social codes that relegate indigenous populations to an inferior status is plain to anyone who looks. Yet, colonization also exacts a psychological toll on the colonized as the oppressed comes to know himself through the eyes of the oppressor. Thus, the identity of the subjugated is defined through the discourse, ideas, and theories of the subjugator, who alone is able to declare what it means to "be" the colonized *other*. As the colonized subject internalizes the "civilization" of the colonizer, he comes to know himself as half-human, inferior, and despicable—ideas sustained by the oppressor to legitimate the act of colonization.

In an attempt to distance himself from this psychologically destructive sense of being, the black subject adopts a "white mask." However, this only serves to further alienate himself from his own identity and consciousness. Moreover, wearing a white mask does not enable the colonized to escape the daily experience of racism, because while he may see himself as "white"—that is, equal—he is forever seen by the colonizer as both "less than" and dangerous. The only sure path to exterminate this oppressive relationship is "absolute violence," a total revolution that destroys colonial rule and, with it, the categories or identities of black and white. This alone will achieve the necessary "collective catharsis" that will allow subjugated populations to purge themselves of the dehumanizing colonial culture that reduces them to the status of animals. While many find Fanon's insistence on violence to be unnecessary, if not abhorrent, he was nevertheless convinced that militant action was the only means to national and psychological liberation from colonial oppression.

While serving as a diplomat in Ghana, Fanon developed leukemia. He died less than a year later in a Bethesda, Maryland, hospital where he was being treated. He was thirty-six years old. Fanon's insights into the psychology of colonialism would inspire a generation of scholars, including Edward Said. Yet, Fanon's influence extends beyond the boundaries of postcolonial studies. For instance, Immanuel Wallerstein, whose world-systems analysis we discussed earlier, also counts Fanon as an important influence on his intellectual development and his political sympathies with those engaged in anticolonial struggles.

That the Hutus would eventually come to govern Rwanda raises an important point. In Rwanda and elsewhere around the world, colonized and enslaved populations have always fought to resist or subvert the domination of their oppressors. Such resistance can take the form of "hybridization," in which the colonized combine or mix their native language and ways of understanding with those imposed by the colonizers, thus creating sites of potential challenge to the authority of their rulers. In expressing their ideas, interests, and desires in words and forms not sanctioned by the colonial powers, the colonized are able to create a sense of identity that is neither identical to their past nor wholly owned by the colonizers. Hybridization takes place not only through literary works written by authors subjected to colonial rule, but also through cultural practices such as religion, in which native beliefs and rituals are combined with those imposed by colonialist Christian missionaries.

Having explored some of the broad themes that inform postcolonial studies, we now turn to a discussion of the work of Edward Said. It was Said's 1978 publication, *Orientalism,* that marked the birth of the field. In this book, Said details how Western scientists, philosophers, novelists, poets, and politicians created the reality of the "Orient" through their writings. Of course, to understand Said's position we first need to know what he means by the Orient. At its most inclusive, the Orient refers to the non-West. It is the Other to the West (or Occident) that encompasses the Near East (more commonly called the Middle East) and the Far East (Asia). That these regions are known or understood in relation to Europe is obvious enough from their names, "Near" and "Far," which take for granted that Europe serves as the central point from which the identities of other regions are designated. And it is Said's most basic contention that the Orient, "the place of Europe's greatest and richest and oldest colonies, the source of its civilizations and languages," was in an important sense invented by the West through the process of **Orientalism** (Said 1978:1). Moreover, in inventing the Orient, Orientalism also helped Europe to define itself by creating a contrasting, inferior Other against which its own identity could be formed. In this light, Europe and the Orient (as well as the rest of the world) are less products of nature than they are products of social invention, or what Said terms **imaginative geography**.

Orientalism has three dimensions to it. First, it refers to all the scientific and academic disciplines whose purpose is to study Oriental cultures and customs. Thus, the Orientalist, whether he or she is a sociologist, an anthropologist, or a historian, is one whose expertise lies in teaching, researching, or writing about the Orient (ibid.:2). (For example, one might think of a scholar who focuses on the history and culture of Southwest Asia, China, or Egypt.) Such academic interests in the Orient began in earnest during the early nineteenth century as Britain and France entered into a competition for colonial expansion. It was around this time that Orientalism was institutionalized as a field of study within universities across these two countries. In translating Oriental texts, explaining the history of the Orient, and deciphering the culture of the Orient, academics were (are) producing a "manifest" form of Orientalism that, in speaking about the region, interpreted its otherwise obscure, inscrutable meaning. Not coincidentally, many academic experts on the Orient served as advisers to colonial governments, dispensing knowledge on how to best "handle" their subject populations. Today in the United States, specialists in "area studies" provide government officials with policy advice on international affairs through their association with a number of think tanks and research institutes such as the RAND Corporation, the American Enterprise Institute, the Brookings Institution, and the Heritage Foundation. For its part, the Middle East Institute (MEI), a Washington, D.C.–based policy and research center, states that its mission is to "strive to increase knowledge of the Middle East among our own citizens and to promote understanding between the peoples of the Middle East and America." Perhaps recognizing the ill effects that stem from a history of "orientalizing" the region and its people, the Institute's president notes, "Now more than ever, we must replace stereotyped and simplistic notions about the Middle East with detailed, objective understanding."[13] Yet, knowledge of the Orient has always been guided by scholars' stated aims to produce a "detailed, objective understanding." In the end, MEI's critical self-reflection can only reproduce what has always been the essence of Orientalism: The Orient is "Orientalized" by the scholar whose judgments about his subject define the "true" Orient. Those living within the Orient are rendered silent, unable to contest the power and validity of the pronouncements made *about* them by the experts whose studies literally create them.

A second dimension refers to Orientalism in a more general sense as a "style of thought," the "ideological suppositions, images, and fantasies about a region of the world called the Orient" (Said 2000:199).

---

[13]http://www.mideasti.org/

This dimension expresses a "latent" form of Orientalism, an "almost unconscious," and thus taken for granted, understanding of the Orient, its culture, and its people. The unchanging certainty that underlies latent Orientalism is derived from a racial/biological determinism that renders Orientals and their culture as singularly backward, degenerate, uncivilized, and morally corrupt. Oriental men were thus understood to be inferior, weak, and effeminate, while the women were portrayed as exotic, sensual, and willing to be dominated by European males. This dimension of Orientalism incorporates not only academic scholarship but also all accounts of the Orient that are rooted in a fundamental distinction between it and the West. Here we find novels and poems, the letters and diplomacy of colonial administrators, and all manner of philosophical and political theories that in their accounts and explanations of the Orient and its people construct the very subject they describe.

The third dimension speaks of Orientalism as a source of power for "dominating, restructuring, and having authority over the Orient" (Said 1978:3). As a mode of discourse that encompasses a specific vocabulary and set of images, Orientalism spoke the "truth" about the Orient and Orientals. This truth provided the justification for the West's imperialist ambitions. Through the knowledge and descriptions it produced on its subject, Orientalism ruled both a place and its people while simultaneously shaping the identity of the West as a rightly dominant, benevolent ruler. Said sums up the "achievements" of Orientalism as follows:

> [T]he British and the French saw the Orient as a geographical—and cultural, political, demographical, sociological, and historical—entity over whose destiny they believed themselves to have traditional entitlement. The Orient to them was no sudden discovery, no mere historical accident, but an area to the east of Europe whose principal worth was uniformly defined in terms of Europe—European science, scholarship, understanding, and administration—the credit for having made the Orient what it was now. And this had been the achievement . . . of modern Orientalism. (ibid.:221)

As illustrations of the three dimensions of Orientalism and its "achievement," Said examined the discourse of a number of individuals including two distinguished English politicians and diplomats, Arthur James Balfour and Evelyn Baring, First Earl of Cromer. Among his many political posts, Balfour was a longtime member of the British Parliament and served as Chief Secretary to Ireland, Secretary to Scotland, and Prime Minister of England. In 1910, Balfour, then leader of the Conservative (Tory) Party, addressed England's House of Commons in an effort to justify his country's continued occupation of Egypt in the face of weakening domestic support for the campaign and a growing Egyptian nationalist movement. Imploring his fellow parliamentarians to uphold their "duty" to govern backward, subject races, Balfour argued that "it is a good thing" for Egypt to be ruled by Britain because the Egyptians "have got under it far better government than in the whole history of the world they ever had before, and which not only is a benefit to them, but is undoubtedly a benefit to the whole of the civilised West" (quoted in Said 1978:33). Balfour rested his claim not on Britain's superior military or economic power but rather on his country's allegedly superior knowledge of Egypt and its people:

> We know the civilization of Egypt better than we know the civilization of any other country. We know it further back; we know it more intimately; we know more about it. It goes far beyond the petty time span of the history of our race, which is lost in the prehistoric period at a time when the Egyptian civilization had already passed its prime. (quoted in ibid.:32)

For Balfour, what this superior knowledge confirms is that Egypt is incapable of "self-government." Thus, the "civilized world [has] imposed" upon Britain, "the dominant race," the "great task" of ruling over Egypt. And without the unwavering support of British politicians and the broader public, Balfour feared that the Egyptians would "lose all that sense of order which is the very basis of their civilization, just as our officers lose all that sense of power and authority, which is the very basis of everything they can do for the benefit of those among whom they have been sent" (ibid.:33).

Another influential figure who sought to maintain Britain's imperialist ambitions was Evelyn Baring, First Earl of Cromer. Lord Cromer served as a colonial administrator in India and in Egypt, where he held the position of British Controller-General, which gave him control over Egypt's finances, before being appointed in 1883 to British Consul-General, an office that effectively anointed him ruler of Egypt. Lord Cromer spent the next twenty-four years as Consul-General. With decades of experience as a colonial ruler, and with all the accumulated knowledge gained through his position, Cromer, as much as anyone, could speak as an authoritative expert on all things Egyptian. Thus, in 1908, one year after resigning from his position, Lord Cromer published his two-volume work, *Modern Egypt,* in which, based on firsthand experience, he explained the need for continued British occupation:

> The European is a close reasoner; his statements of fact are devoid of any ambiguity; he is a natural logician . . . he is by nature sceptical and requires proof before he can accept the truth of any proposition; his trained intelligence works like a piece of mechanism. The mind of the Oriental, on the other hand . . . is eminently wanting in symmetry. His reasoning is of the most slipshod description. Although the ancient Arabs acquired in a somewhat higher degree the science of dialectics, their descendants are singularly deficient in the logical faculty. They are often incapable of drawing the most obvious conclusions from any simple premises of which they may admit the truth. Endeavor to elicit a plain statement of facts from any ordinary Egyptian. His explanation will generally be lengthy, and wanting in lucidity. He will probably contradict himself half-a-dozen times before he has finished his story. He will often break down under the mildest process of cross-examination. (quoted in ibid.:38)

In short, chief among their faults is "the fact that somehow or other the Oriental generally acts, speaks, and thinks in a manner exactly opposite to the European" (quoted in ibid.:39). Because Orientals are "deficient" in their "slipshod" reasoning and "devoid of energy and initiative," England, compelled to act out of its immense beneficence and superior knowledge, must decide what "is best for the subject race."

In emphasizing the connection between knowledge and power as it relates to imaginative geography, Said's indebtedness to the work of Antonio Gramsci and Michel Foucault is particularly apparent.[14] As we noted earlier, central to Gramsci's perspective is the concept of hegemony. Recall that hegemony refers to a mode of domination in which the rulers secure their dominance not through force, but through the "spontaneous consent" of the ruled. This consent is dependent on the subordinated classes adopting as their own the values, beliefs, and attitudes that serve the interests of the ruling class. As a result, the ideas propagated by the ruling class take on the appearance of universality and common sense. For Said, Orientalism is an instance of hegemony in which scholars, missionaries, colonial administrators, and literary authors offer a picture of the Orient and its people that prescribes for them an inferior position relative to the West. Europe's superiority over the "backward" Orient becomes a taken-for-granted truth that then justifies the colonial exploitation of the Other.

Said's discussion of Orientalism also draws on Foucault's notion of power/knowledge (see Chapter 15). For Foucault, power and knowledge are two sides of the same coin: Power is exercised through knowledge, while knowledge is an exercise of power. Knowledge is constructed and communicated through discourse—words that declare a state of being while simultaneously declaring how things are not. For example, when a person says or writes that Palestinians are terrorists, he or she is constructing an identity that at the same time excludes other possible identities. Yet, Palestinians are no more any one "thing" than are Israelis, Americans, Germans, Mexicans, or Koreans. Nevertheless, such a claim often is offered as "knowledge," and to the extent that it gains credibility, it also becomes infused with the power to produce a reality that does not exist outside of the discourse that constitutes it. In this way, Orientalism is a "world of power and representations, a world that came into being as a series of decisions made by writers, politicians,

---

[14]The range of Said's intellectual influences is vast. In addition to drawing from Gramsci and Foucault, his work is inspired by the Hungarian Marxist Georg Lukács (1885–1971), the German philosopher Friedrich Nietzsche (1844–1900), the Marxian literary critic Raymond Williams (1921–88), the French philosopher Maurice Merleau-Ponty (1908–1961), the novelist Joseph Conrad (1857–1924), the Italian historian and philosopher Giambattista Vico (1668–1744), Theodor Adorno (see Chapter 10), and Frantz Fanon (see Significant Others, page 795).

philosophers to suggest or adumbrate one reality and at the same time efface others" (Said 2000:563). This approach to knowledge calls attention to the politics of Orientalism as an academic pursuit (and purportedly objective, impartial academic scholarship more generally), as the production of knowledge invariably involves "interests" on the part of the writer that reflect his or her life circumstances. It is impossible to entirely escape one's own beliefs, attitudes, or social position when researching and writing on a given subject. And in the case of Orientalism, this means, at its most basic level, that the scholar "comes up against the Orient as a European or American first, as an individual second. . . . [This] means being aware, however dimly, that one belongs to a power with definite interests in the Orient, and more important, that one belongs to a part of the earth with a definite history of involvement in the Orient" (Said 1978:11). In this way the Orient *exists* because of, and for, the West, not because of its own internal reality.

We end this section with a brief discussion of contemporary forms of Orientalism, particularly as they are expressed through American views of Arabs and Islam. The Middle East, home to much of the world's Arab population, has occupied a central position in regard to both government policy and popular culture since the 1950s. Much of the American involvement in, and understanding of, this region has been framed by the conflict between the Arabs and Israelis on which the government, corporate media, and academic establish-

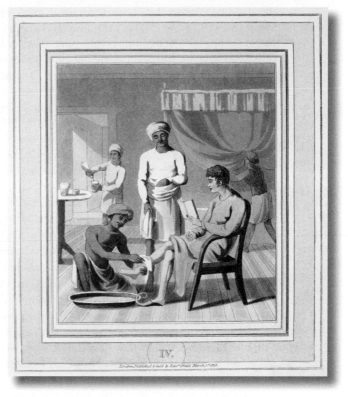

Photo 17.4 Visual orientalism
The inferiority of Orientals was depicted in pictures as well as in words. Above is an 1813 print entitled "A Gentleman Dressing, Attended by His Head Bearer and Other Servants." Drawn by Charles D'Oyly, it was published in *The European in India*. Such illustrations were commonplace throughout the nineteenth century.

ment have adopted a pro-Israeli position—a position that casts Arabs as "evil, totalitarian, and terroristic." The attention focused on the Middle East and its people has only intensified in the aftermath of the September 11 terrorist attacks. Moreover, this intensity is itself in large measure a product of the immediacy with which information and images are disseminated through the electronic media. Well before the attacks on America, however, the media portrayed Arabs in simplistic, stereotypical terms. In television, films, and cartoons, the Arab was a nomadic camel-jockey, an oversexed savage, a treacherous if clever marauder, or an oil sheik who, despite his obvious inferiority, is able to hold the West hostage by controlling the world's energy supply. As for the latter, because Arabs are believed to lack the intelligence and moral qualifications required for possessing such a valuable resource, it is only fitting and right that their oil fields be seized—if necessary, through American military force. This view has taken on renewed significance given the critics of George W. Bush's administration, who claimed that the principal motive for invading Iraq in 2003 was to control that nation's oil supply. Likewise, Said's comments regarding the depiction of Arabs in the news ring as true today as they did thirty years ago, when he wrote,

In newsreels or newsphotos, the Arab is always shown in large numbers. No individuality, no personal characteristics or experiences. Most of the pictures represent mass rage and misery, or irrational (hence hopelessly eccentric) gestures. Lurking behind all of these images is the menace of jihad. Consequence: a fear that the Muslims (or Arabs) will take over the world. (1978:287)

It is important to point out, however, that tension between Islam and the West is by no means confined to the United States. Amidst rising concerns over the preservation of their national identities, Swiss voters in 2009 supported a referendum to ban the building of minarets on mosques, while the French government in 2004 banned the wearing of head scarves in public schools. More recently, a French parliamentary panel recommended banning the wearing of Muslim burqas and niqabs (both are full-face coverings) in certain public facilities, including hospitals, post offices, and banks and on public transportation, and suggested that lawmakers should pass a resolution condemning the garments. Yet, according to the country's interior ministry, in a Muslim population estimated to be between 5 and 6 million—the largest such population in western Europe—fewer than two thousand women wear the burqa (Erlanger, 2010).

⁘ SAID'S THEORETICAL ORIENTATION

Said's work is primarily devoted to exploring a single concept, Orientalism; yet, this is by no means a simple theoretical concept. On the contrary, "Orientalism" is a multidimensional concept that speaks to each of the quadrants in our action/framework, as shown in Figure 17.3. Because of its theoretical multidimensionality and flexibility, this concept can be used to shed light on a range of real-life issues. Consider first Orientalism's individualist/rational dimension. This dimension speaks to the work of specific scholars and colonial administrators who construct the meaning of the Orient in specific ways that serve their interests. This is the manifest form of Orientalism that calls our attention to the interconnection between knowledge and power and, in this case, how knowledge is used by individuals in their attempt to subjugate populations. For the Western Orientalist, the Orient serves "as a kind of culture and intellectual proletariat useful for [his] grander interpretative activity, necessary for his performance as a superior judge . . ." (ibid.:208).

Figure 17.3   Said's Basic Concepts and Theoretical Orientation

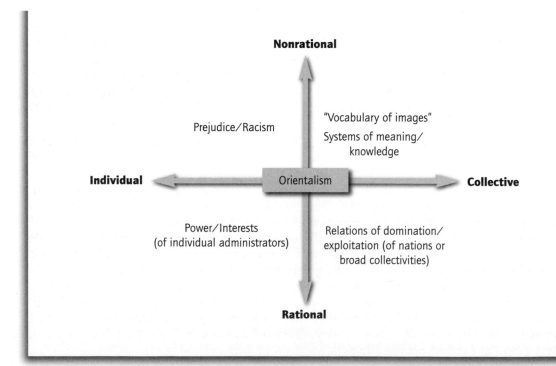

The collectivist/rational aspects of Orientalism become apparent when we move from a focus on individuals to one on groups. Thus, we find Said continually speaking of broad collectivities—East and West, or Orient and Occident—as well as nations. On this level, the interactions that occupy Said's analysis are those taking place between, for instance, Britain and India or between France and Egypt. The emphasis here is on the relations of domination and exploitation that exist between nations and regions of the globe, relations that are strategically sustained by those in power in order to secure their continued geopolitical and economic advantages.

Orientalism operates on a nonrational level as well. This is suggested by Said's notion of latent Orientalism, discussed earlier. As a concept that captures the individual/nonrational dimension, Orientalism refers to an individual's racist attitudes and practices that construct an inferior Other. Thus, individuals use or "mobilize" the dogma of Orientalism in an "almost unconscious" way as they go about the business of describing the world. What an Orientalist (or any person, for that matter) writes and says can betray an underlying racist scheme that positions the Oriental as inherently, and thus undeniably, uncivilized.

Turning to the collectivist/nonrational dimension, Orientalism is that which is mobilized. It is a set of background, taken-for-granted assumptions projected onto the Orient and its people—in short, a hegemonic dogma. Such assumptions are never tested against the reality of the empirical world; they simply exist as an unconscious lens through which the Orient is seen and understood. Orientalism provides a vocabulary of images that "represent or stand for a very large entity, otherwise impossibly diffuse, . . . [that] comes to exert a three-way force, on the Orient, on the Orientalist, and on the Western 'consumer' of Orientalism" (ibid.:66, 67). In this way, Orientalism possesses structural or collectivist properties that give it a life of its own, separate from those individuals who deploy it.

Reading

Introduction to *Orientalism*

In this article, Said begins by defining Orientalism as a field of academic study aimed at discovering and understanding all things "Oriental"—that is, Middle Eastern. Since the early fourteenth century, scholars within the field have produced knowledge covering a vast geographical territory and range of issues including the languages, histories, religions, laws, and economic and political life of the Orient. However, the discourse and images used by academics to understand the Orient have led to the creation of lands and peoples entirely foreign and distinct—though "known"—to the West and Westerners. In other words, knowledge of the Orient and Orientals has produced an "other" in opposition to which we are able to know and understand "us." Through the writings of the Orientalists, the Orient represents the savage and the depraved. So identified, not only is the Orient understood, but in juxtaposition the noble and civilized West is also created at the same time. Yet, this is not solely an academic issue; it is an issue about truth. The Orient is "Orientalized" by the scholar whose judgments about his subject define the "true" Orient. Those living within the Orient are rendered silent, unable to contest the power and validity of the pronouncements made *about* them by the experts whose studies literally create them.

While the Orient refers most generally to the eastern half of the world, Said centers his analysis on the imaginative geography practiced by Orientalists in their attempt to uncover the truth about Islam. From the seventh century until today, Islam has been a geographic, cultural, military, and political provocation to the Christian West. For their part, Orientalists, in an effort to "handle" these provocations and fix the meaning of Islam for the West, have employed a vocabulary that depicts the Islamic world as "militant" and "heretical." Islam is exposed to be a "fraudulent" religion and its prophet Mohammed is an "imposter" and "apostate" who represents the epitome of debauchery and lecherous evil. For Said, it is essential to recognize, however, that such representations are to the actual Orient, or Islam, . . . as stylized costumes are to characters in a play. . . . In other words, we need not look for correspondence between the language used to depict the Orient and the Orient itself, not so much because the language is inaccurate as because it is not even trying to be accurate. What it is trying to do . . . is at one and the same time characterize the Orient as alien and incorporate it schematically on a theatrical stage whose audience, manager, and actors are *for* Europe, and only for Europe. (1978:184–85, emphasis in the original)

In the second portion of this selection, Said turns his attention to Napoleon's invasion of Egypt in 1798. The invasion marked the beginning of the modern era of Orientalism in which the academic expertise offered by Orientalists was an integral part of the larger project to control the Muslim world, less so by military means than by scientific texts. Through knowledge and accessibility, the Orient would be incorporated into the West, transforming it from an obscure and hostile place and people into a submissive "partner." Although Said's analysis here pertains to events that occurred more than two hundred years ago, we suggest that, regardless of one's political persuasion, it should provide the reader with reason for pause to consider the parallels between the period he discusses and the project of democracy building currently under way in Iraq, in Afghanistan, and throughout the Arab Middle East.

Orientalism (1978)[i]

Edward Said

I. IMAGINATIVE GEOGRAPHY AND ITS REPRESENTATIONS: ORIENTALIZING THE ORIENTAL

Strictly speaking, Orientalism is a field of learned study. In the Christian West Orientalism is considered to have begun its formal existence with the decision of the Church Council of Vienne in 1312 to establish a series of chairs, as R. W. Southern notes, in "Arabic, Greek, Hebrew and Syraic at Paris, Oxford, Bologna, Avignon, and Salamanca." Yet any account of Orientalism would have to consider not only the professional Orientalist and his work, but also the very notion of a *field* of study based on a geographical, cultural, linguistic and ethnic unit called the Orient. Fields, of course, are made. They acquire coherence and integrity

in time because scholars devote themselves in different ways to what seems to be a commonly agreed-upon subject matter. Yet it goes without saying that a field of study is rarely as simply defined as even its most committed partisans—usually scholars, professors, experts, and the like—claim it is. Besides, a field can change so entirely, in even the most traditional disciplines like philology, history or theology, as to make an all-purpose definition of subject matter almost impossible. This is certainly true of Orientalism for some interesting reasons.

To speak of scholarly specialization as a geographical "field" is, in the case of Orientalism, fairly revealing, since no one is likely to imagine a field symmetrical to it called Occidentalism. Already the special, perhaps even eccentric, attitude of Orientalism becomes apparent. For

[i]Whereas "Orientalism" usually refers to the study of the whole Orient (including the civilizations of China, Japan, India and the Muslims), it is used in this essay mainly to refer to the Near Orient, that is, the lands of Islam, or the Arabs, or both. Until the eighteenth century the "Orient" was considered in Europe to be Islam, or Turkey, or the lands of the Saracens. After the discovery of large new portions of Asia during the second half of the eighteenth century, the "Orient" expanded accordingly, but in order to retain the coherence of the traditional idea of the Orient, "Orientalism" is treated here as Western attention to the Near East, an attention that includes academic study, imaginative literature, commerce, and attempts at geo-political domination.

although many learned disciplines imply a position taken towards, say, *human* material (a historian deals with the human past from a special vantage point in the present), there is no real analogy for taking a fixed, more or less total geographical position towards a wide variety of social, linguistic, political, and historical realities. A classicist, a Romance specialist, even an Americanist focuses on a relatively modest portion of the world, not on a full half of it. But Orientalism is a field with considerable geographical ambition. And since Orientalists have traditionally occupied themselves with things Oriental (a specialist in Islamic law, no less than experts in Chinese dialects and Indian religions, is considered to be an Orientalist by people who call themselves Orientalists), we must learn to accept enormous, indiscriminate size plus an almost infinite capacity for subdivision as one of the chief things about Orientalism—one of the chief things about its confusing amalgam of imperial vagueness and precise detail.

All of this describes Orientalism as an academic discipline. The "ism" in Orientalism serves to insist on the distinction of this discipline from every other kind. The rule in its historical development as an academic discipline has been its increasing scope, not its greater selectiveness....

Such eclecticism had its blind spots nevertheless. Academic Orientalists for the most part were interested in the classical periods of whatever language or society it was that they studied. Not until quite late in the [nineteenth] century, with the single major exception of Napoleon's Institut d'Egypte, was there much attention given to the academic study of the modern, or actual, Orient. Moreover, the "Orient" studied was a textual universe by and large; the impact of the Orient was made through books and manuscripts, not, as in the impress of Greece on the Renaissance, through plastic artifacts like sculpture and pottery. Even the rapport between an Orientalist and the Orient was textual, so much so that it is reported of some of the early nineteenth-century German Orientalists that their first view of an eight-armed Indian statue cured them completely of their Orientalist taste. When a learned Orientalist traveled in the country of his specialization it was always with abstract unshakeable maxims about the "civilization" which he had studied; rarely were Orientalists interested in anything except proving the validity of these musty "truths" by applying them without great success to uncomprehending, hence degenerate, natives. Finally, the very power and scope of Orientalism produced not only a fair amount of exact positive knowledge about the

Orient, but also a kind of second-order knowledge—lurking in such places as the "Oriental" tale, the mythology of the mysterious East, notions of Asian inscrutability—with a life of its own, what V. G. Kiernan has aptly called "Europe's collective daydream of the Orient." . . .

Today an Orientalist is less likely to call himself an Orientalist than he was at almost any time up to the Second World War. Yet the designation is still useful as when universities maintain programs or departments in Oriental languages or Oriental civilizations. There is an Oriental "faculty" at Oxford and a Department of Oriental Studies at Princeton. As recently as 1959, the British Government empowered a commission "to review developments in the Universities in the fields of Oriental, Slavonic, East European and African studies . . . and to consider, and advise on, proposals for future development." The Hayter report, as it was called when it appeared in 1961, seemed untroubled by the broad designation of the word "Oriental" which it found serviceably employed in American universities as well. For even the greatest name in modern Anglo-American Islamic studies, H. A. R. Gibb, preferred to call himself an Orientalist rather than an Arabist. Gibb himself, classicist that he was, could use the ugly neologism "area study" for Orientalism as a way of showing that area studies and Orientalism after all were interchangeable geographical titles. But this, I think, ingenuously belies a much more interesting relationship between knowledge and geography than is really the case. I should like to consider that relationship briefly.

Despite the distraction of a great many vague desires, impulses, and images, the mind seems persistently to formulate what Lévi-Strauss has called a logic of the concrete. A primitive tribe, for example, assigns a definite place, function and significance to every leafy species in its immediate environment; many of these herbs and flowers have no practical use; but the point Lévi-Strauss makes is that mind requires order, and order is achieved by discriminating and taking note of everything, placing everything of which the mind is aware in a secure, re-findable place, therefore giving things some role to play in the economy of objects and identities that make up an environment. This kind of rudimentary classification has a logic to it, but the rules of the logic by which a green fern in one society is a symbol of grace and in another is considered maleficent are neither predictably rational nor universal. There is always a measure of the purely arbitrary in the way the distinctions between things are seen. And with these distinctions go values whose history, if one could unearth it completely, would probably show the same measure of arbitrariness. This is evident enough in the case of fashion. Why do wigs, lace collars and high buckled shoes appear, then disappear, over a period of decades? Part of the answer has to do with utility, and part with the inherent beauty of the fashion. But if we agree that all things in history, like history itself, are made by men, then we will appreciate how possible it is for objects, or places, or times, to be assigned roles and given meanings that acquire objective validity only *after* the assignments are made. This is especially true of relatively uncommon things, like foreigners, mutants, or "abnormal" behavior.

Obviously, some distinctive objects are made by the mind, and these objects, while appearing to exist objectively, have only a fictional reality. A group of people living on a few acres of land will set up boundaries between their land and its immediate surroundings on the one hand, and on the other, a land beyond theirs which they call "the land of the barbarians." In other words, this universal practice of designating in one's mind a familiar space which is "ours" and an unfamiliar space beyond "ours," which is "theirs," is a way of making geographical distinctions that *can be* entirely arbitrary. I use the word arbitrary here because imaginative geography of the "our land/barbarian land" variety does not require that the barbarians acknowledge the distinction. It is enough for "us" to set up these boundaries in our own minds; "they" become "they" accordingly, and both their territory and their mentality is designated as different from "ours." To a certain extent modern and primitive societies seem thus to derive a sense of their identities negatively. A fifth-century Athenian was very likely to feel himself in a negative sense to be a non-barbarian as much as in a positive sense he felt himself to be Athenian. The geographic boundaries I have been discussing accompany the social, ethnic, and cultural ones in expected ways. Yet very often the sense in which someone feels himself to be not-foreign is based on a very unrigorous idea of what is "out there," beyond one's own territory. All kinds of suppositions, associations, fictions appear to crowd the unfamiliar and strange space outside one's place. . . .

Almost from earliest times in Europe the Orient was something more than what was empirically known about it. At least until the early eighteenth century, as R. W. Southern has so elegantly shown, European understanding of one kind of Oriental culture, the Islamic, was ignorant, but complex. For certain associations—not quite ignorant, not quite informed—always seem to have gathered around the notion of an Orient. . . .

Consider how the Orient, and in particular the near Orient, which is the main focus of this essay, became known in the West, since antiquity, as its great complementary opposite. There were the Bible and the rise of Christianity; there were travellers like Marco Polo who charted the trade routes and patterned a regulated system of commercial exchange, and after him Ludovico and Pietro della Valle; there were fabulists like Mandeville; there were the redoubtable conquering Eastern movements, principally Islam of course; there were the militant pilgrims, chiefly the Crusades. Altogether an internally structured archive is built up from the literature that belongs to these experiences. Out of this comes a restricted number of typical encapsulations: the journey, the history, the fable, the stereotype, the polemical confrontation. These are the lenses through which the Orient is experienced, and they shape the language, perception, and form of the encounter between East and West. What gives the immense number of encounters some unity, however, is the vacillation I was speaking about earlier. Something patently foreign and distant acquires, for one reason or another, a status more rather than less familiar. One tends to stop judging things either as completely novel or as completely well-known; a new median category emerges, a category that allows one to see new things, things seen for the first time, as versions of a previously known thing. In essence such a category is not only a way of receiving new information, it is also a method for controlling what seems to be a threat to some established view of things. If, for example, the mind must suddenly deal with what it takes to be a radically new form of life—as Islam appeared to Europe in the early Middle Ages—the response on the whole is conservative and defensive. Islam is judged to be a fraudulent new version of some previous experience, in this case Christianity. The threat is muted, familiar values impose themselves, and in the end the mind reduces the pressure upon it by accommodating things to itself as either "original" or "repetitious." Islam thereafter is "handled." Its novelty and its suggestiveness are brought under control so that relatively nuanced discriminations, that would have been impossible to make had the raw novelty of Islam been left unattended, are now made. The Orient at large, therefore, vacillates between the West's familiarity of contempt and its shivers of novel delight (or fear).

There is nothing especially controversial or reprehensible about such domestications of the exotic; they take place between all cultures certainly, and between all men. My point, however, is to emphasize the truth that the Orientalist, as much as any one in the European West who thought about or experienced the Orient, performed the kind of mental operation I have been discussing. But what is more important still is the limited vocabulary and imagery which imposes itself as a consequence. The reception of Islam in the West is a perfect case in point, and has been admirably studied by Norman Daniel. One constraint acting upon Christian thinkers who tried to understand Islam was an analogical one; since Christ is the basis of Christian faith, it was assumed—quite incorrectly—that Mohammed was to Islam as Christ was to Christianity. Hence, among other results, the polemic name *Mohammedanism* given to Islam, and the automatic epithet "imposter" applied to Mohammed. Out of such and many other misconceptions, as Daniel remarks, "there formed a circle which was never broken by imaginative exteriorisation. . . . The Christian concept of Islam was integral and self-sufficient." Islam became an *image*—the word is Daniel's but it seems to have remarkable implications for Orientalism in general—whose function was not so much to represent Islam as it was in itself, as to represent Islam *for* the medieval Christian.

> The invariable tendency to neglect what the Qur'an meant, or what Muslims thought it meant, or what Muslims thought or did in any given circumstances, necessarily implies that Qur'anic and other Islamic doctrine was presented in a form that would convince Christians; and more and more extravagant forms would stand a chance of acceptance as the distance of the writers and public from the Islamic border increased. It was with very great reluctance that what Muslims said Muslims believed was accepted as what they did believe. There was a Christian picture in which the details (even under the pressure of facts) were abandoned as little as possible, and in which the general outline was never abandoned. There were shades of difference, but only with a common framework. All the corrections that were made in the interests of an increasing accuracy were only a defence of what had newly been realised to be vulnerable, a shoring up of a weakened structure. Christian opinion was an erection which could not be demolished, even to be rebuilt.

This rigorous Christian picture of Islam was intensified in innumerable ways, including—during the Middle Ages and early Renaissance—a large variety of poetry, learned controversy, and popular superstition. By this time the Near Orient had been all but included in the common world-picture of Latin Christianity. . . . By the middle of the fifteenth century, as R. W. Southern has brilliantly shown, it became apparent to serious European thinkers "that something would have to be done about Islam," which had turned the situation around somewhat by itself having militarily arrived in eastern Europe. Southern recounts a dramatic episode between 1450 and 1460 when four learned men, John of Segovia, Nicholas of Cusa, Jean Germain, and Aeneas Silvius (Pius II), attempted between them to deal with Islam through *contraferentia,* or "conference." The idea was John of Segovia's: it was to have been a staged conference with Islam with Christians attempting the wholesale conversion of Muslims. "He saw the conference as an instrument with a political as well as a strictly religious function, and in words which will strike a chord in modern breasts he exclaimed that even if it were to last ten years it would be less expensive and less damaging than war." There was no agreement between the four men, but the episode is crucial for having been a fairly sophisticated attempt—part of a general European attempt from Bede to Luther—to put a representative Orient in front of Europe, to *stage* the Orient and Europe together in some coherent way, the idea being for Christians to make it clear to Muslims that Islam was just a misguided version of Christianity. Southern concludes,

> Most conspicuous to us is the inability of any of these systems of thought [European Christian] to provide a fully satisfying explanation of the phenomenon they had set out to explain [Islam]—still less to influence the course of practical events in a decisive way. At a practical level, events never turned out either so well or so ill as the most intelligent observers predicted; and it is perhaps worth noticing that they never turned out better than when the best judges confidently expected a happy ending. Was there any progress [in Christian knowledge of Islam]? I must express my conviction that there was. Even if the solution of the problem remained obstinately hidden from sight, the statement of the problem became more complex, more rational, and more related to experience. . . . The scholars who labored at the problem of Islam in the Middle Ages failed to find the solution they sought and desired; but they developed habits of mind and powers of comprehension which, in other men and in other fields, may yet deserve success.

The best part of Southern's analysis, here and elsewhere in his brief history of Western views of Islam, is how he takes it as not surprising that it is finally

Western ignorance that becomes more refined and complex, not some body of positive Western knowledge which increases in size and accuracy. For fictions have their own logic, and their own dialectic of growth or decline. Thus, onto the character of Mohammed in the Middle Ages was heaped a number of attributes, as Daniel has shown, that corresponded to the "character of the [twelfth century] prophets of the 'Free Spirit' who did actually arise in Europe, and claim credence and collect followers." Similarly, since Mohammed was viewed as the disseminator of a false Revelation, he became as well the epitome of lechery, debauchery, sodomy, and a whole battery of assorted treacheries, all of which derived "logically" from his doctrinal impostures. Thus the Orient acquired representatives, so to speak, and representations—each one more concrete, more internally congruent with some Western exigency, than the ones that preceded it. It is as if, having once settled on the Orient as a locale suitable for incarnating the infinite in a finite shape, Europe could not stop the practice; the Orient and the Oriental, Arab, Islamic, Indian, Chinese or whatever, become repetitious pseudo-incarnations of some great Original (Christ, Europe, the West) that they were supposed to be imitating. Only the source of these rather narcissistic Western ideas about the Orient changed in time, not their character. Thus it was commonly believed in the twelfth and thirteenth centuries that Arabia was, as Daniel says, "on the fringe of the Christian world, a natural asylum for heretical outlaws" and that Mohammed was a cunning apostate, whereas in the twentieth century an Orientalist scholar, an erudite man, has pointed out how Islam is really no more than second-order Arian heresy.[ii]

Our initial description of Orientalism as a learned field now acquires a new concreteness. A field is often an enclosed space. The idea of representation is a theatrical one: the Orient is the stage on which the whole East is confined. On this stage will appear figures whose role it is to represent the larger whole from which they emanate. The Orient then seems to be not an unlimited extension beyond the familiar European world, but rather a closed field, a theatrical stage affixed to Europe. An Orientalist is but the particular specialist of a knowledge for which Europe at large is responsible, in the way that an audience is historically and culturally responsible for (and responsive to) dramas technically put together by the dramatist (the

Orientalist). In the depths of this Oriental stage stands a prodigious cultural repertoire: the Sphinx, Cleopatra, Eden, Troy, Sodom and Gomorrah, Astarte, Isis and Osiris, Sheba, Babylon, the Geni, the Magi, Nineveh, Prester John, Mahomet, and dozens more—settings, in some cases names only, half-imagined, half-known, monsters, devils, heroes, terrors, pleasures, desires. The European imagination was nourished extensively from this repertoire: between the Middle Ages and the eighteenth century such major authors as Ariosto, Chaucer, Milton, Marlowe, Tasso, Shakespeare, Cervantes, the authors of the *Chanson de Roland,* and the *Poema del Cid* drew on the Orient's riches for their productions, in ways that sharpened the outlines of imagery, ideas, and figures populating it. In addition a great deal of what was considered learned Orientalist scholarship in Europe pressed ideological myths into service, even as knowledge seemed genuinely to be advancing. . . .

The Orientalist stage, as I have been calling it, becomes a system of moral and epistemological rigor. As a discipline representing institutionalized Western knowledge of the Orient, Orientalism thus comes to exert a three-way force, on the Orient, on the Orientalist, and on the Western "consumer" of Orientalism. It would be wrong, I think, to underestimate the strength of the three-way relationship thus established. For the Orient ("out there" towards the East) is corrected, even penalized, for lying outside the boundaries of European society, "our" world; the Orient is thus *Orientalized,* a process that not only marks the Orient as the province of the Orientalist, but also forces the uninitiated Western reader to accept Orientalist codifications . . . as the *true* Orient. Truth, in short, becomes a function of learned judgment, not of the material itself which in time seems to owe even its existence to the Orientalist. . . .

This whole didactic process is neither difficult to understand nor difficult to explain. One ought to remember that all cultures impose corrections upon raw reality, changing it from free-floating objects into units of knowledge. The problem is not that conversion takes place. It is perfectly natural for the human mind to resist the assault on it of untreated strangeness, so that cultures have always been inclined to impose complete transformations on other cultures, receiving them not as they are but as, for the benefit of the receiver, they

---

[ii]D. B. MacDonald, "Whither Islam," Muslim World, XXIII (January 1933), 2.

ought to be. To the Westerner, however, the Oriental was always *like* some aspect of the West. . . . Yet the Orientalist makes it his work to be always converting the Orient from something into something else: he does this for himself, for the sake of his culture, in some cases for what he believes is the sake of the Oriental. As I have been suggesting, this process of conversion is a disciplined one—it is taught, it has its own societies, periodicals, traditions, vocabulary, rhetoric, all in basic ways connected to and supplied by the prevailing cultural and political norms of the West. And, as I shall demonstrate, it tends to become more, rather than less, total in what it tries to do, so much so that as one surveys Orientalism in the nineteenth and twentieth centuries the over-riding impression is of Orientalism's insensitive schematization of the entire Orient. . . .

Imaginative geography, . . . legitimates a vocabulary, a universe of representative discourse peculiar to the discussion and understanding of Islam and of the Orient. What this discourse considers to be a fact, that Mohammed is an imposter, for example, is a component of the discourse, a statement which the discourse compels one to make whenever the name Mohammed occurs. Underlying all the different units of Orientalist discourse—by which I mean simply the vocabulary employed whenever the Orient is spoken or written about—is a set of representative figures, or tropes. These figures are to the actual Orient, or Islam which is my main concern here, as stylized costumes are to characters in a play; they are like, for example, the cross that Everyman will carry, or the Harlequin costume worn by Pierrot in a *comedia dell'arte* play, and so forth. In other words we need not look for correspondence between the language used to depict the Orient and the Orient itself, not so much because the language is inaccurate as because it is not even trying to be accurate. What it is trying to do . . . is at one and the same time characterize the Orient as alien and incorporate it schematically on a theatrical stage whose audience, manager, and actors are *for* Europe, and only for Europe. Hence the vacillation between the familiar and the alien: Mohammed is always the imposter (familiar: because he pretends to be like the Jesus we know) and always the Oriental (alien: because although he is in some ways "like" Jesus he is after all not like him).

Rather than listing all the figures of speech associated with the Orient—its strangeness, its exotic sensuousness, etc.—we can generalize about them as they were handed down through the Renaissance. They are all declarative and self-evident; the tense they employ is the timeless eternal; they convey an impression of repetition and strength; they are always symmetrical to, and yet opposed and inferior to, a European equivalent, which is sometimes specified, sometimes not. For all these functions it is frequently enough to use the simple copula "is." Thus, Mohammed *is* an imposter, the very phrase canonized in d'Herbelot's *Bibliothéque,* and dramatized in a sense by Dante. No background need be given; the evidence necessary to convict Mohammed is contained in the "is." One does not qualify the phrase, neither does it seem necessary to say that Mohammed *was* an imposter, nor need one consider for a moment that it may not be necessary to repeat the statement. It *is* repeated, he *is* an imposter, and each time one says it, he becomes more of an imposter and the author of the statement gains a little more authority in having declared it. Thus, Humphrey Prideaux's famous seventeenth-century biography of Mohammed is entitled *The True Nature of Imposture.* Finally, such categories as imposter (or Oriental for that matter) imply, indeed, require an opposite, which is neither fraudulently something else nor endlessly in need of explicit identification. And that opposite is "Occidental" or in Mohammed's case, Jesus.

Philosophically, then, the kind of language, thought, and vision that I have been calling Orientalism very generally is a form of radical realism; anyone employing Orientalism, which is the habit for dealing with questions, objects, qualities and regions deemed Oriental, will designate, point to, fix what he is talking or thinking about, with a word or phrase, which then is considered either to have acquired, or more simply to be, reality. Rhetorically speaking, Orientalism is absolutely anatomical and enumerative: to use its vocabulary is to engage in the particularizing and dividing of things Oriental into manageable parts. Psychologically, Orientalism is a form of paranoia, knowledge of another kind, say, from ordinary historical knowledge. These are a few of the results, I think, of imaginative geography and of the dramatic boundaries it draws. There are some specifically modern transmutations of these Orientalized results, however, to which I must now turn.

II. PROJECTS

It is necessary to examine the more flamboyant operational successes of Orientalism if only to judge how totally opposite to the truth was the grandly menacing idea expressed by Michelet, that "the Orient advances, invincible, fatal to the gods of light by the charm of its dreams, by the magic of its *chiaroscuro*." Cultural, material, and intellectual relations between Europe and the Orient have gone through innumerable phases, even though the line between East and West has made a certain constant impression upon Europe. Yet in general it was the West that moved upon the East, not vice versa. Orientalism is the generic term that I have been employing to describe the Western approach to the Orient; Orientalism is the discipline by which the Orient was (and is) approached systematically, as a topic of learning, discovery, and practice. But in addition I have been using the word "Orientalism" to designate that collection of dreams, images, and vocabularies available to anyone who has tried to talk about what lies East of the dividing line. These two aspects of Orientalism are not incongruent, since by use of them Europe could advance securely and unmetaphorically upon the Orient. Here I should like principally to consider material evidence of this advance.

Islam excepted, the Orient for Europe was until the nineteenth century a domain with a continuous history of unchallenged Western dominance. This is patently true of the British experience in India, the Portuguese experience in the East Indies, China, and Japan, and the French and Italian experiences in miscellaneous regions of the Orient. There were occasional instances of native intransigence to disturb the idyll, as when in 1638–39 a group of Japanese Christians expelled the Portuguese. By and large, however, only the Arab and Islamic Orient presented Europe with an unresolved challenge on the political, intellectual, and, for a time, economic levels. For much of its history, then, Orientalism carries within it the stamp of a problematic European attitude towards Islam, and it is this acutely sensitive aspect of Orientalism around which my interest in this study turns.

Doubtless Islam was a real provocation in many ways. It lay uneasily close to Christianity, geographically and culturally. It drew on the Judeo-Hellenic traditions, it borrowed creatively from Christianity, it could boast of unrivalled military and political successes. Nor was this all. The Islamic sacred lands are adjacent to and even overlap the Biblical lands. Moreover, the heart of the Islamic domain has always been the region of the Orient closest to Europe. Arabic and Hebrew are Semitic languages and together they dispose and re-dispose of material that is vitally important to Christianity. From the end of the seventh century until the battle of Lepanto in 1571 Islam, in either its Arab, Ottoman or North African and Spanish forms, dominated or effectively threatened European Christianity.

That Islam out-stripped Rome cannot have been overlooked by any European past or present, and even Gibbon was no exception, as is evident in the following passage from the *Decline and Fall*:

> In the victorious days of the Roman republic it had been the aim of the senate to confine their councils and legions to a single war, and completely to suppress a first enemy before they provoked the hostilities of a second. These timid maxims of policy were disdained by the magnanimity or enthusiasm of the Arabian caliphs. With the same vigour and success they invaded the successors of Augustus and Artaxerxes; and the rival monarchies at the same instant became the prey of an enemy whom they had so long been accustomed to despise. In the ten years of the administration of Omar, the Saracens reduced to his obedience thirty-six thousand cities or castles, destroyed four thousand churches or temples of the unbelievers, and edified fourteen hundred moschs for the exercise of the religion of Mohammed. One hundred years after his flight from Mecca the arms and reign of his successors extended from India to the Atlantic Ocean, over the various and distant provinces. . . .

When the term Orient was not simply a synonym for the Asiatic East taken as a whole, and taken as generally denoting the distant and exotic, it was most rigorously understood as applying to the Islamic Orient. This "militant" Orient came to stand for what Henri Baudet has called "the Asiatic tidal wave." Certainly this is the case in Europe through the middle of the eighteenth century, the point at which repositories of "Oriental" knowledge like d'Herbelot's *Bibliothèque Orientale* stop meaning primarily Islam, the Arabs, or the Ottomans. Until that time cultural memory gave understandable prominence to such relatively distant events as the fall of Constantinople, the Crusades, and the conquest of Sicily and Spain, but if these signified the menacing Orient they did not at the same time efface what remained of Asia. . . .

Access to Indian (Oriental) riches had always to be made by first crossing the Islamic provinces and by withstanding the dangerous effect of Islam as a system of quasi-Arian belief. And at least for the largest part of the eighteenth century, Britain and France were successful. The Ottoman Empire had long since settled into a (for Europe) comfortable senescence, to be inscribed in the nineteenth century as the "Eastern Question." Britain and France fought each other in India between 1744 and 1748 and again between 1756 and 1763 until, in 1769, the British emerged in practical economic and political control of the sub-continent. What was more inevitable than that Napoleon should choose to harass Britain's Oriental empire by first intercepting its Islamic throughway, Egypt? . . .

[F]or Napoleon, Egypt was a project that acquired reality in his mind, and later in his preparations for its conquest, by experiences that belong to the realm of ideas and myths culled from texts, not from empirical reality. His plans for Egypt therefore became the first in a long series of European encounters with the Orient in which the Orientalist's special expertise was put directly to functional colonial use, for at the crucial instant when an Orientalist had to decide whether his loyalties and sympathies lay either with the Orient or with the conquering West he always chose the latter, from Napoleon's time on. As for the Emperor himself, he only saw the Orient as it had been encoded first by classical texts and then by Orientalist experts whose vision based on classical texts seemed a useful substitute for any actual encounter with the real Orient.

Napoleon's . . . idea was to build a sort of living archive for the Expedition, in the form of studies conducted on all topics by the members of the Institut d'Egypte, which he founded. What is perhaps less well known is Napoleon's prior reliance upon the work of the Comte de Volney, a French traveller whose *Voyage en Egypte et en Syrie* appeared in two volumes in 1787. . . . Volney evidently considered himself as a scientist, whose job it was always to record the "état" of something he saw. The climax of the *Voyage* occurs in the second volume, the account of the religious aspect of Islam. Volney's views about Islam were canonically hostile to it as a religion and as a system of political institutions. Nevertheless, Napoleon found this and Volney's *Considérations sur la guerre actuel de Turcs* (1788) of particular importance. For Volney, after all, was a canny Frenchman, and—like Chateaubriand and Lamartine a quarter century after him—eyed the Near Orient as a likely place for the realization of French colonial ambition. What Napoleon profited from in Volney was the enumeration, in ascending order of difficulty, of the obstacles to be faced in the Orient by any French expeditionary force.

Napoleon refers explicitly to Volney in his reflections on the Egyptian expedition, the *Campagnes d'Egypte et de Syrie 1798–1799: Memoires pour servir à l'histoire de Napoleon,* which were dictated by Bonaparte to General Bertrand at Saint-Helena. Volney, he said, considered that there were three barriers to

French hegemony in the Orient and that any French force would therefore have to fight three wars. One against England, a second against the Ottoman Porte, and the most difficult against the Muslims. Volney's assessment was both shrewd and hard to fault since it was clear to Napoleon, as it would be to anyone who read Volney, that the *Voyage* and the *Considérations* were in effect texts to be used by any European wishing to win in the Orient. Volney's work comprised a handbook for attenuating the human shock a European might feel as he directly experienced the Orient: read the books, seems to have been Volney's thesis, and far from being disoriented by the Orient, you will subject it to you.

Napoleon took Volney almost literally, but in a characteristically subtle way. From the first moment that the Armée d'Egypte appeared on the Egyptian horizon, every effort was made to convince the Muslims that "nous sommes les vrais musulmans," as Bonaparte's Proclamation of July 2, 1798, put it to the people of Alexandria. Equipped with a team of Orientalists (and on board a flagship called *Orient*), Napoleon used Egyptian enmity to the Mamelukes and appeals to the revolutionary idea of equal opportunity for all in order to wage a uniquely benign and selective war against Islam. What impressed the first Arab chronicler of the Expedition, 'Abdal-rahman al-Jabarti, more than anything else was Napoleon's use of scholars to manage his contacts with the natives, that and the impact of watching a modern European intellectual establishment at close quarters. Instead of seeming to fight Islam, Napoleon tried everywhere to prove that he was fighting *for* Islam. Everything he said was translated into Koranic Arabic, and the French Army was urged by its command to remember the Islamic sensibility. When it became obvious to Napoleon that his force was too small to impose itself on the Egyptians, he then tried to make the local Imams, Cadis, Muftis and Ulema interpret the Koran in favor of the Grande Armée. To this end, then, the sixty ulemas who taught at the Azhar were invited to Napoleon's quarters, given full military honors, and then allowed to be flattered by Napoleon's admiration for Islam and Mohammed, and by his obvious veneration for the Koran, with which he seemed perfectly familiar. This worked, and soon the population of Cairo seemed to lose its distrust of the occupiers.[iii] Napoleon later gave his deputy Kléber strict instructions after he left always to administer Egypt through the Orientalists and the Islamic religious leaders whom

---

[iii]Yet Napoleon was not just being cynical. It is reported of him that he discussed Voltaire's *Mahomet* with Goethe and defended Islam.

they could win over. Any other politics was too expensive and foolish. . . .

Such a triumph could only have been prepared *before* a military expedition, perhaps only by someone who had no prior experience of the Orient except what books and scholars provided. The idea of taking along a full-scale academy is very much an aspect of this textual attitude to the Orient. And this attitude in turn was bolstered by specific Revolutionary decrees (particularly the one of 10 Germinal An III, March 30, 1793, establishing an *École publique* in the Bibliothèque Nationale to teach Arabic, Turkish, and Persian), whose object was the rationalist goal of dispelling mystery and institutionalizing even the most recondite knowledge: Thus, many of Napoleon's Orientalist translators were students of Sylvestre de Sacy who, beginning in June of 1796, was the first and only teacher of Arabic at the *École publique des langues orientales.* Later, in fact, Sacy became the teacher of nearly every major Orientalist in Europe, where his students dominated the field for about three quarters of a century. Many of them were politically useful, in the way that several were to Napoleon in Egypt.

But dealings with the Muslims were only a part of Napoleon's project to dominate Egypt. The other part was to render it completely open, to make it totally accessible to European scrutiny. From being a land of obscurity and a part of the Orient hitherto known at second hand through the exploits of earlier travelers, scholars, and conquerors, Egypt was to become a department of French learning. Here too the textual and schematic attitudes are evident. The Institut, with its teams of chemists, historians, biologists, archeologists, surgeons, antiquarians, was the learned division of the Army. Its job was no less aggressive—to put Egypt into modern French; and, unlike the Abbé le Mascrier's 1735 *Déscription de l'Egypte,* Napoleon's descriptive translation was to be a universal undertaking. Almost from the first moments of the occupation Napoleon saw to it that the Institut began its meetings, its experiments, its fact-finding mission, as we would call it today. Most important, everything said, seen, studied, was to be recorded in that great collective appropriation of one country by another, the *Déscription de l'Egypte,* published in 23 enormous volumes between 1809 and 1828.

The *Déscription*'s uniqueness is not only its size and the intelligence of its contributors but its attitude to its subject matter, and it is this attitude that makes it of great interest for the study of modern Orientalist projects. . . .

Because Egypt is saturated with meaning for the arts, sciences, and government, it is to be the stage on which actions of world-historical importance will take place.

By taking Egypt, then, a modern power would naturally demonstrate its strength and justify history; Egypt's own destiny is to be annexed, to Europe preferably. In addition, this power would enter a history stocked with figures no less great than Homer, Alexander, Caesar, Plato, Solon, and Pythagoras, who graced the Orient with their presence. The Orient, in short, existed as a set of values attached not to its modern realities but to a series of valorized contacts it had had with a distant European past. This is an example of the textual, schematic attitude I have been referring to. . . .

To restore a region from its present barbarism to its former classical greatness; to instruct (for its own benefit) the Orient in the ways of the modern West; to subordinate or underplay military intervention in order to aggrandize the project of acquiring priceless knowledge in the process of political domination of the Orient; to formulate the Orient, to give it shape, identity, definition with full recognition of its place in memory, its importance to imperial strategy, and its "natural" role as an appendage of Europe; to dignify all the knowledge collected during colonial occupation with the title "contribution to modern learning" when the natives had neither been consulted nor treated as anything except pre-texts for a text, whose usefulness is not to the natives; to feel oneself, as a European, in command, almost at will, of Oriental history, time, and geography; to institute new areas of specialization; to establish new disciplines; to divide, deploy, schematize, tabulate, index, and record everything in sight (and out of sight); to make out of every observable detail a generalization and out of every generalization an immutable law about the Oriental nature, temperament, mentality, custom, or type; and, above all, to transmute living reality into the stuff of texts, to possess (or think one possesses) actuality mainly because as a European nothing in the Orient seems to resist one's powers: these are the features of Orientalist projection fully realized in the *Déscription de l'Egypte,* itself enabled and reinforced by Napoleon's wholly Orientalist engulfment of Egypt by the instruments of Western knowledge and power. . . .

The *Déscription* thereby displaces Egyptian or Oriental history as a record possessing its own coherence, identity, and sense. Instead, history as recorded in the *Déscription* supplants Egyptian or Oriental history by identifying itself directly and immediately with world history, a euphemism for European history. To save an event from oblivion is in the Orientalist's mind the equivalent of turning the Orient into a theater for his representations of the Orient. Moreover, the sheer power of having described the Orient in modern Occidental terms lifts the Orient from the realm of silent obscurity where it has lain neglected (except for the inchoate murmurings of a vast but undefined sense of its own past) into the clarity of modern European science. . . .

Yet the military failure of Napoleon's occupation of Egypt did not destroy the fertility of its over-all projections for Egypt or the rest of the Orient. Quite literally, the occupation gave birth to the entire modern experience of the Orient as interpreted from within the universe of discourse founded by Napoleon in Egypt, whose agencies of domination and dissemination included the Institut and the *Déscription*. The idea, as it has been characterized by Charles-Roux, was that Egypt, "restored to prosperity, regenerated by wise and enlightened administration . . . would shed its civilizing rays upon all its Oriental neighbors." True, the other European powers would seek to compete in this mission, none more so than England. But what would also happen as a continuing legacy of the common Occidental mission to the Orient—despite intra-European squabbling, indecent competition, and outright war—would be the creation of new projects, new visions, new enterprises combining additional parts of the old Orient with the conquering European spirit. After Napoleon the very language of Orientalism changes radically. Its descriptive realism is upgraded and becomes not merely a style of representation, but a language, indeed a means of *creation*. . . . [T]he Orient is reconstructed, reassembled, indeed, *born* out of the Orientalists' efforts. The *Déscription* became the archetype of all further efforts to bring the Orient closer to Europe, thereafter to absorb it entirely and cancel, or at least subdue and reduce, its strangeness and, in the case of Islam, its hostility. For where the Islamic Orient was concerned, Islam would henceforth appear as a category denoting the Orientalists' power and not the Islamic people as humans nor their history as history. . . .

Discussion Questions

1. Wallerstein argues that the capitalist world-economy is based on a division between core, semiperipheral, and peripheral regions. He also notes that any given nation can move from one tier to another. What factors contribute to a nation's advancement or decline? What nations do you think are advancing/declining and why? What, if anything, do you think is happening to the position of the United States?

2. What are some of the positive benefits that globalization has brought about? What are some of the negative consequences? How has the division of the world into core, semiperipheral, and peripheral regions shaped both the positive and negative effects of globalization? What role do "antisystemic" movements organized around class and nationalism play in shaping the development of globalization?

3. Does the fact that the same television shows can be viewed and the same products bought in numerous countries really mean that the world is becoming "one"? What types of data or evidence would you need to conclude that economic and cultural globalization is fundamentally changing the way people see themselves, interact with others, and interpret the world around them?

4. According to Ritzer, what are some of the leading factors accounting for "grobalization"? Are we moving inevitably, with unstoppable force, into a "grobalized" world order? Why or why not? What factors or conditions are sources of resistance to grobalization? What role does glocalization play in shaping contemporary society? Cite specific examples to illustrate your view.

5. In his discussion of "Orientalism," Said examines how academic knowledge and discourse, as well as popular vocabularies and images, legitimize colonial rule and the subjugation of the colonized. While he focuses on events occurring roughly a century ago, do you think similar processes of Orientalism are taking place today? Do magazines, books, television shows, and movies "orientalize" non-Westerners, or do they typically offer unbiased portrayals? While journalists may claim to be objective in their reporting, does the news orientalize non-Western populations? Provide examples to support your view.

GLOSSARY AND TERMINOLOGY

Accounting Practices (Garfinkel): A process during which individuals attempt to order and make sense of their everyday world by constructing and attributing motives from the past to the present.

Action (question of): What are the factors that motivate individuals or groups to act?

Agency: An individual's capability to act or to intervene in their world.

Alienation (Marx): Dehumanizing consequence of workers estrangement or separation from the means of production, the product of their labor, their own species-being, and humanity as a whole.

Americanization (Ritzer): The propagation of American ideas, customs, social patterns, industry, and capital around the world.

Anomie (Durkheim): Moral disorder; lack of moral regulation.

Archaeology (Foucault): A historical methodology within sociology whereby discursive practices are unearthed like the artifacts of past civilizations.

Backstage (Goffman): The performance normally unobserved by and restricted from members of the audience.

Beneficiaries (Coleman): Individuals who hold the right to control the actions of others and who benefit from the actions prescribed or proscribed by norms.

Bifurcation of Consciousness (Smith): A split or separation between the world as one actually experiences it and the dominant view to which one must adapt (the masculine point of view).

Black Feminist Thought (Collins): A school of thought that is located in the everyday experiences and literary traditions of ordinary black women.

Blasé Attitude (Simmel): an intellectualized approach to life, void of an emotional investment in "differences," that protects the individual from becoming overwhelmed by the intensity of city life.

Bourgeoisie (Marx): Owners of the means of production.

Bracketing (Schutz): Systematic process of narrowing perception and cognition by sidestepping existing assumptions about the external world.

Breaching Experiments (Garfinkel): Experiments designed to study everyday interactions through disrupting normal procedures in order to expose them.

Calling (Weber): The Protestant doctrine asserting that each individual has a "life-task" to be performed as a glorification of God.

Capital (Marx): The raw materials and machinery used in the production of commodities.

Character (Goffman): The self understood as a product of interaction; a dramatic effect or image fabricated in concert with others during an encounter.

Charisma (Weber): The "gift of grace" demonstrated through miracles, acts of heroism, or the pronouncement of divine revelations.

Charismatic Authority (Weber): Dominance based on disciples' trust in and devotion to their leader and their duty to recognize the genuineness of their leader's charisma.

Civil Society: Open spaces of debate relatively free from government control.

Class (Marx): Groups of individuals who share a common position in relation to the means of production.

Class (Weber): People who share life chances or possibilities that are determined by economic interests in the possession of goods and opportunities for income within the commodity and labor markets.

Class Consciousness (Marx): An awareness on the part of the members of a class of their common relationship to the means of production.

Collective/Collectivistic Approach to Order: A theoretical orientation that focuses on the power that the overarching structures of society have on individuals and groups in decision making.

Collective Conscience (Durkheim): Totality of beliefs and sentiments common to average citizens of the same society that forms a determinate system which has a life of its own.

Colonization of the Lifeworld (Habermas): A process in which steering media (money and power) and technical/instrumental logic come to replace the negotiation of shared meanings as the foundation for social integration and the reproduction of the lifeworld.

Color Line (Du Bois): Hierarchically organized attitudes, systems on meaning, and social structures revolving around skin color or "race."

Commodification of Feeling (Hochschild): When one's natural capacity to engage in emotion work is sold for a wage and bought to serve a profit motive. As a result, one's feelings become engineered to further corporate and organizational interests.

Commodity (Marx): Product that is bought and sold.

Commodity Fetishism (Marx): Treating goods as if they have magical powers capable of transforming an individual's capabilities.

Communicative Action (Habermas): The process generated by the evolutionary rationalization of the lifeworld in which individuals come to mutual understanding and consensus through open, noncoercive debate or discussion free from the corrosive effects of money, power, and manipulation. It embodies a critical stance through which the cultural stock of knowledge is transmitted, individuals are integrated into the community, and the formation of personal and social identities is secured.

Computerization of Society (Lyotard): The process whereby computers have come to play a major role in the development, manipulation, storage, and rationalization of knowledge.

Conjoint Norms (Coleman): Norms that apply to the same individuals as both the beneficiaries and the targets.

Content (Simmel): The drives, purposes, interests, or inclinations that lead individuals to interact with each other.

Conversation Analysis: A type of analysis that studies not only the words in conversations, but also things such as silences, cutoffs, and laughter.

Core (Wallerstein): A region of the capitalist world-economy that first emerged in northwestern Europe and now includes the United States, Germany, Japan, and other similarly industrialized nation-states. Together, this "area" controls a vast majority of the world's wealth while producing a highly skilled workforce that is controlled through wage payment.

Costs (Homans): Unavoidable punishments that are experienced even though a behavior also elicits rewards. Costs often have the effect of reducing the frequency of an otherwise rewarding behavior.

Counter-Enlightenment: A conservative reaction to the Enlightenment in the late eighteenth century that was wary of the unabashed embrace of rationality, technology, and progress, and that focused on the importance of nonrational factors (such as tradition) rather than reason.

Cultural Capital (Bourdieu): Nonmaterial goods such as education credentials, types of knowledge and expertise, verbal skills, and aesthetic preferences that can be converted into economic capital or that can otherwise used to one's advantage.

Cultural Hegemony: The legitimization of a set of ideas that reinforces existing relations of domination and subordination.

Cultural System (Parsons): A system made up of the values, norms, and symbols which guide our choices, as well as limit the interaction among persons.

Culture Industry (Adorno): Sectors involved in the creation and distribution of mass culture products: television, film, radio, music, magazines, newspapers, books, and the advertisements that sell them and that churn out a never-ending supply of mass-produced, standardized commodities that aborts and silences criticism.

Deference (Goffman): Honor, dignity, and respect afforded to others through individuals' performances.

Definition of the Situation (Goffman): Preliminary to any self-determined act or behavior, there is always a phase of examination and deliberation, where an individual enters a situation and seeks out information about others.

Demeanor (Goffman): Conduct and dress displayed by individuals during their performances.

Deviance (Merton): Modes of action that do not conform to the dominant norms or values in a social group or society.

Disciplinary Society (Foucault): A type of society that relies on possible surveillance to normalize human activity and create a self-induced discipline.

Discipline (Foucault): Self-regulation.

Discursive consciousness (Giddens): What individuals are explicitly able to say about social conditions and the conditions of their own actions.

Disjoint Norms (Coleman): Norms that reflect a conflict of interest as some of the individuals are beneficiaries while others are targets.

Distributive Justice (Homans): Striking a fair balance between the rewards and costs participants experience in interaction.

Double Consciousness (Du Bois): A split or separation of the self from being both "black" and "American."

Duality (Simmel): Interdependent relationship between the individual and society. The fusing of opposites into a coherent whole.

Duality of Structure (Giddens): Social structures as both the "medium and outcome" of the practices they organize.

Dysfunction (Merton): The unintended consequences that occur because of a disconnect between the social and the cultural realms.

Economic Capital (Bourdieu): The material resources such as land, wealth, or money that one possesses and that can be used to advance one's position.

Elementary Social Behavior (Homans): Face-to-face contact between individuals in which the reward or punishment each gets from the behavior of the other is relatively immediate and direct.

Emotion Work (Hochschild): Efforts to alter or manage the intensity or type of feelings that one is experiencing.

Emotional Labor (Hochschild): When one's deep acting (self-induced effort to produce a real feeling) or emotion work is sold for a wage.

Enlightenment: A period of remarkable intellectual development that occurred in Europe during the late seventeenth and early eighteenth century that emphasized reason.

Epistemes (Foucault): A framework of knowledge that shapes discourse; the collection of linguistic tools, rules, descriptions, and habits of logic that make possible specific understandings of the world.

Epistemology: Philosophic question that concerns how we know what we know, how we decide what is valid knowledge.

Exploitation (Marx): The result of laborers producing more value in the form of commodities than they receive in the form of wages.

Externalization (Berger and Luckmann): The process by which human activity and society attain the character of objectivity (also called "objectivation").

Extrinsic Rewards (Blau): Rewards that are "detachable" from the association in which they are acquired; that is, they provide the means for obtaining other benefits (e.g. money).

Favors (Blau): A type of exchange in which one's expression of gratitude serves as a social reward for others.

Feeling Rules (Hochschild): Shared social conventions that determine what we should feel in a given situation, how intensely we should feel it, and how long we should feel.

Field (Bourdieu): A relatively autonomous arena within which actors and institutions mobilize their capital in an effort to capture the distribution of capital that is specific to it.

Forces (or Means) of Production (Marx): The raw materials, technology, and land necessary for production.

Forms (Simmel): The various "shapes" of interaction through which individuals pursue their interests and satisfy their desires.

Free Rider (Coleman): An individual's rational decision not to participate in group activity if the "joint goods" it produces cannot be denied to the individual and if the joint goods supply is not reduced by others consuming it.

Front (Goffman): Composed of the "setting" and the "personal front," it provides for the intentionally or unintentionally given performance by an individual for an audience.

Game Stage (Mead): A more advanced stage of cognitive development marked by an ability to assume the roles of multiple others simultaneously, and to control one's actions on the basis of abstract "rules of the game."

Genealogy (Foucault): A method of sociohistorical analysis of the impact of power on discourse.

Generalized Other (Mead): The organized community or group to which an individual belongs.

Grobalization (Ritzer): Imperialistic ambitions of nations, corporations, organizations and other entities and their desire—indeed, their need—to impose themselves on various geographic areas.

Habitualization (Berger and Luckmann): The process by which the flexibility of human actions is limited, as repeated actions become routinized.

Habitus (Bourdieu): A mental filter that structure's an individual's perceptions, experiences, and practices such that the world takes on a taken-for-granted, commonsense appearance.

Hegemonic Masculinity (Connell): The pattern of practices that allows men's dominance over women to continue, including the gendered character or bureaucracies, workplaces, and educational institutions, including classroom dynamics and patterns of bullying, as well as the media, for instance, the interplay of sports and war imagery as well as the virtual monopoly of men in certain forms of crime.

Heterosexual Matrix (Butler): Normative system that classifies "proper" men or women as heterosexual.

Hyperreality (Baudrillard): A "reality" that has been already reproduced; a separation of signifiers from the signified.

"I" (Mead): The here-and-now, creative aspect of one's self

Ideal Types (Weber): analytical constructs against which real-life cases can be compared.

Imaginative Geography (Said): Fictional realities constructed by the universal practice of designating in one's mind a familiar space which is "ours" and an unfamiliar space beyond "ours," which is "theirs." Such geographical boundaries form the basis of cultural, social, and ethnic identities.

Imbalanced Exchange (Blau): Exchange based on an inequality in resources that in turn leads to relationships marked by power and submission.

Impression Management (Goffman): The verbal and nonverbal practices we employ in an attempt to present an acceptable image of our self to others

Indexicality (Garfinkel): Words, expressions, and practical actions have different meanings in differing contexts.

Individual/Individualistic Approach to Order: A theoretical orientation that focuses on the autonomy of individual actors in making decisions and their ability to affect the social order.

Individualistic Rationality (Marcuse): An oppositional or critical attitude that allows for negating the status quo and hence critically understanding one's world; to freely develop personal objectives and achieve them through rational methods (similar to Horkheimer's concept of objective reason).

Industrial Revolution: The revolution that occurred in England in the eighteenth century that applied power-driven machinery to manufacturing.

Institutional Ethnography (Smith): A method of elucidating and examining the relationship between everyday activities and experiences and larger institutional imperatives.

Institutionalization (Berger and Luckmann): The process by which there is a reciprocal typification of a routinized or habitualized action by types of actors.

Institutionalization (Parsons): The long-standing processes of communal association that bind actors to particular meanings, found in the form of institutions.

Internalization (Berger and Luckmann): The process by which the individual subjectivity is attained.

Internalization (Parsons): The process by which the individual personality system incorporates a specific interpretation of cultural symbols into need-dispositions.

Interpretation (Blumer): A behavioral-thinking process that entails constructing the meaning of another's action, as well as one's own.

Intersubjectivity (Schutz): Shared stocks of knowledge or a shared consciousness among individuals.

Intrinsic Rewards (Blau): Rewards that are pleasurable in and of themselves, not because they provide the means for obtaining other benefits.

Iron Cage (Weber): Dominance of material acquisition and impersonal, bureaucratic forms of organization resulting in the decay of the human spirit and the disenchantment of the world.

Joint Action (Blumer): Group action constituted by the fitting together of the lines of behavior of the separate participants.

Labor Theory of Value (Marx): Theory stating that the value of a commodity is determined by the amount of labor time it takes to produce it.

Latent Function (Merton): The implicit or unintended purpose of an action.

Legitimacy (Weber): Possessing the ability to "rightfully" exercise domination over others. Synonymous with *authority*.

Lifeworld (Schutz): The world of existing assumptions as they are experienced and made meaningful in consciousness

Manifest Function (Merton): The overt or intended purpose of an action.

Matrix of Domination (Collins): The notion that one's position in society is made of multiple contiguous standpoints rather than a single essentialist standpoint.

McDonaldization (Ritzer): A process of growing standardization by which the principles of the fast-food restaurant are coming to dominate more and more sectors of American society as well as the rest of the world.

"Me" (Mead): The phase of the self of which one is relatively conscious; more or less stable sense of who we are that is created, sustained, and modified through our interaction with others.

Meaning (Mead): A threefold relationship between (1) an individual's gesture, (2) the adjustive response by another to that gesture, and (3) the completion of the social act initiated by the gesture of the first individual.

Metanarratives: Grand narratives that are paradigmatic systems of knowledge that contain established and credible worldviews and describe a total picture of reality.

Merchantilization of Knowledge (Lyotard): The process by which knowledge becomes commercialized and is subjected to the whims of the marketplace, to be bought and sold.

Merchants of Morality (Goffman): Individuals' attempts to "engineer" interaction "profits" through properly enacted performances that live up to the standards of morality by which their performances will be judged.

Mind (Mead): a process that allows for the conscious control of one's actions. An internal conversation of gestures that makes possible the imagined testing of alternative lines of conduct.

Mortification of the Self (Goffman): The process of stripping an individual of his or her multiple selves prior to one's entrance into a total institution to be replaced by one totalizing identity under which the individual exercises little if any control.

Negate: To develop a critique of "objective truths" that enables resistance to the domination of the established social order.

Nonrational Motivation for Action: Action that is motivated primarily by values, norms, morals, traditions, the creation of meaning, and/or emotional states.

Norms (Coleman): Socially defined informal rights to control the actions of others.

Norms (Parsons): Written or unwritten rules that govern specific situations.

"Nothing" (Ritzer): A social form that is generally centrally conceived, controlled, and comparatively devoid of distinctive content.

Objective Reason (Horkheimer): A type of reasoning that considers the relative value of the goals of action and thus provides a basis for determining what is just or right.

Objectivation (Berger and Luckmann): The process by which human activity and society attain the character of objectivity (also called "externalization").

Object Relations Theory (Chodorow): Theory that emphasizes that the psychological life of the individual is created in and through relations with other human beings. Object relations theory replaces Freud's emphasis on pleasure-seeking with an emphasis on "relationship-seeking," contending that humans have an innate drive to form and maintain relationships.

Ontological security (Giddens): Relatively unconscious feelings regarding the continuity of one's self identity and the broader social and physical world. A sense of security rooted in a trust that people and the surrounding world are as they appear.

Order (question of): What accounts for the patterns or predictability of behavior that lead us to experience social life as routine?

Orientialism (Said): (1) Scientific and academic disciplines whose purpose is to study Oriental cultures and customs. (2) The ideological suppositions, images, and fantasies about a region of the world called the Orient. (3) A source of power for dominating, restructuring, and having authority over the Orient.

"Other" (Said): The construction of inferior peoples on which one's own superior identity is predicated.

Panopticon (Foucault): An efficient and rational form of prison architecture and surveillance designed by British philosopher Jeremy Bentham (1748–1832) and analyzed by Michel Foucault. The Panopticon features a central guard tower surrounded by a circular gallery of cells; the inward faces of each cell are left unobstructed so that they are completely visible from a single vantage point in the central tower.

Party (Weber): Group aimed at influencing a communal action no matter what it content may be.

Patriarchal Dividend (Connell): The material (e.g., money) as well as nonmaterial (e.g., honor, prestige) dividends that though not *uniformly* distributed among men are nevertheless universally distributed among them under patriarchy.

Pattern Variables (Parsons): A dichotomous set of five choices that describe alternatives of action, applicable both at the individual and collective levels.

Performativity (Butler): The sustained continual nature of gender performance.

Performer (Goffman): the self understood as a derived from an individual's unique psychobiology; the core of one's personality that lies behind and manages his performances.

Periphery (Wallerstein): A region of the capitalist world-economy exploited for its raw materials which are then exported to the core. Often "possessions" or colonies of core states, the periphery is made up of weak nations lacking the economic, political, and military clout necessary for pressing its own agenda. The workforce in the periphery historically has been controlled through coercion and slavery.

Personality System (Parsons): A system of action organized by organic and emotional need dispositions, at the level of the individual.

Play Stage (Mead): An early stage of cognitive development marked by an ability to assume the attitude of only one particular individual at a time.

Postmodernism: An historical moment (after modernity) and a particular type of theorizing (or anti-theorizing) that has emerged over the same course of time, in which time–space configurations are completely changed and a skepticism about the methods, goals, and ideals of modern society, as well as the promises of modern society, is expressed.

Poststructuralism: A theoretical trend that developed in the 1960s converging on themes such as the fragmentation of meaning, the de-centering of the subject, and the decline of the idea of the Truth.

Power (Blau): A type of inequality in which one individual/group is able to supply rewards to others who are unable to offer benefits in return.

Power (Weber): the chance of a man or a number of men to realize their own will in a social action even against the resistance of others.

Practical consciousness (Giddens): The tacit awareness of how and why we do what we do. What individuals know about social conditions, including the conditions of their own actions, but cannot express in words.

Primary Socialization (Berger and Luckmann): The first socialization that an individual undergoes in childhood.

Profane (Durkheim): That which is of the everyday world; mundane, routine.

Proletariat (Marx): Those who sell their labor for a wage.

Protestant Ethic (Weber): Methodical, rational attitude towards worldly affairs that characterized Protestant asceticism.

Pseudo-Individualization (Adorno): Standardized, mass-produced cultural products endowed with a deceptive halo of free choice.

Public Sphere (Habermas): A network for communicating information and opinions where, ideally, private individuals can publicly congregate and freely debate social, ethical, and political issues in a manner free from the distorting influences of money and power.

Queer Theory: A school of theory which emerged from gay and lesbian studies and considers all sexual behaviors and identities to be social constructs.

Rational Motivation for Action: Action that relies on the "rational" or strategic choices of actors, where individuals and group actions are motivated primarily by the attempt to maximize benefits and reduce costs.

Rationalization (Weber): An ongoing process in which social interaction and institutions become increasingly governed by methodical procedures and calculable rules.

Rational-Legal Authority (Weber): Domination based on the rule of impersonal laws. Obedience is owed to the office or position in which authority is vested.

Recipes (Schutz): Innate knowledge from experiences or shared stocks of knowledge that are implicit instructions for everyday life.

Reciprocity (Blau): The mutual paying of costs and receiving of benefits on which all social interaction is based.

Reification (Berger and Luckmann): The extreme step in the process of externalization (also called objectivation) in which human activity and society attain the character of objectivity to such an extent that the real relationship between humans and their world is reversed in consciousness.

Relations of Production (Marx): The social relations that exist within and between classes and between classes and the means of production

Relations of Ruling (Smith): Seemingly "impersonal" and "neutral" ruling apparatus that conceals a male subtext. Relations of ruling include such forms as bureaucracy, administration, management, professional organization, and media as well as scientific, technical, and cultural discourses and texts that facilitate social control.

Repressive Desublimation (Marcuse): False freedoms offered by the managed, controlled liberalization of sexual behaviors and mores that increase the satisfaction with the existing social order.

Resources (Giddens): The capacity or capability to make something happen. Bases for exercising agency.

Ritual (Durkheim): Highly routinized act, such as taking communion

Role (Parsons): Complementary detailed sets of obligations for an interaction.

Role Set (Parsons): The combination of complimentary interdependent social relationships that people are involved in because they occupy a particular social status.

Rules (Giddens): Generalizable techniques or structures applied in the reproduction or enactment of social practices. Commonsense social recipes that allow us to make sense of the words and conduct of others as well as to form our own words and behavior.

Sacred (Durkheim): That which is extraordinary, set apart, above and beyond, the everyday world.

Sanctions (Coleman): Rewards and punishments applied by the beneficiaries of a norm to induce compliance.

Secondary Adjustments (Goffman): Ways in which individuals attempt to stand apart from the roles and the self that are imposed on them by the institutions to which they are attached.

Secondary Socialization (Berger and Luckmann): Processes of socialization that continue throughout an individual's life and induct an already socialized individual into new sectors of society.

Selective Incentives (Coleman): Benefits that are distributed exclusively to those who bear the costs for providing a collective good.

Self (Mead): Self-conscious thinking during which an individual sees himself or herself as an object by assuming the attitudes of others toward himself.

Semiperiphery (Wallerstein): A region of the capitalist world-economy that occupies a position between the core and the periphery—economically and politically weaker than the former, but stronger than the latter. Semiperipheral areas play an essential role in maintaining the political stability of the modern world-system by deflecting some of the protests and political pressures that groups located in the periphery would otherwise direct toward the core states and the economic interests that are aligned with them.

Sign: A sign is that which is socially designated to represent the meaning of objects and experiences.

Significant gestures (Blumer): Words and actions whose meanings are shared by all of those involved in a social act

Significant Symbols (Mead): Words and gestures that have the same meaning for all of those involved in a social act.

Signified: The idea, experience, or object behind a sign.

Signifier: The representation designated to stand for the signified.

Simulacra (Baudrillard): Copies of objects for which there are no true originals.

Simulation (Baudrillard): A model or reproduction that is often more real than "reality" itself.

Social Attraction (Blau): Force that induces humans to establish social associations on their own initiative.

Social Capital (Bourdieu): Networks of contacts and acquaintances that can be used to secure or advance one's position.

Social Capital (Coleman): Resource derived from the *relations* between individuals, particular those that are part of a "closed" social network in which most individuals either directly or indirectly know one another (as in a "friend of a friend").

Social Facts (Durkheim): Conditions and circumstances external to the individual that, nevertheless, determine one's course of action.

Social System (Parsons): A level of integrated interaction between two or more actors that involves cognizance of the others actors ideas and/or intentions, as well as shared norms or expectations and interdependence.

Socialization (Parsons): The process by which individuals come to regard specific norms as binding.

Solidarity (Durkheim): Feeling of "oneness" of a group.
- ♦ **Solidarity (Mechanical):** Solidarity that is typified by feelings of *likeness*.
- ♦ **Solidarity (Organic):** Solidarity that is a function of *interdependence*.

"Something" (Ritzer): A social form that is generally indigenously conceived, controlled, and comparatively rich in distinctive substantive content.

Species-Being (Marx): The creative essence of the human condition.

Standpoint Epistemology (Collins): The philosophic perspective that what one knows is affected by the position that one has in society.

Standpoint Theory (Smith): A theory that emphasizes that what one knows is affected by where one stands in society.

Status (Weber): a positive or negative social estimation of honor expressed through "styles of life" or patterns of consumption.

Steering Media (Habermas): Mechanisms for self-organization within the state (in the form of power) and the economy (in the form of money).

Stocks of Knowledge (Schutz): Relatively unconscious, taken-for-granted experiences or knowledge that provides actors with rules for interpreting interactions, social relationships, organizations, institutions, and the physical world.

Structure (Giddens): The rules and resources drawn upon by individuals in the production and reproduction of social action.

Subjective Reason (Horkheimer): A type of reasoning that is concerned with the means or the most efficient way of achieving a goal and not whether the goal itself is a "reasonable" pursuit.

Surplus Repression (Marcuse): Level of repression that unnecessarily impedes the gratification of individuals' instinctual desires in order to maintain existing relations of domination.

Surplus Value (Marx): The difference between what workers earn for their labor and the price of the goods they produce.

Surplus Value (Absolute): Increasing productivity of labor by extending the workday.

Surplus Value (Relative): Increasing productivity of labor through the introduction of timesaving procedures, e.g., specialization, mechanization.

Surveillance (Foucault): Technologies of observation.

Symbol (Durkheim): Something that stands for something else.

Symbolic Capital (Bourdieu): Commonly labeled as prestige, honor, reputation, or charisma, symbolic capital is the form in which other forms of capital are perceived and recognized as legitimate.

Symbolic Violence (Bourdieu): Acts leading to the misrecognition of reality or distortion of underlying power relations that, in turn, legitimate existing relations of domination as "natural."

System (Giddens): Patterns of relations between individuals or collectives reproduced across time and space. By stretching beyond the time and space of a given interaction, systems make social life orderly and predictable.

System (Habermas): A society's political and economic structures that are responsible for the organization of power relations and the production and distribution of material resources.

Taking the Attitude of the Other (Mead): An internal conversation during which an individual sees and responds to himself or herself as the individual images others will see and respond to him or her.

Targets (Coleman): Those who give up their rights to control their actions in compliance with a norm.

Technological rationality (Marcuse): A type of rationality that is marked by a scientific approach to all human affairs and that promotes an unquestioned conformity to the external, standardized dictates of efficiency, convenience, and the pursuit of profit.

Theory: A system of generalized statements or propositions about phenomena.

Time–space Distanciation (Giddens): System and social integration linking individuals across large spans of time and space.

Total Institutions (Goffman): Places of residence and work where a large number of like situated individuals, cut off from a wider society for an appreciable length of time, together lead an enclosed, formally administered round of life.

Traditional Authority (Weber): Dominance based on the established belief in the sanctity of immemorial traditions. Subjects owe their obedience to their "master" whose demands for compliance and loyalty are legitimated by sacred, inviolable traditions.

Tragedy of culture (Simmel): Where "objective culture"—the products of human creativity—comes to dominate individual will or "subjective culture."

Transnational Capitalist Class (TCC; Sklair): Social class centrally responsible for shaping transnational political practices. It is made up of those who control the major TNCs and media outlets, as well as an elite segment of politicians and state bureaucrats who see their own interests and/or interests of their social and/or ethnic group as best served by an identification with the interests of the capitalist global system

Transnational Corporations (TNC; Sklair): Institutional foundation for transnational economic practices Those who own and oversee TNCs control the manufacturing of goods and the labor necessary to produce and sell them.

Transnational Culture-Ideology (Sklair): Practices found in consumerism and the meanings attached to the globally marketed goods we purchase. The culture-ideology of consumerism is based on values and attitudes aimed at instilling a "need" for the products produced by the TNCs.

Transnational Practices (TNP; Sklair): Practices that cross state borders but do not originate with state actors, agencies, or institutions.

Trust (Coleman): Cooperation with others based on an individual's rational calculation that the chances of reaping rewards are greater than the chances of incurring costs.

Trust (Giddens): Confidence in the reliability of a system or in an individual with regard to a specific set of events or an ôutcome.

Types (Simmel): An individual's identity derived from her relationship to others.

Typification (Schutz): The process through which actors isolate the generic characteristics that are relevant for their particular interactive role (assigning roles).

Unit Act (Parsons): A hypothetical actor in a hypothetical situation, bound by an array of patterns and conditions, such as norms, goals, or required effort.

Veil (Du Bois): A process in which skin color becomes both an objective and a subjective social distancing mechanism.

***Verstehen* (Weber):** Interpretive understanding.

World-System (Wallerstein): A social system, one that has boundaries, structures, member groups, rules of legitimation, and coherence. Its life is made up of conflicting forces that hold it together by tension and tear it apart as each group seeks eternally to remold it to its advantage.

REFERENCES

Abdo, Wendy and Michael P. Broxterman. 2004. *Physician Statistics* Summary. http://www.phg.com/article_a057.htm

Adams, Bert and R. A. Sydie. 2001. *Sociological Theory.* Thousand Oaks, CA: Pine Forge.

Adorno, Theodor. 1941. "On Popular Music." *Studies in Philosophy and Social Science* 9:17–48.

———. 1950. *The Authoritarian Personality.* New York: Harper.

———. [1967] 1981. *Prisms.* Translated by Samuel and Sherry Weber. Cambridge, MA: MIT Press.

———. 1991. *The Culture Industry.* Edited by J. M. Bernstein. London: Routledge.

Alexander, Jeffrey C. 1987. *Twenty Lectures.* New York: Columbia University Press.

Alsop, Rachel, Annette Fitzsimons, and Kathleen Lennon. 2002. *Theorizing Gender.* Cambridge, UK: Polity Press.

Anderson, Elijah. 2000. "The Emerging Philadelphia African American Class Structure." *The Annals of the American Academy of Political and Social Science* 568:54–77.

Antonio, Robert J. 1995. "Nietzsche's Antisociology: Subjectified Culture and the End of History." *American Journal of Sociology* 101:1–43.

Appadurai, Arjun. 1992. *Modernity at Large: Culture Dimensions of Globalization.* Minneapolis: University of Minnesota Press.

Aron, Raymond. [1965] 1970. *Main Currents in Sociological Thought,* vols. 1 and 2. Translated by Richard Howard and Helen Weaver. New York: Anchor Books.

Ashe, Fidelma. 1999. "The Subject." Pp. 88–110 in *Contemporary Social and Political Theory,* edited by Alan Finlayson Ashe, Moya Lloyd, Iain MacKenzie, James Martin, and Shane O'Neill. Philadelphia: Open University Press.

Ashley, David and David Michael Orenstein. 1998. *Sociological Theory: Classical Statements.* Boston: Allyn and Bacon.

Atkinson, J. Maxwell and John Heritage, eds. 1984. *Structures of Social Action: Studies in Conversation Analysis.* Cambridge, UK: Cambridge University Press.

Bellah, Robert. 1973. "Introduction." Pp. ix–lv in *Emile Durkheim on Morality and Society,* edited by Robert Bellah. Chicago: University of Chicago Press.

Bendix, Richard. 1977. *Max Weber: An Intellectual Portrait.* Berkeley: University of California Press.

Benhold, Karen. 2010. "As China Rises, Conflict with the West Rises Too." *The New York Times.* http://www.nytimes.com/2010/01/27/business/global/27yuan.html (accessed January 26).

Benjamin, Walter. [1936] 1968. "The Work of Art in the Age of Mechanical Reproduction." Pp. 217–52 in *Illuminations,* edited by Hannah Arendt. Translated by Harry Zohn. New York: Harcourt, Brace.

Berger, Peter. 1963. *Invitation to Sociology.* A Humanistic Perspective. New York: Anchor Books/ Doubleday.

Berger, Peter L. 1967. *The Sacred Canopy; Elements of a Sociological Theory of Religion.* Garden City, NY: Anchor Books/Doubleday.

Berger, Peter and Thomas Luckmann. 1966. *The Social Construction of Reality.* Garden City, NY: Doubleday.

Berger, Peter and Thomas Luckmann. 1995. *Modernity, Pluralism and the Crisis of Meaning: The Orientation of Modern Man.* Gütersloh: Bertelsmann Foundation Publishers.

Berlant, Lauren. 2000. "Race, Gender, Nation in *The Color Purple.*" Pp. 3–28 in *Alice Walker's* The Color Purple: *Critical Interpretations,* edited by Harold Bloom. Philadelphia: Chelsea House.

Berry, Mary Frances. 2000. "Du Bois as Social Activist: Why We Are Not Saved." *The Annals of the American Academy of Political and Social Science* 568:100–10.

Bertens, Hans. 1995. *The Idea of Postmodernism.* New York: Routledge.

Best, Steven and Douglas Kellner. 1991. *Postmodern Theory: Critical Interrogations.* New York: Guilford.

Bhabha, Homi K. 1986. "The Other Question: Difference, Discrimination and the Discourse of Colonialism." Pp. 148–72 in *Literature, Politics and Theory,* edited by Francis Barker, Peter Hulme, Margaret Iversen, and Diana Loxley. London: Methuen.

Bierstadt, Robert. 1938. "Is Homo Sapient?" Review of *The Structure of Social Action,* by Talcott Parsons. *The Saturday Review of Literature,* March 12, p. 18.

———. 1981. *American Sociological Theory: A Critical History.* New York: Academic Press.

Blau, Peter. 1964. *Exchange and Power in Social Life.* New York: Wiley.

Blum, Deborah. 1997. *Sex on the Brain: The Biological Differences between Men and Women.* New York: Penguin.

Blumer, Herbert. 1969. *Symbolic Interaction: Perspective and Method.* Berkeley: University of California Press.

Bobo, Lawrence. 2000. "Reclaiming a Du Boisian Perspective on Racial Attitudes." *The Annals of the American Academy of Political and Social Science* 568(March):186–202.

Boden, Diedre. 1990. "The World as It Happens: Ethnomethodology and Conversation Analysis." Pp. 185–213 in *Frontiers of Social Theory,* edited by George Ritzer. New York: Columbia University Press.

Bottomore, Tom. 1984. *The Frankfurt School.* Sussex, UK: Ellis Horword Limited.

Bourdieu, Pierre. 1984. *Distinction.* Cambridge, MA: Harvard University Press.

———. 1990a. *In Other Words.* Palo Alto, CA: Stanford University Press.

———. 1990b. *The Logic of Practice.* Translated by Richard Nice. Palo Alto, CA: Stanford University Press.

———. 1991. *Language and Symbolic Power.* Edited and introduced by John B. Thompson. Translated by Gino Raymond and Matthew Adamson. Cambridge, MA: Harvard University Press.

———. 1993. *The Field of Cultural Production.* Edited and introduced by Randal Johnson. New York: Columbia University Press.

———. 1998. *Practical Reason.* Palo Alto, CA: Stanford University Press.

Branaman, Ann. 1997. "Goffman's Social Theory." Pp. xlv–lxxxii in *The Goffman Reader,* edited by Charles Lemert and Ann Branaman. Oxford, UK: Blackwell.

Bryson, Valerie. 1999. *Feminist Debates.* New York: New York University Press.

Burtt, Katrina. 1998. "Male Nurses Still Face Bias." http://www.nursingworld.org/ajn/1998/sept/issu098f.htm

Butler, Judith. [1990] 1999. *Gender Trouble: Feminism and the Subversion of Identity.* New York: Routledge.

———. 1993. *Bodies That Matter: On the Discursive Limits of "Sex."* New York: Routledge.

———. 1999. *Subjects of Desire.* New York: Columbia University Press.

Byerman, Keith. 2003. "W. E. B. Du Bois and the Construction of Whiteness." Pp. 161–71 in *The Souls of Black Folk: One Hundred Years Later,* edited by Dolan Hubbard. Columbia: University of Missouri Press.

Calhoun, Craig. 1992. "Introduction: Habermas and the Public Sphere." Pp. 1–48 in *Habermas and the Public Sphere,* edited by Craig Calhoun. Cambridge, MA: MIT Press.

———. 1995. *Critical Social Theory.* Oxford, UK: Blackwell Cambridge University Press.

———. 2003. "Robert K. Merton Remembered." *American Sociological Association Footnotes* 31(1).

Calhoun, Craig, Joseph Gerteis, James Moody, Steven Pfaff, Kathryn Scmidt, and Indermohan Virk, eds. 2002. *Classical Sociological Theory.* Malden, MA: Blackwell.

Calhoun, Craig and Loïc Wacquant. 2002. "Everything Is Social." *American Sociological Association* 30(2):5, 10.

Camic, Charles. 1991. "Introduction." In *Talcott Parsons, The Early Essays.* Chicago: University of Chicago Press.

Campbell, Marie and Ann Manicom, eds. 1995. *Experience, Knowledge and Ruling Relations: Explorations in the Social Organization of Knowledge.* Toronto, Ontario, Canada: University of Toronto Press.

Carey, P. 1992. "Catching Up with *Candid Camera.*" *Saturday Evening Post* (Indianapolis)

Carroll, Lewis. [1865] 1960. *Alice's Adventures in Wonderland.* New York: New American Library of World Literature, Inc.

Chodorow, Nancy. 1978. *The Reproduction of Mothering.* Berkeley: University of California Press.

Cicourel, Aaron. 1964. *Method and Measurement in Sociology.* New York: Free Press.

Clayman, Steven. 1993. "Booing: The Anatomy of a Disaffiliative Response." *American Journal of Sociology* 58:110–30.

Coleman, J. S. 1990. *Foundations of Social Theory.* Cambridge, MA: Belknap/Harvard University Press.

Collins, Patricia Hill. 1987. "The Meaning of Motherhood in Black Culture and Black Mother/ Daughter Relationships." *Sage* 4(2):3–10.

Collins, Randall, 1975. *Conflict Sociology.* New York: Academic Press.

———. 1979. *The Credential Society.* New York: Academic Press.

———. 1986. "The Passing of Intellectual Generations: Reflections on the Death of Erving Goffman." *Sociological Theory* 4:106–13.

———. 1994. *Four Sociological Traditions.* New York: Oxford University Press.

———. 2004. *Interaction Ritual Chains.* Princeton, NJ: Princeton University.

Collins, Randall and Michael Makowsky. 1998. *The Discovery of Society.* Boston: McGraw-Hill.

Comte, Auguste. [1830–1842] 1974. *The Positive Philosophy.* Edited and translated by Harriet Martineau. New York: AMS Press.

———. 1974. *The Crisis of Industrial Civilization: The Early Essays of Auguste Comte.* Edited by Ronald Fletcher. London: Heinemann.

———. 1975. *Auguste Comte and Positivism: The Essential Writings.* Edited by Gertrude Lenzer. Chicago: University of Chicago Press.

Connell, R. W. 1997. "Why Is Classical Theory Classical." *American Journal of Sociology* 102(6):1511–57.

Cooley, Charles Horton. [1902] 1964. *Human Nature and the Social Order.* New York: Schocken Books.

———. [1909] 1962. *Social Organization: A Study of the Larger Mind.* New York: Schocken Books.

———. [1918] 1966. *Social Process.* Carbondale: Southern Illinois University Press.

Cooper, Anna Julia. [1892] 1988. *A Voice from the South.* New York: Oxford University Press.

Cornell, R. W. 2002. *Gender.* Cambridge, UK: Polity.

Coser, Lewis A. 1977. *Masters of Sociological Thought.* New York: Harcourt, Brace, Jovanovich.

Craib, Ian. 1992. *Anthony Giddens.* London: Routledge.

Crothers, Charles. 1987. *Robert K. Merton.* Sussex, UK: Ellis Horwood Limited.

Davis, Kingsley and Wilbert E. Moore. 1945. "Some Principles of Stratification." *American Sociological Review* 10:242–49.

Deegan, Mary Jo. 1997. "Introduction: Gilman's Sociological Journey from *Herland* to *Ourland.*" Pp. 1–57 in Charlotte Perkins Gilman, *With Her in Ourland,* edited by Deegan and Michael Hill. Westport, CT: Greenwood Press.

Degler, Carl. 1966. "Introduction." Pp. vi–xxxv in Charlotte Perkins Gilman, *Women and Economics.* New York: Harper and Row.

Denning, Steve. 2004. http://www.stevedenning.com/postmodern.html

Dickens, Charles. 1854. *Hard Times.* London: Bradbury and Evans.

Dixon, Wheeler Winston. 1997. Review of *The Gulf War Did Not Take Place,* by Jean Baudrillard, *Film Quarterly* 50(4):4–6.

Donovan, Josephine. [1985] 2000. *Feminist Theory: The Intellectual Traditions* (3rd ed.). New York: Continuum.

Du Bois, W. E. B. [1883] 1986. "Great Barrington Notes." *New York Globe,* September 29. P. 7 in *Newspaper Columns by W. E. B. Du Bois, v. 1 1883–1944,* compiled and edited by Herbert Aptheker. White Plains, NY: Kraus Thomson Organization Unlimited.

———. [1890] 1985. "Something about Me." Pp. 16–17 in *Against Racism: Unpublished Essay, Papers, Addresses, 1887–1961,* edited by Herbert Aptheker. Amherst: University of Massachusetts Press.

———. 1896. *The Suppression of the African Slave-Trade to the United States of America, 1638–1870.* New York: Longmans and Green.

———. [1898] 2000. "The Study of Negro Problems." *The Annals of the American Academy of Political and Social Science* 568(March):13–27.

———. 1899. *Philadelphia Negro: A Social Study.* New York: Lippincott.

———. [1903] 1989. *Souls of Black Folk.* New York: Penguin Books.

———. [1904] 2003. "Du Bois Responds to Critics." *The Independent,* November 17, 57. P. 1152 in *The Souls of Black Folk: One Hundred Years Later,* edited by Dolan Hubbard. Columbia: University of Missouri Press.

———. [1912] 2000. "The Black Mother." P. 294 in *W. E. B. Du Bois: A Reader,* edited by David Levering Lewis. New York: Henry Holt.

———. [1913] 2000. "Hail Columbia!" Pp. 295–96 in *W. E. B. Du Bois: A Reader,* edited by David Levering Lewis. New York: Henry Holt.

———. [1914] 2000. "The Burden of Black Women." Pp. 102–04 in *Du Bois on Religion,* edited by Phil Zuckerman. Walnut Creek, CA: AltaMira.

———. 1915. *The Negro.* New York: Henry Holt.

———. [1920] 2003a. "The Damnation of Women." Pp. 171–91 in *Darkwater: Voices from within the Veil.* (With an introduction by Joe Feagin.) Amherst, NY: Humanity Books.

———. [1920] 2003b. *Darkwater: Voices from within the Veil.* (With an introduction by Joe Feagin.) Amherst, NY: Humanity Books.

———. [1920] 2003c. "The Servant in the House." Pp. 125–35 in *Darkwater: Voices from within the Veil.* (With an introduction by Joe Feagin.) Amherst, NY: Humanity Books.

———. [1920] 2003d. "The Souls of White Folk." Pp. 55–75 in *Darkwater: Voices from within the Veil.* (With an introduction by Joe Feagin.) Amherst, NY: Humanity Books.

———. [1924] 2007. *The Gift of Black Folk: The Negroes in the Making of America.* Oxford, UK: Oxford University Press.

———. [1928] 1995. *Dark Princess.* Jackson: University Press of Mississippi.

———. [1930a] 2007. *Africa: Its Place in Modern History.* Oxford, UK: Oxford University Press.

———. [1930b] 2007. *Africa: Its Geography, People, and Products.* Oxford, UK: Oxford University Press.

———. [1935] 1962. *Black Reconstruction in America: An Essay toward a History of the Part Which Black Folk Played in the Attempt to Reconstruct Democracy in America, 1860–1880.* New York: Atheneum.

———. 1935. *Black Reconstruction in America, 1860–1880.* New York: Atheneum. Brunswick, NJ: Transaction.

———. [1940] 1984. *Dusk of Dawn: An Essay Toward an Autobiography of the Race Concept.* New Brunswick, NJ: Transaction.

———. 1944. "My Evolving Program for the Negro." Pp. 31–70 in *What the Negro Wants* edited by Rayford W. Logan. Chapel Hill: University of North Carolina.

———. 1945. *Color and Democracy: Colonies and Peace.* New York: Harcourt Brace.

———. [1947] 1995. *The World and Africa.* New York: Viking.

———. 1968. *The Autobiography of W. E. B. Du Bois.* New York: International Publishers.

Dufay, François, and Pierre-Bertand Dufort. 1993. *Les Normaliens.* Paris: Editions Jean-Claude Lattes.

Durkheim, Émile. [1893] 1984. *The Division of Labor in Society.* New York: Free Press.

————. [1895] 1938. *The Rules of Sociological Method*. Translated by Sarah A. Solovay and John H. Mueller. Edited by George E. G. Catlin. New York: Free Press.

————. [1895] 1966. *The Rules of Sociological Method*. New York: Free Press.

————. [1897] 1951. *Suicide: A Study in Sociology*. New York: Free Press.

————. [1898] 1973. "Individualism and the Intellectuals." Pp. 4357 in Émile Durkheim, *On Morality and Society*, edited by Robert Bellah. Chicago: University of Chicago Press.

————. [1912] 1965. *The Elementary Forms of Religious Life*. New York: Free Press.

————. [1912] 1995. *The Elementary Forms of Religious Life*. Translated by Karen Fields. New York: Free Press.

Duvenage, Pieter. 2003. *Habermas and Aesthetics*. Cambridge, UK: Polity.

Edles, Laura Desfor. 2002. *Cultural Sociology in Practice*. Malden, MA: Blackwell.

Edles, Laura Desfor and Scott Appelrouth. 2005/2010. *Sociological Theory in the Classical Era: Text and Readings*. Thousand Oaks, CA: Sage.

Ehrenhalt, Alan. 1999. "Because We Like You." Review of *Celebration USA*, by Douglas Frantz and Catherine Collins and *The Celebration Chronicles* by Andrew Ross. *Los Angeles Times Book Review*, September 12, p. 8.

Eliade, Mircea, and Ioan P. Couliano. 1991. *The Eliade Guide to World Religions*. New York: HarperCollins.

Elliot, Anthony. 2001. "Anthony Giddens," pp. 292–303. In *Profiles in Contemporary Social Theory*, edited by Anthony Elliot and Bryan Turner. Thousand Oaks, CA: Sage.

Engels, Friedrich. [1845] 1987. *The Condition of the Working Class in England*. New York: Penguin.

————. [1884] 1942. *The Origin of the Family, Private Property and the State*. New York: International Publishers.

Erhich, Paul. 2000. *Human Natures*. New York: Penguin Books.

Erlanger, Steven. 2010. "Face-Veil Issue in France Shifts to Parliament for Debate." *The New York Times*, http://www.nytimes.com/2010/01/27/world/europe/27france.html (accessed January 26).

Feagin, Joe. 2003. "Introduction." Pp. 9–24 in *Darkwater: Voices from Within the Veil* by W. E. B. Du Bois. New York: Humanity Books.

Featherstone, Mike. 1995. *Undoing Culture*. London: Sage.

Fernandez, Ronald. 2003. *Mappers of Society: The Lives, Times, and Legacies of Great Sociologists*. Westport, CT: Praeger.

Fish, M. Steven. 2003. "Muslim World: Repressing Women, Repressing Democracy." *Los Angeles Times*, October 12, M3.

Fletcher, Ronald. 1966. "Auguste Comte and the Making of Sociology." Speech delivered on November 4, 1965, at the London School of Economics and Political Science.

Ford, Daniel. 2002. "The Horror of the Human Bomb-Delivery System." Review of Albert Axell and Hideaki Kase, *Kamikaze: Japan's Suicide Gods*. *Wall Street Journal*, September 10, D8.

Frank, Andre Gunder. 1967. *Capitalism and Underdevelopment in Latin America*. New York: Monthly Review Press.

Freud, Sigmund. 1961. *Civilization and Its Discontents*. New York: W. W. Norton.

Frisby, David. 1984. *Georg Simmel*. London: Tavistock.

Funt, Allen. 1952. *Eavesdropping at Large: Adventures in Human Nature with Candid Mike and* Candid Camera. New York: Vanguard Press.

Gamson, Joshua. 1998. *Freaks Talk Back: Tabloid Talk Shows and Sexual Nonconformity*. Chicago: University of Chicago Press.

Garfinkel, Harold. 1967. *Studies in Ethnomethodology*. Englewood Cliffs, NJ: Prentice Hall.

Garfinkel, Harold and Harvey Sacks. 1970. "On formal structures of practical actions." Pp. 337–66 in *Theoretical Sociology: Perspectives and Developments*, edited by John C. McKinney and Edward Tiryakian. New York: Appleton-Century-Crofts.

Gerhardt, Uta. 2002. *Talcott Parsons: An Intellectual Biography*. Cambridge, UK: Cambridge University Press.

Gerhenson, Geoffrey and Michelle Williams. 2001. "Nancy Chodorow." Pp. 281–91 in *Profiles in Contemporary Social Theory*, edited by Anthony Elliot and Bryan Turner. Thousand Oaks, CA: Sage.

Giddens, Anthony. 1971. *Capitalism and Modern Social Theory*. London: Cambridge University Press.

————. 1979. *Central Problems in Social Theory*. Berkeley: University of California Press.

————. 1981. "Agency, Institution, and Time-Space Analysis." Pp. 161–74 in *Advances in Social Theory and Methodology*, edited by K. Knorr-Cetina and A. V. Cicourel. London: Routledge and Kegan Paul.

————. 1984. *The Constitution of Society*. Berkeley: University of California Press.

————. 1990. *Consequences of Modernity*. Palo Alto, CA: Stanford University Press.

Gilman, Charlotte Perkins. [1892] 1973. "The Yellow Wallpaper." Originally published as a short story in *New England Magazine* (January). Printed in book form in 1899 by the Feminist Press: *The Yellow Wallpaper*. New York: City University of New York Feminist Press.

————. [1898] 1998. *Women and Economics: A Study of the Economic Relation between Men and Women as a Factor in Social Evolution*. Mineola, NY: Dover.

————. 1900. *Concerning Children*. Boston: Small, Maynard & Co.

————. 1903. *The Home: Its Work and Influence.* New York: McClure, Phillips & Co.

————. 1904. *Human Work.* New York: McClure, Phillips & Co.

————. [1911] 1970. *The ManMade World, or Our Androcentric Culture.* New York: Charlton Co.; NY: Source Book Press.

————. 1923. "Is America Too Hospitable?" *Forum* 70(October):1983.

————. [1935] 1972. *The Living of Charlotte Perkins Gilman: An Autobiography.* New York: D. Appleton-Century.

Goffman, Erving. 1959. *The Presentation of Self in Everyday Life.* New York: Anchor Books.

————. 1961. *Asylums: Essays on the Social Situation of Mental Patients and Other Inmates.* New York: Anchor Books.

————. 1967. *Interaction Ritual.* New York: Anchor Books.

————. 1974. *Frame Analysis: An Essay on the Organization of Experience.* New York: Harper & Row.

————. 1983. "The Interaction Order." *American Sociological Review* 48:1–17.

Golden, Catherine and Joanna Schneider Zangrando. 2000. "Introduction." Pp. 11–22 in *The Mixed Legacy of Charlotte Perkins Gilman,* edited by Golden and Zangrando. Newark: University of Delaware Press.

Goodwin, Jeff, and James M. Jasper. 2004. *The Social Movements Reader.* Malden, MA: Blackwell.

Gramsci, Antonio. 1971. *Prison Notebooks.* Edited and translated by Auintin Hoare and Geoffrey Nowell Smith. New York: International Publishers.

Grenfell, Michael. 2004. *Pierre Bourdieu: Agent Provocateur.* London: Continuum.

Griffin, Erica, compiler. 2003. "Reviews of *The Souls of Black Folk.*" Pp. 18–33 in *The Souls of Black Folk: One Hundred Years Later,* edited by Dolan Hubbard. Columbia: University of Missouri Press.

Griffin, Farah Jasmine. 2000. "Black Feminists and Du Bois: Respectability, Protection and Beyond." *The Annals of the American Academy of Political and Social Science* 568(March):28–40.

Gubrium, Jaber and James Holstein. 1997. *The New Language of Qualitative Method.* New York: Oxford University Press.

Habermas, Jürgen. 1962. *The Structural Transformation of the Public Sphere: An Inquiry into a Category of Bourgeois Society.* Cambridge, MA: MIT Press.

————. 1979. "Toward a Reconceptualization of Historical Materialism." Pp. 130–77 in *Communication and the Evolution of Society.* Boston: Beacon Press.

————. 1984. *The Theory of Communicative Action. Vol. 1, Reason and the Rationalization of Society.* Boston: Beacon Press.

————. 1987. *The Theory of Communicative Action. Vol. 2, Lifeworld and System: A Critique of Functionalist Reason.* Boston: Beacon Press.

————. 1989. *Jürgen Habermas on Society and Politics.* Edited by Steven Seidman. Boston: Beacon Press.

————. 1996. *Between Facts and Norms.* Cambridge, MA: MIT Press.

Hale, Alison. 1998. *My World Is Not Your World.* Glenville, NY: Archimedes Press.

Halperin, David. 1995. *Saint Foucault: A Gay Hagiography.* New York: Oxford University Press.

Harding, Sandra. 1986. *The Science Question in Feminism.* Ithaca, NY: Cornell University Press.

————. ed. 2004. *The Feminist Standpoint Reader.* New York: Routledge.

Hartsock, Nancy. 1990. "Foucault on Power: A Theory for Women?" Pp. 157–75 in *Feminism/Postmodernism,* edited by Linda Nicholson. New York: Routledge.

Hassan, Ihab. 1987. *The Postmodern Turn: Essays in Postmodern Theory and Culture.* Columbus: Ohio State University Press.

Hechter, Michael. 1987. Principles of Group Solidarity. Berkeley: *University of California Press.*

Hedges, Elaine. 1973. "Afterword." Pp. 37–65 in Charlotte Perkins Gilman, *The Yellow Wallpaper.* New York: City University of New York Feminist Press.

Heidegger, Martin. [1927] 1962. *Being and Time.* Translated by John Macquarrie and Edward Robinson. New York: Harper & Row.

Heritage, John. 1984. *Garfinkel and Ethnomethodology.* New York: Polity Press.

————. 1998. "Harold Garfinkel." Pp. 175–88 in *Key Sociological Thinkers,* edited by Rob Stones. London: Macmillan.

Heritage, John and David Greatbatch. 1986. "Generating Applause: A Study of Rhetoric and Response at Party Conferences." *American Journal of Sociology* 92:110–157.

Hill, Mary. 1989. "Charlotte Perkins Gilman: A Feminist's Struggle with Womanhood." Pp. 31–50 in *Charlotte Perkins Gilman: The Woman and Her Work,* edited by Sheryl L. Meryering. Ann Arbor, MI: UMI Research Press.

Hill-Collins, Patricia. 1987. "The Meaning of Motherhood in Black Culture and Black Mother/ Daughter Relationships." *Sage* 4(2):3–10.

————. [1990] 2000. *Black Feminist Thought: Knowledge, Consciousness, and the Politics of Empowerment.* London: HarperCollins.

————. 1998. *Fighting Words: Black Women and the Search for Justice.* Minneapolis: University of Minnesota.

————. 2004. *Black Sexual Politics.* New York: Routledge.

————. 2006. From *Black Power to Hip Hop: Racism, Nationalism, and Feminism.* Philadelphia: Temple University Press.

Hochschild, Arlie Russell. 1983. *The Managed Heart.* Berkeley: University of California Press.

———. 2003. *The Commercialization of Intimate Life.* Berkeley: University of California Press.

Homans, Casper. 1958. "Social Behavior as Exchange." *American Journal of Sociology* 63(6):597–606.

———. 1961. *Social Behavior: Its Elementary Forms.* New York: Harcourt Brace.

———. 1984. *Coming to My Senses.* New Brunswick, NJ: Transaction Books.

Horkheimer, Max. 1941. "The End of Reason." Pp. 366–88 in *Studies in Philosophy and Social Science,* vol. 9. New York: Institute of Social Research.

———. 1947. *The Eclipse of Reason.* New York: Oxford University Press.

———. 1972a. "Traditional and Critical Theory." Pp. 188–243 in *Critical Theory: Selected Essays,* translated by Matthew J. O'Connell, et al. New York: Herder and Herder.

———. 1972b. "Postscript." Pp. 244–52 in *Critical Theory: Selected Essays,* translated by Matthew J. O'Connell, et al. New York: Herder and Herder.

Horkheimer, Max and Theodor Adorno. [1944] 2002. *Dialectic of Enlightenment.* Palo Alto, CA: Stanford University Press.

Horowitz, Irving Louis, ed. and introduction. 1963. *Power, Politics, and People: The Collected Essays of C. Wright Mills.* Oxford, UK: Oxford University Press.

Horster, Detlef. 1992. *Habermas: An Introduction.* Philadelphia: Pennbridge.

Hugo, Victor. [1862] 1879. *Les Miserables.* New York: Carleton Publishing Company.

Huntington, Samuel. 1993. "The Clash of Civilizations?" *Foreign Affairs* 72:3.

Husserl, Edmund. [1913] 1982. *Ideas; General Introduction to Pure Phenomenology.* The Hague, Netherlands: M. Nijhoff.

International Labor Organization. 2006. "ILO Annual Jobs Report Says Global Unemployment Continues to Grow, Youth Now Make Up Half Those Out of Work." http://www.ilo.org/global/About_the_ILO/Media_and_public_information/Press_releases/langen/WCMS_065176/index.htm (accessed July 27, 2009).

James, William. [1890] 2007. *The Principles of Psychology.* New York: Cosimo Classics.

———. [1892] 2001. "The Stream of Consciousness." Pp. 18–43 in *Psychology.* Mineola, NY: Dover.

———. [1902] 1958. *The Varieties of Religious Experience.* New York: New American Library.

Jaspers, Karl. [1957] 1962. *Socrates, Buddha, Confucius, Jesus* (The Great Philosophers, vol. 1). New York: Harcourt Brace.

Jefferson, Gail. 1978. "Sequential Aspects of Storytelling in Conversation." Pp. 219–48 in *Studies in the Organization of Conversational Interaction* edited Jim Schenkein. New York: Academic Press.

Kaspersen, Lars Bo. 2000. *Anthony Giddens.* Oxford, UK: Blackwell.

Kellner, Douglas. 2000. "Habermas, the Public Sphere, and Democracy: A Critical Intervention." Pp. 259–88 in *Perspectives on Habermas,* edited by Lewis Hahn. Chicago: Open Court Press.

King, Martin Luther, Jr. 1970. "Honoring Dr. Du Bois." Pp. 176–85 in *Black Titan: W. E. B. Du Bois,* edited by John Henrik Clarke, Esther Jackson, Ernest Kaiser, and J. H. O'Dell. Boston: Beacon Press.

Kirby, Barbara L. 2001/2005. *The OASIS Guide to Asperger Syndrome.* New: York: Crown.

———."What Is Asperger Syndrome?" http://www.udel.edu/bkirby/asperger/aswhatisit.html

Klein, Stephen and William Leiss. 1978. "Advertising, Needs, and Commodity Fetishism." *Canadian Journal of Political and Social Theory* 2(1)

Lalonde, Marc. 1999. *Critical Theology and the Challenge of Jürgen Habermas.* New York: Peter Lang. Lemert, Charles, ed. 1993. *Social Theory.* Boulder, CO: Westview Press.

———. 1997. "Goffman." Pp. ix–xlii in *The Goffman Reader,* edited by Charles Lemert and Ann Branaman. Oxford, UK: Blackwell.

Lane, Ann J. 1990. *To Herland and Beyond: The Life and Work of Charlotte Perkins Gilman.* New York: Pantheon.

Lemert, Charles, ed. 1993. *Social Theory.* Boulder, CO: Westview Press.

Lemert, Charles. 1997. *Postmodern Is Not What You Think.* Malden, MA: Blackwell.

———. 1998. "Anna Julia Cooper: The Colored Woman's Office." In Charles Lemert and Esme Bhand, eds., *The Voice of Anna Julia Cooper.* Lanham, MD: Rowman and Littlefield Publishers.

Levine, Donald. 1971. "Introduction." Pp. ix–lxv in *Georg Simmel: On Individuality and Social Forms.* Chicago: University of Chicago Press.

Lévi-Strauss, Claude. 1969. *The Raw and the Cooked.* Chicago: University of Chicago Press.

Lewis, David Levering. 2000. *W. E. B. Du Bois.* New York: Henry Holt and Company.

Lipsitz, George. 1994. *Dangerous Crossroads: Popular Music, Postmodernism, and the Poetics of Place.* New York: Verso.

Lloyd, Stephen. 2003. *The Sociology of Anthony Giddens.* London: Pluto Press.

Lorber, Judith. 1998. *Gender Inequality.* Los Angeles: Roxbury Press.

Lukács, Georg. [1923] 1972. *History and Class Consciousness.* Translated by Rodney Livingstone. Cambridge, MA: MIT Press.

Lukes, Steven. 1985. *Emile Durkheim: His Life and Work.* London: Allen Lane, Penguin Press.

Lyotard, Jean-François. [1979] 1984. *The Postmodern Condition: A Report on Knowledge.* Translated by Geoff Bennington and Brian Massumi; Foreword by Fredric Jameson. Minneapolis: University of Minnesota Press.

Mannheim, Karl. [1936] 2001. *Man and Society in an Age of Reconstruction.* London: Routledge and Kegan Paul.

Marable, Manning. 1986. *W. E. B. Du Bois: Black Radical Democrat.* Boston: Twayne.

March, James G. and Herbert A. Simon. 1958. *Organizations.* New York: Wiley.

Marcuse, Herbert. 1941. "Some Implications of Modern Technology." *Studies in Philosophy and Social Science* 9:414–39.

———. 1955. *Eros and Civilization: A Philosophical Inquiry into Freud.* Boston: Beacon Press.

———. 1964. *One Dimensional Man.* Boston: Beacon Press.

———. 1970. *Five Lectures: Psychoanalysis, Politics and Utopia.* Boston: Beacon Press.

Martin, James. 1999. "The Social and the Political." Pp. 155–77 in *Contemporary Social and Political Theory,* edited by Fidelma Ashe, Alan Finlayson, Moya Lloyd, Iain MacKenzie, James Martin, and Shane O'Neill. Buckingham, UK: Open University Press.

Martindale, Don. 1981. *The Nature and Types of Sociological Theory.* 2nd ed. Boston: Houghton Mifflin.

Martineau, Harriet. 1837. *Society in America.* New York: Saunders and Otley.

———. 1838. *How to Observe Manners and Morals.* London: Charles Knight.

———. 1853. *The Positive Philosophy of Auguste Comte.* New York: Calvin Blanchard.

Marx, Karl. [1844] 1977. *The Economic and Philosophic Manuscripts of 1844.* Edited by Dirk J. Struik and translated by Martin Milligan. New York: International.

———. [1852] 1978. "The Eighteenth Brumaire of Louis Bonaparte." Pp. 594–617 in *The Marx-Engels Reader,* edited by Robert C. Tucker. New York: W. W. Norton.

———. [1859] 1978. "Preface to a Contribution to the Critique of Political Economy." Pp. 3–6 in *The Marx/Engels Reader,* edited by Robert C. Tucker. New York: W. W. Norton.

———. [1867] 1978. *Capital,* vol. 1, translated by S. Moore and E. Aveling. New York: International.

Marx, Karl and Friedrich Engels. [1846] 1977. *The German Ideology.* Edited by C. J. Arthur. New York: International.

———. [1846] 1978. "The German Ideology." Pp. 146–201 in *The Marx/Engels Reader,* edited by Robert C. Tucker. New York: W. W. Norton.

———. [1848] 1978. "The Communist Manifesto." Pp. 469–500 in *The Marx-Engels Reader,* edited by Robert C. Tucker. New York: W. W. Norton.

———. [1852] 1978. "The Eighteenth Brumaire of Louis Bonaparte." Pp. 594–617 in *The Marx-Engels Reader,* edited by Robert C. Tucker. New York: W. W. Norton.

———. [1859] 1978. "Preface to A Contribution to the Critique of Political Economy." Pp. 3–6 in *The Marx-Engels Reader,* edited by Robert C. Tucker. New York: W. W. Norton.

———. [1867] 1967. *Capital,* vol. 1. Translated by S. Moore and E. Aveling. New York: International.

———. 1978. *The Marx-Engels Reader.* Edited by Robert C. Tucker. New York: W. W. Norton.

Matustik, Martin. 2001. *Jürgen Habermas: A Philosophical-Political Profile.* Lanham, MD: Rowman and Littlefield.

McCarthy, Thomas. 1978. *The Critical Theory of Jürgen Habermas.* Cambridge, MA: MIT Press.

McGuire, Meredith. 1997. *Religion: The Social Context.* 4th ed. Belmont, CA: Wadsworth.

McMahon, Darrin. 2001. *Enemies of the Enlightenment.* Oxford, UK: Oxford University Press.

Mead, George Herbert. [1934] 1962. *Mind, Self and Society.* Edited by Charles W. Morris. Chicago: University of Chicago Press.

———. [1936] 1964. "Mind Approached Through Behavior" Pp. 65–82 in *George Herbert Mead on Social Psychology,* edited by Anselm Strauss. Chicago: University of Chicago Press.

———. 1938. *The Philosophy of the Act.* Chicago: University of Chicago Press.

———. 1964. "Mind Approached Through Behavior" Pp. 65–82 in *George Herbert Mead on Social Psychology,* edited by Anselm Strauss. Chicago: University of Chicago Press.

Merton, Robert. [1949] 1957. *Social Theory and Social Structure.* Toronto, Ontario, Canada: Free Press.

———. 1994. *A Life of Learning.* American Council of Learned Societies, Occasional Paper, n. 25.

———. 1996. *On Social Structure and Science.* Edited and with an introduction by Piotr Sztompka. Chicago: University of Chicago.

Miller, James. 1959. *Sociological Integration.* New York: Oxford University Press.

———. 1993. *The Passion of Michel Foucault.* New York: Simon & Schuster.

Mills, C. W. 1951. *White Collar.* New York: Oxford University Press.

———. 1958. *Power Elite.* New York: Oxford University Press.

———. 1959. *The Sociological Imagination.* New York: Oxford University Press.

———. 1963. *Power, Politics and People.* Edited by Irving Louis Horowitz. New York: Ballantine Books.

Monteiro, Anthony. 2000. "Being an African in the World of Du Boisian Epistemology." *The Annals of the American Academy of Political and Social Science* 568(March): 220–234.

Morgan, Lewis. [1877] 2000. *Ancient Society.* Piscataway, NJ: Transaction Publishers.

Münch, Richard. 1994. *Sociological Theory.* New York: Nelson-Hall.

National Assembly of France. 1789. "La Déclaration des Droits de l'Homme et du Citoyen" ("The Declaration of Rights of Man and of the Citizen.") http://www.aidh .org/Biblio/Text_fondat/FR_02.htm, https://www .college.columbia.edu/core/students/cc/settexts/nafman 89.pdf (accessed July 27, 2009).

National Association of Home Builders. 2003. *Los Angeles Times,* September 14, K4.

Nazir, Sameena. 2005. "Challenging Inequality: Obstacles and Opportunities Toward Women's Rights in the Middle East and Northern Africa." *Journal of the Institute of Justice and International Studies* 5:31–42.

Nietzsche, Friedrich. [1886] 1917. *Beyond Good and Evil.* Translated by Helen Zimmern, introduction by Willard Huntington Wright. New York: Boni & Liveright.

———. [1866] 1966. *Beyond Good and Evil.* Translated by Walter Kaufmann. New York: Vintage Books.

———. [1882] 1974. *The Joyful Wisdom.* Translated by Walter Kaufmann. New York: Vintage Books.

———. [1887] 1967. *On the Genealogy of Morals.* Translated by Walter Kaufmann, edited with commentary by Walter Kaufmann. New York: Vintage.

———. [1887] 1969. *On the Genealogy of Morals.* Translated by Walter Kaufmann. New York: Vintage Books.

Osborne, Peter and Lynne Segal. 1993. Extracts from *Gender as Performance: An Interview with Judith Butler.* http:// www.theory.org.uk/but-int1.htm

Parsons, Talcott. [1937] 1967. *The Structure of Social Action.* New York: McGraw-Hill.

———. [1943] 1954. "Sex Roles in American Kinship System," selection from "The Kinship System of the Contemporary United States" in *Essays in Sociological Theory.* New York: Free Press.

———. 1961. "An Outline of the Social System." In *Theories of Society,* edited by Parsons et al. New York: Free Press.

———. 1969. *Politics and Social Structure.* New York: Free Press.

———. 1971. *The System of Modern Societies.* Englewood Cliffs, NJ: Prentice Hall.

Parsons, Talcott and Edward A. Shils. [1951] 2001. *Toward a General Theory of Action,* introduction by Neil Smelser. New Brunswick: Transaction Publishers.

Patterson, Orlando. 1998. *Rituals of Blood.* Washington, DC: Civitas/Counterpoint.

Pichanick, Valerie K. 1980. *Harriet Martineau, the Woman and Her Work, 1802–76.* Ann Arbor: University of Michigan Press.

Pollner, Melvin. 1987. *Mundane Reason.* Cambridge, UK: Cambridge University Press.

Poster, Mark. 1989. *Critical Theory and Poststructuralism: In Search of a Context.* Ithaca, NY: Cornell University Press.

Putnam, Robert. 2001. *Bowling Alone: The Collapse and Revival of American Community.* New York: Simon & Schuster.

Rawls, Anne. 2002. "Editor's Introduction." In *Ethnomethodology's Program: Working Out Durkheim's Aphorism,* edited by H. Garfinkel. Oxford, UK: Rowman & Littlefield.

Ritzer, George. 2000a. *Classical Sociological Theory.* 3rd ed. Boston: McGraw-Hill.

———. 2000b. *The McDonaldization of Society.* Thousand Oaks, CA: Pine Forge Press.

Ritzer, George and Douglas J. Goodman. 2004. *Modern Sociological Theory.* 6th ed. New York: McGraw-Hill.

Robbins, Derek. 2000. *Bourdieu and Culture.* London: Sage.

Robertson, Roland. 1970. *The Sociological Interpretation of Religion.* New York: Schocken.

———. 1992. *Globalization.* London: Sage.

Robinson, William I. 2008. "Theories of Globalization." In *The Blackwell Companion to Globalization,* edited by George Ritzer. London/New York: Blackwell.

Rocher, Guy. [1972] 1974. *Talcott Parsons and American Sociology.* London: Nelson.

Rogers, Mary. 1998. *Contemporary Feminist Theory.* Boston: McGraw-Hill.

Rothman, Paul. 1995. *Full Circle* [film]. New York: Filmakers Library.

Rousseau, Jean-Jacques. 1762. *The Social Contract.* Translated by G. D. H. Cole, public domain. http:// www.constitution .org/jjr/socon.htm (accessed July 27, 2009).

Rudwick, Elliot. [1960] 1982. *W. E. B. Du Bois: Voice of the Black Protest Movement.* Urbana: University of Illinois Press.

Sacks, Harvey, Emmanual Schegloff, and Gail Jefferson. 1974. "A Simplest Systematics for the Organization of Turn-taking in Conversation." *Language* 50:396–735.

Said, Edward. 1978. *Orientalism.* New York: Vintage Books.

———. 1980. "Islam through Western Eyes." *The Nation,* April 26.

———. 2000. *Reflections on Exile and Other Essays.* Cambridge, MA: Harvard University Press.

Sartre, Jean-Paul. 1947. *Existentialism.* Translated by Bernard Frechtman. New York: Philosophical Library.

Scharnhorst, Gary. 2000. "Historicizing Gilman: A Bibliographer's View." Pp. 65–73 in *The Mixed Legacy of Charlotte Perkins Gilman,* edited by C. Golden and J. Zangrando. Newark: University of Delaware Press.

Scheler, Max. 1970. *The Nature of Sympathy.* Translated by Peter Heath. General introduction by W. Stark. London: Routledge & Kegan Paul.

Schutz, Alfred. 1964–67. *Collected Papers.* Edited and introduced by Maurice Natanson; Preface by H. L. van Breda. The Hague, Netherlands: M. Nijhoff.

———. 1967. *The Phenomenology of the Social World.* Translated by George Walsh and Frederick Lehnert. Evanston, IL: Northwestern University Press.

———. 1970. *On Phenomenology and Social Relations: Selected Writings.* Edited and Introduction by Helmut R. Wagner. Chicago: University of Chicago Press.

Schutz, Alfred and Thomas Luckmann. 1973. *The Structures of the Life-World.* Translated by Richard M. Zaner and H. Tristram Engelhardt. Evanston, IL: Northwestern University Press.

Seidman, Steven. 1989. "Introduction." Pp. 1–25 in *Jürgen Habermas on Society and Politics,* edited by Steven Seidman. Boston: Beacon Press.

———. 1994. *Contested Knowledge.* Malden, MA: Blackwell.

Simmel, Georg. [1900a] 1978. *The Philosophy of Money.* Translated by Thomas Bottomore and David Frisby. London: Routledge and Kegan Paul.

———. [1900b] 1971. "Exchange." Pp. 43–69 in *On Individuality and Social Forms,* edited by Donald N. Levine. Chicago: University of Chicago Press.

———. [1903] 1971. "The Metropolis and Mental Life." Pp. 324–339 in *On Individuality and Social Forms,* edited by Donald N. Levine. Chicago: University of Chicago Press.

———. [1904] 1971. "Fashion." Pp. 294–323 in *On Individuality and Social Forms,* edited by Donald N. Levine. Chicago: University of Chicago Press.

———. [1908a] 1955. *Conflict and the Web of Group Affiliations.* Translated by Kurt H. Wolff and Reinhard Bendix. New York: Free Press.

———. [1908b] 1971. "Conflict." Pp. 70–95 in *On Individuality and Social Forms,* edited by Donald N. Levine. Chicago: University of Chicago Press.

———. [1908c] 1971. "The Stranger." Pp. 143–49 in *On Individuality and Social Forms,* edited by Donald N. Levine. Chicago: University of Chicago Press.

———. [1908d] 1971. "The Problem of Sociology." Pp. 23–35 in *On Individuality and Social Forms,* edited by Donald N. Levine. Chicago: University of Chicago Press.

———. [1908e] 1971. "How Is Society Possible." Pp. 6–22 in *On Individuality and Social Forms,* edited by Donald N. Levine. Chicago: University of Chicago Press.

———. [1908f] 1971. "Subjective Culture." Pp. 227–234 in *On Individuality and Social Forms,* edited by Donald N. Levine. Chicago: University of Chicago Press.

———. [1910] 1971. "Sociability." Pp. 127–40 in *On Individuality and Social Forms,* edited by Donald N. Levine. Chicago: University of Chicago Press.

———. [1917] 1950. "Fundamental Problems of Sociology (Individual and Society)." Pp. 3–84 in *The Sociology of Georg Simmel,* edited and translated by Kurt Wolff. New York: Free Press.

———. 1950. *The Sociology of Georg Simmel,* edited and translated by Kurt Wolff. New York: Free Press.

———. 1971. *On Individuality and Social Forms.* Edited by Donald N. Levine. Chicago: University of Chicago Press.

———. 1984. *Georg Simmel: On Women, Sexuality, and Love.* Edited and translated by Guy Oakes. New Haven, CT: Yale University Press.

Sitton, John. 2003. *Habermas and Contemporary Society.* New York: Palgrave Macmillan.

Sklair, Leslie. 2002. *Globalization: Capitalism and Its Alternatives.* 3rd ed. Oxford, UK: Oxford University Press.

Smelser, Neil. 2001. "Introduction." Pp. vii–xix in *Toward a General Theory of Action,* by Talcott Parsons and Edward Shils. New Brunswick, NJ: Transaction Publishers.

Smith, Adam. [1776] 1990. *An Inquiry into the Nature and Causes of the Wealth of Nations.* Chicago: Encyclopedia Britannica.

———. [1842] 1993. "The Proper Sphere of Government." Originally printed in *The Nonconformist;* republished in *Spencer: Political Writings,* by Herbert Spencer, Author; John Offer, Editor. Cambridge: Cambridge University Press.

Smith, Dorothy. 1977. *Feminism and Marxism.* Vancouver, Canada: New Star Books.

———. 1987. *The Everyday World as Problematic.* Boston: Northeastern University Press.

———. 1990a. *The Conceptual Practices of Power: A Feminist Sociology of Knowledge.* Boston: Northeastern University Press.

———. 1990b. *Texts, Facts, and Femininity: Exploring the Relations of Ruling.* London: Routledge.

———. 2005. *Institutional Ethnography: A Sociology for People.* Oxford, UK: AltaMira.

Spencer, Herbert. [1873] 1961. *The Study of Sociology.* Edited by Talcott Parsons. Ann Arbor: University of Michigan Press.

———. [1899] 1969. *The Principles of Sociology* (3 vols.). Edited by Stanislav Andreski. New York: Macmillan.

———. 1972. *Herbert Spencer on Social Evolution.* Edited by J. D. Y. Peel. Chicago: University of Chicago Press.

Stack, Carol. 1974. *All Our Kin: Strategies for Survival in a Black Community.* New York: Harper and Row.

Stryker, Sheldon. 1980. *Symbolic Interactionism: A Social Structural Version.* Menlo Park, CA: Benjamin/Cummings.

———. 2003. "Whither Symbolic Interaction? Reflections on a Personal Odyssey." *Symbolic Interaction* 26:95–109.

Swartz, David. 1997. *Culture and Power: The Sociology of Pierre Bourdieu.* Chicago: University of Chicago Press.

Swartz, David and Vera Zolberg. 2004. *After Bourdieu.* Dordrecht, Netherlands: Kluwer.

Sztompka, Piotr. 1986. *Robert K. Merton: An Intellectual Profile.* London: Macmillan.

Thomas, William I. 1923. *The Unadjusted Girl.* Boston: Little, Brown.

Thomas, William I. (with Dorothy Swaine Thomas). 1928. *The Child in America.* New York: Alfred A. Knopf.

Toynbee, Arthur. 1957. *A Study of History.* New York: Oxford University Press.

Tomlinson, John. 1999. *Globalization and Culture.* Chicago: University of Chicago Press.

Tong, Rosemarie. 1998. *Feminist Thought.* Boulder, CO: Westview Press.

Tönnies, Ferdinand. [1887] 1957. *Community and Society.* Translated by Charles P. Loomis. East Lansing: Michigan State University Press.

———. [1935] 1963. *Gemeinschaft und Gesellschaft.* Darmstadt: Wissenshaftliche Buchgesellschaft.

Tripp, Anna. 2000. *Gender: Readings in Cultural Criticism.* New York: Palgrave.

Tucker, Robert C., ed. 1978. *The Marx-Engels Reader.* New York: W. W. Norton.

Turner, Bryan. 2001. "Peter Berger." Pp. 107–16 in *Profiles in Contemporary Social Theory,* edited by Anthony Elliot and Bryan S. Turner. Thousand Oaks, CA: Sage.

Turner, Jonathan H., Leonard Beeghley, and Charles H. Powers. 2002. *The Emergence of Sociological Theory.* Belmont, CA: Wadsworth.

Veblen, Thorstein. [1899] 1934. *The Theory of the Leisure Class.* Edited by Stuart Chase. New York: Macmillan.

———. [1904] 1965. *Theory of Business Enterprise.* New York: A. M. Kelley.

———. 1919. *The Place of Science in Modern Civilization.* New York: W. B. Huebsch.

Venturi, Robert, Denise Scott Brown, and Steven Izenour. 1977. *Learning from Las Vegas: The Forgotten Symbolism of Architectural Form.* Cambridge, MA: MIT Press.

Wagner, Helmut R. 1973. "The Scope of Phenomenological Sociology." In *Phenomenological Sociology,* edited by George Psathas. New York: Wiley.

Wagner-Martin, Linda. 1989. "Gilman's 'The Yellow Wallpaper': A Centenary." Pp. 51–64 in *Charlotte Perkins Gilman: The Woman and Her Work,* edited by Sheryl L. Meryering. Ann Arbor, MI: UMI Research Press.

Wallerstein, Immanuel. 1974. *The Modern World-System: Capitalist Agriculture and the Origins of the European World-Economy in the Sixteenth Century.* New York: Academic Press.

———. 1979. *The Capitalist World Economy.* Cambridge, UK: Cambridge University Press.

———. 1987. "World-Systems Analysis." Pp. 309–24 in *Social Theory Today.* Palo Alto, CA: Stanford University Press.

———. 1990. "Culture as the Ideological Battleground of the Modern World-System." Pp. 31–56 in *Global Culture,* edited by Mike Featherstone. London: Sage.

Walsh, George. 1967. "Introduction." Pp. xv–xxiv in *The Phenomenology of the Social World,* by Alfred Schutz. Bloomington, IN: Northwestern University Press.

Washington, Mary Helen. 1988. "Introduction." In Anna Julia Cooper, *A Voice From the South.* New York: Oxford University Press. Waters, Malcolm. 2001. *Globalization.* 2nd ed. London: Routledge Press.

Watson, John B. [1924] 1966. *Behaviorism.* Chicago: University of Chicago Press.

Weber, Marianne. 1907. *Marriage and Motherhood in the Development of Law. Ehefrau und Mutter in der Rechtsentwicklung.* Tübingen: J. C. B. Mohr.

———. [1926] 1975. *Max Weber: A Biography.* New York: Wiley.

———. 1935. *Women and Love. Frauen und Liebe.* Koonigestein in Taunus: K. B. Langewissche.

Weber, Max. [1903–1917] 1949. *The Methodology of the Social Sciences.* Edited and translated by Edward A. Shils and Henry A. Finch. New York: Free Press.

———. [1904–1905] 1958. *The Protestant Ethic and the Spirit of Capitalism.* Translated by Talcott Parsons. New York: Charles Scribner's Sons.

———. [1915] 1958. "The Social Psychology of the World Religions." Pp. 267–301 in *From Max Weber: Essays in Sociology,* edited and translated by H. H. Gerth and C. Wright Mills. New York: Oxford University Press.

———. [1919] 1958. "Science as a Vocation." Pp. 129–56 in *From Max Weber: Essays in Sociology,* edited and translated by H. H. Gerth and C. Wright Mills. New York: Oxford University Press.

———. [1925] 1978. *Economy and Society.* Edited by Guenther Roth and Claus Wittich. Berkeley: University of California Press.

———. [1925a] 1978. *Economy and Society,* vols. 1 and 2, edited by Guenther Roth and Claus Wittich. Berkeley: University California Press.

———. [1925b] 1978. "Status Groups and Classes." Pp. 926–939 in *Economy and Society,* vols. 1 and 2, edited by Guenther Roth and Claus Wittich. Berkeley: University California Press.

———. [1925c] 1978. "The Types of Legitimate Domination." Pp. 212–301 in *Economy and Society,* vols. 1 and 2, edited by Guenther Roth and Claus Wittich. Berkeley: University California Press.

———. [1925d] 1978. "Bureaucracy." Pp. 956–1005 in *Economy and Society,* vols. 1 and 2, edited by Guenther Roth and Claus Wittich. Berkeley: University California Press.

———. 1947. *The Theory of Social and Economic Organization.* Edited by Talcott Parsons and translated by A. M. Henderson and Talcott Parsons. New York: Oxford University Press.

———. 1958. *From Max Weber: Essays in Sociology.* Edited and translated by H. H. Gerth and C. Wright Mills. New York: Oxford University Press.

———. 1958. "Politics as a Vocation." Pp. 77–128 in *From Max Weber,* edited by Hans Gerth and C. Wright Mills. New York: Oxford University Press.

Wiggershaus, Rolf. 1994. *The Frankfort School: Its History, Theories, and Political Significance.* Translated by Michael Robertson. Cambridge, MA: MIT Press.

Williams, Christine. 1993. "Psychoanalytic Theory and the Sociology of Gender." In *Theory on Gender, Feminism on Theory,* edited by Paula England. New York: Aldine de Gruyter.

Williams, Jr., Robin. 1980. "Talcott Parsons: The Stereotypes and the Realities." *The American Sociologist, 15,* in Peter Hamilton, ed., *Talcott Parsons: Critical Assessments v. 1,* Routledge.

Wrong, Dennis H. 1994. *The Problem of Order.* Cambridge, MA: Harvard University Press.

Zimbardo, Philip. 1985. "Laugh Where We Must, Be Candid Where We Can." *Psychology Today.*

Zimmerman, Don. 1988. "On Conversation." Pp. 406–32 in *Communication Yearbook,* edited by J. Anderson. Beverly Hills, CA: Sage.

Zimmerman, Don and D. Lawrence Wieder. 1970. "Ethnomethodology and the Problem of Order." Pp. 285–98 in *Understanding Everyday Life,* edited by Jack Douglas. Chicago: Aldine.

Zuckerman, Phil, ed. 2000. *Du Bois on Religion.* Walnut Creek, CA: AltaMira.

———, ed. 2004. *The Social Theory of W. E. B. Du Bois.* Thousand Oaks, CA: Pine Forge.

PHOTO CREDITS

Illustration, page 1. Illustration by John Tenniel (1820–1914).

Photo, page 20 (Marx). Library of Congress.

Photo 2.1, page 23. Library of Congress.

Photo 2.2, page 23. Reprinted with permission from the National Labor Committee, www.nlc.org.

Photo 2.3, page 25. © Bettmann/Corbis.

Photo, page 77 (Durkheim). © Betttmann/Corbis.

Photo 3.1a, page 93. © Dr. Klaus Dierks, www.klausdierks.com.

Photo 3.1b, page 93. "Asa Norte" (North Wing), by Augusto C. B. Areal; Copyright 1997. Used by permission.

Photo 3.2, page 102. © Associated Press/Malcolm Browne.

Photo 3.3a, page 115. © Gould/Corbis.

Photo 3.3b, page 115. © Tony Aruzza/Corbis.

Photo, page 125 (Weber). Library of Congress.

Photo 4.1a, page 168. Library of Congress.

Photo 4.1b, page 169. Library of Congress.

Photo 4.1c, page 169. Library of Congress.

Photo, page 186 (Gilman). Library of Congress.

Photo 5.1, page 190. © Lynn Goldsmith/Corbis.

Photo 5.2, page 194. © MGM Studios/Getty Images.

Photo, page 218 (Simmel). The Granger Collection, New York.

Photo 6.1, page 224. © Kodner Gallery.

Photo 6.2, page 227. © Corbis.

Photo 6.3, page 246. Library of Congress.

Photo 6.4, page 246. © Scott Appelrouth. Printed with permission.

Photo 6.5a, page 247. Courtesy of the Samuel Courtauld Trust, Courtauld Institute Art Gallery, London.

Photo 6.5b, page 247. © iStockPhoto.

Photo, page 256 (Du Bois). Library of Congress.

Photo 7.1, page 258. © Bettman/Corbis.

Photo 7.2, page 262. Temple University Libraries, Urban Archives, Philadelphia Pennsylvania.

Photo 7.3, page 263. Department of Special Collections and Archives, W. E. B. Du Bois Library, University of Massachusetts Amherst.

Photo 7.4, page 265. Library of Congress.

Photo 7.5, page 267. Temple University Libraries, Urban Archives, Philadelphia, Pennsylvania.

Photo, page 289 (Mead). The Granger Collection, New York.

Photo 8.1, page 295. UN/DPI. Reprinted with permission.

Photo 8.2, page 308. © Scott Appelrouth. Used with permission.

Photo 8.3a, page 310. © Scott Appelrouth. Used with permission

Photo 8.3b, page 310. © Laura Desfor Edles. Used with permission

Photo page 324 (Parsons). American Sociological Association. Reprinted with permission.

Photo, page 324 (Merton). © 1995 Jim Graham/GSI. Courtesy of Harriet Zuckerman.

Photo 9.1a, page 334. © Wendy Stone/Corbis.

Photo 9.1b, page 334. © Lester Lefkowitz/Corbis.

Photo 9.2a, page 335. USDA Natural Resources Conservation Service.

Photo 9.2b, page 335. Used with permission.

Photo 9.3a, page 358. © Corbis.

Photo 9.3b, page 358. © Corbis.

Photo 9.4a, page 359. © Thinkstock/comstock.

Photo 9.4b, page 359. © Bettmann/Corbis.

Photo, page 375 (Horkheimer). Photographer Kurt Bethke. Copyright © Hessischer Rundfunk.

Photo, page 375 (Adorno). Imagno/Hulton Archive/Getty Images.

Photo, page 375 (Marcuse). © Bettmann/Corbis.

Photo 10.1, page 376. © Scott Appelrouth: Used with permission.

Photo 10.2, page 385. U. S. Marine Corp.

Photo 10.3, page 387. © Scott Appelrouth. Used with permission.

Photo 10.4, page 388. © Scott Appelrouth. Used with permission.

Photo 10.5, page 391. © Scott Appelrouth. Used with permission.

Photo, page 413 (Homans). American Sociological Association. Used with permission

Photo, page 413 (Blau). American Sociological Association. Used with permission

Photo, page 413 (Coleman). American Sociological Association. Used with permission

Photo 11.1, page 416. © Scott Appelrouth. Used with permission.

Photo, page 463 (Goffman). American Sociological Association. Used with permission.

Photo, page 463 (Hochschild). Used with permission of Arlie Hochschild.

Photo 12.1, page 464. © Scott Appelrouth. Used with permission.

Photo 12.2a, page 474. © Scott Appelrouth. Used with permission

Photo 12.2b, page 474. © Thinkstock/Bananastock.

Photo 12.2c, page 474. © Scott Appelrouth. Used with permission.

Photo 12.3a, page 506. © Bettmann/Corbis.

Photo 12.3b, page 506. © Corbis.

Photo, page 518 (Schutz). Used with permission of Evelyn Schutz Lang.

Photo, page 518 (Berger). Used with permission of Peter Berger.

Photo, page 518 (Luckmann). #169 Waseda Schutz Archive.

Photo 13.1, page 524. © Corbis.

Photo 13.2, page 552. © Frank Micelotta/Getty Images/Getty Images Entertainment.

Photo, page 556. (Smith). Reprinted with permission of Dorothy Smith.

Photo, page 556. (Collins). Reprinted with permission.

Photo, page 556. (Connell). Photo by Dianne Reggett.

Photo, page 556 (Butler). Photo by Aimee Friberg.

Photo 14.1, page 558. © YASSER AL-ZAYYAT/AFP/Getty Images.

Photo 14.2, page 600. © Tom Gates/Archive Photos/Getty Images.

Photo, page 607 (Foucault). © Corbis.

Photo, page 607 (Baudrillard). © Eric Fougere/VIP Images/Corbis.

Photo 15.1, page 609. © Bettman/Corbis.

Photo 15.4, page 638. Courtesy of Adbusters Media Foundation.

Photo 15.5, page 640. Photo taken by Bobak Ha'Eri, February 23, 2006.

Photo, page 651 (Bourdieu). Alain Nogues/Sygma/Corbis

Photo, page 651 (Habermas). Copyright © Inamori Foundation. Reprinted with permission.

Photo, page 651 (Giddens). Used with permission of Anthony Giddens.

Photo 16.1, page 657. Library of Congress.

Photo 16.2, page 658. © Thinkstock.

INDEX

"The Absolute in History" (Habermas), 685
Absolute Idea or Spirit, 382
Achievement/ascription, 333 (table), 334–335, 343
Action
    affective, 128
    agency role in, 719–723
    causality to establish probability or elective
        affinity of, 130
    communicative, 690
    critical theorists on motivating forces of, 382–386
    Durkheim's perspective of suicide, 78, 100–113 (table)
    Giddens's stratification model of, 720 (figure)–721
    Habermas's theory of rational, 689–691
    habitualization limiting, 534–535
    how religious beliefs drive, 114–124
    ideal types of, 128–129
    instrumental-rational, 127–128
    latent function and manifest function of, 356–358,
        360–366
    Mead on meaning of, 299–300, 304–307
    nonrational versus rational, 12–15 (table)
    Parsons and Shils on orientation of, 339–342
    Parsons's model of social, 328–329 (figure)
    Simmel's duality regarding subjective/objective culture
        on, 232 (table)
    social facts determining individual's, 80, 83,
        84 (figure), 85–92
    sociological theory focus on, 11–18
    subjective meaning (*gemeinter Sinn*) of, 521
    traditional, 128
    unit act referring to, 328–329
    value-rational, 128
    Verstehen (interpretive understanding) of, 521
    Weber's distinguishment of four types of, 127–128
    Weber's legitimate domination for, 169 (table)
    Weber's notion of class, status, and party dictating,
        160 (table)
    why do people stop at red traffic lights?, 15 (table)–16
    See also Agency; Sanctions; Social behavior

Action systems
    Parsons's AGIL (four functional requirements)
        of, 336 (table)
    Parsons's model on social systems and, 328–332
    See also Social systems
Adaptation
    AGIL scheme at social system level, 337 (table)
    as functional requirement of action system, 336 (table)
    Merton's typologies of, 368–373
Adaptation typologies
    conformity, 368, 369
    innovation, 368, 369
    overview of, 368–369
    rebellion, 368, 372–373
    retreatism, 368, 371–372
    ritualism, 368, 369–371
Addams, J., 291
Adorno, T.
    The Authoritarian Personality by, 379
    basic theoretical orientation of, 392 (figure)–393
    biographical sketch, 377
    "The Culture Industry Reconsidered" by, 400–404
    Dialectic of Enlightenment by Horkheimer and, 379–380,
        400, 685
    Institute for Social Research role by, 378–380
    intellectual influences and core ideas of, 380–392
    introduction to, 8
Advertising
    culture industry fabricating world through, 387 (photo)
    of happiness, 388 (photo)
    offering "convenient" repression, 391 (photo)
    See also Consumption
Affective action, 128
Affectivity/affective-neutrality, 332, 333 (table), 334, 343
*Africa—Its Geography, People, and
    Products* (Du Bois), 262
Africa—Its Place in Modern History (Du Bois), 262
African Americans
    Du Bois on double consciousness of, 269, 270, 284

Note: Illustrated figures are indicated by (figure); photographs by (photos); and tables by (tables).